Human Biology

45-

Human Biology

Clinton L. Benjamin
Lower Columbia College
Gregory R. Garman
Centralia College
James H. Funston

WCB
McGraw-Hill

Boston, Massachusetts BurrRidge, Illinois Dubuque, Iowa
Madison, Wisconsin New York, New York San Francisco, California St. Louis, Missouri

WCB/McGraw-Hill

A Division of The **McGraw-Hill** *Companies*

Human Biology

2 3 4 5 6 7 8 9 0 VNH VNH 9 0 9 8 7

ISBN 0-07-022907-4

This book was set in Caslon by York Graphic Services, Inc.
The editors were Denise Schanck, Holly Gordon, Jack Maisel and Sharon Geary;
the designer was Joan Greenfield; the design manager was Joseph A. Piliero;
the production supervisor was Kathryn Porzio.
The photo editors were Nancy Dyer and Kate Ross;
the photo researchers were Lana Berkovich and Elyse Rieder.
Von Hoffmann Press, Inc., was printer and binder.
Illustrations and illustration concepting: J/B Woolsey Associates

Library of Congress Cataloging-in-Publication Data

Benjamin, Clinton L.
Human biology/Clinton L. Benjamin, Gregory R. Garman, James H. Funston.
p. cm.
Includes bibliographical references and index.
ISBN 0-07-022907-4 (hardcover)
1. Human biology. 2. Human ecology 3. Ecology. I. Garman, Gregory R. II. Funston, James H. III. Title.
QP36.B39 1997
612—dc20 95-45015
CIP

INTERNATIONAL EDITION

When ordering this title, use ISBN 0-07-114050-6.

About the Authors

Clinton L. Benjamin received his Ph.D. from The Ohio State University with an emphasis on vertebrate physiology. He earned a B.S. in biology from St. Mary's College, California, and an M.S. in biology from California State University, Humboldt. Dr. Benjamin has taught human anatomy and physiology and general microbiology at Lower Columbia College in Longview, Washington, for over 20 years, where he has been honored with the Teacher of the Year Award for classroom instruction. He has been awarded an NSF grant for the improvement of undergraduate education and has coauthored a study guide in microbiology. He is an active member of the Human Anatomy and Physiology Society (HAPS) and the Northwest Biology Instructors Organization.

Gregory R. Garman has taught general biology, anatomy, physiology, microbiology, and human biology at Centralia College in Centralia, Washington, for more than 22 years. Prior to his teaching experience, he earned a B.S. in biology at California Polytechnic State University, San Luis Obispo, where his studies focused on field and marine biology. He earned an M.S. in zoology, with an emphasis on comparative physiology and biochemistry, at Oregon State University. He has also studied natural history and education at The Evergreen State College and St. Martins College in Olympia, Washington. He has coauthored research articles on fish ecology and nutritional toxicology and a student study guide in microbiology. In 1992 Mr. Garman was selected as Faculty Scholar of the Year by the students of his college and received an inaugural Exceptional Faculty Award from his peers. He is an active member of the Human Anatomy and Physiology Society and the Northwest Biology Instructors Organization.

James H. Funston received his formal biological education from three institutions: Earlham College (B.A.) in Richmond, Indiana; the Marine Biological Laboratory in Woods Hole, Massachusetts; and Brandeis University (M.A.) in Waltham, Massachusetts. As a graduate student, he taught biology to nonscience students. Later, as a member of the Biology Department of the College of the Holy Cross in Worcester, Massachusetts, he taught biochemistry and cell biology to premedical students. Following his teaching career, he spent nine years in the publishing industry, acquiring, editing, and publishing college textbooks in the life sciences. For more than a decade he has worked as a freelance editor of college textbooks in biology, chemistry, and biochemistry.

To the memory of Louis, Madeline, Loretta, and Jim who were always supportive of my educational pursuits. To my wife, Kathy, for her unending patience and encouragement throughout this project.

—CLB

To Linda, my wife and best friend, for her loving patience and sacrifice and to my parents, parents-in-law, sister, children, and students for their continued interest, support, and encouragement.

—GRG

To my father, James Arthur, who taught me to value words. To Bruce from whom I learn the words to value in the practice of transparency.

—JHF

Contents in Brief

Contents

PART ONE
BUILDING THE BODY: CELLS AND TISSUES

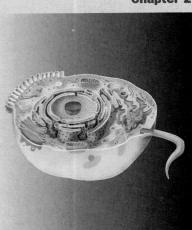

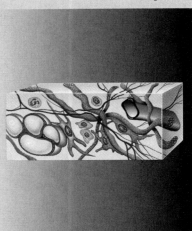

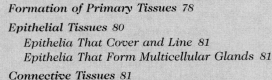

PART TWO
PROTECTING, SUPPORTING, AND MOVING THE BODY

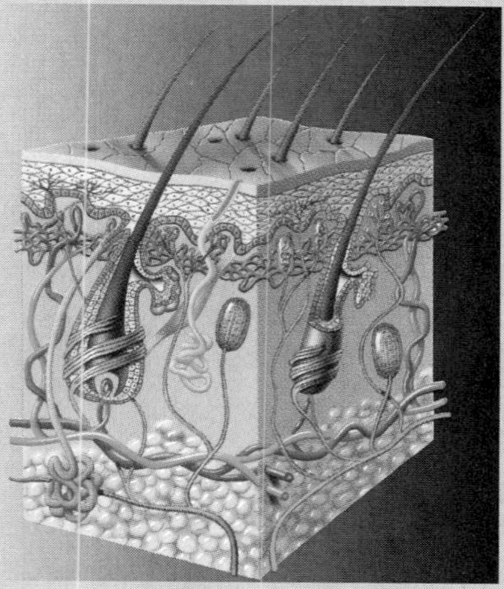

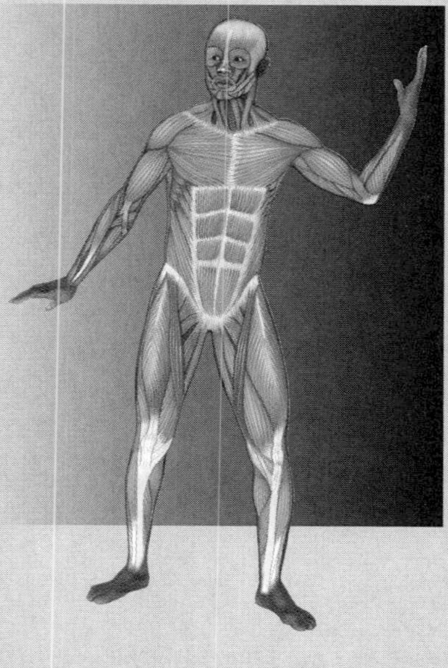

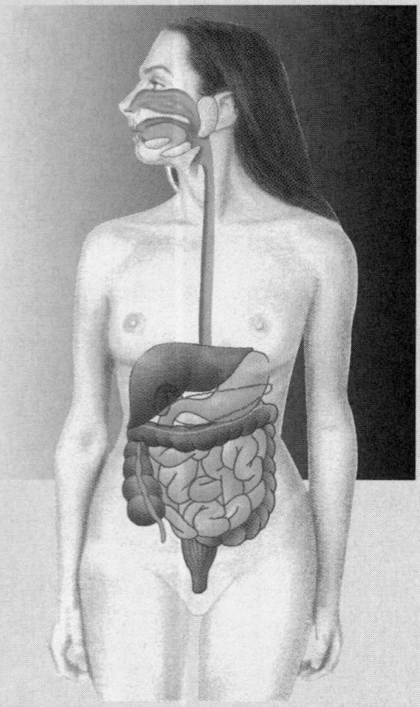

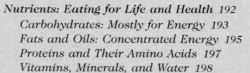

MAYO

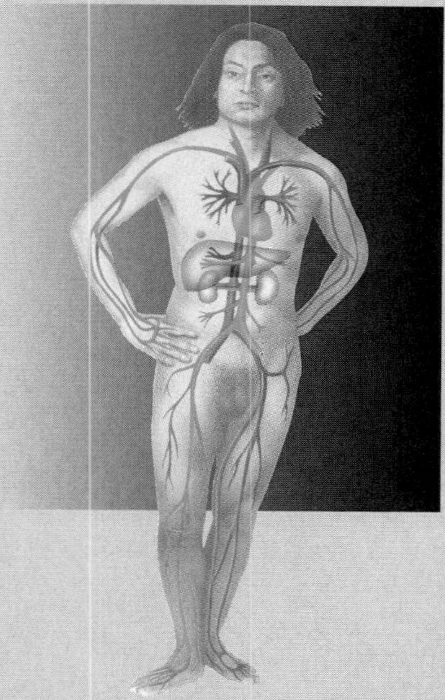

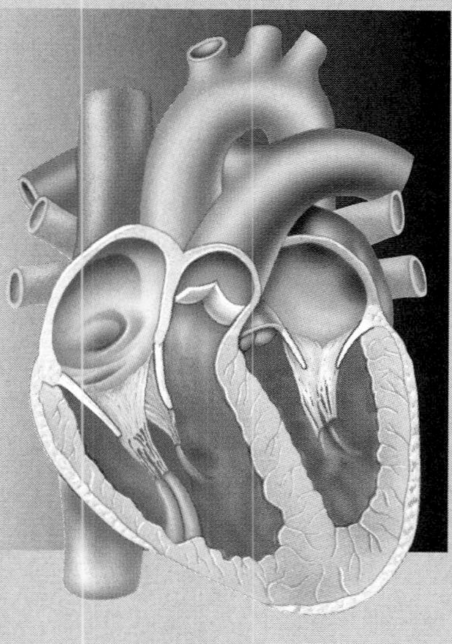

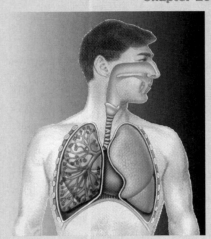

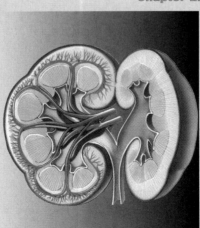

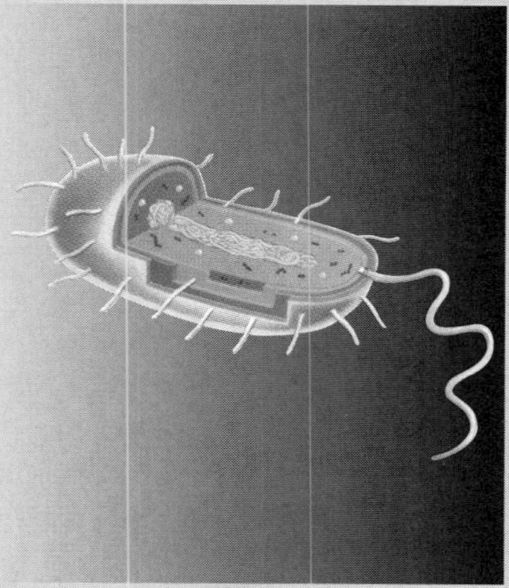

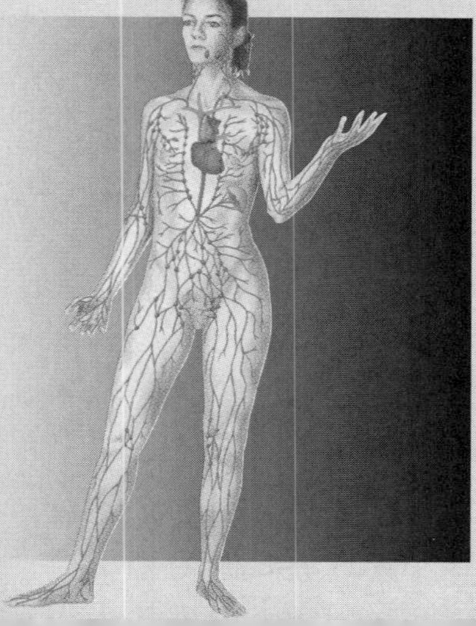

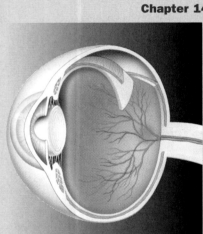

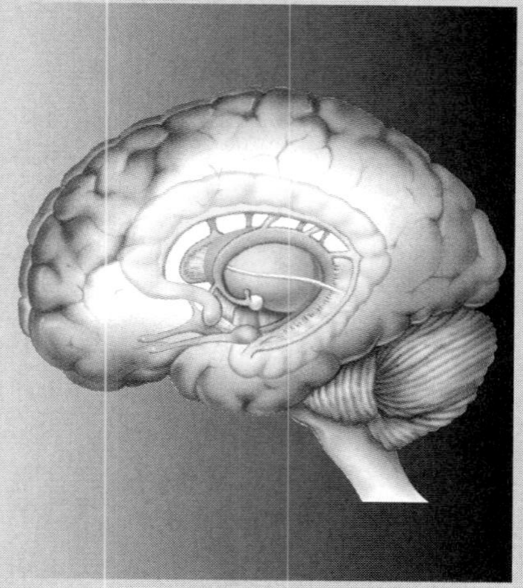

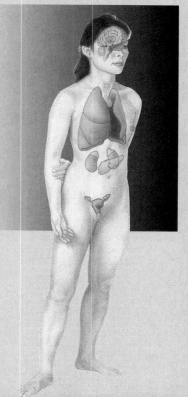

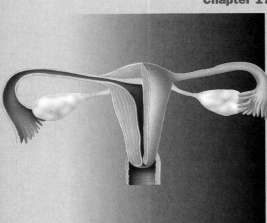

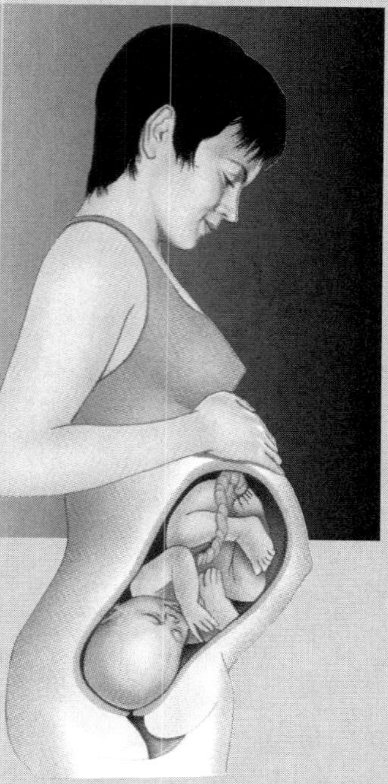

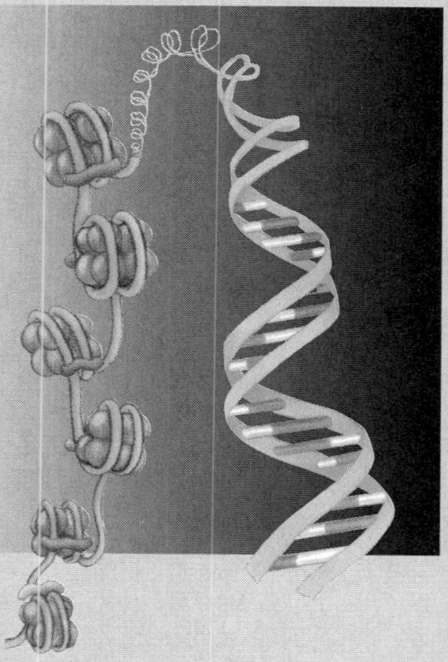

PART EIGHT
LIVING WITH NATURE

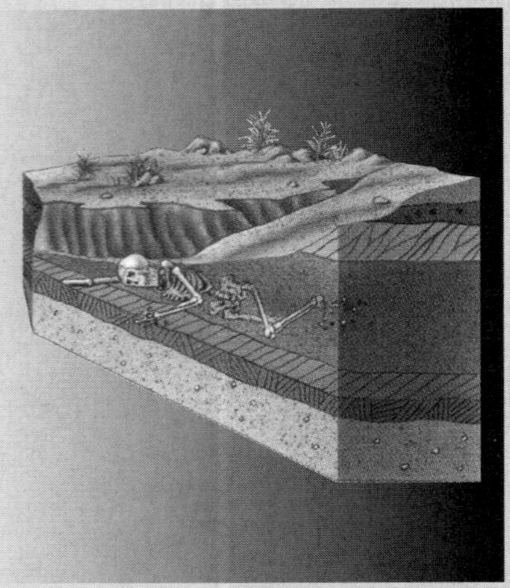

Preface

Who is this "human" we seek to study?

> "What a piece of work," exclaims Shakespeare. "How noble in reason! How infinite in faculty! In form, in moving how express and admirable! In action how like an angel! In apprehension how like a god."

These magnificent characteristics are rooted in matter and energy, and this rooting constitutes the biology of being human.

While our bodies are with us for life, they are often strangers to us. At times we stand in awe and admiration (the birth of an infant, the growth of a child, the vigor and grace of a dancer). At other times we are perplexed and offended by our bodies' pains, limitations, and deterioration. In all conditions, human bodies deserve better care, and the study of the human body helps make such care both effective and appropriate over the entire life span.

To provide human bodies with appropriate care, we must recognize that humans do not exist apart from their physical and biological environment. The history of human life on Earth and the environmental conditions that sustain life are featured in *Human Biology.*

Scope and Approach

Human Biology is written for nonscience students who have little or no background in the sciences. It takes a *structure-and-function approach* that emphasizes the normal operations of each organ system and the role of *homeostasis*.

In addition, special attention is given to selected *diseases and disorders* of each organ system, applying the concepts of normal human biology to help students understand what "goes wrong." The discussion of each organ system closes with a brief description of its *changes over the human life span.*

Since the external environment interacts directly and indirectly with every organ system, each chapter includes environmental considerations as it identifies the regulatory activities that maintain internal conditions. The final section of the book, Living with Nature, explores the broader temporal and environmental forces that both nurture and challenge all forms of life, including humans.

Organization

Human Biology consists of a prologue and 22 chapters covering background information (biological diversity, the nature of science, chemistry, cells and tissues); the organ systems; related topics (infectious diseases and genetics); and evolution, ecology, and human environmental impact.

Human Biology presents the less complex organ systems early (see the table of contents), leaving until later the more complex systems. This approach helps students build study skills, confidence, and knowledge before tackling more challenging topics such as the nervous and immune systems. However, alternative sequences are possible for most chapters.

Related chapters are grouped together to emphasize eight major conceptual themes:

1. Building the Body: Cells and Tissues (cells, tissues, body organization, and aging)
2. Protecting, Supporting, and Moving the Body (homeostasis, skin, skeleton, and muscles)
3. Processing, Absorbing, and Converting Raw Materials (digestion, nutrition, and metabolism)
4. Transporting Materials and Maintaining Internal Conditions (circulation, respiration, and excretion)
5. Threats and Body Defenses (infectious diseases, foreign substances, and immune defenses)
6. Regulation and Integration (nerves, signal processing, and hormones)

7. Reproduction, Development, and Heredity (sex, reproduction, intrauterine development, and genetics)
8. Living with Nature (evolution, human evolution, ecology, and environmental impact)

After each group of chapters there is a special short feature, the Spotlight on Health. These spotlights provide additional information and employ the concepts learned in the preceding chapters.

- Cancer: Causes and Treatment
- Healing Bones and Joints
- Read Right, Eat Well, Live Better: Interpreting Food Labels
- Exercise: Pursuing Fitness
- AIDS: The Disease, the Epidemic, and the Virus
- Alcohol and Drug Use among College Students
- Sexually Transmitted Diseases
- Health and Reproduction: Threats from Pollutants

Chapter Organization

There are three major components in each organ system chapter: structure and function, diseases and disorders, and life span changes.

Structure and function. Each organ system chapter begins by emphasizing the structures of normal component organs, tissues, and cells. With this foundation, the normal functions of the system are presented, emphasizing the relationship between structure and function.

Diseases and disorders. Information on diseases and disorders is presented in two ways: (1) Within the presentation of normal body operations, brief references are made to relevant pathologies. (2) Common pathologies receive more explicit elaboration in the section titled Diseases and Disorders. This discussion applies concepts of normal body biology to help students understand "how things can go wrong" and provides useful information about the symptoms, treatments, and prevention of common diseases.

Life span changes. At the close of each organ system chapter, a section titled Life Span Changes summarizes the major changes that take place in an organ system over the normal human life span from birth to old age.

Friendly Features Aid Learning

The book uses an interpretive approach. This means that where appropriate we "tell a story" and attempt to interpret the significance of scientific facts for the reader. We do not present science merely for the sake of science; instead, we relate the science to familiar human situations, functions, and dysfunctions.

Numerous heads and subheads are used to identify content and divide the material into short "chunks" to aid learning.

Paragraphs are short to encourage effective reading, retention, reflection, and understanding.

Technical terminology has been limited but certainly not avoided. Memory and the proper use of terminology help carry concepts while at the same time empowering the reader and building confidence. All key terms are emphasized with boldface or italic type. Technical terms are explicitly defined when first used, and are redefined when used in later chapters. A glossary is included at the back of the book.

Phonetic pronunciation is provided for unfamiliar terms to encourage students' use of the vocabulary of science, biology, and medicine. Correct pronunciation builds confidence and helps students better understand lectures and class discussion.

Chapter and section introductions consistently prepare the reader to identify the topics that will follow. With such preparation, the reader is never surprised or confused by the scope of a discussion. This aids concentration and contributes to the pleasure and effectiveness of learning.

In-text summaries follow each major chapter section and emphasize the most important points discussed.

End-of-chapter pedagogical aids emphasize important concepts and terms. Included are (1) a chapter summary that is organized around major chapter heads; (2) a selection of review activities that ask students to identify, list, describe, distinguish, or compare and contrast topics, using their own words; and (3) self-quiz with answers.

Illustrations: Linking Image and Information

In *Human Biology,* brief explanations or descriptions are often placed directly on illustrations. This establishes a direct link between the components of the illustration and the explanation, making both more significant and accessible to the reader. Combining graphics with relevant description makes the illustrations more self-instructive and encourages students to spend more time learning from the figures.

Supplements

Instructor's Manual/Test Bank

Greg Garman, Centralia College
The *Instructor's Manual/Test Bank* offers the instructor a complete package of teaching tools. The Chapter Overview, Chapter Concepts, and Chapter Outline sections focus on detailing the major learning objectives of each chapter. Student and Laboratory Activities provide creative ways to involve students in hands-on assignments.

The Classroom Enrichment section provides suggestions for speakers, interactive exercises, debates, and writing activities to help enliven the classroom setting. A listing of commercial videotapes to support each chapter is also provided.

The Appendices feature multimedia ancillary listings, videotape source listings, and pertinent Internet addresses and World Wide Web sites for additional reference materials as well as a listing of overhead transparencies available with the text. Answers to the Review Activities from the textbook can also be found in the Instructor's Manual.

The Test Bank portion of the *Instructor's Manual/Test Bank* provides instructors with a resource of over 1300 questions from which to develop exam material. Among the questions included for each chapter are multiple-choice, true/false, matching, completion, and short answer. Each question is categorized by the type of learning being tested (factual recall, comprehension, or application) and by the level of difficulty.

Computerized Test Bank

The printed test bank is also available in the following computerized formats: IBM $5\frac{1}{4}$, IBM $3\frac{1}{2}$, and Macintosh.

Student's Study Manual

John Capeheart, and Deanna McCullough, University of Houston
The *Student's Study Manual* features Chapter Overview and Learning Objectives sections that briefly introduce students to the major concepts in the chapter while providing specific learning goals. The extended Chapter Outline and subsequent Completion Exercise allow students to test their ability to organize chapter information. Students will be shown how to diagram a concept map to visually make relational connections between concepts. For each chapter, students will be asked to create their own concept maps.

Self-Examination sections include key-term matching, multiple-choice questions, comparative analogies, and essay and critical thinking questions. To aid students in understanding the "big picture," the *Student's Study Manual* contains end-of-part reviews that tie together central topics in the previous section. Case studies are presented to relate biological concepts to everyday life. Critical thinking questions help students analyze and comprehend major text concepts.

Overhead Transparencies

Two hundred and fifty full-color transparencies of key illustrations from the text are available to qualified adopters. Images have been increased in size, when possible, and all labels have been increased in size and boldness for easier use in lectures.

Gratitude for Help Received

If there is any secret to converting unwilling learners into willing learners, it is probably as much a secret of the heart as one of the rational mind. As authors, we have truly enjoyed the many demanding tasks of creating this book, and we hope some of that enthusiasm and pleasure is evident on the pages and will infect the reader.

We gratefully acknowledge the constructive criticisms and helpful suggestions provided by the reviewers who slogged through the early drafts of chapters. Their enthusiasms and comments were valuable in molding and honing the manuscript into its final form. The reviewing panel for this edition included the following individuals.

Master Reviewers

Sheldon R. Gordon
Oakland University
Michael Stewart
The Open University–England

Reviewers

G. Samuel Alspach, Jr.
Western Maryland College
Edmund E. Bedecarrax

City College of San Francisco
Thomas L. Beitinger
University of North Texas
David Brumagen
Morehead University
Gloria M. Caddell
University of Central Oklahoma
Vic Chow
City College of San Francisco
Barbara J. Clarke
The American University
Lisa Danko
Mercyhurst College
Michelle Green
State University of New York College of Technology at Alfred
Martin E. Hahn
William Paterson
N. Gail Hall
Trinity College
Ronald K. Hodgson
Central Michigan University
Carolyn K. Jones
Vincennes University
Myron Cran Lucas
Louisiana State University
Mara L. Manis
Hillsborough Community College
Patricia Matthews
Grand Valley State University
Carol Morris
Tompkins-Cortland Community College
Lloyd M. Pederson
San Joaquin Delta College
Joel Piperberg
Millersville University
Carl Roush
Lower Columbia College
Marc M. Roy
Beloit College
Lynette Rushton
South Puget Sound Community College
Dennis Shaw
Lower Columbia College
Doris M. Shoemaker
Dalton College
Robert J. Sullivan
Marist College
Robin Tyser
University of Wisconsin–LaCrosse

We especially appreciate the high-quality illustrations developed by J/B Woolsey Associates. John Woolsey and Patrick Lane were a pleasure to work with, and their vivid artwork put additional life into our words and helped dramatize the basic concepts of human biology.

We are also indebted to the editorial and production staff of The McGraw-Hill Companies for its guidance and assistance in completing this project. We want to thank our former editor, Kathi Prancan, who brought us together. She shared the vision, set the tone, and encouraged our work. Sharon Geary expertly coordinated the completion of the supplementary materials. Holly Gordon and Jack Maisel confidently coordinated editorial details, and Eric Lowenkron did a superb job copyediting our sometimes idiosyncratic prose. Many thanks to Nancy Dyer and her photo research staff for their persistence in finding just the right photos. As publisher, Denise Schanck has overseen the project from its inception, offering suggestions and guidance.

Special thanks are due Pam Barter, who meticulously read and commented on draft chapters and illustrations. Her comments provided a valuable perspective that kept us aware of our student readers as we wrote and revised.

We are eager to hear about your experiences with *Human Biology*. Please write or e-mail us care of our publisher,

biology_college@mcgraw-hill.com
The McGraw-Hill Companies, Inc.
College Division, 27th Floor
Biology Publisher
1221 Avenue of the Americas
New York, NY 10020

For more information on McGraw-Hill, feel free to browse our website at http://www.mhcollege.com

Clinton. L. Benjamin
Gregory R. Garman
James H. Funston

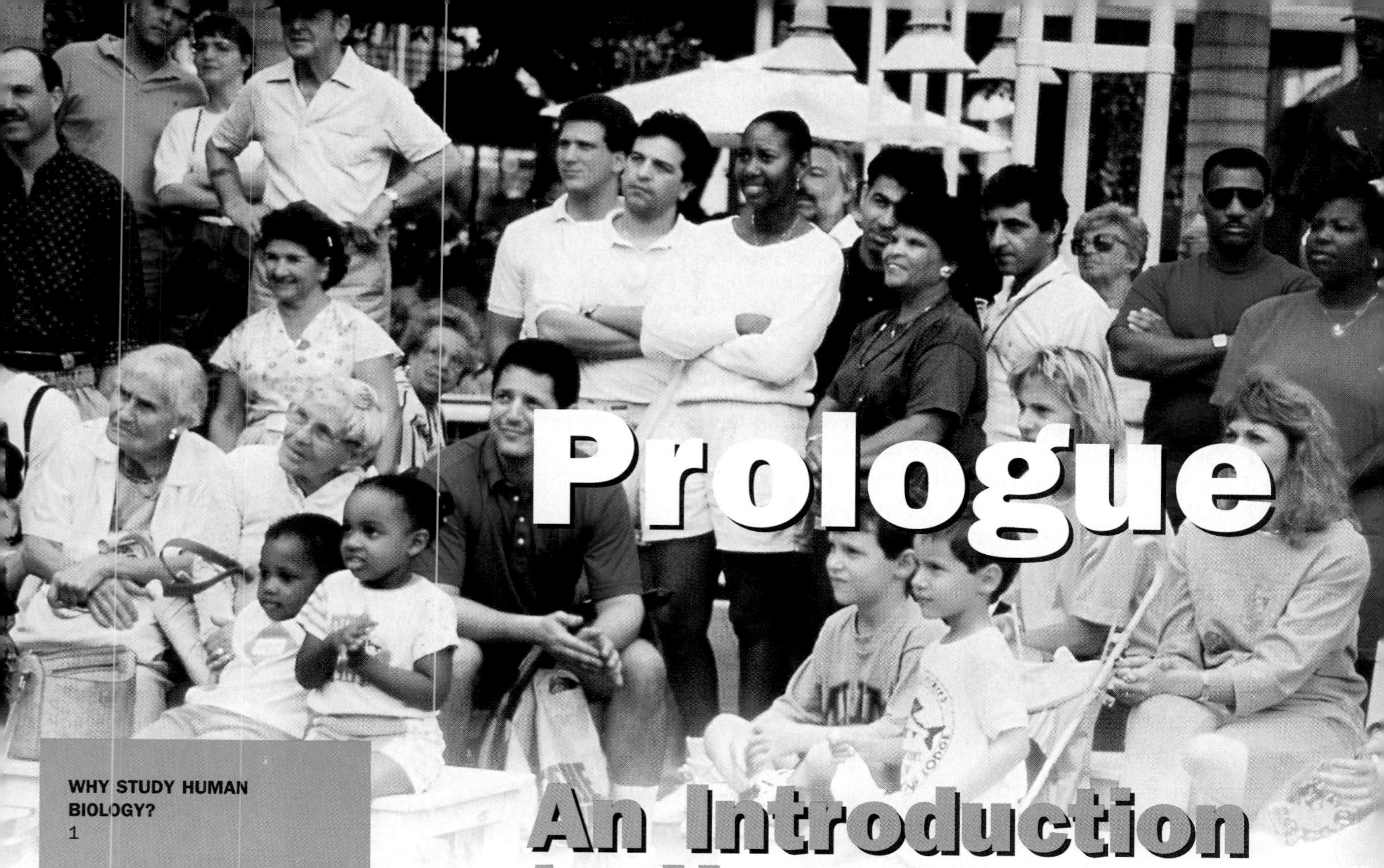

Prologue

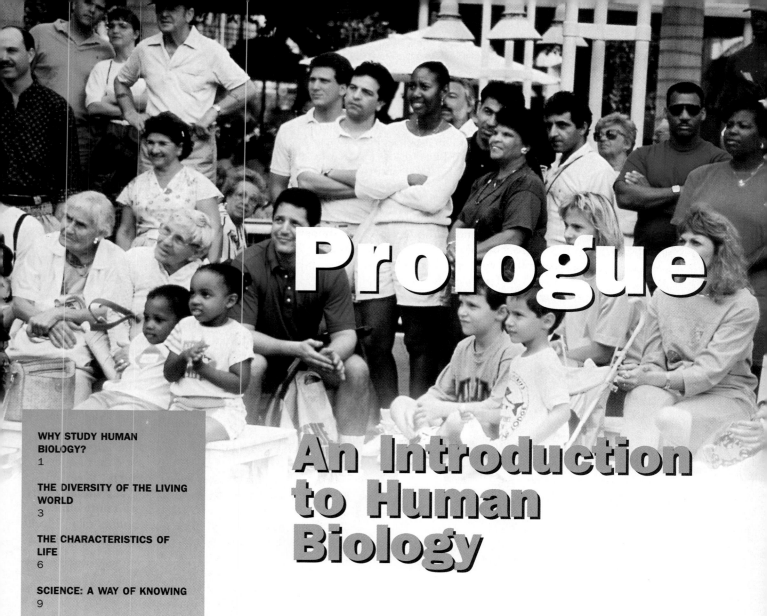

An Introduction to Human Biology

Biology is the study of life (*bio,* "life"; *logy,* "to study"). Human biology is the study of the structure and operation of the human body.

Why Study Human Biology?

Many different interests, circumstances, and requirements helped you decide to take this course and read this book. Before suggesting some answers to the question above, we want you to consider a larger and more personal question: What is the knowledge most worth having?

Knowledge is organized, reliable information, and all knowledge is worth having. In this increasingly complicated world, we need to know a lot of different things (Figure P.1). It is useful to know about electronics, computers, and automobiles; stocks, bonds, and videotape; taxes, laws, and real estate; and plumbing, banking, and building. However, there is more valuable knowledge that is more central to your life. For yourself, how would you answer the question, What is the knowledge *most* worth having?

After some reflection, many people answer, "Self-knowledge is the knowledge most worth having." When you have knowledge of yourself, all other knowledge and information takes on the appropriate significance in your life. While the scope of "self-knowledge" is both too vast and too personal to address in this book, we believe that an understanding of your body helps you understand yourself. It helps you understand your role in the world and the destiny of human life on this small, wet planet 93 million miles from the sun. This is a very practical form of knowledge.

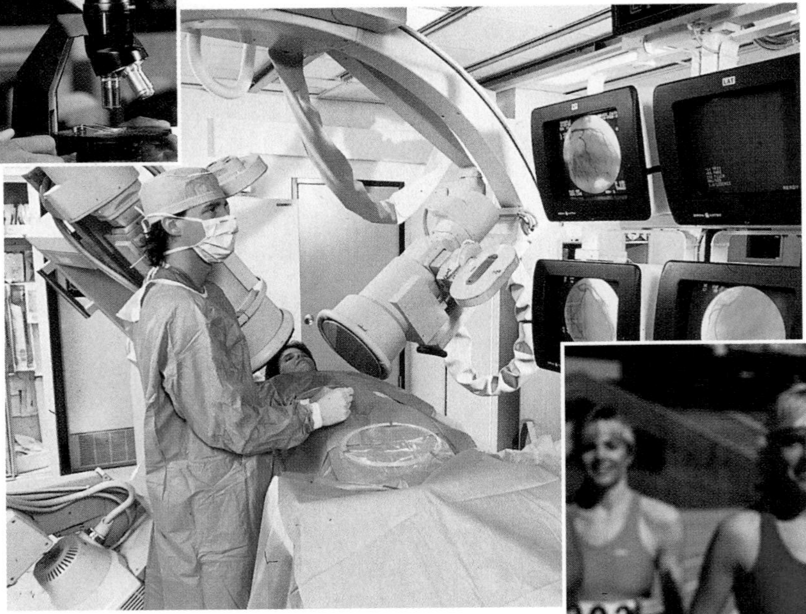

FIGURE P.1 Knowledge lets people understand and accomplish things.

In today's complicated technological world, there are tangible benefits to knowing how the human body functions and how it changes in health, disease, and aging. More than at any other time in the history of human life on Earth, individuals are faced with personal choices that directly affect the quality and length of their lives (Figure P.2). How does exercise benefit your body? What is "cardiovascular fitness"? To keep your heart healthy, how much fat should you eat? What can be done to control cancer? How can a bacterium "eat flesh"? How can you protect yourself against the virus that causes AIDS? Why have antibiotics become less effective against some bacteria? What is natural aging? How do drugs alter the mind and the emotions? How do threats to biodiversity affect your life and your future? How can birth defects be avoided? In this book you'll find knowledge about the human body that helps you answer these questions.

307

The human body is a **multicellular** (*multi,* "many"; *cellular,* "consisting of cells") system that consists of trillions of living cells (Figure P.3). **Cells** are the smallest living units, and they will be examined in detail in Chapter 2. There are hundreds of different types of cells in the human body. Groups of similar cells are associated together to form tissues.

Tissues are organized to form organs. **Organs** are larger structures, such as the heart and the stomach, which perform a specific function, such as the pumping of

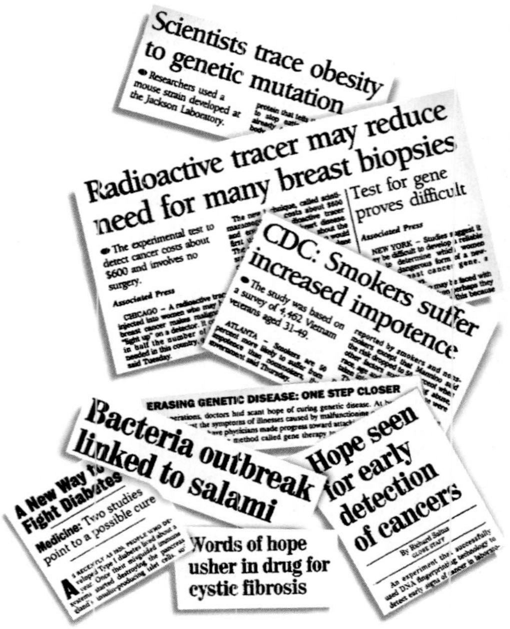
Scientists trace obesity to genetic mutation
Radioactive tracer may reduce need for many breast biopsies
Test for gene proves difficult
CDC: Smokers suffer increased impotence
ERASING GENETIC DISEASE: ONE STEP CLOSER
Bacteria outbreak linked to salami
Hope seen for early detection of cancers
A New Way to Fight Diabetes
Words of hope usher in drug for cystic fibrosis

FIGURE P.2 Headlines from recent newspapers and magazines testify to the significance of biological knowledge in our lives.

blood and digestion. Several organs with closely coordinated functions form an **organ system,** for example, the digestive system. Figure P.4 presents these and additional levels of biological organization.

Understanding the components of each organ system contributes to an understanding of how the body operates normally, how it operates in disease, and how it ages. We'll discuss each organ system in the following chapters. However, before considering the operation of the body, we want to consider three general topics: (1) the diversity of the different forms of life on the Earth, (2) the characteristics that distinguish all forms of life from nonliving things, and (3) the scientific method that generates and verifies our ideas about the natural world (living and nonliving things).

> Knowledge about the human body in health, disease, and aging has tangible benefits. The many cells of the human body are organized into tissues, organs, and organ systems.

The Diversity of the Living World

We do not live by ourselves or only for ourselves. The human body interacts with the outside environment, which includes both living and nonliving things. Each human is part of a **population** of humans that interacts with populations of other living things to form a **community.** Communities are organized into **ecosystems** (see Figure P.4).

Every minute of our lives we are dependent on other living things and on the physical environment for our well-being. All the food we eat is directly and indirectly a product of plants and the chemical processes by which plants use sunlight to construct food from water and the gas carbon dioxide.

Because humans are only one species out of the approximately 1.5 million species scientists have discovered, we will take a brief look at the diversity of life by

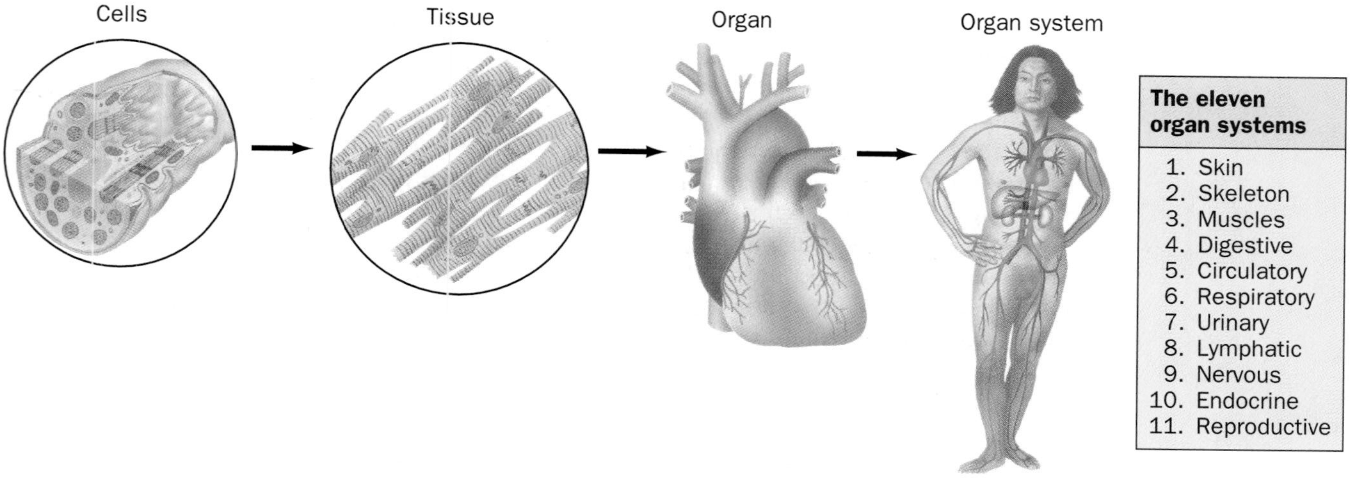
Cells
Tissue
Organ
Organ system
The eleven organ systems
1. Skin
2. Skeleton
3. Muscles
4. Digestive
5. Circulatory
6. Respiratory
7. Urinary
8. Lymphatic
9. Nervous
10. Endocrine
11. Reproductive

FIGURE P.3 The human body consists of distinct interactive components. Organ systems contain several organs, and each organ is made of tissues, which are composed of cells. Cells are the smallest living unit, and there are trillions of living cells in the human body.

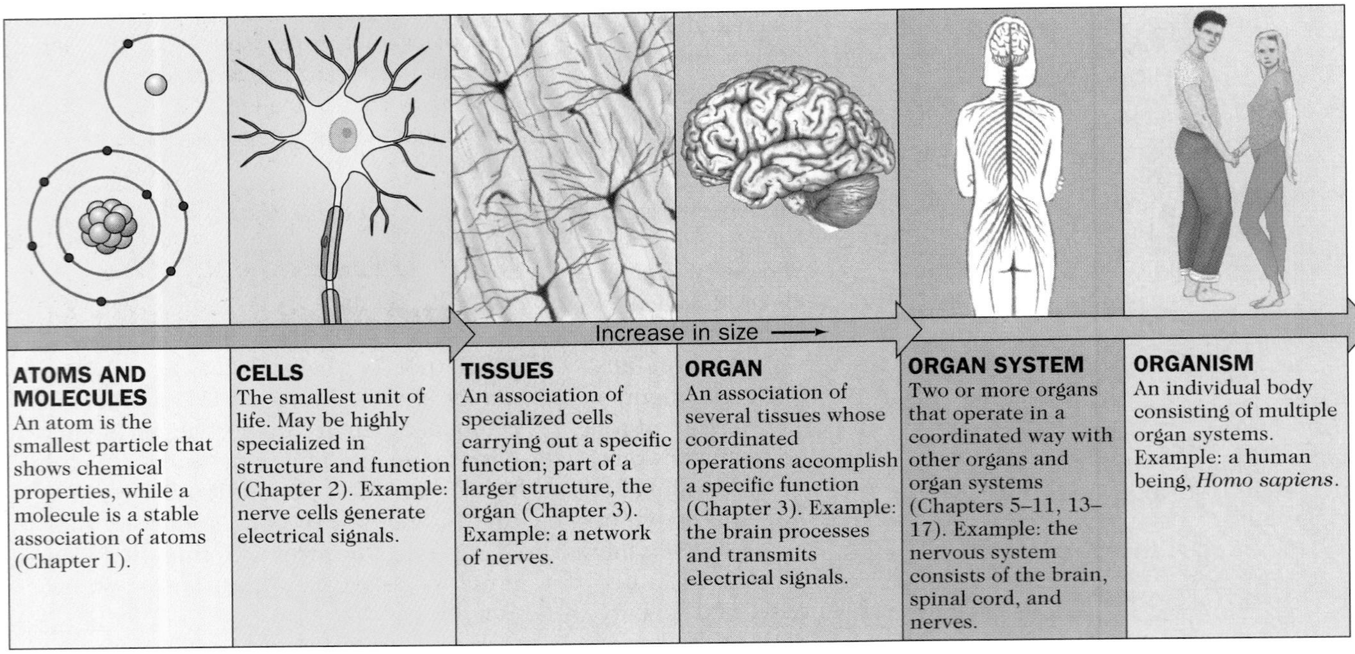
Increase in size
ATOMS AND MOLECULES
An atom is the smallest particle that shows chemical properties, while a molecule is a stable association of atoms (Chapter 1).
CELLS
The smallest unit of life. May be highly specialized in structure and function (Chapter 2). Example: nerve cells generate electrical signals.
TISSUES
An association of specialized cells carrying out a specific function; part of a larger structure, the organ (Chapter 3). Example: a network of nerves.
ORGAN
An association of several tissues whose coordinated operations accomplish a specific function (Chapter 3). Example: the brain processes and transmits electrical signals.
ORGAN SYSTEM
Two or more organs that operate in a coordinated way with other organs and organ systems (Chapters 5–11, 13–17). Example: the nervous system consists of the brain, spinal cord, and nerves.
ORGANISM
An individual body consisting of multiple organ systems. Example: a human being, *Homo sapiens*.

FIGURE P.4 Biological organization. The human body is formed from smaller structures. However, the individual human is also part of larger structures: populations, communities, and ecosystems.

identifying the five kingdoms into which all living things are usually classified.

After studying the extraordinary diversity of organisms, scientists have distinguished five major groups of organisms called **kingdoms** (Figure P.5). The five kingdoms are (1) Monera, (2) Protista, (3) Fungi, (4) Plantae (plants), and (5) Animalia (animals). Each kingdom contains thousands of distinguishable types of organisms called **species.** Although the different species in each kingdom show extraordinary diversity in both size and form, they all share certain characteristics that distinguish them from species in the other kingdoms.

The **Monera** are the smallest, simplest, and oldest living things on Earth. The Monera include the bacteria, whose beneficial and harmful effects we'll discuss further in Chapter 12. Most of the Monera absorb food molecule by molecule directly from their environment. They usually exist as single cells with very simple internal structures. This pattern of a simple internal structure distinguishes the Monera from the cells of the other four groups.

The Protista, Fungi, plant, and animal kingdoms all contain cells with more internal structure than the Monera. We'll discuss internal structures further in Chapter 2. These four kingdoms are believed to have arisen much later than did the Monera, and they contain a great many multicellular forms.

The **Protista** include single cells that usually obtain their food by preying on other living cells, but a few species use the energy of sunlight to manufacture food by means of photosynthesis. A few species cause human diseases. For example, the parasites that cause malaria and the form of pneumonia associated with HIV infection are Protista (Figure P.6). Some Protista, such as amoebas (*ah-me-bahs*) and ciliates (***sill**-ee-ates*), are capable of locomotion.

The **Fungi** include molds, yeasts, and mushrooms. Some live as single cells (yeasts), while others are multicellular (molds and mushrooms). By living off dead and decaying material, Fungi contribute to the recycling of raw materials in the environment. Other species of Fungi (the yeasts) are useful in baking,

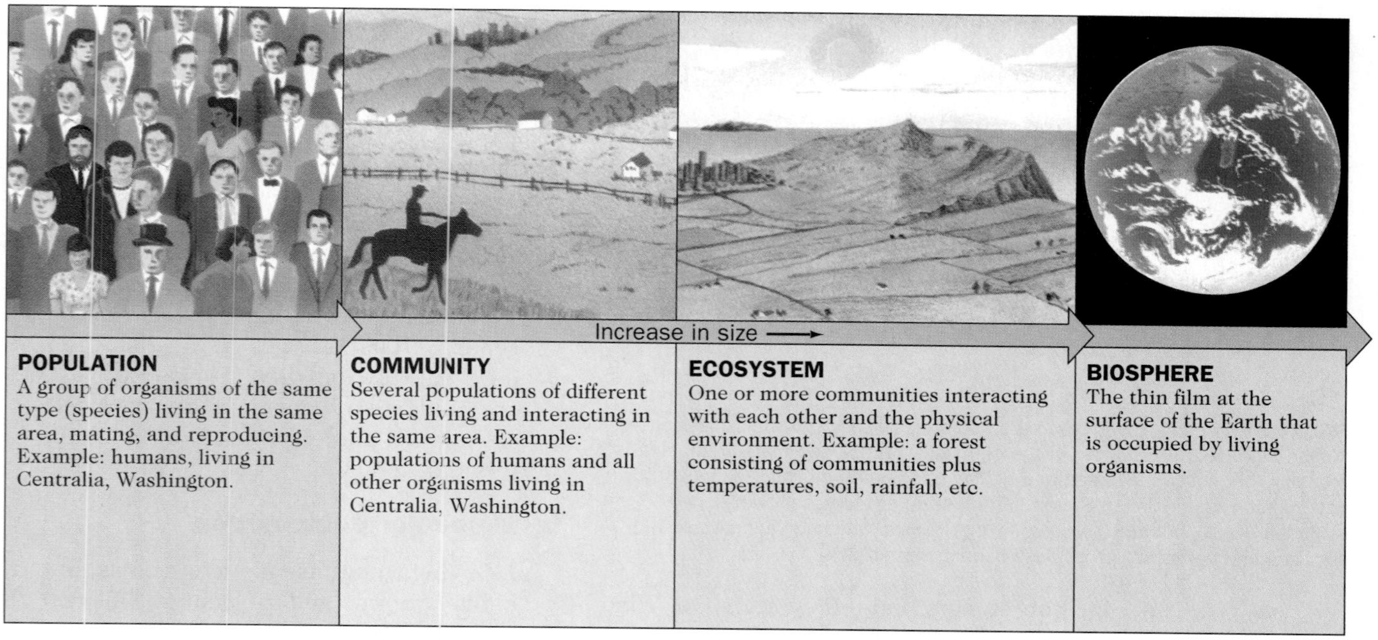
Increase in size →
POPULATION
A group of organisms of the same type (species) living in the same area, mating, and reproducing. Example: humans, living in Centralia, Washington.
COMMUNITY
Several populations of different species living and interacting in the same area. Example: populations of humans and all other organisms living in Centralia, Washington.
ECOSYSTEM
One or more communities interacting with each other and the physical environment. Example: a forest consisting of communities plus temperatures, soil, rainfall, etc.
BIOSPHERE
The thin film at the surface of the Earth that is occupied by living organisms.

FUNGI
PLANTS
ANIMALS
PROTISTA
MONERA

FIGURE P.5 The diversity of living systems and the five kingdoms. More than 1.5 million species have been identified and described. Each species is different in some way from every other species. These different species are classified into five kingdoms.

brewing, and wine production. Some species of molds and yeasts can cause diseases in humans (Chapter 12).

Plants vary greatly in size and complexity, but all contain the green pigment chlorophyll and carry out **photosynthesis.** Photosynthesis is a chemical process by which the energy of sunlight is captured in chemical substances synthesized from carbon dioxide and water. The substances synthesized and stored by plants serve as sources of energy and raw materials for organisms in the other four kingdoms. Our lives are a gift from green plants and sunlight. Because plants produce nutrients, they are referred to as **producers,** and all the other groups are referred to as **consumers** (Figure P.7).

Animals vary tremendously in size and form, but all are multicellular. As consumers, some animals feed directly on plants, while others live off plants indirectly by feeding on other animals (prey) that have fed on plants. Animals usually eat food in large pieces and use specialized internal structures to digest, absorb, and transport food molecules throughout their bodies. Nearly all animals have specialized structures for movement and responsiveness to the environment.

As different in size, shape, and lifestyle as these species are, they all have some-

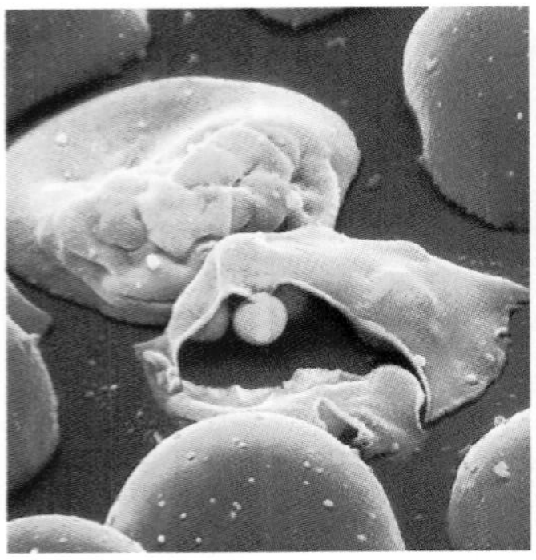

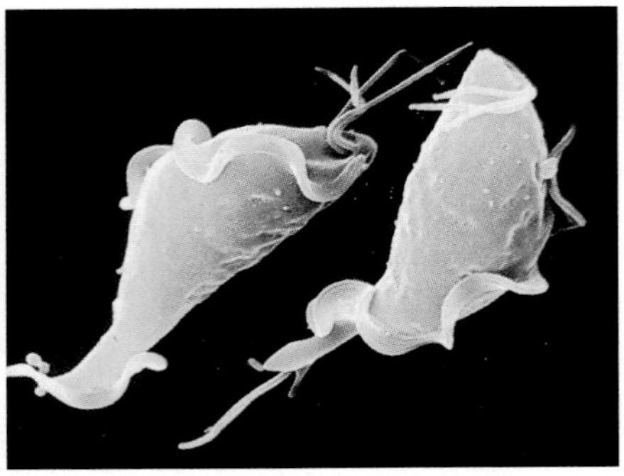

(a) (b)

FIGURE P.6 Among the Protista are some single-cell organisms that cause human diseases. (a) A photo of the organism that causes malaria. Malaria is probably the most widespread and debilitating human disease, although it is not common in the United States and most other industrially developed countries. (b) A photo of the parasite that causes pneumonia in people with poor immune defenses, particularly people with AIDS.

thing in common—they are alive. What does it mean to be alive? Can we define "life"? What are the characteristics all living things possess?

> There are five kingdoms. Among them, Monera have the smallest and simplest cells. Plants are producers and perform photosynthesis. Animals are multicellular consumers.

The Characteristics of Life

It is impossible to give a simple one-sentence definition of life that applies to all examples. Life is easier to identify than to define. For that reason, it makes more sense to list several characteristics that are always or nearly always present when we identify something as living. Seven general characteristics distinguish the living from the nonliving (Figure P.8). These are fundamental concepts that you will see applied and elaborated in the chapters that follow.

Molecular Composition

All living things are made of the same kinds of atoms and molecules. This fact underlies the relatedness of all forms of life. The same kinds of molecules (proteins, carbohydrates, lipids, and nucleic acids) are found in all forms of life from the smallest bacterium to the largest plant (the giant sequoia tree), from the flea to the human. While the kinds of molecules are the same, the individual variations in molecules account for the

FIGURE P.7 Plants use water, carbon dioxide, and the energy in sunlight to produce food. This stored food is what animals, including humans, use to sustain their lives. Some animals (herbivores) eat plants directly, while others (carnivores) live off plants indirectly by feeding on other animals.

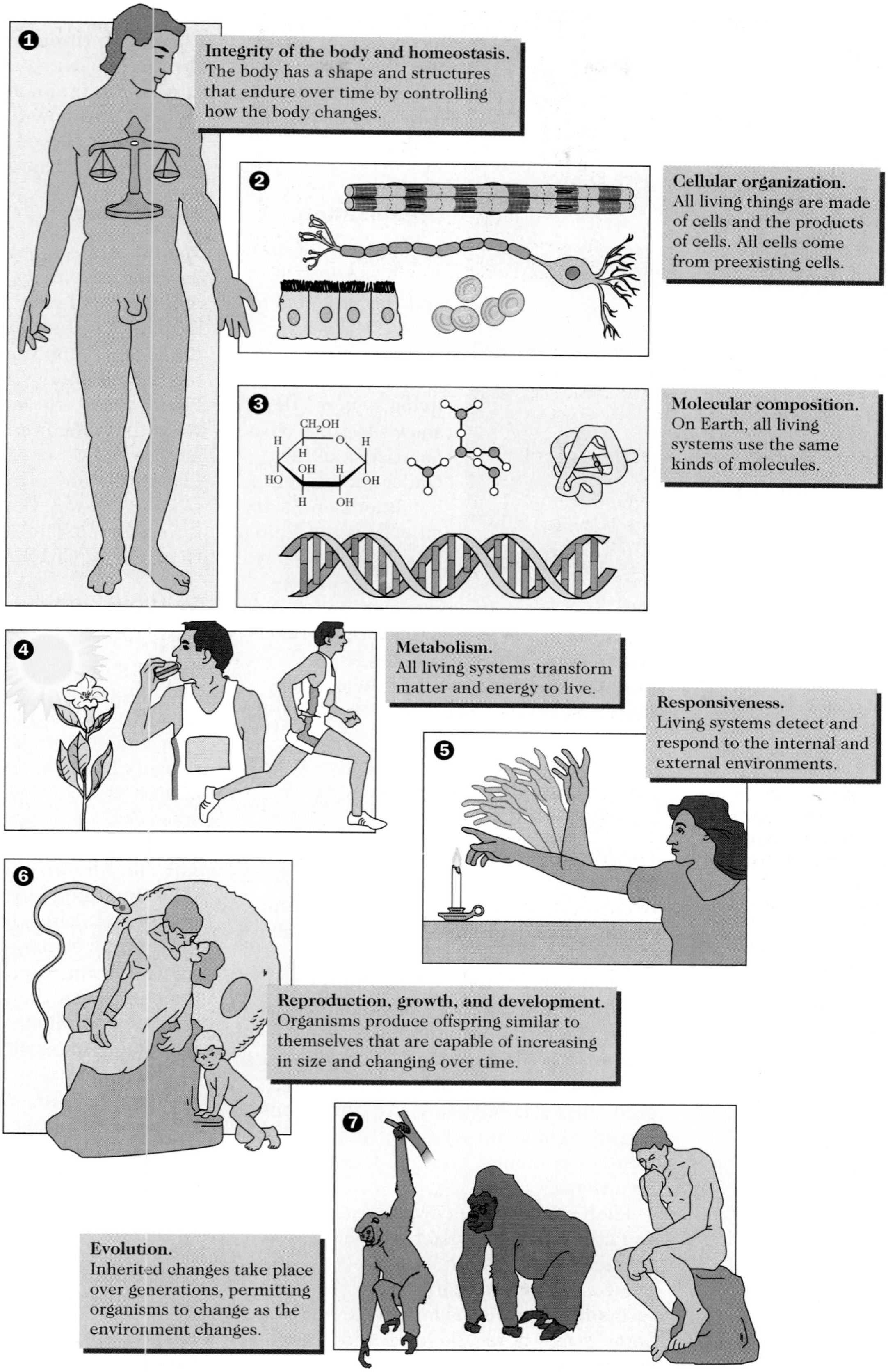
1
Integrity of the body and homeostasis.
The body has a shape and structures that endure over time by controlling how the body changes.
2
Cellular organization.
All living things are made of cells and the products of cells. All cells come from preexisting cells.
3
CH2OH
H
O
H
H
OH
H
HO
OH
H
OH
Molecular composition.
On Earth, all living systems use the same kinds of molecules.
4
Metabolism.
All living systems transform matter and energy to live.
Responsiveness.
Living systems detect and respond to the internal and external environments.
5
6
Reproduction, growth, and development.
Organisms produce offspring similar to themselves that are capable of increasing in size and changing over time.
7
Evolution.
Inherited changes take place over generations, permitting organisms to change as the environment changes.

FIGURE P.8 The characteristics of life. In general, living systems show most of these characteristics. However, individual organisms do not show the seventh characteristic: evolution. Evolution occurs over many generations and is not observed in the life span of an individual human being. Furthermore, within the human body, some specialized cells do not reproduce yet show the other five characteristics.

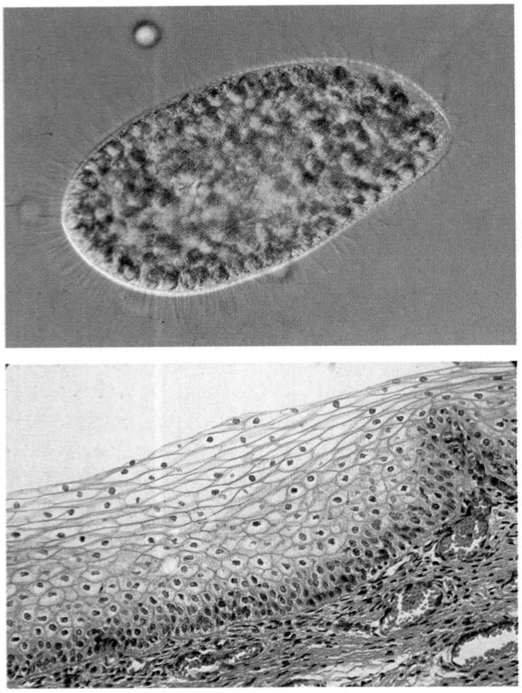

FIGURE P.9 All living things are made of cells. Some organisms are single cells, while other are multicellular, consisting of many cells specialized for certain functions.

diversity among the forms of life. Chapter 1 will focus on the atomic and molecular foundations of life.

Cellular Organization

All living things are composed of cells and the products of cells (Figure P.9). Cells are the smallest structures in living systems that show the other characteristics of life. Under the present conditions on Earth, all cells come from preexisting cells. Some organisms exist as single cells, but many large organisms, such as humans, are **multicellular.** Chapters 2 and 3 focus on the structure and functions of human cells and describe how they associate, forming tissues and organs.

Integrity of the Body and Homeostasis

The living body has structures that endure conditions that change over time. Whether the body is a single cell or multicellular, all living things have a surface barrier that protects the interior and helps maintain interior conditions that are different from those in the outside environment. **Homeostasis** (*home-ee-oh-**stay**-sus*) is the capacity of living systems to maintain stable internal conditions even when the outside environment undergoes significant changes. Chapter 4 focuses on human skin and its role in homeostasis.

Each organ has its own size, shape, and characteristics. Each also has its own balanced state, which contributes to the homeostasis of the entire body. We'll describe the structures, functions, and homeostasis of the skeleton and muscles, digestive system, circulatory system, respiratory system, and urinary system in Chapters 5 through 11. Then we'll look at the direct agents of homeostasis—the nerves, the brain, and secreted hormones—and how these agents operate in controlling reproduction, development, and birth in Chapters 14 through 17.

Metabolism

Metabolism is a collective term used to describe all the chemical transformations of matter and energy that occur in cells to maintain their integrity. All living things must continuously obtain raw materials and energy from the environment. Plants absorb the energy of sunlight and store that energy in the food they manufacture. All other forms of life ultimately live off the chemical energy stored by plants. The way the human body performs these chemical transformations is the focus of Chapters 6 and 7.

Responsiveness

Responsiveness is the capacity to detect and respond to changes in the internal and external environments. Step on a tack or touch a hot stove and you will respond rapidly to protect yourself from dangers in the external environment. Of course, there are also more subtle responses: In a cool breeze, you may put on a sweatshirt; at the smell of food, you salivate. All forms of life are responsive to their environments, avoiding danger and identifying mates, food, and shelter.

In multicellular organisms such as humans, responsiveness extends to detecting changes in the internal environment and the coordinated responses that maintain stable internal conditions. The roles of the human sense receptors, nervous system, and endocrine system are explored in Chapters 14 through 16. Responses to infectious agents and foreign materials are accomplished by the lymphatic system and immunity, as discussed in Chapter 13.

Reproduction, Growth, and Development

Living things undergo many changes that have been programmed in the hereditary material, the DNA. Organisms grow in size. They change in form, functions, and behavior. Living things reproduce, and in this process "like always reproduces like." Cats produce kittens, dogs produce puppies, and humans produce infant humans. This capacity to reproduce is coded into the DNA that is passed between generations. Human reproduction, development, and genetics are explored in Chapters 17 through 19.

Evolution

The forms of life change over generations. The DNA carried in sperm and eggs links the generations. Variation in DNA produces changes that may enhance survival and lead to successful reproduction. The environment acts on this variation to select the forms most adapted for survival and reproduction. In over 3.5 billion years, this process has created the abundance and diversity of life on Earth. Evolution provides a scientific explanation for the origin of humans. It offers explanations for our similarities to other forms of life as well as for our differences.

While all seven of these characteristics are necessary to describe living organisms, they do not all apply to every living cell in the human body. For example, human nerve cells are alive but do not reproduce or evolve.

> Seven characteristics describe living systems. Most, but not all, of these characteristics apply to the cells that make up living bodies. Homeostasis is the capacity to maintain a stable internal environment when the outer environment changes.

Science: A Way of Knowing

We have defined biology as the study of life. Now we'll ask: How is this "study" carried out? What general method do we apply to the study of the human body? How are ideas about the body verified? What is the scientific method, and what are its limits? You will discover that critical thinking is not only for scientists. In fact, the scientific method provides a practical tool for making many decisions.

The Nature of Science

Science is both a collection of organized information and a system of attitudes and methods that produce that information. Most of what you will read in this book is science in the first sense: organized information. However, you also need to understand the scientific method and the personal attitudes that support it. First, let's look at some of the personal attitudes and mental activities that contribute to critical thinking.

Thinking critically. The process of science rests on the **critical thinking skills** of the human beings we call scientists. These skills include reason, an open mind, measurement, and the making of comparisons and contrasts. Using **reason,** scientists seek causes for phenomena observed in the material world, and those **causes** must also belong to the material world. It is reason that requires people to seek the simplest and most direct causes. Reason rejects the use of obscure and complicated causes that cannot be measured or tested.

Scientists need to keep an **open mind** toward all observations whether or not they fit in with traditional, authoritative, or preconceived ideas. To operate as a scientist, one must be willing to abandon ideas that are contradicted by new observations, measurements, and experiments. With **measurement,** scientists use numbers to describe, compare, and contrast what is observed in the natural world. "How much?" is as important to scientific thinking as is "What kind?"

In making **comparisons and contrasts,** the scientist tries to see how things are similar (compare) and how they differ (contrast). This activity is required for descriptions and classification. Making comparisons and contrasts plays a central role in the effort to understand the diversity of living forms on Earth.

In addition to these skills, a scientist needs imagination and intuition. **Imagination** is the ability to form a mental image.

It is the capacity to "see" what might be true. **Intuition** is the capacity to understand something without reasoning. On first consideration, these might not seem to be appropriate skills for a scientist, because they appear to relate to a nonmaterial mental world. However, when we read scientists' accounts of their discoveries, we hear about the roles of imagination, intuition, and inspiration. Of course, these skills are balanced by critical thinking and the application of the scientific method. A scientist must be ready to abandon his or her intuition when it is contradicted by facts.

The scientific method. Stated in the simplest terms, the scientific method is a method of obtaining proof through observation, questioning, hypothesis formation, and hypothesis testing (often through experimentation).

All scientific investigation begins with an **observation** of the natural world (Table P.1). On the basis of this observation, the scientist asks a **question.** The scientist then proposes a **tentative answer** to the question. This tentative answer is the **hypothesis,** and it must be tested. **Hypothesis testing** is done by making further observations or conducting experiments. The results of such testing either support or contradict the hypothesis. If it is contradicted, the hypothesis is discarded and a new one is generated (see Table P.1).

Testing a hypothesis. An **experiment** is a test of a hypothesis under controlled conditions chosen or created by a scientist. The experiment usually involves an examination of two groups or situations which differ in only one way. This single difference is the factor being tested. Examine Figure P.10 carefully to see how an experiment is carried out, observations are made, and conclusions are drawn.

Sometimes it is not possible to do an experiment with human beings; in this case other observations are the only way to test a hypothesis. To see how this method works, let's apply it to a consideration of how the virus that causes AIDS (HIV) is transmitted (Figure P.11).

Starting with the knowledge that a virus causes AIDS (our initial "observation"), we can ask a question: How is the virus transmitted from person to person? One hypothesis could be that *the virus is transmitted by casual contact,* by touching an infected person or contacting surfaces and solid objects used by an infected person (for example, handshakes, doorknobs, pencils, money). How can we test this hypothesis?

It would be unethical to conduct an experiment that risked infecting healthy people by exposing them to the HIV virus, and so an alternative route must be considered.

We can test this hypothesis by asking another question: Is the pattern of relationships among HIV-infected people different from the pattern of a viral disease

TABLE P.1 The Scientific Method

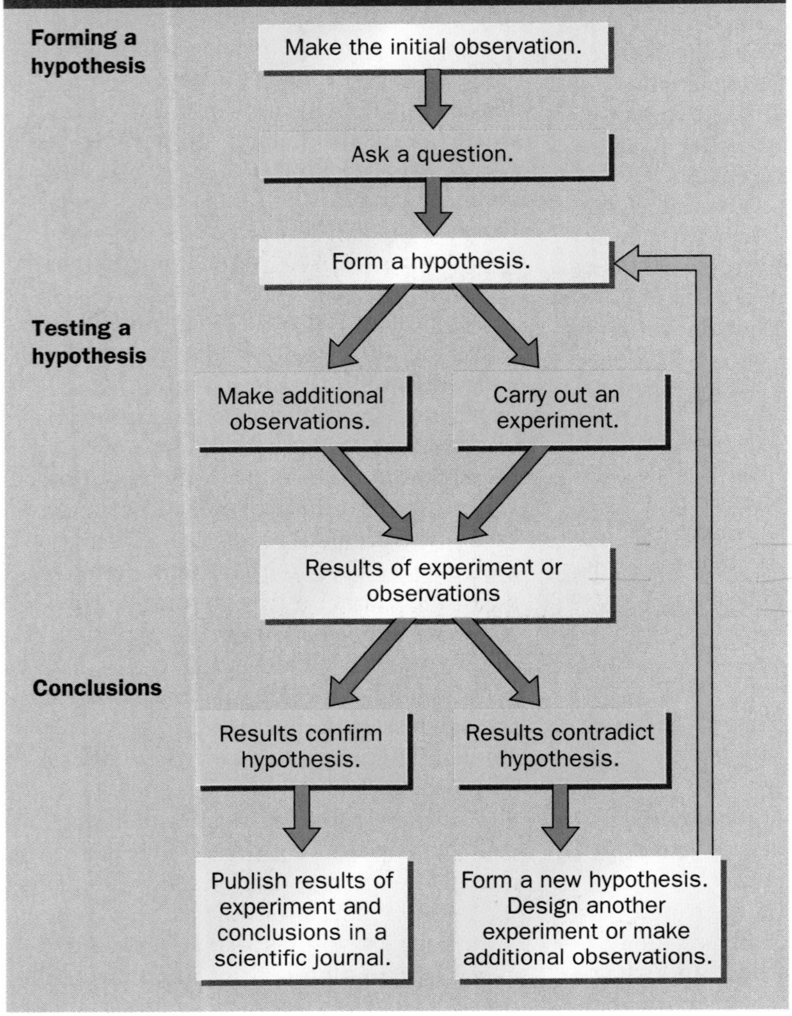
Forming a hypothesis
Make the initial observation.
Ask a question.
Form a hypothesis.
Testing a hypothesis
Make additional observations.
Carry out an experiment.
Results of experiment or observations
Conclusions
Results confirm hypothesis.
Results contradict hypothesis.
Publish results of experiment and conclusions in a scientific journal.
Form a new hypothesis. Design another experiment or make additional observations.

(a)

An experiment is a carefully designed test of the hypothesis. An experiment begins with a single difference between two otherwise identical situations; an unaltered "control" is contrasted to the "experiment" that has been altered in only one way.

At the end of the experiment, the two situations are examined for distinguishing changes that have occurred. Any observed changes can be attributed to the single alteration that initially distinguished the "control" from the "experiment."

In a well-designed experiment, the changes that occur either confirm or deny the hypothesis.

Initial observation:
Some drugs inhibit reproduction in some animal viruses.

Ask a question:
"Can drug X inhibit the reproduction of the human immunodeficiency virus (HIV)?"

Form a hypothesis (tentative answer):
"Drug X inhibits HIV reproduction in human white blood cells, the T4 cells."

Perform an experiment:
White blood cells containing the virus are collected from a patient, and identical numbers are placed in two identical tubes. The drug is added to the tube labeled "experiment" (E). No drug is added to the "control" (C).

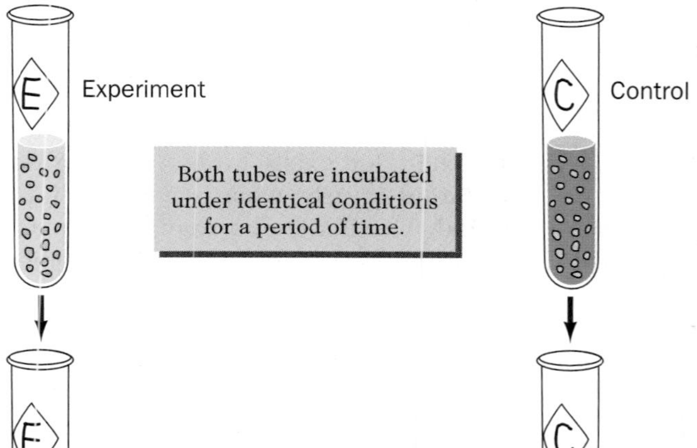
E
Experiment
C
Control
Both tubes are incubated under identical conditions for a period of time.
E
C
After incubation, the cells are collected and the virus particles are extracted and counted.

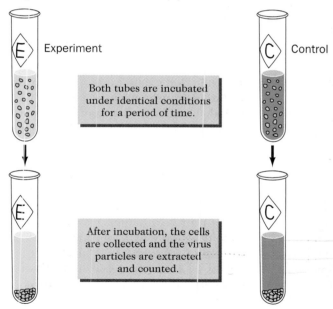

Results:
There are fewer HIV in the experiment tube than in the control tube.

Conclusion:
Drug X inhibits HIV reproduction in these cells. The hypothesis is confirmed.

New hypothesis:
"Drug X prevents HIV reproduction in whole animals, not just cells isolated in a test tube."

(b)

You can use the scientific method every day. For example, when your computer doesn't respond, you ask, "Why didn't my computer turn on when I pushed the switch?" This initial question leads you to pose a tentative answer (the hypothesis): "The computer is unplugged." You then proceed to test this hypothesis.

The results of this observation may confirm the hypothesis ("the computer is unplugged") or may contradict the hypothesis, requiring you to form a second hypothesis.

Initial observation:
Nothing happens when the switch is turned on.

Ask a question:
"Is the electricity reaching the computer?"

Form a hypothesis:
"The computer is unplugged."

Perform an experiment by making an observation:
Compare the location of the computer plug to the location of the plug for the lamp which is operating.

Results:
The computer is not unplugged.

Conclusion:
The hypothesis is contradicted. Failure to operate is not due to the computer plug. You need to form and test a new hypothesis.

New hypothesis:
The switch is broken.

FIGURE P.10 Making and interpreting an experiment. (a) An experiment may be conducted in the laboratory under controlled conditions. (b) The scientific method can be applied to everyday problems and situations.

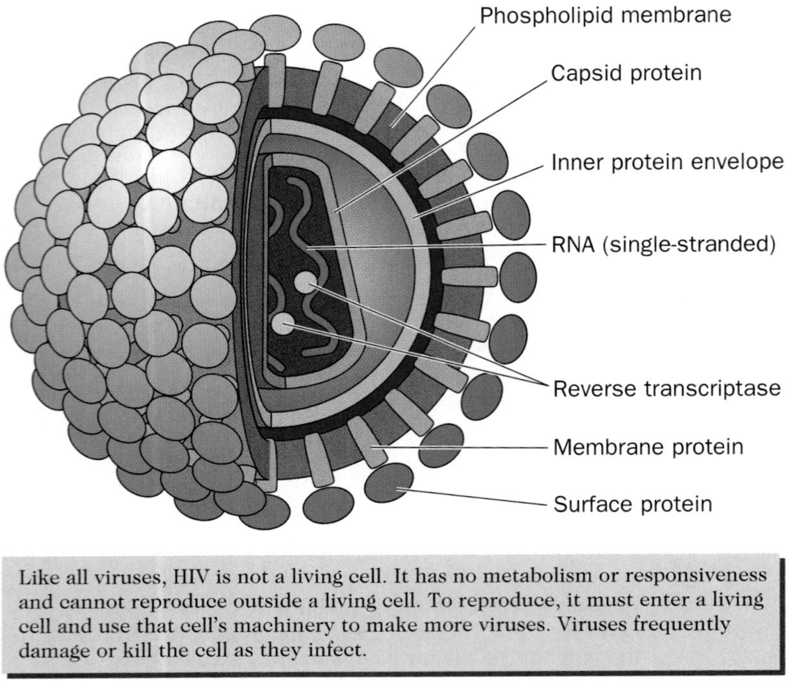
Phospholipid membrane
Capsid protein
Inner protein envelope
RNA (single-stranded)
Reverse transcriptase
Membrane protein
Surface protein
Like all viruses, HIV is not a living cell. It has no metabolism or responsiveness and cannot reproduce outside a living cell. To reproduce, it must enter a living cell and use that cell's machinery to make more viruses. Viruses frequently damage or kill the cell as they infect.

FIGURE P.11 The human immunodeficiency virus (HIV) causes the disease AIDS (Acquired Immune Deficiency Syndrome).

that is transmitted by casual contact, for example, the disease pattern for cold viruses?

Cold viruses can be picked up from surfaces contaminated by an infected person. Cold symptoms appear randomly and widely among people who have not had direct contact with each other. We'll call this the *random pattern* of infection. This pattern differs sharply from the pattern observed for HIV infection. HIV infection and AIDS show a *selected pattern.* The AIDS disease appears in groups of people who have had sexual contact with each other or have shared blood in some way, for example, through transfusions or unclean hypodermic needles.

From these observations, we can discard the hypothesis of transmission by casual contact and form a new hypothesis: *The virus that causes AIDS (HIV) is transmitted by direct contact with body fluids, usually blood, semen, vaginal fluid, or breast milk.* This new hypothesis can be tested in a couple of ways: (1) by comparison of the patterns of infection with blood-borne viruses such as the hepatitis virus and (2) by direct examination of vaginal fluid, blood, and semen for the HIV virus. Both tests support the second hypothesis. [AIDS is discussed more completely in Spotlight on Health: Five (page 354).]

Everyone can use the scientific method. The attitudes and methods of science are the professional requirements for natural scientists such as biologists, chemists, and physicists. The scientific method is also used in one form or another by nearly all academic disciplines, including anthropology, psychology, sociology, history, and economics. In business, aspects of the scientific method are applied when one does "market research" or "test marketing" for a new product.

You can apply the scientific method every day to some parts of your personal life (see Figure P.10). We can all benefit by critical thinking. Do not believe everything you hear and read. Ask questions, find answers, and test the answers. However, although the scientific method is useful and productive, it does have limits.

The limits of science. Neither the body of scientific information nor the methods of science can answer certain questions. Science cannot determine whether God exists, whether something is beautiful or ugly, what is moral or immoral. These are all areas that depend on personal experience and value judgment. Science cannot say it is immoral to destroy vast rain forests or abort a human fetus. However, science can identify changes and predict consequences, and humans can form personal judgments about those consequences. Because science deals only with repeatable phenomena, brief, unpredictable, and unrepeated events such as "miracles" and "ghosts" cannot be investigated well by science. Therefore, we have to deal with these events in other ways.

The Benefits and Perils of Science and Technology

Although science has limits, the application of its methods has contributed to the transformation of human society and culture and the physical world in the last 300 years. While the products of science are information and understanding, science also gives people the power to effectively predict and control parts of the natural world. This application of knowledge is usually referred to as technology, but the distinction between science and technology is not always clear.

The application of science to practical problems has resulted in thousands of tangible benefits for humans. For example, we have seen improvements in the production and preservation of food for human consumption. Certainly improvements in medical care and sanitation (public health) have greatly extended the longevity of individuals in the more developed countries (Figure P.12). However, there are dark clouds on the horizon.

The many "gifts" of science and technology to humankind often come with unforeseen costs and consequences. High levels of production and consumption generate high levels of pollution, and while the production is controlled privately, the pollution is all too public: It affects all of us through the air we breath, the water we drink, and the food we eat. Improvements in the average life span and quality of life for humans (and other living things) are endangered by overpopulation, air pollution, and water pollution.

Some thinkers say that science and technology hold the solutions to these environmental problems. Others say that the solutions require a change in values and a shift away from the high-consumption and high-technology lifestyle that has generated many of these problems. Perhaps both science and changed values are required. It does seem clear that short-term solutions need to give way to long-term plans. Humans need to give up some short-term benefits in favor of longer-term benefits. And we must remember that we do not control nature. We are part of nature and must respect it and work with it.

While the many problems of technology and society are beyond the scope of this book, we believe that the study of human biology will make you better informed, enriching your personal life and the life of future generations.

In Chapter 1, we will begin with a brief examination of the properties of the atoms and molecules that make up living systems and the principles that govern their behavior. We cannot see these atoms and molecules, but they constitute us. An understanding of the dance of the atoms and molecules is one step toward an appreciation of how we are part of nature and how the body operates.

> Critical thinking uses an open mind, reason, and measurement and makes comparisons and contrasts. The scientific method consists of an initial observation, a question, a tentative answer (the hypothesis), and a test of the hypothesis by further observation or experimentation.

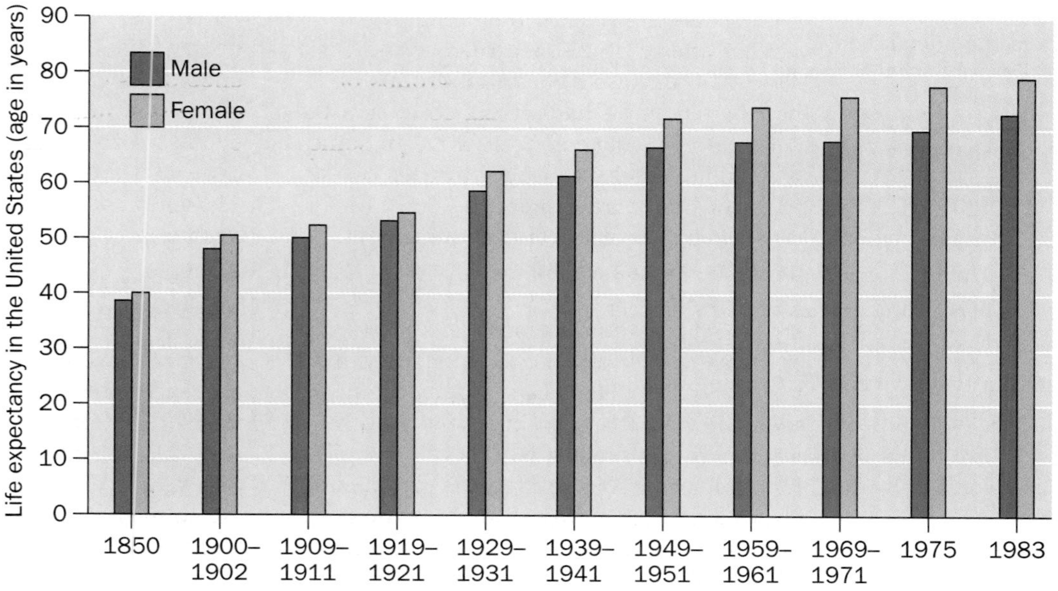
Life expectancy in the United States (age in years)
90
80
70
60
50
40
30
20
10
0
Male
Female
1850
1900–1902
1909–1911
1919–1921
1929–1931
1939–1941
1949–1951
1959–1961
1969–1971
1975
1983

FIGURE P.12 Life expectancy for males and females. Over the last 100 years there has been a slow but significant increase in the life expectancy of people in the United States and other developed countries. This increase is attributed to better public health, better nutrition, and generally improved living conditions. In the next century, will increasing levels of pollutants begin to reduce life expectancy?

Prologue Summary

Why Study Human Biology?

- Knowledge about the human body contributes to self-knowledge and benefits an individual's decision making.
- The human body is made up of trillions of tiny cells. Cells are associated into tissues, and tissues form organs. An organ system usually consists of several organs.

The Diversity of the Living World

- Populations of a living organism form communities with other populations. Communities form ecosystems.
- The diversity of life is classified into five kingdoms: Monera, Protista, Fungi, plants, and animals.
- In their internal structure, the Monera are simpler than and distinct from the other four kingdoms.
- Photosynthesis by producers provides the chemical energy and raw materials on which consumers depend.

The Characteristics of Life

- All living organisms are made of the same kinds of molecules.
- All living organisms are composed of cells.
- All living organisms maintain stable internal conditions (homeostasis).
- All living organisms chemically transform matter and energy (metabolism).
- All living organisms detect and respond to changes in the internal and external environments (responsiveness).
- All living organisms produce a new generation that changes to reach reproductive maturity (reproduction, growth, and development).
- Changes in DNA and its expression take place over many generations (evolution).

Science: A Way of Knowing

- Critical thinking requires an open mind and uses reason to identify causes.
- Measurements help us make comparisons and contrasts.
- The scientific method begins with an observation and a question.
- A hypothesis is a tentative answer. It must be tested by making further observations or conducting an experiment.
- In an experiment, an unaltered situation ("control") is contrasted to a second situation that has been altered in only one way ("experiment").
- At the end of the experiment, any observed changes can be attributed to the single alteration that distinguished the "control" from the "experiment."
- In a well-designed experiment, the changes that occur either confirm or contradict the hypothesis.

Selected Key Terms

cells (p. 2)	homeostasis (p. 8)	organ system (p. 3)	responsiveness (p. 8)
consumers (p. 5)	hypothesis (p. 10)	photosynthesis (p. 5)	scientific method (p. 10)
ecosystems (p. 3)	metabolism (p. 8)	population (p. 3)	species (p. 4)
experiment (p. 10)	organs (p. 2)	producers (p. 5)	

Review Activities

1. What role does knowledge of human biology play in your life?
2. List five tangible benefits to your life that come from a knowledge of human biology.
3. Identify and distinguish the seven characteristics of life. Which five of these characteristics describe nerve cells in the human body?
4. Describe what is meant by the term "homeostasis."
5. Describe and distinguish the five kingdoms into which all living things are classified.
6. Describe what is involved in critical thinking.
7. Describe the steps in the scientific method.
8. Describe the two ways in which hypotheses can be tested.
9. Describe the limits of science and the scientific method.
10. Identify some of the perils of science and technology.

Self-Quiz

Matching Exercise

___ 1. An association of cells, often showing specialized functions
___ 2. The maintenance of stable internal conditions
___ 3. Two characteristics of all living cells
___ 4. The kingdom with the simplest cells
___ 5. A tentative answer to a question
___ 6. One method of testing a hypothesis
___ 7. Two things necessary for critical thinking

A. Hypothesis
B. Reason
C. Homeostasis
D. Metabolism
E. Experiment
F. Multicellularity
G. Monera
H. Open mind

Answers to Self-Quiz

1. F; **2.** C; **3.** C, D; **4.** G; **5.** A; **6.** E; **7.** (*any order*) B, H

Chapter 1

Chemistry and the Human Body

Living is a great adventure, and to be alive is to change. One way to change is by learning more about yourself and about your body and its relationship to the environment. Although you feed it, exercise it, and sleep with it, your body is probably an intimate stranger. In the chapters that follow we'll help you change that. We'll ask questions and look for answers. We'll help you become more aware of how your body is constructed, how it operates, and how it changes.

Nothing is more characteristic of life than change (Figure 1.1). Chemistry is the science that explores changes in matter and energy, and the language of chemistry is used to describe and explain some of the changes that occur in your body. Knowing a little about chemistry will help you understand and sometimes control these changes. In this chapter, we'll present a

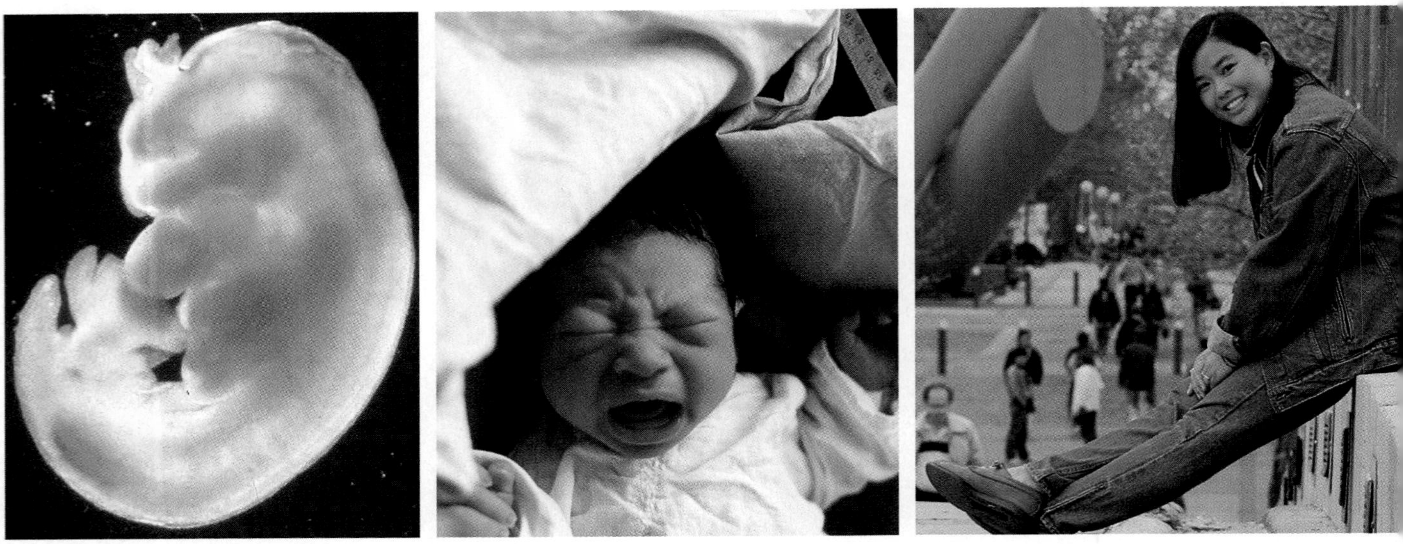

FIGURE 1.1 Chemistry can be used to describe the changes in matter and energy that take place in the human body.

few chemical concepts necessary for an understanding and appreciation of the operation of your body and its relationship to the environment.

Matter, Energy, and the Dance of Life

You are part of the physical universe. You are the stuff of rocks and wind and salt and sea. Like the stars, you are matter and energy. Scientists define **matter** as anything that has mass and takes up space. Because of the Earth's gravity, people experience mass as weight. There are different kinds of matter, and matter takes different forms. A solid rock, a flowing river, and a blowing wind represent different forms of matter—solids, liquids, and gases.

Energy can be defined as the capacity to make some change in matter. We experience energy in the sunlight that warms our skin and the heat of a fire that turns liquid water into a gas. We have some experience of energy when we act to accomplish a task: to move something or make something. Science understands that energy cannot be created or destroyed; this fundamental physical principle is called the **conservation of energy.** However, energy can be converted from one form to another. The activities of the human body carry out some of these conversions.

Converting Energy and Making Changes

For the scientist, energy exists in two forms: potential energy and kinetic energy. **Potential energy** is stored energy; examples include the chemical energy in food and the electrical energy stored in a battery or a nerve. Familiar examples of potential energy are a boulder at the top of a hill, the water stored behind a dam, and a diver on a platform. As each moves down, its potential energy is converted into kinetic energy (Figure 1.2).

Kinetic energy is the energy of motion. As water flows from the dam, potential energy is converted into kinetic energy and work can be done. If the flow of water is coupled to an electrical generator, electricity can be used to light homes and drive machinery. More familiar to you is the potential energy in a match that can be converted to kinetic energy in the form of heat and light (see Figure 1.2). In the human body, the controlled release of the potential energy from food is used for the work of the body. This includes walking, talking, reading, and eating, but it also includes less obvious activities in your cells, for example, the secretion of substances, the production of nerve impulses, and the manufacture of DNA and proteins.

The release of energy within the body is essential to life. Although energy may be more difficult to point to, it is no less

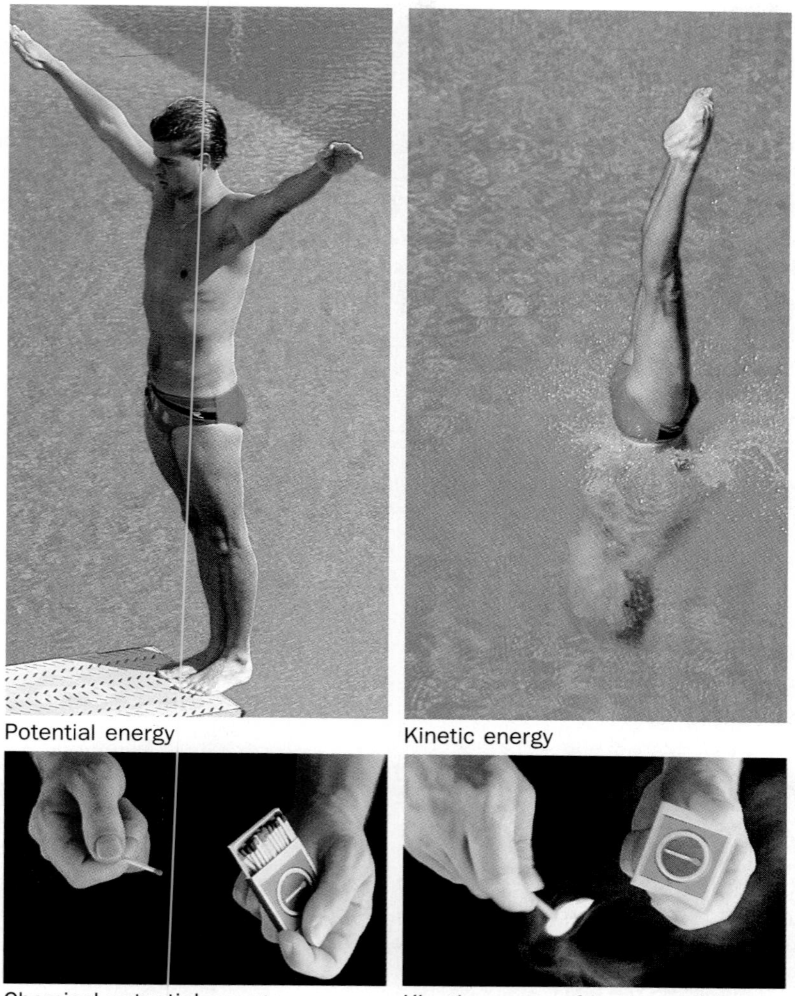
Potential energy
Kinetic energy
Chemical potential energy
Kinetic energy of heat and light

FIGURE 1.2 Potential energy can be converted to kinetic energy and used to do work.

real than matter. Like matter, energy can be measured, and these measurements can be used to understand changes in matter.

Rules for the dance of life. The structure and operation of our bodily organs in both health and disease involve changes in the particles of matter called atoms. Atoms are always in motion—vibrating, rotating, and moving from place to place if they are not held to other atoms. Atoms are like tireless dancers; they gyrate, collide, couple, and uncouple. Although their individual motions are random, there are rules to the dances they perform with other atoms. To understand the lives of your cells and the activities of your body, you need to know a little about the rules of the dance, some principles of chemistry.

Before continuing, consider these suggestions: (1) Fear of chemistry is not required. It is optional, and you don't need it. Look at chemistry as if it were merely a matter of dancers and their dance. The dancers represent matter and energy, while the dance represents the rules of chemistry: the laws and principles of chemistry. (2) As in all your studying, don't be overwhelmed. Read and study in small bites. Pause frequently to review what you've read. Make notes and ask questions. (3) Make the unfamiliar familiar. Memorize the definitions of key terms and apply this understanding when you encounter those terms again.

When you apply these suggestions, both the chemical principles and their relevance will become clearer. Remember, energy makes things change. You have the energy; all you have to do is direct it. We'll begin our discussion of chemistry with a consideration of atoms, their structures, and their associations.

> Energy is the capacity for change, while matter is what changes. Energy can exist in different forms. Potential energy can be converted to kinetic energy and captured to do work.

Atoms: Particles of Matter

Matter exists as particles called atoms. An **atom** is the smallest unit of matter that can undergo chemical change. Atoms are so small that they cannot be seen in detail even with the most powerful electron microscope. The largest atom is about 0.00000004 cm in diameter, or to look at it another way, to span the distance of 1 centimeter (cm) [——] would require 400 million of the largest atoms. There are 109 different kinds of atoms (Figure 1.3a). However, only a few of these atoms are found in living systems. Indeed, only six different kinds of atoms make up 98 percent of the human body (Figure 1.3b). It will be easier to understand what you read if you memorize the names and symbols for the 11 atoms listed in Table 1.1, they will be used throughout the book. Each kind of atom has significantly different characteristics or properties.

Properties can be considered the "personality traits" of an atom. They permit

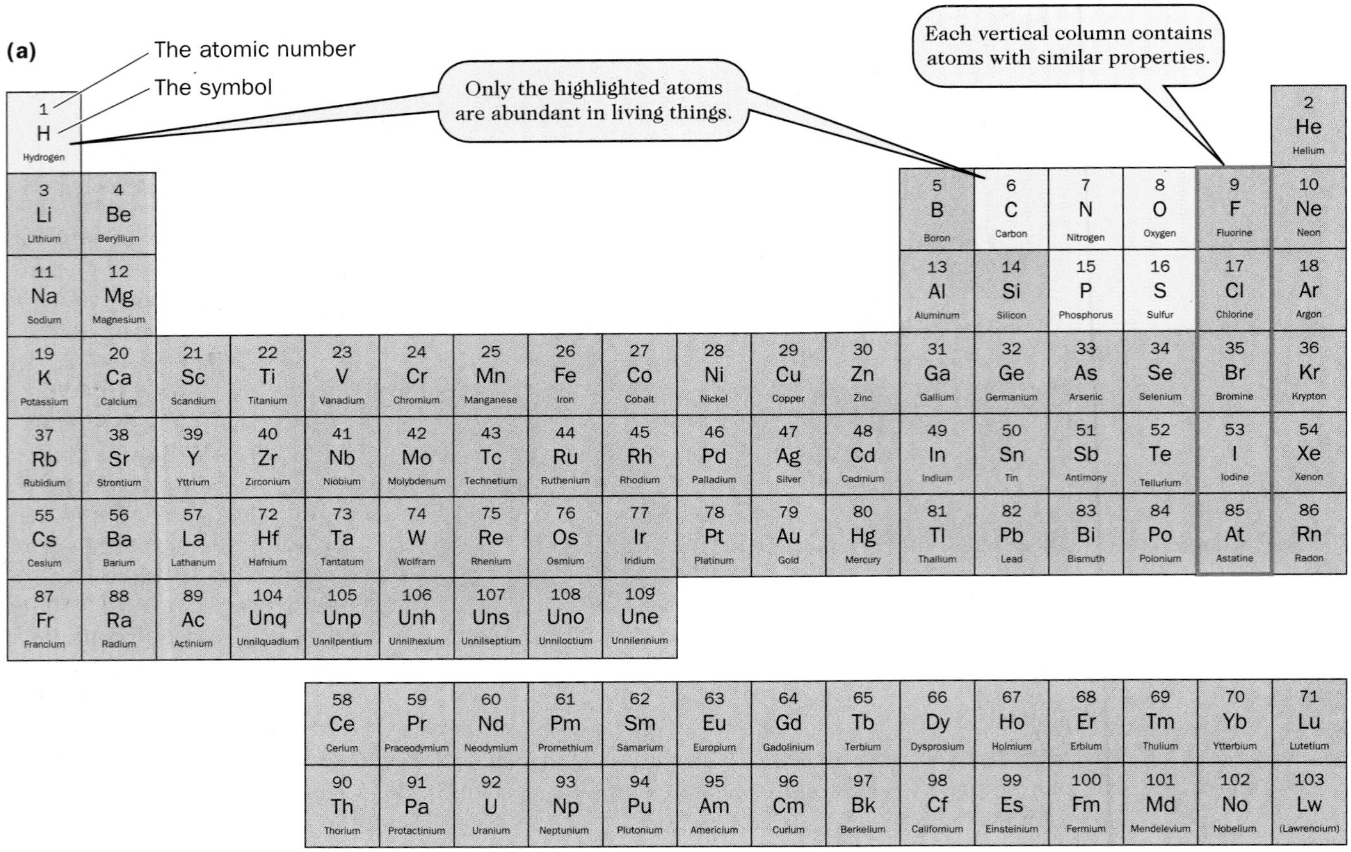
(a)
The atomic number
The symbol
Only the highlighted atoms are abundant in living things.
Each vertical column contains atoms with similar properties.
1 H Hydrogen
2 He Helium
3 Li Lithium
4 Be Beryllium
5 B Boron
6 C Carbon
7 N Nitrogen
8 O Oxygen
9 F Fluorine
10 Ne Neon
11 Na Sodium
12 Mg Magnesium
13 Al Aluminum
14 Si Silicon
15 P Phosphorus
16 S Sulfur
17 Cl Chlorine
18 Ar Argon
19 K Potassium
20 Ca Calcium
21 Sc Scandium
22 Ti Titanium
23 V Vanadium
24 Cr Chromium
25 Mn Manganese
26 Fe Iron
27 Co Cobalt
28 Ni Nickel
29 Cu Copper
30 Zn Zinc
31 Ga Gallium
32 Ge Germanium
33 As Arsenic
34 Se Selenium
35 Br Bromine
36 Kr Krypton
37 Rb Rubidium
38 Sr Strontium
39 Y Yttrium
40 Zr Zirconium
41 Nb Niobium
42 Mo Molybdenum
43 Tc Technetium
44 Ru Ruthenium
45 Rh Rhodium
46 Pd Palladium
47 Ag Silver
48 Cd Cadmium
49 In Indium
50 Sn Tin
51 Sb Antimony
52 Te Tellurium
53 I Iodine
54 Xe Xenon
55 Cs Cesium
56 Ba Barium
57 La Lathanum
72 Hf Hafnium
73 Ta Tantatum
74 W Wolfram
75 Re Rhenium
76 Os Osmium
77 Ir Iridium
78 Pt Platinum
79 Au Gold
80 Hg Mercury
81 Tl Thallium
82 Pb Lead
83 Bi Bismuth
84 Po Polonium
85 At Astatine
86 Rn Radon
87 Fr Francium
88 Ra Radium
89 Ac Actinium
104 Unq Unnilquadium
105 Unp Unnilpentium
106 Unh Unnilhexium
107 Uns Unnilseptium
108 Uno Unniloctium
109 Une Unnilennium
58 Ce Cerium
59 Pr Praceodymium
60 Nd Neodymium
61 Pm Promethium
62 Sm Samarium
63 Eu Europium
64 Gd Gadolinium
65 Tb Terbium
66 Dy Dysprosium
67 Ho Holmium
68 Er Erbium
69 Tm Thulium
70 Yb Ytterbium
71 Lu Lutetium
90 Th Thorium
91 Pa Protactinium
92 U Uranium
93 Np Neptunium
94 Pu Plutonium
95 Am Americium
96 Cm Curium
97 Bk Berkelium
98 Cf Californium
99 Es Einsteinium
100 Fm Fermium
101 Md Mendelevium
102 No Nobelium
103 Lw (Lawrencium)

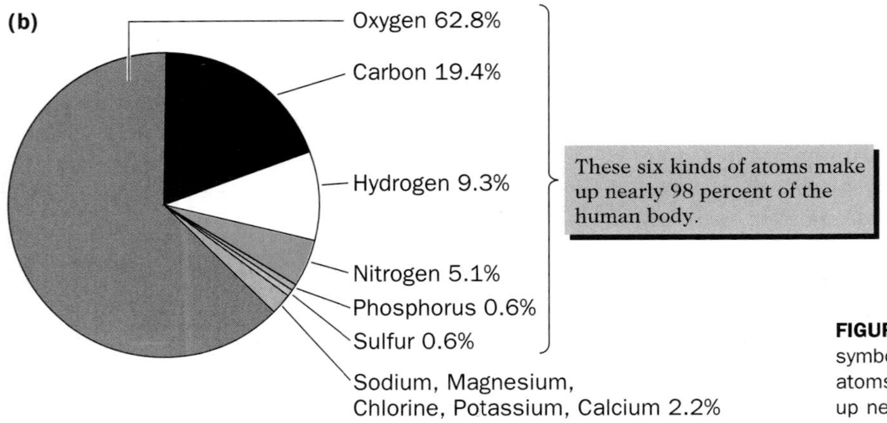
(b)
Oxygen 62.8%
Carbon 19.4%
Hydrogen 9.3%
Nitrogen 5.1%
Phosphorus 0.6%
Sulfur 0.6%
Sodium, Magnesium, Chlorine, Potassium, Calcium 2.2%
These six kinds of atoms make up nearly 98 percent of the human body.

FIGURE 1.3 (a) The periodic table showing the symbols and names for more than 100 kinds of atoms. (b) Notice that only six kinds of atoms make up nearly 98 percent of the human body.

TABLE 1.1 Important Atoms in the Human Body

Name	Symbol
Carbon	C
Hydrogen	H
Nitrogen	N
Oxygen	O
Phosphorus	P
Sulfur	S
Sodium	Na
Magnesium	Mg
Chlorine	Cl
Potassium	K
Calcium	Ca

us to identify, measure, and understand atoms and predict how they will behave in association with other atoms. Properties include things such as mass (weight) and combining capacity. Each atom has a different mass or weight. Collectively, human weight is the sum of the weights of all the atoms in the body.

Atoms also have characteristic **combining capacities** that indicate the ratios in which a specific kind of atom will bond with other atoms to form stable associations called **molecules.** Molecules are represented by **chemical formulas** that indicate the kind and number of atoms present. For example, water has the formula H_2O, indicating that 2 atoms of hydrogen (H) are bonded to 1 atom of oxygen (O).

The sugar glucose, which is present in human blood, has the formula $C_6H_{12}O_6$, indicating 6 atoms each of carbon (C) and oxygen (O) and 12 atoms of hydrogen (H). While these formulas are useful, they don't indicate the actual structure of molecules. They don't tell us which atoms are bonded to which other atoms and how all the atoms are arranged in three-dimensional space. Later in this chapter we will speak more about the structure and shape of molecules.

In discussing the composition and function of the human body, we will often speak of different kinds of molecules. Some of these molecules are small, consisting of only 2 or 3 atoms, such as oxygen gas (O_2) and water (H_2O). Other molecules are larger, with dozens of atoms; table sugar, or sucrose, is an example of such a molecule ($C_{12}H_{22}O_{11}$). Still other molecules are even larger, with hundreds or even thousands of atoms; the proteins that make up enzymes, muscles, and hair are examples.

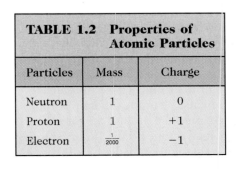

TABLE 1.2 Properties of Atomic Particles

Particles	Mass	Charge
Neutron	1	0
Proton	1	+1
Electron	$\frac{1}{2000}$	−1

To understand the changes that take place in the body, we need to examine in more detail the composition and characteristics of atoms. We'll ask and answer four questions: (1) What are atoms made of? (2) What is the structure or arrangement common to all atoms? (3) How do atoms differ from one another? (4) How does the structure of atoms help us understand how atoms bond together?

> Atoms are the smallest units of matter that undergo chemical change. Only six different kinds of atoms make up 98 percent of the human body. Each kind of atom has characteristic properties and can combine with other atoms.

What Are Atoms Made Of?

Atoms are made up of three tiny particles, the subatomic particles: protons, neutrons, and electrons. Each kind of atom has a characteristic number of protons, neutrons, and electrons. Each **proton** has a mass of 1 and an electrical charge of positive one (+1). Each **neutron** also has a mass of 1 but no electrical charge; neutrons are said to be neutral in charge. The **electron** has a very tiny mass (about 1/2000, or 0.00054, of a proton's mass) and an electrical charge of negative one (−1) (Table 1.2). As a general principle, different electrical charges attract each other, while the same electrical charges repel each other.

What Is the Structure or Arrangement Common to All Atoms?

Positively charged protons are clustered together with all the neutrons in the center of the atom, forming a structure called the **atomic nucleus** (plural, *nuclei*) (Figure 1.4). The atomic nucleus has all the positive charges and all the mass (weight) of an atom but takes up a very small amount of space. The diameter of an atom is about 10,000 times greater than the diameter of its nucleus.

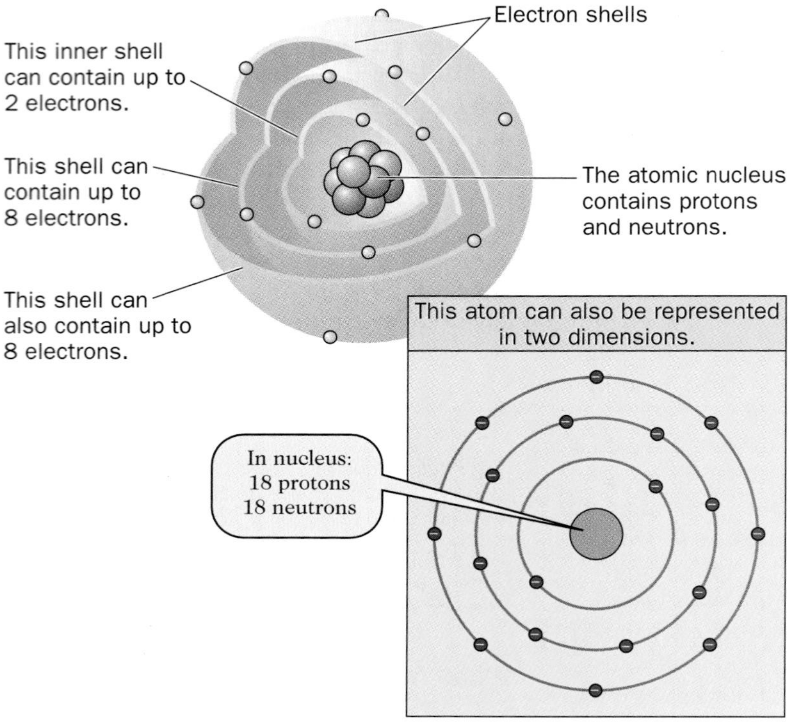
Electron shells
This inner shell can contain up to 2 electrons.
This shell can contain up to 8 electrons.
This shell can also contain up to 8 electrons.
The atomic nucleus contains protons and neutrons.
This atom can also be represented in two dimensions.
In nucleus: 18 protons 18 neutrons

FIGURE 1.4 The structure of atoms.

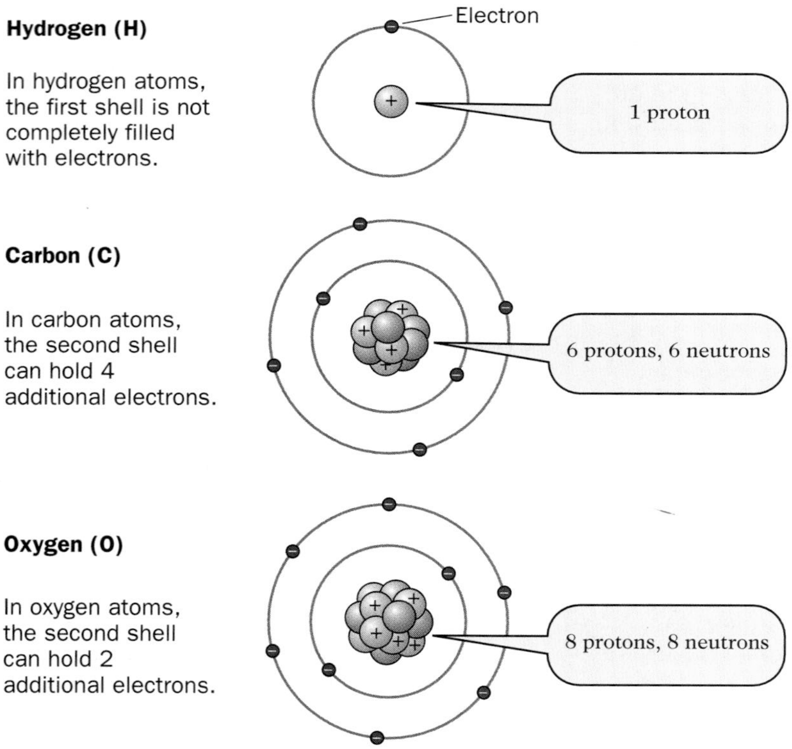
Hydrogen (H)
In hydrogen atoms, the first shell is not completely filled with electrons.
Electron
1 proton
Carbon (C)
In carbon atoms, the second shell can hold 4 additional electrons.
6 protons, 6 neutrons
Oxygen (O)
In oxygen atoms, the second shell can hold 2 additional electrons.
8 protons, 8 neutrons

FIGURE 1.5 The atomic structures of hydrogen, carbon, and oxygen atoms.

Most of the space in an atom is occupied by electrons that move around the nucleus at an extremely high speed. An atom's electrons are not just randomly distributed around the nucleus. They are distributed around the nucleus in a series of **shells** or **energy levels.** Each shell can hold only a limited number of electrons. For the atoms of living cells, this number is usually 8, and this represents a stable condition (see Figure 1.4). For hydrogen — the smallest atom—the stable number of electrons is 2. Within shells, electrons are grouped as pairs, and each pair moves in a defined region of space called an **orbital.**

The number of electrons in the outer shell is important because attaining a stable number of electrons causes atoms to link with other atoms, forming molecules.

How Do Atoms Differ from One Another?

Different atoms have different numbers of protons, neutrons, and electrons. For any single atom, the number of protons is always the same, and there are always as many electrons as there are protons. The positive charges neutralize the negative charges, and so there is no net charge; each atom is neutral in electrical charge. The number of protons in an atom is called the **atomic number** (Table 1.3). Hydrogen, the smallest atom, with an atomic number of 1, has only 1 proton, no neutrons, and 1 electron. Carbon, with an atomic number of 6 has 6 protons, 6 neutrons, and 6 electrons (Figure 1.5). Each kind of atom represents a more or less stable form of matter and energy. However, changes do occur in atoms, sometimes in the nucleus but more often in the distribution of electrons around the nucleus.

Isotopes: different numbers of neutrons. Although the number of protons is always the same for a particular kind of atom, rare forms of hydrogen and carbon (and other atoms as well) contain varying numbers of neutrons. These different forms are called **isotopes.** Some isotopes are stable and, like atoms, do not change their nuclear composition. Other isotopes are not stable and are called **radioactive**

TABLE 1.3 The Components and Masses (Weights) of Common Atoms in the Human Body

Element	Symbol	Number of Electrons in Outer Shell	Number of Protons (Atomic Number)	+ Number of Neutrons	= Atomic Mass
Hydrogen	H	1	1	0	1
Carbon	C	4	6	6	12
Nitrogen	N	5	7	7	14
Oxygen	O	6	8	8	16
Sodium	Na	1	11	12	23
Phosphorus	P	5	15	16	31
Sulfur	S	6	16	16	32
Chlorine	Cl	7	17	18	35
Potassium	K	1	19	20	39
Calcium	Ca	2	20	20	40

isotopes, for example, ^{14}C (carbon 14) and ^{131}I (iodine 131).

Radioactive isotopes give off radiation (a form of energy) and particles of matter from their nuclei to reach a more stable and lower energy state. This emission of radiation can be dangerous, but it can also be put to use. Scientists can "tag" and follow the fate of molecules in the human body by using radioactive isotopes. The condition of your thyroid gland can be examined by injecting a small amount of ^{131}I into the blood (Figure 1.6). The radiation from another radioactive isotope, thallium 201 (^{201}Th), can be used to locate damaged regions of the heart.

The nuclei of the vast majority of atoms are stable and do not change. However, even stable atoms undergo changes. These changes involve the number and distribution of the electrons outside the nucleus. We'll discover this when we consider the fourth question.

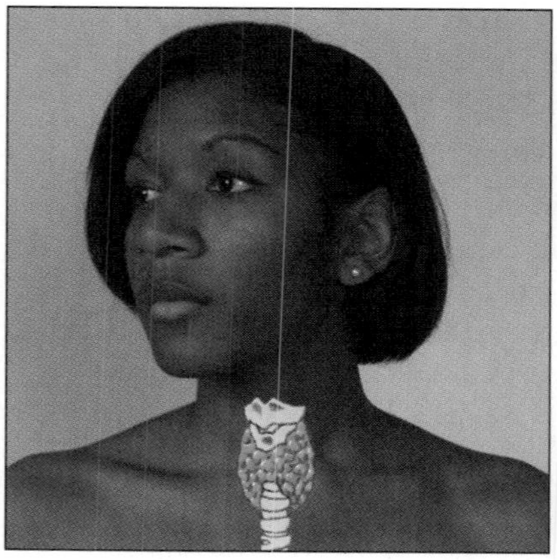

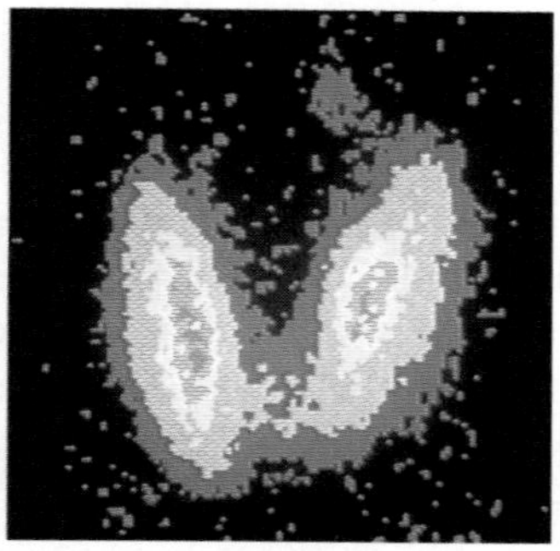

(a) Normal Thyroid

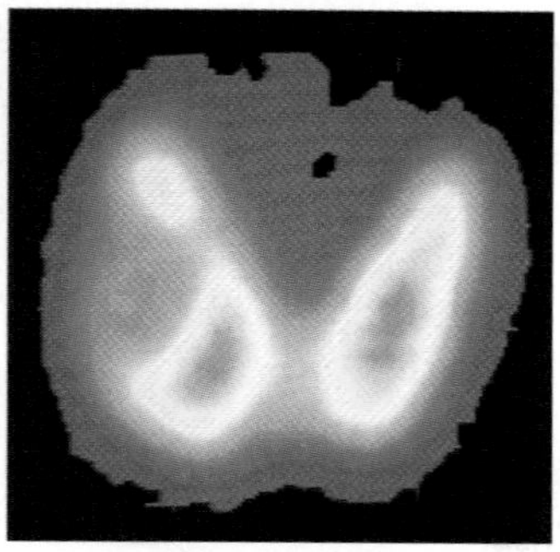

(b) Cancerous Thyroid

FIGURE 1.6 Radioactive atoms behave chemically just as their nonradioactive cousins do, except that they give off radiation that can be used to identify their presence and location.

How Does the Structure of Atoms Help Us Understand How Atoms Bond Together?

Atoms do not usually exist separate and alone. Instead, they bond together to form molecules. The capacity of atoms to bond with each other is directly dependent on the electrons that whirl around the nucleus in the orbitals of the outermost shell. The number and distribution of these electrons determine the combining capacity of each kind of atom (Table 1.4). The *combining capacity* is the number of single linkages or bonds an atom can form with other atoms. Hydrogen can form only one bond with another atom, while oxygen can form two and carbon can form four. These combining capacities are fairly constant and permit us to determine the specific structure of molecules by figuring out which atoms are bonded to which other atoms.

In addition to determining the combining capacity of atoms, the distribution of electrons in the outermost orbitals determines the shape of a molecule, or the **molecular structure.** Molecules have three-dimensional shapes that result from

TABLE 1.4 The Electrons in an Atom s Outer Shell Determine the Atom s Combining Capacity

Atom	Number of Electrons in Outer Shell	Total Electrons Permitted in Outer Shell	Combining Capacity	Example
Hydrogen	1	2	1	H_2O
Oxygen	6	8	2	H_2O, CO_2
Nitrogen	5	8	3	NH_3
Carbon	4	8	4	CH_4
Phosphorus	5	8	3	PO_4^{3-}
Sulfur	6	8	2	H_2S

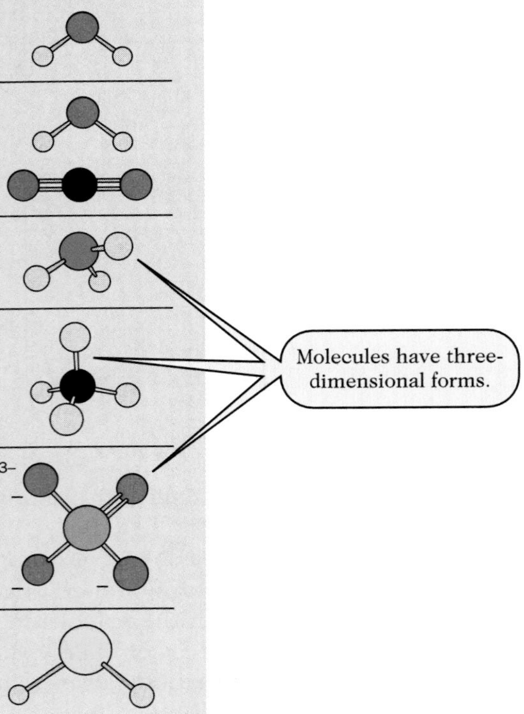
Molecules have three-dimensional forms.

the orientation of their orbitals in space (Figure 1.7). The orbitals of oxygen atoms have different orientations in space than do those of carbon atoms. Consequently, when oxygen is part of a molecule, it contributes a molecular structure different from that of carbon. The structure of both small and large molecules is very important to our understanding of the structure and functioning of the body and its cells.

Molecules form because they represent more stable energy states for atoms and their electrons. However, no molecule is stable under all conditions and circumstances. Molecules change (chemist's say that molecules "react") when their atoms lose, gain, or rearrange electrons to become more stable.

To understand the human body, we need to know how atoms form so many different kinds of molecules. We need to know something about the structure of these molecules, and we need to know why they change (react) in certain predictable ways. First, we'll look at the bonds that hold atoms together.

> Each atom consists of a nucleus of protons and neutrons, with electrons taking up most of the space around the nucleus. Electrons are organized into shells and orbitals. The electrons in the outer shell are responsible for the combining capacity of each atom.

Ions, Bonds, and Molecules

The different electron configurations of atoms permit them to interact with each other in a variety of ways. These interac-

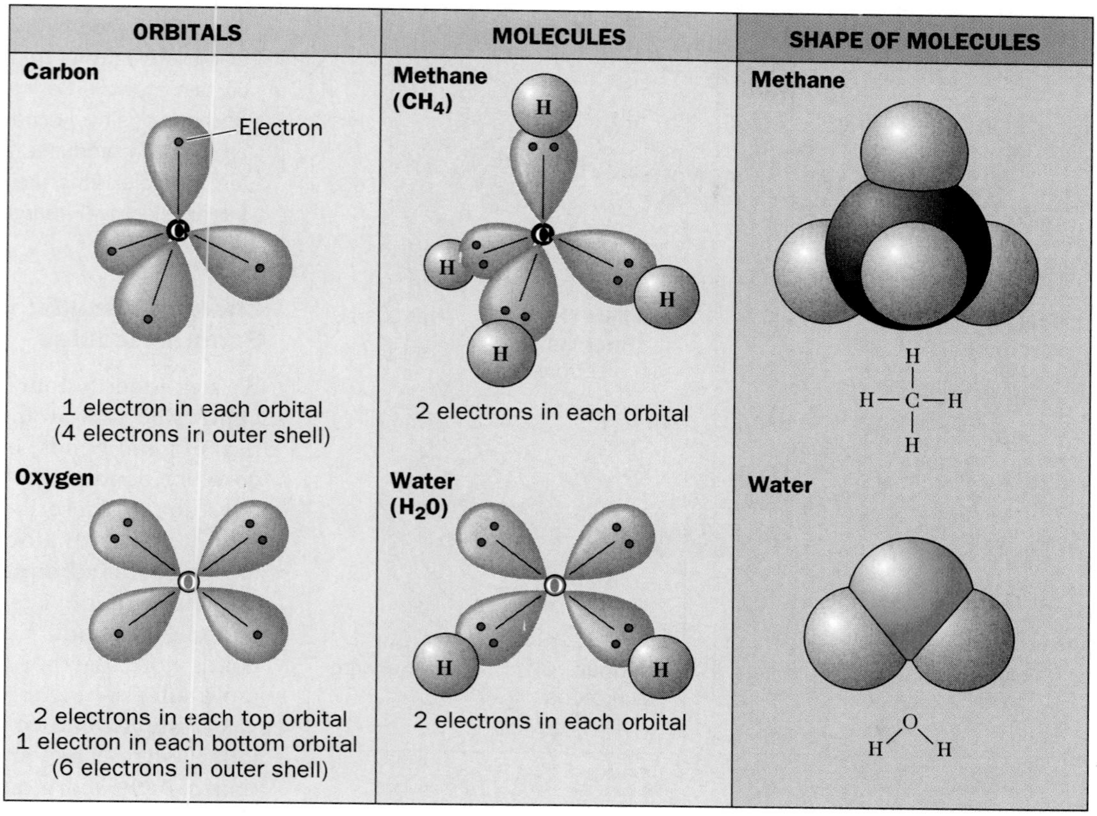

ORBITALS
MOLECULES
SHAPE OF MOLECULES
Carbon
Electron
1 electron in each orbital
(4 electrons in outer shell)
Methane
(CH_4)
2 electrons in each orbital
Methane
Oxygen
2 electrons in each top orbital
1 electron in each bottom orbital
(6 electrons in outer shell)
Water
(H_2O)
2 electrons in each orbital
Water

FIGURE 1.7 The orientation of orbitals in space gives molecules their three-dimensional shapes. This applies to large molecules as well as the small molecules of methane (CH_4) and water (H_2O) depicted here.

tions produce the linkages, called **bonds,** that bind atoms together into molecules. First we'll examine how atoms become electrically charged and how these charged particles are bonded together. Then we'll discuss bonds formed by the sharing of a pair of electrons between atoms.

Ions Are Electrically Charged Atoms

Some atoms establish a more stable electron distribution by losing or gaining one or more electrons to form an electrically charged particle called an **ion.** For example, an atom of sodium (Na) has only 1 electron in its outer shell. It attains a more stable situation by losing this electron and becoming a positively charged particle called the sodium ion, designated as Na^+ (Figure 1.8). The positive charge results from the fact that there is one more proton in the nucleus than there are electrons outside the nucleus. When hydrogen loses its 1 electron, it becomes the hydrogen ion (H^+). There are also negatively charged ions. The chloride ion (Cl^-) is formed when an atom of chlorine acquires a single additional electron (see Figure 1.8). The additional electron gives the ion the more stable 8 electrons in its outer shell.

Groups of atoms may associate together in a stable configuration and form **complex ions.** For example, phosphate is an important complex ion that consists of 1 atom of phosphorus and 4 atoms of oxygen. It carries three negative charges (PO_4^{3-}, symbolized as ℙ). As a stable association of atoms, a phosphate group can attach to and detach from larger molecules. As you will learn later, the "dance" of phosphate molecules is very important to life.

Positive and negative ions attract each other and often are associated together. For example, common table salt (NaCl) is a crystalline solid consisting of sodium and chloride ions packed together in three dimensions (Figure 1.8). The hard structure of bone is due to solid crystals

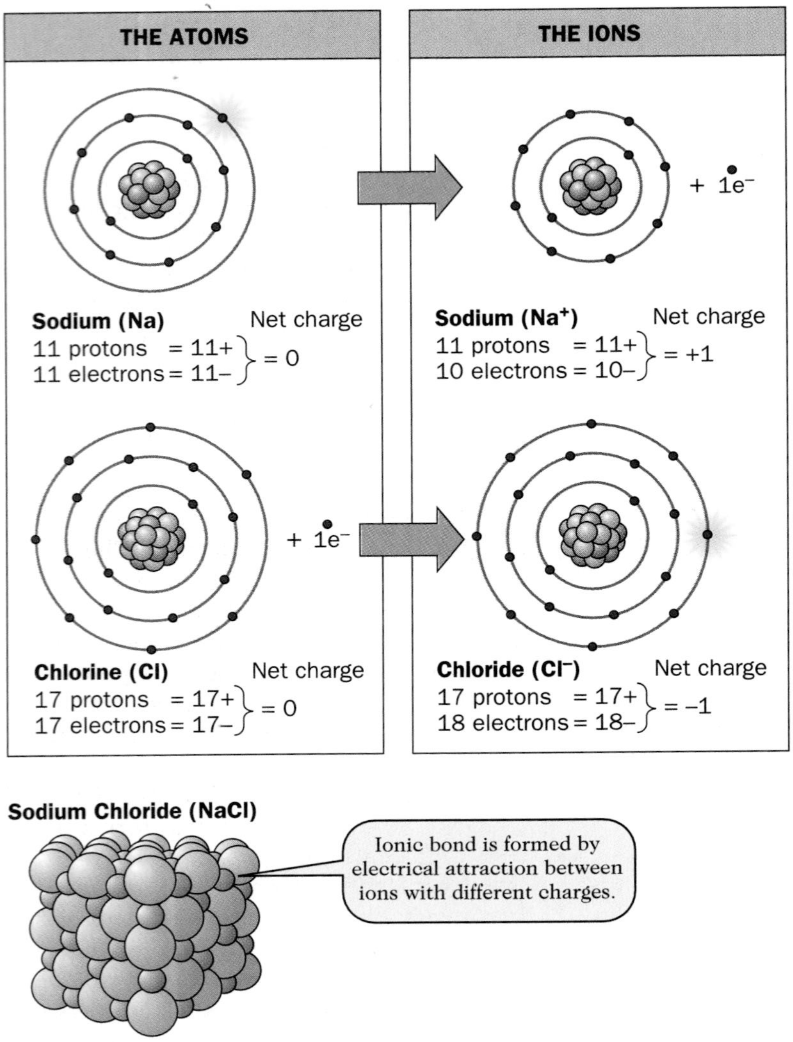
THE ATOMS
THE IONS
+ 1e⁻
Sodium (Na)
Net charge
11 protons = 11+
11 electrons = 11–
= 0
Sodium (Na⁺)
Net charge
11 protons = 11+
10 electrons = 10–
= +1
+ 1e⁻
Chlorine (Cl)
Net charge
17 protons = 17+
17 electrons = 17–
= 0
Chloride (Cl⁻)
Net charge
17 protons = 17+
18 electrons = 18–
= –1
Sodium Chloride (NaCl)
Ionic bond is formed by electrical attraction between ions with different charges.

FIGURE 1.8 Ions and ionic bonds. When electrons are lost or gained, ions are formed from atoms. Ions carry either a positive (+) or a negative (–) electrical charge. Ionic bonds are electrical attractions between ions with opposite charges. Ions with the same charge repel each other.

of calcium phosphate, $Ca_3(PO_4)_2$. As we will see soon, positive and negative ions also can attract and repel each other when they are in a liquid such as water. Our blood contains many ions, including Na^+, K^+, Cl^-, H^+, and Ca^{2+}.

In addition to ions, there is a kind of fragmented molecule called a **free radical.** These molecules have unpaired electrons, are very unstable, and can damage other molecules. Free radicals are produced by sunlight and are found in cigarette smoke and spoiled foods. They may contribute to cancer and the aging process.

Ions are formed by the loss or gain of one or more electrons from the outer shell of certain atoms. The loss of an electron produces a positively charged ion. The gain of an electron produces a negatively charged ion. Complex ions are groups of atoms bearing an electrical charge.

Chemical Bonds: Linking Atoms to Form Molecules

We will identify four types of chemical bonds that link atoms together in living tissue: ionic bonds, covalent bonds, polar covalent bonds, and hydrogen bonds. We have already described ions that are formed when an atom gains or loses an electron. **Ionic bonds** are the electrical attraction that exists between positive and negative ions. For example, ionic bonds hold together the ions in a grain of table salt (see Figure 1.8).

Covalent bonds are very different from ionic bonds. They are formed by atoms that equally share one or more pairs of electrons in their outer shells so that each atom attains the stable 8 electrons (Figure 1.9a). In the case of hydrogen, stability is attained by sharing 2 electrons. The shared pair of electrons, called *bonding electrons,* constitute the covalent bond. When covalent bonds are formed, the atoms are linked together to form a molecule that is more stable. The atoms will stay linked together until enough energy is available to break the bonds.

Covalent bonds are represented by a long dash connecting the letter abbreviations for the atoms, for example, C—H and C—C. Under certain conditions, double covalent bonds are formed between some atoms; for example, oxygen gas (O_2) is double bonded: O═O. Double bonds form carbon dioxide O═C═O (CO_2), and also are found in other molecules. Pure covalent bonds such as C—H and C—C have no net electrical charge.

Intermediate between covalent and ionic bonds are **polar covalent bonds.** These are distorted covalent bonds that result from an unequal sharing of the bonding electrons between two atomic nuclei. Polar covalent bonds have a small net electrical charge, with one end slightly negative and the other end slightly positive (Figure 1.9b). Examples include the O—H and N—H bonds.

The bonds between the oxygen atom and the hydrogen atoms of each water molecule (H_2O) are polar covalent bonds. As a result, water is considered a **polar molecule,** with one end being slightly positive and the other end being slightly negative. Polar molecules such as water are attracted to ions or other polar molecules or functional groups but are not attracted to molecules with pure covalent bonds, such as fats (Figure 1.9c).

Some stable associations of atoms, called **functional groups,** are parts of many different larger molecules. The atoms in functional groups are covalently bonded together in characteristic relationships and arrangements (Table 1.5). Some functional groups, such as phosphate (PO_4^{2-}), carry an electrical charge. Functional groups contribute certain properties to the molecules they are part of.

The fourth type of bond, called the **hydrogen bond,** is a bond that forms between molecules, not within molecules (see Figure 1.9c). Hydrogen bonds are weak electrical attractions that form between molecules with polar covalent bonds. A slightly positive atom in one molecule is attracted to a slightly negative atom in another molecule. Hydrogen bonds are represented by three or more dots, for example, OH···OH. We'll learn more about hydrogen bonds by considering the structure and interactions of water molecules.

Chemical bonds link atoms together to form molecules. Ionic bonds link atoms bearing different electrical charges. Covalent bonds are formed by sharing electrons. Polar covalent bonds have small positive and negative charges. Hydrogen bonds are weak electrical attractions between molecules.

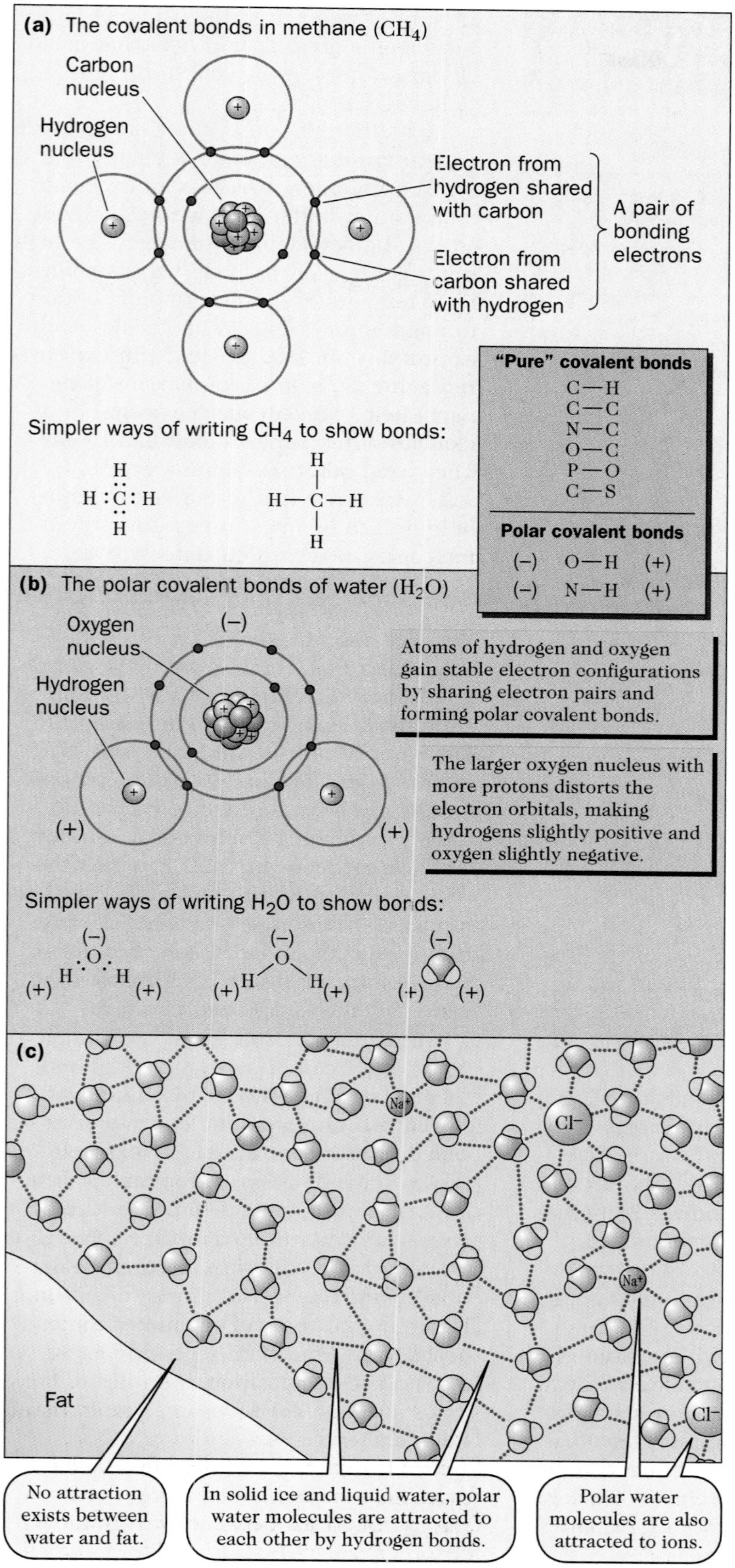

(a) The covalent bonds in methane (CH4)
Carbon nucleus
Hydrogen nucleus
Electron from hydrogen shared with carbon
Electron from carbon shared with hydrogen
A pair of bonding electrons
"Pure" covalent bonds
C—H
C—C
N—C
O—C
P—O
C—S
Polar covalent bonds
(–) O—H (+)
(–) N—H (+)
Simpler ways of writing CH4 to show bonds:
(b) The polar covalent bonds of water (H2O)
Oxygen nucleus
Hydrogen nucleus
Atoms of hydrogen and oxygen gain stable electron configurations by sharing electron pairs and forming polar covalent bonds.
The larger oxygen nucleus with more protons distorts the electron orbitals, making hydrogens slightly positive and oxygen slightly negative.
Simpler ways of writing H2O to show bonds:
(c)
Na+
Cl–
Fat
No attraction exists between water and fat.
In solid ice and liquid water, polar water molecules are attracted to each other by hydrogen bonds.
Polar water molecules are also attracted to ions.

FIGURE 1.9 Atoms can share electrons to fill their outer shells and become more stable. When this happens, a covalent bond is formed between the atoms, and they are held together as a unit. A hydrogen bond forms between molecules with polar covalent bonds.

TABLE 1.5 Some Common Functional Groups and Their Properties

Name/ Symbol	Structure	Properties
Hydroxyl group (OH)	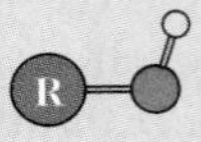	Polar, attracted to water molecules, involved in hydrogen bond formation
Carboxyl group (COOH)	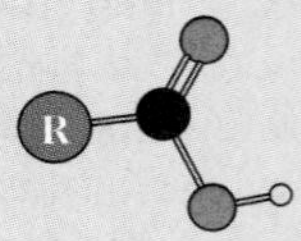	Weak acid; can donate H^+ to become $R–COO^-$
Amine group (NH_2)	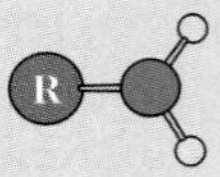	Weak base; can accept H^+ to become $R–NH_3^+$
Phosphate group	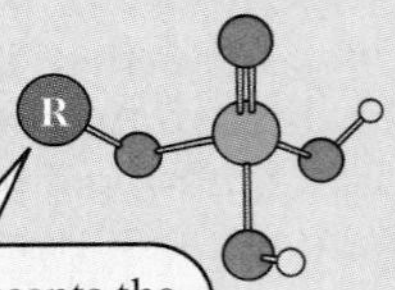	Acidic; can donate two H^+ to become $R–PO_4^{2-}$

The "R" represents the rest of the molecule to which the functional group is bonded.

Water Molecules: How They Interact

Water constitutes 70 percent of the human body and 90 percent of the blood. Water molecules are abundant and important. Everything that goes on in the human body involves water in one way or another. A molecule of water has 2 hydrogen atoms and 1 oxygen atom: H_2O. The hydrogen atoms are linked to the oxygen atom by polar covalent bonds (see Figure 1.9b). The small electrical charges of hydrogen and oxygen permit water molecules to form many hydrogen bonds with other water molecules and with any ions or other polar substances that are present.

Although each hydrogen bond is a weak bond, there are so many of them that they give water many exceptional properties that are beneficial to all living things, including humans. For example, it takes a great deal of energy to break so many hydrogen bonds and convert liquid water into the gas, water vapor. This means that the evaporation of a small amount of water from the skin and lungs carries off a great deal of heat that could be dangerous to your health if it accumulated.

In addition, because water must absorb an extraordinary amount of energy to increase in temperature, it is an excellent temperature buffer both within the body and in the outside environment. The high water content of the body helps stabilize body temperature, preventing disruption to temperature-sensitive molecules such as proteins and nucleic acids. In the environment, the high water content of the Earth helps prevent more extreme changes of temperature between seasons. These and other special properties of water are attributed to the large number of hydrogen bonds formed between water molecules in the liquid state.

Solutions and Their Characteristics

Water is also an excellent solvent. Because water molecules form hydrogen bonds, they can interact with charged substances such as ions. For example, liquid water *dissolves* solid table salt, NaCl (Figure 1.10). In this process, water operates as a solvent to dissolve the solute (NaCl), producing a solution. A **solution** is a homogeneous mixture in which the solutes (ions or entire molecules) are uniformly distributed in a solvent. A **solute** is any substance that dissolves in a solvent to form a solution. A **solvent** is the dissolving medium (usually water).

Operating as a solvent in the human body, water can dissolve many different substances. The amount of solute dissolved in a given volume of water is called the **concentration.** For example, we can dissolve 25 grams (g) of table salt in a liter of water, yielding a concentration of 25 grams per liter (25 g/l). Such a solution is more concentrated than a solution containing only 10 grams of salt in a liter (10 g/l). You will encounter the concept of concentration frequently as we discuss the composition and roles of body fluids such as blood plasma, lymph, tissue fluid, and urine.

One solvent, many solutes. The human body uses only one solvent, water (H_2O), for all its fluids. However, we will

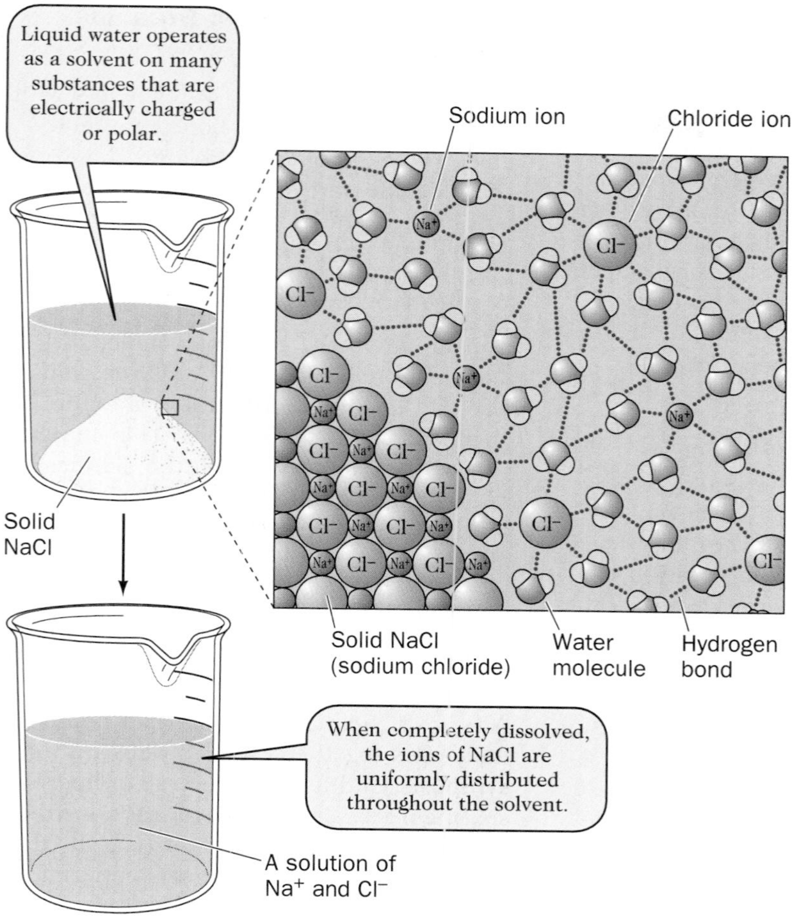
Liquid water operates as a solvent on many substances that are electrically charged or polar.
Sodium ion
Chloride ion
Solid NaCl
Solid NaCl (sodium chloride)
Water molecule
Hydrogen bond
When completely dissolved, the ions of NaCl are uniformly distributed throughout the solvent.
A solution of Na+ and Cl−

FIGURE 1.10 Water dissolves solids, forming solutions.

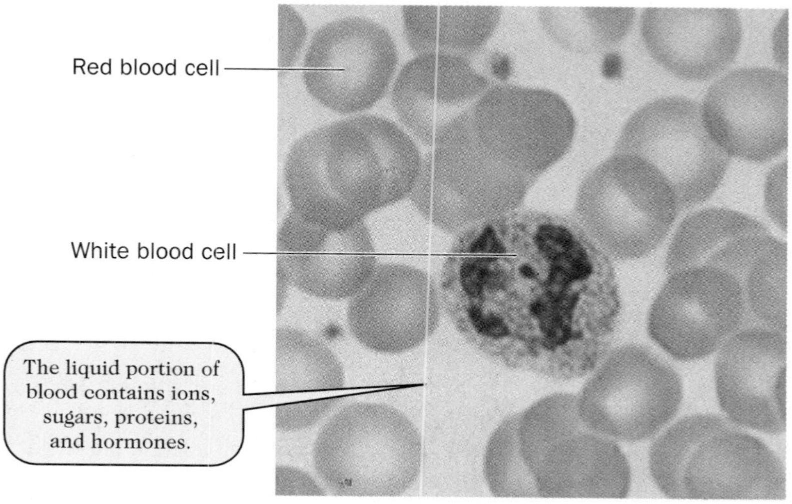
Red blood cell
White blood cell
The liquid portion of blood contains ions, sugars, proteins, and hormones.

FIGURE 1.11 The liquid portion of blood is a complex solution that contains many different ions, sugars, proteins, and hormones.

encounter many different solutes. Some solutes are small ions, such as H^+, Na^+, K^+, Ca^{2+}, and Cl^-. Other solutes are larger molecules, such as the sugars glucose (blood sugar) and sucrose (table sugar), vitamin C, and amino acids. Still other solutes are enormous proteins, such as the proteins in human blood (albumin).

When many different substances dissolve in water, a *complex solution* is formed. Body fluids are all complex solutions in which each of the many different solutes plays a special role (Figure 1.11). Whether they are ions or molecules or are large or small, all these solutes dissolve in water because they have an electrical charge that permits them to be attracted to the polar water molecules.

When sugar, salt, or vitamin C dissolves in water, the solid matter changes form but does not change in terms of atomic composition or properties. No chemical reaction takes place, and we can recover unchanged solutes from the water solvent. However, when a chemical reaction takes place, there are changes in atomic composition and properties. Chemical reactions are necessary to create and maintain life.

> Water is a polar molecule that forms many hydrogen bonds with other water molecules in the liquid state. Water serves as a temperature buffer. It also can dissolve ionic and polar substances, forming a solution in which the solute is uniformly distributed. Body fluids are complex solutions.

Chemical Reactions, Chemical Changes

A chemical reaction is a change. We can define a **chemical reaction** as a change in the atomic composition and properties of molecules. We start with molecules called **reactants.** These molecules undergo changes, and we end up with altered molecules called **products.** To undergo a reaction, reactant molecules dissolved in water must collide with each other. When they collide with sufficient kinetic energy (the energy of motion), bonds are broken and new bonds are formed. The products of the reaction are different from the reactants.

A chemical reaction is represented by a **chemical equation** in which reactants are symbolized on the left. An arrow represents the process of collision and rearrangement, and the product or products are symbolized to the right of the arrow. For example, molecules A and B collide and react, forming products C and D:

$$\underset{\text{Reactants}}{A + B} \longrightarrow \underset{\text{Products}}{C + D}$$

In such a reaction, matter is neither created or destroyed; only its form is changed. This is a fundamental physical principle known as the **conservation of matter.** It means that there are always the same number of atoms in the products as there were in the reactants.

There are many types of chemical reactions, but they all involve a change in the composition and properties of the reactants. Many chemical reactions release energy as reactants are converted to products. Other reactions require energy for the reactants to be converted to products. However, in any chemical reaction, energy is neither created nor destroyed. This is the principle of the **conservation of energy** that was mentioned earlier.

Both of these principles—the conservation of matter and the conservation of energy—apply to all the chemical changes that take place in the living cells of the human body. Cells do not create or destroy matter or energy. However, transformations of matter and energy from one form to another take place all the time in living cells. Such transformations supply energy for growth, reproduction, and responses to the environment.

> If molecules collide with sufficient energy, a chemical reaction takes place. The products of a reaction have compositions and properties that are different from those of the reactants. Neither matter nor energy is created or destroyed, but energy may be consumed or released during a reaction.

Hydrolysis Reactions: Breaking Bonds with Water

In **hydrolysis reactions,** water reacts with and breaks a particular bond in a molecule. The components of water, H and OH, become part of the product molecules generated by the hydrolysis reaction (*hydro,* "water"; *lysis,* "breaking"):

$$A—B + H_2O \longrightarrow A—H + HO—B + \text{energy}$$

Depending on the particular molecules and the conditions, hydrolysis reactions also can liberate considerable amounts of energy for the body's needs. This energy can be used to accomplish the synthesis of other molecules (Figure 1.12a). Some of these molecules are enormous polymers such as proteins, polysaccharides, and nucleic acids (DNA and RNA).

Dehydration Synthesis: Linking Molecules Together

A type of chemical reaction that bonds molecules together is called **dehydration synthesis,** and it is the reverse of hydrolysis (Figure 1.12b). In dehydration synthesis, the components of water (H and OH) are extracted as a bond forms between the reactants (*de,* "loss of"; *hydration,* "water"). This kind of chemical reaction is used to bond together similar small molecules called **monomers** into very long chains called **polymers.** For any particular polymer, the monomers are all of a single type, such as amino acids (A) or sugars, but there may be individual differences (A_1, A_2, A_3, etc.).

Polymers may be hundreds or thousands of monomers long. Proteins, polysaccharides, and nucleic acids are the most important biological polymers we will encounter as we discuss the structures and functions of human organs. We will learn more about them later in this chapter.

> Body functions involve many different kinds of chemical reactions. In hydrolysis reactions, water acts to break apart a molecule. Hydrolysis reactions liberate energy. Dehydration synthesis is the reverse of hydrolysis and bonds monomers into polymers.

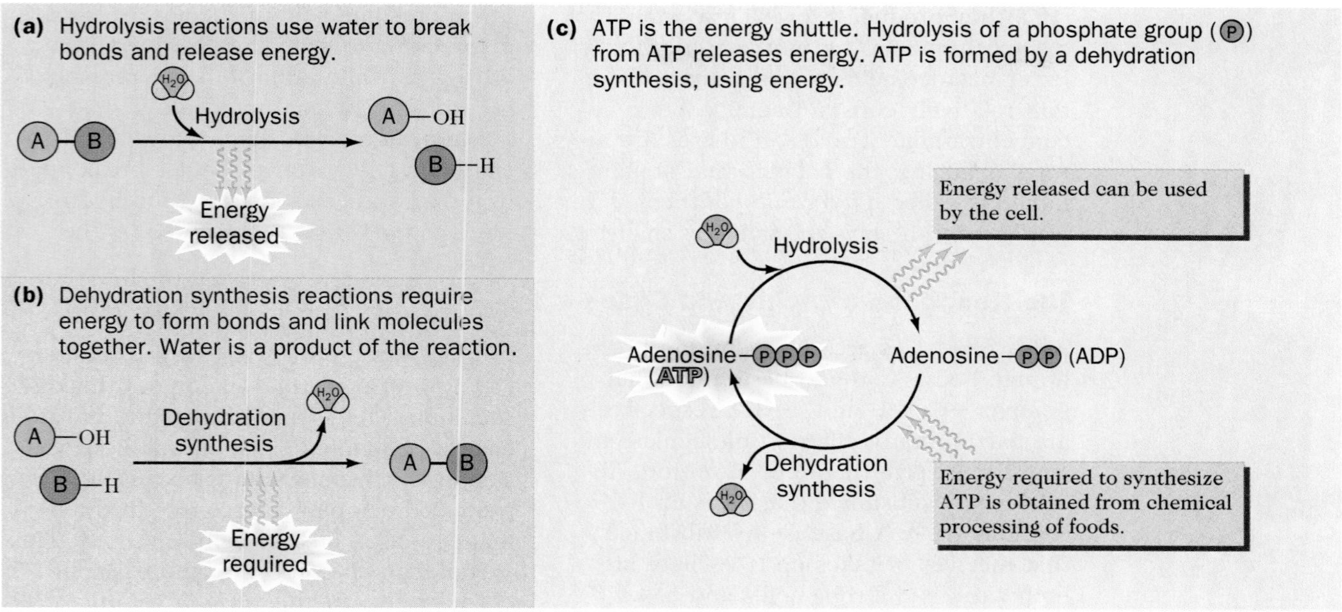

(a) Hydrolysis reactions use water to break bonds and release energy.
H2O
Hydrolysis
A
B
A
OH
B
H
Energy released
(b) Dehydration synthesis reactions require energy to form bonds and link molecules together. Water is a product of the reaction.
A
OH
B
H
Dehydration synthesis
H2O
A
B
Energy required
(c) ATP is the energy shuttle. Hydrolysis of a phosphate group (P) from ATP releases energy. ATP is formed by a dehydration synthesis, using energy.
Energy released can be used by the cell.
H2O
Hydrolysis
Adenosine—PPP (ATP)
Adenosine—PP (ADP)
Dehydration synthesis
H2O
Energy required to synthesize ATP is obtained from chemical processing of foods.

FIGURE 1.12 Breaking molecules apart and linking them together. Hydrolysis and dehydration synthesis reactions and the role of energy.

ATP: Providing Energy Where It Is Needed

The synthesis of polymers requires energy. This energy is supplied by other chemical reactions and is made available to polymerization reactions in a convenient form. This convenient form is the energy-rich molecule ATP (spoken by naming the letters: A-T-P). **ATP** stands for **adenosine triphosphate,** which is a molecule of adenosine with a tail of three phosphate groups: ADENOSINE—ℙ—ℙ—ℙ. ATP operates as an "energy shuttle," transferring energy from storage forms such as sugars and fats to the sites where it is needed for synthesis or movement (Figure 1.12c).

Energy is captured in ATP when ATP is synthesized by a dehydration synthesis reaction between ADP (adenosine diphosphate) and a phosphate group (PO_4^{3-}, or ℙ):

$$\text{ADP} + \mathbb{P} + \text{energy} \longrightarrow \text{ATP} + H_2O$$

Energy is released from ATP when it undergoes a hydrolysis reaction:

$$\text{ATP} + H_2O \longrightarrow \text{ADP} + \mathbb{P} + \text{energy}$$

The energy-capturing and energy-releasing reactions of ATP can be depicted as a cycle (see Figure 1.12c). However, it is important to realize that ATP is formed at certain sites and moves to other sites where its energy is released as it is needed. You will encounter ATP and its energy-providing role frequently in the chapters that follow.

> ATP serves as an energy shuttle. Energy is captured in ATP when ATP is synthesized from ADP and ℙ. ATP can then move to a reaction where energy is needed. Upon hydrolysis of the third ℙ, energy is released and can be used to sustain living processes.

Acids and Bases: Reactions in Equilibrium

Both acids and bases are important substances for the human body. Acids taste sour. They digest food in your stomach, cause heartburn, and dissolve tooth enamel. These examples may make them sound dangerous, but that is not entirely true. Our bodies and cells could not function without acids and the hydrogen ions

(H^+) they produce. However, too much of a good thing is as harmful as too little. Bases balance acids in the body to maintain relatively constant acidity. Because concentrations of acids and bases are always changing, the balance that is maintained is called a *dynamic balance* or *equilibrium*. It stays the same by changing.

The Reactions of Acids and Bases

Acids and bases are very common in the human body. Acid and base functional groups (—COOH and —NH_2, respectively) are part of many different molecules, and so we must give them our attention. An **acid** is any substance that gives up hydrogen ions (H^+). A **base** is any substance that receives or takes up H^+. There are both weak and strong acids and bases. When a **strong acid** such as HCl (hydrochloric acid) dissolves in water, it completely **dissociates.** This means that all the HCl molecules separate into H^+ and Cl^-:

$$HCl \longrightarrow H^+ + Cl^-$$

A **strong base** such as NaOH (sodium hydroxide) also dissociates completely into Na^+ and OH^- ions when it dissolves in water.

Weak acids behave somewhat differently: They do not completely dissociate. Only a few of these acid groups (or acid molecules) undergo dissociation, giving up only a few hydrogen ions. All the reactant molecules are not converted to products. In a water solution, the weak acid acetic acid (vinegar) undergoes such a partial dissociation:

$$\underset{\text{Many acetic acid molecules}}{CH_3\text{—}COOH} \rightleftharpoons \underset{\text{A few hydrogen ions}}{H^+} + \underset{\text{A few complex ions}}{CH_3\text{—}COO^-}$$

In a water solution, all three forms are present: the acid, the H^+, and the complex ion.

In the equation, the arrows point both ways to indicate that the reaction can operate in either direction. In fact, the complex ion can serve as a **weak base,** combining with H^+ to re-form the molecule acetic acid. Since it can operate in either direction, the reaction is an *equilibrium* in which some acid molecules are always dissociating to form the positive and negative ions while other ions are reassociating to form the acid molecule.

Water is a weak acid. In water dissociation, a few water molecules break apart to form hydrogen ions (H^+) and hydroxide ions (OH^-), as represented by the equation

$$H_2O \rightleftharpoons H^+ + OH^-$$

In pure water, only 1 in every 10 million (10^7) water molecules undergoes this dissociation. This is a small amount, but it cannot be ignored. This partial dissociation of water is important because the H^+ produced can play a large role in the normal operation of your cells and body. This partial dissociation is also the basis for the pH scale, the measure of acidity.

pH: Measuring Acidity

In any sample of pure water, the concentration of H^+ is tiny and can be expressed as 0.0000001, or 10^{-7} (where the -7 is the negative exponent). Small as this number may seem, it is important and significant, forming the basis of the **pH scale,** which is a measure of the acidity of any water solution. The pH is defined as the negative exponent of the hydrogen ion concentration, and pure water has a pH of 7, which is considered neutral. If the pH values are lower than 7, the solution is acidic. If they are higher than 7, the solution is basic or alkaline (Figure 1.13). Blood and many other body fluids which are mostly water have pH values of about 7.4, and so blood is slightly alkaline.

Things can happen to water to change its pH value. The addition of a strong acid such as hydrochloric acid (HCl) increases the number of hydrogen ions and lowers the pH (the pH numbers get smaller). Tasting food initiates the secretion of HCl by the stomach, lowering the pH to aid digestion (Chapter 6). However, acid secretion must be regulated. If acid is secreted in the absence of food, it causes pain, and an ulcer may form. Another example involves the addition of acid-producing substances to the atmosphere. These substances have caused precipita-

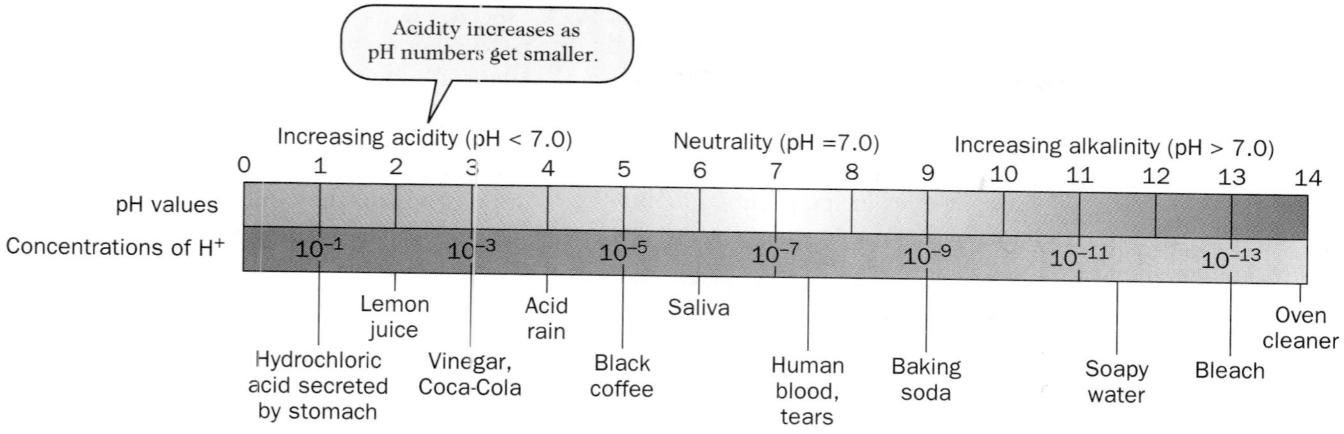
Acidity increases as pH numbers get smaller.
Increasing acidity (pH < 7.0)
Neutrality (pH =7.0)
Increasing alkalinity (pH > 7.0)
0 1 2 3 4 5 6 7 8 9 10 11 12 13 14
pH values
Concentrations of H+
10-1 10-3 10-5 10-7 10-9 10-11 10-13
Lemon juice
Acid rain
Saliva
Oven cleaner
Hydrochloric acid secreted by stomach
Vinegar, Coca-Cola
Black coffee
Human blood, tears
Baking soda
Soapy water
Bleach

FIGURE 1.13 The pH scale measures acidity and alkalinity.

tion of acid rain and snow, with devastating consequences for the environment (Figure 1.14).

Buffers: Limiting pH Changes

In the human body, changes in the pH of fluids can be dangerous and life-threatening. For example, if the pH of blood falls below 7.0 or rises above 7.8, nerves do not function properly and a coma or convulsions may occur. Such pH changes are limited by buffers. A **buffer system** is a solution made from a weak acid and a weak base. The acid and the base operate together to "soak up" or release hydrogen ions when they are added or removed by other processes. By operating in this manner, buffers limit the change in pH that occurs when H^+ is added or removed. The body uses many different buffer systems. One of them involves bicarbonate (a weak base) and carbonic acid (a weak acid):

$$\underset{\text{Carbonic acid}}{H_2CO_3} \rightleftharpoons \underset{\text{Hydrogen ion}}{H^+} + \underset{\text{Biocarbonate ion}}{HCO_3^-}$$

This reaction can shift back and forth to "soak up" or release hydrogen ions as they are added to or removed from the solution. This process helps maintain a constant hydrogen ion concentration (pH).

FIGURE 1.14 Acid rain has a pH of about 4. It profoundly alters the environment, killing many plants and animals.

> An acid gives up H^+, while a base takes up H^+. There are strong and weak acids and bases. Weak acids and weak bases partially dissociate. The pH scale measures the hydrogen ion concentration. Buffers limit the change in pH when hydrogen ions are added to or removed from a solution.

Biological Molecules and Their Roles

As we mentioned in the Prologue, all living things, including the human body, are made up of the same kinds of molecules. There are four major classes of these molecules: proteins and their amino acids,

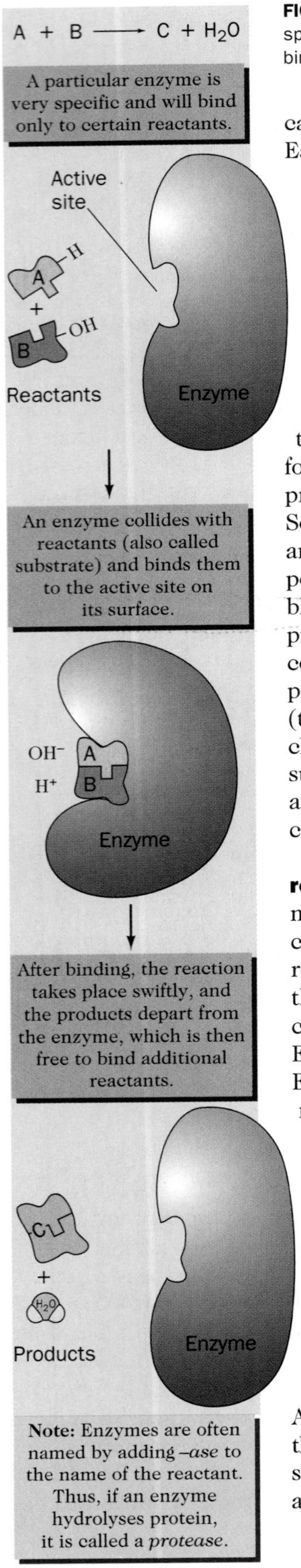
A + B → C + H_2O
A particular enzyme is very specific and will bind only to certain reactants.
Active site
A
H
+
B
OH
Reactants
Enzyme
An enzyme collides with reactants (also called substrate) and binds them to the active site on its surface.
OH^-
H^+
A
B
Enzyme
After binding, the reaction takes place swiftly, and the products depart from the enzyme, which is then free to bind additional reactants.
C
+
H_2O
Products
Enzyme
Note: Enzymes are often named by adding *–ase* to the name of the reactant. Thus, if an enzyme hydrolyses protein, it is called a *protease*.

FIGURE 1.15 Enzymes are large proteins that speed up chemical reactions in cells by combining temporarily with reactants.

carbohydrates, lipids, and nucleic acids. Each type of molecule plays many different roles in the normal functioning of the body.

Proteins: Support, Protection, and Regulation

Proteins are polymers that are made by linking together different amino acid monomers in specific sequences. There are thousands of different proteins in the human body, and they perform many different functions. Some proteins are large, and some are small. Some are long and thin, while others are tight and compact. Some are transport proteins (hemoglobin protein of blood); others provide structure and protection (the keratin of skin and the collagen of connective tissue). Some proteins are responsible for movement (the actin and myosin proteins of muscle); others function as hormones (insulin) and carriers of materials entering and leaving the cell. One important class of proteins is the enzymes.

Enzymes: speeding up chemical reactions. Enzymes are catalysts. This means that they increase the speed of a chemical reaction by interacting directly with the reactants. However, they themselves are not permanently changed by their participation. Enzymes can be used over and over. Enzymes operate by first (1) binding to reactants, which are referred to as substrates, then (2) breaking and forming bonds, and finally (3) releasing the products (Figure 1.15):

$$\text{Enzyme} + \text{substrate}$$
$$\downarrow$$
$$\text{enzyme-substrate complex}$$
$$\downarrow$$
$$\text{enzyme} + \text{product(s)}$$

After releasing the product or products, the enzyme is free to again bind to a substrate and repeat the reaction. Enzymes are very specific for particular molecules and reactions; an enzyme for one molecule and reaction will not operate on another molecule. For this reason, the human body makes thousands of different enzymes.

Enzymes do not make impossible reactions happen; they only speed up reactions. However, this has significant consequences. Enzymes make life as we know it possible. Without enzymes, most of the chemical reactions in the living body could not take place fast enough for survival in a changing environment.

Enzymes are so important that the body has found ways to manipulate the amount of an enzyme present and adjust the speed with which a particular enzyme operates. This permits the **regulation** of a particular reaction, and by this means body functions are also regulated. We will encounter enzymes and their functions frequently in the chapters that follow.

Amino acids and peptide bonds. The units that make up proteins are called **amino acids.** Twenty different amino acids are used to make all the proteins in the human body. Different amounts and sequences of these 20 amino acids produce all the thousands of different kinds of proteins. Each amino acid consists of a central carbon atom to which four different groups are bonded. Three of these groups are identical in all amino acids. The fourth group is different and provides the distinctive structure and properties of an amino acid.

All amino acids have an amino functional group ($—NH_2$) and a carboxyl functional group ($—COOH$) with acid properties. These are the source of the name "amino acids." The third group common to all amino acids is a sole hydrogen atom. The fourth and distinguishing group is identified as R:

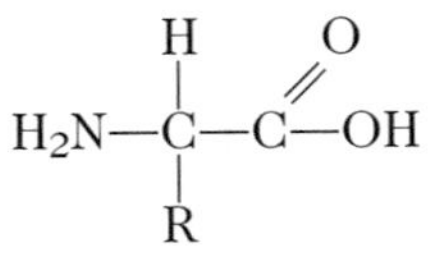
H
O
$H_2N—C—C—OH$
R

The 20 different R groups have different chemical properties. Some of these R groups are acidic, some are basic, others are polar, and still others are nonpolar (Table 1.6).

TABLE 1.6 The Four R Groups of Amino Acids

Properties of R Group	Number of Amino Acids
Nonpolar	9
Polar	6
Basic	3
Acidic	2
Total	20

When amino acids are bonded together to form a protein, the abundance and location of the different R groups determine the distinctive structure and properties of that protein. R groups are also essential to the function of all proteins. In enzymes, particular R groups determine which molecules the enzyme will bind to and what reaction it will promote.

In a protein, the bonds between the amino acids are called **peptide bonds** and are formed by a dehydration synthesis reaction between a carboxyl group and an amino group:

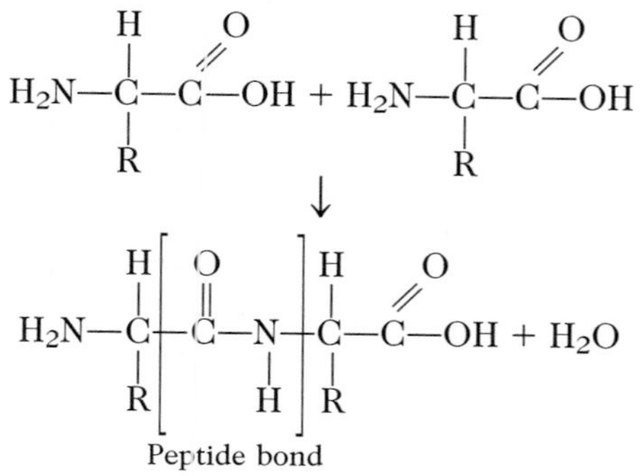
Peptide bond

Repetition of this reaction using different amino acids produces a sequence of linked amino acids called a **polypeptide** or **protein.**

The four levels of protein structure. Each kind of protein has a specific amino acid sequence and a regular three-dimensional geometry that is a product of coiling, bending, and folding. We'll refer to these as the four levels of protein structure: primary, secondary, tertiary, and quaternary. Examine Figure 1.16 carefully to learn essential information about the three-dimensional structure of proteins. Pay particular attention to the role of hydrogen bonds in maintaining the secondary structure.

Protein function depends on a three-dimensional structure. The specific three-dimensional structure of a protein is necessary for its proper function. If a single amino acid in the primary structure is deleted or changed, the normal function of the protein may be lost. This is what happens in the disease **sickle-cell anemia.** A single alteration in one of the hundreds of amino acids in hemoglobin (the O_2-carrying protein) alters the way hemoglobin functions. This small change produces all the symptoms of the disease: abnormal blood cells, fatigue, shortness of breath, obstructed blood flow to organs, and stroke.

Heating and certain chemicals also can alter the three-dimensional structure of a protein and destroy its normal function. This change is called **denaturation.** When proteins are heated, the weak hydrogen bonds that maintain a protein's structure are disrupted, and the proteins assume new unusual structures that do not function (Figure 1.17). Egg white is almost pure protein, and you are familiar with what happens when it is boiled: It changes into a white solid. Similar changes can happen to some of your proteins if your body becomes overheated by excessive exposure to high temperatures, high fever, and/or poor body cooling.

> Proteins are made from amino acids linked by peptide bonds. Each protein has a specific three-dimensional structure that is maintained by hydrogen bonds. Normal protein functions require the maintenance of this structure. Protein denaturation destroys the regular three-dimensional structure and functions.

Carbohydrates: Energy Storage and Support

Like proteins, carbohydrates play many roles in the human body. They store energy, transport energy, provide structure and strength, and serve as signals. The general formula $(CH_2O)_n$ applies to many different kinds of carbohydrates. Some carbohydrates are small and simple, while others are large and complex. Some contain only carbon, oxygen, and hydrogen, while others contain small amounts of other atoms as well (nitrogen, sulfur).

Small carbohydrates, which are called **monosaccharides** or **simple sugars**, have

FIGURE 1.16 Four different levels of structure contribute to the three-dimensional structure of proteins. Different proteins have different amino acid compositions and different three-dimensional structures.

(a) Primary structure consists of the specific sequence of different amino acids bonded to each other like links in a chain.

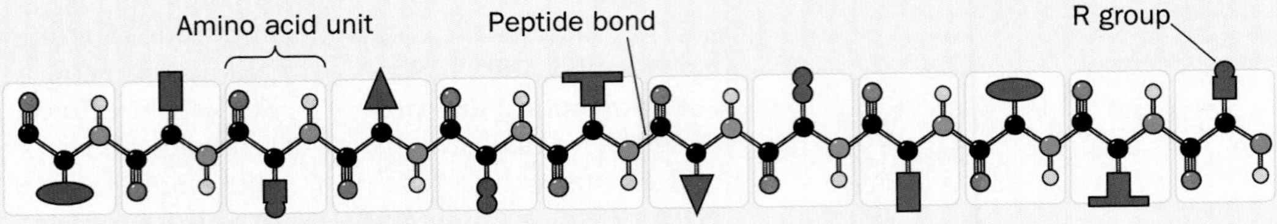
Amino acid unit
Peptide bond
R group

(b) Secondary structure may be the **alpha-helix** (α-helix) which is maintained by hydrogen bonds between atoms associated with the peptide bond.

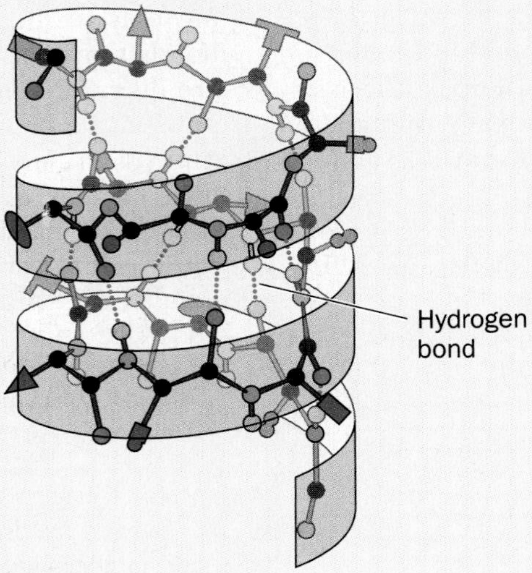
Hydrogen bond

(c) Tertiary structure is formed by bending and folding of the α-helix as determined by the size and properties of the R groups of specific amino acids. The folded structure is maintained by hydrogen bonds, ionic bonds, and covalent bonds between different R groups (not shown here).

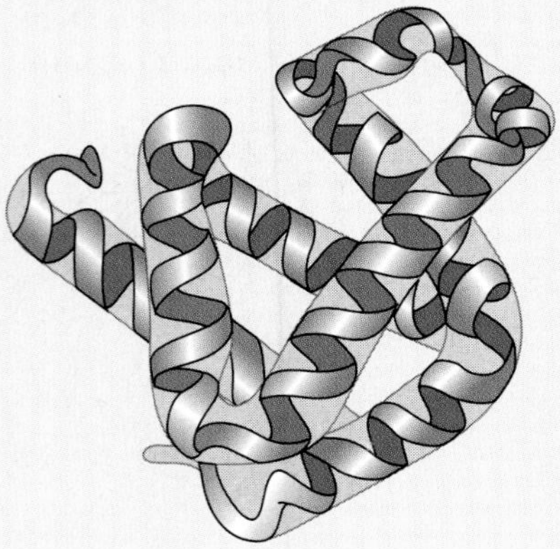

(d) Quaternary structure consists of the intimate association of separate proteins. In such an association, the individual proteins are called **subunits** and are held together by hydrogen bonds. The protein hemoglobin shown here has four such subunits.

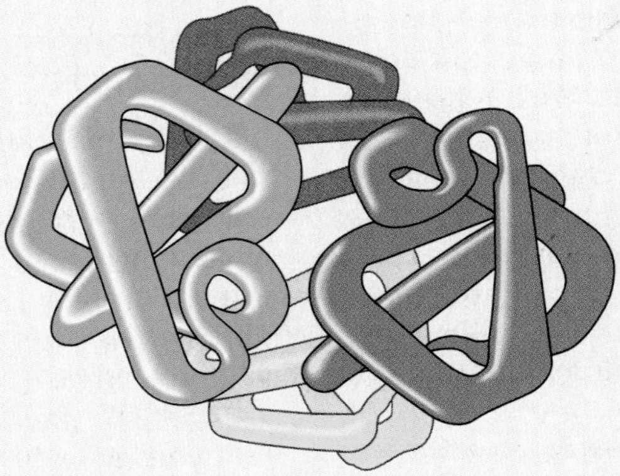

(e) A three-dimensional model of human hemoglobin containing four polypeptide chains (two blue, one pink, and one yellow).

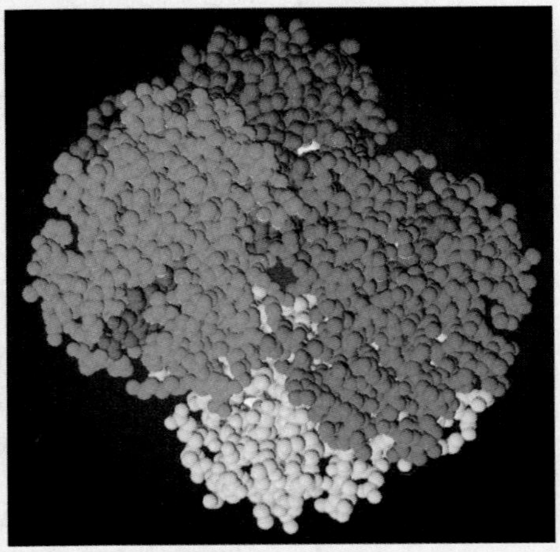

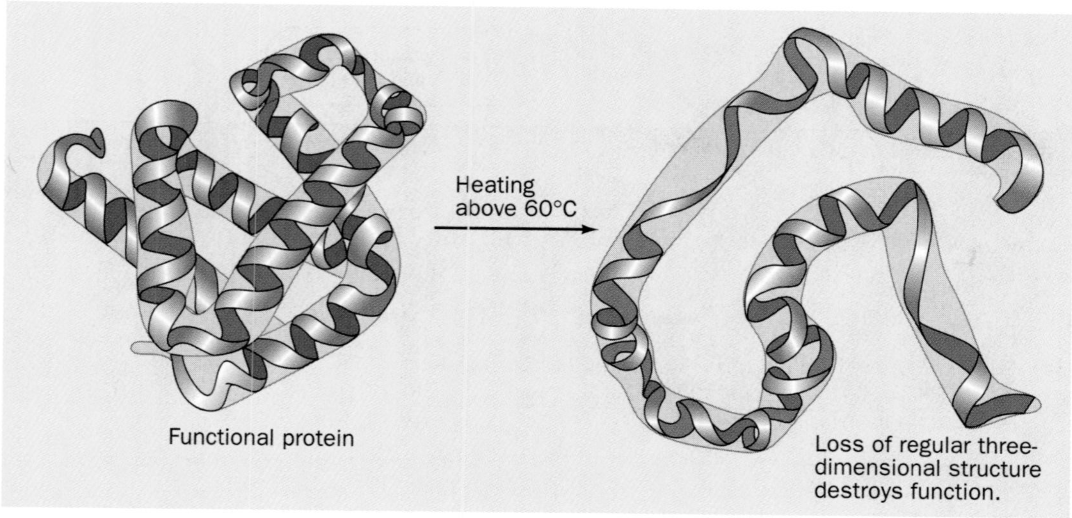
Heating
above 60°C
Functional protein
Loss of regular three-
dimensional structure
destroys function.

FIGURE 1.17 Protein structure and function. The operation of proteins requires the maintenance of the three-dimensional protein structure.

between 3 and 7 carbon atoms (C) bonded to each other and to hydroxyl groups (OH) in specific ways. The abundance of polar OH groups allows carbohydrates to dissolve in water. Examine Figure 1.18 for similarities and differences between glucose and ribose, two simple sugars.

If two simple sugars are covalently bonded together, they form a **disaccharide.** For example, sucrose is table sugar from plants (glucose-fructose), and lactose is the sugar in milk (glucose-galactose).

Large polymers of simple sugars are called **polysaccharides.** When thousands of glucose molecules are bonded together, they form the polysaccharide called **starch** in plants and **glycogen** in animals (see Figure 1.18). Both starch and glycogen store energy. Glycogen is synthesized and stored by cells of your liver and muscles when there is excess glucose in the blood. When your blood glucose declines, the bonds between the units in glycogen are broken by hydrolysis reactions and glucose molecules are liberated into your blood. These free glucose molecules are transported by the blood to all your cells, where they are broken down to CO_2 and H_2O; the released energy is used to synthesize ATP from ADP and ℙ (Chapter 7).

The polysaccharide **cellulose** provides strength to plant cells but is not found in humans. Like starch and glycogen, cellulose is a polymer of glucose, but in cellulose the glucose molecules are bonded together in a different way. Humans do not have a digestive enzyme that can break the bonds in cellulose, and so people cannot liberate and use its glucose for energy. When we eat plants, cellulose passes through our digestive systems as undigested fiber. Al though it cannot be used for energy, this fiber may help prevent colon cancer.

> Monosaccharides (simple sugars) have 3 to 7 carbons. Disaccharides consist of two monosaccharides bonded together. Polysaccharides are large polymers of monosaccharides used to store energy.

Lipids: Energy Storage, Support, and Regulation

The **lipids** are a diverse group of molecules consisting mostly of carbon (C) and hydrogen (H) linked by covalent bonds (C—C, C—H). Because of these nonpolar bonds, lipids are mostly insoluble in water; common examples include cooking oils, butter, and the fat on meat. Although insoluble in water, lipids dissolve in other solvents such as olive oil.

Like carbohydrates and proteins, lipids play several different roles in the body. Some lipids (the neutral fats) store energy—in fact, lipids store twice as much energy per gram as carbohydrates do. Other lipids (the phospholipids) are essential structural molecules in all cellular membranes (Chapter 2). Still other lipids function as regulating signals (steroid hormones, prostaglandins, and some vitamins).

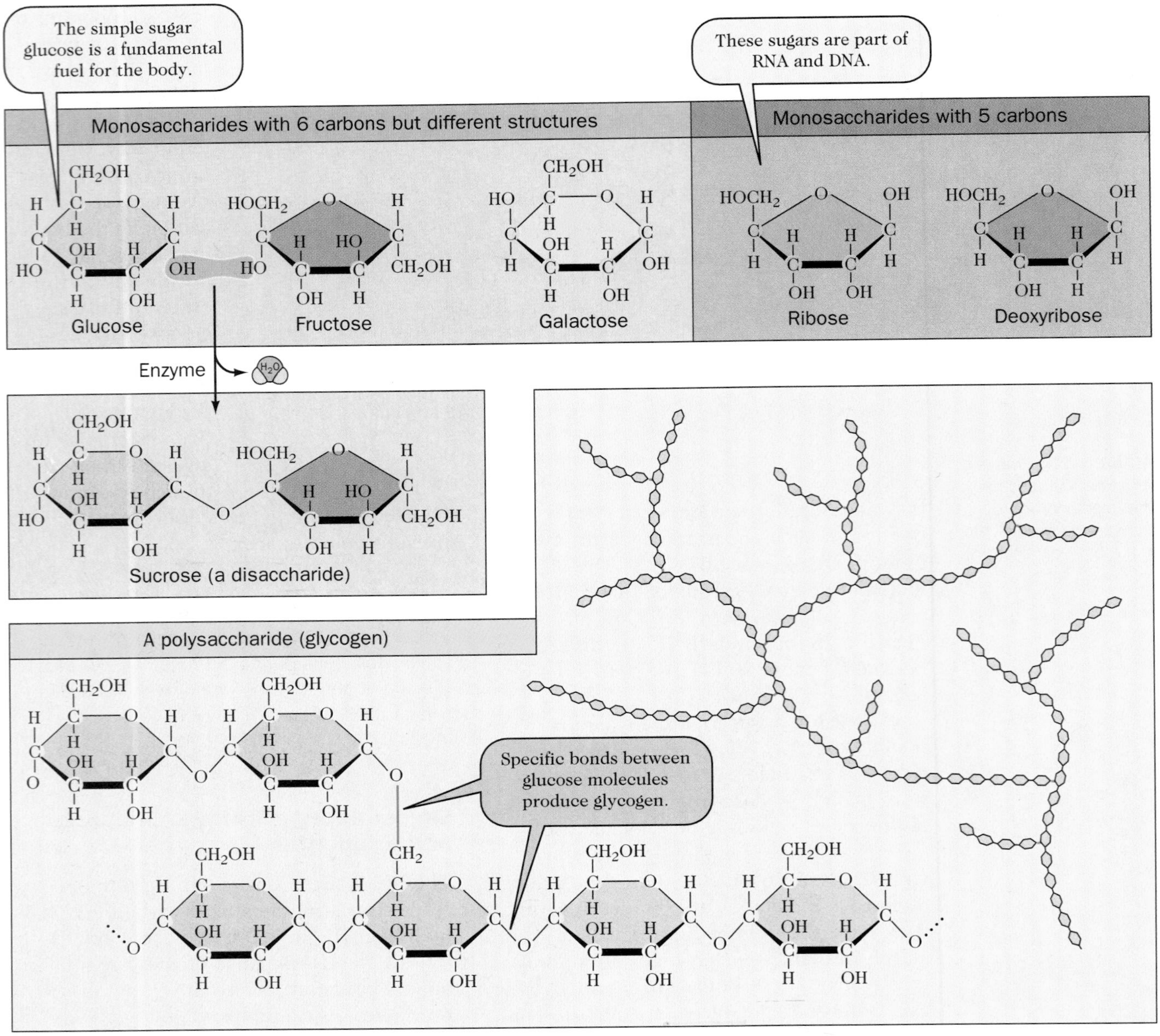

The simple sugar glucose is a fundamental fuel for the body.
These sugars are part of RNA and DNA.
Monosaccharides with 6 carbons but different structures
Monosaccharides with 5 carbons
Glucose
Fructose
Galactose
Ribose
Deoxyribose
Enzyme
Sucrose (a disaccharide)
A polysaccharide (glycogen)
Specific bonds between glucose molecules produce glycogen.

FIGURE 1.18 Carbohydrates differ greatly in size, but all contain carbon, hydrogen, and oxygen. Small carbohydrates such as glucose are soluble in water and are transported in the blood. Enormous polymers such as glycogen are not readily soluble in water and are used to store energy in the liver and muscles.

Fatty acids and neutral fats and oils. **Fatty acids** are long chains of carbon atoms with attached hydrogens. At one end of the chain is a carboxyl (acid) group (Figure 1.19a). Fatty acids vary in length and in the number of carbons, but most have 16 or 18 carbons. They may be either saturated or unsaturated. **Saturated fatty acids** have no double bonds between carbons, and so all the available bonds are occupied by hydrogen atoms. Saturated fatty acids are common in butter and beef fat. **Unsaturated fatty acids** are common in plant oils and have one or more double bonds in the carbon chain. **Polyunsaturated fatty acids** have several double bonds. Studies have shown that high amounts of saturated fats in the diet increase the possibility of developing coronary heart disease, a topic we'll discuss further in Chapter 9.

Most fatty acids do not exist free in the body. Instead, they are linked to **glycerol** (***gliss**-err-all*) molecules, forming **triglyc-**

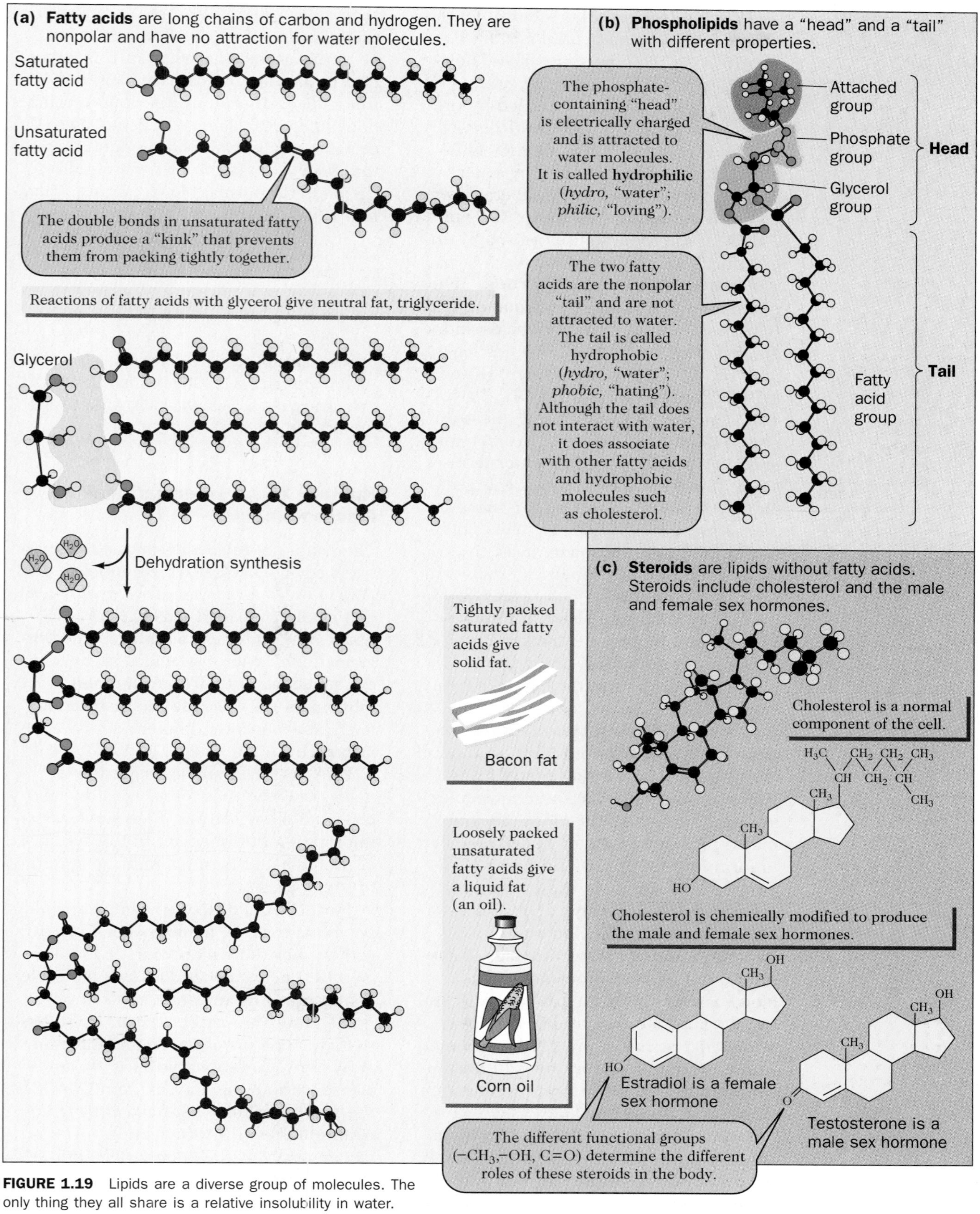

(a) Fatty acids are long chains of carbon and hydrogen. They are nonpolar and have no attraction for water molecules.
Saturated fatty acid
Unsaturated fatty acid
The double bonds in unsaturated fatty acids produce a "kink" that prevents them from packing tightly together.
Reactions of fatty acids with glycerol give neutral fat, triglyceride.
Glycerol
Dehydration synthesis
Tightly packed saturated fatty acids give solid fat.
Bacon fat
Loosely packed unsaturated fatty acids give a liquid fat (an oil).
Corn oil
(b) Phospholipids have a "head" and a "tail" with different properties.
The phosphate-containing "head" is electrically charged and is attracted to water molecules. It is called hydrophilic (hydro, "water"; philic, "loving").
Attached group
Phosphate group
Glycerol group
Head
The two fatty acids are the nonpolar "tail" and are not attracted to water. The tail is called hydrophobic (hydro, "water"; phobic, "hating"). Although the tail does not interact with water, it does associate with other fatty acids and hydrophobic molecules such as cholesterol.
Fatty acid group
Tail
(c) Steroids are lipids without fatty acids. Steroids include cholesterol and the male and female sex hormones.
Cholesterol is a normal component of the cell.
Cholesterol is chemically modified to produce the male and female sex hormones.
Estradiol is a female sex hormone
Testosterone is a male sex hormone
The different functional groups (–CH3,–OH, C=O) determine the different roles of these steroids in the body.

FIGURE 1.19 Lipids are a diverse group of molecules. The only thing they all share is a relative insolubility in water.

eride (*try-**gliss**-er-ride*) or **neutral fat** (see Figure 1.19a). Some neutral fats are liquids (oils), while others are solids. However, all neutral fats store energy.

When stored energy is needed by the body, fatty acids are detached from glycerol by hydrolysis. Then the individual fatty acids are broken down by other chemical reactions, and the liberated energy is used to form ATP from ADP and ℙ. This ATP can then be used to supply energy for the processes of life.

Phospholipids: dual properties. Phospholipids are an important group of lipids that also are built from fatty acids and glycerol. However, phosphoplipids have only two fatty acids per glycerol (Figure 1.19b). Bonded to the third OH of glycerol is a phosphate-containing "head" group. While fatty acid tails are nonpolar and have no attraction for water molecules, this phosphate group is charged and has a strong attraction for water.

The presence of both fatty acid tails and the charged phosphate head gives phospholipids dual properties. They can interact with nonpolar substances such as other fatty acids and also can interact with water and ions. In Chapter 2, we will explore how these dual properties equip phospholipids to form the cell membranes critical to the function of all human cells.

Steroids and cholesterol: structure and regulation. Steroids and cholesterol are lipids that do not have fatty acids. However, like all lipids, these molecules are largely insoluble in water. They are made from carbon atoms linked together to form four fused rings (Figure 1.19c). Small modifications in these structures generate a whole family of substances with important roles. Cholesterol plays a structural role in cell membranes (Chapter 2), and its overabundance in the blood can lead to coronary heart disease (Chapter 9). Steroids, such as testosterone and estrogen, are sex hormones with powerful regulatory roles in human development, behavior, and reproduction (Chapters 15 through 17).

Some vitamins are lipids. The lipid vitamins A, D, E, and K are important regulators of body functions. We will explore them more fully in Chapter 7, and their roles will be discussed in many of the following chapters. Recently it has been recognized that vitamin E is important in limiting the damage caused by free radicals. Because this vitamin can reduce molecular damage, it is added to certain commercial food products to retard spoilage. Vitamin K is required by the body for normal blood clotting (Chapters 7 and 9).

> All lipids have little or no solubility in water. Triglyceride is a storage lipid consisting of three fatty acids bonded to glycerol. Phospholipids have two fatty acid "tails" and a phosphate "head" group which can interact with water and ions. Steroids function as regulating molecules. Vitamins A, D, E, and K are also lipids.

Nucleic Acids: Informational Macromolecules—DNA and RNA

This family of molecules consists of two types of polymers: **deoxyribonucleic acid (DNA)** (*dee-ox-ee-**rye**-bow-new-clay-ick*) and **ribonucleic acid (RNA)** (***rye**-bow-new-clay-ick*). These acids are called **informational macromolecules** because in the sequence of their four different monomers they encode the instructions for the sequence of different amino acids in proteins.

DNA is the hereditary material. It makes up the genes found in all living cells and many viruses. RNA serves as an information intermediate in the synthesis of proteins and is the hereditary material of some viruses.

Both DNA and RNA are long chainlike polymers that are made by bonding together nucleotide monomers. Each **nucleotide** (***new**-clee-oh-tide*) is a molecule consisting of three components: a phosphate group, a 5-carbon sugar, and a base that contains nitrogen (Figure 1.20). In DNA the sugar is **deoxyribose** (which is ribose without one oxygen atom). In RNA the sugar is **ribose.** We have already encountered one RNA nucleotide—ATP—which serves as the energy-rich shuttle.

DNA uses four bases: two double-ringed molecules—**adenine** (***ad**-din-een*) and

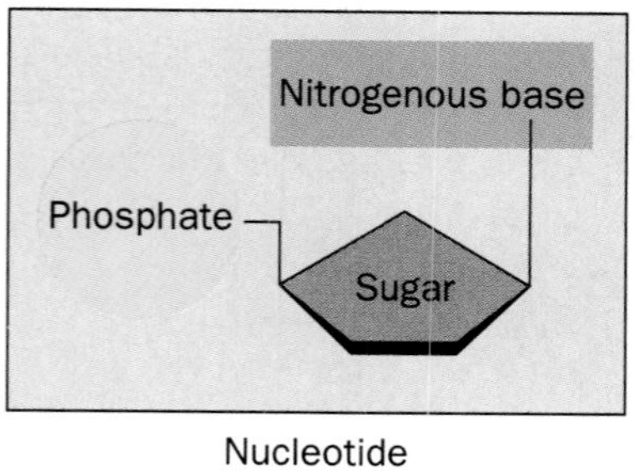
Nitrogenous base
Phosphate
Sugar
Nucleotide

FIGURE 1.20 The nucleotides of DNA and RNA. Illustrated here are the chemical structures of the different molecular units—the nucleotides—that are bonded together to make up the polymers of DNA and RNA.

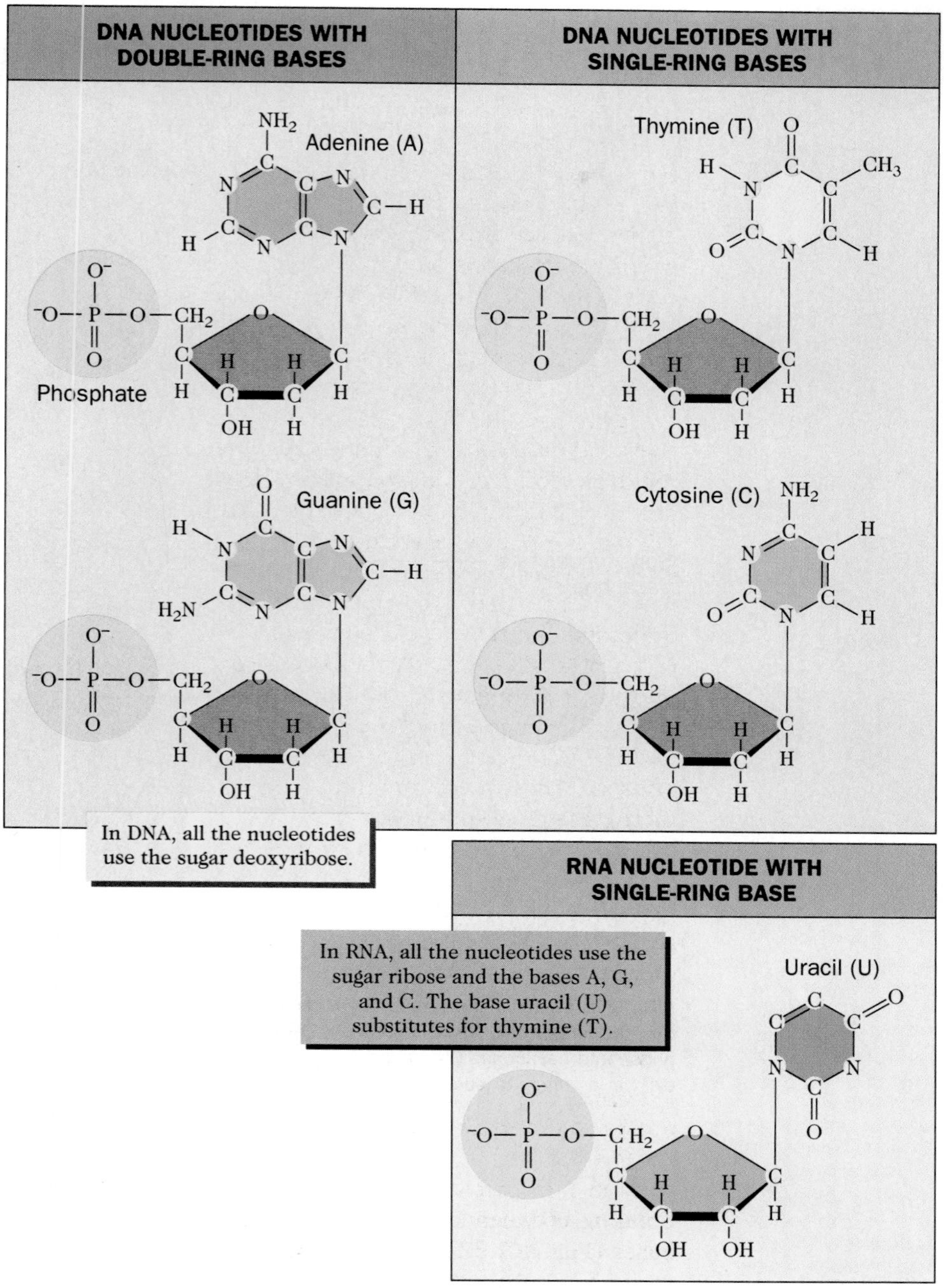
DNA NUCLEOTIDES WITH DOUBLE-RING BASES
DNA NUCLEOTIDES WITH SINGLE-RING BASES
Adenine (A)
Phosphate
Guanine (G)
Thymine (T)
Cytosine (C)
In DNA, all the nucleotides use the sugar deoxyribose.
RNA NUCLEOTIDE WITH SINGLE-RING BASE
In RNA, all the nucleotides use the sugar ribose and the bases A, G, and C. The base uracil (U) substitutes for thymine (T).
Uracil (U)

guanine (***gwah**-neen*)—and two single-ringed molecules—**thymine** (***thigh**-meen*) and **cytosine** (***sigh**-toe-seen*). For convenience, when writing about nucleic acids, we'll use the abbreviations **A, G, T,** and **C** to refer to these bases. In addition to using the sugar ribose, RNA differs from DNA in using the base **uracil** (***your**-a-sill*) instead of thymine (see Figure 1.20).

The polymers DNA and RNA are formed by covalent bonds between the nucleotides. When bonded together, the nucleotides form a long threadlike molecule called a strand. The "backbone" of the strand consists of the sugar and phosphate groups, with the bases extending out to one side (Figure 1.21). This structure permits the bases to interact with other bases.

RNA is usually single-stranded. DNA is a double-stranded helix, and the two strands are held together by hydrogen

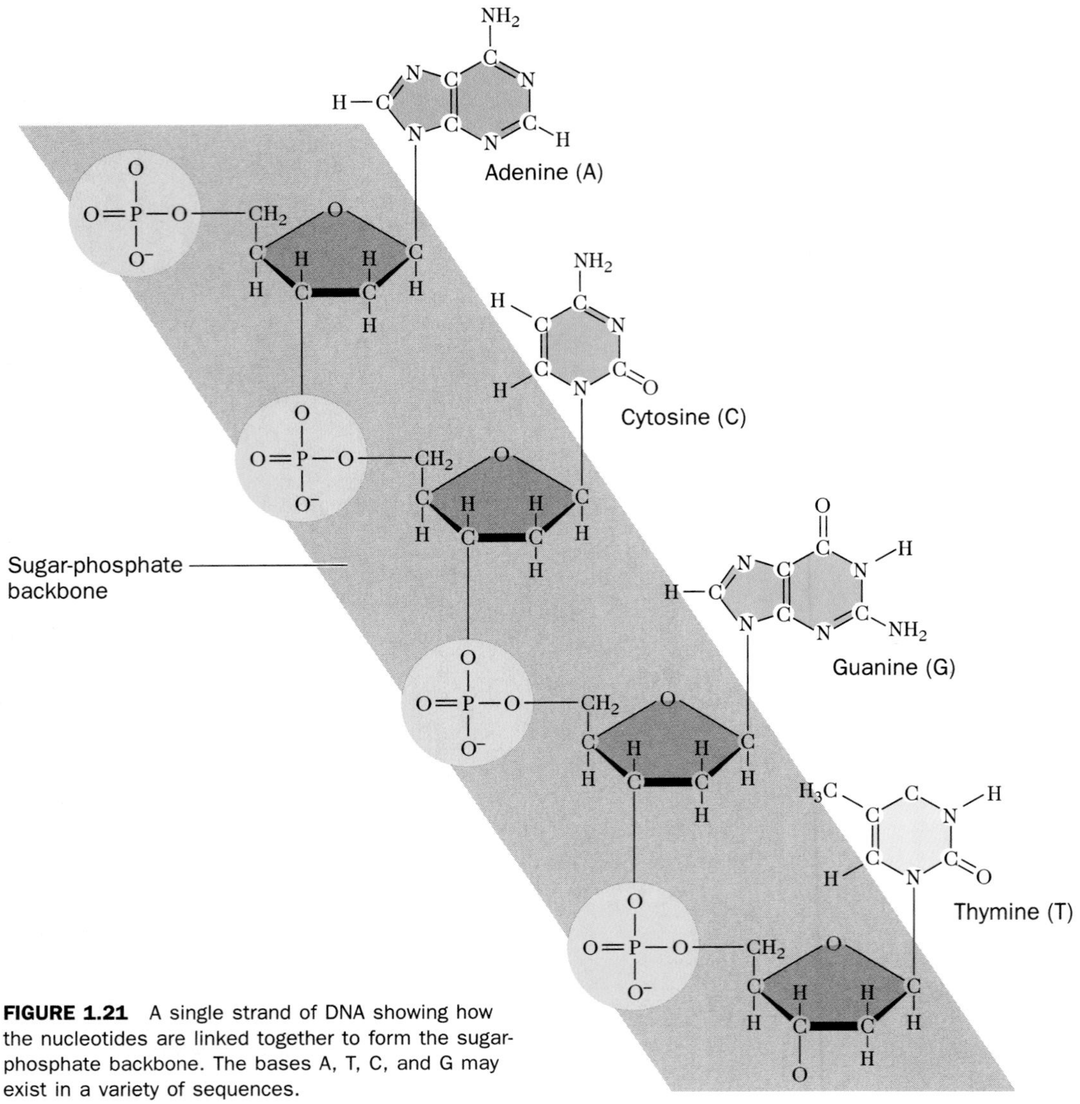
Adenine (A)
Cytosine (C)
Guanine (G)
Thymine (T)
Sugar-phosphate backbone

FIGURE 1.21 A single strand of DNA showing how the nucleotides are linked together to form the sugar-phosphate backbone. The bases A, T, C, and G may exist in a variety of sequences.

bonding between specific complementary bases (Figure 1.22). Thymine always pairs with adenine (A=T), and guanine always pairs with cytosine (G≡C). In a single strand of DNA the bases are present in a specific sequence, and this forms the basis for their coded information.

Because of complementary base pairing, the sequence of nucleotide bases in one DNA strand determines the sequence in the other strand. This is the system the cell relies on to transmit hereditary information (specific sequences of bases) from parent cell to progeny (offspring) cells during cell reproduction (Chapter 2). In addition to passing information between generations (parents to progeny), the sequence of bases in DNA determines the sequence of amino acids in proteins (Figure 1.23).

This is accomplished by using a three-base code for each of the 20 amino acids and employing an RNA intermediate to carry the message from DNA to the sites of protein synthesis. The base sequence of a specific region of a single strand of DNA is copied into a complementary strand of RNA called **messenger RNA**

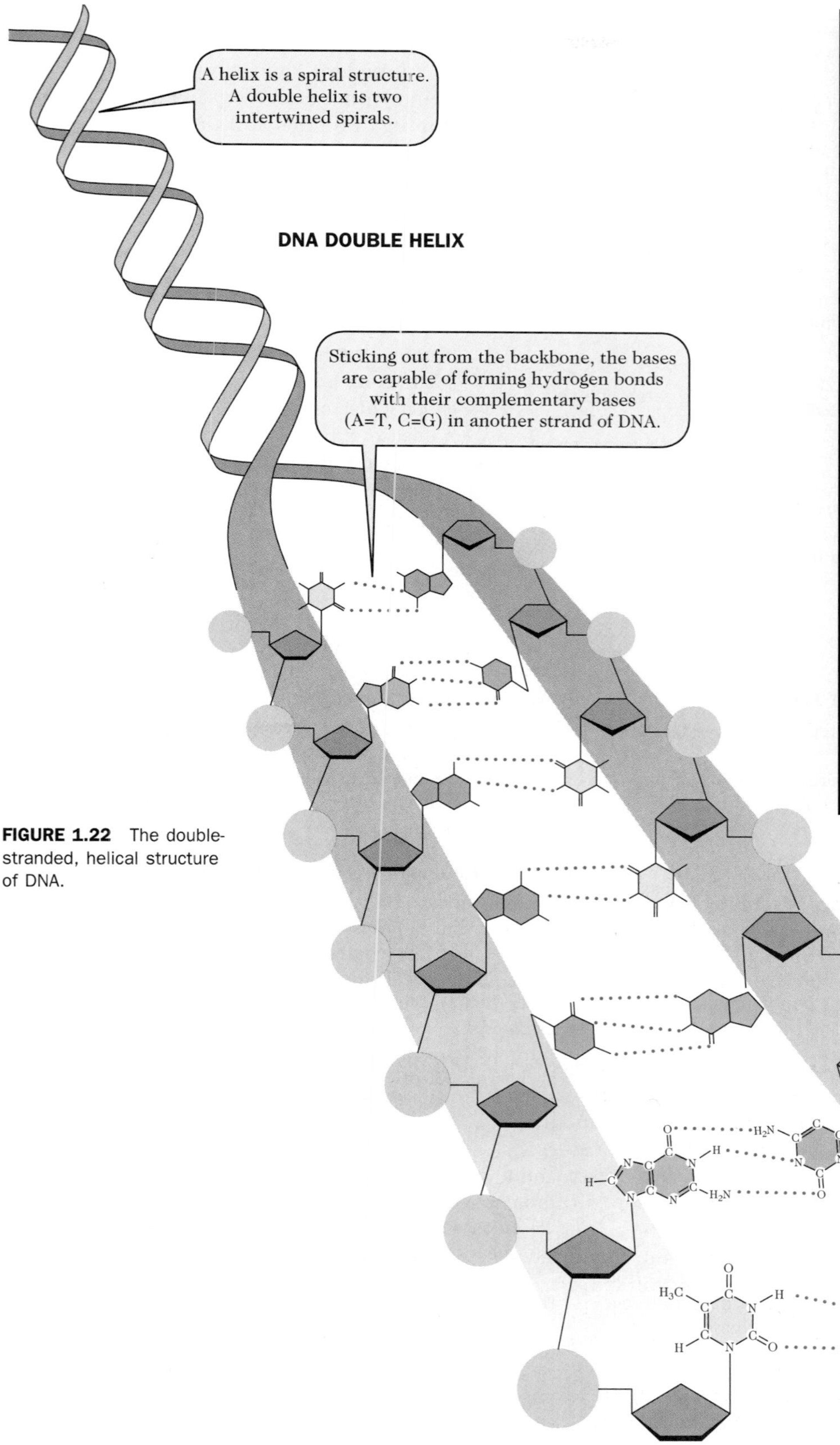
A helix is a spiral structure. A double helix is two intertwined spirals.
DNA DOUBLE HELIX
Sticking out from the backbone, the bases are capable of forming hydrogen bonds with their complementary bases (A=T, C=G) in another strand of DNA.

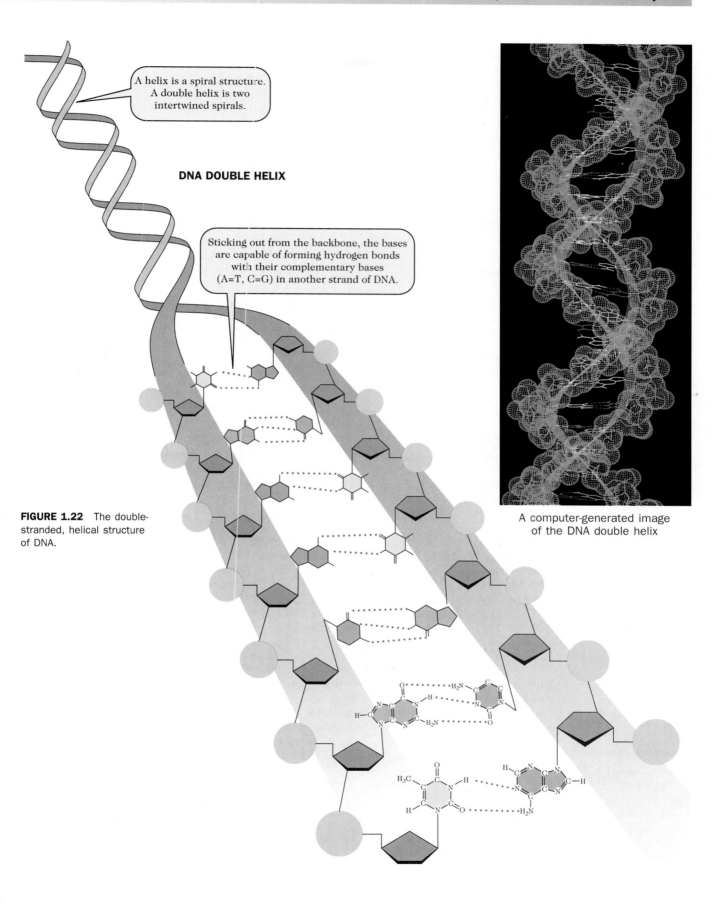

FIGURE 1.22 The double-stranded, helical structure of DNA.

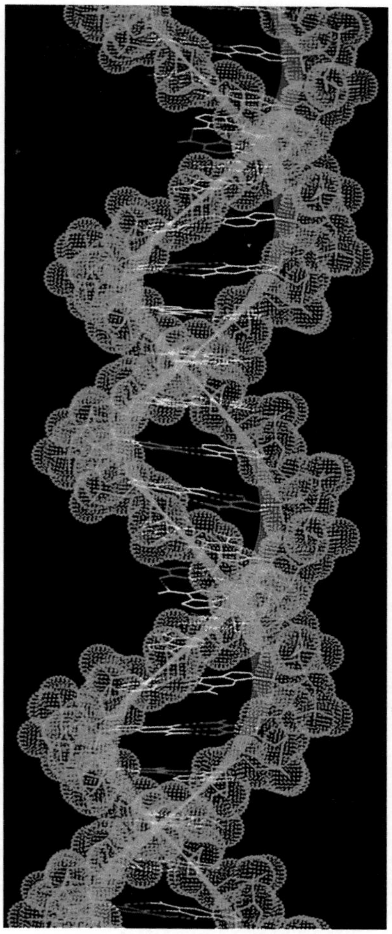

A computer-generated image of the DNA double helix

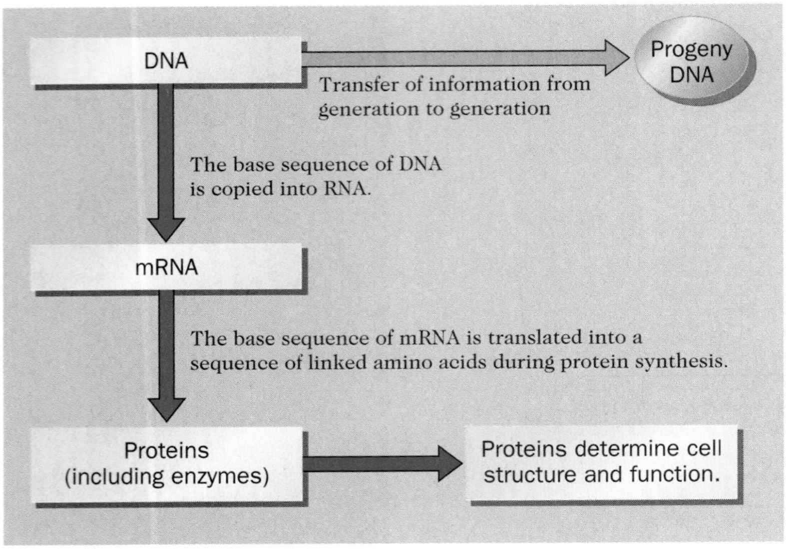
DNA
Progeny DNA
Transfer of information from generation to generation
The base sequence of DNA is copied into RNA.
mRNA
The base sequence of mRNA is translated into a sequence of linked amino acids during protein synthesis.
Proteins (including enzymes)
Proteins determine cell structure and function.

FIGURE 1.23 Information transfer by DNA and RNA. The sequence of bases in DNA codes information for the synthesis of identical DNA (to reproduce a new generation) and the synthesis of proteins (to build and maintain bodies).

(mRNA). This mRNA is then used to synthesize a protein (Chapter 2).

In modern biology, proteins and protein synthesis are very important. It is through proteins and the actions of enzymes that the instructions in the DNA control the moment-to-moment life of cells.

> Nucleic acids (DNA and RNA) are polymers of nucleotides. Each nucleotide consists of a phosphate group, a 5-carbon sugar, and a base. DNA has a double-stranded helical structure. Both RNA and DNA are informational macromolecules in which the sequence of nucleotides codes for the sequence of amino acids in proteins.

Chapter Summary

Matter, Energy, and the Dance of Life

- Neither matter nor energy is created or destroyed, but both can exist in different forms.
- Different kinds of atoms have different numbers of protons (+), electrons (−), and neutrons (no charge).

Ions, Bonds, and Molecules

- Chemical bonds link atoms together to form molecules. Ions are electrically charged atoms. Ionic bonds are formed between oppositely charged ions.
- Covalent bonds are formed when atoms share one or more electron pairs. Polar covalent bonds have small positive and negative charges.
- Hydrogen bonds are weak electrical attractions between molecules whose atoms are linked by polar covalent bonds.
- Functional groups are stable associations of atoms with specific properties that may be attached to a variety of larger molecules.

Chemical Reactions, Chemical Changes

- In a chemical reaction, reactants (A and B) are converted to products (C and D). In all chemical reactions, matter is neither created nor destroyed but merely changes form.
- ATP is an energy-carrying molecule that shuttles energy from storage forms to reactions where it is needed. Energy is trapped in ATP when it is formed from ADP and a phosphate group ℙ. Energy is released from ATP when it breaks down to ADP and ℙ.

Acids and Bases: Reactions in Equilibrium

- Acids are substances that give up hydrogen ions (H^+). Bases are substances that take up H^+.
- The pH scale measures the degree of acidity. Buffers limit pH changes and protect living cells from damage by strong acids and bases.

Biological Molecules and Their Roles

- Proteins consist of amino acids linked by peptide bonds. Enzymes are proteins that speed up chemical reactions.
- Protein function requires the maintenance of the specific three-dimensional shape of proteins, which is established by coiling and folding and is maintained by hydrogen bonding.
- Carbohydrates are primarily involved with the storage and transport of energy.
- Lipids are used for storage (triglycerides), structure (phospholipids), and regulation (steroids).
- In all cells, DNA is the hereditary material of the genes and RNA carries information from DNA to the sites of protein synthesis. Both DNA and RNA are made of nucleotides. RNA nucleotides use the sugar ribose; in DNA, the sugar is deoxyribose. In DNA, the bases are adenine (A), thymine (T), cytosine (C), and guanine (G). In RNA, the base uracil (U) substitutes for T.
- DNA consists of two strands held together in a double helix by hydrogen bonding between complementary bases (A═T, G═C). The sequence of bases in one strand of DNA determines the sequence in the other stand.
- Using RNA as an intermediate, the sequence of bases in DNA determines the sequence of amino acids in proteins.

Selected Key Terms

acid/base (p. 30)
atom (p. 17)
ATP (p. 29)
chemical bonds (pp. 24–25)
DNA (p. 38)
electron (p. 19)
energy (p. 16)
enzyme (p. 32)
functional group (p. 26)
ion (p. 23)
lipid (p. 35)
matter (p. 16)
molecule (pp. 19, 24)
protein (p. 32)
proton (p. 19)
RNA (p. 38)

Review Activities

1. Identify the subatomic particles, their properties, and their locations in an atom.
2. How do the numbers of subatomic particles differ in different kinds of atoms?
3. What are the bonding capacities of the six most abundant kinds of atoms in the human body?
4. Describe what happens to matter and energy during a chemical reaction.
5. Describe the chemical reactions of ATP and its energy role.
6. Distinguish between an acid and a base.
7. What is pH, and how are changes in pH limited?
8. Describe the composition and function of an enzyme.
9. Describe the size, composition, and structural levels of proteins.
10. How do monosaccharides, disaccharides, and polysaccharides differ?
11. Describe the different roles of triglycerides, phospholipids, and steroids in the human body.
12. Identify the roles of DNA and RNA.
13. Describe the chemical and structural differences between DNA and RNA.
14. Describe how the two strands of DNA are held together in a helix.

Self-Quiz

Matching Exercise

___ 1. A change in the atomic composition and properties of molecules
___ 2. A substance that releases hydrogen ions
___ 3. The type of chemical bond that holds the two strands of DNA together in a double helix
___ 4. A class of proteins that speeds up chemical reactions
___ 5. The molecule that forms the hereditary material, the genes
___ 6. A kind of lipid with both polar and nonpolar parts
___ 7. The bond that links amino acids together to form proteins
___ 8. The kind of RNA involved in protein synthesis
___ 9. The energy shuttle molecule
___ 10. The measure of acidity

A. Enzymes
B. DNA
C. ATP
D. Acid
E. pH
F. Chemical reaction
G. Peptide bond
H. Phospholipid
I. Messenger RNA
J. Hydrogen bond

Answers to Self-Quiz

1. F; 2. D; 3. J; 4. A; 5. B; 6. H; 7. G; 8. I; 9. C; 10. E

PART ONE

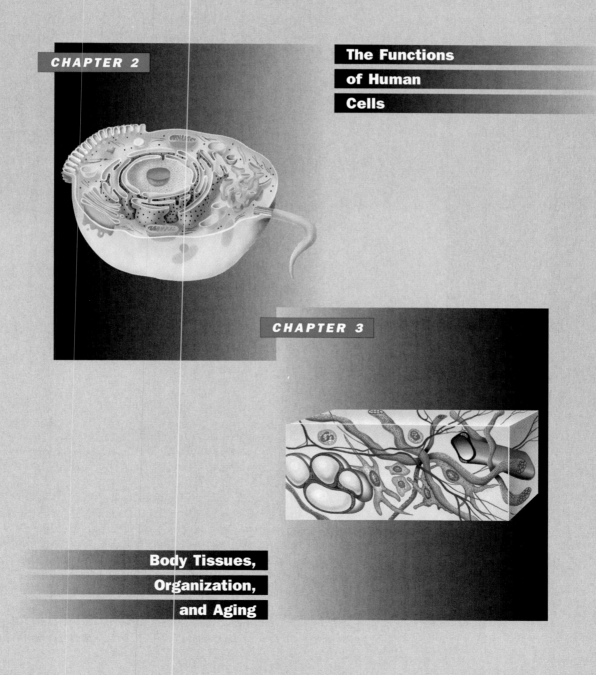

Building the Body: Cells and Tissues

CHAPTER 2

The Functions of Human Cells

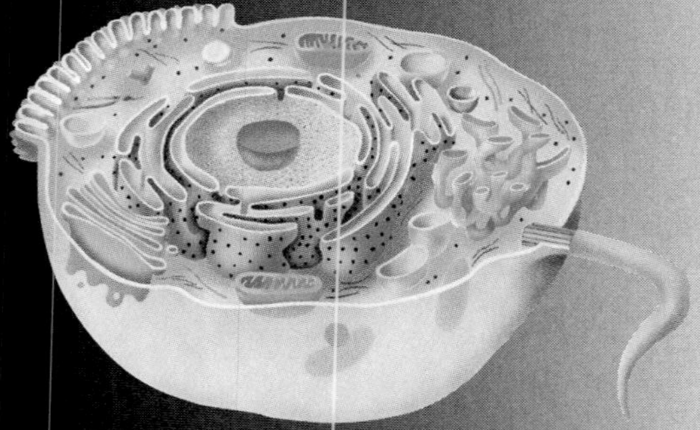

CHAPTER 3

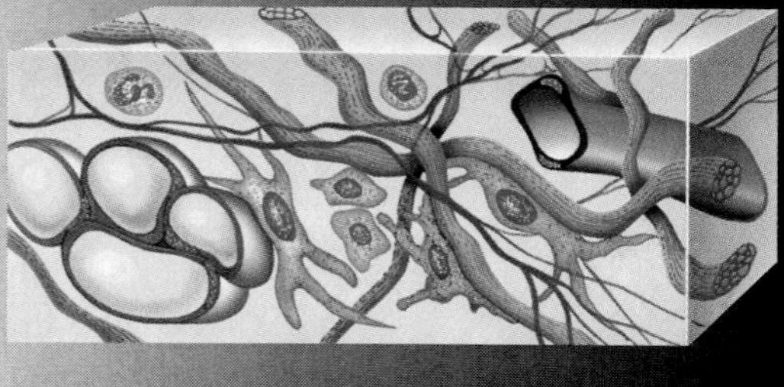

Body Tissues, Organization, and Aging

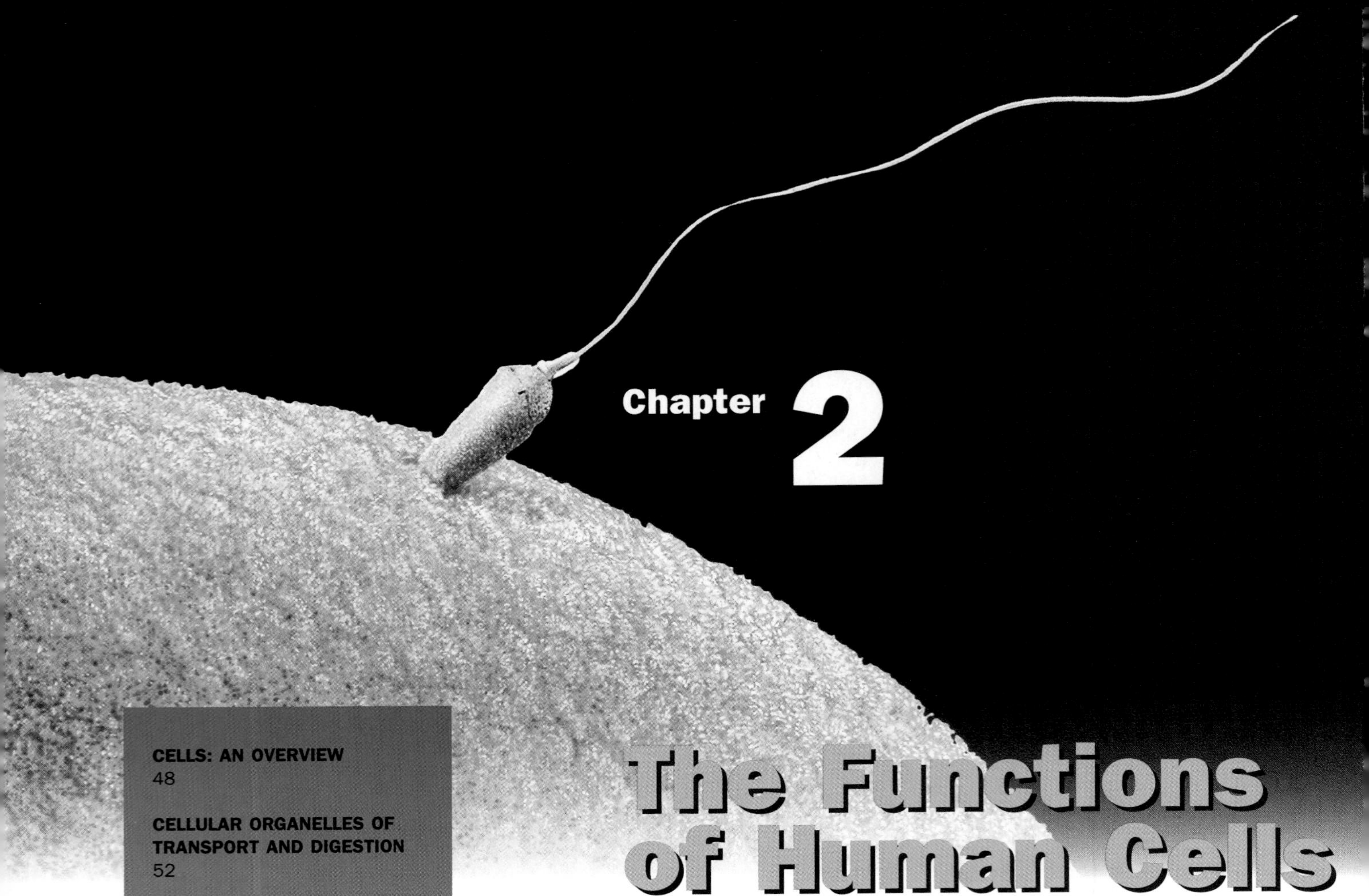

Chapter 2

The Functions of Human Cells

The cells of the human body are too small to see and too numerous to count, but each one is important. The lives of our cells give us life. Their living and dying are our living and dying.

Like all living things, the human body is composed of tiny boxlike units called cells that are too small to see with the unaided eye. However, we know of their existence through the use of microscopes that magnify their images, revealing their structures and lives. From the year 1674, when living cells were first observed, to the present day, millions of observations and experiments have confirmed the hypothesis that (1) all living things are composed of cells or cell products, (2) the cell is the smallest unit that shows all the characteristics of life, and (3) all cells come from preexisting living cells. These three generalizations are called the **cell doctrine.** Let's examine the significance of each statement for the human body.

When we observe any part of the human body with a microscope, we always find living cells and their products. However, not all cells are identical. Some cells are larger than others, and some have different shapes (Figure 2.1). The human body contains about 200 different cell types that perform different functions and use different cell structures. **Structure** refers to the materials used to construct cells and the way those materials are put together. **Function** refers to how cells operate: the kind of work they do and the way they do it.

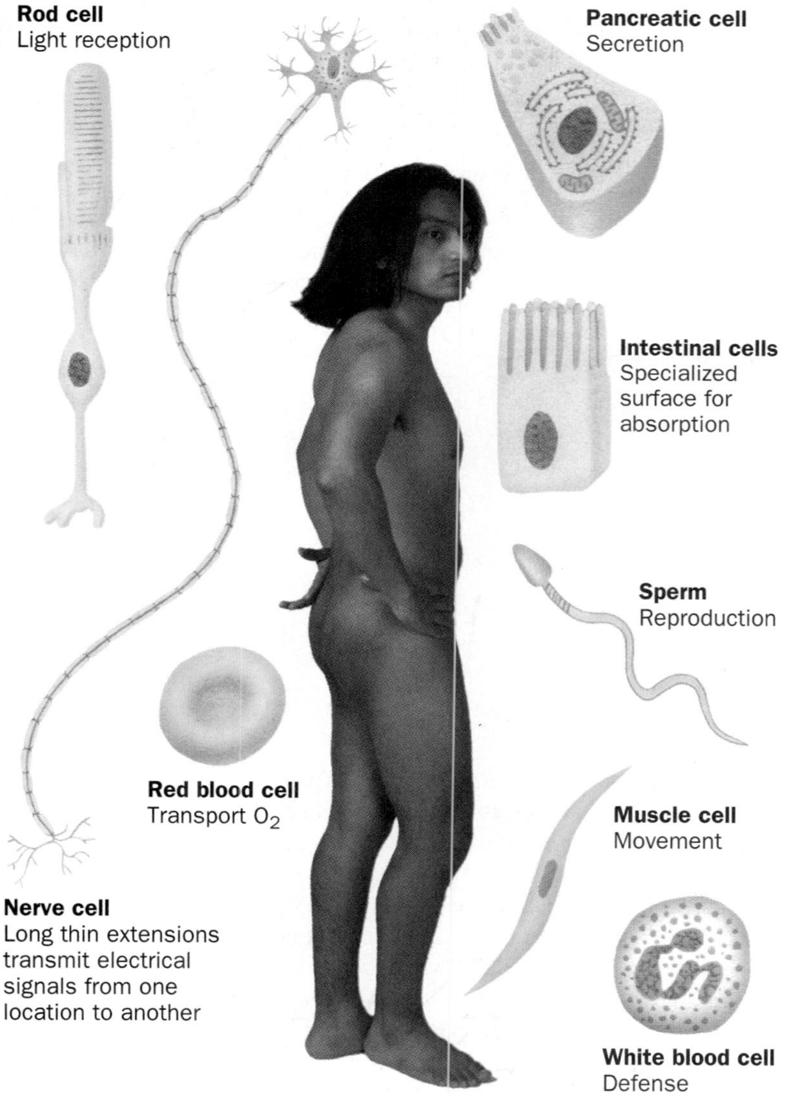
Rod cell
Light reception
Pancreatic cell
Secretion
Intestinal cells
Specialized surface for absorption
Sperm
Reproduction
Red blood cell
Transport O_2
Muscle cell
Movement
Nerve cell
Long thin extensions transmit electrical signals from one location to another
White blood cell
Defense

FIGURE 2.1 The human body is made up of many types of cells that have different sizes, shapes, structures, and functions. The functions a cell performs are related to its structure and shape.

No matter how much cells differ in appearance and function, they all show most of the characteristics of living systems. They take in raw materials, transform matter and energy, excrete waste materials, and respond to their environment. And what about cell reproduction? Can all the cells of your body reproduce themselves? The answer is yes and no. In general, all but the most specialized cells are capable of reproduction.

All the cells in the human body are derived from a single cell—the fertilized egg. This cell undergoes a duplication process and divides into two identical cells. The processes of duplication and cell division are repeated over and over, eventually producing all the cells in the body. However, your cells are not all identical. Cell proliferation is accompanied by a process called development.

During **development**, changes take place in the structure and functions of cells and the cells become more specialized. Developmental changes in cells are not accidental but instead are encoded in the hereditary material of the cell, the DNA. For example, during development some cells become muscle cells with specialized proteins that produce movement. Other cells produce proteins that act as enzymes to digest food. Still other cells become skin cells, containing quantities of a tough, waterproof protein. Although the instructions are found in DNA, the story of cell function is the story of cell proteins. Different kinds of cells perform different functions because of the different kinds of proteins they contain.

In this chapter, we will emphasize the structures and functions common to most of the cells in the human body. We will use the power of modern electron microscopes to gaze inside the cell and examine its component structures, observing how those structures interact with each other so that the cell functions. In later chapters, we will give more emphasis to the characteristics of the specialized cells that are the products of development.

Cells: An Overview

All living cells are enclosed by a surface structure called the **plasma membrane.** This structure surrounds the **cytoplasm,** which is mostly water but contains a variety of chemicals, enzymes, and the cellular machinery needed to maintain the life of the cell.

Human cells are **eukaryotic cells** (*you-care-ee-ot-tick*), as are the cells of all other animals, plants, Fungi, and Protista (four of the five kingdoms described in the Prologue). This means that all these cells share a certain pattern of internal structure and organization that is distinct from that of **prokaryotic cells** (*pro-care-ee-ot-tick*), which are the bacteria (kingdom Monera) (Figure 2.2).

The cytoplasm of eukaryotic cells contains many small compartments called **organelles.** Most organelles are separated from the cytoplasm by a thin membrane. One of these organelles—the **nucleus**—contains the hereditary material **DNA** (Figure 2.3). In fact, the word "eukaryotic" means having a "true" or formed nucleus (*eu,* "true"; *karyote,* "nucleus").

In prokaryotes (*pro,* "before"; *karyote,* "nucleus"), there is no separate compartment to contain the DNA, and these cells do not have a nucleus. The cytoplasm of prokaryotes also lacks many of the organelles found in eukaryotes (see Figure 2.2b). The simpler prokaryotic cells is believed to have evolved before the eukaryotic cell.

In this book, we will focus mostly on eukaryotic cells. However, we will also discuss the life of prokaryotes to help you understand how bacteria make us sick and how antibiotics can kill bacteria but leave our cells unharmed.

The Sizes of Cells: Measuring What Is Tiny

We opened the chapter by saying that cells are small, but how small is small? To measure and compare the sizes of cells, scientists use the metric system of measurement. In this book, we'll usually give dimensions in metric units and usually provide a rough equivalent in the more familiar English system (Table 2.1). However, you should become familiar with the metric system and the names of its units for length, weight, and volume.

In the metric system, the standard unit of length is the **meter** (abbreviated **m**), and it measures about 1 yard (actually 39.4 inches). One meter consists of 100 **centimeters** (*centi,* "hundred"), abbreviated **cm.** Each centimeter is divided into 10 parts called **millimeters (mm).** The prefix *milli* means

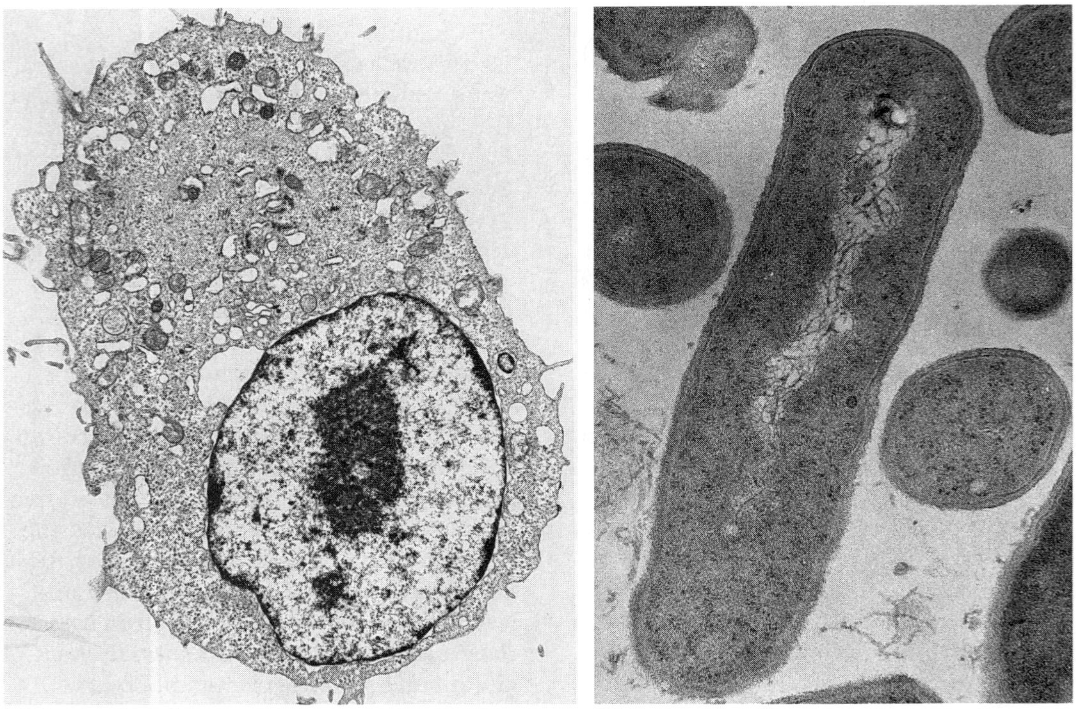

(a) Eukaryotic cell **(b)** Prokaryotic cell

FIGURE 2.2 The structure of eukaryotic cells and prokaryotic cells. As these two photographs taken with the electron microscope show, eukaryotic cells (a) have a great deal more internal structure than do prokaryotic cells (b). These photographs present only a two-dimensional view of the cell interior.

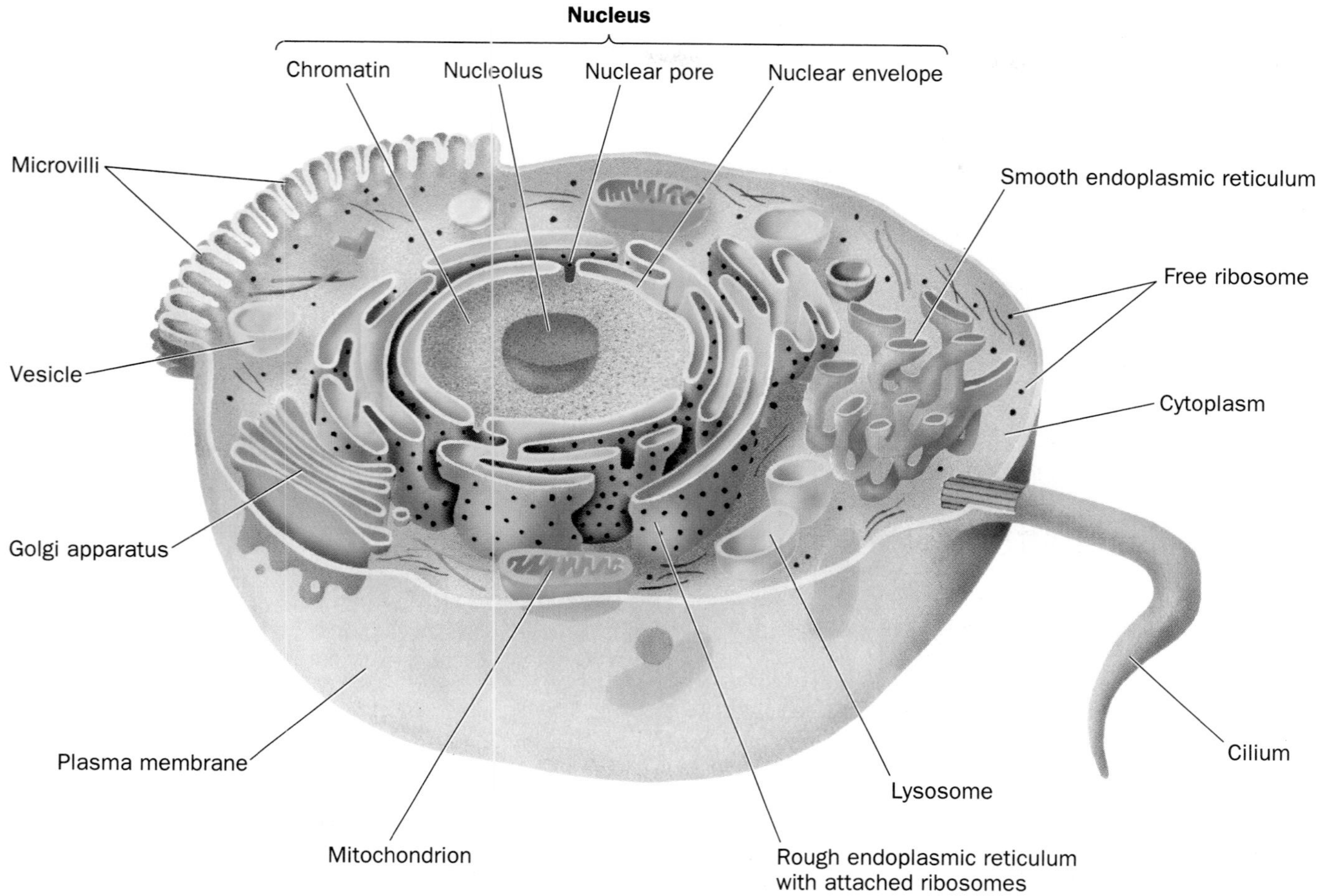
Nucleus
Chromatin
Nucleolus
Nuclear pore
Nuclear envelope
Microvilli
Smooth endoplasmic reticulum
Free ribosome
Vesicle
Cytoplasm
Golgi apparatus
Cilium
Plasma membrane
Lysosome
Mitochondrion
Rough endoplasmic reticulum
with attached ribosomes

FIGURE 2.3 A eukaryotic cell. This three-dimensional drawing shows the many different organelles in the cytoplasm of the eukaryotic cell. The most prominent eukaryotic organelle is the nucleus, which contains the hereditary material.

"thousand," and there are 1000 millimeters in a meter. A millimeter is a small unit of measurement, but with normal eyesight you can easily see things a millimeter in size—about the diameter of a paper clip wire. However, cells are smaller.

To measure most cells, we use the unit called a **micrometer (μm)**. There are 1000 micrometers in 1 millimeter, and things this size are not visible to the unaided eye. Most human cells range in size from 10 to 100 μm. Prokaryotic cells are smaller, ranging from 1 to 3 μm. Although cells are small, their internal structures and molecules are even smaller (Figure 2.4). To measure these structures, we use the unit called a **nanometer (nm).** A micrometer contains 1000 nanometers. Small molecules such as glucose and the amino acids have dimensions of only a few nanometers, while the much larger proteins may be 10 nm or more in diameter.

Why are cells so small? Why is the human body made up of trillions of cells rather than just one large cell? Finding the answer requires us to look at the relative sizes of the cytoplasm and the cell surface. The cytoplasm contains all the cellular machinery. The larger the volume of cytoplasm, the more manufacturing activity. To support this activity, raw materials and energy must enter the cell and wastes must leave, and these exchanges must occur across the cell surface. A larger surface area permits more exchange. As Figure 2.5 shows, when a volume of cytoplasm is divided into many small cells, it has a greater total surface area than does the same volume of cytoplasm in the form of a single large cell.

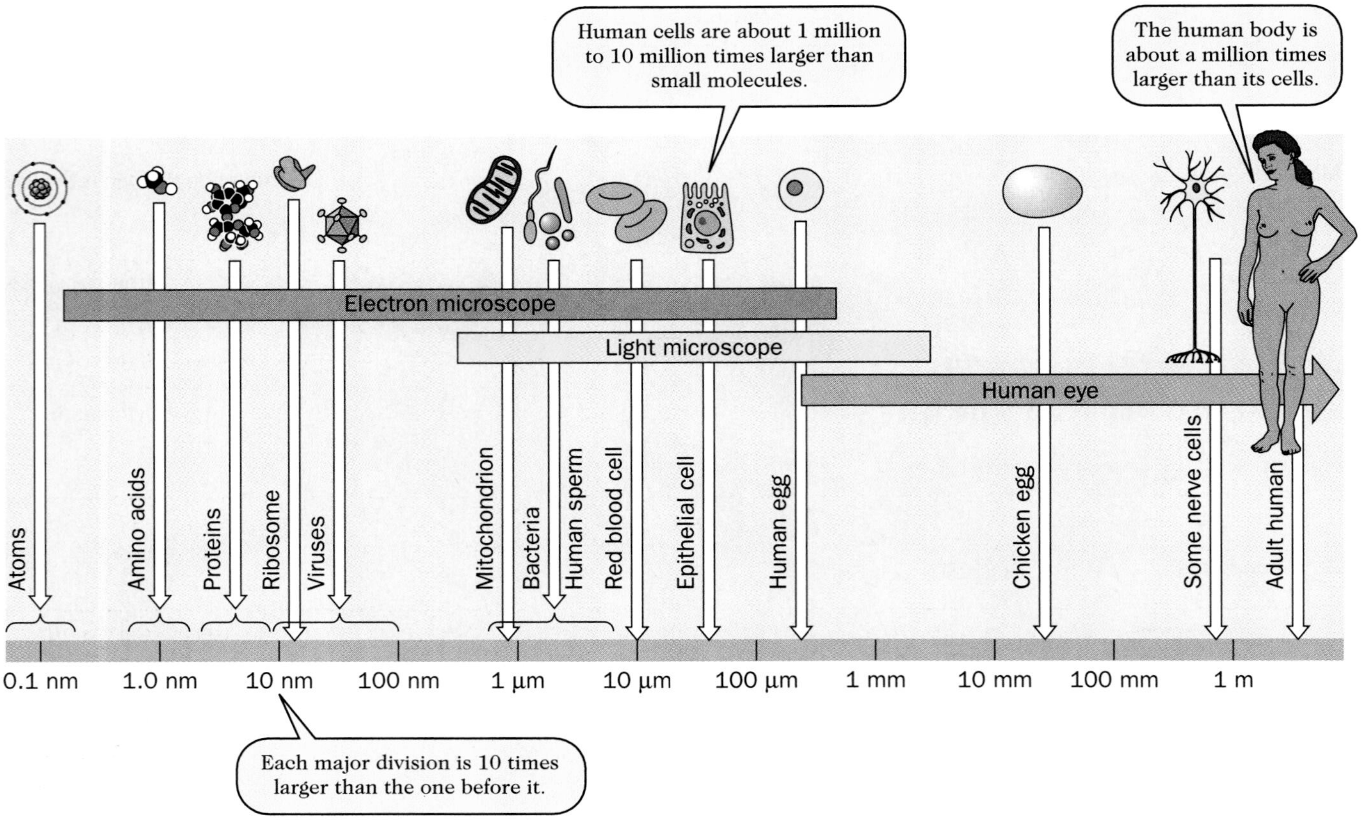
Human cells are about 1 million to 10 million times larger than small molecules.
The human body is about a million times larger than its cells.
Electron microscope
Light microscope
Human eye
Atoms
Amino acids
Proteins
Ribosome
Viruses
Mitochondrion
Bacteria
Human sperm
Red blood cell
Epithelial cell
Human egg
Chicken egg
Some nerve cells
Adult human
0.1 nm
1.0 nm
10 nm
100 nm
1 µm
10 µm
100 µm
1 mm
10 mm
100 mm
1 m
Each major division is 10 times larger than the one before it.

FIGURE 2.4 The sizes of the human body and its cells, organelles, and molecules.

Cells are small because their surface area must be large relative to the volume of their cytoplasm.

In the chapters that follow you will see how important it is for cells to exchange materials with their environments. In fact, some cell surfaces have folds or tubelike structures to increase the surface area over which materials can be exchanged.

stop here

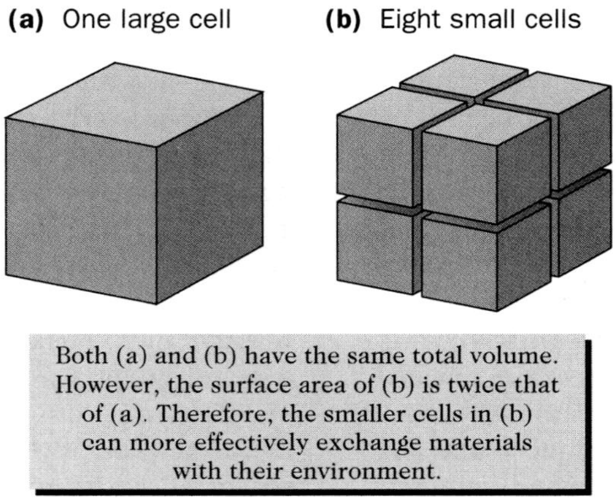
(a) One large cell
(b) Eight small cells
Both (a) and (b) have the same total volume. However, the surface area of (b) is twice that of (a). Therefore, the smaller cells in (b) can more effectively exchange materials with their environment.

FIGURE 2.5 "Why are cells so small?" The surface areas and volumes of cells.

All human cells are eukaryotic cells and have organelles that are not found in prokaryotic cells. Cells are small so that the volume of their cytoplasm is balanced by an adequate surface area over which to exchange materials with the environment.

Revealing the Invisible: The Electron Microscope

As Figure 2.3 shows, cells are not merely bags of molecules dissolved in water. Quite the contrary. Within human cells, molecules of protein, nucleic acid, and phospholipid are organized to form the organelles that perform necessary functions. Pictures taken through the electron microscope reveal the shapes and structures of organelles (see Figure 2.2a). This instrument was invented in 1932 and has been in widespread use only since the early 1950s. Since that time, the instrument has been improved, but the principles of its operation have remained unchanged.

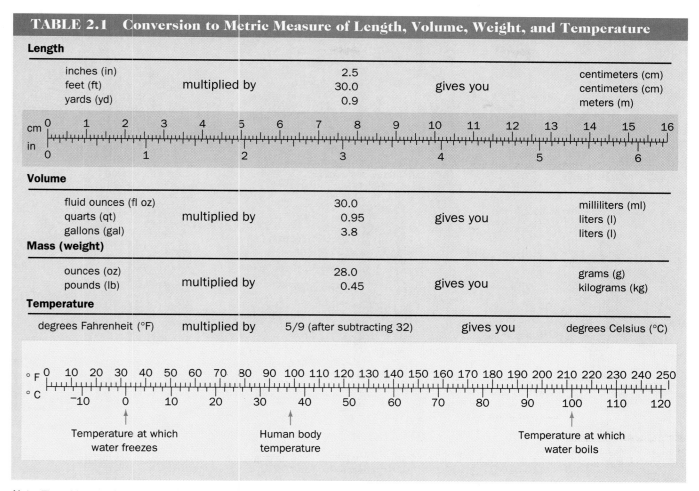

TABLE 2.1 Conversion to Metric Measure of Length, Volume, Weight, and Temperature

Length

inches (in)	multiplied by	2.5	gives you	centimeters (cm)
feet (ft)	multiplied by	30.0	gives you	centimeters (cm)
yards (yd)	multiplied by	0.9	gives you	meters (m)

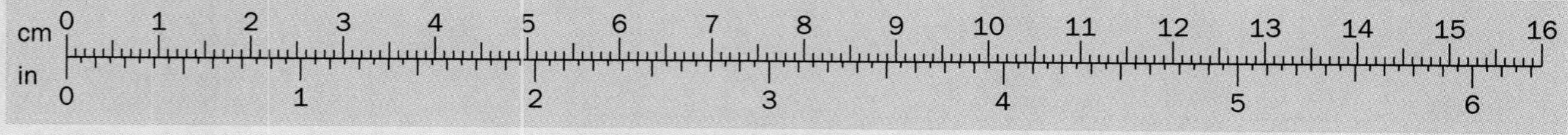
cm
0 1 2 3 4 5 6 7 8 9 10 11 12 13 14 15 16
in
0 1 2 3 4 5 6

Volume

fluid ounces (fl oz)	multiplied by	30.0	gives you	milliliters (ml)
quarts (qt)	multiplied by	0.95	gives you	liters (l)
gallons (gal)	multiplied by	3.8	gives you	liters (l)

Mass (weight)

ounces (oz)	multiplied by	28.0	gives you	grams (g)
pounds (lb)	multiplied by	0.45	gives you	kilograms (kg)

Temperature

degrees Fahrenheit (°F)	multiplied by	5/9 (after subtracting 32)	gives you	degrees Celsius (°C)

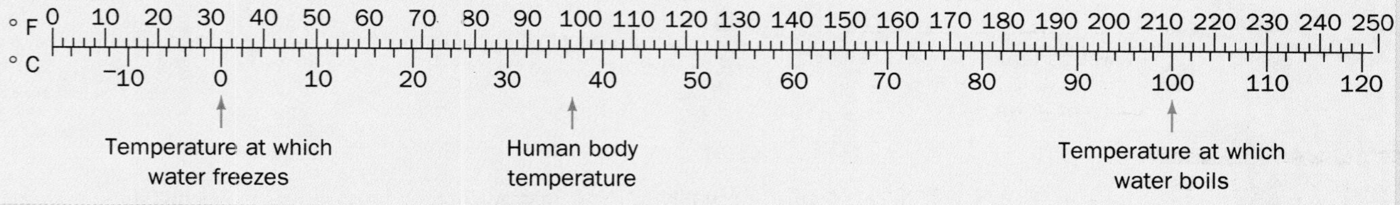
°F 0 10 20 30 40 50 60 70 80 90 100 110 120 130 140 150 160 170 180 190 200 210 220 230 240 250
°C −10 0 10 20 30 40 50 60 70 80 90 100 110 120
Temperature at which water freezes
Human body temperature
Temperature at which water boils

Note: The table specifies multiplication by a factor to convert English units to metric units. Metric units can be converted to English units by division, using the same factor.

The **electron microscope** uses a focused beam of electrons to generate a detailed image of a biological specimen placed in the beam. It is far more effective at revealing detail than is the **light microscope,** which has been employed for about 300 years and uses visible light (Figure 2.6a). The electron microscope can magnify cells up to 100,000 times, compared with a limit of about 1000 times for the light microscope. However, even more important than magnification is the electron microscope's greater resolving power.

Resolving power is the capacity to distinguish two points that are close together. With the light microscope, two points closer than about 300 nm always appear as a single point. However, with the electron microscope, two points in a cell as close together as 2 nm can be distinguished. This greater resolving power is the source of the fine structural detail that is revealed in the photographic images, called *electron micrographs,* produced by the electron microscope (Figure 2.6b).

In both light microscopy (*my-**cross**-co-pea*) and what is called transmission electron microscopy, the beams of light or electrons usually pass through the biological sample. What we focus on is a two-dimensional image of a thin slice of tissue, and we often draw cells as two-dimensional. However, you must remember that the cell and its internal structures are three-dimensional. The three-dimensional shapes of cells are revealed by a special type of electron microscope called the **scanning electron microscope (SEM)** (Figure 2.6c).

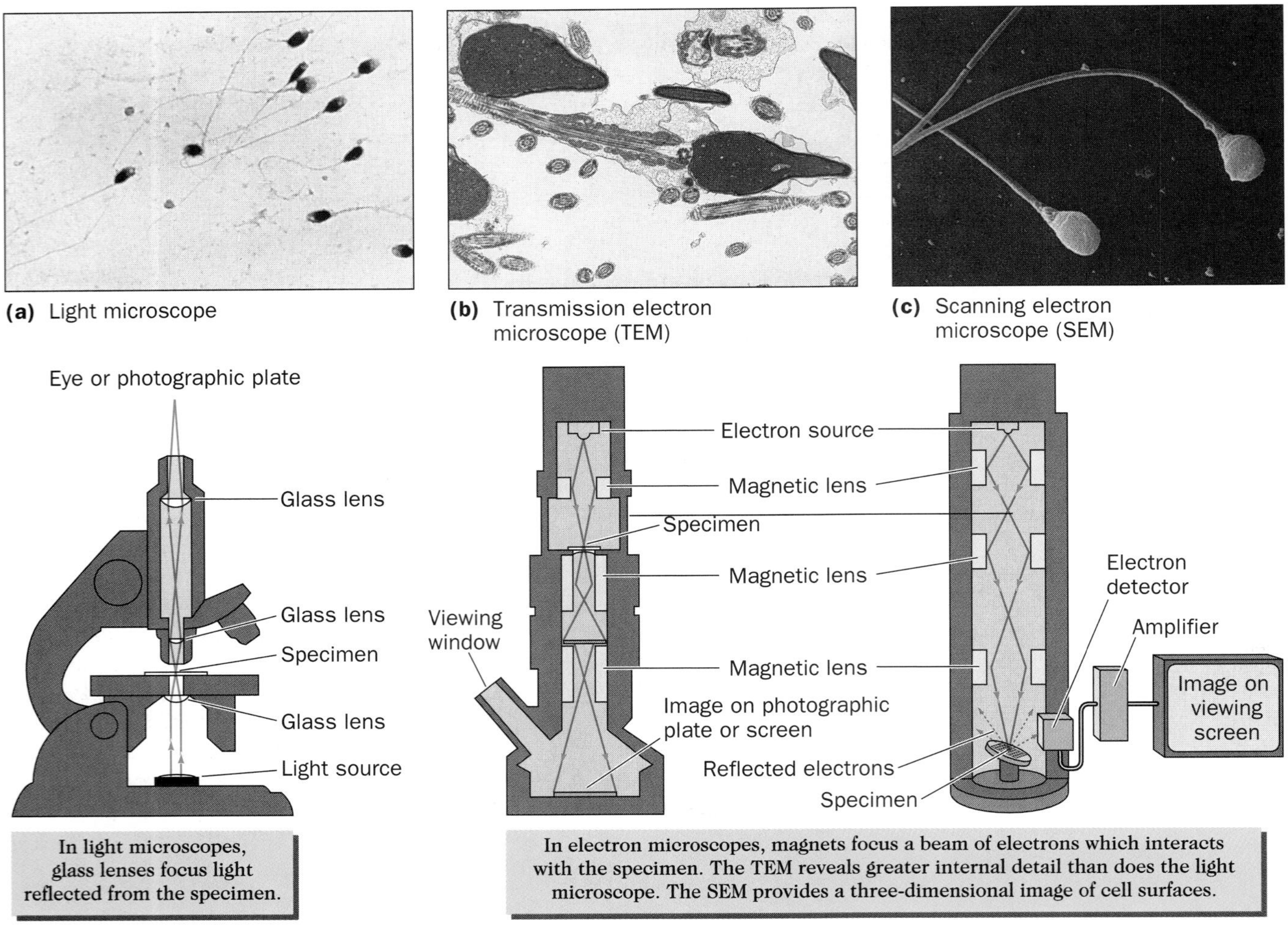
(a) Light microscope
(b) Transmission electron microscope (TEM)
(c) Scanning electron microscope (SEM)
Eye or photographic plate
Glass lens
Glass lens
Specimen
Glass lens
Light source
Viewing window
Electron source
Magnetic lens
Specimen
Magnetic lens
Magnetic lens
Image on photographic plate or screen
Reflected electrons
Specimen
Electron detector
Amplifier
Image on viewing screen
In light microscopes, glass lenses focus light reflected from the specimen.
In electron microscopes, magnets focus a beam of electrons which interacts with the specimen. The TEM reveals greater internal detail than does the light microscope. The SEM provides a three-dimensional image of cell surfaces.

FIGURE 2.6 Both light and electron microscopy generate enlarged images of biological specimens.

> Electron microscopes reveal cellular structures that are not visible in light microscopes. The images from transmission electron microscopy are two-dimensional. The scanning electron microscope gives a three-dimensional image of cell surfaces.

Cellular Organelles of Transport and Digestion

To understand body functions, we need to understand cells, and to understand cells, we need to understand their internal structures. How do these structures function, and how do they contribute to the life of the cell? In the pages that follow we'll describe the structures and functions of the organelles found in most human cells. We will start with the surface of the cell and the plasma membrane and discuss how things get into and out of cells. In examining these processes, we'll pay particular attention to the role and significance of proteins.

The Plasma Membrane and Transport

The **plasma membrane** is a very thin structure about 7 to 9 nm in width that completely surrounds the cytoplasm, separating it from the environment (see Figure 2.3). This membrane assures that the inside of the cell maintains a chemical composition and organization different from the environment outside the cell. The integrity of this membrane is essential to cell life. If it is ruptured, the cell will die. The plasma membrane is composed primarily of phospholipids and proteins but also includes other molecules, such as cholesterol and carbohydrates (Figure 2.7). Indeed, in the plasma membranes of some cells, cholesterol makes up 20 to 25 percent of the lipids.

Membrane phospholipids. Recall from Chapter 1 that **phospholipids** are molecules with both hydrophobic fatty acid "tails" and a hydrophilic "head" group containing the phosphate group (see Figure 1.19b). In a water environment, these dual properties cause phospholipids to form a double layer of molecules called a **bilayer** (*bi,* "two").

In this bilayer, the fatty acid tails interact with each other, forming a **hydrophobic core** (see Figure 2.7). The polar, or charged, head groups are oriented outward, interacting with the polar water molecules in the cytoplasm and in the fluid outside the cell. The lipid cholesterol is also present in the hydrophobic core of the membrane. It adds strength and rigidity to the otherwise flexible and fluidlike behavior of the fatty acids.

Membrane proteins. In addition to these lipid molecules, the plasma membrane includes many proteins that are embedded in the phospholipid bilayer and span its width (see Figure 2.7). There are many different types of membrane proteins. Some are free to move around in the phospholipid bilayer, and some span the bilayer with one end exposed to the cytoplasm and the other end exposed to the outside environment. This general structure of phospholipids and proteins applies to all cellular membranes and is called the **fluid mosaic model.**

Membrane proteins are very significant for the functioning of cells. Indeed, one of the molecular features that distinguishes different cell types (liver cells, blood cells, brain cells) is the presence of different proteins in their plasma membranes. These different proteins perform different functions.

The exposed regions of some membrane proteins help hold cells together. Other membrane proteins function as surface markers, enzymes, or receptors, while still others act as channels and carriers to transport substances into and out of the cell. **Surface markers** are proteins bonded to short chains of carbohydrates. These chains are molecular "name tags" that distinguish your body's cells from foreign cells and diseased cells. They ensure that your body defenses attack dangerous cells, not healthy cells.

Other exposed membrane proteins operate as **enzymes,**

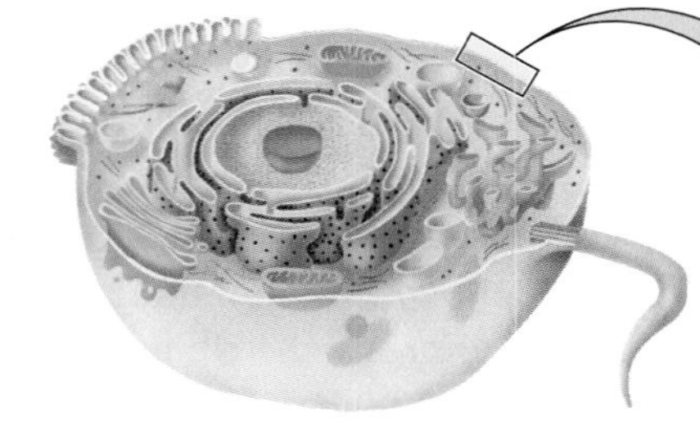

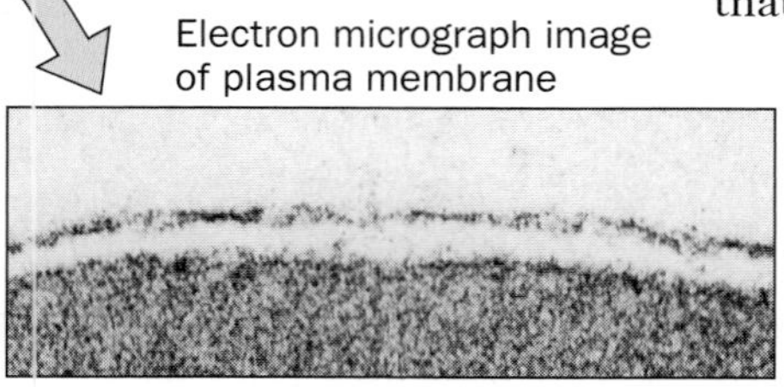
Electron micrograph image of plasma membrane

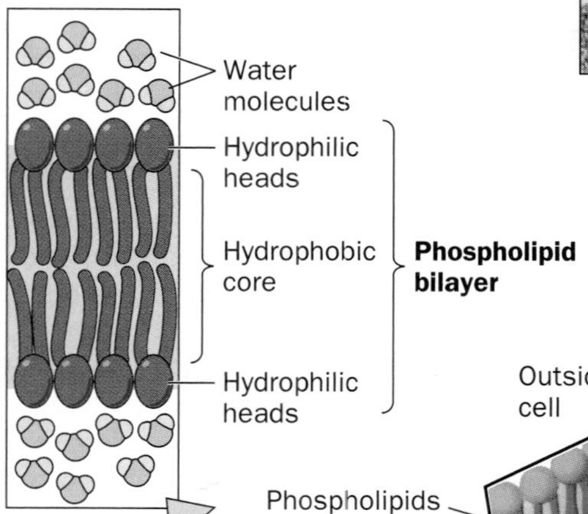
Water molecules
Hydrophilic heads
Hydrophobic core
Phospholipid bilayer
Hydrophilic heads

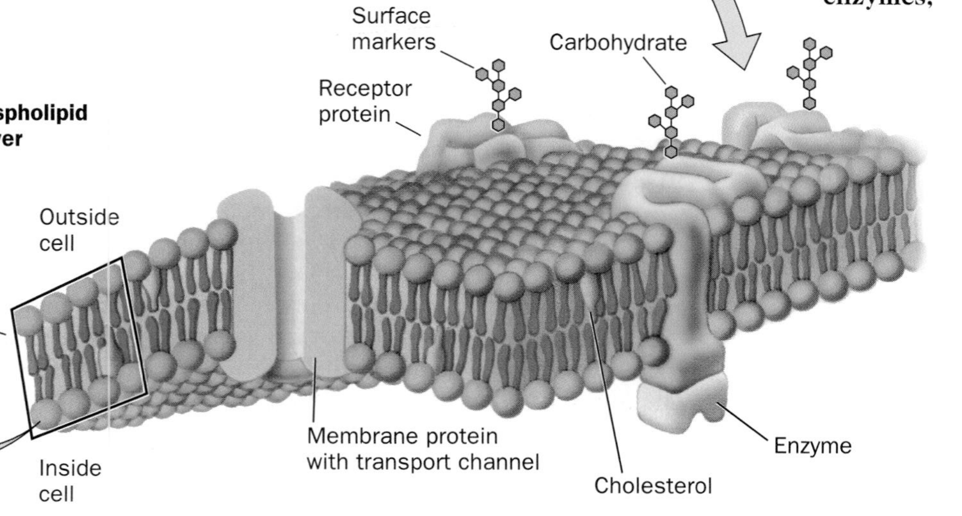
Surface markers
Carbohydrate
Receptor protein
Outside cell
Phospholipids
Inside cell
Membrane protein with transport channel
Cholesterol
Enzyme

FIGURE 2.7 The plasma membrane surrounds the cytoplasm and determines what enters and leaves the cell.

speeding up chemical reactions that are essential to the cell (Chapter 1). As **receptors,** the outer exposed regions can bind to specific molecules in the environment. Such binding causes a change in the protein that is transmitted through the bilayer and into the cytoplasm, where it causes an alteration in cell function.

There are many different kinds of receptor proteins protruding from the outer surface of cells, and each kind is very specific in regard to the molecules to which it will bind. For example, low-density lipoprotein (LDL) receptors bind only to the blood protein LDL. Molecules of LDL transport cholesterol to all the cells in the body. When LDL binds to its receptor, a series of events is initiated by which the cell engulfs the entire LDL particle and eventually extracts the needed cholesterol. We'll describe these events and the extraction of cholesterol later in this chapter.

In addition to functioning as surface markers, enzymes, and receptors, membrane proteins operate as channels or as carriers that determine which molecules and ions enter and leave the cell.

> The plasma membrane separates a cell's interior from the exterior environment. The plasma membrane consists of a phospholipid bilayer in which proteins are embedded. Membrane proteins function as surface markers, enzymes, receptors, and transport channels.

Passive Transport across the Plasma Membrane

The plasma membrane is said to be **selectively permeable.** This means that it permits the transport of substances such as water and carbon dioxide but restricts or controls the movement of other substances. In general, the hydrophobic core of the membrane restricts the movement of larger polar or charged substances such as ions, sugars, and amino acids. These substances cross the plasma membrane by means of channel proteins and carrier proteins.

Channel proteins function like pores, allowing substances to move across the plasma membrane. **Carrier proteins** carry—think of a shuttle—specific substances across the membrane. No matter how different molecules are, they all are transported across the plasma membrane by either passive transport or active transport.

Passive transport does not require the expenditure of cellular energy. In passive transport, substances enter and leave a cell by means of a physical phenomenon called **diffusion.** Diffusion is the movement of substances from a region of higher concentration to a region of lower concentration. Any difference in concentrations establishes what is called a **concentration gradient,** and substances are said to diffuse *down* their concentration gradients, from high to low. Diffusion happens because of the constant random motion of all molecules.

Diffusion may take place through a gas such as air or through a liquid such as water. For example, put a drop of food coloring in a cup of water and observe what happens over time. Initially, the color stays in one place and a concentration gradient exists. However, if you wait long enough, the colored molecules will diffuse throughout the entire cup and the color will become uniformly distributed. When this happens, the concentration gradient is gone and no further changes can be observed.

Diffusion also can take place across a solid membrane if the membrane is permeable to the substance that is diffusing. The body relies heavily on passive transport. For example, water (H_2O) and oxygen gas (O_2) enter cells by diffusion, and the gas carbon dioxide (CO_2) leaves cells the same way.

Some substances, such as CO_2, O_2, and the fat-soluble vitamins, may diffuse directly across the phospholipid bilayer of the membrane. However, ions and most molecules are insoluble in this hydrophobic core and require channel or carrier proteins to cross the membrane (Figure 2.8). For example, chloride ions (Cl^-) require a specific channel protein to diffuse out of the cell. If an altered and defective

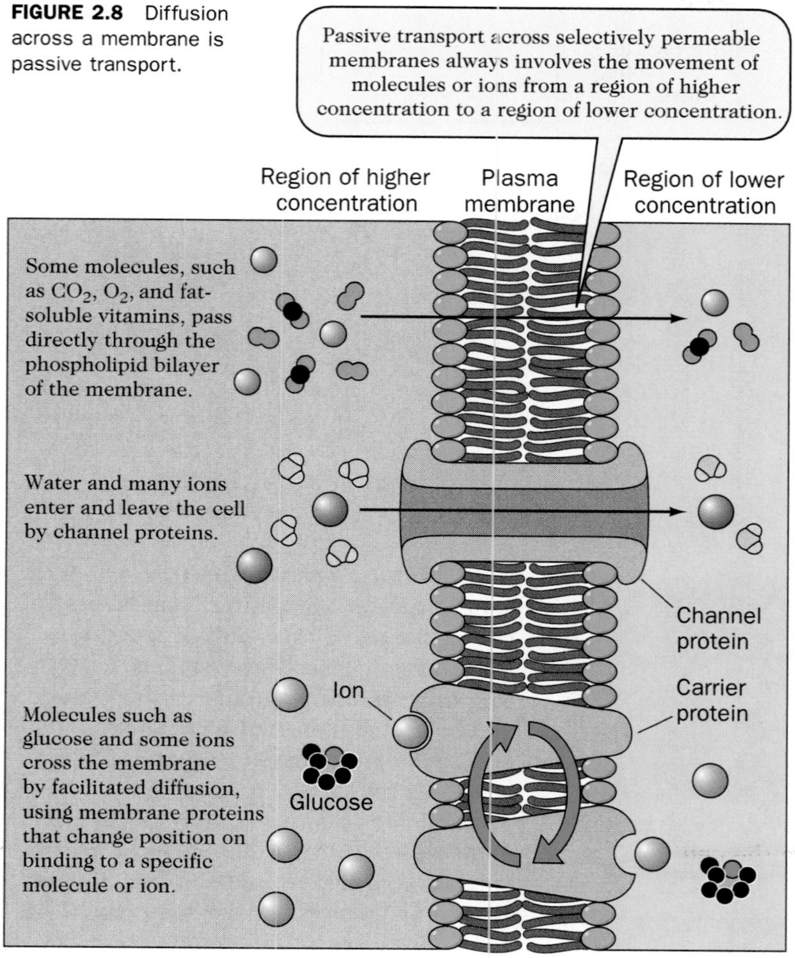
Passive transport across selectively permeable membranes always involves the movement of molecules or ions from a region of higher concentration to a region of lower concentration.
Region of higher concentration
Plasma membrane
Region of lower concentration
Some molecules, such as CO_2, O_2, and fat-soluble vitamins, pass directly through the phospholipid bilayer of the membrane.
Water and many ions enter and leave the cell by channel proteins.
Channel protein
Carrier protein
Ion
Glucose
Molecules such as glucose and some ions cross the membrane by facilitated diffusion, using membrane proteins that change position on binding to a specific molecule or ion.

FIGURE 2.8 Diffusion across a membrane is passive transport.

Passive transport across selectively permeable membranes is accomplished by diffusion down a concentration gradient. Diffusion may take place across the lipid bilayer or through channel proteins and carrier proteins that span the bilayer.

form of this protein exists in the plasma membrane, Cl^- cannot leave the cell normally. The inherited disease **cystic fibrosis** is believed to be caused by a defective channel protein.

Molecules such as glucose diffuse rapidly across the plasma membrane with the aid of carrier proteins in a process called **facilitated diffusion.** This process involves the following steps: (1) The molecule binds to a carrier protein on one side of the membrane. (2) The protein changes shape to carry the molecule across the membrane. (3) The molecule is released from the carrier protein on the other side (see Figure 2.8). Facilitated diffusion is a passive process because the molecules move down a concentration gradient and cellular energy is not required. Such a system speeds the entry of glucose into the cells.

Osmosis: the diffusion of water. **Osmosis** is the diffusion of water across a selectively permeable membrane. The movement of water into and out of cells is directly related to the amount of solute the water contains (see Chapter 1 for a discussion of solutes and solutions).

Because osmosis is an entirely physical-chemical phenomenon, it can be demonstrated with simple laboratory equipment which does not involve living cells. Such a demonstration uses two water-filled compartments separated by an artificial membrane. The membrane must be permeable to water and impermeable to a solute (dissolved substance) that is present in a greater concentration in one of the compartments (Figure 2.9).

The presence of solute in one compartment lowers the concentration of water in that compartment. (The solute takes up space that water molecules could occupy.) Therefore, in the system depicted in Figure 2.9 water moves from the high-concentration compartment into the compartment with the lower water concentration. The same thing happens in your body.

The movement of water molecules into a confined compartment such as a cell creates a pressure called **osmotic pressure.** This pressure is a force that is exerted by the liquid against the entire inner surface of the plasma membrane. In Figure 2.9 the magnitude of the osmotic pressure is evident by the height to which the water rises in the tube connected to the compartment on the right. In a cell, this osmotic pressure may be sufficient to rupture the plasma membrane, killing the cell.

To avoid rupturing, our cells must be constantly bathed in a fluid that has about the same concentration of solutes as does their cytoplasm. Such a fluid is called an **isotonic solution** (*iso,* "the

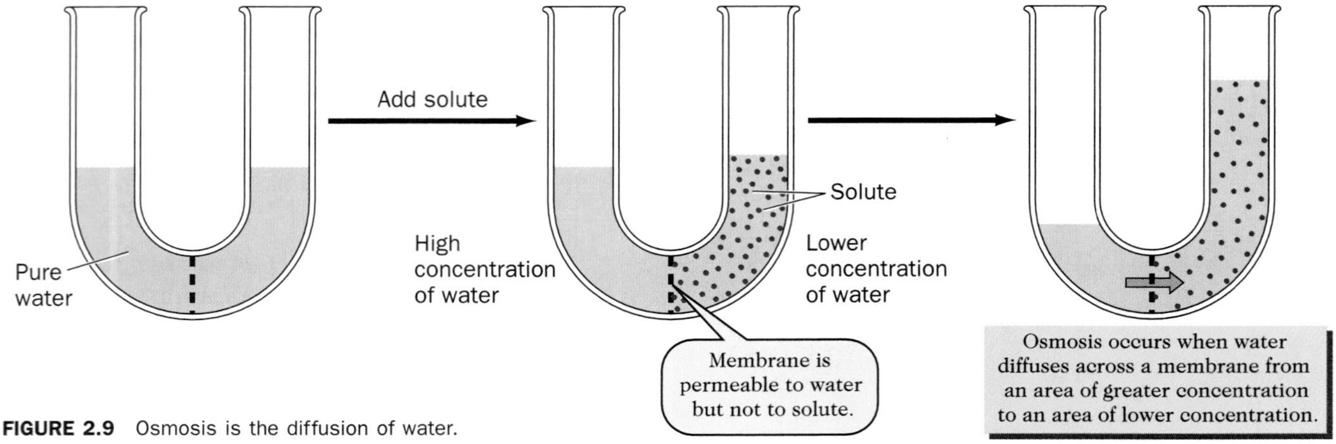
Add solute
Pure water
High concentration of water
Solute
Lower concentration of water
Membrane is permeable to water but not to solute.
Osmosis occurs when water diffuses across a membrane from an area of greater concentration to an area of lower concentration.

FIGURE 2.9 Osmosis is the diffusion of water.

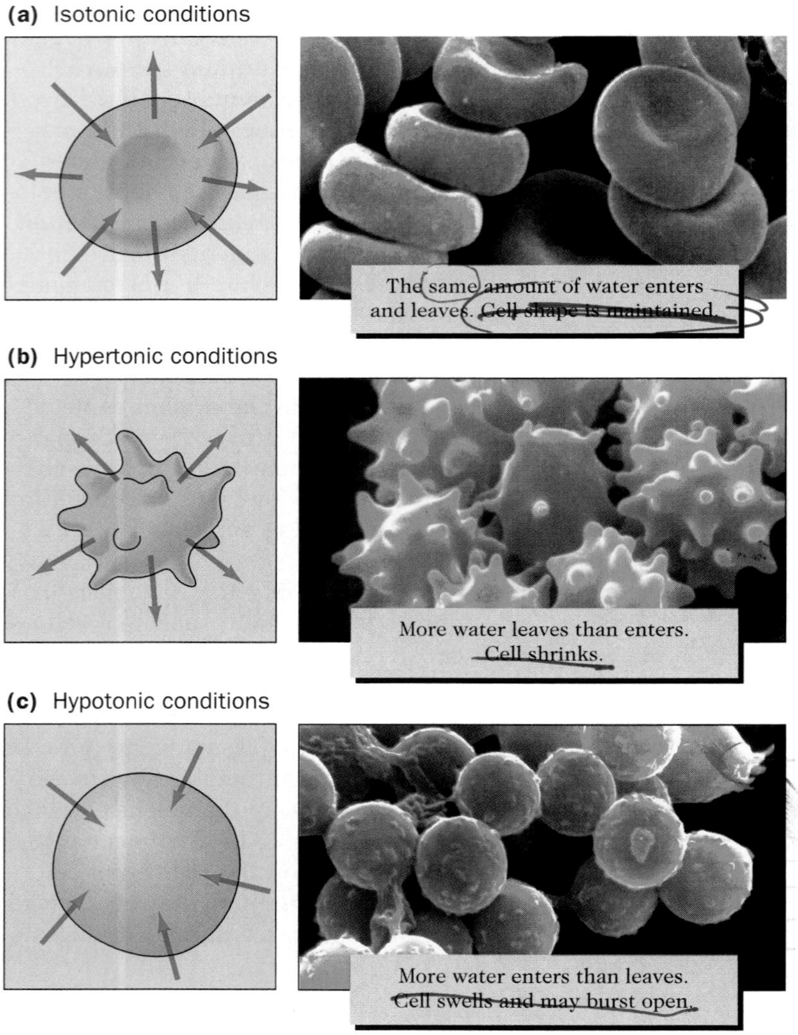
(a) Isotonic conditions
The same amount of water enters and leaves. Cell shape is maintained.
(b) Hypertonic conditions
More water leaves than enters. Cell shrinks.
(c) Hypotonic conditions
More water enters than leaves. Cell swells and may burst open.

FIGURE 2.10 The effect of solute concentration on the diffusion of water (osmosis) into and out of cells.

same"; *tonic,* "strength"). In an isotonic solution water both enters and leaves the cell at the same rate, and so a cell neither gains or loses water (Figure 2.10a). The human kidney regulates the water and solute concentrations of the blood and thus maintains an isotonic fluid surrounding the cells.

If a solution has a higher concentration of solutes than does the cell's cytoplasm, it is called a **hypertonic solution** (*hyper,* "above or greater"). What happens if a cell is placed in a hypertonic solution? Because water is more concentrated inside the cell, it will diffuse out of the cell. If enough water is lost, the cell will shrink in size, as can be seen in the red blood cell shown in Figure 2.10b. Such a loss of water profoundly alters cell functioning and can cause cell death.

What happens if a cell is placed in a very dilute solution called a **hypotonic solution** (*hypo,* "below or lesser")? In a hypotonic solution, there is a lower solute concentration and a higher water concentration than there is in the cytoplasm. Consequently, water diffuses into the cell. In fact, so much water enters that the cell swells and may burst open (Figure 2.10c).

To function properly, cells must maintain a certain water content. Since the solute concentration of the outer environment can affect water composition, the body tries to control the concentrations of solutes dissolved in the fluids of the body. One way this is accomplished is through active transport, our next topic.

> Osmosis is the diffusion of water across a selectively permeable membrane. Water moves from a region of higher water concentration (hypotonic solution) to a region of lower water concentration (a hypertonic solution). The movement of water into a confined space generates pressure.

Active Transport across the Plasma Membrane

Active transport uses cellular energy to move specific molecules from regions of lower concentration to regions of higher concentration. This movement against a concentration gradient is not a process that happens spontaneously. In other words, you do not observe perfume molecules dispersed throughout a room reentering an open container. That would require the expenditure of a considerable amount of effort and energy.

Instead of reducing a concentration gradient, as happens in passive transport, active transport creates or maintains a concentration gradient. To do this, the cell uses energy released by the hydrolysis of ATP ($ATP + H_2O \longrightarrow ADP + \mathbb{P} +$ energy). This energy is used to change the shape or position of a carrier protein, moving substances from one side of the plasma membrane to the other side (Figure 2.11). Active transport is used to maintain the concentration of certain ions within cells and to move larger molecules, such as amino acids, into cells.

The cellular concentrations of two important ions, sodium (Na^+) and potassium (K^+) ions, are maintained by an active transport called the **Na^+/K^+ pump** (sodium/potassium pump). Through the operation of this pump, Na^+ is transported out of the cell and K^+ is simultaneously transported into the cell by a carrier protein that uses energy from the breakdown of ATP.

The operation of this Na^+/K^+ pump maintains a low internal Na^+ and a high internal K^+, while outside the cell Na^+ is high and K^+ is low. The Na^+/K^+ pump operates in all the cells of the human body, but it is particularly significant for the electrical operation of nerves and muscles.

About 30 percent of cellular energy is used for active transport processes (Table 2.2). This large expenditure of energy emphasizes how important it is for the cell to obtain raw materials such as amino acids. It also emphasizes the need to maintain the proper internal concentrations of ions for the operation of cellular machinery.

Stop here

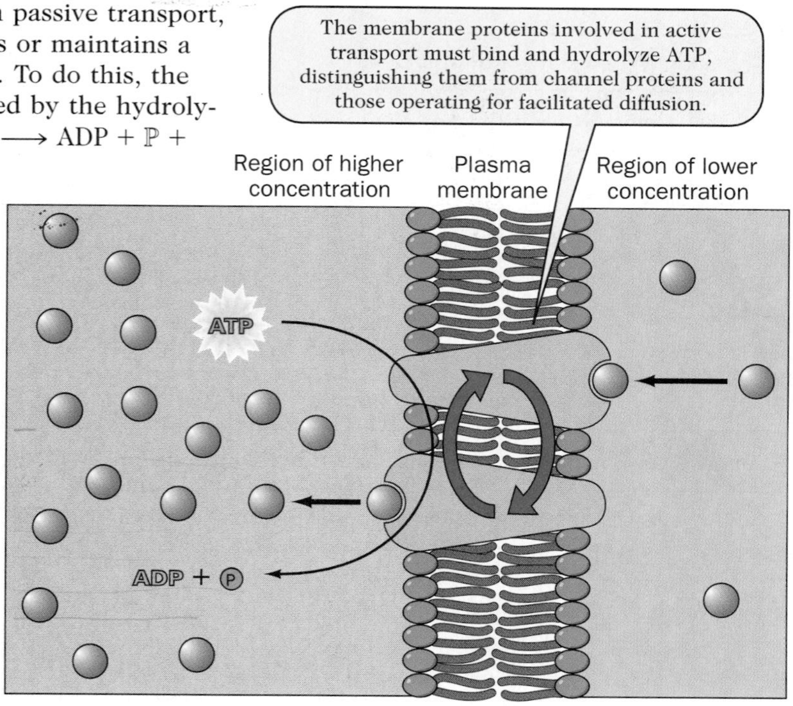
The membrane proteins involved in active transport must bind and hydrolyze ATP, distinguishing them from channel proteins and those operating for facilitated diffusion.
Region of higher concentration
Plasma membrane
Region of lower concentration
ATP
ADP + P

FIGURE 2.11 Active transport across a membrane occurs against a concentration gradient and requires cellular energy, often from ATP.

TABLE 2.2 Passive and Active Transport Related to Concentration Gradients
Passive transport (in direction of arrow) High concentration ⟶ low concentration
Active transport (in director of arrow) Low concentration ⟶ high concentration

Note: Passive transport does not require cellular energy. Active transport requires cellular energy, often from the hydrolysis of ATP.

> Active transport uses cellular energy and specific carrier proteins to transport substances against a concentration gradient (from regions of low concentration to regions of high concentration). The Na^+/K^+ pump is an example of active transport.

Active Transport by Vesicle Formation

The mechanism of active transport we have just discussed transports molecules one by one across the plasma membrane. However, there is another process by which a cell moves materials across its plasma membrane. This is called **cytosis,** and it also requires cellular energy. Cytosis always moves large particles or quantities of matter, millions of molecules at a time. **Exocytosis** (*exo,* "outward or outside") moves large molecules such as proteins out of the cell and is sometimes called secretion. We'll discuss exocytosis later in this chapter. **Endocytosis** (*endo,* "inward or within") moves materials into the cell (Table 2.3).

Endocytosis is a sequence of coordinated events. Initially, there is an infolding of a small region of the plasma membrane (Figure 2.12a). This infolding develops into a small pocket that engulfs material from outside the cell. Eventually this pocket closes, forming a membrane-bounded sac called an **endocytotic vesicle** (*endoe-sigh-***tot***-ic*). This vesicle detaches from the plasma membrane and moves into the cytoplasm. Although they are within the cytoplasm, the contents of an endocytotic vesicle are still isolated from digestive enzymes, but eventually the enzymes will arrive.

There are several different kinds of endocytosis (Figure 2.12b). One form, called **phagocytosis** (*phago,* "eating"), involves cell extensions that surround and then engulf large particles, including bacterial cells. Not all cells show phagocytosis, but it is used extensively by certain white blood cells to engulf and destroy invading bacteria and dead human cells. Another form of endocytosis is called **pinocytosis** (*pino,* "drinking"). In this process, small droplets of fluid with their dissolved solutes are engulfed. For example, the maturing human egg cell uses pinocytosis to take up nutrients before fertilization.

TABLE 2.3 Passive and Active Transport Processes: How Things Enter and Leave Cells

Transport Process	Action in Cells
Passive Transport	**Does Not Require Cellular Energy**
Simple diffusion	Movement of substances from a higher concentration to a lower concentration. May or may not use channel proteins.
Osmosis	A special form of simple diffusion involving movement of water from a higher to a lower concentration.
Facilitated diffusion	Movement of substances from a higher to a lower concentration using a shuttle molecule to speed (facilitate) transport.
Active Transport	**Requires Cellular Energy, Often from ATP**
Active transport	Movement of substances from a lower to a higher concentration. Uses specific membrane transport proteins.
Endocytosis	Moves large particles or quantities of fluid into the cell by means of vesicle formation from the plasma membrane.
Exocytosis	Moves materials out of the cell (secretion) by means of fusion of secretion vesicles with the plasma membrane.

A third form of endocytosis is called **receptor-mediated endocytosis.** This process is similar to phagocytosis but is more specific. It requires binding between a plasma membrane receptor protein and a specific particle. Earlier, we described the binding of the cholesterol-carrying protein LDL to a specific receptor protein. This binding triggers the formation of an endocytotic vesicle that engulfs the LDL particle and carries it into the cytoplasm. After digestion by enzymes, the free cholesterol diffuses out of the vesicles and into the cytoplasm for use by the cell. The way in which this digestion is accomplished is our next consideration.

Vesicle Fusion, Digestion, and the Role of Lysosomes

The large materials engulfed by endocytosis require digestion. This occurs when an endocytotic vesicle containing the engulfed particle fuses with another kind of vesicle called a lysosome. **Lysosomes** are large-diameter (0.4 μm) membrane-bounded vesicles that contain a variety of powerful digestive enzymes.

When the membrane of a lysosome makes contact with the membrane of an endocytotic vesicle, the two membranes fuse, forming a large **digestive vesicle** (Figure 2.13). Within the digestive vesicle, lysosomal enzymes break down large molecules into smaller molecules. These smaller molecules can then diffuse across the vesicle membrane and enter the cytoplasm, where they supply energy and raw materials for the cellular machinery.

Defective cellular digestion: Tay-Sachs disease. We get an indication of how important lysosomes are to human health when we learn that more than 40 diseases are attributed to defects in lysosomes. One such disease is **Tay-Sachs disease,** which is most often found among Jews of eastern European ancestry, among whom it affects 1 in 2500 births. This is about 100 times higher than the incidence in any other ethnic group. The symptoms of the disease are evident by age 6 months and include deafness, blindness, paralysis, and seizures. Rapid brain and nerve deterioration follow, and the child usually dies by age 3.

Tay-Sachs is an inherited disease. It results when a child receives two defective genes, one from each parent. These defective genes direct the synthesis of a defective lysosomal enzyme that is used for the degradation of a complex lipid. Without proper degradation, this complex lipid accumulates within the lysosomes of nerve cells,

FIGURE 2.12 Endocytosis moves materials into the cell.

(a) Formation of an endocytotic vesicle

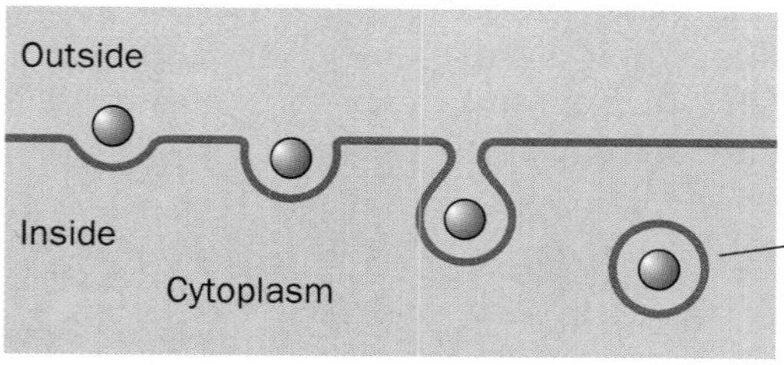
Outside
Inside
Cytoplasm
Plasma membrane
Endocytotic vesicle surrounded by a membrane derived from the plasma membrane

(b) Different forms of endocytosis

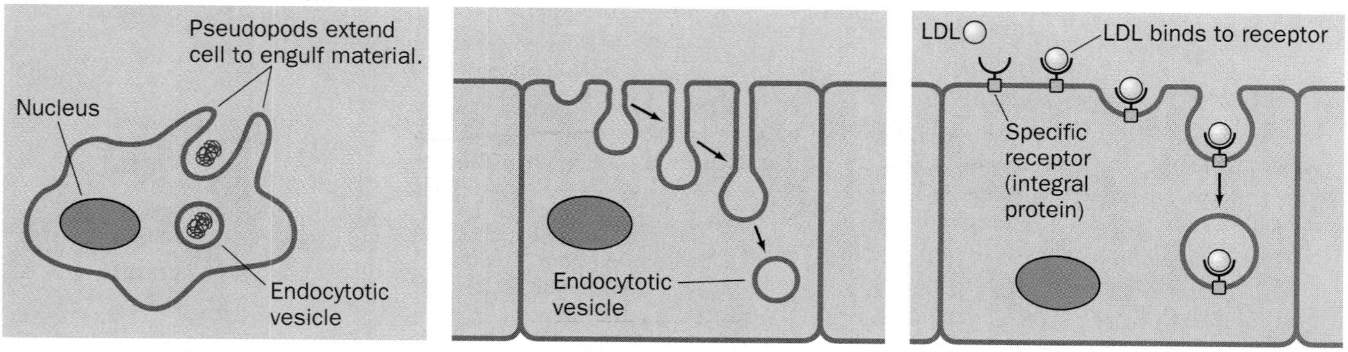
Pseudopods extend cell to engulf material.
Nucleus
Endocytotic vesicle
Endocytotic vesicle
LDL
LDL binds to receptor
Specific receptor (integral protein)

Phagocytosis is used by single cells such as white blood cells to take up material.

Pinocytosis is used by cells to take up fluid and dissolved solutes.

Receptor-mediated endocytosis is used to engulf a specific substance (LDL) that binds to the specific LDL receptor in the plasma membrane.

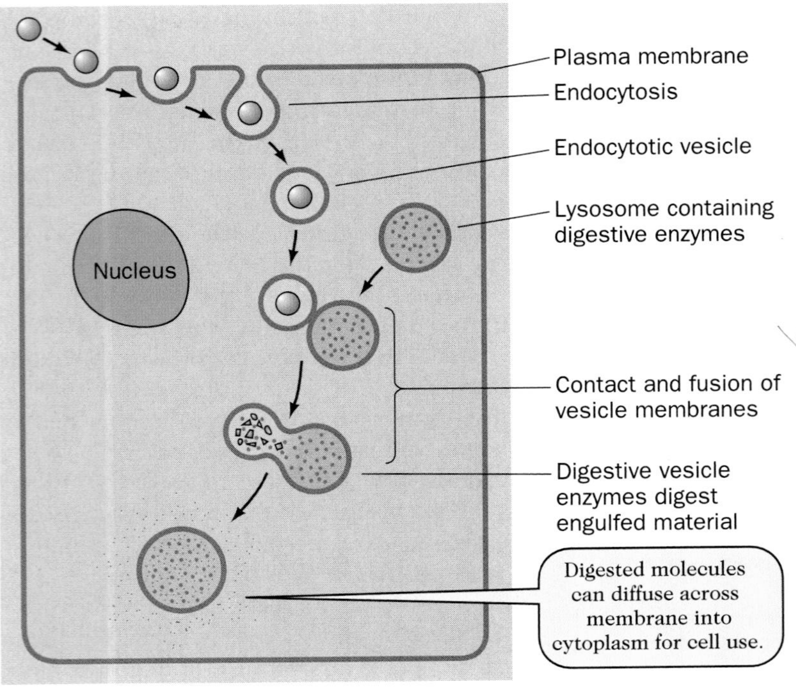
Plasma membrane
Endocytosis
Endocytotic vesicle
Lysosome containing digestive enzymes
Nucleus
Contact and fusion of vesicle membranes
Digestive vesicle enzymes digest engulfed material
Digested molecules can diffuse across membrane into cytoplasm for cell use.

FIGURE 2.13 Cell digestion of engulfed materials occurs when endocytotic vesicles fuse with lysosomes that contain digestive enzymes.

Endocytosis is a process by which the plasma membrane engulfs material outside the cell and forms an endocytotic vesicle that moves into the cytoplasm. Digestion within an endocytotic vesicle is accomplished by its fusion with a lysosome containing powerful digestive enzymes.

disrupting normal functioning. Unfortunately, there is no cure for the condition. However, diagnostic testing can identify parents who carry the defective gene even though they have no symptoms.

The Nucleus and Its DNA

The **nucleus** is the largest organelle in most human cells (5 to 10 μm) and is easily visible under both light and electron microscopy. It contains most of the cellular DNA and is bounded by a double membrane called the **nuclear envelope** (Figure 2.14). The inner and outer membranes of the nuclear envelope are continuous with each other at openings between the nucleus and the cytoplasm called *nuclear pores.* These pores are 70 to 90 nm in diameter and are abundant over the entire surface of the nuclear envelope. However, they are not entirely open passages between the nucleus and the cytoplasm; they are plugged with proteins that control the passage of materials into and out of the nucleus.

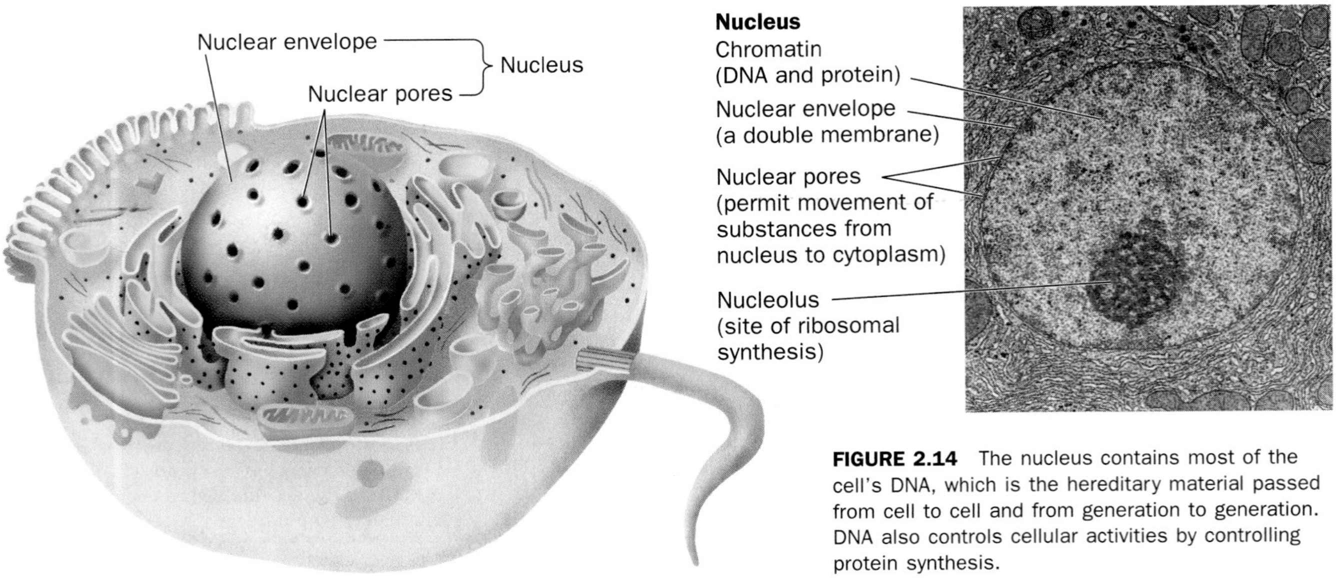
Nuclear envelope
Nuclear pores
Nucleus
Nucleus
Chromatin (DNA and protein)
Nuclear envelope (a double membrane)
Nuclear pores (permit movement of substances from nucleus to cytoplasm)
Nucleolus (site of ribosomal synthesis)

FIGURE 2.14 The nucleus contains most of the cell's DNA, which is the hereditary material passed from cell to cell and from generation to generation. DNA also controls cellular activities by controlling protein synthesis.

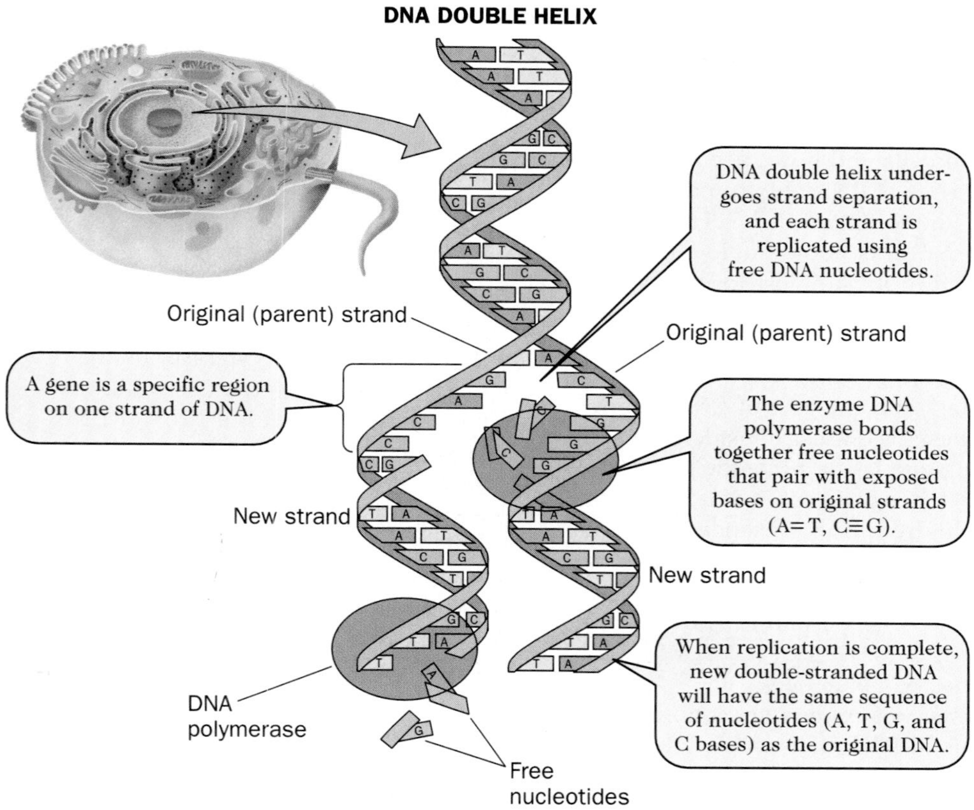
DNA DOUBLE HELIX
DNA double helix undergoes strand separation, and each strand is replicated using free DNA nucleotides.
Original (parent) strand
Original (parent) strand
A gene is a specific region on one strand of DNA.
The enzyme DNA polymerase bonds together free nucleotides that pair with exposed bases on original strands (A=T, C≡G).
New strand
New strand
When replication is complete, new double-stranded DNA will have the same sequence of nucleotides (A, T, G, and C bases) as the original DNA.
DNA polymerase
Free nucleotides

FIGURE 2.15 Duplication of DNA occurs in the nucleus and requires specific enzymes.

In the nucleus, the DNA is bound to a protein, forming long, thin strands called **chromatin.** Consequently, the interior of the nucleus has a diffuse, unstructured appearance most of the time. However, when the cell is preparing to divide, changes take place in the chromatin, and the thin strands condense into shorter, thicker structures called **chromosomes** that are large enough to be observed with the light microscope. We'll return to chromosomes when we discuss cell division later in this chapter.

As the hereditary information, DNA has two essential functions: (1) It carries coded information between the parent cell and the progeny cells produced by cell division, and (2) the coded information in DNA controls cellular activities, determining both the fate and the function of a cell. In the sections that follow we'll examine both of these roles of DNA. First, how does DNA carry information from parent cell to progeny cells?

The Duplication of DNA

As you learned in Chapter 1, DNA is a polymer made up of nucleotides. It is a long, thin double-stranded helical molecule composed of four different types of nucleotides (see Figure 1.20). The sugar and phosphate components of the nucleotides make up the structural backbone of the molecule, while four different bases—adenine (A), thymine (T), guanine (G), and cytosine (C)—are used for coding. These nucleotides can be linked in any order, and the precise sequences of the four bases constitute the coded messages that are the hereditary information.

The two strands of DNA molecules are held together by hydrogen bonding between complementary base pairs (A=T, C≡G). Within the nucleus before cell division, DNA undergoes a process of precise duplication called **replication.** During replication, the two strands of DNA separate, and free DNA nucleotides pair with their complementary bases in the single strands. These separate nucleotides are then bonded together through the action of an enzyme, **DNA polymerase** (Figure 2.15).

When replication has been completed, all the cell's DNA will have been duplicated. With double its usual DNA content, the cell is prepared to distribute equal amounts of DNA to progeny cells (more about this later). The next topic will be how DNA controls cellular activity by controlling the synthesis of proteins.

Stop here

> To duplicate, the DNA double helix unwinds and separates. Free DNA nucleotides base-pair with the complementary nucleotides in the exposed single strands. These free nucleotides are then bonded together into a complementary strand of DNA. The result is the production of two identical double helices of DNA.

DNA, Genes, and Protein Synthesis

DNA determines how cells develop into specialized cells and how specialized cells function. How is DNA able to control the fate and functioning of cells? The answer to this question is both simple and complex. Stated in its simplest form, *DNA controls cells by directing the synthesis of proteins. Proteins control a cell's fate and determine its function.*

To understand the role of DNA in protein synthesis, we must consider the gene. The **gene** is the fundamental unit of heredity. Genes are passed from parents to their children. The presence of a gene in our cells is associated with the appearance of a particular trait or characteristic. For example, with certain genes you may have brown eyes, while a friend with different genes has blue eyes.

Scientists now understand that genes are regions of DNA that consist of specific nucleotide base sequences (see Figure 2.15). The sequence of DNA bases is a code that eventually determines the sequence of amino acids in a protein. For example, the gene for cystic fibrosis (mentioned earlier in this chapter) codes for the channel protein that allows chloride ions (Cl^-) to diffuse out of the cell. Indeed, the idea that one gene codes for one type of protein is fundamental to modern biology. However, DNA does not conduct protein synthesis directly. It controls the synthesis of proteins by using ribonucleic acid (RNA), the other type of nucleic acid.

> Genes are the units of inheritance, and they are composed of DNA. DNA controls cellular activity by controlling the synthesis of proteins. The sequence of bases in DNA specifies the sequence of amino acids in proteins. Different genes code for different proteins, using RNA as an intermediate.

The First Phase: DNA Is Transcribed into mRNA

The first step in DNA's control of protein synthesis involves the synthesis of complementary RNA. This process, which is called **transcription,** involves the faithful copying of a DNA base sequence into the complementary base sequence in a strand of RNA. This process is similar to what happens during DNA replication, except that RNA nucleotides participate, and the base uracil (U) substitutes for thymine (T) (see Figure 1.20).

In the nucleus, free RNA nucleotides form base pairs with their complementary bases in an exposed strand of DNA. Then an enzyme, **RNA polymerase,** bonds the RNA nucleotides together, forming the complementary strand of RNA. A DNA base sequence of ACGCCT is faithfully copied into an RNA sequence of UGCGGA (Figure 2.16).

When transcription of the entire gene is complete, the RNA molecule, which is called the **primary transcript,** separates from the DNA and undergoes chemical editing. Why is editing necessary? As it turns out, all eukaryotic genes, including human genes, contain two kinds of base sequences: introns and exons. Only the **exons** code for proteins and are used in protein synthesis. The **introns** are edited out of the primary transcript and play no further role.

This editorial processing produces what is called **messenger RNA,** or **mRNA** (see Figure 2.16). In mRNA, sequences of three bases, called **codons,** code for specific amino acids. The sequence of these codons in the mRNA determines the sequence of amino acids in a particular protein. Bearing the coded information from DNA, the mRNA moves through the nuclear pores and into the cytoplasm. In the cytoplasm, a specific mRNA directs the synthesis of a specific type of protein.

> A gene segment of DNA is transcribed into a primary RNA transcript that is then edited to provide mRNA. The mRNA will be used in the synthesis of proteins. The mRNA uses a sequence of three bases—a codon—to code for each amino acid and specify its sequence in the protein.

The Second Phase: mRNA Is Translated into Protein

You will recall that proteins are long chains of amino acids bonded together

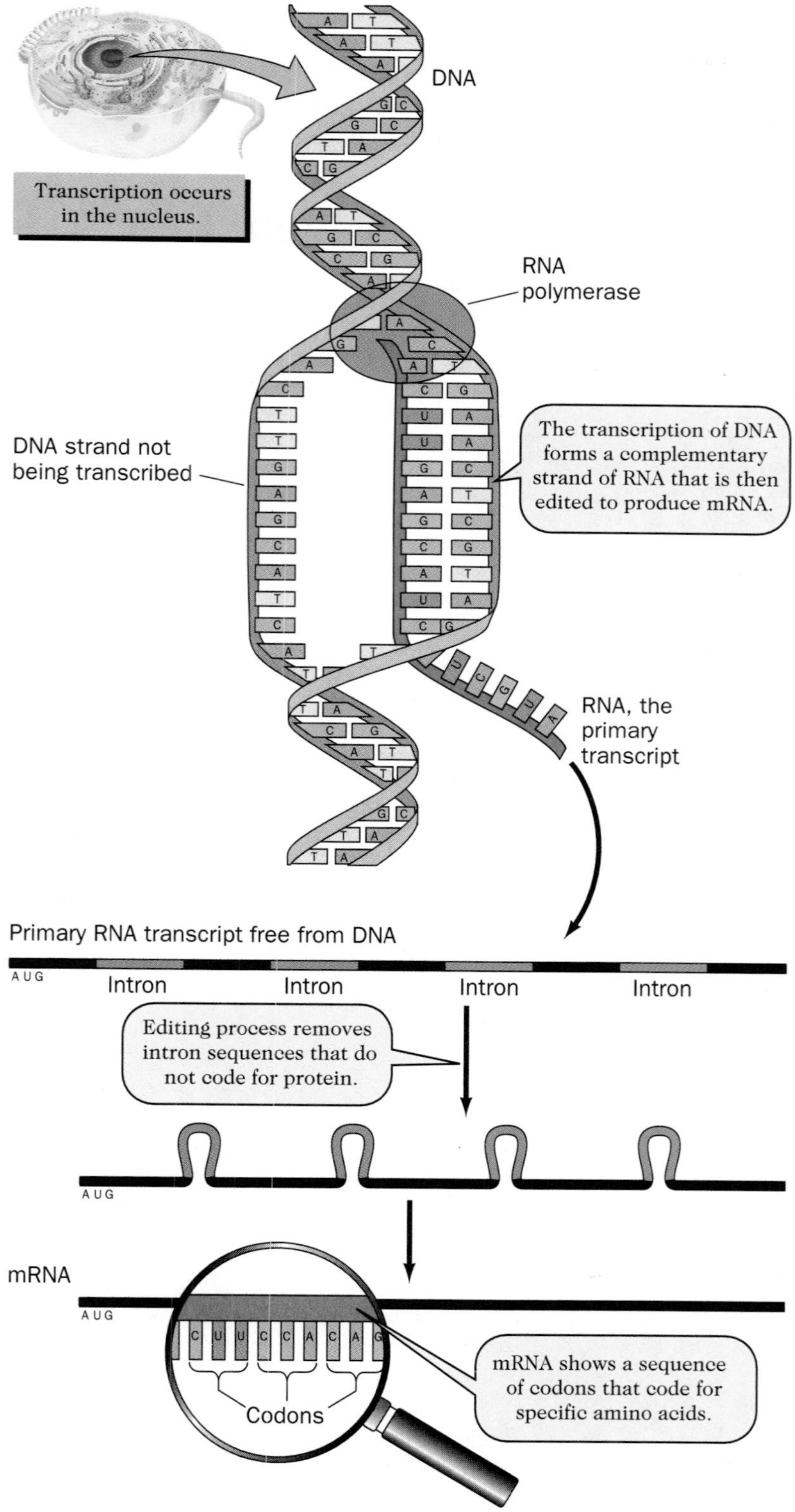
DNA
Transcription occurs in the nucleus.
RNA polymerase
DNA strand not being transcribed
The transcription of DNA forms a complementary strand of RNA that is then edited to produce mRNA.
RNA, the primary transcript
Primary RNA transcript free from DNA
AUG
Intron
Intron
Intron
Intron
Editing process removes intron sequences that do not code for protein.
AUG
mRNA
AUG
Codons
mRNA shows a sequence of codons that code for specific amino acids.

FIGURE 2.16 The transcription of DNA.

through linkages called peptide bonds. Operating as enzymes, receptors, support structures, and hormones, proteins directly control the fate and function of the cell. They are real important.

The synthesis of proteins occurs in the cytoplasm and requires large structures called ribosomes. **Ribosomes** are composed of a special type of RNA called *ribosomal RNA (rRNA)* and several different kinds of *ribosomal proteins.* Ribosomes are about 25 nm in diameter and can be observed with the electron microscope as small dark dots some of which are free in the cytoplasm while others are attached to cellular membranes (see Figure 2.3). There are thousands of identical ribosomes in every cell. Each ribosome is the site where amino acids are linked together one by one as instructed by the mRNA.

In the cytoplasm, each mRNA attaches to a ribosome (Figure 2.17). During protein synthesis, the strand of mRNA moves across the ribosome and provides a stabilizing structure that correctly positions amino acids for linkage into the growing protein molecule. To accomplish these tasks, the ribosome-mRNA complex needs (1) energy from ATP and (2) a means of directly associating amino acids with their codons in mRNA. The second requirement is met by another type of RNA, transfer RNA.

Each of the 20 amino acids used in protein synthesis can be covalently bonded to a distinct type of RNA called **transfer RNA (tRNA).** Each amino acid has its own type of tRNA. The chemical reaction between the amino acid, ATP, and tRNA produces what we'll call a *loaded tRNA,* which is a tRNA that is bonded to its amino acid.

How is tRNA able to directly associate a specific amino acid with a specific site in mRNA? This is accomplished by complementary base pairing between an mRNA codon and a complementary 3-base sequence in the tRNA called the **anticodon** (see Figure 2.17).

The process in which mRNA, ribosomes, and loaded tRNA interact in the synthesis of proteins is called **translation.** It consists of translating a nucleic acid

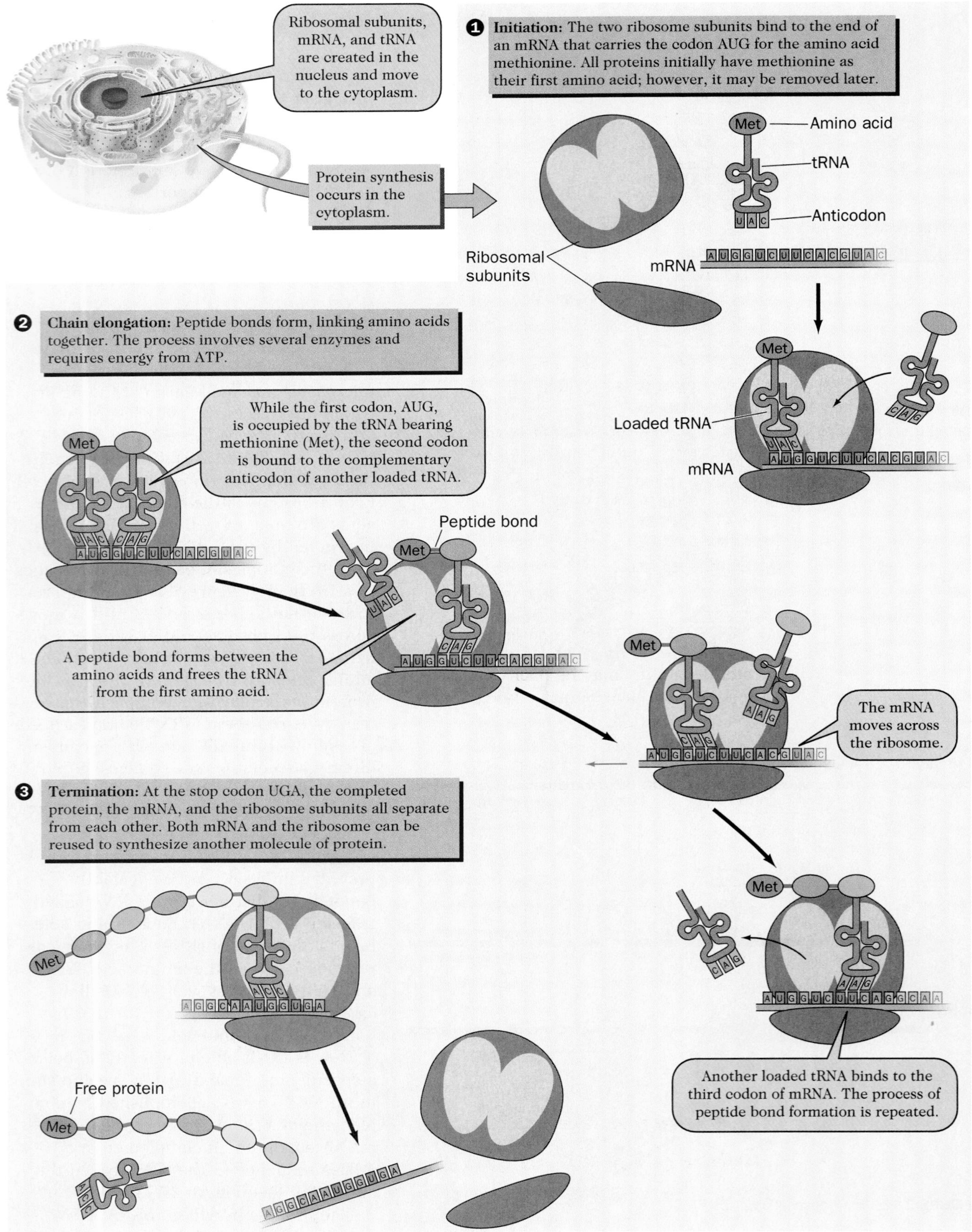
Ribosomal subunits, mRNA, and tRNA are created in the nucleus and move to the cytoplasm.
Protein synthesis occurs in the cytoplasm.
❶ Initiation: The two ribosome subunits bind to the end of an mRNA that carries the codon AUG for the amino acid methionine. All proteins initially have methionine as their first amino acid; however, it may be removed later.
Met
Amino acid
tRNA
Anticodon
UAC
Ribosomal subunits
mRNA
AUGGUCUUCACGUAC
Loaded tRNA
mRNA
CAG
❷ Chain elongation: Peptide bonds form, linking amino acids together. The process involves several enzymes and requires energy from ATP.
While the first codon, AUG, is occupied by the tRNA bearing methionine (Met), the second codon is bound to the complementary anticodon of another loaded tRNA.
Peptide bond
A peptide bond forms between the amino acids and frees the tRNA from the first amino acid.
The mRNA moves across the ribosome.
AAG
❸ Termination: At the stop codon UGA, the completed protein, the mRNA, and the ribosome subunits all separate from each other. Both mRNA and the ribosome can be reused to synthesize another molecule of protein.
Another loaded tRNA binds to the third codon of mRNA. The process of peptide bond formation is repeated.
AUGGUCUUCAGGCAA
ACC
AGGCAAUGGUGA
Free protein

FIGURE 2.17 Protein synthesis by translation of mRNA occurs in three steps: initiation, elongation, and termination.

base sequence into an entirely different type of molecule, a sequence of amino acids. Translation is like going from Spanish to English, while transcription is like writing the same language in a different script.

Translation consists of three steps: initiation, chain elongation, and termination. Carefully examine Figure 2.17 to gain an understanding of the sequence of events involved in these processes. Step by step, as the ribosome moves along the mRNA, the amino acids are linked to each other and the protein grows in length.

The fate of new proteins. The completed protein may have one of several different fates, depending on its properties and site of synthesis. If it was synthesized on free ribosomes in the cytoplasm, the protein will remain in the cytoplasm and function there. However, as we will see in the next section, some ribosomes are attached to membranes in the cytoplasm. Proteins that are synthesized on these ribosomes may become membrane proteins, or they may be packaged into a lysosome or packaged into a vesicle for exocytosis.

To understand how proteins are packaged into vesicles, we need to examine two membranous organelles of the cell: the endoplasmic reticulum and the Golgi apparatus.

> Protein synthesis (translation) requires mRNA, ribosomes, energy, and loaded tRNAs. Peptide bonds between amino acids are formed as the mRNA moves across the ribosome. This process synthesizes a protein whose amino acid sequence is determined by the base sequence in the mRNA.

Organelles for Packaging, Processing, and Secreting

The organelles involved in the synthesis, packaging, and secretion of proteins include the endoplasmic reticulum, the Golgi apparatus, and vesicles (see Figure 2.2). All these organelles are bounded by a membrane composed of a phospholipid bilayer and proteins, similar to the plasma membrane. The membranes of these organelles separate their inner compartment, called the **lumen**, and its contents from the surrounding cytoplasm of the cell.

The Endoplasmic Reticulum

The **endoplasmic reticulum** (*ray-tick-you-lum*), or **ER**, is a network of membranes that forms interconnecting flattened sacs, tubules, and vesicles (Figure 2.18). The membranes are continuous with one another and form a single compartment that is separate from the cytoplasm.

Two types of endoplasmic reticulum are distinguished from each other by their appearance in the electron microscope: smooth ER and rough ER. The **smooth ER** lacks ribosomes and usually appears as a network of tubules (Figure 2.18). It is involved in the synthesis of fats, phospholipids, cholesterol, and steroid hormones, including the male and female sex hormones. In the human liver, the smooth ER is also the site where poisonous chemicals such as alcohol and pesticides are chemically modified to reduce their toxic effects on the body.

The **rough ER** often appears as extensive flattened sacs with numerous ribosomes bound to the surface facing the cytoplasm (Figure 2.18). It is involved in the synthesis of proteins destined for (1) secretion from the cell, (2) association with membranes, or (3) containment within lysosomes.

Proteins synthesized on the ribosomes of the rough ER pass through pores in the membrane and enter the lumen of the ER. There they may undergo additional processing. For example, some terminal amino acids may be removed or carbohydrates may be added before the proteins enter vesicles that bud off from the ER. These protein-containing vesicles move through the cytoplasm and fuse with another membranous organelle: the Golgi apparatus.

FIGURE 2.18 Acting together, the endoplasmic reticulum and Golgi apparatus are involved in the synthesis, processing, packaging, and secretion of proteins.

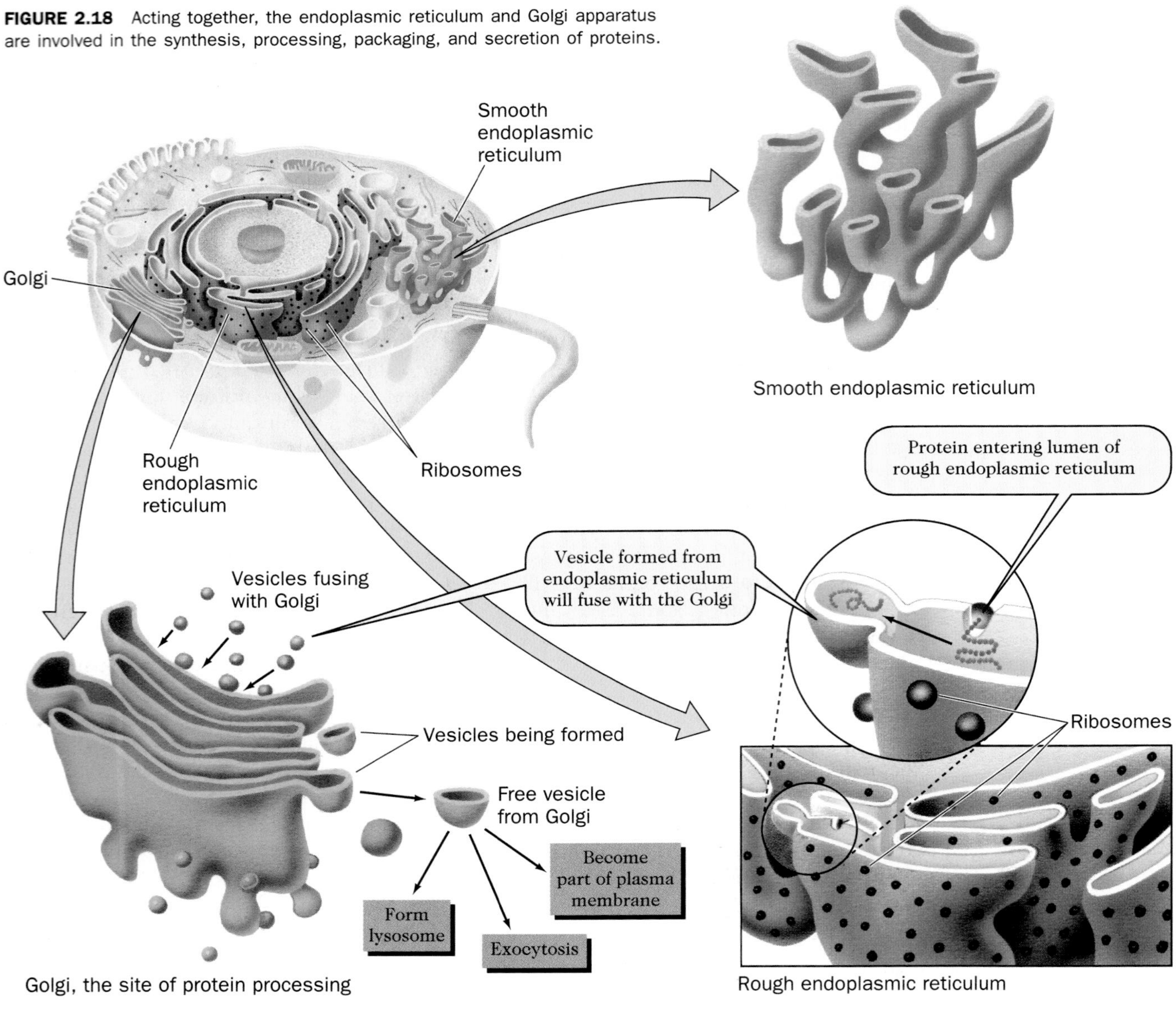

Smooth endoplasmic reticulum
Golgi
Rough endoplasmic reticulum
Ribosomes
Smooth endoplasmic reticulum
Protein entering lumen of rough endoplasmic reticulum
Vesicle formed from endoplasmic reticulum will fuse with the Golgi
Vesicles fusing with Golgi
Vesicles being formed
Ribosomes
Free vesicle from Golgi
Become part of plasma membrane
Form lysosome
Exocytosis
Golgi, the site of protein processing
Rough endoplasmic reticulum

> The endoplasmic reticulum (ER) is a network of tubes and flat sacs that form a large compartment separate from the cytoplasm. Proteins synthesized on the ribosomes of the rough ER enter the lumen of the ER and may be secreted, retained in membranes, or end up in lysosomes.

The Golgi Apparatus: Packaging and Processing

The **Golgi apparatus**, or **Golgi** (***goal**-gee*), is named after the Italian scientist who identified it in 1898, and therefore the word is capitalized. The Golgi is a series of smooth, flattened membrane sacs that are stacked closely together (Figure 2.18). It receives proteins from the ER and processes, concentrates, and packages those proteins. One function of the Golgi is to prepare proteins for secretion from the cell.

Vesicles formed from the ER fuse with the Golgi, adding their contents to the Golgi's interior compartment, where they are processed by chemical modification. When processing is complete, the modified proteins are packaged into new vesicles that form from the Golgi membrane. Some of these Golgi vesicles carry their contents to the cell surface, where the vesicle membrane fuses with the plasma membrane (Figure 2.19). When this happens, the contents of the vesicle are secreted to the outside of the cell in the process called **exocytosis.**

The coordinated activities of the ER, vesicles, and Golgi form the **secretion pathway,** which terminates in exocytosis. Through this process, pancreatic cells secrete the hormone insulin into the blood while other cells in the pancreas secrete digestive enzymes into the small intestine. Plasma cells of the body's immune system secrete antibodies to attack foreign cells and molecules, and liver cells secrete proteins such as albumin into the blood.

Not all Golgi-formed vesicles participate in secretion. Some vesicles contain proteins that are inserted into the plasma membrane, and other vesicles become the lysosomes that were discussed earlier in this chapter.

> Proteins are processed in the Golgi and then isolated into vesicles. The Golgi produces lysosomes and vesicles destined for exocytosis. During exocytosis, vesicles loaded with proteins fuse with the plasma membrane, and the proteins are secreted outside the cell.

Mitochondria: Cellular Power Plants

The **mitochondria** (*my-toe-***con***-dree-ah*) are rather large, membranous bean-shaped organelles that measure about 1 μm by 5 to 10 μm (Figure 2.20). This organelle operates like a power plant by taking one form of chemical energy (fuel molecules like glucose) and converting it to another form (ATP) for use in the cell. Mitochondria are the source of most of a cell's ATP. Enzymes in the mitochondria coordinate with enzymes in the cytoplasm to release all the energy from glucose ($C_6H_{12}O_6$) by completely degrading

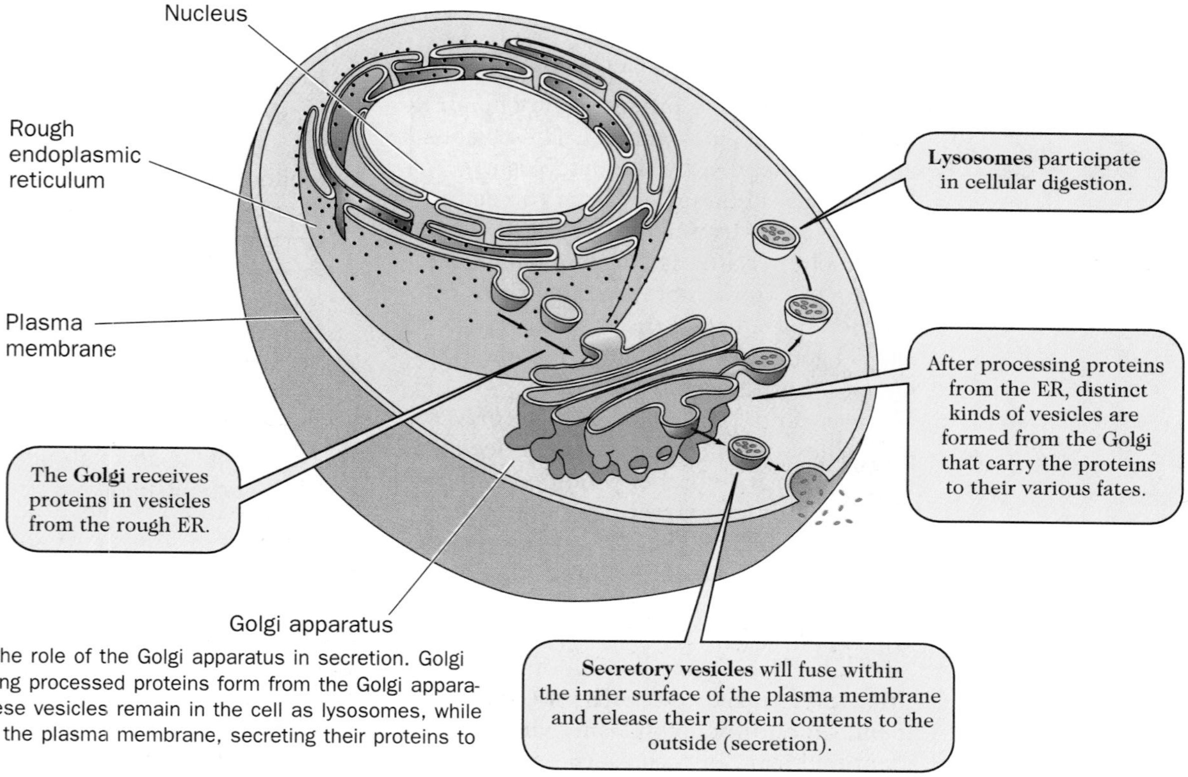

Nucleus
Rough endoplasmic reticulum
Plasma membrane
The Golgi receives proteins in vesicles from the rough ER.
Golgi apparatus
Lysosomes participate in cellular digestion.
After processing proteins from the ER, distinct kinds of vesicles are formed from the Golgi that carry the proteins to their various fates.
Secretory vesicles will fuse within the inner surface of the plasma membrane and release their protein contents to the outside (secretion).

FIGURE 2.19 The role of the Golgi apparatus in secretion. Golgi vesicles containing processed proteins form from the Golgi apparatus. Some of these vesicles remain in the cell as lysosomes, while others fuse with the plasma membrane, secreting their proteins to the outside.

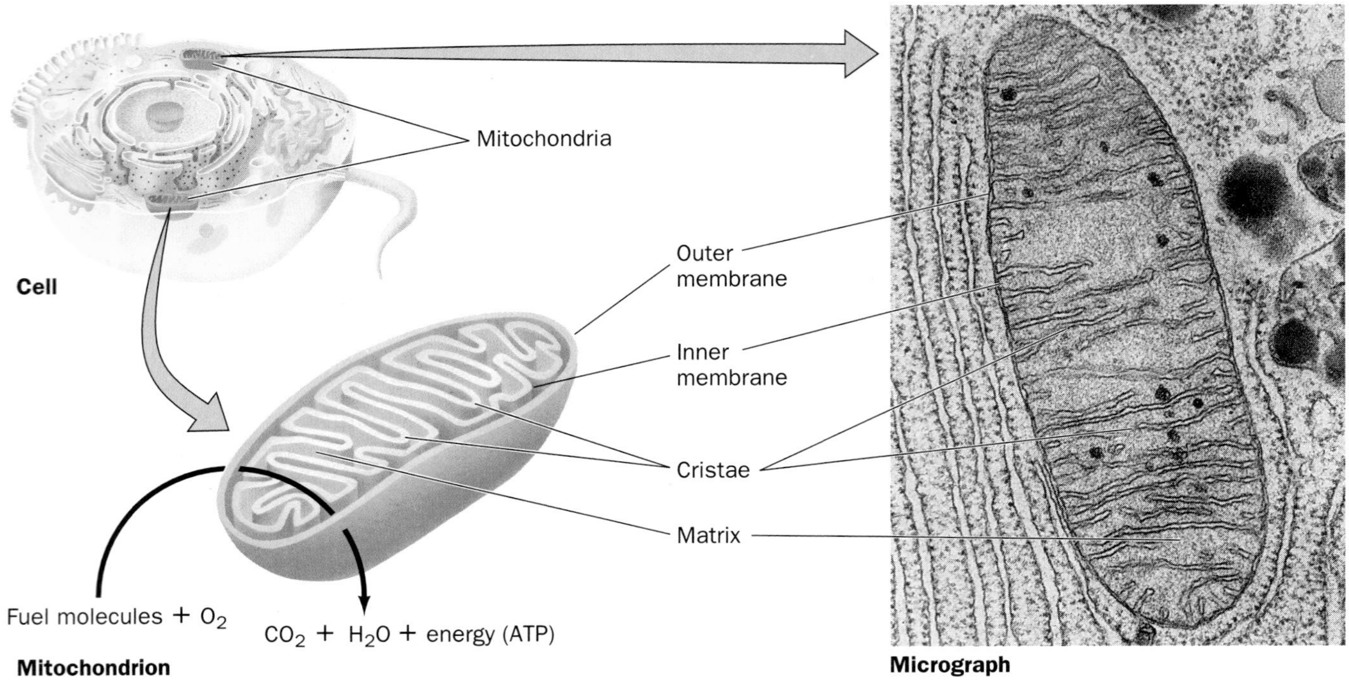
Mitochondria
Cell
Outer membrane
Inner membrane
Cristae
Matrix
Fuel molecules + O_2
CO_2 + H_2O + energy (ATP)
Mitochondrion
Micrograph

FIGURE 2.20 Mitochondria are cellular power plants that convert one form of energy (fuel molecules) to another form (ATP).

it to CO_2 and H_2O (Chapter 7). This process, which is called **cellular respiration,** uses O_2 and has many steps that are summarized by the equation

$$\text{Glucose} + O_2 \xrightarrow[\text{enzymes}]{\text{cellular}} CO_2 + H_2O + \text{ATP}$$

To accomplish this process, mitochondria need fuel molecules and O_2. These molecules are supplied by the cytoplasm, which must obtain them from the cell's immediate environment.

Mitochondria (singular, *mitochondrion*) consist of two membranes enclosing an interior space called the **matrix,** where CO_2 is generated. The outer membrane is smooth, but the inner membrane is arranged into folds called **cristae** (***cris-tea***). These folds increase the surface area of the inner membrane, which contains the enzymes for extracting energy from the fuel molecules. These enzymes utilize O_2, generate H_2O, and capture the energy needed to synthesize ATP from ADP and $\mathbb{P}$ (see Figure 2.20).

After ATP is formed within mitochondria, it is transported across the inner membrane to the cytoplasm, where it can participate in most of the energy-requiring chemical reactions of the cell. ATP provides energy when it is hydrolyzed to ADP and $\mathbb{P}$. Mitochondria are found in nearly all cells, but their number varies greatly. Less active cells such as some white blood cells have only a few mitochondria, while more active cells such as liver and kidney cells may have 500 to 1000 in each cell.

Among the organelles of human cells, mitochondria are unique. They contain their own DNA and their own ribosomes, which are different from cytoplasmic ribosomes. This has led to the hypothesis that very early in cellular evolution mitochondria originated from primitive bacteria that were engulfed by other primitive cells. Rather than be digested, the primitive bacteria proved to be beneficial to their host cells, and an enduring relationship was established.

stop here

In mitochondria, cellular respiration uses O_2 to convert fuel molecules to CO_2 and H_2O, while releasing energy. This energy is used to synthesize ATP from ADP and $\mathbb{P}$. ATP is transported out of the mitochondria to provide energy for the chemical reactions that sustain the life of the cell.

The Cytoskeleton, Cilia, and Flagella

In recent years, improvements in electron microscopy have revealed the presence in the cytoplasm of a network of long thin protein fibers called the **cytoskeleton** (Figure 2.21a). Components of the cytoskeleton can form attachments to each other, to membrane proteins, and to organelles. The entire network operates to strengthen the cell, maintain cell shape, and integrate cellular activities. Components of the network move organelles within the cell, distribute chromosomes during mitosis, and pinch the cell in two at the close of cell division.

The cytoskeleton is made up of three different types of protein fibers: microfilaments, intermediate filaments, and microtubules. Each of these fibers has a distinctive protein composition, structure, and function. **Microfilaments** (*my-crow-fill-ah-mints*) are very thin (Figure 2.21a). They have a diameter of about 7 nm and are formed from units of the protein actin. These structures are not permanent. They can lengthen and shorten by means of the addition and loss of actin units. The assembly and disassembly of microfilaments from actin units permit some cells, such as the white blood cells, to change shape and move.

Microfilaments are involved in many types of cell movement. In the **amoeboid movement** of white blood cells, microfilaments produce localized cell extension called a *pseudopod.* Such movements contribute to the capacity of white blood cells to perform phagocytosis and pass between cells to fight foreign invaders. Microfilaments also pinch cells apart during cell division and make muscles contract.

The second component of the cytoskeleton is the **intermediate filaments,** so named because their 10-nm diameter is intermediate between that of the microfilaments and that of the microtubules (see Figure 2.21a). Intermediate filaments are quite strong and provide a permanent scaffolding to maintain cell shape and the distribution of organelles. Unlike both microfilaments and microtubules, they do not generally disassemble into their component proteins.

Microtubules (*my-crow-***tube***-you-ills*) are long hollow tubes with a diameter of about 25 nm. Like the other components of the cytoskeleton, they are variable in length. Microtubules are formed from units of the compact protein **tubulin.** Depending on cellular conditions, these proteins can assemble into microtubules or disassemble into separate proteins (see

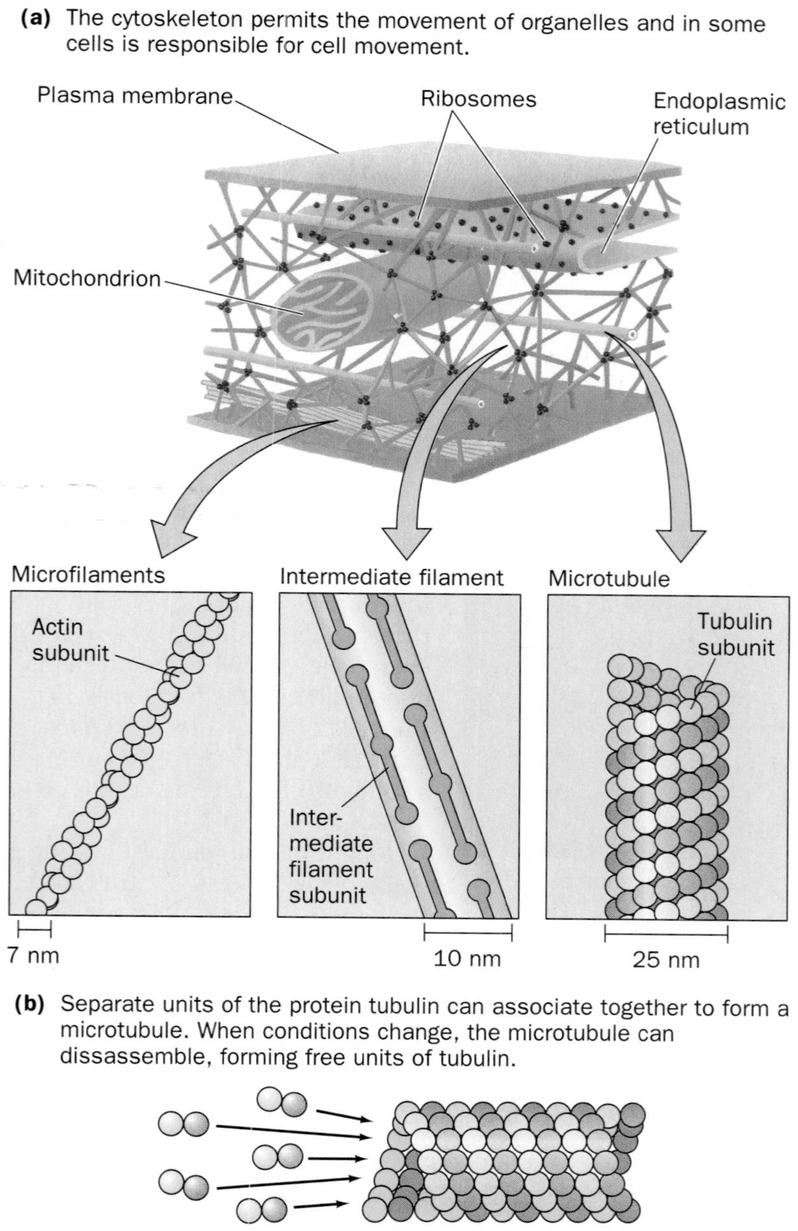
(a) The cytoskeleton permits the movement of organelles and in some cells is responsible for cell movement.
Plasma membrane
Ribosomes
Endoplasmic reticulum
Mitochondrion
Microfilaments
Intermediate filament
Microtubule
Actin subunit
Inter-mediate filament subunit
Tubulin subunit
7 nm
10 nm
25 nm
(b) Separate units of the protein tubulin can associate together to form a microtubule. When conditions change, the microtubule can dissassemble, forming free units of tubulin.

FIGURE 2.21 The cytoskeleton is a network of protein fibers that extends throughout the cell. It is capable of both maintaining and changing cell shape.

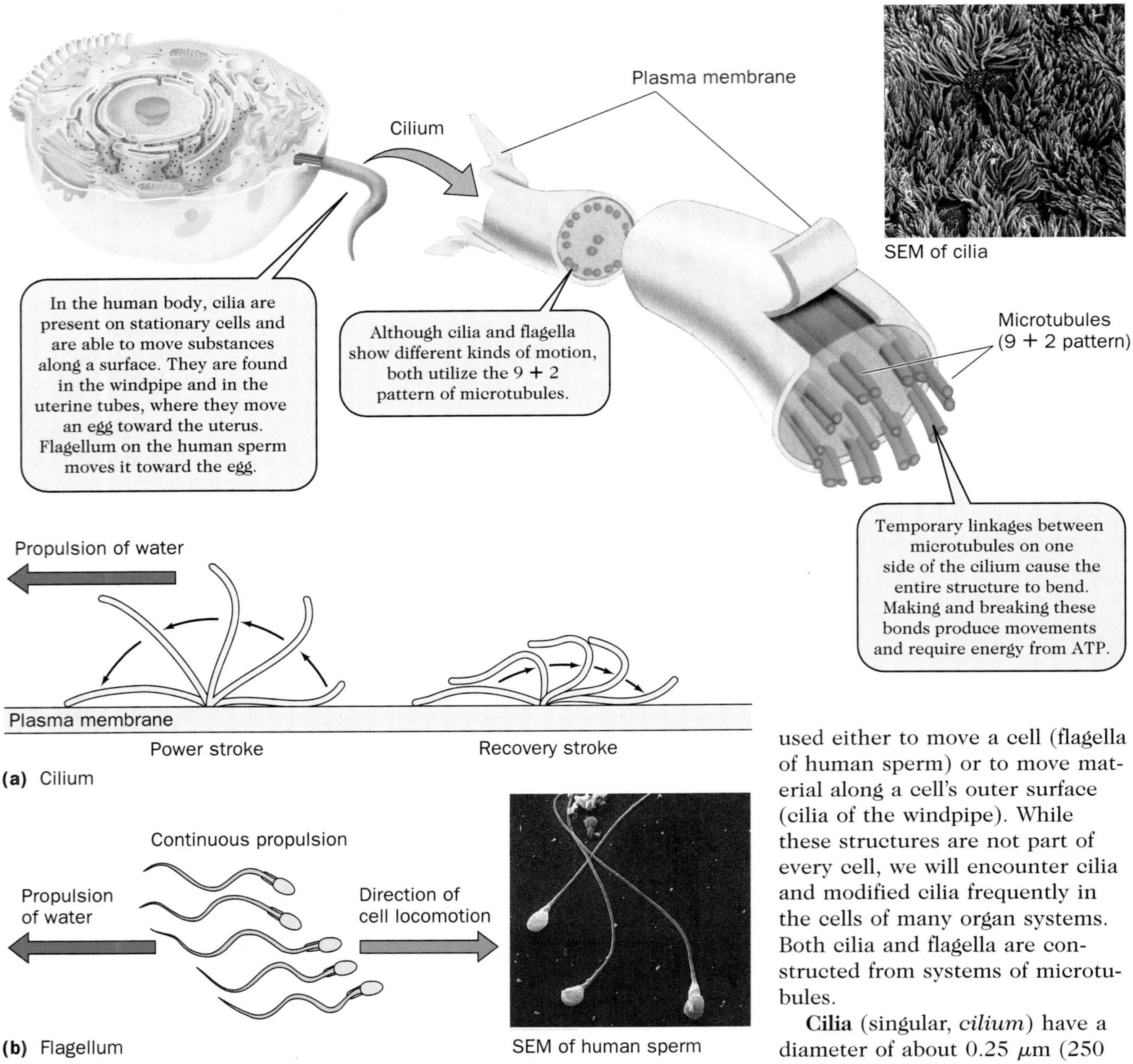
Plasma membrane
Cilium
SEM of cilia
Microtubules (9 + 2 pattern)
In the human body, cilia are present on stationary cells and are able to move substances along a surface. They are found in the windpipe and in the uterine tubes, where they move an egg toward the uterus. Flagellum on the human sperm moves it toward the egg.
Although cilia and flagella show different kinds of motion, both utilize the 9 + 2 pattern of microtubules.
Temporary linkages between microtubules on one side of the cilium cause the entire structure to bend. Making and breaking these bonds produce movements and require energy from ATP.
Propulsion of water
Plasma membrane
Power stroke
Recovery stroke
(a) Cilium
Continuous propulsion
Propulsion of water
Direction of cell locomotion
(b) Flagellum
SEM of human sperm

FIGURE 2.22 Cilia and flagella are movable extensions of the cell.

Figure 2.21b). The addition or loss of tubulin from one end on the microtubule permits the structure to lengthen or shorten. This process contributes to changes in cell shape and to the movement of chromosomes during mitosis.

The Cilia and Flagella

Cilia (*sill-e-ah*) and flagella (*flah-gel-ah*) are structures that protrude from the surfaces of some cells (see Figure 2.22a and b). They are capable of movement and are used either to move a cell (flagella of human sperm) or to move material along a cell's outer surface (cilia of the windpipe). While these structures are not part of every cell, we will encounter cilia and modified cilia frequently in the cells of many organ systems. Both cilia and flagella are constructed from systems of microtubules.

Cilia (singular, *cilium*) have a diameter of about 0.25 μm (250 nm) and may be 2 to 10 μm long. Covered by an extension of the plasma membrane, they protrude in large numbers from the part of the cell surface exposed to an open space. Cilia move in a coordinated manner somewhat like a group of oars, with a power stroke and a recovery stroke (see Figure 2.22a). Heavy inhalation of cigarette smoke damages the cilia of the windpipe and prevents their motion. This prevents the body from clearing accumulated mucus from the windpipe and results in the severe morning cough many smokers experience.

Flagella (singular, *flagellum*) have the same diameter as cilia but are much longer, perhaps 200 μm in length. Sperm are the only human cells to have flagella. Flagellar movement is different from ciliary movement. A bend occurs at one end of a flagellum and moves along its entire length. This creates a force parallel to the flagellum and moves the sperm in that direction (see Figure 2.22b).

The internal structures of flagella and cilia are similar. In both, microtubules extend for the entire length and are arranged in a characteristic 9 + 2 pattern. An outer circle contains nine microtubule doublets with an additional two single microtubules in the center (Figure 2.22). During movement, some of the outer doublets slide past other stationary doublets, causing bending. This process involves the making and breaking of temporary bonds between adjacent doublets in the outer circle. The energy for these events comes from the hydrolysis of ATP to ADP and ℙ. Table 2.4 summarizes the functions of flagella and cilia as well as those of the organelles of the cell.

TABLE 2.4 Structures and Organelles of Eukaryotic Cells

Structure or Organelle	Functions in Human Cells
Transport and Protection	
Plasma membrane	Controls what enters and leaves cells
Heredity and Manufacture	
Nucleus	DNA and RNA synthesis; assembly of ribosomes in a region called the nucleolus
Ribosomes	Site of protein synthesis in cytoplasm
Smooth ER	Lipid synthesis
Rough ER	Using attached ribosomes, synthesis of proteins destined for secretion or isolation in lysosomes
Golgi apparatus	Processing, storage, and transport of proteins destined for secretion; production of lysosomes and secretion vesicles
Secretion and Digestion	
Lysosomes	Contain digestive enzymes that digest the contents of exocytotic vesicles
Secretion vesicles	Formed from the Golgi; contain special proteins destined for secretion by exocytosis
Endocytotic vesicles	Formed by endocytosis; contain materials being transported into the cell
Energy Conversion	
Mitochondria	Chemical conversion of energy in fuel molecules to energy in ATP; site for the production of most of a cell's ATP
Support and Movement	
Cytoskeleton	Maintains cells shape, anchors organelles, and moves cells, using microfilaments, intermediate filaments, or microtubules
Cilia	Movement of materials along a cell surface
Flagella	Propulsion of sperm cells

> The cytoskeleton is an internal network of protein fibers that provides strength, support, and movement for the cell. Cilia and flagella project from the surfaces of some cells and provide movement when their microtubules slide past one another, using ATP as the energy source.

Cell Growth and Cell Division

You began as a single cell. That cell reproduced over and over, producing more and more cells. All the cells in the body are produced by the division of the nucleus followed by cell division. Although all these cells are not identical in structure or function, they all have the same amount and kind of DNA. We will examine the process by which this precise distribution of parental DNA to progeny cells takes place. In doing this, we will recognize that cell division and growth are closely linked phenomena.

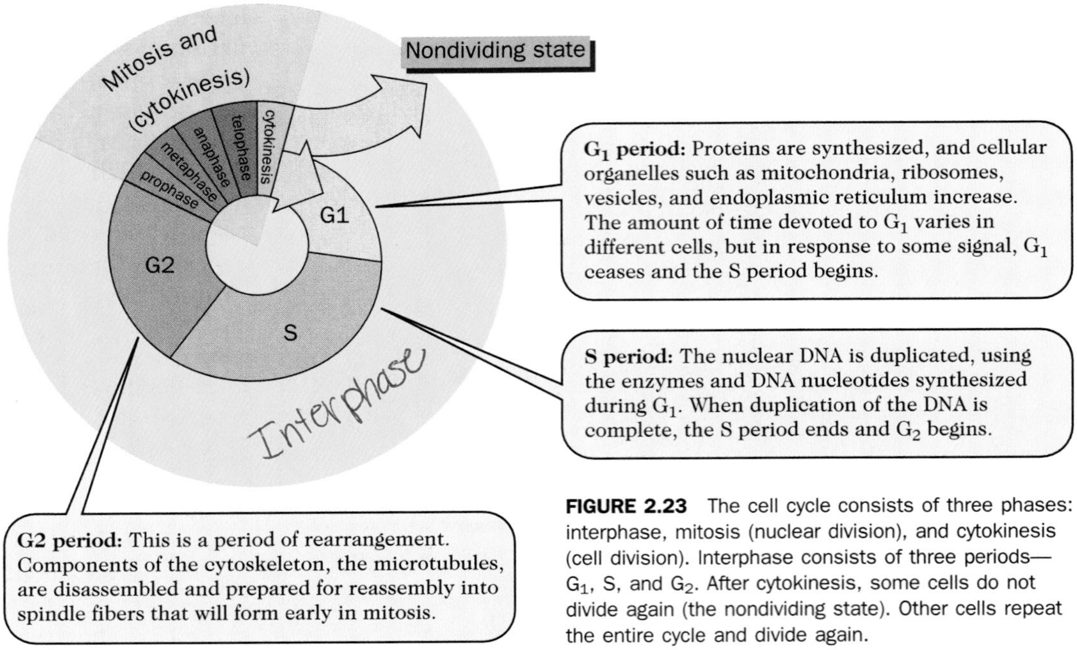
Mitosis and (cytokinesis)
Nondividing state
prophase
metaphase
anaphase
telophase
cytokinesis
G1
G2
S
Interphase
G_1 period: Proteins are synthesized, and cellular organelles such as mitochondria, ribosomes, vesicles, and endoplasmic reticulum increase. The amount of time devoted to G_1 varies in different cells, but in response to some signal, G_1 ceases and the S period begins.
S period: The nuclear DNA is duplicated, using the enzymes and DNA nucleotides synthesized during G_1. When duplication of the DNA is complete, the S period ends and G_2 begins.
G2 period: This is a period of rearrangement. Components of the cytoskeleton, the microtubules, are disassembled and prepared for reassembly into spindle fibers that will form early in mitosis.

FIGURE 2.23 The cell cycle consists of three phases: interphase, mitosis (nuclear division), and cytokinesis (cell division). Interphase consists of three periods—G_1, S, and G_2. After cytokinesis, some cells do not divide again (the nondividing state). Other cells repeat the entire cycle and divide again.

The **cell cycle** consists of growth followed by reproduction. It can be divided into three general phases: interphase, mitosis, and cytokinesis (Figure 2.23). We will examine each of these phases, identifying and describing their particular events and seeing how they contribute to the life of the cell and the human body.

Interphase: A Period of Synthesis and Growth

In the life of the cell, **interphase** can be a relatively long period. When we observe a living cell with a light microscope, little appears to be happening during interphase. However, chemical analysis of cells reveals that all cellular materials are manufactured during interphase. Large molecules are synthesized from raw materials taken up by the cell, and organelles are assembled from molecules. Interphase can be divided into three distinct periods: G_1, S, and G_2. Carefully examine Figure 2.23 for a description of what occurs during G_1, S, and G_2.

The cell cycle consists of interphase, mitosis, and cytokinesis. Interphase is divided into G_1, S, and G_2. During G_1, protein synthesis and assembly occur. During the S period, the cell's nuclear DNA is duplicated. During G_2, there are changes in the cytoskeleton, preparing the cell for mitosis.

Mitosis: Producing Two Identical Nuclei

Mitosis is nuclear division. It is a precise process that assures that the DNA duplicated during the S period of interphase is divided equally between each of the two progeny nuclei that will eventually be produced. Although the events of mitosis occur in the nucleus of the cell and concern the DNA, the process involves the entire cell, and major structural changes can be observed with the light microscope (Figure 2.24, pages 74–75). After nuclear division, the cytoplasm is divided and two separate but identical cells are produced.

To understand mitosis, scientists have distinguished four different phases: prophase, metaphase, anaphase, and telophase. The descriptions that accompany Figure 2.24 refer to each of these phases. Read them carefully as you examine the pictures so that you will understand what happens during mitosis.

Cytokinesis: Producing Two Identical Cells

Cytokinesis (*sigh-toe-kin-ee-sis*) is the process of physically dividing the cytoplasm and all its contents into two separate cells. The distribution of organelles during cytokinesis is less precise and more random than is the distribution of chromosomes during mitosis.

In the cells of animals and humans, cytokinesis is accomplished by the contraction of a belt of microfilaments that encircle the cell just inside the plasma membrane in the center of the cell. As these microfilaments shorten, a cleavage furrow forms and deepens, eventually pinching the cell in two and forming the two separate cells (see Figure 2.24, pages 74–75). We will call these separate cells **progeny cells.** Each progeny cell contains a nucleus with the same kind and amount of DNA contained in the parent cell. In Chapter 18 we will describe nuclear division by meiosis. Meiosis produces gametes (eggs and sperm) with one-half the amount of DNA of the parent cell.

> Mitosis is nuclear division and accomplishes the precise division of nuclear DNA. Mitosis consists of prophase, metaphase, anaphase, and telophase. Cytokinesis follows mitosis and divides the cytoplasm, creating two distinct progeny cells. Mitosis and cytokinesis produce two cells with the same amount and kind of DNA as the parent cell.

Chapter Summary

Cells: An Overview

- All organisms are composed of cells. All specialized human body cells are derived from a fertilized egg through the processes of cell division and development.
- Eukaryotic cells have a DNA-containing nucleus and other organelles that are not present in prokaryotes (bacteria).
- Nutrients and wastes move across the plasma membrane at the cell surface.

Cellular Organelles of Transport and Digestion

- The plasma membrane consists of a phospholipid bilayer in which are suspended mobile proteins that function as receptors, channel proteins, carrier proteins, and enzymes.
- Passive transport operates by diffusion and does not use cellular energy.
- Active transport uses cellular energy (often from ATP) to transport molecules against a concentration gradient.
- In endocytosis, vesicles formed by the plasma membrane transport materials into cells. Digestion occurs within fused vesicles. Exocytosis transports materials out of cells.

The Nucleus and Its DNA

- The nucleus contains most of the cell's DNA and is bounded by the nuclear envelope, which has pores through which materials move to the cytoplasm.
- DNA carries coded hereditary information from parent cell to progeny cells during cell reproduction. Before cell reproduction, the DNA is duplicated.

DNA, Genes, and Protein Synthesis

- Genes are DNA regions that code for the sequence of amino acids in different kinds of proteins.
- In transcription, a complementary stand of mRNA is formed from a strand of DNA.
- In translation, the coded information in mRNA is used to specify the sequence in which amino acids are bonded together to form a protein.
- Some proteins remain in the cell. Others are secreted out of the cell by exocytosis.

Organelles for Packaging, Processing, and Secreting

- The rough endoplasmic reticulum (ER) has attached ribosomes for the synthesis of proteins that enter the lumen of the ER for additional processing and packaging into vesicles.
- Vesicles from the ER fuse with the Golgi membrane and empty their protein contents into the lumen of the Golgi, where further processing takes place.
- Some vesicles from the Golgi fuse with the plasma membrane, secreting their contents outside the cell (exocytosis).

Mitochrondria: Cellular Power Plants

- Mitochrondria are organelles that use O_2 and enzymes to degrade fuel molecules, producing CO_2 and H_2O and releasing energy which is captured as ATP.
- ATP is transported out of each mitochondrion and can provide energy for cellular activities.

The Cytoskeleton, Cilia, and Flagella

- The cytoskeleton is a network of long thin cytoplasmic proteins that maintain cell shape, anchor organelles, participate in cell movements, and divide cells in two.
- Both cilia and flagella are constructed from microtubules using the 9 + 2 pattern. Using energy from ATP, the microtubules slide past one another, causing cilia and flagella to move.

(Continued on page 76)

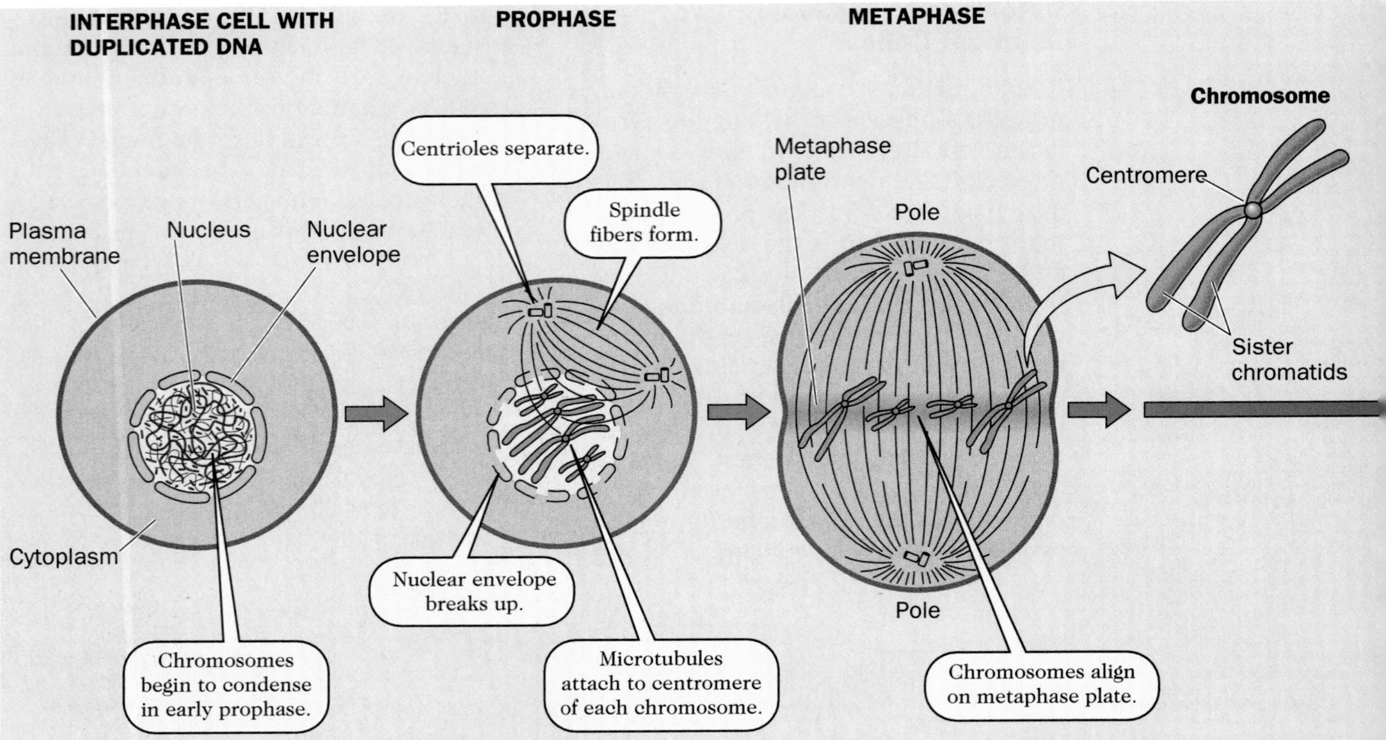
INTERPHASE CELL WITH DUPLICATED DNA
PROPHASE
METAPHASE
Chromosome
Centrioles separate.
Metaphase plate
Centromere
Spindle fibers form.
Pole
Plasma membrane
Nucleus
Nuclear envelope
Sister chromatids
Cytoplasm
Nuclear envelope breaks up.
Pole
Chromosomes begin to condense in early prophase.
Microtubules attach to centromere of each chromosome.
Chromosomes align on metaphase plate.

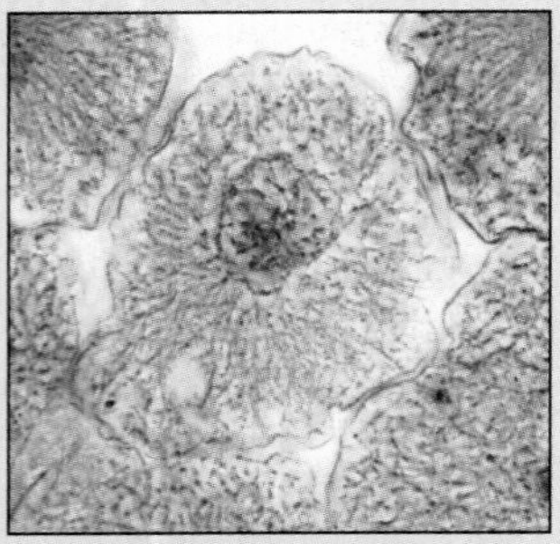

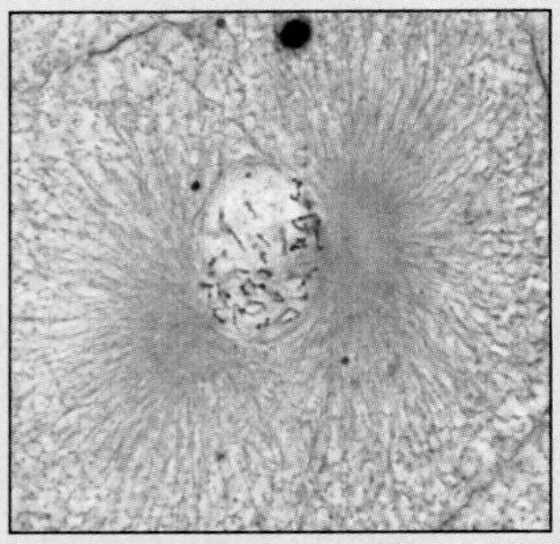

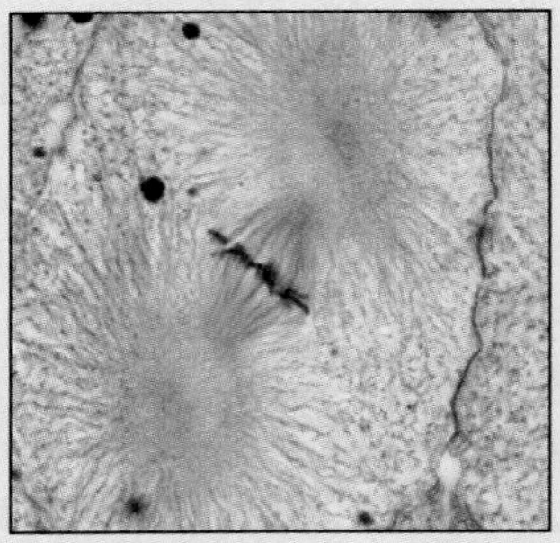

Prophase–During prophase, three things happen:

1. The diffuse chromatin of the interphase nucleus forms structures called chromosomes. In normal human cells undergoing mitosis, there are 46 chromosomes containing all the cell's nuclear DNA. Because the DNA content of the nucleus was doubled during interphase, each chromosome actually consists of two identical structures called chromatids. These chromatids are attached at a constricted site called the centromere.

2. Simultaneously with the changes in the nucleus, there are changes in the cytoplasm. The paired structures, the centrioles, separate and migrate to opposite ends or poles of the cell. Centrioles may play a role in the formation of microtubule spindle fibers from tubulin. Eventually, spindle fibers stretch from one end of the cell to the other. Some of these spindle fibers will be involved in the movements that distribute the chromatids of each chromosome to opposite poles of the cell.

3. Toward the end of prophase, the nuclear envelope breaks down, and a spindle fiber from each pole attaches to the centromere region of each chromosome.

Metaphase

During metaphase, each chromosome is moved to the center of the cell by its attached spindle fibers. The site of chromosome assembly is termed the metaphase plate.

FIGURE 2.24 The stages of mitosis: prophase, metaphase, anaphase, and telophase. The role of mitosis is to precisely distribute identical amounts and kinds of DNA to the two progeny cells, produced by cytokinesis.

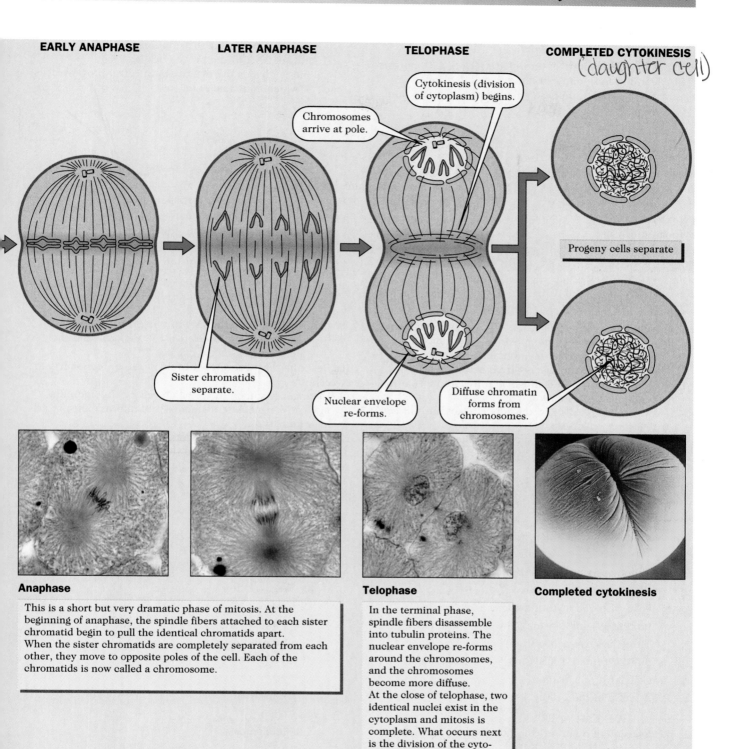

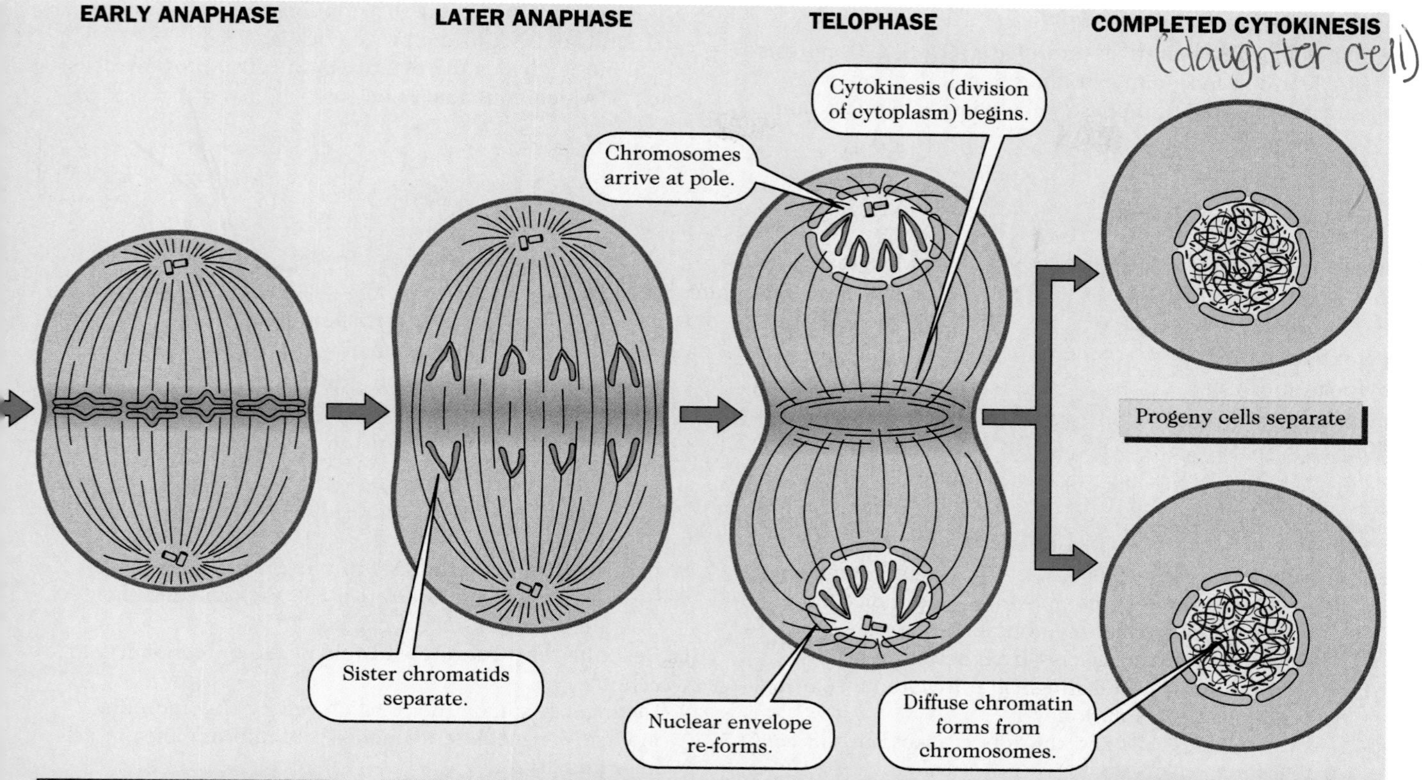
EARLY ANAPHASE
LATER ANAPHASE
TELOPHASE
COMPLETED CYTOKINESIS
(daughter cell)
Cytokinesis (division of cytoplasm) begins.
Chromosomes arrive at pole.
Progeny cells separate
Sister chromatids separate.
Nuclear envelope re-forms.
Diffuse chromatin forms from chromosomes.

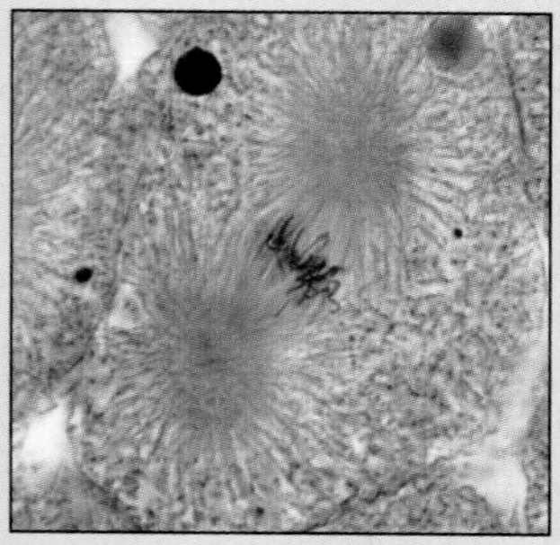
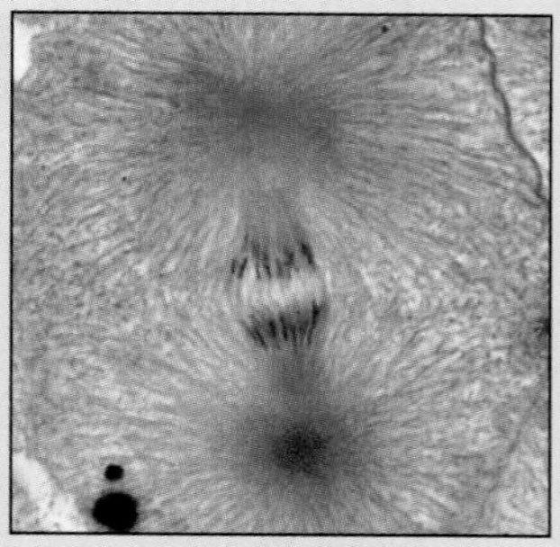

Anaphase

This is a short but very dramatic phase of mitosis. At the beginning of anaphase, the spindle fibers attached to each sister chromatid begin to pull the identical chromatids apart. When the sister chromatids are completely separated from each other, they move to opposite poles of the cell. Each of the chromatids is now called a chromosome.

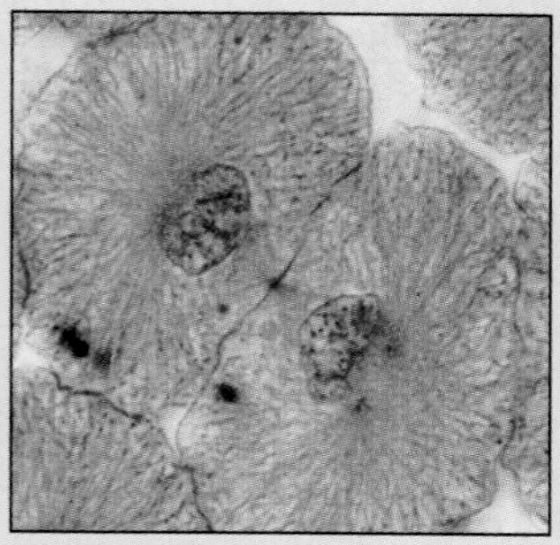

Telophase

In the terminal phase, spindle fibers disassemble into tubulin proteins. The nuclear envelope re-forms around the chromosomes, and the chromosomes become more diffuse.
At the close of telophase, two identical nuclei exist in the cytoplasm and mitosis is complete. What occurs next is the division of the cytoplasm and its contents into two separate cells.

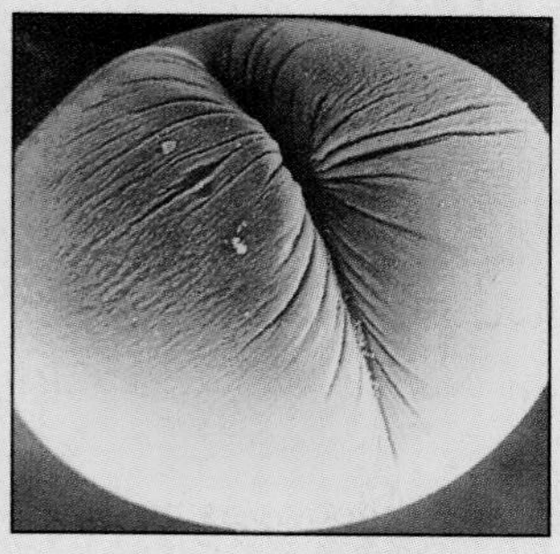

Completed cytokinesis

Cell Growth and Cell Division

- The cell cycle consists of interphase (G_1, S, G_2), mitosis (nuclear division), and cytokinesis (cell division).
- Mitosis consists of prophase, metaphase, anaphase, and telophase.
- Mitosis accomplishes the distribution of equal amounts and kinds of DNA to the progeny cells.
- Cytokinesis divides the cytoplasm into two progeny cells, each of which contains a nucleus.

Selected Key Terms

active transport (p. 57)
cell cycle (p. 72)
chromosome (pp. 61, 74)
cytokinesis (p. 73)
cytoplasm (p. 48)
cytoskeleton (p. 69)
diffusion (p. 54)
endocytosis (p. 58)
endoplasmic reticulum (ER) (p. 65)
exocytosis (pp. 58, 67)
Golgi apparatus (p. 66)
mitochondria (p. 67)
mitosis (pp. 72, 74–75)
nucleus (p. 60)
organelles (p. 48)
osmosis (p. 55)
passive transport (p. 54)
plasma membrane (p. 52)
replication (p. 61)
transcription (p. 62)
translation (p. 63)

Review Activities

1. How do eukaryotic cells differ from prokaryotic cells?
2. Describe the function, composition, and molecular structure of the plasma membrane.
3. Distinguish passive transport from active transport.
4. Describe osmosis and distinguish a hypotonic solution from a hypertonic solution.
5. Describe endocytosis and the role of lysosomes in cellular digestion.
6. Describe the cell nucleus and the duplication of DNA.
7. Describe the role of DNA in protein synthesis.
8. Describe the role of mRNA in protein synthesis.
9. Describe the role of the rough ER, vesicles, and the Golgi in secretion by exocytosis.
10. Describe the structure and role of the mitochondria in cell life.
11. Distinguish among the roles played by the microfilaments, intermediate filaments, and microtubules in cellular activities.
12. Identify the sequential events that take place during interphase and mitosis and describe their activities.

Self-Quiz

Matching Exercise

___ 1. Diffusion of water across a membrane
___ 2. Transport against a concentration gradient
___ 3. Endocytosis performed by white blood cells
___ 4. The formation of mRNA
___ 5. Fuse with endocytotic vesicles for digestion
___ 6. Protein assembly using RNA and ribosomes
___ 7. Three-base sequence in mRNA
___ 8. Site for protein processing before secretion
___ 9. Cellular power plants
___ 10. A phase of the cell cycle consisting of G_1, S, and G_2
___ 11. The first stage of mitosis
___ 12. Division of a parent cell into two progeny cells

A. Interphase
B. Lysosomes
C. Phagocytosis
D. Transcription
E. Golgi
F. Prophase
G. Osmosis
H. Cytokinesis
I. Active transport
J. Translation
K. Codon
L. Mitochondria

Answers to Self-Quiz

1. G; **2.** I; **3.** C; **4.** D; **5.** B; **6.** J; **7.** K; **8.** E; **9.** L; **10.** A; **11.** F; **12.** H

Chapter 3

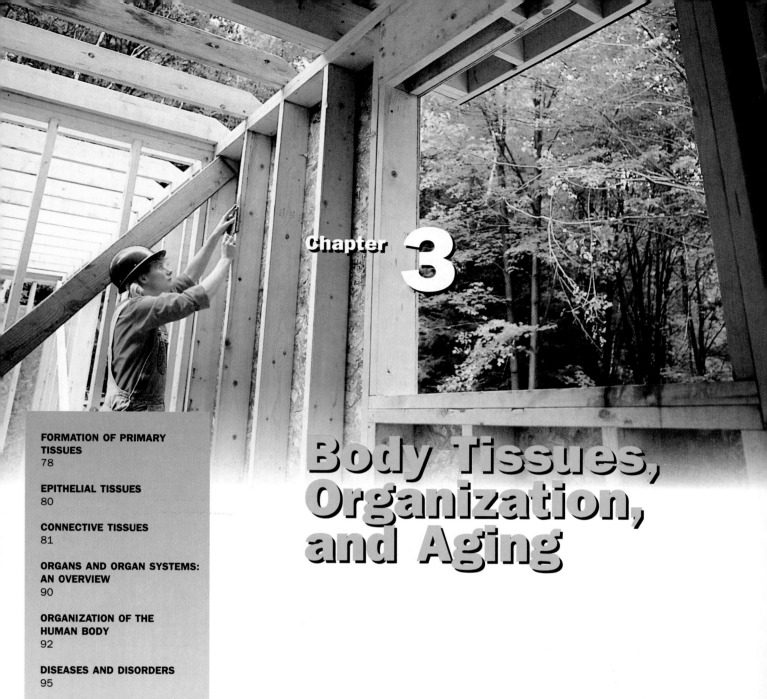

Body Tissues, Organization, and Aging

Boards, beams, bricks, pipes, and wires—these construction materials may be more familiar to you than the structures in your body. Like a house, your body is assembled from simpler building blocks that are organized into specific structures to accomplish specific functions. As you learned in Chapter 1, atoms form molecules (carbohydrates, amino acids, proteins, lipids, and nucleic acids) and ions (including many minerals), and out of those substances all living cells are constructed. As we stated in Chapter 2, cells are the fundamental units of all living systems, and there may be as many as 70 trillion cells in the adult human body.

The human body, however, is not just a random collection of cells; it is organized in very specific ways. Specific cells with different shapes, structures, and functions are organized into tissues. A **tissue** is a group of cells that have a common developmental origin and perform a limited number of specialized functions. There are roughly 200 different kinds of cells in the human body, and those cells can be organized into about 20 types of tissues. Two or more of these tissues combine to form each organ.

Organs usually have a distinct shape and perform one or a few specific functions. For example, the heart is an organ that pumps blood, while the kidneys remove wastes and control the body's water content. Several organs that perform related and coordinated functions are called an **organ system.** Cells make up tissues, tissues make up organs, and organs make up organ systems (Figure 3.1). You have 11 major organ systems: the integumentary (skin), skeletal, muscular, digestive, circulatory, respiratory, urinary, lymphatic (or immune), nervous, endocrine, and reproductive systems.

As an **organism,** you are the product of the coordinated functioning of all your systems. For example, when you breathe, your *muscular* and *skeletal systems* help the *respiratory system* obtain molecular oxygen (O_2) from the air. The *circulatory system* transports O_2 to your body cells, which use it to release energy from food molecules that are processed and absorbed by your *digestive system*. Waste material from cellular activities is removed from your body by the *urinary system*. These functions are coordinated by your *nervous* and *endocrine systems*.

In this chapter, we will describe the major characteristics of tissues and survey the organ systems and the organization of the body. We will then explore the nature of diseases and the aging process. In subsequent chapters, we will focus on individual organ systems and their coordinated interactions in health, disease, and aging.

Formation of Primary Tissues

A mature human body is made up of about 20 different tissues that can be assigned to four basic categories called **primary tissues.** The four primary tissues are the epithelial, connective, muscular, and nervous tissues. The formation of these primary tissues begins very early in our development from a fertilized egg (Figure 3.2). Cell division by mitosis leads

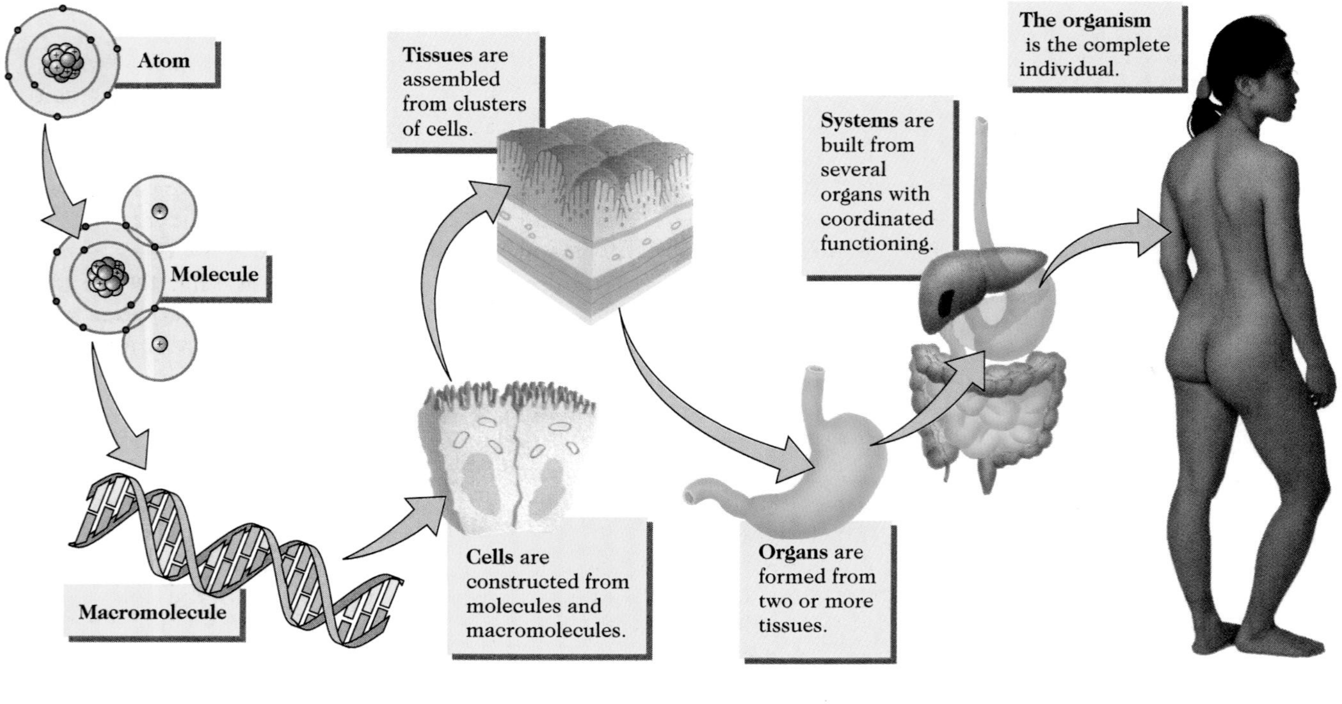
Atom
Molecule
Macromolecule
Cells are constructed from molecules and macromolecules.
Tissues are assembled from clusters of cells.
Organs are formed from two or more tissues.
Systems are built from several organs with coordinated functioning.
The organism is the complete individual.

FIGURE 3.1 Levels of assembly in humans.

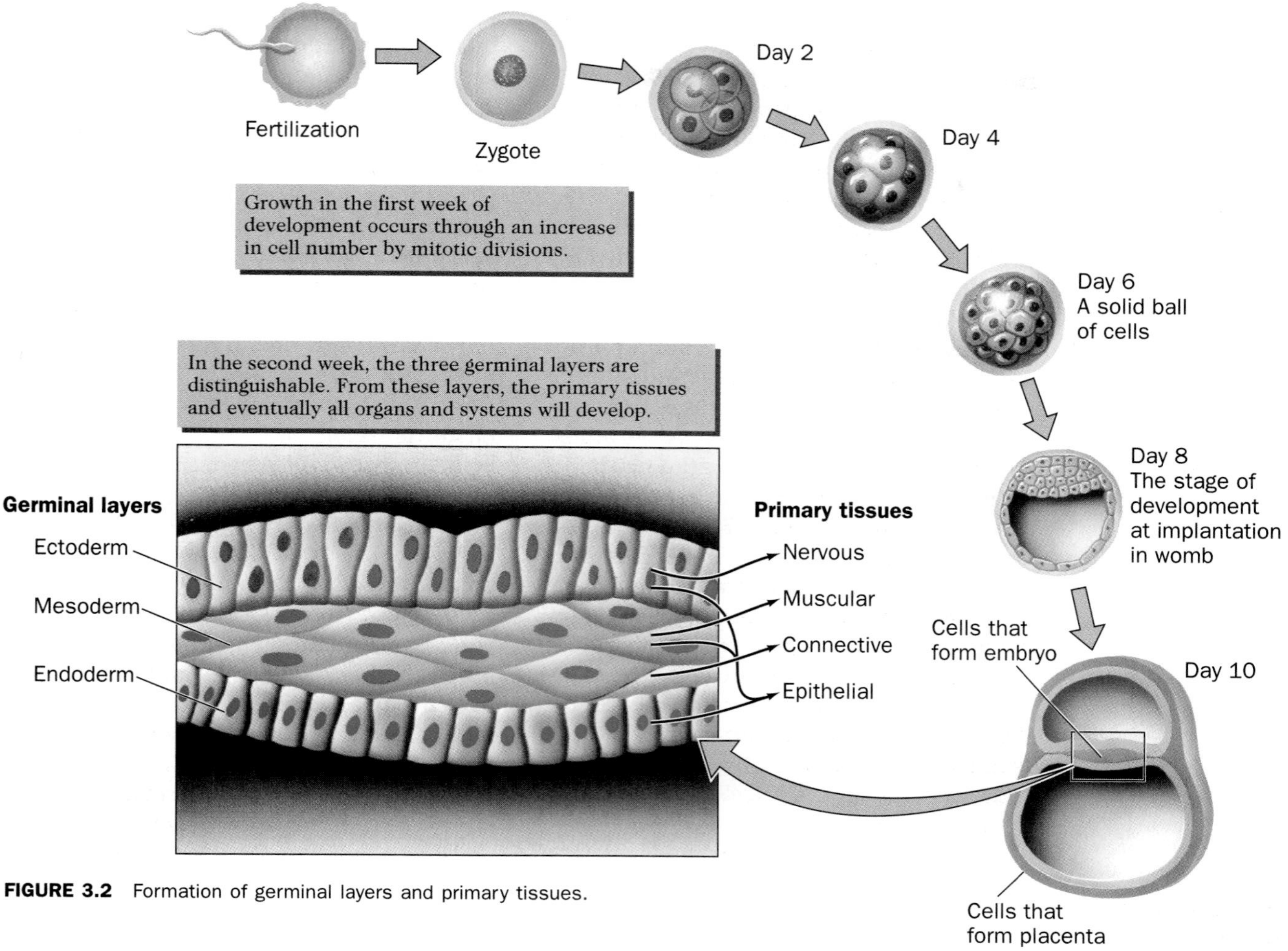

Fertilization
Zygote
Day 2
Day 4
Growth in the first week of development occurs through an increase in cell number by mitotic divisions.
Day 6
A solid ball of cells
In the second week, the three germinal layers are distinguishable. From these layers, the primary tissues and eventually all organs and systems will develop.
Day 8
The stage of development at implantation in womb
Germinal layers
Ectoderm
Mesoderm
Endoderm
Primary tissues
Nervous
Muscular
Connective
Epithelial
Cells that form embryo
Day 10
Cells that form placenta

FIGURE 3.2 Formation of germinal layers and primary tissues.

to cell proliferation and an organized mass of cells called an **embryo.** As the embryo grows in size, it changes and eventually produces all the specialized cells of human tissues and organs. By the second week of development three layers of cells can be distinguished. These layers are called the **germinal layers,** and from them all tissues and organs will eventually form.

The three germinal layers are named for their relative positions in the embryo: **ectoderm** (*ek-toe-derm*), the outermost layer; **mesoderm** (*mez-oh-derm*), the middle layer; and **endoderm** (*en-doe-derm*), the innermost layer. Initially the cells of each germinal layer are identical, but soon they begin to change in shape and function, forming the more specialized primary tissues. Nervous tissue develops from the ectoderm. Mesoderm forms the muscles and the connective tissues that support, protect, insulate, and store energy. Epithelial tissues and most of the other organs are produced from two or all three germinal layers.

In this chapter, we will examine in detail the epithelial and connective tissues. The other, more specialized tissues will be described in later chapters, along with the relevant organ systems.

> The human body is made of molecules, cells, tissues, organs, and organ systems. Four primary tissues (epithelial, connective, muscular, and nervous) form early in embryonic development from three distinct layers of cells: the ectoderm, mesoderm, and endoderm.

Epithelial Tissues

An **epithelium** (*eh-pih-**thee**-lee-um*; plural, *epithelia*) is composed of closely packed, regularly shaped, and often tightly connected cells (Figure 3.3). Because of those tight connections, there is very little extracellular (*extra*, "outside of") space or material between epithelial cells.

Epithelial cells sit on a thin noncellular layer called the **basement membrane,** which helps connect them to an underlying connective tissue. The basement membrane consists of proteins and other substances secreted by both the epithelium and the connective tissue. It is a firm but open network that does not prevent the diffusion of small molecules, and it should not be confused with the plasma membrane that surrounds all cells (Chapter 2).

No blood vessels pass directly through an epithelium. For this reason, all epithelial cells depend on diffusion from the blood vessels in connective tissue for a continuous supply of nutrients and for waste removal. Though they lack blood vessels, epithelia may contain nerves and sensory receptors connected to nerves.

Because epithelial cells form a protective surface layer and are usually subjected to mechanical stress, they are connected to each other by specialized structures (see Figure 3.3). A layer of *intercellular cement* (*inter*, "between") binds epithelial cells to each other and to the basement membrane. *Gap junctions* connect neighboring cells and allow them to communicate through protein channels. *Tight junctions* narrow the extracellular space and prevent material from diffusing between the epithelial cells that line a cavity. *Spot desmosomes* (***dez**-moh-somes*) consist of disks on the inner surface of the plasma membrane connected by filaments to the cytoskeleton and to the plasma membrane of an adjacent cell. Since the intercellular filaments of adjacent desmosomes interlock, this connecting structure resembles a Velcro fastener between neighboring cells.

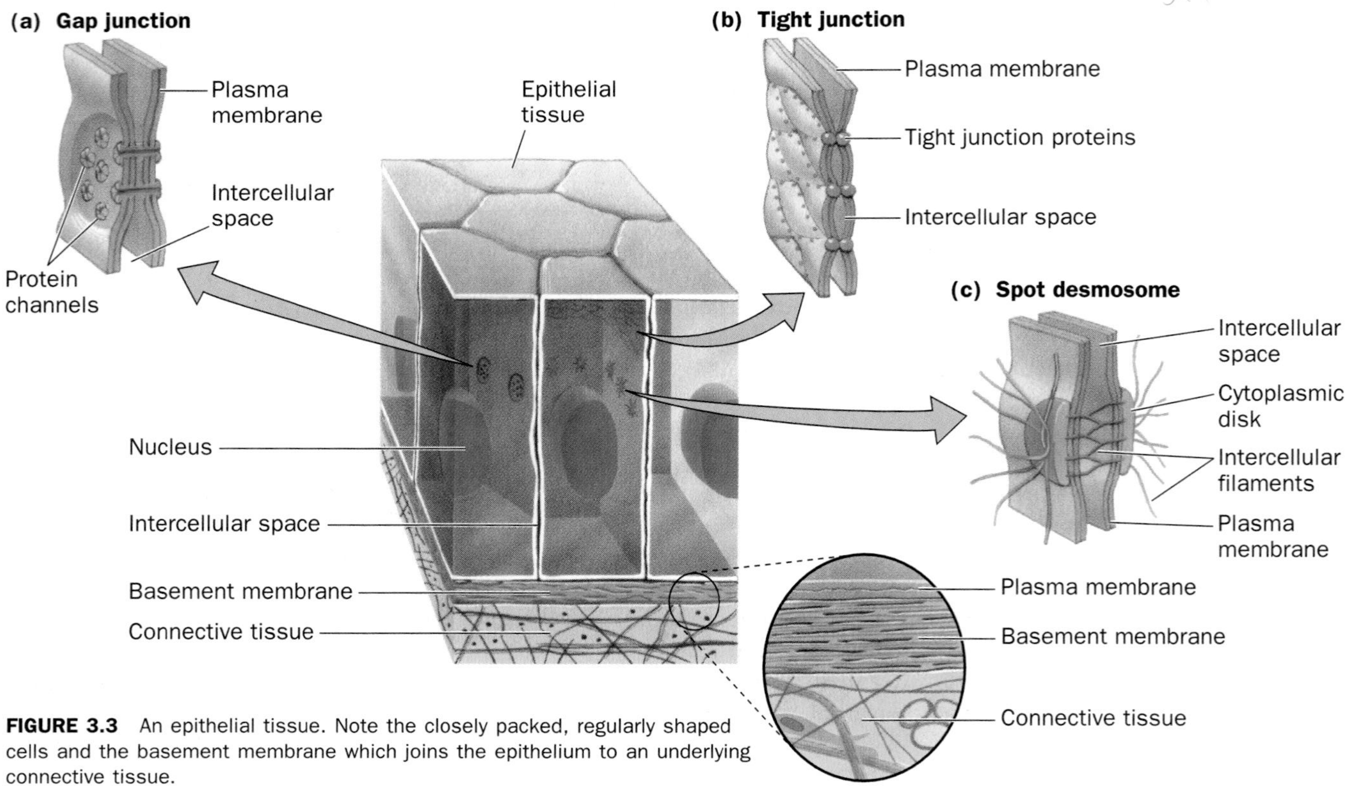
(a) Gap junction
Plasma membrane
Intercellular space
Protein channels
Epithelial tissue
(b) Tight junction
Plasma membrane
Tight junction proteins
Intercellular space
(c) Spot desmosome
Intercellular space
Cytoplasmic disk
Intercellular filaments
Plasma membrane
Nucleus
Intercellular space
Basement membrane
Connective tissue
Plasma membrane
Basement membrane
Connective tissue

FIGURE 3.3 An epithelial tissue. Note the closely packed, regularly shaped cells and the basement membrane which joins the epithelium to an underlying connective tissue.

Epithelial tissues lack blood vessels and obtain nutrients by diffusion from neighboring tissues. The cells of an epithelium are closely packed and tightly connected by intercellular cement, gap junctions, tight junctions, and spot desmosomes.

Epithelia That Cover and Line

Epithelia can be grouped by function into two major categories: (1) covering and lining epithelia and (2) glandular epithelia involved in secretion. The **covering and lining epithelia** cover body surfaces (such as the skin), line passageways (such as the throat, intestines, blood vessels, and sweat gland ducts), line body cavities (such as the abdominal cavity), cover organs, and form the hair and nails (Figure 3.4a). There is always a cavity or open space above covering and lining epithelial cells. A fluid often occupies the open space.

There are several different kinds of covering and lining epithelia, and their classification is based on (1) the number of cell layers and (2) the cell shape. Epithelia consisting of only a single thin layer of cells are called **simple epithelia,** and they function in filtration, secretion, and absorption. Epithelia which contain two or more layers of cells are known as **stratified epithelia,** and this thicker layer functions as a protective layer (for example, the skin).

Three basic cell shapes are found in both simple and stratified arrangements. These cell shapes are squamous (flat), cuboidal (cubelike), and columnar (columnlike). Cilia (Chapter 2) are frequently found on the exposed surfaces of columnar epithelial cells. In certain body locations, such as the respiratory and digestive tracts, some columnar epithelial cells are specialized to secrete enzymes, hormones, or mucus. Other columnar cells are specialized to absorb molecules from passageways and cavities. We will explore these functions in later chapters.

Epithelia That Form Multicellular Glands

The second functional category of epithelia—the **glandular epithelia**—form the secretory portions of multicellular glands. There are two general types of these glands: (1) those with a secretory tube or duct and (2) those without a duct (Figure 3.4b). Glands with a duct are called **exocrine glands** (*ex-oh-krin*). Their secreted products pass through ducts to the surface of the body or into an internal cavity or passageway. Examples of exocrine glands include sweat, tear, salivary, mammary, and digestive glands.

Glands without ducts are called **endocrine glands** (*en-doe-krin*). Their secreted products diffuse into blood vessels and are distributed throughout the body by the circulation of the blood. Examples of endocrine glands include the pituitary, thyroid, and adrenal glands. The secreted products of endocrine glands are called hormones. **Hormones** are chemical signals that regulate and coordinate the functions of the different organ systems. The following chapters will identify the roles of specific hormones in their discussions of the regulation and coordination of individual organs.

Epithelial tissues that cover surfaces or line passageways are arranged as single layers of cells (simple epithelia) or multiple layers of cells (stratified epithelia). Glandular epithelia form multicellular glands with or without secretory ducts (exocrine and endocrine glands).

Connective Tissues

Connective tissues have great variety in terms of their structure, composition, and function and are abundantly distributed in the body. Like the epithelia, they can be divided into different categories (Table 3.1). However, this classification is more convenient than natural. The diversity in the form and function of tissues within

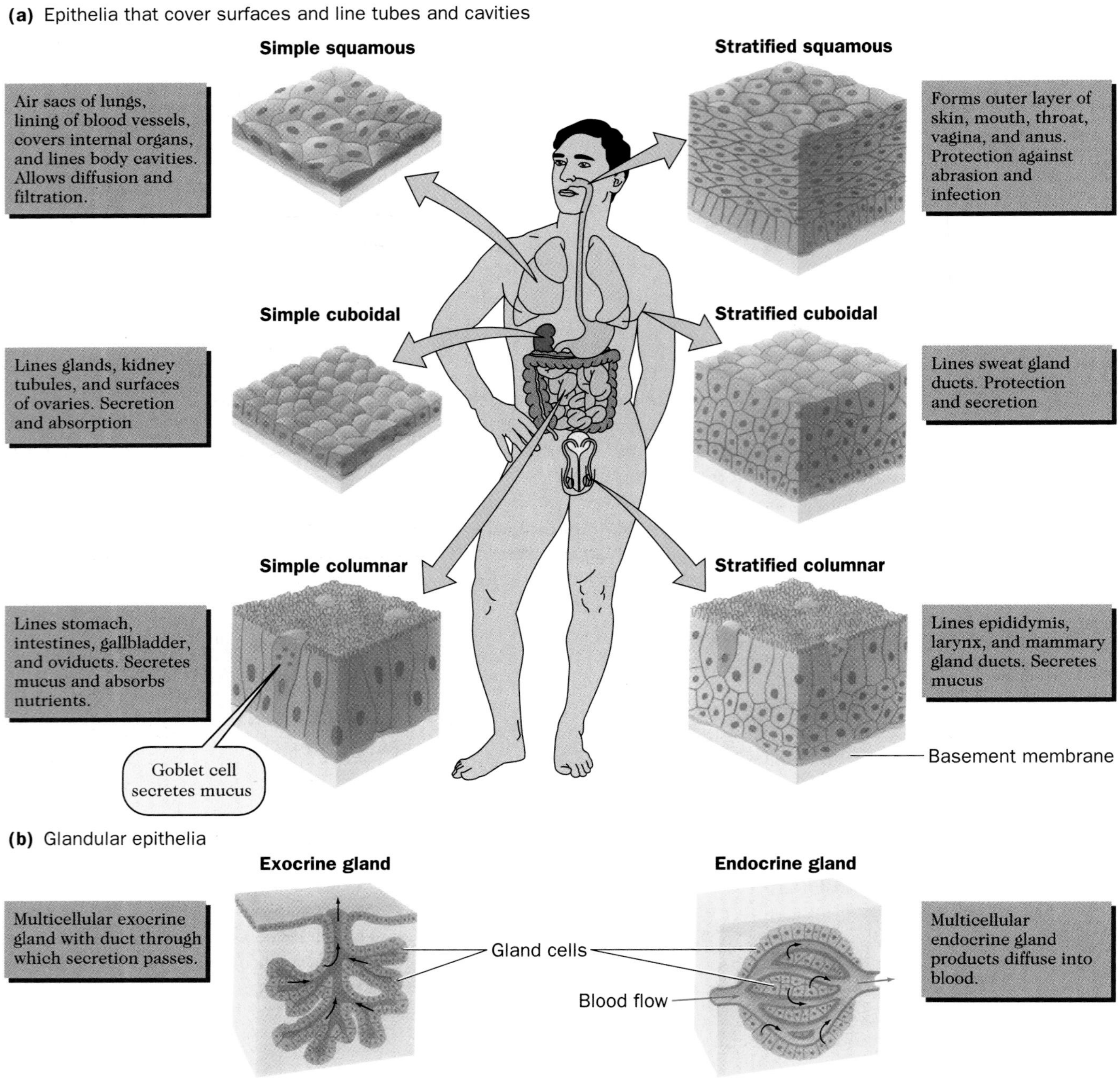
(a) Epithelia that cover surfaces and line tubes and cavities
Simple squamous
Air sacs of lungs, lining of blood vessels, covers internal organs, and lines body cavities. Allows diffusion and filtration.
Stratified squamous
Forms outer layer of skin, mouth, throat, vagina, and anus. Protection against abrasion and infection
Simple cuboidal
Lines glands, kidney tubules, and surfaces of ovaries. Secretion and absorption
Stratified cuboidal
Lines sweat gland ducts. Protection and secretion
Simple columnar
Lines stomach, intestines, gallbladder, and oviducts. Secretes mucus and absorbs nutrients.
Goblet cell secretes mucus
Stratified columnar
Lines epididymis, larynx, and mammary gland ducts. Secretes mucus
Basement membrane
(b) Glandular epithelia
Exocrine gland
Multicellular exocrine gland with duct through which secretion passes.
Gland cells
Blood flow
Endocrine gland
Multicellular endocrine gland products diffuse into blood.

FIGURE 3.4 Covering and lining and glandular epithelia.

each major group is as great as the differences between the groups.

Some types of connective tissues bind, support, and protect (bone and cartilage), other types insulate and store (fat or adipose tissue), while still others function in transportation and defense (blood and lymph). Perhaps the only thing common to all types of connective tissue is their origin from the mesoderm during the development of the embryo. In contrast to most epithelial tissues, connective tissues are never located on exposed body surfaces. Instead, they are buried within the body, where they (1) give internal support to soft organs, (2) surround organs and hold them in place, and (3) attach organs to the body wall.

Different connective tissues vary in the numbers of blood vessels they contain.

TABLE 3.1 A Classification of Connective Tissue

I. Fibrous connective tissue
   - A. Loose collagenous tissue
   - B. Dense collagenous tissue
   - C. Elastic tissue
   - D. Reticular tissue

II. Specialized connective tissue
   - A. Adipose (fat storage) connective tissue
   - B. Cartilage
     1. Hyaline cartilage
     2. Fibrocartilage
     3. Elastic cartilage
   - C. Bone
   - D. Blood

Note: A summary of the appearances, functions, and general locations of the major connective tissues is given in Figure 3.6.

Some adult connective tissues have an abundant supply of blood vessels and thus have ready access to the nutrients the blood transports. Other connective tissues contain few blood vessels and take longer to heal when they are injured. In contrast to epithelial tissue, the cells of connective tissues usually are not in direct contact with each other; instead, they are held apart by a nonliving extracellular matrix which they secrete.

Connective Tissue Matrix

Nonliving extracellular **matrix** consists of a ground substance, tissue fluid, and fibers (Figure 3.5). **Ground substance** is a clear, semifluid gel that is built from complexes of polysaccharides and proteins. In some connective tissues, such as cartilage, the ground substance is rubbery. To get an idea of its texture, think of the gristle on the joint end of a chicken drumstick. In other connective tissues the ground substance may be firm and flexible, or it may be soft and stretchable. In bone, it is hard as rock.

In contrast to the ground substance, **tissue fluid** is clear and watery. It is derived from the liquid portion of the blood by filtration through the walls of blood vessels. Respiratory gases (CO_2 and O_2), nutrients, and wastes diffuse through the tissue fluid as they move between cells and blood vessels.

The **fibers** in the matrix are long, thin proteins that provide strength, a supportive framework, and elasticity to connective tissues.

The three types of fibers in the matrix. The matrix contains collagen fibers, reticular fibers, and elastic fibers (see Figure 3.5). The most abundant fibers are **collagen fibers** (***kol**-uh-jen*). Collagen is a ropelike protein that can bend but not stretch. Collagen fibers provide strength to connective tissue, playing a role somewhat similar to that of steel reinforcing bars placed in concrete. These fibers are abundant in tendons, which join muscles to bone, and ligaments, which join bones together. In the genetic disorder called Ehlers-Danlos syndrome, or "elastic skin," weakened collagen fibers in ligaments cause the bones of body joints to dislocate easily.

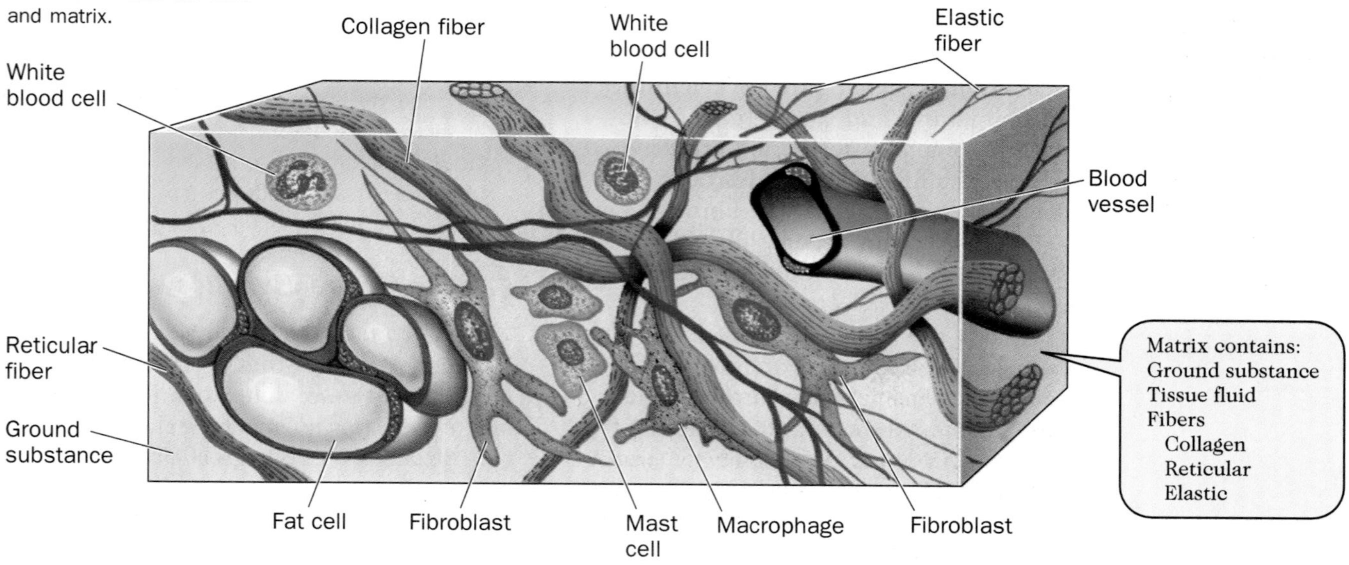

Collagen fiber
White blood cell
Elastic fiber
White blood cell
Blood vessel
Reticular fiber
Ground substance
Fat cell
Fibroblast
Mast cell
Macrophage
Fibroblast
Matrix contains:
Ground substance
Tissue fluid
Fibers
Collagen
Reticular
Elastic

FIGURE 3.5 A connective tissue. Note the cells and matrix.

A second type of fiber is called **reticular fiber** (*reh-**tick**-you-lahr*). Reticular fibers form a thin branching network made mostly of collagen. They provide an internal framework for soft organs such as the spleen, liver, and lymph nodes. They are an important part of the basement membrane that was described earlier in this chapter. Reticular fibers function like the chicken wire used to support a papier-mâché model.

Elastic fibers are the third type of fiber present in connective tissue matrix. These fibers consist of a randomly coiled protein which can stretch without being permanently deformed. The skin, the lungs, and the walls of the blood vessels gain resiliency from the elastic fibers they contain. After filling with air, the lungs are able to deflate and quickly exhale because of elastic fibers.

As is true of their blood vessels, connective tissues differ greatly in the number and kinds of matrix fibers they contain.

Cells Present in Connective Tissue

In many connective tissues there are two types of cells: (1) those involved in the secretion and maintenance of the matrix and (2) specialized accessory cells that defend against foreign substances and infection and aid in wound healing. The cells that secrete and maintain the matrix are called **fibroblasts** (see Figure 3.5). The suffix *blast* indicates a cell that actively produces and secretes the components of a developing tissue. When the tissue matures and stops growing, the cells become less active and are denoted with the suffix *cyte* (as in **fibrocyte**). If the tissues are injured, the fibrocytes are activated to again become fibroblasts and assist in the regeneration that occurs with healing.

Among the accessory cells of the connective tissue are macrophages, mast cells, and plasma cells. All these cells help defend the body and initiate the healing of wounds. **Macrophages** (***mack**-row-fahj-ez*) are a type of white blood cell capable of phagocytosis (see Figure 3.5). Phagocytosis permits them to engulf and digest dead human cells and combat bacteria, viruses, and other foreign cells that may cause disease. Macrophages may be fixed in place by attachments to matrix fibers or may wander through the matrix by means of amoeboid movement (Chapter 2).

Mast cells secrete substances involved in defense against foreign material or cells. One of these substances (heparin) prevents blood clotting. Another secreted substance (histamine) causes blood vessels to enlarge in diameter and become leaky. Both of these substances encourage white blood cells to move out of the blood vessels and into the tissues to fight infection. **Plasma cells** secrete defensive proteins called *antibodies* that attach to and inactivate specific foreign proteins.

All these cells are part of what is called the *inflammation response* to an injury. People experience inflammation as heat, pain, and swelling in the region of an injury. These discomforts are outward manifestations of the defensive actions operating within connective tissues. We will discuss these defenses further in Chapter 13.

The amount and arrangement of fibers, the composition of the ground substance, and the type of cells present can all be useful in distinguishing and classifying connective tissues. In the next two sections, we will explore the two categories of connective tissue: fibrous connective tissues and specialized connective tissues.

> Connective tissues are found throughout the body. They function to bind, support, protect, insulate, store, or transport. Cells in connective tissues are separated by an extracellular matrix which consists of ground substance, tissue fluid, and protein fibers.

Fibrous Connective Tissues

Fibrous connective tissue is a group of connective tissues that includes loose connective tissue, dense connective tissue, elastic tissue, and reticular tissue (Figure 3.6). There is great diversity among these tissues, but they all contain a matrix with some collagen fibers. Blood vessels are found in all types of fibrous

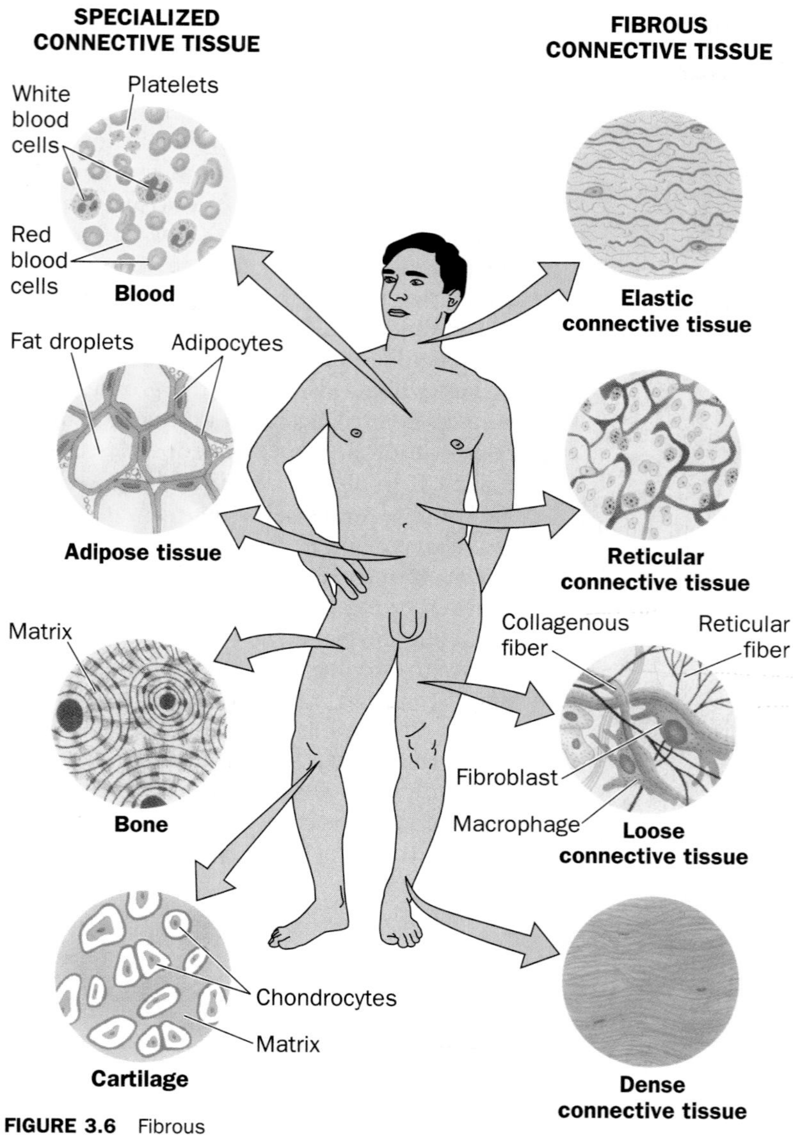
SPECIALIZED CONNECTIVE TISSUE
White blood cells
Platelets
Red blood cells
Blood
Fat droplets
Adipocytes
Adipose tissue
Matrix
Bone
Chondrocytes
Matrix
Cartilage
FIBROUS CONNECTIVE TISSUE
Elastic connective tissue
Reticular connective tissue
Collagenous fiber
Reticular fiber
Fibroblast
Macrophage
Loose connective tissue
Dense connective tissue

FIGURE 3.6 Fibrous and specialized connective tissues.

connective tissues, although dense connective tissue has fewer vessels than do the other types.

Loose connective tissue. Loose connective tissue is the most widely distributed connective tissue in adults. It forms the wrapping around internal organs, blood vessels, muscles, and nerves. In a syrupy ground substance, loose connective tissue contains only a few collagen and elastic fibers lying in all directions. These fibers give it flexibility in all directions but not great strength.

Dense connective tissue. Dense connective tissue contains a higher density of collagen fibers than does loose connective tissue. Dense connective tissue functions more for strength than for flexibility. Some types of dense connective tissues have bundles of collagen fibers lying in all directions that provide strength and flexibility. A dense connective tissue like this is found in the lower layer of the skin, the dermis. You can pinch and twist your skin in different directions without tearing it.

Other types of dense connective tissue have bundles of collagen fibers that lie more or less in one direction. These fibers provide great strength and protect against stress in the direction of the fibers. Tendons and ligaments are formed of such tissue and are quite strong when pulled in one direction. However, they may tear if stress comes unexpectedly from a different direction. Tendons and ligaments have fewer blood vessels than do loose connective tissues; therefore, their injuries often take longer to heal.

Elastic connective tissue. Elastic connective tissue has numerous bundles of elastic fibers which impart their characteristics to the entire tissue, allowing it to stretch without deforming permanently. Elastic tissue is found in the walls of body organs that regularly change shape, such as the stomach, the lungs, the blood vessels, and even the heart. Elastic tissues are also found in the vocal cords and in the ligaments that connect the bones of the spine, where both stretch and support are needed.

Reticular connective tissue. Reticular connective tissue has a finely branched network of reticular fibers (which are actually a very thin form of collagen fibers). It provides an internal framework for the cells that perform the functions of soft organs, for example, the liver, spleen, lymph nodes, and tonsils.

> The four types of fibrous connective tissue bind, protect, and support. However, they differ in their degrees of strength, flexibility, and elasticity and are found at different sites in the body. All fibrous connective tissues contain blood vessels.

Specialized Connective Tissues

Specialized connective tissue is a structurally and functionally diverse group of tissues. It includes adipose (fat storage) tissue, cartilage, bone, and blood (see Figure 3.6). Among these tissues, only cartilage and bone have abundant fibers.

Adipose tissue. Adipose tissue functions for storage. It differs from fibrous connective tissues in that it has very little ground substance and few fibers. It contains cells called **adipocytes** (*add-uh-po-sites*), which are specialized to store fats. Molecules of fat have a great deal of energy (almost twice as much as carbohydrates do). If food intake is reduced, this stored fat provides reserves of chemical energy for cellular activities. Adipose tissue is found under the skin, where it provides insulation. It also forms a protective cushion around many internal organs, most notably the kidneys.

The exact sites in your body where adipose tissue will be deposited are determined in part by your sex, your hormonal balance, and the genes you inherited from your parents (Figure 3.7). Unfortunately, when we form adipocytes, they are with us for life. As we gain weight, the fat content of adipocytes increases and the individual cells increase in size. During weight loss, these cells become smaller as the stored fats are used for energy. However, with weight loss, the number of adipose cells does not decrease. In recent years a controversial surgical procedure called *liposuction* has been used to remove adipose tissues and slenderize the body.

Cartilage. Like dense connective tissue, **cartilage** has a high concentration of collagen fibers. However, the ground substance of cartilage has a unique chemical composition and special properties that permit it to absorb and bind large amounts of water. This bound water makes cartilage resistant to compression. Collagen and elastic fibers may be present in varying proportions. These components give cartilage strength and the capacity to provide firm support, flexibility, and a tough smooth surface for the movements of bones against each other.

Cartilage lacks blood vessels, nerves, and lymphatic vessels in its extracellular matrix. The absence of blood vessels means that injuries to cartilage are slow to heal.

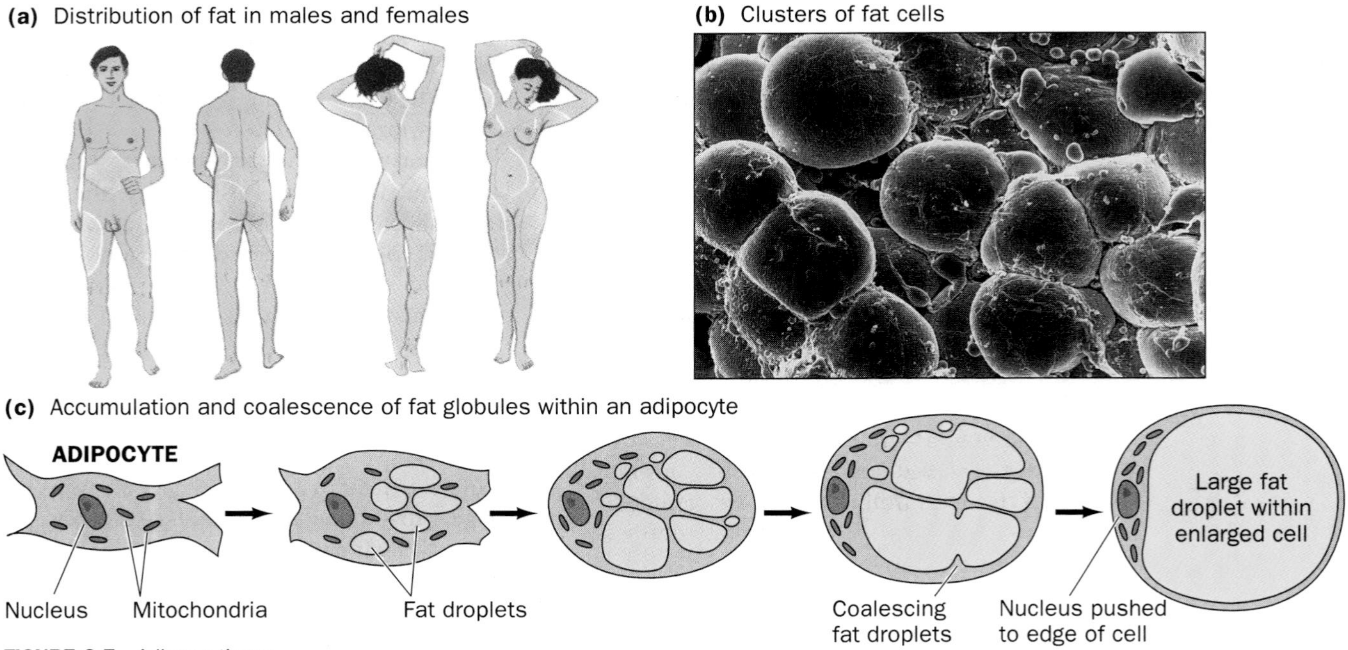
(a) Distribution of fat in males and females
(b) Clusters of fat cells
(c) Accumulation and coalescence of fat globules within an adipocyte
ADIPOCYTE
Large fat droplet within enlarged cell
Nucleus
Mitochondria
Fat droplets
Coalescing fat droplets
Nucleus pushed to edge of cell

FIGURE 3.7 Adipose tissue.

The cells that secrete the cartilage matrix are called **chondroblasts** (***kon**-droh-blasts*). As the entire matrix grows, the chondroblasts become trapped in small spaces within it. These trapped cells, which are called **chondrocytes** (***kon**-droh-sights*), do not actively secrete matrix (see Figure 3.6).

Because cartilage has no blood vessels, chondrocytes receive their nutrients and exchange gases and wastes by slow diffusion through the surrounding matrix. The blood vessels that nourish cartilage are all located in a dense connective tissue called the **perichondrium** (*pair-ih-**kon**-dree-um*) that is wrapped around most cartilage (except on joints).

Three types of cartilage. There are three types of cartilage in the body: hyaline cartilage, fibrocartilage, and elastic cartilage. **Hyaline cartilage** (***hi**-uh-lin*) is both flexible and strong. It is abundant in the body and consists of a mat of collagen fibers in a rubbery matrix. The front part of your nose is made of hyaline cartilage. You can observe its flexibility by holding the tip of your nose and bending it gently from side to side. Hyaline cartilage covers the ends of many bones, providing a slippery surface for the movement of joints. It also joins the ribs to the breastbone (sternum), providing flexibility so that the entire rib cage can rise and fall as you breathe.

Fibrocartilage is tougher and less flexible than hyaline cartilage. Its collagen fibers are thicker and are arranged in dense bundles. Fibrocartilage joins together bones in areas where considerable stress may occur. Pads of fibrocartilage form the disks between the bones of the spine (vertebrae) and join together the bones of the pelvic girdle. Both the spine and the pelvic girdle bear great mechanical stresses as we stand erect and shift our weight from one leg to the other during walking or running.

Elastic cartilage can be slightly stretched without deforming. It has many elastic fibers in its matrix in addition to collagen. Elastic cartilage provides internal support for the ear. Twist your ear in a variety of directions and note how quickly it snaps back into shape and position.

Bone: a hard connective tissue. Bone is very hard and dense because of the minerals, mostly calcium and phosphate, deposited in its matrix. The microscopic structure of bone is described in Chapter 5. Here we will focus on comparing it with cartilage. Like cartilage, bone cells are trapped in an extracellular matrix that they secrete. Unlike cartilage, bone has an abundant supply of blood vessels, and the nourishment from those vessels permits it to heal faster than cartilage does.

Blood: a liquid connective tissue. Blood consists of cells suspended in a liquid matrix called *plasma* (***plaz**-muh*). Unlike other connective tissues, the matrix is not actually produced by the cells suspended in it; instead, it is produced by several organ systems. Furthermore, the chemical composition of the plasma matrix changes as gases, nutrients, and wastes are added or removed during circulation of the blood through the various tissues and organs. The plasma also contains dissolved hormones which circulate to the organs, regulating and coordinating their functions.

The cells of blood include red blood cells, white blood cells, and platelets (see Figure 3.6). The major function of *red blood cells* is the transport of molecular oxygen (O_2). *White blood cells* are involved in defending the body against disease-causing microorganisms. *Platelets* help blood clot, limiting the loss of blood from a wound.

As you can see, blood performs very important functions for all the tissues and organs through which it circulates. We'll devote an entire chapter to blood (Chapter 8). The organ systems are illustrated in Figure 3.8 and discussed in the following section.

Specialized connective tissues include adipose tissue (storage), blood (transport), and cartilage and bone (support). The properties of cartilage permit it to reduce friction and cushion impacts in the joints between bones. Of all the connective tissues, cartilage has the fewest blood vessels.

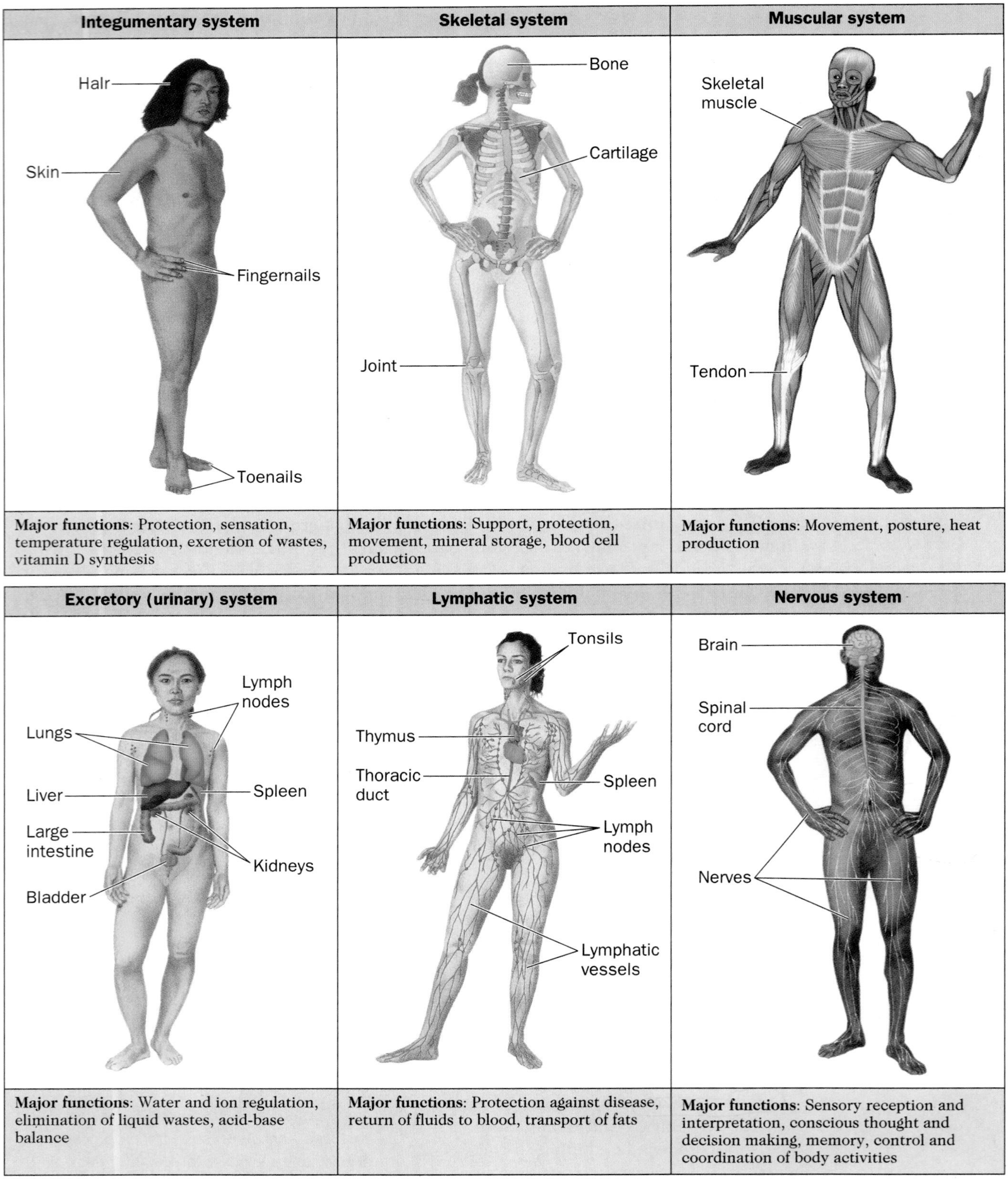
Integumentary system
Halr
Skin
Fingernails
Toenails
Major functions: Protection, sensation, temperature regulation, excretion of wastes, vitamin D synthesis
Skeletal system
Bone
Cartilage
Joint
Major functions: Support, protection, movement, mineral storage, blood cell production
Muscular system
Skeletal muscle
Tendon
Major functions: Movement, posture, heat production
Excretory (urinary) system
Lymph nodes
Lungs
Liver
Spleen
Large intestine
Kidneys
Bladder
Major functions: Water and ion regulation, elimination of liquid wastes, acid-base balance
Lymphatic system
Tonsils
Thymus
Thoracic duct
Spleen
Lymph nodes
Lymphatic vessels
Major functions: Protection against disease, return of fluids to blood, transport of fats
Nervous system
Brain
Spinal cord
Nerves
Major functions: Sensory reception and interpretation, conscious thought and decision making, memory, control and coordination of body activities

FIGURE 3.8 Organ systems of humans.

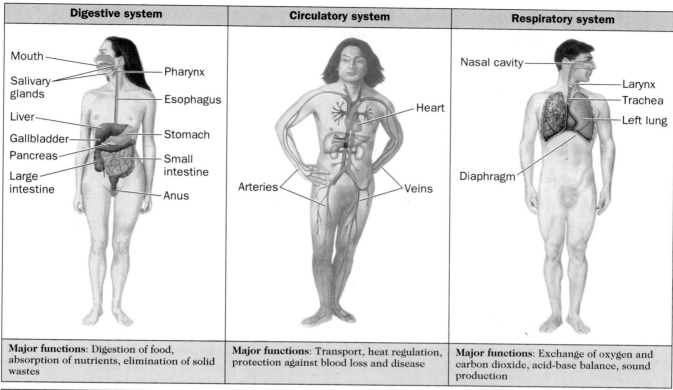

Digestive system

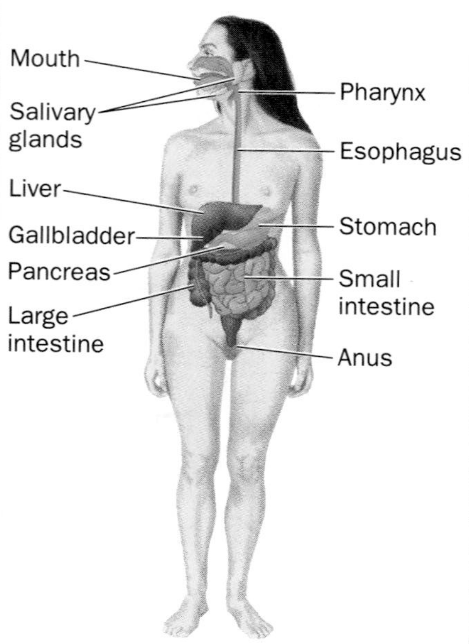
Mouth
Salivary glands
Liver
Gallbladder
Pancreas
Large intestine
Pharynx
Esophagus
Stomach
Small intestine
Anus

Major functions: Digestion of food, absorption of nutrients, elimination of solid wastes

Circulatory system

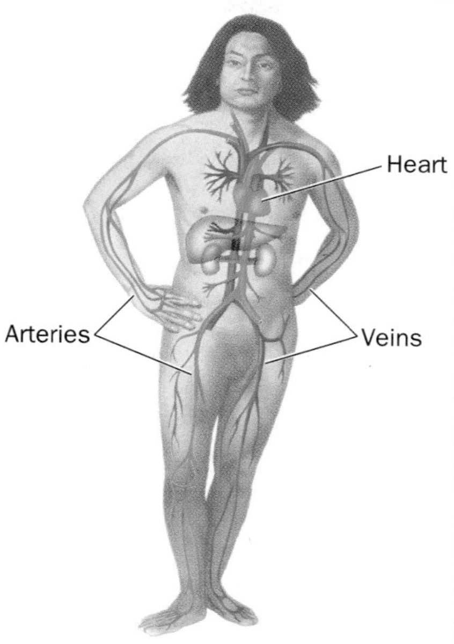
Heart
Arteries
Veins

Major functions: Transport, heat regulation, protection against blood loss and disease

Respiratory system

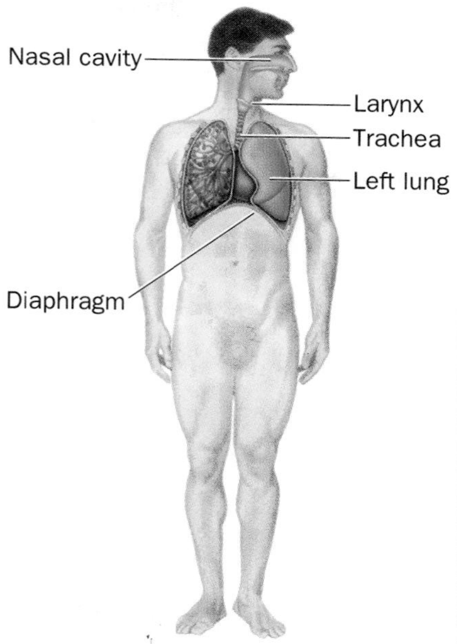
Nasal cavity
Larynx
Trachea
Left lung
Diaphragm

Major functions: Exchange of oxygen and carbon dioxide, acid-base balance, sound production

Endocrine system

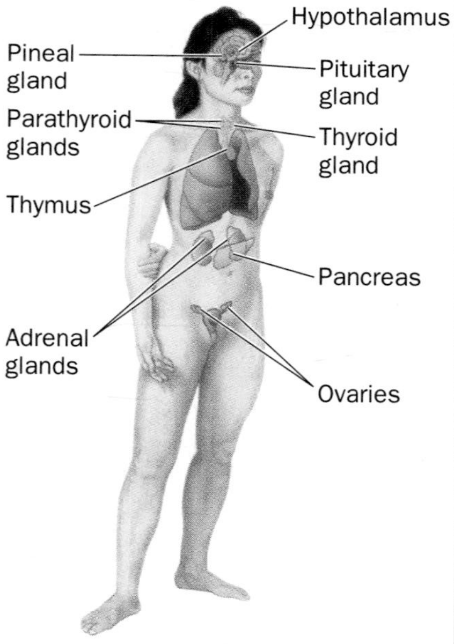
Hypothalamus
Pineal gland
Pituitary gland
Parathyroid glands
Thyroid gland
Thymus
Pancreas
Adrenal glands
Ovaries

Major functions: Growth, metabolism, reproduction, control and coordination of body activities

Male reproductive system

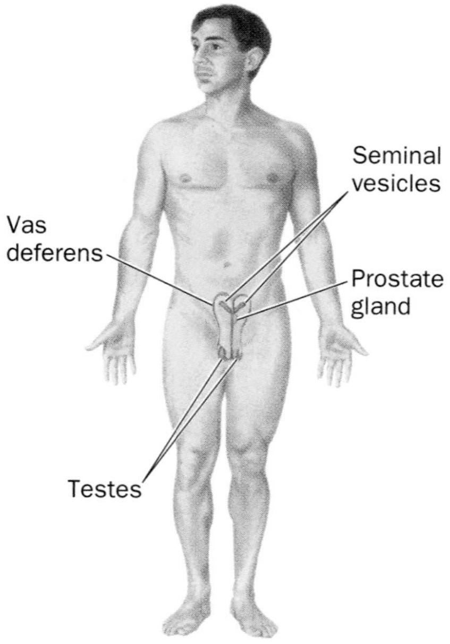
Seminal vesicles
Vas deferens
Prostate gland
Testes

Major functions: Production and transport of reproductive cells

Female reproductive system

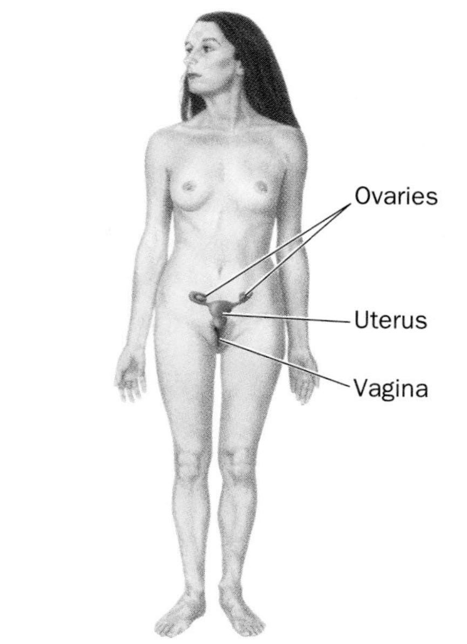
Ovaries
Uterus
Vagina

Major functions: Production and transport of reproductive cells; production, protection, and nourishment of offspring

Organs and Organ Systems: An Overview

Cells form tissues, and tissues form organs. Through the coordinated actions of its tissues, an organ performs a specific function or a group of related functions. Examples of all four primary tissues often can be found in the same organ. For instance, in the heart, an *epithelium* forms a smooth lining for the chambers through which blood flows, and a *connective tissue* helps produce strong, flexible valves to control the movement of blood. *Muscle tissue* provides the pumping action needed to keep blood flowing, and *nervous tissue* regulates the rate of action. When the heart rate is regulated, the supply of blood to the entire body can be regulated. Each component tissue contributes to the functioning of the organ. When two or more organs perform coordinated functions, they constitute an organ system.

Eleven **organ systems** can be distinguished in the human body. Figure 3.8 provides a brief summary of the components, location, and functions of these organ systems.

Although distinguishing separate organ systems provides a convenient way to describe body structures and functions, it is important to realize that there is a great deal of interaction and interdependence among different organ systems. No single organ system can function entirely independently of the others. Each system has its own functions and contributes to the body's welfare, but the proper functioning of all 11 organ systems gives you a healthy life.

In the following chapters, we will focus on each organ system. However, in focusing on a single organ system, we will have to mention the roles of the other interacting organ systems, particularly the nervous and endocrine systems. These two systems play essential roles in regulating organs, coordinating systems, and maintaining stable conditions in the body, and both will be encountered in every organ system. What follows is a preview of the nervous and endocrine systems. These brief descriptions will prepare you to understand their regulating roles in the organ systems that are discussed in the next few chapters.

The Nervous System and Coordination of Organ Systems

The nervous system is the most intricate and complicated organ system, yet it consists of only two types of cells: neurons and glial cells. **Neurons** are the impulse-conducting cells of the nervous system, while **glial cells** play maintenance and support roles. The impulses transmitted by neurons are rapid but short-lived electrical signals that can stimulate or inhibit cells, tissues, and organs. Individual neurons vary in size and shape, but each one has long microscopic extensions. These extensions act somewhat like wires in a telephone system to relay messages. The long extensions of many neurons are bundled together to form *nerves*.

The nervous system is divided into the central nervous system and the peripheral nervous system (Figure 3.9). The **central nervous system (CNS)** consists of the brain and spinal cord, which constitute the control and processing centers of the body. The CNS receives and sends information by means of the peripheral nervous system. In addition to receiving and processing information from all parts of the body, the brain is involved in learning and memory.

The **peripheral nervous system (PNS)** has two components: the somatic division and the autonomic division. The **somatic division** of the PNS contains **sensory nerves** that connect with special receptors on the body surface (for example, in the skin, eyes, and ears). These receptors allow people to sense the external environment. The sensory nerves relay information to the CNS, where it is processed, stored, and interpreted. From the CNS, **motor nerves** of the somatic division carry signals to the muscles that move the body. The somatic division is sometimes called the "voluntary" nervous system because it can execute conscious decisions. However, this does not mean that all its actions are under direct conscious control.

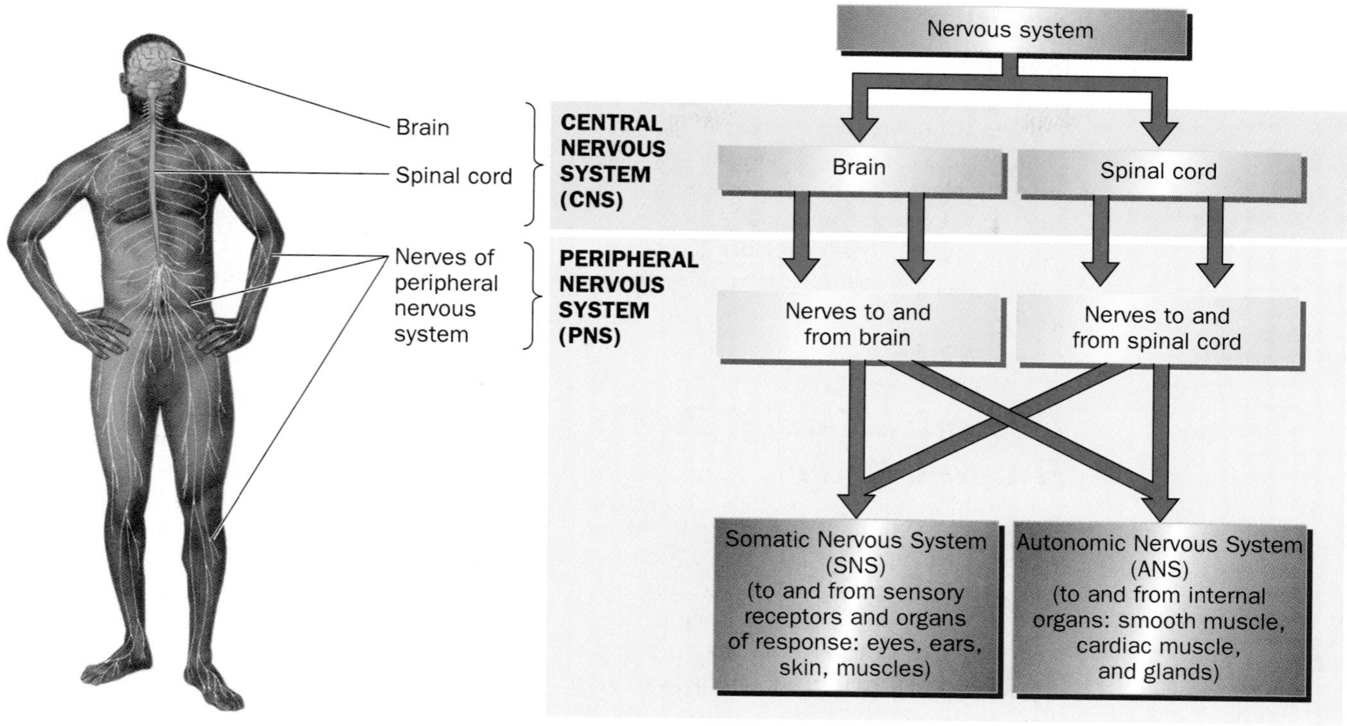
Brain
Spinal cord
Nerves of peripheral nervous system
CENTRAL NERVOUS SYSTEM (CNS)
PERIPHERAL NERVOUS SYSTEM (PNS)
Nervous system
Brain
Spinal cord
Nerves to and from brain
Nerves to and from spinal cord
Somatic Nervous System (SNS) (to and from sensory receptors and organs of response: eyes, ears, skin, muscles)
Autonomic Nervous System (ANS) (to and from internal organs: smooth muscle, cardiac muscle, and glands)

FIGURE 3.9 The organization of the nervous system.

The **autonomic division** of the PNS also contains sensory and motor neurons, but it is involved with regulating the internal environment and controlling the operation of internal organs and glands. It can both stimulate and inhibit the actions of those structures. Heartbeat, digestive activities, responses to temperature changes, and contractions of the urinary bladder are controlled wholly or in part by the autonomic division.

The autonomic division is sometimes called the "involuntary" nervous system because it acts without conscious decision making. The autonomic division is extensively involved in maintaining stable internal conditions within the body (homeostasis). Chapters 14 and 15 present a more complete description of the nervous system.

> The nervous and endocrine systems coordinate body functions. The CNS is the master of integration and the site of consciousness, learning, and memory. The somatic division of the PNS regulates interactions with the external environment, while the autonomic division regulates internal organs.

The Endocrine System and Coordination of Organ Systems

The other major controlling system of the body is the **endocrine system,** which consists of small glands that secrete hormones directly into the blood. **Hormones** are slow-diffusing but long-lasting chemical signals that regulate functions at body sites that are distant from the gland of secretion. While there are discrete endocrine glands (such as the thyroid, pituitary, and adrenals), hormones also are secreted by single epithelial cells in the stomach and intestines and by organs with additional functions (such as the thymus, the pancreas, and the reproductive organs).

In addition to glands formed from epithelium, some of the major endocrine glands are formed by modification of nervous tissue. Indeed, a portion of the brain called the **hypothalamus** (*hi-po-***thal***-ah-muss*) is considered an endocrine gland. This "brain gland" secretes hormones that control the **pituitary gland,** which in turn controls many other glands and systems and even the growth rate of the entire body. Later, we will devote an entire chapter to the endocrine system and the hormones it secretes (Chapter 16).

> The endocrine system consists of glands that secrete chemical signals (hormones) into the blood. These hormones regulate and coordinate the functions of all the body's organs. There is a close functional relationship between the nervous system and the endocrine system.

Organization of the Human Body

The organ systems identified and described in Figure 3.8 all have different sizes, shapes, and locations. In this section, we shall take a closer look at the body cavities in which organs are located and the tissue membranes that attach, lubricate, and protect the surfaces of some organs and their cavities. However, before discussing these structures, we must review some useful terminology that will help us describe structures and locate organs.

The Language of Anatomy: Locating Regions and Structures

Standing before your reflection in a full-length mirror, you quickly realize that the left side of your body resembles the right side. You can imagine a midline running from the top of your head through your nose and belly button. This imaginary line gives your body **bilateral symmetry** (Figure 3.10a) and is useful in describing the locations of certain internal organs. When we describe the location of an organ as *medial,* we mean that it is toward the middle or midline of the body. Organs or structures located away from the midline are described as having a *lateral* position.

Figure 3.10b and c shows that the body can be divided into major surface regions. The names of these regions are useful in locating the internal structures of a particular region. Indeed, the names of some muscles, bones, blood vessels, and nerves include the names of surface regions. If you know the name of a surface region, you can often determine where a particular structure is located by

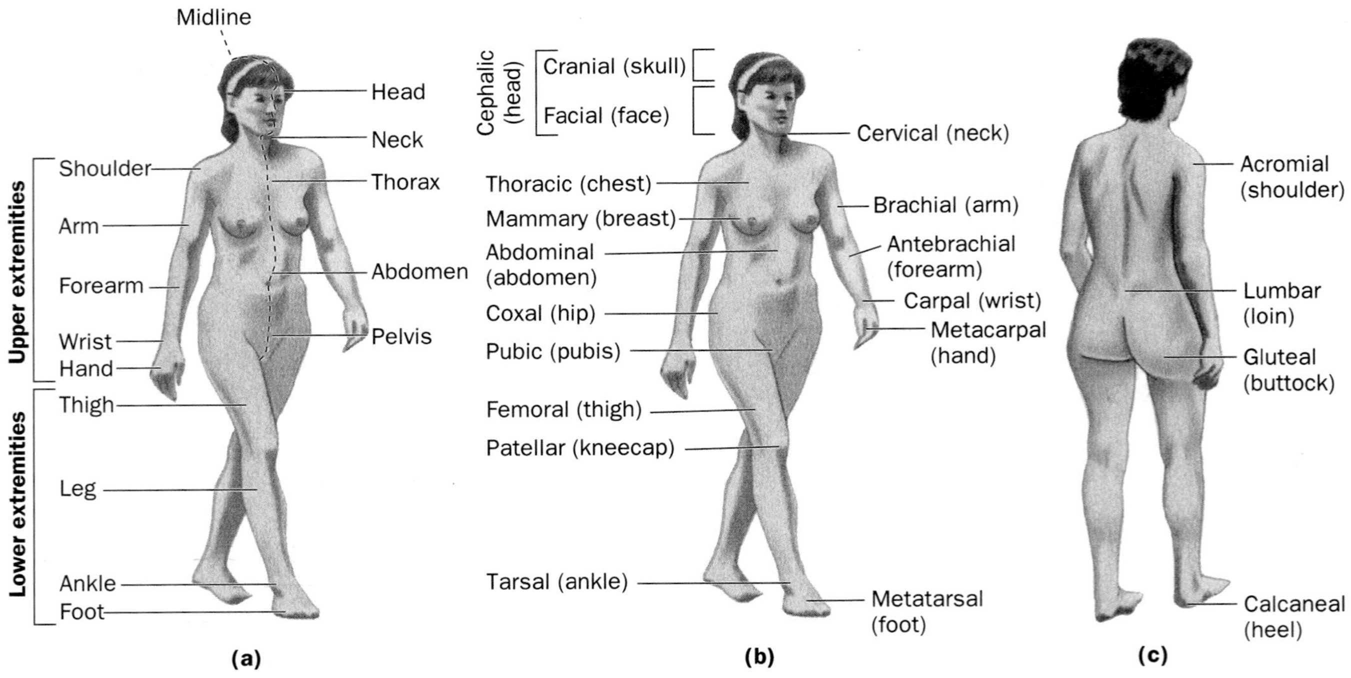

Midline
Head
Neck
Thorax
Abdomen
Pelvis
Upper extremities
Shoulder
Arm
Forearm
Wrist
Hand
Lower extremities
Thigh
Leg
Ankle
Foot
(a)
Cephalic (head)
Cranial (skull)
Facial (face)
Cervical (neck)
Thoracic (chest)
Mammary (breast)
Abdominal (abdomen)
Coxal (hip)
Pubic (pubis)
Femoral (thigh)
Patellar (kneecap)
Tarsal (ankle)
Brachial (arm)
Antebrachial (forearm)
Carpal (wrist)
Metacarpal (hand)
Metatarsal (foot)
(b)
Acromial (shoulder)
Lumbar (loin)
Gluteal (buttock)
Calcaneal (heel)
(c)

FIGURE 3.10 Body surface regions. (a) The midline and bilateral symmetry of the body. (b) Anatomic regions on the anterior of the body. (c) Anatomic regions on the posterior of the body.

the name alone. For example, the upper leg is called the *femoral region*. Within this region lies a large bone known as the *femur;* a *femoral* artery, vein, and nerve; and a large muscle known as the *quadriceps femoris*. Unless you extend your study of human biology by taking anatomy and physiology courses, it is unlikely that you will need to use this regional terminology. However, it is helpful to be aware of this pattern of naming.

A specialized vocabulary is also useful in describing the location of a structure relative to another structure. This vocabulary consists of pairs of terms with opposite meanings: superior versus inferior, anterior versus posterior, distal versus proximal. *Superior* means the upper part of the body or toward the upper part; we say that the nose is superior to the mouth. *Inferior* means toward the lower part of the body; we say that the nose is inferior to the eyes. When we refer to the inferior portion of the spine, we mean the lower portion of the spine.

Another convenient pair of terms is "anterior" and "posterior." *Posterior* means the back or toward the back side, while *anterior* means the front or toward the front side. We say that the spine is posterior to the lungs. The term "distal" means away from the attached portion of a structure, while "proximal" means close to or toward the attached portion of a structure. For example, the hand is distal to the elbow, while the wrist is proximal to the fingers.

> The body has bilateral symmetry. Specific terminology exists to identify body regions and the specific structures found in a region. Pairs of terms with opposite meanings often are used to indicate the relative locations of body structures or their parts.

Body Cavities: Spaces Occupied by Internal Organs

Body cavities are internal body spaces that are protected by the body wall (Figure 3.11). These cavities contain the body's internal organs. There are two major cavities: the posterior cavity and the anterior cavity. The **posterior cavity** contains the brain and the spinal cord. The large **anterior cavity** is divided into an upper **thoracic cavity** (*tho-***rass***-ick*) and a lower **abdominal cavity** by a sheet of muscle called the **diaphragm.**

The thoracic cavity can be divided further into two *pleural cavities* (**plur***-ull*) which surround each of the lungs and a single *pericardial cavity* (*pair-uh-***kar***-dee-ull*) which surrounds the heart. The abdominal cavity surrounds most of the digestive system. In later chapters, we will refer to these cavities and their organs.

FIGURE 3.11 Major body cavities.

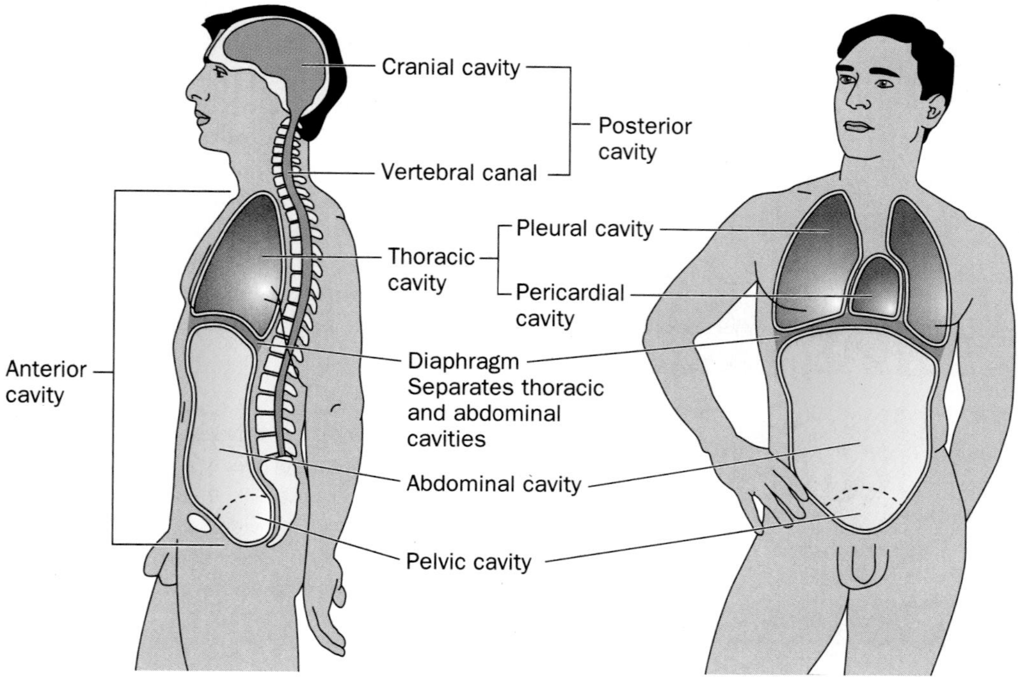

Cranial cavity
Vertebral canal
Posterior cavity
Pleural cavity
Thoracic cavity
Pericardial cavity
Anterior cavity
Diaphragm Separates thoracic and abdominal cavities
Abdominal cavity
Pelvic cavity

Tissue Membranes: Support, Protection, and Lubrication

The combination of an epithelium and supportive connective tissue constitutes a **tissue membrane.** These structures provide support, protection, and lubrication for organs. There are four major tissue membranes in the body: serous membranes, cutaneous membranes, mucous membranes, and synovial membranes.

Don't be confused. This is the third different use of the word "membrane" you have encountered so far in this book. It is important to emphasize that a tissue membrane is made of cells and is quite different from a basement membrane (described earlier in this chapter) and plasma and cellular membranes (described in Chapter 2).

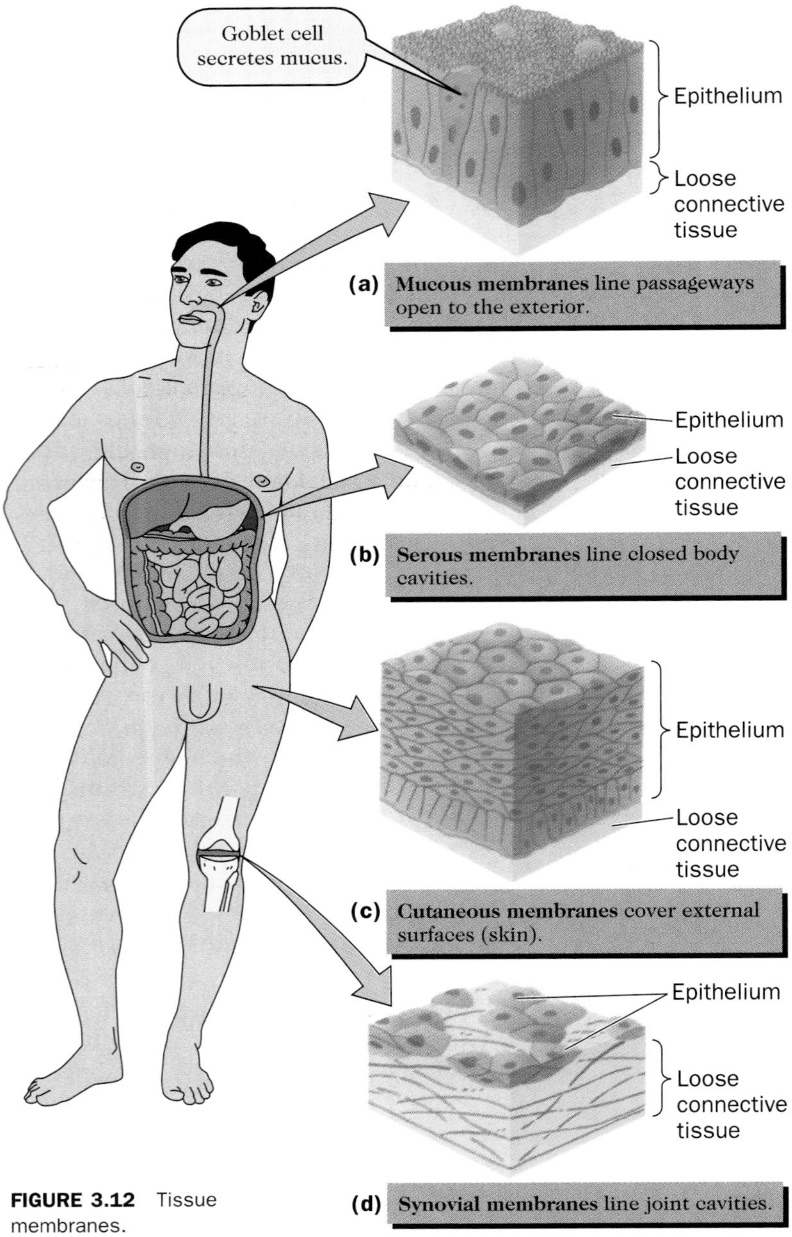
Goblet cell secretes mucus.
Epithelium
Loose connective tissue
(a) **Mucous membranes** line passageways open to the exterior.
Epithelium
Loose connective tissue
(b) **Serous membranes** line closed body cavities.
Epithelium
Loose connective tissue
(c) **Cutaneous membranes** cover external surfaces (skin).
Epithelium
Loose connective tissue
(d) **Synovial membranes** line joint cavities.

FIGURE 3.12 Tissue membranes.

The **serous membranes** (*sear-us*) are moist lubricating surfaces that line body cavities and permit internal organs to move over each other with little friction or sticking (Figure 3.12). The pleural cavities surrounding the lungs are lined by a serous membrane called the **pleura** (***plur**-uh*). The **pericardium** (*pair-uh-**kar**-dee-um*) is the serous membrane which forms the pericardial cavity that surrounds the heart. The abdominal cavity is lined by another serous membrane, the **peritoneum** (***pair**-uh-toe-knee-um*).

The serous membranes may become infected and inflamed. **Pleurisy** is a painful inflammation of the pleura, while inflammation of the peritoneum and the pericardium produces **peritonitis** and **pericarditis**, respectively. In addition to being life-threatening diseases, these conditions are very painful because reduced lubrication of the serous membranes causes internal organs to stick together.

In addition to serous membranes, there are three other types of tissue membranes (see Figure 3.12). A single **cutaneous membrane** is found on the outer surface of the body as the two layers of the skin. **Mucous membranes** line body passageways that are open to the exterior, such as the respiratory, digestive, urinary, and reproductive passages. Unicellular glands in the epithelium of this membrane secrete a thick, sticky fluid called **mucus. Synovial membranes** line cavities around movable joints such as the shoulder and the hip. We will encounter these tissue membranes in later chapters on the skin, muscles and bones, and digestive system.

> The internal organs are housed in two major cavities: the posterior cavity and the anterior cavity. These cavities are lined by serous membranes, which provide support, protection, and lubrication. Cutaneous, mucous, and synovial membranes have highly specialized functions.

Diseases and Disorders

The human body functions for health, but it does not always stay healthy. Any disruption in the normal structure and function of cells, tissues, or organ systems is considered a **disease.** The study of diseases is called **pathology** (*patho*, "ill"; *logy*, "to study").

In the chapters that follow, we will describe the most important diseases of each organ system. In doing this, we will take an objective, scientific, and sensitive approach to diseases. We want you to discover two things: (1) how knowledge of normal structure and function is applied to understanding the abnormal and (2) how reliable general information can be useful in your personal life, family life, and career.

The applied science of medicine and the experience of illness are often separate and distinct. At times one may appear to be inconsiderate of the other. As an applied science, the practice of medicine tries to be objective, logical, and quantitative. It examines things not readily visible to the eye or directly experienced by the patient. It must understand abnormalities in terms of things that can be measured. For example, the number of blood cells, the degrees of fever, and the amount of unusual chemicals in the urine are all **objective symptoms.**

Objective symptoms are distinct from the more **subjective symptoms** we all experience when we are ill: nausea, discomfort, pain, dizziness, and headache. Although these are subjective symptoms, they are no less real and no less important than what can be measured. They are also valuable guides to both diagnosis and treatment. A characteristic combination of objective and subjective symptoms is called a **syndrome.**

In their early stages some diseases lack subjective symptoms, and you may experience no discomfort or disability. Such a disease is said to be **asymptomatic.** However, with the proper examination, it may be possible for medical science to detect early signs of the disease. With some diseases, particularly cancers, such early detection often makes the difference between a cure and death. Because it is based on both objective and subjective symptoms, **diagnosis** is both an art and a science that allows doctors to distinguish one disease from another. Correct diagnosis is the basis for all successful treatment.

In the sections that follow, we will explore some the general phenomena associated with diseases. In Spotlight on Health: One (page 103), we'll explore the nature of cancer, applying your knowledge of cells and tissues.

The Many Ways of Being Ill: A Descriptive Classification of Diseases

While health seems to be a single condition, there are many diseases—many ways for the body to be ill. We will start our consideration of diseases with a general discussion of the kinds of diseases modern medicine recognizes. Medical science has discovered a specific cause for some diseases. This may be an infectious organism, a mutated gene, a poison, a trauma, or an accident of birth. However, some illnesses do not have an easily identifiable cause. These diseases are termed *idiopathic* (*id-ee-oh-**path**-ik*).

Some diseases are **chronic,** which means that they persist over a long period of time. Others are **acute,** meaning that they last for a short and often severe period. Some diseases are **local,** affecting only a single organ or organ system, while others are **systemic,** affecting the entire body. A brief description of the general causes of the diseases follows. As you will discover, some diseases fit into more than one category.

Infectious diseases. Infectious diseases are caused by microscopic viruses and organisms called **pathogens** that invade the body, disrupting cells and tissues. The symptoms of infectious diseases are caused by the tissue destruction and poisons produced as the foreign cells grow. The body can defend itself against pathogens, but its defenses may contribute to the discomfort that is experienced. Measles and chicken pox are viral diseases, while strep throat is caused by bacteria. Ringworm infections of the skin are caused by a fungus, and the pathogen for malaria is a protozoan. Figure 3.13 presents the characteristic stages that occur during the progress of an infectious disease.

Most but not all infectious diseases are communicable (contagious) and can be passed from one person to another. For example, cold viruses are easily communicated by touch, clothing, surfaces, and coughing. In contrast, the virus that causes AIDS (HIV) is less easy to communicate. HIV requires moisture to retain its capacity to infect. It is communicated only through body fluids such as blood, semen, breast milk, and vaginal fluid.

Inherited or genetic diseases. Inherited diseases are associated with the genes and chromosomes people inherit from their parents. Normally, at the moment of conception, each parent contributes one-half of the total number of chromosomes. If an accident occurs during the production of sperm or eggs, they may carry too many or too few chromosomes. When such eggs and sperm unite at fertilization, a genetic disease may result.

Even when the chromosome number is correct, some of the genes that are present on chromosomes may be defective or mutant. The proteins coded for by mutant genes do not operate properly. Hemophilia and cystic fibrosis are genetic diseases that are due to changes in single genes. Hemophilia is a failure of the blood to clot after an injury, and in cystic fibrosis exocrine glands do not secrete properly.

Congenital diseases. Congenital diseases are disorders that are present at birth. They result from chromosomal abnormalities or exposure of the embryo to drugs, toxic chemicals, antibiotics, malnutrition, radiation, or infectious agents. These agents cause abnormal tissue development, leading to organ malfunction after birth.

Heavy alcohol consumption by a pregnant woman may cause **fetal alcohol syndrome** in her child. The symptoms of fetal alcohol syndrome include low birthweight, abnormalities of the face and skull, learning disabilities, mental retardation, and defects in the genitalia and the circulatory system.

Inflammatory diseases. Inflammatory diseases occur when the body responds to the presence of a foreign material with either a local or a systemic inflammation. The symptoms of a localized inflammation are redness, swelling, heat, and pain. A systemic inflammatory response, such as a severe pollen allergy, may cause congestion, swollen and itching eyes, and sneezing. **Anaphylactic shock** is a life-threatening systemic inflammation that results when antibodies on mast cell surfaces react with foreign material. Other inflammatory diseases are associated with the formation of free antibodies.

Antibodies are protective proteins that normally are produced and secreted by plasma cells in response to foreign substances. Antibodies combine with the foreign substances to prepare them for destruction by phagocytosis by the white blood cells. Sometimes antibodies mistakenly recognize the body tissues as foreign and react with them. This autoimmunity (*auto*, "self") produces a group of dis-

FIGURE 3.13 A time line for the progress of infectious disease.

TIME →

Exposure to agents of disease: viruses, bacteria, fungi, protozoa, and larger parasites.	**Incubation:** Organisms or their toxic products accumulate over hours, days, weeks, or months. No symptoms.	**Onset of early symptoms:** fever, headache, general discomfort. Usually 8–24 h.	**Acute illness:** Specific symptoms appear as tissue damage occurs. Death if body responses or medical treatments fail. Days to weeks to rest of life.	**Convalescence:** recovery as tissue damage is repaired. Days to rest of life.

eases known as **autoimmune diseases.** Rheumatic fever is an autoimmune disease that results when antibodies produced against bacteria mistakenly attack heart valves and muscle. Rheumatoid arthritis is an inherited autoimmune disease that causes inflammation and a crippling deformity of the joints.

Degenerative diseases. Degenerative diseases are usually associated with aging, when tissues break down more rapidly than they can be repaired. **Osteoarthritis,** for example, is a painful and chronic joint inflammation that involves a progressive degeneration of the hyaline cartilage over the bone surfaces in movable joints. Another example is the lens of an aging eye that may lose elasticity and fail to focus or fail to admit enough light for clear vision.

Metabolic diseases. Metabolic diseases result from a failure to chemically process raw materials or waste products. They are often caused by a defective protein, perhaps a defective enzyme, receptor, or hormone. An example is **diabetes mellitus** (*dye-uh-beat-eez mell-eye-tuss*), in which cells have difficulty transporting glucose from the blood for energy production.

The metabolic disease **phenylketonuria** (*fen-ell-key-toe-nur-ia*) **(PKU)** is caused by a defective enzyme that normally converts the amino acid phenylalanine into another key amino acid, tyrosine. This blocked conversion causes an accumulation of phenylalanine, resulting in mental retardation unless the patient is treated with a diet low in phenylalanine. You may have noticed the warning concerning the phenylalanine content of artificially sweetened diet sodas.

Mental and emotional diseases. Mental and emotional diseases are as difficult to define as they are sometimes difficult to treat. Three general categories can be distinguished: (1) Internal disorders such as schizophrenia, depression, and compulsions. Although little is known about them, these disorders may have genetic and metabolic causes. (2) Disorders related to external substances (alcohol or drugs) or ongoing external influences (such as physical or emotional abuse). These disorders include addictions to harmful substances, behaviors, and situations. (3) Disorders from identified or unidentified conflicts and past emotional traumas.

Cancers and neoplastic diseases. The word **"neoplasm"** (***knee**-oh-plaz-um*) means "new or abnormal growth." This group of diseases results from the abnormal and uncontrolled growth of cells. Often cell proliferation leads to the formation of a mass of cells called a **tumor.** Neoplasms may be either benign or malignant. We will explore the nature, causes, prevention, and treatments of cancer in Spotlight on Health: One (page 103).

Life Span Changes

How long can a person live? For Americans born today, average **longevity,** or life expectancy, is approximately 72 years for men and 79 years for women. In this country life expectancy has more than doubled in the last 200 years. However, our maximum longevity (the age at death of the longest-lived individuals) has not changed. It appears to have remained at about 120 years. Advances in health care, housing, and sanitation and the availability of nutritious foods are allowing more people to reach the upper limits of human longevity. If these factors can alter the aging process, can more people live to the age of 120 years? This is a good question, and there are other fascinating questions about aging.

Why do we age? When do we begin to age? What controls the aging process? Are aging and death "diseases" that should be treated and cured? How can we prolong our lives? The study of aging is called **gerontology** (*jair-on-**tall**-oh-gee*), and those who study the processes of aging are called gerontologists.

In this section, we will consider, if not answer, some of these questions. First we'll distinguish the different kinds of changes that take place over the normal human life span. Then we will examine the many theories that attempt to explain the aging process. In subsequent chapters, we will discuss the life span changes in each organ system.

Development, Aging, and Senescence: Changes over the Life Span

It is convenient to divide the **life span** into three stages: embryological development, growth and maturation, and adulthood. **Embryological development** occurs between the time of fertilization and the time of birth (Figure 3.14). During this 38-week period, cells proliferate and undergo profound changes. They acquire specialized functions, allowing tissues, organs, and systems to form.

Growth and maturation encompasses the periods of infancy, childhood, adolescence, and early adulthood (from birth to the mid-twenties). This is a dramatic stage of the life span. There are increases in body size and height and great changes in the structure and functions of tissues. The body becomes capable of reproduction (puberty) during this stage, and many human qualities are acquired by learning: reasoning ability, use of language, moral judgment, and social interaction, to name a few. In spite of growth and change, there is a general decline in the *rate of growth* during this stage (Figure 3.15).

We enter **adulthood** when growth in height ceases. Adulthood is marked by a sustained capacity of the body to repair and maintain itself. However, significant changes occur in some organs and systems during this period. For example, the thymus gland decreases in size after helping to establish the body's immune defenses. Late adulthood is associated with a decline in these defensive, repair, and maintenance capacities.

If we exclude the body changes associated with illness and healing, we can distinguish three different kinds of change that take place in the human body: developmental changes, aging changes, and senescence.

Developmental changes produce a fully grown and sexually mature male or female who is capable of independence and a social life. These changes are dramatically evident from conception to maturity 20 years later.

Aging involves what usually are slow changes in the structure and function of the body, altering one's appearance and reducing one's capacity to survive. Aging is a natural process that begins very early in the life of an individual. Usually the aging process is slow, progressive, and cumulative, and people do not experience incapacity until late in the life span. Severe incapacity associated with the end of the life span is called senescence. **Senescence** involves a reduced capacity of the body to defend itself, repair itself, and maintain constant internal conditions. **Death** is marked by the complete inability of the body to regulate and maintain itself.

Aging shows a great deal of variation. Over the entire life span, the rate of aging is not uniform within the same individual

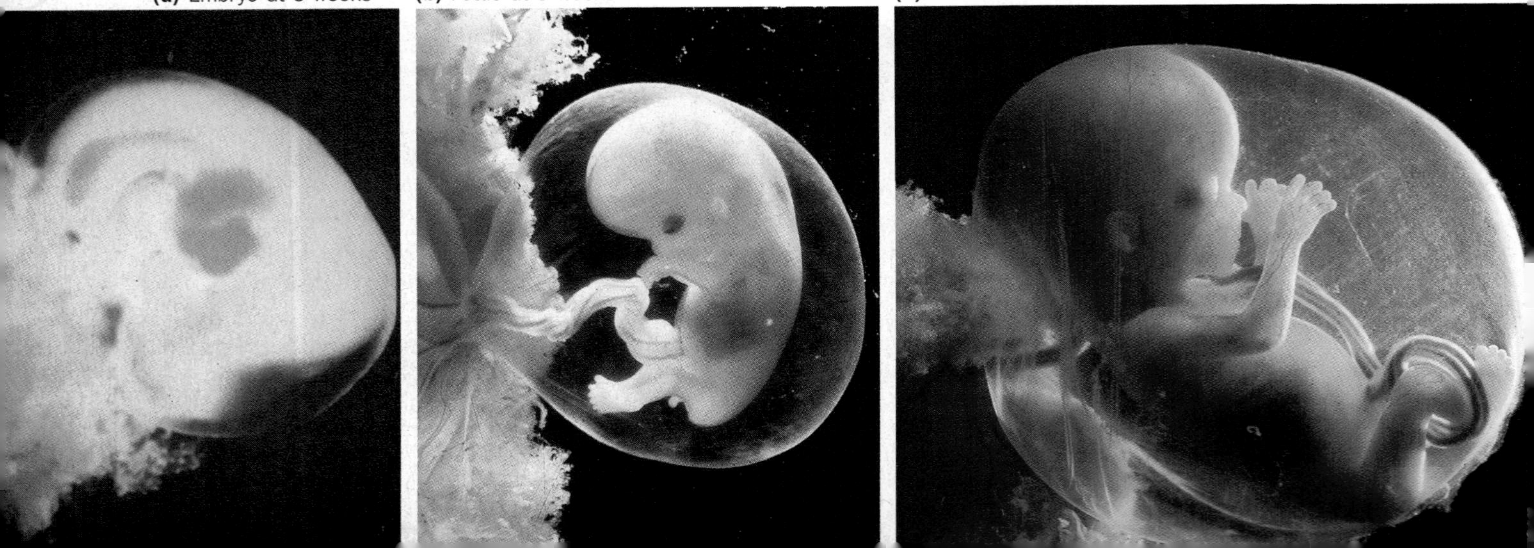

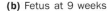

FIGURE 3.14 Human embryological development. **(a)** Embryo at 5 weeks **(b)** Fetus at 9 weeks **(c)** Fetus at 12 weeks

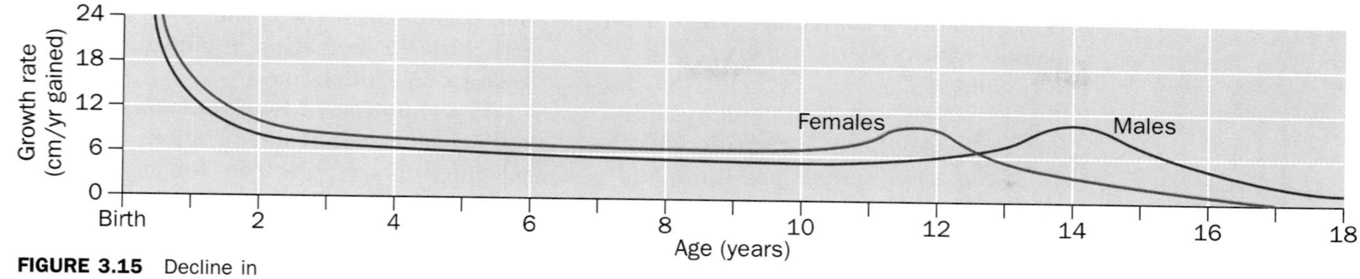
Growth rate (cm/yr gained)
24
18
12
6
0
Birth
2
4
6
8
10
12
14
16
18
Age (years)
Females
Males

FIGURE 3.15 Decline in the growth rate and changes during the growth and maturation stage of human development.

or between different individuals. If we could watch the different organs of one person over time and monitor the changes that occur, we would find that each organ ages at a different rate. If we could compare the same organ in different individuals, we would find that these organs do not age at the same rate. The "biological" age of an organ is therefore not necessarily the same as the "chronological" age of an individual. In humans, the rate of aging is influenced by genetic, psychological, social, and economic factors.

Genes and the Environment in Aging and Senescence

Gerontology research suggests that if your grandparents live past age 80, you will probably live 4 years longer than will someone whose grandparents did not live past age 60. A genetic influence on longevity is also supported by studies of identical twins, who have been found to die at very nearly the same age. Because identical twins have identical genetic material, this observation strongly suggests a genetic control of aging, senescence, and death.

Observations of people with genetic disorders also suggest that genes play a role in aging and senescence. For example, people with *Down syndrome* (*trisomy-21*) have an extra chromosome, and at early ages they show body changes that resemble the changes seen in normal elderly people. They also have a high incidence of diseases characteristic of the elderly including cancer and cardiovascular disease. In these individuals the aging process seems to have been accelerated by their genetic inheritance, and they have shortened life spans.

The rare disease *progeria* suggests that there are biological controls of the aging and senescence processes. Progeria is characterized by premature old age. It begins to appear in children during infancy and progresses until a child shows many of the characteristics of old age, such as gray hair, baldness, thin limbs, and sagging skin (Figure 3.16). Also present is atherosclerosis (fat deposits in blood ves-

FIGURE 3.16 Children showing premature aging (progeria).

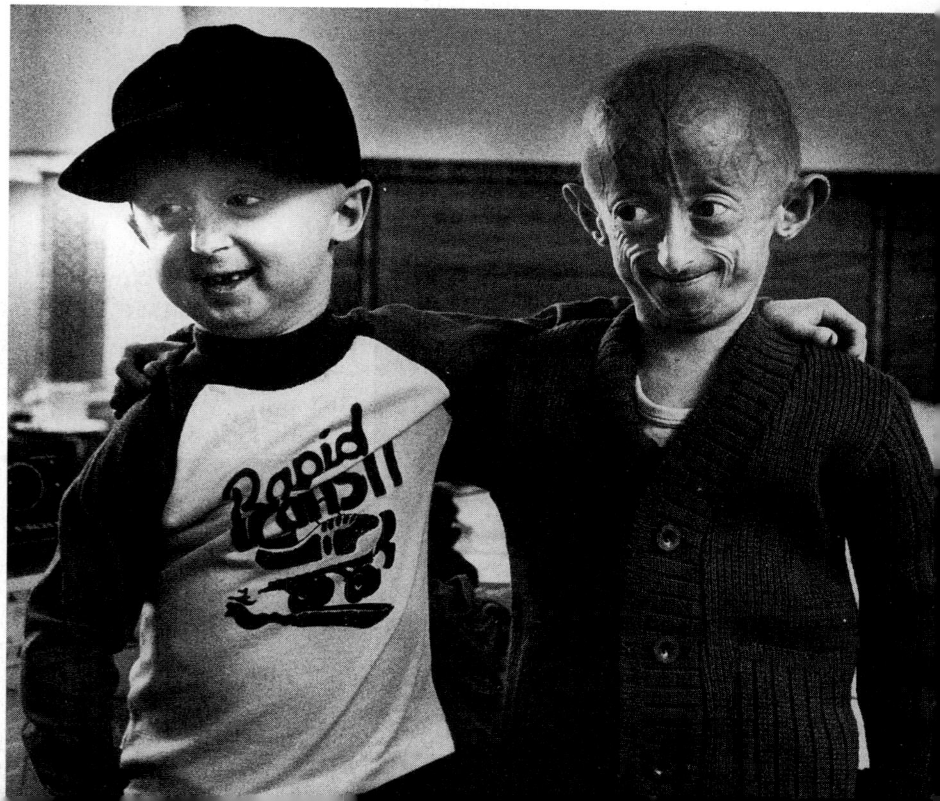

sels), a condition common in the elderly. Children with progeria usually die in their teens.

There are also environmental influences on longevity. Poor nutrition, exposure to radiation and air pollution, and high levels of physical, mental, or emotional stress appear to reduce longevity. The need for social contact between humans is emphasized by studies which show that persons who are married or living together tend to live longer than do those who are alone.

Theories to Explain Aging and Senescence

What are the biological causes of aging and senescence? There are many theories to explain aging. They all suggest that people age because irreversible changes take place in the cells and tissues. All the theories briefly described here may be valid to a greater or lesser extent. No single theory by itself provides an adequate explanation for aging.

Genetic mutations. As we stated earlier, mutated genes code for altered and often defective proteins. Mutations in genes are passed on to all the descendent body cells. Although a single mutation may have only a minimal effect on cell function, an accumulation of these defects spells trouble. The longer an individual lives, the more mutations accumulate in the DNA of his or her body cells. If these are cells that must divide to repair and maintain the body, altered DNA eventually leads to more and more malfunctioning cells.

Cross-linkage of proteins. As cells grow older, their separate proteins may become increasingly cross-linked with each other. Such cross-linking decreases the ability of enzymes to operate properly, and the operation of the cytoskeleton is impaired. Extracellular proteins such as the collagen of connective tissues also may become cross-linked. As a result, the tissue becomes stiffer, reducing the flexibility and movement of tissues. Cross-linkages between collagen fibers can also reduce the diffusion of nutrients and wastes through the matrix and interfere with the migration of white blood cells to fight infection.

Accumulation of free radicals. Free radicals are chemical substances that possess an unpaired electron, for example, superoxide (O_2^-). This makes them very reactive, and they can damage other molecules, including proteins, DNA, and the lipids in cell membranes. In a cell, free radicals may form as a result of certain food additives, chemicals in tobacco smoke, and normal cell processes in which O_2 participates.

Cells have only a limited ability to protect themselves from damage through the use of *free radical scavengers* such as selenium (a trace mineral), vitamin E, vitamin C, and beta-carotene. These scavengers neutralize free radicals. However, over time, more free radicals may accumulate than can be neutralized. This accumulation damages cell structures, alters tissue functions, and contributes to aging.

Accumulation of waste molecules. The accumulation of certain chemical substances within cells interferes with normal cellular activities. One such substance, called *lipofuscin,* begins to accumulate in certain brain cells by age 3 months. By 30 years of age, 84 percent of these cells contain so much lipofuscin that their nuclei have moved out of the usual position. Lipofuscin also accumulates in heart muscle cells. With the passage of time, continued accumulation may contribute to the reduced functioning of these tissues.

Development of autoimmunity. Normally, our immune defenses are able to distinguish our own proteins, cells, and tissues from foreign cells and proteins. We normally react against substances that belong to another individual or species but not to our own tissues. However, as we age, our immune defenses may become "confused" and react with our own proteins. This autoimmunity damages tissues and may contribute to general aging as well as certain specific diseases.

The ticking of a biological clock. Evidence suggests that there is a "clock" running inside cells or organs, determining how long they live. When fibroblasts are grown outside the body, they pass through about 50 cell divisions and then stop dividing. If such cells are allowed to divide 30 times and then are frozen for a few years, when they are thawed, they divide 20 more times and stop. It is as if they had a memory or biological clock that keeps track of the number of divisions. Such a clock may also operate in organs to determine the aging of the entire body. Some research suggests that the hypothalamus in the brain may be such a biological clock. If it were identified and understood, could such a clock be reset to postpone aging?

Chapter Summary

Formation of Primary Tissues

- Three germinal layers in the embryo—ectoderm, mesoderm, and endoderm—produce the four primary tissues—epithelial, connective, muscular, and nervous—from which all adult organs and systems develop.

Epithelial Tissues

- Epithelial tissues lack blood vessels and have squamous, cuboidal, or columnar cells.
- Covering and lining epithelia cover body surfaces, line body passageways and cavities, and form hair and nails; glandular epithelia form exocrine and endocrine glands.
- Simple epithelia (single layer of cells) function in secretion, absorption, and filtration; stratified epithelia (multiple layers of cells) function in protection.

Connective Tissues

- Connective tissues bind organs together, support and insulate the body, and store energy.
- Fibroblasts secrete a nonliving matrix which contains collagen, reticular, or elastic fibers.
- Cartilage (hyaline cartilage, elastic cartilage, and fibrocartilage) has a compression-resistant matrix that lacks blood vessels and repairs slowly. Chondroblasts are trapped in the matrix they secrete.

Organs and Organ Systems: An Overview

- The 11 organ systems are assembled from the primary tissues.
- The nervous system includes a central system—the brain and spinal cord—and a peripheral system—the somatic division (controls body movement) and the autonomic division (coordinates internal activities).
- Endocrine glands use hormones for communication and to control internal activities.

Organization of the Human Body

- The body exhibits bilateral symmetry and has anterior and posterior cavities. The diaphragm subdivides the anterior cavity into the thoracic and abdominal cavities.
- Tissue membranes (an epithelium and a connective tissue) include serous (closed cavities), mucous (open passageways), cutaneous (body surface), and synovial (joint cavity) membranes.

Diseases and Disorders

- A disease is a disruption in the structure or function of cells, organs, or tissues.
- The symptoms of illness are objective (measurable) or subjective (not quantifiable).
- Diseases may be idiopathic (without an identifiable cause), chronic (long-lasting), acute (brief and intense), local (limited to a body region), or systemic (affecting the entire body).
- Categories of disease include infectious, genetic, congenital, inflammatory, degenerative, and mental diseases. Neoplasms are abnormal cell growths, including benign and malignant (cancer) tumors.

Life Span Changes

- The life span includes embryological development, growth and maturation, and adulthood.
- Aging is natural and progressive; individuals and organs age at independent rates.
- Longevity, or life expectancy, is influenced by genetic and environmental factors.
- Senescence is a reduced capacity for defense, repair, and homeostasis.
- Aging may result from an accumulation of genetic mutations, free radicals, or waste products as well as cross-linking of proteins, development of autoimmunity, and the presence of an internal biological clock.

Selected Key Terms

adipose tissue (p. 86)
aging (p. 98)
basement membrane (p. 80)
cartilage (p. 86)
connective tissue (p. 81)
disease (p. 95)
epithelial tissue (p. 80)
fibers (p. 83)
germinal layers (p. 79)
life span (p. 98)
longevity (p. 97)
matrix (p. 83)
organ systems (p. 90)
primary tissues (p. 78)
tissue membrane (p. 93)

Review Activities

1. List the three germinal layers of the embryo and the primary tissues which develop from each layer.
2. Describe five structural or functional differences between epithelial and connective tissues.
3. List the basic categories of covering and lining epithelia and summarize their locations and functions.
4. Explain the difference between exocrine and endocrine glands.
5. List the three types of connective tissue fibers and describe their functions.
6. List the major categories of fibrous connective tissue and describe their location and functions.
7. Summarize the location and function of adipose connective tissue.
8. Distinguish between the three types of cartilage on the basis of structure, location, and function.
9. List the 11 organ systems of the human and briefly summarize their functions.
10. Explain the difference between the central and peripheral nervous systems and between the somatic and autonomic divisions.
11. List the two major body cavities and the organs found in them.
12. Distinguish between serous, mucous, synovial, and cutaneous membranes.
13. List the major categories of human disease.
14. List and describe the three stages of the human life span.
15. List and summarize six theories that attempt to explain the causes of aging.

Self-Quiz

Matching Exercise

___ 1. Germinal layer which forms connective tissue	A. Endocrine system
___ 2. Noncellular layer below epithelial cells	B. Stratified squamous
___ 3. Cells that produce the matrix of connective tissue	C. Dense fibrous
___ 4. Cells that produce the matrix of cartilage	D. Basement membrane
___ 5. Type of epithelium in the outer layer of skin	E. Mesoderm
___ 6. Type of cartilage in the rib cage and the tip of the nose	F. Central nervous system
___ 7. Type of connective tissue specialized for fat storage	G. Hyaline
___ 8. Type of connective tissue in tendons and ligaments	H. Chondroblasts
___ 9. Body system associated with hormone production	I. Adipose
___ 10. Segment of the nervous system containing the brain and spinal cord	J. Fibroblasts

Answers to Self-Quiz

1. E; 2. D; 3. J; 4. H; 5. B; 6. G; 7. I; 8. C; 9. A; 10. F

Cancer: Causes and Treatments

Cancer is not a new disease. The fossilized remains of animals and ancient human bones indicate that it has been around for thousands of years. Only in this century has fundamental scientific research provided the genetic, cellular, chemical, and technological means to detect, diagnose, and treat cancers. Because of improvements in treatment, the frequency of deaths from cancers of the uterus, colon and rectum, and stomach have declined over the last two decades.

At current rates, over 1.5 million Americans will develop cancer annually. Over 500,000 die each year; that is one death per minute. In most cases, cancer is a disease of middle and later life. Many cancer deaths, especially those resulting from sunlight, smoking, and alcohol consumption, can be prevented through changes in personal behavior. Many more can be prevented by means of early recognition and treatment.

What Is Cancer?

Cancer is actually a group of diseases that all involve the uncontrolled division of cells. These cells also lose their attachments to other cells and are capable of moving around. These properties permit the spread of cancerous cells throughout the body.

In the body, rapid cell growth produces a cluster of cells called a **tumor.** These cells don't perform any specialized functions; they just proliferate. Tumors interfere with normal tissue structure and function and can eventually cause death. There are two kinds of tumors: benign and malignant (Figure 1). Their study is known as *oncology*.

Benign tumors have slow-growing cells contained within a layer of connective tissue. Benign tumors can be removed effectively with surgery. They usually do not reappear.

Malignant tumors contain rapidly growing, actively moving cells that tend to invade other tissues. The movement of malignant cells from their original site to other body areas is called **metastasis.** When metastasizing cells gain access to a blood vessel, the circulating blood carries those cells throughout the body. When they lodge at distant body sites, the cells continue to grow, forming new tumors (Figure 1).

What Causes Cancer?

There is no single cause of cancer. This fact has vastly complicated cancer research and treatment for decades. Today we consider environmental factors (toxic chemicals and radiation) and viruses to be the agents that *transform* normal cells to cancerous cells (*transformed cells*). These agents all act on cellular DNA, causing changes that alter crucial genes.

Chemicals and radiation that can transform cells are called **carcinogens** (*car*-**sin**-*oh-jens*). Chemical carcinogens can be found in polluted air that contains automobile and industrial emissions, in tobacco smoke, and in some preservatives, solvents, and saturated dietary fats. Various forms of radiation can also be carcinogenic, including ultraviolet (UV) light, the radiation from radioactive atoms (for example, radon), and x-rays. All types of radiation are forms of energy and can cause chemical changes in DNA that result in gene mutation.

The Role of Genes

Changes (mutations) in certain genes leads to cancer. In their normal state, these genes code for proteins that perform essential cell functions, for example, control of cell division and holding cells together. When they mutate, they become **oncogenes,** or genes that cause cancer (Figure 2). Oncogenes produce defective proteins that contribute to

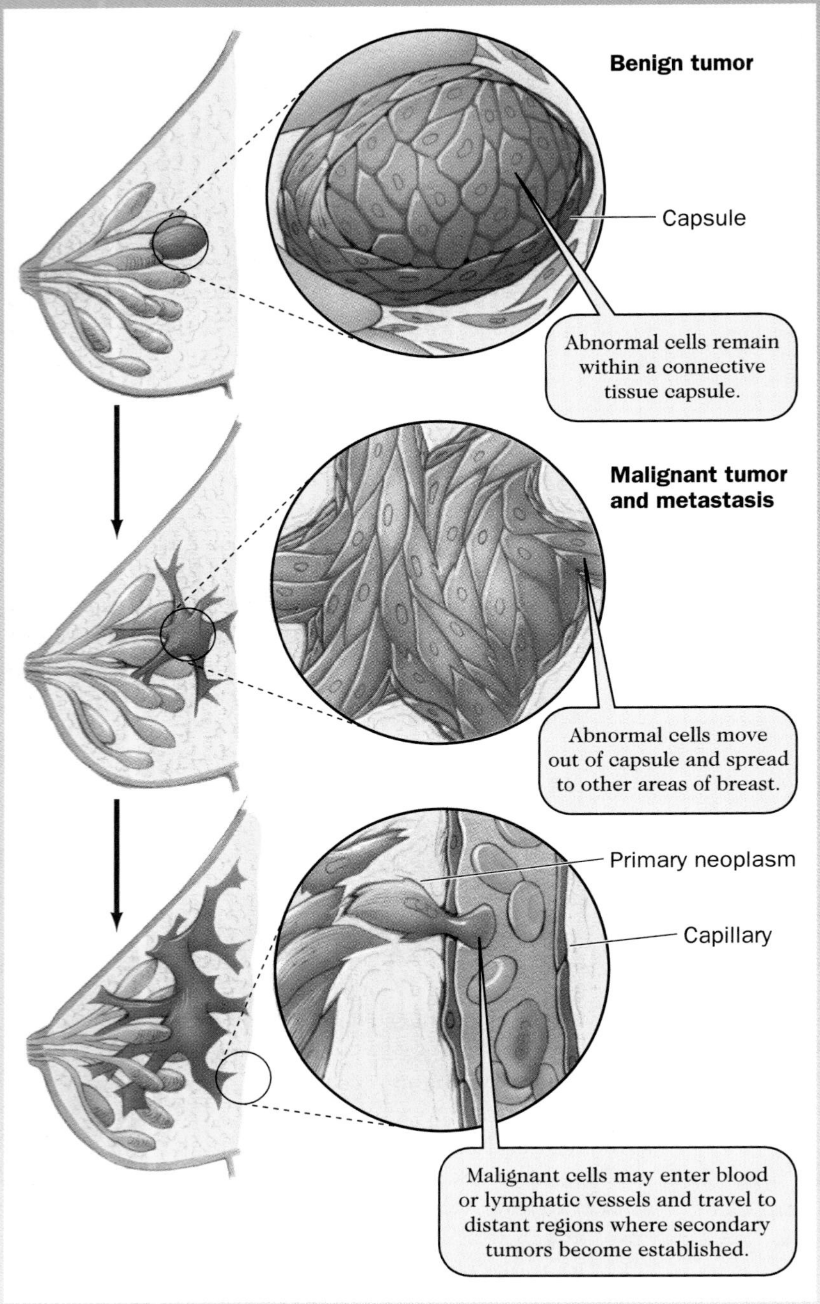
Benign tumor
Capsule
Abnormal cells remain within a connective tissue capsule.
Malignant tumor and metastasis
Abnormal cells move out of capsule and spread to other areas of breast.
Primary neoplasm
Capillary
Malignant cells may enter blood or lymphatic vessels and travel to distant regions where secondary tumors become established.

FIGURE 1 Growth and metastasis of a malignant tumor.

provide only a weak defense against transformed cells.

Several different oncogenes are involved in the sequence of events that lead to cancer of the colon and rectum (Figure 3). Cancers of the bladder, bone, brain, breast, cervix, lung, and ovary also have been linked to oncogenes.

In addition to mutation and inheritance, you may generate an oncogene when unusual processes delete a portion of one chromosome and insert it at a different chromosomal site. Such a move is called a *translocation*. Translocations of genes may occur as a result of mistakes during mitosis, or they may be caused by carcinogens or by viral infection.

Viruses: Agents of Cell Transformation

Research has revealed that viruses can transform cells, and several types of viruses have been associated with cancer in humans (Table 1). To understand how viruses play this role, we need to consider what viruses are.

Viruses are not living cells, although they have some of the properties of cells. In their simplest form, viruses consist of nucleic acid (either RNA or DNA) packaged in a specific protein coat (Figure 4, page 107). Some viruses contain a phospholipid membrane in addition to the protein coat and nucleic acid. The nucleic acid is the hereditary material of the virus. After entering a living cell, the nucleic acid codes for the reproduction of the virus (protein coat and nucleic acid). There are hundreds of different viruses. Some infect bacteria, and others are specific for plant or animal cells.

Although viruses do not have cytoplasm, organelles, or enzymes for energy transformations, they can replicate. To replicate, a virus must use a living cell and its resources. After entering a cell, the viral nucleic acid has two alternative fates: (1) It may produce new virus particles and destroy the cell. (2) It may integrate into the cellular DNA. The second fate may transform a normal cell into a cancer cell by deleting a portion of a gene (DNA) or by moving DNA from one site to another on the chromosome (Figure 4).

the high rates of cell division and the loss of attachments to other cells.

In addition to mutation, you can also inherit oncogenes from your parents. Breast, colon, rectal, and prostate cancers seem to occur more frequently in some families. Of course, no one actually inherits cancer. One inherits genes, and defective genes give one the tendency to form cancerous growths or may

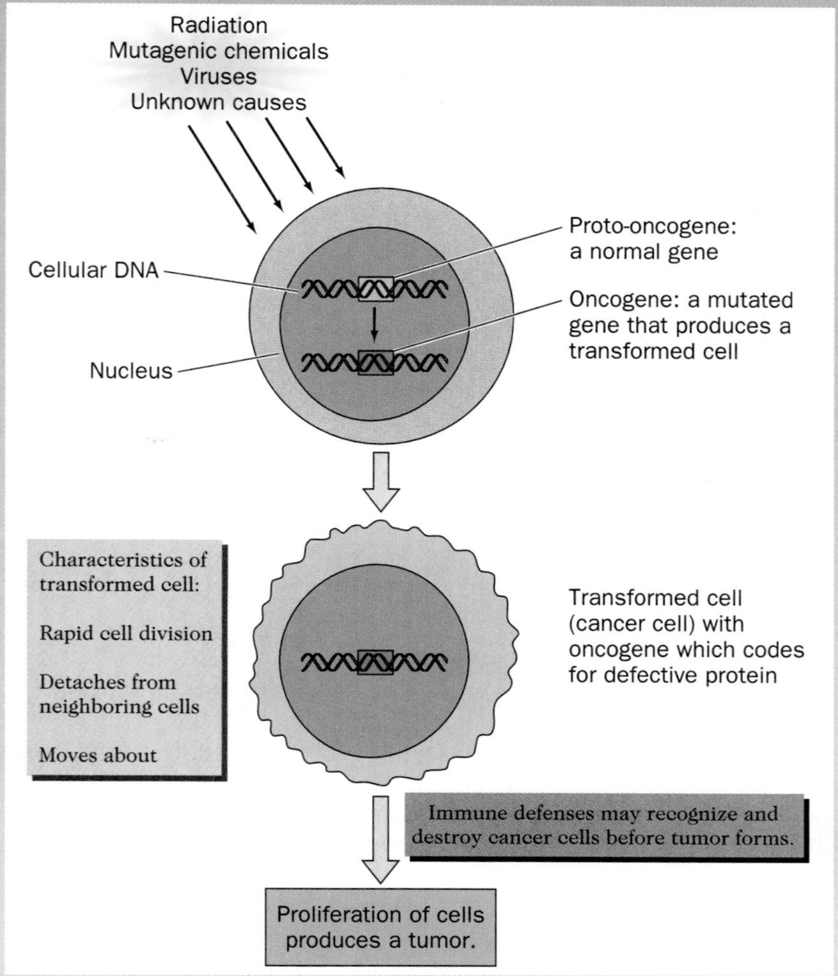

Radiation
Mutagenic chemicals
Viruses
Unknown causes
Cellular DNA
Nucleus
Proto-oncogene: a normal gene
Oncogene: a mutated gene that produces a transformed cell
Characteristics of transformed cell:
Rapid cell division
Detaches from neighboring cells
Moves about
Transformed cell (cancer cell) with oncogene which codes for defective protein
Immune defenses may recognize and destroy cancer cells before tumor forms.
Proliferation of cells produces a tumor.

FIGURE 2 The transformation of normal cells into cancer cells.

The Body's Natural Defense against Cancer

Most cancers are slow to develop. Fortunately, the body has natural protections against cancer cells when they first arise. As part of their transformation from normal cells, cancer cells acquire unusual surface marker proteins (see Figure 2.7). Our immune defenses recognize these proteins as a foreign threat, and an attack is mounted. Thousands of cancer cells are probably destroyed each day by the normal operation of our immune defenses. This is the good news. Unfortunately, there is also bad news.

The bad news is that our natural immune defenses don't catch all of these abnormal cells. If some cells evade detection and form a tumor, our natural defenses are overwhelmed. The other bad news is that some diseases, including AIDS, reduce the effectiveness of our immune defenses, leaving the entire body more vulnerable to cancer cells. You will read more about the operation of the immune system and its diseases in Chapter 13.

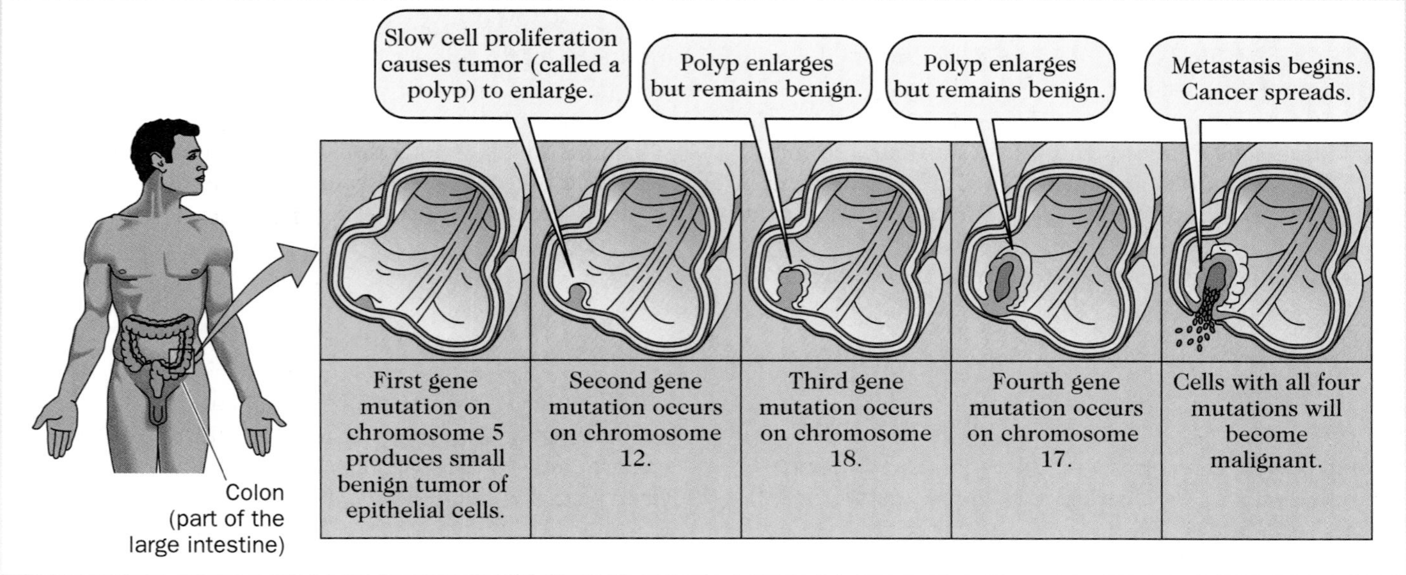

Slow cell proliferation causes tumor (called a polyp) to enlarge.
Polyp enlarges but remains benign.
Polyp enlarges but remains benign.
Metastasis begins. Cancer spreads.
First gene mutation on chromosome 5 produces small benign tumor of epithelial cells.
Second gene mutation occurs on chromosome 12.
Third gene mutation occurs on chromosome 18.
Fourth gene mutation occurs on chromosome 17.
Cells with all four mutations will become malignant.
Colon (part of the large intestine)

FIGURE 3 The development of colorectal cancer involves four genetic changes.

TABLE 1 Oncogenic Viruses That Infect Humans

Type of Virus	Name of Virus	Cancers in Which Virus Has Been Implicated
DNA	Herpes simplex I	Cancers of the lip (also causes cold sores)
	Herpes simplex II	Cancer of the cervix (also causes genital herpes)
	Epstein-Barr	Carcinoma of the nose and throat Burkitt's lymphoma Hodgkin's disease (a lymphoma) (also causes mononucleosis)
	Papillomaviruses	Cancer of the cervix (also causes warts)
	Hepatitis B Hepatitis C	Liver cancer
RNA (retroviruses)	Human T-cell lymphotropic viruses	
	HTLV-I	Adult T-cell leukemia
	HTLV-II	Hairy cell leukemia

Diagnosis and Treatment of Cancer

The traditional medical methods for treating cancers are quite direct and aggressive: Malignant neoplasms are removed surgically, burned with radiation, or poisoned chemically. *Surgery* is most effective for small, localized tumors. However, surgery carries the risk of infection and the burden of recovery for an already ill patient.

Radiation therapy is effective because cancer cells, with their high rate of cell division, are more vulnerable to radiation injury than are normal cells. Radiation can destroy metastasizing cells that a surgeon cannot find or remove. However, normal slowly dividing body cells also may be damaged.

Chemotherapy uses toxic chemicals to kill rapidly dividing cells. Many of these drugs are altered forms of the nucleotides found in DNA. During DNA synthesis these drugs can be incorporated into the cancer cell's DNA, where they prevent further DNA replication or cause defective proteins to be produced (Chapter 2). Without DNA replication, the cancer cells cannot reproduce, and they become more vulnerable to the body's natural defenses. Defective proteins also weaken or kill cancer cells. Unfortunately, both radiation and chemotherapy injure normal cells and may cause side effects such as nausea and hair loss.

New diagnostic technologies have focused on noninvasive imaging of tumors that are still quite small and localized. These methods allow for early detection and guide surgical removal. For example, *magnetic resonance imaging* (*MRI*) uses an electromagnet to detect tumors. *Computed tomography CT*) uses x-rays to show cross-sectional views of the body and locate early tumors. *Positron emission tomography* (*PET*) can reveal the altered metabolic activity of cancers and differentiate them from normal tissue.

Thanks to early detection and new forms of treatment, the 5-year survival rate for all cancers is now about 51 percent. Fifty years ago, it was only 20 percent.

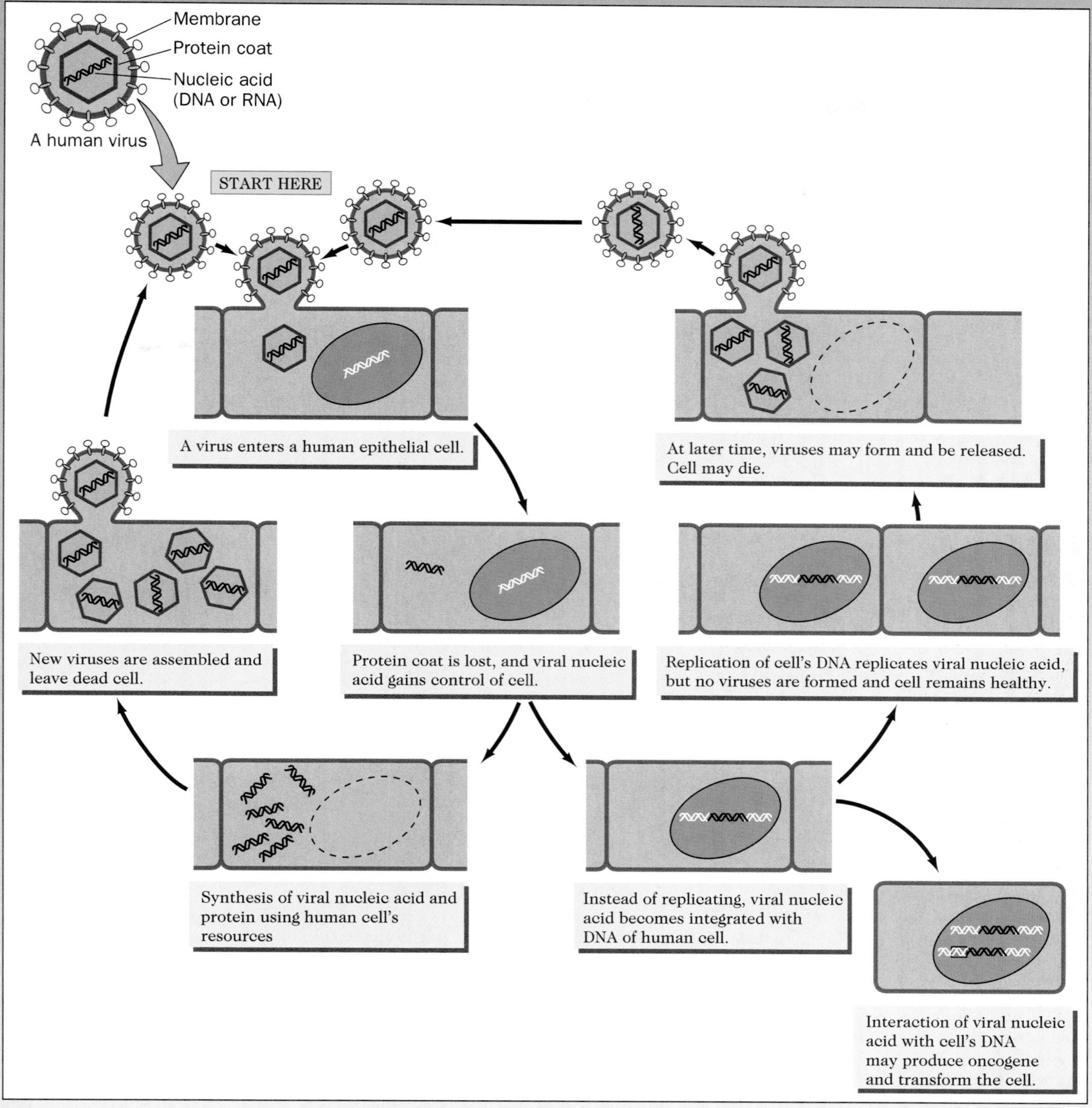
Membrane
Protein coat
Nucleic acid
(DNA or RNA)
A human virus
START HERE
A virus enters a human epithelial cell.
At later time, viruses may form and be released.
Cell may die.
New viruses are assembled and leave dead cell.
Protein coat is lost, and viral nucleic acid gains control of cell.
Replication of cell's DNA replicates viral nucleic acid, but no viruses are formed and cell remains healthy.
Synthesis of viral nucleic acid and protein using human cell's resources
Instead of replicating, viral nucleic acid becomes integrated with DNA of human cell.
Interaction of viral nucleic acid with cell's DNA may produce oncogene and transform the cell.

FIGURE 4 Viruses and their reproduction in human cells (see discussion on page 104).

PART TWO

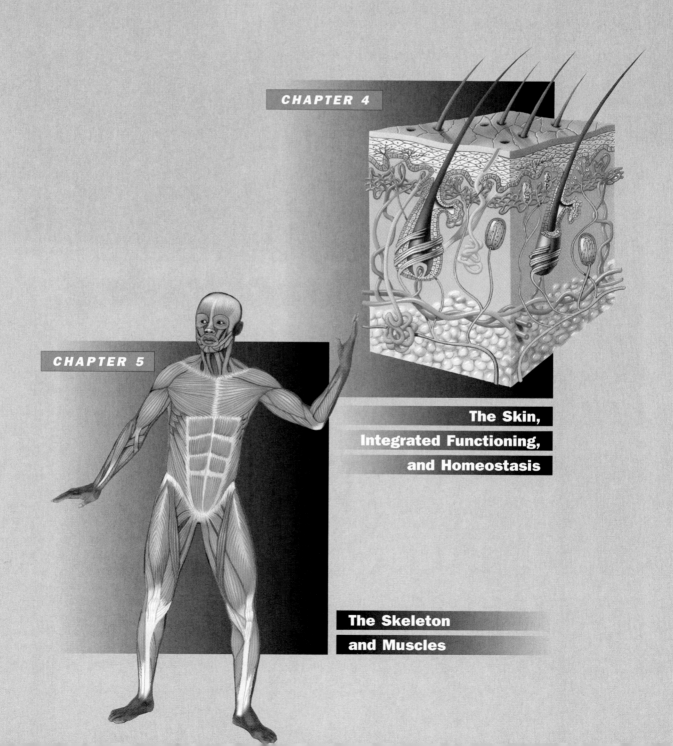

Protecting, Supporting, and Moving the Body

CHAPTER 4

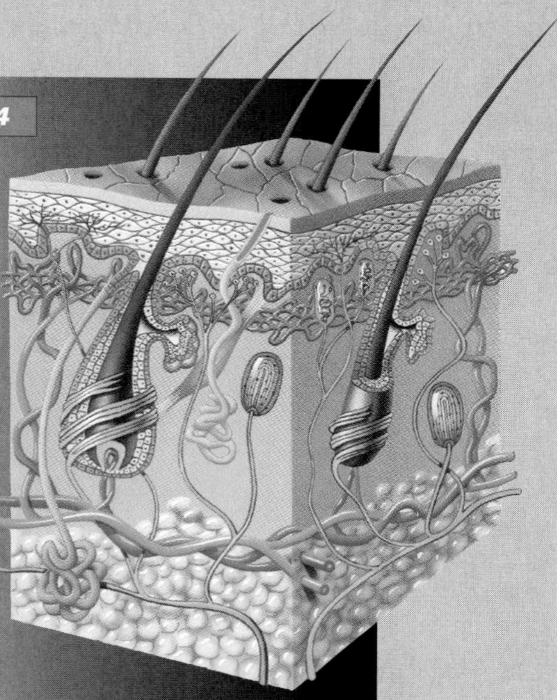

The Skin, Integrated Functioning, and Homeostasis

CHAPTER 5

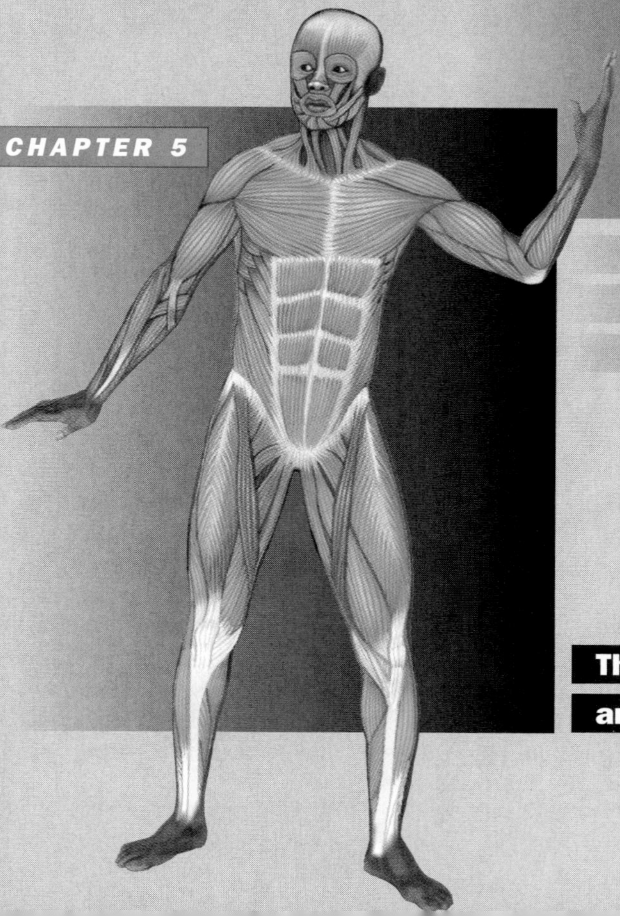

The Skeleton and Muscles

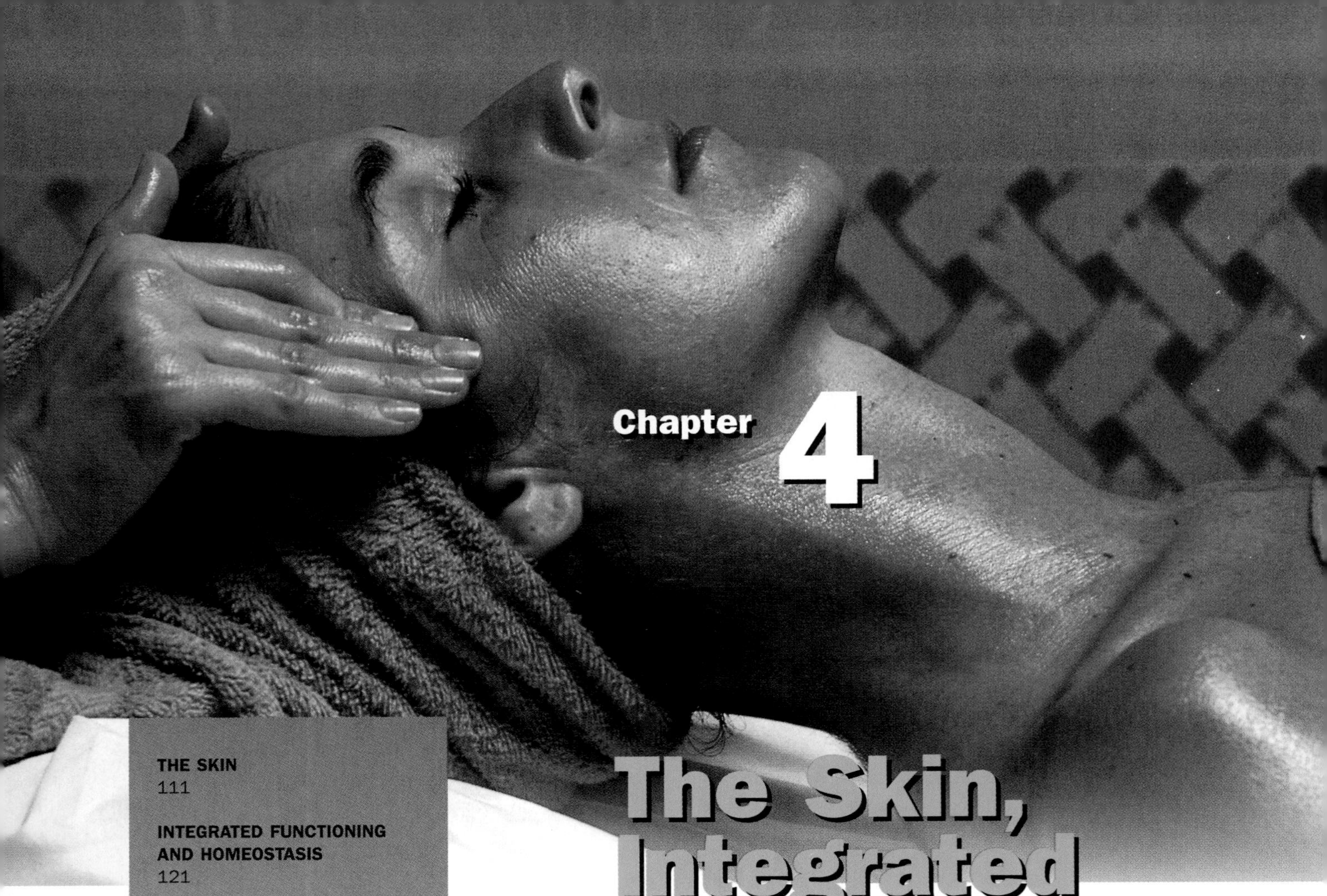

Chapter 4

The Skin, Integrated Functioning, and Homeostasis

We bathe it, shave it, and caress it. The mirror tells the story. Each day, when we inspect our faces, we are examining a portion of our skin, or **integumentary system** (*in-teg-you-**men**-tah-ree*). The skin is actually the body's largest single organ. It makes up about 7 percent of an adult's weight and covers about 15 to 20 square feet (1.4 to 1.7 square meters) of surface area.

In our concern with our skin, we worry about lines, blemishes, rashes, infections, the presence or absence of hair, and signs of aging. These concerns are rooted in the complex structure and multiple functions of the skin. The skin includes all four types of tissues discussed in Chapter 3, and these tissues are organized into specialized structures that carry out the many functions of the skin.

In this chapter, you will learn about the structure and function of the skin. You will learn about appendage structures derived from the skin (hairs and nails) and structures buried within the skin (glands and sensory receptors). You will also learn about some of the common diseases of the skin and the way the skin changes over the normal life span from infancy to old age.

Because the skin is a complex organ that includes different tissues and structures, its activities must be coordinated with those of other body systems, especially the circulatory, nervous, and endocrine systems. Such coordination is called **integrated functioning,** and it enables the body to maintain a stable internal environment for the healthy operation of all cells and systems, even when conditions outside or inside change.

This maintenance of stable internal conditions is called **homeostasis.** Homeostasis is a concept that is essential to understanding the functioning of the human body, and we will return to it frequently in the chapters that follow.

The Skin

As the body's outermost covering, the skin interacts directly with the environment and has several important functions. It limits the entry and exit of materials, provides sensory awareness of our surroundings, and is able to repair and regenerate itself when it is injured. As the following summary of its functions makes clear, the skin is not merely a passive wall around the body.

Protection and defense. The skin is an effective barrier against mechanical injury and the absorption of dangerous chemicals. It blocks penetration by all but the sharpest objects. It is the first line of defense against foreign agents, such as bacteria and viruses, which cause disease.

Sensation. The skin is an early warning system for the body. Specialized sensory organs in the skin continuously monitor the sensations of touch, pressure, pain, warmth, and cold.

Water balance. The outermost layer of the skin contains a water-resistant protein called **keratin.** This waterproofing reduces the loss of the body's precious internal water.

Temperature regulation and excretion of wastes. In humans, the skin plays a direct role in limiting heat loss if the body is too cool and cooling the body if it becomes too warm. To achieve cooling, sweat glands secrete water onto the skin. As it evaporates, this water carries with it heat energy, cools the skin, and lowers body temperature. The sweat glands also secrete salts and small amounts of waste products such as urea and ammonia.

Synthesis of vitamin D. Exposure to the ultraviolet wavelengths of sunlight causes the skin to produce small quantities of vitamin D which supplement the amount supplied in the diet. This vitamin is essential to the construction of bones and teeth from the minerals calcium and phosphate.

> The skin protects against mechanical injury, defends against disease organisms, senses changes in the environment, restricts the loss of body water, regulates internal temperature, excretes waste, and produces vitamin D.

Tissue Structure of the Skin

The functions of the skin that were just described are accomplished by two relatively thin layers of tissue: an outer **epidermis** (*ep-ih-**dur**-mis*; *epi,* "over") and an inner **dermis** (Figure 4.1). The thinner epidermis is constructed from stratified squamous epithelial tissue, while the thicker dermis is a dense connective tissue (Chapter 3). The skin ranges in thickness from 0.5 mm over the eyelids to 6 mm (about $\frac{1}{4}$ in.) or more on areas of the hands and feet that receive heavy wear and tear. Beneath the dermis is a layer called the **hypodermis** (*hi-poh-**dur**-mis*; *hypo,* "below or under") or the subcutaneous (*sub-kew-**tay**-nee-us*) layer. While this is not considered part of the integumentary system, it helps support the two layers of the skin by anchoring them to underlying structures.

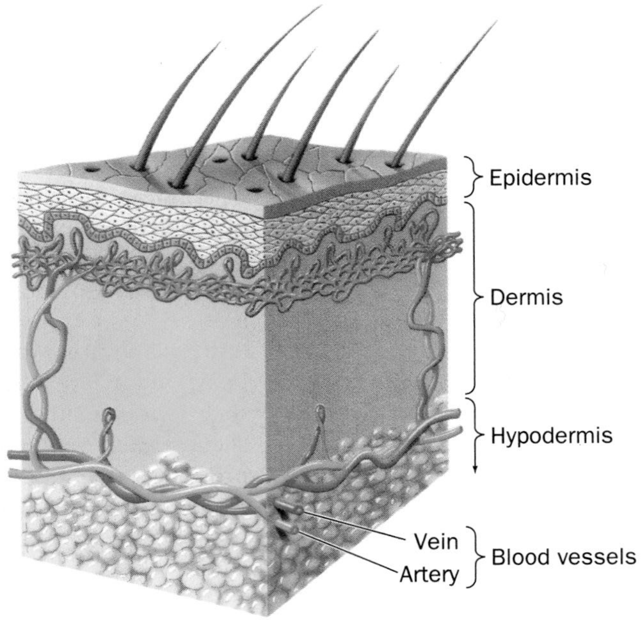

Epidermis
Dermis
Hypodermis
Vein
Artery
Blood vessels

FIGURE 4.1 The organization of the skin.

The epidermis: a thin outer layer. The epidermis consists of up to five discrete layers, or **strata** (singular, *stratum*). These five layers can be distinguished in Figure 4.2, but we will discuss only two layers: the innermost layer, called the **basal stratum** and the outermost layer, called the **cornified stratum.** The epidermis varies in thickness, and all five strata are present only in thick-skinned areas on the palms and the soles of the feet.

The epidermis constantly undergoes growth and renewal. In regions of repeated pressure and friction, production of new cells is stimulated. If your skin has ever formed a **callus** (***kal**-us*) as a result of pressure from tools, shoes, or a guitar string, you know that the epidermis can form thick, tough layers to protect the dermis from abrasion. New cells are produced by mitotic cell division in the basal stratum. As new cells are produced, the older cells on top of them move upward toward the outer surface of the skin, the cornified stratum.

As these cells move upward, they undergo distinctive changes, the most significant of which is the synthesis of massive amounts of the protein keratin, which eventually fills the cells. The cells that make keratin are called **keratinocytes** (***kare**-ah-tin-oh-sites*) and account for about 95 percent of all the cells in the epidermis. In addition to waterproofing, keratin contributes great strength and toughness to the skin and its appendage structures, the nails and hair.

Melanocytes (*mel-**an**-oh-sites*) are a second type of cell found in the epidermis. They are specialized to produce a dark pigment called **melanin** (***mel**-ah-nin*). Melanin protects the DNA of the dividing cells in the basal stratum from damage by ultraviolet wavelengths of sunlight. Changes in basal cell DNA can lead to skin cancer, and this melanin screen pro-

FIGURE 4.2 The epidermis and dermis of the skin.

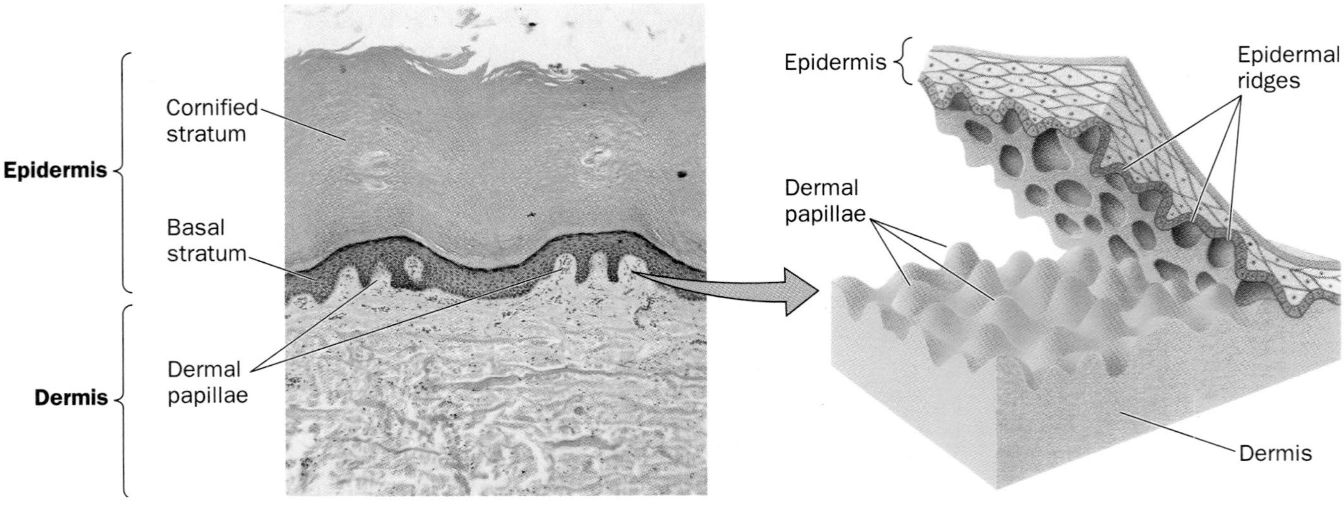

Epidermis
Cornified stratum
Basal stratum
Dermis
Dermal papillae
Epidermis
Epidermal ridges
Dermal papillae
Dermis

vides some protection. Skin cancer can be a serious consequence of extensive exposure to sunlight. We'll say more about this later in the chapter.

As with all epithelial tissues, the epidermis contains no blood vessels. Cells in the lower strata can obtain oxygen (O_2) and nutrients only by diffusion from the dermis, which is well supplied with blood vessels. As the cells reach the upper strata, however, diffusion of nutrients is diminished and the cells die.

The surface of the skin we admire and care for consists of dead cells. Each day millions of dead skin cells are flaked away, to be replaced with new cells from below. Normally it takes about 4 to 6 weeks for cells to move from the basal stratum to the cornified stratum, where they are lost. Some estimates suggest that 80 percent of the dust in our homes is actually dead epidermal cells. Excessive flaking of cells produces **dandruff.**

The dermis: a thicker inner layer. The dermis lies below the epidermis and is substantially thicker than the epidermis. It is a layer of connective tissue that consists of an extracellular matrix with abundant collagen, elastic, and reticular fibers.

FIGURE 4.3 Fingerprints.

(a) A photograph showing the friction ridges that create fingerprints.

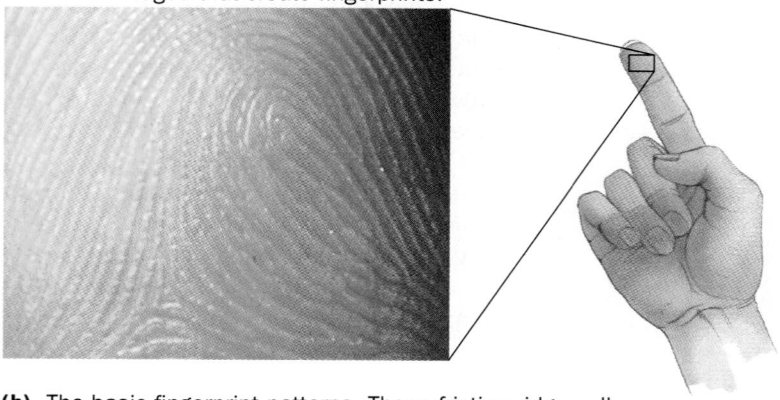

(b) The basic fingerprint patterns. These friction ridges allow firm grasping of objects. The patterns are different for each individual, even identical twins.

Arch
Whorl
Loop
Combination

The most abundant cells in the dermis are fibroblasts and macrophages, but mast cells and white blood cells are also present (Chapter 3).

The dermis contains many blood vessels that supply O_2 and nutrients while removing cellular wastes. When you cut yourself deeply enough to bleed, you have cut into the dermis. The dermis also contains hair follicles, several types of glands, smooth muscle cells, and sensory receptors with their associated nerves (we will say more about these structures later in this chapter).

The uppermost layer of the dermis contains many tiny projections known as **papillae** (*pah-****pil****-ee*) (see Figure 4.2). Papillae help anchor the epidermis to the dermis and are largest and most numerous in the areas of the skin that receive the most wear and tear: the palms and soles. The papillae produce the ridges and lines that are evident in these regions and on the fingertips. On the fingertips, these friction ridges make it easier for us to grasp and hold objects and form the fingerprints that are unique to each individual (Figure 4.3).

It is within the cells of the dermis that vitamin D is formed. Ultraviolet (UV) light that is not absorbed by the melanin screen of the epidermis enters the dermis and participates in vitamin D formation. The UV light in sunlight chemically converts a precursor molecule into an inactive form of vitamin D. This inactive molecule is then transported by the blood from the dermis to the liver and kidneys, where it is chemically converted to active vitamin D.

In this active form, vitamin D is transported throughout the body and supports the intestinal absorption of the calcium (Ca^{2+}) needed for many body processes, particularly the normal growth of bones and teeth and the operation of muscles. The synthesis of vitamin D is an example of the integrated interaction between the skin and other body systems. As we further examine the functions of the skin and other organs, we will see many more examples.

The hypodermis: beneath the dermis. The hypodermis consists of loose connective tissue that anchors the skin to underlying structures, but permits the skin to slide over bones and joints as they move. The hypodermis also contains abundant adipose cells that store energy as fats. Adipose cells also provide a layer of insulation, making it easier for people to retain heat and maintain a stable body temperature. These cells also help cushion fragile structures under the skin. About half the body's fat is stored in the hypodermis, with the exact amount and location determined by one's sex, age, and heredity.

As part of their secondary sex characteristics, women tend to accumulate more adipose tissue in the breasts and over the hips and buttocks. Men store fat over the neck, upper arms, lower back, and buttocks. Your exact sites and amounts of fat deposition are controlled by the genes inherited from your parents (see Figure 3.7). The body shapes of your parents and grandparents provide an indication of what lies ahead for you.

The hypodermis is also important clinically as a site for the injection and slow absorption of medications. This method is called subcutaneous injection, and you will recognize the other name for the site of injection from the name of the instrument that is used—a *hypodermic* needle.

FIGURE 4.4 The diversity of skin coloration in humans.

The skin consists of two layers: an outer epidermis composed of keratin-rich epithelium and an inner dermis built from dense connective tissue. Beneath the skin, the hypodermis consists of loose connective tissue and serves to anchor the skin to underlying structures.

Skin Colors

Worldwide, the human family displays a wide range of skin colors, ranging from pink to yellow to red to brown to black (Figure 4.4). Although skin color is determined largely by heredity, environmental factors play a limited role.

Genes determine the number of blood vessels in the dermis, sensitivity to sunlight, the thickness of the cornified stratum, and responsiveness to certain hormones. Genes also code for three pigments that influence skin color: hemoglobin, carotene, and melanin. **Hemoglobin** (*he-mow-**globe**-in*) is the pigment in red blood cells which binds and transports O_2. Oxygenated blood flowing through small vessels in the skin imparts a pinkish color which is most easily seen in light-skinned white people. If these vessels dilate (increase in diameter) as a result of strong emotions or high body temperature, the skin turns a brighter shade of red and we say that a person is blushing or is flushed.

Carotene (***kare**-oh-teen*) is a yellowish-orange pigment that is related to vitamin A. Carotene is found in cells in the epidermis and adipose cells in the hypodermis. The coloration of many Asians is due in part to carotene, though a third pigment, melanin, also plays a role.

Melanin is the brownish-black pigment produced by the melanocytes that we mentioned earlier. As Figure 4.5 shows, melanocytes have branches that protrude among the keratinocytes of the lower strata and carry the pigment into them. Once inside the keratinocytes, melanin forms a protective cap over the nucleus containing the DNA. Melanocytes are not always uniformly distributed over the surface of the epidermis, and clusters of these cells produce freckles and moles.

All humans have nearly the same number of melanocytes; variations in the amount of melanin they produce result in the range of human skin colors. **Albinos** (*al-**buy**-nose*) are people who lack melanin pigment in the skin. They have inherited defective genes from both parents and are unable to make an enzyme required for melanin formation. Although they lack the pigment, albinos have a normal number of melanocytes.

Unlike albinos, who lack pigment over the entire body, people with ***vitiligo*** (*vit-uh-**lye**-go*) lose melanin from irregular patches of the skin, usually on the backs of the hands and behind the knees and sometimes on the face. These patches remain surrounded by normally pigmented skin. This condition is not considered dangerous, and little research has been conducted into its cause or treatment.

In addition to genes, environmental factors play a role in determining skin color. An individual's skin color normally darkens after exposure to sunlight. This is the familiar tanning reaction, and it is most evident in lightly pigmented individuals. After significant exposure to the UV light in sunlight, a hormone called *melanocyte stimulating hormone (MSH)* is released from the pituitary gland at the base of the brain. Transported by the blood, this hormone reaches the melanocytes, causing an increase in their activity and a darkening of the skin.

Tanning is an attempt by the body to minimize genetic and other damage associated with prolonged unprotected exposure to sunlight. Many people associate a suntan with good health and devote time and money to the pursuit of the perfect tan, but a dark tan indicates that the skin has suffered injury.

Diseases and trauma also can affect skin color. Infections of the liver can reduce the normal removal of the yellow-colored chemical bilirubin from the blood. Bilirubin is formed from the breakdown of hemoglobin from old red blood cells. If bilirubin accumulates in the blood, it causes a yellowing of the skin and eyes known as **jaundice** (***jawn**-diss*). Bruises can also discolor the skin.

Bruises form when blood leaks from damaged blood vessels in the dermis. They temporarily color the skin blue, though greens and yellows

FIGURE 4.5 Melanocytes and melanin production.

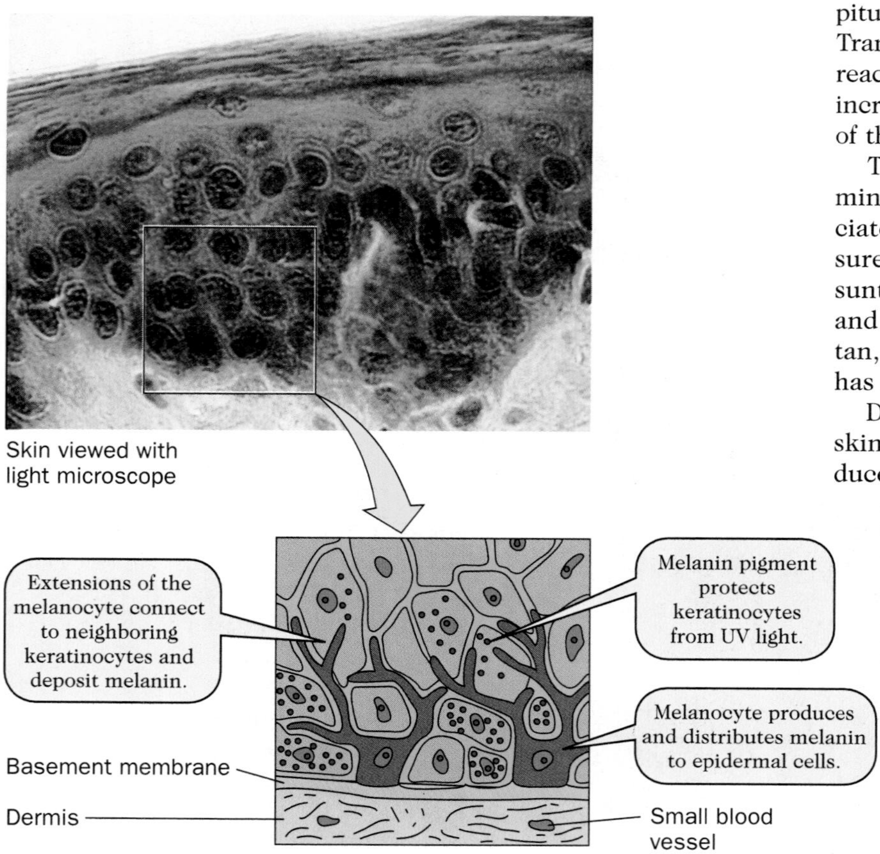
Skin viewed with light microscope
Extensions of the melanocyte connect to neighboring keratinocytes and deposit melanin.
Melanin pigment protects keratinocytes from UV light.
Melanocyte produces and distributes melanin to epidermal cells.
Basement membrane
Dermis
Small blood vessel

are also common. As part of the healing process, white blood cells engulf and digest the broken cells and spilled blood, slowly returning the skin to its normal structure and color.

> Normal skin coloration represents the combined effect of three pigments: reddish hemoglobin, yellowish carotene, and blackish melanin. Color is influenced by heredity and environment as well as disease and trauma. The ultraviolet light in sunlight stimulates melanin production and leads to a tan.

Skin Appendages: Hair and Nails

While the health and appearance of the skin are of great concern, people also lavish attention to the hair and nails. These appendages of the skin are specialized outgrowths of the epidermis, and both are made of keratin.

Hair is found on most body surfaces except the palms, soles, and nipples. There are an estimated 5 million hairs on the body, with over 100,000 on average in the scalp. Blondes average the most scalp hairs (140,000), while redheads have the fewest (90,000).

Although we have nearly as many hairs as other mammals do, human hair is relatively short and thin and plays only a limited role in insulation and protection. Scalp hair offers some protection from injury and sunlight, while eyebrows, eyelashes, and nasal hair trap dust and small insects. Hair also helps provide sensation (touch and air movement over the skin) and is an important element in sexual attraction.

Hair: structure and growth. Each hair consists of a **shaft** which projects from the skin surface and a **root** that extends beneath the surface deep into the dermis. Surrounding the root is a complex structure called a **hair follicle** (*fol-ih-kul*) constructed of epithelial and connective tissues. The follicle is formed from epidermal cells that grow deep into the dermis. It consists of two major parts: a protective sheath and an expanded bulb (Figure 4.6).

At the base of the bulb is a region rich in blood vessels that bring nourishment

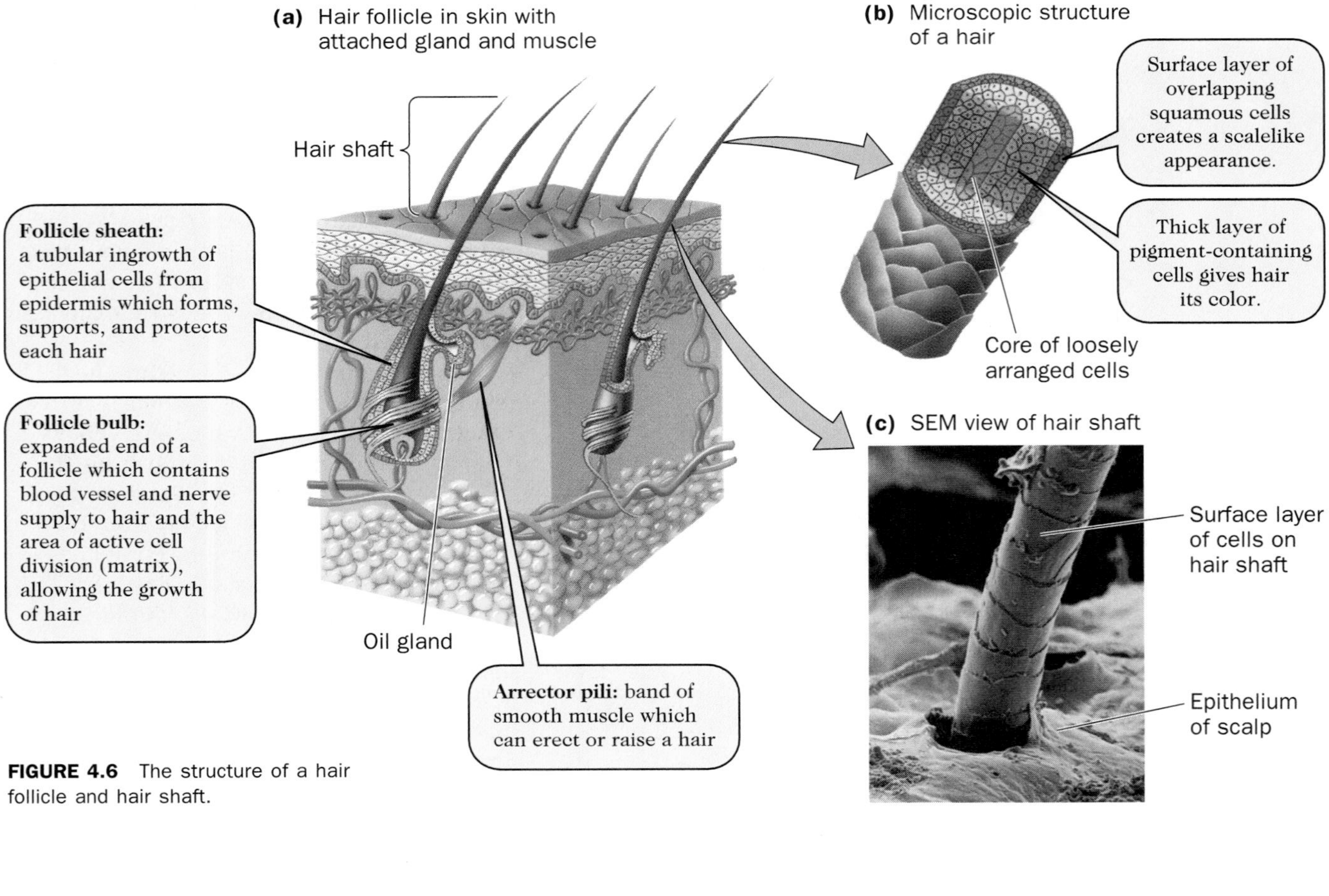
(a) Hair follicle in skin with attached gland and muscle
Hair shaft
Follicle sheath: a tubular ingrowth of epithelial cells from epidermis which forms, supports, and protects each hair
Follicle bulb: expanded end of a follicle which contains blood vessel and nerve supply to hair and the area of active cell division (matrix), allowing the growth of hair
Oil gland
Arrector pili: band of smooth muscle which can erect or raise a hair
(b) Microscopic structure of a hair
Surface layer of overlapping squamous cells creates a scalelike appearance.
Thick layer of pigment-containing cells gives hair its color.
Core of loosely arranged cells
(c) SEM view of hair shaft
Surface layer of cells on hair shaft
Epithelium of scalp

FIGURE 4.6 The structure of a hair follicle and hair shaft.

to a group of rapidly dividing epidermal cells. The cells produced here become **keratinocytes,** which develop abundant amounts of the keratin from which hair is made. As new cells are formed, older cells are pushed up the root farther from the nourishing blood supply. As in the epidermis, these displaced keratinocytes eventually die, but their keratin remains to form the hair shaft. The shaft that emerges from a pore on the skin consists of dead keratinocytes among which are dead melanocytes that contain pigment.

As with skin color, hair color is genetically controlled and is produced by varying amounts and forms of melanin. Red hair, for example, is created by a combination of melanin and iron atoms. As we get older, the amount of melanin in hair is reduced gradually until its absence produces white hairs. Gray hair is a mixture of hairs, each with some, little, or no melanin. As people age, air spaces tend to occur more often between the layers of hair shafts, and this makes hair more brittle.

Attached to the base of most hair follicles is one end of a smooth muscle called the **arrector pili muscle.** The other end of this muscle is attached to the papillary layer of the dermis. Usually the hair follicle is oriented at an acute angle to the surface of the skin, and the hair shaft thus lies against the skin. When the arrector pili muscle contracts, the follicle assumes a more perpendicular orientation and the hair stands straight up, while the skin forms dimples known as "goose bumps."

Since the cells of the shaft are dead and contain no sensory receptors, we can receive a haircut without feeling pain. After cutting, hair continues to grow as long as the growth center in the bulb remains alive and active. If the bulb is damaged, the follicle may die and the hair loss will become permanent. Such permanent hair loss is the goal of electrolysis treatment, in which electric needles are used to damage the bulb and follicle.

Each hair follicle shows a distinctive pattern of growth and resting. After a period of sustained growth, a follicle enters a resting phase during which the hair is lost. When the resting phase is over, cell proliferation begins and a new hair eventually emerges. While the rate of hair growth depends on age and sex, it averages about 2 mm ($\frac{1}{12}$ in.) a week. Hair growth is fastest from puberty to about age 40. After that age, hair growth begins to slow and an increasing number of follicles cease operating altogether. Hair loss may be accelerated by a protein-deficient diet, high fever, certain drugs, and the chemical and radiation therapy given to combat cancer.

Hair loss and pattern baldness. While health and environmental factors can cause hair loss, 90 percent of baldness is pattern baldness, the onset and character of which are programmed by genes. Although both men and women may carry the genes for pattern baldness, the expression of the genes and the loss of hair are controlled by the male sex hormones. For this reason baldness is more common among men than it is among women; it occurs in about 1 of 25 males, but only in 1 of 625 females.

Pattern baldness is due to the programmed degeneration of the hair follicles on the top of the head but not the sides, where hair growth can be abundant. The problem seems to lie with the hair follicles, not with more general conditions of the dermis or epidermis such as infection, dryness, and muscle tension. If healthy hair follicles are transplanted into the scalp from other areas of the body, they grow and produce hair on top of the head. In fact, such transplants constitute a treatment to reverse balding.

Nails: structure and growth. Fingernails and toenails are the other appendages to the skin. Nails offer protection against injury, make it easier to grasp objects, and help people relieve a nagging itch by scratching it. Like hair, **nails** are produced by modified epidermal cells and are made of dead keratinocytes (Figure 4.7).

There are three parts to a nail: the free edge of the nail, the body of the nail with edges embedded in skin, and the root of the nail. At the base of the body of the nail is a thickened opaque region called the *lunula* (**loon-you-la**). Beneath the lunula is a patch of epidermis that is responsible for nail growth, the **matrix.**

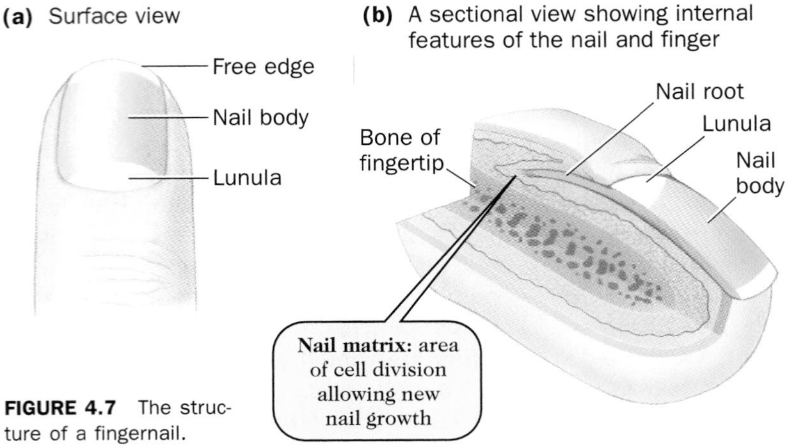
(a) Surface view
Free edge
Nail body
Lunula
(b) A sectional view showing internal features of the nail and finger
Nail root
Lunula
Bone of fingertip
Nail body
Nail matrix: area of cell division allowing new nail growth

FIGURE 4.7 The structure of a fingernail.

Each nail grows about 1 mm per week, adding up to nearly 5 cm (2 in.) of total nail growth per year, though toenails grow more slowly than fingernails do. Severe trauma to the matrix may prevent further growth and result in permanent loss of the nail.

> Hair follicles produce new cells which fill with keratin and gradually die. The dead keratinocytes form the hair root and shaft and are pushed upward toward the surface by the production of new cells. Nails are produced by cells that fill with keratin and die, forming the flat plate of the nail.

Glands of the Skin

Keeping the skin soft and the body cool are functions of the millions of tiny glands distributed throughout the skin. Figure 4.8 shows the three major types of skin glands: eccrine (***ek**-rin*), apocrine (***ap**-oh-krin*), and sebaceous (*see-**bay**-shus*). These exocrine glands produce and secrete substances which reach the surface of the skin. Like hair follicles, these glands develop from epidermal cells that grow down into the dermis and undergo a process of specialization.

Eccrine glands: the source of sweat. The fluid we call sweat or perspiration is produced by eccrine glands. Each **eccrine gland** consists of a coiled tubule which secretes sweat and a long duct which opens at a pore on the surface of the skin. Nearly 3 million eccrine glands are found on the body, except for the lips and genitalia. They are most numerous in the forehead and in the palms and soles, where there may be up to 465 glands per square centimeter (3000 per square inch).

Eccrine glands secrete water that contains small amounts of ions (Na^+, K^+, Cl^-), nitrogenous wastes (ammonia, urea, and uric acid), lactic acid, ascorbic acid, antibodies, and a tiny amount of sugar. Normally, up to 1 liter (about 1 quart) of sweat is produced per day. The evaporation of this water cools the body and leaves a residue of salt which helps limit the growth of bacteria and fungi on the skin surface. An additional inhibitor of bacterial growth is the enzyme called *lysozyme,* which is a constituent of sweat that kills bacteria by disrupting their cell walls.

Eccrine glands and the sweat they produce are essential to cooling the body. An increase in body temperature signals eccrine glands to increase their secretion of sweat. In addition to being stimulated by high body temperature, sweating on the palms and soles can be provoked by emotions. We have all experienced sweaty palms while enduring the emotional stress of job interviews, blind dates, and biology exams. (We'll talk more about sweating and body temperature later in this chapter.)

Apocrine glands: human scent glands? Apocrine glands are located primarily in the armpit and near the genitalia and anus. **Apocrine glands** are very similar to eccrine glands, but the duct passes into a hair follicle rather than terminating at a surface pore (see Figure 4.8).

Apocrine glands secrete an oily substance that is not odorous. However, it contains many nutrients on which skin bacteria can live. As bacteria use these nutrients and grow, they produce waste products that have a smell. Commercial deodorants inhibit the growth of bacteria and mask the odor with a more pleasant scent. Antiperspirants reduce the moisture level on the skin by closing pores and causing eccrine ducts to swell. With less water available, bacterial growth is inhibited and odor is reduced. Apocrine glands begin to function after people reach sexual maturity (puberty) in response to sex hormones circulating in the blood.

It has been suggested that apocrine glands may be human scent glands that play an undetected role in producing sexual and social signals used in people's interactions with others. The role of such scents is well established in the animal world, but in humans it is controversial. In women the activity of apocrine glands is correlated with the menstrual cycle. It is also known that women living together over a period of time (as in a dormitory) tend to synchronize the onset of their menstruation. Is this phenomenon mediated by scents? We don't know.

Sebaceous glands: care for hair. Sebaceous glands are abundant over the entire body except for the palms and the soles of the feet. **Sebaceous glands** are smaller than the other skin glands and release their product into a hair follicle through a short duct (see Figure 4.8). The product of the gland is an oily material called **sebum** (***see-bum***) which waterproofs the skin and lubricates hairs, preventing brittleness. Excessive washing or shampooing reduces the amount of sebum present and may result in brittle hair or dry skin.

In response to increased levels of sex hormones after puberty, sebaceous glands increase their secretion and may cause problems. An overactive gland may block its duct and follicle with a plug of sebum. This may encourage the growth of bacteria, causing local swelling and discomfort. If the follicle wall ruptures, bacteria spill into the surrounding dermis, spread the infection, and cause a pimple.

Other specialized glands. There are other specialized skin glands in certain regions. **Mammary glands** are modified apocrine glands in the breast. In females they produce and secrete milk when they are stimulated by hormones from the pituitary gland and ovary. These glands remain undeveloped and inactive in males. In the ear canal modified sweat glands produce earwax. Glands on the inner surface of the eyelids and at the base of eye-

FIGURE 4.8 Major skin glands.

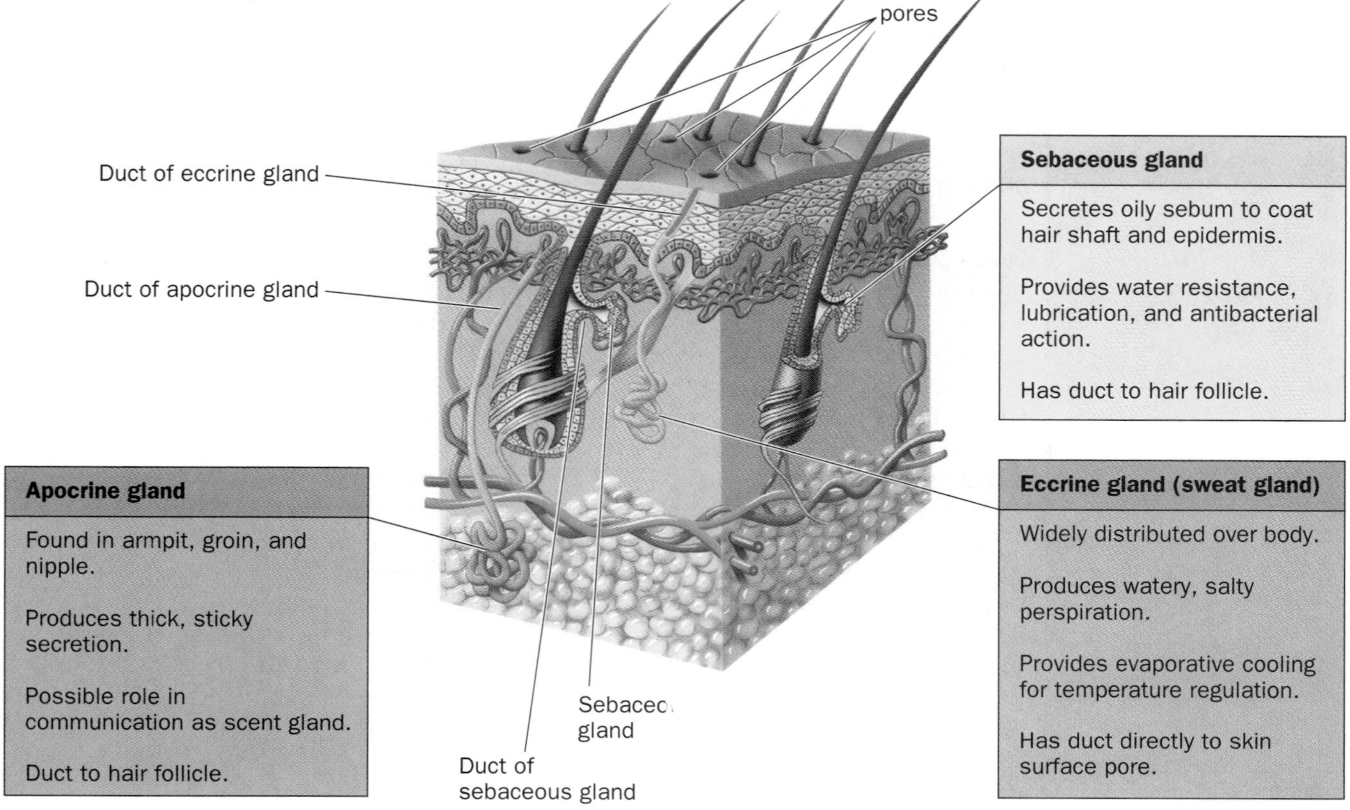
Sweat pores
Duct of eccrine gland
Duct of apocrine gland
Sebaceous gland
Secretes oily sebum to coat hair shaft and epidermis.
Provides water resistance, lubrication, and antibacterial action.
Has duct to hair follicle.
Apocrine gland
Found in armpit, groin, and nipple.
Produces thick, sticky secretion.
Possible role in communication as scent gland.
Duct to hair follicle.
Eccrine gland (sweat gland)
Widely distributed over body.
Produces watery, salty perspiration.
Provides evaporative cooling for temperature regulation.
Has duct directly to skin surface pore.
Sebaceo
gland
Duct of sebaceous gland

lashes produce secretions that prevent the eyelids from sticking together. Bacterial infections of these glands produce what we call a *sty*.

> The eccrine glands produce watery sweat which cools the skin as it evaporates. Apocrine glands may function in scent production. Skin bacteria produce odorous compounds. Sebaceous glands produce sebum, which lubricates hairs and the skin surface.

Sensory Receptors of the Skin

As you lie on the beach, basking in the warmth of the sun, an insect crawls over your arm. You may move to brush it off or, before you respond, you may feel the pain of its bite. If that happens, you may soothe your offended dignity, if not the pain, with a quick plunge into cooling water. When something like this happens, reflect for a moment on the roles of sensory receptors in your skin. **Sensory receptors** are specialized structures that warn your body about changes in the external environment (Figure 4.9).

The several different kinds of sensory receptors in the skin can distinguish among different stimuli: light touch, pressure, vibration, cold, heat, and pain. These specialized structures are located in different layers of the epidermis and dermis. Some receptors are branched endings of nerve cells, while others are modified epithelial cells. Still other receptors are elaborate layered structures. Although they respond to different stimuli, all these receptors produce an electrical signal.

As you learned in Chapter 3, these sensory receptors are connected to a series of nerve cells which transmit electrical signals from a sensory receptor to the spinal cord and brain, where they are processed. This processing functions to prevent injury and maintain a stable internal environment in the face of external changes.

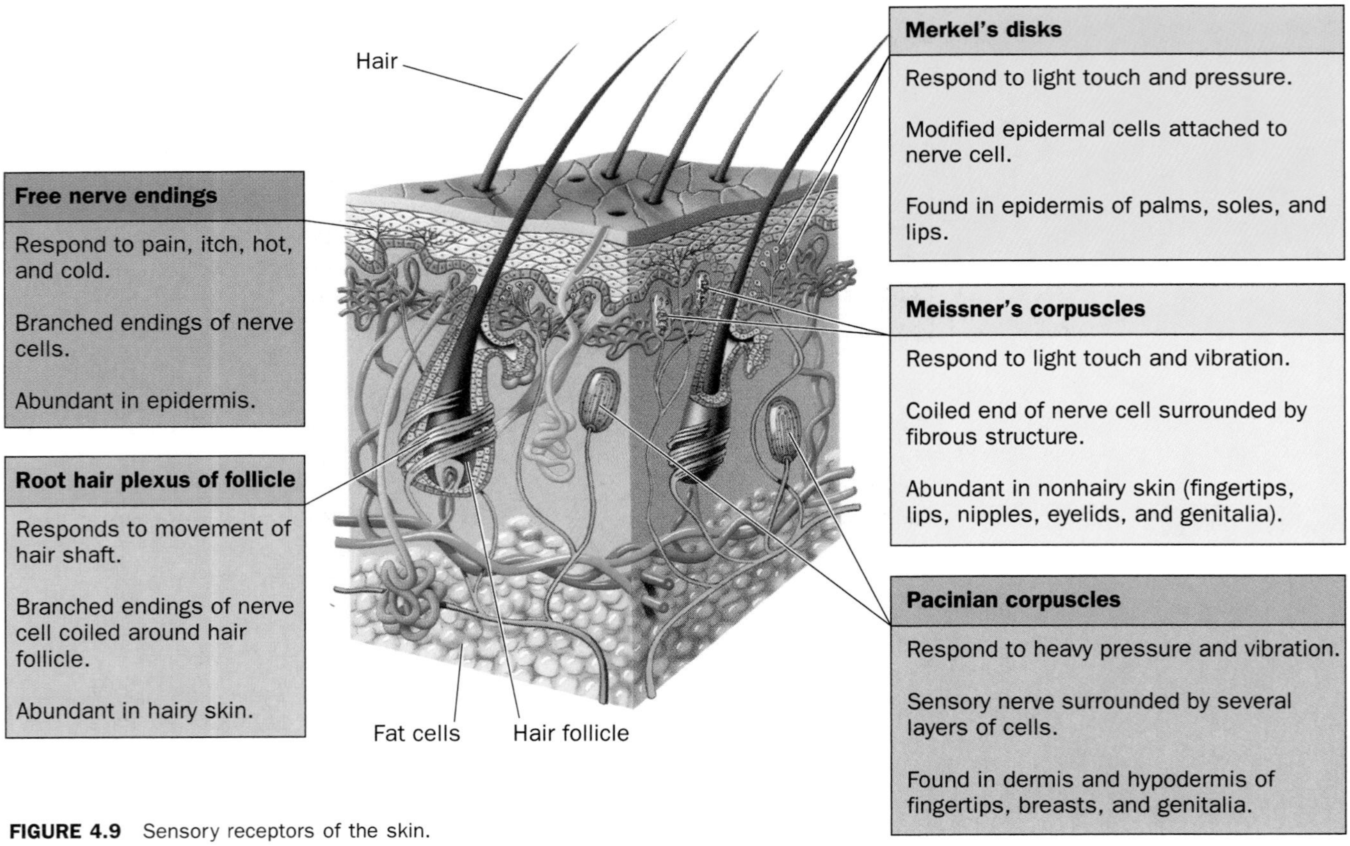
Hair
Free nerve endings
Respond to pain, itch, hot, and cold.
Branched endings of nerve cells.
Abundant in epidermis.
Root hair plexus of follicle
Responds to movement of hair shaft.
Branched endings of nerve cell coiled around hair follicle.
Abundant in hairy skin.
Fat cells
Hair follicle
Merkel's disks
Respond to light touch and pressure.
Modified epidermal cells attached to nerve cell.
Found in epidermis of palms, soles, and lips.
Meissner's corpuscles
Respond to light touch and vibration.
Coiled end of nerve cell surrounded by fibrous structure.
Abundant in nonhairy skin (fingertips, lips, nipples, eyelids, and genitalia).
Pacinian corpuscles
Respond to heavy pressure and vibration.
Sensory nerve surrounded by several layers of cells.
Found in dermis and hypodermis of fingertips, breasts, and genitalia.

FIGURE 4.9 Sensory receptors of the skin.

Some of our **touch receptors** (Meissner's corpuscles and Merkel's disks) are located near the papillae of the dermis. The slight depression of the skin caused by the uninvited insect above was registered by these receptors. When the insect crawled, it moved hairs on the skin, and this was registered by other touch receptors (the root hair plexus) associated with the hair follicle in the dermis. Buried deeper in the dermis are the **pressure receptors** (Pacinian corpuscles), which require substantial pressure for activation. The pressure of a watchband, tight shoes, or a necktie is sensed by these receptors.

Pain receptors are the ends of nerve cells and are located in both the lower epidermis and the dermis, where they respond to cutting, tearing, or burning, alerting you to avoid further injury if possible. An itch or a tickle is actually a very low level pain sensation. Some free nerve endings also respond to heat and cold. There seems to be more cold receptors than heat receptors.

> The skin's sensory receptors respond to stimuli such as touch, pressure, pain, and changes in temperature. Sensory receptors generate electrical signals that are carried by nerves to the spinal cord and brain for processing.

Integrated Functioning and Homeostasis

The human body is made up of trillions of cells, most of which lie below the skin. The skin acts as a tough, unbroken water-resistant covering over the body, helping to shield and protect internal tissues and organs from changes in the external environment. Human cells have only a limited capacity to tolerate physical and chemical changes and can die if external conditions change too much or too quickly.

In a healthy human body, cells have an environment that is relatively stable. You will recall that the maintenance of a stable internal environment is called **homeostasis.** Cells are bathed, buffered, and nourished by blood and extracellular fluids which are both chemically different from external fluids and more stable in their composition. When conditions change outside the body, these fluids do not change much.

Homeostasis depends on the integrity of the skin. Burns and cuts can destroy the protective surface and allow exposure to dangerous external materials or microorganisms as well as loss of internal fluids. A familiar example of the operation of homeostasis occurs when you move from a cool room into a sauna. In the sauna, you can survive temperatures much higher than your body temperature. In response to the high temperature, your skin produces sweat. The evaporation of water from the sweat cools the body and prevents overheating.

Homeostasis requires the **integrated functioning** of many different cells, tissues, and organs. Your body is a complex society of interacting cells whose individual efforts contribute to the common welfare by performing specialized functions and stabilizing the internal body environment.

The concept of homeostasis is important in helping us understand how and why the body functions as it does. It is useful to consider homeostasis as the goal of all our cells, tissues, organs, and systems. Homeostasis is also important in understanding the nature of diseases and dysfunctions. A dysfunction or disease represents a challenge to homeostasis and a deviation from normal functioning. Natural body processes attempt to correct any internal imbalance. Of course, there are limits to the body's capacity to achieve homeostasis in the face of certain challenges. In these cases, the body fails and dies. We can look at the aging process that the human body undergoes as a gradual and progressive inability to maintain stability within body tissues.

> Homeostasis is the maintenance of a stable internal environment. Homeostasis is accomplished through the integrated functioning of the body. This involves the coordinated activity of different cells, tissues, organs, and organ systems.

Feedback Control Systems

Although the body's internal environment is normally stable, small changes are occurring constantly. Internal conditions are rarely the same from moment to moment, and body organs strive to maintain an average value for each physical or chemical condition. This average value can be considered a *set-point* for a particular condition (Figure 4.10). Examples of set-points include the glucose concentration in the blood (90 mg per 100 ml) and the value of tissue fluid pH (7.35).

As the body moves into different external environments, eats different foods, and encounters disease-producing microorganisms, imbalances are created; these are deviations from the set-point value for the normal operation of the body. Such deviations are called **stresses**, and the factors that cause stress are called **stressors.** Stressors operating outside the body include high humidity, the wind-chill factor on a ski slope, low O_2 at a high altitude, loud noise in a factory, and carbon monoxide (CO) on busy city streets. Stressors operating inside the body include low blood sugar, fever, dehydration, high blood pressure, and an acidic blood pH.

To achieve homeostasis, the body needs to do two things: (1) recognize internal and external stressors that cause changes in physical and chemical conditions and (2) make timely, accurate, and appropriate adjustments to reestablish normal conditions (see Figure 4.10). There are, of course, limits to the magnitude and duration of the adjustments the body can make. Chronic stress may start the body on a decline which will make tissues more susceptible to disease and death.

The way in which the body recognizes and responds to stressors involves a combination of chemical and cellular activities that are called **feedback control systems.** Feedback control utilizes information about the current operation of a system to modify its future operation.

FIGURE 4.10 Homeostatic responses to maintain a set-point operation.

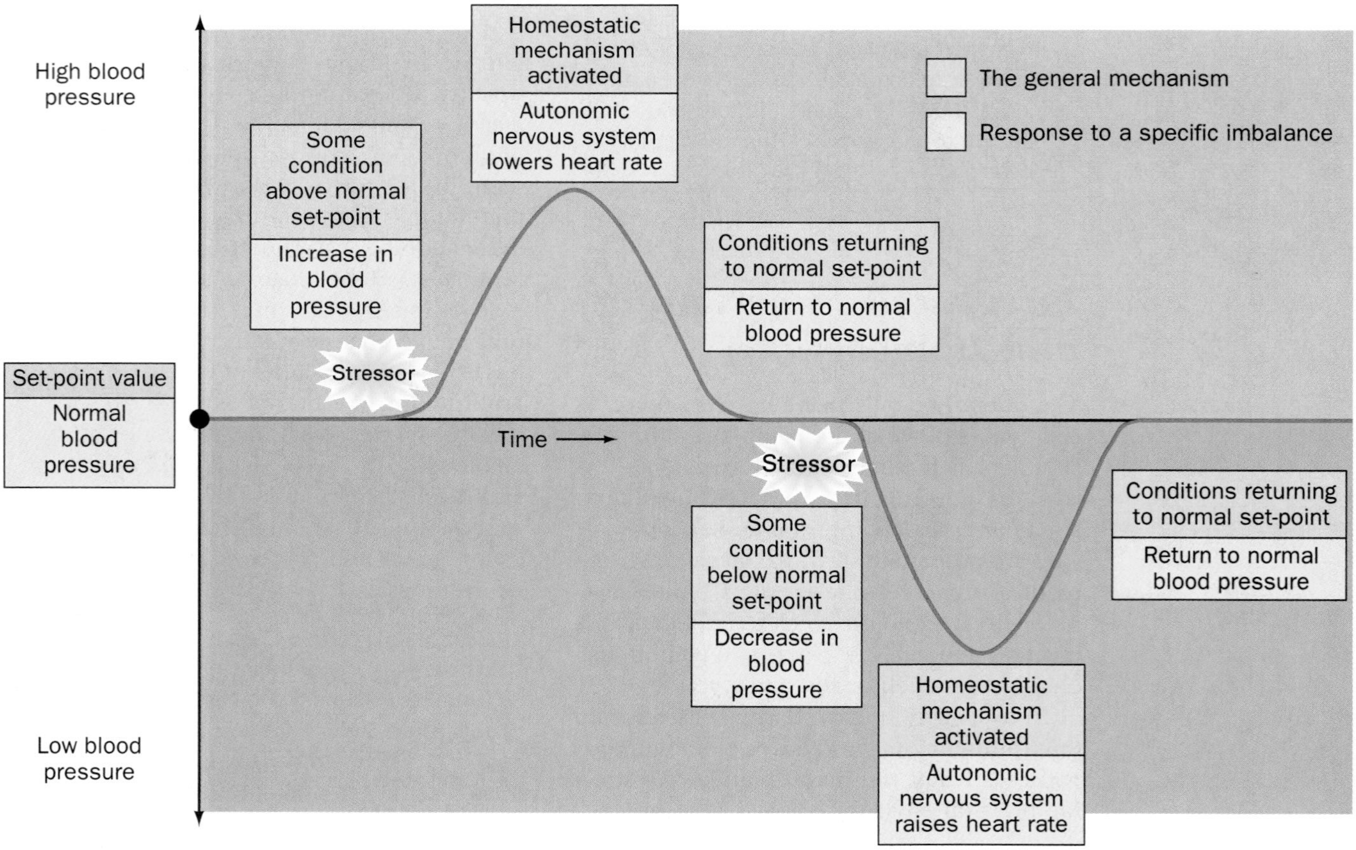
High blood pressure
Low blood pressure
Set-point value
Normal blood pressure
Time
Stressor
Some condition above normal set-point
Increase in blood pressure
Homeostatic mechanism activated
Autonomic nervous system lowers heart rate
Conditions returning to normal set-point
Return to normal blood pressure
Stressor
Some condition below normal set-point
Decrease in blood pressure
Homeostatic mechanism activated
Autonomic nervous system raises heart rate
Conditions returning to normal set-point
Return to normal blood pressure
The general mechanism
Response to a specific imbalance

While we usually think of information in terms of words or written messages, in biological systems it may consist of the increased concentration of a chemical, the presence of different hormones, or an electrical signal passed along the nerves. If the operation of a system slows, this is sensed by a receptor, and a signal is passed to the body's "engines" to speed up—if it goes too fast, a signal is passed to slow it down.

All feedback systems consist of three fundamental components. (1) A **receptor** detects stressors. (2) A **control center** receives and routes information. (3) An **effector** takes action, generating a response to the stressor (Figure 4.11). One type of feedback control system is a special series of nerve cells called a **reflex arc.**

Reflex arcs allow automatic responses to stimuli. For example, when you accidentally touch a hot stove and quickly pull back, you are exhibiting a withdrawal reflex to prevent injury. Receptors in the skin detect the placement of your hand and its increased temperature. An electrical signal is transmitted by your nerves to the spinal cord (control center) and from there to muscles (effectors), which move the hand from the hot stove. Other reflexes help us keep our posture, hold our balance, and maintain internal conditions such as blood pH and body temperature (we will discuss reflexes further in Chapter 14).

In the operation of the human body, some feedback routes or pathways involve nerves. Other feedback systems use the blood as a route to transport chemical signals such as hormones. Conscious thought is not required for feedback or reflex activity. For example, the slowing of the heart rate when the blood pressure rises too high and the breakdown of liver glycogen when blood sugar is too low do not require conscious thought. However, after reflex responses, your conscious brain is informed by signals it receives from sensory nerves. After processing these signals, your brain may initiate additional voluntary actions. For example, you may sit down and rest if you feel flushed and warm as a result of high blood pressure signals, or you may decide to eat an apple in response to the hunger signals associated with low blood sugar.

FIGURE 4.11 The components of a feedback control system.

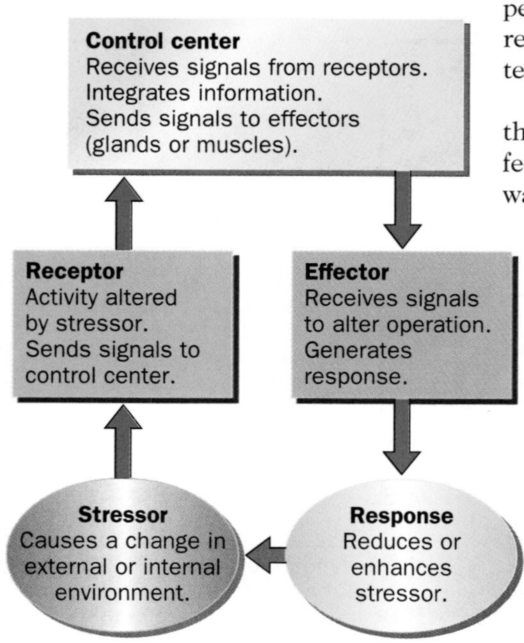
Control center
Receives signals from receptors.
Integrates information.
Sends signals to effectors (glands or muscles).
Receptor
Activity altered by stressor.
Sends signals to control center.
Effector
Receives signals to alter operation.
Generates response.
Stressor
Causes a change in external or internal environment.
Response
Reduces or enhances stressor.

Feedback control systems are distinguished on the basis of how the response of the effector organ influences an imbalance created by stressors. *Positive feedback control systems* intensify an imbalance. Positive feedback occurs during childbirth as the pressure of the infant's head against the exit from the womb stimulates stretch-sensitive receptors. These receptors signal for the release of a hormone from the brain that intensifies labor contractions. The contractions cause the release of additional hormone and continue until stretching is stopped by the infant's birth. Positive feedback systems are relatively rare in the human body. Negative feedback systems are more common. *Negative feedback control systems* reduce an imbalance to restore homeostasis. Examples include the actions that occur in the skin to help maintain the set-point value for body temperature in the face of external warming or cooling.

> Feedback control systems can recognize changes in the external and internal environments and bring about adjustments that help maintain homeostasis. Positive feedback intensifies an imbalance. Negative feedback reduces an imbalance.

Homeostasis and the Regulation of Body Temperature

When asked for the value of "normal" body temperature, most people respond 98.6°F (37°C), which is the set-point around which fluctuations occur. In a normal healthy person, the oral temperature ranges between 97 and 99°F (36 and

37.5°C) throughout the day. Internal temperature, called **core temperature**, is a bit higher, with internal organs working most efficiently at 100°F. The core of the body includes the organs within the thoracic and abdominal cavities, the central nervous system, and the skeletal muscles. It is this core temperature which the body monitors and regulates to maintain stable internal conditions.

Core temperature normally varies about 2°F (about 1°C) during the day. It is highest in the early morning and lowest in the late afternoon. Of course, there are individual variations in the core temperature, and physical activity can change it. Heavy exercise raises the core temperature as metabolic heat is released from active muscles. Women typically experience a slightly higher core temperature during the second half of the monthly menstrual cycle. Not surprisingly, temperature at the body surface varies more widely than does oral or core temperature. Skin temperature may range from 68 to 104°F (20 to 40°C) without damaging cells.

Skin temperature is one of the variables that can be adjusted to maintain core temperature. Skin temperature can be changed by controlling the amount of warm blood that flows from the body core through the dermis. When the blood vessels in the dermis constrict, the amount of blood flowing to the skin is reduced. This reduces the amount of heat lost and maintains or increases the core temperature. When blood vessels in the dermis dilate, more blood flows to the skin and more heat can be lost, reducing the core temperature.

The actions of both eccrine glands and blood vessels are under the control of a region of the brain known as the **hypothalamus** (see Figure 15.6). The hypothalamus acts as the body's thermostat (Figure 4.12). It receives sensory input from specialized temperature receptors both in the skin and in internal organs. Through complex and coordinated reflexes, the hypothalamus initiates changes which result in warming or cooling of the body.

As with other reflexes, conscious decision making by the brain is not required for hypothalamic adjustments in body temperature. However, in addition to such unconscious regulation of body temperature, your conscious mind may undertake modifications in your behavior. You may seek a restful position in the shade when your

FIGURE 4.12 Control of body temperature.

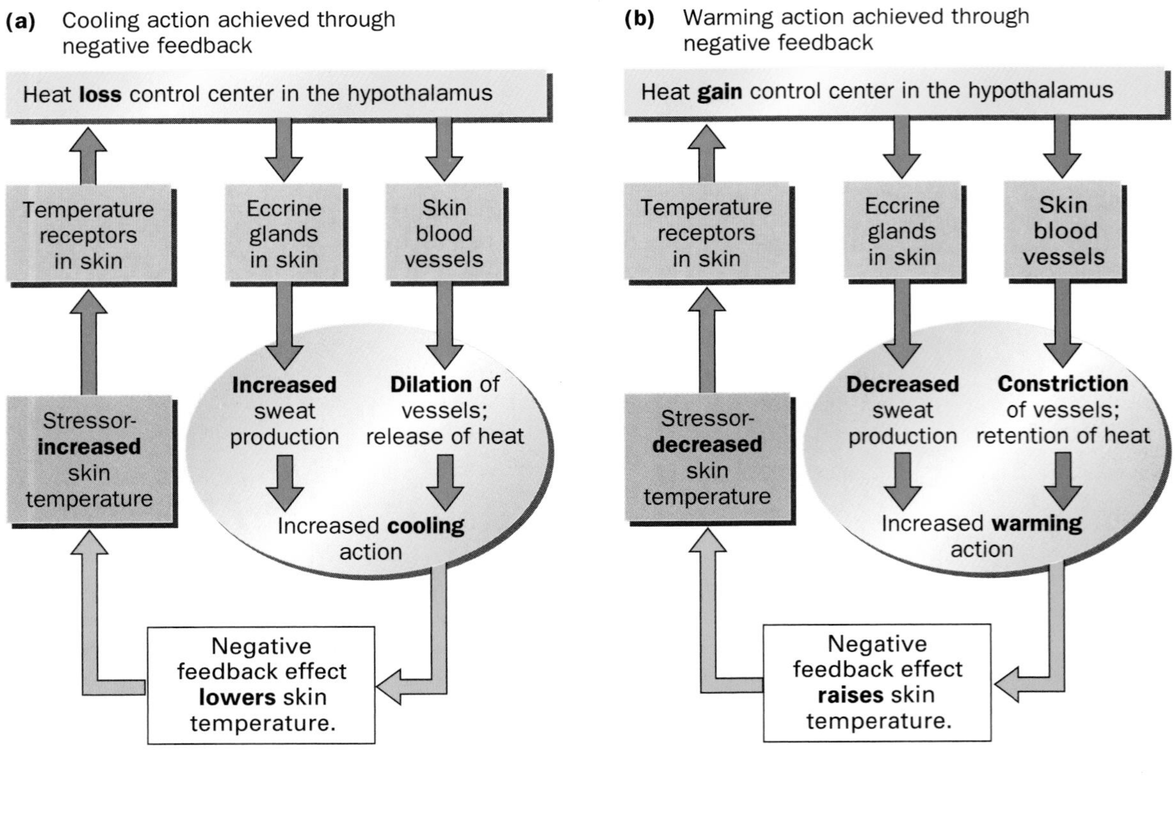
(a) Cooling action achieved through negative feedback
Heat loss control center in the hypothalamus
Temperature receptors in skin
Eccrine glands in skin
Skin blood vessels
Stressor-increased skin temperature
Increased sweat production
Dilation of vessels; release of heat
Increased cooling action
Negative feedback effect lowers skin temperature.
(b) Warming action achieved through negative feedback
Heat gain control center in the hypothalamus
Temperature receptors in skin
Eccrine glands in skin
Skin blood vessels
Stressor-decreased skin temperature
Decreased sweat production
Constriction of vessels; retention of heat
Increased warming action
Negative feedback effect raises skin temperature.

body temperature rises or build a fire and put on a sweater when it begins to drop.

While you can make some conscious decisions to help your body, other decisions may be life-threatening. Alcohol consumption on a cold day produces a false warming sensation. Alcohol actually dilates blood vessels, causing an increased loss of body heat from the skin. Consumption of alcohol on a very cold day can hasten the onset of a life-threatening decrease in core temperature called **hypothermia.** The old cartoon of a Saint Bernard dog with a cask of brandy around its neck for lost Alpine skiers actually represents a false and dangerous remedy. Hugging the warm dog would be safer than drinking the alcohol.

> Body temperature is regulated by negative feedback. Temperature receptors send messages to the hypothalamus in the brain, which then signals eccrine glands to increase or decrease sweating. The hypothalamus also signals blood vessels to dilate or constrict, increasing or decreasing heat loss.

Cuts, Abrasions, and Burns: Healing the Skin

It's six o'clock on Friday evening, and you are busily slicing vegetables for a stir-fry. Suddenly the knife slips, cutting your finger. Blood flows, and you feel pain and anger. However, the bleeding will stop and the wound will heal. Fortunately, the body has a remarkable capacity to repair cuts and abrasions. While you bandage your finger, the healing process has begun. Healing is an example of homeostasis and integrated functioning. Many systems interact to protect the body from damage and maintain stable life-supporting functions.

Healing a Cut or Abrasion

There are many ways in which the skin can be wounded, but its healing always proceeds in four sequential but overlapping phases: clot formation and inflammation, the proliferation of new cells, removal of debris and infection, and completion of dermal healing (Figure 4.13).

Phase 1: clot formation and inflammation. Immediately after your injury, muscle tissue in the walls of damaged blood vessels contracts, reducing the loss of blood. Blood cells called **platelets** release chemicals that start a series of reactions that cause a protein in the blood fluid to form a network of long fibers. These fibers trap blood cells, forming a **blood clot** that plugs the torn blood vessel. In the dermis, injured mast cells release **histamine**, which attracts phagocytic white blood cells and initiates an **inflammation response.** Inflammation is part of the healing process and helps combat microorganisms which have entered the dermis.

Phase 2: proliferation and new cells and tissues. Within a few hours after an injury, cells of the basal stratum of the epidermis and cells of the dermis begin to divide rapidly. The new cells slowly spread over the injured area, replacing lost tissue. New blood vessels are produced to replace injured ones. As the clot dries, it forms a scab and contracts, drawing the edges of the wound together and reducing the gap that will be filled with new tissue.

Phase 3: removal of debris and infection. Within several days, more phagocytic white blood cells migrate into the wound to engulf and digest infection-causing microorganisms and debris from the damaged tissue. Under the protective scab, new epidermal cells seal the skin's surface and stop dividing, but tissue replacement in the dermis continues.

Phase 4: completion of dermal healing. Within 1 to 2 weeks the dried scab falls off, revealing a shallow depression caused by absent dermal tissue. Continued addition of dermal cells normally fills this depression and completes the healing process.

The regenerative powers of the skin are remarkable. However, depending on the extent and depth of the wound, repair of the dermis may be incomplete and **scar tissue** may form. Scar tissue has more collagen fibers and fewer blood vessels than does the normal dermis and is usually less flexible. Depending on the sever-

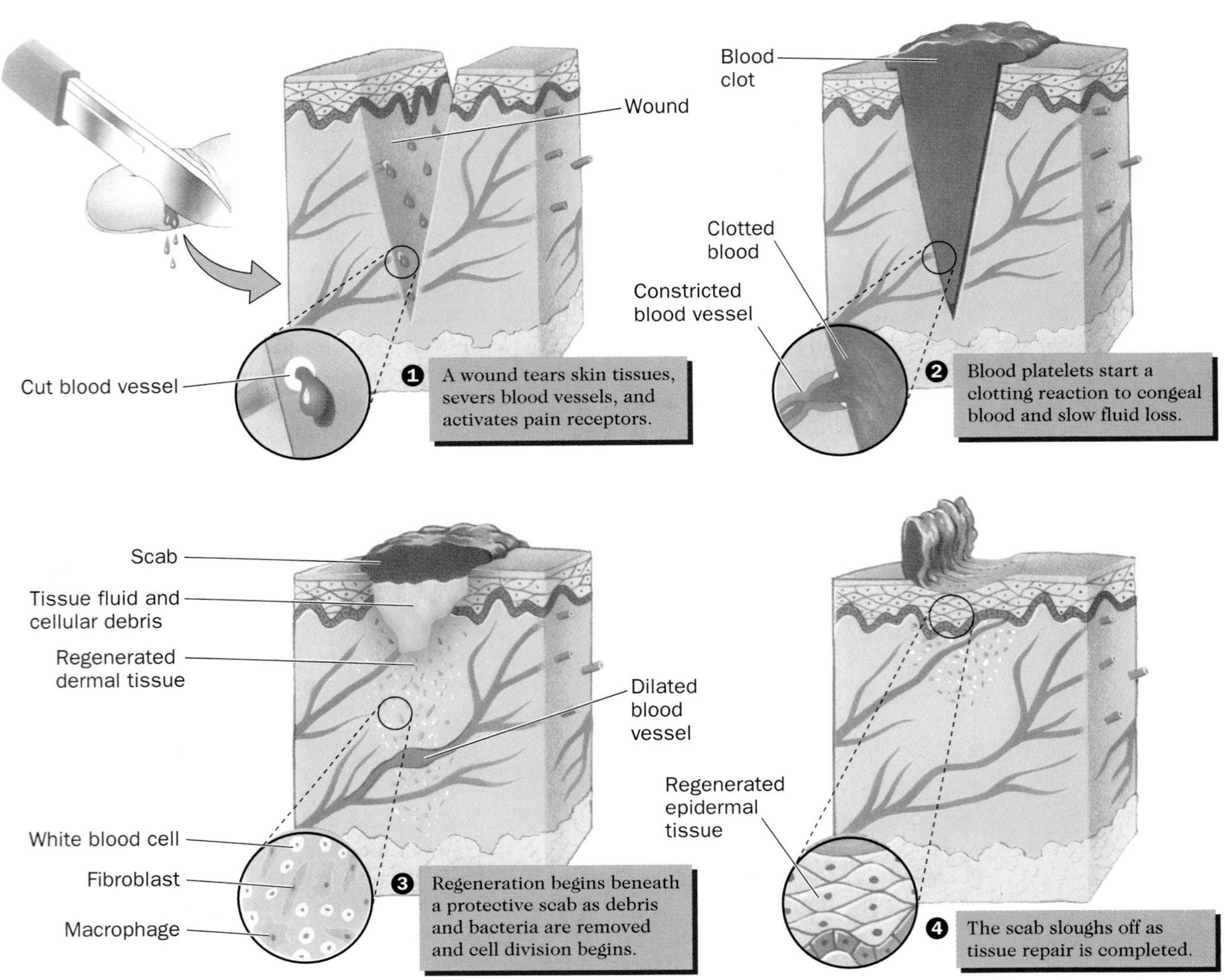
Wound
Blood clot
Clotted blood
Constricted blood vessel
Cut blood vessel
1 A wound tears skin tissues, severs blood vessels, and activates pain receptors.
2 Blood platelets start a clotting reaction to congeal blood and slow fluid loss.
Scab
Tissue fluid and cellular debris
Regenerated dermal tissue
Dilated blood vessel
Regenerated epidermal tissue
White blood cell
Fibroblast
Macrophage
3 Regeneration begins beneath a protective scab as debris and bacteria are removed and cell division begins.
4 The scab sloughs off as tissue repair is completed.

FIGURE 4.13 The process of wound healing.

ity of the injury, scar tissue may lack hair follicles, glands, and sensory receptors.

> When the skin is torn down to the dermis, a blood clot forms in damaged vessels to minimize blood loss. White blood cells migrate into the wound to fight infection. Protected by the clot and later by the scab, the epidermis and dermis produce new tissue to repair the wound.

Burns and Their Healing

Skin burns can be caused by heat, steam, caustic chemicals, radiation, or a strong electrical current. The severity of a burn is determined by the size of the surface area involved and the depth of skin damage. The three categories of burns are based on how severely the epidermis and dermis are damaged.

First-degree burns are limited to the epidermis. They are characterized by the redness, pain, and mild swelling seen in a common sunburn. The surface layers of cells may be shed and replaced in a few days, but healing will occur without scarring. First-degree burns require no special treatment, though gentle rinsing with cold water reduces the pain and swelling.

Second-degree burns involve the entire epidermis and at least a portion of the dermis. Symptoms include redness, considerable pain, and often blister formation as the epidermis and dermis separate and tissue fluid collects between them. Later, rupture of the blisters may provide a route for skin bacteria to enter and cause infection. The hair follicles, glands, and sense organs may be damaged in some second-degree burns, reducing the capacity for repair. Scarring may result. After initial treatment with cold water, soaking the affected area in dilute salt water aids healing.

Third-degree burns involve damage to the epidermis, dermis, hypodermis, and sometimes the underlying structures. The burned area may be bright red, white, or a charred black color, and there is extensive fluid loss. At the center of the burn there may be little pain because the sensory receptors have been destroyed. However, at the margins, where the burn grades out into the second and first degrees, the pain is very intense.

Third-degree burns are easily infected. When they are extensive, they represent a severe challenge to body homeostasis, causing a loss of water and blood. Such third-degree burns are life-threatening, and healing is difficult. Extensive scar tissue may form after healing and distort the skin, making movement difficult. Successful treatment of serious third-degree burns usually requires antibiotics, fluid replacement, and surgical skin grafts.

> First-degree burns involve only the surface epidermal layers. Second-degree burns extend into the dermis, causing separation from the epidermis and leakage of tissue fluid. Third-degree burns damage both layers of the skin and extend into the hypodermis and underlying tissue.

Diseases and Disorders

We are not alone in our bodies. Large numbers of bacteria, viruses, and fungi inhabit our skin, hiding under dead keratinized cells, in the pores and ducts of eccrine glands, in hair follicles, and even within skin cells. You may be carrying up to 3 million bacteria per square centimeter of skin surface. The skin actually presents a rather inhospitable environment, and only a few species are adapted for growth on its exposed, dry, salty landscape.

Some species of microorganisms are permanent residents of the skin and usually do not cause damage. Others are transients that we pick up as we encounter environmental surfaces. Unless the surface of the skin is torn, worn, or cracked, most microorganisms cannot gain entrance to the dermis. However, when surface microorganisms penetrate into the dermis, they encounter a moist environment that is well supplied with nutrients. Here the microorganisms can grow, causing infection.

Infections are not the only assaults on the health of the skin. Environmental conditions, nutrition, and inherited genes can also cause skin problems. Our discussion in this section will be limited to two types of disorders. We will examine some infectious diseases of the skin and then describe skin cancers, which are noninfectious diseases.

Infectious Skin Diseases

Because of its exposure, the skin is vulnerable to many kinds of infections by viruses, bacteria, and fungi. Viral infections include chicken pox, cold sores, warts, and shingles. Bacterial infections include acne, boils, cellulitis, and impetigo. Fungal diseases include ringworm infections such as athlete's foot.

Acne vulgaris. The acne which affects most teenagers and many adults is an infection of the sebaceous glands caused by the bacterium *Proprioniobacterium acnes,* a normal resident on the skin. The infection is most notable after puberty, when high amounts of sex hormones stimulate increased sebum production. Sebum provides moisture and nutrients for bacterial growth in hair follicles, where bacterial waste products can accumulate, causing irritation and redness. If the bacteria spread to the dermis, cysts may form. Cysts are fluid-filled pockets in the skin which may lead to pitting and scarring.

An association between food and acne has not been firmly established, and contrary to popular opinion, peanut butter, oily foods, and chocolate do not promote acne. Treatments for acne include thorough washing and scrubbing to remove bacteria, antibiotics such as tetracycline to limit bacterial growth, and a synthetic form of vitamin A known as Accutane which reduces the activity of sebaceous glands.

Boils. Boils are caused by the bacterium *Staphylococcus aureus,* another normal skin resident. A boil is a large, serious infection of a hair follicle in which an abscess forms. An *abscess* is a pocket of inflamed tissue that contains pus (bacteria, white blood cells, and tissue fluid). This pocket of tissue isolates the infection from the body's normal defenses and from antibiotics. A health care professional may have to surgically open a boil and drain off the pus before the administration of antibiotics.

Impetigo. Impetigo is caused by a different bacterium, *Streptococcus pyogenes,* though *Staphylococcus* bacteria also may be present. The disease appears as raised, round sores with a crusty surface. Impetigo is most common in school-age children. It is very contagious and is spread by direct contact with an infected person. Antibiotics such as penicillin and erythromycin are used for its treatment.

Warts. Warts are localized growths of skin. They are caused by a group of viruses called human papillomaviruses (HPV). These viruses are transmitted by direct contact with an infected individual and usually enter the skin at the site of a cut or abrasion. The virus infects epidermal cells, causing them to divide rapidly and create a raised growth.

Warts most often occur on the fingers or other sites subjected to injury. Genital warts are dangerous and have been implicated in cervical cancer in women. Sometimes warts disappear spontaneously. If they do not, they can be removed by freezing with liquid nitrogen or by burning with an electrical current or an acid. In general, antibiotics do not provide much help against the viral infections that cause warts.

Noninfectious Skin Disorders

Disorders, unlike infectious diseases, cannot be passed from person to person. In most disorders, the kinds of changes that take place in the tissues and cells are usually more complicated and difficult to treat than are those seen in most infectious diseases. Disorders of the skin may be associated with exposure to physical conditions such as sunlight, contact pressure, and caustic chemicals.

Disorders also may be caused by the genes you inherited from your parents (for example, the presence of moles, autoimmunity, and the number and distribution of sweat glands or sensory receptors). A common skin disorder, called *psoriasis,* is characterized by inflamed patches of scaly skin. While the cause of psoriasis is unknown, this disorder seems to appear in families and may have a genetic component. In our discussion, we will focus on cancers of the skin, and you should review Spotlight on Health: One (page 103).

Skin cancers. Skin cancer is the most common human cancer. It is most often associated with increased exposure to the UV wavelengths of sunlight. People who are at risk for skin cancer include those with fair skin (low melanin production), those exposed every day to many hours of sunlight, those who had several blistering sunburns early in life, and those who are over age 50. Over 450,000 Americans are diagnosed with skin cancer each year. The best prevention for skin cancer is to reduce sunlight exposure with clothing or a hat and to use a sunscreen ointment with an SPF rating of 15 or higher on exposed areas.

Cancers of epithelial tissues are called *carcinomas.* The most common and least dangerous form of skin cancer is **basal-cell carcinoma.** As its name implies, this cancer begins with the transformation of cells in the basal stratum. The transformed cells cease keratin production, lose contact with neighboring cells, and invade the dermis and hypodermis, producing slow-growing nodules. On the surface of the skin, the abnormal tissue develops a central, scaly ulcer with a beaded margin, as shown in Figure 4.14a. This form of cancer rarely spreads (metastasizes) and can be treated effectively with surgery.

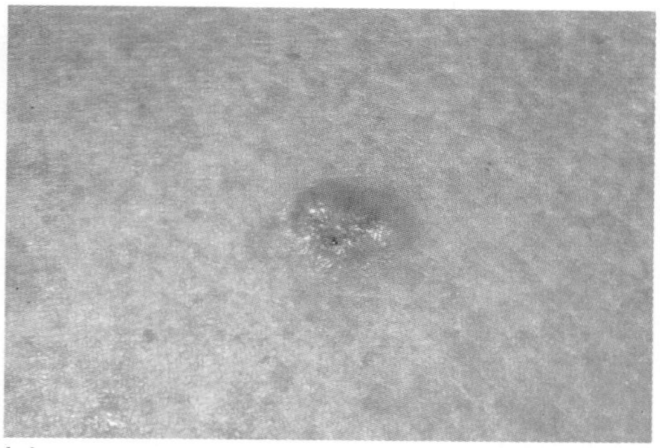

(a)

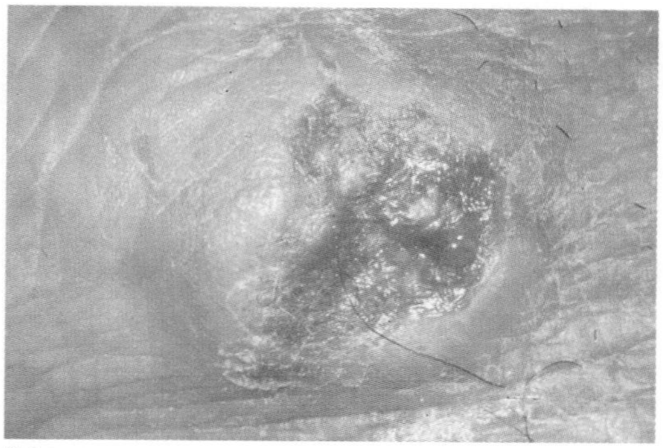

(b)

FIGURE 4.14 Skin cancer. (a) Basal-cell carcinoma. (b) Squamous-cell carcinoma.

Another type of skin cancer is **squamous-cell carcinoma,** which develops in sun-exposed keratinocytes that then form raised, hard, scaly red nodules. This cancer, as shown in Figure 4.14b, grows rapidly and may spread to the dermis and underlying tissues. Early surgery and radiation therapy are effective treatments.

Malignant melanoma (***mel**-ah-no-mah*) is the most life-threatening skin cancer. It develops in melanocytes, often in a preexisting mole, and is most common in fair-skinned individuals with red or blond hair and lightly pigmented eyes. The cancer quickly metastasizes to neighboring lymph nodes and internal organs.

To protect themselves, all individuals should periodically examine their skin, even the scalp and between the toes, and apply the **ABCD rule** for spotting melanoma (Figure 4.15):

A = **asymmetry:** The two sides of the lesion are different.

B = **border irregularity:** The borders are irregularly scalloped.

C = **color varied:** The color of the lesion varies from one region to another and may include shades of brown, black, tan, red, white, or even blue.

D = **diameter:** The patch is larger in diameter than a pencil eraser, usually 6 mm or greater.

Early recognition of melanoma can lead to successful removal by means of surgery and radiation therapy. After metastasis, which spreads the fast-growing cancer cells to other parts of the body, treatment is difficult and death may follow. According to the American Cancer Society, 22,000 people develop melanoma each year and 5500 die from the disorder. The incidence continues to increase, and soon 1 in 100 Americans will develop melanoma annually.

Life Span Changes

Like all body organ systems, the skin changes over the human life span. A healthy baby's skin is soft, moist, and elastic. However, over the years, the mirror tells a story. As we grow older, we see the assaults of time that accompany even the most healthy lives (Figure 4.16).

In regard to these changes, it is difficult to separate normal aging from the cumulative effect of sun, wind, and other environmental or occupational factors. However, those factors only speed up a normal process. Less obvious than wrinkling are reductions in the skin's ability to combat disease and repair itself after normal wear and tear.

FIGURE 4.15 The ABCD rule applied to malignant melanoma. A normal mole is included for comparison.

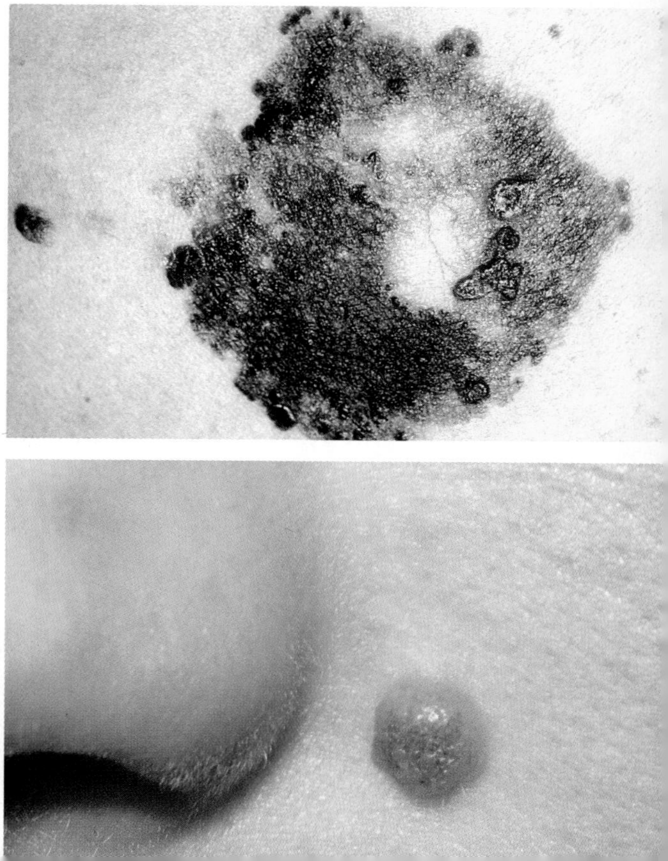

FIGURE 4.16 Aging of the skin. Both internal aging and environmental exposure contribute to aging of the skin.

Princess Elizabeth at age 6 to 9 months.

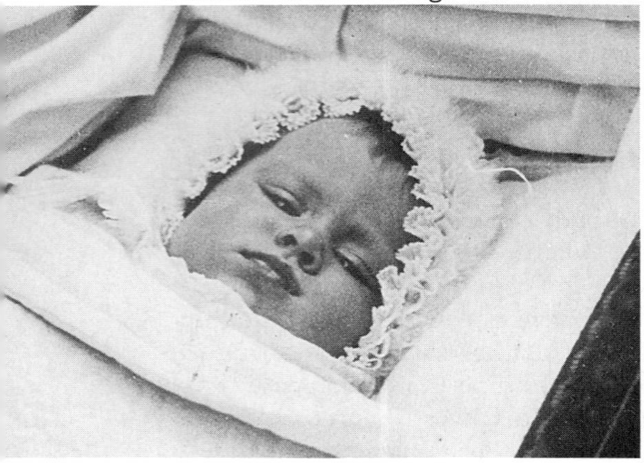

Princess Elizabeth at age 4 years.

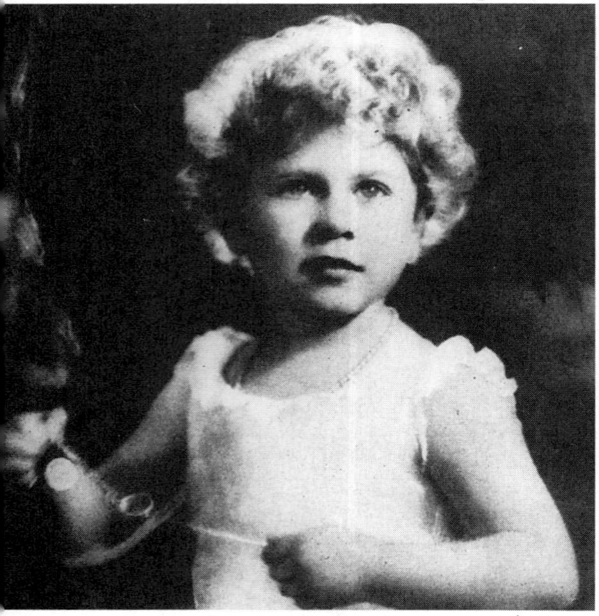

Princess Elizabeth at age $14\frac{1}{2}$ years.

Her Majesty, Queen Elizabeth II, at age 27 years.

Her Majesty, Queen Elizabeth II, at age 65 years.

By themselves, most of these changes in the skin are not life-threatening. However, they damage our protective and sensitive outer envelope and threaten our ability to maintain a stable internal environment.

Changes in the Epidermis and Dermis

As we grow older, keratinocytes in the epidermis adhere less strongly to each other. As a result, the epidermis becomes more vulnerable to penetration by foreign substances. In addition, there is a decrease in the production of new cells by the basal stratum. This slows wound healing and enhances the chance of developing an infection (Figure 4.17). After age 30, the number of melanocytes in the epidermis decreases by about 20 percent each decade, making painful sunburn more likely in light-skinned people. The melanocytes that remain become larger and gradually cluster more closely together, producing dark spots on the skin that are often called "liver spots."

Below the epidermis, the dermis also changes with age, becoming thinner and less elastic. Less active fibroblast cells decrease the amount of collagen and elastic fibers in the extracellular matrix. This substantially reduces the flexibility and elasticity of the skin. You can observe this by pinching and releasing a bit of skin on the back of your hand. In a younger individual, the skin quickly settles back into place. In an elderly person, the skin remains peaked and settles slowly.

The number of blood vessels in the dermis also declines with age, and there is a general reduction in blood flow. The remaining vessels are less permeable, and the exchanges of nutrients, O_2, CO_2, water, and wastes are reduced. Fewer white blood cells migrate into the tissues to fight infections and remove debris. The inflammation and blood-clotting responses are slowed, and in the elderly even a slight bump produces significant bruising and discoloration.

FIGURE 4.17 Summary of age-related changes in the skin.

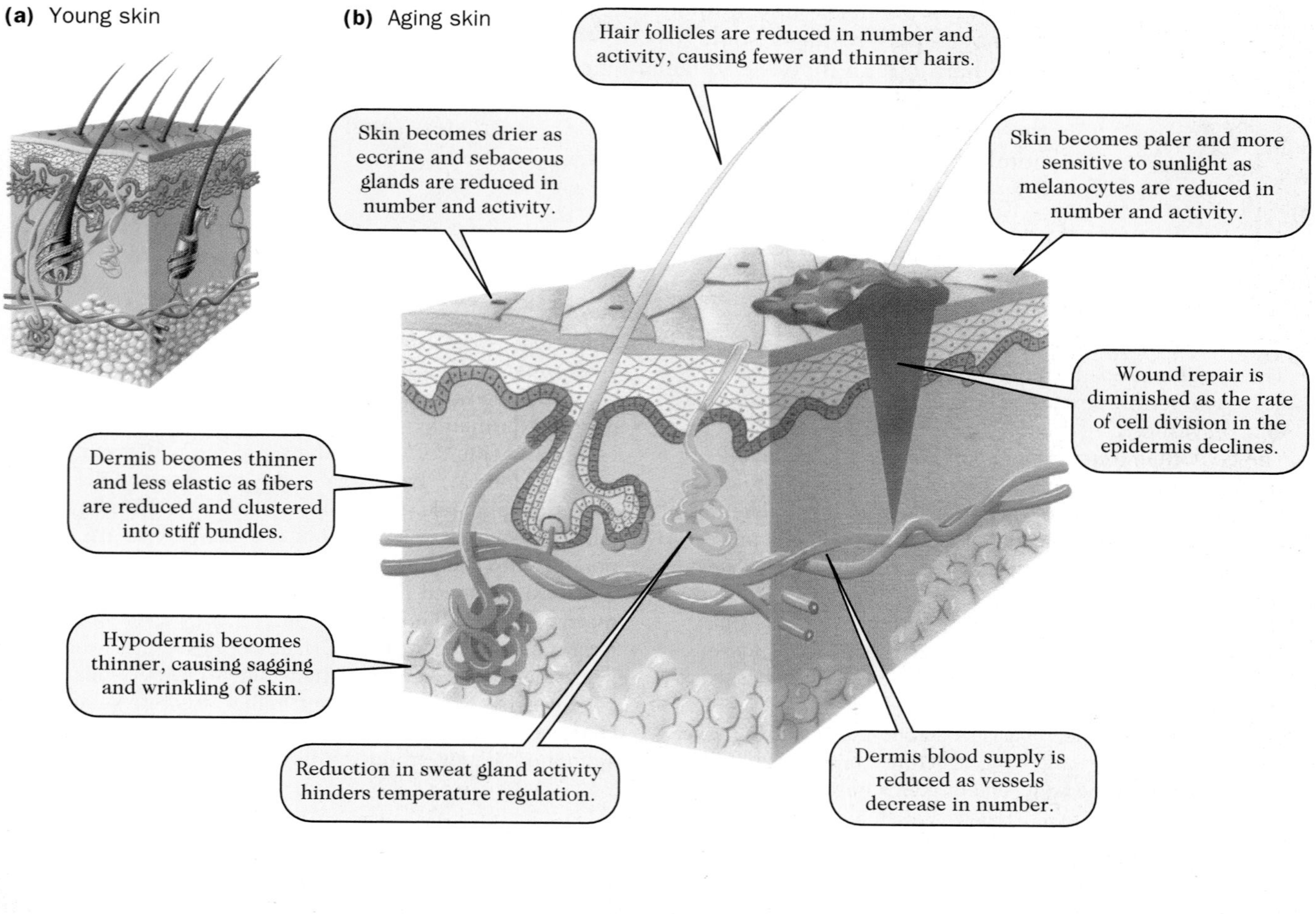
(a) Young skin
(b) Aging skin
Hair follicles are reduced in number and activity, causing fewer and thinner hairs.
Skin becomes drier as eccrine and sebaceous glands are reduced in number and activity.
Skin becomes paler and more sensitive to sunlight as melanocytes are reduced in number and activity.
Wound repair is diminished as the rate of cell division in the epidermis declines.
Dermis becomes thinner and less elastic as fibers are reduced and clustered into stiff bundles.
Hypodermis becomes thinner, causing sagging and wrinkling of skin.
Reduction in sweat gland activity hinders temperature regulation.
Dermis blood supply is reduced as vessels decrease in number.

The reduced flow of blood to the dermis also reduces service to the epidermis, which has no vessels and relies on diffusion to and from the dermis. Because of the reduced blood flow, older skin is paler in color and functions less effectively in temperature regulation.

Changes in Glands, Hair, and Sensory Receptors

As people grow older, there are decreases in sweat glands, hair follicles, hair, and hair pigment. The decrease in the number, size, and activity of eccrine sweat glands makes cooling of the body more difficult and temperature homeostasis less effective. The decline in the activity of apocrine and sebaceous glands leads to drier skin that is more susceptible to infection.

In men, an early sign of passing years is thinning of the hair on the scalp. The number of hair follicles is reduced, and the remaining follicles are less active, producing hair more slowly. Women normally experience only minimal hair loss through middle age. However, after menopause, changes in a woman's hormones may result in some loss of scalp hair and an increase in facial hair.

In both men and women, graying of the hair is caused by a decrease in the number of melanocytes in hair follicles, less melanin pigment, and trapped air pockets within hairs. The age at which graying starts may be anywhere from the early twenties to the mid-forties, and its onset is controlled by the presence and activity of genes.

Changes in the Hypodermis

Difficulty in maintaining body temperature is associated with a decrease in the thickness of the body's layer of fatty insulation in the hypodermis. As people age, fat storage shifts from the skin to the body cavities. This reduction in fat content below the skin, together with abnormal cross-linkages between collagen proteins, contributes to skin wrinkling. In some individuals changes in fat distribution may even alter body proportions. There is a general thickening of the trunk and a thinning of the appendages, influencing the comfort and grace with which we perform our daily activities. These changes, along with cosmetic changes in the skin, may affect the way we view ourselves and interact with others.

Inevitable as these changes are, they do not doom us, and for many people the aging process is a source of satisfactions, revealing capacities that were overlooked or undervalued in the younger years. The greatest satisfaction remains a life well lived.

Chapter Summary

The Skin

- The skin functions in protection, sensation, water balance, temperature regulation, waste excretion, and vitamin D synthesis.
- The epidermis (outer layer) contains up to five layers of cells. Keratinocytes produce keratin to waterproof and strengthen the skin, and melanocytes produce melanin for protection against ultraviolet (UV) light. Cells produced in the basal stratum migrate to the cornified stratum, die, and slough off.
- The dermis (inner layer) contains connective tissue that provides support, elasticity, and nutrients to the epidermis. Vitamin D formation begins in the dermis after exposure to UV light.
- The hypodermis, which consists of adipose connective tissue, anchors the skin to underlying structures, provides an energy reserve, insulates the body, and cushions fragile organs.
- Skin color is determined genetically but is influenced by environmental factors. Pigments that contribute to skin color include hemoglobin, carotene, and melanin.
- A hair is formed by keratinocytes and consists of a shaft (above the skin surface) and a root (below the skin surface in a follicle). Epidermal cells grow into the dermis to form the follicle. Hair color results from different amounts and forms of melanin as melanocytes are incorporated into the growing hair.
- Nails are produced from keratin-rich epidermal cells; the matrix is the nail growth region.
- Eccrine glands (duct to a surface pore) secrete sweat. Apocrine glands (duct to a hair follicle) secrete oils. Sebaceous glands (duct to a hair follicle) secrete sebum. Modified sweat glands produce wax in the ear canal. Mammary glands produce milk in the breast.

- Sensory receptors include Meissner's corpuscles (touch), Merkel's disks (touch), root hair plexuses (touch), Pacinian corpuscles (pressure), and free nerve endings (pain, heat, cold).

Integrated Functioning and Homeostasis

- Homeostasis is the maintenance of a stable internal environment and requires the integrated functioning of all body systems. Disease represents an imbalance in homeostasis.
- Feedback control systems (a receptor, a control center, and an effector) adjust internal conditions. Positive systems intensify imbalances; negative systems reduce imbalances and restore homeostasis.
- Skin temperature is controlled by the hypothalamus through a negative feedback system.

Cuts, Abrasions, and Burns: Healing the Skin

- Wounds heal in four sequential phases: clot formation and inflammation, proliferation of new cells, removal of debris and infection, and completion of dermal healing.
- Blood cells called platelets initiate clot formation.
- Mast cells release histamine to begin an inflammation response.
- First-degree burns are limited to the epidermis, second-degree burns injure the epidermis and dermis, and third-degree burns penetrate to the hypodermis and below.

Diseases and Disorders

- Skin infections are caused by microorganisms. Resident microorganisms cause infection when they penetrate below the skin. Nonresident microorganisms are obtained from environmental surfaces.
- Bacterial infections include acne, boils, and impetigo. Viruses produce warts and cold sores. Athlete's foot and ringworm result from fungal infections.
- Basal-cell carcinoma rarely metastasizes and is the most common and least dangerous form of skin cancer. Malignant melanoma readily metastasizes and is the most life-threatening form. Melanoma is recognized from the ABCD rule: *a*symmetry, *b*order irregularity, *c*olor varied, *d*iameter large.

Life Span Changes

- With age, the epidermis becomes less able to repair itself and is more susceptible to infection and sunburn.
- The dermis is thinner, less elastic, and wrinkled with age because of decreased fiber production.
- The skin's infection-fighting and temperature-regulating abilities are reduced with age.
- Sweat glands, hair follicles, and hair pigments decrease in abundance with age.
- Changes in fat distribution shift fat storage from the hypodermis to the body cavities, altering body proportions and causing wrinkling and sagging of the skin.

Selected Key Terms

apocrine gland (p. 118)
arrector pili muscle (p. 117)
dermis (p. 111)
eccrine gland (p. 118)
epidermis (p. 111)
feedback control system (p. 122)
histamine (p. 125)
homeostasis (p. 121)
hypodermis (p. 111)
inflammation response (p. 125)
integumentary system (p. 110)
keratin (p. 112)
malignant melanoma (p. 129)
melanin (p. 112)
sebaceous gland (p. 119)

Review Activities

1. List the general functions associated with the skin.
2. Describe three differences between the epidermis and the dermis.
3. Compare the protective functions of keratinocytes and melanocytes and their products.
4. Describe three functions associated with the hypodermis.
5. List and describe three pigments which contribute to skin color.
6. Diagram and label a hair and its follicle. Include the associated glands and muscles (see Figure 4.6).
7. Summarize the cycle of growth within a hair follicle.
8. In what way are nails similar to hairs?
9. Compare the structure, function, and location of eccrine, apocrine, and sebaceous glands.
10. What is the general function of sensory receptors in the skin? List several examples of receptors and the sensations they monitor.
11. Define homeostasis.
12. Explain the concept of integrated functioning.
13. List the three general components of a feedback system.
14. Compare the actions of positive and negative feedback control systems and give an example of each. Which type is more common in the body?
15. Summarize the role of the hypothalamus and the skin in temperature regulation.

Self-Quiz

Matching Exercise

__ 1. Connect epidermis to dermis	A. Epidermis
__ 2. Synthesized in skin with the help of UV light	B. Dermis
__ 3. Innermost layer of skin	C. Hypodermis
__ 4. Adds strength to skin, hair, and nails	D. Melanin
__ 5. Produce sebum to lubricate hairs	E. Keratin
__ 6. Protects DNA of skin cells from UV light	F. Vitamin D
__ 7. Outermost layer of skin	G. Papillae
__ 8. Produce sweat for cooling	H. Eccrine glands
__ 9. Cushions, insulates, and stores energy	I. Apocrine glands
__ 10. Possible scent glands	J. Sebaceous glands

Answers to Self-Quiz

1. G; 2. F; 3. B; 4. E; 5. J; 6. D; 7. A; 8. H; 9. C; 10. I

Chapter 5

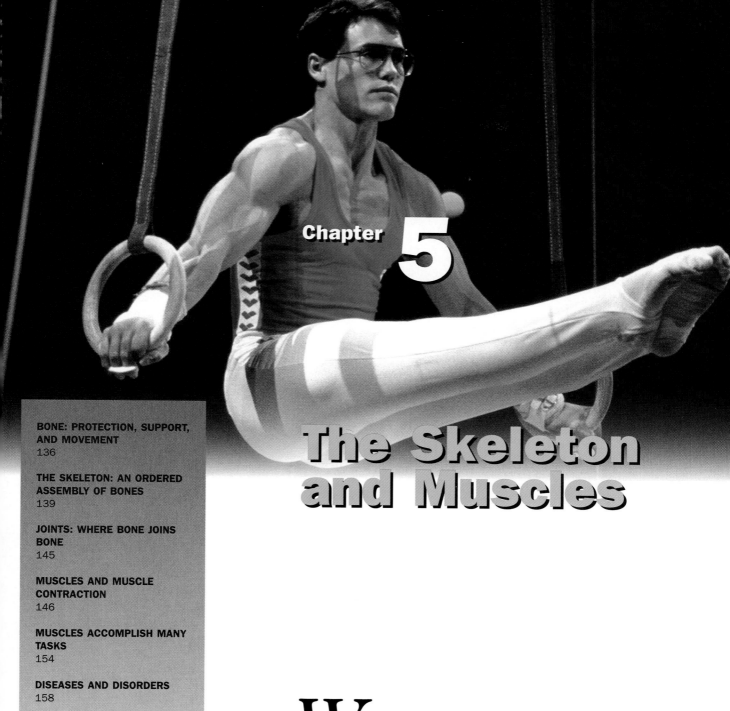

The Skeleton and Muscles

We jump, we turn, we dance. Nothing is more characteristic of life than movement, and humans are capable of an astonishing diversity of precise movements from doing a high jump to threading a needle. Two organ systems are directly involved in accomplishing these movements: the bones of the skeleton and the muscles that are attached to them.

The adult human skeleton consists of 206 bones linked in a specific order. As we will see, many of those bones are joined in ways that permit movement. Actual movement of the bones is accomplished by means of the contraction and relaxation of the muscles attached to them. Muscles also operate in less visible ways to stabilize and support the skeleton. A gymnast, poised and motionless before leaping into action, is using muscles to

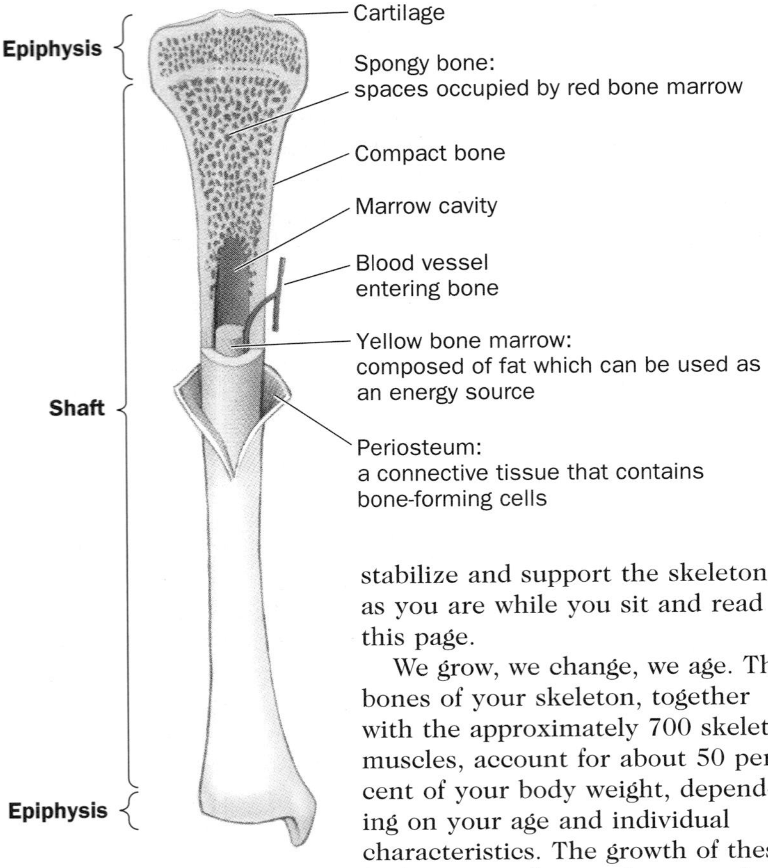
Epiphysis
Shaft
Epiphysis
Cartilage
Spongy bone:
spaces occupied by red bone marrow
Compact bone
Marrow cavity
Blood vessel
entering bone
Yellow bone marrow:
composed of fat which can be used as an energy source
Periosteum:
a connective tissue that contains bone-forming cells

FIGURE 5.1 A partially cut tibia showing the general features of long bones.

stabilize and support the skeleton, as you are while you sit and read this page.

We grow, we change, we age. The bones of your skeleton, together with the approximately 700 skeletal muscles, account for about 50 percent of your body weight, depending on your age and individual characteristics. The growth of these body systems begins during early development and continues until about age 20. However, changes in the bones and muscles do not end then. Like all other organ systems, bones and muscles undergo alterations during the entire life span, from infancy to old age. Even when bones break, a programmed series of changes permits them to heal.

We will begin this chapter with an examination of the structure and function of bone. Then we will see how bone is fashioned into the skeleton and how the skeleton is held together and operated by the muscles. At the close of this chapter we will look more closely at certain diseases and disorders of these systems and discuss how these systems change over the life span.

Bone: Protection, Support, and Movement

Bone is a living connective tissue composed of a hard matrix material and bone cells (Chapter 3). In a museum or teaching laboratory, however, it is easy to misinterpret bone as being a rocklike nonliving material. In preserved specimens, you see only the solid nonliving matrix, which is composed mostly of calcium phosphate. In such specimens, the cells have been removed during preparation.

The hardness and strength of this nonliving matrix make it easy to appreciate how bone can support the body, anchor the muscles for movement, and protect internal organs. Less obvious, however, are the other functions associated with living bone cells. These cells give bone the capacity to grow, change shape, and repair itself.

As bones grow in children and teenagers, they are remodeled continuously to maintain correct proportions. Even adult bones undergo some remodeling in response to changes in weight and activity. This remodeling requires the coordinated action of several different types of bone cells and is supported by the integrated functioning of other body systems. In this section, we will examine the structure and organization of bone, study the way bone grows in length, and discuss the way it changes through remodeling.

Bone Tissue: Structures Visible to the Unaided Eye

To better understand the general anatomy of bone, let us examine a long bone from the lower leg, the tibia. This bone consists of a cylindrical *shaft* with an internal cavity and enlarged knoblike ends each of which is called an *epiphysis* (*ee-**piff**-uh-sis*). In Figure 5.1 a partially cut tibia shows the features associated with most long bones.

The shaft is composed of a rather dense type of bone called **compact bone.** Compact bone also covers each end of a long bone. Although it appears to be rock-solid to the naked eye, compact bone actually has many microscopic canals running through it. The shaft of a living long bone contains a cylindrical cavity filled with *yellow bone marrow,* which is stored fat that can be used as an energy source.

Internally, each epiphysis is composed of **spongy bone,** which, as its name implies, is less dense than compact bone.

Spongy bone consists of tiny bars and plates that form an irregular gridlike pattern. In living bone, the spaces between the bars and plates contain *red bone marrow,* which produces most of the red and white blood cells.

Where the end of a long bone forms a movable joint with another bone, its surface is covered with *cartilage,* a tough, smooth connective tissue that reduces friction between moving surfaces. Covering the outer surface of the shaft is the *periosteum* (*pair-ee-**oss**-tea-um*), a thin, tough covering of connective tissue whose inner layer contains many specialized bone-forming cells. The periosteum is well supplied with nerves and blood vessels, some of which pass into the compact bone.

> Long bones consist of a shaft and an epiphysis at each end. The shaft is made of dense compact bone and contains yellow marrow. The interior of each epiphysis is composed of spongy bone and contains red bone marrow.

Bone Tissue: Structures Observed with the Microscope

With the microscope, we can observe the cellular components of bone and focus on the less obvious functions associated with those cells. We find that compact bone and spongy bone differ in terms of microscopic organization.

In compact bone, the numerous bone cells, or **osteocytes** (***oss**-tea-oh-sites*), are embedded in a solid nonliving matrix of calcium phosphate that is deposited on collagen fibers. Within this matrix individual bone cells appear to be isolated from one another, but they are in fact connected by slender cellular extensions that pass through tiny channels in the hardened matrix (Figure 5.2).

To sustain life and perform their specialized functions, osteocytes must have access to blood vessels to obtain nutrients and excrete wastes. To provide this access, osteocytes are organized into circular patterns around microscopic canals called Haversian canals. Each **Haversian canal** (*Ha-**ver**-shun*) contains blood vessels and a nerve (see Figure 5.2). The osteocytes closest to the blood vessels obtain nutrients by diffusion and pass them to more distant osteocytes by means of their slender cellular extensions. In a similar fashion, waste products produced in distant osteocytes diffuse through a series of cells until they reach the blood vessels in a Haversian canal.

Spongy bone also contains osteocytes, but they are not organized around Haversian canals. The bony bars and plates of spongy bone are no wider than a

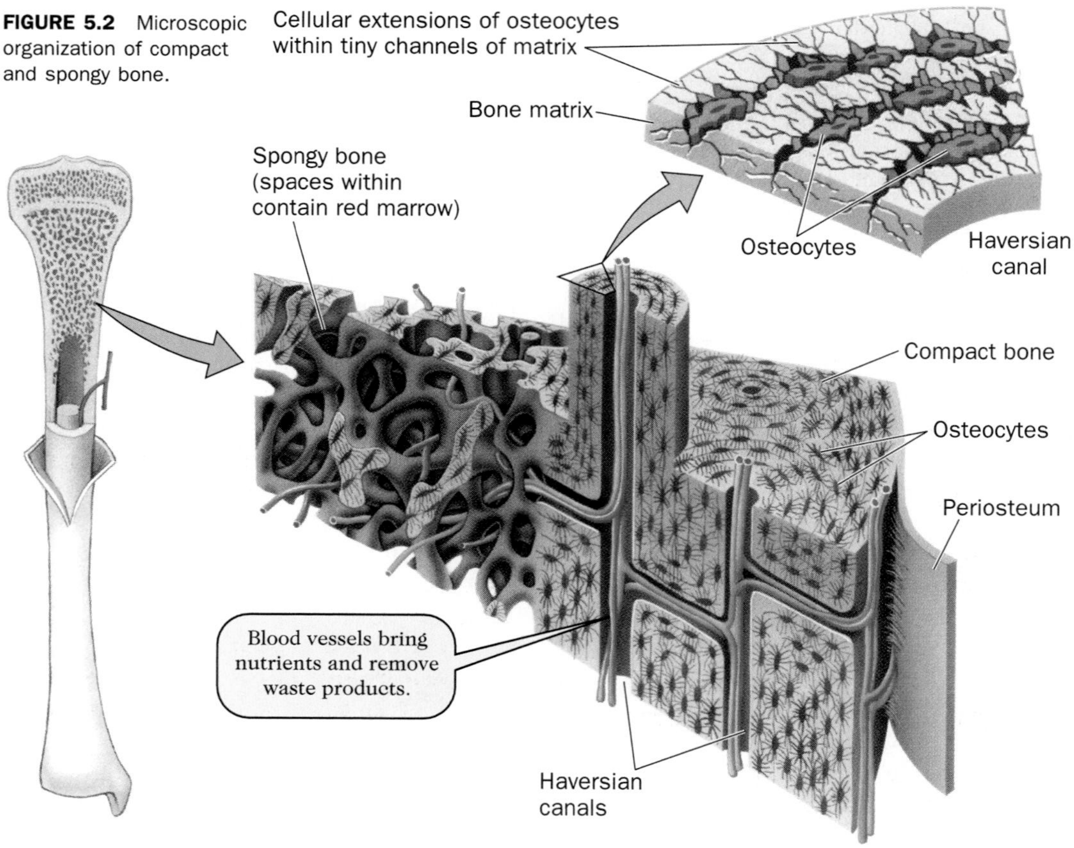

Cellular extensions of osteocytes within tiny channels of matrix
Bone matrix
Spongy bone (spaces within contain red marrow)
Osteocytes
Haversian canal
Compact bone
Osteocytes
Periosteum
Blood vessels bring nutrients and remove waste products.
Haversian canals

FIGURE 5.2 Microscopic organization of compact and spongy bone.

few cells; therefore, each osteocyte has intimate access to the red marrow for the exchange of nutrients and wastes (see Figure 5.2).

While osteocytes maintain bone, two other types of cells form new bone and remove bone. Bone-forming cells are called **osteoblasts,** and they are located within the periosteum. Osteoblasts remove soluble calcium (Ca^{2+}) and phosphate (PO_4^{3-}) ions from the blood and deposit them as calcium phosphate on thin collagen fibers of the bone matrix. During bone development, osteoblasts become entrapped in the matrix as it forms. Once they have been entrapped, they are transformed into osteocytes.

Bone-removing cells are called **osteoclasts.** Osteoclasts are located on the surface of bone and remove bone by secreting enzymes and acids that dissolve the solid calcium phosphate. As a result, Ca^{2+} and PO_4^{3-} are released into the blood when insufficient dietary Ca^{2+} is consumed.

> The osteocytes of compact bone are organized around Haversian canals that contain blood vessels. Spongy bone does not have Haversian canals. Osteocytes maintain bone, osteoblasts build bone, and osteoclasts remove bone by dissolving the solid calcium phosphate.

Bone Development, Growth, and Remodeling

Most skeletal bones begin as a cartilage model that consists of hyaline cartilage. This model does not change into bone. Instead, it is gradually dissolved and replaced by bone. This process is called **ossification** (Figure 5.3).

Ossification begins in the second- or third-month old fetus. Initially, a bone such as the femur is defined only by its cartilage model. The cartilage cells undergo a programmed death, and as they degenerate, they leave many tiny spaces. Blood vessels grow into those spaces, car-

FIGURE 5.3 Events associated with bone development and growth.

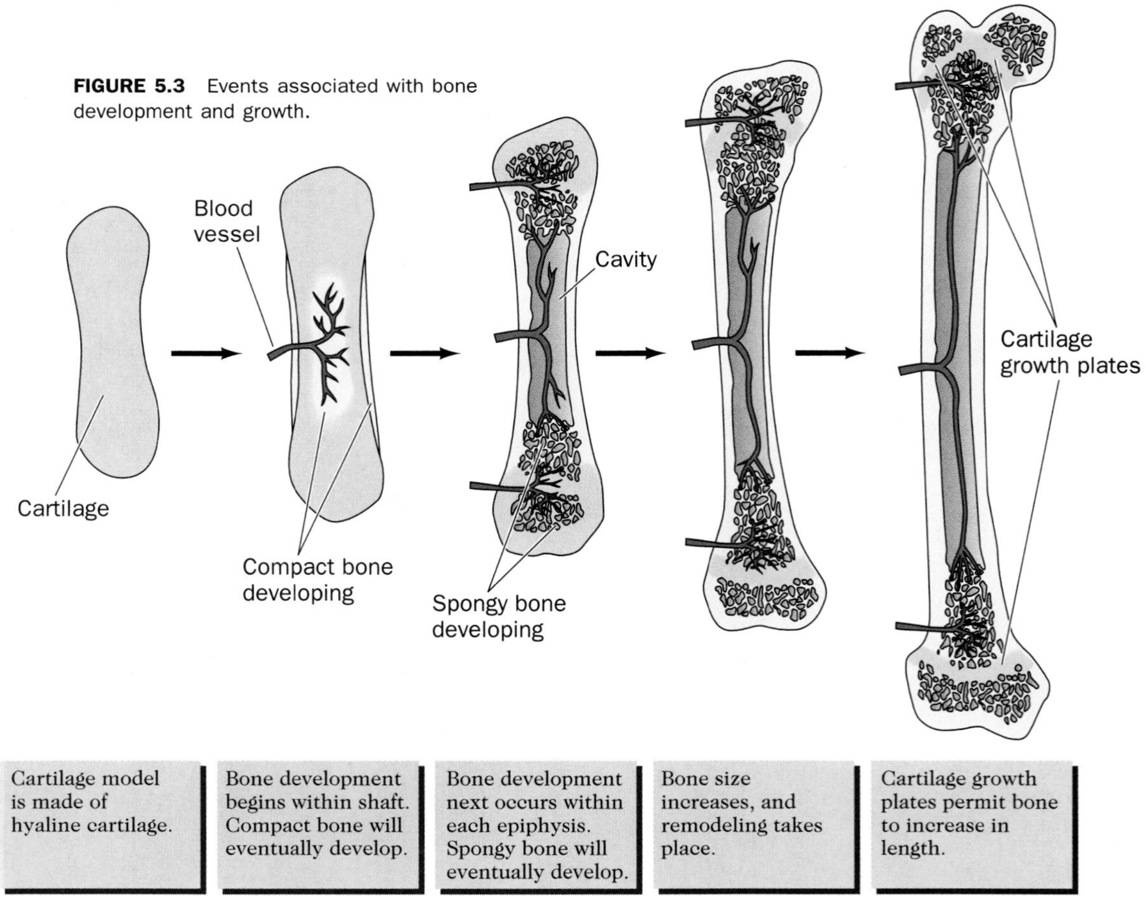
Blood vessel
Cavity
Cartilage growth plates
Cartilage
Compact bone developing
Spongy bone developing
Cartilage model is made of hyaline cartilage.
Bone development begins within shaft. Compact bone will eventually develop.
Bone development next occurs within each epiphysis. Spongy bone will eventually develop.
Bone size increases, and remodeling takes place.
Cartilage growth plates permit bone to increase in length.

rying with them osteoblasts that take up residence and begin the deposition of solid matrix.

Before birth, ossification occurs primarily within the bone shaft. However, at around the time of birth, ossification begins within the cartilage of each epiphysis. As ossification proceeds in each epiphysis, there remains a narrow strip of cartilage known as the **growth plate.** This plate is responsible for the increase in the length of long bones that continues for the first two decades of life.

The growth plate accomplishes bone elongation as cartilage cells on one side of the growth plate undergo continued cell division (Figure 5.4). As this occurs, the cartilage tissue on the other side of the plate undergoes ossification. As a result of both processes, bone length increases. The entire process is regulated by hormones from the endocrine system. By the time you reach your late teens, these hormones cause the cartilage growth plate to cease functioning and be replaced by bone. Once this occurs, your long bones do not grow in length.

Although adult bones do not continue to grow in length, they can be repaired and remodeled throughout adult life. For example, in response to mechanical forces such as running and weight lifting, bones increase their density and thickness. This is accomplished by the beautifully orchestrated interplay between osteoblasts and osteoclasts. Bone may be removed from one area by osteoclasts and then deposited in another area by osteoblasts. The balance of these activities in a healthy adult is a major homeostatic accomplishment that is necessary for survival. See Spotlight on Health: Two (page 163) for more on the interactions of osteoblasts and osteocytes.

Most bones develop in areas previously occupied by hyaline cartilage models. The cartilage cells die and are replaced by bone cells that deposit the bone matrix in a process called ossification. Ossification in long bones continues at the growth plate until a person's late teens.

The Skeleton: An Ordered Assembly of Bones

When you examine the human skeleton in Figure 5.5, you will see that bones vary in shape and size. It is helpful to distinguish four categories of bone shape: long bones, short bones, flat bones, and irregular bones.

Long bones are bones that are longer than they are wide; examples include the bones of the limbs, such as the femur and the humerus, and the bones of the fingers. *Short bones* are approximately as wide as they are long; examples include the small bones of the wrist. *Flat bones,* not surprisingly, tend to be broad, thin, and flattened; examples include the skull bones, sternum, and scapula. *Irregular bones* are bones with rather complex shapes which do not easily fit into any of the other categories; exam-

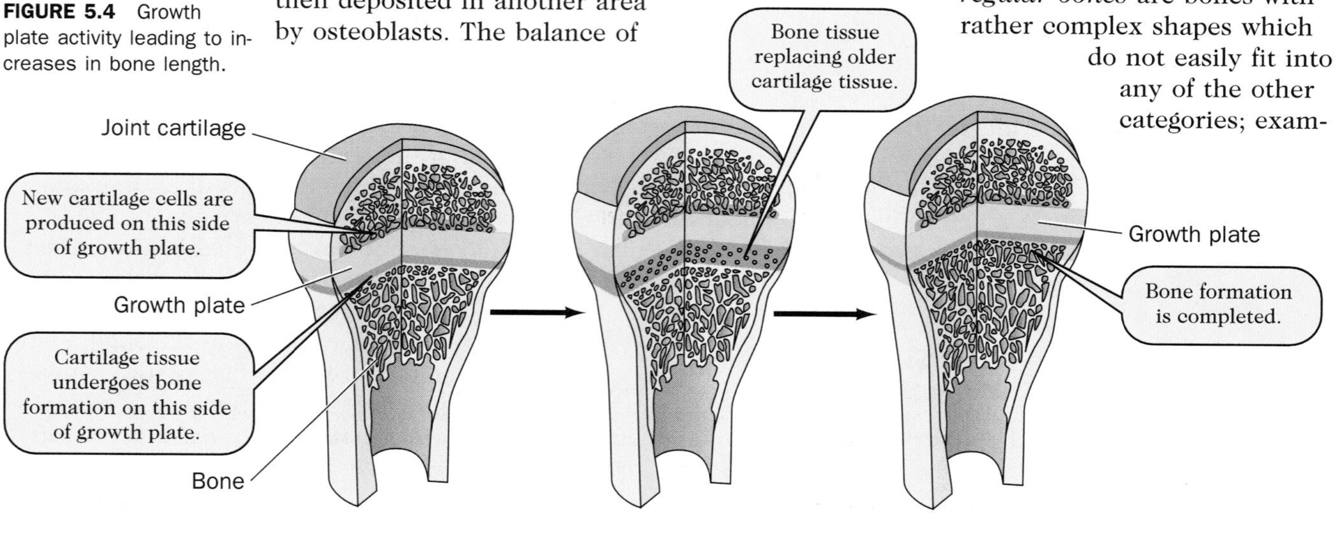
Bone tissue replacing older cartilage tissue.
Joint cartilage
New cartilage cells are produced on this side of growth plate.
Growth plate
Cartilage tissue undergoes bone formation on this side of growth plate.
Bone
Growth plate
Bone formation is completed.

FIGURE 5.4 Growth plate activity leading to increases in bone length.

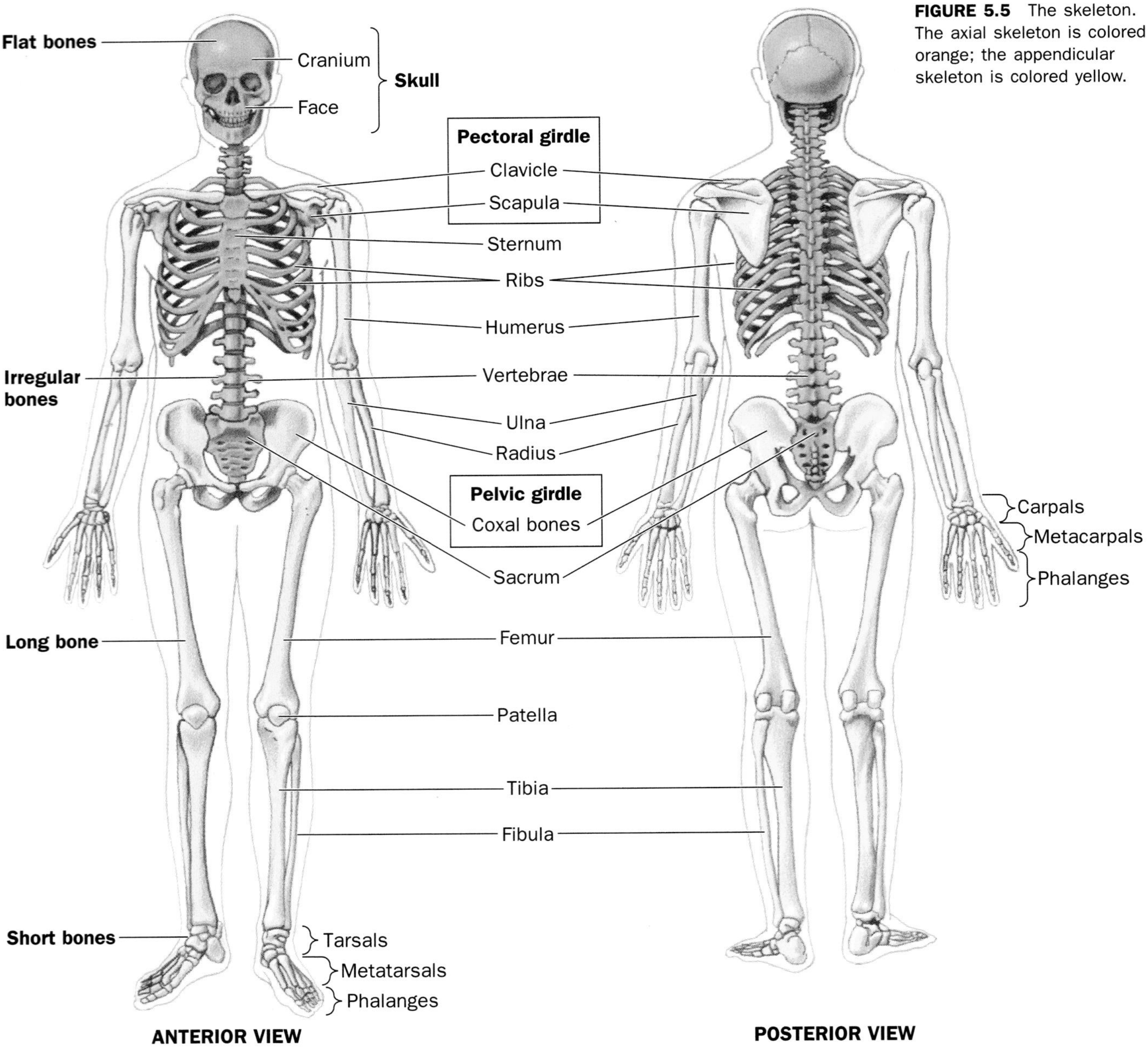

Flat bones
Cranium
Face
Skull
Pectoral girdle
Clavicle
Scapula
Sternum
Ribs
Humerus
Irregular bones
Vertebrae
Ulna
Radius
Pelvic girdle
Coxal bones
Sacrum
Carpals
Metacarpals
Phalanges
Long bone
Femur
Patella
Tibia
Fibula
Short bones
Tarsals
Metatarsals
Phalanges
ANTERIOR VIEW
POSTERIOR VIEW

FIGURE 5.5 The skeleton. The axial skeleton is colored orange; the appendicular skeleton is colored yellow.

ples include the vertebrae and several facial bones.

Let us begin our study of the skeleton by distinguishing its two major divisions: the axial skeleton and the appendicular skeleton.

The Axial Skeleton: Focus on the Skull

The **axial skeleton** includes the bones that form the "axis," or midline, of the body: the skull, vertebral column, and rib cage (see Figure 5.5). We will describe each of these structures, beginning with the skull, which consists of both cranial bones and facial bones.

Cranial bones. Cranial bones are flat bones that surround and protect the brain (Figure 5.6). The *frontal bone* forms the entire front portion of the skull, including the forehead. Just posterior to this bone are the two *parietal bones* (*pa-**rye**-eh-tal*) that constitute the upper left and right sides of the skull. Below the parietals are the two *temporal bones* (***temp***-*ore-al*) that form the lower left and right sides of the skull. Each temporal bone has an external opening leading into the ear canal through which sound waves pass on their way to the eardrum.

Attached to the frontal, parietal, and temporal bones is the *sphenoid bone* (***sfee***-*noid*). This sphenoid is centrally located within the skull and helps connect many of the cranial and facial bones (see Figure 5.6). The *occipital bone* (*ox-**sip**-ih-tal*) forms the back and base of the skull, and its lower portion has a large opening, the *foramen magnum* (*foe-**ray**-men*), which in Latin means "large opening." Through this opening the spinal cord enters the skull and connects with the brain. On either side of the foramen magnum are two small bony projections that rest on the first vertebra and support the entire skull while permitting a nodding motion of the head.

FIGURE 5.6 The skull.

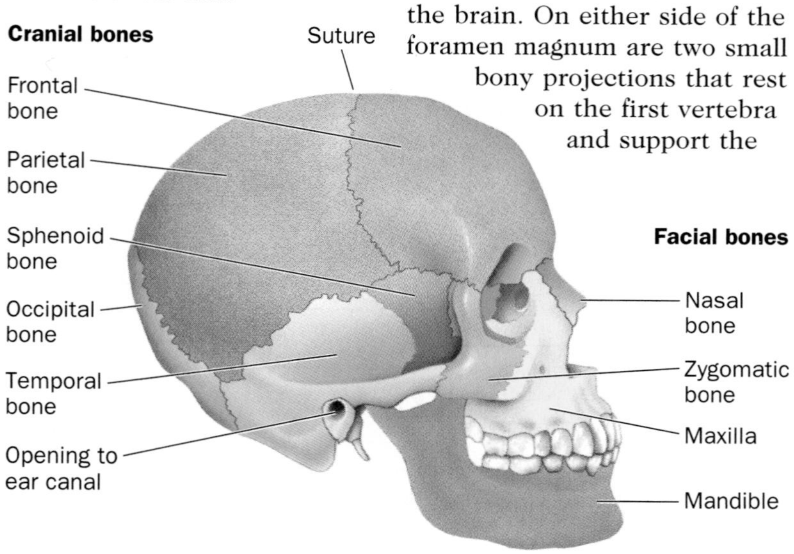
Cranial bones
Suture
Frontal bone
Parietal bone
Sphenoid bone
Occipital bone
Temporal bone
Opening to ear canal
Facial bones
Nasal bone
Zygomatic bone
Maxilla
Mandible

LATERAL VIEW

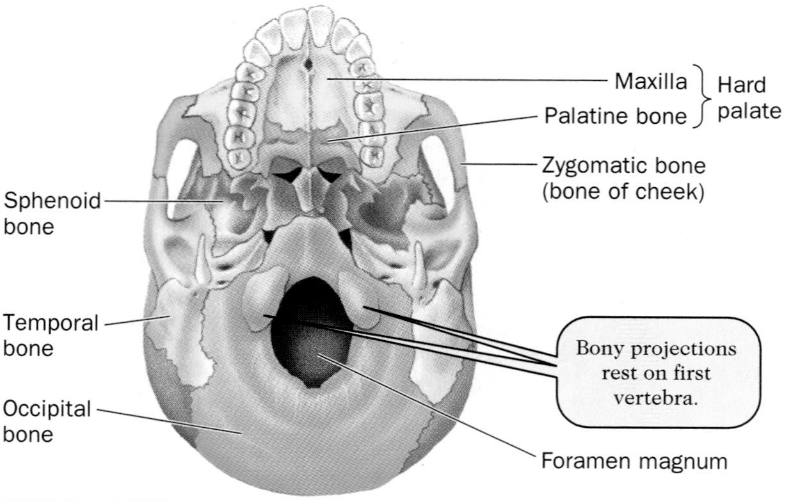
Maxilla
Palatine bone
Hard palate
Zygomatic bone (bone of cheek)
Sphenoid bone
Temporal bone
Occipital bone
Bony projections rest on first vertebra.
Foramen magnum

INFERIOR VIEW

Facial bones. The facial bones make up the front of the skull (see Figure 5.6). The two *maxilla bones* (***macks***-*illa*) are situated on either side of the nose and help form the orbits of the eyes and the upper part of the mouth (the hard palate). They also contain the sockets that secure the upper teeth. The hard palate is also formed by the two *palatine bones.* Between the maxillary bones, at the level of the eyes, are the two short, narrow *nasal bones.* These bones form only the upper bridge of the nose and are much smaller than the nose that is visible on the face. What we call the nose is mostly cartilage and is not considered part of the skeletal system. The maxilla bones and the nasal bones partially define a space called the *nasal cavity.*

One of the defining features of the face is provided by the two *zygomatic bones* (*zye-go-**mat**-ic*). These bones join with narrow projections from the two temporal bones, and together they provide the bony structure that makes up the cheeks. Another defining feature of the face is provided by the lower jaw, the **mandible.** This is a single U-shaped bone that is hinged with the temporal bone of the cranium and contains sockets for the lower teeth.

Sinuses are spaces. Several of the cranial and facial bones have air spaces within them (Figure 5.7). These spaces, which are called **sinuses,** lighten the skull and help give the voice its distinctive resonant qualities. Each sinus is lined with a mucus-secreting epithelium and is connected to the nasal cavity by small tubes through which mucus normally drains. During a sinus infection, the inflamed epithelium swells, preventing the tubes from draining properly. A sinus headache results from pressure caused by fluid accumulation in the sinuses.

> The axial skeleton consists of the skull, vertebrae, and rib cage. The skull includes the cranial and facial bones. Some of these bones have sinuses.

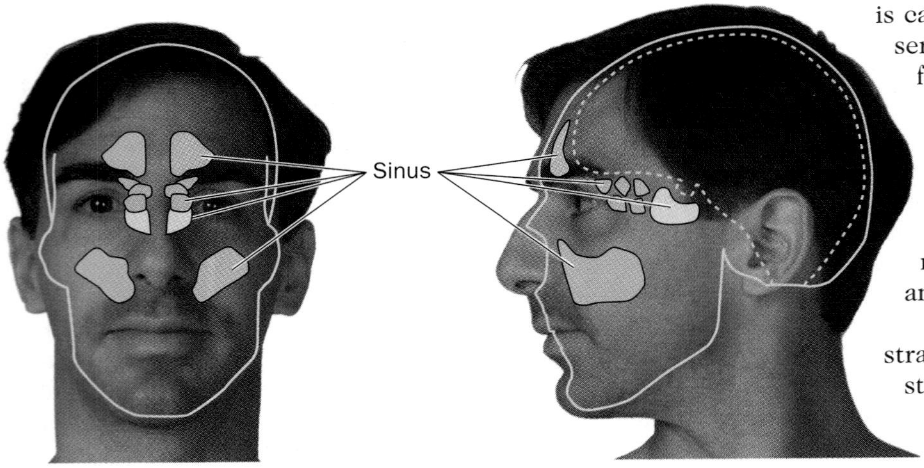
Sinus

FIGURE 5.7 The location of sinuses within the skull.

The Axial Skeleton: Focus on the Vertebrae and Rib Cage

The **vertebral column** is also known as the backbone or spine. In an adult, it consists of 26 irregular bones called **vertebrae** (***ver**-teh-bray*); a single spinal bone is called a vertebra (***ver**-teh-bra*). This series of interacting bones extends from the skull to the pelvis (Figure 5.8). Together these bones have three major functions. They (1) provide flexible support for the head and trunk, (2) protect the delicate spinal cord, and (3) permit the passage of spinal nerves to and from the spinal cord.

The vertebral column is not straight like a series of spools on a string but instead is curved. The curvatures seen in Figure 5.8 are due to the different structure of each vertebra and the support given by the attached ligaments and muscles. Together these structures give the entire column strength and flexibility. This is an example of the integrated functioning of the skeletal and muscular systems.

Structurally, the vertebral column can be divided into five regions (see Figure

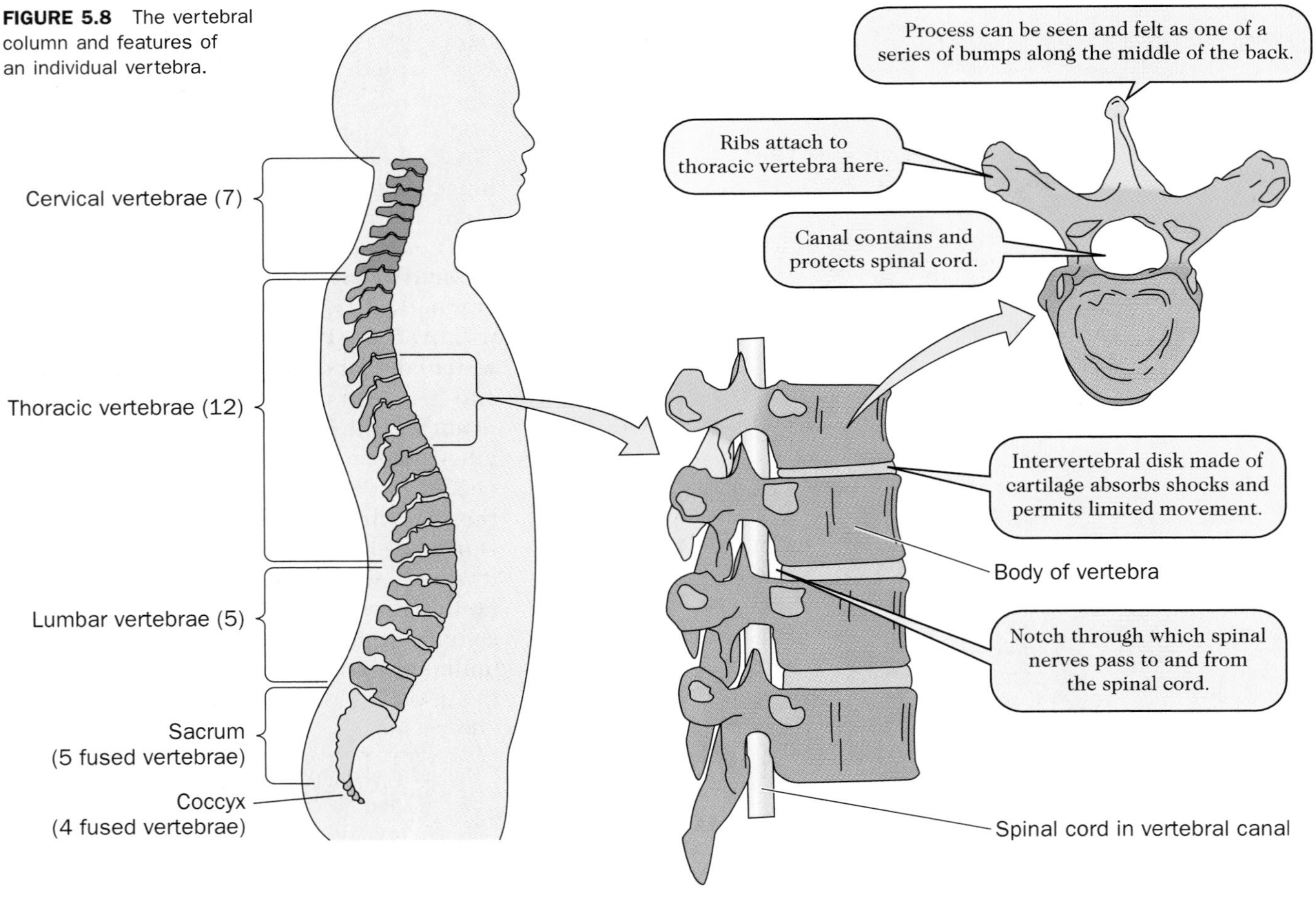
Cervical vertebrae (7)
Thoracic vertebrae (12)
Lumbar vertebrae (5)
Sacrum (5 fused vertebrae)
Coccyx (4 fused vertebrae)
Process can be seen and felt as one of a series of bumps along the middle of the back.
Ribs attach to thoracic vertebra here.
Canal contains and protects spinal cord.
Intervertebral disk made of cartilage absorbs shocks and permits limited movement.
Body of vertebra
Notch through which spinal nerves pass to and from the spinal cord.
Spinal cord in vertebral canal

FIGURE 5.8 The vertebral column and features of an individual vertebra.

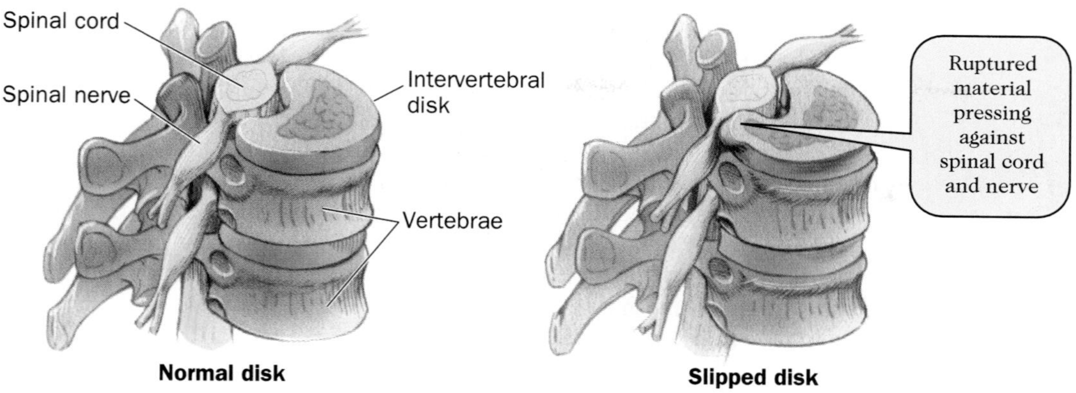
Spinal cord
Spinal nerve
Intervertebral disk
Vertebrae
Normal disk
Ruptured material pressing against spinal cord and nerve
Slipped disk

FIGURE 5.9 A slipped disk pressing against the spinal cord and spinal nerve.

5.8): the *cervical* (***serve**-ick-ul*), *thoracic* (*tho-**rass**-ick*), *lumbar, sacral* (***say**-kral*), and *coccygeal* (*cock-**sij**-ee-al*) regions. These terms can also refer to specific vertebrae and to the curvatures associated with each region; thus, one can speak of the cervical vertebrae and the cervical curvature.

Although the vertebrae from different regions of the column differ from one another, they all share two features: (1) A bony canal for the spinal cord (see Figure 5.8), and (2) a notch on each side that forms an opening which permits nerves to pass through the vertebral column to and from the spinal cord (see Figure 5.8).

Adjacent vertebrae are separated from each other by a pad of cartilage known as the **intervertebral disk** (*inter-**ver**-teh-bral*). These disks act as cushions, protecting the vertebrae from the shocks associated with activities such as walking, running, and jumping. They also provide a certain degree of flexibility, which we experience when we swing a golf club or touch our toes.

FIGURE 5.10 The rib cage.

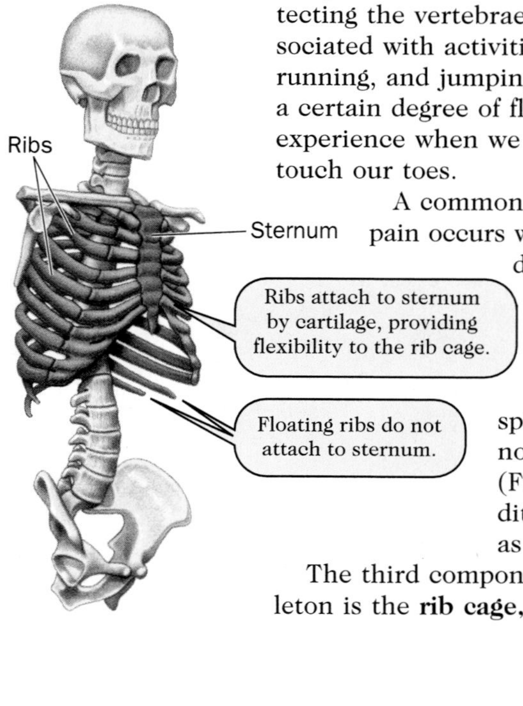
Ribs
Sternum
Ribs attach to sternum by cartilage, providing flexibility to the rib cage.
Floating ribs do not attach to sternum.

A common cause of severe back pain occurs when an intervertebral disk ruptures and presses against the spinal cord or against nerves that enter and leave the spinal cord through the notch in the vertebra (Figure 5.9). This condition is referred to as a **slipped disk.**

The third component of the axial skeleton is the **rib cage,** which in both men and women consists of 12 paired ribs that form a protective cage for some internal organs. The interior of this cage is known as the *thoracic cavity,* and it houses the heart and lungs. One end of each pair of ribs joins with one of the 12 thoracic vertebrae (Figure 5.10). The other ends of the upper 10 pairs of ribs are attached to the **sternum** (***stir**-num*), or *breastbone,* by cartilage. As Figure 5.10 shows, the lower two pairs of ribs do not attach to the sternum and are called *floating ribs.*

The vertebral column and rib cage are parts of the axial skeleton. The vertebral column consists of 26 vertebrae divided into five groups. Each vertebra is separated by intervertebral disks made of cartilage. The ribs and the sternum form the thoracic cavity.

The Appendicular Skeleton: The Limbs and Their Support

The **appendicular skeleton** is the second division of the human skeleton. It includes the two supportive frames called the pectoral and pelvic girdles and their attached limbs, the arms and legs. Each girdle is a supportive bony structure that is secured to the axial skeleton by ligaments and muscles.

Pectoral girdle. The pectoral girdle forms the bony portion of each shoulder and consists of right and left **scapulas** (***scap**-you-lahs*) and **clavicles** (***clav**-ick-uls*) (Figure 5.11). You know the scapula as the shoulder blade. It is a triangular flat bone that forms a movable joint with the upper arm bone. The clavicle is also known as the collarbone. It is a curved bone that joins with the scapula and acts as a brace helping to hold the scapula in position.

Review Figures 5.5 and 5.11 carefully and you will see that the scapula and

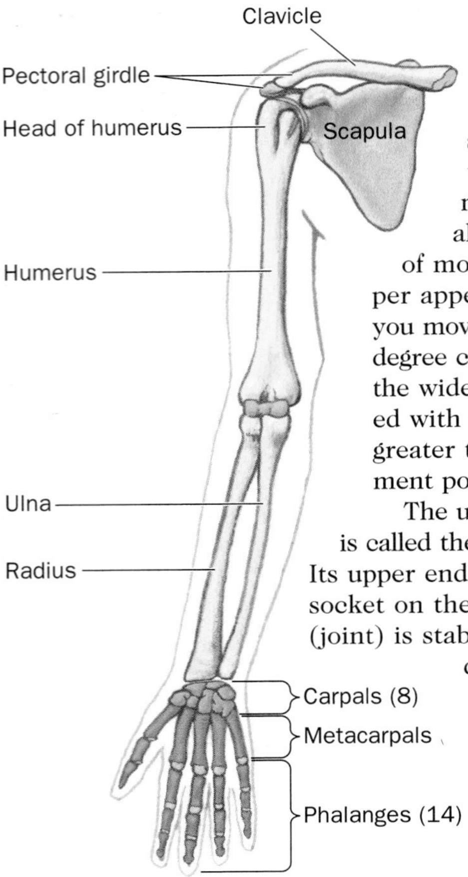

Clavicle
Pectoral girdle
Head of humerus
Scapula
Humerus
Ulna
Radius
Carpals (8)
Metacarpals
Phalanges (14)

FIGURE 5.11 Anterior view of the right pectoral girdle and arm and hand bones.

clavicle are not well secured to each other or to the rib cage. The pectoral girdle is actually stabilized and secured to the rib cage by a group of muscles. This arrangement allows a considerable range of movement in the entire upper appendicular skeleton. When you move your upper arm in a 360-degree circle, you can appreciate the wide range of motion associated with the pectoral girdle. This is greater than the range of movement possible with any other joint.

The uppermost bone of the arm is called the **humerus** (***hume***-*er-us*). Its upper end fits into a saucer-shaped socket on the scapula. This attachment (joint) is stabilized by muscles and tendons and permits a wide range of movement. The lower end of the humerus joins with two bones of the lower arm, the **ulna** and **radius,** forming what is called the elbow. When you accidentally bump your elbow at the "crazy bone," you may feel a tingling sensation. The tingling sensation results from hitting the ulnar nerve, which passes along the back of the elbow.

The lower end of the ulna and radius is attached to a group of bones called the *carpal bones* (***car***-*pul*) that form the wrist. The palm of the hand contains the *metacarpal bones.* The fingers, or *phalanges* (*fuh*-**lan**-*gees*), consist of bones that join with the metacarpal bones.

Pelvic girdle. The pelvic girdle consists of two large bones called the **coxal bones,** commonly known as the hipbones (Figure 5.12). Linked together by cartilage in the front (the pubic symphysis), the two hipbones also join with the vertebral column in back, forming a ringlike structure called the *pelvis.* The pelvises of adult males and females differ somewhat in shape and size; this accounts for the visible differences in the shape and size of the hips in men and women (Figure 5.13).

The female pelvis is wider, shallower, and rounder than the male pelvis and provides a larger opening for the passage of the infant's head during birth. Differences between the male and female pelvises begin to develop during puberty, when the production of sex hormones initiates bone remodeling that specializes the female pelvis for pregnancy and birth.

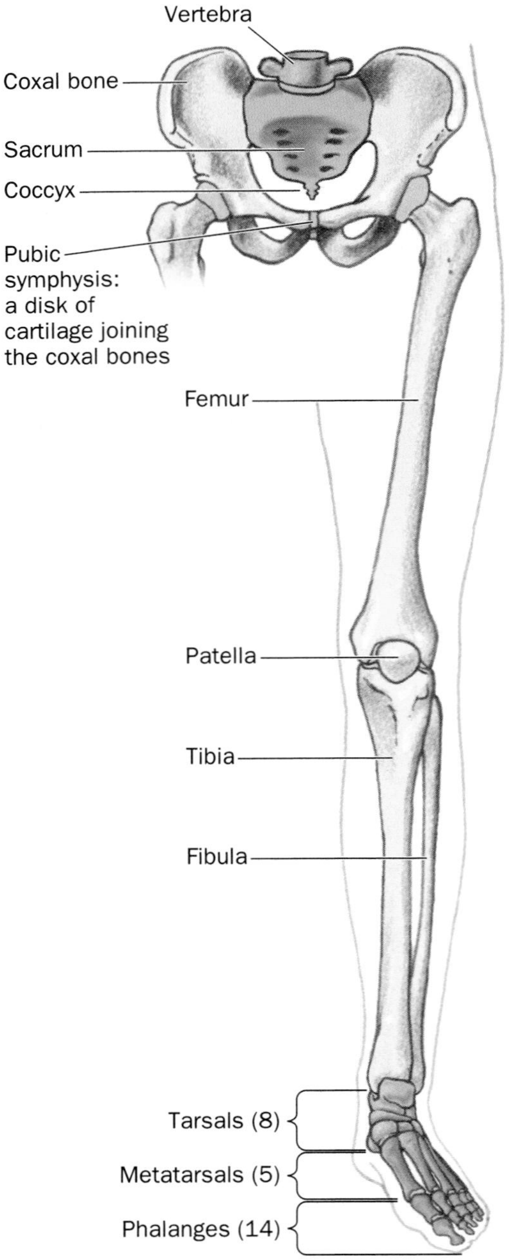

Vertebra
Coxal bone
Sacrum
Coccyx
Pubic symphysis: a disk of cartilage joining the coxal bones
Femur
Patella
Tibia
Fibula
Tarsals (8)
Metatarsals (5)
Phalanges (14)

FIGURE 5.12 Anterior view of the pelvic girdle with left leg and foot bones.

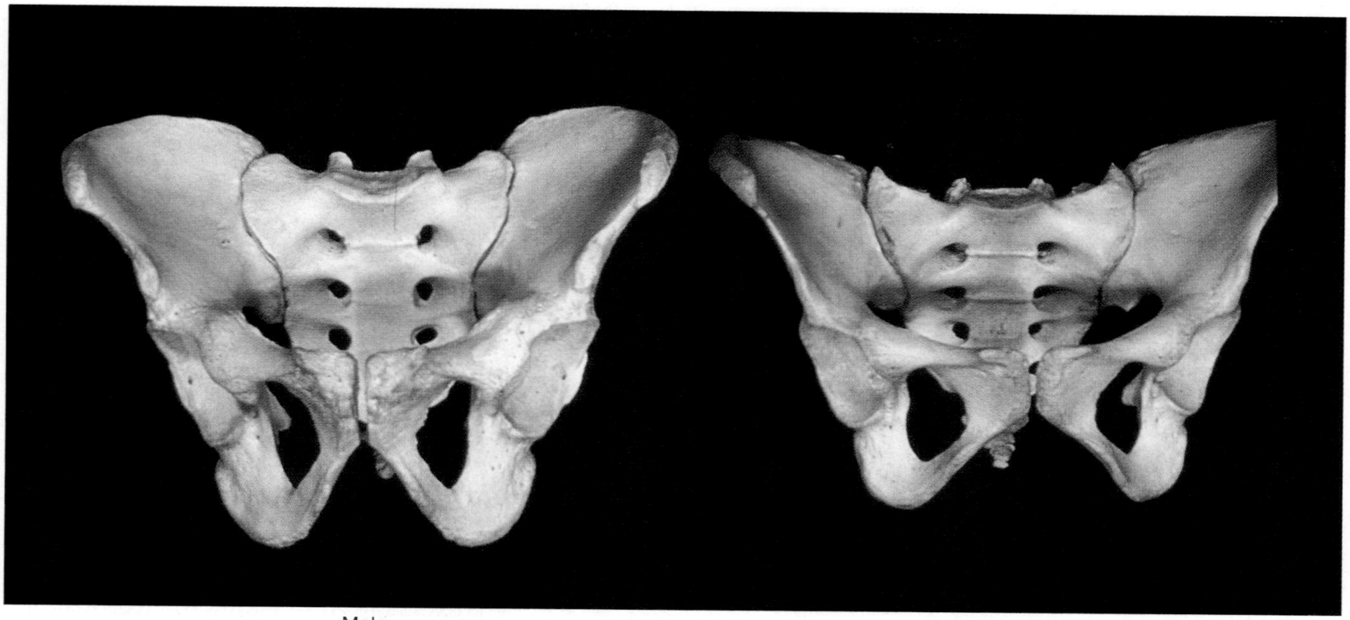
Male
Female

FIGURE 5.13 Photograph comparing male and female pelvises. The lower portion of the female pelvis is wider which allows for the passage of the infant during childbirth.

The upper leg bone is called the **femur** (*fee-mur*) or thighbone (see Figure 5.12). Its upper end is formed into a ball-shaped unit that fits into a well-defined socket of a coxal bone. This ball-and-socket arrangement results in a stable load-bearing joint that is capable of a great range of movement. At the knee, the lower end of the femur meets two bones of the lower leg: the larger **tibia** (***tib**-ee-ah*) and the smaller **fibula** (***fib**-you-lah*). The knee joint forms where the femur joins directly with the tibia. In front of each knee joint is a small triangular bone, the *patella,* or kneecap, which protects the knee joint and acts as a pulley for the upper thigh muscles.

The tibia and fibula join with a number of small *tarsal* (***tar**-sul*) bones to form the ankle. The tarsal and *metatarsal bones,* along with the *phalanges* of the toes, form the foot (see Figure 5.12).

> The appendicular skeleton includes the pectoral and pelvic girdles The pectoral girdle includes the scapulas and clavicles. The pelvic girdle includes the coxal bones. The pectoral and pelvic girdles attach the appendicular skeleton to the axial skeleton.

Joints: Where Bone Joins Bone

Where two or more bones join together, a **joint,** or **articulation,** is formed. All your body movements occur at joints. Many joints are held together and stabilized by bands of fibrous connective tissue called **ligaments** (Chapter 3). Some joints permit little or no movement between adjoining bones (for example, the sutures of the skull), while others allow a wide range of movement (for example, the hip joint). In this section we will briefly describe both types.

Joints Permitting Little or No Movement

The bones of the skull join together in a tight-fitting interlocking joint called a **suture.** This immovable joint provides stability to the skull. In a newborn infant, the bones of the skull are still developing and there are relatively wide spaces between them (Figure 5.14). These spaces contain tough sheets of connective tissue that connect the bones and are called *fontanels* (*fon-ta-***nells**). The "soft spot" on top of an infant's skull is a fontanel. Fontanels allow a certain amount of

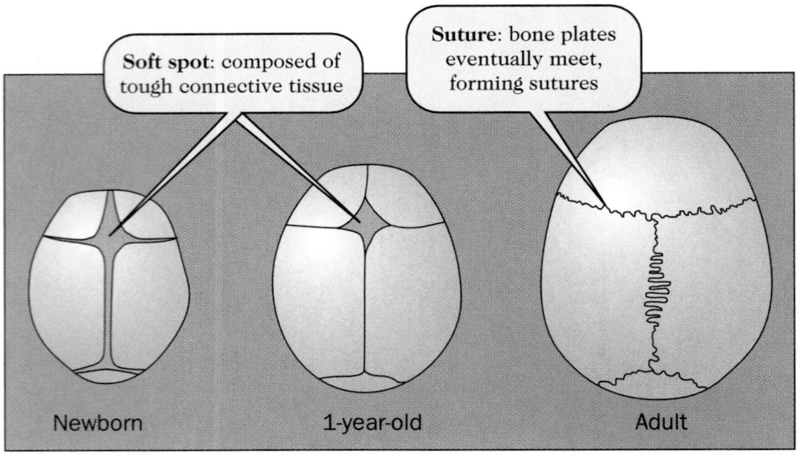
Soft spot: composed of tough connective tissue
Suture: bone plates eventually meet, forming sutures
Newborn
1-year-old
Adult

FIGURE 5.14 Suture development.

"give" in the skull as it passes through the mother's pelvic opening during birth, and they also permit brain growth. After birth, as the infant grows, the connective tissue sheet is replaced by bone, forming the sutures found between adult cranial bones.

Some joints allow limited movement between two adjoining bones. Examples include the cartilage joints between the hipbones (pubic symphysis), those between the sternum and the ribs, and those between the vertebrae (see Figures 5.8, 5.10, and 5.12). Such joints provide firm support but allow a degree of flexibility.

Joints Permitting Movement

The many joints that permit movement between adjoining bones are called **synovial joints** (*sin*-**oh**-*vee-al*), and they are found at the shoulder, elbow, wrist, hip, knee, phalanges, and jaw, to name a few places. Synovial joints have four common characteristics (Figure 5.15): (1) They have a *joint cavity* between the articulating bones. (2) *Cartilage* covers the articulating ends of bones. (3) A fibrous connective tissue forms a *capsule* that encompasses the entire joint. (4) A delicate tissue membrane called the *synovial membrane* lines the inside of this capsule and secretes *synovial fluid.* Synovial fluid coats the inner surfaces, reducing friction. Figure 5.16 shows the variety of body movements permitted by synovial joints.

> Joints are specialized structures that form where bones meet and are secured to each other. Joints may be immovable or movable. Synovial joints permit the most movement.

Muscles and Muscle Contraction

Our muscles are in use during every waking and sleeping moment. Some muscles are under conscious voluntary control, while others operate without conscious control. All muscle tissues have three characteristics in common: (1) Muscles can contract and shorten in length. (2) After contraction, muscles relax and return to their former length. (3) Muscles are excitable, responding to electrical or chemical signals (stimuli) from the nervous and endocrine systems. Muscle is the only tissue in the body that has all these properties.

There are three types of muscle: skeletal muscle, smooth muscle, and cardiac muscle. Each type is specialized in terms of structure, function, and location in the body. In this chapter we will focus primarily on skeletal muscle.

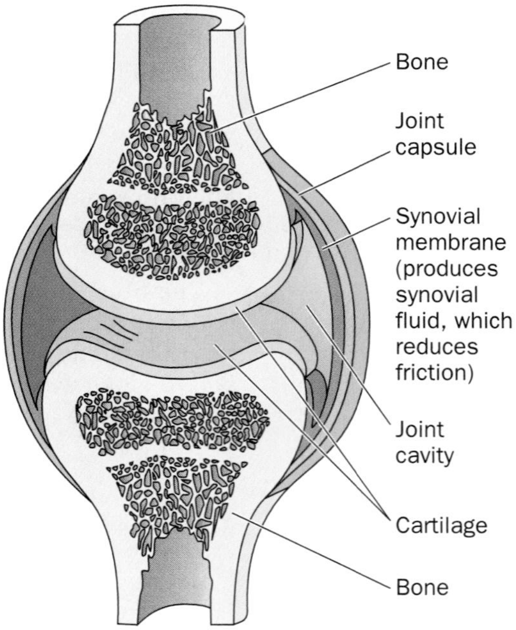
Bone
Joint capsule
Synovial membrane (produces synovial fluid, which reduces friction)
Joint cavity
Cartilage
Bone

FIGURE 5.15 Anatomic structures characteristic of synovial joints.

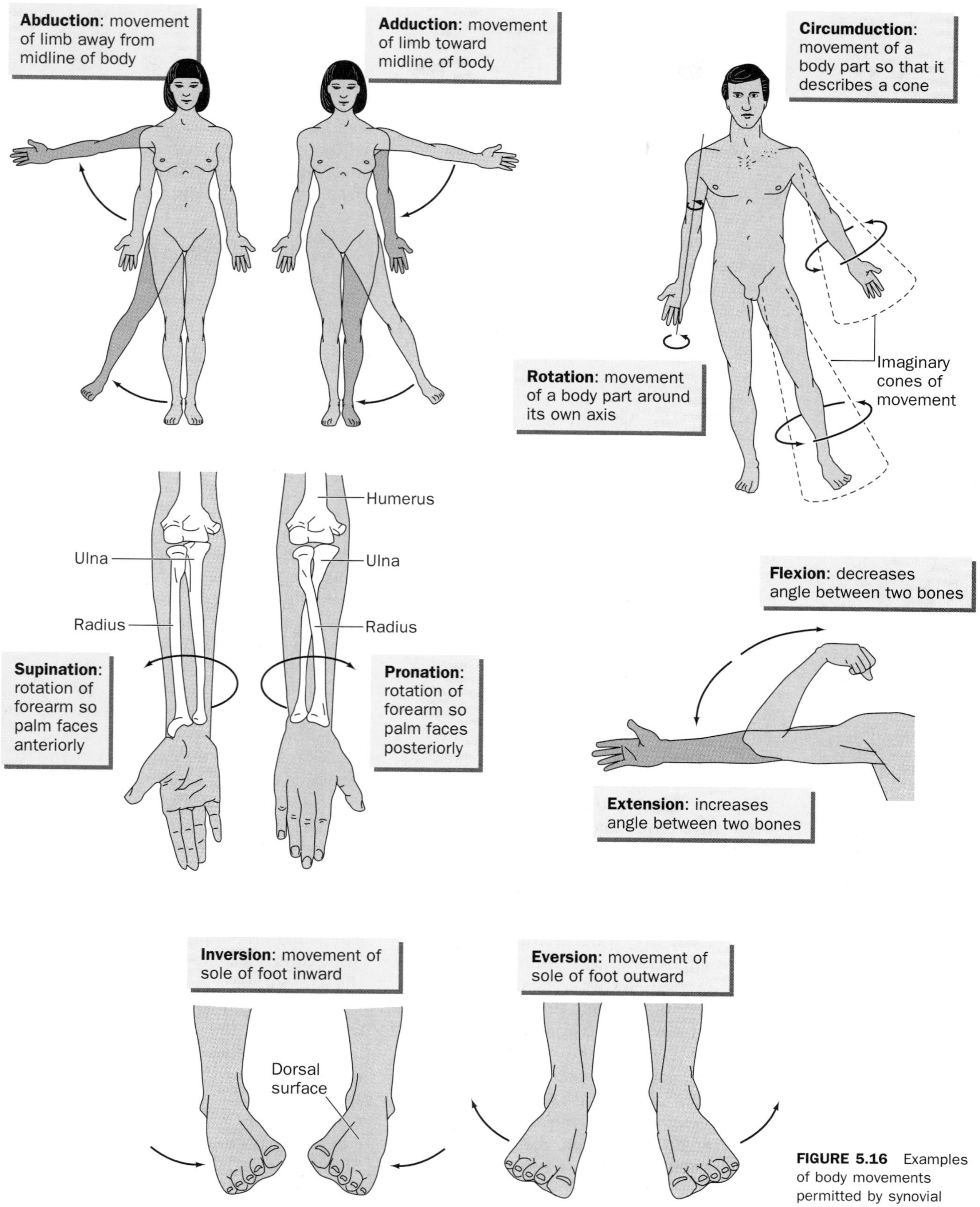
Abduction: movement of limb away from midline of body
Adduction: movement of limb toward midline of body
Circumduction: movement of a body part so that it describes a cone
Rotation: movement of a body part around its own axis
Imaginary cones of movement
Humerus
Ulna
Ulna
Radius
Radius
Supination: rotation of forearm so palm faces anteriorly
Pronation: rotation of forearm so palm faces posteriorly
Flexion: decreases angle between two bones
Extension: increases angle between two bones
Inversion: movement of sole of foot inward
Dorsal surface
Eversion: movement of sole of foot outward

FIGURE 5.16 Examples of body movements permitted by synovial joints.

Skeletal Muscle: Structures Visible to the Unaided Eye

Skeletal muscles attach to bones and move the skeleton. They also attach to other muscles and to the facial skin. The approximately 700 skeletal muscles in the human body differ greatly in size and shape, as shown in Figure 5.17.

Some skeletal muscles are very large, such as the gastrocnemius, the major muscle forming the calf in the lower leg. Others are very small, such as the muscles of the eyelid. Muscle shapes vary. Some muscles are triangular (deltoid), while others are rectangular (rectus abdominis) or trapezoidal (trapezius). Many are spindle-shaped (biceps brachii), narrow at the ends where they attach to bone and larger in the middle or body of the muscle.

Each skeletal muscle is composed of parallel bundles of individual *muscle cells* (Figure 5.18). Each muscle cell is surrounded by delicate connective tissue. Bundles of these cells, called *fascicles* (***fas***-*i-kuls*) are wrapped in another thin layer of connective tissue. Many fascicles make up a single muscle. Each muscle is wrapped by another sheath of connective tissue called a *fascia* (***fash***-*ee-ah*), which supports and holds the entire muscle together. These connective tissue wrappings are actually continuous with each other at the end of the muscle, where they unite to form an extremely tough fibrous connective tissue material called a **tendon.** Tendons attach muscle to bone.

Muscles usually are attached to the skeleton at two locations. One end of the muscle attaches to a stationary bone and is called the **origin** of the muscle. The other end attaches to a bone that moves when the muscle contracts and is called the **insertion.** Figure 5.19 shows the origin and insertion of two muscles in your arm.

Moving your forearm. Consider the roles of the biceps and triceps muscles when your forearm is moved up and down (Figure 5.19). Your forearm is flexed when your *biceps* contracts (shortens). To move the forearm back to its original position, the biceps relaxes, and another muscle—the *triceps*—contracts.

These muscles function as antagonistic pairs and move your forearm in opposite directions. If both muscles contracted at the same time, no movement would take place, and so coordination of their contractions and relaxations is essential. When the biceps contracts, the triceps must relax. Most skeletal muscles are arranged in similar antagonistic pairs.

> A muscle is composed of many muscle cells that are bundled into fascicles. Connective tissues covering individual muscle cells, fascicles, and a muscle are continuous at the ends of the muscle, where together they form a tendon that connects muscle to bone.

Skeletal Muscle: Structures Observed with the Microscope

To understand how a muscle performs its actions, we must look more closely at the structure and operations of an individual muscle cell. What are its internal structures and important proteins? How are they organized to accomplish movement?

Individual skeletal muscle cells have a tubular shape, contain many nuclei per cell, and appear striated or striped. These muscle cells are relatively large compared with other human cells. Their length ranges between 1 mm and 30 cm with a diameter of 10 to 100 μm. Long muscles, like those of the thigh, have longer muscle cells.

Muscle cells contain the same cellular components or organelles found in most other human cells, but some components have specialized functions or are unique. The smooth endoplasmic reticulum forms into parallel saclike compartments called the **sarcoplasmic reticulum** (***sar***-*koh-plas-mick ree-****tick****-you-lum*), or **SR** (Figure 5.20). On the outer surface of the muscle cell's plasma membrane are the openings of tiny tubes called **transverse tubules,** or **T tubules.** These tubules are extensions of the plasma membrane that run deep into the cell and lie close to the SR. They relay the electrical signals that trigger contraction.

Within each muscle cell and running its entire length are numerous protein

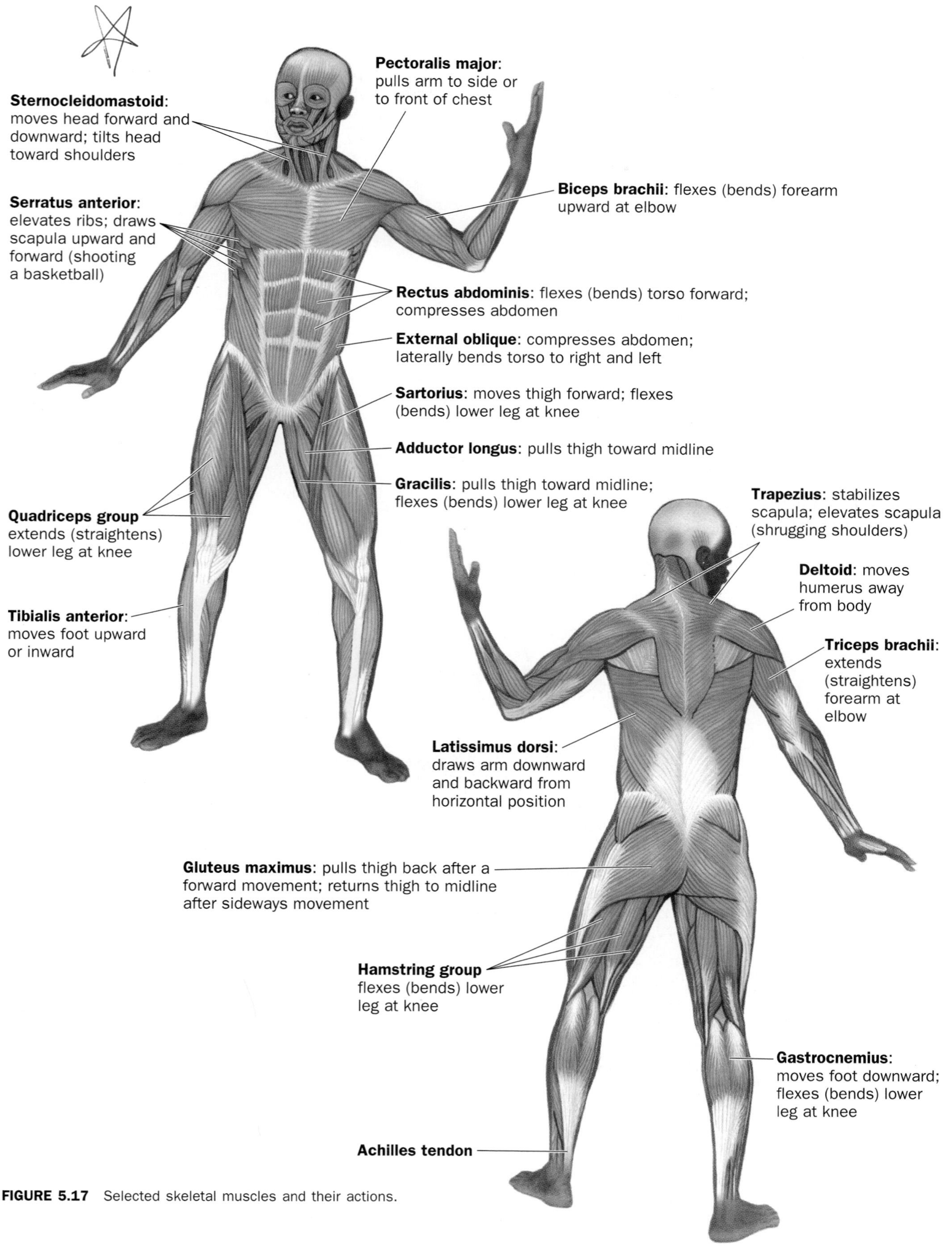
Pectoralis major: pulls arm to side or to front of chest
Sternocleidomastoid: moves head forward and downward; tilts head toward shoulders
Biceps brachii: flexes (bends) forearm upward at elbow
Serratus anterior: elevates ribs; draws scapula upward and forward (shooting a basketball)
Rectus abdominis: flexes (bends) torso forward; compresses abdomen
External oblique: compresses abdomen; laterally bends torso to right and left
Sartorius: moves thigh forward; flexes (bends) lower leg at knee
Adductor longus: pulls thigh toward midline
Gracilis: pulls thigh toward midline; flexes (bends) lower leg at knee
Quadriceps group extends (straightens) lower leg at knee
Trapezius: stabilizes scapula; elevates scapula (shrugging shoulders)
Deltoid: moves humerus away from body
Tibialis anterior: moves foot upward or inward
Triceps brachii: extends (straightens) forearm at elbow
Latissimus dorsi: draws arm downward and backward from horizontal position
Gluteus maximus: pulls thigh back after a forward movement; returns thigh to midline after sideways movement
Hamstring group flexes (bends) lower leg at knee
Gastrocnemius: moves foot downward; flexes (bends) lower leg at knee
Achilles tendon

FIGURE 5.17 Selected skeletal muscles and their actions.

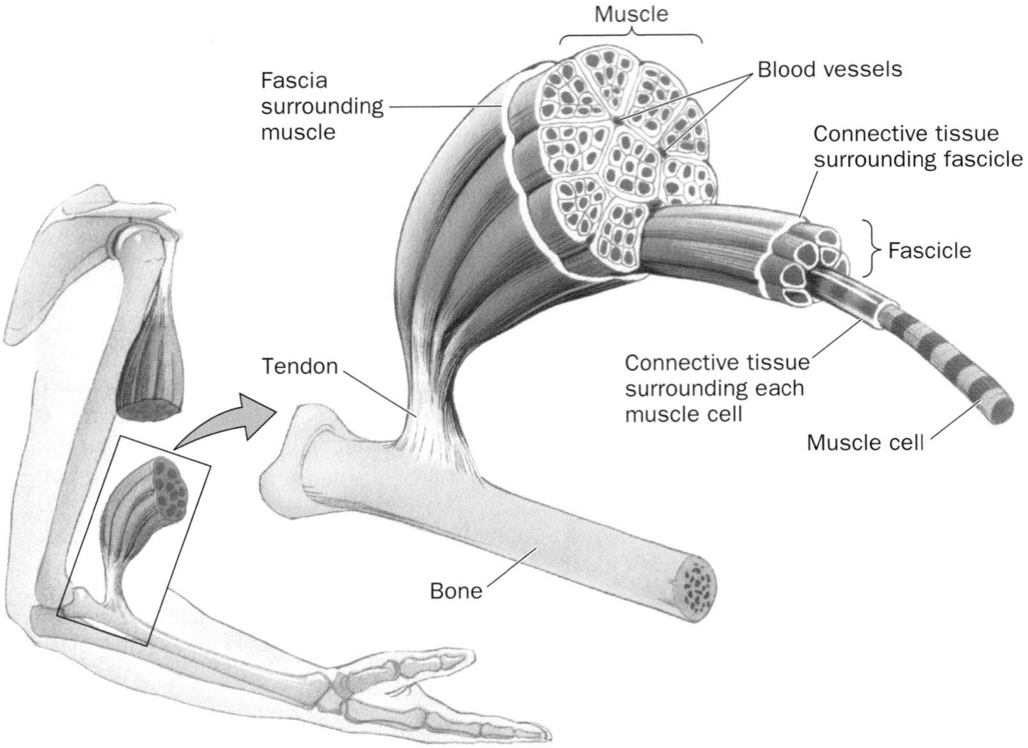
Muscle
Blood vessels
Fascia surrounding muscle
Connective tissue surrounding fascicle
Fascicle
Tendon
Connective tissue surrounding each muscle cell
Muscle cell
Bone

FIGURE 5.18 Relationship of a skeletal muscle with its muscle cells and connective tissue wrappings.

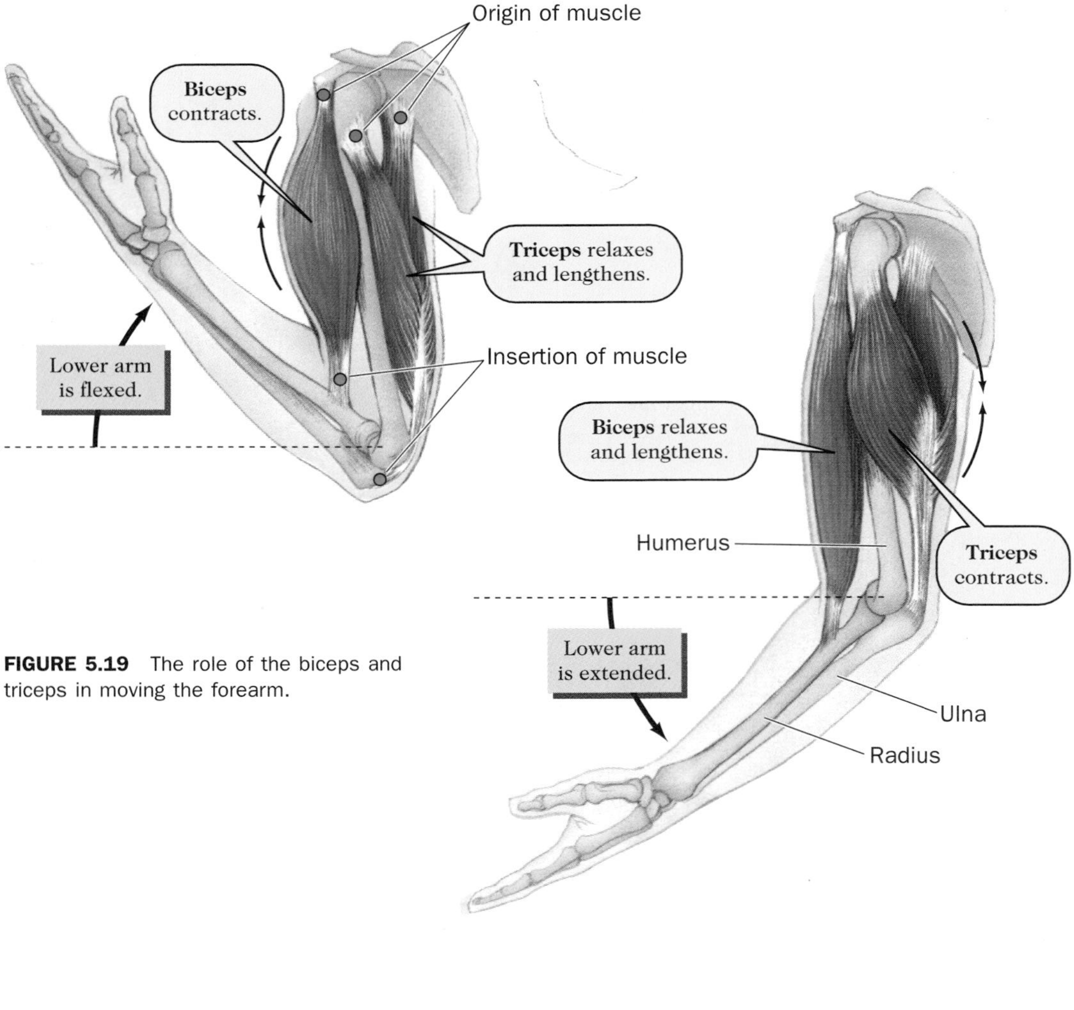
Origin of muscle
Biceps contracts.
Triceps relaxes and lengthens.
Lower arm is flexed.
Insertion of muscle
Biceps relaxes and lengthens.
Humerus
Triceps contracts.
Lower arm is extended.
Ulna
Radius

FIGURE 5.19 The role of the biceps and triceps in moving the forearm.

fibers called **myofibrils** (*my-o-***fye***-brills*). Myofibrils are the structures responsible for contraction. Each myofibril consists of a series of units called **sarcomeres** (***sar****-koh-mears*). The orderly arrangement of sarcomeres gives a striated appearance to skeletal muscle cells (see Figure 5.20).

Each sarcomere is made of a rather complex arrangement of several kinds of protein. Two of these proteins are **actin** and **myosin.** Thousands of actin molecules link together to form long threadlike macromolecules called *actin filaments.* Similarly, thousands of myosin molecules form the somewhat thicker *myosin filaments* (see Figure 5.20). As we will see, the molecular interaction between actin and myosin filaments contributes directly to muscle contraction.

> Each skeletal muscle cell contains many nuclei and specialized organelles, including the sarcoplasmic reticulum, the transverse tubules, and numerous myofibrils that run the length of the cell. Each myofibril contains many sarcomere units made of the proteins actin and myosin.

FIGURE 5.20 A cutaway view of an individual muscle cell, revealing its internal structure.

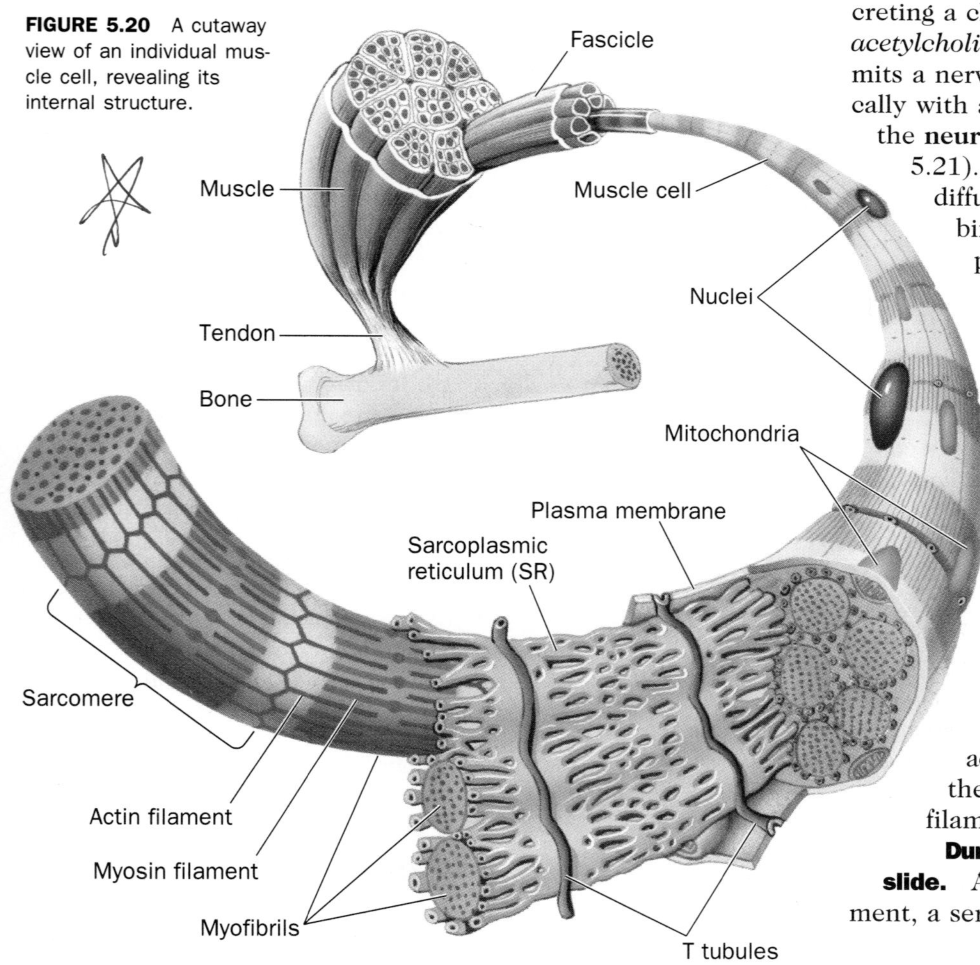

Fascicle
Muscle
Muscle cell
Nuclei
Tendon
Bone
Mitochondria
Plasma membrane
Sarcoplasmic reticulum (SR)
Sarcomere
Actin filament
Myosin filament
Myofibrils
T tubules

Contraction of a Skeletal Muscle Cell

Before you can understand how an entire muscle contracts, you need to know how an individual muscle cell contracts. How is contraction triggered and controlled? How does the molecular interaction between actin and myosin result in shortening? What follows is a description of the events that occur within muscle cells before, during, and after contraction.

For contraction, a stimulus is needed. Before a muscle cell can contract, it must be stimulated. In the human body, stimulation is provided by nerve impulses that pass along nerve cells (Figure 5.21). A nerve cell stimulates a muscle cell by secreting a chemical substance called *acetylcholine (ACh).* Acetylcholine permits a nerve cell to communicate chemically with a muscle cell at a site called the **neuromuscular junction** (see Figure 5.21). Upon its release, acetylcholine diffuses across the junction and binds to specific receptors on the plasma membrane of the muscle cell. This binding initiates an electrical impulse which travels down the T tubules into the interior of the cell.

When the impulse reaches the SR, Ca^{2+} stored in the SR is released. The released Ca^{2+} forms bonds with the *troponin-tropomyosin protein complex* which is associated with the actin filaments (Figure 5.22). When Ca^{2+} forms this bond, the troponin-tropomyosin complex shifts position, exposing sites on the actin filaments for attachment to the head portions of the myosin filaments.

During contraction, actin and myosin slide. After exposing sites for attachment, a series of events cause the short-

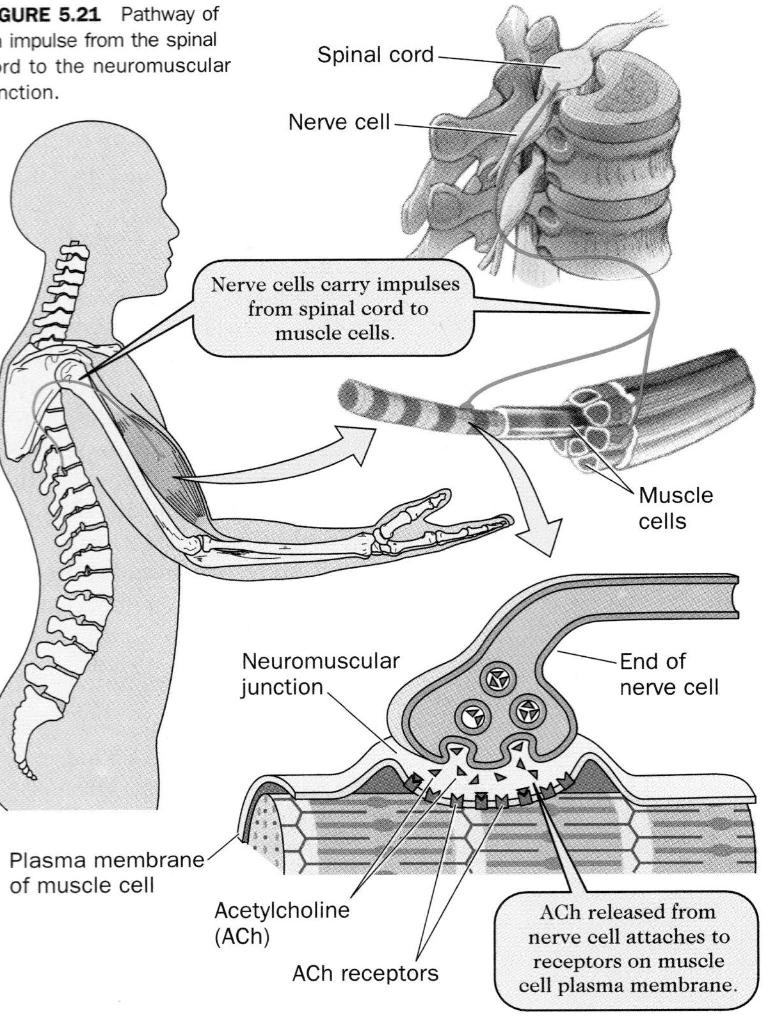
Spinal cord
Nerve cell
Nerve cells carry impulses from spinal cord to muscle cells.
Muscle cells
Neuromuscular junction
End of nerve cell
Plasma membrane of muscle cell
Acetylcholine (ACh)
ACh receptors
ACh released from nerve cell attaches to receptors on muscle cell plasma membrane.

FIGURE 5.21 Pathway of an impulse from the spinal cord to the neuromuscular junction.

ening or contraction of a muscle cell (see Figure 5.22). Each myosin head projecting from a myosin filament bonds with an actin filament. After forming these bonds, the myosin heads change shape and pull the actin filaments toward the center of the sarcomere.

This reaction requires energy, which comes from the ATP attached to each myosin head. Myosin acts as an enzyme (when Ca^{2+} is present) and breaks the ATP into ADP + ℙ + energy. This energy is used to change the shape of the myosin head, which causes the actin filaments to slide a little toward the center of the sarcomere.

After sliding, a new ATP attaches to the myosin head, breaking the myosin-actin bond. When this bond is broken, the myosin head is released and returns to its original shape. Immediately, the myosin head bonds to another site on the actin filament. Then it changes shape again, and the filaments slide a little farther past one another.

The sequence of myosin-actin bonding, sliding, and release is repeated rapidly over and over until the muscle has shortened sufficiently. This is the **sliding filament mechanism** for muscle contraction, and it consumes a great deal of ATP.

The sliding filament mechanism can be compared to walking on a nonmotorized treadmill. Each time your foot (myosin head) lands on the treadmill's belt (actin filament), your foot changes position a little and the belt moves a short distance. A series of rapid steps will move the belt a longer distance.

How does the sliding of actin and myosin filaments cause the contraction of a muscle cell? When an individual muscle cell is stimulated, the actin filaments in every sarcomere unit slide toward the middle of the sarcomere. As a result, the myofibrils shorten. A muscle cell shortens (contracts) when all its myofibrils shorten. An entire muscle contracts when many of its muscle cells contract simultaneously.

Ending contraction. Contraction ends when the muscle cell ceases to be stimulated by the nerve cell. If an impulse no longer travels down the T tubules to the cell's interior, Ca^{2+} is actively transported back into the SR. Without Ca^{2+}, the troponin-tropomyosin protein complex shifts back into its original position, masking the sites on the actin filaments and preventing further bonding by the myosin heads. When this happens, the muscle relaxes and returns to its original shape.

> Upon stimulation of a muscle cell, calcium ions (Ca^{2+}) are released from the SR. In reacting with a protein complex, Ca^{2+} unmasks myosin-binding sites on the actin filaments. As a result the actin filaments slide toward the center of the sarcomere.

The Energy for Muscle Contraction

Muscle contraction requires a large number of ATP molecules. Running, for exam-

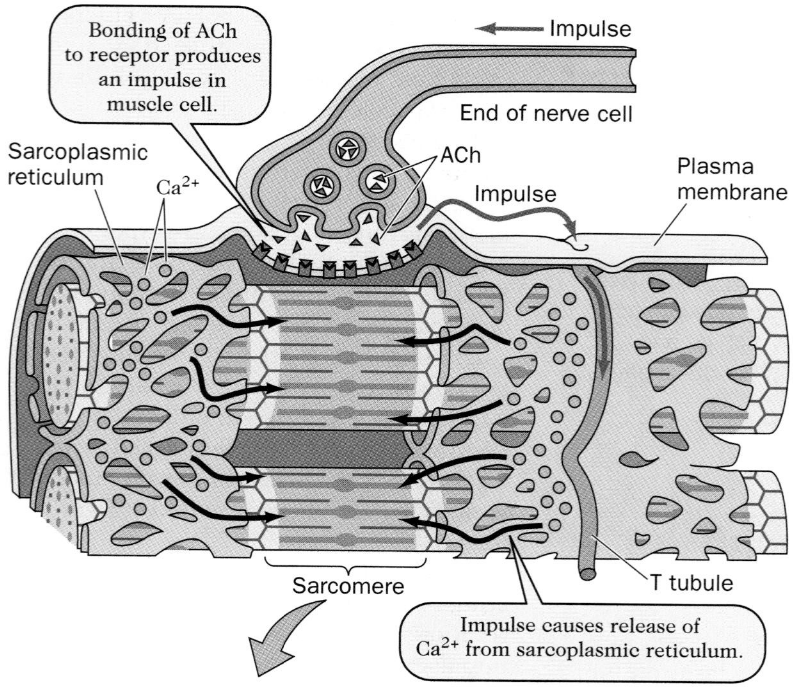

Bonding of ACh to receptor produces an impulse in muscle cell.
Impulse
End of nerve cell
Sarcoplasmic reticulum
Ca^{2+}
ACh
Impulse
Plasma membrane
Sarcomere
T tubule
Impulse causes release of Ca^{2+} from sarcoplasmic reticulum.

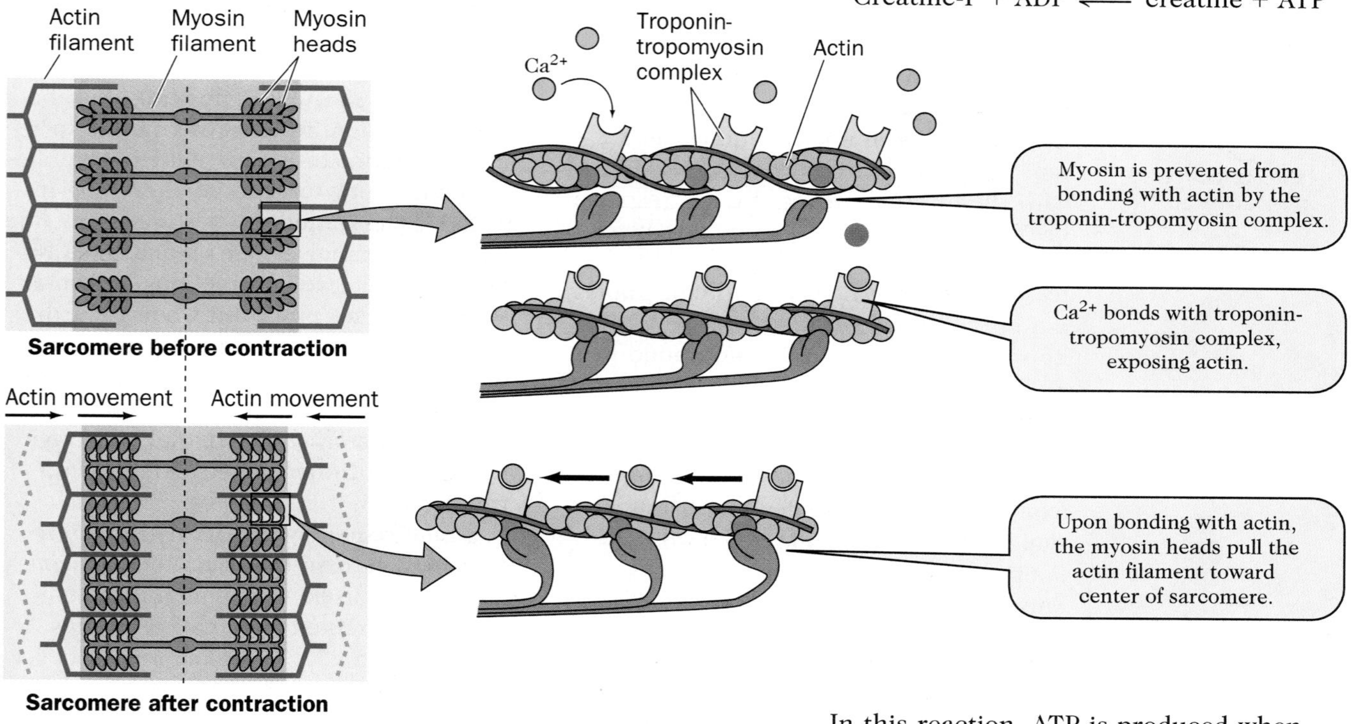

Actin filament
Myosin filament
Myosin heads
Sarcomere before contraction
Actin movement
Actin movement
Sarcomere after contraction
Ca^{2+}
Troponin-tropomyosin complex
Actin
Myosin is prevented from bonding with actin by the troponin-tropomyosin complex.
Ca^{2+} bonds with troponin-tropomyosin complex, exposing actin.
Upon bonding with actin, the myosin heads pull the actin filament toward center of sarcomere.

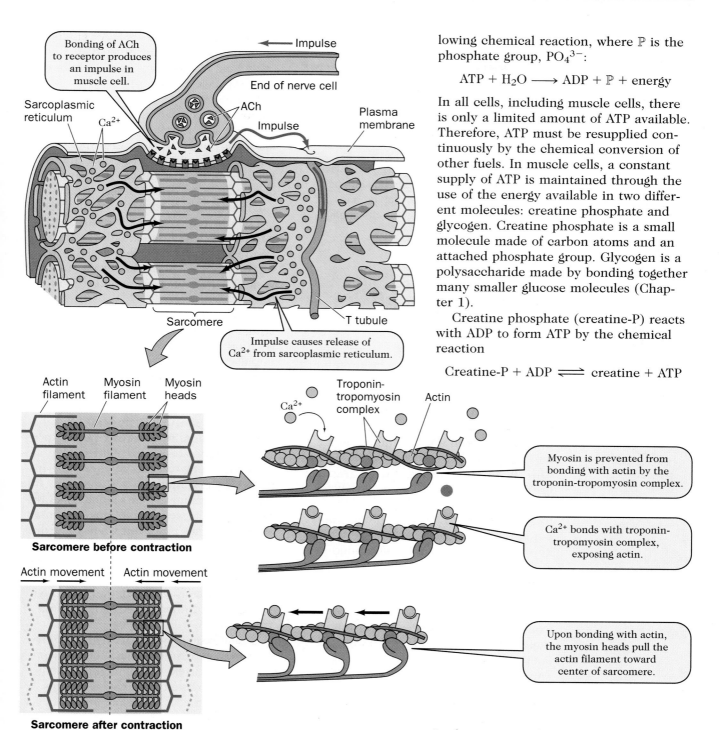

FIGURE 5.22 The interactions between ACh, Ca^{2+}, actin, and myosin that lead to muscle contraction.

ple, requires the use of large leg muscles that are composed of thousands of individual muscle cells. The rapid contraction of all these cells requires a large amount of energy from ATP.

How do cells obtain sufficient ATP? Recall from Chapter 1 that ATP stores energy. This energy is liberated by the following chemical reaction, where $\mathbb{P}$ is the phosphate group, PO_4^{3-}:

$$ATP + H_2O \longrightarrow ADP + \mathbb{P} + \text{energy}$$

In all cells, including muscle cells, there is only a limited amount of ATP available. Therefore, ATP must be resupplied continuously by the chemical conversion of other fuels. In muscle cells, a constant supply of ATP is maintained through the use of the energy available in two different molecules: creatine phosphate and glycogen. Creatine phosphate is a small molecule made of carbon atoms and an attached phosphate group. Glycogen is a polysaccharide made by bonding together many smaller glucose molecules (Chapter 1).

Creatine phosphate (creatine-P) reacts with ADP to form ATP by the chemical reaction

$$\text{Creatine-P} + ADP \rightleftharpoons \text{creatine} + ATP$$

In this reaction, ATP is produced when the phosphate and the stored energy in creatine phosphate are transferred to ADP. The double arrow in this equation indicates that the reaction is reversible; it can proceed in either direction. When ATP is not used for muscle contraction, it begins to accumulate, and any excess is used to resynthesize creatine phosphate, which then is stored.

During contraction, when stores of creatine phosphate have been depleted, glucose molecules are removed from the glycogen stored in the muscle cell. The energy of glucose is used for the synthesis of ATP, using a series of chemical reactions that will be discussed in Chapter 7.

> The immediate source of energy for muscle contraction is ATP. During strenuous muscle activity, stored creatine phosphate and glycogen can supply energy for the chemical synthesis of more ATP.

Muscles Accomplish Many Tasks

The muscles in our bodies all work by means of contraction and relaxation, but they perform more than a thousand different functions. The 700 or more skeletal muscles move bones, skin, and other muscles. In addition to skeletal muscle, there are smooth muscle and cardiac muscle. Smooth muscles are associated with the internal organs and are responsible for moving food along the digestive system, changing the diameter of arteries to regulate blood flow, and forcing babies out of the uterus. Cardiac muscle is the muscle of the heart and is not found in any other organ. Its contractions pump blood through the blood vessels to nourish the entire body.

In the sections that follow, we will examine some of the characteristics of skeletal muscle that allows it to fill so many of the body's movement needs. Finally, we will briefly describe smooth muscle and cardiac muscle.

Responses of Skeletal Muscle

Have you ever been annoyed by a persistent muscle twitch? Perhaps you have watched a series of jerky muscle contractions take place under your skin. Fundamentally, skeletal muscles function by twitching. But how can our most graceful movements be the product of "twitches"? Our task in this section is to describe how muscles produce smooth controlled movements that result from the "twitches" of individual muscle cells and entire muscles.

In a laboratory, it is possible to electronically record and measure the strength of a muscle's contraction, which is called muscle tension, in response to different stimuli. The recording that relates a stimulus to muscle tension is called a *myogram.* If a single stimulus (a single impulse) is applied to a muscle for only an instant, that muscle will quickly contract and then relax, and the myogram will record a single **muscle twitch.** Analysis of the myogram permits us to identify three different periods (Figure 5.23).

The *latent period* is a brief period from stimulation to the start of muscle contraction. It is during this period that the impulse travels to the SR, calcium ions are released, and the myosin heads bond with the actin filaments. The *contraction period* is the time during which the muscle actually shortens. The *relaxation period* is the time it takes the muscle to relax and return to its original length.

A muscle twitch can be altered by increasing the frequency of stimulation (the number of impulses per second) (see Figure 5.23). Muscle contractions known as **summation** are produced when the frequency of stimulation is so rapid that a muscle does not have time to relax completely after each impulse. Summation produces a greater strength of contraction by adding together many separate contractions.

If the frequency of stimulation is further increased so that there is no relaxation, the muscle will remain in a completely contracted state called **tetanus** (see Figure 5.23). These laboratory observations illuminate the properties and capabilities of skeletal muscle, but how do they relate to the contraction of muscle in the normal human body?

Muscle contractions that produce movements such as eating, walking, and lifting an object result from a rapid series of stimulations that originate in the brain and spinal cord and pass to the muscle through the nerves. The strength of a

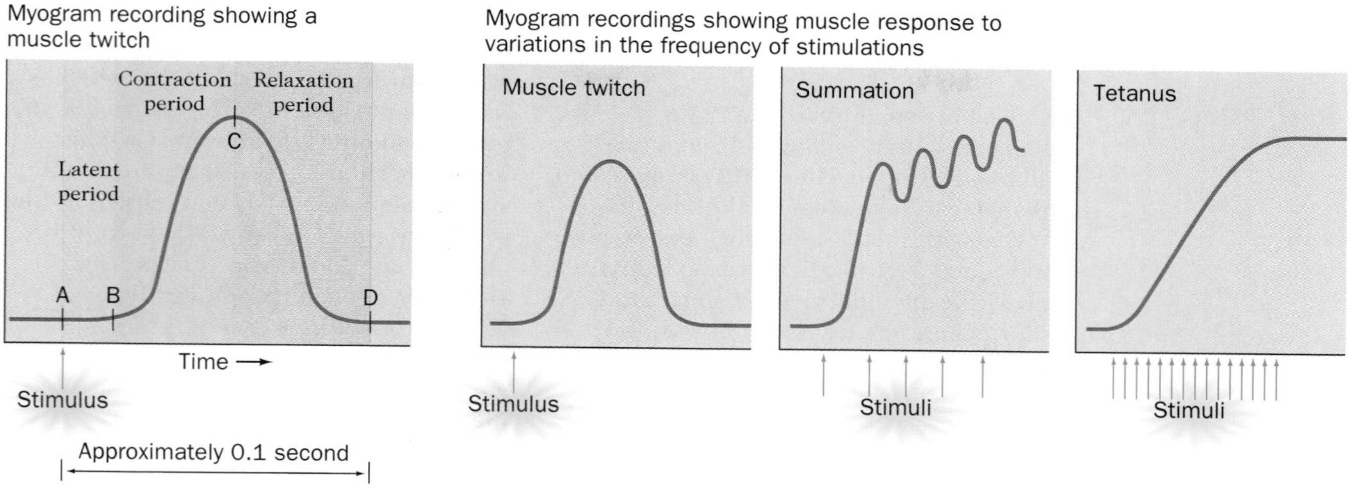
Myogram recording showing a muscle twitch
Contraction period
Relaxation period
Latent period
C
A
B
D
Time
Stimulus
Approximately 0.1 second
Myogram recordings showing muscle response to variations in the frequency of stimulations
Muscle twitch
Summation
Tetanus
Stimulus
Stimuli
Stimuli

FIGURE 5.23 Myogram showing a muscle twitch and other responses to stimulations.

muscle contraction depends on the strength of the stimulus applied; however, not all the muscle cells in a muscle respond to the same stimulus strength.

The thresholds for different muscle cells. A muscle cell contracts only when the stimulus it receives is of sufficient strength. The smallest strength of stimulus that causes contraction is called the *threshold stimulus.* When a muscle cell receives a threshold stimulus, it contracts with maximum strength. Even if a muscle cell receives a stimulus greater than the threshold, it will contract with the same maximum contraction strength. This is called the **all-or-none law**, and it states that a muscle cell will contract maximally or not at all.

The all-or-none law applies to individual muscle cells but not to an entire muscle. You know from personal experience that an entire muscle does not contract completely with each movement. A whole muscle produces varying strengths of contraction and matches its effort to the resistance it encounters. This can be explained by understanding that the muscle cells in a muscle do not all have the same threshold value.

If a stimulus is threshold for only 25 percent of a muscle's cells, only those cells will contract and a small muscle tension will be produced. However, if a greater stimulus is applied that is at or above the threshold for 75 percent of the muscle cells, the contraction of all those cells will produce greater tension. The tension or force of contraction produced by 25 percent of the muscle's cells is less than that produced by 75 percent of its cells. This explains how an arm that picks up a feather can also pick up a textbook; more muscle cells contract when picking up a book.

Maintaining stabilization. In addition to muscle contractions that produce movements such as walking, certain skeletal muscles also stabilize the bones of the skeleton so that we can sit, stand, or maintain our posture. These groups of muscles receive prolonged and rapid stimuli from the nervous system, and this produces tetanus in these muscles.

These stabilizing muscles do not fatigue easily even though they are in a state of tetanus because not all of their muscle cells contract at the same time. Instead, only some of the muscle cells contract, while the others relax. As the contracting cells start to fatigue, they relax and a different group of cells will contract. This cycling between contraction and relaxation allows the work of the muscle to be shared among many cells while maintaining the strength of contraction.

> The events of a single muscle twitch form the basis for all muscle contraction. When a muscle cell receives a threshold stimulus, it will contract with maximum strength. This is called the all-or-none law.

Skeletal Muscle and the Benefits of Exercise

Exercise is popular with many people. We undertake activities such as swimming, walking, jogging, cycling, and aerobics to enhance cardiovascular health, maintain body weight, increase physical endurance, and reduce psychological stress. Exercise usually means movement of some kind, and it is appropriate to ask how muscles respond to repeated and extended use.

Isotonic and isometric muscle contractions are used in certain exercise activities. **Isotonic** means "same strength" (constant tension). In an isotonic contraction, the strength of the contraction remains the same but the muscle shortens. In an exercise such as weight lifting, the muscle contracts and exerts a constant tension that is sufficient to move the weight. Any exercise activity that moves part of the skeleton involves isotonic contractions.

Isometric means "same length." In an isometric contraction, the strength of the contraction increases, but there is practically no shortening of the muscle. This type of contraction does not result in body movement. Sitting upright in a chair while contracting your abdominal muscles is an example of an isometric contraction. Isometric contractions are important during exercise because they stabilize the skeleton. The performance of most exercises typically involves a combination of isotonic and isometric contractions.

Muscle cells contract at different rates. Each skeletal muscle is made up of a combination of three types of muscle cells. *Slow-twitch* muscle cells contract more slowly, are more resistant to fatigue, and contain relatively more mitochondria. *Fast-twitch* muscle cells contract faster, are less resistant to fatigue, and have relatively few mitochondria. *Intermediate-twitch* muscle cells lie somewhere between the fast and slow categories.

The percentage of slow-, intermediate-, and fast-twitch cells in a muscle varies from individual to individual and is an important factor in determining an individual's athletic capabilities. For example, tests have shown that the leg muscles of marathon runners contain a high percentage of slow-twitch muscle cells, whereas the same muscles in sprinters have a high percentage of fast-twitch muscle cells. The proportion of the different cells in your skeletal muscle is determined by the genes you inherited. While you cannot change your genes, you can increase the efficiency of each type of cell through physical conditioning and training.

Exercise builds muscle protein. Exercising muscles can alter their shape and the contours of the body. When you use maximum tension to exercise your muscles, as in weight training, your muscles respond by increasing in size and strength. Such a response occurs because maximum tension exercises stimulate your muscle cells to synthesize more actin and myosin protein filaments. Muscles enlarge, therefore, because of an increase in muscle cell size, not an increase in the number of cells.

If a muscle is not used for an extended period, *muscle atrophy* occurs. This is a decrease in muscle size and strength caused by loss of the actin and myosin contractile proteins. If inactivity is prolonged over several months, some muscle cells will completely degenerate and will be replaced by connective tissue. When this happens, the loss will be permanent. To prevent muscle atrophy, physical therapy is necessary when an individual (or a single limb) is immobilized for an extended period.

Anabolic steroids and athletics. Muscle growth can be stimulated by the consumption of synthetic hormones called **anabolic steroids.** These steroids are chemically similar to the male sex hormone testosterone and can stimulate the production of muscle proteins (actin and myosin). In patients with muscle atrophy, these substances have been prescribed to help reestablish muscle tissue and strength. However, these drugs are powerful and dangerous, particularly when taken in high doses without medical supervision.

In recent years some male and female athletes have taken as much as 9 times the prescribed dose of anabolic steroids in an attempt to increase muscle growth. The resulting increase in muscle mass may improve strength and competitive performance. However, the relationship between high doses and increased performance varies greatly among individuals, and there are dangerous side effects.

To maintain their effects, anabolic steroids must be used throughout an athlete's competitive career. Such long-term use can lead to mental and emotional disorders, liver damage, liver cancer, and blood and circulatory problems. In women, large doses disrupt the menstrual cycle and may stimulate a general masculinizing of features, including the growth of facial and body hair. Infertility may occur in both men and women.

Exercise, O_2, and muscle fatigue. During strenuous exercise, we have all had experiences that relate directly to the use of energy by the skeletal muscles. Imagine that you are running. As you run, your muscle contractions increase and you start to breathe faster. Faster breathing increases the amount of oxygen (O_2) available to your cells, so glucose can be more efficiently utilized to produce ATP. Such O_2-requiring activities are known as **aerobic exercise.**

As muscle contraction becomes more intense, O_2 cannot be supplied fast enough and the muscle cells start to break down glucose less efficiently. A byproduct of this less efficient breakdown of glucose is *lactic acid.* During strenuous exercise, lactic acid accumulates in the cells, causing an irritation which is partly responsible for the muscle soreness you may experience after intense exercise.

When you stop running, you continue breathing rapidly for a period of time. During this period, the extra breathing supplies the additional O_2 needed to break down the lactic acid that has accumulated.

Reducing body fat. Exercise can be used to reduce body fat. With increased exercise and muscle contractions, your need for energy may exceed your daily intake of energy from food. When this happens, the energy eventually will be supplied by stored fat. Exercise programs designed to reduce body fat usually involve large muscle groups of the legs, torso, and arms. These muscles use more energy than do smaller muscles.

Smooth Muscle and Cardiac Muscle: General Characteristics

Smooth muscle is associated with the internal organs, such as those of the digestive tract (esophagus, stomach, intestines), respiratory passages, urinary bladder, and female and male reproductive structures. Blood vessels and the iris of the eye contain smooth muscle, as do the arrector pili muscles of the skin.

We generally have no conscious or voluntary control over the contraction of smooth muscles, which are often called *involuntary* muscles. This is in contrast to the skeletal muscles, which are called *voluntary* muscles because of the great deal of conscious control we exert over their actions. As we will see in later chapters, smooth muscle contraction can be stimulated by the brain, hormones, and local changes in oxygen (O_2) and carbon dioxide (CO_2) concentrations.

There are fundamental differences between smooth muscle and skeletal muscle (Figure 5.24). Smooth muscle cells are shorter and have tapered ends, and each contains one nucleus. Their actin and myosin filaments have a random orientation instead of the ordered arrangement seen in the myofibrils of skeletal muscle. As a result, they do not appear striated, and that is why they are called smooth muscle cells. The cellular characteristics of smooth muscle cells give these muscles their functional properties.

Smooth muscle generally has slower contraction and relaxation rates than does skeletal muscle. This provides for the slow rhythmic contractions used in the intestines to move and mix food during digestion. Smooth muscle can be stretched greatly yet continue to maintain a relatively constant tension. This quality is well adapted for a structure such as the urinary bladder.

Cardiac muscle is the third type of muscle in the body. It is found only in

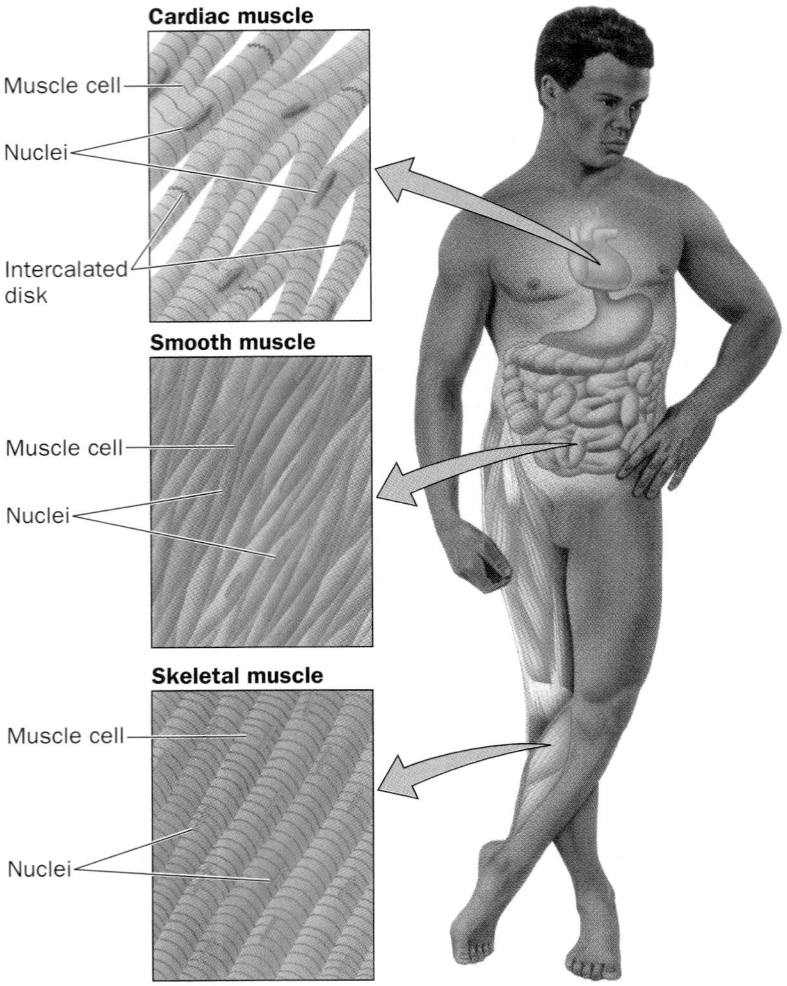
Cardiac muscle
Muscle cell
Nuclei
Intercalated disk
Smooth muscle
Muscle cell
Nuclei
Skeletal muscle
Muscle cell
Nuclei

FIGURE 5.24 Summary of skeletal, smooth, and cardiac muscle characteristics.

the heart, where its rhythmic contractions move blood throughout the body. Cardiac muscle cells are striated and have one or two nuclei per cell. Each cell has branches that form tight junctions called *intercalated disks* with other cardiac muscle cells. These disks are specialized to permit the rapid transfer of electrical signals for contraction from one cardiac muscle cell to another. Cardiac muscle is considered involuntary muscle because we have no conscious control over its contractions. We will encounter cardiac muscle again in Chapter 9.

> Smooth muscle cells are involuntary and are associated with the internal organs. They lack striations because their actin and myosin are oriented randomly. Cardiac muscle is found only in the heart. It is also involuntary but has striations.

Diseases and Disorders

There are many disease and disorders of bone and muscle tissues, including inherited conditions caused by malfunctioning genes, malnutrition, improper use, infection, and tumors. In the following sections, we will focus on three topics: (1) common disorders resulting from the misuse of bone and muscle, (2) muscular dystrophy, which is a genetic disorder, and (3) degenerative disorders of bone and joints. The healing of bone fractures is described in Spotlight on Health: Two (page 163).

Disorders of Bones and Muscles

Shinsplints. People who regularly run, jog, or perform other aerobic exercises often develop a painful condition called shinsplints. The symptoms include pronounced pain or soreness along the anterior surface of the tibia (commonly called the shinbone). This condition often results from poor conditioning, inadequate warm-up exercises, and/or athletic shoes that give poor support.

The pain emanates from either the tibia or the lower leg muscles. Pain from the tibia may be due to small fractures or inflammation of its periosteum. Leg muscle pain can result from an inflammation of the tendon that attaches the tibialis anterior muscle to the tibia and from swelling of the lower leg muscles. Swelling causes pain by putting pressure on the connective tissue (fascia) that wraps around these muscles. These symptoms usually disappear after a week or two of rest.

Muscle cramps. Less serious but more common than shinsplints is the prolonged contraction of a skeletal muscle referred to as a muscle cramp. Cramps have a variety of causes, but in an active, healthy individual they most often follow a period of exercise or exertion. In such cases the cramp appears to be caused by changes in the muscle cells: a depletion of ATP, dehydration, reduced ion concentrations, and/or an accumulation of lactic acid. Cramps can be alleviated by gentle massage and stretching.

Pulled muscles. A common disorder that is associated with vigorous and often unaccustomed exercise is called a pulled muscle. This happens when a muscle is stretched too far and some of the muscle cells are torn away from the others. This results in bleeding from ruptured blood vessels and may cause swelling and pain.

Tendinitis. A condition called **tendinitis** results from the inflammation or actual tearing of the tendons that connect muscles to bones. It is most common in the knees, heels (Achilles tendon), shoulders, and elbows ("tennis elbow"). Because a tendon is composed of connective tissue with few blood vessels, healing and recovery from tendinitis is a slow process that usually requires reduced use of the affected region for a period of time.

Muscular Dystrophy

Muscle cramps, pulled muscles, and tendinitis are common and painful conditions, but they are not life-threatening. Other muscle disorders are more serious. One of these disorders is **muscular dystrophy** (***dis**-tro-fee*), which results from the inheritance of a defective gene. In one form of the disease, the muscle cells atrophy and are replaced by connective tissue. Disease symptoms begin as early as age 2, and muscle atrophy continues until death occurs by the mid-twenties from respiratory or cardiac muscle dysfunction. Intense research in recent decades has shown that the defective gene interferes with muscle contraction by causing Ca^{2+} to continuously leak out of the sarcoplasmic reticulum.

Degenerative Disorders of Bone and Joints

Osteoporosis. Osteoporosis (*oss-tea-oh-pore-**oh**-sis*) is a bone disorder that results from the loss of calcium and phosphate from the bone matrix (Figure 5.25). As you learned earlier, throughout your life bone is continuously deposited and removed by osteoblasts and osteoclasts. In osteoporosis, osteoblast activity is decreased and new bone is deposited more slowly than it is removed by osteoclasts.

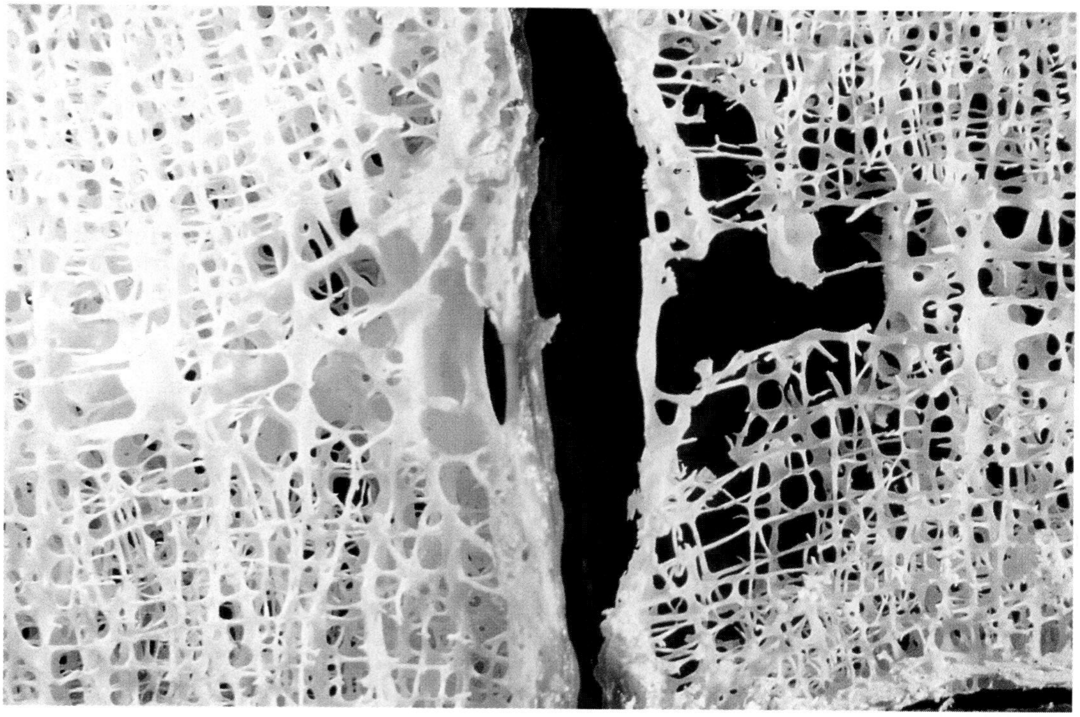

FIGURE 5.25 Spongy bone. Normal condition *(left)* and early osteoporosis condition *(right)* with thinner bars and plates.

Osteoporosis primarily affects postmenopausal women. Menopause is a time (around age 50) when there is a decline in a woman's production of the female sex hormone estrogen, one function of which is the stimulation of osteoblast activity. After menopause, some women may lose 30 percent or more of their bone calcium. As a result, the skeleton is weakened and bone fractures can occur even with normal activity. Over time, vertebrae may become compressed, contributing to the hunch-backed condition known as "dowager's hump."

To help delay the onset of osteoporosis, all women should increase their dietary intake of calcium and maintain a regular schedule of weight-bearing exercise. Medically supervised estrogen replacement therapy may also be needed.

Osteoarthritis. Osteoarthritis (*oss-tea-oh-ar-**thry**-tis*) is an inflammation of the large weight-bearing joints. To some extent it affects everyone over age 60. However, severe cases are 3 times more frequent among women than among men. It is believed that osteoarthritis results from the gradual wearing of the surface cartilage of a joint. As the cartilage is worn, the underlying bone is exposed and may thicken, reducing the space within the joint capsule. This can restrict movement and cause chronic pain.

Although osteoarthritis often accompanies normal aging, its onset and severity can be accelerated by injury, infection, and misuse of a joint. There is some evidence that genes play a role in the severity of the disease, but their specific effects are unknown. Medication may be prescribed to reduce inflammation and pain, and in some cases surgical replacement of a joint (particularly the hip joint) may be required.

Rheumatoid arthritis. Rheumatoid (***room**-a-toid*) arthritis is a degenerative disease of the joints that occurs with equal frequency in men and women. It is more serious and debilitating than osteoarthritis. The symptoms are swollen, painful, and stiff joints. These symptoms can occur at any age, even in young children, although they usually develop later in life.

This disease causes inflammation of the synovial membrane and disruption of the cartilage surface of joints. As it progresses, the cartilage surface of the joint may be completely destroyed, causing friction, pain, and immobility. Bones may even fuse together. The small bones of the hands and feet are most often involved, but it is common for the disease to progress to larger joints.

Rheumatoid arthritis is an autoimmune disease in which the body's antibodies mistakenly attack one's own tissues, causing their inflammation. Heredity may play a role in this disorder, but this is unclear. There is no cure, and treatment is aimed at reducing inflammation and pain. In severe cases, the deformed joint may be replaced with an artificial one.

Life Span Changes

From birth to old age, there are significant changes in the structure and function of the skeleton and musculature of the body. Let's examine some of these changes, starting with infancy and continuing to old age.

Bone development begins before birth and is not complete until the late teens. Our bones grow in length through infancy and childhood, and we grow taller. During adulthood our bones are continuously remodeled as we place new stresses on them through exercise, weight changes, and postural changes. Genes, injuries, and environmental conditions also play a role.

There are two primary degenerative effects of aging on bone. The first involves the loss of calcium and phosphate, and the other involves a decreased rate of protein synthesis in bone cells. When bone cells make fewer protein fibers for the matrix, bones become less flexible and more brittle. This explains why the bones of elderly people fracture so easily.

As a child's skeleton grows and changes, so do the child's muscles. We are born with all our muscle cells. As we grow to maturity, our muscle cells increase in size, permitting entire muscles to increase in both size and strength. In addition, there is increased coordination and control of muscles as the nervous system matures. As nerve cells grow, they make more contacts with muscles and produce and release more acetylcholine, the chemical that functionally links nerve cells to muscle cells.

Starting at around age 30, skeletal muscles begin to lose some of their mass. Muscle cells become somewhat smaller in diameter because of fewer myofilaments, and some cells die. As we age, there is also a reduction in the number of nerve cells controlling a muscle and a reduction in acetylcholine production. The result of these changes is a gradual loss of muscle size, strength, and endurance.

Research shows that a regular program of exercise can delay many of these muscular changes and maintain the strength of our bones. We must adapt a "use it or lose it" philosophy for the long-term health of our muscles and bones.

Chapter Summary

Bone: Protection, Support, and Movement

- Bone is a living connective tissue that is made of a hard calcified matrix and bone cells.
- The shaft of a long bone contains yellow bone marrow; each epiphysis contains red bone marrow.
- The osteocytes of compact bone are organized around Haversian canals.
- Osteocytes maintain bone, osteoblasts form new bone, and osteoclasts remove bone.
- Bone development begins from a cartilage model before birth.

The Skeleton: An Ordered Assembly of Bones

- The skeleton is organized into two divisions: the axial and appendicular skeletons.
- The axial skeleton includes the skull, vertebral column, and rib cage.
- The appendicular skeleton includes the pectoral and pelvic girdles and their attached limbs.

Joints: Where Bone Joins Bone

- A joint is formed when two or more bones join together.
- Immovable joints, such as suture joints, permit little or no movement.
- Movable joints, such as those of the shoulder and knee, are called synovial joints.

Muscles and Muscle Contraction

- There are three types of muscle tissue: skeletal, smooth, and cardiac.
- Skeletal muscles attach to bones or to facial skin. Each skeletal muscle is composed of parallel bundles of muscle cells wrapped by connective tissue. Tendons attach muscle to bone.
- Skeletal muscle cells are packed with tiny fibers called myofibrils. Each myofibril contains a series of sarcomere units. Each sarcomere contains actin and myosin filaments that bond with each other when a muscle cell contracts (sliding filament mechanism).
- Nerve cells release acetylcholine, which attaches to receptors on a muscle cell's plasma membrane.
- When a muscle cell is stimulated, Ca^{2+} is released and attaches to the troponin-tropomyosin protein complex. As a result, myosin bonding sites are exposed, permitting bonding between myosin and actin.
- A muscle cell contracts when the sarcomeres along each myofibril shorten.

Muscles Accomplish Many Tasks

- A threshold stimulus is the minimum stimulus needed to stimulate a muscle cell to contract.
- The all-or-none law states that a muscle cell contracts maximally or not at all.
- Smooth muscle is involuntary and is found in the internal organs. Its cells are spindle-shaped, contain one nucleus, and are not striated.
- Cardiac muscle is found in the heart. Its cells are branched, contain one or two nuclei, and are striated.

Diseases and Disorders

- Shinsplints result from poor conditioning, inadequate warming up, and/or improper athletic shoes.
- Muscle cramps often result when muscle cells lack sufficient ATP, are dehydrated, have reduced ion concentrations, or are irritated by lactic acid.
- In osteoporosis, calcium is lost from the bone matrix. This disorder primarily affects postmenopausal women.
- Rheumatoid arthritis is an autoimmune disease that causes inflammation of joint capsules.
- Osteoarthritis results from the gradual wearing of the surface cartilage of a joint.

Life Span Changes

- Bone development begins before birth and continues until the late teens.
- During aging, bone loses calcium and phosphate and contains fewer protein fibers.
- During aging, skeletal muscle loses mass; this decreases muscle size, strength, and endurance.

Selected Key Terms

actin (p. 151)
appendicular skeleton (p. 143)
axial skeleton (p. 140)
cardiac muscle (p. 157)
growth plate (p. 139)
Haversian canal (p. 137)
joint (p. 145)
muscle twitch (p. 154)
myosin (p. 151)
neuromuscular junction (p. 151)
osteocyte (p. 137)
sarcomere (p. 151)
skeletal muscle (p. 148)
sliding filament mechanism (p. 152)
smooth muscle (p. 157)

Review Activities

1. Describe six anatomic structures of a long bone.
2. How does the microscopic structure of compact bone differ from that of spongy bone?
3. Describe the major events in bone development.
4. Explain how growth plate activity increases bone length.
5. Name and locate nine bones in your skull.
6. Name and locate the five regions of your vertebral column.
7. Locate the scapula, clavicle, and coxal bones on your body.
8. Name and locate the bones of your arms and legs.
9. Sketch a synovial joint and label its parts. Identify three synovial joints in the body.
10. Name and locate 15 muscles of the body.
11. Differentiate between myofibrils, sarcomeres, actin filaments, and myosin filaments.
12. Describe how the interaction between Ca^{2+} and the troponin-tropomyosin complex permits actin and myosin to bond during muscle contraction.
13. Draw and explain the latent, contraction, and relaxation periods of a muscle twitch.
14. Using the concept of threshold stimulus and the all-or-none law, explain how a muscle produces varying strengths of contraction.
15. Identify cellular differences between skeletal, smooth, and cardiac muscle.

Self-Quiz

Matching Exercise

___ 1. The type of vertebrae that form the neck region
___ 2. The cartilage pad positioned between vertebrae
___ 3. Bone-forming cells
___ 4. The type of bone marrow that produces blood cells
___ 5. The bones that form the fingers
___ 6. The bone type that contains Haversian canals
___ 7. Chemical in muscle cells that reacts with ADP to make ATP
___ 8. Minimum stimulus required for a muscle cell to contract
___ 9. The bone that forms the lower jaw
___ 10. Unit along a myofibril that contains actin and myosin

A. Threshold
B. Sarcomere
C. Compact
D. Red
E. Intervertebral disk
F. Cervical
G. Osteoblasts
H. Phalanges
I. Creatine phosphate
J. Mandible

Answers to Self-Quiz

1. F; 2. E; 3. G; 4. D; 5. H; 6. C; 7. I; 8. A; 9. J; 10. B

SPOTLIGHT ON HEALTH: TWO

Healing Bones and Joints

Bones are extraordinary things. Their strength is legendary, yet injuries occur. Automobile and sports accidents produce broken bones, as well as dislocations that displace a bone from its joint, and sprains that stretch or tear ligaments.

A **bone fracture** is a serious condition requiring immediate medical assistance. Most fractures occur because of traumatic injuries that twist, shatter, or crack a bone (Figure 1). There are two main categories of bone fractures: simple fractures and compound fractures. In *simple fractures,* the broken bone does not puncture the skin and infection is rare. In *compound fractures,* a bone penetrates the surface of the skin and provides an opening for microorganisms to enter the body, causing an infection.

Proper healing begins after the broken ends of a bone have been realigned. This usually requires expert medical assistance, and in complicated fractures it may have to be done under general anesthesia that renders the patient safely unconscious. After realignment, the broken bone must be immobilized by a cast or another device to maintain alignment as healing proceeds.

Within hours after a bone is broken, a large blood clot encircles the area of the fracture (Figure 2). Osteoblasts from the periosteum migrate into the fracture area and begin to deposit bone. At the same time, osteoclasts remove bone fragments. Over the next 2 weeks, the blood clot slowly dissolves and

FIGURE 1 Types of fractures.

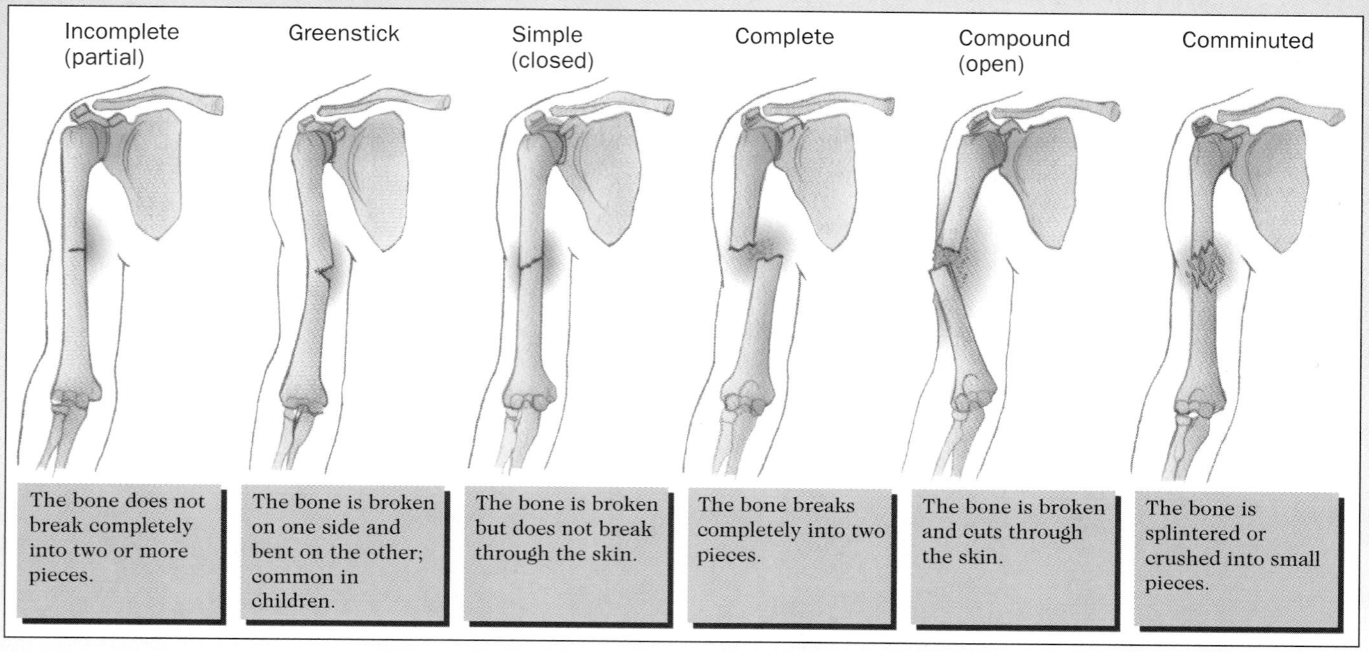
Incomplete (partial)
The bone does not break completely into two or more pieces.
Greenstick
The bone is broken on one side and bent on the other; common in children.
Simple (closed)
The bone is broken but does not break through the skin.
Complete
The bone breaks completely into two pieces.
Compound (open)
The bone is broken and cuts through the skin.
Comminuted
The bone is splintered or crushed into small pieces.

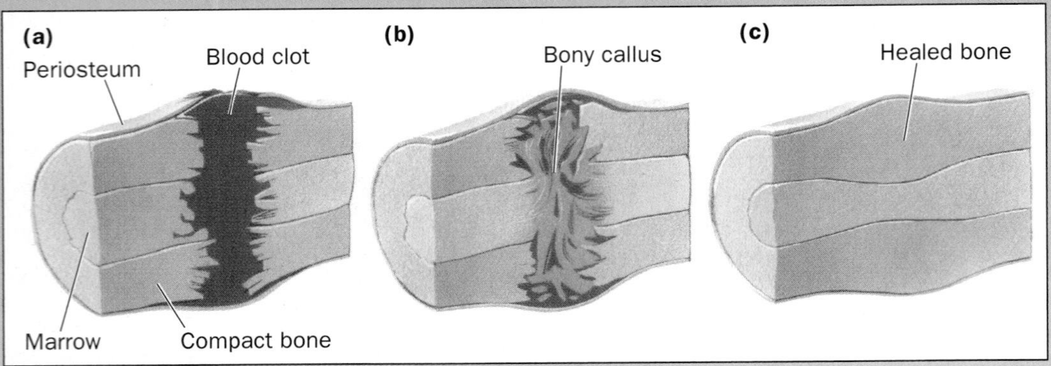
(a)
Periosteum
Blood clot
Marrow
Compact bone
(b)
Bony callus
(c)
Healed bone

FIGURE 2 Stages in the repair of a simple bone fracture.

new bone tissue is deposited in the fractured area, eventually forming a bony callus around the fracture.

Over 1 or 2 months a stable bony union develops, and remodeling occurs. After bone remodeling has been completed, there may be little visual evidence that a fracture occurred. At the end of the repair process, the bone is capable of withstanding the normal mechanical stresses of its attached muscles. Because of the abundant blood supply and living cells in bone tissue, healing occurs surprisingly fast under most conditions.

Although fairly common, fractures are not the only injuries that befall bone. In strenuous athletics or after severe bodily trauma, two bones may become separated at their joint so that the joint no longer operates properly. Such a condition is called a **dislocation.** In addition to displaced bone, there may be torn ligaments and damage to the connective tissue capsule surrounding the joint. The surrounding muscles, blood vessels, and nerves also may be damaged, resulting in considerable pain, swelling, and discoloration.

Usually the dislocated bones must be repositioned under general anesthesia, and healing may require immobilization of the joint in a cast or sling for several weeks. Healing is often slow because of the limited blood supply to the connective tissues of the ligaments and tendons.

Sprains are less severe than dislocations and result from the stretching or partial tearing of the ligaments that support and stabilize a joint. Joints in the knees, ankles, and fingers are the ones most often sprained. The pain is related to the extent of the damage, and healing is faster if the affected joint is not used.

P A R T   T H R E E

Processing, Absorbing, and Converting Raw Materials

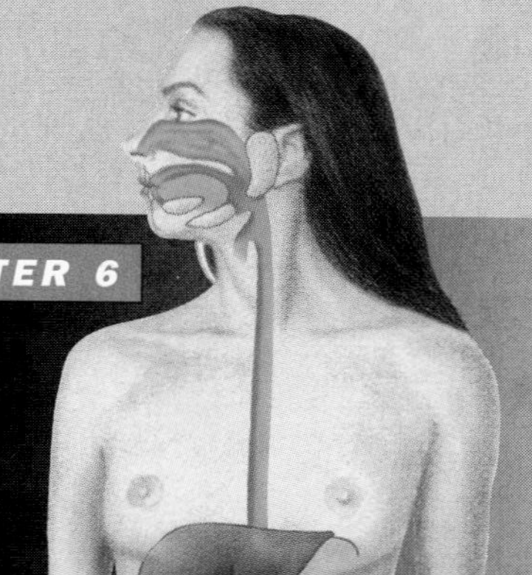

CHAPTER 6

The Digestive System

MAYO

CHAPTER 7

Nutrition and Metabolism

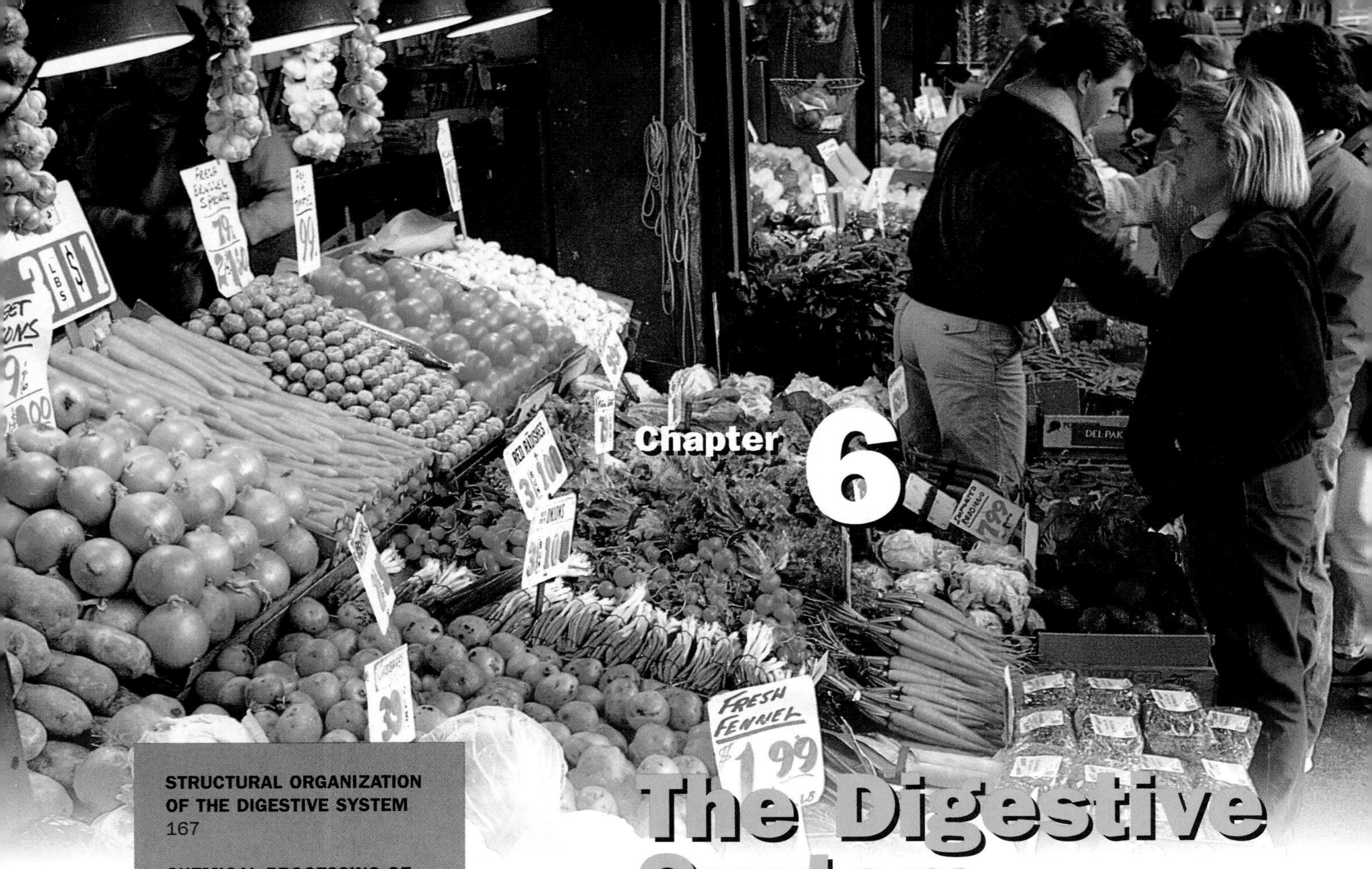
FRESH FENNEL
1.99
DEL PAK

Chapter 6

The Digestive System

Whether it is filet mignon, a caesar salad, or a cheeseburger, food sustains our life. Humans treat food with much ceremony and ritual. We spend time, effort, and money growing, preserving, marketing, and preparing food. However, after we put food into our mouths, what happens to it? Certainly we don't look like the food we eat. For example, the proteins in our bodies are different from the proteins in the meats and vegetables we eat.

We can live on a variety of foods because our digestive systems process the foods we eat. This processing involves both mechanical and chemical operations which break down the larger structures and molecules in food to smaller components. For example, from proteins come amino acids, and from polysaccharides come individual sugars. These amino acids and sugars are among the raw materials used by our cells for energy, growth, and repair.

In this chapter, we will examine the structures of the digestive system and their functions. Then we will describe the chemical breakdown of food and the absorption of the breakdown products into the blood. At the end of the chapter, we'll look at certain common diseases and disorders of the digestive system and summarize the normal changes that occur over a life span.

Structural Organization of the Digestive System

The digestive system consists of an extraordinary tube called the gastrointestinal (GI) tract or alimentary canal and its attached accessory organs (Figure 6.1). The **alimentary canal** (*al-ih-men-tar-ry*) is approximately 25 feet long and is located mostly in the abdominal cavity. The major sections of this tube are the mouth, pharynx, esophagus, stomach, small intestine, large intestine, rectum, and anus. Each section is specialized to perform distinct mechanical and/or chemical functions in order to digest and absorb the nutrients in food.

Accessory organs are attached by ducts to the alimentary canal and include the salivary glands, liver, gallbladder, and pancreas. These structures manufacture enzymes and other chemicals which contribute to the chemical processing of food.

In this section, we will describe the tissues found throughout the alimentary canal. Then we will discuss the anatomy and mechanical operation of the specialized regions of the canal and its accessory organs.

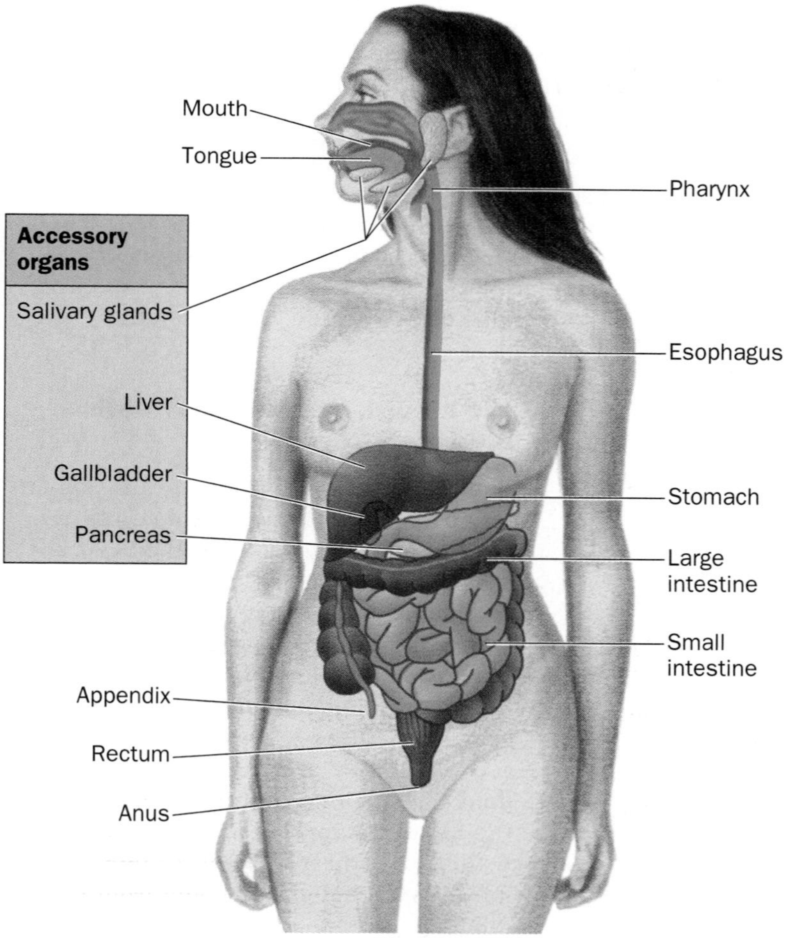

Mouth
Tongue
Pharynx
Accessory organs
Salivary glands
Esophagus
Liver
Gallbladder
Stomach
Pancreas
Large intestine
Small intestine
Appendix
Rectum
Anus

FIGURE 6.1 The alimentary canal and its attached accessory organs.

Four Tissue Layers of the Alimentary Canal

Except in the mouth, the wall of the alimentary canal consists of four layers. Starting from the innermost layer and proceeding outward, they are the mucosa, submucosa, muscularis, and serosa. The interior space of the tube—the place occupied by the food being processed—is referred to as the **lumen** (Figure 6.2).

The **mucosa** is a moist epithelial tissue that forms the inner lining of the alimentary canal from the mouth to the anus. Different types of epithelia are found in the different sections of the canal. Lining the mouth and esophagus is stratified squamous epithelium, which protects those areas from the normal abrasion caused by the passage of food. Lining the stomach, small intestine, and large intestine is simple columnar epithelium, which is specialized for the absorption of nutrients, the secretion of digestive enzymes, or the secretion of protective mucus.

The **submucosa** consists of connective tissue that binds the mucosa to the next layer, the muscularis. It contains many blood vessels, lymph vessels, and nerves which supply the tissues of the digestive tract.

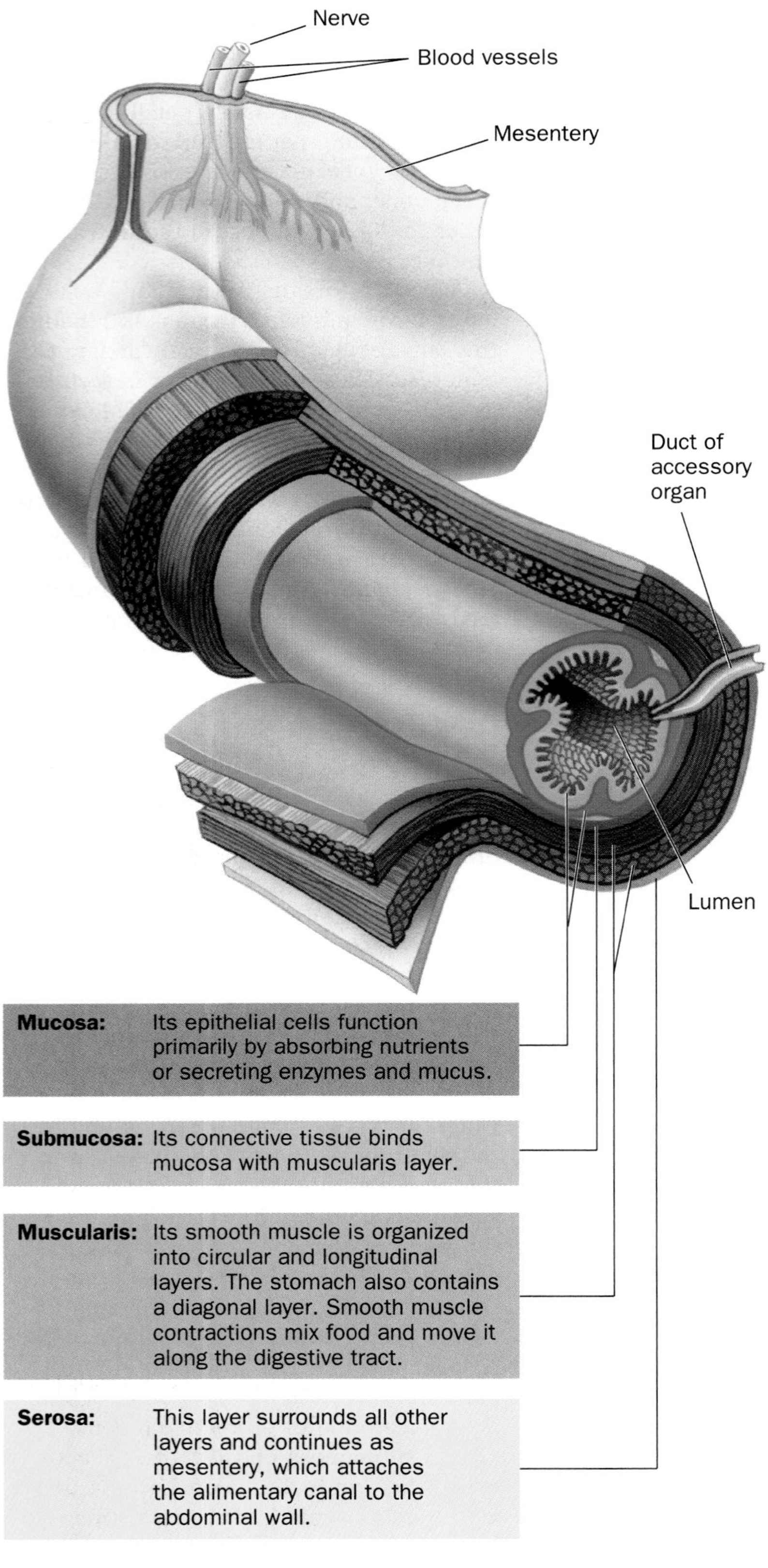
Nerve
Blood vessels
Mesentery
Duct of accessory organ
Lumen
Mucosa: Its epithelial cells function primarily by absorbing nutrients or secreting enzymes and mucus.
Submucosa: Its connective tissue binds mucosa with muscularis layer.
Muscularis: Its smooth muscle is organized into circular and longitudinal layers. The stomach also contains a diagonal layer. Smooth muscle contractions mix food and move it along the digestive tract.
Serosa: This layer surrounds all other layers and continues as mesentery, which attaches the alimentary canal to the abdominal wall.

FIGURE 6.2 Tissue layers of the alimentary canal.

The **muscularis** (*mus-cue-**lair**-iss*) contains mostly smooth muscles. These muscles produce the involuntary rhythmic contractions called **peristalsis** (*pair-iss-**tall**-sis*) which mix food and move it along the canal. In the upper third of the esophagus there is skeletal muscle, which gives us voluntary control of the swallowing process.

Along the remaining sections of the canal only smooth muscle is found, and it is organized into two or three sublayers. The first sublayer encircles the canal (circular muscle), and the second sublayer runs the length of the canal (longitudinal muscle). A third sublayer of smooth muscle encircles the stomach on a diagonal (oblique muscle). This layer adds a churning motion to the stomach that is not found in other parts of the canal.

The **serosa** (*ser-**oh**-sah*) is a connective tissue layer that forms the outermost covering of the alimentary canal. Special supportive structures called *mesenteries* are extensions of the serosa that attach it to the peritoneal tissue that lines the abdominal cavity (see Figure 6.2).

> The alimentary canal consists of four layers. The mucosa absorbs, secretes, or protects. The submucosa binds the mucosa to the muscularis. The muscularis is mostly smooth muscle that mixes and moves food. The serosa is the outermost layer.

The Mouth, Tongue, and Teeth: Digestion Begins

The **mouth** is the entrance to the alimentary canal (see Figure 6.1). With its tongue, teeth, and salivary glands, it is specialized for tasting, moistening, and mechanically digesting food. The mouth is where chemical digestion begins.

Outside the alimentary canal and near the mouth are three pairs of accessory glands called the **salivary glands** (see Figure 6.1). These glands produce a watery fluid called *saliva* that is transported into the mouth by ducts. Saliva moistens food, helping to form it into a manageable mass for chewing and swallowing. It also contains the enzyme *amylase* (***am**-ill-ase*),

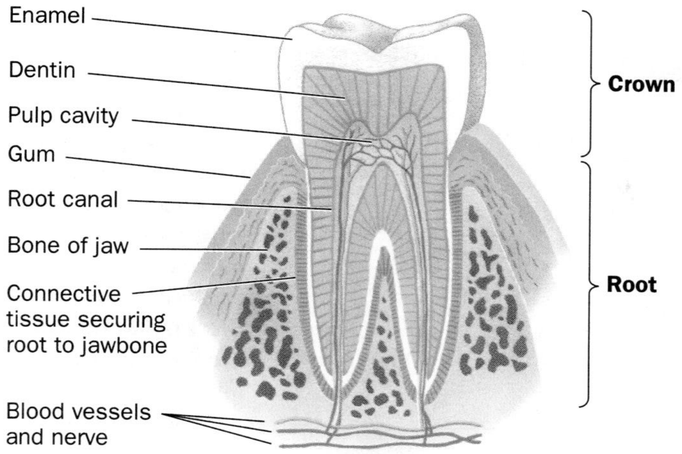
Enamel
Dentin
Pulp cavity
Gum
Root canal
Bone of jaw
Connective tissue securing root to jawbone
Blood vessels and nerve
Crown
Root

FIGURE 6.3 The anatomy of a tooth.

which functions to chemically break down starch, the polysaccharide found in foods such as potatoes, rice, and pasta.

The **tongue** is a muscular and sensory structure that is anchored to the floor of the mouth. The surface of the tongue contains sensory receptors called taste buds that permit us to enjoy food and detect spoiled or dangerous foods. Because the tongue is composed of skeletal muscle, we can consciously control its movement. This allows us to manipulate food during chewing and swallowing.

Teeth: biting, chewing, and grinding. The **teeth** mechanically digest food by reducing its size and preparing it for swallowing. Each tooth has two regions: the crown and the root (Figure 6.3). The **crown** is the exposed portion above the gums, and the **root** is embedded in the jawbone and secured to it by connective tissue. Most of the crown and the root is composed of a living bonelike material called *dentin.* Although it is as hard as bone, dentin cannot withstand the continual wear from chewing and the acidity of some foods, and so the dentin of the crown is protected by enamel. *Enamel* is an extremely hard nonliving substance that is made of calcium and phosphate.

The interior of the tooth contains a space called the *pulp cavity.* This cavity contains nerves and blood vessels which nourish and maintain the dentin. Nerves and blood vessels enter the pulp cavity through a small opening at the bottom of the root.

There are four types of teeth: incisors, canines, premolars, and molars (Figure 6.4a and Table 6.1). The incisors and canines are used for biting and tearing food, while the premolars and molars grind and mash food. During your life, two separate sets of teeth form. The *deciduous teeth,* or baby teeth, are completely developed by about 2 years of age. They are gradually lost and replaced by the *permanent teeth* by the end of adolescence. The one exception is the wisdom teeth (third molars), which typically appear between 17 and 25 years of age.

Everyone is familiar with the tongue, teeth, and saliva. Less obvious are the bacterial residents of the mouth (Figure 6.4b). Our mouths are hosts to large numbers of bacteria, and we usually experience their presence through their harmful actions on the teeth. These bacteria live on the sugars in our diet and release acid which over time can produce tiny holes in the enamel called cavities or *dental caries.* If they are not treated, cavities may progress into the dentin and pulp cavity, causing a toothache.

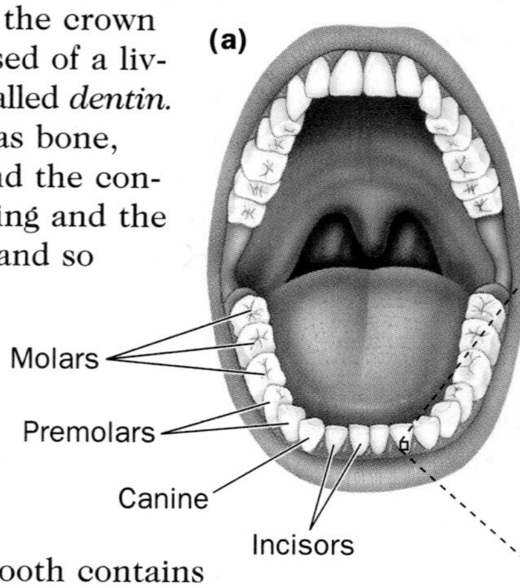
(a)
Molars
Premolars
Canine
Incisors

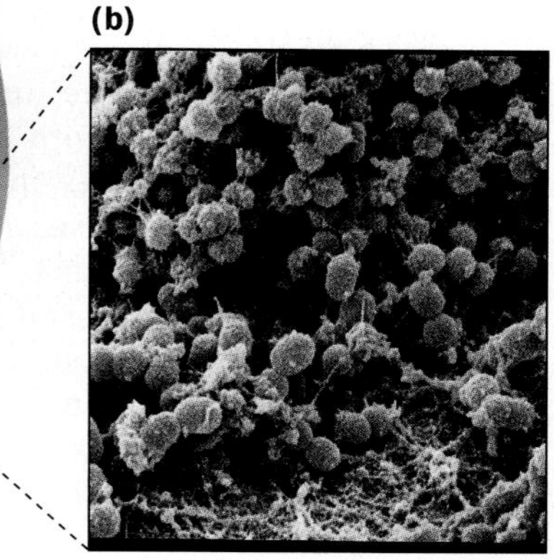
(b)

FIGURE 6.4 (a) The four types of teeth and their location in the mouth. (b) A highly magnified photograph of bacteria adhering to a tooth surface.

TABLE 6.1 Deciduous and Permanent Teeth

Teeth			Total Number	
Tooth Type	**Shape**	**Function**	**Deciduous**	**Permanent**
Incisor	 Chisel	Cutting	8	8
Canine	 Conical	Grasping, piercing, tearing	4	4
Premolar	 Broad surface	Crushing, grinding	0	8
Molar	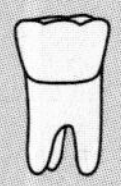 Broad surface	Crushing, grinding	8	12
		Total	20	32

There are four types of adult teeth. Each tooth has a crown and root consisting of dentin and a pulp cavity. The crown is covered by enamel. Salivary glands secrete saliva, which contains the enzyme amylase.

The Pharynx and Esophagus: To Swallow or Not to Swallow

At the posterior end of the mouth there is a passageway called the **pharynx** (*fair-inks*) (see Figure 6.1). Often referred to as the throat, it is a common passageway for both air and food. This dual use requires special structures and muscle responses so that when you swallow, food passes to the stomach, not into the windpipe and lungs.

The act of swallowing is so familiar that we take it for granted until something goes wrong and we choke. When that happens, the blockage of the windpipe (trachea) can cause death. What are the coordinated movements that occur every time you swallow? First, as food enters the pharynx the soft palate moves up and prevents liquids and solids from entering the nasal cavity (Figure 6.5a and b). Second, the upper portion of the windpipe, called the voice box or *larynx,* moves up and the tongue moves back. As a result, a flap of cartilage at the base of the tongue called the *epiglottis* (*ep-ee-**glot**-tis*) covers the opening of the windpipe.

These coordinated reflexes of the tongue, larynx, and epiglottis prevent food from entering the lungs. When food or liquid accidentally enters the windpipe, we cough. A cough is a reflex response that consists of a forceful expulsion of air from the lungs to clear material from the windpipe.

Esophagus: a last chance not to swallow. The **esophagus** (*eh-**sof**-ah-gus*) is a relatively short (25 centimeters, or 10 inches), narrow tube that passes through the diaphragm and connects the pharynx to the stomach (see Figures 6.1 and 6.5c). After swallowing, food passes from the pharynx to the esophagus. Because of the presence of skeletal muscle in the wall of the pharynx and the upper part of the esophagus, we can voluntarily retrieve food that has entered this region. Recall the last time you almost lost a piece of chewing gum to an inadvertent swallow but were able to save it before it went too far down your throat.

In the remaining part of the esophagus, the muscularis consists of smooth muscle and peristalsis moves food to the stomach (see Figure 6.5c). The body does not rely on gravity to move food and liquids. Peristalsis transports these substances even when you are in a horizontal position (eating while lying down) or beyond the Earth's gravitational force (traveling in space).

The pharynx is the common route for both food and air. The esophagus joins the pharynx to the stomach. Peristalsis moves food along the esophagus to the stomach.

The Stomach: Churning for Digestion

The **stomach** is an enlarged J-shaped region of the alimentary canal that is located in the upper left region of the abdominal cavity (Figures 6.1 and 6.6). We often locate the stomach by pointing to the area of the naval (belly button), but this is incorrect. The stomach is actually located higher in the abdominal cavity. It is the site of both mechanical digestion and chemical digestion, after which food passes into the small intestine.

The passageways into and out of the stomach are opened and closed by specialized circular bands of smooth muscle called **sphincters** (***sfink**-ters*) (see Figure 6.6). The *esophageal sphincter* (*eh-sof-ah-**gee**-al*) is located at the junction of the esophagus and the stomach, and the *pyloric sphincter* (*pie-**lore**-ick*) is located at the junction of the stomach and the small intestine. When food enters the esophagus, the esophageal sphincter relaxes, allowing the food to enter the stomach. When food leaves the stomach, the pyloric sphincter relaxes so that material can pass into the small intestine. When both of these sphincters are contracted, the stomach forms a closed compartment capable of holding 1 to 3 liters (about 1 to 3 quarts) of food.

After a large meal the esophageal sphincter may fail to close completely, and gastric fluids can enter the esophagus, causing a burning sensation known as heartburn. The name for this uncomfortable sensation emphasizes how close the lower portion of the esophagus is to the heart. Indeed, heartburn can sometimes be mistakenly identified as a heart attack.

Glands and secretions. The stomach mucosa contains millions of microscopic

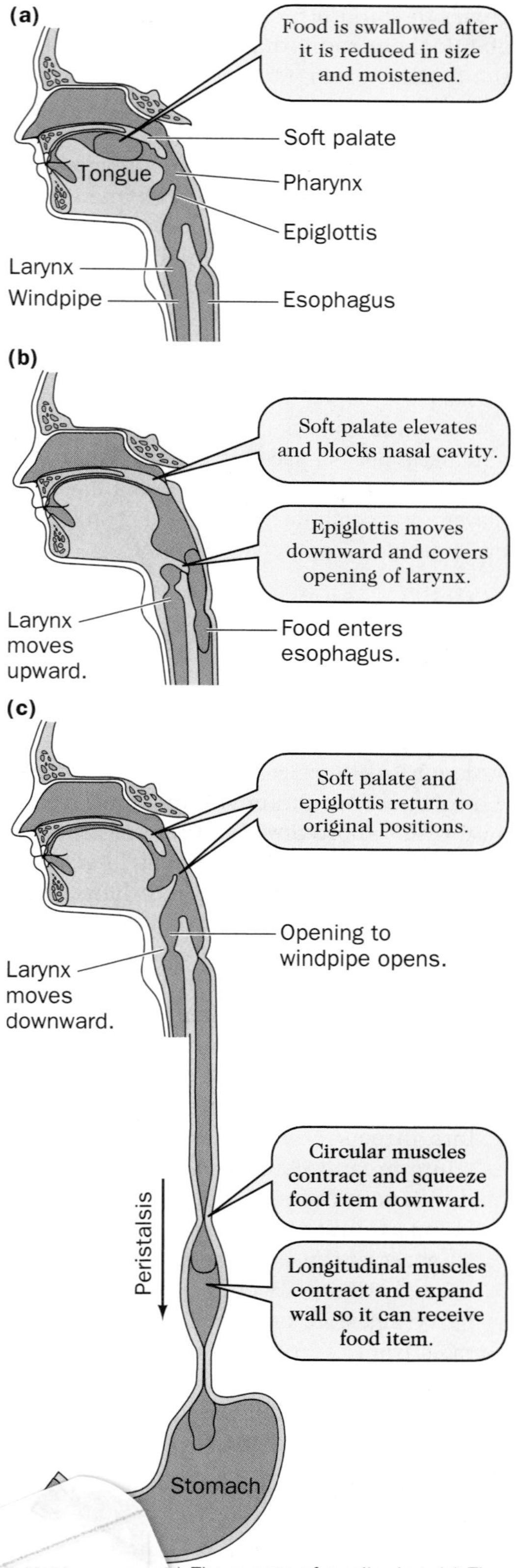

(a)
Food is swallowed after it is reduced in size and moistened.
Soft palate
Tongue
Pharynx
Epiglottis
Larynx
Windpipe
Esophagus
(b)
Soft palate elevates and blocks nasal cavity.
Epiglottis moves downward and covers opening of larynx.
Larynx moves upward.
Food enters esophagus.
(c)
Soft palate and epiglottis return to original positions.
Opening to windpipe opens.
Larynx moves downward.
Circular muscles contract and squeeze food item downward.
Peristalsis
Longitudinal muscles contract and expand wall so it can receive food item.
Stomach

) The events of swallowing. (c) The ... along the esophagus by peristalsis.

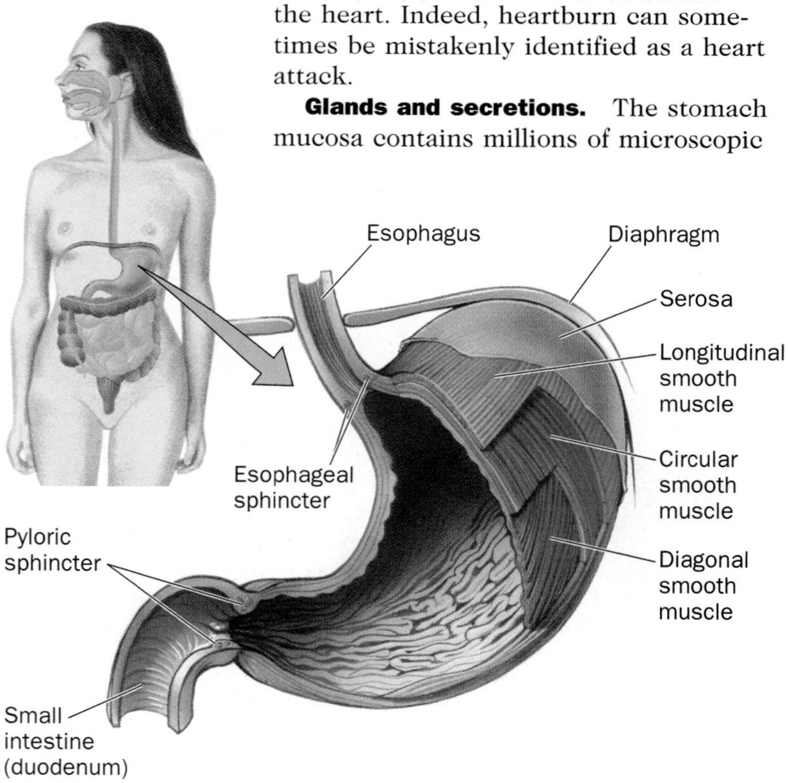

Esophagus
Diaphragm
Serosa
Longitudinal smooth muscle
Circular smooth muscle
Diagonal smooth muscle
Esophageal sphincter
Pyloric sphincter
Small intestine (duodenum)

FIGURE 6.6 The anatomy of the stomach.

pitlike depressions called **gastric pits** (Figure 6.7). Each gastric pit contains columnar epithelial cells known as **gastric glands.** Some of these cells secrete hydrochloric acid (HCl) and other substances that are collectively called *gastric juice.* Other gastric gland cells secrete the hormone *gastrin,* which helps regulate digestion. We will discuss more about gastric juice and gastrin again later in this chapter.

The stomach's churning action. Because the stomach has a third sublayer of smooth muscle that is oriented at a diagonal, it can contract in a special fashion. The coordinated contraction of the three layers of smooth muscle produces a mechanical churning that mixes food with the gastric juices. Mixing and churning usually continue for 2 to 4 hours, depending on the type and quantity of food consumed. Eventually, a solid meal is converted to a semiliquid called **chyme** (*kime*). In this form, food leaves the stomach and enters the small intestine for further chemical digestion and absorption.

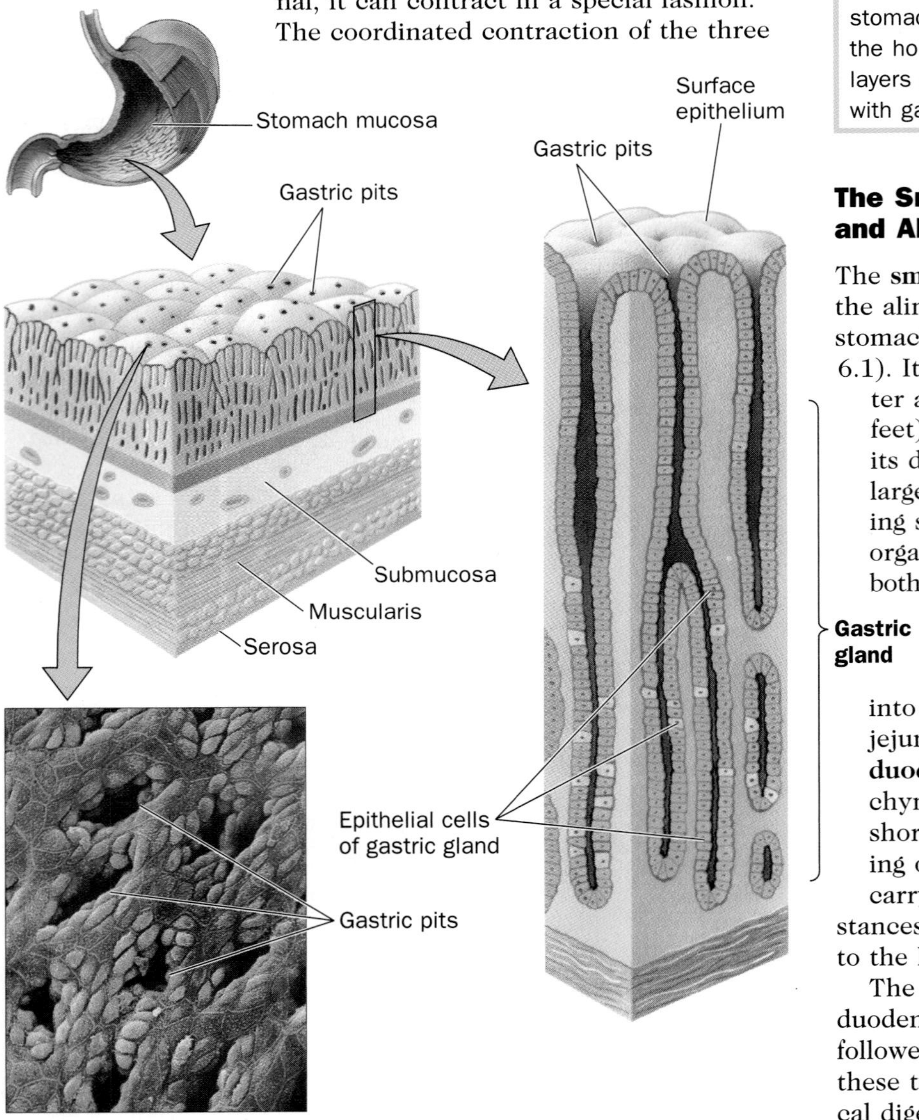

Stomach mucosa
Gastric pits
Submucosa
Muscularis
Serosa
Surface epithelium
Gastric pits
Gastric gland
Epithelial cells of gastric gland
Gastric pits

Magnified photo of stomach mucosa

FIGURE 6.7 Enlarged view of stomach mucosa showing gastric pits and epithelial cells of a gastric gland.

> Sphincters control the entrance and exit of materials to and from the stomach. The stomach mucosa secretes gastric juice and the hormone gastrin. The stomach's three layers of smooth muscle mix and churn food with gastric juice, producing chyme.

The Small Intestine: Digestion and Absorption

The **small intestine** is the next portion of the alimentary canal, extending from the stomach to the large intestine (see Figure 6.1). It is about 2.5 cm (1 in.) in diameter and approximately 6 meters (20 feet) long. It is called "small," because its diameter is smaller than that of the large intestine. However, there is nothing small about the importance of this organ. The small intestine is the site of both mechanical and chemical processes that prepare nutrients for absorption.

The small intestine is divided into three sections: the duodenum, jejunum, and ileum (Figure 6.8). The **duodenum** (*dew-oh-**dee**-num*) receives chyme from the stomach. It is the shortest of the three sections, measuring only about 25 cm (10 in.). Ducts carry digestive enzymes and other substances from the pancreas and gallbladder to the lumen of the duodenum.

The small intestine continues from the duodenum as the **jejunum** (*jeh-**june**-um*), followed by the **ileum** (***ill**-ee-um*). It is in these two regions that most of the chemical digestion and absorption of nutrients occur. Both digestion and absorption are made more efficient by structural modifications that increase the total surface area of the intestinal mucosa. The larger the surface area is, the more rapidly absorption occurs.

The surface area for absorption. The structural modifications that increase the surface area for absorption include circular folds, villi, and microvilli (Figure 6.8). The large **circular folds** project into the lumen of the intestine for a distance of about 1 cm. Besides increasing the surface area, they help mix chyme with the secretions from the pancreas and gallbladder.

The mucosa of the small intestine forms small fingerlike projections called **villi** (***vill-eye***). A single projection is a villus (***vill-us***). Each villus is about 1 millimeter in height and surrounds small blood vessels called *capillaries* and a lymphatic vessel called a *lacteal.* These numerous small villi give the intestinal lining a velvetlike texture.

Each villus is covered with columnar epithelial cells which are directly involved in absorbing nutrients from the lumen and passing them to the capillaries and lacteals. To further increase the surface area, the exposed surface of each columnar cell is covered by hundreds of delicate projections of the plasma membrane called **microvilli** (Figure 6.8). Together, the folds, villi, and microvilli increase the surface area of the mucosa about 600-fold. As a result, you absorb nutrients over a surface area almost as large as the surface area of a singles tennis court.

Mixing and churning. During mechanical digestion, chyme is mixed and churned in a process called **segmentation.** During segmentation, the circular smooth muscles contract so that the intestine is divided into a series of segments resembling a chain of sausages (Figure 6.9). Within a short time (10 seconds), these circular muscles relax as other muscles contract and subdivide the "sausages." This continues along the entire length of the small intestine, thoroughly mixing the chyme and assuring that it comes into contact with the columnar cells where absorption takes place.

> The small intestine is divided into three regions. The surface area of the mucosa is increased by circular folds, villi, and microvilli. Contractions of the smooth muscle produce segmentation and peristalsis to mix and move chyme.

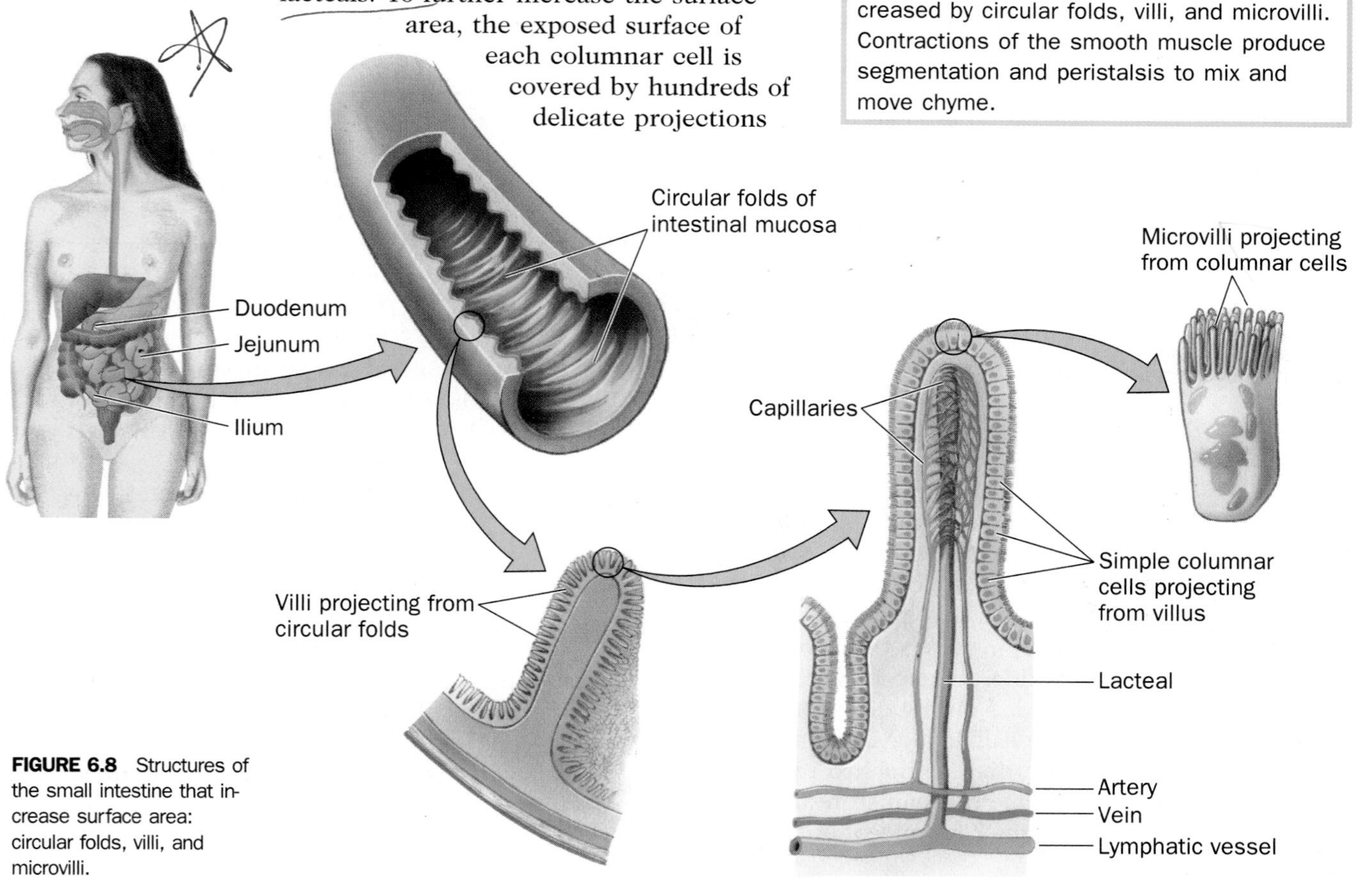
Duodenum
Jejunum
Ilium
Circular folds of intestinal mucosa
Villi projecting from circular folds
Capillaries
Microvilli projecting from columnar cells
Simple columnar cells projecting from villus
Lacteal
Artery
Vein
Lymphatic vessel

FIGURE 6.8 Structures of the small intestine that increase surface area: circular folds, villi, and microvilli.

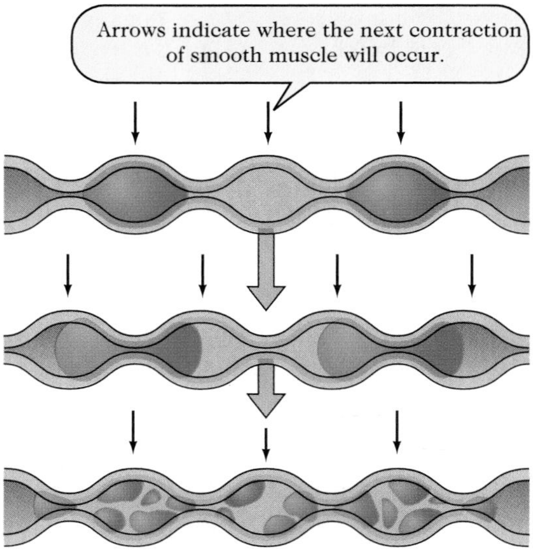
Arrows indicate where the next contraction of smooth muscle will occur.

FIGURE 6.9 A series of segmentation contractions showing how chyme in the small intestine is mixed.

The Accessory Organs: Agents of Digestion

As we mentioned earlier in this chapter, the **accessory organs** of the digestive system include the salivary glands, the liver and gallbladder, and the pancreas. These organs supply secretions for the chemical digestion of food. Before discussing these secretions, we will describe the anatomic relationships of these structures with the small intestine. Earlier in this chapter we located the salivary glands and described their role.

The liver: processing fats and doing other things. The **liver** is a large organ situated in the upper right region of the abdominal cavity, immediately below the diaphragm (see Figure 6.1). The liver manufactures and secretes a liquid substance called **bile,** which is involved in the processing of fats. In addition to its role in digestion, the liver has a number of other important functions (Table 6.2).

Once bile has been secreted by the liver, it travels through ducts to the **gallbladder,** where it is stored until it is needed during digestion (Figure 6.10). The gallbladder is a small saclike structure that is attached to the underside of the liver. It can store up to 70 milliliters (about 2.5 ounces) of bile. When it is stimulated, the gallbladder releases its stored bile into a duct that empties into the lumen of the duodenum.

The pancreas: two functions. The **pancreas** is an elongated organ that is about 15 cm (6 in.) long and is positioned next to the stomach and duodenum (see Figure 6.10). It has both exocrine and endocrine functions. As an exocrine gland, the pancreas manufactures and secretes (1) sodium bicarbonate ($NaHCO_3$) to neutralize stomach acid and (2) various enzymes that chemically digest food. These secretions are carried to the duodenum by ducts. As an endocrine gland, the pancreas secretes hormones that regulate the amount of sugar in the blood (Chapter 16).

> The liver produces bile, which aids in fat digestion. The gallbladder stores bile and releases it to the duodenum. The secretions of the pancreas include sodium bicarbonate and digestive enzymes which are released into the duodenum.

TABLE 6.2 Nondigestive Functions of the Liver

The Liver
• **Stores** glycogen, vitamins (A, D, E, and B_{12}), and some minerals, such as iron (Fe) and copper (Cu).
• **Detoxifies** many harmful substances that have been consumed or have otherwise entered the body, such as alcohol and pesticides.
• **Interconverts** certain nutrients. Glucose and amino acids may be converted to fat. Amino acids and lactic acid may be converted to glucose.
• **Maintains** normal concentrations of glucose, amino acids, and fats in the blood.
• **Manufactures** blood proteins. Some of these proteins are required for blood clotting, while others help maintain the proper osmotic balance between blood and tissues.

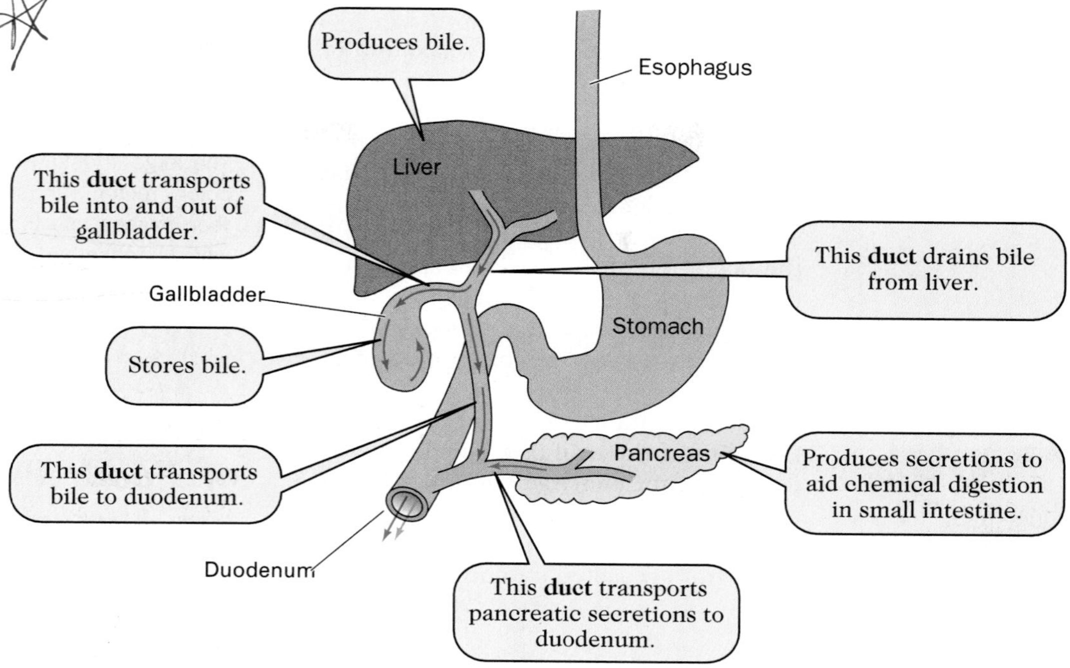
Produces bile.
Esophagus
Liver
This duct transports bile into and out of gallbladder.
This duct drains bile from liver.
Gallbladder
Stomach
Stores bile.
Pancreas
Produces secretions to aid chemical digestion in small intestine.
This duct transports bile to duodenum.
Duodenum
This duct transports pancreatic secretions to duodenum.

FIGURE 6.10 Anatomic and functional relationships between the liver, gallbladder, pancreas, and duodenum.

The Large Intestine: Absorbing Water and Storing Wastes

Any undigestible food residue from the small intestine passes through the **large intestine** before being eliminated from the body. In the large intestine, water is absorbed from undigested food and feces are formed and stored before their elimination from the body. The large intestine is about 1.5 m (5 ft) long and 6.25 cm (2.5 in.) in diameter and is divided into the cecum, colon, rectum, and anal canal (Figure 6.11).

The **cecum** (***seek**-um*) is a short pouchlike compartment that receives chyme from the small intestine. The passage of chyme into the cecum is controlled by the operation of a flaplike valve. This valve permits chyme to move in one direction, from the small intestine to the cecum. Attached to the cecum is a narrow pouch about as long as your ring finger that is called the appendix.

The function of the *appendix* is unknown, but this pouch is significant because it may become inflamed or infected, resulting in *appendicitis.* This is a serious condition since a severely inflamed appendix may break open. If this happens, the contents of the appendix, including bacteria, spill into the abdominal cavity, where they can cause a life-threatening infection.

After the cecum comes the largest portion of the large intestine, the **colon** (***coal**-un*), which consists of four different parts distinguished by position or shape: the ascending colon, transverse colon, descending colon, and sigmoid colon (see Figure 6.11). Following the sigmoid colon is the rectum (***rek**-tum*), which stores feces until they can be eliminated. During defecation, feces pass from the rectum through the *anal canal* and out an opening, the **anus.** The wall of the anal canal contains two sphincter muscles that play a role in defecation.

Controlling elimination. Elimination of feces is controlled by both an involuntary reflex called the **defecation reflex** and a learned process that is under conscious control. As feces enter the rectum, the rectal wall is stretched and nerves are stimulated, carrying signals to the spinal cord and back to the rectum, where the smooth muscle of the rectum is stimulated to contract. The nerves from the spinal cord also signal the *internal anal* *sphincter* to relax (see Figure 6.11). Contraction of the rectum and relaxation of the internal sphincter force feces out of the body.

In addition to the internal sphincter, there is an *external anal sphincter* that contains skeletal muscle and is under conscious control (see Figure 6.11). It is the operation of this external sphincter that a developing child learns to control. This control prevents defecation from being entirely a reflex action in a normal healthy person.

In adults, increased blood pressure in the veins of the anal canal sometimes causes them to enlarge and bulge into the canal, forming *hemorrhoids*, or piles. Increased pressure within these veins can result from repeated straining during defecation or during pregnancy, when the weight of the uterus and fetus puts pressure on veins in the pelvic region.

> The large intestine includes the cecum, colon, rectum, and anal canal. Undigestible food passes into the large intestine, where water is absorbed and feces are formed and stored before defecation. Defecation involves both unconscious and conscious control.

Chemical Processing of Food: Preparation for Absorption

In the preceding sections we focused primarily on the general anatomy of the digestive system and the mechanical processing of food. In this section we will examine how substances secreted by both the mucosa and the accessory organs accomplish chemical digestion of the carbohydrates, proteins, fats, and nucleic acids in food. These processes take place in the mouth, stomach, and small intestine; the large intestine is not involved. A summary of mechanical and chemical digestion processes and sites is presented in Table 6.3.

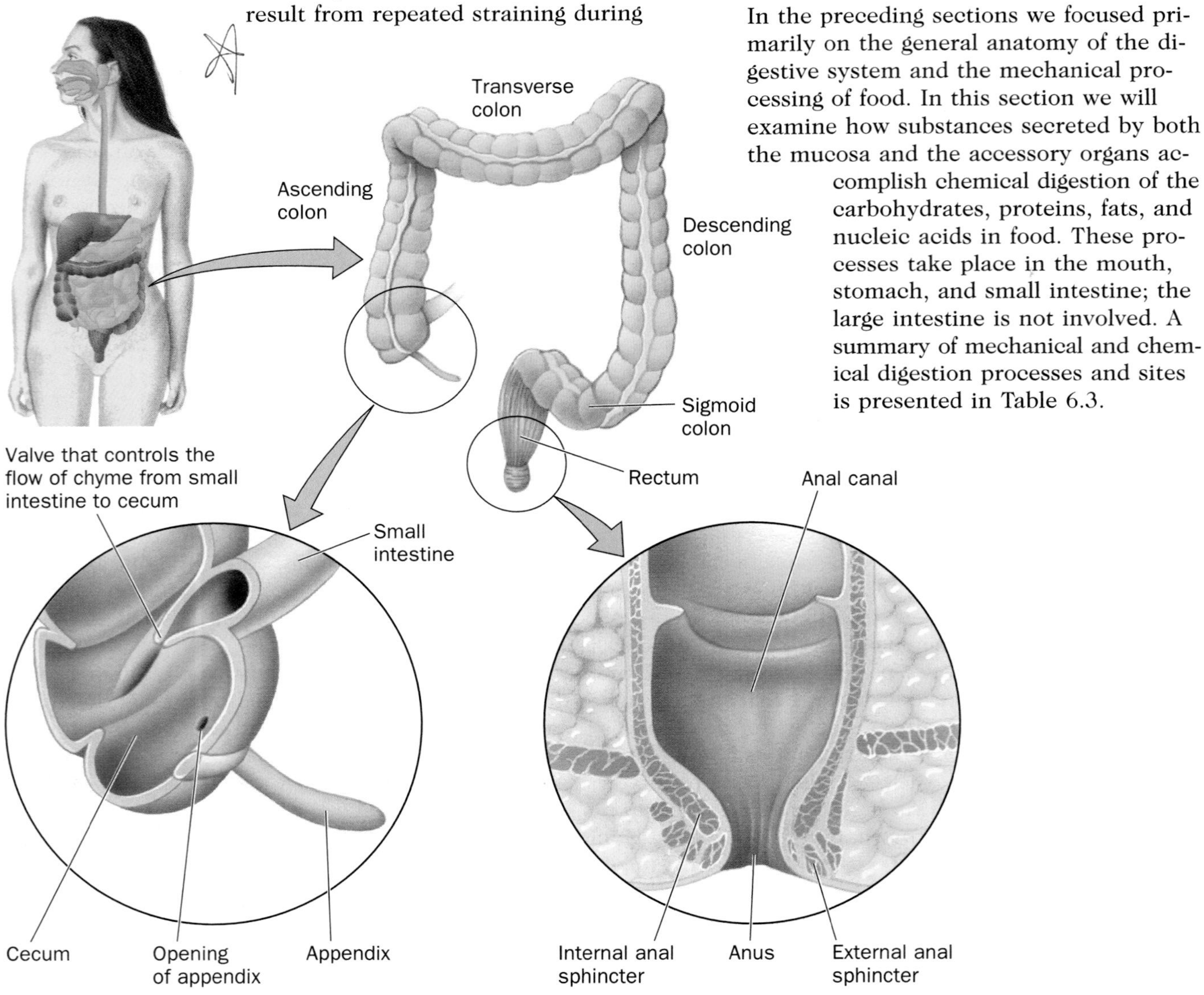
Transverse colon
Ascending colon
Descending colon
Sigmoid colon
Rectum
Anal canal
Valve that controls the flow of chyme from small intestine to cecum
Small intestine
Cecum
Opening of appendix
Appendix
Internal anal sphincter
Anus
External anal sphincter

FIGURE 6.11 The anatomy of the large intestine.

TABLE 6.3 Summary of Mechanical and Chemical Digestion

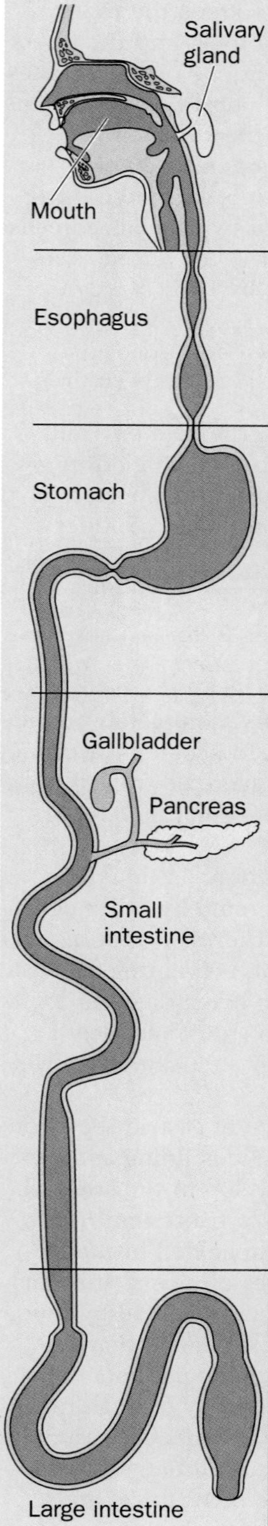

	Mechanical Digestion	Chemical Digestion: Carbohydrates	Chemical Digestion: Proteins	Chemical Digestion: Fats
Mouth, Salivary gland	Chewing reduces size of food and mixes it with saliva.	**Salivary glands:** Secrete amylase, which begins starch digestion. Starch → (*Amylase*) → Disaccharide, maltose	No action	No action
Esophagus	Peristalsis moves food to stomach.	No action	No action	No action
Stomach	Peristalsis churns food and mixes it with gastric secretions, producing a semiliquid called chyme.	No action	HCl softens food and activates pepsinogen to pepsin. Pepsin begins protein digestion. Protein → (*Pepsin*) → Short chains of amino acids (peptides)	No action
Small intestine (Gallbladder, Pancreas)	Peristalsis moves chyme along intestine. Segmentation mixes chyme with secretions from gallbladder and pancreas.	**Pancreas:** Releases amylase into intestine, which continues starch digestion. Starch → (*Amylase*) → Disaccharide, maltose **Intestinal mucosa:** Disaccharidases released, which break down disaccharides. Disaccharides → (*Disaccharidases*) → Monosaccharides	**Pancreas:** Releases enzymes such as trypsin into intestine, which continues protein digestion. Protein → (*Trypsin*) → Short chains of amino acids (peptides) **Intestinal mucosa:** Peptidases released, which break down short chains of amino acids to individual amino acids. Short chain of amino acids → (*Peptidase*) → Individual amino acids	**Pancreas:** Releases lipase, which digests fat. Fat → (*Lipase*) → Fatty acid + Glycerol **Gallbladder:** Releases bile, which emulsifies large fat globules. Large fat globules → (*Bile*) → Small fat globules
Large intestine	Peristalsis moves chyme along intestine.	Resident bacteria in colon digest material in chyme not previously digested in small intestine. Vitamin K and some B-complex vitamins are produced and released by bacteria.		

The Mouth and Saliva: Beginning Starch Digestion

With an empty stomach, we eagerly anticipate eating. The sight, smell, and taste of food activate reflexes that stimulate the salivary glands to secrete saliva. Saliva not only moistens food but contains the enzyme **amylase,** which chemically digests starch. Starch is a polysaccharide consisting of millions of glucose molecules bonded together (Chapter 1). The human digestive system cannot absorb large starch molecules, and so they must be broken down into smaller glucose units which can be absorbed.

Salivary amylase begins this process by converting starch into maltose, a disaccharide sugar that consists of two glucose molecules bonded together. Because food usually remains in the mouth briefly, less than 10 percent of the starch you eat is converted to maltose in your mouth. After swallowing, the digestion of starch is stopped by the acid condition of the stomach which inactivates amylase.

> Saliva contains amylase, which breaks down starch, forming maltose. Amylase is inactivated by the acidity of the stomach.

The Stomach: Beginning Protein Digestion

The sight, smell, and taste of food stimulate the secretion of gastric juice into the stomach. Gastric juice contains water, intrinsic factor, hydrochloric acid (HCl), pepsinogen, and mucus. How does each of these substances contribute to the processing of food?

Intrinsic factor for vitamin B_{12}. The protein called *intrinsic factor* is not directly involved in digestion but is needed for the absorption of vitamin B_{12} from the small intestine. Vitamin B_{12} is required for the manufacture of red blood cells and must be obtained from the food we eat. An insufficient quantity of vitamin B_{12} results in the disorder called *pernicious anemia.*

Hydrochloric acid. Hydrochloric acid is a very strong acid that is secreted in large amounts by specialized cells in the gastric pits. Stomach acidity (about pH 2) is entirely normal and is important in several ways: (1) HCl kills many microorganisms, such as bacteria, helping to protect the digestive tract from any disease-causing organisms that enter with food or fluids. (2) This acid softens the tough connective tissues in meat and the fibers in vegetables and fruits. (3) HCl activates pepsinogen, forming pepsin, which begins protein digestion in the stomach.

Activating pepsinogen to digest protein. *Pepsinogen* is an inactive enzyme that is secreted by specialized cells in the gastric pits. It is converted to its active form, **pepsin,** by HCl:

$$\underset{\text{Inactive enzyme}}{\text{Pepsinogen}} \xrightarrow{\text{HCl}} \underset{\text{Active enzyme}}{\text{pepsin}}$$

Pepsin works well in the acid environment of the stomach, breaking down large protein molecules into shorter chains of amino acids called peptides:

$$\underset{\text{Long chains of amino acids}}{\text{Protein}} \xrightarrow{\text{pepsin}} \underset{\text{Short chains of amino acids}}{\text{peptides}}$$

For absorption and use by the cells, proteins must be digested completely to individual amino acids. To accomplish this, additional enzymes secreted from the pancreas and the intestinal mucosa are necessary.

Mucus for protection. *Mucus* is a thick slimy fluid secreted by other cells in the gastric pits. It forms a coating over the stomach mucosa, preventing the wall of the stomach from being digested by its own HCl and pepsin. You can digest a thick steak, but your stomach lining will be untouched.

An overabundance of gastric secretions can penetrate the mucus lining and cause sores (ulcers) to develop in the stomach mucosa. Recently, the bacterium *Helicobacter pylori* was implicated in the formation of some types of ulcers. It seems that this bacterium is protected by mucus and releases a toxic chemical that destroys the mucosa. Some patients with ulcers have been cured by taking oral antibiotics that destroy this bacterium.

Only protein is substantially digested in the stomach. The mechanical and chemical processes of the stomach reduce solid chunks of food to the semiliquid chyme. Chyme passes into the small intestine in small volumes as the pyloric sphincter periodically relaxes. Depending on the volume and type of food consumed, a typical meal may take up to 4 hours to be processed and moved completely into the duodenum. It is in the small intestine that the chemical digestion of food is completed.

> Gastric juice contains intrinsic factor, HCl, pepsinogen, and mucus. Intrinsic factor is necessary for vitamin B_{12} absorption. HCl converts pepsinogen to the active enzyme pepsin, which digests proteins to peptides. Mucus protects the stomach mucosa.

The Small Intestine: Completing Digestion

The small intestine carries out chemical digestion by using bile released by the gallbladder and a variety of enzymes secreted by the pancreas and the intestinal mucosa. The pancreas also secretes the alkaline substance sodium bicarbonate, which neutralizes the acidity of the chyme that enters from the stomach (pH 2). This neutralization is necessary because the intestinal enzymes operate at neutral or slightly alkaline conditions (pH values of 7 to 8).

Bile: breaking up fat globules. The *bile* secreted by the gallbladder contains chemicals called bile salts that reduce the size of fat globules. Everyone is familiar with the fact that fats and oils do not mix with water. In water, they form into layers or large globules. You may have seen such globules floating on the top of your coffee after you dunk a cookie. In your intestine bile breaks up large fat globules into much smaller fat droplets in a process called *emulsification*. This is important because the enzyme that digests fat operates more efficiently on small droplets:

$$\text{Larger fat globules} \xrightarrow{\text{bile}} \text{smaller fat droplets}$$

Pancreatic enzymes: digesting "everything." Released into the duodenum by a duct, the pancreatic enzymes include lipase (to digest fats), amylase (to digest starch), trypsinogen (to digest proteins), and nucleases (to digest nucleic acids).

Lipase acts on individual fat molecules, breaking their bonds to yield glycerol and fatty acids:

$$\text{Fat molecules} \xrightarrow{\text{lipase}} \text{fatty acids} + \text{glycerol}$$

Amylase from the pancreas continues the breakdown of starch which began in the mouth:

$$\text{Starch} \xrightarrow[\text{amylase}]{\text{pancreatic}} \text{maltose}$$

Trypsinogen is an inactive form of the enzyme **trypsin.** After being activated in the small intestine, trypsin continues the breakdown of proteins which began with pepsin in the stomach:

$$\text{Protein} \xrightarrow{\text{trypsin}} \text{peptides}$$

Nucleases digest DNA and RNA into individual nucleotide molecules:

$$\text{RNA or DNA} \xrightarrow{\text{nucleases}} \text{individual nucleotides}$$

After they are broken down, nucleotides provide the body with pentose sugars, nucleotide bases, and phosphate. These components can be used to duplicate your unique cellular DNA before the division of your cells and to manufacture RNA for protein synthesis (Chapter 2).

Through the combined actions of bile and pancreatic enzymes, large molecules of fat, protein, starch, and nucleic acid are digested to smaller molecules. However, the process is not yet complete. Before they can be absorbed, peptides and disaccharides must be broken down further by enzymes secreted by the mucosal cells of the small intestine.

Enzymes from the intestinal mucosa. The epithelial cells of the intestinal mucosa manufacture and secrete two types of digestive enzymes: disaccharidases and peptidases. These enzymes are not released into the lumen of the intestine but remain closely associated with the microvilli of the epithelial cells. *Disaccharidases* break down disaccharide sugars into monosaccharide sugars.

Sometimes digestion is incomplete and some disaccharides are not broken down and absorbed. For example, some people have a condition called **lactose intolerance** which results when epithelial intestinal cells fail to produce the enzyme lactase. Normally, this enzyme breaks down the sugar lactose. Since lactose cannot be absorbed into the body, it passes into the large intestine, where the resident bacteria use it. During this process the bacteria release gas that accumulates in the large intestine, causing discomfort, bloating, flatulence, and in severe cases diarrhea. To alleviate the problem, one can avoid milk products with a high lactose content or eat lactase tablets before consuming milk products.

The *peptidases* of the intestinal mucosa complete the digestion of proteins. Peptidases break down peptides into their individual amino acids. These amino acid molecules then are absorbed by the microvilli of the epithelial cells and enter the blood. Table 6.3 summarizes mechanical and chemical digestion.

> Bile emulsifies large fat globules. Pancreatic secretions neutralize stomach acid and break down starch, proteins, fat, DNA, and RNA. Intestinal mucosa enzymes complete the digestion of disaccharides and peptides.

The Absorption of Nutrients

You cannot benefit from digested food until it is absorbed. Absorption is the process by which nutrients are moved out of the lumen of the intestine and into the blood or lymph for circulation throughout the body. Sugars, amino acids, fats, vitamins, water, and ions are absorbed by the small intestine. Water and vitamins also are absorbed in the large intestine.

The Small Intestine: Moving Nutrients to Blood or Lymph

In the small intestine, small nutrient molecules enter the intestinal mucosa cells. From these cells, the nutrients follow one of two routes, entering either the blood capillaries or the lacteals of the villi. Small water-soluble nutrients such as amino acids, simple sugars (monosaccharides), water-soluble vitamins, and ions pass into the blood capillaries of the villi (Figure 6.12). After passing through the liver, these nutrients are distributed to the tissues of the body.

The other route involves the *lymph vessels* and *lymph fluid,* which are part of the *lymphatic system* and are separate and distinct from the blood and blood vessels. Substances that are not soluble in water, such as fats, cholesterol, and fat-soluble vitamins, move into the lymph capillaries (lacteals) of the villi and enter the lymph fluid for transport. Lymph eventually flows into blood vessels in the shoulder area, and these fatty substances are then distributed to the entire body by the general circulation (Figure 6.12).

The liver: regulating, storing, and detoxifying. Water-soluble nutrients are carried from the small intestine to the liver by the *hepatic portal vein* before being distributed throughout the body (see Figure 6.12). This permits the liver to regulate nutrient concentrations in the blood and chemically modify any toxic substances that have been absorbed.

For example, the liver regulates the amount of glucose that enters the general circulation after it is absorbed from the small intestine. Excess glucose is removed from the blood and stored in the liver as *glycogen,* a polysaccharide. As a result, blood leaving the liver carries the concentration of glucose needed to fuel current body activities. Such regulation contributes to body homeostasis.

> From intestinal mucosa cells, amino acids and simple sugars pass into blood vessels of the villi and travel to the liver. Fatty materials enter the lacteals of the villi and travel to blood vessels in the shoulder.

Passive and Active Absorption of Nutrients

The movement of nutrients from the lumen of the small intestine into the epithelial cells of the mucosa is accom-

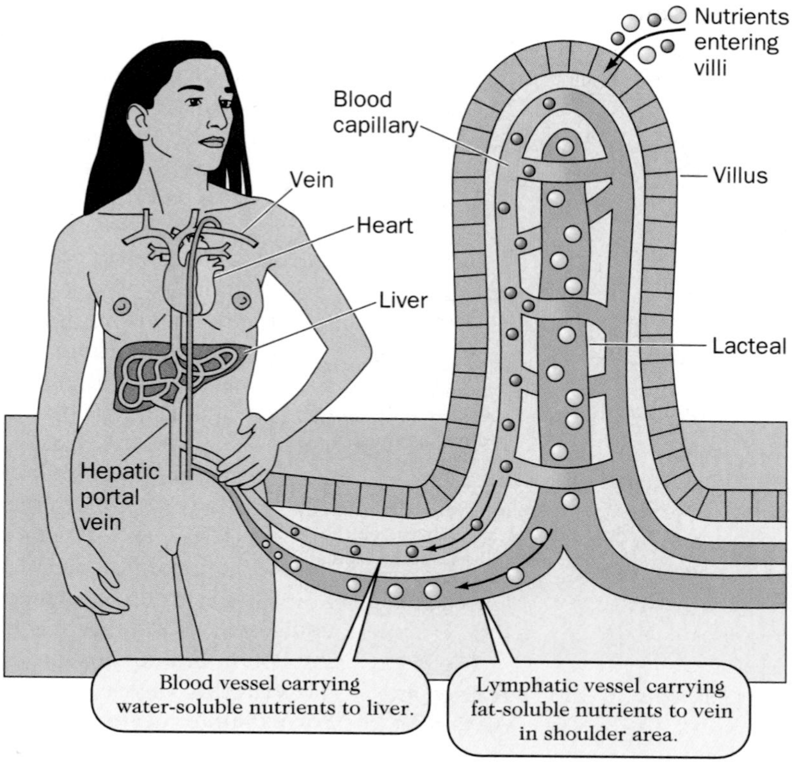
Nutrients entering villi
Blood capillary
Vein
Heart
Liver
Villus
Lacteal
Hepatic portal vein
Blood vessel carrying water-soluble nutrients to liver.
Lymphatic vessel carrying fat-soluble nutrients to vein in shoulder area.

FIGURE 6.12 Transport of nutrients away from the villi.

plished by different processes for different nutrients. The monosaccharides *glucose* and *galactose* enter intestinal mucosa cells by active transport, while *fructose* enters by facilitated diffusion. The 20 *amino acids* that are the products of protein digestion enter the intestinal mucosa cells by active transport (Figure 6.13). Monosaccharides and amino acids move out of the mucosa cells and into the capillaries of the villi by diffusion.

Water, mineral ions, and vitamins are also absorbed. Water enters the epithelial cells by osmosis. Mineral ions (electrolytes) such as sodium, chloride, iron, potassium, and calcium move into the intestinal mucosa cells by diffusion or active transport. The entry of calcium ions requires an active form of vitamin D. Fat-soluble vitamins (vitamins A, D, E, and K) dissolve in dietary fats and move by passive diffusion into the intestinal mucosa cells along with those fats. Most of the water-soluble vitamins (B-complex and C) are absorbed by passive diffusion from the small intestine.

Absorbing fats: breakdown and reassembly. Most of the fats you eat are *triglycerides* and *cholesterol.* Digestive processes in the small intestine break down triglyceride molecules into fatty acids and glycerol. Fatty acids, glycerol, and cholesterol diffuse into the epithelial cells. Once inside these cells, glycerol and fatty acids are reassembled into triglycerides (see Figure 6.13).

After reassembly, triglycerides and cholesterol are wrapped in a protein, forming fat-cholesterol-protein complexes (called chylomicrons). These complexes prevent fat molecules from forming a giant oil slick in the water of the blood and lymph. The fat-cholesterol-protein complexes leave the epithelial cells and enter the lacteals, then the lymph vessels, and eventually the blood (see Figures 6.12 and 6.13).

Figure 6.14 provides a summary of the digestive system's functions: mechanical digestion, chemical digestion, and absorption.

> Monosaccharides are passively or actively transported into the intestinal mucosa cells. Amino acids are actively transported. Water enters by osmosis, while mineral ions enter passively or by active transport. Fatty acids, glycerol, and cholesterol enter by diffusion.

The Large Intestine: Water, Benevolent Bacteria, and Gas

By the time chyme reaches the large intestine, both digestion and nutrient absorption are nearly complete. So why does a person have about 5 feet of colon? It is in the large intestine that additional water is absorbed and bacterial action occurs, contributing to our health.

As peristaltic contractions move chyme along the colon, water is absorbed by the epithelial cells and transported to the bloodstream. As water is absorbed, chyme becomes a semisolid mass called feces. As much as 1.5 liters (about 1.5 quarts) of chyme enters the large intestine each day, and about 90 percent of its water is absorbed. This conserves body water and reduces the volume of feces to be eliminated, making storage and control of defecation more efficient. Feces consist of

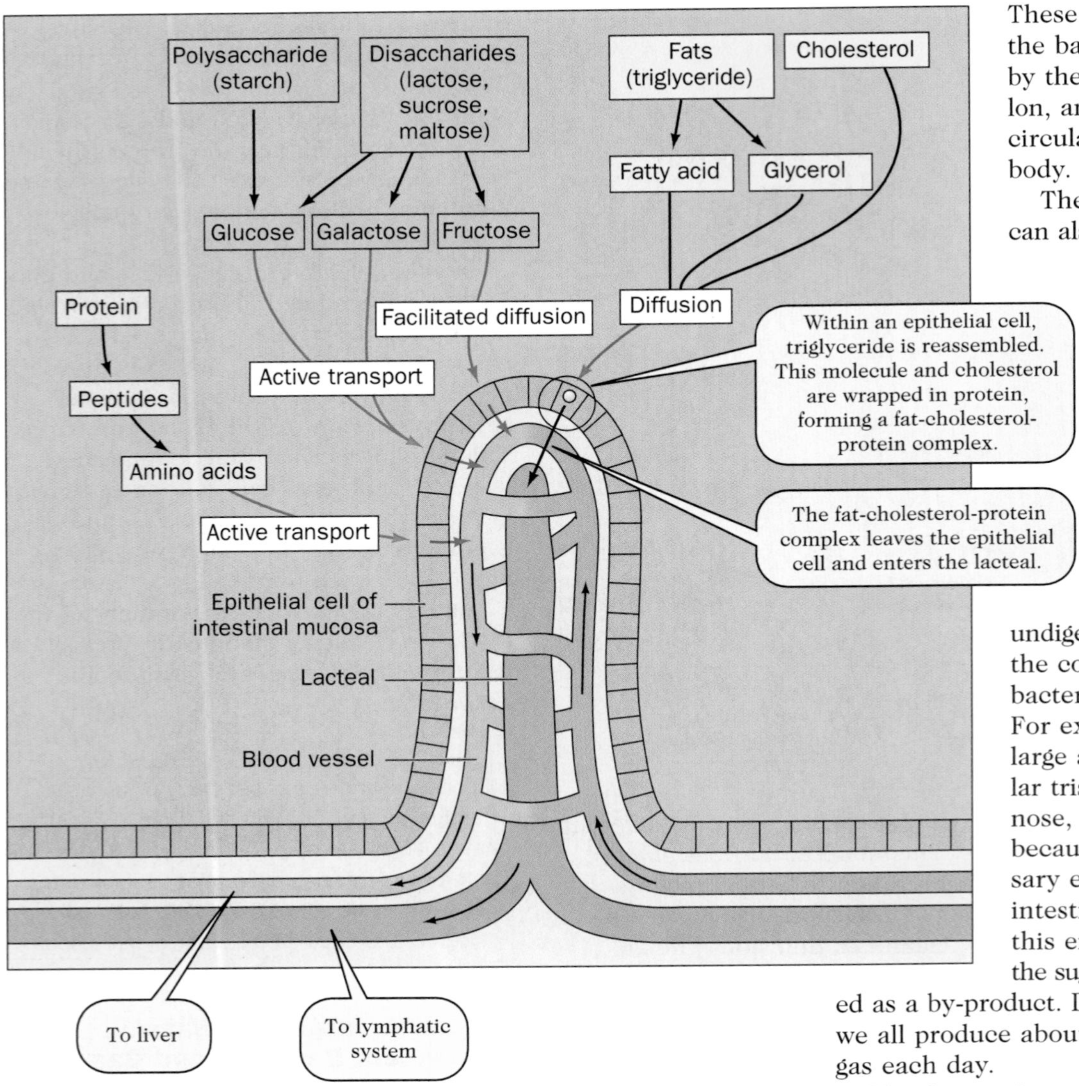

Polysaccharide (starch)
Disaccharides (lactose, sucrose, maltose)
Fats (triglyceride)
Cholesterol
Fatty acid
Glycerol
Glucose
Galactose
Fructose
Protein
Facilitated diffusion
Diffusion
Within an epithelial cell, triglyceride is reassembled. This molecule and cholesterol are wrapped in protein, forming a fat-cholesterol-protein complex.
Active transport
Peptides
Amino acids
The fat-cholesterol-protein complex leaves the epithelial cell and enters the lacteal.
Active transport
Epithelial cell of intestinal mucosa
Lacteal
Blood vessel
To liver
To lymphatic system

FIGURE 6.13 How nutrients are absorbed into the villi and transported.

undigestible materials, water, and bacteria.

The large intestine is host to numerous bacteria that are usually harmless. Our relationship with these bacteria is mutually beneficial. The bacteria benefit because they have a physical environment and nutrients for growth and reproduction. We benefit because the bacteria can manufacture certain substances that we require but that our cells are unable to make.

Our resident bacteria give off valuable by-products such as several B-complex vitamins and vitamin K. Indeed, bacterial synthesis is our major source of vitamin K, which is required for bloodclotting. These vitamins seep out of the bacteria, are absorbed by the epithelium of the colon, and enter the blood for circulation throughout the body.

The actions of bacteria can also be annoying. Intestinal gas is a common by-product of the bacterial breakdown of certain foods, such as beans, cabbage, and brussels sprouts. Gases such as carbon dioxide (CO_2), methane (CH_4), and hydrogen sulfide (H_2S) are produced when undigested substances reach the colon and encounter bacteria that digest them. For example, beans contain large amounts of a particular trisaccharide sugar, raffinose, that we cannot digest because we lack the necessary enzyme. However, our intestinal bacteria produce this enzyme. They digest the sugar, and gas is produced as a by-product. Depending on our diet, we all produce about 1 liter (1 quart) of gas each day.

Diarrhea and constipation. Our digestive organs release as much as 9 liters (about 9 quarts) of secretions into the alimentary canal every day. Much of this is water, and it represents more than 3 times the amount of water we usually drink or consume with food. Most of this water is reabsorbed into the blood from the colon.

Diarrhea is a condition in which there is frequent defecation of liquid feces (loose stools). It occurs when the colon becomes irritated. As a result, peristalsis increases and chyme moves rapidly through the colon, preventing adequate water absorption. Diarrhea can challenge the homeostasis of the body. Not only is water lost, causing dehydration, you also

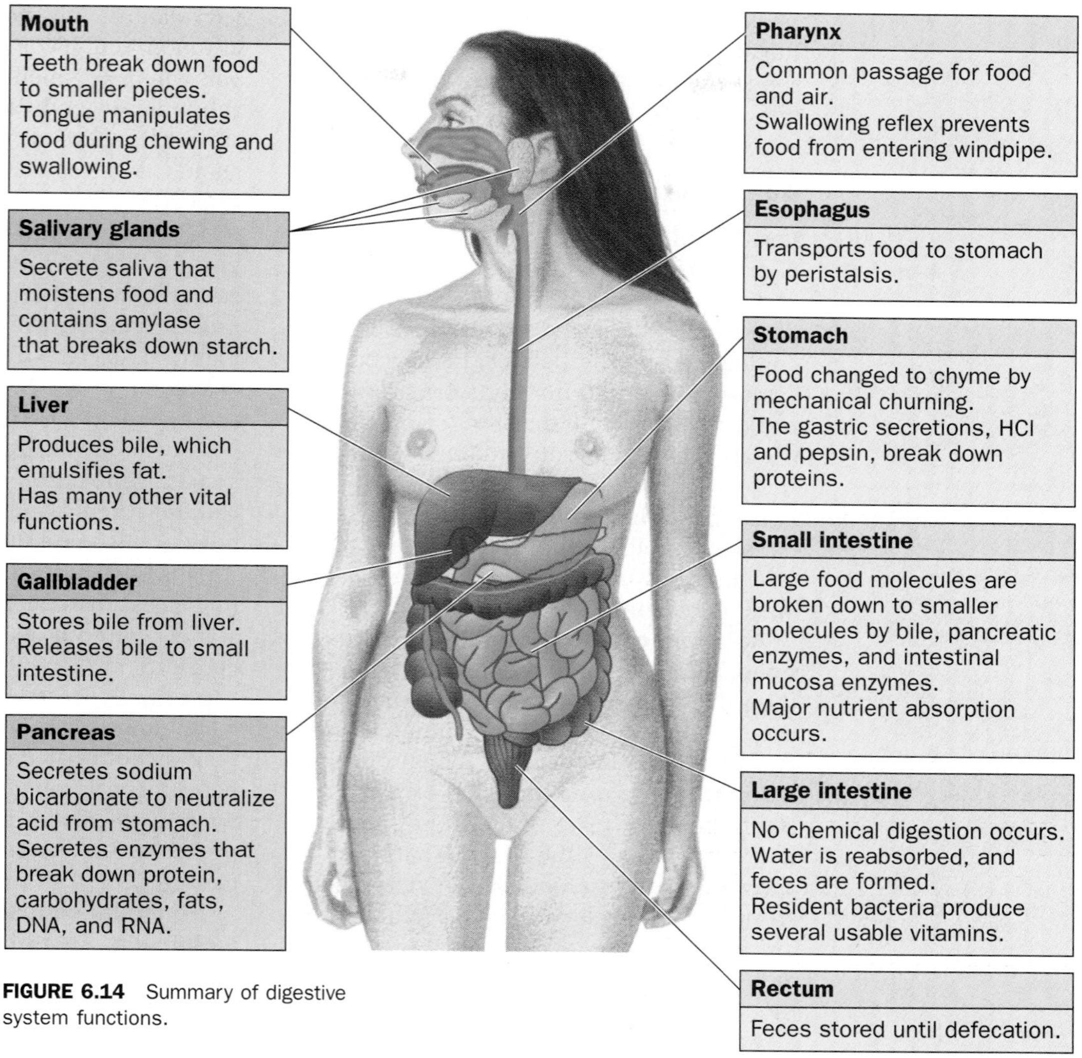
Mouth
Teeth break down food to smaller pieces. Tongue manipulates food during chewing and swallowing.
Salivary glands
Secrete saliva that moistens food and contains amylase that breaks down starch.
Liver
Produces bile, which emulsifies fat. Has many other vital functions.
Gallbladder
Stores bile from liver. Releases bile to small intestine.
Pancreas
Secretes sodium bicarbonate to neutralize acid from stomach. Secretes enzymes that break down protein, carbohydrates, fats, DNA, and RNA.
Pharynx
Common passage for food and air. Swallowing reflex prevents food from entering windpipe.
Esophagus
Transports food to stomach by peristalsis.
Stomach
Food changed to chyme by mechanical churning. The gastric secretions, HCl and pepsin, break down proteins.
Small intestine
Large food molecules are broken down to smaller molecules by bile, pancreatic enzymes, and intestinal mucosa enzymes. Major nutrient absorption occurs.
Large intestine
No chemical digestion occurs. Water is reabsorbed, and feces are formed. Resident bacteria produce several usable vitamins.
Rectum
Feces stored until defecation.

FIGURE 6.14 Summary of digestive system functions.

lose important ions such as Ca^{2+} (calcium), Na^+ (sodium), HCO^{3-} (bicarbonate), and K^+ (potassium). Prolonged diarrhea is a serious condition. If it persists for more than 48 hours in an adult, medical attention should be sought. Diarrhea can be life-threatening to infants and small children. Worldwide, over half the deaths of children under age 4 are directly caused by the dehydration that results from diarrhea.

Constipation is basically the opposite condition: the slow passage of feces through the large intestine. Slower passage means more time for water absorption. As a result, the feces are drier and therefore more difficult to eliminate.

> The large intestine functions to absorb water, form feces, and store feces before defecation. Resident bacteria digest nutrients that are not processed in the small intestine and in turn release B-complex vitamins and vitamin K.

Regulating Digestion

As food is processed in the digestive system, regulatory mechanisms coordinate the release of digestive secretions with the arrival of food. This prevents the waste of energy and body resources that would occur if digestive secretions were released into an empty alimentary canal. Regulation also assures that in the absence of food, the mucosa of the alimentary canal is not damaged by digestive secretions.

Dual Control of Secretion

Both the nervous system and the endocrine system contribute to the regulation and coordination of the entire digestive process. *Nervous control* is accomplished by the part of the nervous system that we do not consciously control, the autonomic division (also called the autonomic nervous system—ANS). *Endocrine control* is accomplished by hormones released from specialized endocrine cells. Although hormones circulate throughout the body, they act only on specific cells.

Autonomic control of salivary secretions. Regulation and coordination of the salivary glands are accomplished by the autonomic division. Hormones do not participate in this process. When our sensory receptors are stimulated by the sight, smell, or taste of food, they send nerve impulses to the brain. The brain activates the autonomic division, which stimulates the salivary glands. When you are not stimulated by food, a small amount of saliva is released to keep your mouth moist.

Autonomic and hormonal control of stomach secretions. As with the salivary glands, the brain acts through the autonomic division to stimulate the stomach when we smell, view, taste, or even think about food. Before food enters the stomach, nerve stimulation causes gastric glands to release HCl, pepsin, and mucus (Figure 6.15).

When the stomach wall is stretched or partially digested protein is present, specialized endocrine cells in the stomach mucosa release the hormone gastrin (see Figure 6.15). **Gastrin,** like all hormones, circulates in the blood. When it reaches the stomach, gastrin stimulates the mucosa cells to release more HCl and pepsinogen. Certain substances, such as caffeine, alcohol, cinnamon, and chili powder, also stimulate gastrin release. This is why individuals with ulcers cannot ingest foods containing these substances without suffering stomach discomfort.

Normally, gastrin release is inhibited when stomach acidity reaches pH 2. As a result, the further release of HCl and pepsinogen is reduced or stopped. This negative feedback mechanism normally prevents the excessive accumulation of damaging HCl in the stomach.

Hormonal control of pancreatic and gallbladder secretions. Although pancreatic secretion is partially controlled by the autonomic division, the more significant regulation is achieved by two hormones: secretin and cholecystokinin

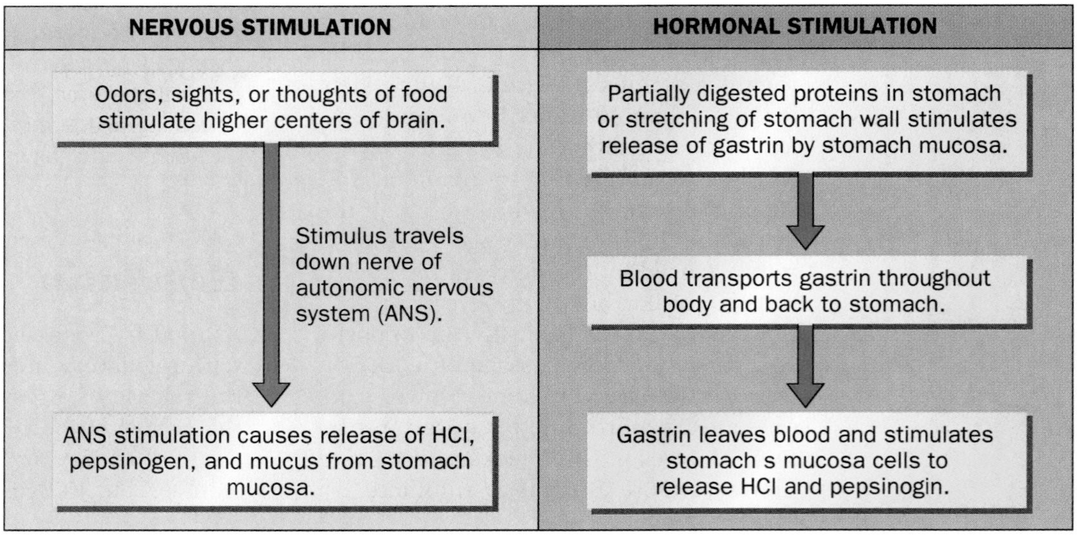
NERVOUS STIMULATION
HORMONAL STIMULATION
Odors, sights, or thoughts of food stimulate higher centers of brain.
Stimulus travels down nerve of autonomic nervous system (ANS).
ANS stimulation causes release of HCl, pepsinogen, and mucus from stomach mucosa.
Partially digested proteins in stomach or stretching of stomach wall stimulates release of gastrin by stomach mucosa.
Blood transports gastrin throughout body and back to stomach.
Gastrin leaves blood and stimulates stomach s mucosa cells to release HCl and pepsinogin.

FIGURE 6.15 Nervous and hormonal regulation of stomach secretions.

(CCK). The presence of acid chyme in the duodenum stimulates the release of secretin (Figure 6.16). **Secretin** stimulates the pancreas to secrete a watery fluid that is rich in sodium bicarbonate into the duodenum. The sodium bicarbonate neutralizes the HCl from the stomach. **Cholecystokinin (CCK)** (*coal-lee-sis-toe-kie-nin*) is released in response to partially digested proteins and fats entering the duodenum (see Figure 6.16). CCK stimulates the pancreas to secrete digestive enzymes and causes the gallbladder to release bile.

When chyme has passed out of the duodenum and no more enters from the stomach, the release of these two hormones ceases. Without these hormones, secretions from the pancreas and gallbladder also cease.

Salivary glands are stimulated by the ANS. Gastric glands are stimulated by the ANS and gastrin. The pancreas and gallbladder are regulated primarily by secretin and CCK.

Diseases and Disorders

The digestive system is exposed to the external environment in ways that make it vulnerable to a variety of diseases and disorders. With every meal, there is a risk of ingesting toxic substances and disease-producing microorganisms. For example, a number of deadly infectious diseases, such as cholera, typhoid fever, and dysentery, attack the human digestive system.

Even in the absence of infectious organisms, we can abuse the digestive system by what we eat, how we eat, and how much we eat. In the following sections we will explore several common problems of the digestive system that are associated with the alimentary canal and its accessory organs.

Disorders of the Alimentary Canal

A **peptic ulcer** involves localized destruction of the mucosa in the stomach (gastric ulcer) or the first part of the duodenum (duodenal ulcer). This destruction

FIGURE 6.16 Hormonal regulation of the pancreas and gallbladder by CCK and secretin.

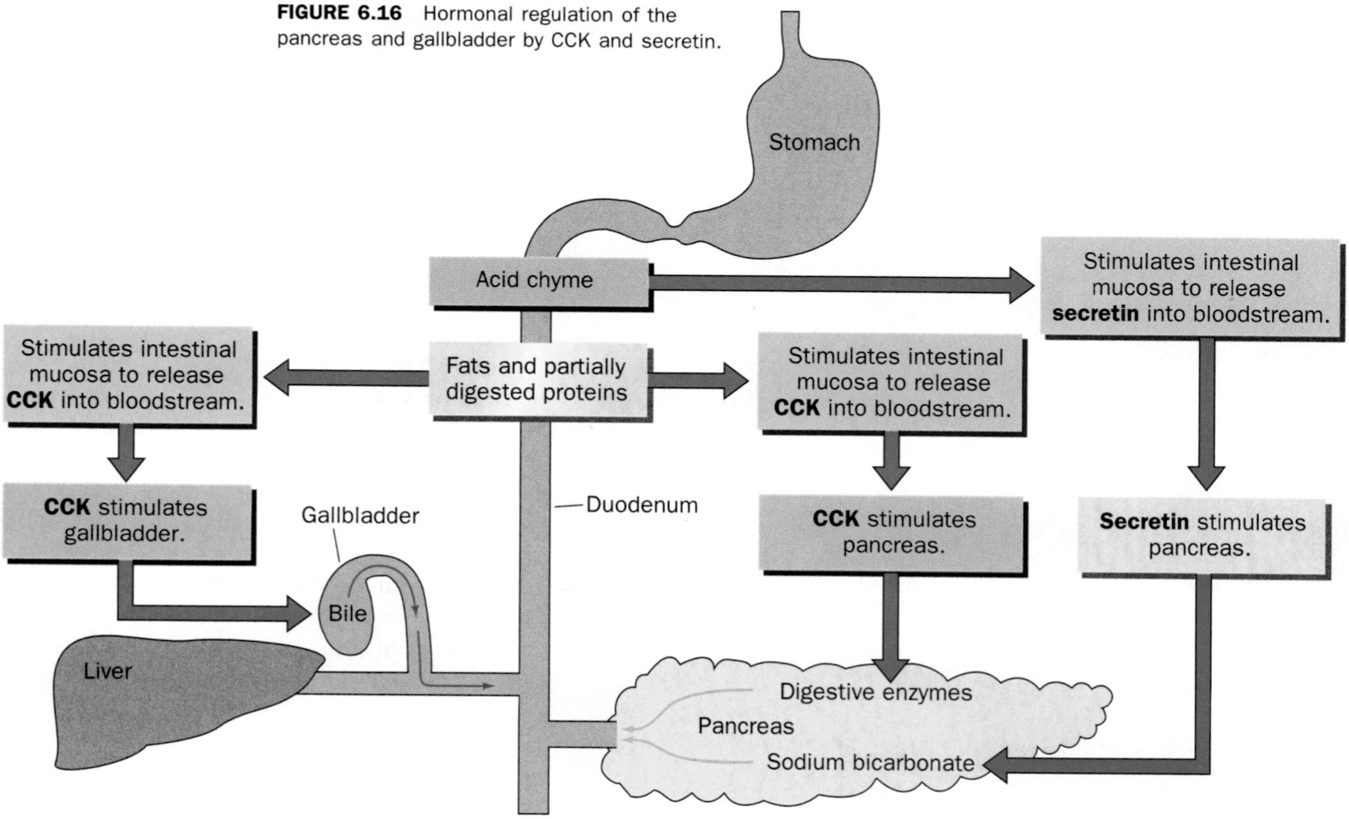
Stomach
Acid chyme
Stimulates intestinal mucosa to release secretin into bloodstream.
Stimulates intestinal mucosa to release CCK into bloodstream.
Fats and partially digested proteins
Stimulates intestinal mucosa to release CCK into bloodstream.
CCK stimulates gallbladder.
Gallbladder
Duodenum
CCK stimulates pancreas.
Secretin stimulates pancreas.
Bile
Liver
Digestive enzymes
Pancreas
Sodium bicarbonate

produces an open sore that exposes nerves and blood vessels of the submucosa to food, acid, and enzymes (Figure 6.17). Peptic ulcers develop because of overproduction of gastric HCl or inadequate secretion of protective mucus. If you have a peptic ulcer, you usually experience a persistent burning pain in the upper abdomen and lower chest.

A number of conditions are known to increase the occurrence of peptic ulcers, including persistent emotional stress and conflict, cigarette smoking, and the consumption of substances that increase digestive secretions (alcohol, caffeine, and certain spices). Some drugs, such as aspirin and ibuprofen, irritate the stomach mucosa and may result in gastric ulcers when they are used for extended periods. The genes you inherit may also increase your chances of developing a peptic ulcer. As we noted earlier in this chapter, there is a bacterium that appears to cause some ulcers.

Ulcers can develop into a serious health problem if tissue destruction progresses or damaged blood vessels bleed continuously. If an ulcer is not treated, a hole may form through the entire wall of the stomach or intestine, resulting in a *perforated ulcer*. This is dangerous, because the contents of the alimentary canal, including bacteria, may enter the abdominal cavity, causing life-threatening infections. Fortunately, if an ulcer is diagnosed early, there are a number of drugs that reduce HCl secretion (Tagamet and Zantac) or neutralize acid (oral antacids).

FIGURE 6.17 Highly magnified photograph showing an ulceration in the stomach lining.

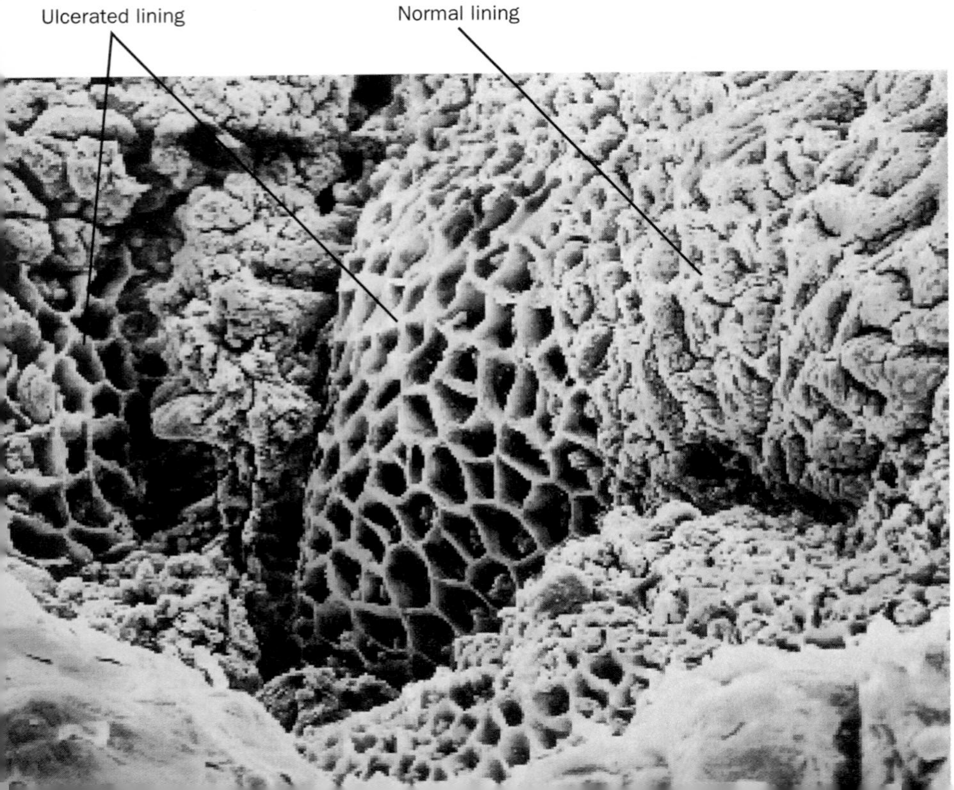
Ulcerated lining
Normal lining

Traveler's diarrhea is most commonly caused by ingesting unfamiliar but not necessarily disease-producing bacteria. You will recall that the human colon contains large populations of several types of normally harmless bacteria, including *Escherichia coli* (*esh-er-***eek***-ee-ah* **koh***-lye*). These populations are referred to as normal flora. Slightly different forms or strains of these bacteria constitute the normal flora of people living in different parts of the world. When we travel abroad, we may ingest food and water containing an unfamiliar strain of intestinal bacteria common to the local area (Table 6.4).

These unfamiliar bacteria have characteristics slightly different from the characteristics of those already residing in our intestines. For example, some strains of *E. coli* produce by-products that cause short-term irritation to the intestinal mucosa. Such irritation may cause both excessive water secretion into the intestinal lumen and impaired water absorption. These conditions result in the watery feces of diarrhea.

If you stay for an extended time in a foreign country, the local bacterial strains will become part of your normal flora and irritation of your intestinal mucosa will cease. Thus, you will cease to suffer from this form of diarrhea.

A **hiatal hernia** is a condition in which the upper stomach protrudes into the thoracic cavity at the opening (the hiatus) where the esophagus passes through the diaphragm (Figure 6.18). Under these circumstances, the esophageal sphincter gets less support from the diaphragm and consequently may not close completely. This permits the acidic contents of the stomach to enter the esophagus, irritating its mucosa and causing a painful condition known as *heartburn*.

Cancer of the colon or rectum is one of the leading causes of cancer deaths in the United States. The cause of colon-rectal cancer is under investigation. It is known that heredity plays a role, because studies have shown that one is at increased risk if one's family members have had the disease. Recently, a specific human gene has been identified that appears to contribute directly to about 15 percent of these cancers.

Diet is also thought to play a role, particularly a diet high in fat and low in fiber. No one knows exactly how fat and fiber are involved with colon cancer, but it has been suggested that dietary fat stimulates the release of bile. The bile acids in bile are chemically similar to certain cancer-causing chemicals (carcinogens). Dietary fiber from fruits, vegetables, and grains may interact with bile acids or effectively dilute them, reducing the exposure that the intestinal mucosa receives.

Several methods are used to detect this cancer. One method detects small amounts of blood in the feces, since precancerous and cancerous growths called polyps bleed easily (Figure 6.19). Sigmoidoscopy and colonoscopy are two methods for viewing the interior of the colon, allowing the identification and removal of polyps without the need for surgery. Early detection and treatment increase the survival rate of this cancer to 90 percent.

TABLE 6.4 Traveler's Diarrhea and Its Prevention

Chances of Contracting Traveler's Diarrhea, %	When Traveling In
40	Less developed countries
10	European or Mediterranean countries
2–5	North America

Tips to reduce the risk of contracting traveler's diarrhea while visiting countries with inadequate water treatment facilities:

- Never drink or brush your teeth with tap water; use only bottled water. Avoid ice in drinks, as freezing does not kill all bacteria.
- Water can be treated by boiling for 15 minutes or by adding chlorine bleach (5–10 drops per liter or quart of water).
- Eat only cooked vegetables or fruits that must be peeled, for example, bananas and oranges.

FIGURE 6.18 Normal anatomic position of the stomach and diaphragm compared with the abnormal hiatal hernia condition.

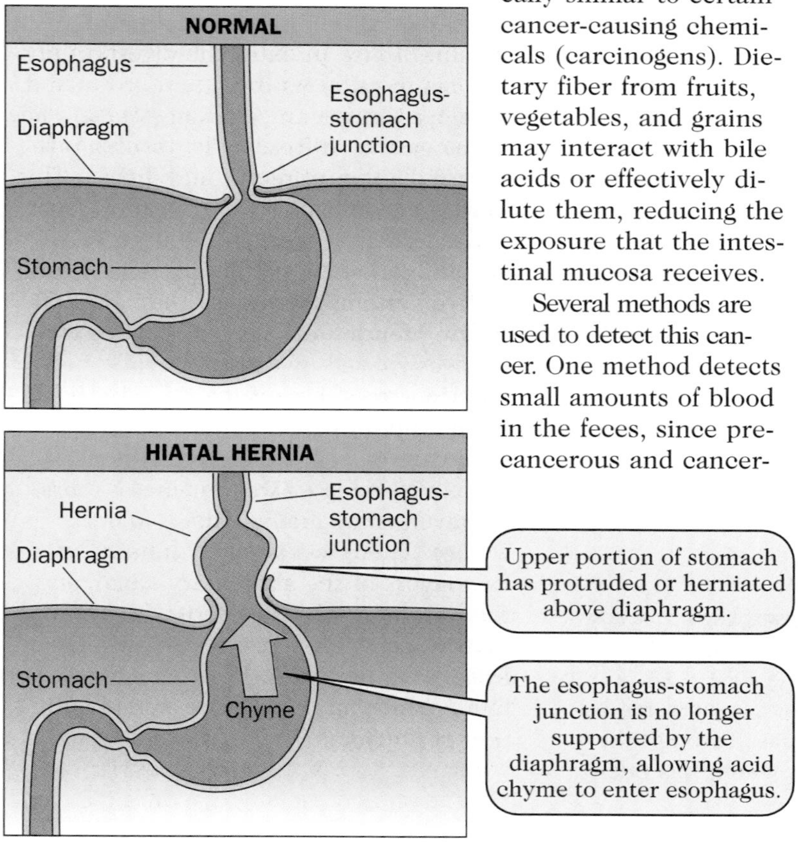
NORMAL
Esophagus
Diaphragm
Esophagus-stomach junction
Stomach
HIATAL HERNIA
Hernia
Diaphragm
Esophagus-stomach junction
Stomach
Chyme
Upper portion of stomach has protruded or herniated above diaphragm.
The esophagus-stomach junction is no longer supported by the diaphragm, allowing acid chyme to enter esophagus.

Diseases and Disorders of the Accessory Organs

Hepatitis is the general term for a liver inflammation that can be caused by viruses, chemicals, or drugs. These agents interact with the cells of the liver, destroying their structural integrity. Because the liver performs so many important functions (see Table 6.2), hepatitis can be a very serious disease.

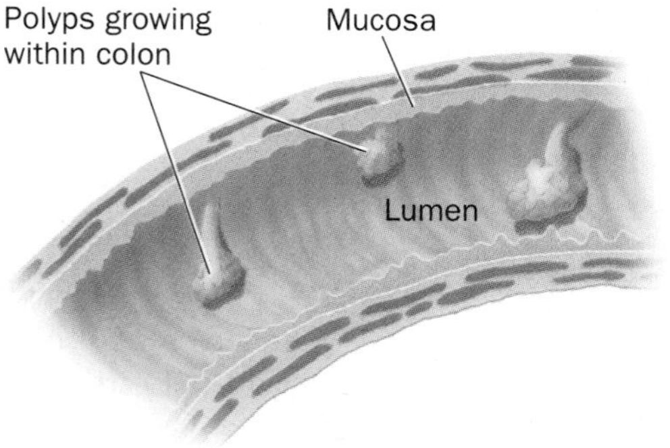
Polyps growing within colon
Mucosa
Lumen

FIGURE 6.19 Polyps growing in the colon.

Cirrhosis of the liver is a condition in which the liver slowly deteriorates. As liver cells die, they are gradually replaced by nonfunctional scar tissue. There are several causes of cirrhosis, including malnutrition, chronic hepatitis, parasites, and toxic chemicals.

The most common cause of cirrhosis of the liver is prolonged alcohol abuse. The liver detoxifies many chemicals, including alcohol, and in this process some liver cells die. In itself, this is not dangerous, because the liver has the capacity to replace those cells. However, chronic use of alcohol overwhelms the replacement process, and fibrous connective tissue and fat are deposited instead. Cirrhosis resulting from alcohol abuse can be corrected by terminating alcohol consumption, which is difficult for a person addicted to alcohol. If alcohol abuse continues, an increasing load is placed on fewer and fewer surviving liver cells, and liver failure and death are likely.

Gallstones are hard pebble-sized structures that form in the gallbladder. They develop when excess cholesterol in the bile precipitates out of solution, trapping calcium and bile salts. Small gallstones are common and often produce no pain or other symptoms. Gallstones become a medical problem when they become large enough to obstruct the ducts that carry bile to the duodenum.

Treatments for problem gallstones include surgical removal of the gallbladder or drug therapy to dissolve the stones. A nonsurgical procedure focuses high-energy sound waves on the stones to pulverize them, preventing duct blockage. Women are 3 times more likely than men to develop gallstones, and a diet high in fat increases one's chance of developing gallstones.

Life Span Changes

Changes within our digestive systems begin during birth as we pass through the birth canal. The womb, which has been our home during 9 months of development, is normally sterile. That is, the mother's body serves as a protective barrier against microorganisms. It is during birth that we encounter our first microorganisms. Thus begins a lifelong relationship between our bodies and bacteria, viruses, protozoa, and fungi.

Bacteria from the environment enter an infant's digestive tract through the mouth. They pass into the intestine, where they become established as normal flora. Breast-fed infants receive natural antibodies from the mother's milk. These antibodies favor the growth of "friendly" bacteria in the infant's intestine and discourage the growth of "unfriendly" bacteria.

Whether through the mother's milk or in a commercial "formula," the infant receives a liquid diet rich in carbohydrates, fats, and proteins for energy and growth. In addition to pepsinogen and HCl, the infant's stomach also secretes the enzymes rennin and lipase. *Rennin* acts to curdle milk, making milk proteins more accessible to the digestive action of pepsin. *Lipase* initiates the digestion of the fats in milk. As the infant grows and develops, programmed changes take place in the gastric mucosa cells, turning off the production of rennin and lipase. This change accompanies the transition from liquid food to more solid food.

Although changes take place in our digestive systems as we age, they are difficult to identify and distinguish from the changes caused by an individual's diet, disease, stress, loss of teeth, and alcohol consumption. It seems that if we take good care of it, the digestive system is capable of serving us well for many years. However, some gradual changes occur as we age. There is a general thinning of all four layers of the alimentary canal, and this results in reduced peristalsis, reduced gastric secretion, and reduced absorption. In general, however, these changes do not hinder the overall digestion and absorption of nutrients.

Chapter Summary

Structural Organization of the Digestive System

- The alimentary canal wall has four layers: mucosa, submucosa, muscularis, and serosa.
- Smooth muscle in the muscularis layer produces rhythmic contractions called peristalsis that mix and move food along the alimentary canal.
- Major sections of the alimentary canal include the mouth, pharynx, esophagus, stomach, small intestine, large intestine, and anus.
- The mouth contains the tongue and teeth and receives saliva from the salivary glands.
- The pharynx is the common passageway for air and food.
- The esophagus connects the pharynx to the stomach.
- The stomach forms a closed compartment when its two sphincters contract.
- The small intestine mucosa is organized into circular folds, villi, and microvilli.
- Accessory glands include the salivary glands, liver, gallbladder, and pancreas.
- The large intestine includes the colon, rectum, anal canal, and anus.

Chemical Processing of Food: Preparation for Absorption

- Salivary glands release amylase into the mouth and begin the digestion of starch.
- Stomach mucosa cells release the inactive enzyme pepsinogen, HCl, and mucus.
- The small intestine receives secretions from the gallbladder, the pancreas, and its mucosa cells.
- The gallbladder releases bile, which emulsifies fat.
- The pancreas releases lipase, amylase, trypsinogen, nuclease, and sodium bicarbonate.
- Mucosa cells of the small intestine release enzymes, disaccharidases, and peptidases.
- The large intestine absorbs water from chyme and contains bacteria which produce some vitamins.

The Absorption of Nutrients

- Most nutrients are absorbed by the small intestine.
- Water-soluble nutrients include monosaccharides, amino acids, minerals, and water-soluble vitamins. These nutrients enter blood capillaries of the villi and pass directly to the liver.
- Fatty acids, glycerol, cholesterol, and fat-soluble vitamins enter the lacteals of the villi and then pass into blood vessels in the shoulder area.

Regulating Digestion

- The autonomic nervous system (ANS) regulates the release of saliva from the salivary glands.
- The ANS and the hormone gastrin regulate the release of gastric juice from the stomach.
- The ANS and the hormones secretin and cholecystokinin (CCK) regulate the release of pancreatic and gallbladder secretions. Secretin stimulates the pancreas to release sodium bicarbonate. CCK stimulates the release of pancreatic enzymes and bile from the gallbladder.

Diseases and Disorders

- Peptic ulcers are sores of the stomach or intestinal mucosa.
- Traveler's diarrhea occurs when new strains of bacteria enter the intestines.
- A hiatal hernia occurs when the upper stomach partially bulges through the diaphragm.
- Colon-rectal cancer may result from hereditary factors and a diet high in fat and low in fiber.
- Hepatitis is an inflammation of the liver caused by viruses, chemicals, and drugs.
- Cirrhosis of the liver involves a deterioration of the liver in which liver tissue is replaced by connective tissue and fat.
- Gallstones are pebble-sized structures that form in the gallbladder.

Life Span Changes

- Intestinal bacteria are acquired from the environment after birth.
- Rennin and lipase are released by the stomach mucosa of infants.
- As we age, there is a gradual thinning of the alimentary canal's four layers.

Selected Key Terms

accessory organs (p. 174)
alimentary canal (p. 167)
amylase (pp. 178, 179)
bile (p. 179)
cholecystokinin (p. 185)
gastrin (p. 184)
lacteal (p. 173)
large intestine (p. 175)
pepsin (p. 178)
peristalsis (p. 168)
secretin (p. 185)
small intestine (p. 172)
stomach (p. 171)
trypsin (p. 179)
villus (p. 173)

Review Activities

1. Name and locate the major sections of the alimentary canal.
2. Name and locate the four accessory organs and describe their functions.
3. How does the mucosa of the mouth and esophagus differ from the rest of the alimentary canal mucosa?
4. Draw a tooth and include the crown, root, enamel, dentin, and pulp cavity.
5. Locate the incisors, canines, premolars, and molars in your mouth.
6. Describe the events of swallowing which prevent food and drink from entering the airway.
7. Describe the chemical components of gastric juice and explain the function of each one.
8. Draw a villus and include the columnar cells, microvilli, lacteal, and blood capillary.
9. Explain the difference between peristalsis and segmentation.
10. Describe the chemical digestion of carbohydrates in the small intestine.
11. Describe the chemical digestion of proteins in the small intestine.
12. Describe the chemical digestion of triglycerides in the small intestine.
13. Explain how the defecation reflex controls defecation.
14. Identify which nutrient molecules from carbohydrate, protein, and fat digestion are absorbed by the mucosa of the small intestine.
15. Describe the role of gastrin, secretin, and CCK in digestion.

Self-Quiz

Matching Exercise

___ 1. Sphincter of stomach that regulates release of chyme	A. Sodium bicarbonate
___ 2. The inactive enzyme secreted by stomach mucosa	B. Lipase
___ 3. Folds in the plasma membrane of small intestine mucosa cells	C. ANS
___ 4. Opening through which feces pass out of alimentary canal	D. Anus
___ 5. Enzyme from pancreas that digests fat	E. Pyloric
___ 6. Acid-neutralizing chemical released by the pancreas	F. Small intestine
___ 7. Chemical from gallbladder that emulsifies fat	G. Pepsinogen
___ 8. Enzymes from small intestine mucosa that digest maltose and sucrose	H. Disaccharidases
___ 9. The part of the alimentary canal where minerals are absorbed	I. Microvilli
___ 10. Part of nervous system that regulates digestion	J. Bile

Answers to Self-Quiz

1. E; **2.** G; **3.** I; **4.** D; **5.** B; **6.** A; **7.** J; **8.** H; **9.** F; **10.** C

SPECIAL SANDWICHES
CHICKEN FILLET
FISH FILLET
HAM 'N CHEESE SANDWICH
SIDE ORDERS
TOSSED SALAD
COTTAGE CHEESE N FRUIT
FRENCH FRIES
ONION RINGS
CHILI
BEVERAGES
COCA COLA
TAB
COFFEE
MILK
SANKA
TEA, HOT CHOCOLATE
BIG W 'N CHEESE
HAMBURGER
CHEESEBURGER
DOUBLE CHEESEBURGER
TODAY'S SPECIAL
STEAK SANDWICH
SOUP OF THE DAY
BEEF NOODLE
TOMATO

Chapter 7

Nutrition and Metabolism

Let's have lunch. The local fast-food restaurant offers sandwiches, salads, beverages, and desserts. I order a grilled chicken sandwich (breast of chicken, tomato, lettuce, and bun), a garden salad (lettuce, tomatoes, cauliflower, carrots, broccoli, and cucumbers with no cheese and no salad dressing), and water. This lunch contains the nutrients my body needs for energy, cell maintenance, and growth (Figure 7.1).

I could have made other choices. For example, yesterday I ordered a quarter-pound hamburger with mayonnaise, a large order of French fries, a cola, and a chocolate chip cookie (Figure 7.1). This meal also contains the nutrients my body needs, but it has unhealthy amounts of certain nutrients. In recent years, research has shown that the foods we eat affect our health, moods, and productivity as well as how long we live. I

Lunch 1 (Today's Lunch)				
Food	Calories	Protein	Fat	Carbohydrates
Grilled chicken breast	99 C	18 g	3 g	-
Bun	171 C	5 g	3 g	31 g
Green salad:	40 C	1 g	-	9 g
Lettuce	8 C	-	-	2 g
Tomatoes	12 C	-	-	3 g
Cauliflower	4 C	-	-	1 g
Carrots	8 C	-	-	2 g
Broccoli	4 C	-	-	1 g
Cucumbers	2 C	-	-	-
Water	0 C	0 g	0 g	0 g
Totals	310 C	24 g	6 g	40 g

MAYO

Lunch 2 (Yesterday's Lunch)				
Food	Calories	Protein	Fat	Carbohydrates
$\frac{1}{4}$-pound hamburger	184 C	19 g	12 g	-
Bun	171 C	5 g	3 g	31 g
Mayonnaise	63 C	-	7 g	-
French fries (6 oz)	470 C	6 g	22 g	62 g
Cookie (1)	285 C	3 g	13 g	39 g
Cola (8 oz)	100 C	0 g	0 g	25 g
Totals	1273 C	33 g	57 g	157 g

Note: g = gram(s); - = less than 1 g.

FIGURE 7.1 A comparison of the nutrients in two lunches from a fast-food restaurant.

am trying to make healthier food choices to maintain my health and lengthen my life.

As we try to inform ourselves about healthy food choices, words scream at us from ads and food-packages: "high-fiber," "cholesterol-free," "vitamin-enriched," "low-fat." What do these words mean, and why are they important? Evaluating such information and determining its application to your food choices require an understanding of nutrients. What are nutrients, and how are they important to normal growth and health? How much do we need? What foods contain the nutrients we need? What happens when we get too much of certain nutrients? How do we get energy from food, and how is it used in the body?

The answers to these questions are presented in this chapter. We will begin by discussing nutrients and their utilization. Then we will explore how some nutrient molecules can be chemically converted to other nutrients. Finally, we will discuss selected nutritional disorders and nutritional changes that occur over the life span.

Nutrients: Eating for Life and Health

Nutrition is the study of nutrients and their uses in our bodies. **Nutrients** are chemical substances needed by the body for growth, maintenance, repair, and reproduction. Nutrients are usually obtained from the food we eat, and six categories of nutrients can be distinguished: carbohydrates, fats, proteins, vitamins, minerals, and water. Most foods have several of these nutrients.

On the basis of animal experimentation and observations of humans, the Food and Nutrition Board of the National Research Council (U.S.) has established a *recommended dietary allowance* (*RDA*)

TABLE 7.1 The Recommended Daily Intake of Carbohydrates, Fats, and Proteins Based on 2000 Calories per Day

	Carbohydrates	Fats	Proteins
Percentage of Total Calories Recommended	60%	30%	10%
Number of Calories Recommended Based on 2000 Calories per Day	1200 Calories	600 Calories	200 Calories
Grams Recommended per Day Based on 2000 Calories	300 grams (10.5 ounces)	67 grams (2.3 ounces)	50 grams (1.75 ounces)
How to Calculate Grams from Calories	1 gram = 4 Calories 1200 Cal ÷ 4 Cal/gram = 300 grams	1 gram = 9 Calories 600 Cal ÷ 9 Cal/gram = 67 grams	1 gram = 4 Calories 200 Cal ÷ 4 Cal/gram = 50 grams

for most nutrients (Table 7.1; see also Tables 7.6 and 7.7). Similar guidelines have been established by the Canadian government. RDAs are defined as "the levels of intake considered to be adequate to meet the nutritional needs of practically all healthy persons." RDAs are not always an appropriate guide to proper nutrition, however, because circumstances may increase or decrease the need for certain nutrients. For example, age, sex, physical activity, pregnancy, breast-feeding, and illness can all influence your personal nutritional needs.

Obtaining an adequate supply of nutrients is made complicated because the foods we eat contain different quantities of nutrients. Consequently, eating foods without regard to their nutrient content can result in nutritional deficiencies and excesses. These conditions can lead to functional disorders, diseases, and even death. In the sections that follow, we will present each nutrient category and discuss its dietary sources, uses in the body, and dietary requirements for a healthy life.

Carbohydrates: Mostly for Energy

Plants are the source of most of our dietary carbohydrates. As you read in Chapter 1, there are small carbohydrate molecules called **simple sugars,** such as glucose, fructose, lactose, maltose, and sucrose. There are also much larger molecules called **complex carbohydrates** or polysaccharides, such as starch, glycogen, and plant fiber. Polysaccharides are actually long chains of simple sugars linked together by chemical bonds (see Figure 1.18).

Simple sugars are found naturally in fruits and vegetables (Figure 7.2). Processed plant materials that contain simple sugars include molasses, table sugar, corn syrup, and honey (thanks to honeybees).

Starch is a complex carbohydrate that is found in many plants. It is particularly abundant in potatoes, grains (rice, wheat, oats), and legumes (beans). Processed forms of starch include flours made from grains such as corn, wheat, and rye. My sandwich bun was made from wheat, and the French fries I ate yesterday were made from potatoes, which are mostly starch. Little of our dietary carbohydrate comes from animal products. However, meat (skeletal muscle) and liver supply small amounts of the complex carbohydrate glycogen, and milk products supply the simple sugar lactose.

Uses of carbohydrate in the body. Carbohydrates are used primarily as an energy source by the body. One gram of any carbohydrate contains about 4 Calories of energy. A Calorie is a measure of energy and will be discussed later in this chapter. Our cells remove glucose from

FIGURE 7.2 Fruits contain simple sugars which make them sweet.

the blood and use it as an energy source for the production of adenosine triphosphate (ATP). If there is more carbohydrate in the diet than your cells need, the excess is stored as glycogen or converted to fat. Glycogen and fat are reserves of energy than can be used if we eat less or are more active.

Plant fiber is a complex carbohydrate that we cannot utilize. The human digestive system does not produce the enzymes needed to digest it, and it passes through the alimentary canal undigested. My lunch salad of lettuce, tomatoes, cauliflower, carrots, broccoli, and cucumbers contained fiber. All of our dietary fiber is divided into two categories on the basis of whether it dissolves in water: insoluble fiber and soluble fiber (Table 7.2).

Insoluble fiber includes cellulose and other structural components of plants. Sources include bran (the outer layer of grains) and the skins and fibrous parts of most fruits and vegetables. Insoluble fiber aids the movement of materials through the alimentary canal. It is also thought to reduce colon cancer by binding carcinogens and increasing their rate of movement through the colon. **Soluble fiber** is abundant in grains (oats, barley), legumes (beans), and many fruits and vegetables. Soluble fiber forms a gel-like substance in the digestive tract and is known to benefit humans by slowing the absorption of glucose and cholesterol from the small intestine.

Dietary requirements for carbohydrates. It is recommended that most adults obtain about 60 percent of their daily Calories from carbohydrates. Thus, if a person requires 2000 Calories each day, that person should consume as many as

TABLE 7.2 Characteristics of Insoluble and Soluble Fiber

Fiber Type	Major Forms	Solubility in Water	Food Sources
Insoluble	Cellulose, lignin, insoluble hemicellulose	Does not dissolve in water but can absorb water and swell	Outer covering (bran) of grains such as wheat, corn, rice Pulp of fruits, vegetables, grains, and legumes Skins of fruits and root vegetables
Soluble	Pectin, plant gums, soluble hemicellulose	Dissolves in water and forms a gel-like solution	Grains such as oats, barley, and psyllium Legumes such as kidney, navy, and pinto beans Pulp of fruits such as apples and oranges

1200 Calories or 300 grams (10.5 ounces) of carbohydrates daily (see Table 7.1). Complex carbohydrates (starch) should constitute the major portion of these 300 grams.

The bun from my grilled chicken sandwich supplied about 124 Calories, and my garden salad supplied another 36 Calories (Table 7.3). Thus, with today's lunch I have consumed 160 carbohydrate Calories. It is also recommended that 20 to 35 grams (about 1 ounce) of fiber be consumed daily, and this should include both insoluble and soluble types. My salad contained fiber as well as other nutrients, such as vitamins.

> A gram of carbohydrate contains about 4 Calories of energy. Excess blood glucose can be stored as glycogen or fat. Fiber is not digested or absorbed. Approximately 60 percent of our daily Calories should come from carbohydrates.

Fats and Oils: Concentrated Energy

Fats, oils, cholesterol, and phospholipids are all **lipids.** Our discussion in this section will focus on fats and oils, which are also known as triglycerides. A **triglyceride** molecule is made up of three **fatty acid** molecules covalently bonded to a **glycerol** molecule (Chapter 1). Fatty acids differ in the number of carbon atoms and the number of double bonds present. We will focus on two categories of fatty acids: saturated fatty acids and unsaturated fatty acids (see Figure 1.19). Saturated fatty acids do not have double bonds between their carbon atoms and contain the maximum number of hydrogen atoms. Unsaturated fatty acids have one or more double bonds between their carbon atoms and contain fewer hydrogen atoms.

Saturated fats can be identified because they remain solid at room temperature. They are common in foods such as meat and dairy products as well as in

TABLE 7.3 The Nutrient Content of Two Lunches from a Fast-Food Restaurant

Lunch 1 (Today's Lunch)	Carbohydrate (4 Calories/gram)		Fat (9 Calories/gram)		Protein (4 Calories/gram)	
	Calories	Grams	Calories	Grams	Calories	Grams
Grilled chicken breast	0	0	27	3	72	18
Bun	124	31	27	3	20	5
Green salad (no dressing or cheese)	36	9	0	0	4	1
Total Calories = 310	160	40	54	6	96	24

Percent of Calories from carbohydrates, fats, and proteins

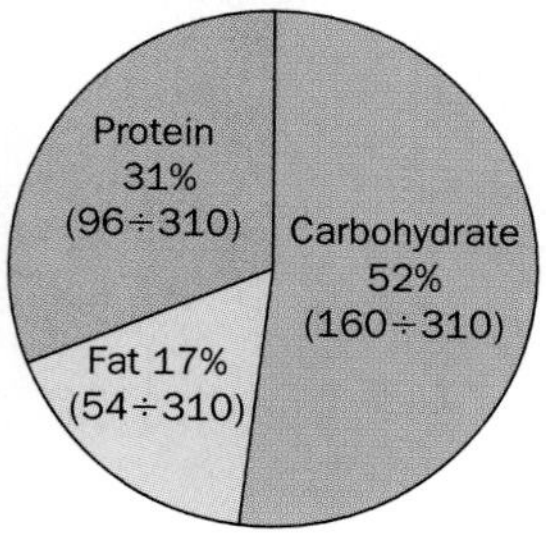
Protein
31%
(96÷310)
Carbohydrate
52%
(160÷310)
Fat 17%
(54÷310)

Lunch 2 (Yesterday's Lunch)	Carbohydrate (4 Calories/gram)		Fat (9 Calories/gram)		Protein (4 Calories/gram)	
	Calories	Grams	Calories	Grams	Calories	Grams
$\frac{1}{4}$-pound hamburger	0	0	108	12	76	19
Bun	124	31	27	3	20	5
Mayonnaise	0	0	63	7	-	-
French fries	248	62	198	22	24	6
Cookie	156	39	117	13	12	3
Cola	100	25	0	0	0	0
Total Calories = 1273	628	157	513	57	132	33

Percent of Calories from carbohydrates, fats, and proteins

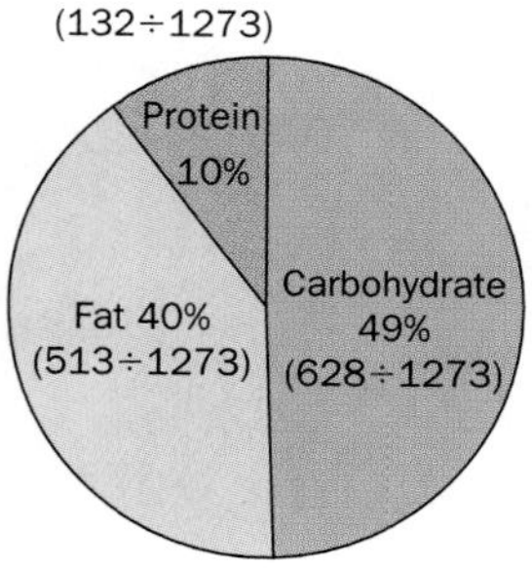
(132÷1273)
Protein
10%
Carbohydrate
49%
(628÷1273)
Fat 40%
(513÷1273)

Note: -= less than 1 g.

some plant oils, such as those from coconuts and palm seeds (Figure 7.3). Diets high in saturated fats have been implicated in atherosclerosis and heart disease. **Unsaturated fats** remain liquid at room temperature and are commonly called oils. They are found in seeds and nuts (olives, corn, walnuts, peanuts). In general, diets high in unsaturated fats seem to protect us against some types of heart disease.

Food processing can convert unsaturated fats to saturated fats by adding hydrogen atoms, a process known as *hydrogenation*. Margarine and peanut butter are two common products in which hydrogenation is used to convert plant oils to a more solid form.

Cholesterol is a lipid with a chemical structure different from that of fat (Chapter 1). It is found only in foods of animal origin; plants lack cholesterol. Consequently, cholesterol is found in muscle meats (beef, pork, fish), body organs (liver, brain), and animal products (eggs, milk products). In addition to being present in these foods, cholesterol is manufactured by the human liver.

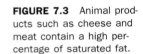

FIGURE 7.3 Animal products such as cheese and meat contain a high percentage of saturated fat.

An overabundance of cholesterol in the blood has been linked to the development of fatty deposits called **plaques** in the arteries. We will discuss the relationship between cholesterol and cardiovascular disease in Chapter 9.

Fat can be synthesized in the body from dietary carbohydrates and proteins, and so fat is not a nutrient that must be acquired from the diet. However, a particular fatty acid called *linoleic acid* must be obtained from the food we eat, because it cannot be synthesized by the body. Deficiencies of this acid are rare—most diets supply adequate amounts of linoleic acid. As little as a tablespoon of plant oil, such as soybean, corn, sunflower, or safflower oil, can supply our daily requirement of linoleic acid.

Uses of lipids in the body. Fats and other lipids have several uses within the human body: (1) Fats store energy. One gram of fat contains about 9 Calories, or more than twice the amount of energy contained in carbohydrates. (2) Fats insulate the body, reducing heat loss. (3) Internal organs such as the kidneys and eyes are cushioned by fat. (4) Phospholipids are structural components of all cell membranes. (5) The lipid cholesterol is a component of all cell membranes, but it is also used for the synthesis of bile and the sex hormones, testosterone and estrogen.

Dietary requirements for fat. It is recommended that in an adult diet, fat should make up 30 percent or less of the total Calories. For someone requiring 2000 Calories per day, this would amount to about 600 Calories or 67 grams (2.2 ounces) (see Table 7.1). Cholesterol intake should be limited to 250 milligrams per day because of the role cholesterol plays in cardiovascular diseases.

My grilled chicken sandwich was relatively low in fat, containing 54 Calories or 6 grams (Table 7.3). Since I ate my salad without cheese or dressing, it contained less than 1 gram of fat, which we'll consider to be 0 Calories. The 54 Calories from fat represent 17 percent of the total of 310 Calories in my lunch. This contrasts with the 513 Calories from fat in yesterday's lunch of a hamburger, French fries, and a cookie. In yesterday's lunch, fat represented 40 percent of the total Calories consumed (Table 7.3). It also represents almost all the fat I should consume in a day (57 grams of the 67 grams recommended). Yesterday's lunch represents a less healthy choice. If a lunch like this were a rare event, there would be no long-term health risk. However, if I eat such a high percentage of fat with every meal, there can be unhealthy consequences.

> Saturated fats are solid at room temperature; unsaturated fats are liquid at room temperature. Each gram of fat contains about 9 Calories of energy. Fats should make up 30 percent or less of our daily Calories.

Proteins and Their Amino Acids

Animal and plant products contribute to our daily intake of protein. Remember, however, that it is not the protein but the amino acids in the protein that are required by our cells. We need a diet that provides the correct types and amounts of amino acids.

In Chapter 2 you learned that all our cells use 20 different types of amino acids to produce their required protein molecules. Some of these amino acids can be synthesized by the cells and do not need to be included in the diet. They are called **nonessential amino acids;** that is, it is not essential to have them in the diet (Table 7.4). Other amino acids, called **essential amino acids,** cannot be synthesized by the cells, and they must be included in the diet. The meaning of these two terms has nothing to do with whether an amino acid is necessary (essential) for making protein; all 20 amino acids are needed for protein synthesis.

Dietary proteins that contain all 10 essential amino acids are called **complete proteins.** The proteins in animal products (meats, milk products, eggs) are complete proteins. Dietary proteins that contain some, but not all, of the essential amino acids are called **incomplete proteins.** Most plant proteins (beans, grains, fruits, nuts) lack one or more essential amino acid and are considered to be incomplete proteins.

If you are a vegetarian and consume no animal products, you must eat a variety of different plant products every day to assure that you get all the essential amino acids (Table 7.5). This is known as *protein complementing*. Protein deficiency diseases occur when people, particularly children, must depend entirely on a single plant source for all their food.

Uses of protein in the body. Proteins serve a number of diverse functions within the body, including *structural support* for cells (membrane and cytoskeleton proteins) and tissues (collagen, keratin), *contraction* of muscles (actin and myosin proteins), *catalysis* of chemical reactions (enzymes), and *regulation* of many body functions (protein hormones). A number of different types of protein have other functions, such as *transport* (hemoglobin and membrane proteins) and *defense* (antibodies).

Proteins can also be used to supply energy for the body and, when utilized, provide the same amount of energy that carbohydrates supply. Protein is not the body's first choice of energy, however. Only when one is fasting, dieting, or starving will the body use its own muscle protein to supply energy needs.

Dietary requirements for protein. It is recommended that protein should account for about 10 percent of the total Calories consumed daily by adults. For someone requiring 2000 Calories, this would amount to 200 Calories or 50 grams (1.75 ounces) of protein per day (see Table 7.1). The grilled chicken sandwich provided about 92 Calories or 23 grams (0.8 ounce) of protein, while the

TABLE 7.4 Essential and Nonessential Amino Acids

Essential Amino Acids (Not Synthesized by Our Cells)	Nonessential Amino Acids (Synthesized by Our Cells)
1. Isoleucine	1. Alanine
2. Leucine	2. Asparagine
3. Lysine	3. Aspartic acid
4. Methionine	4. Cysteine
5. Phenylalanine	5. Glutamic acid
6. Threonine	6. Glutamine
7. Tryptophan	7. Glycine
8. Valine	8. Proline
9. Arginine*	9. Serine
10. Histidine*	10. Tyrosine

*Essential only during childhood because they cannot be synthesized in sufficient quantities during periods of rapid growth.

TABLE 7.5 Essential Amino Acids Missing from Selected Plant Foods and Protein Complementing Using Wheat and Beans

Plant Food	Essential Amino Acids Missing from Plant	Examples of Plant Combinations Supplying All Essential Amino Acids
Grains (wheat, rice, barley, millet)	Lysine, threonine	Corn + green peas
		Rice + beans
(Corn)	Lysine, tryptophan	Wheat + beans
Legumes (kidney, lima, pinto beans, soybeans, green peas)	Methionine	Corn + beans

Essential Amino Acids Found in Wheat		Essential Amino Acids Found in Beans
Isoleucine Leucine Lysine Methionine Phenylalanine Threonine Tryptophan Valine Arginine Histidine	Eaten together, wheat and beans provide all 10 essential amino acids	Isoleucine Leucine Lysine Methionine Phenylalanine Threonine Tryptophan Valine Arginine Histidine

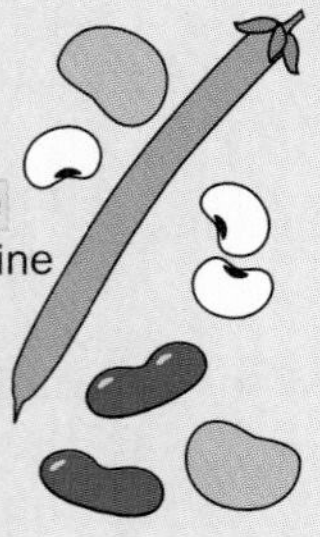

☐ = Amino acid in low quantities or missing

quarter-pound hamburger with a bun I had yesterday had about 96 Calories or 24 grams. Either one of these meats would have provided about 30 percent of the protein recommended for a 2000 Calorie per day diet.

> Plant and animal products supply dietary protein. There are 10 essential amino acids that the body cannot manufacture. Each gram of proteins contain 4 Calories. About 10 percent of daily Calories should be derived from protein.

Vitamins, Minerals, and Water

Most **vitamins** are not produced by our cells and must be obtained from the diet. Compared with carbohydrates and proteins, only tiny amounts of vitamins are needed. Vitamins are organic molecules that function as parts of enzymes or as direct participants in enzyme-catalyzed reactions. The National Research Council has established an RDA for most vitamins (Table 7.6).

Although only small amounts are needed, no single food can supply all the vitamins we require. It is necessary to eat a variety of foods to obtain all these vitamins. Insufficient vitamins can lead to deficiency diseases such as night blindness (vitamin A), scurvy (vitamin C), beriberi (vitamin B_1, or thiamine), rickets (vitamin D), and pellagra (niacin). Vitamin supplements, including vitamin-enriched foods, have greatly reduced the frequency of these diseases in developed countries. Because of their rapid growth and depen-

TABLE 7.6 Water-Soluble and Fat-Soluble Vitamins

Vitamin	Primary Function in the Body	Major Dietary Source	Symptoms from Prolonged Deficiency
		Water-Soluble	
B_1 (thiamin)	A coenzyme in carbohydrate metabolism; required for changing pyruvic acid to acetyl	Meats, eggs, grains, legumes	Beriberi: damage to heart tissue, nerve damage leading to muscle weakness, pain, and paralysis
B_2 (riboflavin)	A coenzyme in carbohydrate metabolism; part of FAD	Meat, eggs, grains, dairy products, green leafy vegetables	Scaly skin, cracks at corner of mouth, purple-red tongue
B_6 (pyroxidine)	A coenzyme in amino acid and fatty acid metabolism	Meat, eggs, grains, green leafy vegetables	Nerve and muscular irritability, skin lesions, anemia
B_{12}	A coenzyme used in DNA/RNA synthesis and formation of red blood cells	Meat, eggs, dairy products	Pernicious anemia, nerve damage
Niacin	A coenzyme in carbohydrate metabolism; part of NAD	Meat, eggs, grains	Pellagra: diarrhea, skin damage, red and sore tongue
C (ascorbic acid)	Collagen synthesis, antioxidant	Fruits, vegetables	Scurvy: bleeding gums, poor wound healing, loosened teeth, muscle weakness
Folic acid	A coenzyme in DNA/RNA synthesis and formation of red blood cells	Leafy green vegetables, legumes, grains	Anemia, intestinal disturbances
Biotin	A coenzyme used in the metabolism of fat, glycogen, and some amino acids	Meat, eggs, dairy products, grains, legumes	Drying of skin, muscular weakness and pain
Pantothenic acid	Part of coenzyme A	Meat, eggs, dairy products, grains, legumes	Intestinal disturbances, fatigue, anemia
		Fat-Soluble	
A	Vision, especially at night, and development of bones and teeth	In dark green and yellow vegetables, eggs	Night blindness, dry skin
D	Promotes absorption of calcium and phosphorus from intestine	Fish oil, fortified milk	Rickets, osteomalacia
E	Antioxidant	Plant oils	Anemia
K	Participates in blood clotting	Meat, green vegetables (also produced by intestinal bacteria)	Abnormal bleeding

TABLE 7.7 The Major Minerals and Selected Trace Minerals of the Body

Mineral	Source	Primary Function
	Major Minerals	
Calcium (Ca)	Dairy products, meat	Maintenance of bones and teeth, blood clotting, muscle and nerve cell function
Phosphorus (P)	Dairy products, meat	Maintenance of bones and teeth, acid-base balance of blood, ATP and DNA/RNA formation
Magnesium (Mg)	Cereals, green leafy vegetables	Enzyme function, muscle and nerve cell function
Sulfur (S)	Meat, dairy products	Part of some amino acids and vitamins
Sodium (Na)	Table salt, processed foods	Muscle and nerve cell function, maintenance of water balance (osmosis)
Potassium (K)	Meats, cereals, fruits, vegetables	Muscle and nerve function, maintenance of water balance (osmosis)
Chloride (Cl)	Table salt, processed foods	Acid-base balance of blood, production of gastric HCl
	Trace Minerals	
Iron (Fe)	Meats, eggs, legumes, cereals	Part of hemoglobin molecule
Iodine (I)	Seafood, iodized salt	Part of thyroid hormone
Fluoride (F)	Seafood, drinking water	Provides extra strength for teeth
Zinc (Zn)	Meats, seafood, cereals	Part of several enzymes, important for growth and fertility
Manganese (Mn)	Nuts, cereals, vegetables	Part of several enzymes
Selenium (Se)	Seafood, meats, cereals	Part of several enzymes, functions with vitamin E
Copper (Cu)	Seafood, nuts, legumes	Synthesis of hemoglobin molecule

dence on adults, children are frequently vulnerable to vitamin deficiencies. Government programs such as WIC (**W**omen, **I**nfants, and **C**hildren) have largely prevented severe childhood deficiency diseases among poor families in this country.

Water-soluble vitamins. Depending on their solubility, vitamins can be divided into two categories: water-soluble and fat-soluble (see Table 7.6). Vitamins B_1, B_2, B_6, and B_{12}; niacin; vitamin C; pantothenic acid; biotin; and folic acid all

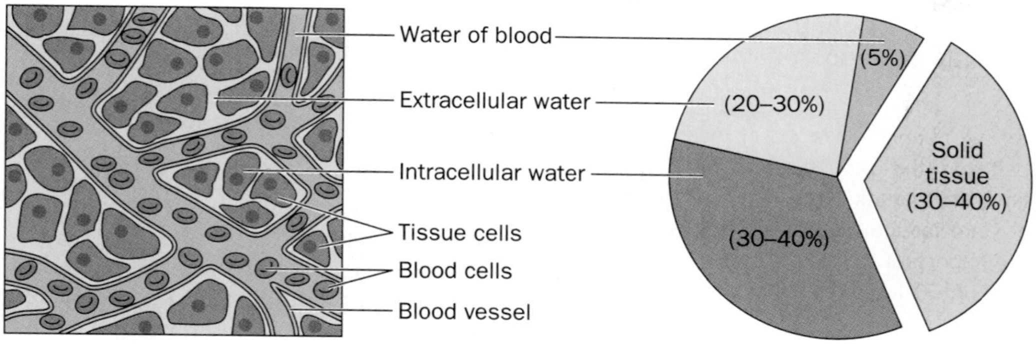
Water of blood
Extracellular water
Intracellular water
Tissue cells
Blood cells
Blood vessel
(5%)
(20–30%)
(30–40%)
Solid tissue (30–40%)

FIGURE 7.4 Location of water within tissue and the percentages of solid tissue and water that make up total body weight.

dissolve in water and are absorbed with water from the small intestine. Only a limited amount of water-soluble vitamins can be stored. Any excess that is consumed is excreted in the urine, and this is why most of these vitamins should be eaten daily. Among the water-soluble vitamins, only niacin can be synthesized by the body.

Fat-soluble vitamins. Vitamins A, D, E, and K dissolve in fat and are absorbed with fat from the small intestine. Fat-soluble vitamins are stored in the body, particularly in the liver. Among the fat-soluble vitamins, only vitamin D can be synthesized by the body (Chapter 4).

The best sources of most vitamins are fresh fruits and vegetables and meats (see Table 7.6). Although good food choices should supply adequate amounts of vitamins, some people take vitamin supplements. Whether necessary or not, these supplements can be expensive, and it is possible to take too much of some vitamins. The fat-soluble vitamins are toxic at high doses, as is the water-soluble vitamin niacin.

Vitamins as health protectors. Several vitamins are gaining publicity as antioxidants, that is, chemicals that disarm cell-damaging molecules called free radicals. **Free radicals** are unstable and highly reactive compounds that are normal by-products of cell metabolism. Our cells produce chemicals that neutralize free radicals, but some damage can still occur to cell proteins (enzymes, membrane proteins) and nucleic acids (DNA, RNA). Over time, such damage accumulates and may contribute to diseases associated with aging, such as cancer, hardening of the arteries, and cataracts.

Recent research has indicated that vitamin E, vitamin C, and beta-carotene (used by the body to make vitamin A) may neutralize free radicals. To help prevent free radical damage, some health experts suggest that we consume more fruits and vegetables rich in these vitamins.

Minerals. The minerals we need are a diverse group of chemical substances that include calcium, phosphorous, potassium, sulfur, sodium, chloride, and magnesium plus many others that are needed in tiny amounts (trace minerals), such as iron and iodine (Table 7.7 on page 200). Minerals participate in a variety of cellular and body functions. Some minerals form solid structures (calcium and phosphorous in bones and teeth), some are associated with proteins (iron in hemoglobin, iodine in thyroid hormone), and some are required for enzyme activity. Other minerals are present as ions in body fluids, maintaining the electrical activity of nerves and muscles (sodium, potassium, and calcium ions: Na^+, K^+, Ca^{2+}).

Water. Water is a major constituent of the body, contributing up to 60 to 70 percent of body weight. It is found within cells (intracellular) and between cells (extracellular) and forms a major portion of the blood (Figure 7.4). Water has a variety of functions. It serves as a solvent into which chemicals dissolve, transports chemicals throughout the body, and provides a medium in which chemical reactions take place. Its molecules participate in some chemical reactions, and it is important in temperature regulation during evaporative cooling (Chapter 4).

On average, we need between 2 and 3 liters (about 2 and 3 quarts) of water a day to replace that which is lost from the body in urine and from general evaporation (breathing, perspiration). Water from the consumption of food and liquids provides our daily water needs. Increases in water consumption are necessary when water is lost from the body through prolonged periods of vomiting, diarrhea, or excessive perspiration.

> Vitamins, minerals, and water must be obtained from the diet. The B vitamins and vitamin C are water-soluble. Vitamins A, D, E, and K are fat-soluble. Minerals have a variety of important functions in the body. Water makes up 60 to 70 percent of body weight.

Metabolism: The Body's Chemical Transformations

The term **metabolism** (*me-**tab**-ho-liz-um*) refers to the chemical reactions that occur in the cells of the body. Metabolism can be differentiated into two types of processes: catabolism and anabolism.

Catabolism (*kah-**tab**-oh-liz-um*) includes the chemical reactions that break down nutrient molecules to release energy (Figure 7.5). The energy released during catabolic reactions is captured (conserved) by the synthesis of ATP from ADP and ℙ (phosphate):

$$\text{ADP} + \mathbb{P} + \begin{matrix}\text{energy from}\\ \text{catabolism}\end{matrix} \longrightarrow \text{ATP}$$

ATP is a high-energy molecule that is the immediate source of energy for cellular anabolic reactions.

Anabolism (*an-**ab**-oh-liz-um*) includes the chemical reactions that synthesize complex molecules from simpler nutrient molecules (Figure 7.5). For these chemical reactions to occur, energy is required in the form of ATP. This energy is made available when the high-energy ATP molecule is chemically broken down into ADP and ℙ (phosphate).

$$\text{ATP} \longrightarrow \text{ADP} + \mathbb{P} + \begin{matrix}\text{energy for}\\ \text{anabolism}\end{matrix}$$

In the sections that follow, we will discuss the metabolism of carbohydrates, lipids, and proteins. We will show what happens to each of these nutrients during catabolism and anabolism. By its nature, this topic requires us to apply some of the chemistry presented in Chapter 1.

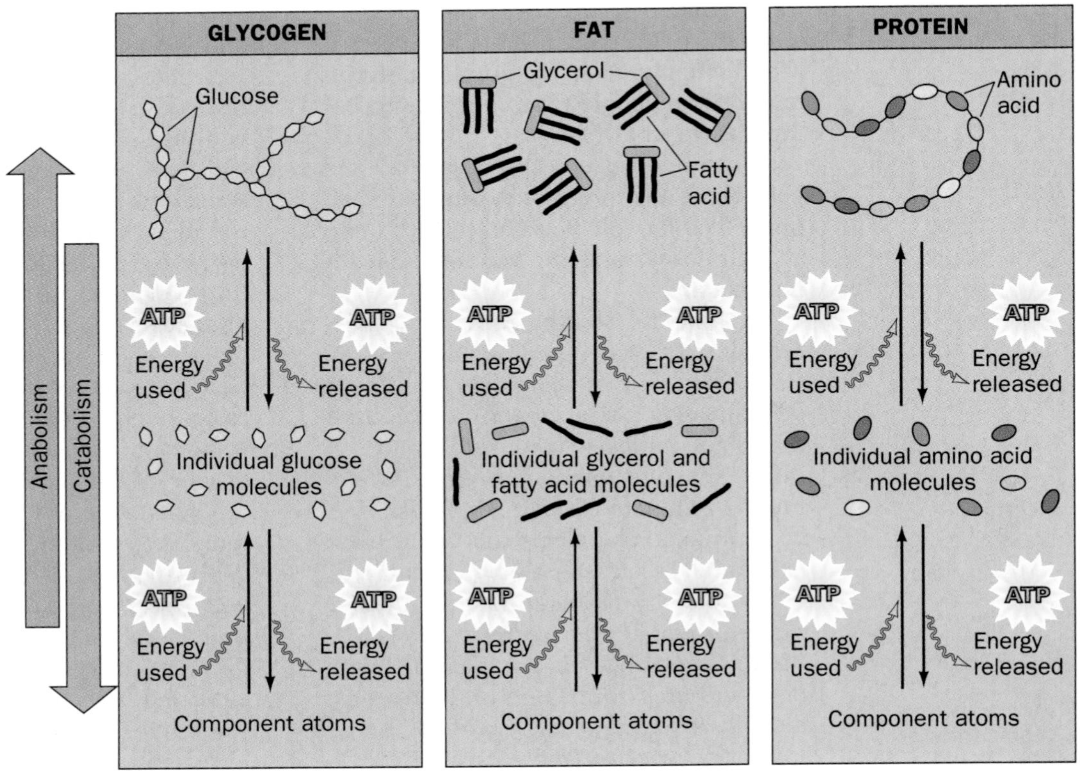
Anabolism
Catabolism
GLYCOGEN
Glucose
ATP
Energy used
Energy released
Individual glucose molecules
Component atoms
FAT
Glycerol
Fatty acid
Individual glycerol and fatty acid molecules
PROTEIN
Amino acid
Individual amino acid molecules

FIGURE 7.5 The catabolism and anabolism of glycogen, fat, and protein.

Learning this chemistry will enable you to better apply the fundamentals of nutrition to decisions about your life and health.

Carbohydrate Catabolism: Breaking Down Glucose for Energy

Carbohydrate catabolism is accomplished through a sequence of reactions that release the chemical energy stored in glucose. Recall that glucose is the carbohydrate most commonly used by cells for energy. The energy-releasing reactions of glucose catabolism can be divided into three distinct stages: glycolysis, the Krebs cycle, and the electron transport system. Together, these three stages extract much of the energy contained in the glucose molecule and use it to produce ATP. An overview of glycolysis, Krebs cycle, and the electron transport system is presented in Figure 7.6. Further details of these three stages are provided in Figures 7.7 through 7.9.

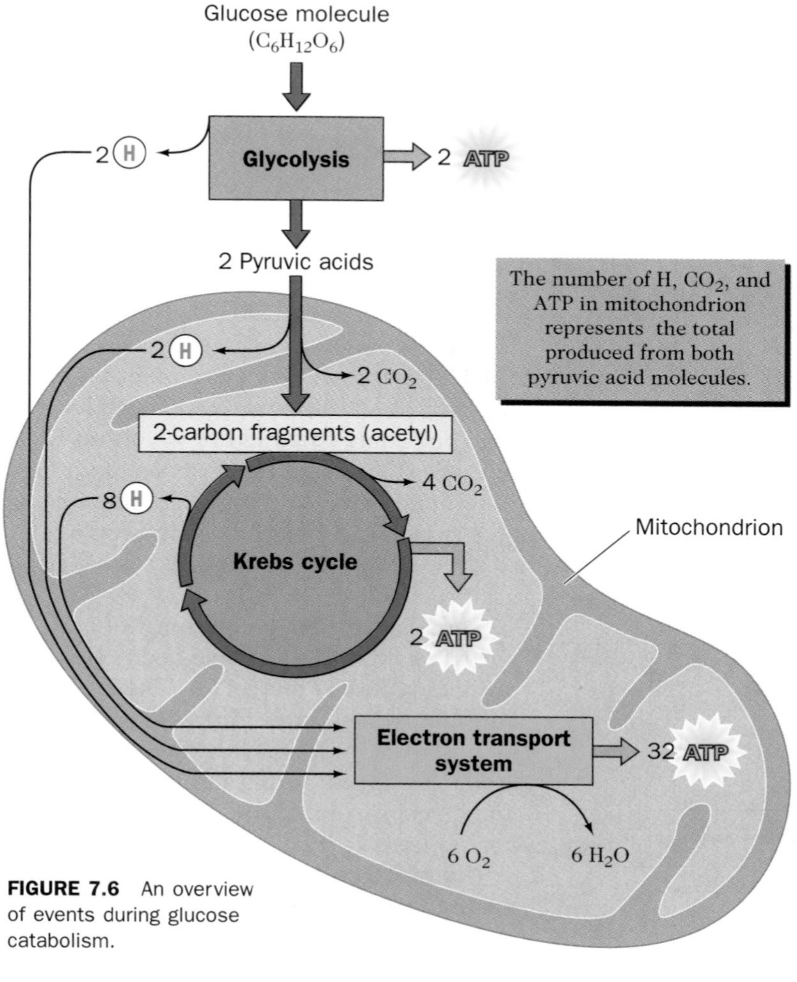

Glucose molecule
($C_6H_{12}O_6$)
2 H
Glycolysis
2 ATP
2 Pyruvic acids
The number of H, CO_2, and ATP in mitochondrion represents the total produced from both pyruvic acid molecules.
2 H
2 CO_2
2-carbon fragments (acetyl)
4 CO_2
8 H
Mitochondrion
Krebs cycle
2 ATP
Electron transport system
32 ATP
6 O_2
6 H_2O

FIGURE 7.6 An overview of events during glucose catabolism.

Glycolysis: converting glucose to pyruvic acid. The chemical reactions of glycolysis occur in the cytoplasmic fluid of a cell, not within the organelles. During a series of nine chemical reactions, a 6-carbon *glucose* molecule is split in half, ultimately forming two 3-carbon molecules called *pyruvic acid* (see Figure 7.6). The reactions of glycolysis also release 2 ATP molecules and 2 energy-rich hydrogen atoms. As you will see, the energy in these hydrogen atoms is extracted and used to make additional ATPs.

Pyruvic acid is further processed by either an anaerobic pathway or an aerobic pathway. During periods of insufficient oxygen (O_2) supply, the **anaerobic pathway** operates (Figure 7.7). Pyruvic acid is converted to lactic acid by reactions called *lactic acid fermentation*. You will recall that during strenuous muscle activity, skeletal muscle cells may not receive sufficient O_2, and as a consequence, lactic acid accumulates, causing fatigue and muscle soreness (Chapter 5).

Under normal operating conditions, however, the body has sufficient O_2 and the **aerobic pathway** operates. The aerobic pathway takes place in the mitochondria and begins with the Krebs cycle, the second stage of the breakdown process.

The Krebs cycle: processing pyruvic acid. The Krebs cycle involves a series of chemical reactions in the mitochondria that further process pyruvic acid so that more energy can be extracted from it.

After entering the mitochondria, each pyruvic acid molecule is processed in the following way before entering the Krebs cycle (Figure 7.8). First, an energy-rich hydrogen atom is released. Second, 1 carbon atom is lost as CO_2. The remaining 2-carbon fragment (an acetyl group) joins with a carrier compound called *coenzyme A*, forming a molecule called *acetyl coenzyme A* (acetyl CoA). The 2-carbon acetyl part of each acetyl CoA then enters the reactions of the Krebs cycle. Coenzyme A brings the 2-carbon acetyl into the Krebs cycle, after which it detaches and is reused.

The Krebs cycle reactions are referred to as a cycle because they begin and end

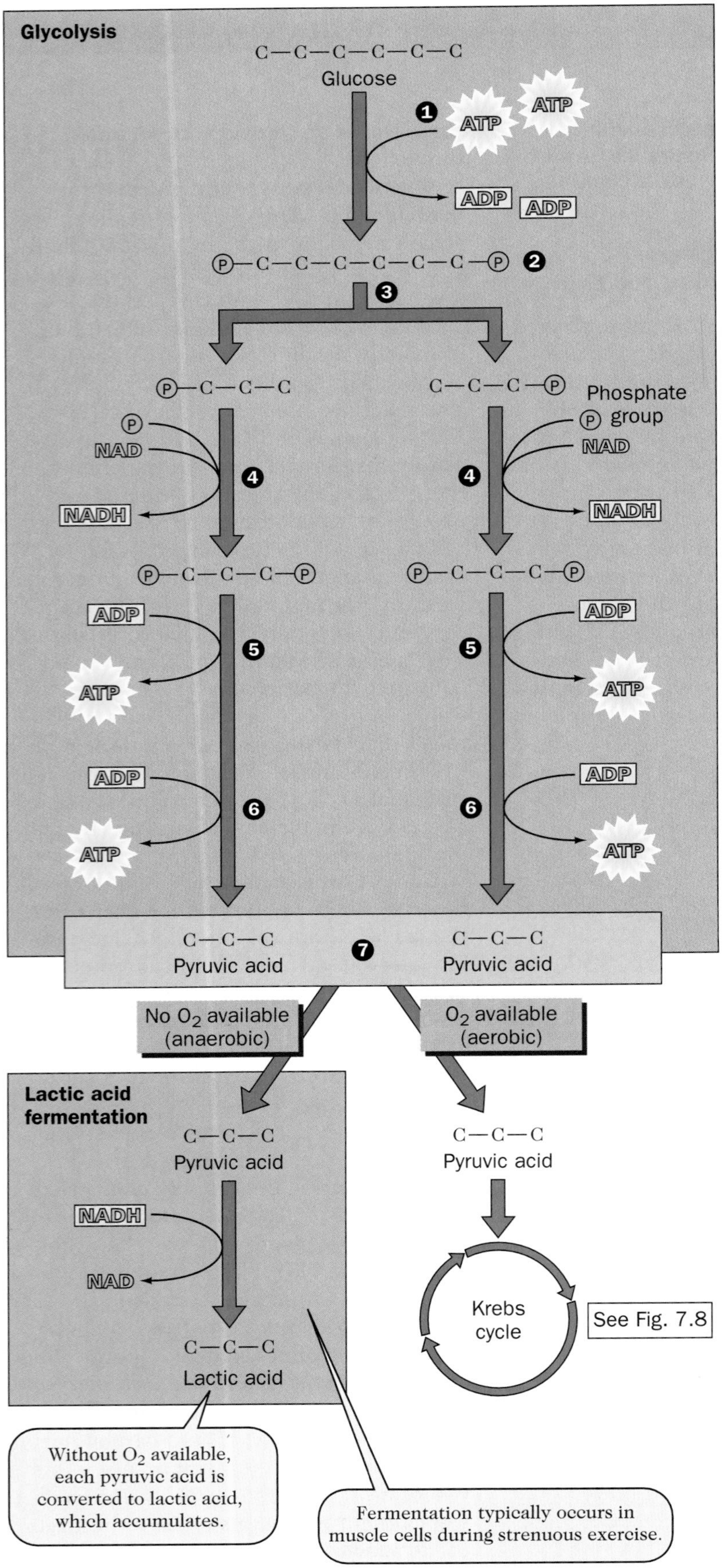
Glycolysis
C—C—C—C—C—C
Glucose
ATP
ATP
ADP
ADP
P—C—C—C—C—C—C—P
P—C—C—C
C—C—C—P
Phosphate group
P
NAD
NADH
P—C—C—C—P
ADP
ATP
C—C—C
Pyruvic acid
No O_2 available (anaerobic)
O_2 available (aerobic)
Lactic acid fermentation
NADH
NAD
Lactic acid
Krebs cycle
See Fig. 7.8
Without O_2 available, each pyruvic acid is converted to lactic acid, which accumulates.
Fermentation typically occurs in muscle cells during strenuous exercise.

FIGURE 7.7 The major chemical reactions of glycolysis and lactic acid fermentation. Only the carbon atoms are depicted; oxygen and hydrogen atoms are not shown.

1. Two ATPs required to energize glucose.
2. A phosphate from each ATP attaches to glucose.
3. Energized glucose molecule splits in half, forming two 3-carbon molecules.
4. An inorganic phosphate (P) from cytoplasm attaches to each 3-carbon molecule as each molecule releases hydrogen.
5. 3-carbon molecule releases a phosphate to ADP, forming ATP.
6. The second phosphate is released, forming another ATP.
7. End product of glycolysis is two pyruvic acid molecules.

with the same molecule. The following chemical events occur during each cycle: (1) The 2-carbon acetyl fragment is dismantled, releasing two CO_2 molecules. (2) Four energy-rich hydrogen atoms are released. (3) A single ATP molecule is generated. Because the Krebs cycle operates twice for every glucose molecule, 4 CO_2, 8 energy-rich hydrogens, and 2 ATP are produced.

Thus far, the 3 carbon atoms of each pyruvic acid molecule have been released, producing 3 CO_2 molecules. These molecules are transported to the lungs by the bloodstream and exhaled. The hydrogen atoms produced during the Krebs cycle and glycolysis enter the electron transport system.

The electron transport system: producing lots of ATP. During glycolysis and the Krebs cycle, less than 10 percent of the energy contained in the original

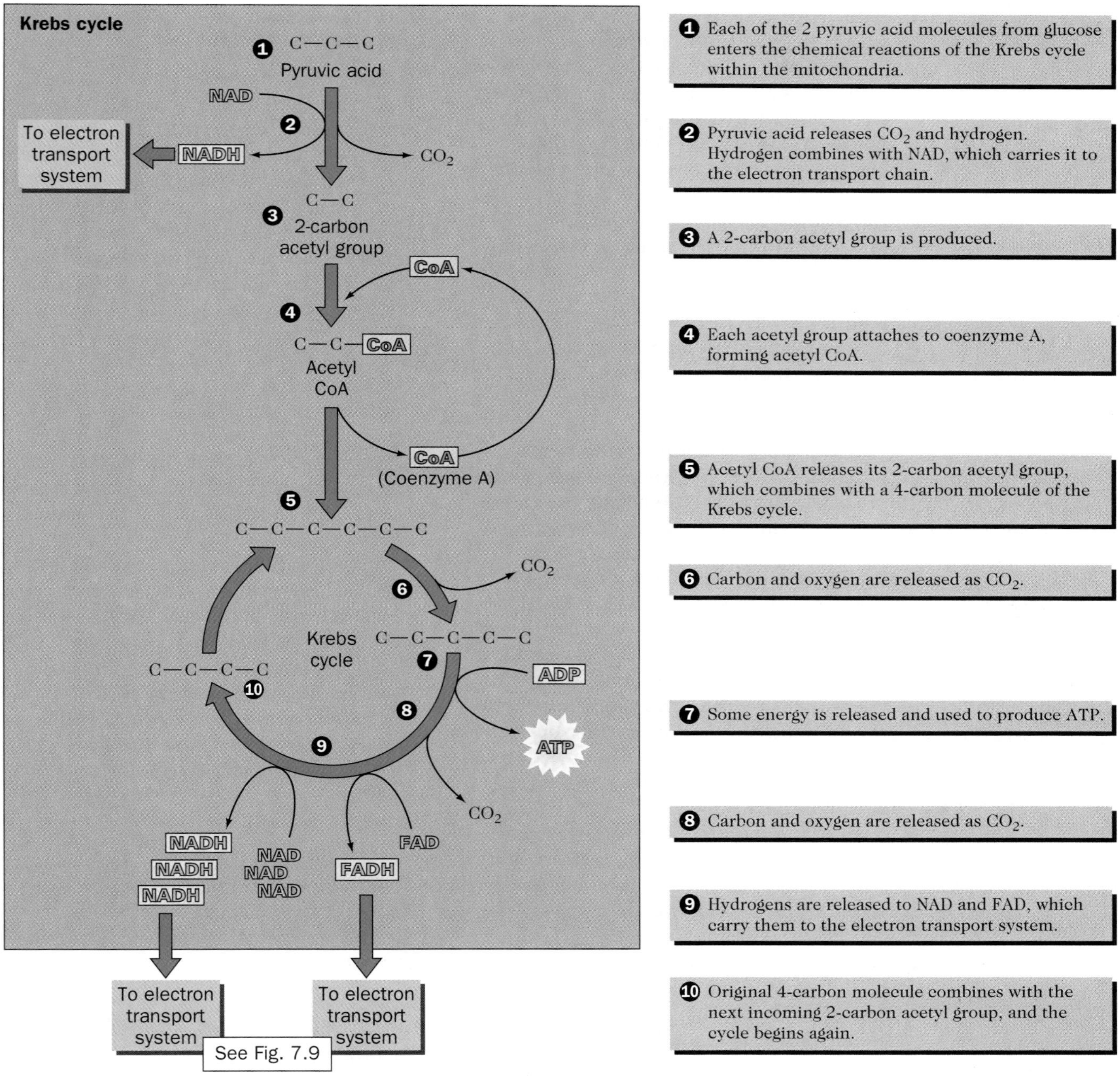
Krebs cycle
1 C—C—C
Pyruvic acid
NAD
2
To electron transport system
NADH
CO_2
C—C
3 2-carbon acetyl group
CoA
4
C—C—CoA
Acetyl CoA
CoA
(Coenzyme A)
5
C—C—C—C—C—C
CO_2
6
Krebs cycle
C—C—C—C—C
7
ADP
C—C—C—C
10
8
ATP
9
CO_2
NADH
NADH
NADH
NAD
NAD
NAD
FADH
FAD
To electron transport system
To electron transport system
See Fig. 7.9
1 Each of the 2 pyruvic acid molecules from glucose enters the chemical reactions of the Krebs cycle within the mitochondria.
2 Pyruvic acid releases CO_2 and hydrogen. Hydrogen combines with NAD, which carries it to the electron transport chain.
3 A 2-carbon acetyl group is produced.
4 Each acetyl group attaches to coenzyme A, forming acetyl CoA.
5 Acetyl CoA releases its 2-carbon acetyl group, which combines with a 4-carbon molecule of the Krebs cycle.
6 Carbon and oxygen are released as CO_2.
7 Some energy is released and used to produce ATP.
8 Carbon and oxygen are released as CO_2.
9 Hydrogens are released to NAD and FAD, which carry them to the electron transport system.
10 Original 4-carbon molecule combines with the next incoming 2-carbon acetyl group, and the cycle begins again.

FIGURE 7.8 The major chemical reactions that process pyruvic acid into and through the Krebs cycle. Only the carbon atoms are depicted; oxygen and hydrogen atoms are not shown.

glucose molecule is released for ATP formation. Through the electron transport system, much more energy is released and used to generate many ATP molecules. These reactions, which occur along the inner mitochondrial membrane, extract the energy from the energy-rich hydrogen atoms produced during both glycolysis and the Krebs cycle (see Figure 7.6).

The energy-rich hydrogen atoms are transported to the electron system by a carrier molecule known as **NAD** (**n**icotinamide **a**denine **d**inucleotide, pronounced *nick-oh-**tin**-am-eyed **add**-in-neen di-**nuke**-lee-oh-tide*). As a result, NAD changes to NADH (Figures 7.7 and 7.8). A related hydrogen carrier called **FAD** (**f**lavin **a**denine **d**inucleotide) also carries some of the hydrogens from the Krebs cycle and changes to FADH. Both of these hydrogen carriers are made from B vitamins. NAD is derived from niacin, and FAD is derived from riboflavin (B_2).

During processing by the electron transport system, electrons are removed from each hydrogen atom, producing a hydrogen ion (H^+) (Figure 7.9). The electrons are then passed along a series of reactions that extract their energy. It is this energy that is used to produce ATP molecules from ADP and ℙ. Once energy extraction has occurred, the electrons are transferred to O_2, which in turn bonds with the H^+ to form water. A total of 32 ATP molecules are produced through the electron transport system's processing of all the high-energy hydrogens derived from a single glucose molecule.

A summary of energy production. A total of 36 ATPs are produced per glucose molecule during its catabolism (see Figure 7.6). The greatest number of these molecules (32 ATPs) are generated when the high-energy hydrogen atoms are processed in the oxygen-dependent reactions of the electron transport system. Now you know why we need to breath: Our cells require a continuous supply of O_2 so that the electron transport system can produce the ATP needed by our cells. Only four of the 36 ATPs are directly produced during glycolysis (net of 2 ATPs) and the Krebs cycle (2 ATPs).

Knowing the number of ATPs generated helps us understand the amount of energy released from glucose. It is estimated that about 40 percent of the chemical energy contained in 1 glucose molecule is captured through the reactions of the aerobic pathway and used to make ATP. This may appear to be a rather low level of efficiency, but it is much higher than that of the most efficient automobile, which operates at about 20 percent efficiency. The remaining 60 percent of glucose energy is lost as heat and is used to maintain body temperature or is released from the body.

> Glucose catabolism releases the energy in glucose. In the aerobic pathway, glucose is processed by glycolysis, the Krebs cycle, and the electron transport system. Of the 36 ATPs produced, 32 are derived from the electron transport system.

Carbohydrate Anabolism: Glycogen from Glucose

Carbohydrate anabolism includes the chemical reactions that synthesize glycogen from glucose or synthesize glucose molecules from noncarbohydrate molecules such as amino acids and glycerol. *Glycogen synthesis* occurs in liver and muscle tissue. These reactions combine individual glucose molecules together to form glycogen (see Figures 7.5 and 7.10). Glycogen is the storage form of glucose. An adult liver synthesizes and stores about 100 grams (about $\frac{1}{4}$ pound) of glycogen. Our skeletal muscles can also synthesize and store from 300 to 400 grams (about $\frac{2}{3}$ of a pound) of glycogen. Together, liver glycogen and skeletal muscle glycogen can provide enough glucose to support a few hours of strenuous activity.

Fats and proteins can be used in *glucose synthesis* when glycogen stores are depleted during periods of fasting, dieting, or starvation (Figure 7.10). Under these conditions, body fat is broken down to glycerol and fatty acids and skeletal muscle proteins are broken down to amino acids. Glycerol and amino acids are brought to the liver, where they are converted to glucose (fatty acids are not converted to glucose).

> Carbohydrate anabolism produces glycogen from glucose or from noncarbohydrate sources such as fat and protein. Glycogen is stored in the liver and skeletal muscles.

Fat Catabolism and Anabolism: Storing and Using Fats

Stored fat, called triglycerides, constitutes about 98 percent of the body's energy reserves. Glycogen accounts for only about 1 percent. Adipose cells (fat cells) are distributed throughout the body, and their stored triglyceride molecules can be broken down by fat catabolism to provide energy.

During *fat catabolism,* triglyceride breaks down to glycerol and three fatty acids (see Figures 7.5 and 7.10). Glycerol is converted to pyruvic acid, which is then processed in the Krebs cycle and the electron transport system to release its chemical energy and produce ATP. Each fatty acid molecule is a chain of 16 or 18 carbon atoms with attached hydrogen atoms (see Figure 1.19). Fatty acids are disassembled into individual 2-carbon acetyl groups. Each acetyl group is processed by the Krebs cycle and the electron transport system, and ATP molecules are produced.

Fat anabolism produces fat molecules (triglycerides) from excess carbohydrates and proteins consumed in the diet. In other words, you can accumulate fat by eating too much carbohydrate and protein. After glycogen storage has been completed, any excess glucose is converted into two different molecules: glycerol and the 2-carbon acetyl group (see Figure 7.10). Excess amino acids are also converted to 2-carbon acetyl groups. The acetyl groups derived from glucose and amino acids are used to synthesize fatty acids. Triglyceride (fat) molecules are formed when glycerol and fatty acid molecules combine. Fats are stored in adipose cells until they are needed for ATP production (fat catabolism).

During fat catabolism, fatty acids and glycerol are processed by the Krebs cycle and the electron transport system. Fat anabolism converts excess carbohydrates or amino acids to fat molecules.

Protein Catabolism and Anabolism

Protein catabolism is very different from protein digestion, which takes place in the stomach and small intestine. *Protein catabolism* breaks down cellular protein molecules into their individual amino acids, which can be used as a source of energy during dieting or starvation. The amino acid molecules enter different reactions of glycolysis and the Krebs cycle (see Figure 7.10). Once this happens, they are catabolized and ATP is produced, as was described earlier for carbohydrates.

One important point about protein catabolism must be mentioned here. Amino acids contain an amino group with a nitrogen atom ($—NH_2$). Before catabolism can occur, the amino group of each amino acid must be removed. This removal produces ammonia (NH_3), which is very toxic. To protect cells and the body as a whole, the liver rapidly converts ammonia to a less toxic chemical known as *urea,* which is carried to the kidneys for excretion in the urine (Chapter 11).

Protein anabolism consists of reactions that synthesize nonessential amino acids and reactions that are associated with protein synthesis (Chapter 2). Recall that nonessential amino acids can be made by the body (see Table 7.4). During the synthesis of nonessential amino acids, certain molecules from the

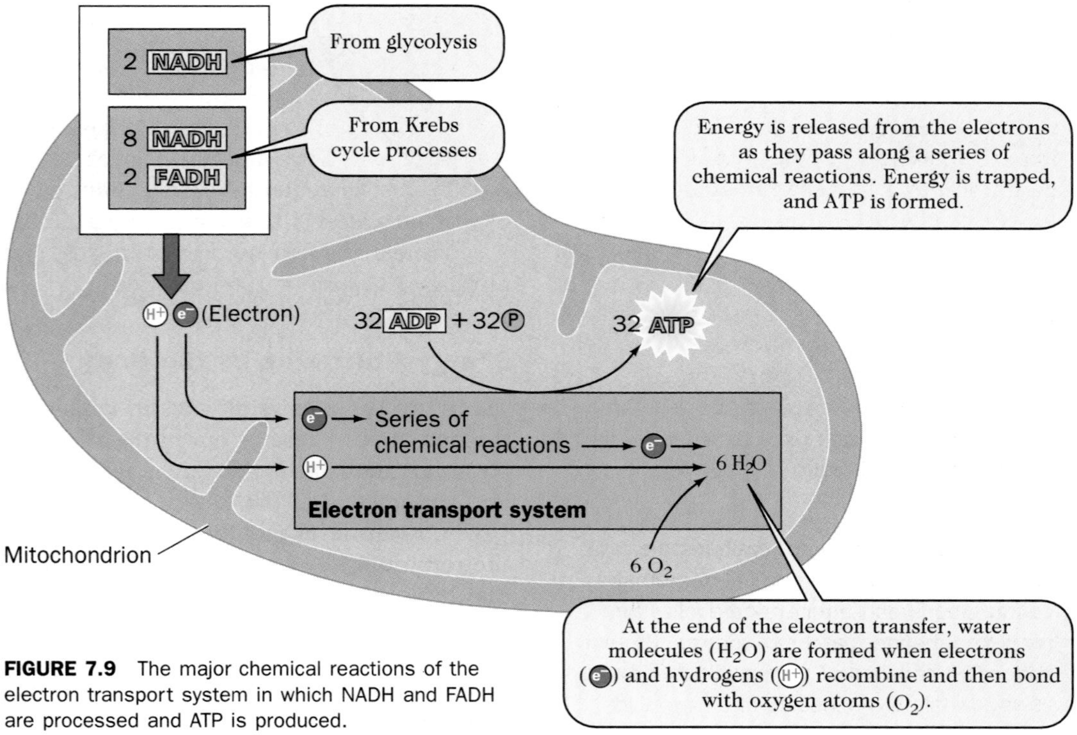
2 NADH
From glycolysis
8 NADH
2 FADH
From Krebs cycle processes
Energy is released from the electrons as they pass along a series of chemical reactions. Energy is trapped, and ATP is formed.
H+ e (Electron)
32 ADP + 32 P
32 ATP
e → Series of chemical reactions → e →
H+ → 6 H2O
Electron transport system
Mitochondrion
6 O2
At the end of the electron transfer, water molecules (H2O) are formed when electrons (e) and hydrogens (H+) recombine and then bond with oxygen atoms (O2).

FIGURE 7.9 The major chemical reactions of the electron transport system in which NADH and FADH are processed and ATP is produced.

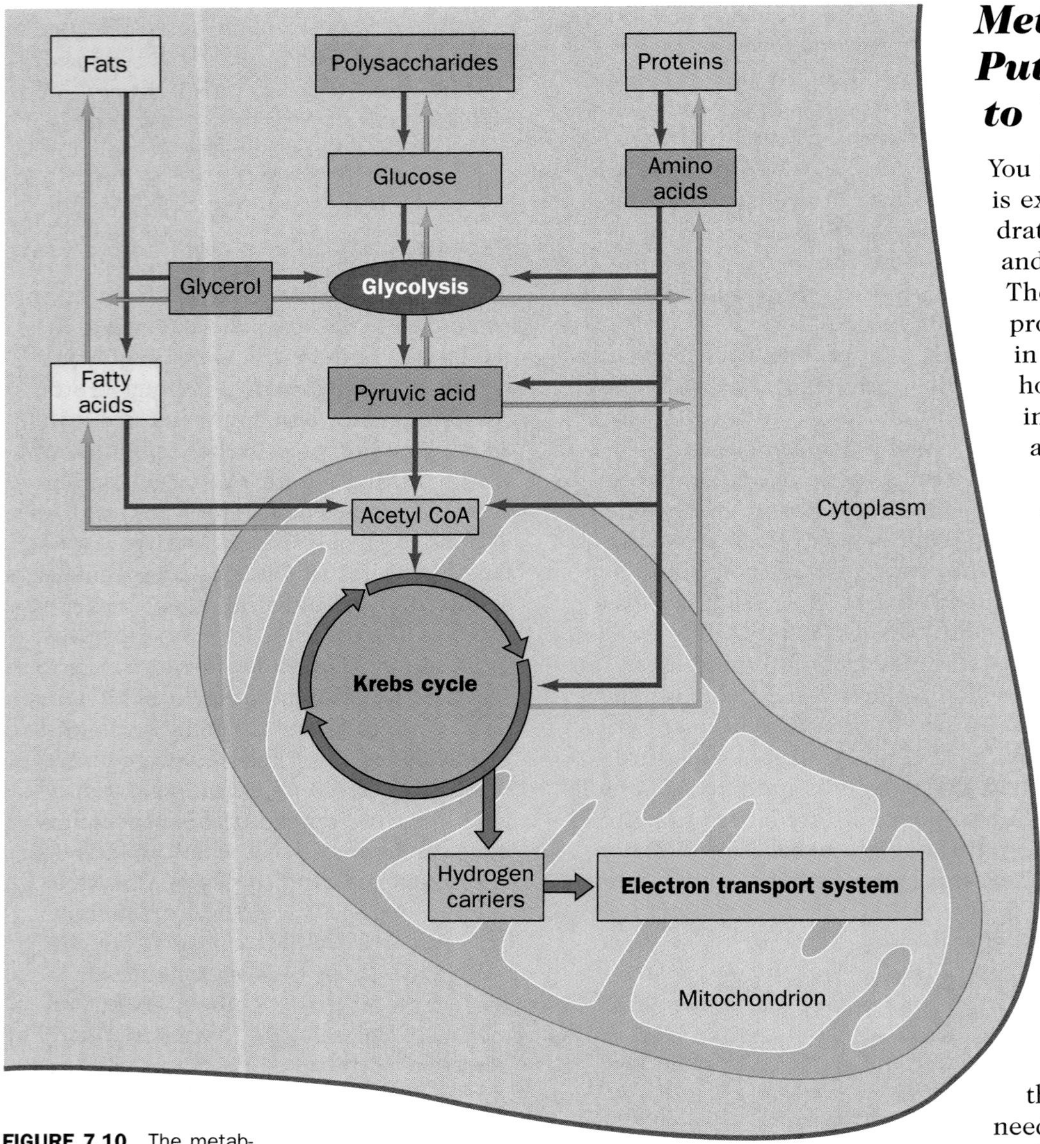

Fats
Polysaccharides
Proteins
Glucose
Amino acids
Glycerol
Glycolysis
Fatty acids
Pyruvic acid
Acetyl CoA
Cytoplasm
Krebs cycle
Hydrogen carriers
Electron transport system
Mitochondrion

FIGURE 7.10 The metabolism of carbohydrates, fats, and proteins, showing how these nutrients can be interconverted.

Krebs cycle are siphoned off, receive an amino group, and then are further modified into an amino acid.

> During protein catabolism, proteins are broken down to amino acids which lose their amino groups and then are catabolized in the same manner as carbohydrates. Protein anabolism consists of protein synthesis and the synthesis of nonessential amino acids.

Metabolism: Putting Energy to Work

You have learned how energy is extracted from carbohydrates, fats, and proteins and used to produce ATP. These energy extraction processes are very similar in all body cells no matter how different the cells are in size, shape, structure, and function.

The amount of energy contained in nutrients can be measured in Calories [also called kilocalories (kcal)]. A **Calorie** is defined as the amount of heat needed to raise 1000 grams of water 1° Celsius. As we saw in the early part of this chapter, this is the measurement used in nutrition (and on food labels) to designate the energy in food. Don't confuse this with the calorie, spelled with a small "c." A calorie is the amount of heat needed to raise 1 gram of water 1° Celsius and is the unit of measurement used in chemistry and physics: 1 Calorie = 1000 calories.

Energy Utilization by the Body

There are three general ways in which the body uses energy from nutrients: for basal metabolism, for physical activity, and for processing the food we eat.

The term **basal metabolism** refers to the minimum energy expended to maintain basic life processes. It represents the energy required for activities such as active transport, heartbeat, respiration, maintenance of body temperature, muscle tone, and brain function. The amount of

TABLE 7.8 Factors That Influence the Basal Metabolic Rate (BMR)

Factor	Effect on BMR
Age	BMR decreases with age due to a decline in muscle mass
Sex	BMR in females is lower, except during pregnancy
Body composition	BMR increases as muscle mass increases and decreases as fat tissue increases
Thyroid hormone	BMR increases as thyroid hormone levels increase
Body temperature	BMR increases when body temperature increases and decreases when body temperature decreases
Environmental temperature	BMR increases in hot or cold environments
Pregnancy	BMR increases
Disease conditions and fever	BMR increases 7% for each degree Fahrenheit of fever
Stress	BMR increases

energy required to maintain basic life processes for a given period is called the **basal metabolic rate (BMR).** On a daily basis, the BMR is quite high and can account for at least 60 to 70 percent of all the energy expended by the body. For example, if 2000 Calories is required to maintain an individual's daily activities, 1200 to 1400 of those Calories will be used for basal metabolism.

The BMR is most commonly measured by determining the amount of O_2 consumed while one is resting after a 10-hour fast. The Calories can be calculated from the O_2 consumed. The BMR varies between individuals and can even vary in one individual when health or physical conditions change. Factors that affect the BMR include age, sex, body composition, thyroid hormone levels, body temperature, environmental temperature, pregnancy, and illness (Table 7.8).

Physical activity utilizes the next highest amount of energy. It involves the action of skeletal muscles such as those associated with walking, housecleaning, swimming, cycling, chopping wood, dancing, and playing the piano. It is not unusual for physical activities to account for 20 to 30 percent of the Calories used by the body each day. The amount of energy used during physical activity is influenced by the type of activity, its intensity, and its duration. Under most conditions, an activity using larger muscles will expend more energy than will one using smaller muscles.

Food processing by the digestive system also uses energy. The muscles of the digestive tract require energy for chewing and peristalsis, the accessory organs require energy to produce their secretions, and many nutrients are absorbed into the intestinal cells by active transport, another energy-requiring process.

Table 7.9 describes how to calculate your personal energy needs. As you perform these calculations, note that the Calories used to maintain BMR, physical activity, and food processing must be replaced by the Calories consumed in the diet. To maintain a constant body weight, there must be a balance between the energy used and the energy consumed (Figure 7.11). If the amount of energy con-

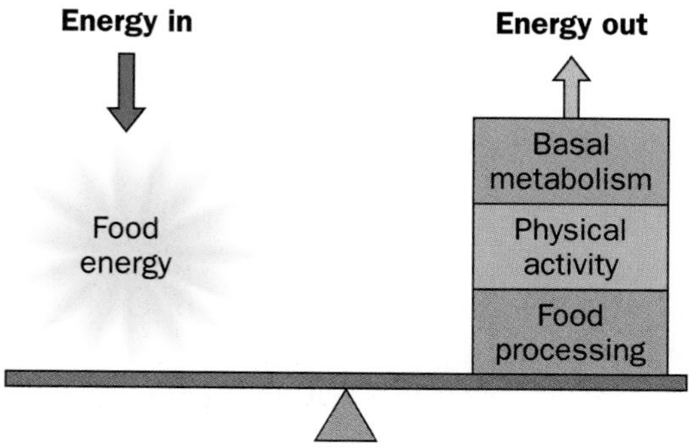
Energy in
Energy out
Food energy
Basal metabolism
Physical activity
Food processing

FIGURE 7.11 An energy balance occurs when the amount of food energy consumed equals the amount of energy used.

TABLE 7.9 Estimating Your Daily Energy Needs

By following the example below, you can estimate your daily energy requirements by determining the Calories you need for basal metabolism, physical activity, and food processing.

> Jane D. is a 130-pound, full-time college student. Each weekday she walks several miles to and from classes, studies for several hours, and works part-time in the library. She does not perform regular aerobic exercises. How many Calories does she need to maintain her basal metabolism, physical activity, and food processing?

1. Estimating Calories needed for basal metabolism per day (BMR)
To estimate the BMR, multiply body weight by 10 for females and by 11 for males. For Jane this would be

Weight in Pounds	X	BMR Factor for a Female	=	Calories for BMR
130	X	10	=	1300

2. Estimating Calories needed for physical activity per day
To estimate the Calories needed for physical activity, use the accompanying Physical Activity Level Chart. Multiply the BMR by the estimated activity level. If Jane's activity level is 40 percent, then

BMR	X	Activity Level	=	Calories for Physical Activity
1300 Calories	X	40%	=	520

3. Estimating Calories needed for food processing per day
To estimate the Calories needed for food processing, add the Calories for BMR and physical activities and then multiply this sum by 10 percent:

BMR Calories	+	Physical Activity Calories	=	Total Calories for BMR and Physical Activity
1300	+	520	=	1820

BMR + Physical Activity Calories	X	% of Calories for Food Processing	=	Calories Required for Food Processing
1820	X	10%	=	182

Jane needs 2002 Calories per day to maintain her BMR, physical activity, and food processing (1300 + 520 + 182).

Physical Activity Level Chart for Estimating Calories

Activity	Activity level, %
Light activity Sitting or standing much of the workday (driving, typing)	30–40
Moderate activity Workday includes some physical activity (housecleaning, factory work, aerobic exercise)	40–50
Heavy activity Strenuous physical activity during much of the workday (laborer, athletic training)	50–100

Calculate your personal energy needs

1. ______ X ______ = ______
   Your weight; BMR factor; BMR Calories

2. ______ X ______ = ______
   BMR calories; Activity level estimate; Physical activity Calories

3. 10% X (______ + ______) = ______
   BMR Calories; Physical activity Calories; Food processing Calories

______ BMR Calories
+ ______ Physical activity Calories
+ ______ Food processing Calories

= ______ Your estimated Calorie needs.

sumed is less than the amount of energy used, the body will use its energy reserves (glycogen, fat, protein), resulting in weight loss. However, if the amount of energy consumed is more than the amount of energy used, the body will store the excess energy as fat, resulting in weight gain. We will discuss the implications of this concept in the next section.

> Energy is used by the body for basal metabolism, physical activity, and food processing. For daily energy expenditures, basal metabolism accounts for 60 to 70 percent, physical activity for 20 to 30 percent, and the remainder is used for food processing.

Applying Nutritional and Metabolic Concepts

Not everyone is satisfied with his or her body weight. Some people want to lose weight; others want to gain weight. Accomplishing either task relates directly to the information presented in this chapter, for weight control fundamentally involves an understanding of food requirements (nutrition) and the way food energy is used and stored by the body (metabolism).

Food intake is regulated by an area of the brain called the hypothalamus (Chapter 15). In the hypothalamus, a region called the **hunger center** regulates the need to eat and a **satiety center** indicates when to stop eating. The way in which this regulation occurs is not well understood, but it appears that the need to eat and the decision to stop are partially influenced by blood glucose levels. For example, low blood glucose levels stimulate the hunger center, and we become hungry and eat. An increase in the blood glucose level stimulates the satiety center, and we become satiated and stop eating.

There is, however, a psychological component of eating that is called **appetite.** Appetite can override regulation by the hypothalamus. Your appetite is in effect when, after a full meal, you eat a serving of dessert even though you are satiated from the large meal. Conversely, when you are hungry, a stressful situation may make you lose your appetite and not eat. Hunger, satiety, and appetite are all interrelated with behavior and greatly influence body weight.

In the following section, we will explore how the metabolism of nutrients causes changes in body weight during periods of dieting or overconsumption of food.

Weight Control: Energy Consumed versus Energy Used

If the energy consumed from the diet equals the energy used by the body, then an equilibrium has been reached and there will be no change in body weight (see Figure 7.11). Body weight will increase, however, if more food energy is consumed than is required by the body to support basal metabolism, physical activity, and food processing. This concept is fundamental to every weight loss or weight gain program. Before we continue, let us review a few key elements concerning the metabolism of nutrients.

- Fat molecules are energy-rich and contain 9 Calories per gram compared with 4 Calories per gram for carbohydrate and protein molecules.
- Carbohydrates, proteins, and fats consumed in excess of what is required by the metabolic processes of the body are converted to fat molecules and stored in fat cells.
- When the body needs extra energy to support metabolic processes, it can break down stored fat, glycogen, or the protein of skeletal muscle.

Every pound of body fat contains 3500 Calories. We can use this information to understand weight control. To lose a pound of fat, a person must consume 3500 fewer Calories than that person uses. Conversely, to gain a pound of fat, the opposite has to occur.

Gradual weight loss can be accomplished by consuming 500 fewer Calories per day than is actually used by the body. As a result, 1 pound of body weight will be lost each week (500 Calories per day $\times$ 7 days = 3500 Calories). The body will simply use its stored energy reserves, primarily fat, to supply the needed 3500 Calories, and in so doing a pound of fat will be lost. Conversely, an increase of 500 Calories per day will increase body fat by 1 pound a week. This is the essence of weight control. It seems simple, but as many of us know, it is not easy.

Weight can accumulate slowly over an extended period if one eats only a few extra Calories each week. For example, recall the student whose daily energy requirement we calculated to be 2002 Calories in Table 7.9. What would happen if Jane D. consumed an extra 100 calories each day for 35 days? One hundred Calories would be the equivalent of a cup of low-fat yogurt or half a doughnut—not extravagant by any standard.

At the end of 5 weeks Jane will have gained 1 pound. Hardly noticeable, you say, but after 10 months Jane will have gained over 8 pounds:

$$30 \text{ days/mo.} \times 10 \text{ mo.} = 300 \text{ days}$$
$$300 \text{ days} \times 100 \text{ Cal/day} = 30{,}000 \text{ Cal}$$
$$30{,}000 \text{ Cal} \div 3500 \text{ Cal/lb} = 8.6 \text{ lb}$$

Over a period of several years, the accumulated body weight from this small but consistent indulgence will be significant. The same concept can be used to help a person lose body weight. If Jane was 8 pounds overweight, she could easily lose those 8 pounds in 10 months by omitting the 100 Calories per day in snacks.

How could Jane enjoy her extra 100 Calories a day and not gain weight? She could increase her physical activity to use the extra Calories. By consulting Table 7.10, you can determine the Calories required for various activities. If Jane walks briskly for 1 hour a day, her body will use an extra 200 Calories of energy. This is more than enough to offset her indulgence.

Another strategy Jane can use is to snack on nutritional foods that contain more carbohydrates and proteins but fewer fats. Recall that fat has about twice as many Calories per gram as do carbohydrates and proteins. Thus she would be able to consume twice as much food volume by eating foods containing only carbohydrates and proteins. Much of the food we eat is a combination of carbohydrate, fats, and proteins, and so to be more realistic, she could choose foods containing proportionally less fat (plain yogurt instead of sour cream) or could reduce the amount of fat she adds to her food (less butter, salad dressing, cheese). In either situation, she would consume fewer Calories per serving of food.

> Every pound of body fat contains 3500 Calories. By consuming 3500 more Calories than are needed for basal metabolism, physical activity, and food processing, we gain a pound of body weight. If one uses 3500 more Calories than are consumed, a pound of body weight will be lost.

TABLE 7.10 Calories Required for Different Types of Activities

Type of Activity	Intensity	Calories per Hour
Walking	2.6 mph (23 min/mile)	200
Jogging	5.3 mph (12 min/mile)	570
Bicycling	10.0 mph (6 min/mile)	300
Swimming	50 yd/min	560
Rowing	20 strokes/min	820
Sawing wood or shoveling snow	Moderate intensity	480

Source: A. J. Vander, J. H. Sherman, and D. S. Luciano, *Human Physiology,* 5th ed. New York: McGraw-Hill, 1990, p. 584. Used by permission.

Diseases and Disorders

There are two types of metabolic disorders: those caused by too many or too few nutrients and those resulting from the inability to metabolize a nutrient. The first disorders are environmental in the broadest sense of the word and are considered **eating or nutritional disorders;** the second group are **genetic disorders** that stem from an abnormal gene coding for an abnormal enzyme that prevents a needed chemical reaction from taking place. Here we will focus on vitamin deficiencies and three eating disorders that are common in industrialized societies: obesity, anorexia nervosa, and bulimia.

Vitamin Deficiency Disorders

With the exception of vitamin D and niacin, the body must obtain all its vitamins from food or from supplements such as vitamin pills. When poverty or famine prevents the consumption of nutritious foods, vitamin deficiency symptoms appear, particularly in children. Table 7.6 provides a brief description of deficiency symptoms for specific vitamins.

The minimal quantities of vitamins needed (RDA) are usually available in economically developed countries. However, even within developed countries, certain populations may be at risk for vitamin deficiencies. These groups include the poor, the homeless, the chronically ill (including alcoholics), institutionalized people, and mentally impaired people living with or without supervision. Because of their dependency, children are always at risk.

TABLE 7.11 Height and Weight Table

Height†	Weight (lb)*			
	19 to 34 Years		35 Years and Over	
	Midpoint	Range	Midpoint	Range
5′0″	112	97–128	123	108–138
5′1″	116	101–132	127	111–143
5′2″	120	104–137	131	115–148
5′3″	124	107–141	135	119–152
5′4″	128	111–146	140	122–157
5′5″	132	114–150	144	126–162
5′6″	136	118–155	148	130–167
5′7″	140	121–160	153	134–172
5′8″	144	125–164	158	138–178
5′9″	149	129–169	162	142–183
5′10″	153	132–174	167	146–188
5′11″	157	136–179	172	151–194
6′0″	162	140–184	177	155–199
6′1″	166	144–189	182	159–205
6′2″	171	148–195	187	164–210
6′3″	176	152–200	192	168–216
6′4″	180	156–205	197	173–222
6′5″	185	160–211	202	177–228
6′6″	190	164–216	208	182–234

Note: The higher weights in the ranges generally apply to men, who tend to have more muscle and bone; the lower weights more often apply to women, who have less muscle and bone. The higher weights for people age 35 and older reflect recent research that seems to indicate that people can carry a little more weight as they grow older without added risk to their health.

*Without clothes.

†Without shoes.

Source: U.S. Department of Agriculture and U.S. Department of Health and Human Services, Home and Garden Bulletin No. 232, *Nutrition and Your Health: Dietary Guidelines for Americans*, 3d ed. Washington, D.C.: U.S. Government Printing Office. 1990.

Obesity: Overweight and Overeating

Before discussing obesity, we should first distinguish between three weight conditions: normal weight, overweight, and obesity. The **normal weights** for men and women are shown in Table 7.11. If an individual's weight is 10 to 20 percent above the midpoint for that person's weight range, then that person is considered **overweight.** If an individual's weight is more than 20 percent above the midpoint, then that person is considered **obese.**

Many factors play a role in determining a person's correct body weight, including age, frame size, and body composition. Body composition refers to the amount of body fat compared with the amount of skeletal muscle. This can be important, for medical research has shown that the amount of body fat is a better indicator of fitness and health than is body weight alone. For proper fitness and health, it is recommended that the percentage of body fat should range between 15 and 18 percent for men and 20 and 25 percent for women.

There seem to be two general components associated with obesity: genetic and environmental. Some evidence for the *genetic origin of obesity* comes from studies of identical twins separated at birth and raised in different environments. Identical twins inherit identical genetic material, and so differences (if any) between separated twins must be due to environmental factors. Studies have shown that identical twins raised in different environments are very similar to each other in weight and height. If one twin is overweight or obese, the other one also is. Therefore, the genes, not the environment (food, learned eating habits, etc.), must be playing a role.

A popular proposal suggests that each person is genetically programmed to weigh a certain amount. This is called the set-point theory. It states that if you lose or gain weight, your body will continually adjust food intake (hunger) or the BMR so that you return to your genetically programmed weight, the so-called set-point. This theory explains why 80 to 90 percent of dieting individuals regain the weight they lose within a year.

The *environmental component of obesity* focuses on eating too much, consuming too much fat, and eating when not hungry as a way to reduce stress or boredom or satisfy one's emotional needs. However, food is not the only environmental component. Societal and cultural attitudes and messages from advertising also can contribute. For example, it is generally more acceptable for a mature man to be overweight than it is for a woman.

A reliable treatment that permanently corrects obesity is not available. Treatment for overweight or obesity should focus on altering eating habits and increasing physical activity levels. Continued success requires a lifelong commitment to such alterations.

TABLE 7.12 Do You Suffer from Anorexia Nervosa?

Diagnostic Criteria for Anorexia Nervosa
• Refusal to maintain body weight over a minimal normal weight for age and height, resulting in a body weight that is 15 percent below that expected
• Intense fear of gaining weight or becoming fat, even though underweight
• Disturbance in the way one's body weight, size, or shape is experienced; example, "feeling fat" or believing that one area of the body is "too fat" even though emaciated
• In females, absense of at least three consecutive menstrual cycles that were otherwise expected to occur

Source: American Psychiatric Association, *Diagnostic and Statistical Manual of Mental Disorders,* 3d ed., Revised. Washington, D.C.: American Psychiatric Association, 1987, p. 67. Used by permission.

Anorexia Nervosa: Refusing to Eat

Anorexia is an eating disorder most typically found in young women from the teenage years to the midthirties. It is estimated that about 1 in every 250 young women between the ages of 12 and 18 suffers from **anorexia nervosa** (*an-oh-**rex**-ee-ah nerve-**oh**-sah*). Those suffering from anorexia nervosa have certain physiological and emotional characteristics (Table 7.12).

Individuals with this disorder severely limit the amount of food they consume even though they are hungry. As a result, the effects of anorexia nervosa are similar to those of starvation and include a reduced BMR, amenorrhea (absence of a monthly menstrual period), emaciation, and weakened heart muscle resulting in lowered blood pressure. If untreated, these conditions can become life-threatening.

Emotionally, anorexia nervosa patients often have a low self-esteem, a poor self-image, and periods of depression. Furthermore, they often believe that food consumption is the only part of life they can control. An anorexia nervosa patient has typically accepted the exaggerated images of the female form that our culture holds up as models. For young women, a powerful cultural and societal message is that "you cannot be too thin" and that "the thinner you are, the more acceptance and approval you will get." In addition to nutritional counseling, treatment includes psychotherapy to identify and resolve the underlying emotional conflicts.

Bulimia: Binging and Purging

Bulimia (*bull-**ee**-me-a*), or binge-purge eating, is characterized by the consumption of large volumes of food (binging) followed by self-induced vomiting (purging). This eating disorder is typically found in young women from the teenage years to early adulthood. It is roughly estimated that 1 to 4 percent of college-age women suffer from bulimia. Those suffering from bulimia have certain behavioral characteristics (Table 7.13).

Bulimics have little control of their binging and purging episodes. In addition to the loss of nutrients, chronic bulimic vomiting of stomach acid erodes tooth enamel and irritates the esophagus and voice box. The loss of electrolytes, particularly potassium (K^+), can lead to abnor-

TABLE 7.13 Do You Suffer from Bulimia?

Diagnostic Criteria for Bulimia
• Recurrent episodes of secretive binge eating, that is, rapid consumption of a large amount of food in a discrete period
• A feeling of lack of control over eating behavior during eating binges
• Regular practice of self-induced vomiting, use of laxatives, strict dieting or fasting, or vigorous exercise to prevent weight gain
• A minimum average of two binge eating episodes a week for at least 3 months
• Persistent overconcern with body shape and weight

Source: American Psychiatric Association, *Diagnosis and Statistical Manual of Mental Diseases,* 3d ed., Revised. Washington D.C.: American Psychiatric Association, 1987, p. 67. Used by permission.

mal heart activity, resulting in sudden death. In addition to nutritional counseling, treatment involves psychotherapy to regain control over food consumption and address problems of self-esteem similar to those of anorexia nervosa patients.

Life Span Changes

Changes in metabolic and nutritional requirements occur over the entire life span. In considering these changes, we will address three stages: prenatal (before birth) and infancy, childhood, and old age. Each stage is represented by specific metabolic and nutritional needs.

The **prenatal and infant** periods of human development involve two types of growth: an increase in cell numbers (proliferation) and then an increase in cell size. The nutrients required to support such growth during these periods are supplied by the mother. During the prenatal period, the fetus receives all its nutrients from the mother's bloodstream. During early infancy, a breast-fed child will continue to receive all its nutrients from its mother's milk for a period of time (say, 4 to 6 months). To ensure the health and well-being of both the mother and the child, it is important that the mother receive adequate nutrition during these periods.

In many less developed countries, a lack of adequate nutrition is an ever-present reality. In those countries, when breast-feeding ceases, a child may not have sufficient nutrients to support proper growth. It is not uncommon for these children to have a high-carbohydrate and low-protein diet. This can have devastating long-term consequences, especially during the first 18 months after birth.

It is during the prenatal period and the first 18 months after birth that all our brain cells are produced. After this time no new brain cells are produced, no matter how long we live. Without sufficient protein from the diet and the essential amino acids they supply, fewer brain cells will be produced during this period. As a result, mental dysfunctions and learning disorders may develop.

During **childhood,** a lack of nutrients can lead to inadequate growth. Again, this is especially true in impoverished countries. Insufficient Calories and/or essential nutrients greatly reduce the growth rate and overall health of children. Malnourished children not only are smaller in stature, they also are much more susceptible to infectious diseases.

Elderly people require the same basic nutrients, but the amounts may differ compared with younger adults. The elderly generally have a reduced BMR and engage in less physical activity, and so they often require fewer Calories. It is recommended that the Calories from protein increase from 10 percent to about 22 percent, while those from fats should be reduced. Also, the need for certain vitamins (C, B_{12}) and minerals (calcium, iron) may increase because of reduced gastric secretion and intestinal absorption.

It is also important to appreciate the fact that older adults often take medications for long periods of time and that these medications can affect their nutrient requirements. For example, diuretics used to reduce excess body water can increase the need for potassium; corticosteroids used as an anti-inflammatory medication decreases zinc absorption and the utilization of calcium; and laxatives result in poor absorption of calcium and potassium.

Chapter Summary

Nutrients: Eating for Life and Health

- The six nutrient categories are carbohydrates, fats, proteins, vitamins, minerals, and water.
- Carbohydrates contain 4 Calories per gram and should account for about 60 percent of our daily Calories.
- Fats are categorized as saturated or unsaturated. Fats contain 9 Calories per gram and should account for 30 percent or less of our daily Calories.
- Complete proteins contain all 10 essential amino acids. We can synthesize nonessential amino acids. Proteins contain 4 Calories per gram and should account for about 10 percent of our daily Calories.
- Vitamins and minerals are required in small amounts and come from the diet. Water typically constitutes up to 60 to 70 percent of body weight.

Metabolism: The Body's Chemical Transformations

- Metabolism refers to chemical reactions in cells. Catabolic reactions break down nutrient molecules for energy; anabolic reactions synthesize more complex molecules.
- Carbohydrate catabolism releases chemical energy from glucose via glycolysis, the Krebs cycle, and the electron transport system when oxygen is present. A total of 36 ATP molecules is produced, with the majority coming from the electron transport system.
- Carbohydrate anabolism synthesizes glycogen from glucose or glucose from noncarbohydrates.
- Fat catabolism releases stored energy from glycerol and fatty acids, producing ATP. Fat anabolism synthesizes fat molecules from excess carbohydrates and proteins.
- Protein catabolism releases stored energy from individual amino acids, producing ATP. Protein anabolism makes new protein through the cellular processes of protein synthesis.

Metabolism: Putting Energy to Work

- A Calorie is defined as the amount of heat needed to raise 1000 grams of water 1°C.
- Calories are used for three processes: basal metabolism, physical activity, and food processing.
- To maintain constant weight, the amount of Calories consumed (energy) must equal the amount used for basal metabolism, physical activity, and food processing.

Applying Nutritional and Metabolic Concepts

- Food intake is regulated by the hunger center and the satiety center in the hypothalamus.
- A pound of fat contains 3500 Calories. A pound of body fat can be lost over a period of time by eliminating 3500 Calories from the diet or expending 3500 Calories in physical activity.

Diseases and Disorders

- Obesity is defined as being 20 percent overweight. Obesity has a genetic component and an environmental component.
- Anorexia nervosa is characterized by an extremely underweight condition that is maintained by limiting food consumption.
- Bulimia is characterized by an extremely underweight condition that is maintained by food binging and self-induced vomiting.

Life Span Changes

- A lack of sufficient nutrients, especially proteins, during the first 18 months after birth can reduce the number of brain cells produced, resulting in mental dysfunctions.
- A lack of sufficient nutrients during childhood can lead to stunted growth.
- The elderly typically require fewer Calories than do younger adults because of reduced BMR and physical activity.

Selected Key Terms

anabolism (p. 202)
basal metabolism (p. 208)
Calorie (p. 208)
catabolism (p. 202)
complete proteins (p. 197)
electron transport system (p. 204)
essential amino acids (p. 197)
fiber (p. 194)
glycolysis (p. 203)
Krebs cycle (p. 203)
metabolism (p. 202)
NAD (p. 205)
saturated fat (p. 195)
unsaturated fat (p. 196)
vitamin (p. 198)

Review Activities

1. Identify the six categories of nutrients.
2. What are the recommended daily requirements for carbohydrates, fats, and proteins?
3. Distinguish between the following: insoluble and soluble fiber, unsaturated and saturated fats, water-soluble and fat-soluble vitamins, essential and nonessential amino acids, complete and incomplete proteins, and catabolism and anabolism.
4. List the number of Calories per gram in carbohydrates, fats, and proteins.
5. What type of molecule is cholesterol, and in what types of food is it found?
6. Calculate the number of Calories in 6 grams of fat (see Table 7.3 for the answer).
7. A meal contains 1273 Calories and includes 57 grams of fat. Calculate the percentage of fat contained in the meal (see Table 7.3 for the answer).
8. Identify three uses of fat in the body.
9. Identify three uses of protein in the body.
10. What is the approximate quantity of water needed daily to replace that which is lost in urine and evaporative cooling?
11. During glucose catabolism, what happens to the carbon, hydrogen, and oxygen atoms of glucose?
12. Where do the following chemical reactions occur in the cell: glycolysis, Krebs cycle, electron transport system?
13. How is coenzyme A important in glucose catabolism?
14. During fat catabolism, what is the fate of glycerol and fatty acid molecules?
15. How many ATPs (net) are produced during glycolysis, the Krebs cycle, and the electron transport system?

Self-Quiz

Matching Exercise

_ 1. Complex carbohydrates that are not digested by enzymes of the digestive tract
_ 2. Produces the minimum energy required to maintain basic life processes
_ 3. Part of brain that regulates food intake
_ 4. Number of Calories in a pound of fat
_ 5. Number of Calories in a gram of protein
_ 6. During fat catabolism, the part of the cell where fatty acids are processed to produce ATP
_ 7. The part of a cell where glycolysis occurs
_ 8. The series of reactions that process NADH and FADH
_ 9. The molecule that forms when carbon and oxygen atoms are released
_ 10. A process that can synthesize fat molecules from excess carbohydrate and protein

A. Cytoplasm
B. Carbon dioxide
C. Electron transport system
D. 3500
E. Basal metabolism
F. 4
G. Fiber
H. Hypothalamus
I. Mitochondria
J. Fat anabolism

Answers to Self-Quiz

1. G; 2. E; 3. H; 4. D; 5. F; 6. I; 7. A; 8. C; 9. B; 10. J

Read Right, Eat Well, Live Better: Interpreting Food Labels

New food label guidelines adopted by the Food and Drug Administration (FDA) in 1993 must be included on all processed foods regulated by the FDA. The new labeling guidelines provide consumers with more information about what is in their food and how it applies to a healthy diet. What follows summarizes how to interpret the information on all food labels.

Human
BIOLOGY
CEREAL
Low in Fat and Cholesterol Free

Human
BIOLOGY
CEREAL

HUMAN BIOLOGY CEREAL
NUTRITION FACTS
Serving size: 3/4 c (28 g)
Servings per container: 14

Amount per serving	
Calories 110 cal	Calories from fat 9
	Percent of Daily Value*
Total Fat 1 g	2%
Saturated fat 0 g	0%
Cholesterol 0 mg	0%
Sodium 250 mg	10%
Total Carbohydrate 23 g	8%
Sugars 10 g	2%
Dietary fiber 1.5 g	6%
Protein 3 g	2%

Vitamin A 25% • Vitamin C 25% • Calcium 2% • Iron 25%

*Percent Daily Values are based on a 2000 calorie diet. Your daily values may be higher or lower depending on your calorie needs.

Nutrient		2000 Calories	2500 Calories
Total fat	Less than	65 g	80 g
Sat Fat	Less than	20 g	25 g
Cholesterol	Less than	300 mg	300 mg
Sodium	Less than	2400 mg	2400 mg
Total Carbohydrate		300 g	375 g
Fiber		25 g	30 g

1 g Fat = 9 calories
1 g Carbohydrates = 4 calories
1 g Protein = 4 calories

INGREDIENTS, listed in descending order of prominence: Corn, Sugar, Salt, Malt flavoring.
VITAMINS and MINERALS: Vitamin C (Sodium ascorbate and Ascorbic acid), Niacinamide, Iron, Vitamin B_6 (Pyridoxine hydrochloride), Vitamin B_2 (Riboflavin),Vitamin A (Palmitate), Vitamin B_1 (Thiamin hydrochloride), Folic acid, and Vitamin D.

Serving size is set by the FDA. They are now standardized and better reflect the portions typically eaten.

Calories from fat allows consumers to know how much fat is contained in each serving. It helps consumers meet dietary guidelines of no more than 30% of daily Calories from fat.

% daily value shows how much of a day's worth of each nutrient is provided per serving. The percentages are based on a daily diet of 2000 Calories.

List of nutrients must include the three major food groups and the two vitamins and minerals shown. Food manufacturers have the option of listing other vitamins and minerals. Nutrients such as cholesterol and sodium are listed so consumers can better access their daily consumption of these health-related nutrients.

Daily values are the same on all food labels. The values are based on the dietary guideline for a 2000-Cal and a 2500-Cal daily diet. A 2000-Cal diet approximates the need for most children, women, and men over 50. A 2500-Cal diet more closely represents the need for teenage boys, younger men, and more active individuals.

Calories per gram states the number of calories in each gram of fat, carbohydrate, and protein.

Ingredients are listed in descending order of concentration by weight. What appears first is present in a larger quantity than what appears second, and so forth.

Garman's
CHUNK LIGHT
TUNA
IN SPRING WATER

Containers with 12 to 40 square inches of surface area are permitted to have less nutritional information.

NUTRITION FACTS
Serving size: 1/4 c (56 g)
Servings: 2.5
Calories: 70
Fat Cal. 10
*Percents Daily Value (DV) are based on a 2,000 calorie diet.

Amount/serving		%Daily Value*	Amount/serving		%Daily Value*
Total Fat	1g	2%	Total Carb	0g	0%
Sat. Fat	0g	0%	Dietary fiber	0g	0%
Cholest.	25mg	8%	Sugars	0g	
Sodium	250mg	10%	Protein	15g	

Vitamin A 0% • Vitamin C 0% • Calcium 0% • Iron 0%

INGREDIENTS: WHITE TUNA, SPRING WATER, VEGETABLE BROTH, SALT, HYDROLYZED SOY PROTEIN, PYROPHOSPHATE.

FUNSTON & BENJAMIN
BOMBERS

Containers with a surface area less than 12 square inches need provide only an address or phone number where nutritional information can be obtained.

PART FOUR

Transporting Materials and Maintaining Internal Conditions

CHAPTER 8

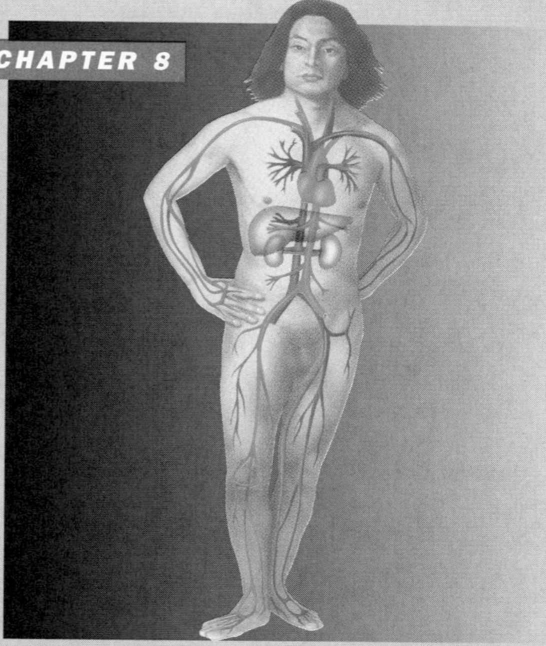

The Circulatory System: Blood

CHAPTER 9

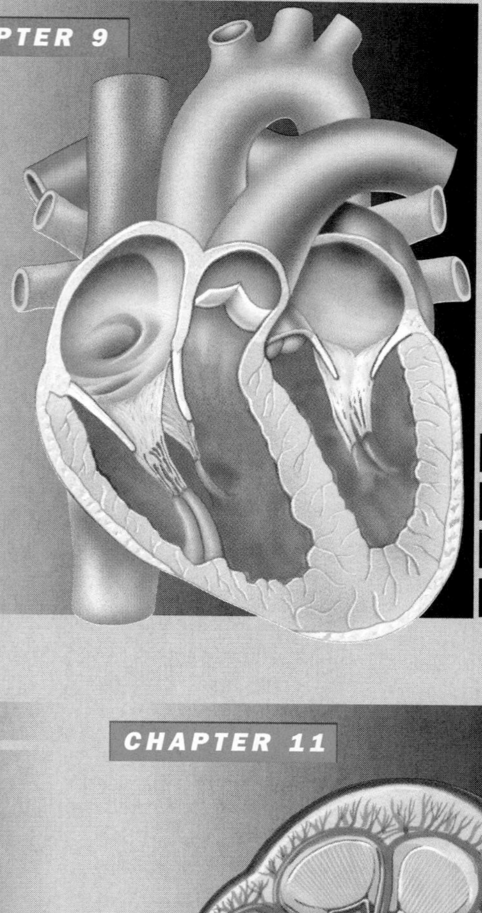

The Circulatory System: Blood Vessels and the Heart

The Respiratory System

CHAPTER 10

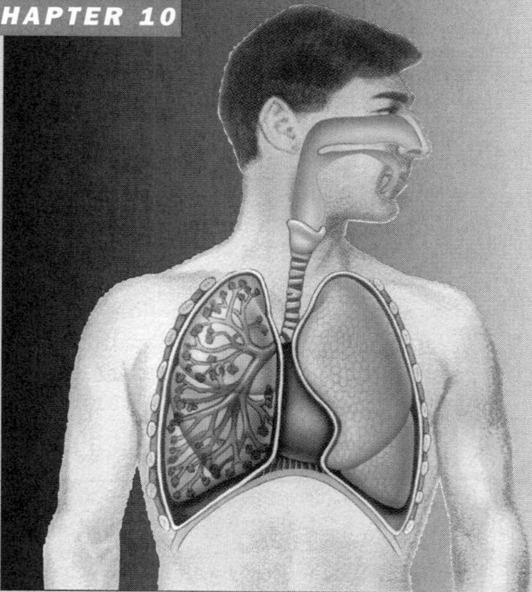

CHAPTER 11

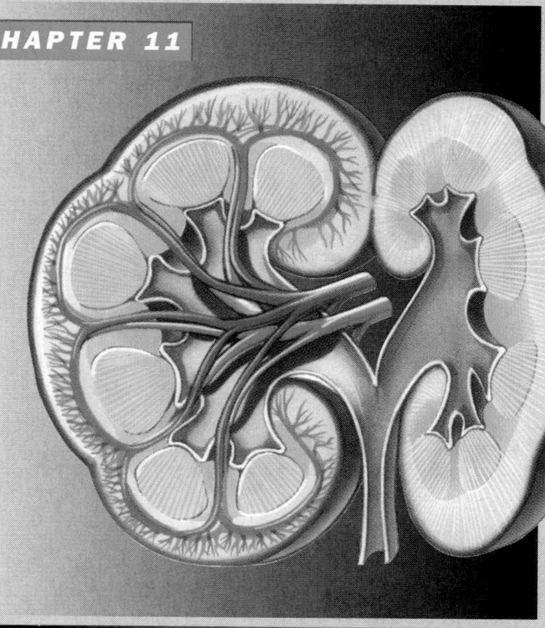

The Urinary System

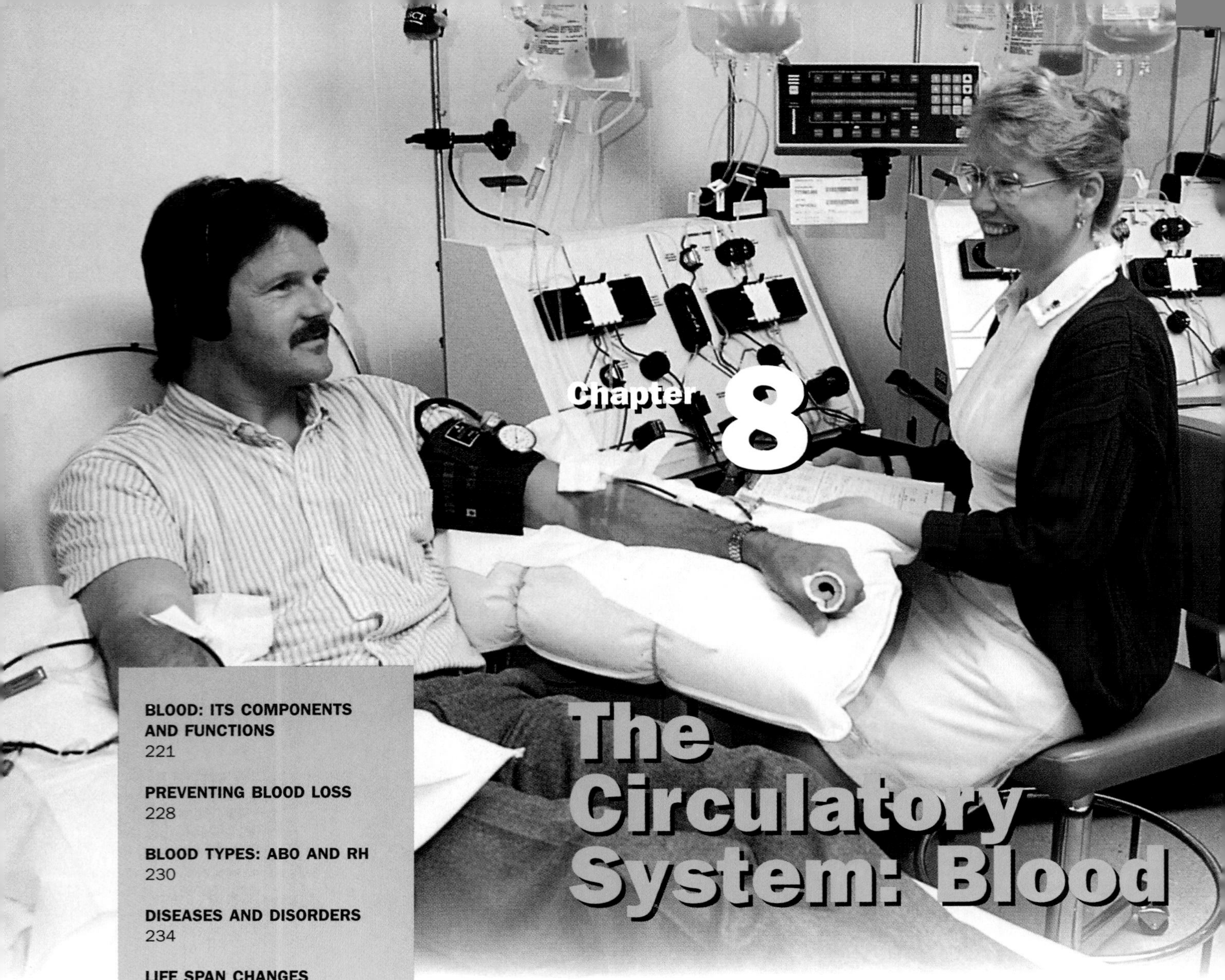

Chapter 8

The Circulatory System: Blood

Blood, sweat, and tears. These words have an emotional power that captures much of the drama of human existence. From ancient times to the present, blood has been thought of as both a vital component of the body and a potent symbol for life, death, conflict, and sacrifice. Sweat is associated with activity, exertion, and effort, while tears accompany strong emotions. Biologically, these three body fluids are related. Sweat and tears, like other body fluids, are derived from blood. It is not an exaggeration to say that blood is central to all body processes.

Blood is a connective tissue that consists of cells suspended in liquid matrix. The **heart** pumps the blood through a series of tubes called **blood vessels**. These three components—blood, heart, and blood vessels—form the circulatory system (Figure

8.1). The movement of blood through this system transports materials to and from our trillions of cells. Our survival depends on this function operating without interruption.

We will discuss the heart and blood vessels in Chapter 9. Here in Chapter 8 we will discuss the many functions of blood. We will describe its molecular and cellular components and show how they perform their life-sustaining actions. Then we will examine several common diseases and disorders of the blood and the changes that normally occur over a life span.

Blood: Its Components and Functions

The average adult has about 5 liters (over 5 quarts) of continuously circulating blood. The movement of blood throughout the body helps maintain homeostasis in three general ways—transportation, regulation, and protection: (1) Blood transports O_2 from the lungs and nutrients from the digestive system. It transports cellular waste products such as carbon dioxide (CO_2) and urea to sites of elimination. It also transports hormones. (2) Blood helps regulate the body's normal acid-base balance, water content, and temperature. (3) Blood protects the body from disease agents such as bacteria and viruses with its specialized defensive cells. The clotting mechanism in blood prevents excessive blood loss from damaged vessels.

To understand and appreciate these functions, we need to distinguish the two major components of blood and consider them separately: the liquid component and the cellular component. The liquid component, called **plasma**, is mostly water in which are dissolved a variety of substances.

Suspended within the plasma is the cellular component, which consists of **red blood cells (RBCs)** for transporting O_2 and a little CO_2, **white blood cells (WBCs)** for defense against disease agents, and **platelets**, which are cellular fragments that help stop bleeding from damaged vessels (Figure 8.2). In the sections that follow, we will describe the chemical composition of plasma and the characteristics of the cellular component, focusing on how these contribute to homeostasis. We'll start with blood plasma.

Blood Plasma: The Liquid Component

Plasma makes up about 55 percent of our blood volume, and the cellular component makes up the remaining 45 percent. By removing a small sample of blood from the body and subjecting it to high-speed rotation (a process called centrifugation), we can separate plasma from the cellular component and study each one separately (Figure 8.3). Without blood cells and platelets, plasma is a pale yellow liquid that is about 90 percent water. The other 10 percent includes a variety of dissolved substances. Recall from Chapter 1 that a dissolved substance (called a solute) is

FIGURE 8.1 An overview of the human circulatory system.

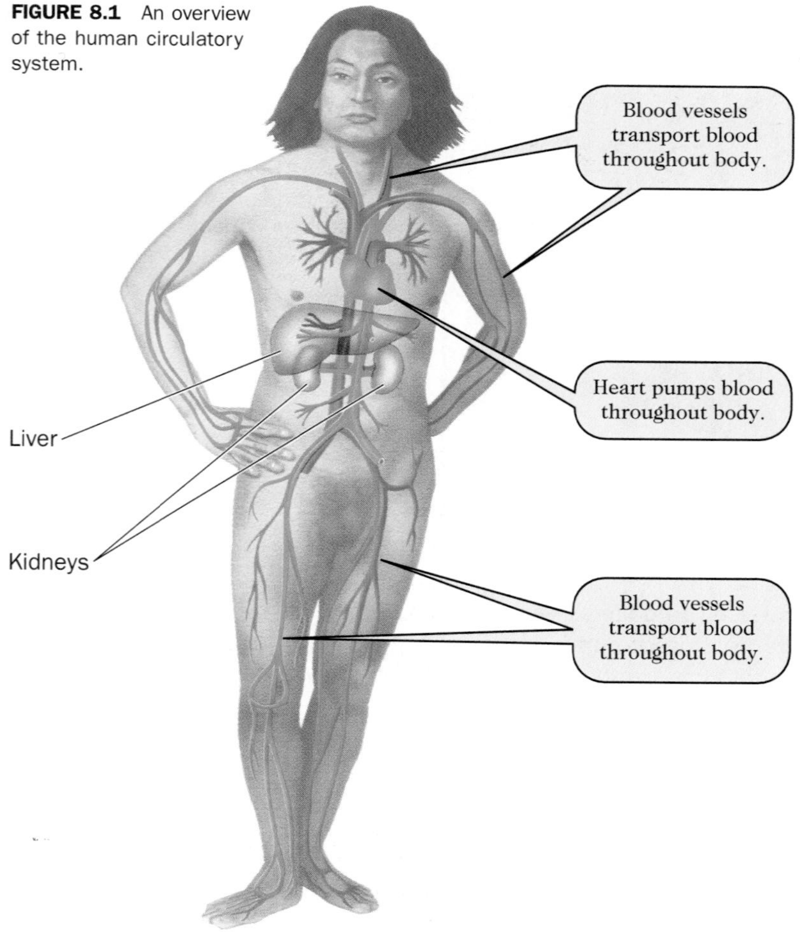
Blood vessels transport blood throughout body.
Heart pumps blood throughout body.
Liver
Kidneys
Blood vessels transport blood throughout body.

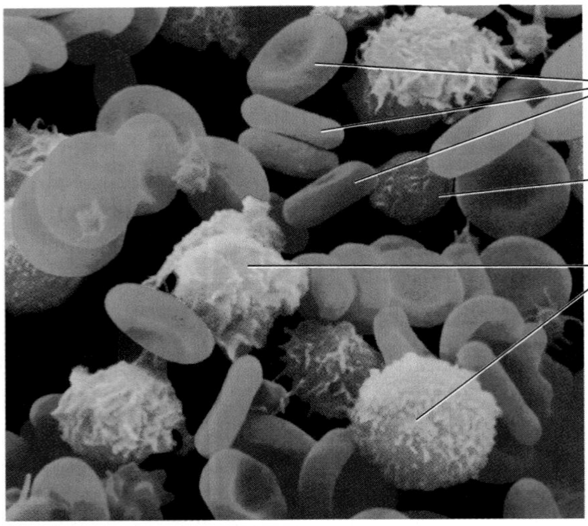
Red blood cells
Platelet
White blood cells

FIGURE 8.2 A scanning electron micrograph of the three types of blood cells.

surrounded by water molecules and tends to be uniformly distributed throughout the water. This is true whether the solute is a small ion or gas molecule or a large protein molecule.

Plasma proteins: a variety of functions. An important group of solutes dissolved in the plasma is referred to collectively as the **plasma proteins**. Do not confuse plasma proteins with dietary proteins. The proteins we eat are not found in the bloodstream because they are digested to amino acids in our intestines. There are three types of plasma proteins: albumins, globulins, and blood-clotting proteins (Figure 8.3). Each type is actually a family of similar proteins with small but important differences. Except for a specific type of globulin (gamma globulin), all plasma proteins are manufactured in the liver.

The **albumins** (*al-***byou***-mins*) are plasma proteins that help maintain the water balance between blood and tissue. Their normal confinement to the blood vessels helps maintain the osmotic relationship between the blood and the fluid that surrounds tissue cells (tissue fluid). In other words, albumins assure that there will not be excessive movement of water from the plasma to tissues outside the circulatory system.

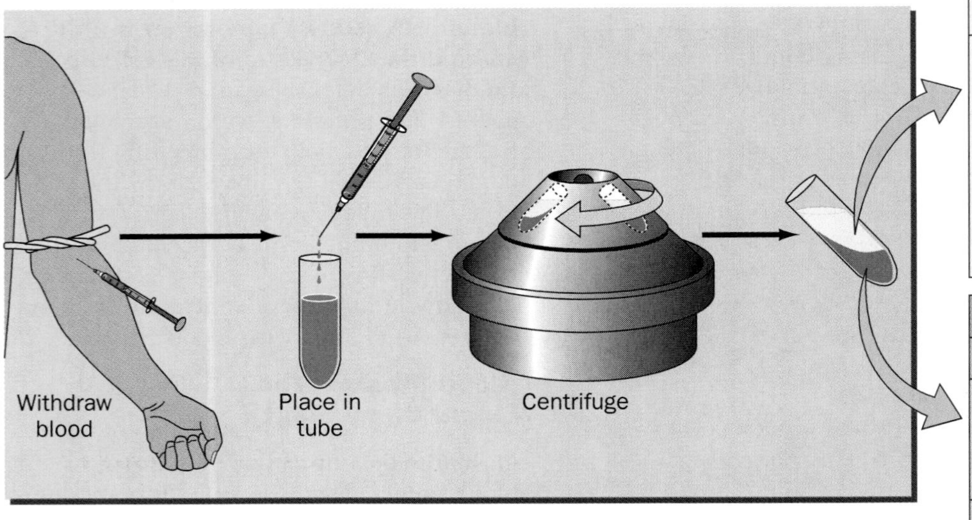
Withdraw blood
Place in tube
Centrifuge

PLASMA 55%	
Components	Functions
Water	Solvent for carrying other substances
Plasma proteins	
Albumin	Osmotic balance
Globulins	Transport and defense
Fibrinogen	Clotting
Other substances transported by plasma Nutrients (e.g., glucose, fatty acids, vitamins) Waste products of metabolism Respiratory gases (O_2 and CO_2) Hormones Electrolytes (Na^+, Ca^{2+}, HCO_3^-, K^+)	

CELLULAR COMPONENTS 45%	
Cell type	Functions
Red blood cells	Transport O_2 and help transport CO_2
White blood cells	Defense and immunity
Platelets	Blood clotting

FIGURE 8.3 The composition of whole blood.

The **globulins** (**glob**-*you-lins*) are a diverse group of plasma proteins. There are three varieties: alpha, beta, and gamma. *Alpha* and *beta globulins* are transport proteins that bind to and transport fat, cholesterol, fat soluble vitamins, and some hormones. These substances are not soluble in water, and their transport would be difficult without the help of water-soluble proteins. For example, lipids such as fat and cholesterol become soluble in plasma after attaching to alpha and beta globulins. The binding of lipid molecules to one of these proteins produces a complex called a *lipoprotein*. The concentration in the blood of two types of lipoproteins—high-density lipoproteins (HDLs) and low-density lipoproteins (LDLs)—is often used as an indicator of potential damage to blood vessels (Chapter 9). The third type of globulin is *gamma globulin*. These proteins are the antibodies that help protect us from invading bacteria and viruses. Gamma globulins are produced by specialized white blood cells called lymphocytes.

The **blood-clotting proteins** participate in blood-clotting reactions that help seal damaged vessels and prevent blood loss. *Fibrinogen* (*fie-**brin**-oh-gin*) is an important blood-clotting protein that we will examine later in this chapter.

Other plasma solutes. In addition to large proteins, plasma contains a variety of smaller molecules, including nutrients and electrolytes, waste products, respiratory gases, and hormones (see Figure 8.3). **Nutrients** are substances that have been absorbed from the digestive tract or derived from metabolic reactions in the body. They include simple carbohydrates (glucose, fructose, and galactose), amino acids, vitamins, and electrolytes such as sodium (Na^+), potassium (K^+), calcium (Ca^{2+}), and bicarbonate (HCO_3^-).

The **waste products** dissolved in plasma include substances produced by metabolic reactions within cells, such as CO_2, lactic acid, and urea. The removal of these substances from the blood is a function of the respiratory and urinary systems, which will be discussed in Chapters 10 and 11.

The **respiratory gases** in plasma include CO_2 and O_2. Most of the CO_2 is dissolved in plasma as the bicarbonate ion (HCO_3^-). Most of the O_2 in blood is carried by the red blood cells. The **hormones** produced by endocrine organs are secreted into the blood plasma and carried by the plasma to their target tissues.

Water and body fluids. As important as these dissolved constituents of plasma are, we must not overlook the role of water. Water is important beyond its function as the transport solvent for solutes. Through osmosis, the water of plasma is made available to all the cells of the body and is the basis for all body fluids: tissue fluid, lymph fluid, saliva, semen, tears and sweat.

In their studies, chemists have learned that water has unique physical properties. For example, it can absorb a large amount of heat without large changes in its temperature, making it a good temperature buffer and transporter of heat. The heat generated by tissue metabolism deep in the body is carried to the skin by the water of blood plasma. As blood flows through the skin capillaries, heat is released to the environment.

Some of the substances dissolved in the water of plasma are small (ions), while others are large (plasma proteins). However, the plasma proteins are small compared with the cellular components. In the three sections that follow, we will describe red blood cells, white blood cells, and platelets.

> Blood plasma consists of water in which nutrients, plasma proteins, waste products, ions, hormones, CO_2, and a little O_2 are dissolved. Plasma proteins consist of albumins, globulins, and blood-clotting proteins.

Red Blood Cells: Transporting O_2

Blood gets its red color from its abundant **red blood cells (RBCs)** or **erythrocytes** (*eh-**rith**-row-sites*). In each cubic millimeter of blood (1 mm equals about $\frac{1}{25}$ inch), there are approximately 5 million red blood cells, making them the most abundant blood cells. To give you an idea of their size, about 60 red blood cells placed edge to edge would span the diameter of the period at the end of this sentence. Their primary function is to transport O_2 (Figure 8.3 and Table 8.1).

Red blood cells are highly specialized to accomplish this task. In fact, they are not really typical cells. Mature red blood cells cannot divide or carry on the many

TABLE 8.1 A Comparison of the Three Blood Cell Types

Blood Cell Type	Description	Average Number of Cells	Major Functions
Red blood cells (erythrocytes)	Biconcave disk without nucleus; contains hemoglobin	5,000,000 per mm^3*	Transports oxygen and a small amount of carbon dioxide.
White blood cells (leukocytes)		8,000 per mm^3	Protects against disease-causing organisms.
Neutrophil	Granular leukocytes (many granules in cytoplasm)	60% of white cells	Engulfs infectious agents and foreign debris.
Eosinophil	Granular leukocytes (many granules in cytoplasm)	1-4% of white cells	Combats parasites and allergy-causing agents.
Basophil	Granular leukocytes (many granules in cytoplasm)	< 1% of white cells	Releases histamine and heparin.
Monocyte	Agranular leukocytes (few granules in cytoplasm)	5% of white cells	Engulfs bacteria, dead cells, and debris.
Lymphocyte	Agranular leukocytes (few granules in cytoplasm)	30% of white cells	Destroys bacteria and cancer cells, regulates immune system, and produces antibodies.
Platelets	Cell fragments	200,000–400,000 per mm^3	Important in blood clotting.

* mm^3 = cubic millimeter.

life processes characteristic of other cells, because during its development each erythrocyte loses its nucleus and organelles (mitochondria, endoplasmic reticulum, and ribosomes). Without these organelles, a red blood cell assumes the shape of a biconcave disk (Figure 8.4). This shape allows RBCs to bend and flex so that they can pass more easily through the tiny blood vessels called capillaries. In addition, without a nucleus or organelles, there is more space for hemoglobin molecules, which are the proteins that bind O_2. About 250 million hemoglobin molecules can be packed into one mature red blood cell.

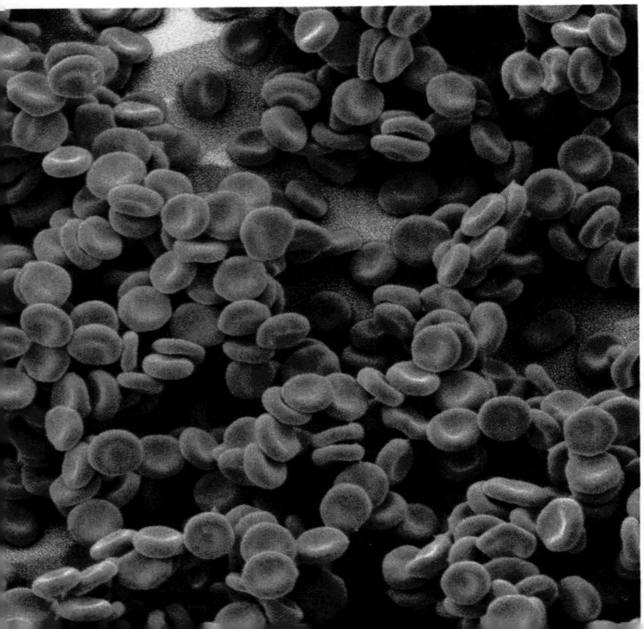

FIGURE 8.4 A highly magnified view of red blood cells.

Hemoglobin (*he-mow-globe-in*) is a large protein molecule which efficiently picks up O_2 molecules in the lungs, where the O_2 concentration is high, and releases them to metabolizing tissues, where the O_2 concentration is low. A hemoglobin molecule consists of two chemical components: an iron-containing component (heme) and a protein component (globin). The **globin** protein is made by binding together four individual polypeptide chains consisting of two alpha chains and two beta chains. One **heme** component is bound to each polypeptide chain (Figure 8.5).

The iron atom in each heme can form a bond with an O_2 molecule. Since there are four heme groups, each hemoglobin molecule can transport four O_2 molecules at a time. It is the iron atoms in hemoglobin that give erythrocytes their reddish color. Because we have so many RBCs, our blood appears red.

RBC production, destruction, and recycling. As we mentioned in Chapter 5, new RBCs are produced in the red marrow of bone. Contained within the red marrow spaces are specialized cells called **stem cells** that undergo mitosis to produce immature cells that develop into mature red blood cells (Figure 8.6). This process takes about 1 week, and it is dur-

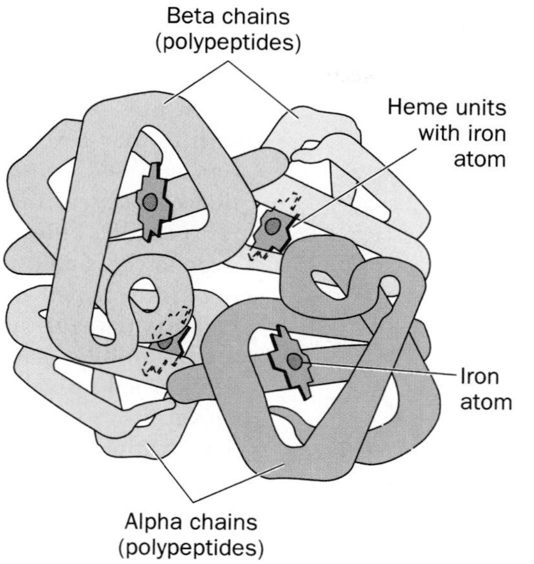
Beta chains
(polypeptides)
Heme units
with iron
atom
Iron
atom
Alpha chains
(polypeptides)

FIGURE 8.5 The chemical structure of hemoglobin.

ing this time that the nucleus and the organelles leave the cell. In a normal, healthy person, the number of mature RBCs in the body remains relatively constant because their production is regulated.

The regulation of RBC production is accomplished by the hormone *erythropoietin* (*eh-rith-row-**poy**-ee-tin*), which is secreted by the kidneys (Figure 8.7). Erythropoietin production increases whenever there is a reduction in the amount of O_2 in the tissues, as can occur during excessive bleeding or when O_2 availability is reduced (high-altitude locations). This hormone circulates to the red bone marrow, where it increases stem cell mitosis and speeds the development of red blood cells. When enough RBCs are again available to supply our tissues with O_2, erythropoietin production is decreased and

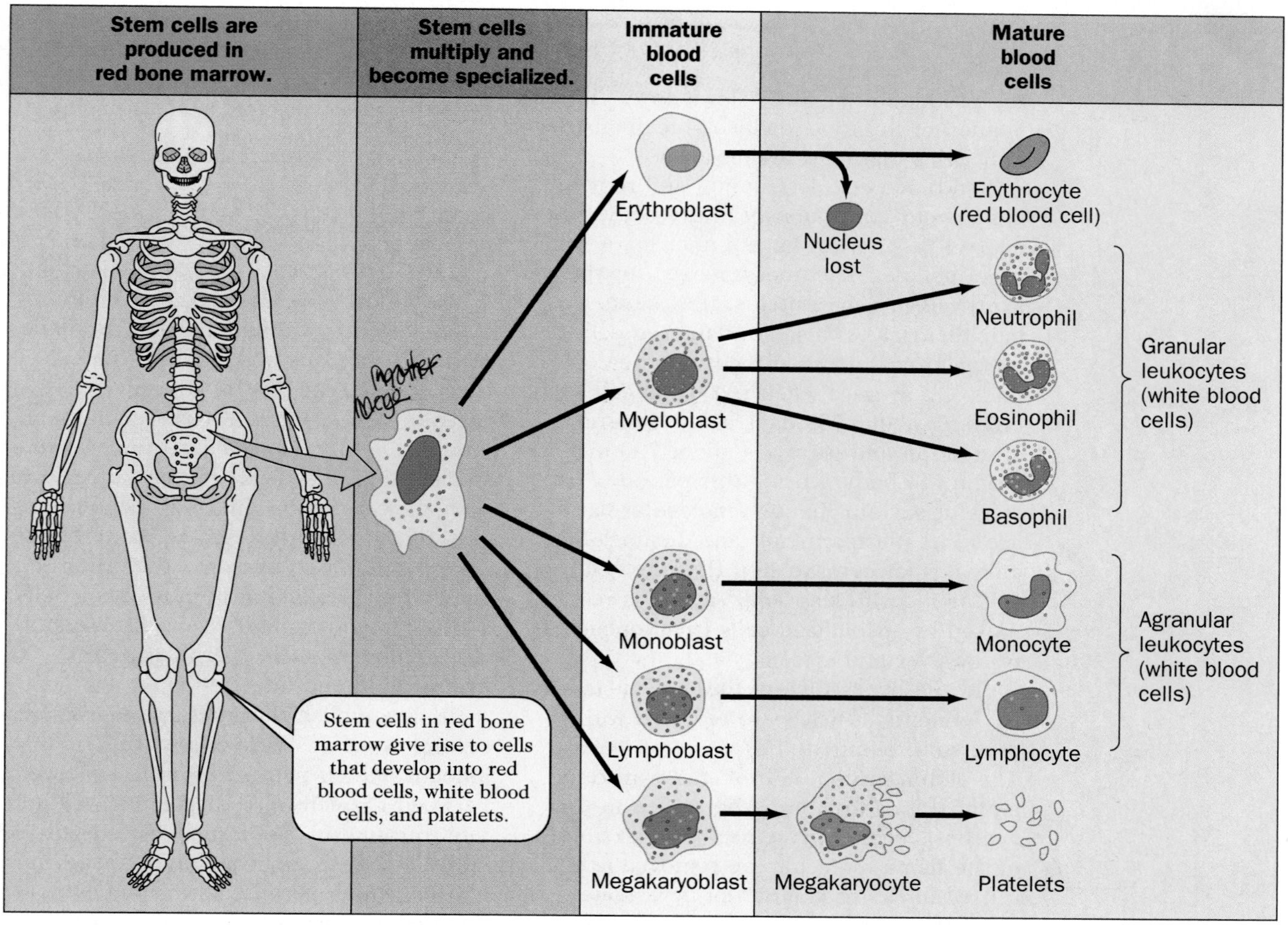
Stem cells are produced in red bone marrow.
Stem cells multiply and become specialized.
Immature blood cells
Mature blood cells
Erythroblast
Nucleus lost
Erythrocyte (red blood cell)
Myeloblast
Neutrophil
Eosinophil
Basophil
Granular leukocytes (white blood cells)
Monoblast
Monocyte
Lymphoblast
Lymphocyte
Agranular leukocytes (white blood cells)
Megakaryoblast
Megakaryocyte
Platelets
Stem cells in red bone marrow give rise to cells that develop into red blood cells, white blood cells, and platelets.

FIGURE 8.6 The production of blood cells.

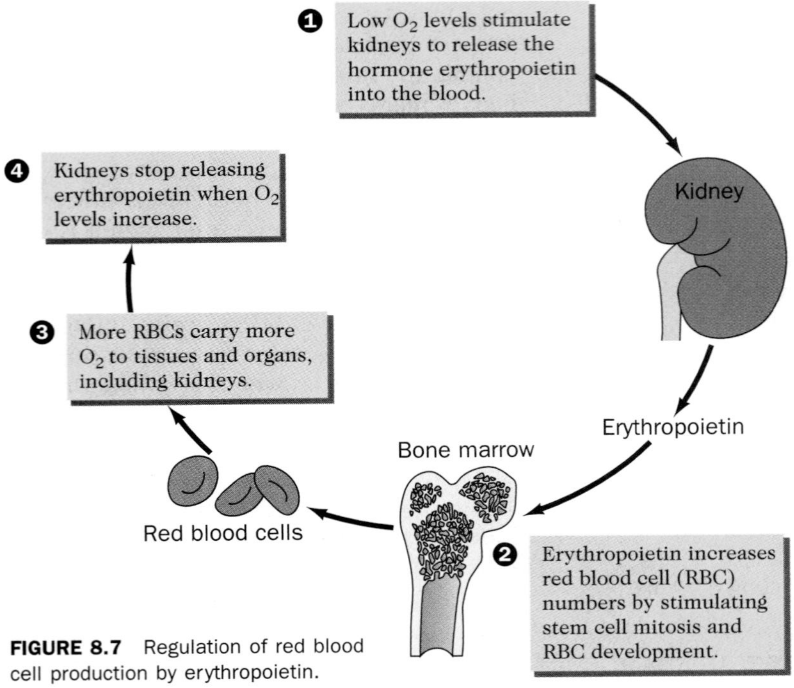

1 Low O_2 levels stimulate kidneys to release the hormone erythropoietin into the blood.
Kidney
Erythropoietin
Bone marrow
2 Erythropoietin increases red blood cell (RBC) numbers by stimulating stem cell mitosis and RBC development.
Red blood cells
3 More RBCs carry more O_2 to tissues and organs, including kidneys.
4 Kidneys stop releasing erythropoietin when O_2 levels increase.

FIGURE 8.7 Regulation of red blood cell production by erythropoietin.

the production of RBCs slows down. This is another negative feedback mechanism that helps maintain homeostasis.

Red blood cell destruction and recycling occur continuously in the body. Because RBCs do not have a nucleus or the other cellular machinery needed for the synthesis of new proteins, they cannot repair themselves. Instead, damaged cells are destroyed and replaced with new ones. On average, each red blood cell survives for about 120 days in the general circulation and it travels about 700 miles within the body. During this period of travel it sustains enough molecular damage to its plasma membrane, hemoglobin, enzymes, or cytoskeleton that the cell can be identified as "old," and it is removed by specialized cells (macrophages) in the liver and spleen.

In the liver, some of the molecular components of RBCs are recycled while others are eliminated as waste products. The globin protein is broken down to its amino acids, which can be reused in the synthesis of new proteins. The iron atoms of the heme molecule are removed and reused in the production of new heme units. The remaining portion of the heme molecule is converted in the liver to a yellow chemical called *bilirubin*. If the liver is damaged or diseased, bilirubin may accumulate in the blood plasma, causing the whites of the eyes and unpigmented skin to turn a yellowish color. This condition is called *jaundice* and is an indication of a severe liver disorder.

Normally, bilirubin is secreted from the liver to the small intestine along with bile. After it passes into the intestine, bacteria convert bilirubin to a yellow pigment (urobilinogen), some of which is reabsorbed into the blood and then removed from the blood by the kidneys. This pigment gives urine its yellowish color. The characteristic color of feces is due to a brownish pigment produced by the further action of the intestinal bacteria on bilirubin.

> RBCs transport O_2 from the lungs to the tissues. Mature RBCs have a biconcave shape and lack organelles. They contain hemoglobin, the O_2-carrying protein. RBC development from stem cells is regulated by the hormone erythropoietin.

White Blood Cells: Defending the Body

White blood cells (WBCs), or **leukocytes** (***luke**-oh-sites*), are larger than and fundamentally different from RBCs. Whereas there is one type of RBC, there are five types of WBCs. While RBCs are simple in structure and are specialized to perform a single function, WBCs are more diverse in both their structure and their cellular operation. In contrast to RBCs, white blood cells contain a nucleus and the other organelles characteristic of most living cells. WBCs protect the body from disease and help remove nonliving cellular debris from wound and infection sites.

To accomplish their defensive mission, several types of WBCs can move like an amoeba, which permits them to squeeze between the cells of the capillary wall and move around in the tissue spaces of the body. They are also capable of phagocytosis, which permits them to engulf, process, and destroy invading organisms, foreign substances, and cellular debris.

White blood cells constitute a much smaller population of blood cells than do erythrocytes. Depending on the state of your health, you have around 8000 WBCs per cubic millimeter of blood, compared with 5 million RBCs. This is referred to as the "normal white cell count" and changes in this number are used to diagnose illness and determine treatment. During infections, the WBC count increases as they are produced to fight the invading organisms. When the invading organisms have been destroyed and cell debris has been cleaned up, the number of WBCs returns to normal.

There are five different types of white blood cells. Depending on the abundance of granules (mostly lysosomes) in their cytoplasm, each type can be distinguished as either granular leukocytes or agranular leukocytes (see Table 8.1). The word "granular" was applied to the appearance of these cells before the electron microscope revealed that the granules were actually numerous vesicles in the cytoplasm.

Granular leukocytes: many granules. The three types of **granular leukocytes** are neutrophils, eosinophils, and basophils. **Neutrophils** (***new**-tro-fills*) are the most abundant, making up about 60 percent of the WBCs circulating in the blood. Their amoeboid movement carries them across capillary walls and through tissue spaces to reach the site of infection or injury. They seek out, engulf (by phagocytosis), and destroy (by lysosomes) foreign cells and substances that have penetrated the body.

Eosinophils (*ee-oh-**sin**-oh-fills*) represent a relatively small percentage of the WBCs circulating in the blood vessels (about 1 percent). Like neutrophils, they are capable of amoeboid movement and can cross the capillary wall to enter tissue spaces. Eosinophils have two major functions. One function is to attack large parasites such as worms (tapeworms, pinworms). These parasites are much too large to be engulfed by phagocytosis. Instead, numerous eosinophils cluster around a large parasite and secrete powerful digestive enzymes to destroy it. The second function is the release of chemicals that control the inflammatory response during allergic reactions initiated in part by another type of WBC, the basophils.

The **basophils** (***bay**-so-fills*) constitute less than 1 percent of the WBCs circulating in the blood. Their cytoplasm granules contain the chemical substance *histamine*, which initiates the inflammation response. When tissue damage occurs, basophils release histamine, causing nearby capillaries to leak plasma into the tissue spaces. This allows nutrients and other compounds important for tissue repair to flow into the injured area.

Since localized clotting would defeat the functions of histamine, basophils also secrete *heparin*, a chemical that prevents blood clotting. The effects of histamine and heparin are believed to enhance the events associated with inflammation. Though often uncomfortable, inflammations are a natural part of the body's defensive and healing processes.

Those of us who suffer from hay fever experience local inflammation of the mucous membranes of the nasal cavity and eyes. This inflammation results when certain plant pollens or other substances stimulate the release of histamine. *Antihistamine* medication prevents this release and the capillary leakage that produces the annoying runny nose and watery eyes. Allergies such as hay fever are exaggerated responses of the body to a foreign substance that usually is not life-threatening (Chapter 13).

Agranular leukocytes: few granules. White blood cells with relatively few granules in their cytoplasm are called **agranular leukocytes**. There are two types: monocytes and lymphocytes (see Table 8.1). **Monocytes** (***mon**-oh-sites*) constitute a small number of circulating WBCs (about 5 percent). They can leave the bloodstream and migrate into tissue spaces by amoeboid movement. They are "supereaters" that actively engulf bacteria, dead cells, and other debris.

Lymphocytes (***lim**-foe-sites*) account for a relatively large number of circulating WBCs (about 30 percent). They are found in the bloodstream and within tissues of the lymph nodes, thymus gland, tonsils, and spleen. There are two major types of lymphocytes: B-cell lymphocytes and T-cell lymphocytes. Both types are associated with what are called the body's immune defenses.

When properly stimulated, *B-cell lymphocytes* manufacture highly specialized proteins called *antibodies* and secrete them into the blood plasma. Antibodies help protect us from foreign cells and substances. *T-cell lymphocytes* recognize and kill specific cells such as bacteria, virus-infected cells, and cancer cells. The role of lymphocytes in immunity will be examined in Chapter 13.

White blood cell production. WBCs are produced in the red bone marrow from the same stem cells that produce RBCs (see Figure 8.6). The stem cells produce immature WBCs that develop into the five different types of leukocytes. All immature WBCs except T-cell lymphocytes mature in the red bone marrow. Immature T-cell lymphocytes leave the bone marrow and are carried by the circulating blood to the thymus gland, lymph nodes, and spleen, where they mature.

Most WBCs do not live long (only hours or days) because of the damage they sustain while defending the body from disease and injury. Dead or damaged WBCs are continuously removed from the circulation by the spleen and liver. Consequently, the supply of new WBCs must be constantly replenished by the red bone marrow. The regulation of WBC production involves a poorly understood series of complex chemical signals from existing leukocytes.

Platelets: Preventing Blood Loss

Platelets (***plate**-lets*) circulating in the blood are small cell fragments, not entire cells (see Table 8.1). These fragments are derived from larger cells called *megakaryocytes* (*meg-ah-**kare**-ee-oh-sites*). They number between 200,000 and 400,000 per cubic millimeter of blood. Like blood cells, megakaryocytes develop from stem cells in the red marrow (see Figure 8.6). As platelets are pinched off, they enter the bloodstream and are available to help prevent blood loss from a damaged vessel. We will explore this process in the next section.

> The five types of WBCs defend the body and are found in the blood and in tissue spaces. All WBCs develop from stem cells in the red bone marrow. Blood also contains platelets that prevent blood loss.

Preventing Blood Loss

In the rush to complete a kitchen task we have all felt the knife slip and watched the blood rush from a wound. When a wound is small, most of us lose little blood because of two processes: platelet plug formation and blood clot formation. Without these processes, even minor damage to blood vessels could result in extensive blood loss.

Platelet Plug Formation

As platelets circulate in the bloodstream, they normally do not clump together. However, when platelets encounter damaged blood vessels, they swell, become sticky, and form spiny extensions. Platelets quickly accumulate, adhere to each other, and plug the opening in the vessel, preventing extensive blood loss (Figure 8.8a). Small tears or openings of blood vessels can be sealed by a platelet plug in several seconds. For larger wounds, the body has another defense: blood clotting.

Blood Clot Formation

The development of a blood clot involves a complex series of chemical reactions that ultimately plug damaged blood vessels with a network of protein fibers (Figures 8.8b and 8.9). We will focus on the final events. In reality, the process is much more complex with many additional steps and at least 12 "clotting factors" involved.

As with platelet plugs, these reactions are strictly regulated and occur only when there is damage to a blood vessel or the nearby tissue. Without such regulation, frequent formation of blood clots would plug capillaries and endanger our lives. How is clot formation regulated? The answer involves successive reactions that convert the soluble plasma protein fibrinogen to insoluble fibers called fibrin.

When blood vessels and their surrounding tissues are damaged, they and nearby platelets release *prothrombin activator* (*pro-***throm***-bin*). In the presence of calcium ions (Ca^{2+}), *prothrombin activator* speeds the conversion of the inactive plasma protein **prothrombin** to the active enzyme **thrombin** (***throm****-bin*).

$$\underset{\text{Inactive plasma protein}}{\text{Prothrombin}} \xrightarrow{\text{prothrombin activator}} \underset{\text{Active enzyme}}{\text{thrombin}}$$

As an enzyme, thrombin speeds the conversion of the soluble plasma protein **fibrinogen** to long fibers called **fibrin** that are insoluble in plasma.

$$\underset{\text{Soluble plasma protein}}{\text{Fibrinogen}} \xrightarrow{\text{thrombin}} \underset{\text{Insoluble protein fibers}}{\text{fibrin}}$$

Because they are long and insoluble, these threadlike fibers can attach to the damaged tissue and to each other, forming a meshwork that traps platelets, blood

FIGURE 8.8 (a) Development of a platelet plug. (b) Development of a blood clot.

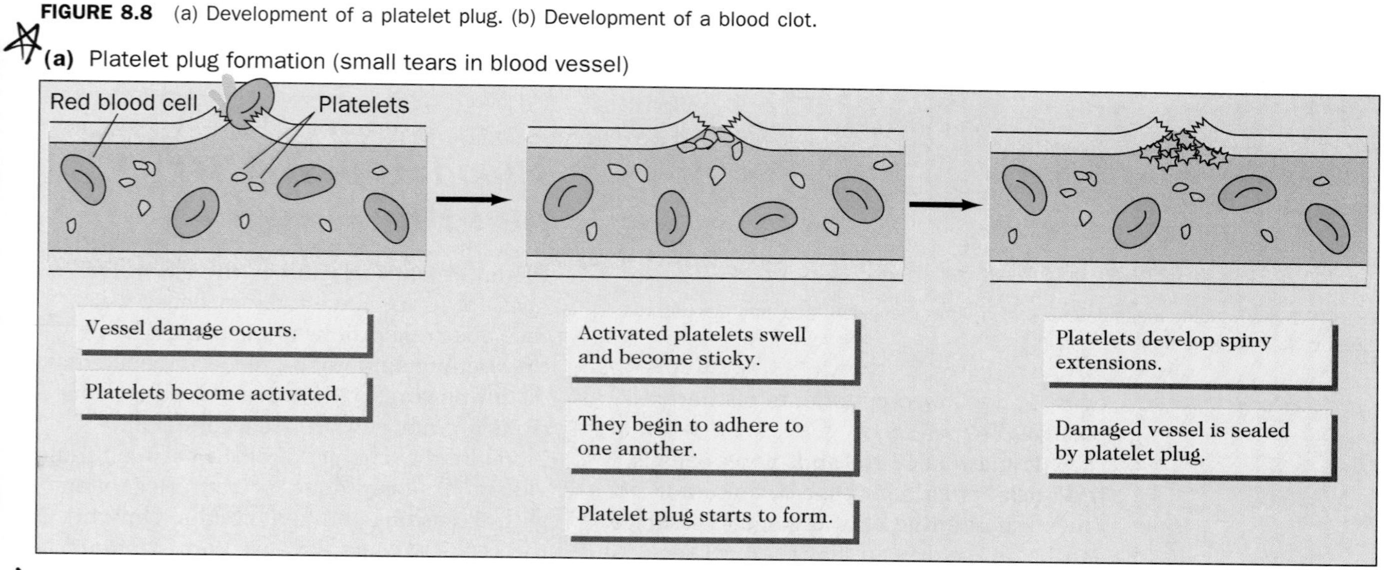

(a) Platelet plug formation (small tears in blood vessel)
Red blood cell
Platelets
Vessel damage occurs.
Platelets become activated.
Activated platelets swell and become sticky.
They begin to adhere to one another.
Platelet plug starts to form.
Platelets develop spiny extensions.
Damaged vessel is sealed by platelet plug.

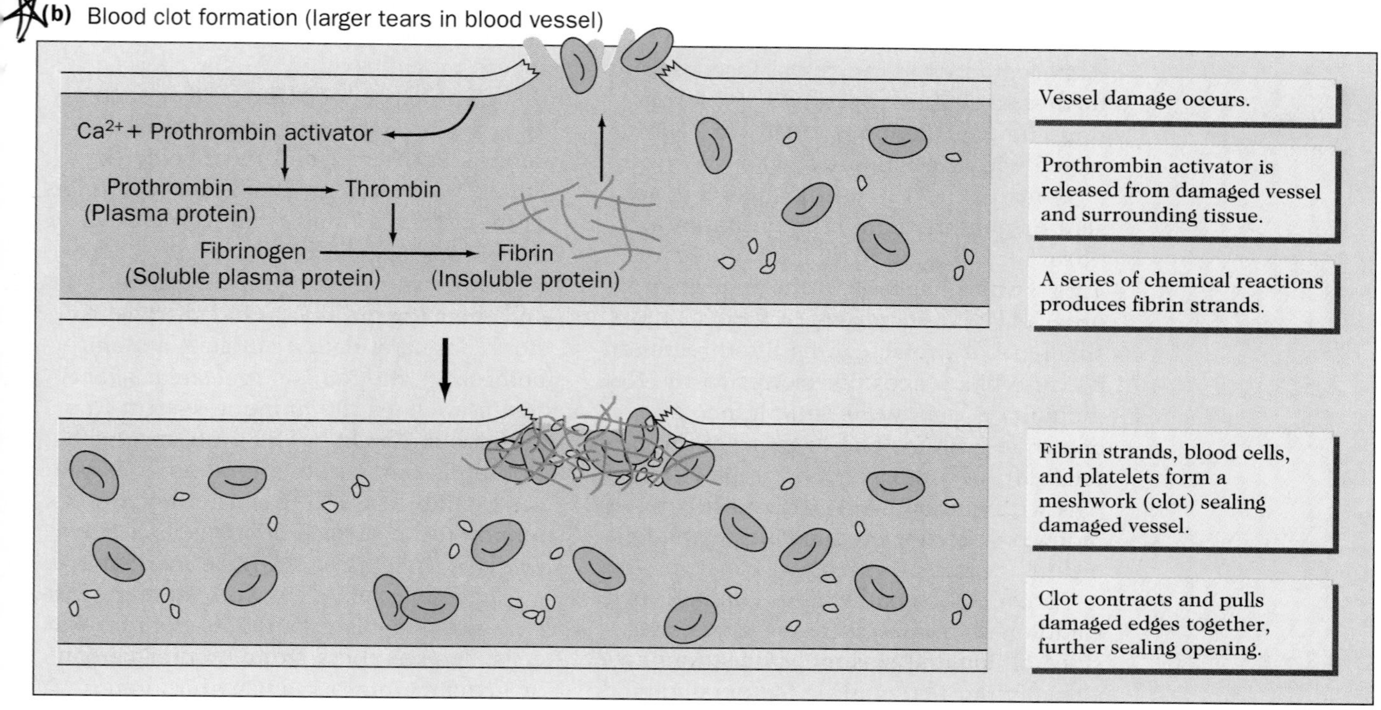

(b) Blood clot formation (larger tears in blood vessel)
Ca2+ + Prothrombin activator
Prothrombin (Plasma protein)
Thrombin
Fibrinogen (Soluble plasma protein)
Fibrin (Insoluble protein)
Vessel damage occurs.
Prothrombin activator is released from damaged vessel and surrounding tissue.
A series of chemical reactions produces fibrin strands.
Fibrin strands, blood cells, and platelets form a meshwork (clot) sealing damaged vessel.
Clot contracts and pulls damaged edges together, further sealing opening.

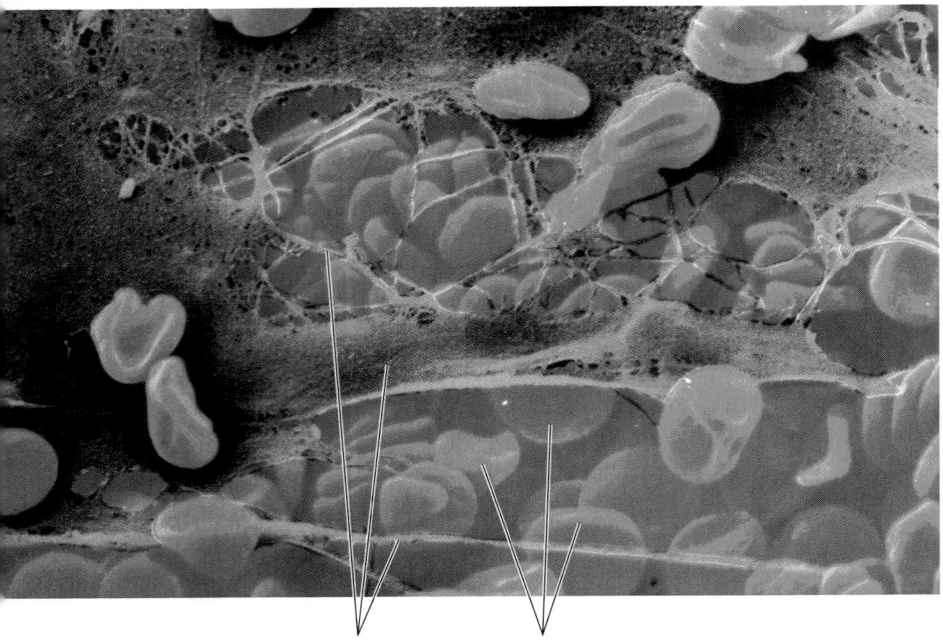
Fibrin
Red blood cells

FIGURE 8.9 A highly magnified view of a fibrin meshwork that has trapped red blood cells and platelets.

cells, and other components of the plasma (Figure 8.9).

This mass of fibrin and blood cells is the initial *blood clot* that covers and plugs the opening of a damaged vessel. It can begin to form in less then 60 seconds. As time progresses, the platelets in the clot start to contract, pulling the damaged edges of the vessel together and further sealing the opening. Clot formation and contraction normally are completed within 60 minutes.

If any of the 12 clotting factors is absent or nonfunctional, the formation of fibrin may be slowed or may not occur. This is what happens in the inherited disorder called *hemophilia,* in which even a small cut or bruise can be life-threatening because blood does not clot properly. Hemophilia is due to the inheritance of a defective gene for one or more of the protein clotting factors. For example, individuals with hemophilia A, the most common type, do not produce clotting factor eight.

To control bleeding, hemophiliacs have traditionally required frequent transfusions of donated plasma or injections of partially purified clotting factors obtained by pooling blood from thousands of donors. Unfortunately, these sources of help have not been entirely free of the viruses for hepatitis and AIDS (HIV). The good news is that genetic engineering with recombinant DNA has produced a pure form of the protein clotting factor for hemophilia, type A.

> Platelet plugs and blood clots limit blood loss from damaged vessels. A plug forms when platelets adhere to each other. A blood clot forms through a series of chemical reactions that convert a soluble plasma protein into insoluble fibrin.

Blood Types: ABO and Rh

Blood transfusion can be "the gift of life." Many of us are alive today because we received transfusions after losing life-threatening amounts of blood in accidents or during surgery. However, blood transfusions are not new. To treat illness and blood loss, exchange of blood between individuals has been done for centuries, often with disastrous and fatal results. Only in the last 100 years have we come to understand the complexities of the circulatory system, allowing transfusions to become successful and routine. We have learned that all blood is not identical. For successful transfusion, the blood "types" of the donor and the recipient must be determined and matched. The **blood type** refers to the particular kind of antigen present in the plasma membrane of RBCs.

Before we can fully explain blood types and their importance in blood transfusions, we must define antigens and antibodies. An **antigen** is a molecular agent that stimulates the immune system to produce antibodies. The proteins attached to the plasma membrane of our cells can act as antigens, and because they are found on the surface of our cells, we can think of them as cell surface markers.

These markers allow your immune system to distinguish your cells from foreign cells, such as those from another person or invading microorganisms. If foreign cells enter your body, their surface markers (antigens) stimulate your immune system to produce antibodies. The surface markers on RBCs are the basis for blood types and will be particularly important when we describe the transfusion of blood from one person to another.

An **antibody** is a protein molecule made by the lymphocytes. An antibody binds to a specific foreign antigen and neutralizes its danger. Antibodies help protect us from disease. You might think of them as being "anti" or against foreign "bodies" or structures that enter our tissues. These antibodies are the gamma globulin class of soluble plasma proteins we described earlier in this chapter. Antigen-antibody binding reactions are very specific; that is, a specific type of antigen reacts only with a specific antibody. The specificity of this reaction is similar to a lock and key mechanism: only a specific key acts to open a specific lock.

ABO Blood Types

For transfused blood to be the "gift of life," we must know the RBC surface antigens for both the donor and the recipient of the blood. Two antigens are responsible for the ABO blood types, and they are designated A and B. Each of us inherits a blood type. That is, we inherit the genes for synthesizing type A or type B antigens. If you inherit only the gene to make type A antigen, you are *blood type A*. If you inherit only the gene to make type B antigen, you are *blood type B*. If you inherit both genes, you are *blood type AB*, and if neither gene is inherited, you are *blood type O*. This is easier to remember if you think of O as a zero, indicating that all type O red blood cells contain zero A and B antigens in their membranes.

The plasma of A, B, and O individuals contains certain associated antibodies (Figure 8.10a). Type A blood contains B antibodies (antibodies that neutralize type B antigens) in the plasma. Type B blood contains A antibodies in the plasma. Type O blood has both A and B antibodies in the plasma. Type AB blood contains neither A nor B antibodies in the plasma because the body does not produce antibodies against its own proteins, in this case, the A and B antigens. This antigen and antibody pattern occurs in all of us.

(a) The red blood cell antigen and associated plasma antibodies found in each blood type

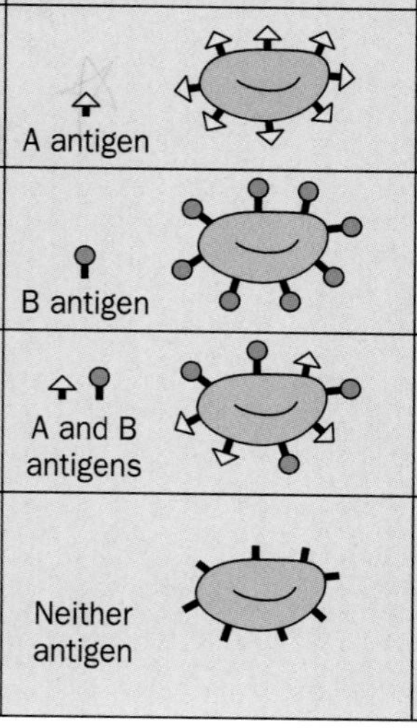

Blood type	Red blood cells showing surface antigen	Plasma antibodies
A	A antigen	Antibody B
B	B antigen	Antibody A
AB	A and B antigens	None
O	Neither antigen	Antibody A, Antibody B

(b) The antigen-antibody reaction that occurs when a donor's type A blood is transfused into a type B recipient

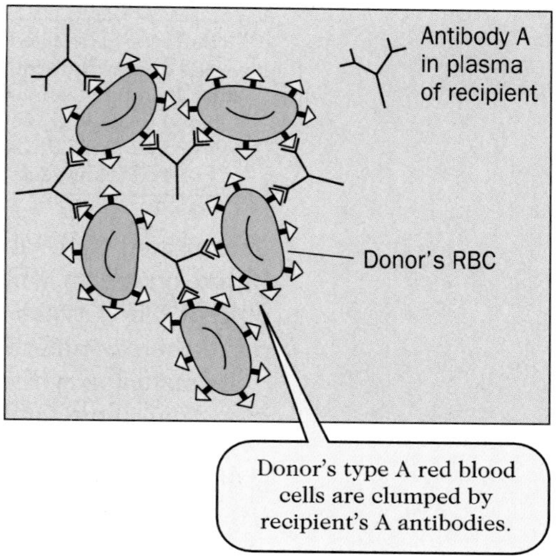
Antibody A in plasma of recipient
Donor's RBC
Donor's type A red blood cells are clumped by recipient's A antibodies.

FIGURE 8.10 (a) The relationship between RBC surface antigens and plasma antibodies. (b) RBC clumping after transfusion of a donor's mismatched blood.

Blood transfusion: matching blood types. A successful blood transfusion is one in which the antigens and antibodies of the donor's blood are compatible with those of the recipient's blood. This can be

FIGURE 8.11 A chart showing whether an antigen-antibody reaction (clumping) will occur when a donor's blood type is mixed with that of a recipient.

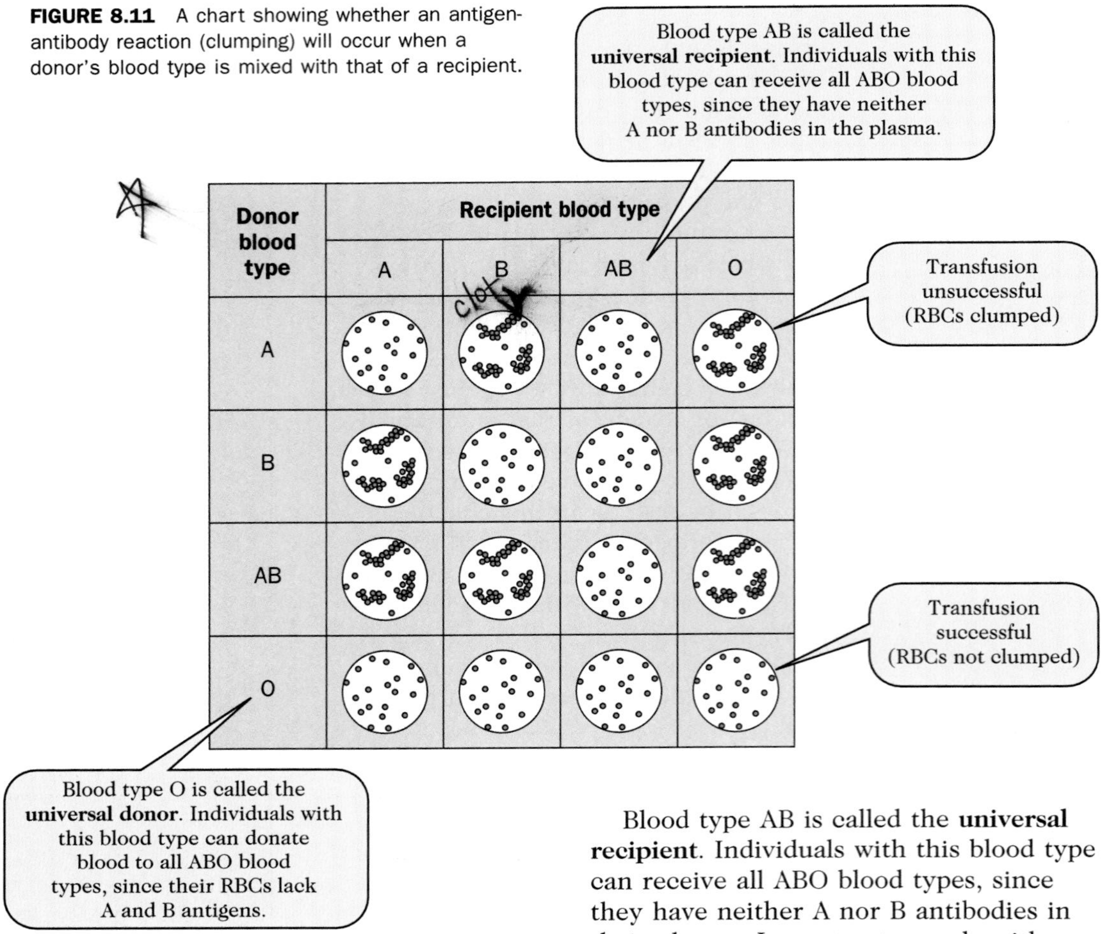
Blood type AB is called the **universal recipient**. Individuals with this blood type can receive all ABO blood types, since they have neither A nor B antibodies in the plasma.
Donor blood type
Recipient blood type
A
B
AB
O
A
B
AB
O
Transfusion unsuccessful (RBCs clumped)
Transfusion successful (RBCs not clumped)
Blood type O is called the **universal donor**. Individuals with this blood type can donate blood to all ABO blood types, since their RBCs lack A and B antigens.

determined by a laboratory process called **blood typing** in which the donor and recipient blood types are matched to determine their compatibility.

Transfusion with incorrectly matched blood can have fatal consequences. For example, incorrect matching occurs when a donor with type A blood (A antigens on RBCs) gives blood to a type B person (A antibodies in plasma). As the donor's type A red blood cells enter the type B recipient, the donor's A antigen will react with the A antibodies in the recipient's plasma (Figure 8.10b). Such an antigen-antibody reaction will cause the donor's RBCs to clump together. As these small clumps move through the circulatory system, they plug the capillaries. Such plugging obstructs normal blood flow and prevents O_2 and nutrients from reaching tissues. Without O_2, tissues are damaged and may die. Figure 8.11 presents examples of successful and unsuccessful blood transfusions.

Blood type AB is called the **universal recipient**. Individuals with this blood type can receive all ABO blood types, since they have neither A nor B antibodies in their plasma. In contrast, people with blood type O are called **universal donors** because they can donate their blood to all ABO blood types. Their red blood cells do not have A and B antigens, and so an antigen-antibody reaction will not occur with A or B antibodies in the blood plasma of the recipient.

> ABO blood types are determined by A or B antigens on red blood cells. Blood types A, B, and AB contain the A, B, and AB antigens, respectively. Blood type O contains neither A nor B antigens. The blood plasma of each blood type contains antibodies to the A and B antigens not found on the red blood cells.

Rh Blood Types

The A and B antigens are not the only protein antigens on the RBC surface that are important in blood transfusions. The *Rh antigen* is another important protein. The abbreviation "Rh" stands for "rhesus

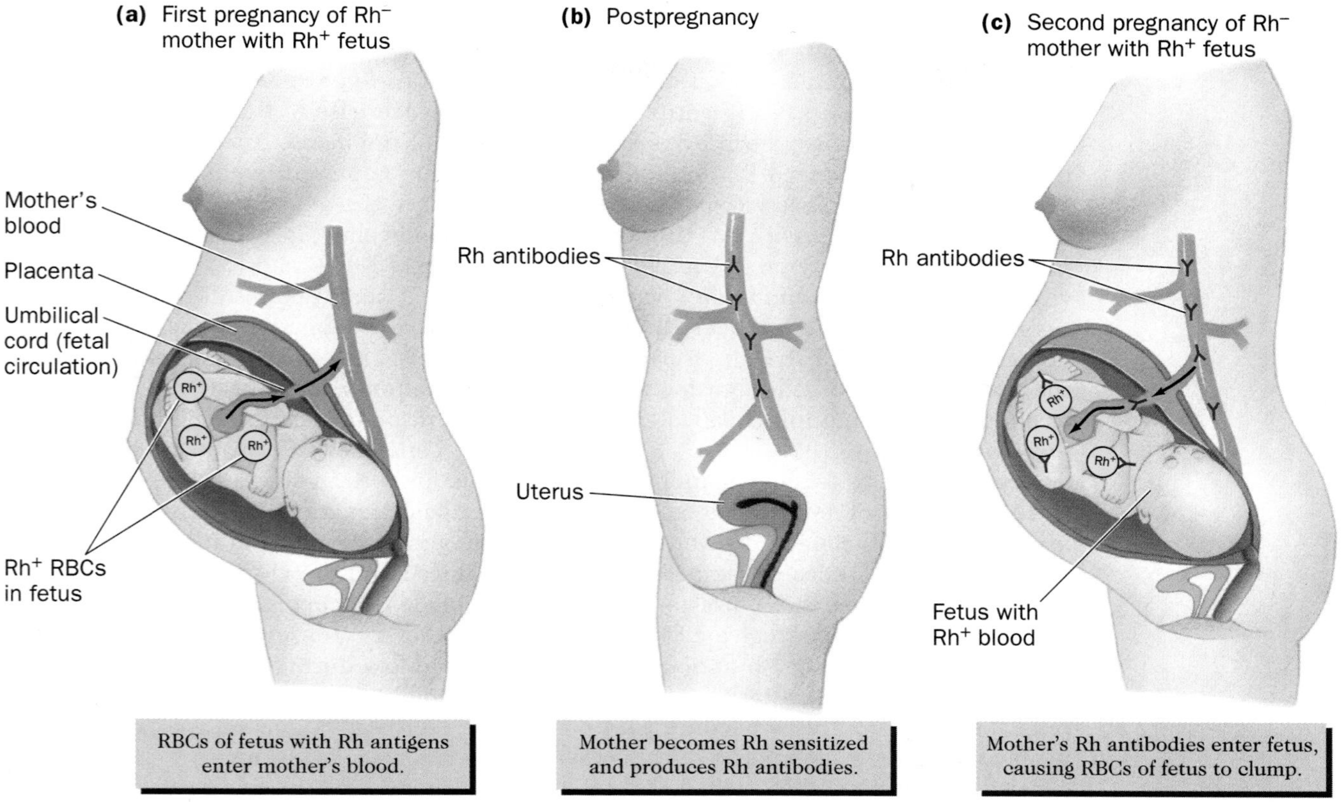
(a) First pregnancy of Rh⁻ mother with Rh⁺ fetus
(b) Postpregnancy
(c) Second pregnancy of Rh⁻ mother with Rh⁺ fetus
Mother's blood
Placenta
Umbilical cord (fetal circulation)
Rh⁺ RBCs in fetus
Rh antibodies
Uterus
Rh antibodies
Fetus with Rh⁺ blood
RBCs of fetus with Rh antigens enter mother's blood.
Mother becomes Rh sensitized and produces Rh antibodies.
Mother's Rh antibodies enter fetus, causing RBCs of fetus to clump.

FIGURE 8.12 The events associated with possible Rh problems during the first and second pregnancies involving an Rh^- mother and an Rh^+ fetus.

monkey," the species used in the initial research on this blood type. A person who inherits the gene to make this protein antigen is said to be *Rh-positive* (Rh^+), while a person without this gene cannot make the Rh antigen and is said to be *Rh-negative* (Rh^-). An individual with an A^+ blood type has red blood cells with the A antigen and the Rh antigen (indicated by the superscript plus sign).

Rh-positive individuals do not produce Rh antibodies, because doing so would lead to an antigen-antibody reaction. Rh-negative individuals do not naturally have Rh antibodies, but they are capable of producing them when stimulated to do so. Stimulation can occur when an Rh-negative individual receives blood from an Rh-positive donor. The Rh antigen on the donor's RBCs will stimulate the recipient's immune system to make Rh antibodies.

An Rh-negative person receiving a first transfusion with Rh-positive blood does not usually have an antigen-antibody reaction because it takes 1 to 2 months for the immune system to produce sufficient Rh antibodies. Within that period, most of the donated RBCs will have been removed by natural processes. Thus, for this first transfusion with incompatible Rh blood, there are no dangerous consequences. However, the recipient is now said to be Rh sensitized, and subsequent Rh^+ transfusions may cause a problem.

Rh sensitization means that Rh antibodies can be synthesized rapidly in large numbers if Rh-positive blood is transfused again. Upon a second transfusion, there will be an immediate antigen-antibody reaction between the donor's red blood cells and the recipient's manufactured Rh antibodies, causing the donor's RBCs to clump. As with ABO antigen-antibody reactions, small blood vessels become blocked, possibly resulting in death.

Rh and pregnancy. Special problems associated with Rh can occur during pregnancy. The embryo and fetus produces its own blood, which does not mix with the mother's blood. However, some blood may be exchanged between the mother and the fetus during birth or through small tears that sometimes develop in the placenta during late pregnancy. The problem arises when an Rh^- mother is carrying an Rh^+ fetus (Figure 8.12).

As in Rh transfusions, the first contact with Rh^+ blood usually does not lead to antigen-antibody problems. Because of the time required to produce antibodies, the child will have been born by the time the mother has produced Rh antibodies. However, if an Rh-sensitized mother becomes pregnant again with an Rh^+ fetus, her Rh antibodies may cross the blood vessels of the placenta and enter the fetus's bloodstream. The mother's Rh antibodies will react with the Rh antigens on the fetus's red blood cells, causing them to clump. This condition is called *erythroblastosis fetalis*.

Erythroblastosis fetalis results in massive destruction of fetal red blood cells, leading to fetal anemia and severe tissue damage and sometimes to the death of the fetus. To prevent such disasters, it is now common practice to give all Rh^- mothers an injection of Rh antibodies (called RhoGam) after a birth, miscarriage, or abortion. This procedure eliminates any Rh^+ red blood cells of the fetus circulating in the mother's bloodstream, preventing her from becoming Rh sensitized. Clumping of fetal RBCs in the mother is less of a threat to her because her cardiovascular system is much larger than the fetal system.

> Individuals with the Rh antigen are designated Rh^+; those without the antigen are designated Rh^-. Rh^- individuals produce Rh antibodies if they receive Rh^+ blood. This leads to Rh sensitization, which may cause problems during a second transfusion or pregnancy.

Diseases and Disorders

Serious health problems can occur when blood cells function improperly or when there are too many or too few of them. A poor diet can also lead to blood disorders, as occurs when insufficient dietary iron limits the production of hemoglobin.

In this section, we discuss several blood disorders.

Anemia: Carrying Too Little O_2

Anemia (*a-knee-me-ah*) is a general term for any decrease in the O_2-carrying capacity of the blood. Anemia may result from three different conditions: (1) a reduction in the amount of hemoglobin in red blood cells, (2) a reduction in the number of RBCs, and (3) production of abnormal hemoglobin. All three conditions have similar symptoms, including fatigue, weakness, and shortness of breath on minimum exertion.

Nutritional anemia results from a dietary deficiency of iron that leads to the depletion of iron stores in the body. Without adequate amounts of iron, insufficient numbers of hemoglobin molecules are produced and the O_2-carrying capacity of red blood cells decreases. Females require more dietary iron than do males because of blood loss during menstruation. It is recommended that women age 20 to 50 receive 18 milligrams of iron per day and that men receive 10 mg.

Hemorrhagic anemia results from excessive blood loss, as can occur during injury or bleeding ulcers. Chronic loss of even small amounts of blood can result in anemia through the depletion of iron stores in the body.

Aplastic anemia results from the inability of the red bone marrow to produce RBCs. White blood cells and platelets also decrease in number, since they are also produced by red marrow stem cells. The normal function of the bone marrow can be destroyed or inhibited by ionizing radiation, certain toxic chemicals, and the effects of leukemia (a form of cancer we'll discuss below).

Pernicious anemia results from the lack of vitamin B_{12} and/or *intrinsic factor* needed for the absorption of B_{12} from the digestive tract. Vitamin B_{12} is required for the cell divisions that result in RBC formation.

Sickle-cell anemia is a genetic disorder that results in the production of abnormal hemoglobin molecules. It occurs because of a defective (mutant) gene that codes for an altered beta polypeptide chain in the globin portion of the hemoglobin (Figure 8.5). Sickle-cell hemoglobin differs from normal hemoglobin by a single amino acid out of the 146 amino acids in each beta chain.

This defect causes hemoglobin molecules to distort after they release O_2 to the tissues. Hemoglobin distortion causes RBCs to be distorted into a sickle shape (Figure 8.13). This damages the RBCs, encouraging their early destruction and causing problems with the passage of these cells through the capillaries. Sickle-cell anemia is more prevalent among Africans and African-Americans than among other racial groups.

Carbon Monoxide Poisoning: Binding to Hemoglobin

Carbon monoxide (CO) is a colorless, odorless, and very poisonous gas that is produced as a by-product of burning carbon-containing materials such as wood, gasoline, and coal. When inhaled, carbon monoxide enters the bloodstream and bonds to the iron atom of the hemoglobin molecule as O_2 does. As more CO attaches to hemoglobin, less O_2 can attach. Unfortunately, carbon monoxide's affinity for iron is about 200 times greater than that of O_2. This means that inhaling even small amounts of CO can interfere with the O_2-carrying capacity of the red blood cells. Prolonged exposure to small quantities of carbon monoxide gas can result in coma and death.

Mild cases of carbon monoxide poisoning can be treated by breathing fresh air.

FIGURE 8.13 Red blood cells from a person with sickle-cell anemia. (a) Shape of red blood cells is normal when oxygen levels are high. (b) Red blood cells become sickle-shaped when oxygen levels are low.

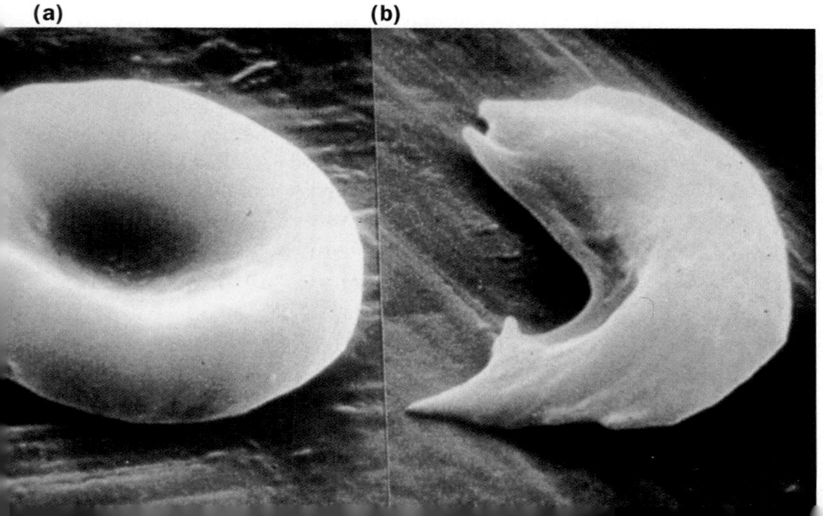
(a)
(b)

Severe cases require the administration of either pure or pressurized O_2 to quickly displace carbon monoxide from hemoglobin.

Leukemia: A Cancer of the White Blood Cells

Leukemia (*lew-**key**-me-ah*) is a type of cancer characterized by the uncontrolled production of immature or abnormal leukocytes. These cells do not form a tumor but, like normal WBCs, are shed into the blood plasma and transported throughout the body.

Leukemia can be categorized as an acute or chronic state depending on the duration of the disease. *Acute leukemia* develops rapidly, and if it is not treated, survival can be measured in months. *Chronic leukemia* develops slowly, and if it is not treated, survival can be measured in years.

Not only does leukemia affect the number and function of white blood cells, it also can interfere with the normal production and formation of red blood cells and platelets. As a result, there can be a variety of symptoms associated with the disease: (1) There is increased susceptibility to infection because of immature or inactive leukocytes. (2) Anemia results from the production of fewer red blood cells. (3) Internal bleeding may result from the production of fewer platelets.

The treatment of leukemia consists of radiation therapy and chemotherapy, which are used to destroy the rapidly dividing cancer cells [see Spotlight on Health: One (page 103)]. Such therapy can be effective in ridding the body of all evidence of the disease, a fortunate outcome called **remission**. Transplantation of bone marrow tissue between individuals may be possible when a donor can be found with cell surface markers (antigens) that will not be interpreted by the recipient's (patient's) body as "foreign."

Mononucleosis: A Viral Infection

Mononucleosis (*mon-oh-nuke-lee-**oh**-sis*), commonly called "mono," is a highly contagious disease caused by the Epstein-Barr virus. It is often transmitted by kissing. The symptoms typically appear 30 to

40 days after the infection and include fever, fatigue, sore throat, and enlarged lymph nodes. Examination of a patient's blood shows an increased number of lymphocytes and monocytes. Many of these lymphocytes are enlarged and abnormal in appearance, resembling monocytes; this is why the disease is called mononucleosis. There is no specific treatment, and complete recovery usually occurs in several weeks with proper rest and nutrition.

Blood Poisoning: Infections of the Blood

Blood poisoning is a common name for an infection of the blood that is also called *septicemia* or *toxemia*. If certain microorganisms, such as bacteria, gain entrance to the blood in substantial numbers, they may overwhelm body defenses and grow rapidly in the nutrient-rich plasma. The poison may be the bacteria themselves or the poisonous substances called toxins excreted by the bacteria as they grow (Chapter 12). Deep puncture wounds, dental extractions, urinary system infections, and major burns may lead to septicemia. Antibiotics are usually very effective against such blood infections.

Life Span Changes

During fetal development and after birth there are dramatic changes in the blood. Unless there is a major blood disease or disorder, however, the blood remains fairly constant over a long life span. This is not surprising, given the blood's extensive homeostatic role. If major changes took place in the blood, life could be shortened.

During embryonic development and after birth, changes occur in blood cell production. Prior to the development of bone marrow, the embryonic liver and spleen produce blood cells. Later, as bone develops, it becomes the site of blood cell production.

Changes also take place in the kind of hemoglobin in red blood cells. Normally, the mother's blood cells do not enter the fetus and the fetus obtains O_2 from the mother's blood through gas exchange. To facilitate this exchange, fetal hemoglobin is slightly different from adult hemoglobin in its amino acid composition and structure. These differences give fetal hemoglobin greater affinity for O_2 compared to the mother's hemoglobin, permitting fetal hemoglobin to successfully obtain O_2 from the mother's hemoglobin. Shortly after birth, the production of fetal-type hemoglobin ceases and regular adult hemoglobin is produced.

As we age, there is a decrease in the ability of red bone marrow to produce blood cells and platelets. Under normal conditions, this decline is not sufficient to reduce the number of circulating blood cells. However, if excessive bleeding occurs, blood cell production may be inadequate to replace all the cells that are lost. This may make an elderly accident victim or surgery patient more vulnerable to blood loss, anemia, and infections.

Chapter Summary

Blood: Its Components and Functions

- Blood can be divided into two components: liquid plasma and blood cells.
- Blood plasma consists of about 90 percent water and 10 percent dissolved substances. The dissolved substances include proteins (albumin, globulins, blood-clotting agents), nutrients, respiratory gases, electrolytes, hormones, and waste products.
- Blood cells include red blood cells, white blood cells, and platelets. All these cells are produced in the red bone marrow.
- Red blood cells, or erythrocytes, are numerous (5 million per cubic millimeter). Mature RBCs lack a nucleus and other organelles and have a biconcave shape. They are filled with hemoglobin molecules which transport oxygen molecules.
- White blood cells, or leukocytes, are less numerous than RBCs (8000 per cubic millimeter). They protect the body from disease. There are two types: granular leukocytes (neutrophils, eosinophils, basophils) and agranular leukocytes (lymphocytes, monocytes).
- Platelets are cell fragments. They help control bleeding.

Preventing Blood Loss

- Platelet plug formation and blood clot formation limit blood loss.

- In platelet plug formation, platelets accumulate and plug small openings in blood vessels.
- In blood clot formation, insoluble fibers of fibrin plug damaged blood vessels.

Blood Types: ABO and Rh

- The plasma membranes of red blood cells contain proteins which can act as antigens during blood transfusions. Two of these proteins are designated A and B.
- Blood type is inherited. Individuals with blood type A inherit the gene for making the A protein; those with blood type B inherit the B gene; those with blood type AB have both A and B genes; and those with blood type O have neither gene.
- Each blood type has related A and B antibodies floating in the plasma. Blood type A has B antibodies; blood type B has A antibodies; blood type AB has neither; and blood type O has both A and B antibodies.
- During a transfusion, the A and B proteins on the RBCs of the donor must be compatible with the recipient's plasma antibodies or the blood cells will clump together.
- Individuals who are Rh^+ have inherited the gene for making the Rh protein on the RBC plasma membrane.
- Rh^- individuals are stimulated to make Rh antibodies if they are exposed to Rh^+ red blood cells.
- Erythroblastosis fetalis can occur when an Rh^- mother is pregnant with an Rh^+ fetus.

Diseases and Disorders

- Anemia can occur because of a decrease in the O_2-carrying capacity of the blood caused by iron deficiency, excessive bleeding, inability of red bone marrow to produce RBCs, lack of vitamin B_{12}, or a defective hemoglobin gene.
- Carbon monoxide poisoning occurs when carbon monoxide binds to the hemoglobin molecule, preventing oxygen from attaching to it.
- Leukemia occurs when there is uncontrolled production of immature or abnormal leukocytes.
- Mononucleosis is a contagious viral infection that causes abnormal-looking lymphocytes.
- Blood poisoning, or septicemia, occurs when microorganisms infect the blood.

Life Span Changes

- The blood of the mother and the blood of the fetus do not mix under normal conditions.
- Before birth, the embryo and fetus produce their own blood cells in the liver and spleen, but blood cells will be produced by bone as the fetus develops.
- The hemoglobin of the fetus has a greater affinity for O_2 than does adult hemoglobin.

Selected Key Terms

agranular leukocytes (p. 227)
blood type (p. 230)
erythrocyte (p. 223)
erythropoietin (p. 225)
fibrinogen (p. 229)
granular leukocytes (p. 227)
hemoglobin (p. 224)
leukocyte (p. 226)
plasma (p. 221)
plasma proteins (p. 222)
platelet (p. 228)
stem cells (p. 224)
thrombin (p. 229)

Review Activities

1. Identify at least five categories of chemicals dissolved in plasma.
2. How many RBCs, WBCs, and platelets are found in blood?
3. Describe the chemical composition of hemoglobin.
4. What is the life span of red blood cells, and how are they recycled?
5. List the three types of granular leukocytes and briefly describe their functions.
6. List the two types of agranular leukocytes and briefly describe their functions.
7. Where are RBCs, WBCs, and platelets produced in the body?
8. Starting with prothrombin activator, describe the sequence of reactions that ultimately produces fibrin.
9. What is the difference between an antigen and an antibody?
10. Could someone with blood type AB successfully donate blood to a recipient with blood type O? Explain.
11. Could someone with blood type A successfully donate blood to someone with blood type B? Explain.
12. Why is a person with blood type AB called a universal recipient?
13. Why will an Rh-positive individual not become Rh sensitized if he or she receives a transfusion of negative blood?
14. Why can erythroblastosis fetalis occur only when an Rh-negative mother is pregnant with an Rh-positive fetus?
15. Explain why erythroblastosis fetalis does not occur with the first pregnancy of an Rh-negative mother and an Rh-positive fetus.

Self-Quiz

Matching Exercise

___ 1. A plasma protein that transports fat, cholesterol, and fat-soluble vitamins	A. Hemoglobin
___ 2. The number of RBCs per cubic millimeter of blood	B. Stem cells
___ 3. O_2-carrying protein found in RBCs	C. 8000
___ 4. The atom in the heme portion of hemoglobin that binds with O_2	D. Thymus gland
___ 5. The specialized cells in red marrow that produce RBCs, WBCs, and platelets	E. Alpha globulin
___ 6. The hormone secreted by the kidneys that regulates RBC formation	F. Eosinophil
___ 7. The number of WBCs per cubic millimeter of blood in a healthy person	G. Erythropoietin
___ 8. The granular leukocyte that attacks large parasites	H. Iron
___ 9. An area of the body where T-cell lymphocytes mature	I. Type O
___ 10. The blood type with A and B antibodies in the plasma	J. 5 million

Answers to Self-Quiz

1. E; 2. J; 3. A; 4. H; 5. B; 6. G; 7. C; 8. F; 9. D. 10. I

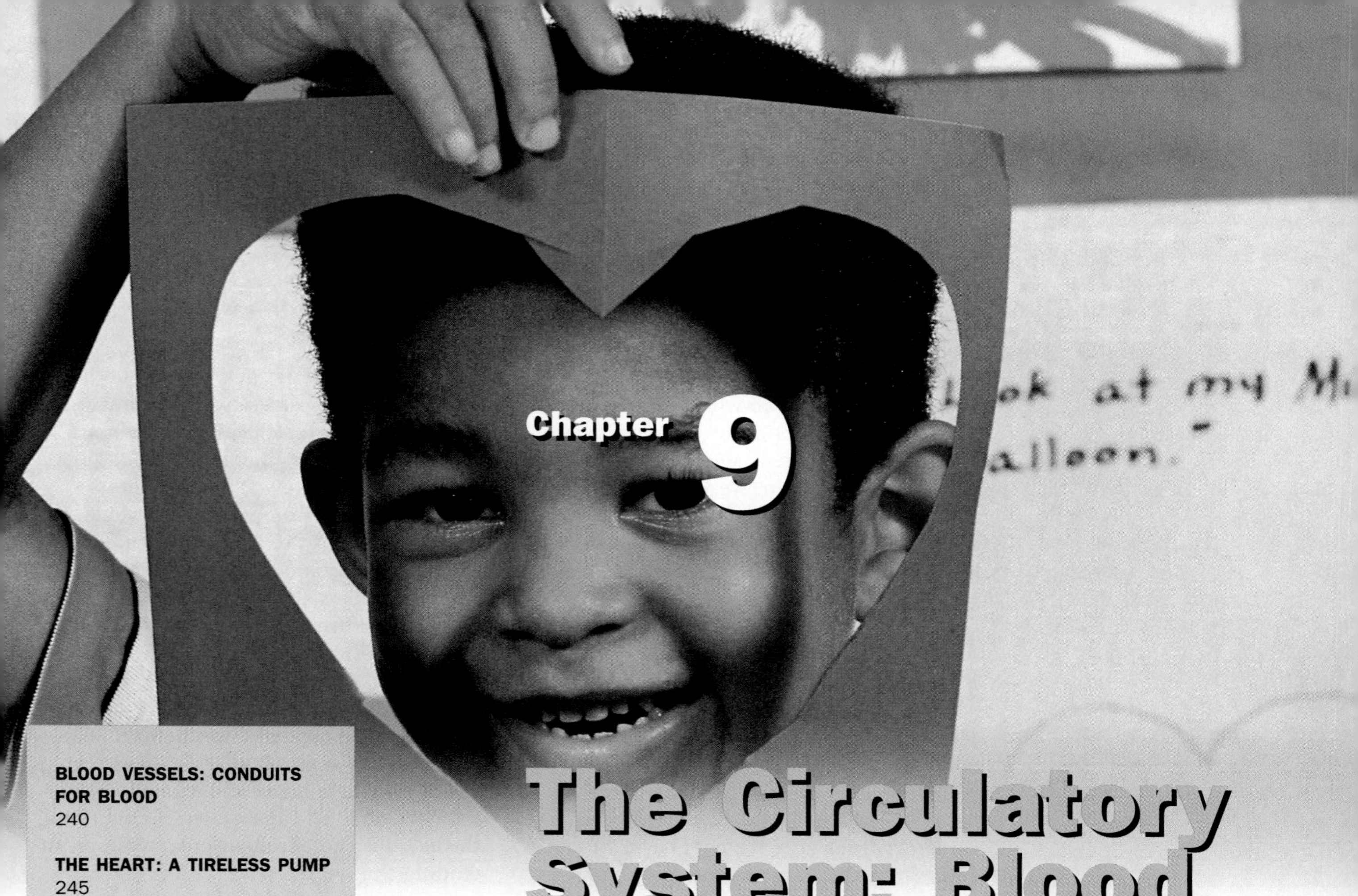

Chapter 9

The Circulatory System: Blood Vessels and the Heart

"Take heart" and have courage. Mourn a loss with a "broken heart." Deep emotion is "heartfelt." Life may end with a "heart attack." From ancient cultures to modern times, the heart has been linked to joy, love, courage, and pain. While these linkages are not expressions of scientific facts, they capture the truth. The heart is the center of your life. Your every move is accompanied by its beating, and it responds with changing rhythms to love, fear, and death.

The heart and blood vessels move about 300 liters (about 80 gallons) of blood throughout the body every day. This is a closed **circulatory system,** meaning that blood normally is confined within vessels and moves in a continuous or "circular" path from the heart, to body tissues, and then back to the heart.

The heart and blood vessels are often identified as the **cardiovascular system.** In addition, there is an associated transport system called the **lymphatic system.** In contrast to the circulatory system, the lymphatic system moves fluid in one direction. It transports excess tissue fluid back to the circulatory system.

In this chapter, we will describe the structure and function of the cardiovascular system and briefly describe the lymphatic system. Then, we will examine some of their common diseases and disorders and the changes that occur over the life span.

Blood Vessels: Conduits for Blood

The 60,000 miles of blood vessels in your body are the conduits for distributing your blood. There are three types of blood vessels: arteries, capillaries, and veins. **Arteries** (***are**-ter-ees*) carry blood away from the heart to the body's tissues (Figure 9.1). En route to the tissues, blood passes into smaller and smaller arteries and eventually enters the smallest arteries, which are called *arterioles* (*are-**teer**-ee-oles*). Arterioles are about the diameter of the period at the end of this sentence, and they connect with even smaller, microscopic vessels called capillaries. **Capillaries** (***cap**-ill-air-ees*) are narrow, thin-walled vessels that branch extensively. This branching ensures that a capillary vessel is near every cell in the body. The structure of capillary walls permits an exchange of nutrients, waste products, oxygen (O_2), and carbon dioxide (CO_2) between the blood and the neighboring cells of the body. Blood leaving a capillary enters tiny veins called *venules* (***ven**-yules*). Venules merge into larger-diameter **veins,** which carry blood back to the heart.

Our blood vessels have three principal functions. They (1) form a continuous set of tubes that distribute blood throughout the body, (2) permit an exchange of materials between the blood and our cells, and (3) help regulate the flow of blood to different organs and tissues. In this section, we will describe the structure of blood vessels and relate those structures to the above three functions.

Arteries: From the Heart to the Tissues

Arteries carry blood away from the heart. The diameter of arteries varies, but it is largest for arteries closest to the heart. Our largest artery, the aorta, has a diameter of approximately 2.5 centimeters (about 1 inch). Compared with veins, the walls of arteries are thick to withstand the increased blood pressure produced when the heart contracts. An artery wall consists of three layers which surround a hollow interior, the lumen, through which blood flows (Figure 9.1).

The **inner layer** is primarily a layer of epithelial cells called endothelium. The *endothelium* (*end-oh-**thee**-lee-um*) consists of a single layer of flattened epithelial cells which forms a continuous lining throughout the cardiovascular system, including the heart.

The **middle layer** of the artery wall consists of smooth muscle and elastic connective tissue. The smooth muscle permits regulation of blood flow. When the smooth muscle contracts, the diameter of the lumen becomes smaller and blood flow decreases, a process called *vasoconstriction.* Relaxation of the muscle has the opposite effect and is called *vasodilation.* The elastic connective tissue permits larger arteries to stretch when blood is pumped with each heartbeat.

Large arteries, such as the aorta, have more elastic connective tissue and less smooth muscle. This arrangement allows them to expand and receive the surge of blood pumped with each heartbeat. Smaller arteries (the diameter of a pencil) farther from the heart contain more smooth muscle and less elastic connective tissue. The smooth muscle permits these vessels to regulate the passage of blood to body tissues.

Vasoconstriction and vasodilation are controlled by the autonomic nervous system (ANS) in response to both external and internal conditions. Vasodilation is partially responsible for the increased blood flow that produces erection of the penis and clitoris during sexual arousal. When we blush in embarrassment, vasodilation has channeled more blood to the skin; when we turn pale with fright, vasoconstriction has restricted the flow of blood to the skin.

The **outer layer** of arteries is composed entirely of connective tissue. This layer acts as a supportive wrapping to protect arteries from damage and stabilize them by means of attachments to the surrounding tissues. A weakening of the middle layer or outer layer can cause a balloon-like bulge in an artery called an *aneurysm*. When this happens, the wall of the artery becomes thin and may rupture.

Arteries carry blood away from the heart. An artery's inner layer consists of endothelium. Its middle layer is composed of smooth muscle and elastic connective tissue, and its outer layer consists of supportive connective tissue.

Capillaries: Vessels for Exchange and Regulation

Arterioles branch into capillaries, which are so narrow that blood cells must pass through them in single file (Figure 9.2). They form a network of vessels called the **capillary network** in the body's tissues. The walls of capillaries are formed by a single layer of endothelial cells, and it is across these walls that exchanges take place between the blood and the tissues.

FIGURE 9.1 The types of blood vessels, their wall structure, and their relationship to the circulation of blood in the body.

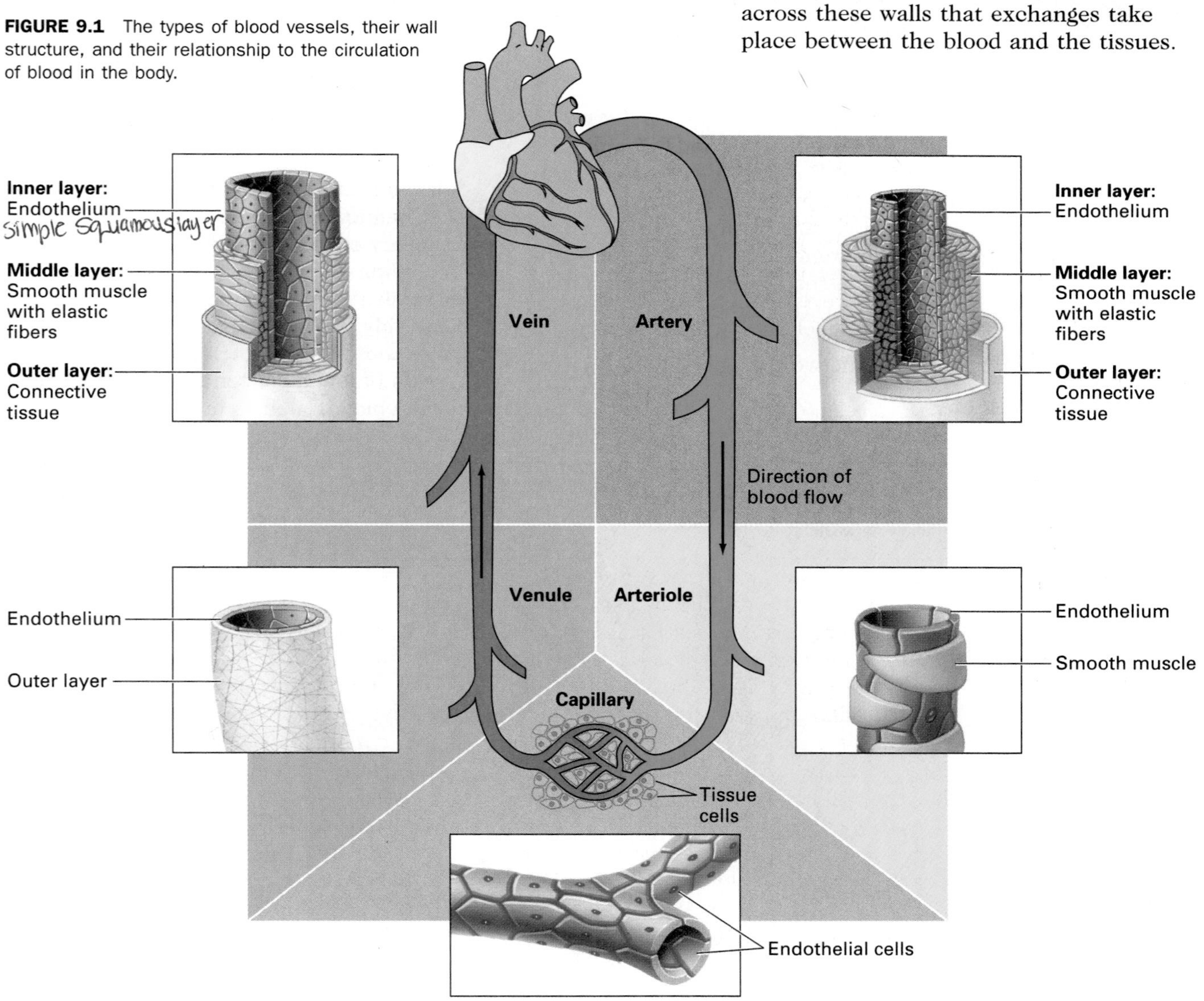
Inner layer:
Endothelium
simple squamous layer
Middle layer:
Smooth muscle with elastic fibers
Outer layer:
Connective tissue
Vein
Artery
Inner layer:
Endothelium
Middle layer:
Smooth muscle with elastic fibers
Outer layer:
Connective tissue
Direction of blood flow
Venule
Arteriole
Endothelium
Outer layer
Endothelium
Smooth muscle
Capillary
Tissue cells
Endothelial cells

I Love you a lot!!

FIGURE 9.2 A highly magnified view of a capillary.

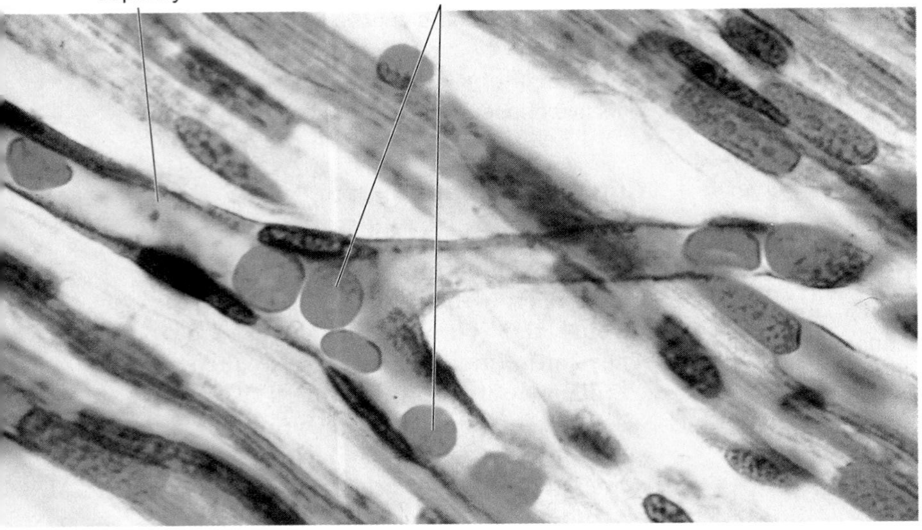
Capillary
Red blood cells

Blood flow through the capillary network is regulated by bands of smooth muscle called **capillary sphincters** (*sfink-ters*) located at the beginning of many capillary vessels. If all the sphincters contract simultaneously, blood will flow directly from the arteriole to the venule (Figure 9.3a). If all the capillary sphincters relax, blood will flow through the capillary network (Figure 9.3b). Contraction and relaxation of capillary sphincters are regulated by local conditions in the area of a particular capillary such as temperature, pH, and the concentrations of O_2 and CO_2. This arrangement provides for the monitoring and regulation of blood flow to meet the metabolic and temperature requirements of local tissues.

Exchanges between capillaries and body tissues normally occur only with substances that can move through the single cell layer of the endothelium. Chemicals that move easily across the capillary wall include nutrients, electrolytes, O_2 and CO_2, hormones, and waste products (Figure 9.4). Molecules such as plasma proteins and red blood cells are too large to diffuse across the endothelial cells of the capillary wall, and they normally remain within the blood vessels. Of course, as we mentioned in Chapter 8, amoeboid movement allows white blood cells to squeeze between the endothelial cells and move out of the capillaries into the tissue spaces.

Chemical substances move across the capillary endothelium by several processes, including diffusion, blood pressure, and osmosis. With **diffusion**, substances move passively across the plasma membrane from a region of higher concentration to a region of lower concentration. Chemicals in the blood, such as nutrients and O_2, diffuse from a capillary into the tissues because they are more concentrated in the blood than they are in the tissues. Similarly, metabolic waste products such as urea and CO_2 diffuse from the tissues, where they are produced, and into the capillaries. However, diffusion is not the whole story. Blood pressure also plays a role in moving water across the capillary wall.

Blood pressure is the force by which

FIGURE 9.3 Regulation of blood flow through capillaries by capillary sphincters.

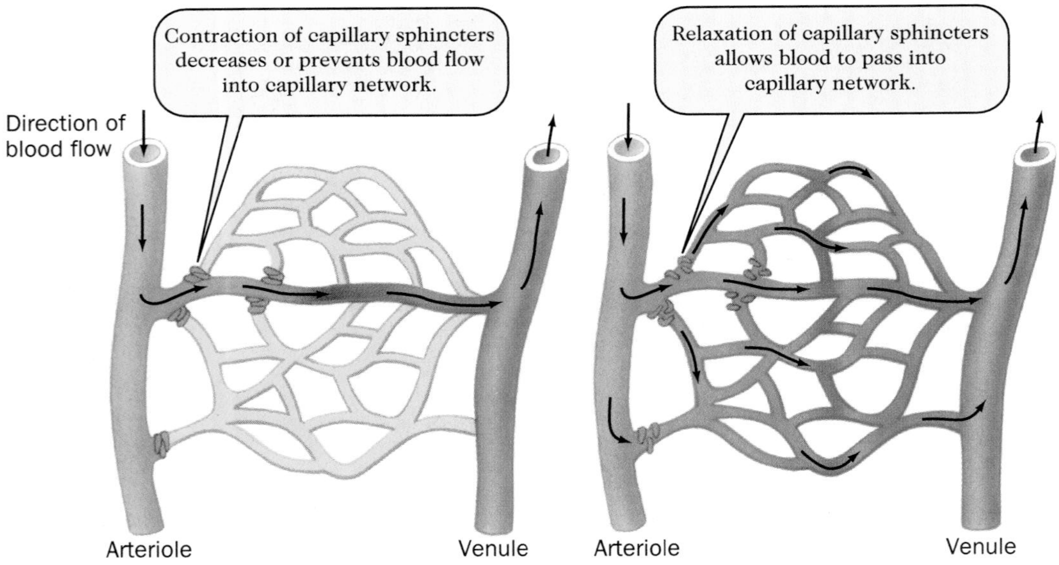
Contraction of capillary sphincters decreases or prevents blood flow into capillary network.
Relaxation of capillary sphincters allows blood to pass into capillary network.
Direction of blood flow
Arteriole
Venule
Arteriole
Venule

(a) Sphincters contracted

(b) Sphincters relaxed

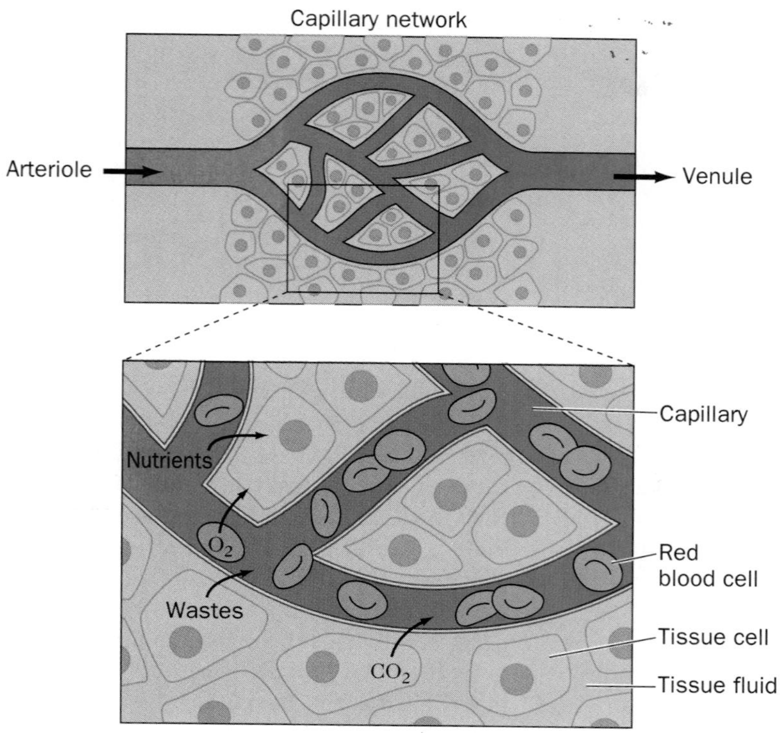
Capillary network
Arteriole
Venule
Capillary
Nutrients
O_2
Red blood cell
Wastes
Tissue cell
CO_2
Tissue fluid

FIGURE 9.4 A capillary network illustrating the exchange of nutrients, waste products, oxygen, and carbon dioxide.

blood is pushed through blood vessels as a result of heart contractions. As blood travels through arteries and arterioles, blood pressure becomes progressively reduced. Consequently, the blood pressure in our capillaries is very low (Figure 9.5). Even so, the pressure within capillaries is greater than the pressure exerted by tissue fluid. As a result, fluid is pushed from the blood plasma through the thin layer of endothelium and into the tissue spaces.

The fluid filtered from the blood by pressure is mostly water and dissolved solutes; blood cells and plasma proteins remain in the capillaries. With the loss of this fluid from the capillaries, the plasma proteins become more concentrated in the blood. These conditions encourage the movement of some water back into the capillaries by the process of **osmosis**. As you may recall from Chapter 2, the diffusion of water across a membrane is called osmosis.

During exchanges between capillary blood and the surrounding tissue, slightly more water leaves the capillaries than returns. What happens to this tissue fluid? It does not accumulate in the tissues but is removed by microscopic vessels called *lymphatic capillaries* which are distributed throughout our tissues. Once in the lymphatic capillaries, the tissue fluid is called **lymph.** Lymphatic capillaries transport lymph to larger lymphatic vessels, which ultimately merge and eventually

FIGURE 9.5 A graph illustrating the reduction in blood pressure as blood travels through arteries, capillaries, and veins.

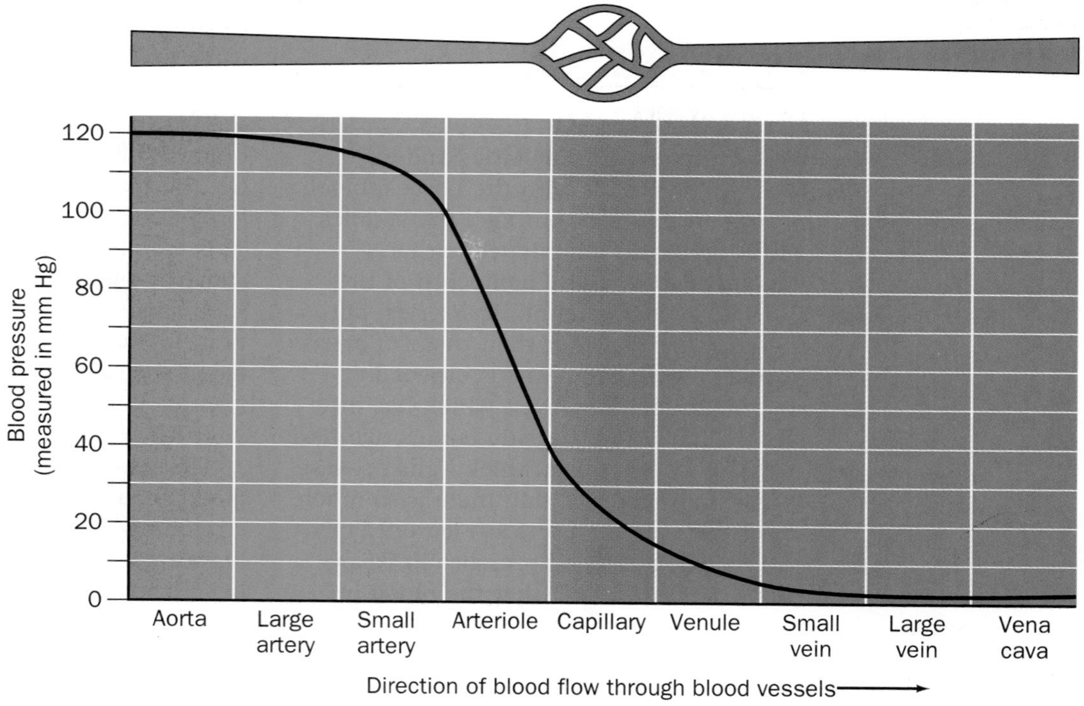
Blood pressure (measured in mm Hg)
120
100
80
60
40
20
0
Aorta
Large artery
Small artery
Arteriole
Capillary
Venule
Small vein
Large vein
Vena cava
Direction of blood flow through blood vessels

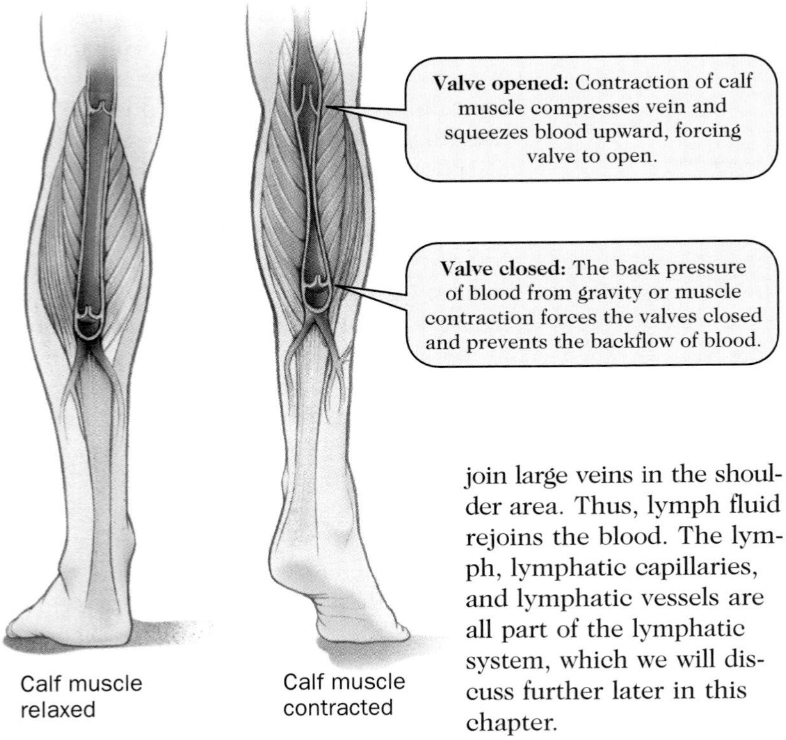

Valve opened: Contraction of calf muscle compresses vein and squeezes blood upward, forcing valve to open.
Valve closed: The back pressure of blood from gravity or muscle contraction forces the valves closed and prevents the backflow of blood.
Calf muscle relaxed
Calf muscle contracted

FIGURE 9.6 One-way valves in veins help move blood toward the heart.

join large veins in the shoulder area. Thus, lymph fluid rejoins the blood. The lymph, lymphatic capillaries, and lymphatic vessels are all part of the lymphatic system, which we will discuss further later in this chapter.

> Blood flow through a capillary network is controlled by capillary sphincters. Water and solutes move across the capillary wall by osmosis, diffusion, or blood pressure. Fluid remaining in the tissue spaces is removed by lymph vessels and rejoins the blood.

Veins: From Capillaries to the Heart

After reaching the capillaries and exchanging materials with the tissues, blood returns to the heart through the venules and veins, as we mentioned earlier. Venules and veins are composed of the three basic layers found in arteries, but the layers are thinner than those of arteries (see Figure 9.1). This is especially true for the middle layer, which has little smooth muscle and elastic connective tissue. Veins do not need thick walls because the blood pressure within them is so much lower than that in arteries (Figure 9.5).

The reduced blood pressure in veins, coupled with the fact that the blood in veins must often move against the force of gravity (as when one is standing), can interfere with the return of blood to the heart. Without assistance, blood tends to accumulate in the veins of the legs and lower torso. To aid the return of blood to the heart, the body uses a combination of three mechanisms: (1) one-way valves within veins, (2) contractions of skeletal muscles, and (3) movements associated with breathing. Since the adequate circulation of blood is so important, we'll examine these mechanisms more fully.

One-way valves and the role of skeletal muscle. The veins in our limbs have valves consisting of small flaps of tissue which extend into the lumen of the vein (Figure 9.6). Their orientation allows them to open, permitting blood to move toward the heart, and then close, preventing backward flow. In our limbs, veins pass between many different skeletal muscles. The contraction and relaxation of these muscles during normal body movements compress veins and squeeze blood toward the heart (Figure 9.6). As blood is squeezed along the veins, the one-way valves prevent backflow. Without the combined action of valves and skeletal muscles, blood slowly accumulates in veins, particularly those in the legs. This can lead to problems.

When blood accumulates in veins, their thin walls stretch, increasing their diameter and allowing even more blood to accumulate. Accumulation of blood in the veins has two consequences. First, there is less blood available for circulation, and the first organ noticeably affected is the brain. Without sufficient blood to supply O_2, the brain does not function well and loss of consciousness (fainting) can occur. This may happen if you are required to stand in one position for long periods, such as a soldier standing rigidly at attention. In such situations, you can prevent blood accumulation in your veins by performing periodic isometric contractions of your leg, buttock, and arm muscles.

The second major consequence of blood accumulation in veins is permanent stretching of veins. This is especially true for surface veins in the leg that do not benefit from the pumping actions of skeletal muscles. Surface veins that have been permanently stretched are called *varicose veins*, and they can often be seen bulging from under the skin of the lower leg. People with jobs that require long periods of standing, such as surgeons and bank tellers, have an increased risk of developing varicose veins.

The pumping action of breathing. The large veins in the walls of the thoracic and abdominal cavities do not have valves, but blood flowing through these veins must still move against gravity. This flow is aided by the actions of breathing that put pressure on the veins, squeezing blood toward the heart. Improper breathing while weight lifting interferes with this process and can lead to a light-headed feeling when insufficient blood is circulated to the brain.

> Veins carry blood from capillaries to the heart. The blood pressure in veins is low. Movement of blood through veins is aided by one-way valves, the pumping actions of skeletal muscles, and the pressure changes associated with breathing.

The Heart: A Tireless Pump

The heart is about the size of a fist and is located in the thoracic cavity, just behind the sternum and between the lungs. As a muscle, the heart is impressive. It usually contracts and relaxes about 75 times a minute to meet normal body needs. This amounts to 4500 beats an hour and more than 100,000 beats a day. This goes on without interruption for your entire life. Furthermore, the heart is capable of increasing its pumping activity to meet the greater demands of the body which occur during physical activity, stress, and danger.

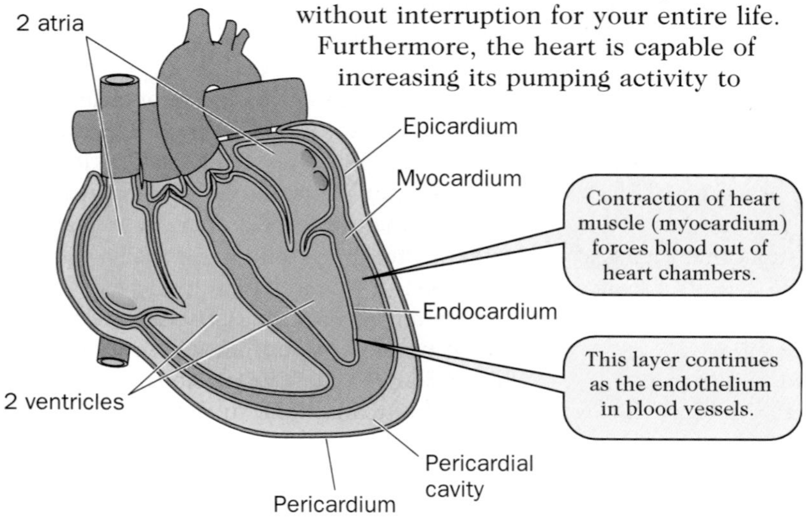
2 atria
Epicardium
Myocardium
Contraction of heart muscle (myocardium) forces blood out of heart chambers.
Endocardium
This layer continues as the endothelium in blood vessels.
2 ventricles
Pericardial cavity
Pericardium

FIGURE 9.7 The two pairs of heart chambers and the layers of the heart wall.

In this section, we will describe the structure of the heart, explain how blood flows through the heart, and discuss how the heart pumps blood through the entire body. Then we will describe how regulation of the heart's rhythm maintains homeostasis as the needs of the body change.

The Heart Wall: A Muscular Pump

The heart wall surrounds four chambers. It is this wall that contracts and relaxes to pump the blood. The heart wall is composed of three different layers: endocardium, myocardium, and epicardium (Figure 9.7). The **endocardium** (*end-oh-**car**-dee-um*) is a thin layer of endothelium that lines the interior of the chambers. The heart's endothelium is continuous with the endothelium of the blood vessels and has the same tissue structure.

The **myocardium** (*my-oh-**car**-dee-um*) is the thick middle layer and consists of cardiac muscle. Each cardiac muscle cell has many short branches that form specialized junctions called *intercalated disks* (*in-**ter**-cal-lat-ed*) with other cardiac muscle cells (Chapter 5). These junctions allow a large number of individual muscle cells to transmit electrical impulses rapidly and to contract as a single coordinated unit. When the myocardium contracts, the chambers of the heart become smaller and blood is ejected from them into attached blood vessels.

The third and outer layer of the heart is called the **epicardium** (*ep-ee-**car**-dee-um*). It is a thin moist layer of epithelium and connective tissue that is firmly attached to the myocardium. An extension of the epicardium forms a loose-fitting sac called the *pericardium* that encloses the entire heart. Between the heart and the pericardium is a space called the *pericardial cavity* (Figure 9.7). This cavity is coated with a lubricating fluid which helps reduce friction between the heart and the pericardium during contractions. An inflammation of the pericardium is called *pericarditis.* In such an inflammation, parts of the pericardium

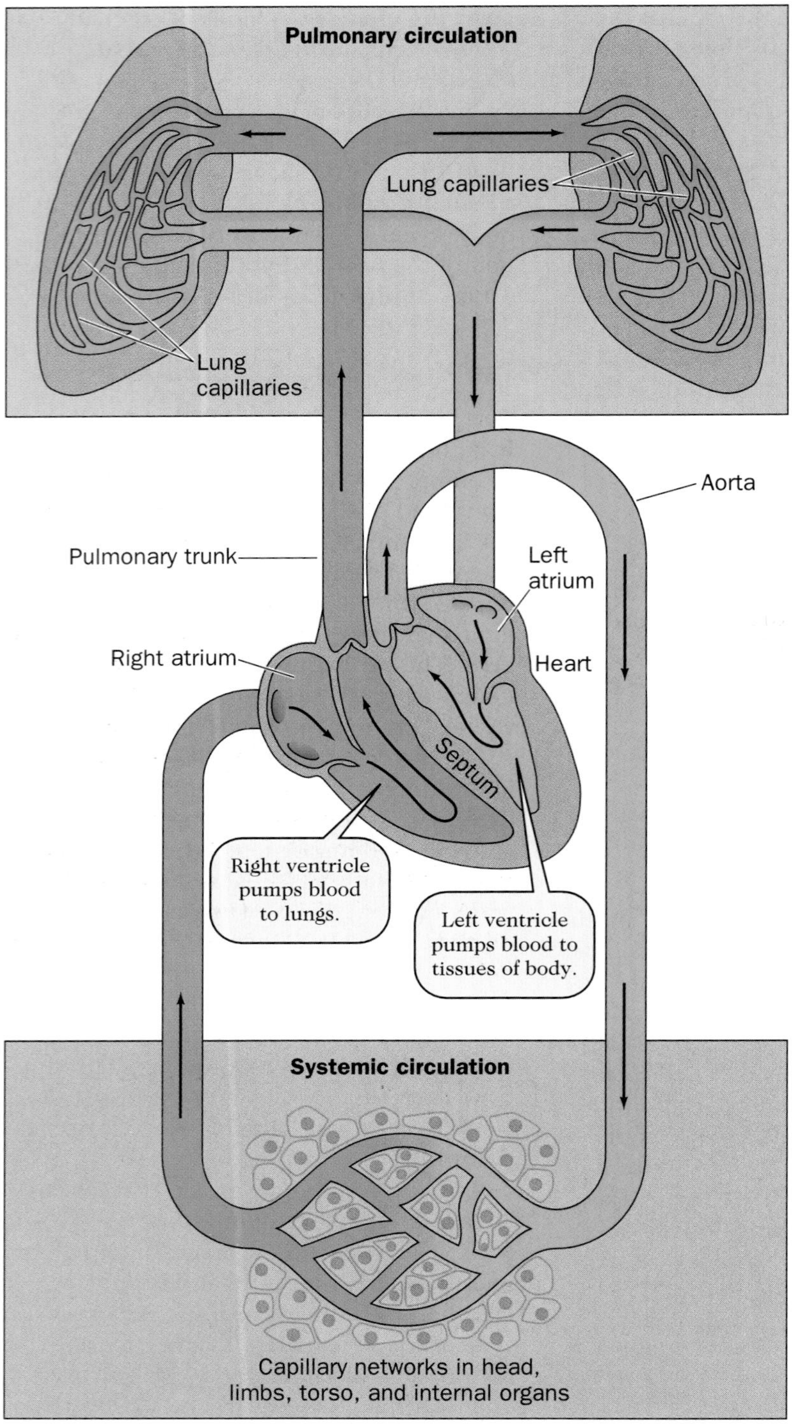
Pulmonary circulation
Lung capillaries
Lung
capillaries
Aorta
Pulmonary trunk
Left
atrium
Right atrium
Heart
Septum
Right ventricle
pumps blood
to lungs.
Left ventricle
pumps blood to
tissues of body.
Systemic circulation
Capillary networks in head,
limbs, torso, and internal organs

FIGURE 9.8 The dual pumping action of the heart, illustrating the pulmonary circulation and systemic circulation pathways.

may stick to the epicardium, causing pain and interfering with the normal operation of the heart.

> The heart wall has three layers. The inner endocardium is continuous with the endothelium of blood vessels. The thicker myocardium is the cardiac muscle. The outer epicardium attaches to the myocardium.

Chambers of the Heart and Blood Flow

The heart consists of four separate chambers: two atria (singular, *atrium*) and two ventricles. Although the heart is a single organ, it functions as two separate pumps. The right side of the heart pumps blood to the lungs, while the left side pumps blood to all the other tissues of the body. By inspecting Figure 9.8, you can identify the four chambers and the path of blood from each chamber to the next.

The **right atrium** receives blood returning from the body, and the **right ventricle** pumps it to the lungs via the pulmonary trunk (an artery). This blood has participated in exchanges with tissue cells, and so it has less O_2 and thus is known as *deoxygenated blood.* In the lungs, blood picks up its O_2 and releases its CO_2. It is now called *oxygenated blood.* The **left atrium** receives oxygenated blood from the lungs, and the **left ventricle** pumps it to all areas of the body via the aorta (an artery). A muscular partition, the **septum,** separates the right and left chambers.

As Figure 9.8 shows, the heart actually pumps blood into two separate circulatory pathways which both begin and end at the heart. The right atrium and right ventricle power the **pulmonary circulation,** which moves blood from the right ventricle to the lungs and back to the left atrium.

The left atrium and left ventricle power the **systemic circulation,** which moves blood from the left ventricle to the tissues of the body and back to the right atrium. The left ventricle has a thicker myocardium than does the right ventricle because it must generate more force to push blood a longer distance and through more blood vessels.

> The four heart chambers function as two separate pumps. The right atrium and ventricle pump blood into the pulmonary circulation, which carries deoxygenated blood to the lungs. The left atrium and ventricle pump oxygenated blood into the systemic circulation, which carries it to the body tissues.

Heart Valves: Ensuring One-Way Flow

Blood must move through the heart in one direction. To ensure that this occurs, the heart has two sets of one-way valves that prevent the backward flow of blood. One set of heart valves, the **atrioventricular valves (AV)** (*ate-tree-oh-ven-**trick**-you-lar*), is situated between the atria and the ventricles (Figure 9.9). The AV valve between the right atrium and right ventricle has three flaps and is called the *tricuspid valve,* while the valve between the left atrium and left ventricle has two flaps and is called the *bicuspid valve* or mitral valve.

AV valves are constructed of thin, tough connective tissue that projects into the ventricles. The lower portion of each valve is loosely anchored by connective tissue cords (chordae tendineae) to muscular projections of the ventricular wall. This arrangement secures the AV valves and prevents them from inverting into the atria during ventricular contraction.

The second set of heart valves, called the **semilunar valves,** is situated in the area where the ventricles connect with the arteries leaving the heart (Figure 9.9). These valves are made of connective tissue and consist of three cup-shaped units that prevent blood from dropping back into the ventricles as they relax. Each cup-shaped unit of a semilunar valve is similar in shape to a half-moon (*semi,* "half"; *luna,* "moon"). One semilunar valve, the *pulmonary semilunar valve,* is located between the right ventricle and the pulmonary trunk. The other semilunar valve, the *aortic semilunar valve,* is located between the left ventricle and the aorta.

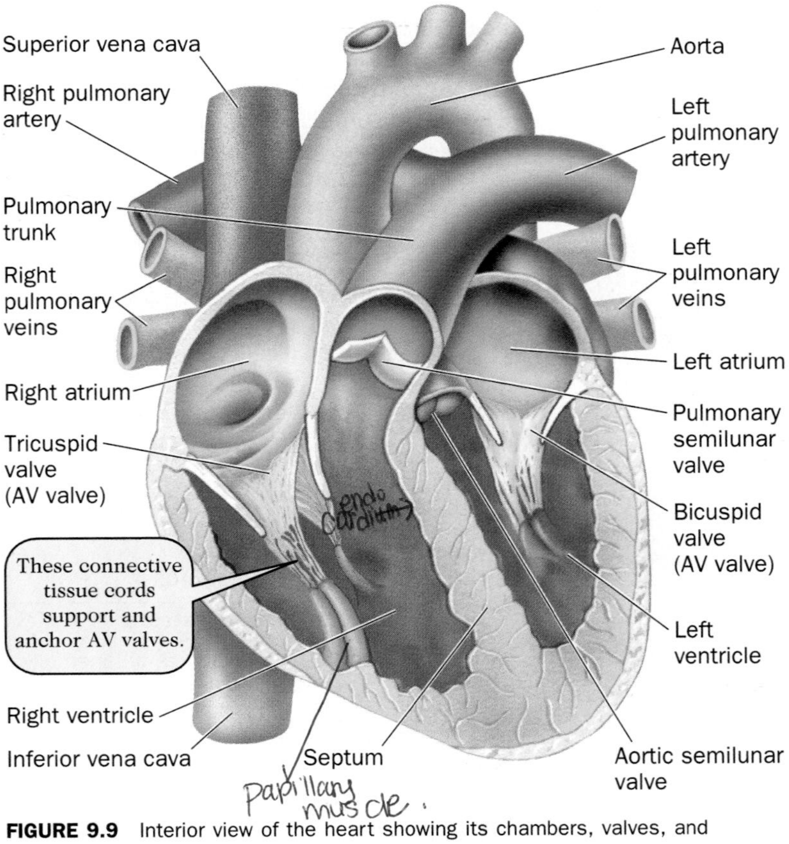
Superior vena cava
Right pulmonary artery
Pulmonary trunk
Right pulmonary veins
Right atrium
Tricuspid valve (AV valve)
These connective tissue cords support and anchor AV valves.
Right ventricle
Inferior vena cava
Septum
Aorta
Left pulmonary artery
Left pulmonary veins
Left atrium
Pulmonary semilunar valve
Bicuspid valve (AV valve)
Left ventricle
Aortic semilunar valve
endocardium
papillary muscle

FIGURE 9.9 Interior view of the heart showing its chambers, valves, and attached blood vessels.

> The heart has two sets of valves to prevent the backflow of blood. The atrioventricular (AV) valves are located between the atria and the ventricles. The semilunar valves are located where the ventricles connect with their attached blood vessels.

The Path of Blood: A Step-by-Step Summary

The right atrium receives deoxygenated blood from two large veins that drain the upper and lower body regions: the superior vena cava and inferior vena cava (see Figure 9.9). When the right atrium contracts, blood is forced through the tricuspid valve into the right ventricle. Contraction of the right ventricle forces blood against the tricuspid valve, which closes. Blood thus exits the right ventricle through the pulmonary semilunar valve to enter the pulmonary trunk and its branches, the pulmonary arteries. As the right ventricle relaxes, the pulmonary semilunar valve closes, preventing blood from reentering the right ventricle.

Oxygenated blood is carried from the lungs to the left atrium by the pulmonary veins. Contraction of the left atrium forces blood through the bicuspid valve and into the left ventricle. As the left ventricle contracts, blood is forced against the bicuspid valve, closing it. Blood leaves the left ventricle through the aortic semilunar valve, enters the aorta, and is distributed throughout the body. As the left ventricle relaxes, the aortic semilunar valve closes, preventing blood in the aorta from reentering the left ventricle.

> The path of blood through the pulmonary pathway follows this sequence: right atrium, right ventricle, pulmonary trunk and arteries, lungs, and left atrium. The path of blood through the systemic pathway follows this sequence: left atrium, left ventricle, aorta, body tissues, and right atrium.

The Cardiac Cycle: Coordinating Contractions

During a complete heartbeat, blood flows through the heart because of a sequence of coordinated contractions known as the **cardiac cycle.** In one cardiac cycle, both atria simultaneously relax and contract, followed by the relaxation and contraction of both ventricles. One cardiac cycle is quickly followed by another, as one heartbeat follows another.

Relaxation of the atria and ventricles is called **diastole** (*dye-as-toe-lee*). During diastole, the atria and ventricles fill with blood (Figure 9.10). Contraction of the atria and ventricles is called **systole** (***sis**-toe-lee*). It is during systole that the atria and ventricles pump blood. If the heart is beating at a rate of 75 times per minute, each cycle takes about 0.8 second to complete (60 seconds ÷ 75 beats per minute). Diastole and systole are generally of equal duration, each lasting about 0.4 second.

Heart sounds: closing valves. As the heart valves snap shut during each heartbeat, distinctive sounds can be detected. These sounds can be heard easily with a simple instrument called a stethoscope. The heart sounds are typically characterized as a series of *lub-dup* sounds. The

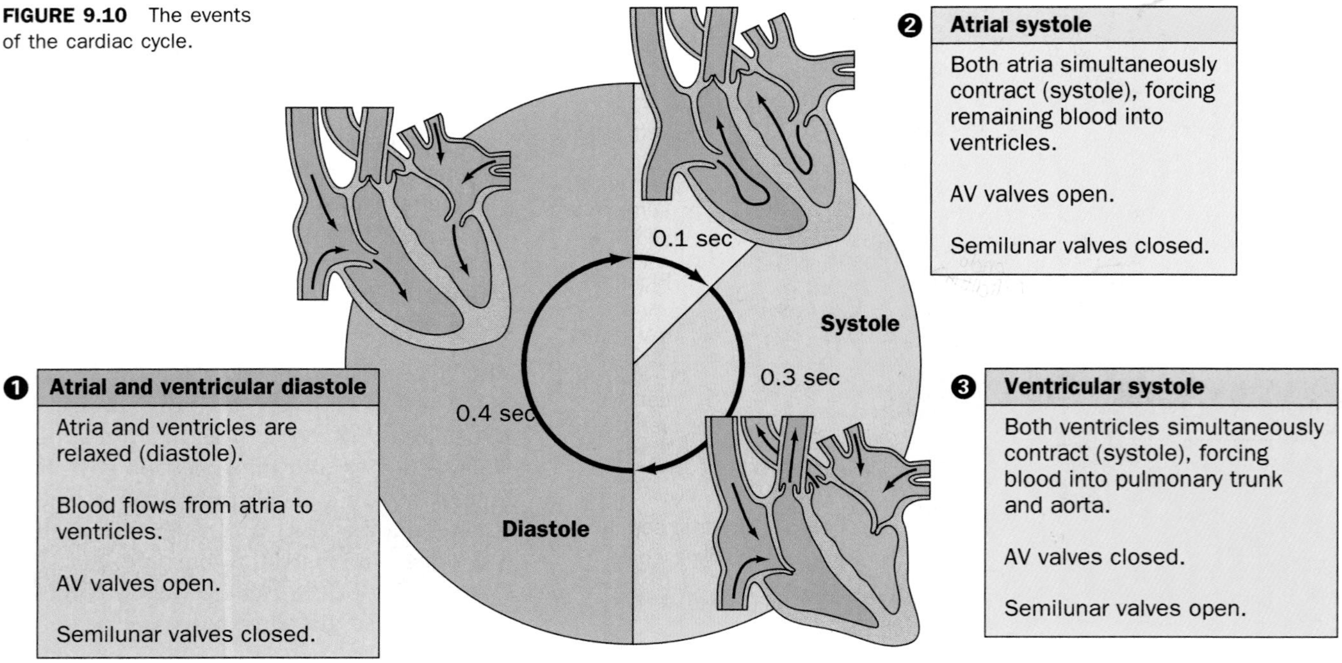

0.1 sec
0.3 sec
0.4 sec
Systole
Diastole
❶ Atrial and ventricular diastole
Atria and ventricles are relaxed (diastole).
Blood flows from atria to ventricles.
AV valves open.
Semilunar valves closed.
❷ Atrial systole
Both atria simultaneously contract (systole), forcing remaining blood into ventricles.
AV valves open.
Semilunar valves closed.
❸ Ventricular systole
Both ventricles simultaneously contract (systole), forcing blood into pulmonary trunk and aorta.
AV valves closed.
Semilunar valves open.

FIGURE 9.10 The events of the cardiac cycle.

lub represents the sound produced when the two AV valves close during ventricular contraction, and the *dup* represents the sound produced when the two semilunar valves close during ventricular relaxation. If any of these valves do not close tightly, some blood will be forced through them, producing a swishing noise called a murmur.

A *heart murmur* results from small deformities of a valve. Severe murmurs may substantially reduce the efficient flow of blood through the heart and hinder a person's ability to sustain normal activity levels. When necessary, heart surgery can replace a defective valve with an artificial one.

> The sequence of events during one heartbeat is called the cardiac cycle. Diastole is the relaxation of the atria and ventricles. Systole is the contraction of the atria and ventricles. Heart sounds are caused by the closing of heart valves.

Systemic Circulation: Blood Flow through the Body

Let us now take a more detailed look at the systemic circulation pathway that pumps oxygenated blood through blood vessels of the body. Arteries carry blood away from the heart, and the largest artery in the body is the *aorta.* Along its length, the aorta gives off branches that are collectively known as **systemic arteries.** It is through systemic arteries that blood is distributed to all regions of the body. Figure 9.11 shows the aorta and the major systemic arteries.

Blood flows from arteries to smaller arterioles and then to the capillary networks. As blood passes through capillaries, gases are exchanged, nutrients are delivered to the tissues, and waste products are removed. Deoxygenated blood leaves the capillaries, enters venules, and passes through **systemic veins** which return it to the heart. Figure 9.11 shows the major systemic veins.

Supplying blood to the heart muscle. The heart muscle's need for O_2 and nutrients is so great that it cannot be supplied merely by blood moving through its chambers. Therefore, the heart muscle has its own blood vessels that continuously supply and remove blood from the heart tissue. These vessels form the **coronary circulation.**

Two coronary arteries—the *right* and *left coronary arteries*—branch from the aorta as it leaves the heart (Figure 9.12). Branches of the right and left coronary arteries eventually form arterioles and capillary

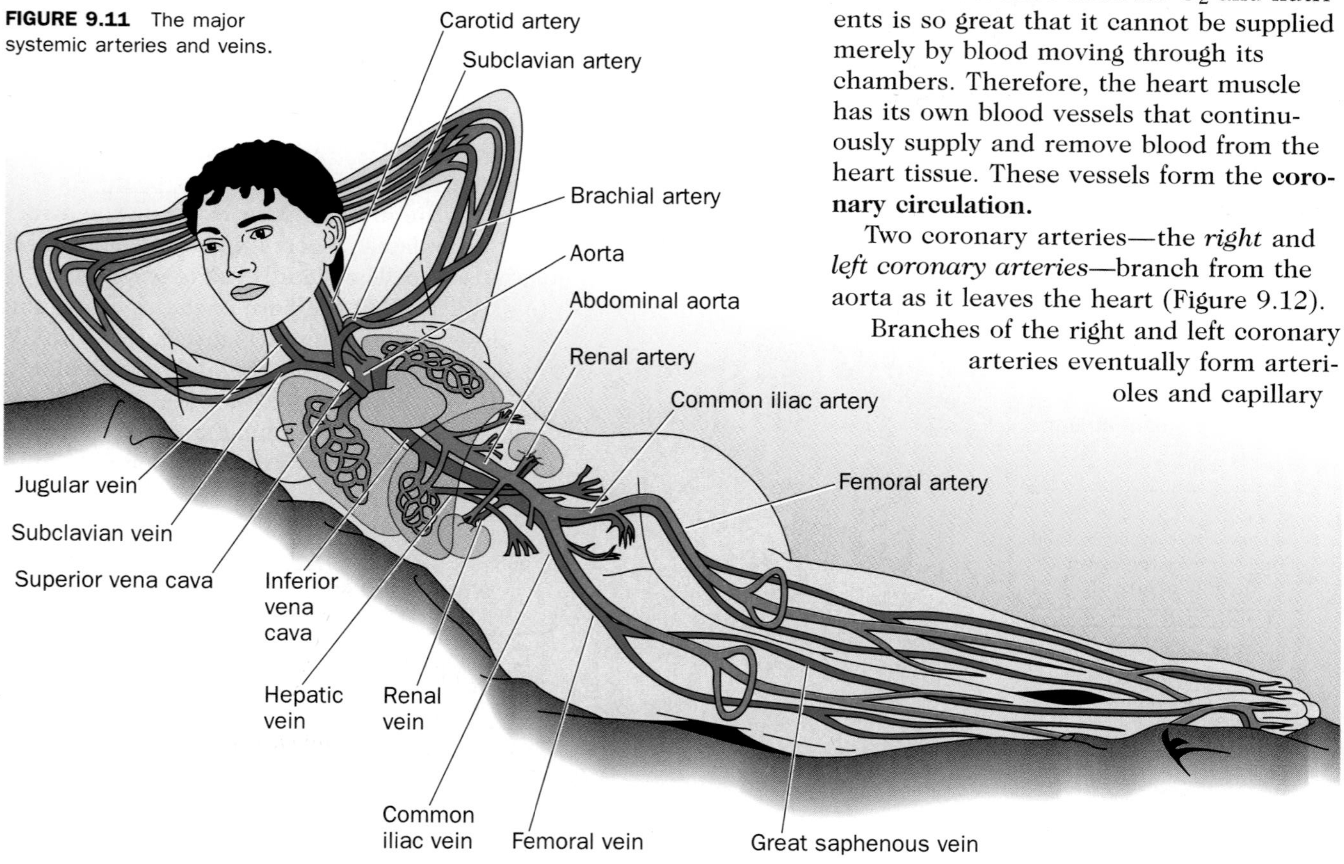
Carotid artery
Subclavian artery
Brachial artery
Aorta
Abdominal aorta
Renal artery
Common iliac artery
Femoral artery
Jugular vein
Subclavian vein
Superior vena cava
Inferior vena cava
Hepatic vein
Renal vein
Common iliac vein
Femoral vein
Great saphenous vein

FIGURE 9.11 The major systemic arteries and veins.

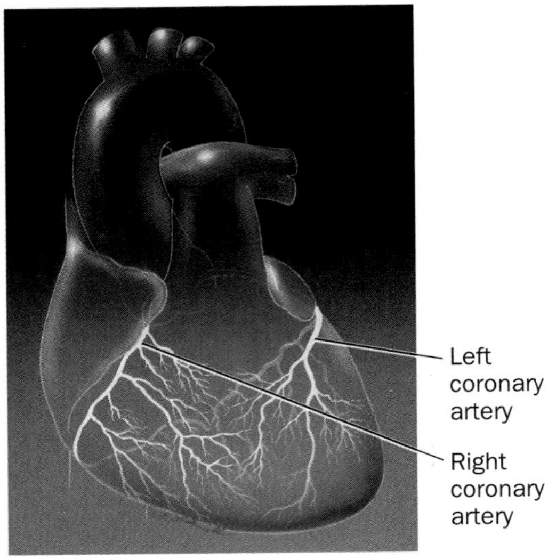
Left
coronary
artery
Right
coronary
artery

FIGURE 9.12 Illustration of a human heart showing the left and right coronary arteries and their branches.

networks, supplying the heart muscle with oxygenated blood and nutrients. *Coronary veins* collect the deoxygenated blood and return it to the right atrium, where it mixes with deoxygenated blood returning from the general systemic circulation. In "Diseases and Disorders" in this chapter, we will discuss what happens during a heart attack when the coronary arteries become blocked.

> Systemic arteries branch from the aorta and carry oxygenated blood to all body regions. Systemic veins carry deoxygenated blood back to the heart. Two coronary arteries supply the heart with oxygenated blood.

FIGURE 9.13 The cardiac conduction system. Arrows represent impulse direction.

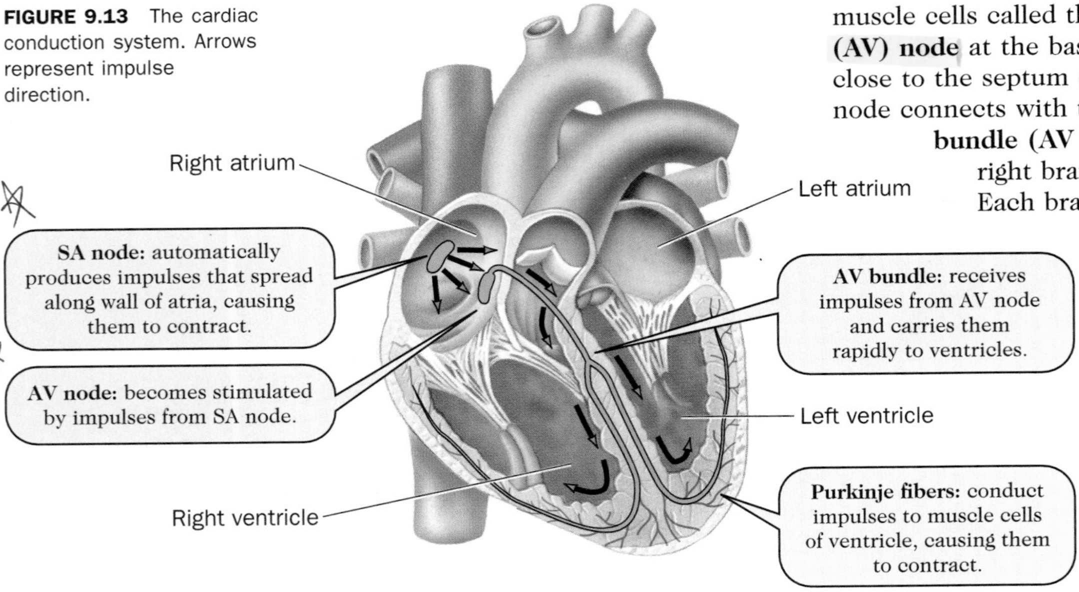
Right atrium
SA node: automatically produces impulses that spread along wall of atria, causing them to contract.
AV node: becomes stimulated by impulses from SA node.
Right ventricle
Left atrium
AV bundle: receives impulses from AV node and carries them rapidly to ventricles.
Left ventricle
Purkinje fibers: conduct impulses to muscle cells of ventricle, causing them to contract.

Regulating the Heart and Circulation

So far we have described the types of vessels that transport blood and have presented the heart as a double pump whose coordinated contractions force blood through vessels. Now we will examine how each heartbeat is produced and regulated. Then we will describe blood pressure and discuss how it is measured.

Initiating and Coordinating the Heartbeat

The heart contains a system of specialized muscle cells called the **cardiac conduction system** that initiates and conducts electrical impulses. These impulses stimulate the heart muscle to contract. The cardiac conduction system consists of four specialized structures: the sinoatrial node, the atrioventricular node, the atrioventricular bundle, and the Purkinje fibers (Figure 9.13).

The **sinoatrial (SA) node** is a small mass of specialized muscle tissue in the upper wall of the right atrium. The SA node automatically initiates an impulse, which travels across the atria, stimulating artrial muscle cells to contract. This node is often called the **pacemaker** because it sets the "pace" of each heartbeat.

As the impulse spreads along the atria, it stimulates a second mass of specialized muscle cells called the **atrioventricular (AV) node** at the base of the right atrium close to the septum (Figure 9.13). The AV node connects with the **atrioventricular bundle (AV bundle)**, which has a right branch and a left branch. Each branch conducts the impulse down to the tip (apex) of the heart and forms additional branches called **Purkinje fibers** (*per-**kin**-gee*). These fibers conduct impulses to the myocardium of the ventricles, stimulating their contraction.

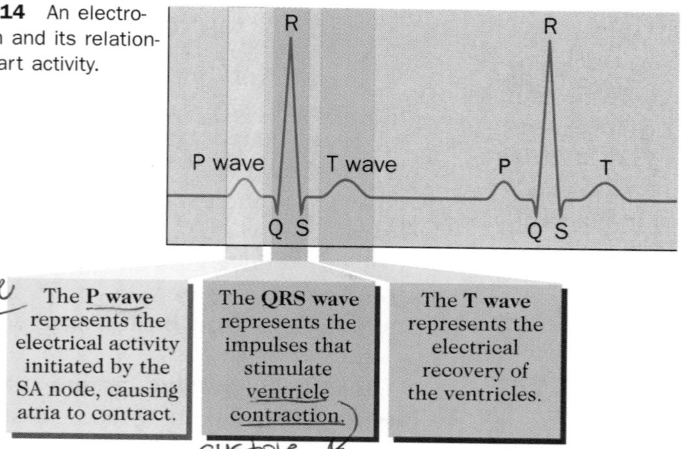
R
R
P wave
T wave
P
T
Q S
Q S
diastole
The P wave represents the electrical activity initiated by the SA node, causing atria to contract.
The QRS wave represents the impulses that stimulate ventricle contraction.
systole
The T wave represents the electrical recovery of the ventricles.

FIGURE 9.14 An electrocardiogram and its relationship to heart activity.

The passage of an electrical impulse through the cardiac conduction system takes about 0.2 second. This sequence of events not only stimulates the myocardium but also ensures coordinated contraction of the atria and ventricles in two ways. First, it ensures that the atria contract before the ventricles do. This occurs because each impulse is momentarily delayed at the AV node, giving both atria time to empty their contents into the ventricles before the ventricles contract. Second, coordination allows the lower portion of the ventricles to contract before the upper portion does. This ensures that blood is pumped in the direction of the pulmonary trunk and aorta.

Recording the heart's electrical impulses. A recording of the electrical activity in the cardiac conduction system is called an **electrocardiogram** (**ECG** or **EKG**). Such a recording is possible because the electrical activity of the heart can be detected by metal electrodes placed on the surface of the skin. The electrodes transmit the impulses to an amplifier so that the signals can be observed and a permanent ECG recording can be made (Figure 9.14).

By observing changes in the wave pattern or in the time intervals between them, trained medical personnel can use an ECG to diagnose the type and severity of damage to the heart muscle or its conduction system (Figure 9.15). As you can imagine, any irregularity in the initiation or conduction of the heart's electrical impulses can affect the operation of the heart. Irregularities in the initiation and conduction of impulses are called *arrhythmias* (*ay-***rith***-me-uz*).

> The cardiac conduction system consists of the SA node, AV node, AV bundle, and Purkinje fibers. This system initiates and coordinates the movement of electrical impulses of the heart.

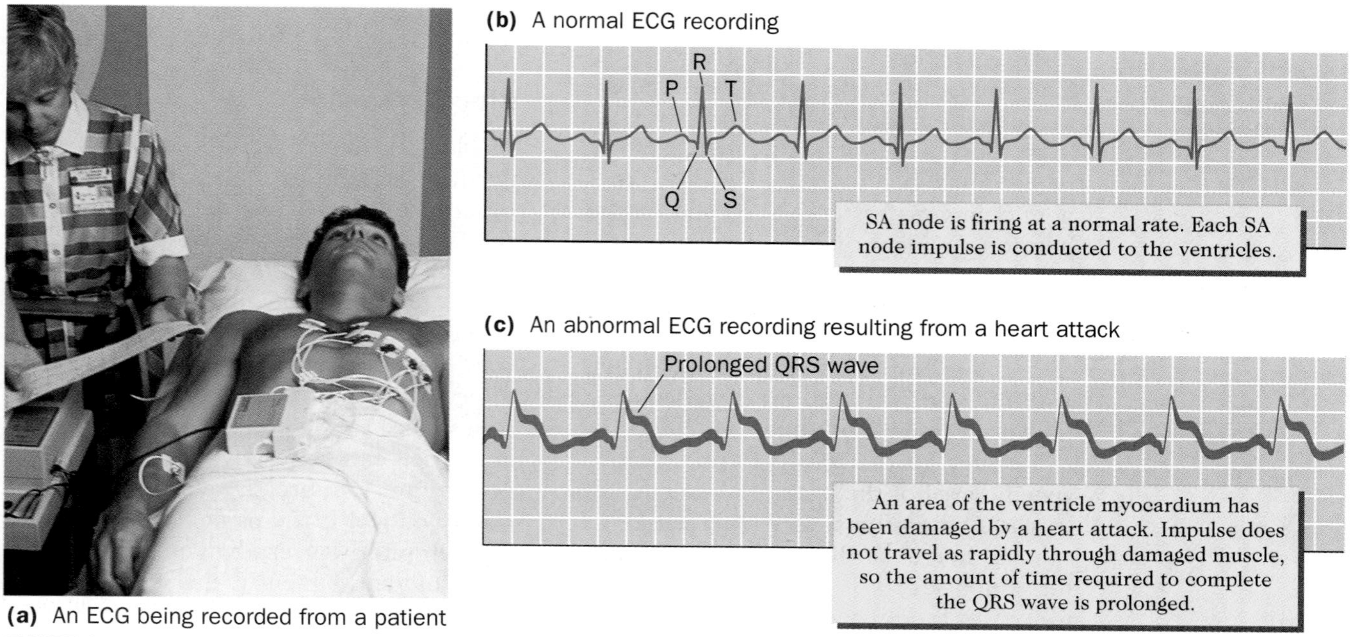
(b) A normal ECG recording
R
P
T
Q
S
SA node is firing at a normal rate. Each SA node impulse is conducted to the ventricles.
(c) An abnormal ECG recording resulting from a heart attack
Prolonged QRS wave
An area of the ventricle myocardium has been damaged by a heart attack. Impulse does not travel as rapidly through damaged muscle, so the amount of time required to complete the QRS wave is prolonged.

(a) An ECG being recorded from a patient

FIGURE 9.15 How an ECG is made and two ECG recordings of patients.

Regulating Heart Activity

The cardiac conduction system establishes a heart rate of about 75 beats per minute. This is sufficient to meet the body's needs while at rest, but it is not sufficient to meet the expanded needs of our tissues during periods of physical activity or stress. During these periods, the heart contracts faster, pumps blood faster, and moves blood more rapidly through the body. This helps maintain homeostasis by supplying more O_2 and nutrients and removing more waste products such as CO_2 and lactic acid.

Cardiac output is the volume of blood pumped by each ventricle in 1 minute. It is determined by two factors: heart rate and stroke volume. The *heart rate* is simply the number of heartbeats per minute. The *stroke volume* is the volume of blood ejected by a ventricle at each contraction. For an adult at rest, the heart rate is about 75 beats per minute and the stroke volume is about 70 milliliters (ml) per contraction. Using these values, the cardiac output would be 5250 ml per minute, or 5.25 liters (about 5 quarts) per minute:

$$\textit{Heart rate} \times \textit{stroke volume} = \textit{cardiac output}$$
$$75 \text{ beats/min} \times 70 \text{ ml/beat} = 5250 \text{ ml/min}$$

Usually the amount of blood ejected per minute by each ventricle equals the amount of blood returning to the heart.

The question now arises: How can cardiac output be increased to meet the demands of the body? One important way is by increasing the heart rate. This can be accomplished by nerves of the autonomic nervous system (ANS) and by the hormone epinephrine (also called adrenaline).

Increasing cardiac output: the role of nerves. Remember that the ANS controls the operation of our internal organs and functions without our conscious control or awareness (Chapter 3). The control center for the ANS is located in a specific area of the brain, the medulla oblongata. When this area is stimulated, specific nerves of the ANS send stimulatory impulses to the SA and AV nodes as well as to the myocardium of the ventricles. As a result, there is an increase in heart rate and the heart muscle contracts more forcefully.

The control center for the ANS also receives nerve signals from parts of the brain associated with the emotions. This is one reason why emotions such as anger, fear, and excitement cause elevated heart activity. For example, when the astronauts touched down on the moon, their heart rates were over 170 beats per minute even though they were confined to their seats.

Increasing cardiac output: the role of a hormone. The hormone *epinephrine* is released into the blood by the adrenal glands above each kidney. When epinephrine reaches the heart, it stimulates the SA node, causing an increase in the heart rate and the force of contraction. One may wonder why epinephrine is needed, since the heart is already stimulated by nerves from the ANS. The answer lies in the fact that epinephrine circulates throughout the body and prepares all areas of the body for emotional responses or physical activity.

> Cardiac output is the volume of blood pumped by each ventricle in 1 minute. It is calculated by multiplying the heart rate by the stroke volume. Heart rate is increased by ANS stimulation and by the hormone epinephrine.

Blood Pressure

As blood is ejected from the ventricles during systole, it exerts a force known as **blood pressure** against the wall of a vessel. When a volume of blood passes through an artery, the pressure exerted by the blood causes the artery to bulge. Because of the elastic quality of the artery wall, the bulge is not permanent, and the artery will recoil to its normal size. In arteries close to the body surface, the bulge and recoil can be felt by placing one's finger on the skin directly over an artery. What you feel is the pulse.

Each **pulse** corresponds to one heartbeat, and consequently, the pulse rate (pulses per minute) is equal to the heart rate. There are several sites on your body

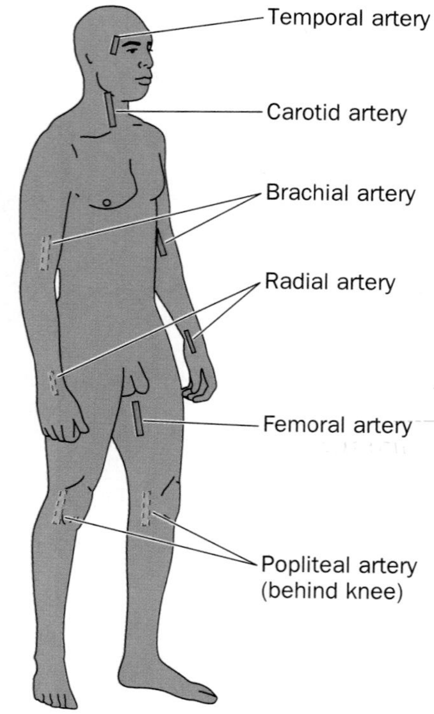
Temporal artery
Carotid artery
Brachial artery
Radial artery
Femoral artery
Popliteal artery (behind knee)

FIGURE 9.16 Several body locations where your pulse can be felt.

where arteries are close to the surface and your pulse can be detected (Figure 9.16). To a skilled health care professional, feeling the pulse can reveal not only the heart rate but the regularity and strength of the heart's contractions.

Maintenance of blood pressure is absolutely necessary to the movement of blood throughout the body. However, if blood pressure becomes too low or too high, it endangers good health. Therefore, it is useful to measure blood pressure. Blood pressure is actually a measure of two values: ventricular systole (contraction) and ventricular diastole (relaxation).

Measuring blood pressure. You can measure blood pressure with a device called a *sphygmomanometer* (*sfig-mow-ma-**nom**-eh-ter*). This device allows an arm cuff to be inflated over the brachial artery while observing the pressure values on a connected gauge (Figure 9.17). Inflating an arm cuff with air pressure compresses the brachial artery, preventing blood from passing through it. Blood pressure is determined in the following manner:

(1) A stethoscope is positioned over the closed artery so that sounds can be heard when blood again begins to flow. (2) Air is released slowly from the arm cuff, lowering the cuff pressure. (3) At some point the blood pressure will slightly exceed the cuff pressure, and blood will begin to flow through the brachial artery. When you hear the sound of blood moving, the reading on the pressure gauge corresponds to the **systolic pressure,** that is, the pressure developed during ventricular contraction. (4) As more air pressure in the cuff is released, there will be a reading on the gauge when sound is no longer heard. At that moment, the artery is no longer constricted and blood is flowing freely. This gauge reading corresponds to the **diastolic pressure,** that is,

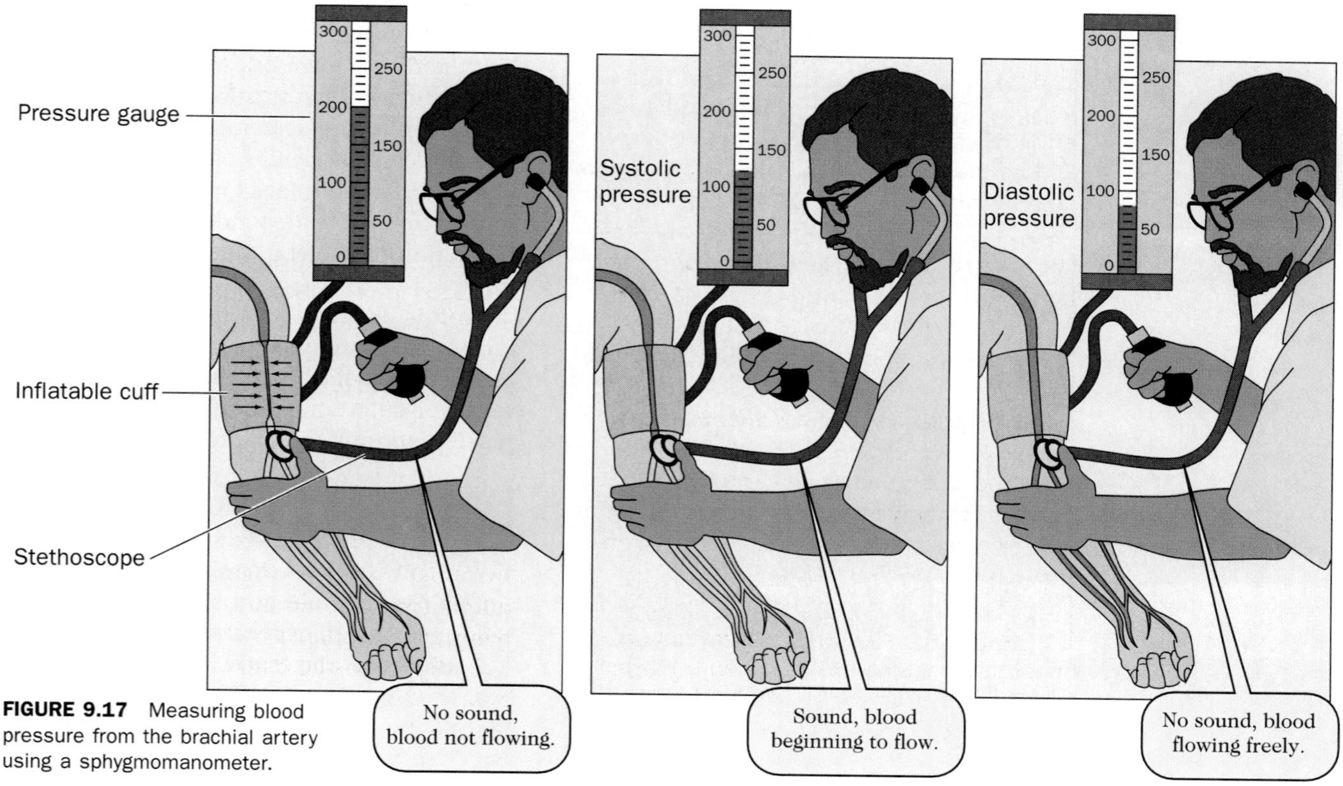
Pressure gauge
Inflatable cuff
Stethoscope
300
250
200
150
100
50
0
Systolic pressure
Diastolic pressure
No sound, blood not flowing.
Sound, blood beginning to flow.
No sound, blood flowing freely.

FIGURE 9.17 Measuring blood pressure from the brachial artery using a sphygmomanometer.

the pressure that exists during ventricular relaxation.

Blood pressure values are usually given in millimeters of mercury (mm Hg). This is the amount of pressure required to raise a column of liquid mercury up a narrow glass tube for a distance measured in millimeters. For a healthy adult, a normal blood pressure is 120 mm Hg for systolic pressure and 80 mm Hg for diastolic pressure. It is expressed as the systolic pressure over the diastolic pressure, or "120 over 80" (120/80).

Individual differences in blood pressure. Although we have identified a blood pressure of 120 over 80 as normal for healthy adults, there is wide individual variation. It has been difficult for medical science to say exactly what a minimum or maximum systolic and diastolic pressure should be for a normal, healthy adult at rest. Table 9.1 lists the different systolic and diastolic pressures used to define high blood pressure for individuals 18 years and older.

TABLE 9.1 Normal and Abnormal Blood Pressures for Individuals 18 Years of Age and Older

Blood Pressure, mm Hg	Blood Pressure Classification
Diastolic	
84 or less	Normal
85–89	High normal
90–104	Mild hypertension
105–114	Moderate hypertension
115 or higher	Severe hypertension
Systolic*	
139 or less	Normal
140–159	Borderline
160 or higher	Severe hypertension

*When diastolic is below 90.
Source: 1988 Report of the Joint National Committee on Detection, Evaluation, and Treatment of High Blood Pressure (NIH Publication No. 88-1088).

This issue is further complicated because personal circumstances can alter one's blood pressure. For example, anxiously awaiting a medical appointment can significantly increase blood pressure, and a second measurement at the close of the appointment may be lower and more accurate. A stress-filled workweek also may raise blood pressure so that it is higher on Friday afternoon than it is on Sunday evening after a relaxing weekend.

> Blood pressure is the pressure of blood against the wall of a blood vessel. It is measured with a sphygmomanometer. A normal blood pressure is a systolic of 120 mm Hg and a diastolic of 80 mm Hg.

The Lymphatic System

Like many systems of the body, the **lymphatic system** (*lim-fat-ick*) has more than one function. It is involved in defending the body, transporting fats absorbed from the intestines, and maintaining the fluid balance of the tissues. The lymphatic system consists of lymph fluid, vessels, and specialized organs. The vessels are distributed throughout the tissues of the body and provide a one-way transportation network for returning excess tissue fluid back to the systemic circulation. Unlike the cardiovascular circulation, the lymphatic system lacks a specialized pump. Instead, it depends on the pumping action of skeletal muscles to move fluid.

In this chapter, we will discuss the lymphatic system as a pathway by which excess tissue fluid is returned to the systemic circulation. We will discuss its defense functions in Chapter 13.

Lymphatic Vessels: Collecting and Moving Lymph

Lymphatic vessels originate as delicate microscopic closed-end vessels known as **lymphatic capillaries** which form a network between the cells of tissues. Lymphatic capillaries are similar to blood capillaries. However, the cell edges of the

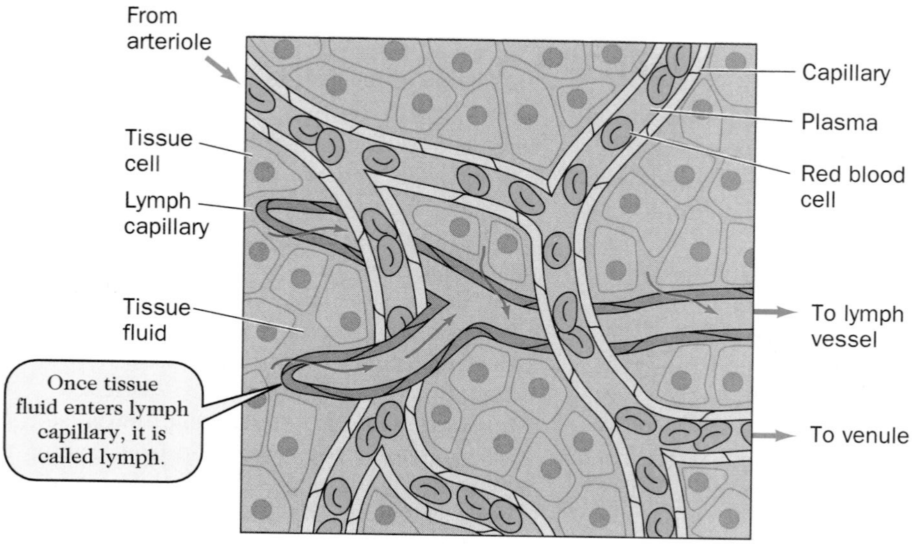
From arteriole
Tissue cell
Lymph capillary
Tissue fluid
Once tissue fluid enters lymph capillary, it is called lymph.
Capillary
Plasma
Red blood cell
To lymph vessel
To venule

FIGURE 9.18 Movement of tissue fluid into a lymph capillary.

lymphatic endothelium overlap, permitting fluid to enter the vessel but not exit (Figure 9.18). Consequently, once it is in those capillaries, the fluid (now called lymph) does not easily move back into the tissue spaces. This arrangement allows for the efficient collection and transportation of fluid away from the tissue. This is a necessary and important activity, for recall that some of the fluid filtered from the capillary at its arterial end is not reabsorbed at the venule end.

On a daily basis, 2 to 3 liters (about 2 to 3 quarts) of fluid leaves our capillaries and accumulates in our tissues. This fluid remains in the tissue spaces unless it is removed by the lymphatic capillaries. Even a minor interference in the removal of this tissue fluid will be noticed as tissue swelling, a condition called *edema* (*eh-**dee**-mah*).

Lymphatic capillaries join together to form larger **lymphatic vessels,** which are similar to veins in structure (Figure 9.19). Their walls are constructed of the same three thin layers, and there are one-way valves along their length to prevent the backflow of lymph. As with veins, the

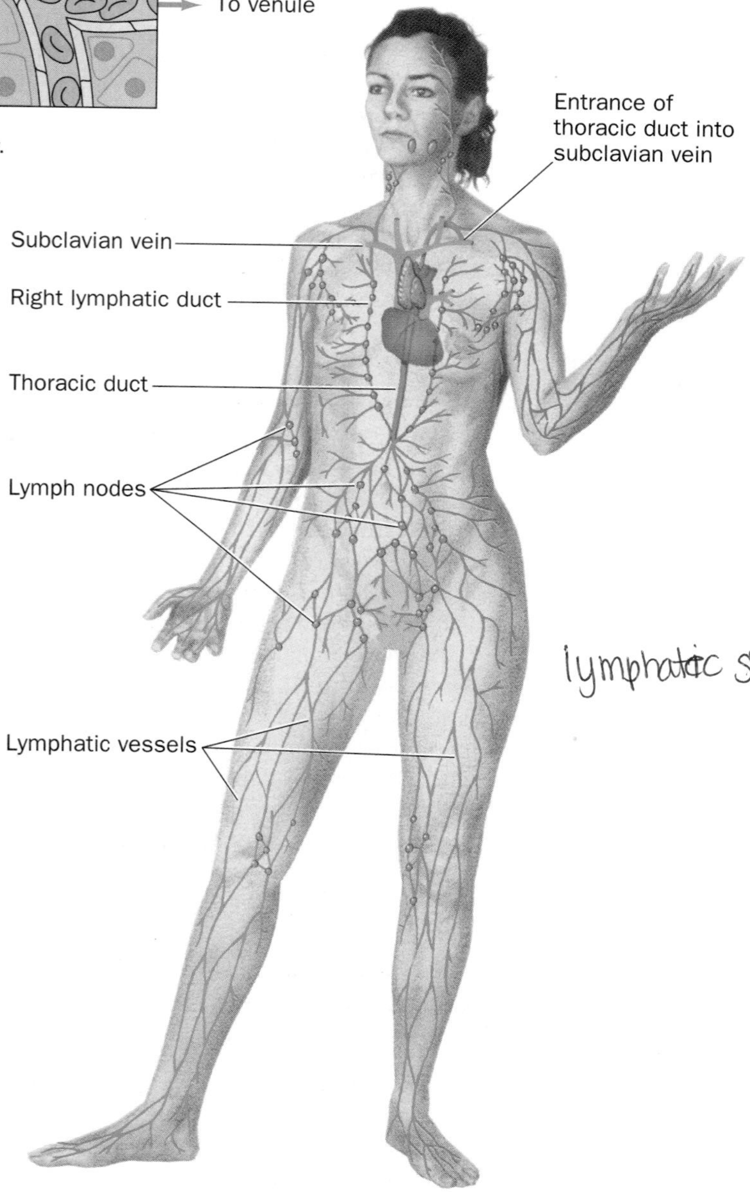
Entrance of thoracic duct into subclavian vein
Subclavian vein
Right lymphatic duct
Thoracic duct
Lymph nodes
Lymphatic vessels

lymphatic system

FIGURE 9.19 The lymphatic vessels.

squeezing action of skeletal muscle contractions forces lymph through the vessel. Along the lymphatic vessels are many small structures called **lymph nodes** that contain macrophages and lymphocytes. These cells destroy bacteria, viruses, and wandering cancer cells as the lymph flows through the nodes.

Lymphatic vessels continue to converge, forming larger vessels, until eventually two major lymphatic ducts are formed: the *thoracic duct* and the *right lymphatic duct.* These ducts empty lymph into the veins (subclavian veins) in the shoulder area, and the lymph is returned to the general circulation (Figure 9.19).

> Lymphatic capillaries are microscopic closed-end vessels that remove excess fluid from the tissue spaces for transport back to the general circulation. Lymph is moved by the actions of skeletal muscles.

Diseases and Disorders

As we have seen, the circulatory system has many different components, all of which must work together to maintain both the system and the body's homeostasis. A defect in a single component will affect the entire system sooner or later.

Some circulatory diseases do not show symptoms until they are well advanced and damage has been done to several organs. For example, high blood pressure and cholesterol deposition in arteries can occur over decades. By contrast, circulatory shock can happen suddenly at any time. Without proper recognition and treatment, it can be immediately life threatening.

Inadequate Blood Flow: Circulatory Shock

Circulatory shock threatens a person's life by reducing blood flow to the tissues. External or internal trauma that causes bleeding and reduces the volume of circulating blood leads to *hypovolemic shock.* With the loss of 15 to 20 percent of the blood volume (about 1 liter), blood pressure drops. Only a weak pulse can be detected, and many functions of the blood are reduced as the patient loses both blood cells and the plasma containing water and electrolytes.

An increase in blood vessel diameter (vasodilation) that permits blood to accumulate in veins also can cause shock. As blood pools, less returns to the heart, reducing cardiac output and blood flow to the body. For example, *anaphylactic shock* is widespread vasodilation caused by toxins (poisons) released from bacteria or from allergenic reactions to bee and wasp stings, penicillin, or other injected medications.

High Blood Pressure: Too Much of a Good Thing

Blood pressure increases during physical exertion, high temperatures, and emotional stress. Usually blood pressure returns to normal when you are at rest and relaxed. However, if your blood pressure remains high, you are said to have **hypertension,** or **high blood pressure.** A *normal blood pressure* range for an adult would be a systolic pressure between 120 and 140 and a diastolic pressure below 85. Table 9.1 compares a normal pressure with mild, moderate, and severe hypertension.

Over 90 percent of high blood pressure cases do not have a single identifiable cause. However, both genetic and environmental factors contribute (Table 9.2). High blood pressure often goes unnoticed. Either there are no symptoms or the symptoms are temporary and are ignored. Prolonged high blood pressure directly damages or increases the possibility of damaging the heart, blood vessels, kidneys, and brain.

High blood pressure can be lowered by reducing sodium intake, reducing weight, stopping smoking, increasing exercise, and improving stress management. If these measures do not reduce blood pressure adequately, prescription drugs are available.

TABLE 9.2 Risk Factors for Hypertension

Risk Factor	Explanation
Heredity	There can be a family history of hypertension.
Age	Hypertension increases between ages 30 and 50.
Race	African-Americans have twice the incidence found in white people.
Obesity	Blood must be pumped through more adipose tissue.
Sodium (Na^+)	Sodium increases water retention, which increases blood pressure.
Smoking	Nicotine from cigarette smoke constricts small arteries, which increases blood pressure.
Prolonged emotional stress	Stress stimulates the ANS, which increases cardiac output and blood pressure.
Inadequate exercise	The role of exercise is not well understood. It may involve a number of factors, including changes in the ANS, reduction in blood lipids, and weight loss.

Cholesterol, Plaques, and Atherosclerosis

There is much in the news about the relationship between high blood cholesterol and heart disease. Cholesterol is both obtained from the diet and manufactured by our cells. It is transported throughout the body by the blood plasma. When cholesterol is measured in the plasma, values above 200 milligrams per 100 milliliters are associated with an increased risk of developing deposits of cholesterol on the inner surface of arteries (Figure 9.20). Such deposits are called *plaques* and, over time, as more cholesterol is deposited, they enlarge. This endangers the circulation in three ways: (1) Plaques narrow an artery's diameter, reduce blood flow, and increase blood pressure. (2) Cellular changes in the artery wall reduce the elasticity of the artery and cause hypertension. (3) Plaques damage the endothelium, permitting the formation of a blood clot called a *thrombus* (***throm**-bus*).

A thrombus can enlarge and further decrease vessel diameter, or it may detach and travel along the artery as an *embolus* (*em-**bowl**-us*). An embolus becomes dangerous when it lodges in a small artery, preventing blood flow. This may lead to a stroke if it occurs in the brain or to a heart attack if it occurs in a coronary artery.

Because heart disease accounts for 50 percent of the deaths in the United States, medical science has tried to discover the exact events that cause cholesterol to be deposited in plaques. In our blood, cholesterol is carried by special proteins, the lipoproteins, which are manufactured in the liver. Two classes of these cholesterol-containing lipoproteins are significant for the heart and blood vessels.

Low-density lipoproteins (*LDL*) contain triglyceride and cholesterol. If cells are in need of cholesterol for plasma membranes or steroid hormone synthesis, they take up LDL. If this process does not occur, the LDL loaded with cholesterol remain in the plasma and deposition of cholesterol as plaques becomes possible.

High-density lipoproteins (*HDL*) are the second group of lipoproteins important to the "cholesterol story." HDL picks up cholesterol from cells and transports it to the liver for disposal. The liver releases the excess cholesterol with the bile into the small intestine.

Today, most measurements of blood cholesterol determine the amounts of HDL and LDL as well as the "total cholesterol." It has been found that higher HDL levels are associated with a *lower risk* of

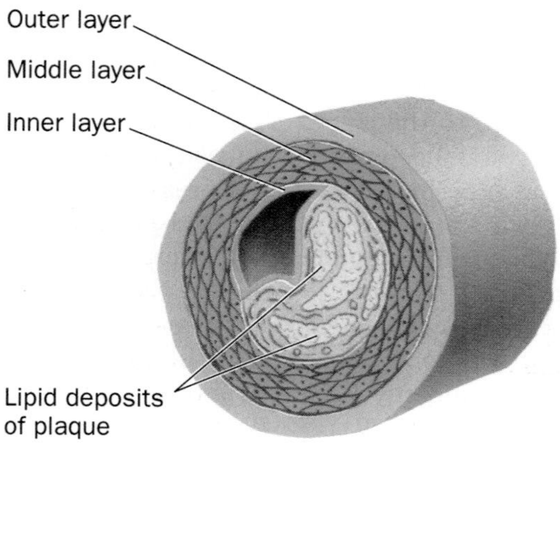
Outer layer
Middle layer
Inner layer
Lipid deposits of plaque

FIGURE 9.20 Plaque deposit in an artery.

TABLE 9.3 The Major Risk Factors for Atherosclerosis
Factors That Can Be Modified by Changes in Behavior
• High blood cholesterol • Smoking • Overweight or obesity • High blood pressure (hypertension)
Other Factors
• Diabetes mellitus • Heredity: family history of heart disease • Gender: under age 45, males are 10 times more at risk than are females

developing atherosclerosis, while high LDL levels are associated with a *higher risk*. This is why HDL has been called "good" cholesterol, and LDL "bad" cholesterol. LDL levels below 130 milligrams per deciliter and HDL levels above 40 mg/dl are considered desirable.

Atherosclerosis (*ath-er-oh-skler-**oh**-sis*), or hardening of the arteries, is a result of both plaque deposition and reduced elasticity of artery walls. Although atherosclerosis is often considered a disorder of the elderly, it begins in childhood and slowly develops, unnoticed, throughout a lifetime. None of us is entirely free of plaque formation in the arteries. Even in our early thirties, some of us have well-developed plaques. The question is not whether you will develop plaques, for you will. The question is how severe and extensive they will be.

What can be done to slow plaque formation and atherosclerosis? We can answer this question by considering the risk factors related to plaque formation (Table 9.3). The first four risk factors can be controlled by modifying one's behavior. Even gradual modifications of behavior and health habits can be helpful.

Starving the Heart and Heart Attack

Coronary heart disease is a condition in which plaque deposits form in the coronary arteries that supply blood to the heart. The coronary vessels may become so blocked that blood flow to the heart does not meet its needs. As a result, the myocardium does not receive sufficient O_2 and nutrients. A heart attack, or *myocardial infarction* (MI) (*my-oh-**car**-dee-uhl in-**fark**-shun*) occurs when blood flow to part of the myocardium is suddenly interrupted by blockage of a coronary artery. Without O_2, cardiac muscle cells begin to die. Once cardiac muscle cells die, they are replaced by connective tissue, not by new muscle cells. Consequently, the more cardiac cells that die in a heart attack, the less functional the heart becomes.

Angina pectoris (*ang-**eye**-nah **peck**-tore-iss*), or simply angina, is a period of pain that occurs when increased circulatory demands are placed on a heart with reduced coronary circulation resulting from coronary atherosclerosis. The reduced blood flow to the heart is typically noticed during periods of exercise or emotional stress. The pain is commonly experienced in the chest area as a heavy, squeezing sensation. Sometimes pain is experienced in the arms, neck, or jaw.

To reduce the pain of angina and, more important, to prevent a heart attack, several treatments are available: (1) Prescription drugs such as nitroglycerin quickly dilate the coronary arteries. (2) The diameter of a blocked coronary artery can be enlarged by *balloon angioplasty.* In this technique, a small plastic tube attached to a balloonlike device is inserted into a coronary artery narrowed by plaque formation. Inflation of the balloon compresses the plaque material and widens the narrowed vessel. (3) In **bypass surgery,** a blocked artery is "bypassed" with a length of vein taken from the leg (Figure 9.21).

Congestive Heart Failure

Normally, the amount of blood pumped out of the heart equals the amount returning to the heart. **Congestive heart failure** results when this balance is disturbed and blood accumulates in veins. This accumulation of venous blood produces a back pressure which increases the pressure in the capillaries. The increased pressure forces fluid into the tissue spaces, causing edema. Congestive heart failure is caused by any circulatory abnormality that reduces cardiac output. Such abnormalities include heart damage, long-term high blood pressure, and disorders of the heart valves.

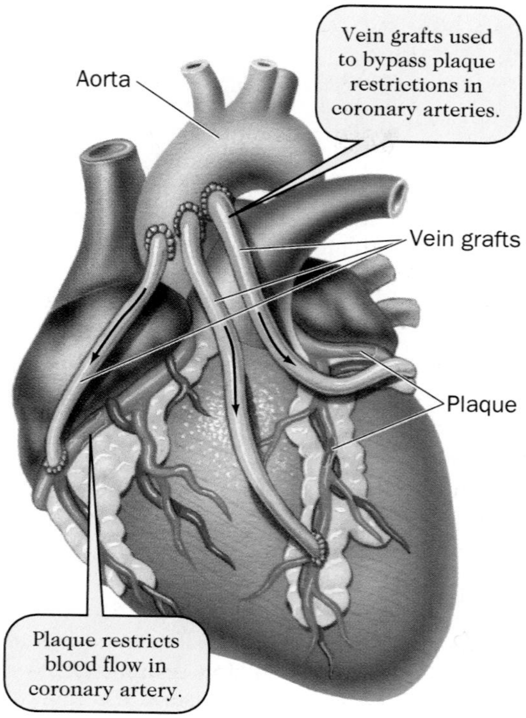
Vein grafts used to bypass plaque restrictions in coronary arteries.
Aorta
Vein grafts
Plaque
Plaque restricts blood flow in coronary artery.

FIGURE 9.21 How coronary bypass surgery increases blood flow to the area beyond a restricted artery.

When reduced cardiac output occurs in the pulmonary circulation, fluid moves from the lung capillaries into the air spaces of the lung. This accumulation of fluid in the lungs is called *pulmonary edema.* When reduced cardiac output occurs in the systemic circulation, fluid is forced from the capillaries into the surrounding tissues. This is called *peripheral edema,* and it is frequently observed as swelling of the feet and ankles.

Life Span Changes

Over the life span, both your heart and your blood vessels undergo changes. Some of these changes are part of normal aging. Other changes are brought on by genetic or environmental factors that reduce the effective operation of the system.

It is not always possible to determine whether these changes in the circulatory system are due to normal aging or are changes that have a genetic or environmental component. For example, plaque deposition and atherosclerosis seem to go on throughout the life span and are part of normal aging. However, genetic inheritance and environmental factors such as diet and exercise may affect the rate and extent of deposition. While your heart and blood vessels may function adequately for many years, changes do take place in both. Plaque formation in the arteries is one of these changes, but there are others.

As the heart ages, changes take place in the protein collagen, which surrounds each cardiac muscle cell. As we grow older, this collagen becomes stiffer and less flexible. In some regions of the heart more collagen may be deposited, changing the ratio of collagen to muscle. In people under 50 years of age, the heart has 46 percent muscle and 17 percent collagen. In hearts over 75 years old, the ratio is 27 percent muscle and 36 percent collagen. Of course, with fewer muscle cells, the action of the heart is weakened.

It has been estimated that cardiac output decreases about 1 percent per year after age 20 because of the gradual reduction in the number of muscle cells. Such decreases do not interfere with the heart's ability to meet normal circulatory demands. However, the heart may not be able to meet the increased demands that come with hot weather or increased physical activity.

The good news is that decreases in cardiac output can be partially reversed by a program of regular exercise designed to gradually increase the circulatory demands on the heart. Cardiac output is then increased as individual cardiac muscle cells become stronger. It is never too late to start, for even sedentary people above age 60 can increase their cardiac output after a program of regular exercise appropriate for their age.

Chapter Summary

Blood Vessels: Conduits for Blood

- Arteries transport blood away from the heart. They are composed of three layers.
- Capillaries are narrow vessels consisting of a single layer of cells. As blood flows through capillaries, nutrients, respiratory gases, and waste products are exchanged between the blood and the tissues.
- Veins transport blood back to the heart. Like arteries, they contain three layers. The blood in veins is moved with the help of one-way valves, skeletal muscle contractions, and breathing movements.

The Heart: A Tireless Pump

- The heart wall is composed of three layers: the endocardium, an inner layer of endothelium; the myocardium, a layer of cardiac muscle; and the pericardium, an outer layer of connective tissue.
- The pulmonary circulation, which is powered by the right atrium and ventricle, pumps blood to the lungs and back to the heart.
- The systemic circulation, which is powered by the left atrium and ventricle, pumps blood to the tissues throughout the body and back to the heart.
- There are two atrioventricular valves: the tricuspid (right side) and the bicuspid (left side). Both prevent blood from entering the atria during ventricular contraction.
- The two semilunar valves prevent blood from flowing back into the ventricles from the pulmonary trunk and aorta during ventricular relaxation.
- During one complete heartbeat, or cardiac cycle, both atria contract simultaneously, followed by both ventricles contracting simultaneously.
- The coronary arteries and veins circulate blood through the heart muscle.

Regulating the Heart and Circulation

- The cardiac conduction system generates and conducts electrical impulses through the heart, causing it to contract. It consists of the SA node, the AV node, the AV bundle, and Purkinje fibers.
- An electrocardiogram (EKG or ECG) is a recording of these electrical events.
- Cardiac output is the volume of blood pumped by each ventricle in 1 minute and equals about 5.25 liters. It is calculated by multiplying heart rate by stroke volume. Cardiac output can be increased by ANS and hormonal (epinephrine) stimulation of the heart.
- Blood pressure is a measure of the pressure in an artery during ventricular systole and ventricular diastole. A normal blood pressure is 120 over 80.

The Lymphatic System

- The lymphatic system consists of a series of vessels and organs located throughout the body.
- An important function of this system is the removal of excess tissue fluid leaked from blood capillaries. This fluid eventually is emptied into veins of the circulatory system in the shoulder area.

Diseases and Disorders

- Circulatory shock occurs when there is an excessive reduction in blood volume or blood pooling.
- High blood pressure, also known as hypertension, can be mild, moderate, or severe.
- Cholesterol is transported in the blood by lipoproteins. Increased levels of low-density lipoproteins (LDL) are associated with an increased risk of developing atherosclerosis.
- Coronary heart disease results when plaque formation in the coronary arteries reduces blood flow to the heart muscle. A heart attack, or myocardial infarction, occurs when blood flow through a coronary artery is interrupted, resulting in damage to the heart.

Life Span Changes

- As we age, changes in the heart muscle reduce its stroke volume and cardiac output. Cardiac output decreases about 1 percent per year after age 20. Such decreases can be offset by regular exercise.

Selected Key Terms

artery (p. 240)
atrioventricular valves (p. 247)
atrium (p. 246)
AV node (p. 250)
capillary (p. 241)
cardiac cycle (p. 248)
cardiac output (p. 252)
myocardium (p. 245)
pulmonary circulation (p. 246)
SA node (p. 250)
semilunar valves (p. 247)
systemic circulation (p. 246)
systole (p. 248)
vein (p. 244)
ventricle (p. 246)

Review Activities

1. Name and describe the functions of the three major types of blood vessels.
2. How do the processes of diffusion, blood pressure, and osmosis participate in the movement of substances across the capillary wall?
3. What is the importance of the lymphatic capillaries to the circulatory system?
4. Describe how one-way valves, skeletal muscle contraction, and breathing action help move blood through our veins. Why are these mechanisms necessary for moving blood in veins?
5. Draw a simple diagram representing the heart and position these structures in it: two atria and two ventricles, septum, epicardium, myocardium, endocardium, pericardial cavity, and tricuspid and bicuspid valves.
6. Trace a drop of blood through the pulmonary circulation, starting in the superior vena cava and ending in the left atrium. Identify the heart chambers, heart valves, and blood vessels through which the drop of blood will pass.
7. Trace a drop of blood through the systemic circulation, starting in the left atrium and ending in the right atrium. Identify the heart chambers, heart valves, and blood vessels through which the drop of blood will pass.
8. What is the function of the atrioventricular valves?
9. Describe the coordinated contractions (systole) and relaxations (diastole) of the atria and ventricles during one cardiac cycle. How long does a cardiac cycle last?
10. Describe how the SA node, AV node, AV bundle, and Purkinje fibers conduct electrical impulses that coordinate the contractions of the heart.
11. What is meant by cardiac output, and how is it calculated?
12. Describe how cardiac output is increased by the ANS and the hormone epinephrine.
13. What is the difference between systolic pressure and diastolic pressure?
14. Name the device used to measure blood pressure.
15. How is an individual's blood pressure taken?

Self-Quiz

Matching Exercise

___ 1. Small veins that receive blood from capillaries	A. Left
___ 2. Muscle layer of the heart	B. Aorta
___ 3. Heart ventricle that pumps oxygenated blood	C. 80 mm Hg
___ 4. Large vessel that receives blood from the right ventricle	D. Lymphatic
___ 5. The large artery from which coronary arteries originate	E. Venules
___ 6. A normal systolic pressure reading	F. 120 mm Hg
___ 7. A normal diastolic pressure reading	G. Heart rate
___ 8. A value that is equivalent to the pulse rate	H. Pulmonary trunk
___ 9. The amount of blood ejected at each stroke volume	I. Myocardium
___ 10. Type of capillary that removes excess fluid from tissues	J. 70 milliliters

Answers to Self-Quiz

1. E; 2. I; 3. A; 4. H; 5. B; 6. F; 7. C; 8. G; 9. J; 10. D

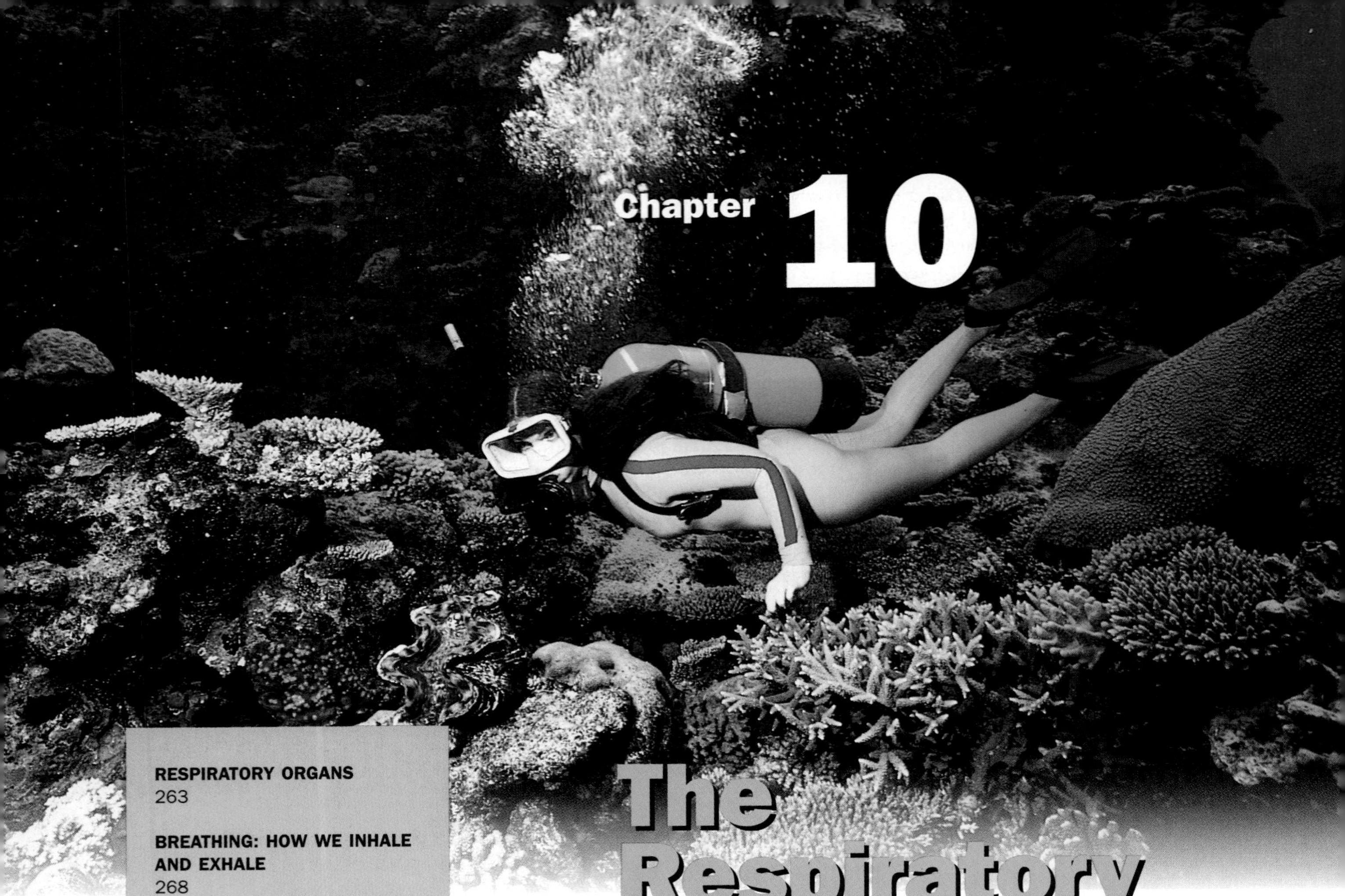

Chapter 10

The Respiratory System

Take a deep breath. Now exhale. This conscious act accomplishes what your body does automatically 20,000 times a day. As you inhale, a mixture of gases in the air enters your body through a series of passageways. The journey begins in your nose, continues down your throat and into your chest, and ends in tiny sacs in your lungs. Together, these structures form the **respiratory system,** which supplies your cells with oxygen (O_2) and removes carbon dioxide (CO_2) from your body.

The function of the respiratory system can be divided into three processes: breathing, external respiration, and internal respiration. **Breathing** moves air into and out of the lungs and requires the coordinated activities of muscles and the nervous system. The second process, **external respiration,** occurs in the lungs and involves gas exchange between the inhaled air and the blood. The third process, **internal respiration,** takes place throughout the body and involves the exchange of O_2 and CO_2 between the blood and the body cells. Both external respiration and internal respiration require the operation of the circulatory system.

Now pause. Take another deep breath and exhale. The O_2 you inhaled with this breath will travel to your cells and participate in chemical reactions that extract energy from nutrient molecules to form ATP. This is the aspect of metabolism we described in Chapter 7. Although it is not considered an activity of the respiratory system, we mention it here because it shows how important the proper exchange of O_2 and CO_2 is to homeostasis. Without sufficient ATP, cells cannot function and begin to die.

We will begin by describing the structure and function of the respiratory organs. Then we'll explore how O_2 and CO_2 are exchanged across membranes and transported by the blood. Regulatory mechanisms that control breathing will complete our discussion of normal respiration. Finally, we will discuss diseases and disorders and life span changes that alter the function of the respiratory system.

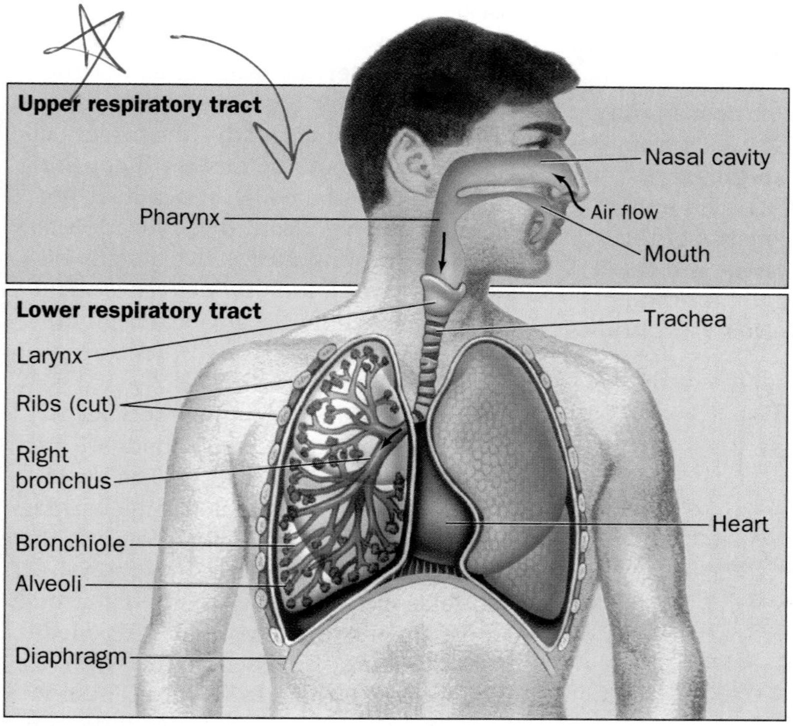
Upper respiratory tract
Nasal cavity
Pharynx
Air flow
Mouth
Lower respiratory tract
Trachea
Larynx
Ribs (cut)
Right bronchus
Heart
Bronchiole
Alveoli
Diaphragm

FIGURE 10.1 Structures that form the passageways of the respiratory system.

Respiratory Organs

The respiratory system includes a series of branched passageways that conduct air into and out of the lungs. For convenience, these passageways can be divided into two regions: the upper respiratory tract and the lower respiratory tract (Figure 10.1). The **upper respiratory tract** consists of the external nose, nasal cavity, and pharynx, while the **lower respiratory tract** consists of the larynx, trachea, bronchi, bronchioles, and alveoli. In this section, we will examine the structures and contributions of each tract by following the path of air from the nose to the alveoli.

Upper Respiratory Tract: Nose, Nasal Cavity, and Pharynx

The *nose* includes an external portion and an internal portion (Figure 10.2). The external portion protrudes from the face and is formed by a framework consisting of two nasal bones and cartilage. The internal portion, called the **nasal cavity,** is a large space directly behind the external nose. Both the external nose and the nasal cavity are separated into left and right channels by a *nasal septum.*

Incoming air enters through the two openings (the nostrils) at the base of the external nose and eventually passes into the nasal cavity. The epithelial tissue lining the nasal cavity is moist, ciliated, and richly supplied with blood vessels. Its moist surface results from mucus secreted by the epithelium. The mucus helps trap tiny particles such as dust, bacteria, pollen, and other debris in incoming air (Figure 10.3). The coordinated beating of numerous cilia continuously move debris-loaded mucus toward the back of the nasal cavity and into the pharynx, where it can be swallowed or expelled through the mouth. In addition to trapping particles, the mucus humidifies the incoming air as it loses water through evaporation. Incoming air

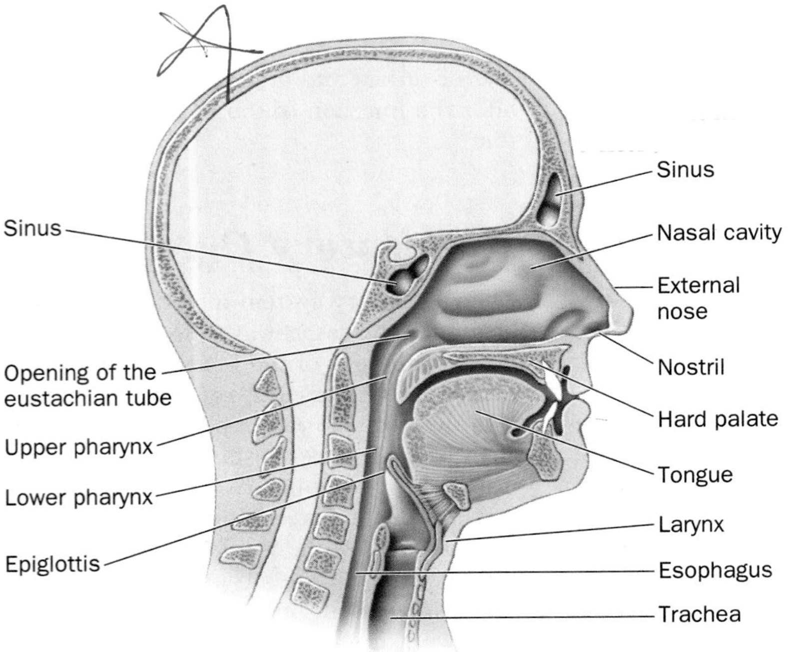
Sinus
Sinus
Nasal cavity
External nose
Nostril
Opening of the eustachian tube
Hard palate
Upper pharynx
Tongue
Lower pharynx
Larynx
Epiglottis
Esophagus
Trachea

FIGURE 10.2 A lateral view of the head and neck showing the structures of the upper respiratory tract.

is also warmed by the abundant blood vessels.

The ducts of our *sinuses* drain into the nasal cavity, as do the right and left *tear ducts.* These tiny ducts drain fluid away from the surface of the eye. When local irritation or powerful emotions cause tears to flow, the excess fluid travels down these ducts into the nasal cavity and we experience a "runny nose."

The pharynx: a common passageway. The **pharynx** (*fair-inks*) extends from the back of the nasal cavity to the larynx, or voice box (see Figure 10.2). The upper portion of the pharynx consists of the area between the nasal cavity and the roof of the mouth. Opening into the upper pharynx are the two *eustachian tubes* from the middle ear chambers. These tubes equalize the air pressure between the middle ear chambers and the outside air (Chapter 14).

The lower portion of the pharynx is often called the throat and can be observed by looking into the back of the mouth. The throat is the common passageway for food passing to the esophagus and air passing to the larynx on its way to the lungs.

FIGURE 10.3 The movement of debris-ladened mucus by the cilia of epithelial cells.

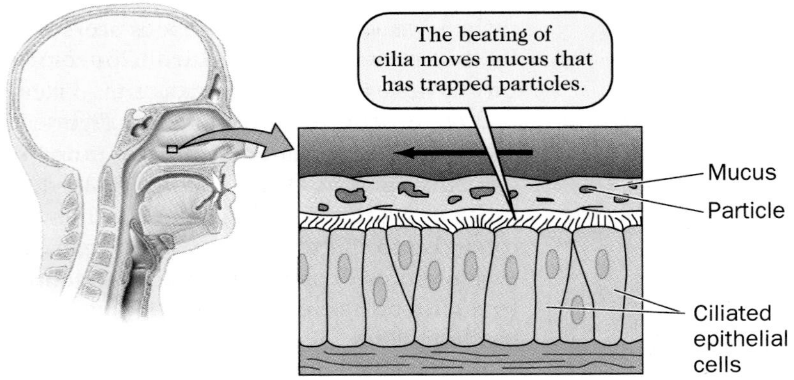
The beating of cilia moves mucus that has trapped particles.
Mucus
Particle
Ciliated epithelial cells

> The upper respiratory tract includes the nose and pharynx. The nose filters, moistens, and warms incoming air. The pharynx extends from the posterior of the nasal cavity to the larynx. It is a common passageway for food and air.

Lower Respiratory Tract: From Larynx to Alveoli

The **larynx** (*lair-inks*) is a relatively short passageway made of cartilage that connects the pharynx with the trachea (Figure 10.4). It includes two important structures: the epiglottis and the vocal cords. The *epiglottis* is a flap of cartilage that covers the opening to the larynx when you swallow. This ensures that liquids and solids enter the esophagus, not the trachea (Chapter 6). However, we sometimes swallow incorrectly, and larger substances can block the airway. If coughing stimulated by the cough reflex does not open the airway, death by suffocation may follow if the substance is not quickly dislodged. The *Heimlich maneuver* is a simple, safe procedure for unblocking an obstructed airway (Figure 10.5). Its correct use has saved many lives.

The larynx also contains the **vocal cords,** which produce the sounds we use for speech (see Figure 10.4). The vocal cords are enclosed within a large cartilaginous structure, commonly known as the Adam's apple. The vocal cords consist of two folds of connective tissue that stretch across the airway. When air is expelled from the lungs, it passes by the vocal cords, causing them to vibrate. These vibrations are transmitted through the air as sound waves.

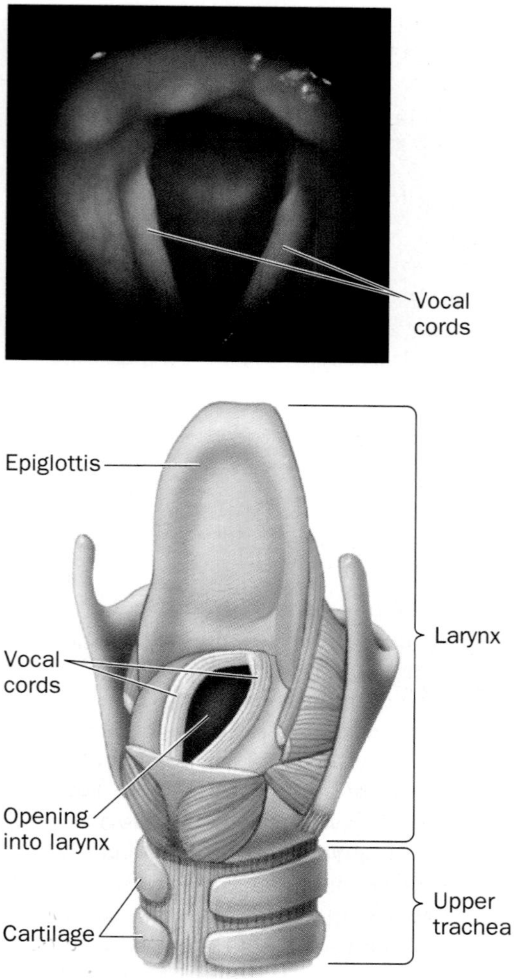
Vocal cords
Epiglottis
Vocal cords
Opening into larynx
Cartilage
Larynx
Upper trachea

FIGURE 10.4 Photograph of the vocal cords in an open position as when one is breathing. Posterior view of the larynx.

To some extent, we can consciously control the loudness and pitch of the sounds we produce. Loudness depends on how much air is forced past the vocal cords. The pitch, or frequency, of the sound is altered by changing the tension on the cords through the contraction and relaxation of muscles in the larynx. When these muscles contract, the two cords are stretched and higher-pitched sounds are produced. Relaxation of these muscles produces lower-pitched sounds.

However, the individual sound qualities that distinguish your voice are not entirely under your conscious control. The relative size and shape of the vocal cords, pharynx, nasal cavity, mouth, nose, and even sinuses contribute to the overtones and resonances that distinguish your voice from others. So distinctive are the sound qualities of individual human voices that they can be used for identification, much as fingerprints are used.

The trachea: our windpipe. After air leaves the larynx, it passes into the **trachea** (***tray-key-ah***), a tube that extends downward from the larynx to the left and right bronchi (Figure 10.6). The wall of the trachea is reinforced by a series of C-shaped rings of cartilage positioned so that the open portion of each ring faces the posterior. The C shape of the rings gives a certain flexibility to the trachea wall so that larger than normal volumes of air can be accommodated during strenuous physical activity.

The epithelial lining of the trachea has mucus-secreting cells and ciliated cells that operate like those of the nasal cavity to trap and move tiny foreign particles upward, where they can be swallowed or ejected from the mouth. Usually the trapping and elimination processes occur without much awareness. However, if irritating or sufficiently large substances enter the trachea, they stimulate the cough reflex. This causes you to cough, which helps force the particles out of the airway.

Tobacco smoke contains tiny irritating particles and chemicals that cause secretion of excessive mucus and decrease activity of the cilia. Without the action of the cilia, mucus and trapped substances accumulate in the airway. In heavy smokers, the cilia of the trachea may be entirely paralyzed so that violent coughing becomes the only way to remove the mucus and irritants from the passageway.

The bronchi enter the lungs. From the trachea, air moves into two passageways called the right and left **bronchi** (***brong-keye***; singular, *bronchus*) (Figure 10.6). Each bronchus enters the lung and then branches into a series of progressively narrower tubes called **bronchioles** (***brong-key-oles***). The walls of the bronchi and the large bronchioles are supported by cartilage, and their epithelial lining secretes mucus that is moved upward by the action of cilia. The smallest bronchioles are 1 millimeter or less in diameter and lack cartilage and cilia. They continue branching into ever narrower tubes until each ends in a terminal sac where gas exchange takes place.

FIGURE 10.5 The Heimlich maneuver.

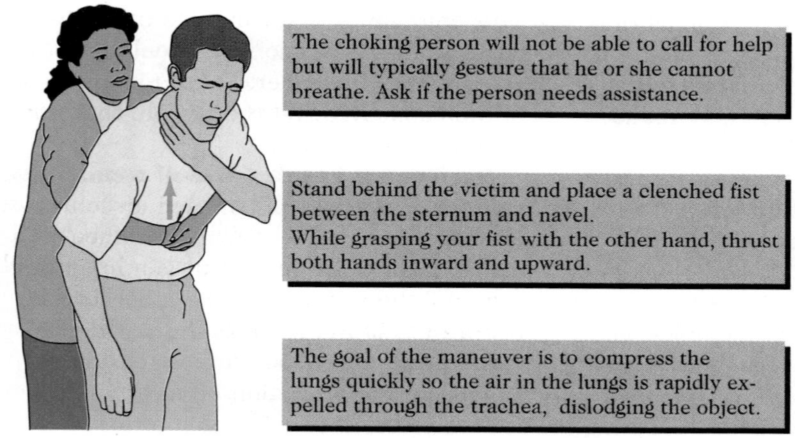
The choking person will not be able to call for help but will typically gesture that he or she cannot breathe. Ask if the person needs assistance.
Stand behind the victim and place a clenched fist between the sternum and navel.
While grasping your fist with the other hand, thrust both hands inward and upward.
The goal of the maneuver is to compress the lungs quickly so the air in the lungs is rapidly expelled through the trachea, dislodging the object.

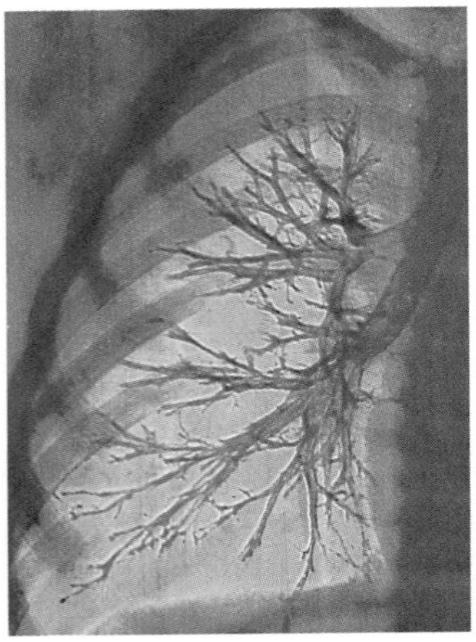

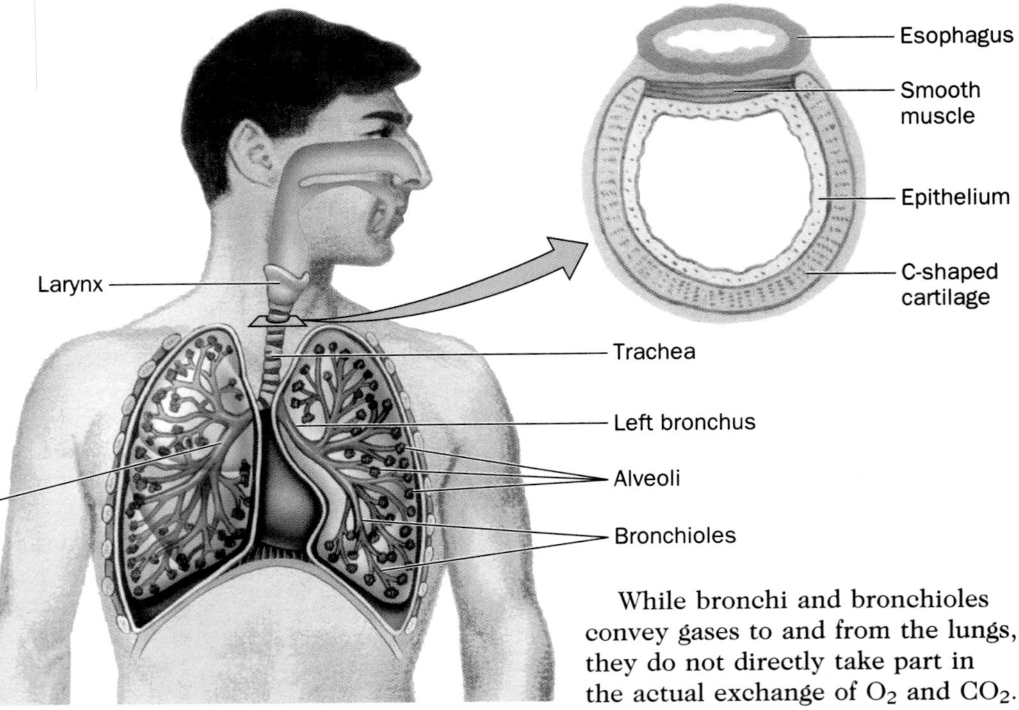
Esophagus
Smooth muscle
Epithelium
C-shaped cartilage
Larynx
Trachea
Left bronchus
Alveoli
Bronchioles
Right bronchus

FIGURE 10.6 Structures of the lower respiratory tract. Photograph of highlighted arteries of the right lung showing the branching pattern of bronchioles.

The lungs, alveoli, and gas exchange. Our two **lungs** consist of supportive tissues surrounding the bronchi, the bronchioles, the terminal sacs, and their blood vessels. The lungs are positioned on either side of the heart and occupy much of the space in the thoracic cavity (Figure 10.7). Each lung is enclosed by two layers of epithelial membranes called **pleural membranes** or **pleura** (***plur**-ah*). One of these membranes is attached directly to the lung, while the other lines the thoracic cavity. These two membranes are separated by a narrow space called the *pleural cavity*. Fluid secreted by the pleura coats the membranes, reducing friction as the lungs inflate and deflate during breathing. An inflammation of the pleural membranes called *pleurisy* (***plur**-ah-see*) may reduce fluid secretion, increase friction, and cause localized adhesions of the pleura.

Each lung is divided into lobes. The right lung has three lobes, and the left lung has two (Figure 10.7). Each lobe is further subdivided into smaller sections with their own bronchi, bronchioles, and blood vessels. These lobes are rather independent, permitting surgical removal of portions of the lung without disrupting the function of the entire organ.

While bronchi and bronchioles convey gases to and from the lungs, they do not directly take part in the actual exchange of O_2 and CO_2. This essential function is performed by tiny terminal sacs called **alveolar sacs** (*al-**vee**-oh-lar*) (Figure 10.8). Each microscopic alveolar sac consists of a cluster of cup-shaped bulges called **alveoli** (*al-**vee**-oh-lie;* singular, *al**ve**olus*). The external appearance of an alveolar sac and its alveoli is frequently described as resembling a cluster of grapes. It is estimated that our lungs contain approximately 300 million individual alveoli that provide a total surface area of approximately 750 to 800 square feet for gas exchange. By comparison, our total skin surface is 15 to 20 square feet.

The wall of each alveolus consists of epithelium that is only one cell thick. This epithelium is surrounded by capillaries from the pulmonary circulation (Figure 10.8). It is here, at the junction of the capillaries and the alveoli, that gas exchange takes place.

Surfactant: keeping alveoli open. Certain epithelial cells in each alveolus are specialized to secrete a substance called **surfactant.** Surfactant is a lipoprotein molecule that coats the inner surface of the alveoli and reduces the surface tension in each alveolus. Surface tension results from the attraction between the water molecules that coat the interior of alveoli. Without surfactant, the tension

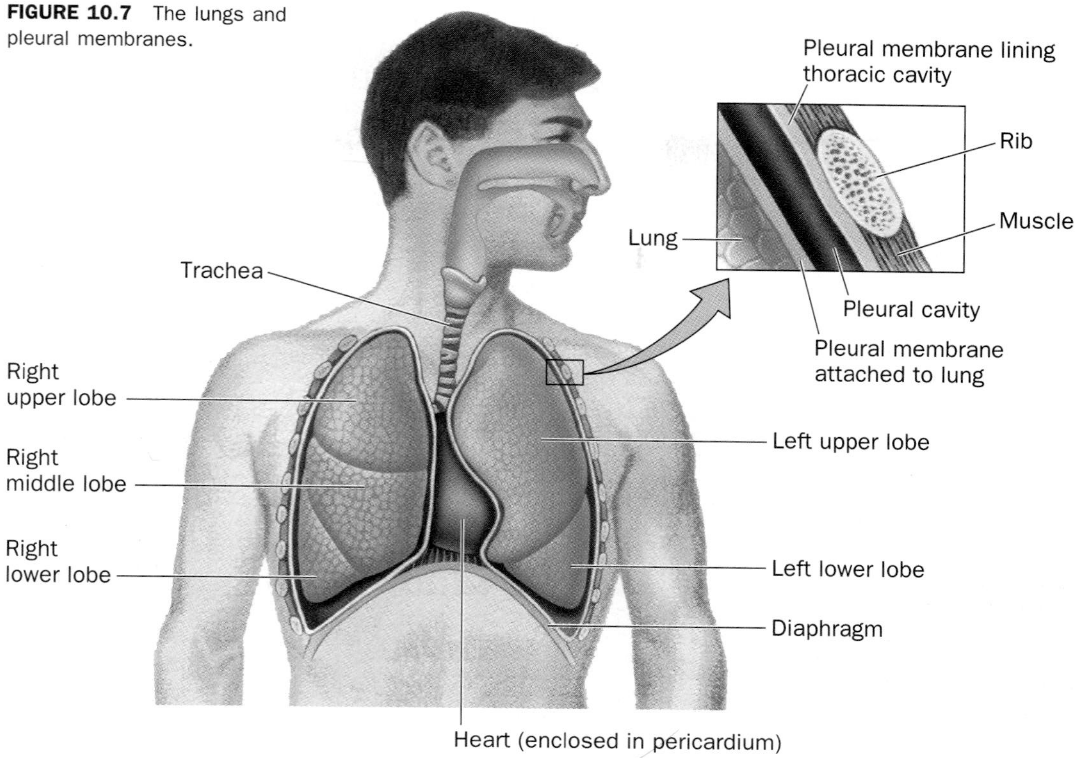
Pleural membrane lining thoracic cavity
Rib
Muscle
Lung
Trachea
Pleural cavity
Pleural membrane attached to lung
Right upper lobe
Right middle lobe
Right lower lobe
Left upper lobe
Left lower lobe
Diaphragm
Heart (enclosed in pericardium)

FIGURE 10.7 The lungs and pleural membranes.

FIGURE 10.8 Relationship between the alveolar sacs, alveoli, and pulmonary capillaries. Photograph is an SEM of alveolar sacs.

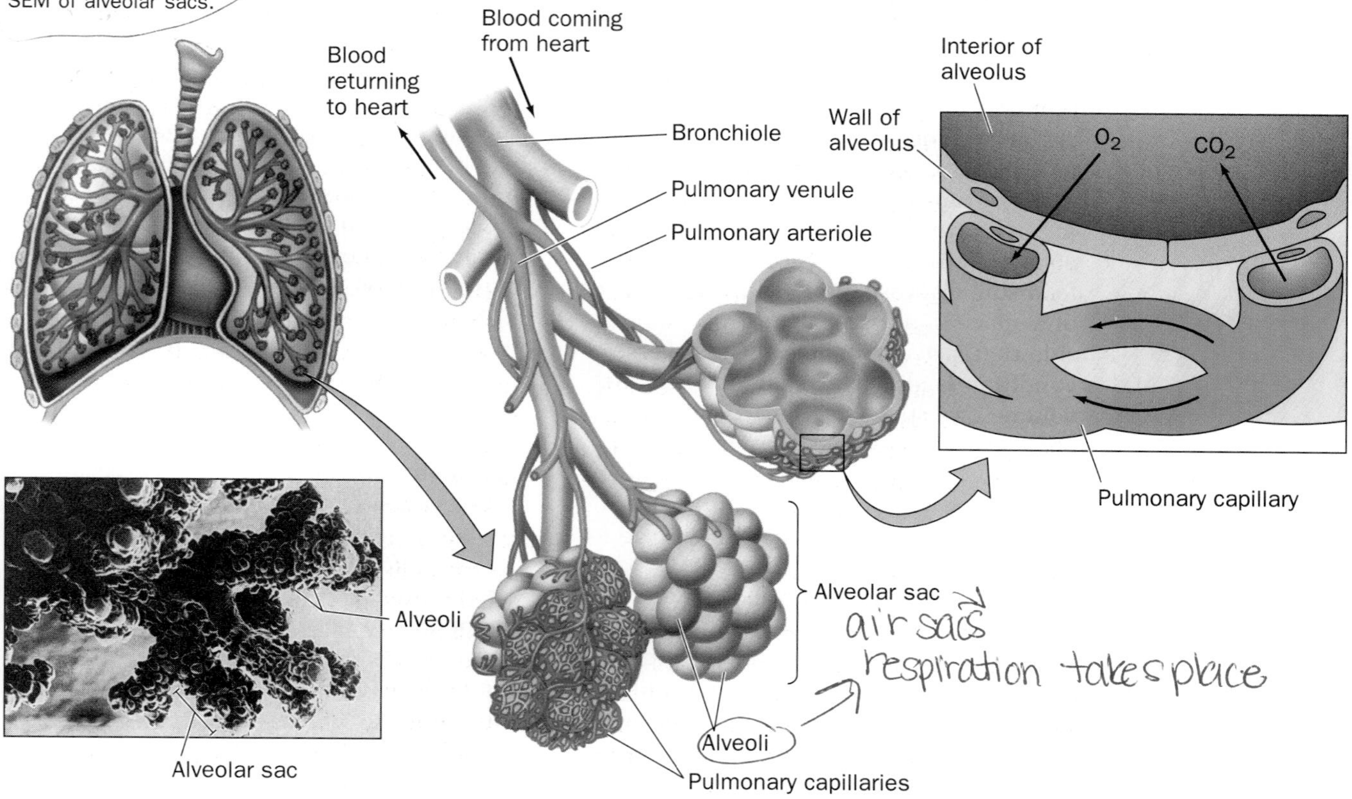
Blood coming from heart
Blood returning to heart
Bronchiole
Pulmonary venule
Pulmonary arteriole
Interior of alveolus
Wall of alveolus
O_2
CO_2
Pulmonary capillary
Alveolar sac
air sacs
respiration takes place
Alveoli
Alveoli
Alveolar sac
Pulmonary capillaries

between the water molecules would be great enough to collapse the alveoli, preventing effective gas exchange. In addition to these specialized secretory cells, the alveoli also have *macrophages* that roam around and engulf inhaled foreign debris such as dust, bacteria, and pollutants.

> The lower respiratory tract includes the larynx, trachea, bronchi, bronchioles, and alveoli. The larynx contains the vocal cords. Gas exchange occurs in the alveoli.

Breathing: How We Inhale and Exhale

So far we have traced the passage of air from outside the body to the sites of gas exchange in the lungs. Let us now consider how the movement of air is accomplished. Using specific muscles coordinated by the nervous system, your body pumps air into and out of your lungs by a mechanical process involving changes in air pressure and air volume. This process is called breathing and will be discussed in the section that follows.

Within the air spaces of the lungs (alveoli) and the chest cavity (thoracic cavity), changes in air pressure and air volume occur as breathing takes place. To help you understand how gases move during breathing, two fundamental physical concepts will be applied throughout the following discussion. (1) When the volume of a space *increases,* the air pressure inside that space *decreases.* The reverse condition is also true; that is, when the volume of a space *decreases,* the air pressure inside that space *increases.* (2) If they are not constrained, all substances move from regions of higher pressure to regions of lower pressure. This is true for gases, liquids, and solids.

Before we proceed, a quick review of thoracic anatomy will be helpful. The thoracic cavity is formed by the ribs, the sternum, the thoracic vertebrae, and their attached skeletal muscles. Two muscles are central to our explanation of breathing: the intercostal muscles and the diaphragm (Figure 10.9). The **intercostal muscles** are located between the ribs, and the **diaphragm** is a broad sheet of muscle positioned so that it separates the thoracic cavity from the abdominal cavity. With this background, we can ask: How are volumes of air moved into and out of our lungs during breathing?

> The mechanics of breathing are based on changes in the volumes of closed cavities. An increase in volume decreases pressure (the reverse is also true). All substances move from regions of higher pressure to regions of lower pressure.

How Air Is Inhaled and Exhaled

We will begin our story a moment before the inhalation of a breath. At this instant the diaphragm is in its relaxed, somewhat elevated position, the chest wall is not expanded, the rib cage is in a lowered position, and the lungs are not inflated (Figure 10.9).

When you **inhale**, muscles contract and produce anatomic changes. First, the diaphragm contracts and moves downward toward the abdominal cavity. Second, the outer intercostal muscles contract and pull the rib cage upward. As a result, the thoracic cavity expands in size.

This is where our discussion of air volume and air pressure relationships becomes relevant. As the thoracic cavity expands, it increases the volume of the closed pleural cavity, and as a result the pressure in the pleural cavity decreases. In response to this lowered pressure, the lungs start to expand. As the volume of the lungs gets larger, the pressure within the lungs decreases. Because the lungs are open to outside air, this decrease in pressure causes outside air to rush into the lungs, and inhalation takes place.

When you **exhale,** the diaphragm and outer intercostal muscles relax. When this happens, the thoracic cavity decreases in size and so do the lungs. Elastic tissue in the lungs allows this decrease in lung volume to occur quite rapidly. As the lungs' volume gets smaller, the air pressure with-

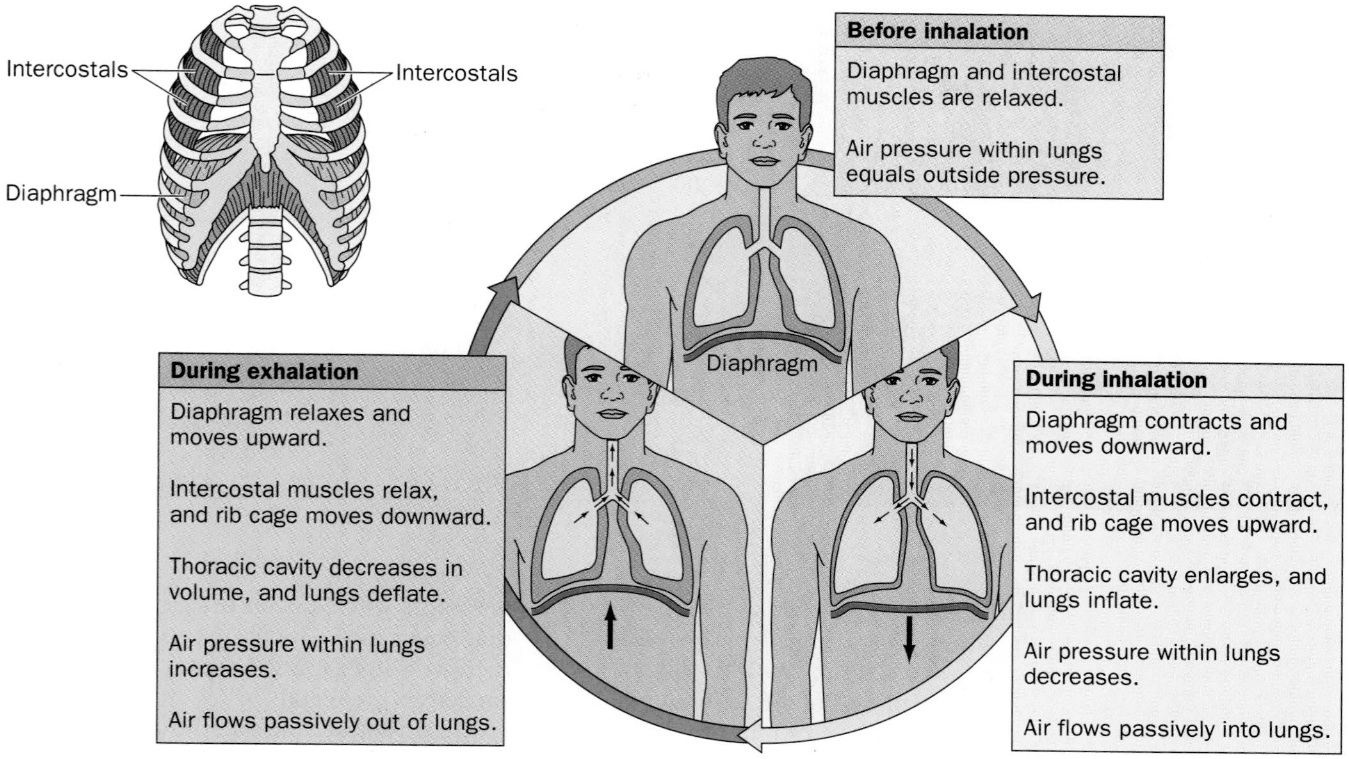
Intercostals
Intercostals
Diaphragm
Before inhalation
Diaphragm and intercostal muscles are relaxed.
Air pressure within lungs equals outside pressure.
Diaphragm
During exhalation
Diaphragm relaxes and moves upward.
Intercostal muscles relax, and rib cage moves downward.
Thoracic cavity decreases in volume, and lungs deflate.
Air pressure within lungs increases.
Air flows passively out of lungs.
During inhalation
Diaphragm contracts and moves downward.
Intercostal muscles contract, and rib cage moves upward.
Thoracic cavity enlarges, and lungs inflate.
Air pressure within lungs decreases.
Air flows passively into lungs.

FIGURE 10.9 The events of breathing.

in them increases. Now, because the air pressure within the lungs is slightly higher than the air pressure outside the body, air rushes out of the lungs, and exhalation takes place.

These are the events which occur during normal quiet breathing, when the body is at rest. However, the deep and labored breathing that occurs during strenuous activity requires the action of several additional muscles. During deep inhalation, the contraction of additional rib cage muscles raises the rib cage higher. During deep exhalation, additional abdominal muscles contract, forcing the diaphragm up higher into the thoracic cavity, and the rib cage is pulled downward by contraction of the inner intercostal muscles.

> Muscle contraction enlarges the thoracic cavity, causing the lungs to expand and lowering the air pressure in them. Outside air under higher pressure moves into the lungs. Exhaling reverses this process.

Lung Volumes and Vital Capacity

As you sit quietly reading these pages, you are inhaling and exhaling about 12 to 15 times per minute. This is called your *respiration rate.* The volume of air entering and leaving your respiratory system with each breath is called your **tidal volume.** On average, your tidal volume equals about 500 milliliters (about 1 pint) of air (Figure 10.10). However, only about 350 ml of this air reaches the alveoli and participates in O_2 and CO_2 exchange. The remaining 150 ml remains in the airways and is known as the *dead air volume,* since it does not participate in gas exchange.

When the diaphragm and intercostal muscles are contracted further, much more air can be inhaled. The additional volume of air inhaled is about 3100 ml (about 3 quarts) and is known as the *inspiratory reserve volume.* The volume of air that can be forcibly exhaled after the tidal volume has been exhaled is about 1200 ml (about 1 quart) and is known as the *expiratory reserve volume.* Even after

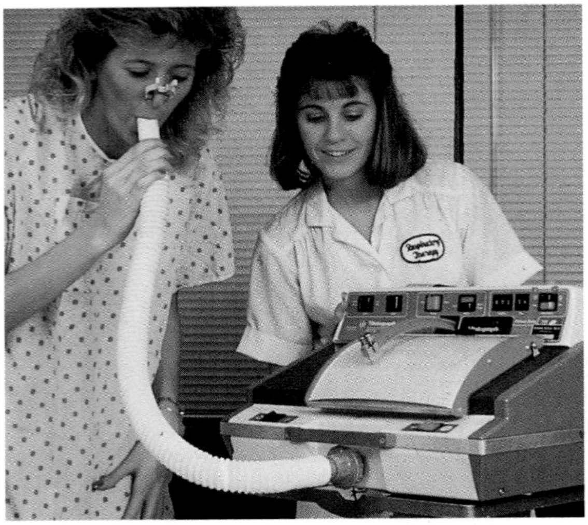

(a)

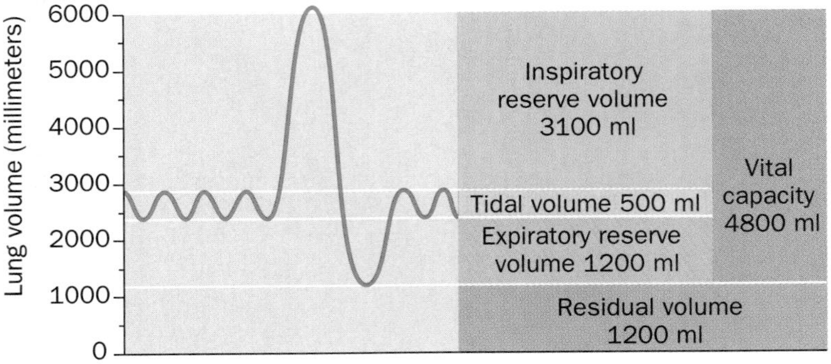
Lung volume (millimeters)
6000
5000
4000
3000
2000
1000
0
Inspiratory reserve volume 3100 ml
Tidal volume 500 ml
Expiratory reserve volume 1200 ml
Residual volume 1200 ml
Vital capacity 4800 ml

(b)

FIGURE 10.10 (a) Lung volumes being recorded on a spirometer. (b) A recording taken by a spirometer and its relationship to lung volumes.

exhaling strenuously, some air remains in the lungs; this is called the *residual volume.*

The total volume of air an individual can move in and out in one breath is called the **vital capacity.** It is the sum of tidal volume, inspiratory reserve volume, and expiratory reserve volume and equals about 4800 ml of air (500 ml + 3100 ml + 1200 ml). Vital capacity is measured by an instrument called a *spirometer.* These measurements are useful in diagnosing respiratory diseases such as lung cancer and emphysema.

> Our respiration rate during quiet breathing averages about 12 to 15 breaths per minute, and the tidal volume is about 500 ml. The vital capacity is the sum of the tidal, inspiratory reserve, and expiratory reserve volumes.

Gas Exchange between the Lungs, Blood, and Tissues

Although the respiratory organs move O_2 and CO_2 into and out of the lungs, it is the circulatory system that transports those gases to all the tissues of the body. A malfunction of either system hinders the movement of O_2 to the cells. The next few pages will focus on the gas exchanges that occur in the lungs and the tissues, and then we will examine how O_2 and CO_2 are carried in the blood.

In our discussion, the diffusion of gases will be central to the explanation of gas exchange. All gases exert pressure against surfaces, and so the diffusion of gases is discussed in terms of pressures. Gases move from areas of high pressure to areas of low pressure. This is the same principle we used earlier to describe the diffusion of solutes in a water solution or across a membrane. In those situations we spoke of "concentrations"; however, "pressure" is a more convenient measure for gases.

If only one kind of gas molecule were present, all the pressure would be due to that single gas. However, in a mixture of gases, the pressure exerted by the mixture is the sum of the pressures exerted by each kind of gas. At sea level, the pressure exerted by atmospheric air on the surface of the Earth is 760 mm Hg. Recall from Chapter 9 that the measure mm Hg represents the distance in millimeters (mm) that an applied pressure pushes liquid mercury (Hg) up a narrow glass tube. The 760 mm Hg pressure is the total pressure produced by the mixture of gases in the air. The air we inhale contains about 78 percent nitrogen (N_2), 21 percent oxygen (O_2), 0.04 percent carbon dioxide (CO_2), and very small amounts of several other gases.

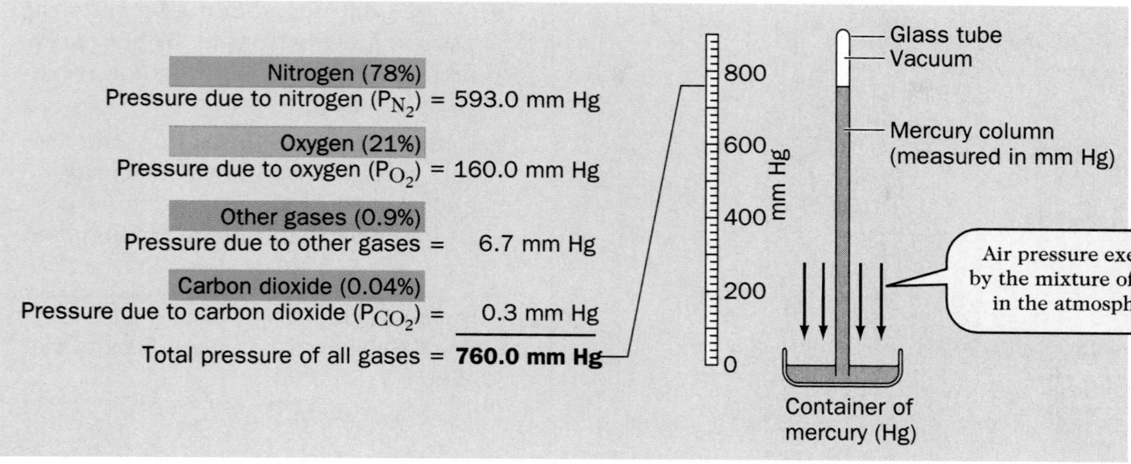
Nitrogen (78%)
Pressure due to nitrogen (P_{N_2}) = 593.0 mm Hg
Oxygen (21%)
Pressure due to oxygen (P_{O_2}) = 160.0 mm Hg
Other gases (0.9%)
Pressure due to other gases = 6.7 mm Hg
Carbon dioxide (0.04%)
Pressure due to carbon dioxide (P_{CO_2}) = 0.3 mm Hg
Total pressure of all gases = 760.0 mm Hg
800
600
400
200
0
mm Hg
Glass tube
Vacuum
Mercury column
(measured in mm Hg)
Air pressure exerted by the mixture of gases in the atmosphere
Container of mercury (Hg)

FIGURE 10.11 Atmospheric gas molecules and their relationship to partial pressure at sea level.

The **partial pressure** of a specific gas represents the "part" it contributes to the total pressure in a mixture of gases. Partial pressure is designated by the capital letter "P" and is measured in mm Hg. Because O_2 represents about 21 percent of the gas mixture in air, the partial pressure of oxygen (P_{O_2}) is 21 percent of 760 mm Hg; that is, P_{O_2} is 160 mm Hg (760 mm Hg × 21 percent = 159.6 mm Hg) (Figure 10.11).

If this seems complicated, it may be helpful to think of partial pressure as another way of representing the concentration of a particular gas in a mixture. Remember, substances always move from regions of higher concentration or pressure to regions of lower concentration or pressure.

External Respiration: Exchanges between Alveoli and Blood

The exchange of O_2 and CO_2 in the lungs occurs by diffusion and depends on the partial pressure of each of these gases in the alveoli and the blood (Figure 10.12). The gases diffuse across two cell layers: one associated with the alveoli and the other associated with the capillary wall. The inhaled air entering the alveoli mixes with the oxygen-poor and carbon dioxide–rich air that is being exhaled. Even with this mixing, the partial pressure of O_2 in the alveoli is still higher than that in the blood, while the partial pressure of CO_2 in the alveoli is lower than that in the blood. As a consequence of these differences in partial pressures, O_2 and CO_2 diffuse across the alveolar and capillary membranes in the following directions:

- Oxygen diffuses from the alveoli to the blood, moving from its higher partial pressure in the alveoli to its lower partial pressure in the blood. Blood leaving the lungs has a higher partial pressure of O_2 compared with blood entering the lungs.
- Carbon dioxide diffuses from the blood to the alveoli, moving from its higher partial pressure in the blood to its lower partial pressure in the alveoli. Blood leaving the lungs has a lower partial pressure for CO_2 compared with blood entering the lungs.

> The exchange of O_2 and CO_2 between the lungs and blood occurs by diffusion, which is determined by the partial pressure of each gas in the alveoli and blood. O_2 diffuses from the alveoli to the blood. CO_2 diffuses from the blood to the alveoli.

Internal Respiration: Exchanges between Blood and Tissues

As with external respiration, the exchange of O_2 and CO_2 between the blood and the body's cells occurs by diffusion from regions of high partial pressure to regions of low partial pressure (Figure 10.12). The gases must pass across two cell layers: one associated with the capillary wall and the other associated with the cell's plasma membrane.

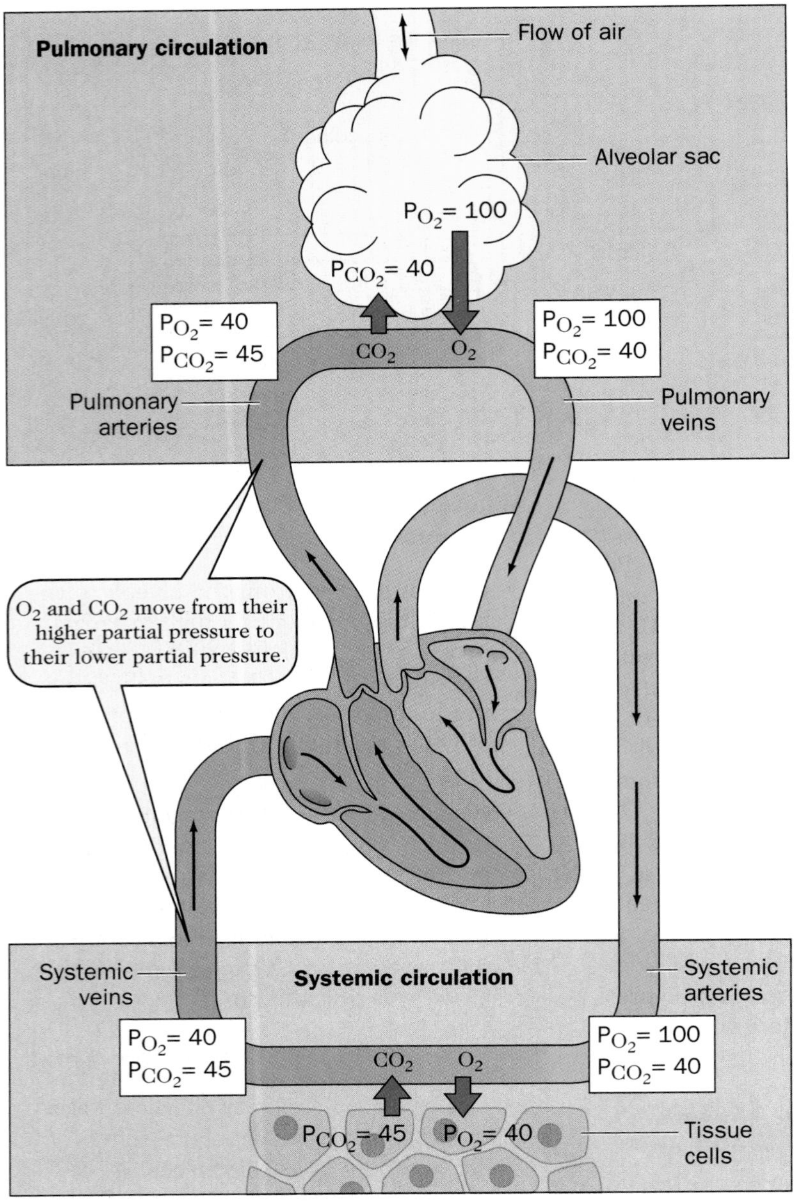

Pulmonary circulation
Flow of air
Alveolar sac
P_{O_2}= 100
P_{CO_2}= 40
P_{O_2}= 40
P_{CO_2}= 45
CO_2
O_2
P_{O_2}= 100
P_{CO_2}= 40
Pulmonary arteries
Pulmonary veins
O_2 and CO_2 move from their higher partial pressure to their lower partial pressure.
Systemic veins
Systemic circulation
Systemic arteries
P_{O_2}= 40
P_{CO_2}= 45
CO_2
O_2
P_{O_2}= 100
P_{CO_2}= 40
P_{CO_2}= 45
P_{O_2}= 40
Tissue cells

FIGURE 10.12 The differences in the partial pressures of oxygen and carbon dioxide in the alveoli, blood, and tissues.

Because our cells continually consume O_2 and produce CO_2, our tissues have a lower partial pressure of O_2 and a higher partial pressure of CO_2 than does the blood in our systemic arteries. As a consequence of these differences in partial pressure, O_2 and CO_2 diffuse in the following directions across the capillary wall:

- Oxygen diffuses from the blood to the tissues, moving from its higher partial pressure in the blood to its lower partial pressure in the tissues.
- Carbon dioxide diffuses from the tissues to the blood, moving from its higher partial pressure in the tissues to its lower partial pressure in the blood.

O_2 diffuses from the blood to the tissues, while CO_2 diffuses from the tissues to the blood.

How O_2 Is Transported in the Blood

Oxygen is transported by the blood in two ways: dissolved in plasma and attached to hemoglobin (Hb) in red blood cells (Figure 10.13). Because O_2 is not very soluble in water, only about 2 percent of the transported O_2 is carried by the blood in the dissolved state. The rest of the O_2 in the blood—about 98 percent—is transported by *hemoglobin.* Recall from Chapter 8 that each hemoglobin molecule is composed of four protein chains and has a nonprotein iron-containing heme group which can form a weak bond with an O_2 molecule. This permits each hemoglobin molecule (Hb) to combine with four O_2 molecules, forming **oxyhemoglobin** (HbO_2):

$$\underset{\text{Hemoglobin}}{Hb} + \underset{\text{Oxygen}}{O_2} \rightleftharpoons \underset{\text{Oxyhemoglobin}}{HbO_2}$$

The reaction between hemoglobin and O_2 is reversible, as shown by the arrows in the chemical reaction above. This means that O_2 combines with hemoglobin under some conditions but detaches under others. The partial pressure of O_2 is one condition that determines whether O_2 attaches to or detaches from hemoglobin. In oxygen-rich environments (high partial pressure), more O_2 and hemoglobin molecules combine to form more oxyhemoglobin. In oxygen-poor environments (low partial pressure), less oxyhemoglobin forms. This explains why O_2 combines with hemoglobin in the lungs, where there is a higher partial pressure of O_2, and why O_2 detaches from hemoglobin in the

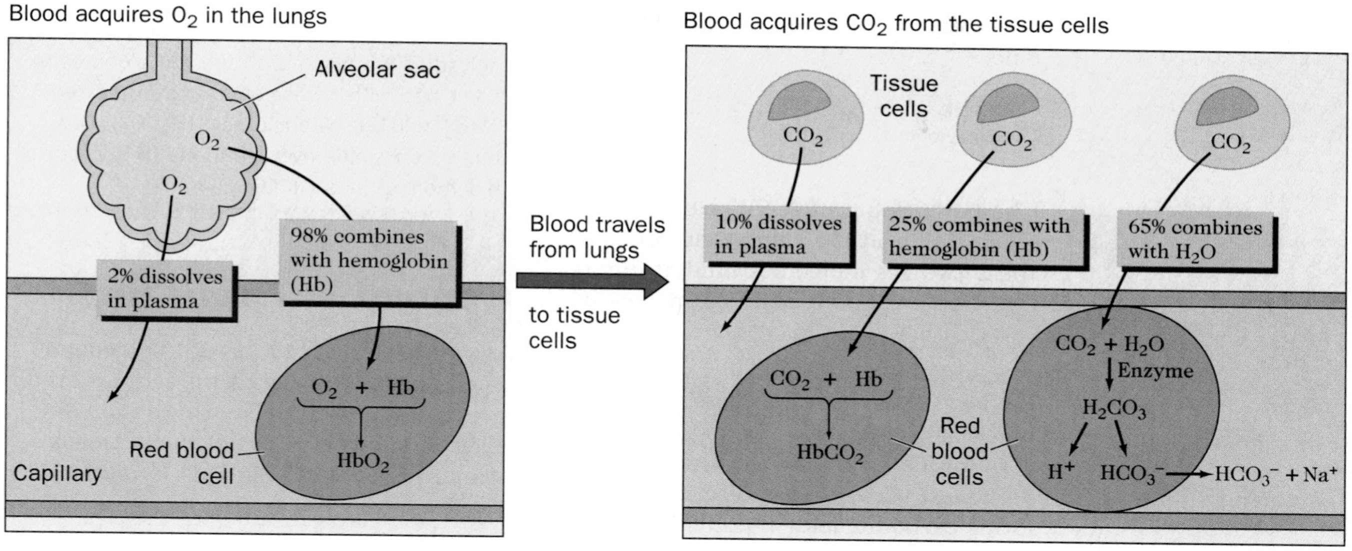
Blood acquires O_2 in the lungs
Alveolar sac
O_2
O_2
98% combines with hemoglobin (Hb)
2% dissolves in plasma
O_2 + Hb
HbO_2
Red blood cell
Capillary
Blood travels from lungs
to tissue cells
Blood acquires CO_2 from the tissue cells
Tissue cells
CO_2
CO_2
CO_2
10% dissolves in plasma
25% combines with hemoglobin (Hb)
65% combines with H_2O
CO_2 + Hb
$HbCO_2$
Red blood cells
$CO_2 + H_2O$
Enzyme
H_2CO_3
H^+ HCO_3^- → $HCO_3^- + Na^+$

FIGURE 10.13 How oxygen and carbon dioxide are transported in the blood.

tissues, where the partial pressure of O_2 is lower. This relationship between the partial pressure of O_2 and HbO_2 formation is presented in graphic form in the *oxygen-dissociation curve* (Figure 10.14).

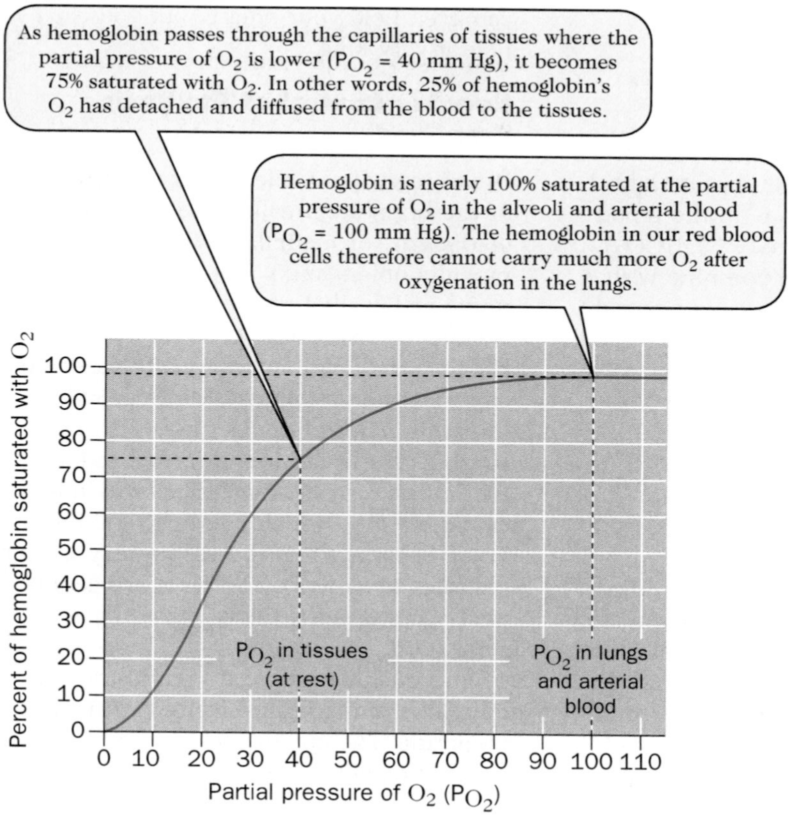
As hemoglobin passes through the capillaries of tissues where the partial pressure of O_2 is lower (P_{O_2} = 40 mm Hg), it becomes 75% saturated with O_2. In other words, 25% of hemoglobin's O_2 has detached and diffused from the blood to the tissues.
Hemoglobin is nearly 100% saturated at the partial pressure of O_2 in the alveoli and arterial blood (P_{O_2} = 100 mm Hg). The hemoglobin in our red blood cells therefore cannot carry much more O_2 after oxygenation in the lungs.
Percent of hemoglobin saturated with O_2
100
90
80
70
60
50
40
30
20
10
0
P_{O_2} in tissues (at rest)
P_{O_2} in lungs and arterial blood
0 10 20 30 40 50 60 70 80 90 100 110
Partial pressure of O_2 (P_{O_2})

FIGURE 10.14 The oxygen-dissociation curve showing the relationship between the partial pressure of oxygen and oxyhemoglobin formation.

The hemoglobin of red blood cells transports 98 percent of the O_2. In the lungs, where the partial pressure of O_2 is high, more oxygen bonds with hemoglobin. In the tissues, where P_{O_2} is low, O_2 detaches from hemoglobin.

How CO_2 Is Transported in the Blood

The operation of cellular metabolism continuously produces the waste product CO_2, which must be eliminated from the body. After its production, this small molecule diffuses from the cell and enters the bloodstream. In the blood it is transported in three ways: dissolved in plasma, combined with hemoglobin, and as bicarbonate ions (see Figure 10.13).

1. A small amount of CO_2 (about 10 percent) dissolves in the plasma. The remainder (about 90 percent) diffuses into red blood cells. It is in the red blood cells that the next two reactions take place.
2. About 25 percent of the CO_2 entering red blood cells combines with the pro-

tein portion of the hemoglobin molecule:

$$\underset{\text{Hemoglobin}}{Hb} + \underset{\text{Carbon dioxide}}{CO_2} \rightleftharpoons HbCO_2$$

The hemoglobin molecule can carry O_2 and CO_2 at the same time because these two gas molecules bind at different sites on the hemoglobin molecule: O_2 with heme and CO_2 with globin.

3. The remaining molecules of CO_2 (about 65 percent) that enter the red blood cells react with water (H_2O) in a reaction catalyzed by the enzyme *carbonic anhydrase.* This reaction rapidly forms carbonic acid (H_2CO_3):

$$\underset{\text{Carbon dioxide}}{CO_2} + \underset{\text{Water}}{H_2O} \xrightleftharpoons{\text{carbonic anhydrase}} \underset{\text{Carbonic acid}}{H_2CO_3}$$

Once carbonic acid forms, it quickly breaks apart into a hydrogen ion (H^+) and a bicarbonate ion (HCO_3^-):

$$\underset{\text{Carbonic acid}}{H_2CO_3} \rightleftharpoons \underset{\text{Hydrogen ion}}{H^+} + \underset{\text{Bicarbonate ion}}{HCO_3^-}$$

The bicarbonate ions diffuse out of the red blood cells, enter the plasma, and combine with sodium ions in the blood to form sodium bicarbonate ($NaHCO_3$). The hydrogen ions (H^+) combine with hemoglobin and thus remain in the red blood cell, where they have an additional role to play. The attachment of H^+ with hemoglobin weakens the bond between hemoglobin and O_2, causing more O_2 to be released. This benefits body cells, for as more CO_2 is produced by metabolically active tissues, more O_2 will be released for cellular respiration.

Carbon dioxide is eliminated from the blood as it passes through the pulmonary capillaries surrounding the alveoli. The chemical reactions discussed above are reversible, and so when carbon dioxide–rich blood (high P_{CO_2}) passes next to the alveoli (low P_{CO_2}), the chemical reactions reverse themselves and CO_2 is released. Carbon dioxide then diffuses into the alveolar space and is exhaled.

> Blood CO_2 is transported in three ways: dissolved in the plasma, in the red blood cells combined with hemoglobin, and in the red blood cells as carbonic acid (H_2CO_3). Carbonic acid yields hydrogen ions (H^+) and bicarbonate ions (HCO_3^-).

Control of Breathing

As you have learned, breathing depends on the actions of the diaphragm and the intercostal muscles. The rhythmic contraction and relaxation of these muscles is maintained and regulated by nerve impulses from the brain.

In this section, we will discuss how the brain and the respiratory muscles interact to maintain quiet breathing. Most of this regulation occurs by means of the nervous system. Although it is an involuntary process, we are not entirely without voluntary (conscious) control of our breathing. You can take a deep breath. You can even hold your breath, but you can't hold it very long.

Nervous Control of Rhythmic Breathing

Breathing is controlled from the region of the brain stem called the *medulla oblongata* (or medulla). Within the medulla oblongata is a group of unique nerve cells called the **respiratory center** (Figure 10.15). These cells are special because they can automatically generate electrical nerve impulses at 2-second intervals. These impulses establish the rhythms for breathing. Nerves conduct the impulses to the diaphragm and intercostal muscles, which contract, and inhalation occurs. When the respiratory center resumes its activity, impulses are generated again and the cycle of contraction-relaxation starts again.

The medulla oblongata is not the only area of the brain involved with breathing. The medulla is connected to another region of the brain stem called the *pons.* Nerve impulses from the pons are transmitted to the respiratory center, where they fine-tune the breathing rhythm.

Without connections from the pons, breathing would occur in gasps.

> The respiratory center is located in the medulla oblongata of the brain. The respiratory center generates nerve impulses that stimulate the diaphragm and intercostal muscles to contract and relax, producing inhalation and exhalation.

Regulating the Rate and Depth of Breathing

Although the rate and depth of breathing are controlled by the respiratory center, they are regulated by factors associated with the changing demands of the body. The most important of these factors is the concentration of three familiar substances: CO_2, H^+, and O_2.

Regulation by CO_2 and H^+. The levels of CO_2 and H^+ are the most important regulators of the rate and depth of breathing. Regulation is accomplished by a group of specialized nerve cells on the surface of the medulla oblongata called a *chemical receptor* (Figure 10.15). These cells monitor increases in CO_2 in the arterial blood. As CO_2 increases, some of the molecules diffuse from the blood into the fluid that surrounds the brain. As CO_2 increases in this fluid, there is a corresponding increase in H^+ because of the reaction

$$CO_2 + H_2O \longrightarrow H_2CO_3 \longrightarrow H^+ + HCO_3^-$$

The chemical receptor cells on the surface of the medulla respond to the H^+ that is produced when CO_2 reacts with water; they do not respond directly to CO_2. As H^+ levels rise, the chemical receptor cells become activated and

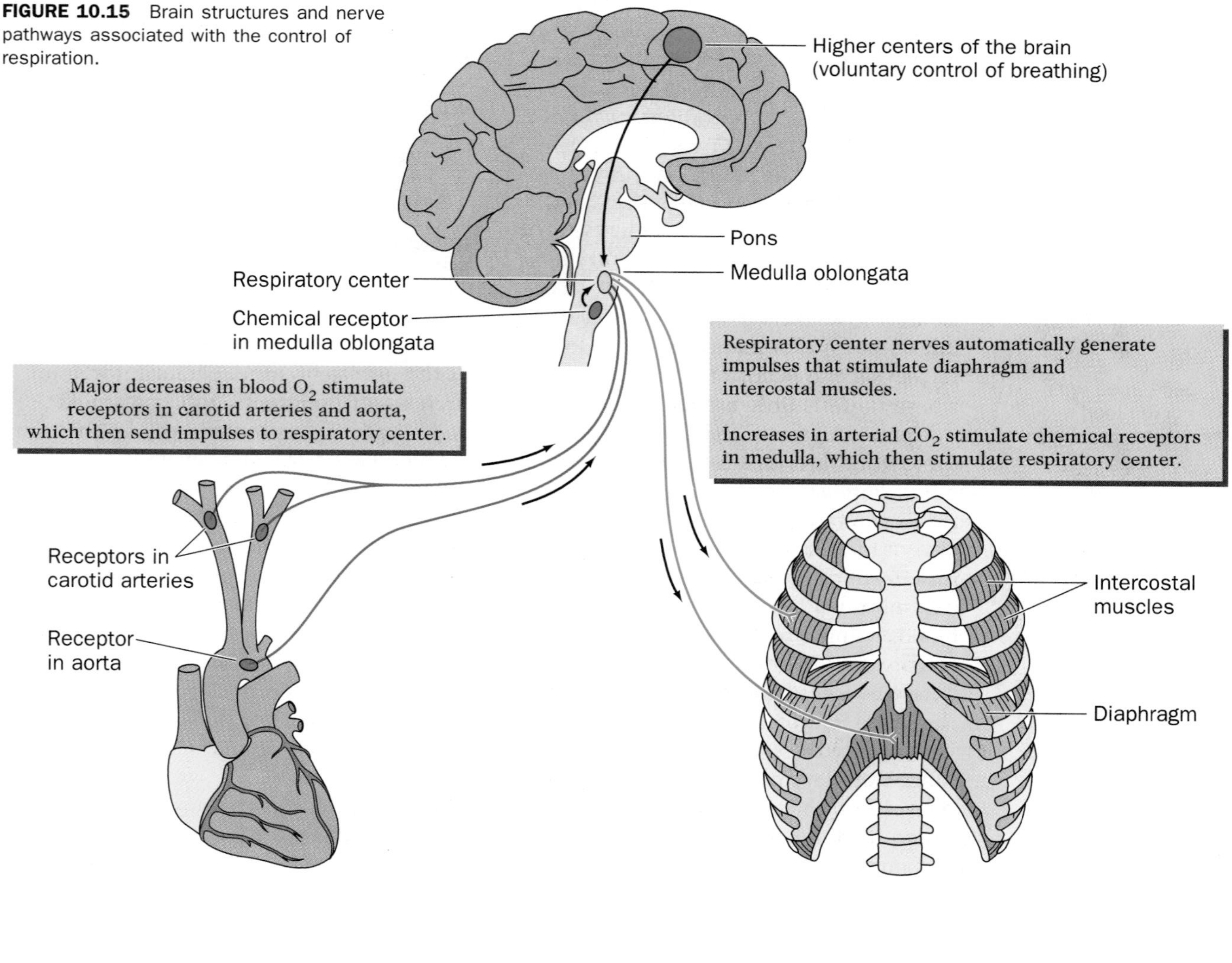
Higher centers of the brain (voluntary control of breathing)
Pons
Medulla oblongata
Respiratory center
Chemical receptor in medulla oblongata
Major decreases in blood O2 stimulate receptors in carotid arteries and aorta, which then send impulses to respiratory center.
Respiratory center nerves automatically generate impulses that stimulate diaphragm and intercostal muscles.
Increases in arterial CO2 stimulate chemical receptors in medulla, which then stimulate respiratory center.
Receptors in carotid arteries
Receptor in aorta
Intercostal muscles
Diaphragm

FIGURE 10.15 Brain structures and nerve pathways associated with the control of respiration.

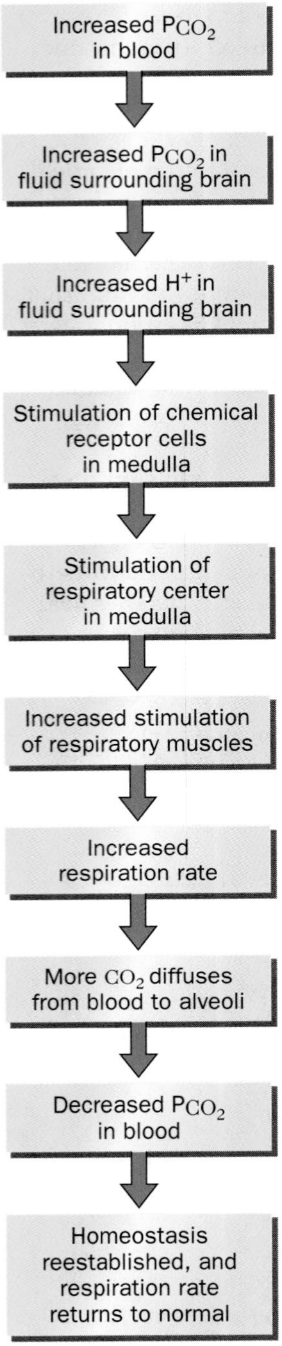
Increased P_{CO_2} in blood
Increased P_{CO_2} in fluid surrounding brain
Increased H^+ in fluid surrounding brain
Stimulation of chemical receptor cells in medulla
Stimulation of respiratory center in medulla
Increased stimulation of respiratory muscles
Increased respiration rate
More CO_2 diffuses from blood to alveoli
Decreased P_{CO_2} in blood
Homeostasis reestablished, and respiration rate returns to normal

FIGURE 10.16 Increased blood P_{CO_2} causes an increased breathing rate and reestablishes homeostasis.

send nerve impulses to the respiratory center in the medulla. After this stimulation, the respiratory center increases the frequency of impulses to the diaphragm and intercostal muscles, and the rate and depth of breathing increase.

As the rate and depth of breathing increase, more CO_2 is exhaled, lowering the level of CO_2 in the arterial blood. With less CO_2 in the blood, CO_2 diffuses out of the fluid surrounding the brain and H^+ levels drop. With a decrease in H^+ the chemical receptor cells in the medulla are no longer stimulated and the respiratory center quiets down. This control mechanism maintains homeostasis during both rest and activity (Figure 10.16).

Regulation by O_2. The rate and depth of breathing are also regulated by the concentration of O_2 in the blood. Receptors in the carotid arteries and aorta monitor the O_2 level of arterial blood (see Figure 10.15). Normally, the partial pressure of O_2 (P_{O_2}) in arterial blood is about 100 mm Hg. However, these O_2 receptors become activated only when the partial pressure of O_2 drops below 60 mm Hg, and this rarely happens. This means that under conditions of normal health and activity, O_2 regulation of breathing is not as important as regulation by CO_2 and H^+. Extremely low levels of arterial O_2 are rare but do occur. For example, low blood P_{O_2} levels can occur at high altitudes where there is little atmospheric O_2.

Upon activation, the O_2 receptors in the carotid arteries and aorta send nerve impulses to the respiratory center in the medulla, which increases the rate of impulses to the diaphragm and intercostal muscles. With faster breathing, more O_2 is made available to the blood, helping to maintain homeostasis (Figure 10.17).

Conscious control of breathing. To a limited extent, we can alter our breathing voluntarily through conscious decision making. Voluntary control of our breathing permits us to do such things as talk, sing, and hold our breath. There are limitations, however, to our ability to alter breathing voluntarily. For example, we cannot hold our breath indefinitely. After several minutes without a breath, we will inhale again. Our voluntary control is overridden by involuntary controls that respond to the increases in CO_2 and H^+ described above.

The rate and depth of breathing are regulated by two chemical systems. The CO_2/H^+ system is the more important. In the second system, O_2 receptors are stimulated when P_{O_2} falls below 60 mm Hg.

Diseases and Disorders

Because the respiratory system interacts directly with the outside environment, it is susceptible to a multitude of irritants, toxins, infectious microorganisms, and obstructions. Any of these elements can interfere with the body's immediate or continued ability to obtain O_2 and release CO_2. In the following sections we will briefly describe how various agents or conditions interfere with the exchange of gases in the lungs, causing respiratory diseases and disorders.

Inflammation by Irritants and Infectious Agents

Our respiratory systems are constantly exposed to chemical and biological agents in the air we breathe. Chemical irritants such as air pollutants, tobacco smoke, and allergens can inflame the epithelium of the airways, causing swelling, excess mucus secretion, tissue damage, and pain. These problems are also caused by infectious agents, such as viruses, bacteria, and fungi, that enter and grow on the respiratory epithelium. Although many types of bacteria normally reside on the epithelium of the upper respiratory tract, they do not cause harm. In fact, the presence of these microorganisms may inhibit the growth of harmful bacteria.

Inflammation of the nasal cavity, which is called *rhinitis* (*rine-**eye**-tis*), may be caused by irritating chemical agents, viruses, or bacteria. The familiar

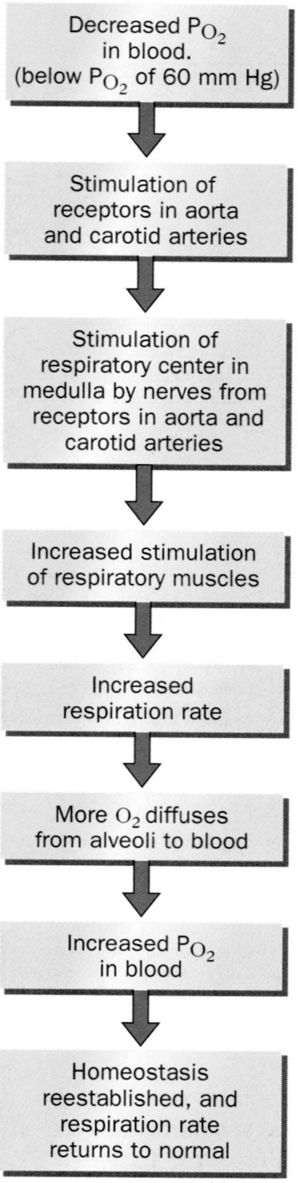
Decreased P_{O_2} in blood. (below P_{O_2} of 60 mm Hg)
Stimulation of receptors in aorta and carotid arteries
Stimulation of respiratory center in medulla by nerves from receptors in aorta and carotid arteries
Increased stimulation of respiratory muscles
Increased respiration rate
More O_2 diffuses from alveoli to blood
Increased P_{O_2} in blood
Homeostasis reestablished, and respiration rate returns to normal

FIGURE 10.17 Decreased blood P_{O_2} causes an increased breathing rate and reestablishes homeostasis.

"runny nose" is caused by the excess fluid produced by inflamed mucous membranes. This fluid also drains into the lower airways, where it may stimulate the cough reflex.

Compounding the effects of irritating and infectious agents, prolonged speaking, shouting, or singing can cause *laryngitis,* which is an inflammation of the larynx and a swelling of the vocal cords. Swelling interferes with vocal cord vibration, and you may become hoarse or even lose your voice. Prolonged cigarette smoking, say, for 20 or more years, also may lead to a hoarse voice from the chronic irritation and damage to the larynx and vocal cords.

Pneumonia (*new-mow-knee-ah*) is an inflammation of the alveoli which causes them to secrete and fill with excess fluid (Figure 10.18). As a result, the exchange of O_2 and CO_2 is greatly inhibited. If the inflammation spreads, affecting many alveoli, suffocation and death may result. Pneumonia is usually associated with infectious agents such as viruses, bacteria, protozoa, and fungi, but inhaled chemical irritants also can inflame the alveoli and cause pneumonia.

Tuberculosis (TB) is caused by the bacterium *Mycobacterium tuberculosis.* These bacteria are usually inhaled with contaminated airborne droplets coughed or sneezed by an infected person. The bacteria enter the lungs, slowly destroy tissue, and damage blood vessels. Bleeding from these damaged vessels produces bloody mucus (sputum), a symptom of TB (Figure 10.18). A simple skin test called the *tuberculin test* can be used to determine whether one has been exposed to the bacterium.

Cases of tuberculosis declined dramatically after the development and introduction of antibiotics in the 1940s. However, since 1986 the number of tuberculosis cases has been increasing in the United States. A combination of factors are responsible for this, including antibiotic-resistant strains of the bacterium, impaired body defenses in HIV-infected people, and declining living standards and health care for the poor, the homeless, and the mentally ill.

Obstructions to Air Flow or Gas Exchange

Chronic respiratory disorders such as asthma, cystic fibrosis, and emphysema obstruct airflow or reduce gas exchange in the alveoli. Asthma occurs both in young people and in adults. Children frequently seem to outgrow the affliction, while adult-onset asthma is more likely to become a chronic condition. Cystic fibrosis affects the young and those under age 30. Emphysema, by contrast, is usually found in those over the age of 60. Although not considered a chronic respiratory disease, lung cancer not only affects the lungs and airways but can spread rapidly, damaging other organs. It occurs more frequently in those over age 40.

Asthma (*as-mah*) is a condition that periodically and severely reduces airflow, causing breathing difficulties, wheezing, and coughing. Asthma attacks can occur without warning and can last from minutes to hours. During an attack, contractions of the smooth muscles encircling the smaller bronchi and bronchioles reduce airflow to the alveoli. Further obstruction may occur if the epithelial tissue becomes inflamed and secretes mucus into the airway.

Asthma affects about 10 percent of all children and 5 percent of all adults. There is no single cause. Common triggers of an asthma attack include allergens, air pollutants, food additives, tobacco smoke, strenuous exercise, and emotional upset. Asthma cannot be cured, but attacks can be controlled by medications that relax bronchial smooth muscles and open the airways.

Cystic fibrosis (CF) is an inherited disorder that affects many organs, including those of the respiratory system. It occurs about once in every 2000 births of white children; the incidence is much lower in other racial groups. Because of a damaged gene, individuals with CF cannot synthesize a specific membrane transport protein. As a result, secreted mucus becomes

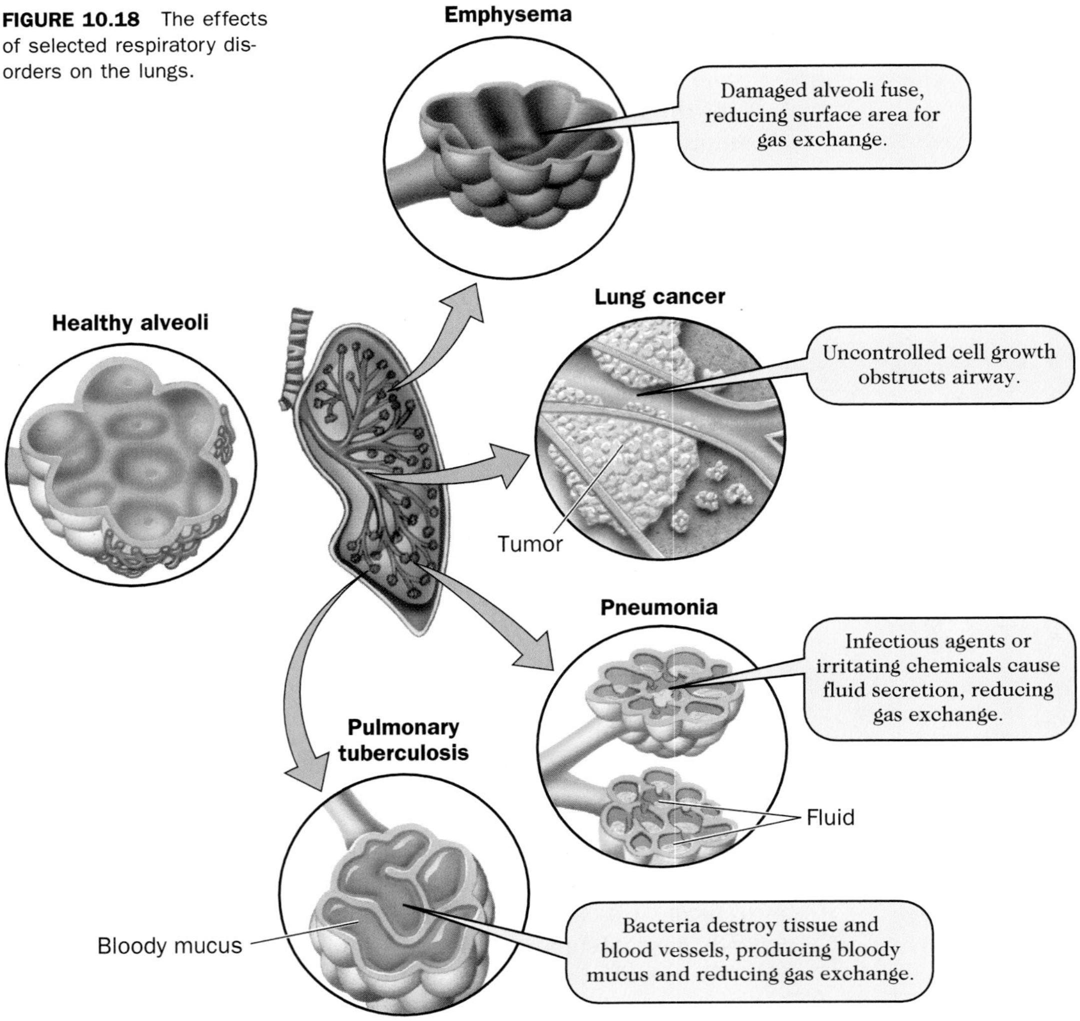

Emphysema
Damaged alveoli fuse, reducing surface area for gas exchange.
Healthy alveoli
Lung cancer
Uncontrolled cell growth obstructs airway.
Tumor
Pneumonia
Infectious agents or irritating chemicals cause fluid secretion, reducing gas exchange.
Fluid
Pulmonary tuberculosis
Bloody mucus
Bacteria destroy tissue and blood vessels, producing bloody mucus and reducing gas exchange.

FIGURE 10.18 The effects of selected respiratory disorders on the lungs.

abnormally thick, causing two problems in the airways. First, the thickened secretions accumulate, reducing airflow and gas exchange. Second, the accumulated mucus provides a growth medium for bacteria. People suffering from CF have frequent bacterial infections of the airways.

The effects of cystic fibrosis on the body can be partially managed with various types of therapy. However, over the years, enough tissue and organ damage occurs to cause death before an individual with CF reaches age 30. However, there is hope. The gene responsible for CF has been identified, isolated, and analyzed. Sooner than was thought possible, CF may be cured by insertion of the correctly engineered gene into patients with the disorder.

Emphysema is a condition in which the epithelium of the alveoli becomes damaged and the alveoli fuse into flattened, shallow chambers (Figure 10.18). The result is a substantial reduction in the surface area for gas exchange. Some degree of emphysema is common in the lungs of older persons and may be part of the normal aging process. However, the severity of the disease increases among those who are heavy smokers, live or work in areas with air pollution, or suffer from chronic asthma. Lung damage from emphysema is irreversible, but treatment with supplemental O_2, medications, and breathing techniques can help.

Lung cancer most often originates in the bronchial epithelium. This disease kills more people than does any other form of cancer in the United States. Lung cancer is most prevalent in persons between 40 and 75 years of age, because it may take years of exposure to cancer-causing agents (carcinogens) to transform normal cells to cancer cells. Furthermore, it may take 5 to 10 years for the cancer cells to become numerous enough to be detected (Figure 10.18). By that time, some cells will have spread to the lymph nodes, liver, brain, and bone. Treatment can include surgical removal of one or more lobes of the lung, radiation therapy, and chemotherapy [see Spotlight on Health: One (page 103) for a general discussion of cancer]. Many of these deaths could be prevented by not smoking cigarettes or other tobacco products.

Many substances are believed to cause lung cancer, but cigarette smoke alone contains as many as 15 known carcinogens. Statistics show that after 20 years of smoking cigarettes, you have a 10 to 20 times greater risk of developing lung cancer than do nonsmokers. Even nonsmokers who regularly breathe exhaled tobacco smoke (called secondhand smoke) have an increased risk of developing lung cancer. However, there is some good news. The risk from smoking diminishes when you stop smoking. After 20 years of smoking, if you stop completely, your risk of lung cancer declines. After about 12 years without smoking, your risk of lung cancer is the same as that of someone who has never smoked.

Neuromuscular Disorders of Respiration

Breathing is affected and death may result when nerve stimulation of the respiratory muscles is impaired or severed. *Brain injury* from trauma, tumors, or stroke may damage the respiratory center in the medulla oblongata. This will interfere with or prevent adequate nerve stimulation to the diaphragm and intercostal muscles. An *upper spinal cord injury* may damage the nerves that transmit impulses from the respiratory center to the diaphragm. When this happens, breathing ceases.

Toxins produced by certain bacteria can prevent the transmission of nerve impulses at the junction of the nerve and muscle. The botulism toxin produced by the bacterium *Clostridium botulinum* can be absorbed from improperly cooked food, causing paralysis of all skeletal muscles, including the diaphragm and intercostal muscles.

Life Span Changes

In this section we will emphasize the programmed changes that are essential to life or are a natural part of aging. These life span changes can be conveniently divided into two categories: those which occur during birth and early infancy and those which begin in early adulthood and are most evident in old age.

Birth and Early Infancy

While developing within the womb, a fetus is totally immersed in a fluid-filled environment. Its airways and lungs are filled with fluid and do not participate in gas exchange. The fetus depends entirely on the placenta for gas exchange with the mother's blood: The mother breathes for the fetus. At birth, the fluid drains from the infant's respiratory system, but breathing does not begin until the partial pressure of CO_2 increases in the infant's blood. This stimulates the respiratory center, and the respiratory muscles are activated. The newborn takes its first breath, and life independent of the mother begins.

In newborns, two respiratory disorders may cut life short. *Infant respiratory distress syndrome* (*IRDS*) occurs when the alveoli of the lungs do not expand sufficiently to achieve adequate gas exchange. Premature babies are most vulnerable to this condition because their lungs do not produce sufficient surfactant. IRDS is treated by using a machine called a respirator to force air into the alveoli so that they remain expanded and functional. Surfactant can also be supplied with inhaled air to aid the development of normal lung expansion. In the United States, approximately 1 in every 100 babies is born with IRDS, which represents more than 40,000 newborns each year.

Another affliction of newborns is *sudden infant death syndrome (SIDS)*, or *crib death.* SIDS is the leading cause of infant mortality in the United States and is responsible for about 5500 deaths annually. Death occurs in healthy infants less than a year old when breathing stops unexpectedly while they are sleeping. We don't know what causes this to happen, but suggested causes include viral infections, an improperly functioning respiratory center, and allergic responses. Some experts propose that the sleeping position may also be a factor and suggest that babies should be placed on their sides when sleeping. Others propose that overheating while sleeping increases the risk of SIDS.

Although these are desperate disorders for both the infant child and the parents, they are relatively rare. Most infants are born with a normal healthy respiratory system.

Changes Later in the Life Span

Over the entire life span, changes take place in the muscular, connective, and epithelial tissues of the respiratory system. These changes are similar to the changes that take place in other organ systems.

Beginning in the middle twenties, airflow to the lungs begins to decrease. Vital capacity declines 40 percent between ages 25 and 85. This usually does not reflect any change in the total volume of the lungs; instead, it reflects changes in the capacity of the airways, the flexibility of the rib cage, and the number and activity of muscle fibers in the diaphragm and intercostal muscles.

Although the elderly have as many alveoli as do the young, changes in the shape of the alveoli reduce the efficiency of gas exchange. Until about age 40 the alveoli have the characteristic cuplike structure. After age 40, the alveoli slowly become more saucer-shaped, reducing the surface area for gas exchange. By age 70, as much as 20 percent of the alveolar surface area can be lost.

Changes also occur in the epithelium of the airways. Ciliated mucous membranes release less fluid, and the activity and number of cilia are reduced. As a result, foreign particles and substances are not trapped and removed as effectively. This increases our susceptibility to infections and inflammations as we get older.

Although the consequences of some of these changes are most apparent and pronounced in the elderly, it is important to emphasize that the changes occur naturally over the entire life span. Of course, there are variations among individuals, and both the onset and pace of these changes are affected by the environment and one's lifestyle.

Chapter Summary

Respiratory Organs

- The upper respiratory tract includes the external nose, nasal cavity, and pharynx.
- The lower respiratory tract includes the larynx, trachea, bronchi, bronchioles, and alveoli.
- The pleura is an epithelial lining that covers the lung surface and the thoracic cavity.
- Surfactant is a lipoprotein that coats the inner lining of alveoli and prevents their collapse.

Breathing: How We Inhale and Exhale

- When the volume of a space increases, the air pressure inside that space decreases. When the volume of a space decreases, the air pressure inside that space increases.
- Inhalation occurs when the diaphragm and outer intercostal muscles contract, producing an increase in thoracic cavity volume. As a result, the lungs expand and outside air rushes into the lungs. The opposite occurs during exhalation.
- The respiration rate is the number of times we inhale and exhale per minute.
- Tidal volume is the amount of air inhaled and exhaled during quiet breathing (500 milliliters).
- Vital capacity is the maximum air volume inhaled and exhaled in one breath (4800 milliliters).

Gas Exchange between the Lungs, Blood, and Tissues

- In a mixture of gases, the partial pressure of a particular gas represents the pressure it contributes to the total pressure of the mixture.

- External respiration is the exchange of gases between the alveoli and the blood. O_2 diffuses from the alveoli to the blood. CO_2 diffuses from the blood to the alveoli.
- Internal respiration is the exchange of gases between the blood and the tissues. O_2 diffuses from the blood to the tissues. CO_2 diffuses from the tissues to the blood.
- About 98 percent of the O_2 transported in the blood is attached to the heme group of hemoglobin.
- Carbon dioxide is transported in the blood in three ways: dissolved in plasma, attached to the globin protein of hemoglobin, and as bicarbonate after reacting with H_2O.

Control of Breathing

- Breathing is controlled by nerve cells of the respiratory center located in the medulla oblongata.
- Nerves from the respiratory center stimulate the diaphragm and intercostal muscles.
- The concentrations of CO_2 and H^+ regulate the rate and depth of breathing.
- Only very low blood O_2 levels stimulate an increase in the rate and depth of breathing.

Diseases and Disorders

- Chemical irritants and infectious agents can inflame the inner epithelial lining of the respiratory airways.
- Conditions that obstruct airflow or gas exchange include asthma, cystic fibrosis, emphysema, and lung cancer.
- Neuromuscular disorders affecting stimulation of the respiratory muscles can result from brain damage, upper spinal cord damage, and toxins from certain bacteria.

Life Span Changes

- While a fetus is in the womb, fluid fills its lungs and gas exchange occurs in the placenta.
- Infant respiratory distress syndrome occurs predominantly in premature babies because of insufficient surfactant.
- Sudden infant death syndrome occurs in healthy infants less than 1 year old. Its cause is unknown.
- During adulthood, gradual changes take place in the tissues of the respiratory airways and alveoli which reduce the vital capacity by as much as 40 percent.

Selected Key Terms

alveolus (p. 266)
bronchiole (p. 265)
bronchus (p. 265)
carbonic anhydrase (p. 274)
diaphragm (p. 268)
external respiration (p. 271)
internal respiration (p. 271)
larynx (p. 264)
oxyhemoglobin (p. 272)
respiration rate (p. 269)
respiratory center (p. 274)
tidal volume (p. 269)
trachea (p. 265)
vital capacity (p. 270)
vocal cords (p. 264)

Review Activities

1. Trace the pathway of inhaled air through the upper respiratory tract.
2. Trace the pathway of inhaled air through the lower respiratory tract.
3. What is the function of surfactant?
4. Describe how increases and decreases in air volume affect air pressure within a space.
5. What are the roles of the diaphragm and the outer intercostal muscles during inhalation and exhalation?
6. What is the difference between tidal volume and dead air volume?
7. Identify the three respiratory volumes that contribute to vital capacity.
8. Give values for the following: percent of atmospheric O_2, percent of atmospheric CO_2, atmospheric air pressure at sea level, P_{O_2} of inhaled air, and P_{CO_2} of inhaled air.
9. Explain why oxygen diffuses from the alveoli to the blood.
10. What is the origin of CO_2 in the blood?
11. Will more oxyhemoglobin be formed under high P_{O_2} or low P_{O_2}? Will oxyhemoglobin release O_2 under high P_{O_2} or low P_{O_2}?
12. Describe the three ways CO_2 is transported in the blood.
13. Describe how the respiratory center in the medulla oblongata regulates quiet breathing.
14. Describe how high CO_2 levels in the blood increase the rate and depth of breathing.
15. Do high or low O_2 levels in the blood cause an increase in the rate and depth of breathing?

Self-Quiz

Matching Exercise

___ 1. Part of respiratory tract that is the common passageway of food and air
___ 2. Part of respiratory tract that contains the vocal cords
___ 3. Another name for the windpipe
___ 4. Saclike structures where gas exchange takes place in the lungs
___ 5. The lipoprotein that reduces surface tension of alveoli
___ 6. The respiratory muscle that separates the thoracic and abdominal cavities
___ 7. The maximum volume of air that can be inhaled and exhaled
___ 8. The number of O_2 molecules that can attach to one hemoglobin molecule
___ 9. The enzyme that increases the reaction rate between CO_2 and H_2O in red blood cells
___ 10. The part of the brain that contains the respiratory center

A. Larynx
B. Medulla oblongata
C. Surfactant
D. Pharynx
E. Diaphragm
F. Four
G. Trachea
H. Vital capacity
I. Alveoli
J. Carbonic anhydrase

Answers to Self-Quiz

1. D; 2. A; 3. G; 4. I; 5. C; 6. E; 7. H; 8. F; 9. J; 10. B

LT 850
United States

Chapter 11

The Urinary System

Water has many moods and plays many roles. It can be soft and solid, firm and flexible, cold and hot, powerful and gentle. In our cells, it is both an actor and the stage on which all the action takes place. It is essential to life, and it rules human lives. We need it, but we can have too much or too little of it.

Because we are terrestrial animals, our problem is most frequently with too little water, and we often are in danger of drying out. There are many ways in which water can be lost from the body. We lose water in our sweat and tears and with each breath we exhale. Our greatest water loss, however, is associated with the dilution, transport, storage, and removal of wastes. The water we lose must be replaced by drinking fluids and eating foods.

FIGURE 11.1 Excretory organs of the body.

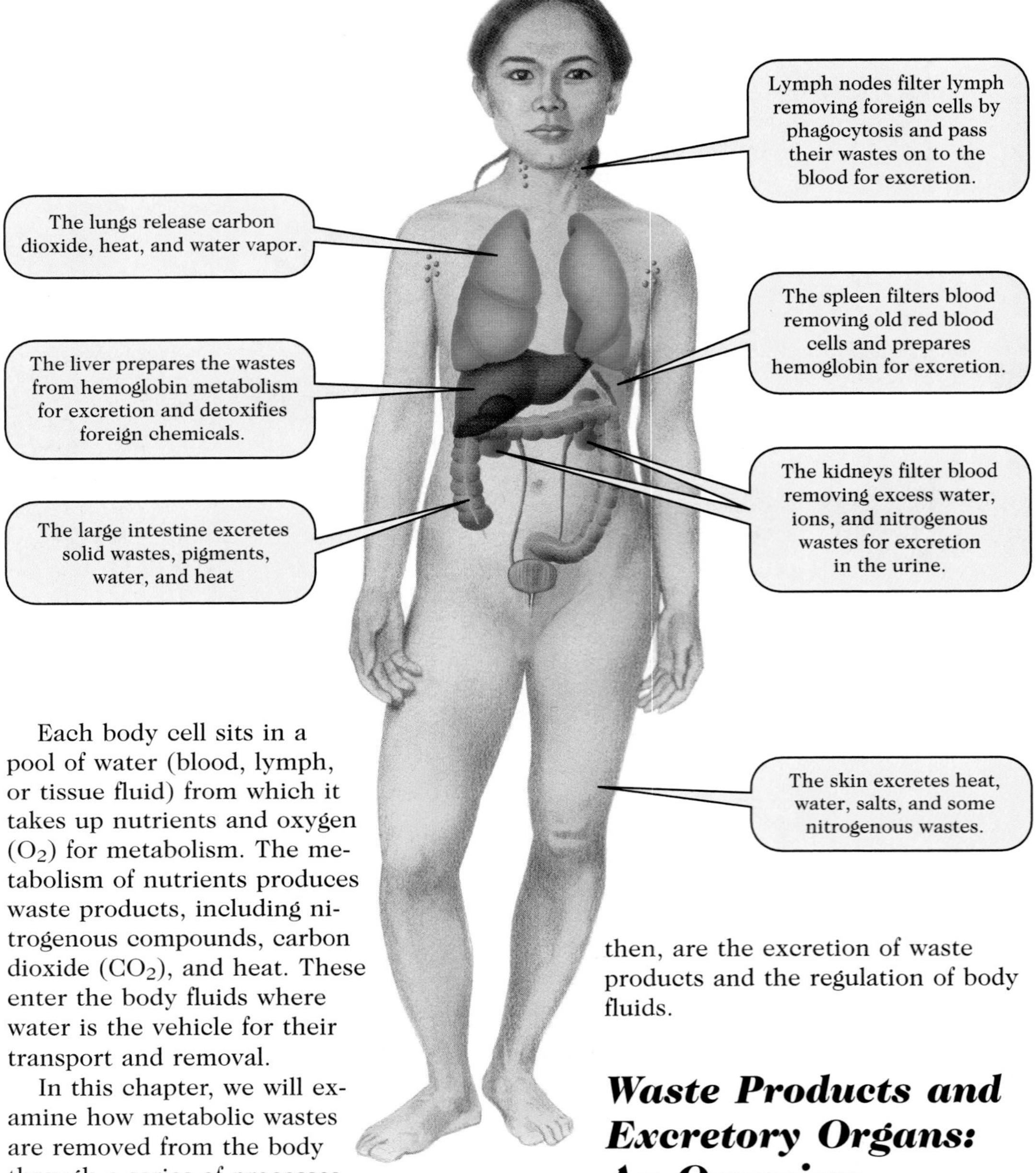
Lymph nodes filter lymph removing foreign cells by phagocytosis and pass their wastes on to the blood for excretion.
The lungs release carbon dioxide, heat, and water vapor.
The spleen filters blood removing old red blood cells and prepares hemoglobin for excretion.
The liver prepares the wastes from hemoglobin metabolism for excretion and detoxifies foreign chemicals.
The kidneys filter blood removing excess water, ions, and nitrogenous wastes for excretion in the urine.
The large intestine excretes solid wastes, pigments, water, and heat
The skin excretes heat, water, salts, and some nitrogenous wastes.

Each body cell sits in a pool of water (blood, lymph, or tissue fluid) from which it takes up nutrients and oxygen (O_2) for metabolism. The metabolism of nutrients produces waste products, including nitrogenous compounds, carbon dioxide (CO_2), and heat. These enter the body fluids where water is the vehicle for their transport and removal.

In this chapter, we will examine how metabolic wastes are removed from the body through a series of processes called *excretion* that involve several **excretory organs,** including those of the **urinary system** (Figure 11.1). However, the story is not only about wastes and their removal. The story of excretion is also about *regulation,* the process of controlling the composition of the body fluids in which our cells live and thrive. To maintain stable internal conditions (homeostasis), the excretory organs also must help regulate internal concentrations of water, H^+, and various solutes, including the ions Na^+ and K^+. The key themes of this chapter, then, are the excretion of waste products and the regulation of body fluids.

Waste Products and Excretory Organs: An Overview

Wastes are produced when we process foods, conduct metabolism, combat disease, and repair damaged structures. A variety of substances can be considered wastes: toxic substances, excess nontoxic substances, and substances that cannot be processed further and used. All such substances must be removed from the body. We often think of wastes as only the undigestible matter (feces) that accumulates in the large intestine and is removed by defecation. However, in this chapter we will focus on metabolic wastes that are soluble in the body fluids and are stored and removed in a liquid state. These substances are toxic or may become so if allowed to accumulate in cells or body fluids. The most potentially toxic metabolic wastes are those which contain nitrogen.

Nitrogenous Wastes

Atoms of nitrogen are essential components of amino acids and nucleotides, which form proteins and nucleic acids (DNA and RNA), respectively. Unfortunately, some forms of nitrogen are toxic, and our cells and organs must handle those substances carefully. The normal metabolic breakdown of amino acids and nucleotides produces smaller molecules, some of which contain nitrogen and are toxic. For example, when amino acids are broken down, the nitrogen-containing amino groups (—NH_2) are removed in a process called *deamination.* The amino groups are chemically converted to **ammonia (NH_3)** and then to a less toxic molecule called urea.

Urea is produced in the liver when two molecules of ammonia are combined with one molecule of CO_2 in a complex series of chemical reactions called the *urea cycle.* Urea, which can be diagrammed structurally as

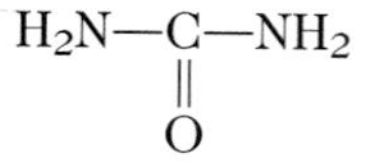

is about 100,000 times less toxic than ammonia. Urea can be transported, stored, and released in a concentrated form. Although urea is less toxic than ammonia, too much urea can be toxic. Failure to excrete sufficient urea from the blood produces *uremia,* a disorder which may cause nausea, vomiting, vertigo, convulsions, and coma.

Other Metabolic Wastes: Too Much of a Good Thing

In the preceding chapters we saw how the body releases excess heat, transports and removes CO_2, and rids itself of bile acids from the breakdown of heme. The body must also excrete the excess **ions** in the food we eat. Although ions are essential for life, when in excess, they are waste. Ions present in body fluids include sodium (Na^+), potassium (K^+), bicarbonate (HCO_3^-), hydrogen (H^+), calcium (Ca^{2+}), chloride (Cl^-), sulfate (SO_4^{2-}), and phosphate (PO_4^{3-}). The most abundant ions (Na^+, K^+, and Cl^-) are important in maintaining the osmotic pressures of fluids and the electrical charges on the plasma membranes of all cells. Other ions (HCO_3^- and HPO_4^{2-}) help maintain the pH of body fluids.

In addition to being essential for life processes, **water** can be considered a waste product if it is present in excess, and so the body's content and distribution of water must be closely regulated. In addition to being ingested in food and drink, water is produced during the catabolism of sugars, proteins, and fats. We manufacture around 200 milliliters (about 1 cup) on an average day. This amount is small compared with the amount we ingest (about 2200 ml or 2.4 quarts) daily. On a typical day our water intake plus our metabolic water production should equal our water output.

> Substances that are poisonous or excessive or that cannot be used by our bodies are excreted as wastes. Waste nitrogen is excreted from the body primarily in the form of urea. Though essential for life, when in excess, water and ions are excreted as wastes.

Water and Urine

Water is the most abundant molecule in our bodies. As infants, humans are about 75 percent water. As we grow older, this percentage declines. An adult male is about 60 percent water, while an adult female is about 55 percent. The difference lies in a woman's slightly higher percentage of body fat.

Although some body water is lost from the lungs and skin, most excess water is excreted as urine by the kidneys. Urine is usually 95 percent water and only 5 percent dissolved wastes. The exact composition of urine varies daily, depending on the food and drink you consume, your activity, and your general health. The volume of urine you produce each day varies between 1 and 2 liters (about 1 and 2 quarts), depending on your consumption of fluids and salt (NaCl) and your blood pressure.

The chemical analysis of urine, which is called **urinalysis,** identifies and measures specific substances in the urine. Such determinations can reveal metabolic changes in the body that are associated with diseases and disorders. Examination of urine samples with the microscope may also reveal abnormal cells, microorganisms, and deposits which may indicate disease states.

> The 1 to 2 liters of urine produced daily is mostly water but also contains nitrogenous wastes and ions which are present in excess in the blood.

Excretory Organs

The excretory organs of the body include the skin, lungs, liver, large intestine, lymph nodes, and spleen, as well as the kidneys and their associated structures. This is a long list, yet each of these organs plays a direct role in the filtration, transport, and excretion of waste materials from the body.

The *skin* is the major organ for dissipating heat through the evaporation of sweat. In addition to excreting water, the *sweat* glands of the skin are a vehicle for the excretion of ions, urea, ammonia, and small amounts of other substances. Although we properly think of the *lungs* as the site for the release of CO_2 from the body, they also release water vapor and heat when we exhale.

The *spleen* and *lymph nodes* filter and cleanse the blood and lymph, respectively, using phagocytic cells to remove debris, microorganisms, and worn-out red blood cells. Wastes from phagocytic cells eventually pass into the blood and are filtered out by the kidneys. The **liver** is another internal organ that plays a role in excretion. It chemically alters foreign and toxic substances absorbed into the blood, reducing their toxicity and making them easier to excrete into the urine.

While these different organs are all involved in waste removal, the **kidneys** are responsible for most of the excretion of nitrogenous wastes and the regulation of chemical conditions in the blood.

> Organs excreting metabolic wastes include the skin, lungs, spleen, lymph nodes, liver, and kidneys. The kidneys are the most important organs for nitrogenous waste excretion.

The Urinary System

The organs of the urinary system include the paired kidneys, the paired ureters, the urinary bladder, and the urethra (Figure 11.2). In this section, we will describe the general location and structure of these organs. Although they may appear to lie in the abdominal cavity, these organs are actually positioned between the peritoneum (the serous membrane lining the abdominal cavity) and the posterior body wall.

Kidneys: Filtering and Regulating

As the primary organs of excretion, the kidneys perform two major functions. They *filter* the blood, removing water, ions, and nitrogenous wastes and forming urine. In doing so, they *regulate* the volume, osmotic pressure, and pH of the blood and body fluids.

The kidneys are each about as large as your fist. They are situated below the diaphragm on the left and right sides, and are partially protected by the eleventh and twelfth pairs of ribs. The right kidney is a bit lower than the left because the liver takes up so much room on that side. Each kidney is surrounded by layers of fibrous connective tissue and a mass of adipose (fat) tissue which cushions it and holds it in place.

Each kidney is bean-shaped and has an indentation called a *hilus* (***high-luss***) where blood vessels, lymphatic vessels, and nerves enter and leave the organ.

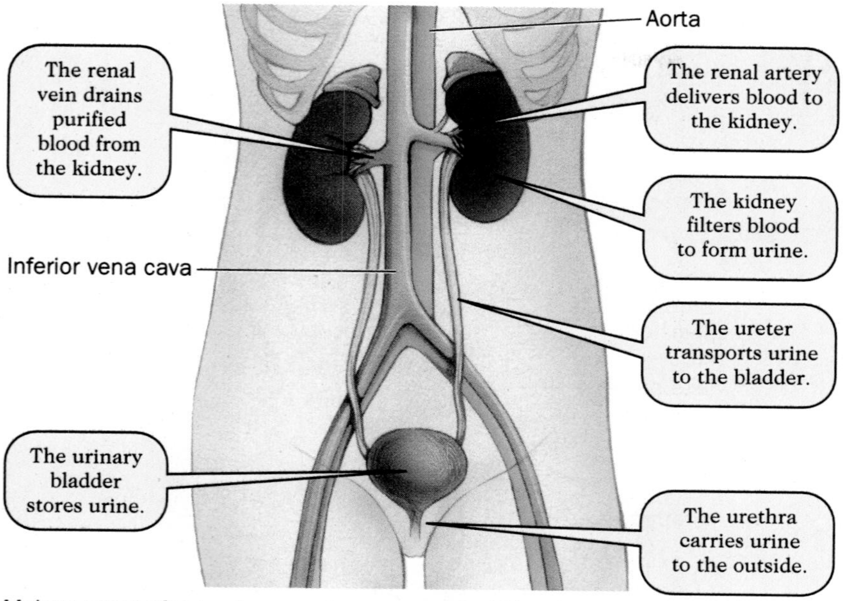
Aorta
The renal vein drains purified blood from the kidney.
The renal artery delivers blood to the kidney.
The kidney filters blood to form urine.
Inferior vena cava
The ureter transports urine to the bladder.
The urinary bladder stores urine.
The urethra carries urine to the outside.

Major organs of the urinary system, associated blood vessels, and their functions

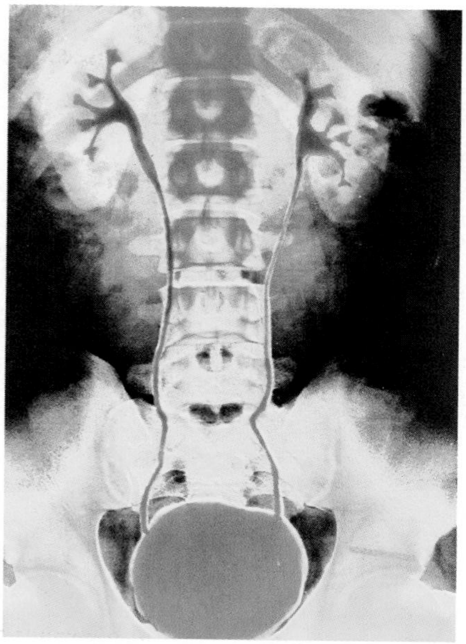

Can you identify the structures in this colored x-ray?

FIGURE 11.2 The urinary system.

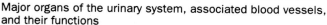

The hilus is also the site where a drainage tube, the ureter, exits each kidney.

Ureters: Transport Tubes

The **ureters** (*you-ree-ters*) are narrow tubes that carry urine from the kidneys down to the urinary bladder for temporary storage. Their walls contain a relatively thick layer of smooth muscle which contracts rhythmically, moving urine to the bladder, especially while you are lying down. The ureters join the bladder on its posterior wall. As the bladder fills with urine, the increased internal pressure closes the openings to prevent the backup of urine into the ureters.

Urinary Bladder: Temporary Storage

The **urinary bladder** is a muscular sac that stores urine until it can be released to the outside. Because of differences in their reproductive systems, the position of the bladder is somewhat different in men and women (Figure 11.3a, b). In women, the overlying uterus compresses the bladder somewhat, causing it to hold a smaller volume of urine than it does in most men. Depending on sex and body size, bladder volume ranges between 0.6 liter and 1 liter (about $\frac{2}{3}$ to 1 quart).

The structure of the bladder fits its role as a storage organ. The lining of the bladder is formed from a specialized stratified epithelium which is designed for stretch and distension. Called *transitional epithelium,* this tissue is found only in the urinary system. The bladder wall contains three layers of smooth muscle, which together are called the **detrusor muscle** (*dee-**true**-sore*). Contractions of the detrusor muscle help expel urine forcefully from the bladder into the urethra.

Urethra: Releasing Urine

The **urethra** (*you-ree-thra*) is a single tube, about as wide as a pencil, which exits from the base of the bladder and carries urine to the outside of the body. Surrounding the urethra where it exits from the bladder is a ring of *smooth muscle,* called the **internal urethral sphincter,** which relaxes and opens for urine release. Below this is a similar ring of *skeletal muscle,* the **external urethral sphincter,** which allows voluntary control of urine release.

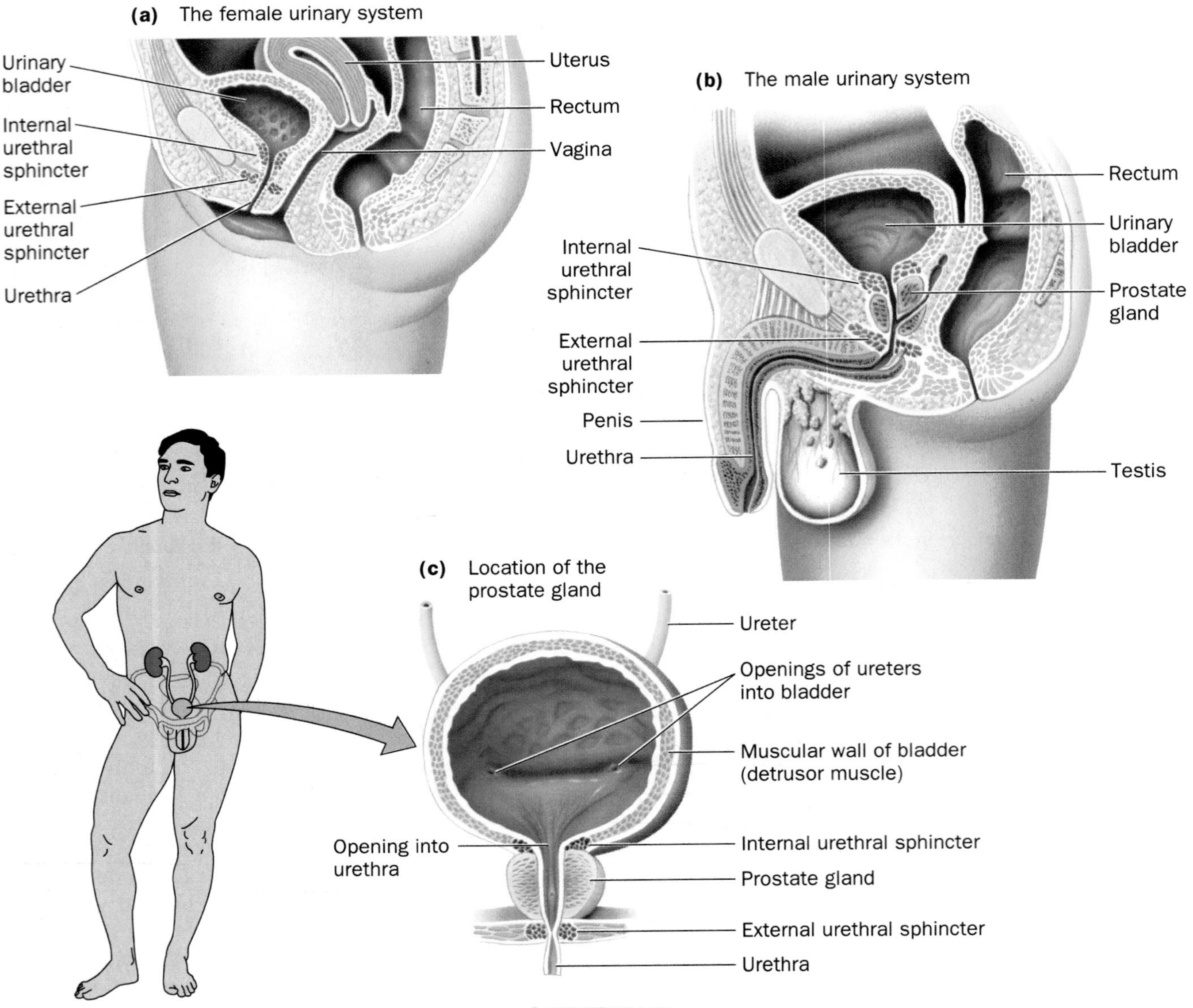
(a) The female urinary system
Urinary bladder
Internal urethral sphincter
External urethral sphincter
Urethra
Uterus
Rectum
Vagina
(b) The male urinary system
Rectum
Urinary bladder
Prostate gland
Internal urethral sphincter
External urethral sphincter
Penis
Urethra
Testis
(c) Location of the prostate gland
Ureter
Openings of ureters into bladder
Muscular wall of bladder (detrusor muscle)
Opening into urethra
Internal urethral sphincter
Prostate gland
External urethral sphincter
Urethra

FIGURE 11.3 Position of the bladder.

The urethra of a woman is shorter than that of a man, only about 4 centimeters ($1\frac{1}{2}$ inches) in length. That means that bacteria and yeast on the body surface have to travel only a short distance to reach the bladder; consequently, bladder infections are more common in women than in men. In women, the exit of the urethra lies just anterior to the entrance into the vagina, the first portion of the female reproductive tract. The female urethra carries only urine. This separation of the urinary and reproductive passageways minimizes the opportunity for infections to spread from the urinary system to the reproductive organs, especially during pregnancy.

A man's urethra is longer, measuring some 20 centimeters (8 inches), because it passes through the penis before it reaches the exterior. At its point of exit from the bladder, the male urethra is encircled by a reproductive gland called the **prostate** (***pross**-tate*) (Figure 11.3c). In older males, this gland may enlarge, compress the urethra, and interfere with urine release. The male urethra fulfills a double function, carrying both urine and reproductive fluids and sperm.

Urine is produced in the kidney and exits through the ureter at the hilus. The ureters carry urine to the bladder for storage. Sphincter muscles control urine release from the bladder into the single urethra, which carries urine to the outside.

Inside the Kidney: Structures and Functions

Your body contains about 5 liters (about 1.4 gallons) of blood. Each day all this blood passes through both kidneys more than 350 times. From this blood the kidneys remove about 180 liters (nearly 50 gallons) of filtered fluid. However, this fluid is not all excreted as urine. Most of it is reabsorbed back into the blood after wastes have been separated away. In fact, over 99 percent of the filtered fluid is reabsorbed, and only about 1 to 2 liters (about 1 to 2 quarts) of urine is produced and released each day. To understand how this remarkable feat is accomplished, we need to look inside the kidney and examine the microscopic filtration units where all the action takes place.

If we carefully sliced through the middle of one kidney and folded back half of it, we would be able to see the internal features labeled in Figure 11.4. A thin, smooth layer at the kidney surface is called the **renal cortex.** A darker central region is called the **renal medulla** (*muh-doo-lah*). Within the medulla are several pyramid-shaped regions called *renal pyramids.* The apex of each pyramid is directed toward a large chamber called the **renal pelvis.** The ureter attaches to the renal pelvis at the hilus of the kidney.

The renal cortex and medulla are the functional regions of each kidney. Together, they contain about 1 million microscopic filtration units called **nephrons** (*nef-rons*). Nephrons are connected to collecting ducts which pass through the renal pyramids and open into the renal pelvis. Urine formed in the nephrons and collecting ducts empties into the renal pelvis and is carried away from the kidney by the ureter.

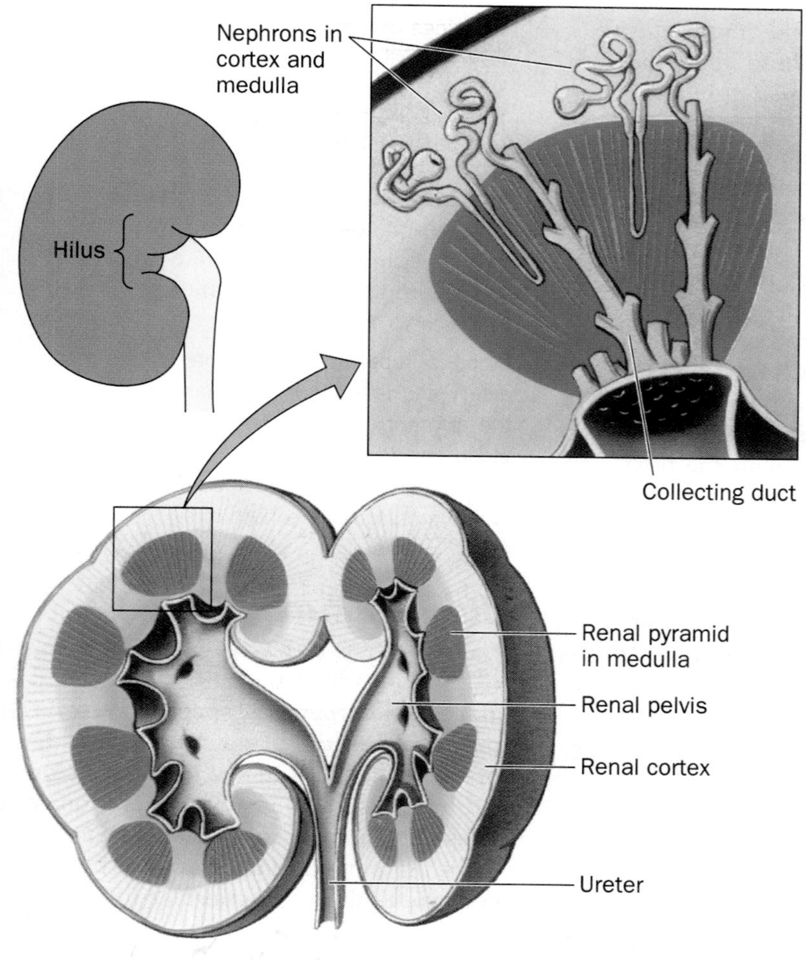
Nephrons in cortex and medulla
Hilus
Collecting duct
Renal pyramid in medulla
Renal pelvis
Renal cortex
Ureter

FIGURE 11.4 Internal anatomy of the kidney.

The Structure of the Nephron

A nephron consists of two parts—a cup-shaped capsule and a long tubule—both made of simple cuboidal epithelium (Figure 11.5). The nephron begins as a double-walled cup called **Bowman's capsule** which surrounds a network of capillaries known as the **glomerulus** (*glow-mare-you-luss*). Together, Bowman's capsule and the glomerulus form the structure where blood is filtered to begin the process of urine formation.

Attached to each Bowman's capsule is a long (3 cm or $1\frac{1}{4}$ inch), thin tubule with three distinct regions. The first region is called the **proximal convoluted tubule.** "Proximal" means that it is near Bowman's capsule, and "convoluted" describes its coiled and looped shape. The proximal

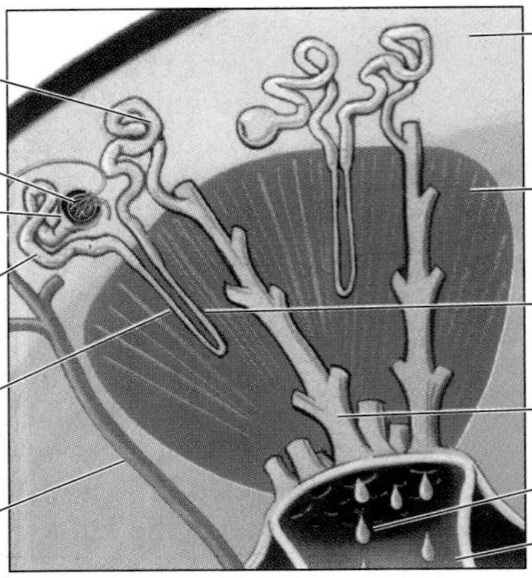

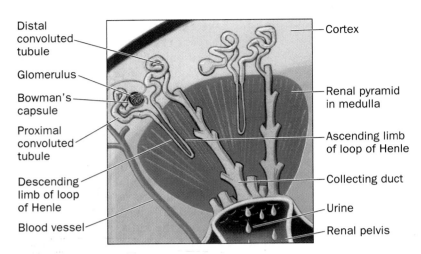

Distal convoluted tubule
Glomerulus
Bowman's capsule
Proximal convoluted tubule
Descending limb of loop of Henle
Blood vessel
Cortex
Renal pyramid in medulla
Ascending limb of loop of Henle
Collecting duct
Urine
Renal pelvis

FIGURE 11.5 A nephron and its collecting duct.

convoluted tubule connects to the second region, the **loop of Henle** (*hen-lee*). Shaped like a hairpin, the loop of Henle dips deeply into the medulla within a renal pyramid and then loops back toward the cortex. Since it is shaped like a hairpin, the loop of Henle is described as having a *descending limb* and an *ascending limb*. We will find that these limbs have different properties and play different roles in urine formation.

The third region of the nephron tubule is called the **distal convoluted tubule.** "Distal" means that it is farther from Bowman's capsule than the other regions. Distal convoluted tubules from many nephrons all connect to a common tube, the **collecting duct,** which empties into the renal pelvis (Figures 11.4 and 11.5).

The collecting duct has important functions in regulating the composition of urine. As you will discover in the next section, water, ions, and nutrients are reabsorbed from the filtrate in the nephron tubules and collecting ducts. This reabsorption prevents the loss of useful nutrients, ions, and water and provides an opportunity for tubule cells to regulate the composition of blood and body fluids.

> Each nephron consists of a glomerulus, a Bowman's capsule, and a long tubule that consists of three regions: proximal convoluted tubule, loop of Henle, and distal convoluted tubule. Several nephrons connect to a single collecting duct which empties into the renal pelvis.

A Nephron's Blood Supply: Two Capillary Networks

There is an intimate association between the blood vessels and the nephrons of the kidney. This association permits both extensive filtration from the blood and selective reabsorption back into the blood. After entering each kidney, the **renal artery** branches repeatedly, forming smaller and smaller arteries, until tiny arterioles reach each of the 1 million nephrons (Figure 11.6).

An **afferent arteriole** delivers blood to the glomerulus capillaries for filtration, and an **efferent arteriole** drains filtered blood away from the same glomerulus (*afferent,* "to bring toward"; *efferent,* "to carry away"). The efferent arteriole connects to a second network of capillaries, the **peritubular capillaries,** which are closely associated with the nephron tubule. It is into these peritubular capillaries that water, ions, and nutrients are reabsorbed from the filtrate in the nephron tubule. After leaving the vicinity of the nephron, blood flows through progressively larger veins until it reaches the *renal vein,* which leaves the kidney and returns blood to the inferior vena cava.

With nephron location, structure, and blood supply clearly in mind, we are ready to examine the cellular processes that accomplish urine formation.

> Blood is supplied to the kidney by the renal artery and reaches the glomerulus through the afferent arteriole. From the glomerulus, blood flows into the efferent arteriole and then the peritubular capillaries, which reabsorb water, ions, and nutrients.

The Formation and Release of Urine

You learned earlier that specific wastes are excreted in the urine. You also read that water and ions in the blood are regulated by kidney actions. The excretion of wastes and the regulation of water and ions in the blood are accomplished by three processes: (1) glomerular filtration, (2) tubular reabsorption, and (3) tubular secretion.

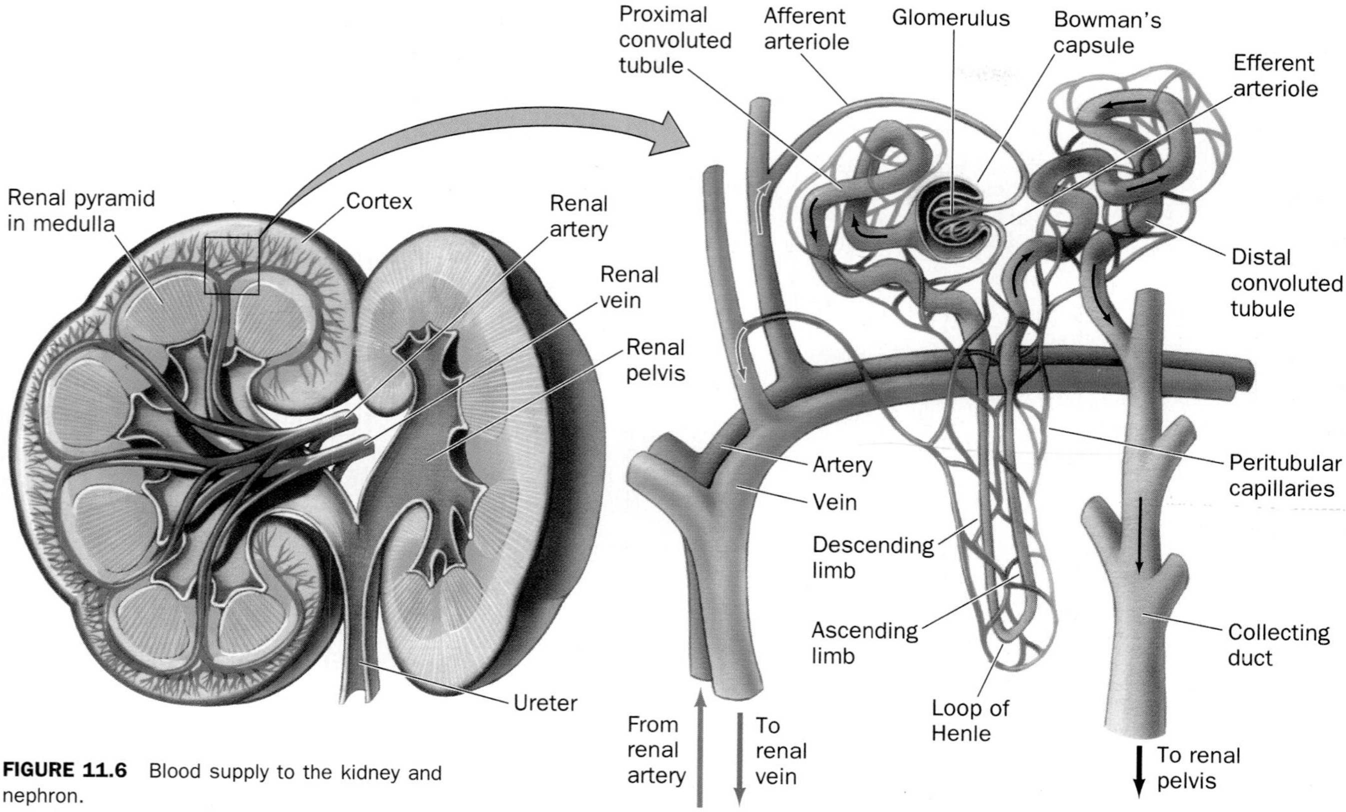

Renal pyramid in medulla
Cortex
Renal artery
Renal vein
Renal pelvis
Ureter
Proximal convoluted tubule
Afferent arteriole
Glomerulus
Bowman's capsule
Efferent arteriole
Distal convoluted tubule
Artery
Vein
Peritubular capillaries
Descending limb
Ascending limb
Collecting duct
Loop of Henle
From renal artery
To renal vein
To renal pelvis

FIGURE 11.6 Blood supply to the kidney and nephron.

Glomerular Filtration of Blood

Glomerular filtration is the first of the three processes that form urine. Urine formation begins with filtration of blood through the epithelial walls of the glomerulus and Bowman's capsule. The capillary wall of the glomerulus is quite porous, over 100 times more permeable than are most capillaries. The fluid portion of the blood, which consists of water, urea, ions, nutrients, and small proteins, is able to move across the capillary wall. Blood cells and larger blood proteins, however, cannot cross and are retained in the blood.

The filtered fluid is called **glomerular filtrate.** Both kidneys produce glomerular filtrate at a rate of about 125 ml/min (4 oz/min). Water and dissolved substances are present in the filtrate at about the same concentrations as they are in the blood. If the glomerular filtrate were excreted from your body unchanged, you would be in constant danger of both dehydration and starvation. You would need to spend most of your life drinking and eating to compensate for water and nutrient losses. Fortunately, you do not excrete the glomerular filtrate. Water and other useful materials are reabsorbed from the filtrate, and only a small volume of concentrated urine is actually formed. You can realize the magnitude of what is accomplished by studying Table 11.1, which contrasts the compositions of blood, filtrate, and urine.

Glomerular filtration occurs because the pressure of the blood flowing in the glomerular capillaries is higher than the pressure of the filtrate in Bowman's capsule. In other words, blood pressure drives glomerular filtration, and because the process takes advantage of a pressure gradient, glomerular filtration does not require the expenditure of energy by kidney cells.

To prevent the rate of glomerular filtration from changing when blood pressure is altered as a result of exercise or other conditions in the body, a certain degree of self-regulation over filtration occurs in the kidney. Specialized cells in the nephron wall sense changes in blood pressure and secrete chemicals that change the

TABLE 11.1 A Comparison of the Amounts of Various Substances in the Blood, Glomerular Filtrate, and Urine, Illustrating the Reabsorptive Powers of the Kidney*

Substance	Total Amount in Blood	Amount Filtered Each Day	Percentage Reabsorbed Each Day	Amount Excreted in Urine Each Day
Water	3 liters	180 liters	99	1–2 liters
Protein	200 g	2 g	95	0.1 g
Sodium (Na^+)	10 g	580 g	99	5 g
Chloride (Cl^-)	11 g	640 g	99	6 g
Potassium (K^+)	0.5 g	30 g	93	2 g
Bicarbonate (HCO_3^-)	5 g	275 g	100	0
Glucose	3 g	180 g	100	0
Urea†	5 g	53 g	53	25 g
Uric acid†	0.2 g	8.5 g	88	1 g
Creatinine†	0.03 g	1.6 g	0	1.6 g

*The actual amounts of these substances vary with age, sex, diet, health, and level of physical activity.
†These wastes contain nitrogen.

diameter of the arterioles connected to the glomerular capillaries. Changing the size of these vessels alters the amount of blood flowing through the glomerulus, maintaining a relatively stable rate of glomerular filtration and urine formation.

> Blood pressure drives filtration in the glomerulus. Cells and large proteins are retained in the blood but water, urea, nutrients, and ions are filtered into Bowman's capsule.

Tubular Reabsorption of Nutrients and Ions

Tubular reabsorption is the second process in the formation of urine from filtrate. As a result of tubular reabsorption, much of the filtrate passes out of the nephron tubule and returns to the blood through the peritubular capillaries. As much as 99 percent of the material in the filtrate is reabsorbed, preventing the loss of water, nutrients, and ions from the body. As a consequence of tubular reabsorption, urine contains mostly waste materials and excess water. Reabsorption occurs within all three regions of the nephron and in the collecting duct, but most of it takes place within the proximal convoluted tubule.

The epithelial cells of the proximal convoluted tubule have numerous microvilli which increase the surface area available for reabsorption (Figure 11.7). During reabsorption, molecules move out of the lumen of the tubule and enter the tubule's epithelial cells. They then pass out of the epithelial cells, cross into the peritubular capillaries, and enter the blood.

Depending on the type of molecule being reabsorbed, movement into and out of epithelial cells occurs by passive transport or active transport. Water and urea, for example, are reabsorbed by passive transport, by which they move from regions of higher concentration to regions of lower concentration (water is reabsorbed by osmosis, and urea by simple diffusion). Glucose and amino acids are reabsorbed by active transport. Recall that active transport requires ATP energy and specific carrier proteins in plasma membranes to move molecules from regions of lower concentration to regions of higher concentration.

The reabsorption of Na^+ occurs by both passive and active transport. Na^+ moves passively by diffusion from the filtrate into tubule cells but is actively transported out of the tubule cells on its way to the peritubular capillaries. The transport of Na^+ will become important to us when we discuss the osmotic movement of water from the filtrate to the peritubular capillaries during the production of concentrated urine.

> Tubular reabsorption uses passive and active transport to recapture up to 99 percent of the material in glomerular filtrate. This prevents the loss of water, nutrients, and ions from the blood. As a result of tubular reabsorption, urine contains mostly waste material.

Tubular Secretion

Certain chemicals in the blood that are not removed by filtration from the glomerular capillaries are removed by a third process of urine formation called **tubular secretion.** These chemicals are moved from the blood in the peritubular capillaries to the nephron tubule by both passive transport and active transport. After entering the proximal or distal convoluted tubules, the chemicals are mixed with the glomerular filtrate and eliminated from the body with the urine. The chemicals moved by tubular secretion include those foreign to the body and the ions and molecules that are toxic at elevated levels.

Ions removed from the blood by tubular secretion include potassium (K^+), hydrogen (H^+), and ammonium ($NH_4{}^+$). The secretion of H^+ is an important way in which the kidneys help control blood pH, as will be discussed later in this chapter. In addition to ions, other molecules, such as food preservatives, some pesticides, medications such as penicillin, and metabolic by-products such as creatinine, are removed from the blood by tubular secretion. The kidneys also rid the body of harmful drugs such as marijuana, cocaine, and heroin by tubular secretion. Urine drug testing, though controversial, has become commonplace for athletes and employees in certain occupations, such as transportation.

> Tubular secretion moves chemical substances from the blood in the peritubular capillaries into the filtrate in the nephron tubules. Ions, metabolic waste, food preservatives, pesticides, and many drugs are removed by this process.

FIGURE 11.7 Tubular reabsorption in the proximal convoluted tubule.

Photomicrograph of epithelial cells in the proximal convoluted tubule

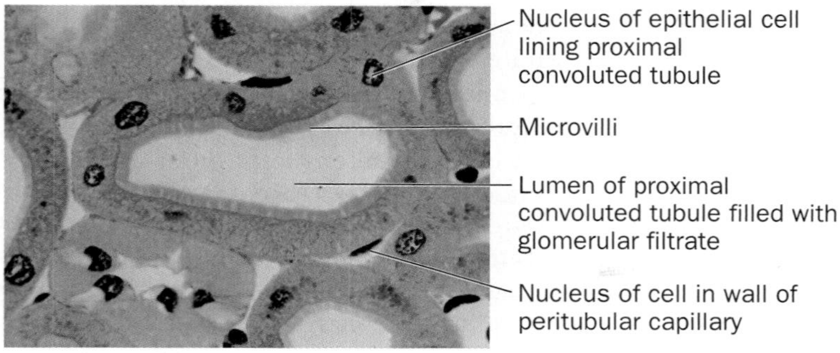
Nucleus of epithelial cell lining proximal convoluted tubule
Microvilli
Lumen of proximal convoluted tubule filled with glomerular filtrate
Nucleus of cell in wall of peritubular capillary

The peritubular capillaries receive water, salts, and nutrients that have been reabsorbed from the glomerular filtrate by epithelial cells.

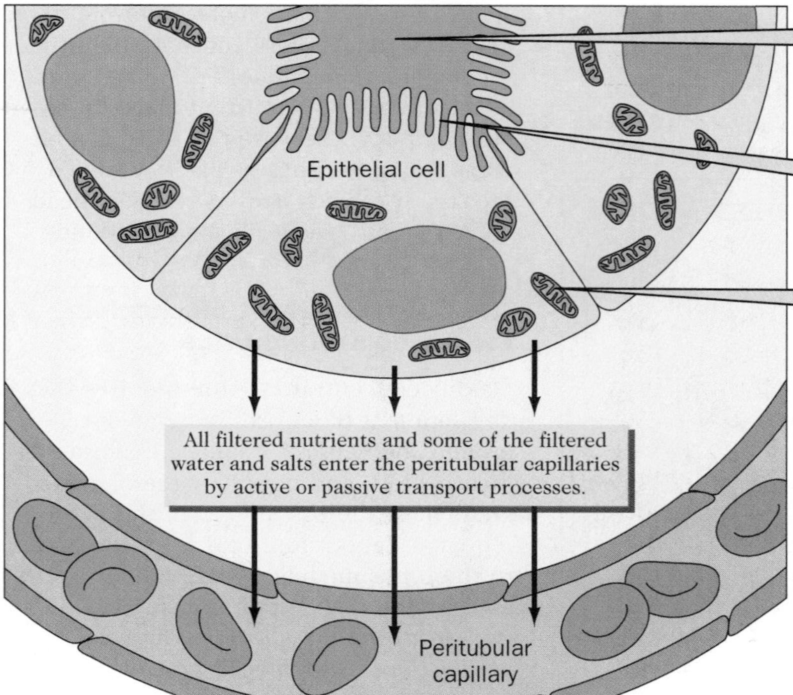
Epithelial cell
All filtered nutrients and some of the filtered water and salts enter the peritubular capillaries by active or passive transport processes.
Peritubular capillary

Lumen of proximal tubule. Water, salts, nutrients, and wastes leave the tubule and enter epithelial cells by active or passive transport processes.

Microvilli increase the surface area of the epithelial cells for reabsorption.

Mitochondria provide ATP energy for active transport processes.

Summarizing Urine Formation

Let us review the processes that lead to the formation of urine. Initially, blood is filtered from the glomerular capillaries by a process called glomerular filtration. The filtrate produced is then altered in volume and concentration by two processes: tubular reabsorption and tubular secretion. These processes and the sites of their actions are summarized in Figure

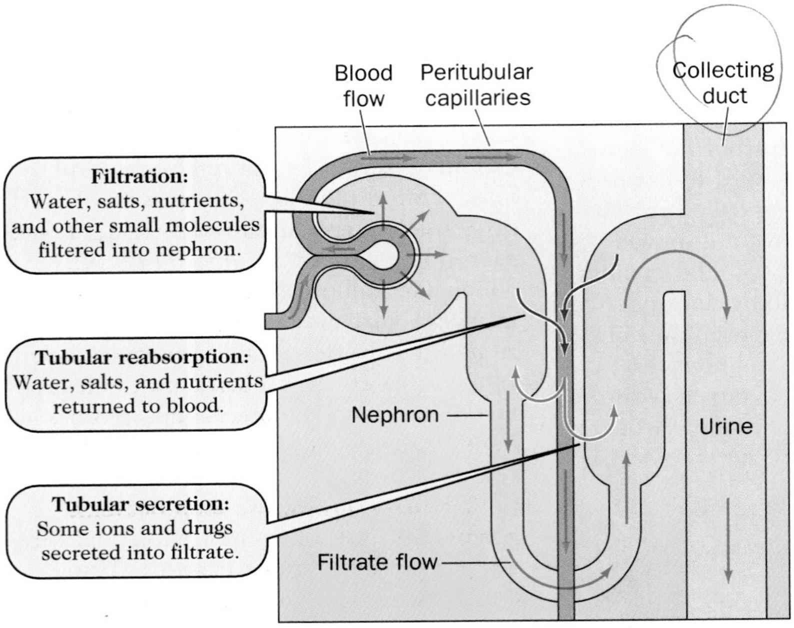
Blood flow
Peritubular capillaries
Collecting duct
Filtration:
Water, salts, nutrients, and other small molecules filtered into nephron.
Tubular reabsorption:
Water, salts, and nutrients returned to blood.
Tubular secretion:
Some ions and drugs secreted into filtrate.
Nephron
Urine
Filtrate flow

FIGURE 11.8 A summary of the processes involved in urine formation.

11.8. On average, only about 1 percent of the original glomerular filtrate appears in the urine released by the kidneys.

Producing Dilute or Concentrated Urine

Dilute blood produces a dilute urine. If you consume a large volume of water over a short time span, water will move into and dilute your blood. This increase in blood volume increases your blood pressure and decreases the concentration of ions in your blood and tissue fluids. This means that there is too much water in the fluid bathing your cells. To avoid osmotic swelling and disruption of cells, you excrete the excess water by producing a large volume of dilute urine.

Making a dilute urine means that you reabsorb less water and more ions from the filtrate in your nephron tubules (Figure 11.9). This action occurs in the ascending limb of the loop of Henle, a region in your nephrons that is impermeable to water. Ions, particularly Na^+, are reabsorbed here, but water is not. The filtrate becomes more dilute and remains so in your distal convoluted tubules and collecting ducts. The urine that you ultimately release can be nearly three times more dilute than your blood.

Concentrated blood produces concentrated urine. Most of the time, our problem is not overconsumption of water but underconsumption. Since we live on dry land, our problem involves losing water more rapidly than it can be replaced. When you sweat excessively you lose water, and your blood volume and pressure decline. As the water content of your tissue fluid also declines, your cells are in danger of dehydrating. Your kidneys respond by reabsorbing more water but fewer ions, excreting a more concentrated urine. Producing a concentrated urine is accomplished by a countercurrent mechanism, our next topic. Because this topic is a bit complicated, we suggest that you take a break. Eat a snack and have a clear head and an empty bladder before continuing.

FIGURE 11.9 The formation of a dilute urine removes excess water from the blood.

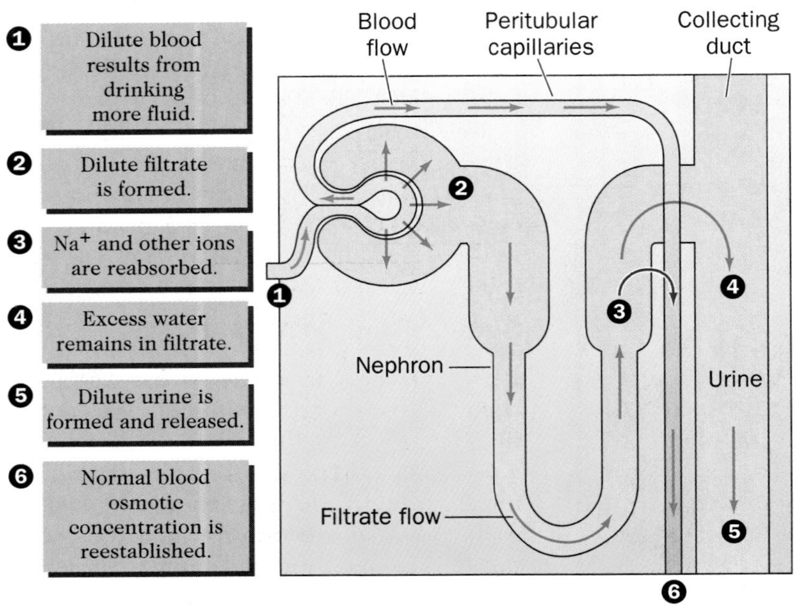
1 Dilute blood results from drinking more fluid.
2 Dilute filtrate is formed.
3 Na+ and other ions are reabsorbed.
4 Excess water remains in filtrate.
5 Dilute urine is formed and released.
6 Normal blood osmotic concentration is reestablished.
Blood flow
Peritubular capillaries
Collecting duct
Nephron
Urine
Filtrate flow

The Countercurrent Mechanism for Concentrating Urine

Making concentrated urine requires the development of an osmotic gradient around the nephron tubules. Examine Figure 11.10 and note how the osmotic gradient of the tissue fluid in the kidney increases (about fourfold) from the cortex to the inner medulla. The gradient is created by the reabsorption of Na^+, Cl^-, and urea from the loop of Henle and the collecting duct. During reabsorption, the ions and urea collect in the tissue fluid

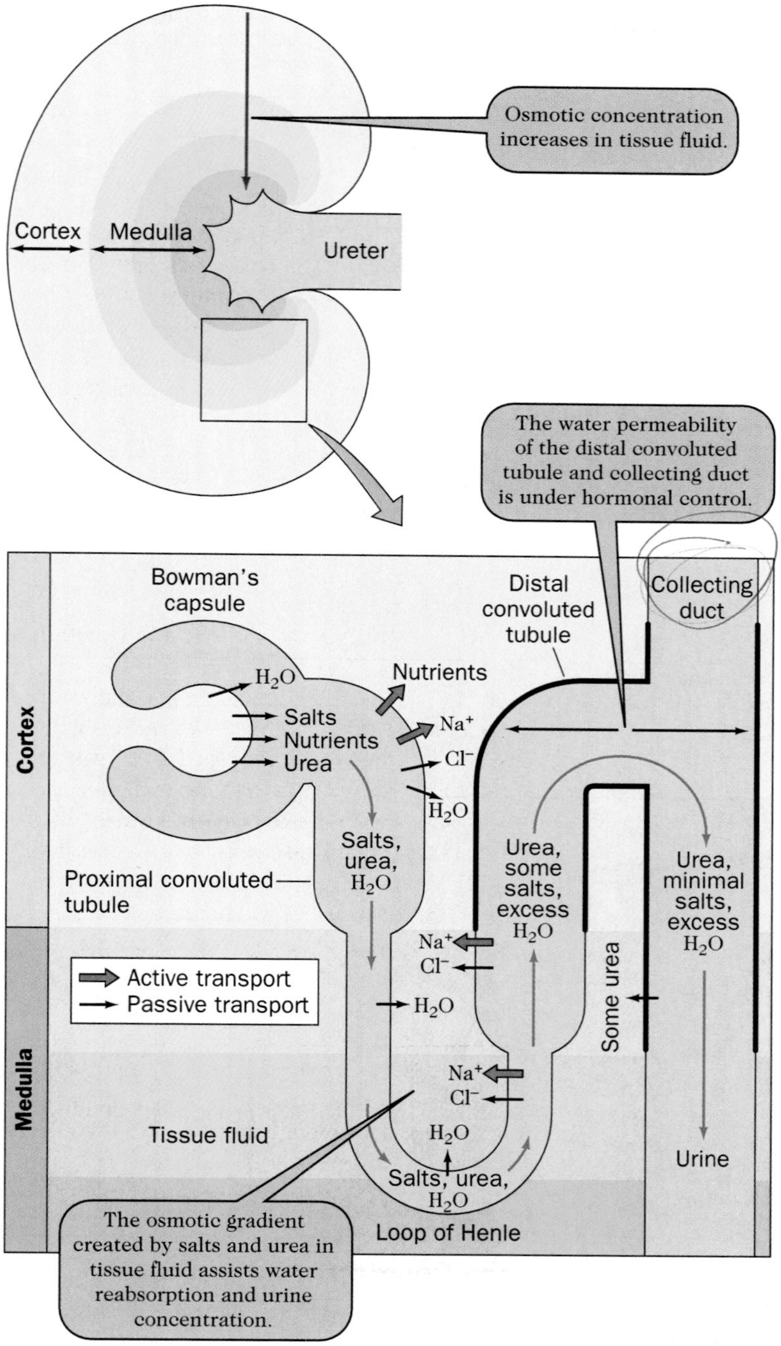

Osmotic concentration increases in tissue fluid.
Cortex
Medulla
Ureter
The water permeability of the distal convoluted tubule and collecting duct is under hormonal control.
Bowman's capsule
Distal convoluted tubule
Collecting duct
H_2O
Nutrients
Salts
Nutrients
Urea
Na^+
Cl^-
H_2O
Salts, urea, H_2O
Urea, some salts, excess H_2O
Urea, minimal salts, excess H_2O
Proximal convoluted tubule
Cortex
Medulla
Na^+
Cl^-
Active transport
Passive transport
H_2O
Some urea
Na^+
Cl^-
Tissue fluid
H_2O
Salts, urea, H_2O
Urine
Loop of Henle
The osmotic gradient created by salts and urea in tissue fluid assists water reabsorption and urine concentration.

FIGURE 11.10 Formation of an osmotic gradient around kidney tubules.

more rapidly than they can be taken into the peritubular capillaries. Their increasing concentration from the cortex to the medulla produces an osmotic gradient which can drive the reabsorption of water, producing a concentrated urine.

The two parallel limbs of the loop of Henle and the surrounding peritubular capillaries help produce and maintain this osmotic gradient in the medulla. Parallel tubules transporting fluids in opposite directions produce what is called a **countercurrent mechanism,** as summarized in Figure 11.11. A countercurrent mechanism allows the fluid flowing in one tubule to influence the conditions in the fluid flowing in the opposite tubule. In this manner, small differences across the wall of a tubule (between the tissue fluid outside the nephron and the filtrate inside) can be magnified into large differences from one end of the tubule to the other end.

The two limbs of the loop of Henle lie parallel to each other, with filtrate moving in opposite directions through the two limbs. The descending and ascending portions of the peritubular capillaries also lie parallel to each other and to the loop of Henle. When filtrate moves through the loop of Henle and blood moves through the peritubular capillaries, both fluids are exposed to an increasing osmotic gradient on their descent toward the medulla and to a decreasing osmotic gradient on their ascent toward the cortex.

When it is exposed to this external osmotic gradient, filtrate moving through the descending limb of the loop of Henle loses water but retains ions. Filtrate moving through the ascending limb retains water but loses ions. This situation develops because the wall of the ascending limb has two characteristics: (1) It actively pumps Na^+ and Cl^- into the surrounding tissue fluid in the medulla, and (2) it is impermeable to water.

Blood moving through the peritubular capillaries also is exposed to the osmotic gradient between the medulla and the cortex. Blood loses water and gains ions in the descending arterial portion but gains water and loses salts in the ascending portion. As a result, the ascending portion of the peritubular capillaries reabsorbs water from the nephron and contributes ions to the osmotic gradient (Figure 11.12).

The collecting ducts also contribute to the production of concentrated urine. The osmotic gradient in the medulla allows water to be reabsorbed from the filtrate in the collecting ducts as it passes

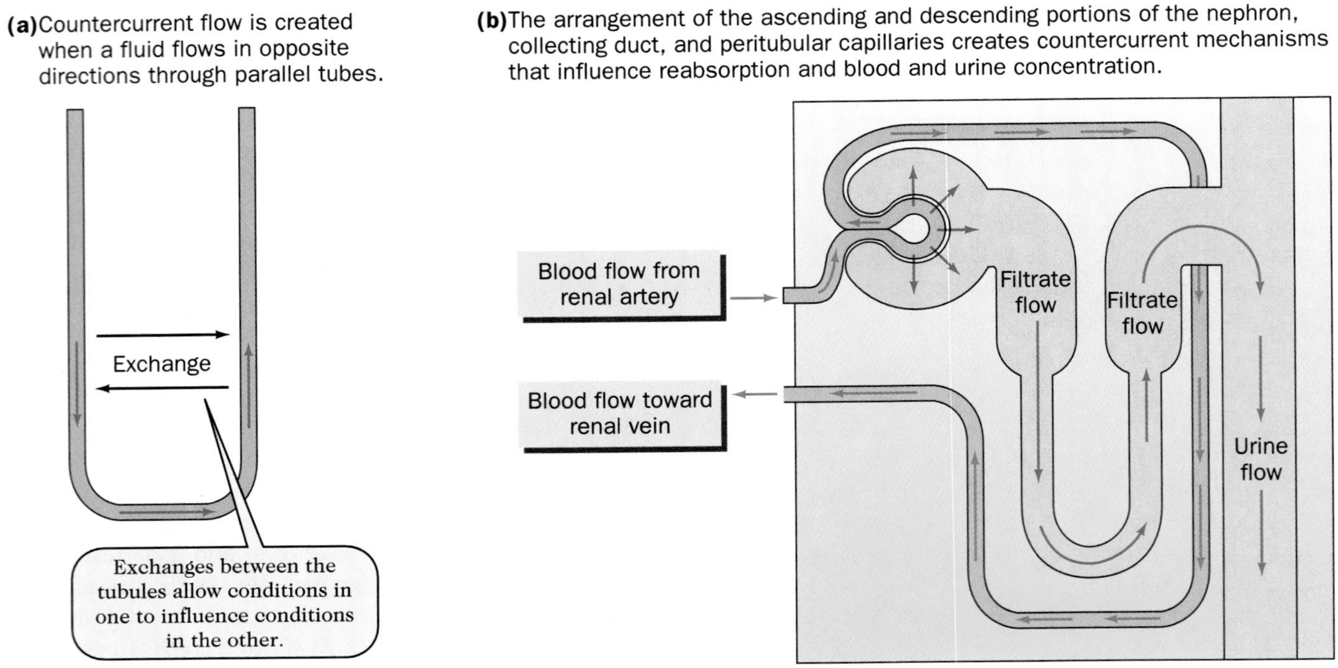
(a) Countercurrent flow is created when a fluid flows in opposite directions through parallel tubes.
Exchange
Exchanges between the tubules allow conditions in one to influence conditions in the other.
(b) The arrangement of the ascending and descending portions of the nephron, collecting duct, and peritubular capillaries creates countercurrent mechanisms that influence reabsorption and blood and urine concentration.
Blood flow from renal artery
Blood flow toward renal vein
Filtrate flow
Filtrate flow
Urine flow

FIGURE 11.11 Countercurrent mechanisms in the nephron and peritubular capillaries.

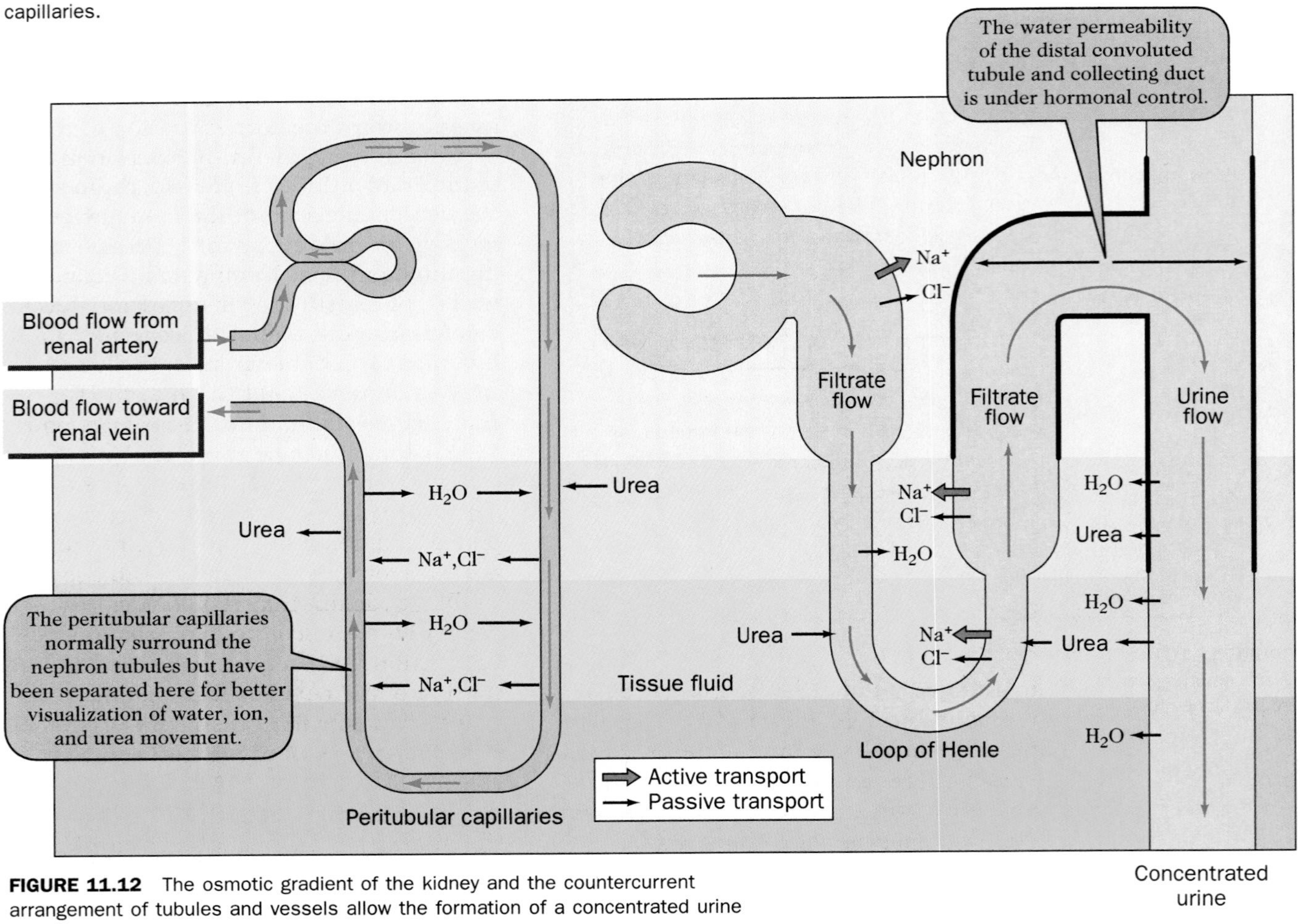
The water permeability of the distal convoluted tubule and collecting duct is under hormonal control.
Nephron
Na+
Cl−
Blood flow from renal artery
Blood flow toward renal vein
Filtrate flow
Filtrate flow
Urine flow
H2O
Urea
Urea
Na+,Cl−
Na+
Cl−
H2O
H2O
Urea
H2O
Na+,Cl−
Urea
Na+
Cl−
Urea
H2O
The peritubular capillaries normally surround the nephron tubules but have been separated here for better visualization of water, ion, and urea movement.
Tissue fluid
Loop of Henle
H2O
Active transport
Passive transport
Peritubular capillaries
Concentrated urine

FIGURE 11.12 The osmotic gradient of the kidney and the countercurrent arrangement of tubules and vessels allow the formation of a concentrated urine as water reabsorbed from the filtrate is returned to the blood.

TABLE 11.2 Summary of Kidney Functions and Urine Formation

1. Initially, blood is filtered from the glomerular capillaries by glomerular filtration.
2. The filtrate produced is then altered in volume and concentration by two processes: tubular reabsorption and tubular secretion. On average, only about 1 percent of the original glomerular filtrate appears in the urine released by the kidneys.
3. Urine formation helps stabilize conditions in the blood and the tissue fluid bathing cells. It contributes to homeostasis by maintaining the body's water and solute concentrations and the pH of body fluids while excreting wastes, excess materials, and toxic substances.

through the gradient on its way to the renal pelvis. The permeability of the collecting ducts to water is regulated by a hormone called antidiuretic hormone (ADH), which will be discussed in the next section. The functions of the kidneys and urine formation are summarized in Table 11.2.

> When blood and body fluids are dilute, the nephron reabsorbs more ions and less water from the glomerular filtrate, producing dilute urine. When blood and body fluids are concentrated, the nephron reabsorbs fewer ions and more water from the filtrate, producing a concentrated urine.

Hormonal Control of Urine Formation

Reabsorption of Na^+ and water in the distal tubule and the collecting duct is controlled by two hormones: aldosterone and antidiuretic hormone. **Aldosterone** (*al-**doss**-ter-own*) is a hormone secreted by the outer layer of the adrenal gland, a gland which sits like a cap above the kidney (Figure 11.13a and b). When aldosterone is present in the blood, all the Na^+ in the filtrate is reabsorbed by the epithelial cells of the collecting duct. When aldosterone is absent, some Na^+ remains in the filtrate and is excreted with the urine. The release of aldosterone is controlled by negative feedback, as shown in Figure 11.13c.

When your blood Na^+ concentration is high, for example, after a late-night snack of pizza with anchovies, aldosterone is withheld, allowing the excess Na^+ to be excreted. When Na^+ is excreted, so is water, and you may feel dehydrated in the morning. By contrast, after several hours of sweaty exercise, your blood Na^+ will be low, and aldosterone is released, causing reabsorption of additional Na^+. Retaining Na^+ raises the osmotic pressure of your blood and reduces water loss from your body.

Antidiuretic hormone (ADH) (*an-tie-die-yoo-**ret**-ick*) increases the reabsorption of water by the distal tubule and collecting duct. The release of ADH from the pituitary gland is controlled by special cells in the hypothalamus of the brain that monitor water concentration of the blood (Figure 11.14). If the water concentration is low, ADH is released, more water is reabsorbed in the collecting duct, and only a small amount of concentrated urine is produced. If the water concentration is high, ADH is withheld, less water is reabsorbed, and a larger volume of dilute urine is produced.

Alcohol inhibits the release of ADH, and caffeine interferes with ADH action and sodium reabsorption; thus, both cause artificially dilute urine to be produced. Drugs called *diuretics* increase the production of dilute urine and prevent the excessive water retention and tissue swelling (edema) that may accompany congestive heart failure, high blood pressure, and other conditions. Furosemide (Lasix), a commonly prescribed diuretic, operates by inhibiting Na^+ reabsorption in all three regions of the nephron tubule.

> The reabsorption of sodium by the nephron is enhanced by the hormone aldosterone, while antidiuretic hormone increases water reabsorption.

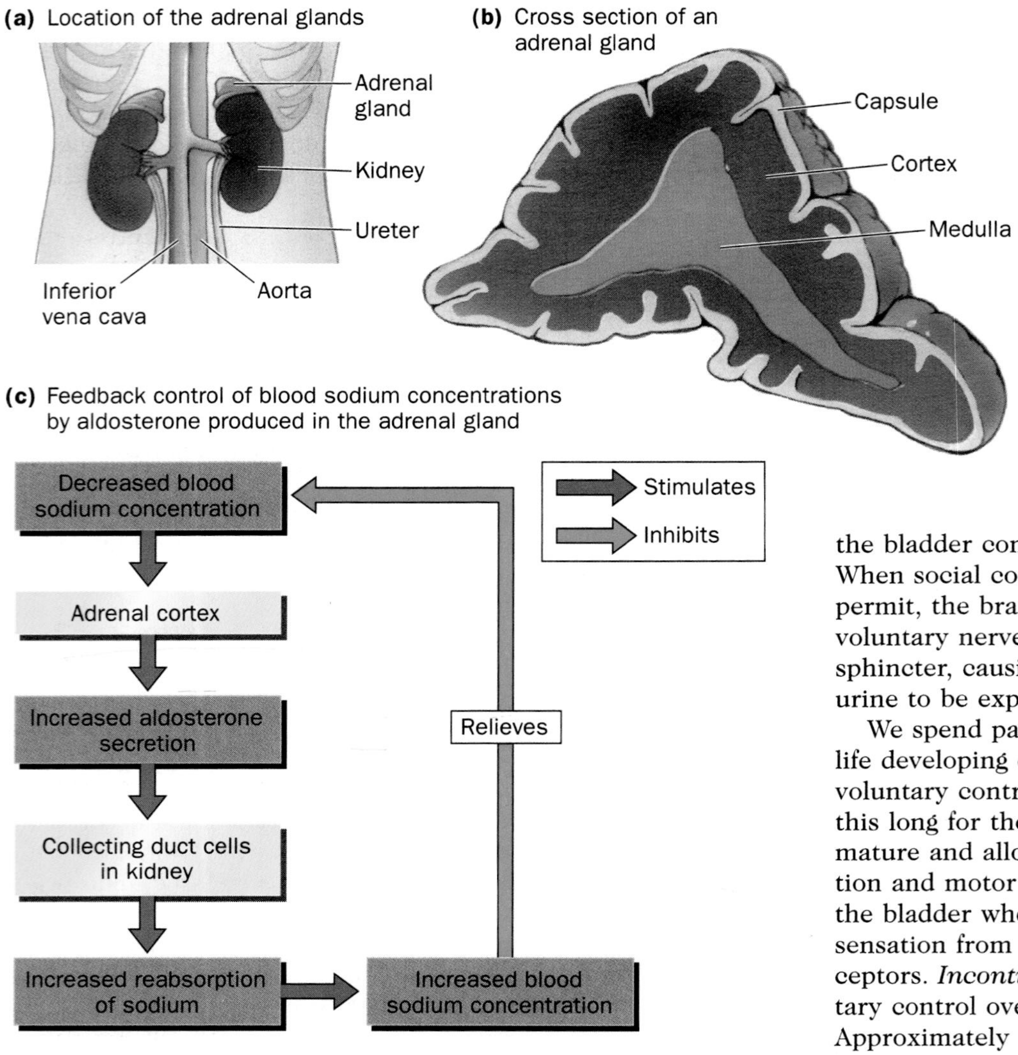
(a) Location of the adrenal glands
Adrenal gland
Kidney
Ureter
Inferior vena cava
Aorta
(b) Cross section of an adrenal gland
Capsule
Cortex
Medulla
(c) Feedback control of blood sodium concentrations by aldosterone produced in the adrenal gland
Decreased blood sodium concentration
Adrenal cortex
Increased aldosterone secretion
Collecting duct cells in kidney
Increased reabsorption of sodium
Increased blood sodium concentration
Relieves
Stimulates
Inhibits

FIGURE 11.13 Location and function of adrenal glands.

Releasing Urine: The Role of Nerves and Learning

Urine stored in the urinary bladder can be expelled through the urethra by a process known clinically as *micturition* or more commonly as *urination.* Once the urinary bladder fills with about 250 milliliters (about 1 cup) of urine, sensory receptors in the wall of the bladder are stretched. Nerve impulses travel from these receptors to the spinal cord and eventually to the brain. From the spinal cord, involuntary nerves of the autonomic nervous system (ANS) carry impulses back to the bladder wall and cause the *internal* urethral sphincter to relax. The nerves of the ANS also stimulate contraction of the detrusor muscle in the bladder, raising the pressure within the bladder. The participation of the sensory receptors, sensory nerves, and nerves of the ANS is called the *micturition reflex.*

Meanwhile, the brain interprets the incoming sensory information as a sensation of fullness, and the urge to urinate begins to intensify as the bladder continues to fill and stretch. When social conditions and surroundings permit, the brain sends impulses along voluntary nerves to the *external* urethral sphincter, causing it to relax and allowing urine to be expelled through the urethra.

We spend part of our first 3 years of life developing **continence,** which means voluntary control over urination. It takes this long for the nerve circuits to fully mature and allow us to coordinate sensation and motor response. Infants empty the bladder when they receive the initial sensation from their bladder stretch receptors. *Incontinence* is the lack of voluntary control over micturition in adults. Approximately 5 percent of all Americans and 30 percent of persons over age 65 experience some degree of incontinence.

Temporary incontinence is common during pregnancy because of the massive weight of the overlying uterus and its pressure against the bladder, which allow a cough or sneeze to trigger the release of urine. After several pregnancies, reduced muscle tension along the front wall of the vagina, which supports the bladder, may interfere with the operation of the urethral sphincters, causing a loss of bladder control.

Urine retention is the failure to void urine and may be caused by bladder or urethra obstructions or by a nerve injury that blocks the sensation to urinate. Older men may suffer urine retention if the prostate gland enlarges and obstructs the passage of urine through the urethra.

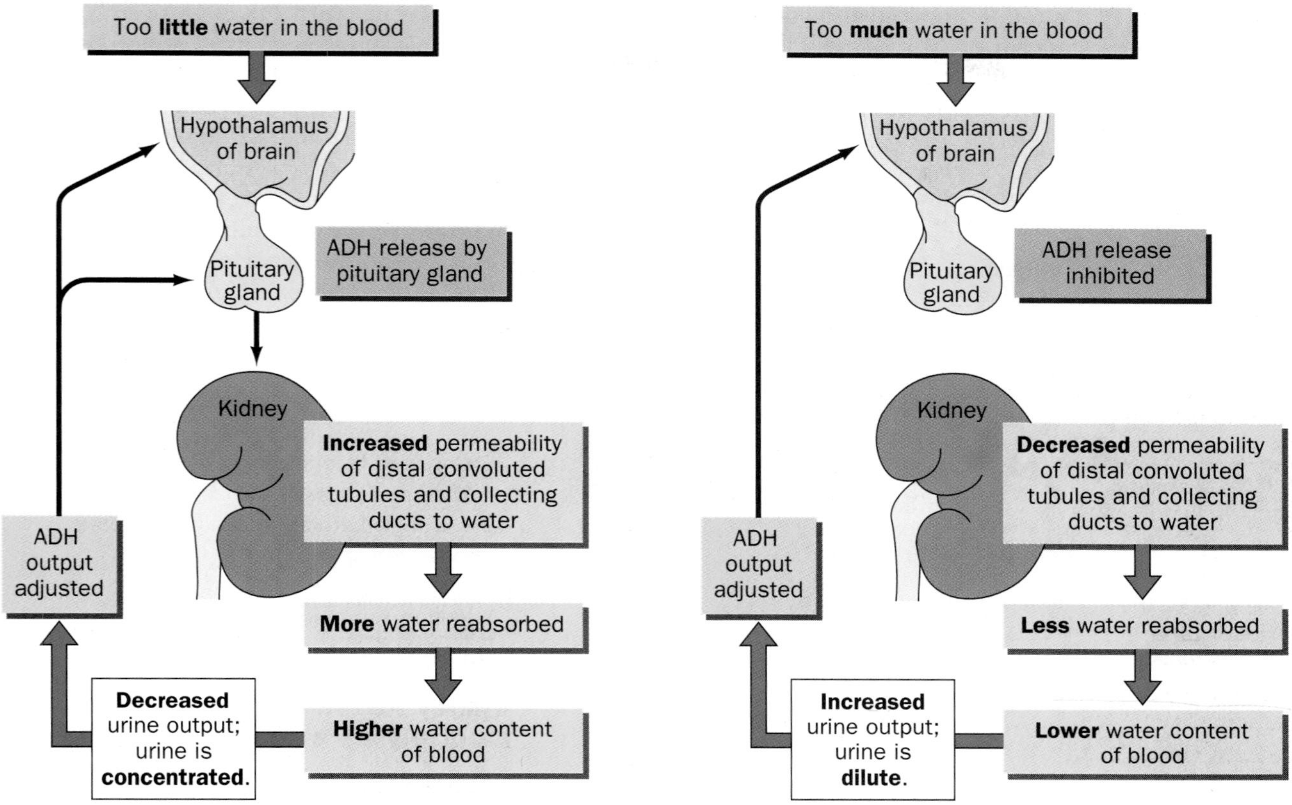
Too little water in the blood
Hypothalamus of brain
Pituitary gland
ADH release by pituitary gland
Kidney
Increased permeability of distal convoluted tubules and collecting ducts to water
More water reabsorbed
Higher water content of blood
Decreased urine output; urine is concentrated.
ADH output adjusted
Too much water in the blood
Hypothalamus of brain
Pituitary gland
ADH release inhibited
Kidney
Decreased permeability of distal convoluted tubules and collecting ducts to water
Less water reabsorbed
Lower water content of blood
Increased urine output; urine is dilute.
ADH output adjusted

FIGURE 11.14 Antidiuretic hormone (ADH) controls the permeability of the nephron to water and regulates the volume and concentration of urine.

Difficulty in urination is frequently an early sign of prostate enlargement. Most growths are benign, but prostate cancer claims more than 30,000 lives per year in the United States.

> The release of urine is controlled by involuntary and voluntary nerve signals. Stretch receptors in the bladder wall signal when the urinary bladder is full.

Homeostatic Functions of the Urinary System

As you read in the preceding sections, the kidneys not only excrete nitrogenous wastes but also regulate the composition of blood and body fluids to maintain homeostasis for the entire body. Since they receive a quarter of the heart's cardiac output, the kidneys are positioned ideally to sense and respond to changes in the blood and influence conditions in other body fluids which are formed from blood. You have already been introduced to some of these regulatory functions and can now appreciate the broader implications of kidney function.

Water Balance and Blood Pressure

The retention or excretion of water determines the fluid volume of the blood. Because the blood circulates in a closed system, an increase in its volume increases the pressure exerted by the blood on all parts of the circulatory system. The control of water balance influences blood pressure. As we have discussed, an increase or decrease in the water content of the blood is regulated by the kidneys under the control of ADH.

In addition to control by ADH, a special group of cells called the *juxtaglomerular apparatus* (***juks***-*tuh-glow-mare-*

you-lahr) operates to control blood volume and pressure (Figure 11.15). The juxtaglomerular apparatus is located at a site where the distal convoluted tubule makes close contact with the afferent arteriole and consists of specialized cells from both. The tubule cells monitor the sodium concentration of the glomerular filtrate, while the cells of the afferent arteriole secrete the enzyme *renin*, which raises blood pressure when blood volume is low (dehydration or excessive bleeding). Renin activates the hormone *angiotensin II* in the blood. Angiotensin II stimulates the adrenal gland to release aldosterone, which increases retention of both Na^+ and water.

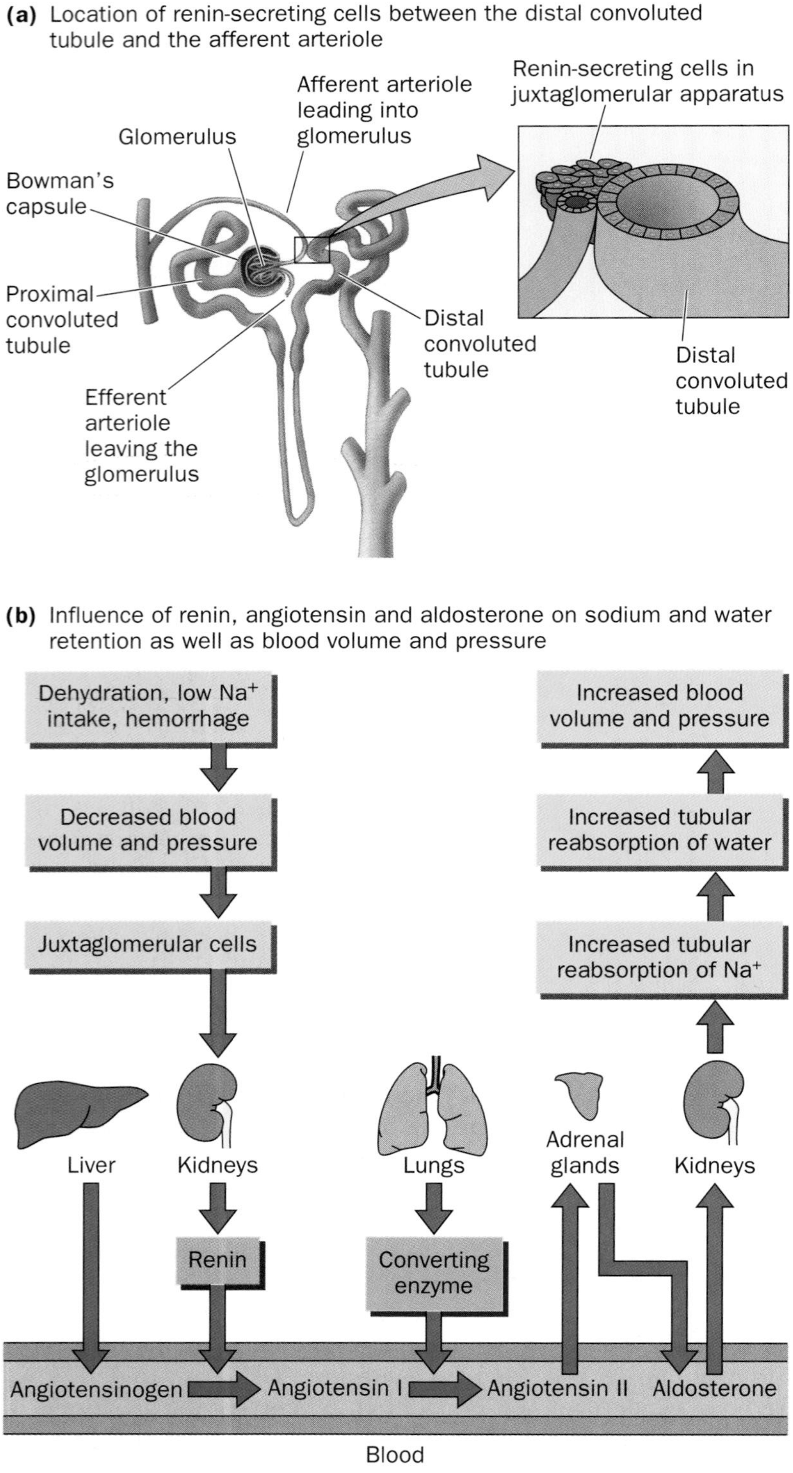

(a) Location of renin-secreting cells between the distal convoluted tubule and the afferent arteriole
Afferent arteriole leading into glomerulus
Renin-secreting cells in juxtaglomerular apparatus
Glomerulus
Bowman's capsule
Proximal convoluted tubule
Efferent arteriole leaving the glomerulus
Distal convoluted tubule
Distal convoluted tubule
(b) Influence of renin, angiotensin and aldosterone on sodium and water retention as well as blood volume and pressure
Dehydration, low Na+ intake, hemorrhage
Decreased blood volume and pressure
Juxtaglomerular cells
Increased blood volume and pressure
Increased tubular reabsorption of water
Increased tubular reabsorption of Na+
Liver
Kidneys
Lungs
Adrenal glands
Kidneys
Renin
Converting enzyme
Angiotensinogen
Angiotensin I
Angiotensin II
Aldosterone
Blood

FIGURE 11.15 The juxtaglomerular apparatus.

> Changes in fluid intake or output alter blood volume and pressure. In addition to ADH, the juxtaglomerular apparatus operates through renin, angiotensin II, and aldosterone to decrease or increase water retention and blood pressure.

Kidney Regulation of Electrolytes and pH

The ions found both inside cells and in the extracellular fluid (blood and tissue fluid) are called **electrolytes** because a solution of ions conducts electricity. You can see in Figure 11.16 that some electrolytes are more abundant outside cells (Na^+, Cl^-, HCO_3^-), while others are more abundant inside (K^+, Mg^{2+}, $H_2PO_4^-$, and negatively charged proteins). The concentrations and distributions of these ions are maintained by active transport pumps in plasma membranes. By regulating the concentrations of electrolytes in the blood and tissue fluid, the kidneys play a direct role in the healthy operation of each cell in the body.

Na^+ and K^+ are abundant and important electrolytes. Any change in their concentrations affects osmosis and the distribution of water in the blood and tissue fluid and can disrupt the operation of heart muscle and neurons. Na^+ and K^+ can be lost from the body during excessive sweating, vomiting, and diarrhea, causing fatigue, muscle cramps, nausea, and cardiac irregularities. Replacement of lost body fluids with pure water does not provide the Na^+ and K^+ needed. To obtain these ions after periods of sweating, vomiting, or diarrhea, you may want to

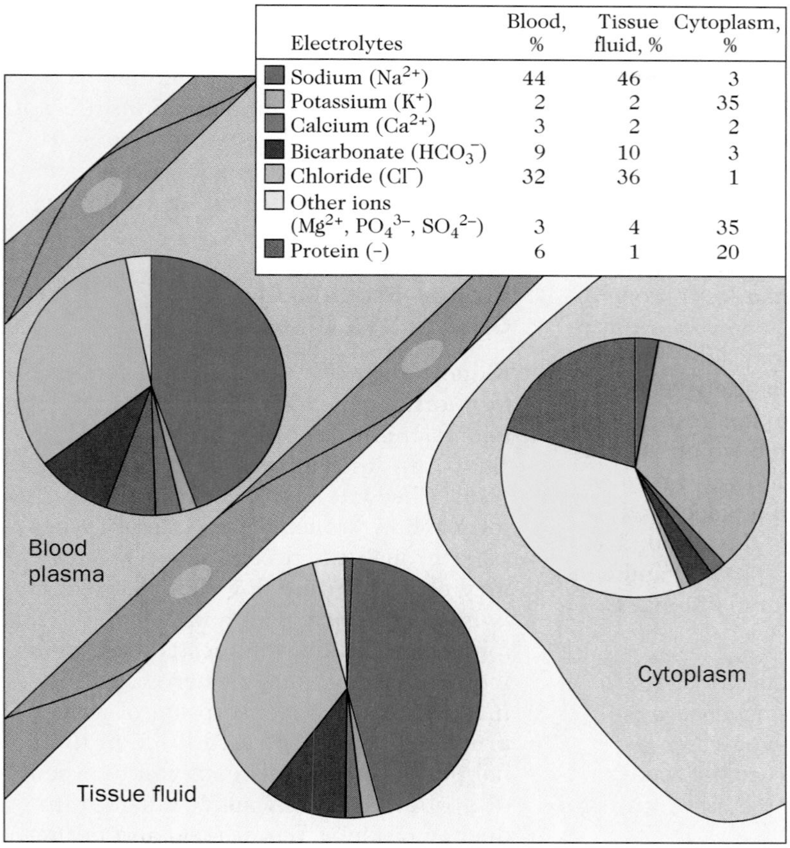
Blood plasma
Tissue fluid
Cytoplasm

Electrolytes	Blood, %	Tissue fluid, %	Cytoplasm, %
Sodium (Na^{2+})	44	46	3
Potassium (K^{+})	2	2	35
Calcium (Ca^{2+})	3	2	2
Bicarbonate (HCO_3^-)	9	10	3
Chloride (Cl^-)	32	36	1
Other ions (Mg^{2+}, PO_4^{3-}, SO_4^{2-})	3	4	35
Protein (-)	6	1	20

FIGURE 11.16 Relative abundance of major electrolytes in extracellular and intracellular fluids. Percentages are based on concentrations per liter (quart) of water.

drink fruit juices or special commercial fluids containing both water and electrolytes, such as Gatorade and All Sports. The levels of Na^+ and K^+ in the blood and urine are controlled by the hormone aldosterone, which promotes the tubular reabsorption of Na^+ and the tubular secretion of K^+.

Blood pH and the secretion of acid. *Acids* are molecules that produce hydrogen ions (H^+), while *bases* are molecules that combine with H^+. There are strong and weak acids and bases. In a water solution, the strength of an acid or base is measured by the pH scale (Chapter 1).

Our cells and body fluids contain numerous different acids and bases, and many of our biochemical processes generate H^+. As a crucial part of homeostasis, the pH of blood is held in the narrow range of 7.35 to 7.45. Variation below this level causes the life-threatening condition *acidosis,* while an increase in pH above 7.5 causes the equally serious *alkalosis.* Buffer systems, which are combinations of weak acids and weak bases, are present in body fluids to prevent large pH fluctuations by combining with or releasing the H^+ produced by metabolism or consumed in food or drink.

Tubular secretion of H^+ by the nephron allows us to produce a urine that is 1000 times more acidic than our blood. The H^+ secreted from the blood into the filtrate has to be buffered to prevent acid damage to urinary system tissue. This buffering is accomplished by the presence of ammonia (NH_3) in nephron tubule cells and the presence of monohydrogen phosphate (HPO_4^{2-}) in glomerular filtrate. As you read earlier, NH_3 is generated from amino acids by deamination and phosphate is formed from the breakdown of ATP. In tubule cells, H^+ joins with ammonia to form ammonium (NH_4^+), which is then secreted into the filtrate and excreted in the urine:

$$\underset{\text{Hydrogen ion}}{H^+} + \underset{\text{Ammonia}}{NH_3} \rightleftharpoons \underset{\text{Ammonium}}{NH_4^+}$$

In the filtrate, H^+ that has been secreted by tubule cells reacts with monohydrogen phosphate filtered from the blood, forming dihydrogen phosphate ($H_2PO_4^-$), which is then excreted with the urine. This buffering action helps prevent the urine from becoming too acidic:

$$\underset{\text{Hydrogen ion}}{H^+} + \underset{\text{Monohydrogen phosphate}}{HPO_4^{2-}} \rightleftharpoons \underset{\text{Dihydrogen phosphate}}{H_2PO_4^-}$$

> The hormone aldosterone increases the reabsorption of Na^+ by the nephron and the excretion of K^+. To help maintain the pH of the blood and body fluids, the kidneys secrete H^+ and NH_4^+. Urine is prevented from becoming too acidic through the use of a phosphate buffer that reacts with H^+.

Kidney Regulation of Red Blood Cells and Vitamin D

In addition to controlling water and solute concentrations in blood and tissue fluids, the kidneys regulate the manufacture of red blood cells and vitamin D. Under conditions of reduced O_2 in the blood, the kidneys release an enzyme that activates the hormone *erythropoietin* (*eh-rith-row-**poy**-ee-tin*). Erythropoietin stimulates the production of additional O_2-carrying red blood cells in the bone marrow.

Vitamin D is essential for absorbing calcium and phosphorus for the construction of bones and teeth. While most of your vitamin D is obtained from food, your body manufactures some in a three-step process involving your skin, liver, and kidneys. When sunlight strikes your skin, a cholesterol precursor in the skin is converted to a molecule which is transported to the liver. In the liver, further chemical conversions take place, and the liver product then enters the blood. The kidneys pick up this liver product and convert it into the active form of vitamin D.

> Low O_2 in the blood stimulates kidney activation of erythropoietin, a hormone which stimulates red blood cell production. The kidneys provide the final chemical conversion for the production of the active form of vitamin D.

Diseases and Disorders

A wide range of disorders may afflict the urinary system, but they are usually centered in the kidneys and bladder. The symptoms may include painful urination, frequent or infrequent urination, incontinence, and blood in the urine. Since we can live well with only one kidney, kidney infections and inflammations are not life-threatening unless both kidneys are affected. Chemical and microscopic analysis of urine can be helpful in confirming a diagnosis of urinary disease.

Kidney Stones: An Obstructive Disorder

Kidney stones, or *renal calculi* (***cal**-cue-lie*), form from calcium oxalate, uric acid, and calcium phosphate (Figure 11.17). They form in the renal pelvis when these solutes become so concentrated that they form solids (stones). When these stones pass through the ureters or urethra, they may obstruct urine flow and their sharp points may cause pain and bleeding. The formation of kidney stones is more common in warm climates where sweating increases kidney reabsorption of water, producing a concentrated urine in the renal pelvis. Individuals confined to a bed or a wheelchair may suffer a higher frequency of kidney stones because of the demineralization of bone associated with reduced physical activity.

When a kidney stone is present, the initial treatment often consists only of increased water intake and pain medication until the stone passes through the urinary system. If this "low-tech" approach does not work, high-frequency sound may be used to shatter the stone into fragments small enough to pass with the urine.

Inflammation of Nephrons

Different parts of the kidney and the nephron can become inflamed by infection, allergens, or toxins. *Glomerulonephritis* (*glow-mare-you-low-nef-**fry**-tus*) is an inflammation of the glomerulus that often follows a bacterial infection elsewhere in the body, for example, strep throat. It is actually an allergic response to the toxins released by the invading bacteria. The glomeruli become inflamed, swollen, and more permeable than normal, allowing large proteins and even red blood cells to enter the filtrate and appear in the urine. Most cases of glomerulonephritis heal once the infection is

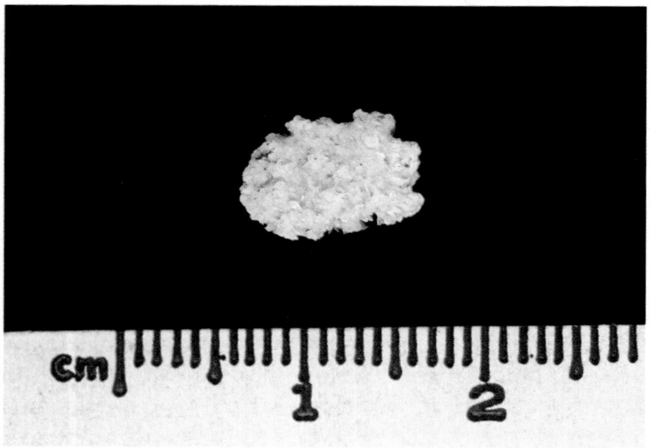
cm
1
2

FIGURE 11.17 Photograph of a kidney stone (renal calculus).

cleared, but a chronic condition called *Bright's disease* sometimes results.

Urinary Tract Infections

Urinary tract infections (UTIs) include infections of the bladder and urethra. They are among the most common human infections. However, they are more common in females because a woman's shorter urethra allows bacteria and yeast cells to spread into the bladder from the body surface, lower digestive tract, or reproductive system. By age 30, at least 20 percent of all women have acquired at least one urinary tract infection. Such infections can best be avoided by increased water intake, frequent urination (especially after sex), and frequent bathing.

Kidney Failure

An abrupt or progressive decrease in glomerular filtration with reduced urine output is called *kidney failure* or *renal failure. Acute renal failure* (*ARF*) can follow an episode of kidney stones or renal tubule disease but also may be observed after severe blood loss or heart disease that reduces cardiac output. *Chronic renal failure* (*CRF*) develops more slowly and may follow long-term glomerulonephritis. The disorder progresses until over 90 percent of the nephrons no longer function and urine production almost ceases. The symptoms include generalized swelling (edema), acidosis, high blood urea and potassium levels, and a decrease in red blood cells resulting from reduced erythropoietin production. People with CRF often require hemodialysis or kidney transplantation.

Hemodialysis for Chronic Renal Failure

You can certainly survive with just one kidney if the other is removed because of disease or organ donation. Indeed, with only a quarter of their tissue functioning, the kidneys can keep you alive and healthy. However, with a further decline in kidney function, the use of an *artificial kidney* becomes necessary to filter the blood (Figure 11.18a). An artificial kidney is a machine consisting of permeable tubes and large volumes of clean fluids to perform *hemodialysis* (*he-mow-die-**al**-eh-sis*), the process of artificially separating wastes from blood cells and plasma proteins.

In hemodialysis, a tube connects the machine to an artery in the patient's arm. Blood passes through a series of small tubules formed from a selectively perme-

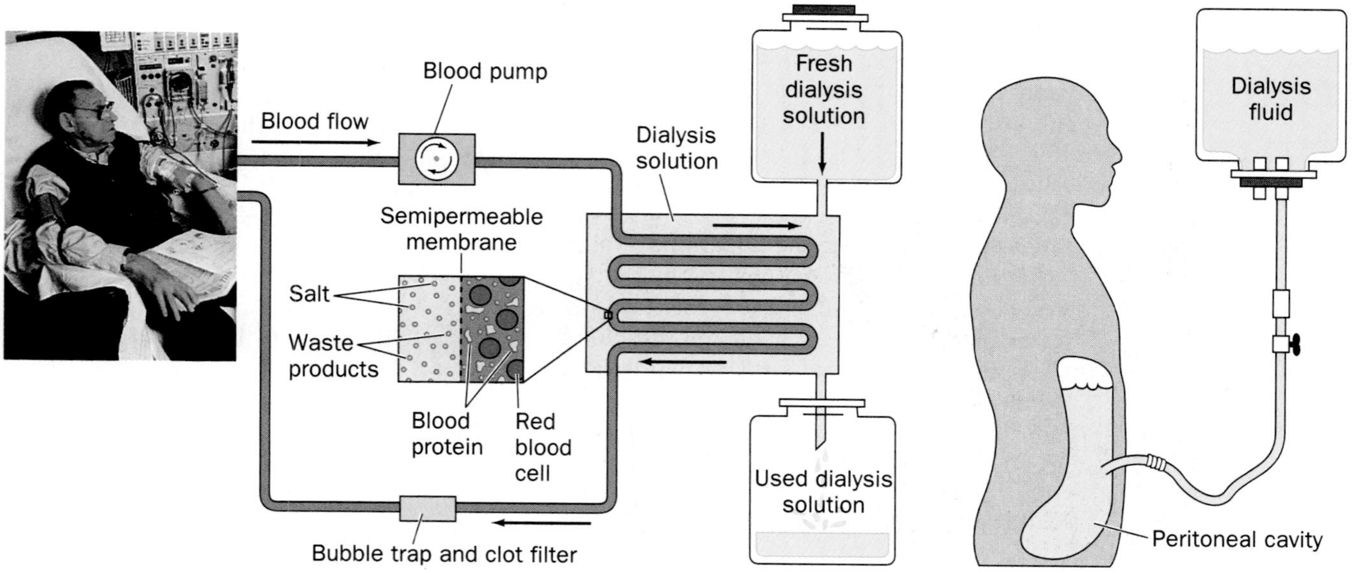
Blood pump
Blood flow
Dialysis solution
Fresh dialysis solution
Semipermeable membrane
Salt
Waste products
Blood protein
Red blood cell
Used dialysis solution
Bubble trap and clot filter
Dialysis fluid
Peritoneal cavity

(**a**) Hemodialysis with an artificial kidney.

(**b**) Continuous ambulatory peritoneal dialysis (CAPD) allows freedom of movement and reduced expense.

FIGURE 11.18 Techniques for replacing kidney function.

able membrane. The pores in the membrane are large enough for waste and water to pass through but too small for blood cells and the larger blood proteins to leave the blood. These tubules are immersed in a dilute water solution called the dialysate (*die-al-**eye**-zate*), and so the blood is continuously exposed to an osmotic gradient. To some extent this reproduces the conditions and operations of the nephron.

After passing through the artificial kidney, blood is returned through a tube to a vein in the arm or leg. Patients with no kidney function may require three or more hemodialysis sessions a week, with each session lasting for several hours. Artificial kidneys are expensive, and considerable technical expertise is required for their operation; therefore, most dialysis sessions occur in a hospital or special clinic.

Artificial devices are not perfect replacements for the kidneys. Blood cells may be damaged as they pass through the machine, and chemicals must be added to prevent blood clotting. The process is slow, patient mobility must be restricted, and there is always a risk of infection. To circumvent the mobility problem, portable units have been developed. Also, a new technique has been developed to utilize body membrane systems for hemodialysis.

Continuous ambulatory peritoneal dialysis (CAPD) is a process that permits a patient to move around while using the peritoneal membranes for hemodialysis (Figure 11.18b). The dialysate fluid is stored externally in a plastic reservoir and is added to the peritoneal cavity (see Chapter 3) through a tube (catheter) permanently inserted in the abdominal wall.

Within the peritoneal cavity, the dialysate accepts the diffusion of waste and water from capillaries, cleansing the blood while the patient conducts normal daily activities. CAPD provides freedom of movement but still carries the risk of infection because of the permanent catheter.

Kidney Transplantation

The solution to chronic renal failure and freedom from artificial hemodialysis may be achieved through the transplantation of a normal kidney. However, this is not always a simple or effective procedure. First, a compatible kidney must be found. To avoid rejection of the transplanted kidney by the recipient's immune system, the donor of the kidney should be a family member. When a biologically related kidney is not available, drugs may be required to suppress the immune response to the foreign tissue (Chapter 13).

However, immune suppressing drugs may lead to complications, and the transplanted kidney may be rejected by the host's body. Several transplants may be necessary before the donated kidney is retained. A shortage of donated organs means that many eligible hemodialysis patients must wait a long time for a chance at transplantation. That is a good reason for checking the organ donor box on your driver's license if you live in a state where anatomical gifts are permitted.

Bladder Cancer

Bladder cancer is the fourth most common cancer in men and the ninth most common in women, striking over 52,000 people annually. Blood in the urine and frequent urination are common symptoms. Bladder tumors tend to grow into the bladder cavity, permitting diagnosis by a *cystoscope* (***sis**-toe-skope*), a narrow tube with a light and lens for visual inspection that is inserted through the urethra into the bladder. Cigarette smoking appears to double the risk of developing bladder cancer. Surgery, often in combination with chemotherapy or radiation therapy, is the normal means of treatment. As with most cancers, early diagnosis and treatment are the keys to recovery and long-term survival.

Life Span Changes

At birth, our kidneys together weigh about 50 grams (1.1 ounces). They grow rapidly, doubling in weight in the first 6 months and nearly tripling by the end of the first year. After that the kidneys grow much more slowly until about age 30, when they reach their maximum adult weight of about 270 g (0.6 lb) each. The cortex of a young kidney is quite thin. A child's glomerular filtration rate is only about 60 ml/min, about half the adult rate.

Before puberty, the pelvic girdle in both boys and girls is quite narrow, causing the urinary bladder to push up against the peritoneum and project into the abdominal cavity. After puberty, the hips widen, more so in girls than in boys, allowing the bladder to drop slightly and remain within the pelvic cavity. As the bladder descends, the ureters lengthen.

During middle and later adulthood, the kidneys gradually decrease in weight, losing 30 percent of their mass by age 90, corresponding to a 50 percent decrease in the number of glomeruli. The loss of glomeruli reduces the rate at which filtrate is formed from about 125 ml/min in a healthy young adult to about 100 ml/min by age 80. Older adults are more likely to exhibit at least a trace of protein in their urine. The renal arteries, which supply blood to the kidney, exhibit an increase in collagen fibers and a decrease in smooth muscle cells, making them less flexible and less able to respond to changes in blood pressure.

An older bladder stores less urine and retains more urine after urination. About 65 percent of adults over age 65 experience sleep disruption from frequent nighttime urination. Older women are more likely to experience incontinence than are men. After menopause, the amount of muscle in the bladder wall, the vagina, and the urethral sphincter may decrease significantly, causing a loss of bladder control. Older men often experience frequent, incomplete urination or interference with urination because of enlargement of their prostate glands.

Chapter Summary

Waste Products and Excretory Organs: An Overview

- Wastes are produced by cell metabolism, processing of food, and the repair of disease or injury.
- Urea forms when NH_3 is combined with CO_2.
- Excretory organs include the skin, lungs, liver, large intestine, lymph nodes, spleen, and the urinary system.

The Urinary System

- The urinary system includes two kidneys (filter wastes from the blood); two ureters (transport urine); a bladder (stores urine); and a urethra (releases urine).
- Transitional epithelium in the bladder allows stretch: the detrusor muscle contracts to expel urine.

Inside the Kidney: Structures and Functions

- The kidneys reabsorb most of the fluid filtered from the blood, producing 1 to 2 liters of urine daily.
- Each kidney has an outer cortex and an inner medulla, where a renal pelvis connects to a ureter.
- Kidney nephrons begin at a Bowman's capsule (surrounds a glomerulus) and continue as a proximal convoluted tubule, a loop of Henle, and a distal convoluted tubule.
- Blood enters the kidney through a renal artery which sends branches to each glomerulus. Peritubular capillaries surround the nephron tubules. Blood drains from the kidney through a renal vein.

The Formation and Release of Urine

- Glomerular filtrate forms when blood pressure forces water, ions, wastes, and nutrients through the wall of the glomerulus.
- Tubular reabsorption removes small nutrients, water, and ions from the glomerular filtrate.
- Tubular secretion rids the blood of excess wastes, ions, acids, medications, and drugs.
- The production of a dilute urine requires reabsorbing more ions and less water.
- The production of a concentrated urine requires an osmotic gradient in the kidney medulla. A countercurrent mechanism assists the formation of the osmotic gradient and the reabsorption of water.
- Antidiuretic hormone increases water reabsorption by a nephron.
- Urine release is controlled by the micturition reflex.

Homeostatic Functions of the Urinary System

- Renin is released by the juxtaglomerular apparatus when blood volume and/or pressure is low. Renin activates angiotensin II, causing the adrenal gland to release aldosterone. Aldosterone increases Na^+ and H_2O reabsorption, raising blood volume and pressure.
- The kidneys influence the osmotic concentration of body fluids; they secrete acid (H^+) to hold blood pH within homeostatic limits; they regulate red blood cell production by releasing erythropoietin; and they contribute an essential step in vitamin D production.

Diseases and Disorders

- Kidney stones form when minerals precipitate out of the urine.

- Glomerulonephritis is an allergic response to a bacterial infection.
- Urinary tract infections are more common in women because of their short urethral length.
- Renal failure is an abrupt or progressive decrease in glomerular filtration and urine formation.
- Bladder cancer is recognized by blood in the urine, frequent urination, and cystoscopic observation.

Life Span Changes

- The kidneys reach maximum size about age 30, then slowly lose weight through late adulthood.
- With aging, the kidneys lose functional glomeruli and the rate of glomerular filtration decreases.
- The older bladder stores less urine. Older men often suffer interference with urination as the prostate gland enlarges.

Selected Key Terms

aldosterone (p. 297)
antidiuretic hormone (p. 297)
countercurrent mechanism (p. 294)
glomerular filtration (p. 291)
kidney (p. 286)
micturition reflex (p. 298)
nephron (p. 289)
renal calculi (p. 302)
renin (p. 300)
tubular reabsorption (p. 292)
tubular secretion (p. 293)
urea (p. 285)
ureter (p. 287)
urethra (p. 287)
urinary bladder (p. 287)

Review Activities

1. Explain how nitrogen is prepared for excretion from the body in a relatively nontoxic form.
2. List the major excretory organs of the body. Which one is responsible for nitrogen excretion?
3. List the organs of the urinary system in the proper order to trace the flow of urine from the site of formation to the point of release.
4. Explain how the structure of the bladder assists its function.
5. Describe the general structural arrangement of the kidney.
6. List the tubules of the nephron in functional order.
7. Describe the blood supply to the nephron. Where do water and useful nutrients return to the blood?
8. Briefly summarize the activities that occur during glomerular filtration, tubular reabsorption, and tubular secretion.
9. Explain how a countercurrent mechanism assists in the formation of a concentrated urine.
10. Compare the actions of aldosterone and ADH.
11. Explain the actions which allow the release of urine.
12. Explain the importance of the juxtaglomerular apparatus to the regulation of blood volume and pressure.
13. Explain how the kidneys help regulate blood electrolyte concentration and pH.
15. Summarize the operation of an artificial kidney during hemodialysis.

Self-Quiz

Matching Exercise

___ 1. Primary form of nitrogen waste
___ 2. The process by which water and small molecules are removed from blood
___ 3. The process by which water and small molecules are removed from filtrate
___ 4. The process by which additional wastes and chemicals are added to filtrate
___ 5. Site of water and nutrient return to blood
___ 6. Site of blood filtration
___ 7. Male reproductive gland around urethra
___ 8. Release of urine
___ 9. Hormone that stimulates Na^+ reabsorption and K^+ excretion
___ 10. Hormone that stimulates water reabsorption

A. Tubular reabsorption
B. Aldosterone
C. Micturition
D. ADH
E. Urea
F. Tubular secretion
G. Glomerular filtration
H. Prostate
I. Glomerulus
J. Peritubular capillaries

Answers to Self-Quiz

1. E; 2. G; 3. A; 4. F; 5. J; 6. I; 7. H; 8. C; 9. B; 10. D

Exercise: Pursuing Fitness

From daily television advertisements for exercise videotapes and equipment to membership in local exercise facilities, we are offered products designed to improve our fitness. How can you decide which of these products best suits your needs? The knowledge that you have acquired thus far about the muscular system, nutrition, the circulatory system, and the respiratory system will be helpful in evaluating answers to this question. In this Spotlight on Health, you will apply your knowledge of the body to make informed decisions about how to reach a level of fitness suited to your personal needs.

What Is Fitness?

Fitness can be defined broadly as the ability to perform physical activities without becoming overly fatigued. Physical activities range from everyday tasks such as walking up a flight of stairs or shoveling snow to the more intense activities performed by trained athletes.

Fitness involves the following body characteristics: flexibility, muscle strength, muscle endurance, and cardiovascular endurance. We all exhibit a certain level of fitness for each of these characteristics. *Flexibility* refers to a joint's range of motion, that is, how far muscles, tendons, and ligaments can be stretched to move the bones of a joint. All of us have some degree of flexibility or we would be incapable of performing simple tasks such as brushing our hair and getting in and out of a car.

Muscle strength refers to the ability of muscles to contract and move objects, as occurs in using a screwdriver, lifting a suitcase, or powering a golf swing. *Muscle endurance,* by contrast, refers to the ability of muscles to perform a task over a period of time without quickly becoming fatigued. How many times can you turn that screwdriver before your forearm muscles become fatigued, and how many stairs can you climb before your leg muscles become fatigued? As with flexibility, we all have some degree of muscle strength and endurance or we would be unable to accomplish the most routine tasks.

Cardiovascular endurance refers to the ability of the heart, blood vessels, and lungs to supply oxygen and nutrients to and remove waste products from the tissues. A person's ability to supply adequate amounts of oxygen and nutrients is directly related to the length of time that person can continue an activity. For example, individuals who are cardiovascularly fit can walk, jog, and swim for longer periods of time than can those who are less fit.

Exercising to Increase Fitness

Regular physical exercise will improve a person's level of fitness. The American College of Sports Medicine recommends an exercise program that includes

Using carotid artery to take pulse

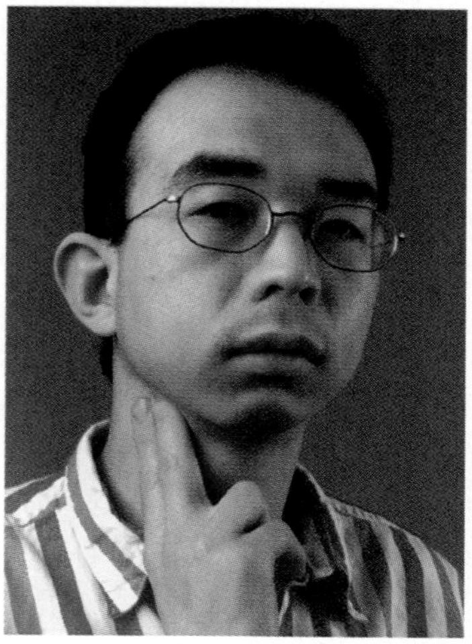

Using radial artery to take pulse

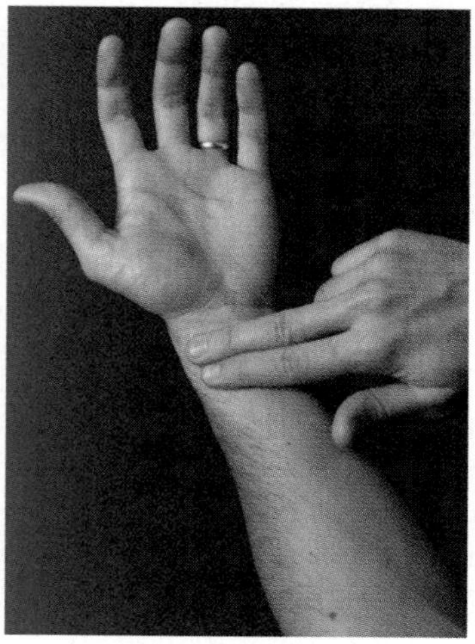

1. To take your pulse, stop exercising for a moment. For accuracy, start counting your pulses immediately. Your heart rate will begin to decrease with rest.
2. Lightly place your fingertips over artery and count the pulses for 15 seconds.
3. To determine your heart rate, multiply the number of pulses counted in 15 seconds by 4.

FIGURE 1 How to determine your heart rate.

strength training as well as aerobic exercise. The components of a balanced exercise program designed to increase fitness are described below.

Flexibility is best achieved through stretching routines that stretch muscles, tendons, and the ligaments of joints with a slow, steady, gently applied force. When you stretch, relax and breathe regularly while holding each stretch position for about 30 seconds.

Muscle strength is best achieved by lifting free weights (barbells) or using weight machines to exercise the major muscle groups. You also can use your own body weight by doing push-ups, sit-ups, and pull-ups. The key to increasing muscle strength is to lift moderately heavy weights about three times a week. Start with weights that can be lifted comfortably a maximum of 10 to 15 times. This is called a set, and a workout should include two or three sets for each muscle exercise. As a guideline to a safe and effective amount of weight for lifting, consider a weight too heavy if you cannot lift it at least 10 times.

The amount of weight that can be lifted will increase as your muscles becomes larger and strength increases. Recall from Chapter 5 that muscle enlargement is related to an increase in the diameter of each muscle cell. Muscle cell diameter enlarges primarily because the amount of actin and myosin increases to meet the new demands placed on the muscle.

Muscle endurance is best achieved by lifting light to moderate weights 15 to 20 times in each set. However, depending on a person's goal, the use of weights may not be required. If the goal is to increase leg muscle endurance, for example, an activity such as cycling or jogging may work equally well.

Muscle endurance increases as muscles become more efficient in obtaining and processing O_2 for the production of ATP. As muscle endurance is elevated, there is a corresponding increase in the number of capillaries supplying the muscle cells and an increase in the number of mitochondria in each cell. As you learned in Chapter 7, mitochondria require the O_2 supplied by blood to generate ATP, and so any increase in the number of mitochondria and capillaries will result in the generation of more ATP.

Cardiovascular endurance is best achieved through walking, jogging, swimming, rowing, aerobic dancing, cross-country skiing, or any other activity that increases the heart rate. These are commonly referred to as aerobic exercises because they require a lot of O_2.

The maximum effect on cardiovascular endurance is achieved by performing an exercise continuously for at least 20 minutes three days a week. An aerobic activity should be intense enough to elevate the heart rate to 70 to 90 percent of its maximum. The maximum rate can be roughly computed by subtracting your age from 220. If you are 20 years old, your maximum heart rate is about 200 beats per minute. The target heart rate falls between 70 and 90 percent of this value, or between 140 and 180 beats per minute. A person's pulse rate is used to determine that person's heart rate. Figure 1 shows how to take your pulse and determine your heart rate.

The cardiovascular system responds in several ways to the demands of such exercises. First, the ventricular myocardium of the heart thickens and the size of the chambers increases. Second, these changes are accompanied by an increase in stroke volume (the volume of blood leaving the heart). Third, the heart rate, both at rest and during exercise, decreases as cardiovascular endurance increases.

Before starting an exercise program, it is prudent to follow the suggestions listed in Table 1.

TABLE 1 Starting an Exercise Program: Guidelines to Follow and Why They Are Important

1. If you are over 35 and have heart disease or another medical condition, check with your doctor before starting an exercise program.
2. Warm up before each workout session by running slowly in place for several minutes. This will allow the skeletal muscles and circulatory system to meet the demands of an exercise program gradually.
3. Stretch before starting your exercise routine. This helps prevent muscles and tendons from developing small tissue tears from the pulling stresses associated with the exercise.
4. Start your exercise program slowly and with modest goals to prevent injury and health complications. This is particularly true for anyone with a sedentary lifestyle.
5. Do not perform the same exercise every day. Include a day of rest after exercising a particular muscle group. Exercising skeletal muscles often produces minute tears in muscle tissue. A day of rest not only allows for tissue repair but also provides time for the production of more actin and myosin in muscles cells.
6. The saying "no pain, no gain" is wrong. Pain is a warning that you are causing damage. An exercise that produces pain should be stopped or reduced in intensity until the pain disappears.
7. Include a gradual cooling-down period after the workout. You can simply repeat what was done during the warm-up period. This will help avoid a sudden drop in blood pressure which could prevent sufficient blood supply to the brain and result in fainting.

PART FIVE

Threats and Body Defenses

CHAPTER 12

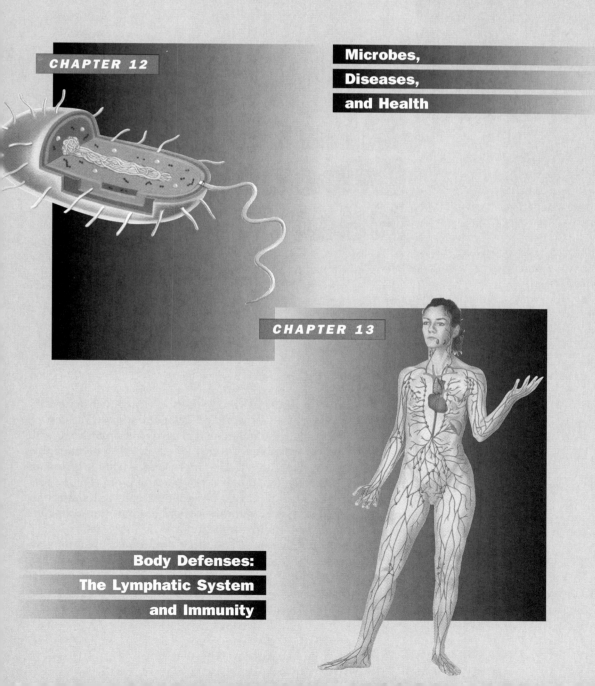

Microbes, Diseases, and Health

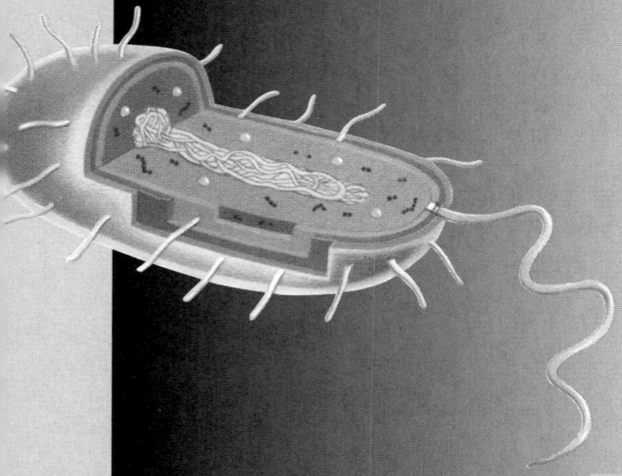

CHAPTER 13

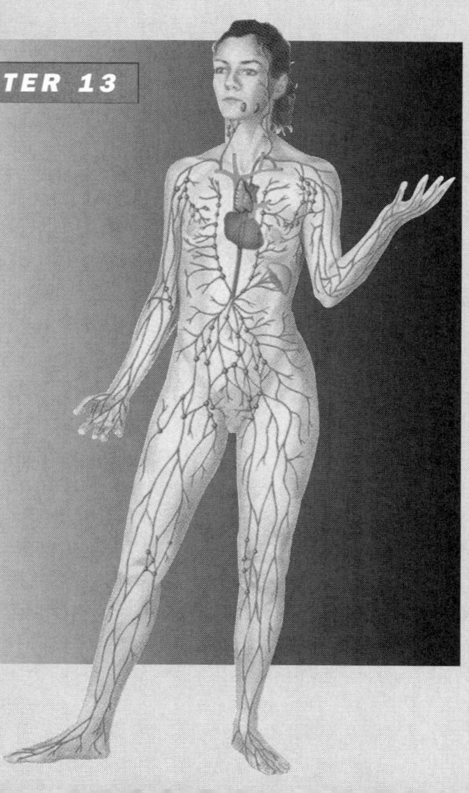

Body Defenses: The Lymphatic System and Immunity

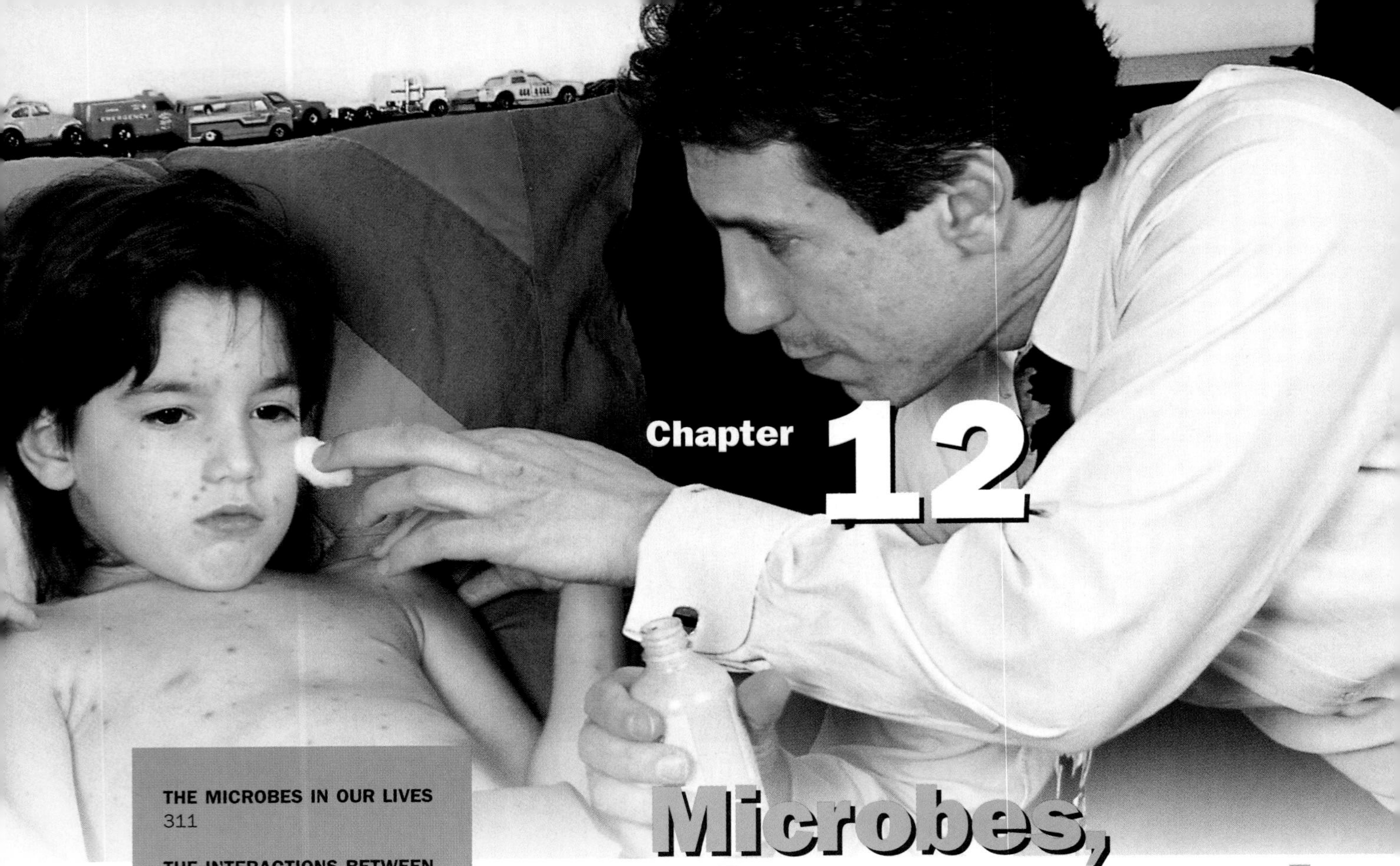

Chapter 12 Microbes, Diseases, and Health

On your skin, in your mouth, and throughout your body there is a world of living creatures that can't be seen with the naked eye. These tiny organisms, which are called **microorganisms** (*micro,* "very small") or **microbes,** are visible only with the aid of a microscope. They are so small, that thousands of them can fit on the point of a pin (Figure 12.1). Each organism usually consists of a single cell, but sometimes individual cells clump together. Although small, microbes are abundant, and they are involved in our everyday lives.

In the preceding chapters we often mentioned infections and diseases caused by microorganisms. In this chapter we will focus on this large and diverse group to better understand its beneficial and detrimental roles in our lives. The study of such very small organisms is appropriately called **microbiology** and includes the

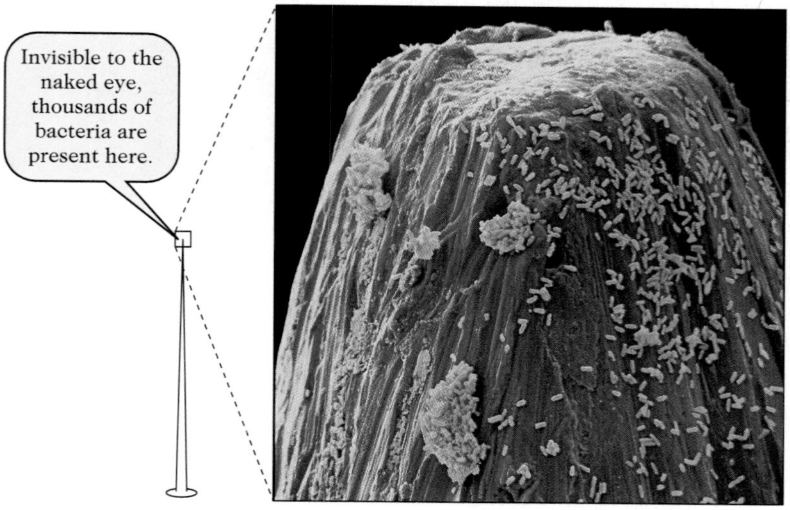
Invisible to the naked eye, thousands of bacteria are present here.

FIGURE 12.1 Scanning electron micrograph showing bacteria on tip of pin.

organisms called bacteria, fungi, and protozoa. Although viruses are not living cells, they are included in microbiology because of their interactions with living cells.

We will begin this chapter with a description of several types of microorganisms. Then we'll ask: How do microbes interact with us in health and disease? How do they actually cause disease? How can we protect ourselves from dangerous microbes? How is the combating of disease threatened by the overuse of antibiotics?

The Microbes in Our Lives

All microbes are small. Most are smaller than human cells (Figure 12.2). Some of the larger microbes are eukaryotes (protozoa and fungi), while the smaller ones are prokaryotes (bacteria) or viruses (Prologue and Chapter 2). Some microbes, such as protozoa, use specialized feeding mechanisms to capture and devour other cells. Others absorb nutrients directly from the environment across the entire cell surface. This rapid absorption of nutrients supports their rapid growth.

Microbial growth consists of proliferation by means of cell division. Rather than getting larger in size, microbes grow by increasing in number. Given nutrients and favorable conditions, most microbes grow rapidly (Figure 12.3). As this figure indicates, some bacteria can divide every 20 minutes. This means that one cell can grow to a thousand cells in about 3.5 hours and to a million cells in about 6.5 hours. Such rapid growth permits microbes to spread quickly.

The rapid growth of these cells also permits beneficial changes in their hereditary material to pass rapidly to progeny cells. This has permitted microbes to adapt successfully to diverse environments, from the depths of the oceans to the highest mountains, from the ice of the Antarctic to sulfur hot springs with temperatures about 50°C (122°F). Microbes have adapted to live on your skin and in your intestines. In animals and plants, microbes can be either beneficial

FIGURE 12.2 Size relationships between a eukaryotic cell, a prokaryotic cell, and viruses.

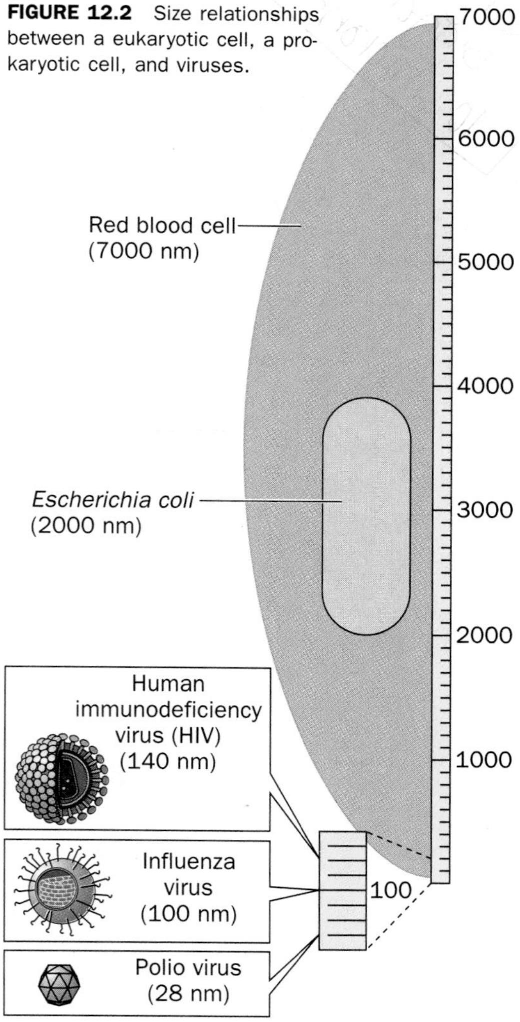
7000
6000
5000
4000
3000
2000
1000
100
Red blood cell (7000 nm)
Escherichia coli (2000 nm)
Human immunodeficiency virus (HIV) (140 nm)
Influenza virus (100 nm)
Polio virus (28 nm)

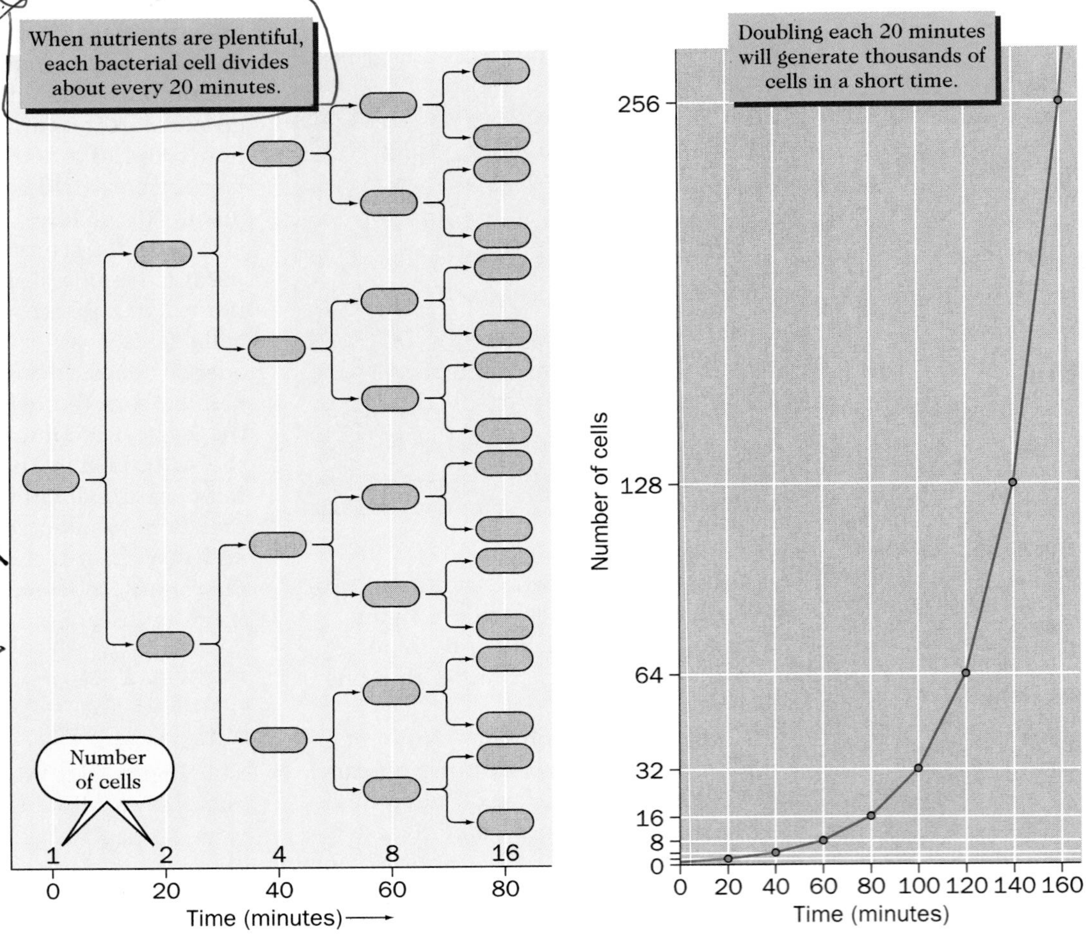
When nutrients are plentiful, each bacterial cell divides about every 20 minutes.
Number of cells
1
2
4
8
16
0
20
40
60
80
Time (minutes)
Doubling each 20 minutes will generate thousands of cells in a short time.
Number of cells
256
128
64
32
16
8
0
0 20 40 60 80 100 120 140 160
Time (minutes)

FIGURE 12.3 The rapid growth of bacteria.

companions or invaders capable of causing disease. When they cause disease, microbes are referred to as **pathogens.** These are the "germs" we are all warned against catching.

To gain a general understanding of the role of microbes in health and disease, we need to examine the characteristics of the different groups. We begin the discussion with bacteria and then proceed to fungi, protozoa, and viruses.

Bacteria: Being Successful by Being Small

Although different types of **bacteria** (singular, *bacterium*) vary in size, most are 5 to 10 times smaller than human cells. What is so special about being small? Although small in size, bacteria have a large surface area relative to the volume of their cytoplasm (see Figure 12.2). This permits bacteria to absorb nutrients rapidly from the environment to fuel biochemical synthesis and rapid growth. Like all cells, bacteria have a plasma membrane surrounding the cytoplasm. However, as prokaryotes, bacteria lack the membranous organelles found in human cells and other eukaryotes. They lack a nucleus, mitochondria, Golgi apparatus, endoplasmic reticulum, and vesicles (Figure 12.4).

The hereditary material of bacteria consists of a molecule of DNA. As in all other cells, bacterial DNA directs protein synthesis. Some antibiotics, such as tetracycline, function by disrupting protein synthesis in disease-causing bacteria while not disrupting it in human cells.

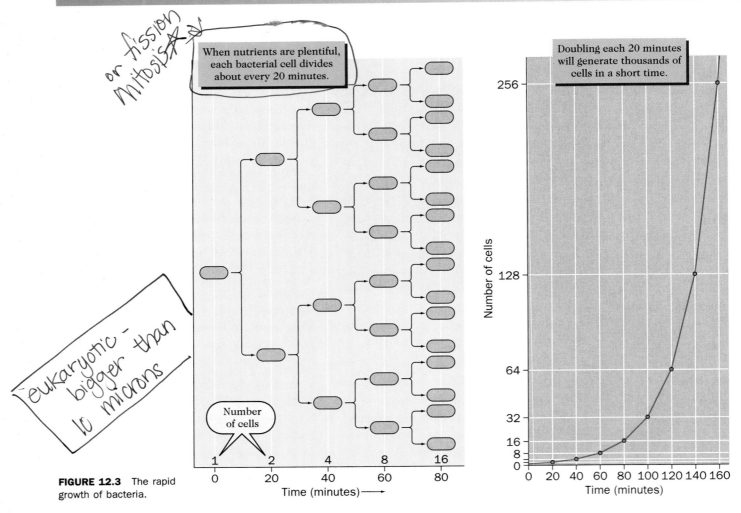

Bacteria are found in several shapes, including spherical, rod-shaped, and curved (Figure 12.5). Spherical bacteria, or *cocci* (singular, *coccus*), can occur

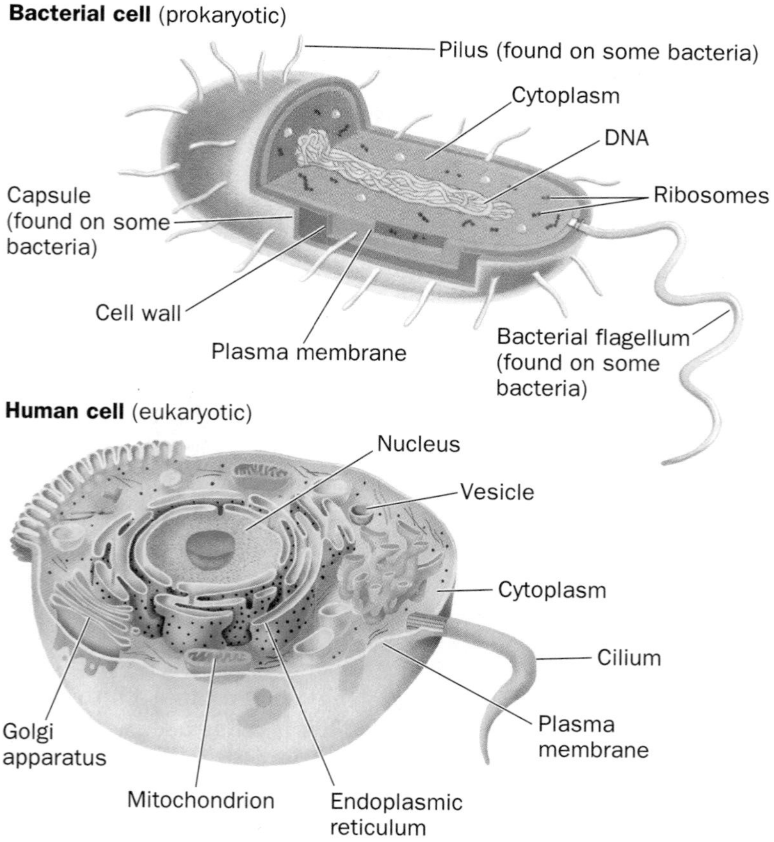
Bacterial cell (prokaryotic)
Pilus (found on some bacteria)
Cytoplasm
DNA
Ribosomes
Capsule (found on some bacteria)
Cell wall
Plasma membrane
Bacterial flagellum (found on some bacteria)
Human cell (eukaryotic)
Nucleus
Vesicle
Cytoplasm
Cilium
Plasma membrane
Golgi apparatus
Mitochondrion
Endoplasmic reticulum

FIGURE 12.4 The structural differences between bacteria and human cells.

FIGURE 12.5 Scanning electron micrographs showing the common shapes of bacteria.

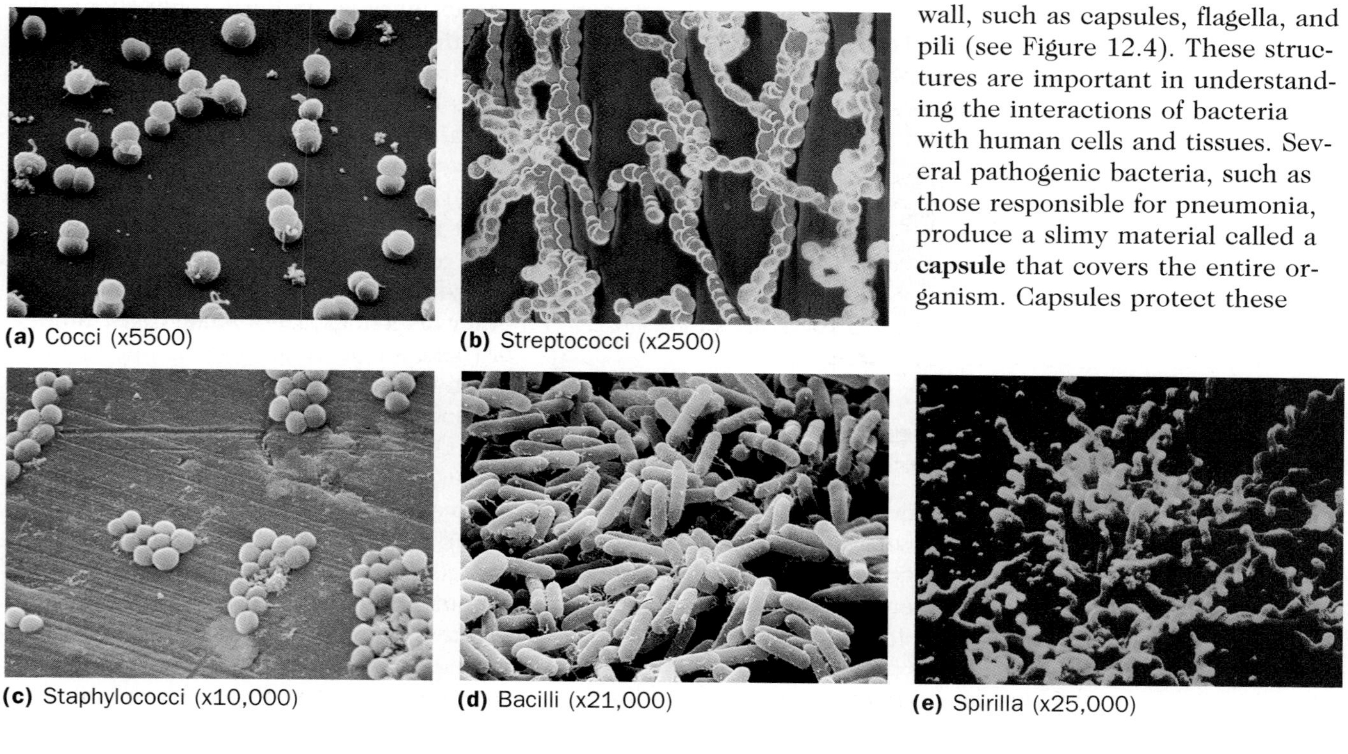

(a) Cocci (x5500)
(b) Streptococci (x2500)
(c) Staphylococci (x10,000)
(d) Bacilli (x21,000)
(e) Spirilla (x25,000)

singly, in chains of many bacteria (streptococci), or in clusters (staphylococci) similar to a stem of grapes. Rod-shaped bacteria, or *bacilli* (singular, *bacillus*), and curve-shaped bacteria, or *spirilla* (singular, *spirillum*), most often occur singly, as separate cells.

These shapes are maintained by a rigid **cell wall** that surrounds the fragile plasma membrane. Some antibiotics, such as penicillin, kill bacteria by interfering with cell wall formation. Without normal cell walls, bacteria burst open and die.

When conditions are unfavorable for growth, the cells of certain types of bacteria may undergo a transformation into dry, thick-walled protective structures called **endospores** (Figure 12.6). Their thick walls and low water content let them endure environmental extremes such as heat, drying, cold, and caustic chemicals. When environmental conditions become more favorable, each endospore develops into a growing bacterial cell (Figure 12.6). Endospores of certain bacteria are directly responsible for wound infections leading to tetanus, gas gangrene, and the food poisoning associated with botulism.

Some bacteria have structures that are located outside the cell wall, such as capsules, flagella, and pili (see Figure 12.4). These structures are important in understanding the interactions of bacteria with human cells and tissues. Several pathogenic bacteria, such as those responsible for pneumonia, produce a slimy material called a **capsule** that covers the entire organism. Capsules protect these

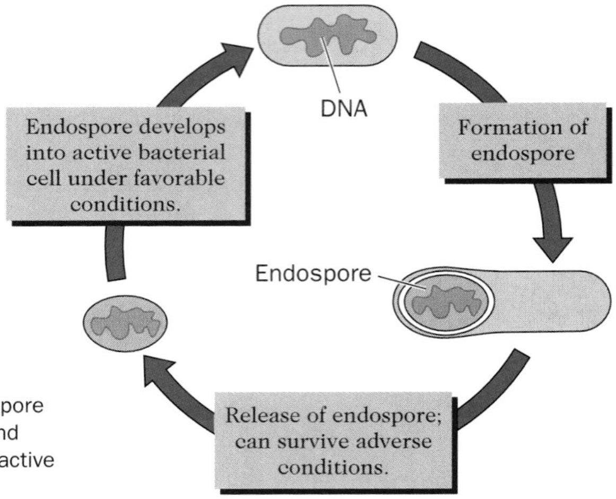
DNA
Formation of endospore
Endospore
Release of endospore; can survive adverse conditions.
Endospore develops into active bacterial cell under favorable conditions.

FIGURE 12.6 Endospore formation, release, and development into an active bacterial cell.

cells from phagocytosis by white blood cells.

Other types of bacteria have long, slender whiplike extensions called **flagella** (*fla-**gel**-ah*) that are used for movement. Flagella help bacteria move toward nutrients and spread through our tissues. Some pathogenic bacteria have shorter and thinner surface extensions called **pili** (***pie**-lee*). These structures are not used for movement but help the invading bacteria attach to tissue cells. For example, pili attach *Neisseria gonorrhoeae* (*nye-**seer**-ee-ah gon-or-**ree**-ee*) to cells of the urinary tract, producing the sexually transmitted disease gonorrhea (*gon-or-**ree**-ah*). The pili prevent these bacteria from being washed away during urination.

> Bacteria are prokaryotes and have a cell wall that maintains their characteristic shape: spherical, rodlike, or curved. Some bacteria form endospores that are resistant to environmental extremes. Some bacteria possess capsules, flagella, or pili.

Some Eukaryotes and Viruses Are Pathogenic

Fungi (***fun**-jii*) (singular, *fungus*) are a diverse group of eukaryotes ranging from microscopic single-celled yeast to much larger molds and mushrooms that can be seen with the naked eye. Fungi have a cell wall which surrounds the plasma membrane. This fungal wall is made of chemicals not found in bacterial cell walls. Consequently, fungi are not vulnerable to the antibiotics that disrupt the cell walls of bacteria.

Except for single-celled yeasts, fungi are multicellular with elongated cells forming long branching structures called **filaments** (Figure 12.7). Whether single-celled or filamentous, mature fungi do not have flagella or other structures that permit movement. Consequently, for long-range dispersal some fungi rely on specialized reproductive cells called **spores.** These spores are produced in large numbers and are dispersed widely by air currents. When inhaled, they cause some of us to suffer allergic inflammations of the nasal cavity and a runny nose.

If fungal spores land on a suitable nutrient, they germinate and develop filaments which secrete enzymes that chemically break down organic material into smaller nutrient molecules. The smaller nutrient molecules then diffuse into the fungal cells. This is what occurs when mold grows on fruit, bread, or leather. The same thing happens when fungi grow on the skin, as in athlete's foot. The athlete's foot fungus releases enzymes that chemically break down the upper layer of dead skin cells.

Of the approximate 100,000 different species of fungi on Earth, only about 100 are human pathogens. Most fungal diseases are contracted not from other humans but by exposure to air, dust, or soil contaminated with pathogenic fungi or their spores. However, the two most familiar fungal diseases—ringworm and yeast infections—can be passed from human to human.

Protozoa (singular, *protozoan*) are another diverse group of eukaryotes. Some of these single-celled organisms may be as tiny as bacteria, while a few are just large enough to be seen with the naked eye. Of the 20,000 species known, only about 30 cause human disease, but some of these are devastating diseases, such as malaria. Although malaria is rare in the United States, it affects 200 million to 300 million people worldwide, primarily in the tropics. In Africa alone, it kills more than 1 million people each year.

Most types of protozoa have some means of movement, including flagella,

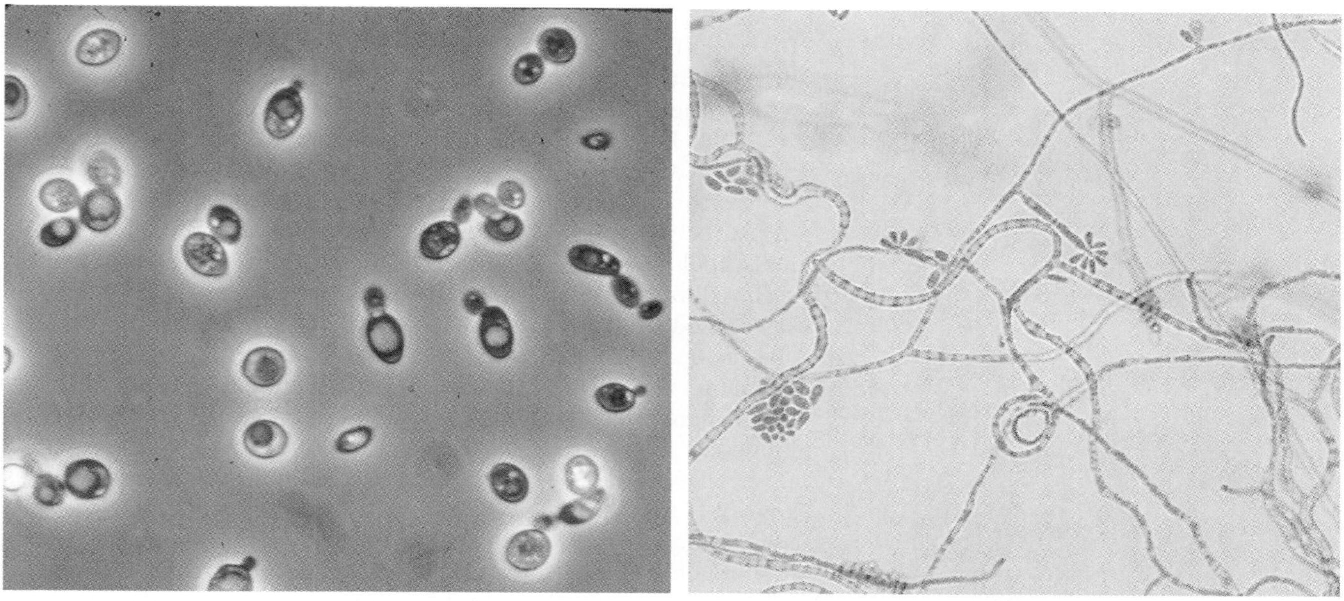

FIGURE 12.7 Two cell arrangements in fungi, single cells and branching filaments.

cilia, and cellular extensions (pseudopodia) similar to those used by white blood cells during phagocytosis. Some protozoa live on dead and decaying organic matter, while others capture and ingest small living organisms such as bacteria, algae, and other protozoa. Pathogenic protozoa obtain nutrients by attaching to, invading, and destroying cells. For example, the malaria parasite enters red blood cells, causing them to rupture.

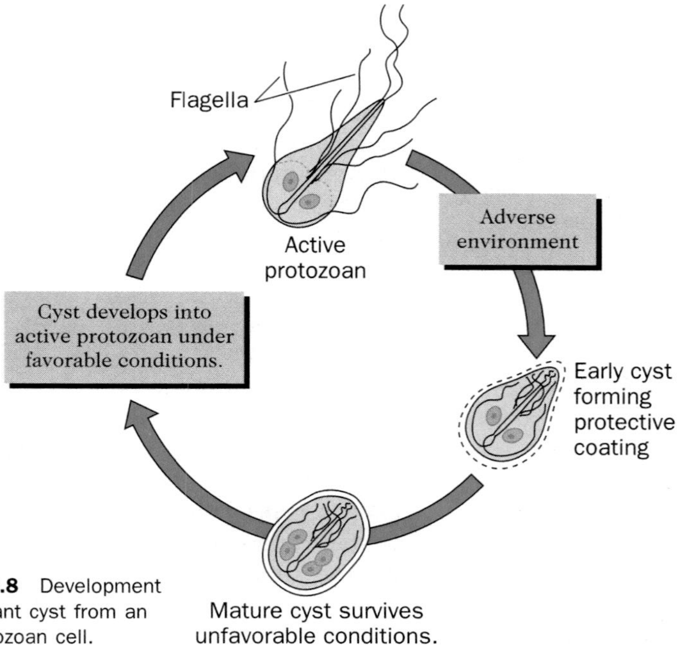
Flagella
Active protozoan
Adverse environment
Early cyst forming protective coating
Mature cyst survives unfavorable conditions.
Cyst develops into active protozoan under favorable conditions.

FIGURE 12.8 Development of a resistant cyst from an active protozoan cell.

Under unfavorable conditions such as extreme heat, cold, and drying, some protozoa are capable of transforming into an inactive protective unit called a **cyst.** Cysts function like bacterial endospores, allowing the protozoa to survive extreme environmental conditions (Figure 12.8). Under more favorable conditions, cysts germinate and become actively reproducing and growing cells. Certain pathogenic protozoa are spread by cysts. For example, cysts transmitted in drinking water and food cause severe infections of the digestive tract, such as amoebic dysentery and giardiasis.

Viruses basically consist of a molecule of nucleic acid surrounded by a protein coat (Figure 12.9a). They lack the organelles and enzymes necessary for energy transformation, protein synthesis, and independent reproduction. To reproduce, viruses must enter a living cell, termed the *host cell,* and take over its organelles and enzymes. Using these structures, the viral nucleic acid directs the production of thousands of new viruses identical to the infecting virus. These new viruses are eventually released from the host cell and are capable of infecting additional cells. Because viruses cannot metabolize and reproduce on their own, they are often considered nonliving and are referred to as *infectious agents* instead of infectious organisms.

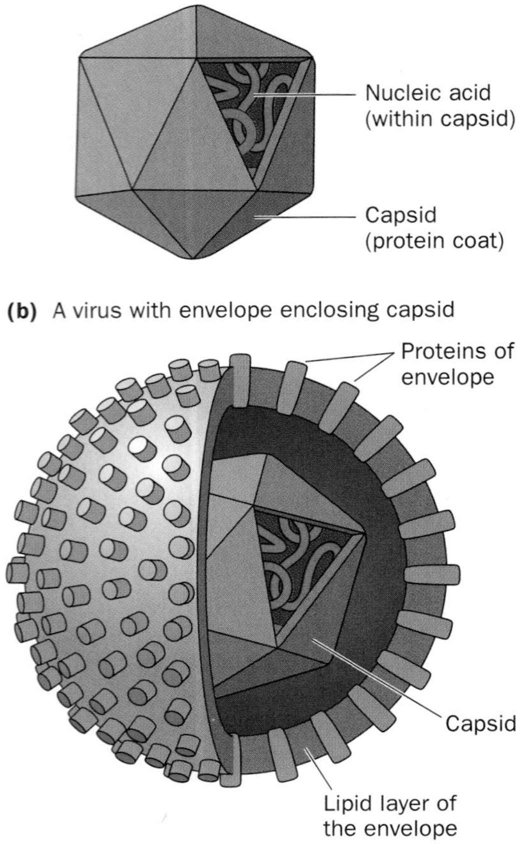
(a) A virus with capsid surrounding nucleic acid molecule
Nucleic acid (within capsid)
Capsid (protein coat)
(b) A virus with envelope enclosing capsid
Proteins of envelope
Capsid
Lipid layer of the envelope

FIGURE 12.9 Two forms of viruses.

Many viruses use DNA for their hereditary material, while other viruses use RNA. Both RNA and DNA viruses can cause diseases in plants and animals, including humans. Human diseases caused by RNA viruses include the common cold, AIDS, mumps, measles, influenza, and polio. DNA viruses are responsible for warts, herpes, hepatitis B, mononucleosis, and chicken pox.

The protein coat that encases the viral nucleic acid is called a **capsid** (Figure 12.9a). A capsid is constructed by bonding hundreds of protein molecules together. In addition, some viruses have a membrane called an **envelope** surrounding the entire capsid (Figure 12.9b). The envelope is acquired from the host cell's plasma membrane as new viruses are released.

> Fungi are eukaryotes with a cell wall. Protozoa are eukaryotes and lack a cell wall. Some form cysts that are resistant to environmental extremes. Viruses consist of nucleic acid (DNA or RNA) in a protein coat.

The Interactions between Microbes and Humans

Although our focus in this chapter is on microorganisms that cause human disease, relatively few microbes are actually pathogenic. Most microbes are harmless to us, and many are even beneficial to our health, our economy, and the environment. A variety of microbes normally inhabit regions of our bodies without causing illness. Indeed, these resident microbes create conditions that help us defend ourselves against invasion by pathogens. We use certain bacteria to make a variety of foods and industrial and medical products. Environmentally, microbes are naturally found in the soil, where they aid in the decay of organic material, enriching the soil for plants and small animals.

Unfortunately, some microbes do infect us and cause disease. Most of these pathogens are not natural inhabitants of our bodies but live in other environments or hosts. By understanding where pathogens normally reside and how they are transmitted to humans, we can protect our health by avoiding contact with them, controlling their proliferation, or limiting their further spread.

In the following discussions, we'll look first at some specific ways in which humans benefit from microbes. Then we'll discuss pathogens: what they are, where they are found, how they are transmitted, and how they cause disease.

Microbes Beneficial to Humans

Everywhere we search, we find beneficial microbes. They are on the human body, at work in industrial processes, and in the environment.

Friendly microbes: the normal flora. Microbes that normally reside on the human body without causing illness are called **normal flora.** Our normal flora includes a variety of bacteria, fungi, and protozoa. The exact types and numbers of these organisms are determined by conditions such as moisture, pH, temperature, and the supply of oxygen (O_2) and nutrients. The normal flora is usually associated with body surfaces directly exposed to the environment. Thus, the skin, the mucous membranes of body openings (mouth, nasal cavity, pharynx, vagina), and the intestines contain normal flora. However, other body locations do not contain microbes; such areas include the internal organs, bone, body cavities, and blood.

You are not born with your normal flora; you acquire it from the environment after birth. The normal flora provides both nutrition and protection. Bacteria residing in our intestines, for example, help to break down undigested foods, and some synthesize the vitamins we need (see Chapter 6). Our normal flora also protects us by secreting substances that hinder the growth of pathogens while encouraging its own growth.

The maintenance of a balanced and diverse normal flora is essential to your health. When conditions change and permit one type of microbe to proliferate, crowding out others, we have problems. For example, the vagina contains a protective normal flora consisting of several types of bacteria and yeast. The numbers of each microbe are kept in check by the conditions they help create. Some of these microbes release acids that make the vaginal pH unsuitable for pathogens. Consequently, excessive douching can alter the vaginal pH and change the composition of the normal flora. This permits unwanted microbes to become established and grow, causing an inflammation of the vagina called *vaginitis*.

The normal flora is stable and effective in maintaining a local environment that prevents pathogenic invasion. In fact, the normal flora is part of the body's homeostasis. It functions in this way precisely because it consists of several different types of interacting microbes. However, for some industrial purposes a mixture of different microbes is not useful.

Microbes working in industry. For many industrial processes, *pure cultures* consisting of only one type of microbe are required. The particular type of microbe is chosen because of its specific biochemical capacities. Medical products derived from microbes include antibiotics, vaccines, hormones, and steroid drugs (cortisone and estrogen). In recent years, an entirely new industry has developed using *genetic engineering* of microbes to manufacture medical products. (We will explore the principles of genetic engineering in Chapter 19.)

The food industry relies on *mixed cultures* consisting of different microbes that function together to produce the desired changes in food. Mixed cultures are used to produce fermented dairy products (yogurt, buttermilk, sour cream, and cheese), breads, vinegar, and fermented vegetable products (sauerkraut, soy sauce, and olives).

Microbes working in the environment. We've said that microbes are everywhere—air, land, and water. Photosynthetic bacteria, along with algae and green plants, act as *producers,* trapping and using light energy to drive the synthesis of carbohydrates from carbon dioxide (CO_2) and water (H_2O). Other microbes (bacteria, fungi, protozoa) act as *decomposers* and break down the organic remains of dead organisms so that carbon, sulfur, nitrogen, and phosphorous can be recycled to support new life. If this did not occur, most of the nutrients on Earth would remain locked up within the bodies of dead organisms and would be unavailable for new life.

> A collection of microbes, the normal flora, inhabit areas of our body and contribute to our health. Specific types of microbes are economically important for the synthesis of chemicals and medicines. Microbes in the environment decompose organic matter.

Pathogenic Microbes

Now we turn to the microbes that do us harm: the pathogens, or pathogenic microbes. These are the microorganisms capable of growing in human tissues and causing disease. Table 12.1 lists some of the common pathogens of bacteria, protozoa, fungi, and viruses.

Pathogens vary in the severity of the diseases they cause. Some produce mild to moderate discomfort (common cold and flu) and are quickly controlled by the body's defenses. If we are generally healthy, they usually do not threaten our lives. Other pathogens, such as those causing rabies, cholera, and plague, overwhelm the body defenses of even healthy people. If treatment is not provided, they kill nearly everyone they infect.

Some microbes infect and cause disease only if an opportunity arises. They

TABLE 12.1 Selected Microbial Diseases, the Scientific Names of Microbes, and the Areas of the Body Infected

Disease	Microbe	Primary Location of Infection
Bacteria		
Botulism	*Clostridium botulinum*	Neuromuscular junction
Cholera	*Vibrio cholerae*	Intestine
Food poisoning	*Staphylococcus aureus*	Intestine
Gas gangrene	*Clostridium perfringens*	Infected tissues
Gonorrhea	*Neisseria gonorrhoeae*	Reproductive tract
Syphilis	*Treponema pallidum*	Reproductive tract
Chlamydia	*Chlamydia trachomatis*	Reproductive tract
Infantile and traveler's diarrhea	*Escherichia coli* (*E. coli*)	Intestine
Salmonellosis	*Salmonella enteritidis*	Intestine
Tetanus	*Clostridium tetani*	Nerves
Tuberculosis	*Mycobacterium tuberculosis*	Lungs
Typhoid	*Salmonella typhi*	Intestine
Protozoa		
Amoebic dysentery	*Entamoeba histolytica*	Intestine
Giardiasis (beaver fever)	*Giardia lamblia*	Intestine
Malaria	*Plasmodium malariae*	Bloodstream
Fungi		
Aspergillosis	*Aspergillus fumigatus*	Lungs
Athlete's foot	*Trichophyton*	Skin
Histoplasmosis	*Histoplasma capsulatum*	Lungs
Candidiasis (yeast infection)	*Candida albicans*	Mouth, intestine, vagina
Viruses		
Influenza	Influenza virus	Upper respiratory tract
Chicken pox	Varicella virus	Skin
Rabies	Rabies virus	Brain, spinal cord
Hepatitis B	Hepatitis B virus	Liver
Hepatitis A	Hepatitis A virus	Intestine, liver
Fever blisters, genital herpes	Herpes simplex	Skin
German measles	Rubella	Skin
Polio	Poliovirus	Spinal cord (paralytic polio)
AIDS	HIV	Immune system

are appropriately called *opportunistic pathogens* and produce *opportunistic infections*. These microbes often infect a host whose health or natural defenses have been diminished in some way. Pneumocystic pneumonia is an opportunistic infection of the lung found in AIDS patients with severely impaired immune defenses.

Even without illness, some beneficial microbes of the normal flora can become opportunistic pathogens if they gain access to a different body location. For example, when beneficial bacteria in the digestive tract called *Escherichia coli* (*esh-er-**eek**-ee-ah **koh**-lye*), or just *E. coli*, accidentally gain access to the urethra and bladder, they can cause a urinary tract infection.

How do pathogens enter the body? Before pathogens can grow and cause disease, they must first gain access to tissues. The sites through which pathogens enter the body are called *portals of entry*. Portals of entry are breaks in the skin or the exposed mucous membranes of the eye or the respiratory, digestive, reproductive, and urinary tracts.

Some pathogens have specific portals of entry, and others do not. For example, cold viruses enter through the mucous membranes of the eyes and upper respiratory tract but not the mucous membranes of the reproductive and urinary tracts. By contrast, the bacteria that cause the sexually transmitted disease gonorrhea can enter through the mucous membranes of the eye, upper respiratory tract, reproductive tract, and anus.

How are pathogens transmitted? Many pathogens are transmitted by passing directly from an infected host, either human or nonhuman, to an uninfected host. Some are transmitted indirectly by contact with contaminated objects. Microbes may not actually grow on these objects but may remain alive long enough to be passed to you upon contact.

Diseases transmitted directly or indirectly between hosts are called *communicable diseases*. There are so many possible ways in which pathogens are transmitted, they are too numerous to mention. Your imagination is the only limitation (Figure 12.10).

It is important to realize that not all pathogens are transmitted in the same way. The viruses for the common cold and hepatitis can be transmitted after exposure to dry conditions on nonliving objects, but this is not the case for other viruses. For example, the virus responsible for AIDS (HIV) is easily inactivated by exposure to air, dryness, and soap and water. For transmission, it requires a moist medium—a body fluid such as blood or semen. This fact explains why the disease profile for AIDS is very different from the disease profile for the common cold (Prologue).

Some human pathogens are not acquired directly or indirectly from an infected person or animal. Such diseases are identified as *noncommunicable diseases*. For example, gas gangrene, botulism, tetanus, and legionnaire's disease (all caused by different bacteria) and the lung infections aspergillosis and histoplasmosis (caused by different fungi) are all noncommunicable diseases. These pathogens normally reside in the soil and water, and it is from these environments that we acquire them.

> Pathogens cause disease. Opportunistic infections occur when normal body defenses are weak. Communicable diseases are transmitted between living hosts. Noncommunicable diseases are transmitted from the environment.

How Do Pathogens Make Us Ill?

If pathogens are not killed by the host's initial defenses, the process of infection begins. During an infection, a microbe enters a tissue area, multiplies in number, and begins to cause tissue damage. The accumulated tissue damage eventually alters body functions and produces the symptoms characteristic of the disease. Each type of pathogen has particular processes which allow it to invade, multi-

Water contaminated with human feces
Respiratory droplets (sneeze or cough)
Fly
Mucus-to-mucus contact
Contaminated inanimate objects
Direct contact
Food
Insect bite

FIGURE 12.10 Examples showing how communicable diseases can be transmitted.

ply, and damage a specific tissue or group of tissues.

In the following sections, we will select representative examples from each microbial group, showing only a few of the processes microbes use to invade and destroy our cells. We will start with bacteria and viruses because they are responsible for the greatest variety of human infections and then conclude with the fungi and protozoa.

How Do Bacteria Cause Disease?

Bacteria damage tissues in many ways. Usually they produce and release enzymes and chemical toxins that help them invade and destroy tissues. As bacteria destroy cells, nutrients are liberated which the bacteria use for growth, making more bacteria. More bacteria lead to more tissue damage. Most types of bacteria proliferate in the space outside cells and do their damage from the outside. However, a few bacteria penetrate into cells, destroying them.

Tissue damage by secreted enzymes. Recall that enzymes are proteins that speed up specific chemical reactions. Through the action of enzymes secreted by bacteria, the structure and chemical composition of tissues can be changed rapidly to benefit the microbe and damage the host (Figure 12.11a).

The bacteria that cause *gas gangrene* secrete an enzyme that allows them to disrupt the molecular "glue" that binds cells together in a tissue. This permits the bacteria to pass between the cells and spread throughout the tissue. These bacteria are often introduced by wounds that penetrate deeply and cause extensive tissue damage.

A variety of other enzymes are secreted by different bacteria for tissue penetration and destruction. Some of these enzymes destroy molecules in the plasma membranes of the host cell, causing its membrane to burst open and resulting in the host cell's death (Figure 12.11a). Other secreted enzymes destroy white

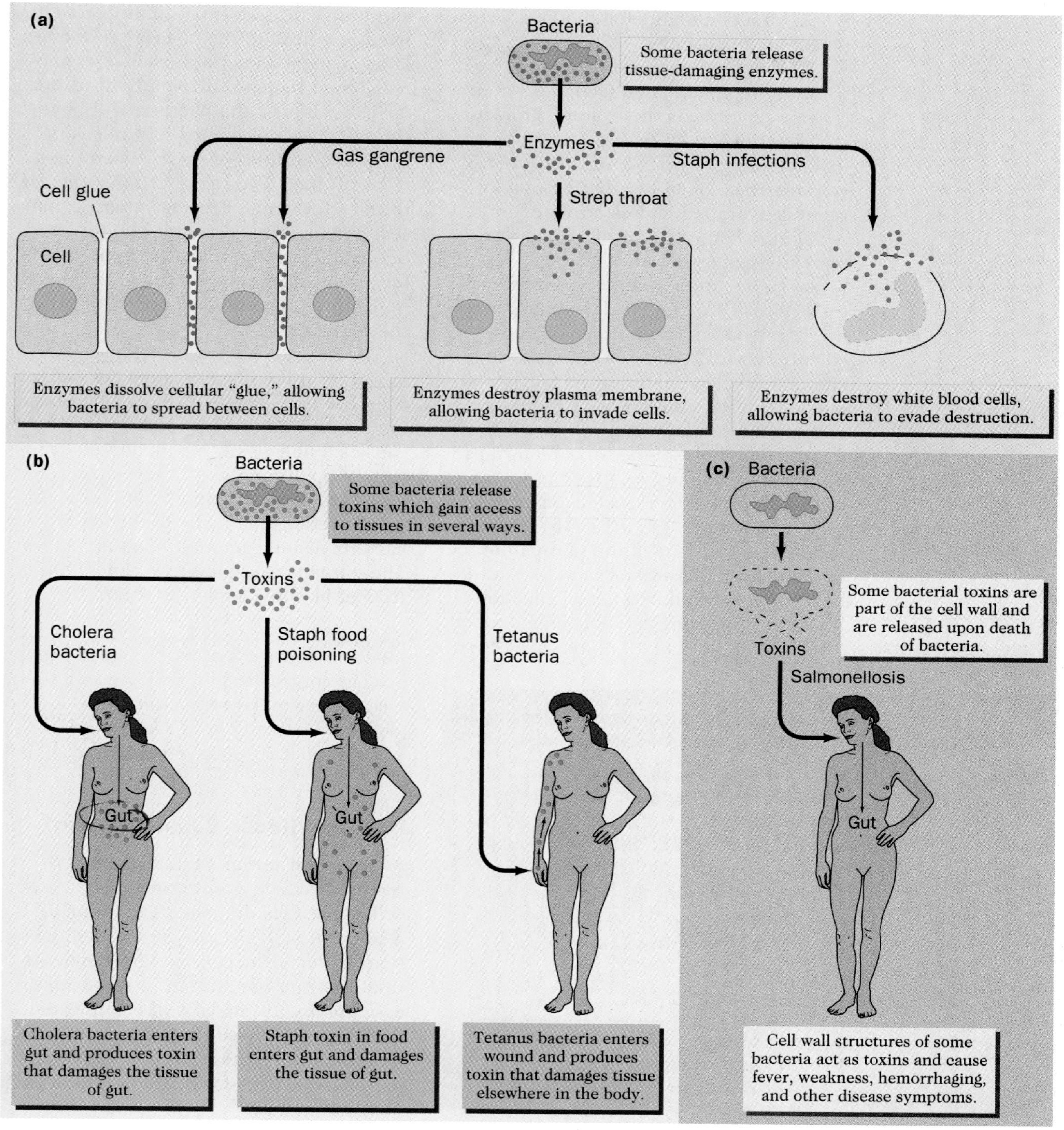

(a)
Bacteria
Some bacteria release tissue-damaging enzymes.
Enzymes
Gas gangrene
Staph infections
Strep throat
Cell glue
Cell
Enzymes dissolve cellular "glue," allowing bacteria to spread between cells.
Enzymes destroy plasma membrane, allowing bacteria to invade cells.
Enzymes destroy white blood cells, allowing bacteria to evade destruction.
(b)
Bacteria
Some bacteria release toxins which gain access to tissues in several ways.
Toxins
Cholera bacteria
Staph food poisoning
Tetanus bacteria
Gut
Gut
Cholera bacteria enters gut and produces toxin that damages the tissue of gut.
Staph toxin in food enters gut and damages the tissue of gut.
Tetanus bacteria enters wound and produces toxin that damages tissue elsewhere in the body.
(c) Bacteria
Some bacterial toxins are part of the cell wall and are released upon death of bacteria.
Toxins
Salmonellosis
Gut
Cell wall structures of some bacteria act as toxins and cause fever, weakness, hemorrhaging, and other disease symptoms.

FIGURE 12.11 Three ways in which bacteria damage cells. (a) Enzymes are released. (b) Toxins are released from living cells. (c) Toxins are released from dead cells.

blood cells and allow captured bacteria to escape phagocytosis.

Tissue damage by bacterial toxins. Poisonous chemicals that damage host tissue are called **toxins.** There are two categories of bacterial toxins: (1) those produced and released by living, growing bacteria and (2) those released from dead, disintegrating bacteria.

The first category consists of toxins produced either by pathogenic bacteria infecting the body (cholera, tetanus) or by bacteria contaminating the food we eat (staph food poisoning).

Cholera bacteria enter the digestive tract with contaminated food and water (Figure 12.11b). As the bacteria grow in our intestines, they release a toxin that irritates the intestinal lining, causing severe diarrhea. Such diarrhea results in rapid dehydration and electrolyte loss.

Tetanus bacteria typically enter the body through a wound (Figure 12.11b). As they grow in the tissues, tetanus bacteria release a toxin which is transported by the blood and causes damage elsewhere. The toxin affects the nerves that stimulate skeletal muscles. As a consequence, muscles do not relax properly and spasms result.

Staph food poisoning, once known as ptomaine poisoning, results from toxins produced by staphylococcus bacteria growing on food (Figure 12.11b). These bacteria can be part of the normal flora of the nasal passages but can also cause boils, pimples, and other skin infections. Therefore, sneezing and coughing on food and preparing food with unprotected hands can lead to the deposition of millions of staph bacteria. If such contaminated food remains at room temperature for a few hours, the bacteria will have time to grow, producing and releasing harmful quantities of toxin. When ingested with food, the toxin irritates the gastrointestinal tract, causing vomiting, nausea, and diarrhea. Summer picnics, food buffets, and potluck meals are notorious for producing food poisoning. For prevention, store prepared food and leftovers in the refrigerator.

The second category of toxins are released from the disintegrating cell wall of dead bacteria (Figure 12.11c). Such toxins are associated with the bacteria that cause *salmonellosis,* a common contaminant of poultry. When these bacteria are ingested, stomach acid and other digestive secretions kill many of them. As the bacteria disintegrate, the toxin is released. These toxins cause fever, weakness, intestinal bleeding, and even shock.

> Bacterial pathogens damage tissues by secreting enzymes and toxins. Enzymes permit bacteria to spread through tissue. Toxins are released by living bacteria or from disintegrating cell walls of dead bacteria.

TABLE 12.2 Different Ways Viruses Are Released from Host Cells and Cause Damage

Methods of Release	Effects on Host Cells
Rapid release 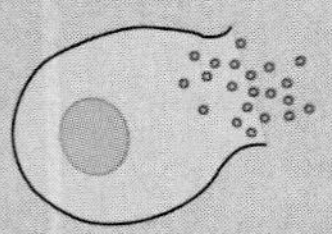	Cell ruptures and dies when viruses are released.
Slow release 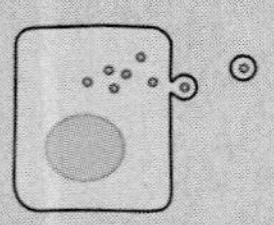	Cell damage and death occur over relatively long periods of time.
Periodic release 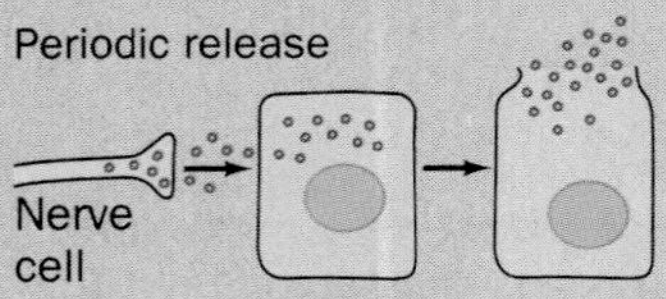 Nerve cell	Little or no damage to nerve cells, where viruses are lifelong residents. Periodic damage to epithelial cells upon release from nerves.
Integration of viral nucleic acid into host DNA 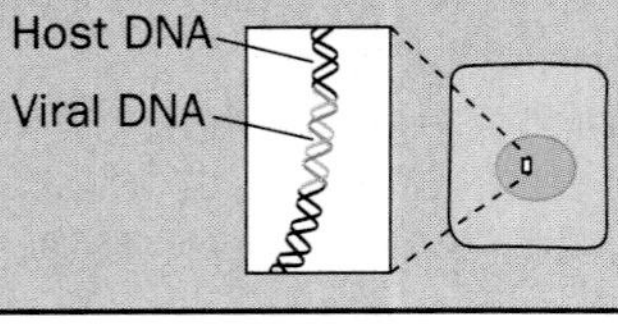 Host DNA Viral DNA	When activated, viral nucleic acid directs production of new viruses. This damages the host cell.

How Do Viruses Cause Disease?

As we noted earlier in this chapter, viruses reproduce by entering a living host cell. Once they are inside, viral nucleic acid (RNA or DNA) reprograms cell operations to serve the goals of viral replication. The specific symptoms of a viral disease are due to the type of cell infected, damaged, and killed. In this section, we'll describe the events of viral infection and the consequences for cells and the human body as a whole.

Viral infection of a host cell is traditionally divided into four sequential phases: attachment, penetration, biosynthesis and assembly, and release (Figure 12.12). After penetration, biosynthesis and assembly, the release of viruses from host cells occur in several different ways, depending on the type of virus (Table 12.2). Viruses lacking an envelope are of-

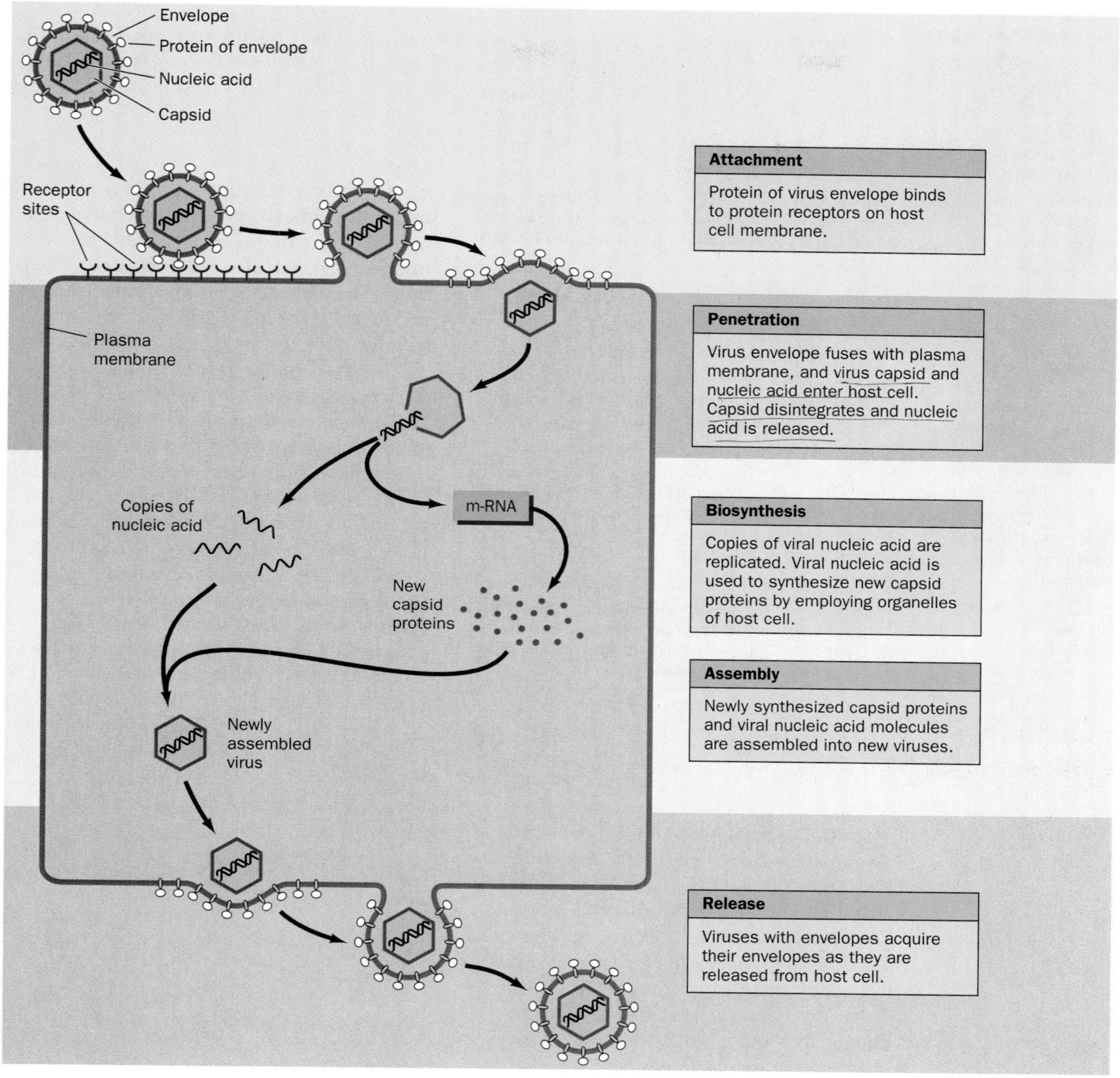

Envelope
Protein of envelope
Nucleic acid
Capsid
Receptor sites
Plasma membrane
Copies of nucleic acid
m-RNA
New capsid proteins
Newly assembled virus
Attachment
Protein of virus envelope binds to protein receptors on host cell membrane.
Penetration
Virus envelope fuses with plasma membrane, and virus capsid and nucleic acid enter host cell. Capsid disintegrates and nucleic acid is released.
Biosynthesis
Copies of viral nucleic acid are replicated. Viral nucleic acid is used to synthesize new capsid proteins by employing organelles of host cell.
Assembly
Newly synthesized capsid proteins and viral nucleic acid molecules are assembled into new viruses.
Release
Viruses with envelopes acquire their envelopes as they are released from host cell.

FIGURE 12.12 Stages and associated events of viral multiplication. The virus shown here has an envelope.

ten all released rapidly as the dead host cell disintegrates. Viruses with an envelope are released slowly over a period of time without immediate disintegration of the host cell. As single viral particles are released, they acquire a little bit of the host's plasma membrane, and this constitutes their envelope. Of course, the host cell eventually disintegrates as a result of the accumulated damage.

In addition to rapid release and slow release, there is a third possibility. In this case, new viruses do not kill the host cell and are not immediately released. Instead, they become lifelong residents of the host cell. During this residency, their periodic release may cause painful symptoms. The herpes simplex virus that causes *fever blisters* and *genital herpes* is an example of this possibility.

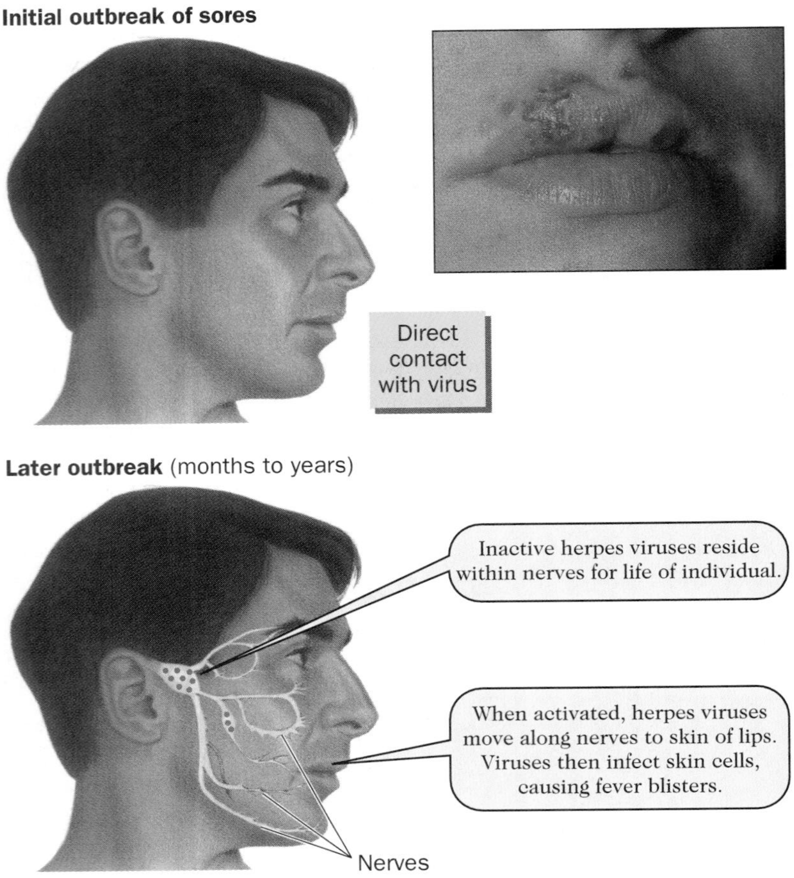
Initial outbreak of sores
Direct contact with virus
Later outbreak (months to years)
Inactive herpes viruses reside within nerves for life of individual.
When activated, herpes viruses move along nerves to skin of lips. Viruses then infect skin cells, causing fever blisters.
Nerves

FIGURE 12.13 Patients with fever blisters have herpes viruses residing in nerves of the face.

Initially, the herpes virus infects and damages epithelial tissue. It then spreads to nerve cells and becomes inactive (Figure 12.13). During the inactive period, the viruses reside in nerves near the skin. Physical and emotional stress may reactivate the viruses. Upon reactivation, they leave the nerve cells (without damaging them) and infect epithelial cells of the skin, where damage occurs. These active and inactive episodes can be periodically repeated throughout a lifetime.

Chicken pox is another lifelong viral disease. After the usual childhood case of chicken pox, this virus remains inactive in a person's nerve cells. If reactivated later in life, it causes the painful skin condition known as *shingles*.

Viral nucleic acid may hide in a cell's DNA. There is yet another strategy for viral survival, reproduction, and damage to host cells. After penetrating a cell, some viruses do not immediately follow the sequence of events outlined above. Instead, before it produces viruses, the viral nucleic acid is inserted into and becomes part of the host cell's DNA, a process called **viral integration** (Table 12.2). The viral nucleic acid will remain integrated within the host cell's DNA for the life of the cell. Every time the host cell's DNA is replicated before mitosis, the viral nucleic acid will also be replicated. When the host cell divides after mitosis, the viral nucleic acid will be passed with the cellular DNA to the two progeny cells.

Viral integration may persist for months or years without damage to the host cells. However, all is not well. Eventually the viral nucleic acid may become activated by chemicals or other microbial infections. Upon activation, the integrated viral nucleic acid produces nucleic acid and viral proteins that are assembled into viruses. This process damages and eventually destroys the host cell. For good reason, such viruses are getting much attention today. The human immunodeficiency virus (HIV) that causes acquired immune deficiency syndrome (AIDS) is an RNA virus, and its integration into the host's DNA chromosomes requires that its RNA be copied as DNA. This is further discussed in Spotlight on Health: Five (page 354).

> After infecting a host cell, some viruses multiply quickly and are released as the host cell dies. Other viruses do not immediately kill and are released slowly. Some viruses become lifelong residents of host cells and are occasionally released.

How Do Fungi and Protozoa Cause Disease?

Fungi have mechanisms similar to those of bacteria for invading and killing human tissues. By releasing enzymes and chemical toxins, pathogenic fungi invade and digest our tissues, using the breakdown products as nutrients for their growth. Pathogenic fungi have two main portals of entry: the respiratory system and the skin. Fungal spores may be inhaled with the air we breathe or contaminate our skin, hair, and nails.

A common respiratory system disease caused by a fungus is *histoplasmosis*. This fungus grows in soils rich in nutrients from bird droppings, and so areas where birds congregate are often places where histoplasmosis infections are prevalent.

The most common fungal infections of the skin affect its upper layer and associated structures. *Ringworm* is the collective name of such fungal skin diseases, which include athlete's foot. These fungi release an enzyme which breaks down the keratin of the skin's upper layer, hair, and nails (Figure 12.14). Yeasts are fungi, and some yeast infections are caused by a pathogenic yeast named *Candida*. These fungi can infect the upper skin surface as well as the mucous membranes of the mouth, the intestinal tract, and the vagina.

Pathogenic protozoa infect a range of organs, including the intestinal tract, brain, liver, and blood. Intestinal tract infections are contracted from food and liquids contaminated with the pathogen. *Amoebic dysentery* and *giardiasis* (*gee-are-***dii***-ah-sis*) are two common intestinal tract infections caused by protozoa. In these two diseases, the active protozoa cells are destroyed by stomach acid before entering the intestines. However, their cysts are resistant to stomach acid, and they enter the intestinal tract, where they develop into active cells. It is these cells that invade the intestinal lining, causing violent attacks of diarrhea, abdominal cramps, and nausea.

Malaria is the world's most widespread disease caused by protozoa. It is transmitted between humans by a mosquito that feeds on human blood. In malaria, the protozoa enter the bloodstream, mature in the liver, and then reenter the blood to infect red blood cells (RBCs). Over a period of weeks, RBCs are destroyed, leading to anemia, capillary obstruction, and general tissue damage. However, with appropriate care, most patients recover.

> Pathogenic fungi usually infect the skin and the respiratory system. Pathogenic protozoa usually infect the intestines, brain, liver, and blood. Cysts of protozoa survive stomach acid and germinate in the intestines.

Control of Microbes

Microbes are everywhere in our environment. There are medical and economic situations in which microbes must be eliminated or their numbers must be greatly reduced. Sterilization destroys all the

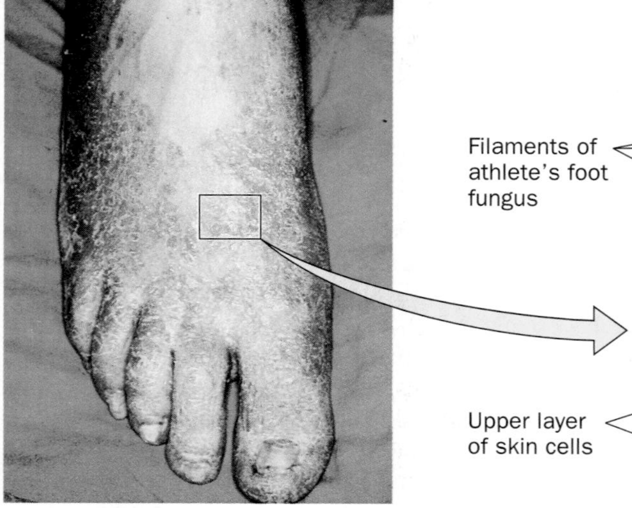

Athlete's foot fungus growing on skin of foot

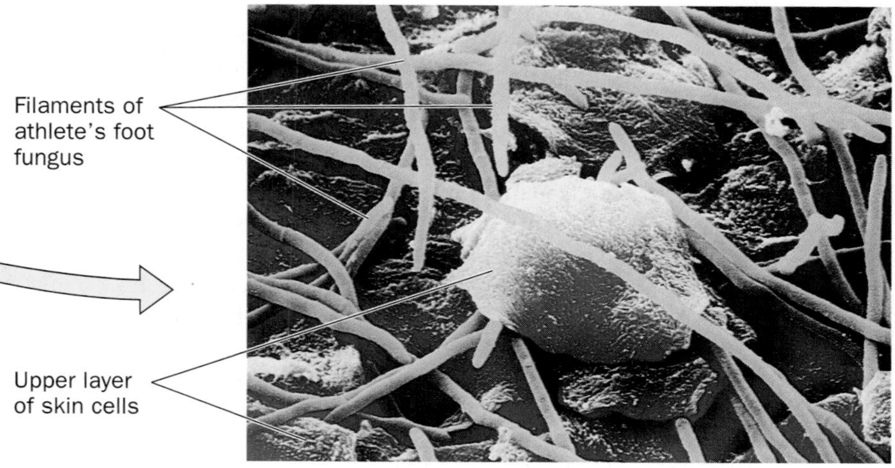
Filaments of athlete's foot fungus
Upper layer of skin cells

SEM showing athlete's foot filaments growing over and under surface of skin cells

FIGURE 12.14 Photographs showing athlete's foot fungus infecting skin.

microbes present, while various other processes, such as pasteurization, greatly reduce their numbers. These processes and methods are of direct personal importance in the prevention of infectious disease and food spoilage. In this section, we will present an overview of how physical methods, chemical agents, and antibiotics destroy microbes or control their numbers. We will also look at the growing problem of pathogen resistance to antibiotic treatment.

Physical Methods of Controlling Microbes

To survive and grow, all microbes require certain minimum conditions of moisture, pH, temperature, and O_2 or CO_2. These physical conditions vary among bacteria, fungi, and protozoa and among the microbes in each group. Such variation means that some microbes are more susceptible to physical damage than are others. Methods such as heating, cooling, drying, and radiation are used to destroy microbes, reduce their numbers, or slow their growth. Table 12.3 summarizes these physical methods and explains how they can be used to control microbes in our environment.

Chemical Methods of Controlling Microbes

A variety of chemical substances either destroy microbes or inhibit their growth. Chemicals that do this on nonliving surfaces such as kitchen equipment and bathroom facilities are known as *disinfectants*. Milder chemicals called *antiseptics* do the same thing on body surfaces without causing irritation.

Disinfectants and antiseptics act by disrupting plasma membranes, cell walls, nucleic acid molecules, or proteins, including enzymes. If applied at appropriate concentrations, chemicals such as alcohol, chlorine, and iodine are effective in destroying microbes (Table 12.3).

Chemicals are also used to purify drinking water. For example, some of us enjoy wilderness backpacking and must rely on environmental sources of water for drinking and cooking. It is wise to purify such water, since it could contain pathogenic microbes. This can be done by adding 2 drops of household bleach (such as Chlorox or Purex) to a liter of water and waiting 30 minutes before drinking. Unfortunately, such treatment may not destroy the cysts of pathogenic protozoa. If the presence of protozoan cysts is suspected, the water must be boiled for at least 30 minutes to destroy the cysts.

> Physical methods that destroy or control microbes include heating, cooling, freezing, drying, and radiation. Chemicals such as alcohols, chlorine, and iodine destroy microbes or inhibit their growth when applied at the proper concentration.

Antibiotics: Wonder Drugs That Save Lives

Antibiotics are natural chemical substances produced by a variety of bacteria and fungi. Antibiotics are effective against bacteria because they interfere with cellular processes and structures unique to bacteria. Protozoa, viruses, fungi, and human cells which lack these structures are not harmed. Antibiotics kill or inhibit bacteria by inhibiting cell wall synthesis and protein synthesis or by damaging the plasma membrane (Figure 12.15).

Antibiotics such as penicillin and bacitracin interfere with the production of bacterial cell walls. A weakened cell wall decreases the bacteria's ability to withstand the osmotic pressure of the surrounding tissue fluid, and the bacterial cell ruptures and dies.

Other antibiotics, such as tetracycline and erythromycin, inhibit bacterial protein synthesis without affecting protein synthesis in eukaryotic cells. This fortunate distinction occurs because bacterial ribosomes, the site of protein synthesis, are slightly different from eukaryotic ribosomes.

Bacterial plasma membranes have the same general molecular structure as the plasma membranes of eukaryotic cells. However, bacterial membranes are composed of different lipids. Antibiotics such

TABLE 12.3 A Summary of Selected Methods Used to Kill Microbes or Inhibit Their Growth

Method of Control	How to Perform Control Method	Suggested Uses	Action on Microbes
Heat			
Incineration	Burn by fire or pass material through open flame	Flammable objects (paper); inflammable objects (knife)	Destroys living microbes, endospores, and cysts
Dry heat	Oven temperatures: 2 hours at 160°C (320°F)	Glassware, metals, soil	Destroys living microbes, endospores, and cysts
Moist heat	Boiling water or steam for 30 minutes	Glassware, metal, food	Destroys living microbes and cysts; does not destroy endospores, some viruses
	Autoclave or pressure cooker (15 minutes at 121°C (250°F)	Glassware, metal, food	Destroys living microbes, endospores, cysts, and viruses
	Pasteurization: 15 seconds at 72°C (162°F)	Foods such as milk and beer	Reduces microbe numbers; destroys pathogens
Cold			
Refrigeration	Temperature at 4°C (40°F)	Foods and other perishables	Slows growth of microbes
Freezing	Temperature below 0°C (32°F)	Foods and other perishables	Destroys some microbes; stops growth of others
Drying			
	Elimination of water from product	Foods	Destroys some microbes; stops growth of others
Radiation			
	Ultraviolet Gamma rays	Surfaces of materials Surface and interior of materials	Radiation destroys living microbes, endospores, cysts, and viruses
Chemicals			
Isopropyl alcohol	Apply at 70–90% concentration	Minor cuts and abrasions; nonliving surface structures	Destroys living microbes ineffective for endospores, cysts, and some viruses
Chlorine	Household bleach (5–10% concentration)	Nonliving surface structures	Destroys living microbes and viruses; slowly destroys endospores and cysts
	Household bleach (2 drops per liter)	Treat questionable drinking water for 30 minutes	Destroys living microbes but ineffective for endospores, cysts, and many viruses
Iodine	Tincture of iodine (2% dissolved in alcohol)	Minor cuts and abrasions	Destroys living microbes, endospores, cysts, and many viruses

as colistin and polymyxin B react with these lipids and damage the membrane. The damaged membrane leaks, and as a result, the bacteria die.

In general, antibiotics have been wonder drugs against specific pathogenic microbes. Many of us are alive today because of them. New antibiotics that have been discovered in nature or synthesized in the laboratory have greatly increased our ability to successfully treat infectious diseases. However, more and more, physicians and scientists are finding pathogenic bacteria that are resistant to antibiotics.

Antibiotic resistance: a threat to the future. Antibiotics kill susceptible cells,

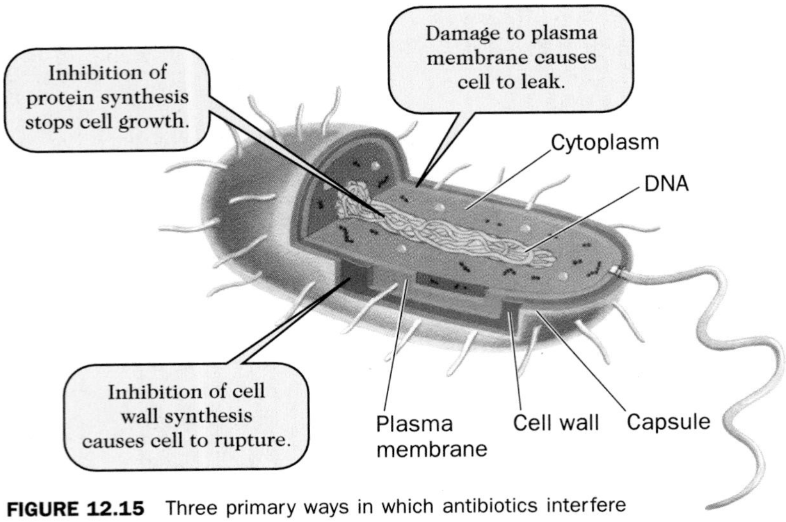
Damage to plasma membrane causes cell to leak.
Inhibition of protein synthesis stops cell growth.
Cytoplasm
DNA
Inhibition of cell wall synthesis causes cell to rupture.
Plasma membrane
Cell wall
Capsule

FIGURE 12.15 Three primary ways in which antibiotics interfere with bacterial function.

but there are always a few cells whose genes make them resistant to antibiotics. Before the widespread use of antibiotics, there were so few resistant cells that they did not present much of a problem. However, with widespread indiscriminate use of antibiotics over 50 years, these resistant cells have survived and flourished. Today, resistant microbes are a major medical concern. Tuberculosis, gonorrhea, syphilis, and pneumonia are examples of bacterial diseases whose pathogens have developed some degree of antibiotic resistance.

A recent and serious example of antibiotic resistance is provided by the bacteria that causes tuberculosis (TB). Tuberculosis is a serious infection of the lungs. It is transmitted through the air and is extremely contagious. Before antibiotics, TB was difficult to treat, debilitating, and deadly. With effective antibiotic treatment, TB has been in decline for the last four decades. With antibiotics, patients with TB could be effectively treated at home without risk to family members, friends, and the general population. However, since 1985, the number of reported TB cases has increased an alarming 20 percent.

This resurgence of TB has occurred for several reasons. Difficult economic times have reduced the money available for medical diagnosis, care, and antibiotics. At greatest risk are the impoverished, the homeless, and drug addicts. A lack of adequate health care among these populations has made it difficult to initiate and complete antibiotic therapy for TB. To successfully treat TB, antibiotics must be taken continuously for 6 to 9 months. If treatment is discontinued before completion, the surviving tuberculosis bacteria are those most resistant to killing by antibiotics. As a consequence, antibiotic-resistant forms of the bacteria proliferate and are spread to others. Because the TB pathogen is spread through the air, everyone is potentially susceptible to infection.

> Antibiotics are produced by a variety of fungi and bacteria. They destroy or inhibit growth in bacteria. Most antibiotics interfere with the cell wall structure, protein synthesis, or the plasma membrane.

Chapter Summary

The Microbes in Our Lives

- Bacteria are prokaryotes. Bacterial shapes (cocci, bacilli, and spirilla) are maintained by a rigid cell wall. External structures found in some bacteria include capsules, flagella, and pili. Some bacteria can survive environmental extremes by developing endospores.
- Fungi are eukaryotes and have a cell wall chemically different from that of bacteria. Fungi are either unicellular (yeast) or multicellular filaments (molds).
- Protozoa are eukaryotes and lack a cell wall. Some protozoa develop protective cysts that are resistant to environmental extremes.
- Viruses must infect a host cell to produce new viruses. A virus consists of nucleic acid (DNA or RNA) surrounded by a capsid of protein. Some viruses have an outer envelope surrounding the capsid.

The Interactions between Microbes and Humans

- Beneficial microbes include our normal flora, pure or mixed cultures used by industry, and environmental microbes that recycle elements.
- Pathogens are microbes that cause disease.
- Opportunistic pathogens usually infect only when there is an opportunity for their growth.
- Communicable diseases are transmitted between individuals either directly or indirectly.

How Do Pathogens Make Us Ill?

- Bacteria produce disease by releasing either enzymes or toxins that damage tissue. Enzymes and many toxins are secreted by living bacteria; some toxins are released from the cell wall after bacteria die.
- Virus infection of a host cell occurs in four stages: attachment, penetration, synthesis and assembly, and release.
- Viral release can happen in four ways: rapidly, killing the host cell; slowly, over time without immediately killing the host cell; periodically, without killing the host cell; and slowly, after integration into host DNA.
- Fungi and protozoa release enzymes and toxins for invading and destroying tissue.

Control of Microbes

- Physical methods for controlling or killing microbes include heating, cooling, drying, and radiation.
- Chemical methods for killing or inhibiting microbes include disinfectants for use on nonliving surfaces and antiseptics for use on body surfaces. Antibiotics are natural chemicals produced by certain bacteria and fungi.
- Antibiotic resistance is a worldwide concern. Tuberculosis is a disease caused by bacteria that are now resistant to many antibiotics.

Selected Key Terms

bacteria (p. 312)
capsid (p. 316)
communicable disease (p. 319)
cyst (p. 315)
endospore (p. 313)
fungus (p. 314)
normal flora (p. 316)
opportunistic pathogen (p. 319)
pathogen (pp. 312, 317)
protozoa (p. 314)
toxin (p. 321)
viral integration (p. 324)
virus (p. 315)

Review Activities

1. How do the cell structures of bacteria, fungi, and protozoa differ?
2. Which microbes—bacteria, fungi, protozoa, viruses—have structures for movement? Identify the structures used.
3. How do endospores of bacteria, spores of fungi, and cysts of protozoa function?
4. Why must viruses infect a living host cell to produce new viruses?
5. Identify at least five ways that communicable diseases can be transmitted between hosts.
6. Which body areas normally have microbes present, and which body areas do not?
7. When is the normal flora acquired?
8. How are normal flora microbes important to us?
9. Identify several ways in which microbes are economically and environmentally important.
10. Explain why a patient infected with the herpes simplex virus will continue to have periodic fever blisters.
11. Identify three diseases caused by bacteria, fungi, protozoa, or viruses.
12. Explain what occurs during each of the four phases of viral infection (attachment, penetration, biosynthesis and assembly, and release).
13. Explain viral integration.
14. Describe the three types of heat which can be used to destroy microbes.
15. Explain how antibiotic resistance has developed through the widespread use of antibiotics.

Self-Quiz

Matching Exercise

___ 1. The bacterial structure resistant to environmental extremes
___ 2. The bacterial structure used for movement
___ 3. The protozoan structure resistant to environmental extremes
___ 4. The protein coat structure enclosing the nucleic acid of a virus
___ 5. Type of bacterial secretion causing staph food poisoning
___ 6. Insertion of viral nucleic acid into host cell's DNA
___ 7. General term for a microbe that causes disease
___ 8. The respiratory disease associated with antibiotic resistance
___ 9. Name given to spherical bacteria
___ 10. An example of a noncommunicable disease

A. Capsid
B. Cocci
C. Endospore
D. Botulism
E. Flagella
F. Toxin
G. Pathogen
H. Cyst
I. Tuberculosis
J. Viral integration

Answers to Self-Quiz

1. C; 2. E; 3. H; 4. A; 5. F; 6. J; 7. G; 8. I; 9. B; 10. D

Chapter 13

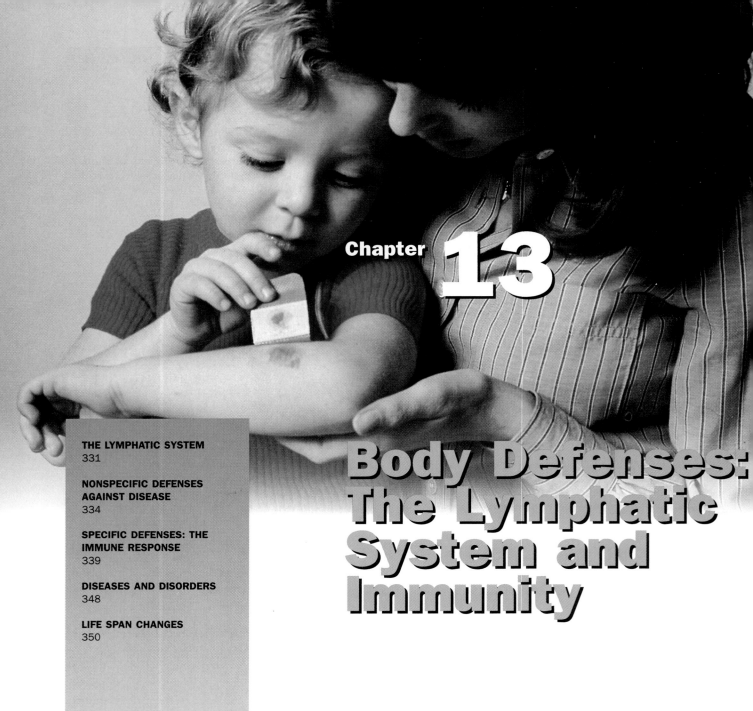

Body Defenses: The Lymphatic System and Immunity

Who are you? How are you different from me? There are several ways to answer these questions. To start with, we can talk about your parents, your family, where you were born, and your experiences growing up. There are also some less obvious answers. We can speak of differences in genes, proteins, and cells that mark you as distinctive from me. In this chapter we'll examine the mechanisms that operate to defend the "you" from all that is "not you."

As we learned in Chapter 12, there are many microbial assaults on your body at every moment. When challenged to maintain the body against invasion by pathogens and foreign substances, your defenses must be able to (1) consistently distinguish the "you" from the "not you" (failing at this task would mean that the body was attacking itself) and (2) attack, kill, or

FIGURE 13.1 The lymphatic system.

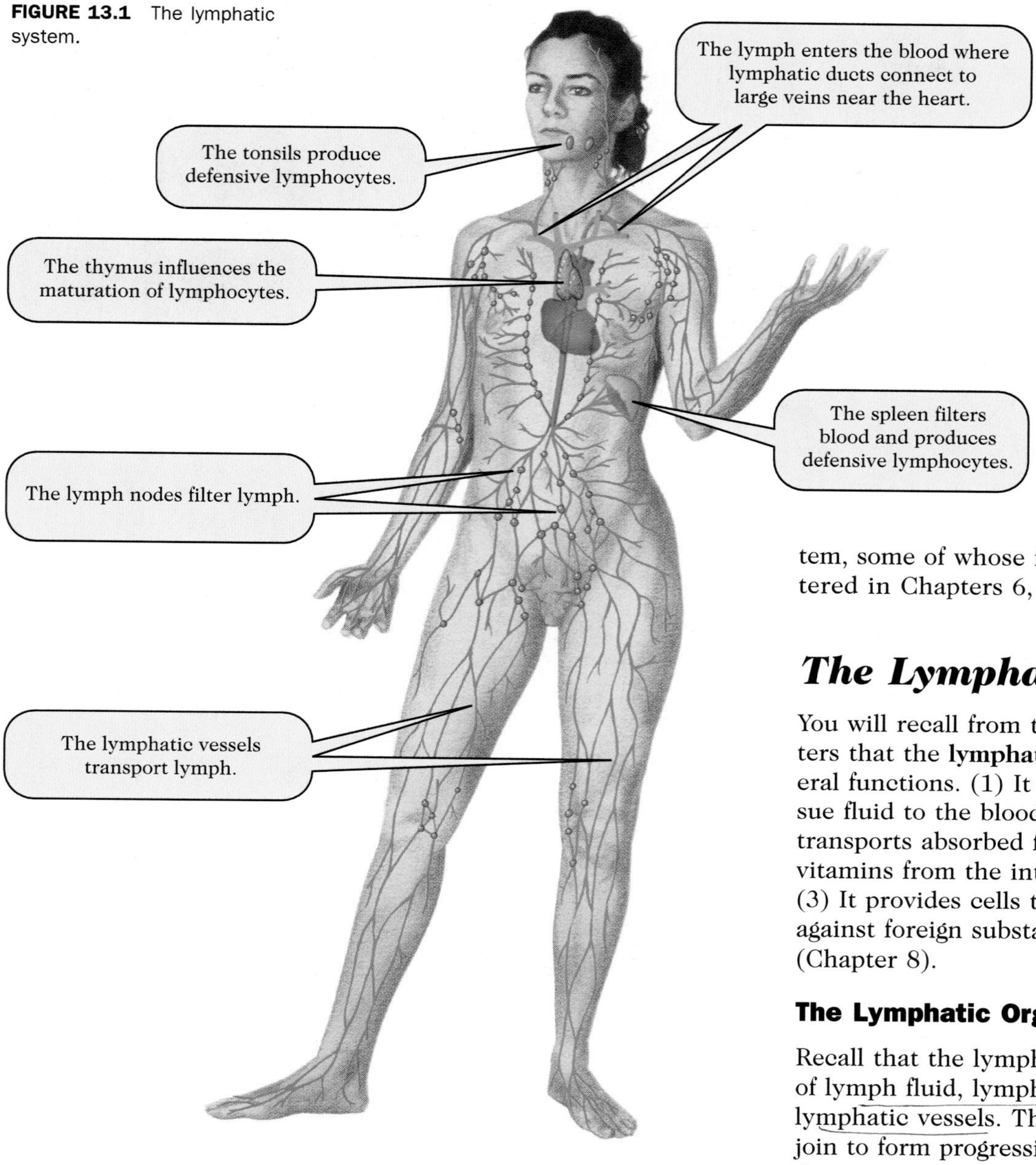

The lymph enters the blood where lymphatic ducts connect to large veins near the heart.
The tonsils produce defensive lymphocytes.
The thymus influences the maturation of lymphocytes.
The spleen filters blood and produces defensive lymphocytes.
The lymph nodes filter lymph.
The lymphatic vessels transport lymph.

alter foreign cells and substances so that they are no threat to your body. Both of these events occur thousands of times daily without our conscious awareness. More often than not, the defenses are successful—we are healthy, and we feel well. This is another example of homeostasis. Our bodies use two types of defense: nonspecific and specific. *Nonspecific defenses* are uniform responses to all invading threats. These defenses include passive barriers, chemical agents, and active combat by phagocytic cells such as neutrophils and macrophages. *Specific defenses* include responses capable of distinguishing between different threats and are collectively known as the *immune response,* which will be the focus of much of this chapter.

It is important to realize that nonspecific defenses and specific defenses operate at the same time and are cooperative. Both involve the lymphatic system, some of whose functions we encountered in Chapters 6, 8, and 9.

The Lymphatic System

You will recall from the preceding chapters that the **lymphatic system** has several functions. (1) It returns excess tissue fluid to the blood (Chapter 9). (2) It transports absorbed fats and fat-soluble vitamins from the intestines (Chapter 6). (3) It provides cells that defend the body against foreign substances and organisms (Chapter 8).

The Lymphatic Organs and Tissues

Recall that the lymphatic system consists of lymph fluid, lymphatic capillaries, and lymphatic vessels. The lymphatic vessels join to form progressively larger lymphatic ducts. Eventually these ducts join two large veins in the shoulder area and the lymph rejoins the blood (Figure 13.1). At various sites, the lymph is filtered by special structures called **lymph nodes.** These nodes contain the combat cells of the system: the phagocytes and lymphocytes.

The **phagocytes** and **lymphocytes** are the frontline soldiers in the defensive operations of the lymphatic system. Phagocytes remove foreign organisms and cell debris, and lymphocytes produce defensive chemicals to combat infection and eliminate cancer cells. Both cell types are abundant in the lymph and the lymph

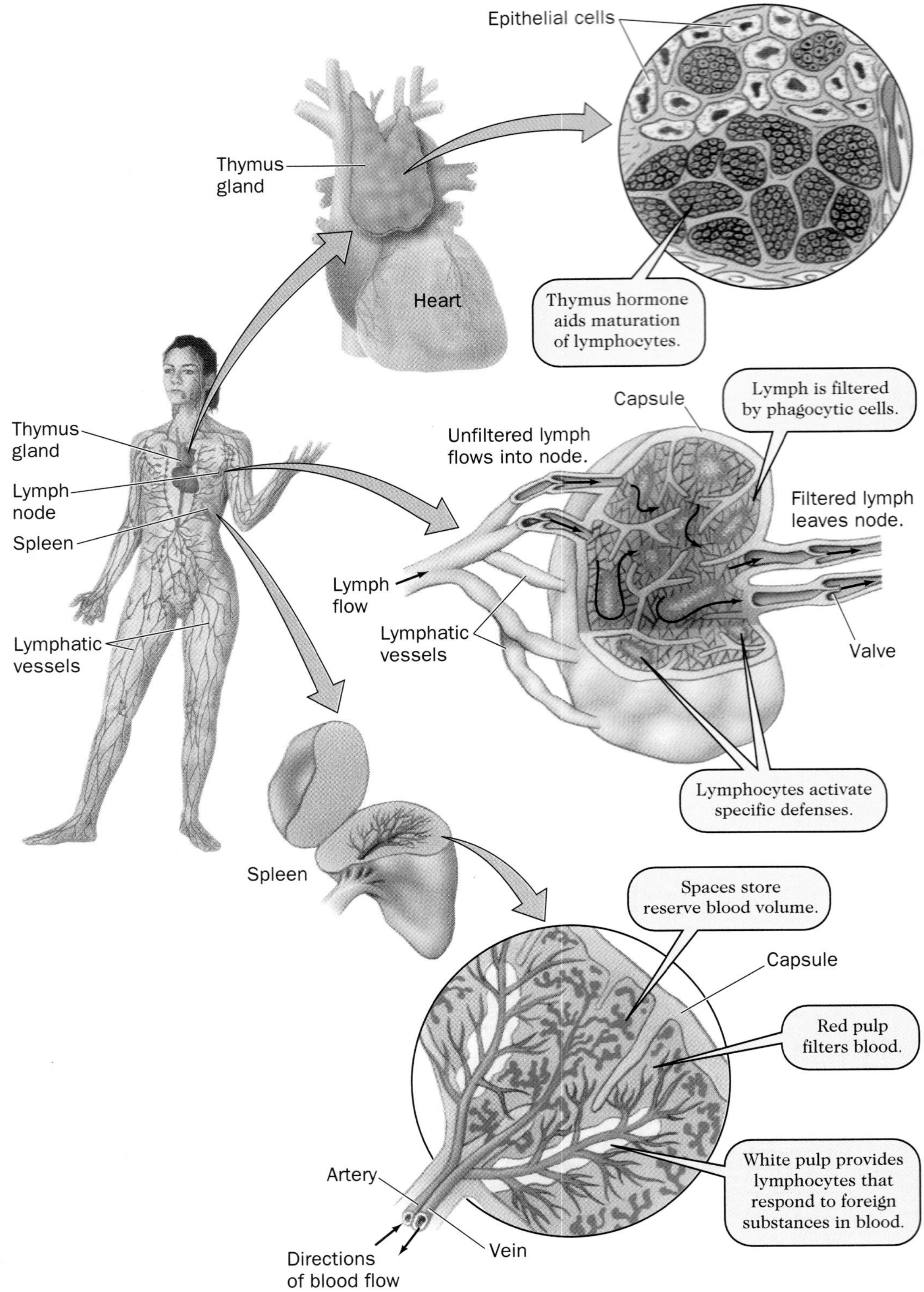
Epithelial cells
Thymus gland
Heart
Thymus hormone aids maturation of lymphocytes.
Capsule
Lymph is filtered by phagocytic cells.
Unfiltered lymph flows into node.
Filtered lymph leaves node.
Thymus gland
Lymph node
Spleen
Lymph flow
Lymphatic vessels
Valve
Lymphatic vessels
Lymphocytes activate specific defenses.
Spleen
Spaces store reserve blood volume.
Capsule
Red pulp filters blood.
White pulp provides lymphocytes that respond to foreign substances in blood.
Artery
Directions of blood flow
Vein

FIGURE 13.2 Structures of lymph nodes, thymus, and spleen.

nodes and other lymphatic organs, such as the thymus, spleen, and tonsils. To get an overview of the lymphatic system, let us now briefly examine these structures.

Lymph nodes: filtering lymph fluid. The lymph nodes are about the size and shape of a lima bean. They are clustered at the groin, armpit, neck, and lower jaw and around the digestive tract. Each node is encased by a connective tissue capsule (Figure 13.2). Internally, a node contains connective tissues, phagocytes, and lymphocytes.

Lymphatic vessels transport lymph to and from each node. Valves within those vessels assure that lymph flows in only one direction through the node. As lymph flows through a node, it comes in contact with both phagocytes that remove cellular debris and lymphocytes that activate specific defenses against microbes, foreign cells, and abnormal cells. Among the various lymphatic organs, only the lymph nodes act to filter lymph.

Thymus gland: maturing lymphocytes. The thymus gland is a double-lobed organ located behind the sternum, just above the heart (Figure 13.2). A connective tissue capsule covers the surface of the organ, while internally it contains packed lymphocytes and epithelial cells.

The thymus actually functions in two body systems—the lymphatic system and the endocrine system—by producing the hormone thymosin. **Thymosin** aids in the maturation of a particular group of lymphocytes which are important in specific defenses against disease. We'll speak more about these cells later in this chapter. The thymus is active and larger while we are young and still developing our specific defenses. It degenerates slowly after adolescence, but by that time our specific defenses are usually well established.

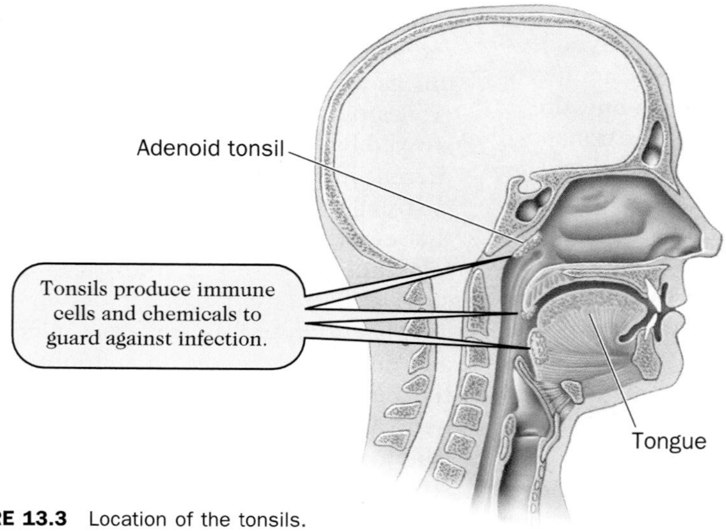
Adenoid tonsil
Tonsils produce immune cells and chemicals to guard against infection.
Tongue

FIGURE 13.3 Location of the tonsils.

Spleen: filtering and storing blood. The spleen is a soft, spongy, fist-sized mass of lymphatic tissue in the upper left portion of the abdominal cavity (Figure 13.2). A dense connective tissue with scattered smooth muscle cells covers the surface of the organ. Internally, there are two distinct masses of tissue: the white pulp and the red pulp. The *white pulp* is packed with lymphocytes which can generate a specific immune response against foreign substances in the blood.

The *red pulp* filters and stores blood. As a filter, it contains phagocytic cells which remove damaged, worn-out red blood cells and platelets. Filtered blood is stored in the red pulp as a reserve blood volume. When you exercise or suffer blood loss from an injury, the smooth muscle cells in the capsule contract to squeeze out the reserve blood and increase your circulating blood volume.

Although it is protected by the rib cage, the spleen is frequently injured by blows to the left side of the body. Because its connective tissue covering is thin, such blows may rupture the spleen, causing massive internal bleeding. Surgical removal of the spleen, called a *splenectomy* (*sple-***neck***-toe-me*), may be necessary to prevent a fatal hemorrhage. Because other organs and tissues can replace its functions, the spleen is one of the few body organs that can be removed entirely without threatening our health and survival.

Tonsils: guarding the throat. The tonsils are clusters of lymphatic tissue that contain lymphocytes and are embedded in the mucous membrane that lines the throat. Five tonsils are arranged in a ring at the junction of the nasal and oral cavities where the pharynx begins (Figure 13.3). Their job is to provide lymphocytes and immune chemicals to guard against infection entering through the mouth or nose.

The single **adenoid** (*add-uh-noid*) tonsil lies in the posterior wall of the pharynx. This tonsil may become enlarged, interfering with nose breathing and causing snoring. A pair of tonsils lie in small side pockets of the pharynx. They may become infected and require surgical removal, a procedure called a *tonsillectomy.* Another pair of tonsils are located at the base of the tongue and also may be removed in a tonsillectomy.

> Lymphatic vessels transport lymph. Lymph nodes filter lymph and combat foreign organisms. The spleen filters and stores blood. The thymus influences lymphocyte development. Tonsils produce lymphocytes.

Nonspecific Defenses against Disease

Our **nonspecific defenses** act as a first line of defense against invaders. Recall that nonspecific defenses are uniform responses to all threats; they do not discriminate between kinds of threats. These defenses react to invading bacteria in the same way they react to invading viruses or fungi, and there is no lasting memory of a particular infection.

Some nonspecific defenses involve one or more components of the lymphatic system. These defenses include the actions of phagocytic cells and the process of inflammation. Other nonspecific defenses do not involve the lymphatic system. These defenses include passive barriers to invasion such as the skin and mucous membranes, secretions, and body processes such as coughing. Nonspecific defenses are summarized in Table 13.1.

Nonspecific Defenses outside the Lymphatic System

We'll begin our discussion of nonspecific defenses by briefly describing those which do not involve the lymphatic system: species barriers, physical barriers, secretions, and body processes.

Species barriers. You do not have to worry about contracting avocado sun blotch disease, potato spindle tuber disease, fish fin rot, or even hog cholera. As a human, you are protected against most of the diseases of other animals and all the diseases of plants. Indeed most pathogens are quite specific for the kinds of cells they can infect. They bind only at certain receptor sites on plasma membranes. If these sites are absent, a pathogen cannot infect. Only a few pathogens, most notably rabies, can be shared by a variety of organisms.

Physical barriers. The keratinized epithelium of the skin provides a protective outer envelope for the body. It is difficult for pathogens to move between the tightly packed epithelial cells. Also, the high level of cell division in the epidermis causes surface cells of the skin to flake away continually, carrying off attached microbes. In addition to the skin, the body passageways that open to the exterior are all lined with mucous membranes. The secretion and movement of mucus make it difficult for microbes to attach and grow there.

Secretions that trap and destroy. In addition to mucus, secretions such as saliva, tears, earwax, sweat, and stomach acid all play roles in trapping and destroying invaders. Saliva helps remove microbes and bits of food from the mouth. It also helps prevent the mucous membrane of the mouth from drying out and forming cracks where pathogens might enter. Microbes that reach the stomach are destroyed by the hydrochloric acid (HCl) present in stomach secretions. This acid is strong enough to destroy most microbes that enter with food and drink. Only certain endospores, cysts, and viruses can survive it.

In addition to washing away microbes and debris, tears moisten and lubricate the surface of the eye. One of the components in tears is the enzyme **lysozyme,** which attacks bacteria. Earwax provides a sticky lining in the external ear canal, discouraging the entrance of microbes, dust, and small insects.

TABLE 13.1 Summary of Nonspecific Defenses

Defense	Function
Species barrier	The unique structure of human cells protects the body against the diseases of most other animals and all plants.
Physical barriers	
Skin and mucous membranes	Provide a mechanical barrier to most microbes.
Secretions	
Mucus, saliva, tears, earwax, gastric juices (HCl), sweat, sebum	Trap microbes and prevent penetration into body. Stomach and skin acids kill microbes or limit their growth. Salt of sweat is hypertonic and inhibits microbial growth.
Body processes	
Urination, defecation, vomiting, coughing, sneezing	Coordinated body actions that periodically or occasionally remove microbes and irritants from the body.
Fever	Increases metabolism; inhibits microbial growth; stimulates white blood cell production.
Inflammation	Dilates blood vessels and increases their permeability to enhance blood flow to infected tissues and assist in tissue repair.
Phagocytic cells	Neutrophils, monocytes, and macrophages engulf and destroy invading microbes and cellular debris.
Natural killer cells	Destroy virus-infected cells and tumor cells.
Interferon	Secreted by virus-infected cells and prevents viral replication in uninfected cells.
Complement	Kills microbial cells; enhances phagocytosis; stimulates inflammation.

In addition to cooling the skin, sweat is slightly salty and creates an inhospitable hypertonic environment on the skin's surface, discouraging the growth of many microbes. The skin is also protected by the secretion of sebum by the glands around hair follicles (Chapter 4). Sebum keeps the skin rather acidic (pH 3.0 to 5.0), and this limits microbial growth.

Body processes. Coordinated body processes such as urination, defecation, vomiting, coughing, and sneezing are other nonspecific defenses that help the body remove dangerous substances and pathogens. During urination, the flow of urine may help remove pathogens that have gained entrance to the urethra. Daily defecation carries away some of the normal flora microbes and occasional intestinal pathogens. When the intestinal lining is irritated by a chemical toxin or pathogen, the secretion of water into the lumen of the intestine and the vigorous contraction of the intestinal smooth muscle produce *diarrhea.* This is one way in which the body rids itself of irritating substances. Vomiting, coughing, and sneezing are all coordinated responses that eject irritating substances from the upper portions of the digestive and respiratory tracts.

Although most people view a fever as a nuisance, fevers below 38°C (100°F) are not dangerous and are actually a sign that the body is responding to an infection. Mild fever increases blood flow, metabolism, and the production of white blood cells as well as inhibiting microbial growth and the assembly of progeny viruses.

> The skin, mucous membranes, and secretions are physical barriers to infection. Coordinated processes rid the body of irritants.

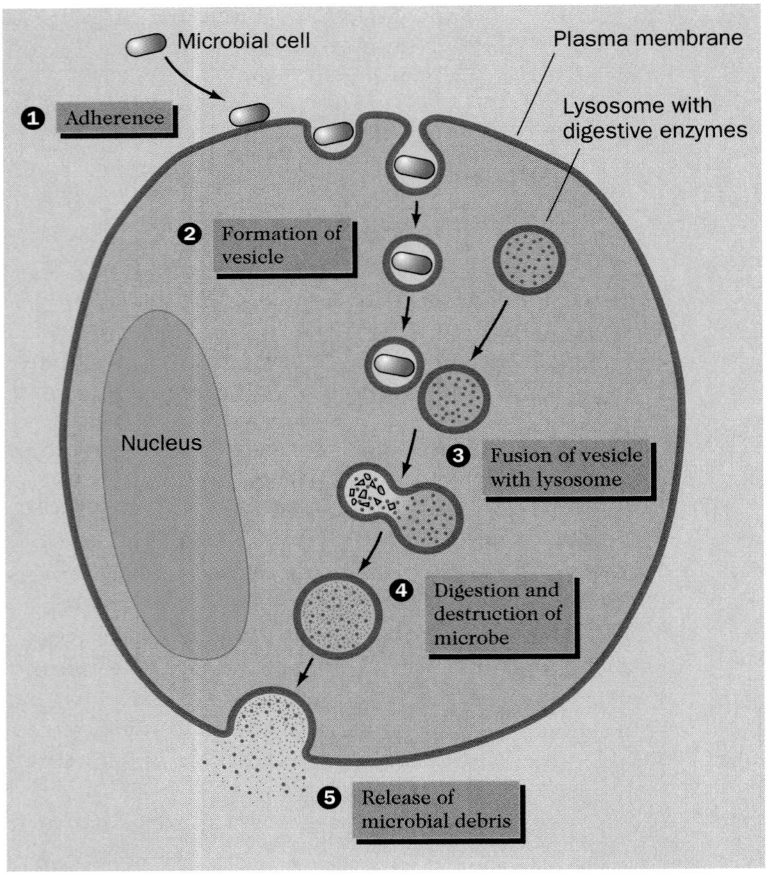
Microbial cell
Plasma membrane
1 Adherence
Lysosome with digestive enzymes
2 Formation of vesicle
Nucleus
3 Fusion of vesicle with lysosome
4 Digestion and destruction of microbe
5 Release of microbial debris

Phagocytosis of a bacterial cell

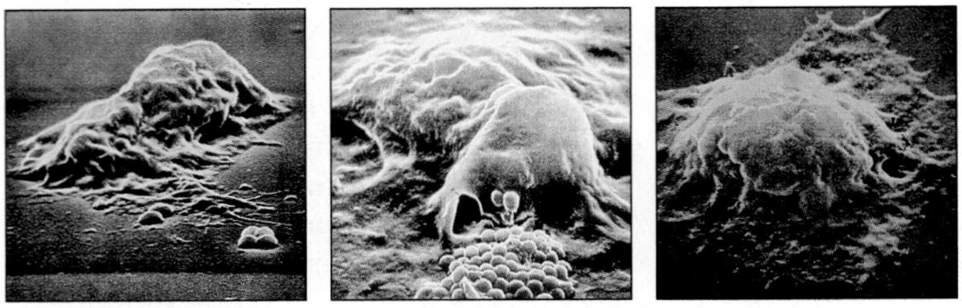

Scanning electron micrographs showing a macrophage phagocytosing bacteria

FIGURE 13.4 Stages in the phagocytosis of a microbial cell.

Nonspecific Defenses of the Lymphatic System

Some nonspecific defenses involve components of the lymphatic system: phagocytic cells, natural killer cells, interferon, complement, and the inflammation response.

Phagocytic cells. Microorganisms that successfully invade the body may encounter a second line of defense that involves cells which can engulf and destroy them. Using the process of **phagocytosis,** these cells (1) engulf foreign cells and debris, (2) contain them within cytoplasmic vesicles, and (3) digest them with enzymes from lysosomes (Figure 13.4).

The body's most numerous phagocytes are the white blood cells called *neutrophils* and *macrophages.* As we explained in Chapter 8, these cells are capable of amoeba-like movement, and can migrate out of blood vessels and into tissues. In the tissue spaces, they are attracted to the site of an infection by chemicals released from damaged and infected cells. Although some phagocytes move around through body tissues, others are more permanently attached to connective tissues in the spleen, lymph nodes, lungs, liver, and brain.

In the absence of an infection, these cells normally live for a few days. However, during an infection, when phagocytes engulf and digest quantities of pathogens and cell debris, these cells survive for only a few hours. The dead phagocytes, debris from damaged body cells, and the tissue fluid that collects at an infection site constitute *pus.* If pus is not expelled to the outside, it may be walled off by connective tissue, forming an *abscess.* Abscesses are common in soft tissue of the breast (mastitis) and gums (dental abscesses). Surgical removal is often necessary.

Natural killer cells. **Natural killer (NK) cells** are modified white blood cells that attack abnormal body cells rather than invading microbes (Figure 13.5). They are able to recognize virus-infected cells and tumor cells because of changes in the plasma membrane proteins on those abnormal cells. NK cells respond by rupturing the abnormal cells, reducing viral infections, or slowing the spread of tumor cells and the development of cancers.

Interferons: infected cells fight back. Lymphocytes, macrophages, and fibroblasts that have been infected by viruses are capable of producing and secreting a special group of proteins called **interferons** (*in-tur-fear-onz*). These proteins diffuse out of the virus-infected cells and

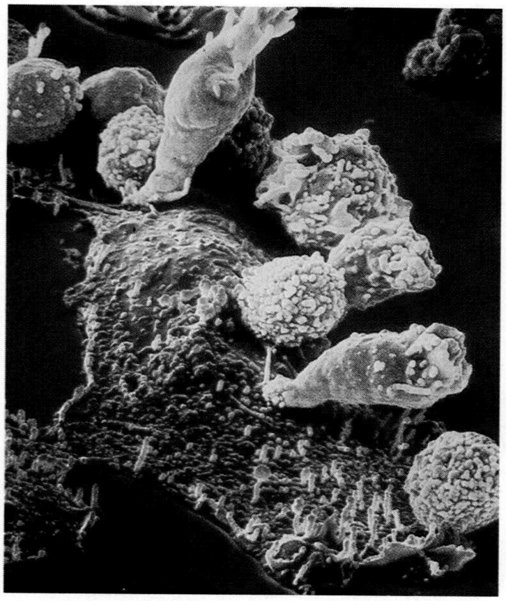

FIGURE 13.5 White blood cells attack a tumor cell.

bind to plasma membrane receptors on neighboring cells. This binding causes neighboring cells to produce enzymes which interfere with the synthesis of viral proteins and the assembly of new viruses. Thus, interferons limit cell damage from viral infection. Several different types of interferon are made by different body cells. In addition to their antiviral action, some interferons enhance the activity of phagocytes and inhibit the development of transformed cells into tumors.

The complement system. The complement system is a group of 20 plasma proteins which enhance, or "complement," other defensive responses. These proteins are normally present in the blood in an inactive state, but when an infection is present, they are activated to accomplish four fundamental goals. (1) They create holes in the plasma membranes of bacteria, allowing the cell contents to spill out, killing the cell. (2) They stimulate the release of histamine from mast cells, initiating the inflammation response. (3) They create a trail of chemical signals that attract phagocytes to an infected site. (4) They attach to pathogens in ways that help phagocytes recognize and engulf the pathogens.

Inflammation: coordinated cellular defenses. Remember the appearance of and the sensations associated with your last infected splinter or hangnail? Redness, pain, warmth, and swelling are the familiar symptoms of inflammation. Since this is a nonspecific defense for the body, the symptoms are the same for a wide variety of injuries, irritations, and infections. Inflammation proceeds in three basic stages: (1) vasodilation and increased leakage from capillaries, (2) migration of phagocytes to the site of infection, and (3) tissue repair (Figure 13.6).

Vasodilation and increased capillary leakage represent one of the first tissue responses after a trauma. The dilation of blood vessels in the vicinity of an injury increases local blood flow, reddens the skin, and causes swelling as fluids leak through the enlarged vessels and into damaged tissues. The increased flow also brings more defensive cells and chemicals to the injury while carrying off toxins, microbes, cellular debris, and dead cells.

Several chemicals contribute to vasodilation, including histamine, kinins, and prostaglandins. *Histamine* is released from injured mast cells in tissues. *Kinins* (***kye*-*ninz***) are plasma proteins activated by lysosomal enzymes during phagocytosis. In addition to vasodilation, kinins attract phagocytes. *Prostaglandins* (*pros-tuh-**glan**-dinz*) are hormonelike fatty acids released from damaged plasma membranes. They intensify the effect of histamine and kinins. Prostaglandins also stimulate phagocyte migration and cause pain, bringing an infection to our conscious awareness.

Phagocyte migration involves neutrophils and macrophages that collect at the scene of an infection, typically within an hour after inflammation starts. They find the site of infection by following a trail of kinin chemical signals. Monocytes arrive within several hours and are transformed into macrophages as they enter the tissue spaces. In addition to engulfing microbes and debris, macrophages remove dead neutrophils from the initial defensive response.

Tissue repair occurs once the invaders and the damaged tissues have been cleared away. New cells that are produced by mitosis replace damaged fibers and other tissue structures.

> Phagocytes digest microbes and debris. Natural killer cells destroy virus-infected cells and tumor cells. Interferon inhibits viral replication. Complement activates inflammation, enhances phagocytosis, and kills bacteria. Inflammation attracts phagocytes to the site of infection.

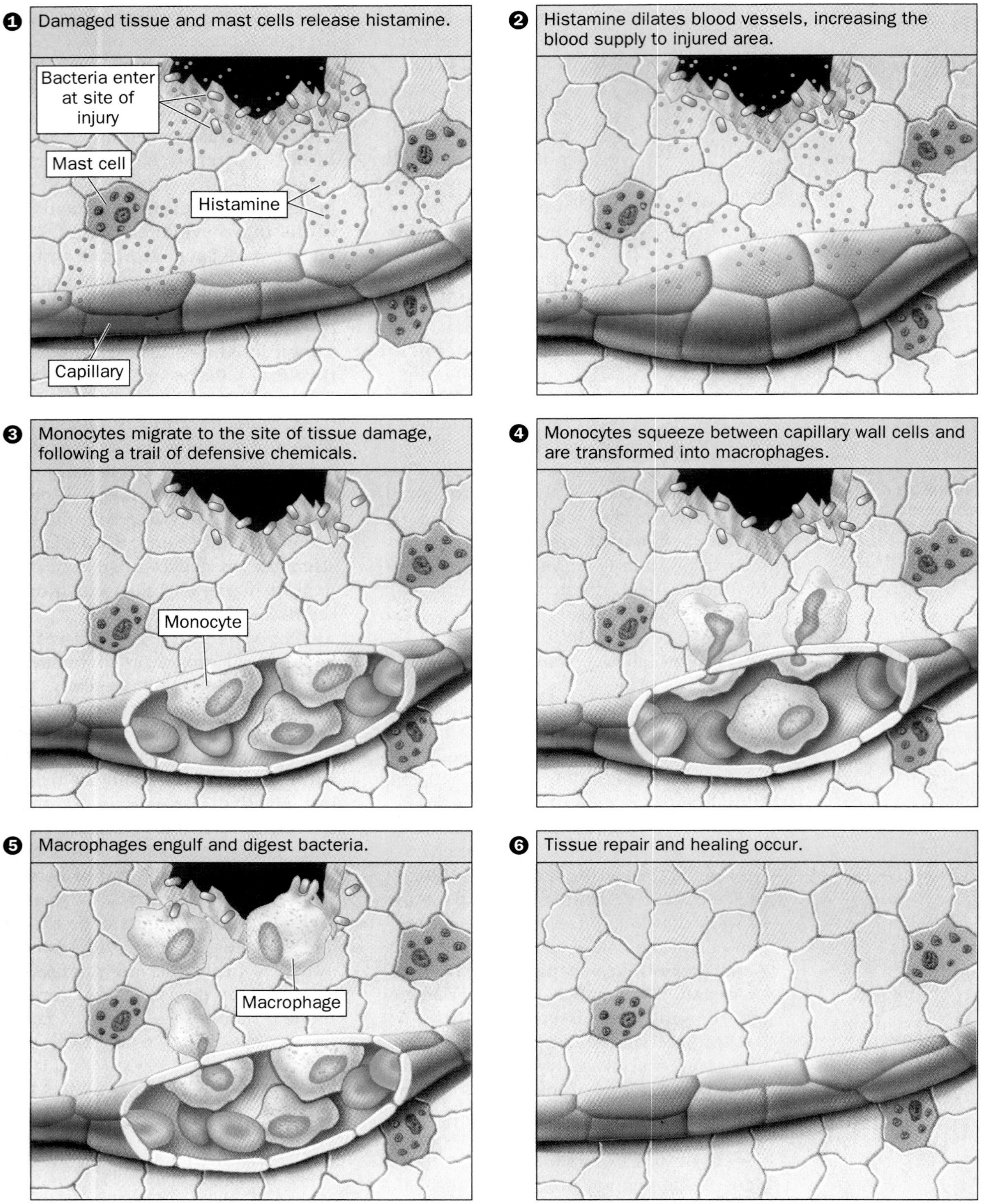
❶ Damaged tissue and mast cells release histamine.
Bacteria enter at site of injury
Mast cell
Histamine
Capillary
❷ Histamine dilates blood vessels, increasing the blood supply to injured area.
❸ Monocytes migrate to the site of tissue damage, following a trail of defensive chemicals.
Monocyte
❹ Monocytes squeeze between capillary wall cells and are transformed into macrophages.
❺ Macrophages engulf and digest bacteria.
Macrophage
❻ Tissue repair and healing occur.

FIGURE 13.6 Inflammation.

Specific Defenses: The Immune Response

When infectious agents overwhelm the nonspecific defenses, the body is further protected by a series of responses collectively called **specific defenses** or the **immune response.** Recall that specific defenses can distinguish between different foreign materials and use a slightly different response to combat each pathogen or foreign substance. Hence, we say that these defenses are specific for each foreign material.

In most cases, when an immune response is mounted, the body retains a memory, called an **immune memory,** of the pathogen so that any later reinfection by the same pathogen results in a more rapid and effective defense. The subsequent defense may be so rapid and successful that we do not become ill. For example, after recovering from measles, you retain an immune memory of the measles virus and thus acquire a lifelong protection, or *immunity,* against this specific pathogen and disease.

The extraordinary feats of the specific defenses are accomplished by the integrated functioning of the lymphatic, circulatory, and endocrine systems. However, the most important actors in this drama are lymphocytes. As we noted in Chapter 8, there are two general types of lymphocytes in the body: T-cell lymphocytes and B-cell lymphocytes. Each is charged with a different aspect of the specific immune response, and their roles are the focus of the next section.

The Lymphocytes and Immunity

T-cell lymphocytes (or **T cells**) are named after the thymus gland, because it is there, under the influence of the hormone thymosin, that these lymphocytes mature into their active form. Transported by the lymph and blood, T cells are found in tissue spaces throughout the body. They are responsible for **cell-mediated immunity (CMI),** which is active against multicellular parasites, fungi, cells infected by bacteria or viruses, cancer cells, and organ or tissue transplants. When T cells arrive at a site of infection, they release special signaling proteins called *lymphokines* (*lim-fo-**kynz***). These proteins help coordinate not only the activities of T cells but also those of B cells and macrophages (Table 13.2).

B-cell lymphocytes (or **B cells**) are named for a lymphatic organ in chickens, the animals in which they were first discovered. B cells are responsible for **antibody-mediated immunity (AMI),** which, as the name implies, involves the production of specific defensive proteins called antibodies. Although they are large proteins, **antibodies** are soluble in the water of blood plasma, lymph, and tissue fluid and move through the body in those fluids. Antibodies react with specific foreign molecules called **antigens** (*ann-teh-jenz*) to inactivate or neutralize their effects on the body. AMI is most effective against bacteria, bacterial toxins, viruses, and soluble foreign molecules.

An essential common feature of both T cells and B cells is their ability to recognize foreign

TABLE 13.2 Summary of Lymphokines Produced by T-Cell Lymphocytes

Lymphokine	Phagocytes	T Cells	Helper T Cells	Killer T Cells	B Cells	Natural Killer Cells
Interleukin-1 (IL-1)	•	✓			✓	
Interleukin-2 (IL-2)			•	✓	✓	✓
Interleukin-4 (IL-4)			•		✓	
Interleukin-5 (IL-5)			•		✓	
Gamma-interferon	✓		•	•		✓
Lymphotoxin				•		
Perforin				•		

Note: ✓ = stimulates; • = secreted by.

cells and cell products as being distinct from human cells. This is the capacity to distinguish the *self* from the *nonself* we spoke about at the outset of this chapter. This recognition is based on antigens found on the surfaces of cells, among other places.

T cells release lymphokines and are responsible for CMI. B cells produce antibodies as a part of AMI. Both B cells and T cells can distinguish your cells from foreign cells and cell products.

FIGURE 13.7 The processing and display of antigen fragments by an antigen-presenting cell (APC).

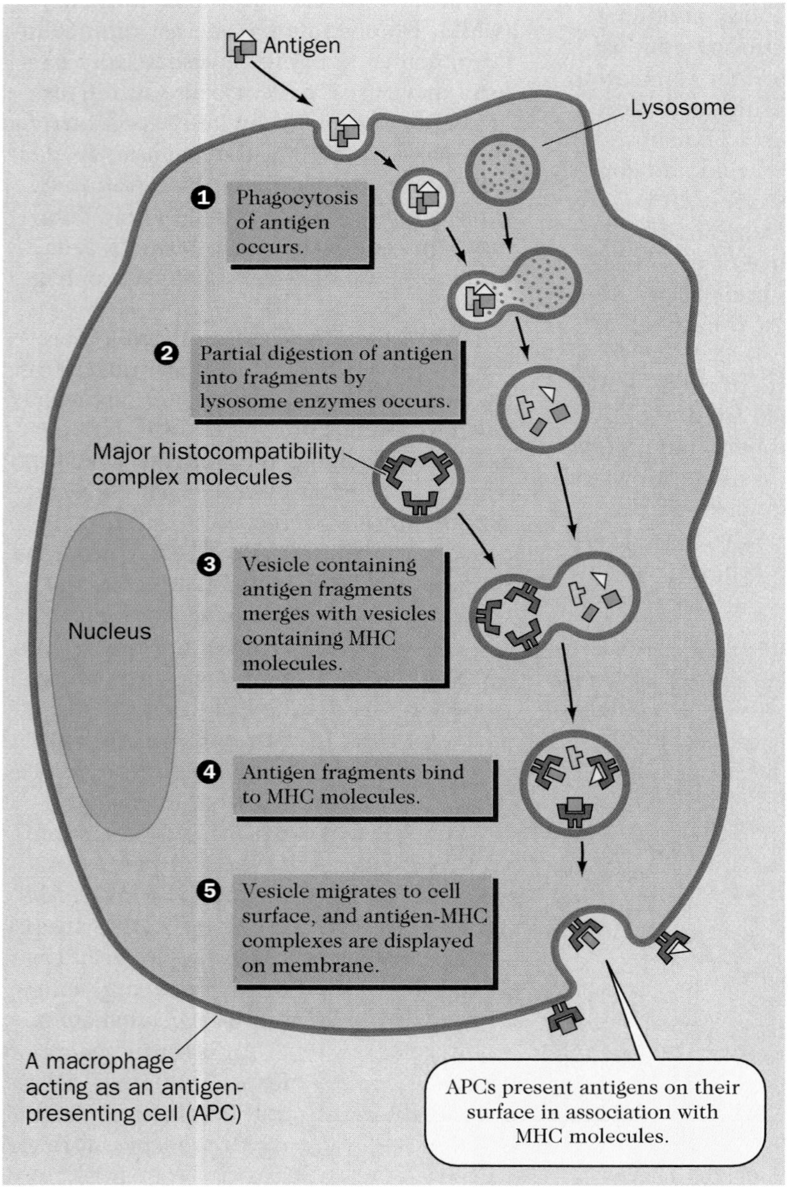
Antigen
Lysosome
1 Phagocytosis of antigen occurs.
2 Partial digestion of antigen into fragments by lysosome enzymes occurs.
Major histocompatibility complex molecules
Nucleus
3 Vesicle containing antigen fragments merges with vesicles containing MHC molecules.
4 Antigen fragments bind to MHC molecules.
5 Vesicle migrates to cell surface, and antigen-MHC complexes are displayed on membrane.
A macrophage acting as an antigen-presenting cell (APC)
APCs present antigens on their surface in association with MHC molecules.

Antigens: Distinguishing Self from Nonself

Antigens are generally large and complex molecules. Most antigens are proteins, but some polysaccharides may serve as antigens. Some plant toxins (from poison ivy, for example) and antibiotics (penicillin) are not by themselves antigenic but, when they combine with human proteins, are capable of functioning as antigens.

Antigens on cell surfaces can serve as recognition markers, providing a kind of name tag identity for cells. The antigens on your cell surfaces are called **self-antigens** and identify your cells as belonging to you. Your self-antigens are a bit different from those of your parents, brothers, and sisters and very different from those of your friends. The more closely two individuals are related, the greater the similarity between their self-antigens.

The membrane proteins on the cells of other humans and on bacteria, fungi, parasites, and even viruses are all considered **nonself antigens** by your body. In addition, such antigens are found on pollen grains and mold spores. If nonself antigens enter your body, your T cells and B cells recognize them as foreign, and an immune response is initiated against them.

On occasion, nonself antigens may be so similar to self-antigens that mistakes are made, and the body initiates an immune response against its own cells. Such disorders are called *autoimmune disorders* and include rheumatoid arthritis, one form of diabetes, and lupus. We'll discuss autoimmune disorders in "Diseases and Disorders" later in this chapter.

How Do B Cells and T Cells React with Antigens?

Before T cells and B cells can generate an immune response to a nonself antigen, they must recognize and react with it. This process is somewhat different in B cells. B cells react to antigens present in lymph, blood plasma, or tissue fluid. The initial T-cell recognition and reactions are a little more involved.

T cells can react only to portions of antigens, and thus T cells require that some *antigen processing* occur before recognition and response take place. Cells, such as macrophages, that conduct antigen processing are called **antigen-presenting cells (APCs).** APCs take up an antigen by phagocytosis, partially digest it, and then display fragments of the antigen on their cell surfaces in conjunction with helper proteins called **major histocompatibility complex (MHC) proteins** (Figure 13.7). T cells can recognize and respond to an antigen displayed by APCs in this manner.

Cell-Mediated Immunity: Proliferation of Antigen-Specific T Cells

Cell-mediated immunity (CMI) is based on the activity of T cells. As shown in Figure 13.8, T cells form when immature lymphocytes leave the bone marrow and are transported by the blood to the thymus gland. Within the thymus gland they mature into inactive T cells with specialized antigen receptors. These receptors permit T cells to recognize and bind with the antigen fragments displayed by APCs.

During their maturation, T cells also receive one of two sets of surface proteins: either CD4 proteins or CD8 proteins. These proteins allow T cells to link up with the MHC proteins on antigen-presenting cells. The presence of either CD4 or CD8 proteins results in the production of two different kinds of T cells with different functions. We'll describe the role of CD4-bearing cells and then that of cells bearing CD8 proteins.

Helper T cells carry CD4 surface proteins. When an APC presents an MHC-bound nonself antigen to a T cell with a CD4 receptor, the T cell develops into what is called a **helper T cell.** The new helper T cell undergoes mitosis, rapidly producing a large number of identical cells called a **clone** (Figure 13.9). The cells of this clone all carry identical surface receptors and are therefore responsive to the same antigen. Most of the cells of this clone immediately secrete lymphokines, especially the substances called interleukins (Table 13.2).

These *interleukins* stimulate the defensive activities of other cells: phagocytes, CD8 cells, natural killer cells, and even B cells. By activating B cells, helper T cells form an important linkage between cell-mediated and antibody-mediated immunity. Other helper T cells are not immediately active; instead, they provide immune memory.

These other helper T cells, which are called **memory T cells,** live in an inactive state for a long time, retaining receptors for the antigen that originally initiated their production. If that antigen appears again, the memory T cells will transform into new helper T cells and proliferate rapidly to attack and destroy the antigen-bearing cells. These memory T cells are an important factor distinguishing specific defenses from nonspecific defenses.

Cytotoxic T cells carry CD8 surface proteins. When a T cell with a CD8 protein on its surface encounters an MHC protein with a nonself antigen, the CD8 cell is activated to produce a clone of cells called **cytotoxic T cells** (also called T8 or killer T cells) (Figure 13.9).

Once they are activated, cytotoxic T cells leave lymphatic tissue and migrate to the site of infection or a tumor, where they secrete substances that are toxic to abnormal cells. One of these substances, a lymphokine called *perforin,* punctures holes in the membranes of infected cells, causing leakage of cytoplasm and death. Another substance, *lymphotoxin,* kills cells by fragmenting their DNA, while the

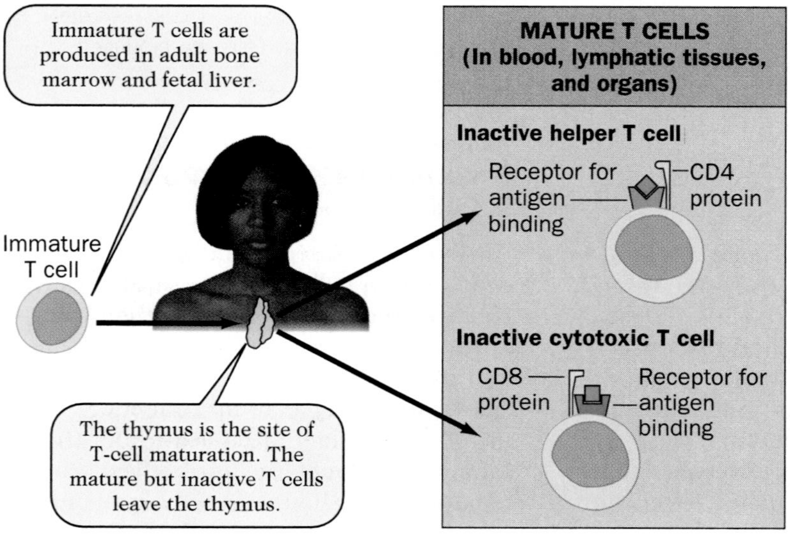

Immature T cells are produced in adult bone marrow and fetal liver.
Immature T cell
The thymus is the site of T-cell maturation. The mature but inactive T cells leave the thymus.
MATURE T CELLS (In blood, lymphatic tissues, and organs)
Inactive helper T cell
Receptor for antigen binding
CD4 protein
Inactive cytotoxic T cell
CD8 protein
Receptor for antigen binding

FIGURE 13.8 Maturation of T-cell lymphocytes in the thymus gland.

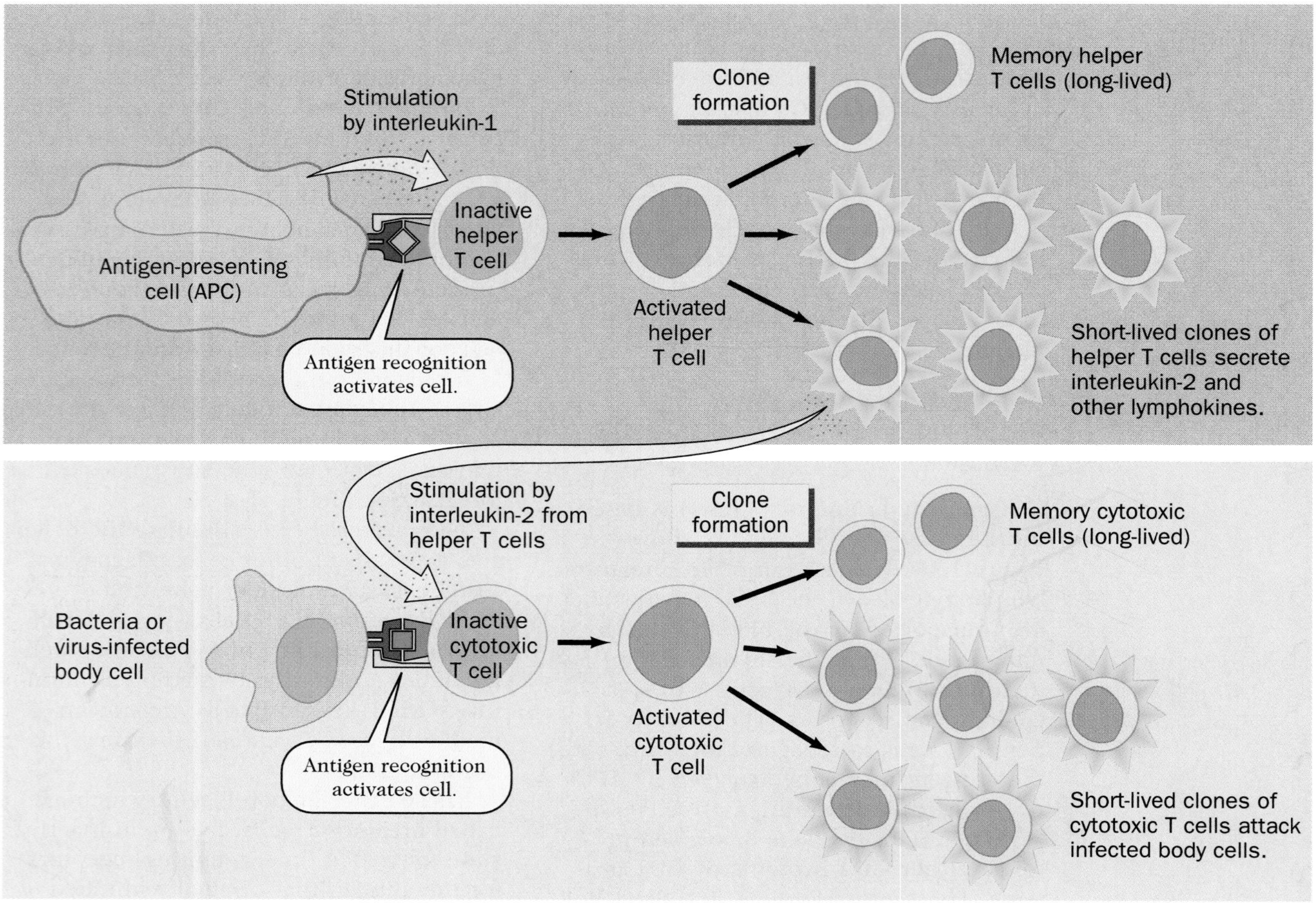

Stimulation
by interleukin-1
Clone
formation
Memory helper
T cells (long-lived)
Antigen-presenting
cell (APC)
Inactive
helper
T cell
Activated
helper
T cell
Antigen recognition
activates cell.
Short-lived clones of
helper T cells secrete
interleukin-2 and
other lymphokines.
Stimulation by
interleukin-2 from
helper T cells
Clone
formation
Memory cytotoxic
T cells (long-lived)
Bacteria or
virus-infected
body cell
Inactive
cytotoxic
T cell
Activated
cytotoxic
T cell
Antigen recognition
activates cell.
Short-lived clones of
cytotoxic T cells attack
infected body cells.

FIGURE 13.9 Cell-mediated immunity: formation of helper T cell and cytotoxic T-cell clones.

release of *gamma-interferon* activates nearby phagocytic and NK cells to assist in cellular destruction (Table 13.2). The cell-killing action of cytotoxic T cells is illustrated in Figure 13.10.

Memory cells can also be produced from this clone of cytotoxic T cells. If the antigen is presented again in the future, these memory cells are activated and proliferate into cytotoxic T cells for an immediate attack.

Suppressor T cells. Some research suggests that there is a third group of T cells called **suppressor T cells.** These cells may stop the immune response when an antigen is no longer present. However, their functions and identity are not well understood. They may actually be a subset of helper T cells rather than a completely separate type of cell. (The functions of the various T-cell lymphocytes are summarized in Figure 13.12.)

> T cells react with MHC proteins on APCs. CD4-bearing T cells become helper T cells. CD8-bearing T cells become cytotoxic T cells. Helper T cells and cytotoxic T cells secrete lymphokines. Memory cells are produced.

Antibody-Mediated Immunity: Proliferation of Antibody-Secreting B Cells

Antibody-mediated immunity (AMI) is a product of B cells. In humans, B cells mature in bone marrow and are then transported by the blood to lymphatic organs and tissues—lymph nodes, spleen, or tonsils—where they take up residence. Once in residence, they rarely leave, but their immune products, the **antibodies,** are distributed throughout the body by lymph, blood plasma, and tissue fluid. B cells in lymphatic tissues are inactive until

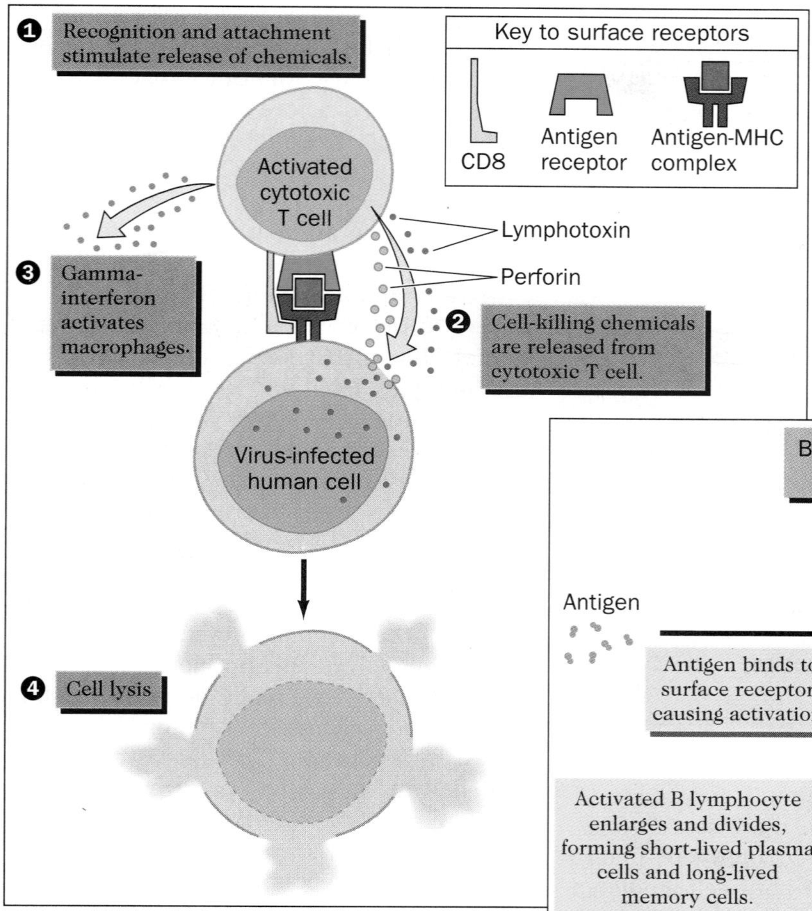

❶ Recognition and attachment stimulate release of chemicals.
Key to surface receptors
CD8
Antigen receptor
Antigen-MHC complex
Activated cytotoxic T cell
Lymphotoxin
Perforin
❸ Gamma-interferon activates macrophages.
❷ Cell-killing chemicals are released from cytotoxic T cell.
Virus-infected human cell
❹ Cell lysis

FIGURE 13.10 Cell-killing action of cytotoxic T cells.

they encounter a foreign antigen. On the plasma membrane, B cells carry specific antigen receptors which can bind to a corresponding specific antigen (Figure 13.11).

Upon binding to a specific antigen, a B cell becomes activated to undergo mitosis, producing a clone of B cells that are all sensitized to the same specific antigen. A few cells of this clone form **memory cells,** which will remain inactive until the same specific antigen appears at a later time. However, most of the B cells in this clone are specialized for producing and secreting a specific type of antibody. (In fact, the word "antigen" is short for *anti*body-*gen*erating molecule.) These specialized B cells are called **plasma cells.** They are the antibody-producing cells of the body's immune response.

When specific antibodies encounter the correspondingly specific antigen molecules, they bind together, forming an **antigen-antibody complex.** The binding of an

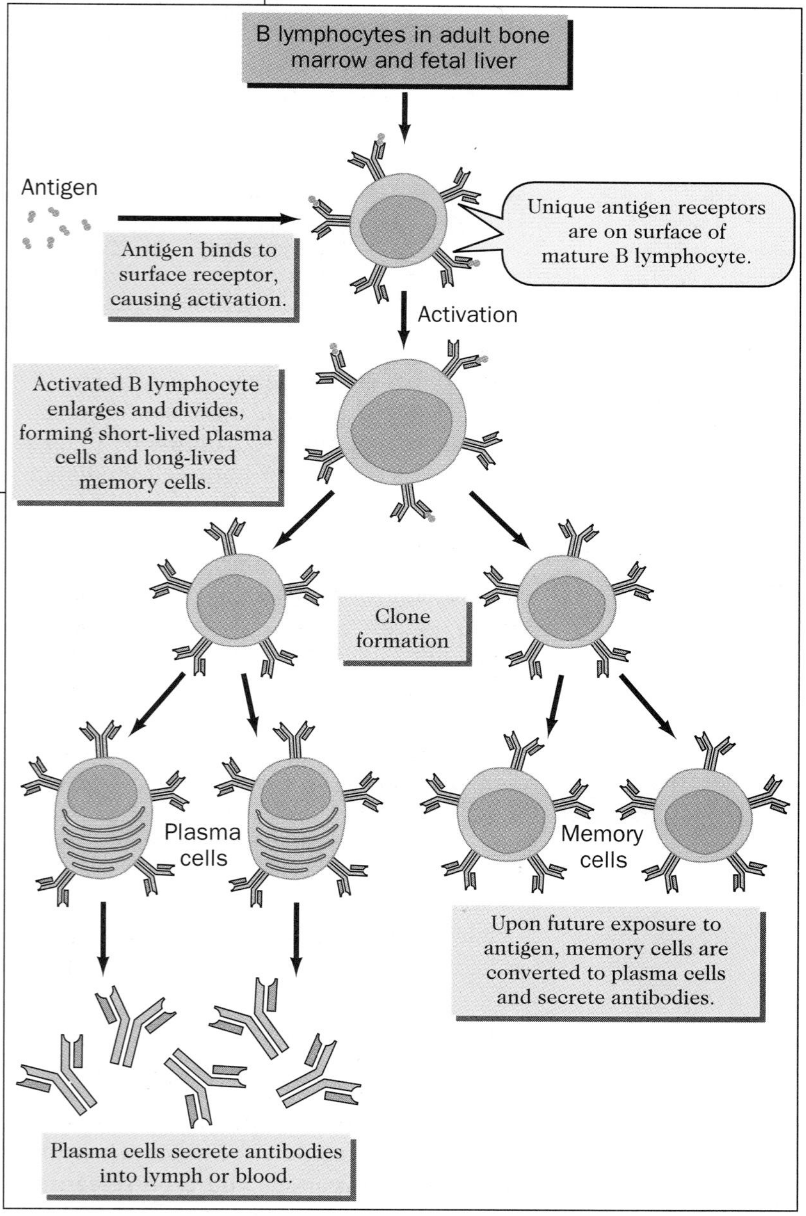

B lymphocytes in adult bone marrow and fetal liver
Antigen
Antigen binds to surface receptor, causing activation.
Unique antigen receptors are on surface of mature B lymphocyte.
Activation
Activated B lymphocyte enlarges and divides, forming short-lived plasma cells and long-lived memory cells.
Clone formation
Plasma cells
Memory cells
Upon future exposure to antigen, memory cells are converted to plasma cells and secrete antibodies.
Plasma cells secrete antibodies into lymph or blood.

FIGURE 13.11 Antibody-mediated immunity: formation of B-cell clones.

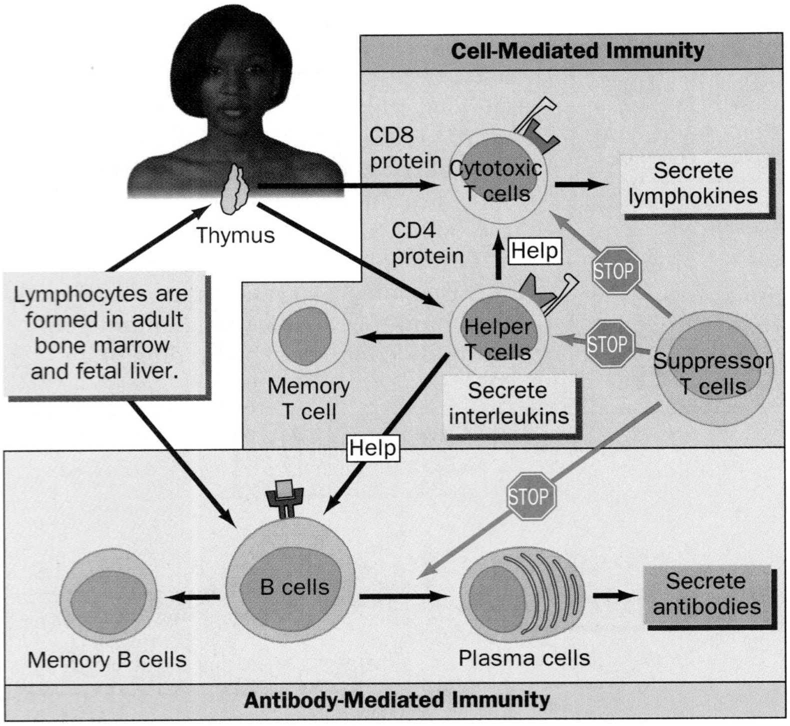
Cell-Mediated Immunity
CD8 protein
Cytotoxic T cells
Secrete lymphokines
Thymus
CD4 protein
Help
STOP
Lymphocytes are formed in adult bone marrow and fetal liver.
Helper T cells
STOP
Suppressor T cells
Memory T cell
Secrete interleukins
Help
STOP
B cells
Secrete antibodies
Memory B cells
Plasma cells
Antibody-Mediated Immunity

FIGURE 13.12 Summary of cell-mediated immunity and antibody-mediated immunity.

antibody to an antigen will inactivate the antigen or make it more susceptible to phagocytosis by neutrophils or macrophages. A summary of both cell-mediated immunity and antibody-mediated immunity appears in Figure 13.12.

Next, we want to take a closer look at how the body can generate so many different specific antigen receptors for the B cells and T cells. This diversity of receptors means that these lymphocytes are able to react to the many different antigens that exist in the world and may invade our bodies.

> B cells are activated and produce a clone of plasma cells that are specialized to secrete antibodies. Some of the clone cells become memory cells. The specific binding of antigen and antibody may inactivate the antigen.

Generating Receptor Diversity and Clonal Selection

This section might be subtitled "Creating an arsenal of weapons and then selecting just the right weapon from the arsenal and making more of it." We will focus on two important questions. First, how is the arsenal of antigen receptors generated for all possible foreign or abnormal antigens? That is, how are different antigen receptors generated before the antigens actually show up? Second, how is a specific lymphocyte selected to attack a specific invader?

Generating antigen receptor diversity. The diversity of receptors on different inactive lymphocytes equips the lymphocytes to react with all possible antigens before the antigens even enter the body. As many as 1 billion (1,000,000,000) different antigen receptors may be required. Since these antigen receptors are proteins, they must be synthesized in the lymphocytes from instructions provided by the genes. However, there are only about 100,000 human genes, not enough for all the antigen receptors needed plus all the other cellular proteins. How has evolution solved this problem?

Fortunately, we really don't need a billion genes to generate a billion different antigen receptors. This task is accomplished by only a few hundred genes. Unlike most of our other genes, immune genes are short sequences that can be rearranged and recombined with each other to generate enormous diversity. This rearrangement allows these gene sequences to code for all the possible antigen receptor proteins. As a result, specific lymphocytes are ready and waiting for a particular antigen when and if it shows up.

If a gene sequence is generated for a receptor to self-antigens, the lymphocyte bearing this gene is quickly destroyed. In this manner, our bodies rid themselves of the lymphocytes that could attack our own cells. In so doing we produce a *tolerance* for self-antigens. By retaining lymphocytes with genes for receptors to nonself antigens, we produce an *intolerance* for foreign and abnormal antigens. This intolerance constitutes our capacity to defend ourselves against these threats.

Clonal selection. We begin life with a very large number of diverse T cells and B cells. These lymphocytes represent an arsenal of specific receptors, ready and waiting for all the various antigens that could show up. If a specific antigen invades the body, the lymphocyte with the correct specific receptor can bind to it (and only to it) and begin the immune response that was described earlier for the B cells and T cells. The binding process does two things: (1) It selects a lymphocyte and (2) serves as a signal for that lymphocyte to proliferate, producing a clone. This process is known as **clonal selection.**

You might wonder, then: If we have a variety of lymphocytes and antigen receptors already available, why do we ever get sick? The answer is that even though an invading antigen is identified by a lymphocyte, it takes some time to produce enough lymphocytes to destroy that antigen. During this period of proliferation, the antigen may make us ill.

> Gene sequences are reshuffled to code for the many different receptor proteins on separate T cells and B cells. An antigen entering the body binds to a specific preexisting receptor on a lymphocyte. This binding stimulates the cell's proliferation.

Structure and Types of Antibodies

As you have seen in our discussion of B cells, antibodies are at the center of our specific defenses. A specific antibody will react with only one type of antigen. Since antibodies are water-soluble, they are rapidly distributed throughout body fluids. Antibodies are classified as part of the *globulin* group of blood plasma proteins, and since they are associated with immunity, they are often referred to as **immunoglobulins (Ig).**

There are five classes of immunoglobulins: IgG, IgA, IgM, IgD, and IgE. Although they are all built on the same plan, each class has a different shape, size, abundance, location, and immune responsibility (Table 13.3). Several types are common in blood and lymph. Other types are more restricted. For example, immunoglobulin A (IgA) is found in body secretions such as tears, saliva, and breast milk. IgE is attached to the surfaces of white blood cells, including mast cells. IgM forms the ABO blood group antibodies. IgG is

TABLE 13.3 Kinds of Immunoglobulins

Name and Shape of Immunoglobulin	Percentage of Total Immunoglobulin	Location	Function
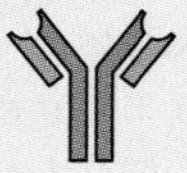 IgG	75	Blood, lymph, tissue fluid, intestines	Protect against bacteria and viruses; enhance phagocytosis; neutralize toxins; abundant in second exposure to antigen; small enough to cross placenta from mother to fetus and give short-term protection after birth.
 IgA	15	Tears, mucus, milk, saliva, blood, lymph, digestive juices	Provide protection on mucous membranes; decrease in abundance during stress, reducing resistance to disease.
 IgM	5–10	Blood, lymph	Abundant in first exposure to antigen; ABO antibodies are of this type; can lyse microbes.
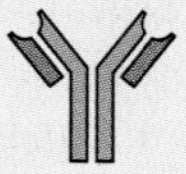 IgD	<1	Blood, lymph, surface of B cells?	Antigen receptors involved in activation of B cells.
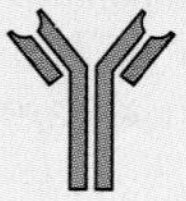 IgE	0.1	B cells, mast cells, basophils	Involved in allergic responses such as asthma and hay fever; trigger release of histamine to cause inflammation.

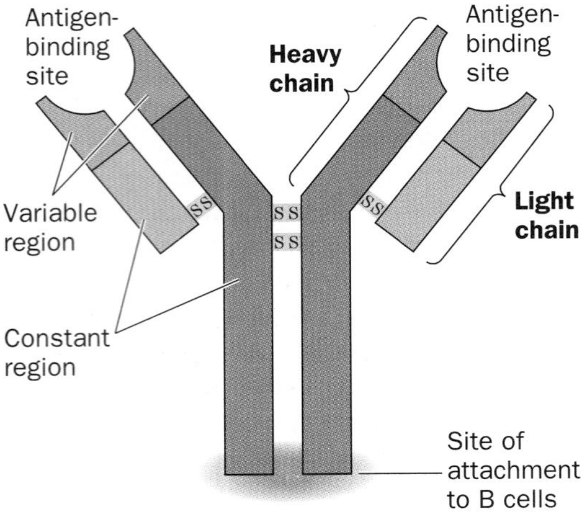
Antigen-
binding
site
Heavy
chain
Antigen-
binding
site
Light
chain
Variable
region
Constant
region
Site of
attachment
to B cells

FIGURE 13.13 Structure of an antibody (IgG).

small enough to cross the placenta from an immune mother to her fetus.

We can use IgG as a model for the basic structure of an antibody. IgG has four polypeptide chains arranged in a Y shape (Figure 13.13). Two of the chains are large, and so they are designated as *heavy chains.* The two smaller chains are called *light chains.* The four chains are joined securely together by covalent bonds between sulfur atoms in amino acids.

A part of each heavy chain and each light chain has the same amino acid sequence in all the antibodies we make no matter which antigen provoked their synthesis. This is called the *constant region.* A smaller portion near the tip of each chain has a different and specific amino acid sequence, depending on the specific antigen that provoked its synthesis. This is called the *variable region.* The shape of the variable region corresponds to the shape of an antigen and is the site of binding between antigen and antibody. This *antigen-binding site* provides the specificity that allows a particular antibody to bind to only one type of antigen.

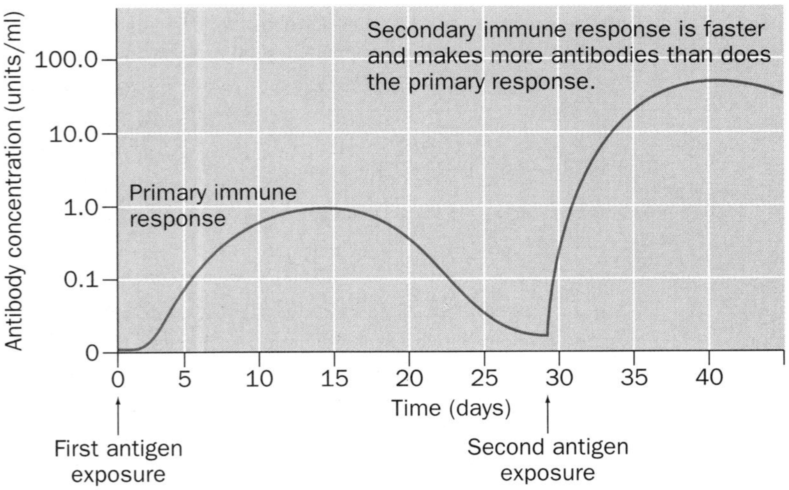
Secondary immune response is faster and makes more antibodies than does the primary response.
Primary immune response
Antibody concentration (units/ml)
100.0
10.0
1.0
0.1
0
0
5
10
15
20
25
30
35
40
Time (days)
First antigen exposure
Second antigen exposure

FIGURE 13.14 Primary and secondary immune responses.

The antigen-binding site is located at the tip of the variable region of each antibody molecule. The shape of this binding site determines the specificity of antibody-antigen reactions.

Immune Memory, Vaccines, and Immunization

As you recall, one of the features which distinguish specific defenses from nonspecific defenses is the capacity for memory. Specialized T cells and B cells retain the memory of an antigen, and this is often a lifelong memory. Examine Figure 13.14. When you are first exposed to an antigen, you slowly generate a **primary immune response** and produce clones of specific cytotoxic T cells, helper T cells, and plasma cells to combat the intruder. During this primary response you also produce memory cells. How many clones of memory cells are present in your body today? There should be a clone for each infection you have ever experienced.

After your initial infection and recovery, your memory cells lie in wait for any return of the same antigen. If it enters the body again, the memory cells bind to it and rapidly proliferate, forming large clones of active lymphocytes that secrete lymphokines (T cells) and antibodies (plasma cells). This proliferation is the **secondary immune response,** and it is more rapid and more massive than the primary response. As a result, the invading pathogen is usually overwhelmed before an infection can cause significant tissue damage and before you feel really ill. At worst you may experience only a vague general discomfort that soon disappears.

Why do we seem to get some diseases, such as the common cold, so many times? There are about 100 different cold viruses. They differ in their antigens, and so you have to experience a primary response for each one separately. Also, since cold viruses have a high rate of mutation, their antigens are rapidly altered, and each new form requires a different antibody. It also appears that our immune memory for some cold viruses is not as long-lasting as it is for other viruses and microbes.

Vaccines and immunization. The secondary response explains the tremendous advantage provided by vaccines and the process of immunization. A **vaccine** is a laboratory-prepared solution of dead or noninfective pathogens or their antigens. For example, the oral form of polio vaccine (Sabin vaccine) is made from noninfective viruses, while the injectible form (Salk vaccine) is produced from chemically inactivated viruses. Vaccines initiate the primary immune response and cause the production of memory cells while sparing you from the painful and disabling symptoms of the disease. Commercial vaccines are available in the United States for about two dozen common diseases.

> The memory cells produced in a primary response rapidly generate a clone of lymphocytes upon a subsequent encounter with the same antigen. The secondary response acts rapidly to prevent disease. A vaccine establishes memory cells without producing disease symptoms.

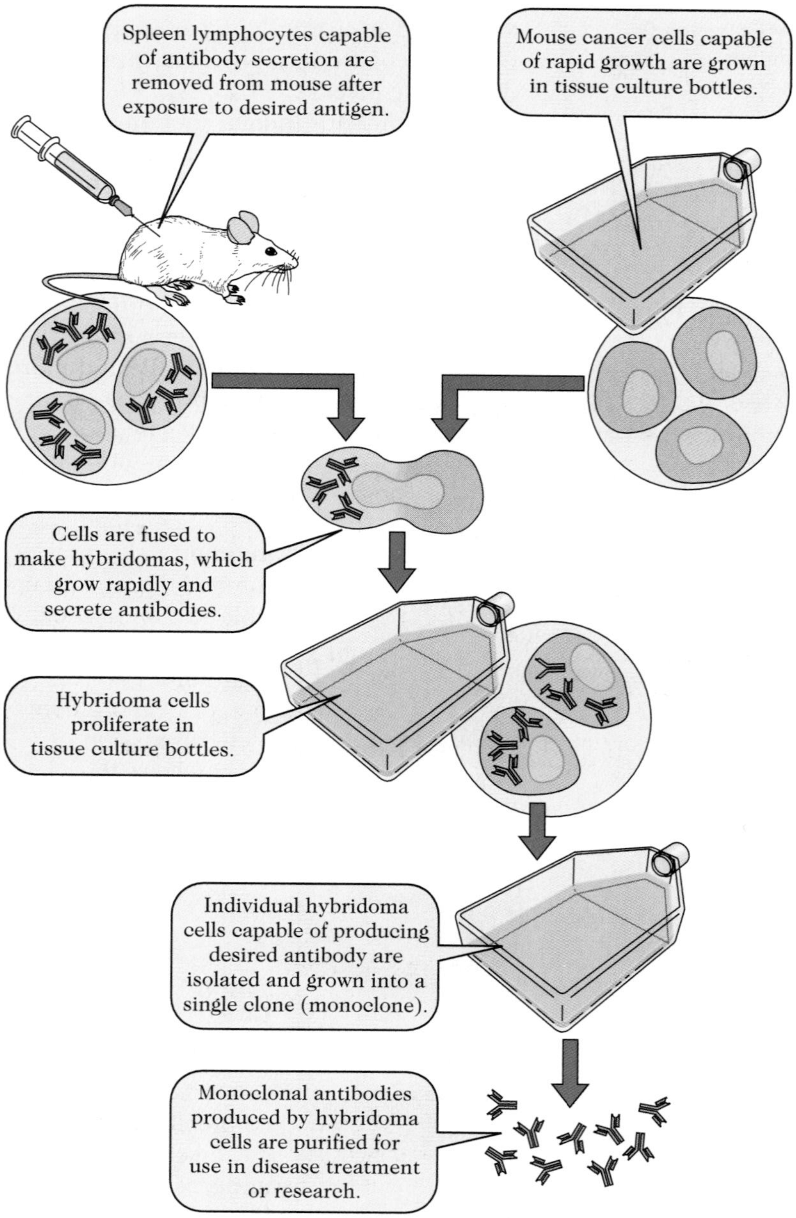
Spleen lymphocytes capable of antibody secretion are removed from mouse after exposure to desired antigen.
Mouse cancer cells capable of rapid growth are grown in tissue culture bottles.
Cells are fused to make hybridomas, which grow rapidly and secrete antibodies.
Hybridoma cells proliferate in tissue culture bottles.
Individual hybridoma cells capable of producing desired antibody are isolated and grown into a single clone (monoclone).
Monoclonal antibodies produced by hybridoma cells are purified for use in disease treatment or research.

FIGURE 13.15 Production of monoclonal antibodies.

Monoclonal Antibodies

In recent years, an exciting technological advance has involved the large-scale laboratory production of antibodies against specific antigens. Antibodies produced in this fashion are called **monoclonal antibodies** (Figure 13.15).

In the laboratory, this procedure usually involves exposing a mouse to a particular antigen and allowing immunity to develop. Eventually, sensitized B cells are removed from the animal's spleen and fused with cancerous mouse lymphocytes called *myeloma* cells. The resulting fused cells are called *hybridomas.* They retain the capacity for antibody production (from the B cell) but also show the high rate of cell division and immortal life span that are characteristic of cancer cells.

Single hybridoma cells are allowed to divide repeatedly to form a population of identical cells called a *monoclone.* All the cells of the clone are capable of producing quantities of a single type of antibody. Such monoclonal antibodies have many scientific, diagnostic, and therapeutic uses.

Monoclonal antibodies are widely used in diagnostic testing. A monoclonal antibody test is available to detect pregnancy within 10 days after conception. Rapid diagnosis of diseases such as hepatitis, influenza, chlamydia, and HIV (AIDS virus) has been made possible by the use of monoclonal antibodies. Monoclonal antibodies chemically bonded to radioactive atoms can be used for early detection of malignancies, especially those which have metastasized in small clusters of cells from the primary tumor. Therapeutically, monoclonal antibodies to which anticancer drugs or radioactive atoms have been attached may permit selective killing by attachment to transformed cells while avoiding damage to normal cells.

> Monoclonal antibodies are produced in the laboratory when single B cells are fused with fast-growing cancer cells to produce hybridoma cells. Hybridoma cells can be used to provide large quantities of a single antibody.

Diseases and Disorders

Because the defense of the biological self is essential to life, any disease or disorder of the lymphatic system and its immune response directly endangers homeostasis and our lives. These disorders may cause (1) reduction or loss of a specific response to infectious agents (a deficient immune response), (2) inappropriate recognition of antigen, (3) exaggerated responses to antigen, or (4) proliferation of cancer cells. In this section, we'll examine selected examples of these four kinds of immune disorders.

AIDS: Acquired Immune Deficiency Syndrome

The world is currently experiencing an epidemic of a communicable immune deficiency disease: AIDS. AIDS is caused by infection with **human immunodeficiency virus (HIV).** HIV initially infects helper T cells but can spread to other body cells, including brain cells.

As the HIV infection progresses, there is a decline in the number of helper T cells. This decline makes the infected person less able to combat HIV and more susceptible to opportunistic infections which may ultimately be fatal (Chapter 12). Loss of cell-mediated immunity also makes the body more susceptible to tumor cells, such as Kaposi's sarcoma and cervical cancer. A more complete discussion of AIDS follows this chapter in Spotlight on Health: Five (page 354).

TABLE 13.4 Selected Autoimmune Disorders

Disorder	Location of Self-Antigens	Description
Addison's disease	Adrenal cortex	Undersecretion of adrenal cortex hormones; weakness, nausea, weight loss, low blood sodium, low blood volume and pressure, darkened skin pigmentation
Diabetes mellitus (type I, or juvenile)	Beta cells of pancreas	Undersecretion of insulin; high blood sugar
Graves' disease	Thyroid cells	Oversecretion of thyroid hormone; high metabolic rate
Hemolytic anemia	Red blood cells (RBCs)	Destruction of RBCs; low RBC count; anemia
Multiple sclerosis	Myelin sheath around nerve cells	Loss of precise muscle control
Myasthenia gravis	Neuromuscular junction	Muscular weakness and fatigue
Pernicious anemia	Stomach cells; intrinsic factor	Low production of intrinsic factor required for absorption of vitamin B_{12}, a substance needed for RBC production; anemia
Rheumatic fever	Heart cells; valves	Self-antigens similar to those of streptococcal bacteria; disorder follows "strep" infections; heart inflammation, valve damage, heart murmur
Rheumatoid arthritis	Immunoglobulin G	Mutant IgM attacks IgG, inflaming joints, destroying cartilage, and fusing bones

Autoimmune Diseases: Antigenic Confusion

Autoimmunities represent a failure of T cells or B cells to distinguish self from nonself antigens, with a resulting attack on normal cells or cell products. Consequently, the

FIGURE 13.16 The allergic response.

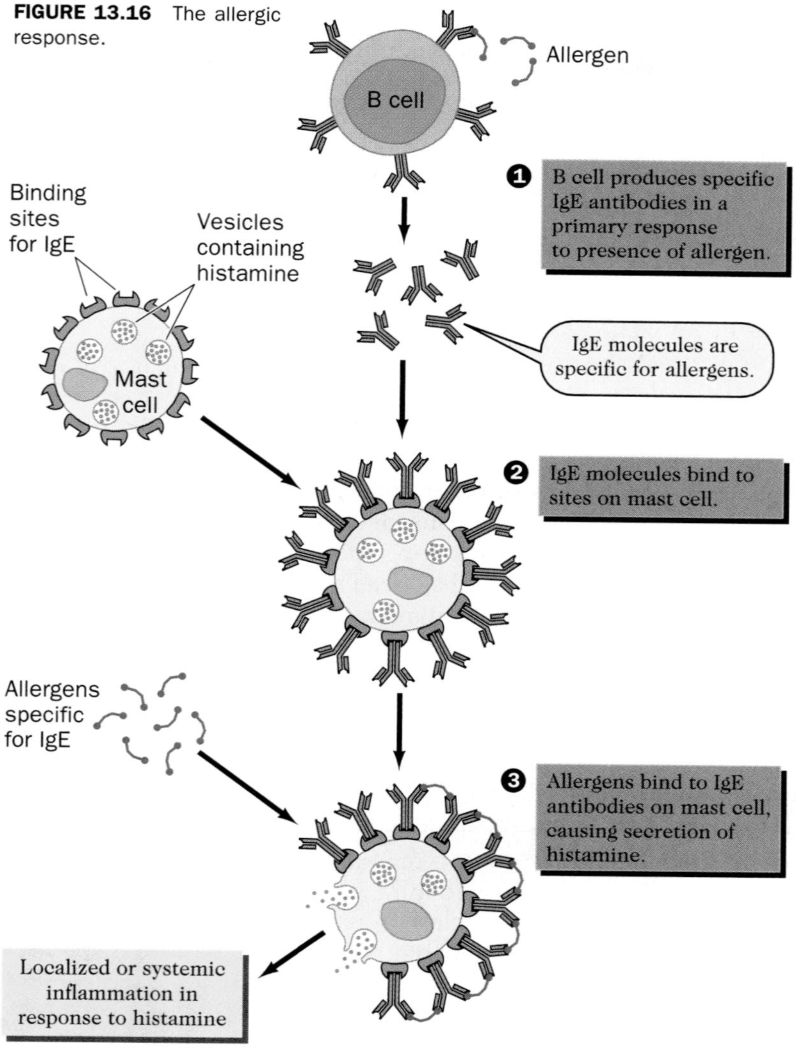
Allergen
B cell
1 B cell produces specific IgE antibodies in a primary response to presence of allergen.
Binding sites for IgE
Vesicles containing histamine
Mast cell
IgE molecules are specific for allergens.
2 IgE molecules bind to sites on mast cell.
Allergens specific for IgE
3 Allergens bind to IgE antibodies on mast cell, causing secretion of histamine.
Localized or systemic inflammation in response to histamine

body's specific defenses attack self-antigens and the cells that bear them. Examples of autoimmune diseases include some forms of diabetes, scleroderma, multiple sclerosis, rheumatic fever, and rheumatoid arthritis (Table 13.4).

One of the more common autoimmunities is *systemic lupus erythematosus* (*loo-pus eh-**rith**-theh-muh-**toe**-sus*), or *SLE.* SLE is a chronic inflammation affecting joints, kidneys, the nervous system, and the skin. Symptoms include arthritislike joint pain, fatigue, enlarged lymph nodes and spleen, weight and hair loss, anemia, a characteristic "butterfly rash" on the face, and ulceration of the skin. The cause of lupus is not well understood, though it appears to be an autoimmune response to connective tissue. The disease occurs more frequently in young women, and renal failure is often a cause of death because antibody-antigen complexes block glomeruli in the kidney nephrons, causing reduced urine formation.

Allergy: An Overactive Immune Response

Allergies represent exaggerated responses of the immune system. The affected individual is overly sensitive—**hypersensitive**—to a particular antigen called an **allergen.** Allergens include dust, mold spores, pollen grains, proteins or sugars in foods, antibiotics, dyes, insect and reptile venoms, cosmetics, and chemicals in plant oils and extracts.

When allergens enter the body, they react with IgE on the surface of mast cells and basophils, causing those cells to release histamine from vesicles in their cytoplasm where it is stored (Figure 13.16). Histamine then causes an inflammation-type response by triggering the dilation of blood vessels, increasing leakage from blood vessels, and causing contraction of smooth muscle.

The response to an allergen may be *localized,* as in the case of *hay fever,* a pollen allergy. The symptoms of watery eyes, nasal congestion, and sneezing are familiar to all who suffer from this common allergic response to pollen entering the nasal cavity. *Hives,* also a localized response, are raised, red, and swollen patches of skin that result from surface contact with an allergen.

When the response to an allergen is widespread and affects several organ systems and body regions, it is called a *systemic* response. Systemic responses occur when allergens such as those from bee or wasp stings enter the blood. Symptoms often include general edema, constricted air passageways, low blood volume, and low blood pressure. In the extreme, suffocation and death may follow. Treatment of severe cases may require the administration of *epinephrine,* a hormone which dilates respiratory passages. Treatment of milder forms of allergy may be managed with *antihistamines,* which reduce the severity of the inflammation response. *Cortisone,* an adrenal cortex hormone, may be given to control inflammation, though it is slower acting than antihistamines.

The allergic responses described so far are called **immediate hypersensitivities** because they occur quickly, usually within a few minutes or hours of allergen exposure. **Delayed hypersensitivities** develop more slowly, usually 1 day to 2 weeks after allergen exposure. They occur when allergens are engulfed by antigen-presenting cells and taken to lymph nodes for presentation to T cells. After interaction, the selected T cells proliferate and migrate to the site of allergen exposure, where they release lymphokines which stimulate inflammation. Exposure to the oils of poison oak or poison ivy, certain chemicals in cosmetics, the tuberculosis skin test, and transplanted tissues are examples of delayed hypersensitivity responses.

Rejection of Transplanted Tissues and Organs

When an injury or disease causes tissue destruction or organ failure, transplants from donors may allow adequate replacement of function. Transplants of tissue from one part of the body to another in the same individual bear no risk of rejection by cell-mediated immunity because the transplanted tissues carry the same antigens as the rest of the body. Also at low risk for rejection are transplants of tissue or organs between closely related individuals. The closer the relationship, the more similar the self-antigen markers are.

Transplants between unrelated individuals are less successful because of differences in self-antigens, though they are still done routinely. Blood transfusions, however, can be successful if the ABO and Rh factors are closely matched. Corneas, which lack blood and lymphatic vessels and thus lack T cells, also can be transplanted between individuals with success. Pig skin has been used successfully as a temporary covering for severe human burns, and pig heart valves have replaced malfunctioning human heart valves.

Skin and organ transplants between unrelated persons or species may require the use of *immunosuppressive drugs* to block the immune response and provide time for tolerance to develop, allowing the eventual acceptance of foreign tissue as "self." One such drug, *cyclosporine,* inhibits the activation of killer T cells by inhibiting the production of interleukin-2 from helper T cells.

Malignancies of the Lymphatic System

Hodgkin's disease is a malignancy of lymphatic tissue characterized by masses of abnormal lymphocytes. It usually begins in a single lymph node but often spreads to other nodes, which become enlarged. The cause is unknown, but it may involve genes or a viral infection. Hodgkin's disease has been successfully treated with bone marrow transplants, radiation therapy, and chemotherapy.

Multiple myeloma (*my-eh-**low**-muh*) is a malignancy of antibody-producing plasma cells which usually arises late in life. Excessive growth of these cells causes anemia as they crowd out blood-forming cells in bone marrow. Such growth may also cause widespread destruction of bone, weakening the skeleton and causing pain. Both the anemia and the inability of immune cells to respond to infections are life-threatening.

Chronic lymphocytic leukemia (*CLL*) is a slowly progressing malignancy that is marked by an overproduction of lymphocytes. The lymphocytes invade the bone marrow, replacing blood-forming cells and creating anemia and fatigue. Reduced numbers of platelets lead to chronic bleeding, and loss of phagocytes reduces defenses against opportunistic infections.

Life Span Changes

From birth to old age, there are dramatic changes in the lymphatic system. Some of these changes occur early in life and prepare the system to function in our defense. Other changes occur later and erode some of the protection offered in earlier years.

Changes in the Lymphatic System

Of all the components of the lymphatic system, the thymus undergoes the most dramatic changes in the first two decades of life. From birth to about the age of 10 years, the thymus attains its maximum size of about 40 grams (1.4 ounces). During these years we experience common childhood diseases and receive vaccinations, thereby establishing general immunities.

Between the ages of 20 and 60, the thymus slowly shrinks, declining to about 10 percent of its maximum weight and reducing its production of thymosin. Both cell-mediated immunity and antibody-mediated immunity are affected by the decline in thymus size and activity. This does not mean that specific defenses are lost, only that fewer new immune defenses are established.

Changes in Specific Defenses

Over the life span, changes in specific defenses result from changes in the numbers of T cells and B cells.

T-cell activity. With the decline in the thymus, there is a decline in the production of mature T cells and a gradual decline in cell-mediated immunity. The number of antigen receptors, their antigen-binding ability, and the motility of T-cell lymphocytes appear to be affected by age. Although an older person retains memory cells established earlier in life, those cells respond less vigorously to antigens.

The decline in cell-mediated immunity may, at least in part, be responsible for the increased incidence of cancer in the elderly. In old age there are fewer T cells to recognize and destroy transformed cells that grow into tumors.

B-cell activity. Over a life span, the total number of B-cell lymphocytes does not change significantly in healthy people, and neither does the total amount of antibodies. However, there are changes in the types of immunoglobulins present. As we grow older, the levels of IgA and IgG increase while those of IgE and IgM decrease.

There is also a decline in our ability to recognize and respond to foreign antigens. For example, after antigen exposure in older people, IgG does not remain in as high a concentration in the blood as it does in younger persons. This decline is due in part to the decline in the thymus and the production of fewer helper T cells, which are essential to the activity of B cells.

Changes in Nonspecific Defenses

Over the life span, there is a general decline in the body's physical and chemical barriers. The epidermis of the skin becomes a less effective barrier as the rate of cell replacement declines with age. There is a decrease in the number of antigen-presenting mast cells in the skin. Wound healing also declines with age. Sweat and sebum production decreases, making the skin less resistant to microbes. The production of saliva, tears, and even gastric acid may decline slowly as life progresses. All these factors make it easier for microbes to enter the body and challenge your specific defenses.

Chapter Summary

The Lymphatic System

- The lymphatic system transports excess tissue fluid and fat-soluble materials to the blood.
- Lymph nodes filter lymph and produce lymphocytes that conduct specific immune responses.
- The thymus produces thymosin and influences the maturation of certain lymphocytes.
- The spleen filters blood, responds to foreign materials, and stores a reserve blood volume.
- Tonsils provide lymphocytes.

Nonspecific Defenses against Disease

- Nonspecific defenses provide uniform responses to disease agents.
- The skin and mucous membranes offer physical barriers.
- Body secretions kill or inhibit microbes.
- Body processes such as sneezing remove microbes.
- Phagocytes digest cellular debris or microbes.
- Natural killer cells attack infected body cells or tumor cells rather than invading microbes.
- Interferon prevents viral assembly in uninfected cells.

- Complement kills foreign cells, stimulates histamine release, and enhances phagocytosis.
- Inflammation (redness, swelling, heat, and pain) involves vasodilation, phagocyte migration, and tissue repair.

Specific Defenses: The Immune Response

- Specific defenses distinguish between disease threats and allow lifelong immunity to a particular disease.
- T cells provide cell-mediated immunity.
- B cells provide antibody-mediated immunity.
- T cells and B cells can distinguish self-antigens from nonself antigens.
- T cells react to antigen fragments displayed by antigen-presenting cells on MHC proteins.
- T cells receiving CD4 receptors become helper T cells.
- T cells receiving CD8 proteins become cytotoxic T cells.
- Memory cells retain a long-lasting response to an antigen.
- Suppressor T cells may stop an immune response when an antigen is no longer present.
- B cells form antibody-secreting plasma cells and memory cells.
- Clonal selection occurs when an antigen binds to a receptor on a lymphocyte which forms a clone of reactive cells.
- Antibodies are classified as IgA, IgD, IgE, IgG, or IgM.
- Antibodies have constant and variable regions and an antigen-binding site.
- Initial antigen exposure causes a primary immune response, while subsequent antigen exposure causes a secondary immune response.
- Vaccines allow an immune response without disease symptoms.
- Monoclonal antibodies are made by hybridoma cells (lymphocytes fused with myeloma cells).

Diseases and Disorders

- AIDS occurs when HIV infection reduces the helper T-cell population, increasing susceptibility to infections.
- Autoimmune disorders represent immune responses to self-antigens.
- Allergies are exaggerated immune responses.
- The lymphatic system is subject to malignancies.

Life Span Changes

- The thymus, which is large during childhood, shrinks during adulthood, reducing CMI and AMI.
- Memory cells formed early in life are retained into adulthood but become less reactive to antigens.
- Nonspecific defenses decline as the skin ages and external glandular activity decreases.

Selected Key Terms

allergy (p. 349)
antibody (p. 339)
antibody-mediated immunity (p. 339)
antigens (p. 339)
autoimmunity (p. 348)
B-cell lymphocytes (p. 339)
cell-mediated immunity (p. 339)
clonal selection (p. 344)
inflammation (p. 337)
lymphokines (p. 339)
nonspecific defenses (p. 334)
primary immune response (p. 346)
secondary immune response (p. 346)
specific defenses (p. 339)
T-cell lymphocytes (p. 339)

Review Activities

1. Distinguish between nonspecific and specific defenses and list several examples of each.
2. List the functions and components of the lymphatic system.
3. Describe the structure and function of lymph nodes, the thymus, the spleen, and tonsils. In what way are lymph nodes different from the other structures?
4. Summarize the mechanism of phagocytosis.
5. Discuss the importance of natural killer cells, interferon, and complement.
6. List the stages of inflammation. Explain how these stages account for the observed symptoms.
7. List the types of T cells and the role they play in cell-mediated immunity.
8. List the types of B cells and the roles they play in antibody-mediated immunity.
9. Explain how antigen receptor diversity is achieved.
10. Summarize the process of clonal selection of lymphocytes by antigens.
11. Sketch and label an IgG antibody.
12. List the five categories of immunoglobulins along with their locations and functions.
13. Explain the difference between primary and secondary immune responses.
14. Relate the importance of a vaccine in the development of an immune response.
15. Summarize the process by which monoclonal antibodies are produced.

Self-Quiz

Matching Exercise

___ 1. Site of lymph filtration	A. T cells
___ 2. Site of blood filtration and storage	B. Autoimmunity
___ 3. Process characterized by vasodilation, phagocyte migration, and tissue repair	C. Spleen
___ 4. Cells responsible for cell-mediated immunity	D. Vaccine
___ 5. Cells responsible for antibody-mediated immunity	E. Inflammation
___ 6. Prevents viral assembly in uninfected cells	F. Lymph nodes
___ 7. Stimulates inflammation response	G. Immunoglobulins
___ 8. Another term for antibodies	H. B cells
___ 9. Immune response against self-antigens	I. Histamine
___ 10. Solution of altered or inactivated pathogens	J. Interferon

Answers to Self-Quiz

1. F; **2.** C; **3.** E; **4.** A; **5.** H; **6.** J; **7.** I; **8.** G; **9.** B; **10.** D

SPOTLIGHT ON HEALTH: FIVE

AIDS: The Disease, the Epidemic, and the Virus

You know that **AIDS** stands for **acquired immune deficiency syndrome.** This name communicates a great deal about the disease. The word "syndrome" indicates that the disease is a collection of symptoms which may vary among different patients. The words "immune deficiency" indicate that the underlying cause of these symptoms is a disorder in the immune defenses. The word "acquired" indicates that this deficiency is due to an infectious agent acquired from the environment; that agent is believed to be a virus.

The **human immunodeficiency virus (HIV)** infects and eventually kills the T cells (helper T cells) of the body's immune system (Chapter 13). Loss of helper T cells destroys the body's capacity to fight foreign agents (cancer cells and infectious bacteria, viruses, protozoa, and fungi) (Chapter 12). HIV is a rather fragile virus that is easily deactivated outside cells and body fluids such as blood, semen, vaginal fluid, and breast milk. AIDS is considered a sexually transmitted disease because HIV can be transmitted during sexual intercourse. However, it is also passed between intravenous (IV) drug users who share blood-contaminated needles. Before HIV was identified, donated blood for transfusions and blood products used by hemophiliacs often were contaminated by HIV.

AIDS: A Historical Profile

The syndrome we now know as AIDS was first described in 1981, after statistics on new cases and deaths began to be compiled carefully in the United States by the Centers for Disease Control and Prevention (CDC).

Since 1981, the disease has spread rapidly and has shown a high death rate. In the United States during the first decade of the disease (1981–1990), a total of 189,153 cases were diagnosed and there were 115,923 deaths. This amounts to a 60 percent mortality rate. These data represent only diagnosed cases of AIDS. The number of people infected but not yet showing symptoms may be nearly 10 times larger!

The global picture is even more grim. By the end of 1992 as many as 14 million adults and 1 million children may have been infected with HIV, with at least 2 million of those individuals in the United States alone (Figure 1). There is the potential for 40 million to 110 million infections worldwide by the year 2000.

Initially, in the United States, AIDS was diagnosed most frequently in male homosexuals, though the rate of new cases in that population has declined. AIDS is now increasing rapidly among IV drug users and their sexual partners. Heterosexual transfer (male to female or female to male) has remained at a low but steadily increasing level in the United States. In Africa, heterosexual transfer of the disease is the most common form of transmission. Women are usually more at risk of being infected by a male than a male is by a female.

Currently, health officials are very much concerned about the rapid increase in HIV infection among 13- to 19-year-olds. In spite of moral instruction, information on HIV, and increased access to condoms, infections in this age group continue to increase because of IV drug use and unprotected sexual intercourse with multiple partners.

The Transmission of HIV

To anyone who has been even mildly attentive to the information media in recent years, the increase in AIDS cases and the cry of alarm are familiar. However, the good news is that infection can be prevented. HIV is a fragile virus. It requires moist conditions for transmission and is fairly easily inacti-

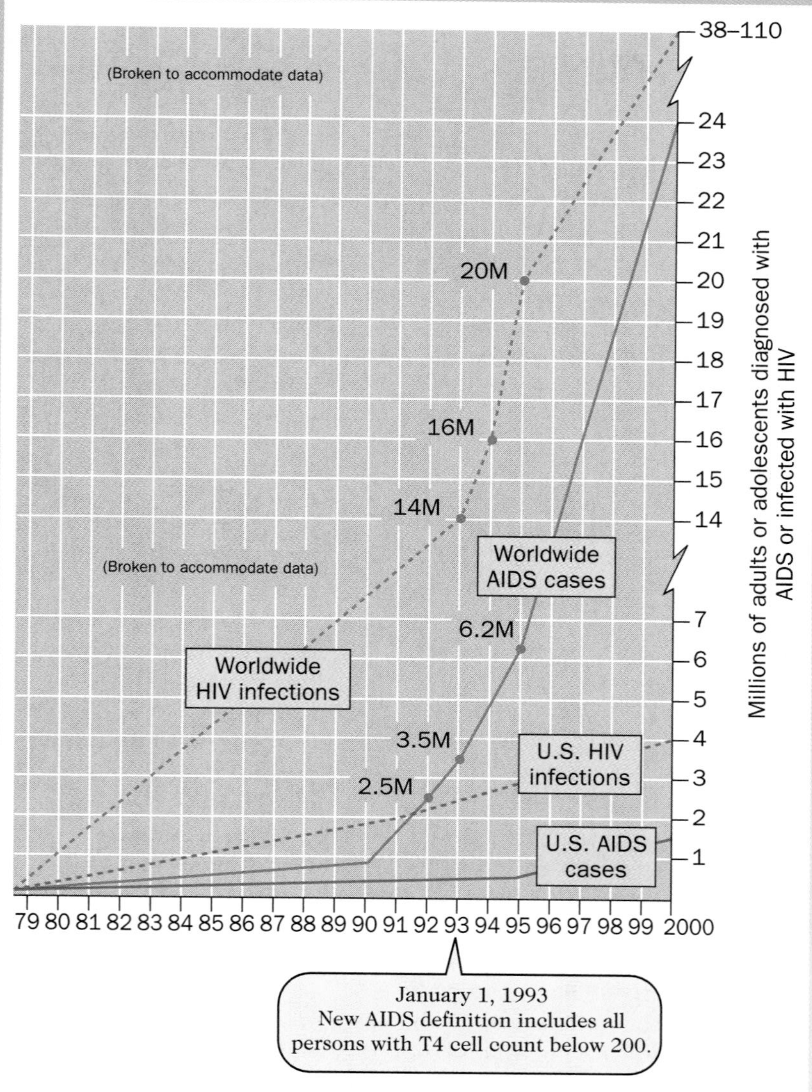
(Broken to accommodate data)
(Broken to accommodate data)
38–110
24
23
22
21
20
19
18
17
16
15
14
7
6
5
4
3
2
1
Millions of adults or adolescents diagnosed with AIDS or infected with HIV
20M
16M
14M
6.2M
3.5M
2.5M
Worldwide AIDS cases
Worldwide HIV infections
U.S. HIV infections
U.S. AIDS cases
79 80 81 82 83 84 85 86 87 88 89 90 91 92 93 94 95 96 97 98 99 2000
January 1, 1993
New AIDS definition includes all persons with T4 cell count below 200.

Sources: Data from the Harvard Global AIDS Policy Commission, 1992, and the World Health Organization, 1993.

FIGURE 1 IV infection and AIDS diagnoses in the United States and the world.

vated. Outside the body, it is killed by exposure to hot water (135°F for 10 minutes) during clothes washing and dish washing. Soap and water and common disinfectants (bleach and Lysol and the chlorine in swimming pools) also can kill the virus.

HIV appears to be transmitted most easily in blood, semen, vaginal fluids, and the breast milk of infected mothers. During sexual contact, the virus gains entry to the body through sores or tiny tears in the mucous membranes which line the vagina, rectum, mouth, and male urethra. Intravenous drug users who share needles and syringes with infected individuals may directly inject virus-contaminated blood into their veins. The virus can be passed from an infected mother to her fetus through the placenta or by exposure to blood and vaginal secretions during birth.

The presence of the virus in other body products, such as tears, saliva, urine, and feces, has been demonstrated, but transmission in this manner is unlikely and no documented cases have been traced to exposure to those materials. There have been very, very few HIV infections among those providing care to AIDS patients, and so casual transmission of the virus through hugging, closed-mouth kissing, handshaking, coughing, or bathing is unlikely.

Protecting Yourself from AIDS

The best protection from HIV infection is abstinence from sexual encounters with anyone who might be HIV-infected. Unfortunately, a person's HIV status is rarely known, and abstinence runs counter to many individuals' emotional needs, peer pressure, and society's mixed messages about sexual liberation. If abstinence is not possible, the barrier protection provided by male and female latex condoms can prevent transmission of the virus. The spermicide nonoxynol-9 is effective against HIV. However, condoms and spermicides must be used consistently and properly, and this requires a level of control that may not be possible, particularly if alcohol use precedes sexual encounters. Strict monogamy with an uninfected person permits both sexual pleasure and freedom from HIV and AIDS.

To prevent infection, you must also avoid contact with the blood of other individuals through needle sharing. Many women have become infected by having unprotected sexual intercourse with an IV drug user who is infected. HIV-infected women should avoid becoming pregnant because the virus can be passed to their unborn children. Fortunately, the blood used in transfusions and blood products has been screened for HIV, and the blood supply is considered nearly 100 percent safe (99.8 percent).

AIDS: The Clinical Picture

The course of HIV infection can be divided into three phases (Figure 2): an initial period after infection, a second period during which the number of T cells steadily declines, and a third phase in which T-cell counts are so low that they cannot protect a

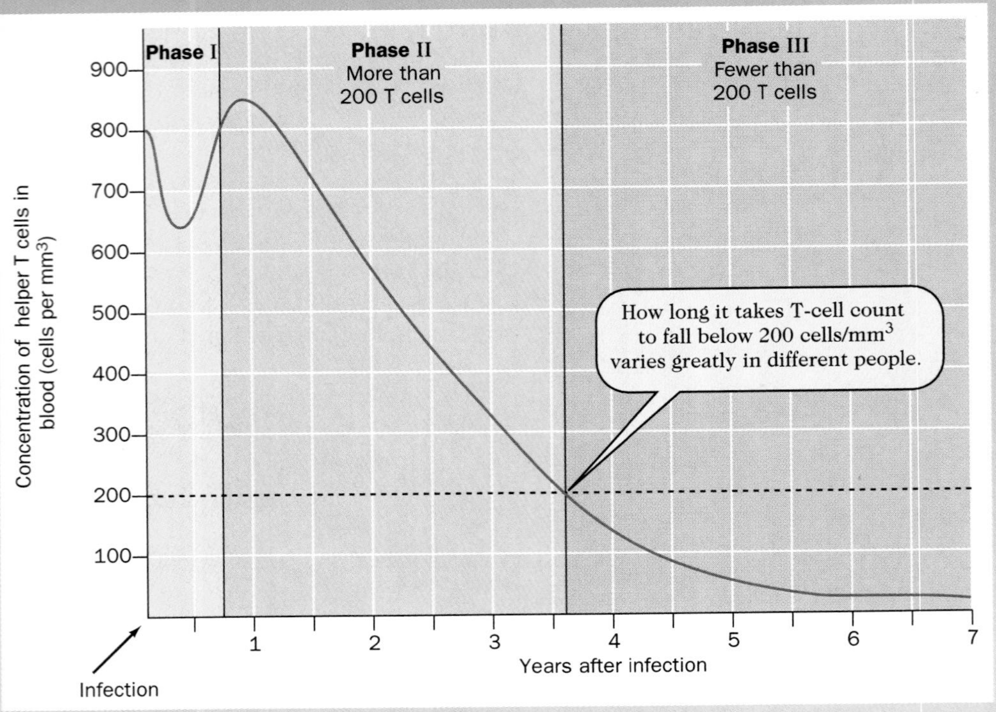
Phase I
Phase II
More than
200 T cells
Phase III
Fewer than
200 T cells
How long it takes T-cell count to fall below 200 cells/mm3 varies greatly in different people.
Concentration of helper T cells in blood (cells per mm3)
900
800
700
600
500
400
300
200
100
1
2
3
4
5
6
7
Years after infection
Infection

FIGURE 2 Changes in the number of T cells after infection by HIV.

of helper T cells may occur in 6 months or take up to 10 years. In addition to recurring infections, the symptoms during this period include persistently swollen lymph nodes in the neck, armpits, and groin.

The Third Phase: A Low T-Cell Count and Life-Threatening Infections

When the number of helper T cells drops below 200 per cubic millimeter, a person is officially diagnosed as having AIDS if one or more of the following are present: pulmonary tuberculosis (TB), recurrent pneumonia, and for women, invasive cervical cancer. Pneumonia caused by *Pneumocystis carinii* is present in about 80 percent of AIDS patients in the United States. However, many other opportunistic infections are encountered in AIDS patients (Table 1).

As infections become more frequent and spread throughout the body, AIDS patients require frequent hospitalization. The symptoms include degeneration of the nervous system, loss of motor control, dementia, and changes in behavior. Complications from multiple infections eventually overwhelm the body, and death follows.

person from opportunistic infections. Strictly speaking, AIDS refers only to the third phase.

The First Phase: HIV Infection without Severe Symptoms

Initially HIV infects white blood cells: helper T cells and macrophages. The early symptoms of HIV infection resemble those of flu and do not last long: enlarged lymph nodes, fatigue, body aches, fever, and chills. The number of helper T cells declines initially but rebounds as antibodies are produced against the virus (Figure 2).

The presence of antibodies in a blood sample is currently the primary means of identifying HIV infection. When antibodies against HIV are present, the person is said to be HIV-positive (HIV+). Unfortunately, these antibodies are not effective defenses, because the virus can hide inside cells.

The Second Phase: HIV Kills the T Cells and Symptoms Appear

In a normal healthy adult, there are about 800 or more helper T cells per cubic millimeter of blood. During the second phase of HIV infection, there is a steady decline in the number of these T cells and an increasing vulnerability to opportunistic infections. Depending on one's general health, the destruction

The Virus, Infection, and Treatment: The Scientific Picture

HIV belongs to a group of viruses known as *retroviruses* that use RNA as their hereditary material. The RNA is wrapped in a protein coat and surrounded by an envelope derived from the plasma membrane of a host cell (Figure 3). Accompanying the RNA is an enzyme called *reverse transcriptase* that permits the viral nucleic acid to integrate into the host cell's DNA, where it may hide for years.

Initially, HIV infects helper T cells and macrophages, but later infection may spread to other cell

TABLE 1 Opportunistic Infections Associated with HIV Infection (Phases 2 and 3)

Disease	Microbial Agent	Symptoms
P. carinii pneumonia	*Pneumocystis carinii* (protozoan)	Pneumonia, difficult breathing, suffocation
Toxoplasmosis	*Toxoplasma gondii* (protozoan)	Fatigue, brain lesions, seizures, cerebral swelling
Cryptosporidiosis	*Cryptosporidium coccidi* (protozoan)	Extreme diarrhea, dehydration, shock
Cryptococcosis	*Cryptococcus neoformans* (fungus)	Pneumonia, piercing headaches, paralysis (meningitis)
Candidiasis	*Candida albicans* (fungus)	White oral patches of fungus, erosion of esophagus
Histoplasmosis	*Histoplasma capsulatum* (fungus)	Pneumonia, lesions of visceral organs, paralysis
Cytomegalovirus disease	Cytomegalovirus (virus)	Pneumonia, liver and kidney disease, impaired vision
Herpes simplex	Herpes simplex virus (virus)	Body sores and blisters
Tuberculosis	*Mycobacterium tuberculosis* (bacterium)	Lesions of lung, difficult breathing, lesions of visceral organs

types, including brain cells. The plasma membranes of all these cells contain protein receptors that permit HIV attachment. After attachment, the virus enters the cell and takes over the cellular functions, producing new RNA viruses and killing the host cell. The released viruses infect more and more cells, and those cells die. This is the first clinical phase.

In some infected cells, instead of reproducing RNA viruses and killing the host cells immediately, the viral RNA is copied as DNA which is integrated into the host cell's DNA (see Spotlight on Health: One, Figure 4, page 103). In this integrated form, the coded information of the viral nucleic acid is reproduced every time the cell reproduces, spreading potential sources of the virus throughout the body.

In its integrated form, the virus may remain inactive for up to 15 years (10 years is the average). However, during this second phase it appears that infected cells are steadily destroyed—how or why is not clear. With the loss of these cells, the body's immune defenses are weakened, and HIV-infected people become more susceptible to opportunistic infections as well as certain kinds of cancer (Kaposi's sarcoma). It is these "complications due to HIV" that end up killing most HIV-infected people.

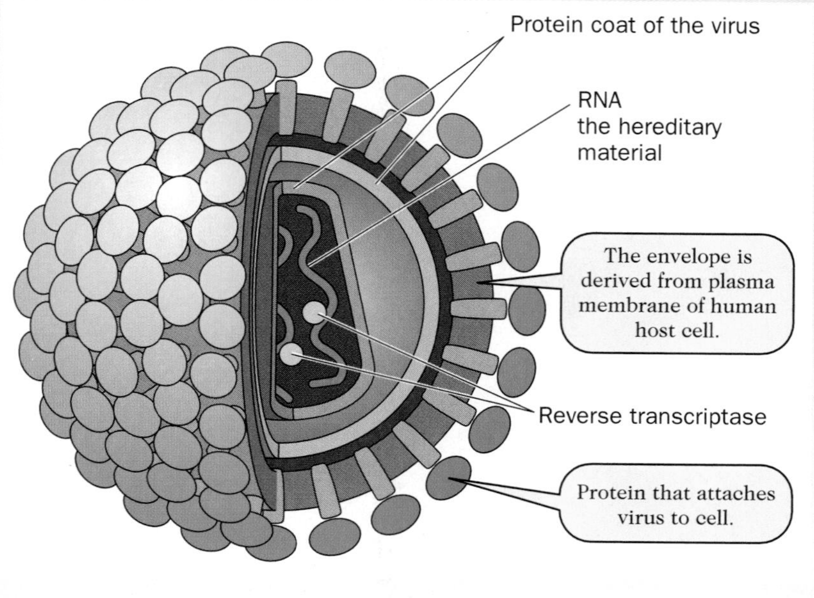
Protein coat of the virus
RNA
the hereditary
material
The envelope is derived from plasma membrane of human host cell.
Reverse transcriptase
Protein that attaches virus to cell.

FIGURE 3 The human immunodeficiency virus (HIV).

Inhibiting the Replication of HIV

Conventional treatment for HIV/AIDS is difficult because of three characteristics of the disease: (1) the dormancy exhibited by HIV, (2) the variety of cells it can infect, and (3) the multiple opportunistic infections that are possible. At present there are very few drugs that can act directly on the replication of the virus. One such drug is AZT (Retrovir).

AZT interferes with the making of viral DNA by the HIV enzyme *reverse transcriptase*. However, it does not interfere with the replication of human DNA, since human cells utilize a different enzyme, DNA polymerase. Unfortunately, AZT has serious side effects, and may not be as effective in prolonging patient life as once believed. Drugs, such as ddI and ddC, and 3TC also inhibit reverse transcriptase. But new drugs that inhibit a protease enzyme necessary for HIV replication have shown great promise. When these protease inhibitors are used in combination with AZT, ddC, or 3TC, the HIV in the blood has gone down dramatically in less than a year, and infection fighting helper T cells have increased in number. Patients using the new combination drugs report overcoming infections, cancers, and exhaustion to return to normal life. This is good news. But it remains to be seen if HIV can fight back.

A Vaccine for HIV?

The development of a vaccine for AIDS has been agonizingly slow. Most experimental vaccines have used purified proteins from the envelope that surrounds HIV. A vaccine prepared from these proteins is not infective but should stimulate the body to produce specific antibodies.

Is HIV the Cause of AIDS?

Before HIV was identified, there was much speculation about the cause of AIDS. Many possible causes were considered, among them the use of so-called recreational drugs, stress, conventional sexually transmitted diseases, and nutritional factors. However, since the identification of the virus in 1984, more and more evidence has accumulated that HIV causes the decline in helper T cells. Consider the following facts: (1) HIV can be found in almost 100 percent of AIDS cases. (2) After HIV-contaminated blood was removed from blood banks in 1985, the number of blood transfusion–associated cases of AIDS declined sharply. (3) In cases of AIDS among recipients of blood transfusions, the donor blood was found to contain HIV. (4) In a study of gay men, the frequency of AIDS increased as the frequency of HIV infection increased. (5) In the laboratory, HIV has been shown to infect and destroy helper T cells.

PART SIX

Regulation and Integration

CHAPTER 14

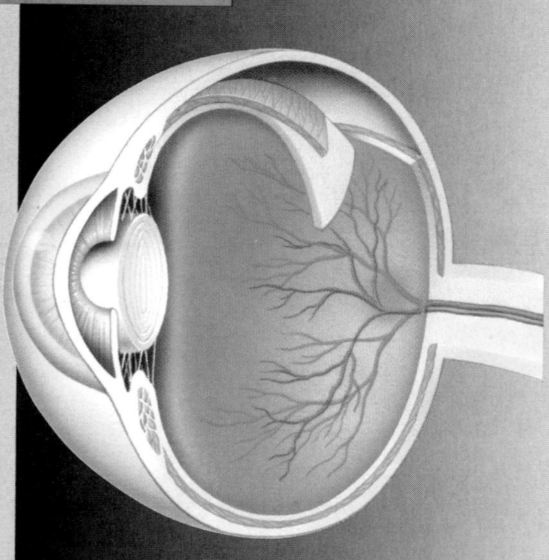

The Nervous System and Senses

CHAPTER 15

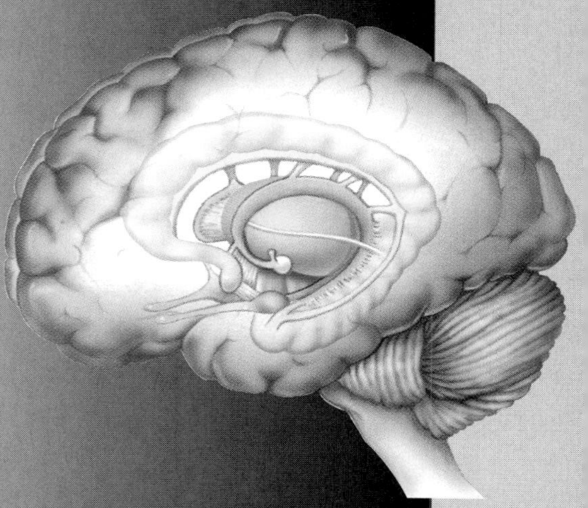

The Brain

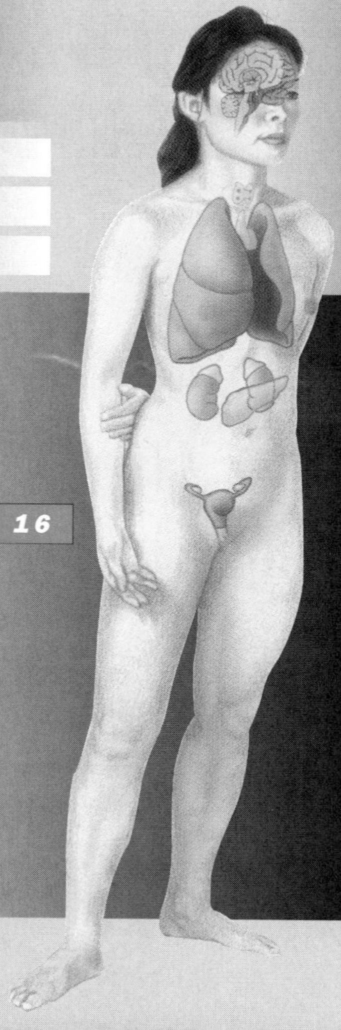

CHAPTER 16

The Endocrine System

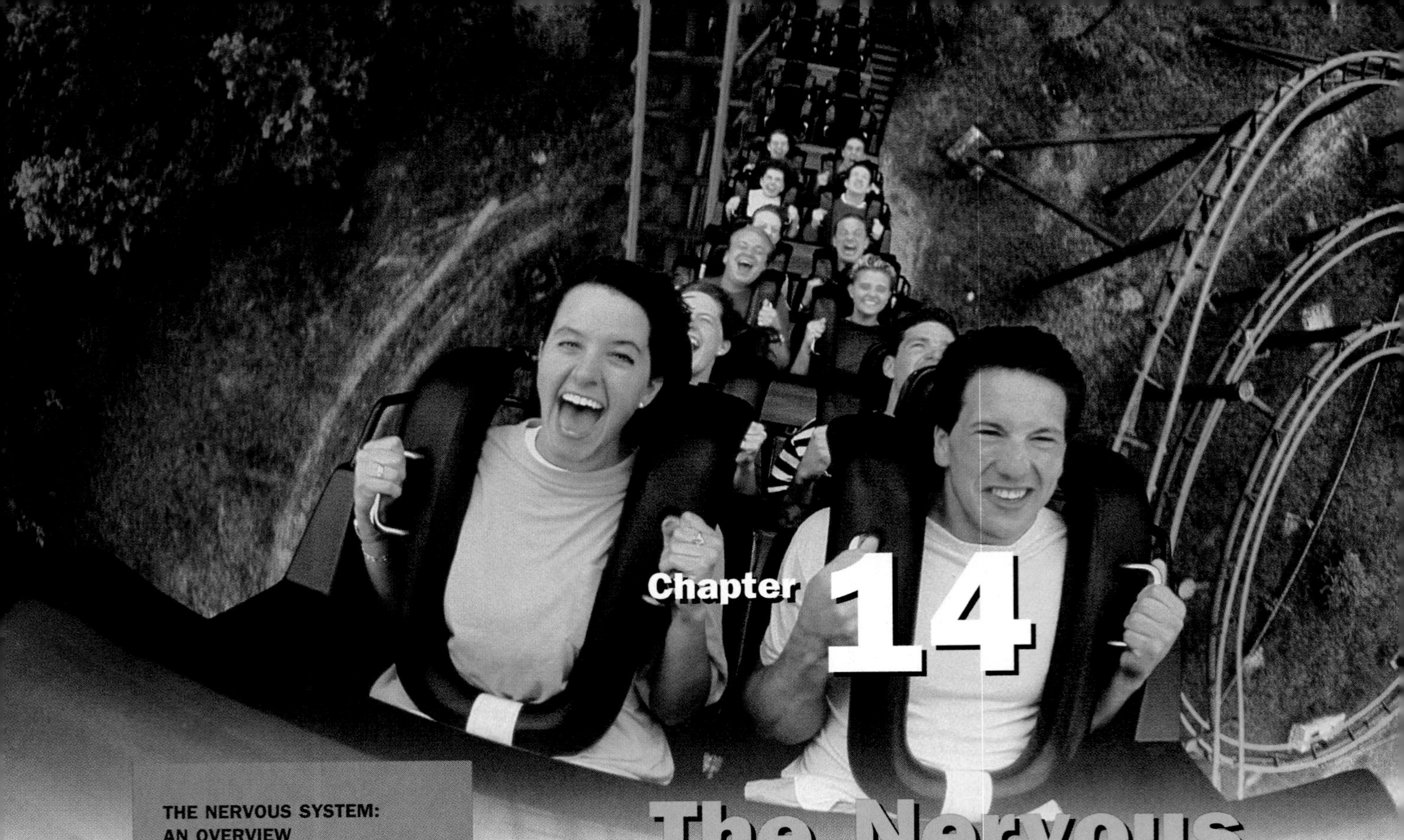

Chapter 14

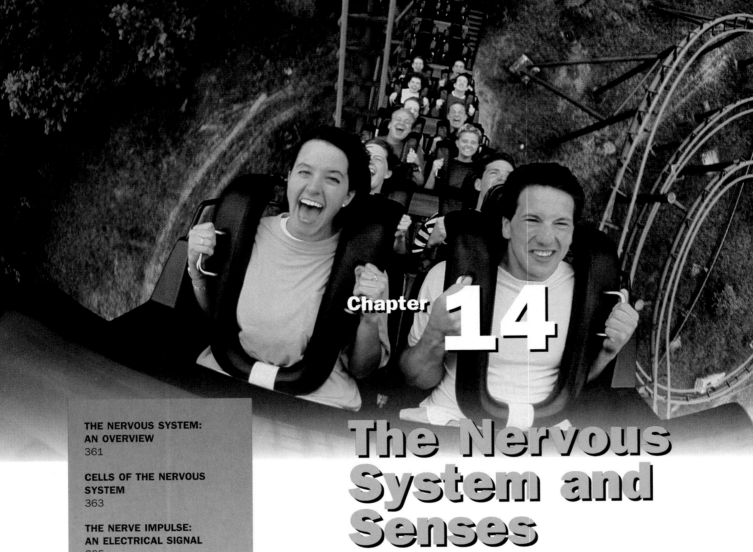

The Nervous System and Senses

To see, to hear, perhaps to dream. As long as we have life, we respond to our environments, both internal and external. When we do this, the awareness we have of ourselves is a product of the nervous system. The nervous system is called our great communicator because it communicates with all our organs to accomplish homeostasis. Through its operation we also perceive the dangers and beauties of the outside world.

Our internal and external environments provide stimuli (singular, *stimulus*). A **stimulus** can be defined as anything that produces a response. What is a response? A **response** may be complex or simple but it always involves a change in molecules, cells, tissues, or organs. Depending on the stimulus, this change may be large or small, visible or invisible, external or internal, immediate or delayed. You may or may not be conscious of the change, but it is always there. Between the stimulus and your response lie the components of your nervous system and their operation.

The complexity of the nervous system is almost beyond belief, but the principles of its operation are fairly simple: reception of stimuli, processing of signals, and generation of a response. The nervous system receives stimuli through *sensory receptors*. Some of these receptors receive internal stimuli; others receive external stimuli, such as sounds and images. Upon receiving a stimulus, receptors produce electrical signals that travel along specialized cells called nerve cells. Nerve cells serve as communication pathways, carrying signals from receptors to central processing units (the brain and spinal cord), where they are integrated with other signals and interpreted. A response signal is then carried by other nerve cells to the appropriate muscles and glands—the effectors—which generate a response.

In this chapter, we will give a very brief overview of the entire nervous system and then focus on the peripheral nervous system and spinal cord, leaving the brain and its functions for the next chapter. Here in Chapter 14, we will focus on three closely related topics: (1) the structure and function of nerve cells, (2) the organization of the nervous system, and (3) the senses. As in the preceding chapters, we'll conclude by examining changes in the nervous system associated with diseases, disorders, and the life span.

The Nervous System: An Overview

The nervous system has two components: the central nervous system and the peripheral nervous system (Figure 14.1). The **central nervous system (CNS)** consists of the brain and spinal cord. The CNS functions as the "central" control unit where information about our internal and external environments is received and processed and appropriate responses are initiated (Table 14.1). The *brain* is also capable of conscious decision making.

The *spinal cord* receives messages from throughout the body via sensory nerves. It functions by (1) processing messages directly, without involving the brain, (2) relaying messages to the brain for processing and integration, or (3) doing both. After processing, messages leave the spinal cord through motor nerves which stimulate muscles or glands.

The **peripheral nervous system (PNS)** consists of nerve cells outside the brain and spinal cord. It is through the sensory and motor nerves of the PNS that com-

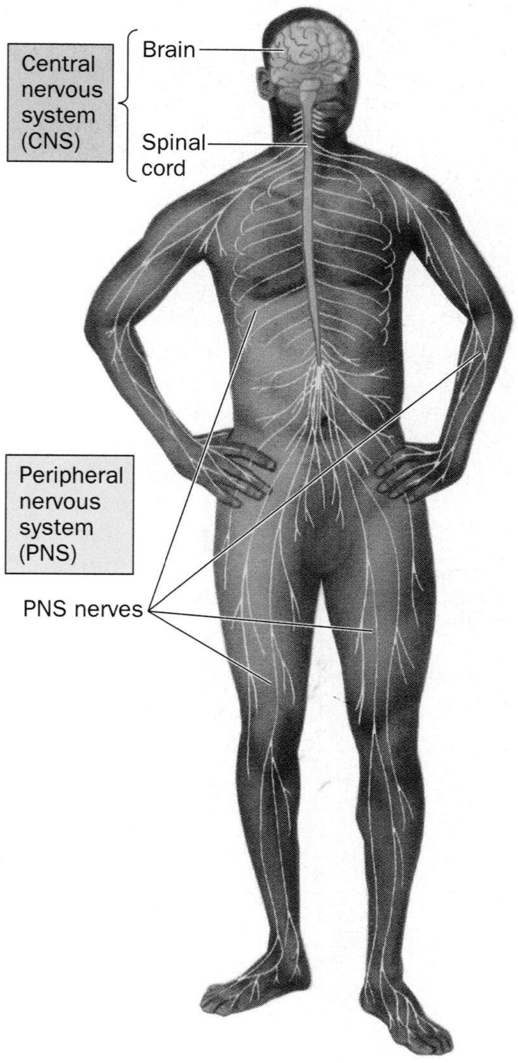
Central nervous system (CNS)
Brain
Spinal cord
Peripheral nervous system (PNS)
PNS nerves

FIGURE 14.1 An overview of the nervous system showing the relationship between the PNS and the CNS.

TABLE 14.1 Organization of the Nervous System

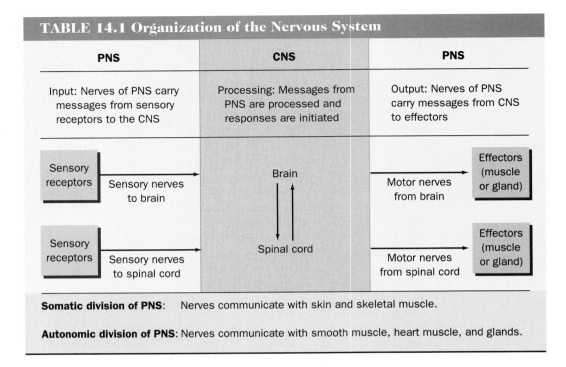

PNS	CNS	PNS
Input: Nerves of PNS carry messages from sensory receptors to the CNS	Processing: Messages from PNS are processed and responses are initiated	Output: Nerves of PNS carry messages from CNS to effectors
Sensory receptors → Sensory nerves to brain	Brain	Motor nerves from brain → Effectors (muscle or gland)
Sensory receptors → Sensory nerves to spinal cord	Spinal cord	Motor nerves from spinal cord → Effectors (muscle or gland)

Somatic division of PNS: Nerves communicate with skin and skeletal muscle.

Autonomic division of PNS: Nerves communicate with smooth muscle, heart muscle, and glands.

munication takes place between our tissues and the CNS (see Table 14.1). Functionally, the PNS is subdivided into the somatic division and the autonomic division. The *somatic division* communicates with the skin and skeletal muscles. The *autonomic division* communicates with smooth muscle, heart muscle, and glands.

Nerve cells of the PNS are organized into larger cordlike structures called nerves. A **nerve** is composed of many nerve cell extensions that are wrapped into parallel bundles by connective tissue (Figure 14.2). Nerves vary in diameter, depending on the number of nerve cells they contain. Some nerves are microscopic in size, while others are easily seen with the unaided eye.

There are three types of nerves: sensory, motor, and mixed. Nerves that transmit impulses from sensory receptors to the CNS are called *sensory nerves* (Table 14.1). Nerves that transmit impulses from the CNS to effectors (muscles and glands) are called *motor nerves*. Some nerves contain both sensory and motor nerve cells and are called *mixed*

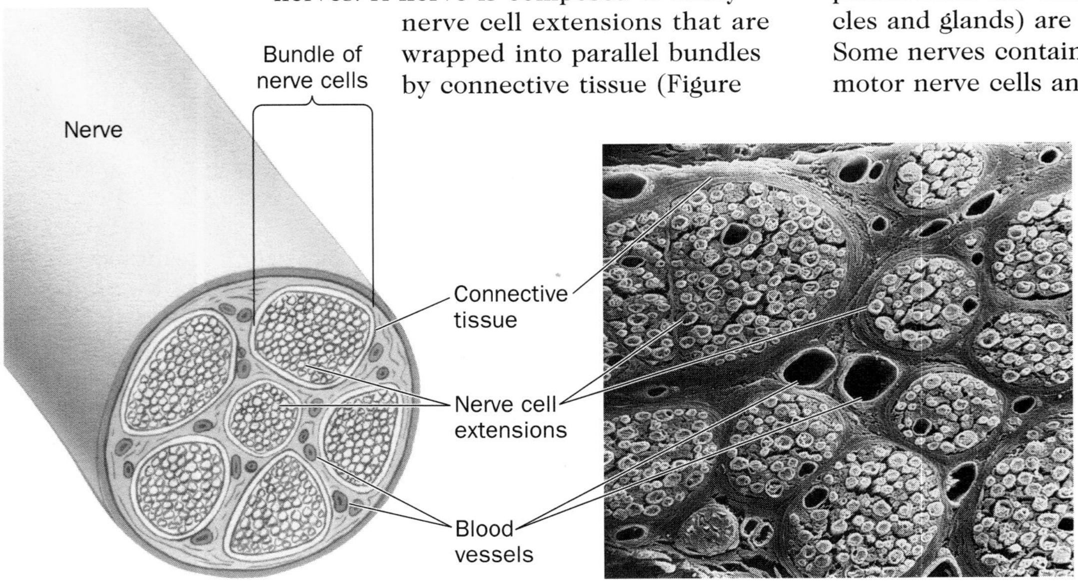
Bundle of nerve cells
Nerve
Connective tissue
Nerve cell extensions
Blood vessels

FIGURE 14.2 An SEM showing a cross section of a nerve. (From *Tissues and Organs: A Text-Atlas of Scanning Electron Microscopy* by Richard G. Kessel and Randy H. Kardon. © 1979 by W. H. Freeman and Company, p. 79.)

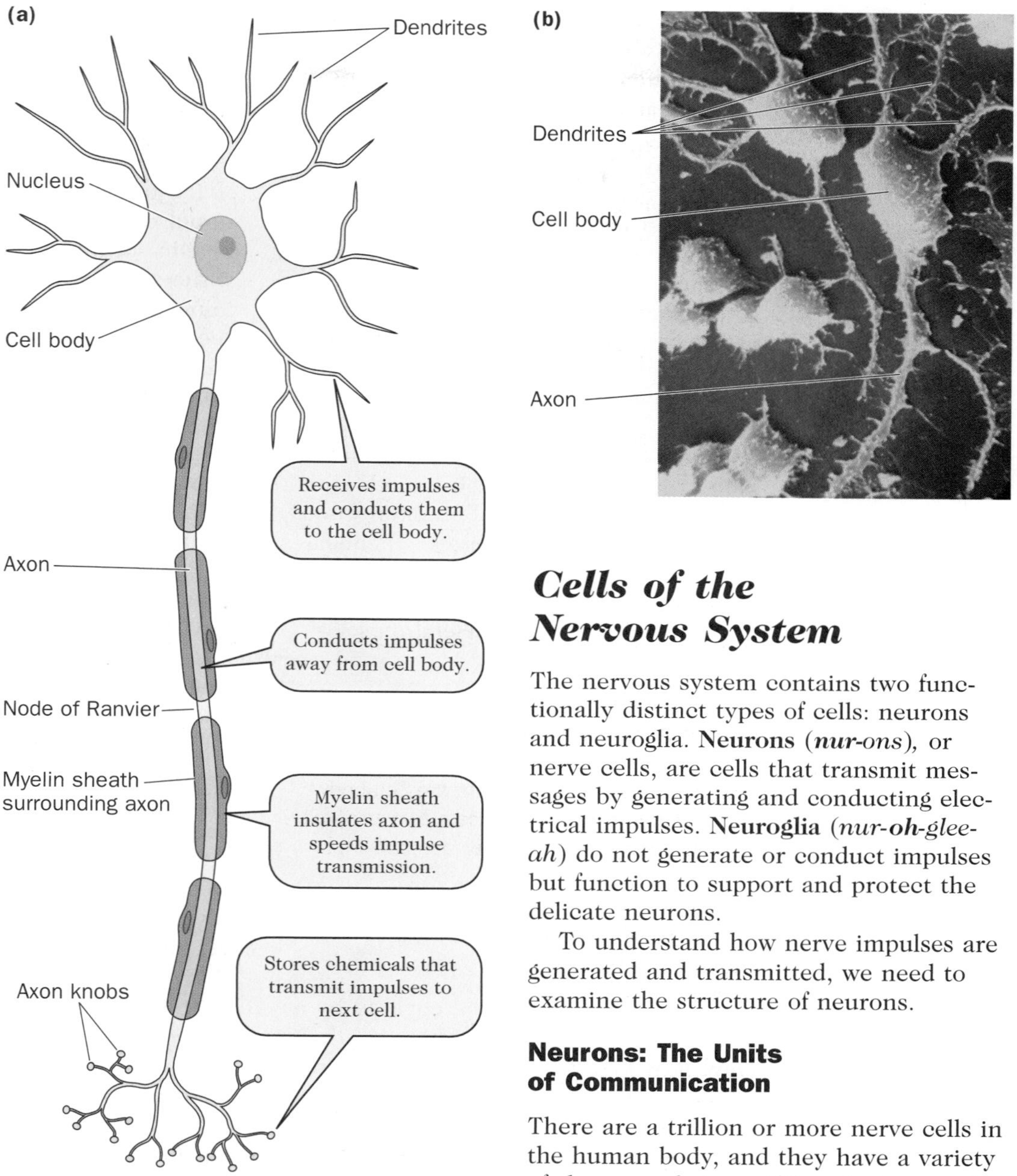
(a)
Dendrites
Nucleus
Cell body
Receives impulses and conducts them to the cell body.
Axon
Conducts impulses away from cell body.
Node of Ranvier
Myelin sheath surrounding axon
Myelin sheath insulates axon and speeds impulse transmission.
Axon knobs
Stores chemicals that transmit impulses to next cell.
(b)
Dendrites
Cell body
Axon

FIGURE 14.3 (a) A motor neuron with a myelin sheath. (b) A magnified view of motor neurons from the spinal cord.

nerves. To understand how nerves carry signals, we need to examine the structure and operation of nerve cells.

> The CNS (brain and spinal cord) receives, processes, and initiates responses. The PNS consists of nerves that communicate with skin and skeletal muscles (somatic) and nerves that communicate with the internal organs (autonomic).

Cells of the Nervous System

The nervous system contains two functionally distinct types of cells: neurons and neuroglia. **Neurons** (*nur-ons*), or nerve cells, are cells that transmit messages by generating and conducting electrical impulses. **Neuroglia** (*nur-oh-glee-ah*) do not generate or conduct impulses but function to support and protect the delicate neurons.

To understand how nerve impulses are generated and transmitted, we need to examine the structure of neurons.

Neurons: The Units of Communication

There are a trillion or more nerve cells in the human body, and they have a variety of shapes and sizes. Although structurally diverse, all nerve cells consist of three principal parts: the cell body and its extensions, the dendrites and axon. In what follows, we will describe a motor neuron as presented in Figure 14.3.

The **cell body** contains the cell structures typically found in other human cells: nucleus, mitochondria, endoplasmic reticulum, ribosomes, and Golgi (Chapter 2). Within the cell body, energy transformations, ATP production, and protein synthesis all occur. If the cell body is damaged, as happens after infection by the polio virus, the entire neuron may cease to function.

The many narrow extensions of the cell body are called **dendrites**. They receive incoming impulses from other neurons and conduct them toward the cell body. Most neurons have a great number of dendrites, permitting them to receive messages from many other neurons.

The **axon** is a single narrow extension of the cell body. It is generally much longer than the dendrites, and there is usually one axon for each neuron. It conducts impulses away from the cell body. Toward its end, an axon forms many thin branches. Each branch ends in a knoblike unit called an *axon knob*. These knobs store and release chemical substances that transmit a nerve impulse from one neuron to another or to an effector organ such as a skeletal muscle.

> All neurons have a cell body and extensions called dendrites and axons. Dendrites conduct impulses toward the cell body. Axons conduct impulses away from the cell body. Axon knobs contain chemicals that transmit an impulse from one cell to another.

Insulating Nerve Cells: Myelination

In the PNS, axons may be covered by a fatty sheath called the **myelin sheath** (***my**-eh-lin*) (see Figure 14.3). During the development of the PNS, the myelin sheath is formed by neuroglia cells called *Schwann cells.* These cells become distributed along the length of the axon, where they wrap around and around forming a compact jelly roll–like layering of their plasma membrane (Figure 14.4a). Between adjacent Schwann cells is a tiny gap known as the *node of Ranvier* (***ron**-vee-ay*) that exposes the surface of the axon. This exposed surface plays an important role in the transmission of nerve impulses in myelinated nerves.

The myelin sheath has three functions: It (1) provides electrical insulation for the axon, (2) speeds the transmission of an impulse along the axon, and (3) guides the nerve cell during axon regeneration. The fatty material in the compacted membrane of the sheath insulates the axon from ions that could short-circuit a nerve impulse. Because of this insulation, the electrical impulses must jump along the axon from one node of Ranvier to the next.

By jumping from node to node, an impulse travels faster in neurons with a myelin sheath than it can in neurons without one. For myelinated neurons the impulse speed is about 120 meters per second (about 270 miles per hour), compared with only 2.3 meters per second (about 5 miles per hour) for neurons without myelin. Damage to the myelin sheath slows impulse transmission and may cause short circuits that interfere with the messages that are sent over neurons.

Schwann cells also play an important role in the regeneration of axons in a severed PNS nerve. When an axon is cut, the severed portion degenerates and nerve function is lost. However, the tubelike channel formed by the Schwann cells remains intact. During regeneration, the cut axon regrows down through the tubelike channel and reconnects with the muscle or gland. Axons grow at a rate of about 1.5 millimeters (nearly $\frac{1}{16}$ inch) per day. Depending on the nerve's length and the severity of damage, completion of the regeneration process may require from several weeks to more than a year.

In the CNS, other neuroglia cells called *oligodendrocytes* (*ol-ig-go-**den**-dro-sites*) form the myelin sheath by sending out cellular extensions that wrap several axons (Figure 14.4b). Unlike the myelin sheath formed by Schwann cells, the sheath formed by oligodendrocytes degenerates after nerve injury. It cannot guide the regeneration of neurons in the brain and spinal cord. For this reason, nerve damage to the CNS cannot be repaired and brain and spinal cord injuries result in a permanent loss of function.

> The myelin sheath of PNS nerve cells is formed by Schwann cells, while in the CNS it is formed by oligodendrocytes. The myelin sheath insulates, speeds impulse transmission, and guides regeneration of PNS axons. CNS axons do not regenerate.

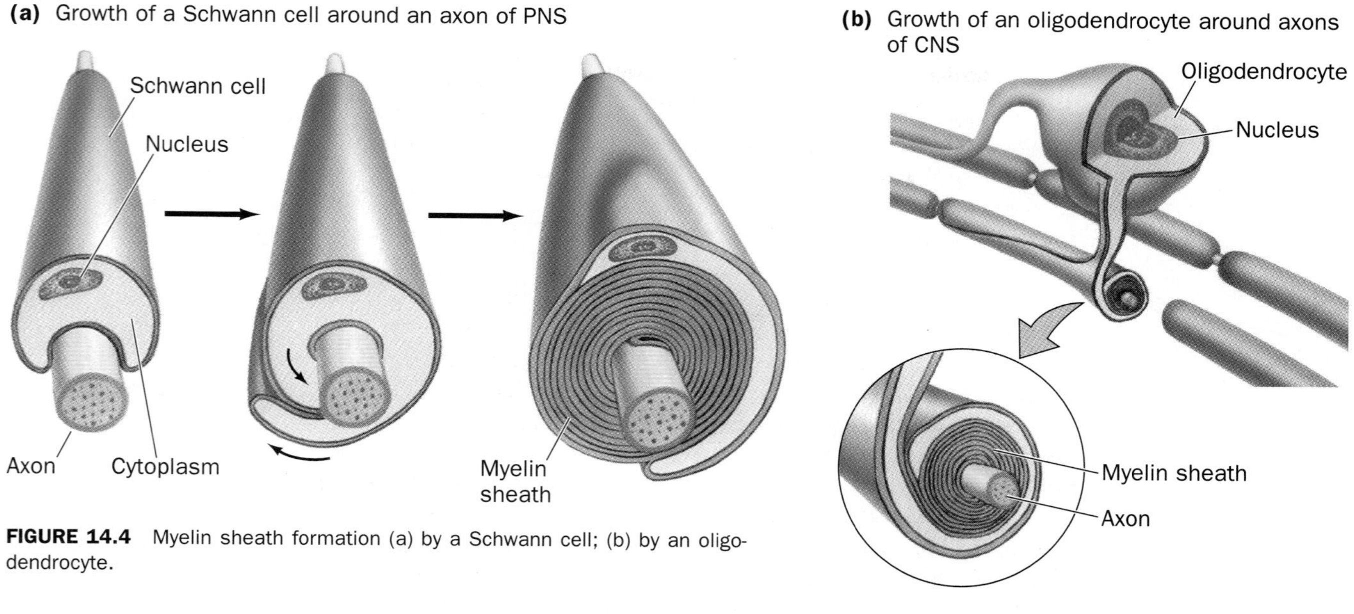
(a) Growth of a Schwann cell around an axon of PNS
Schwann cell
Nucleus
Axon
Cytoplasm
Myelin sheath
(b) Growth of an oligodendrocyte around axons of CNS
Oligodendrocyte
Nucleus
Myelin sheath
Axon

FIGURE 14.4 Myelin sheath formation (a) by a Schwann cell; (b) by an oligodendrocyte.

The Nerve Impulse: An Electrical Signal

The function of the nervous system is communication. The language of this communication consists of the nerve impulse, a small electrical charge generated and transmitted by each neuron. Neurons are specialized to transmit these electrical impulses and communicate with other cells. In this section, we will explore (1) the nature of this electrical impulse, (2) the way it travels along the axon, and (3) the way it travels from cell to cell. We'll start with a little information about the electrical nature of the impulse.

Recall from Chapter 1 that atoms contain positive and negative charges. Like charges (+ and +) repel each other, and unlike charges (+ and −) attract each other. Some atoms, called ions, are charged, for example, Na^+, K^+ and Cl^-. Large molecules such as proteins also may carry a net negative charge.

When given the opportunity, positive and negative charges move toward each other. An *electrical current* is the movement of electrical charges. For example, a flashlight battery has two opposite poles that can be identified as positive and negative. If these two poles are connected by a wire, an electrical current flows from one to the other.

All batteries store energy, and this stored energy can be described as having the potential to do work or make a change. Such a potential is often called an *electrical potential*. **Voltage** is the more common term for electrical potential, and it can be measured by an instrument called a voltmeter.

Any time opposite electrical poles (+ and −) are separated by a switch (as in flashlights and homes) or by a plasma membrane (as in cells), an electrical potential, or voltage, occurs. In neurons this small voltage is called the resting potential.

The Resting Potential: Preparing for Impulse Transmission

Like all cells, a neuron is completely surrounded by a plasma membrane that separates the interior of the cell from the exterior environment. As we discussed in Chapter 2, the plasma membrane is a selective barrier to the movement of molecules and ions between the cytoplasm and its exterior. Some molecules and ions can cross the plasma membrane, while others cannot. The cytoplasm has a composition very different from that of the fluid outside the cell.

The cytoplasm of neurons contains an abundance of negatively charged proteins (Figure 14.5). Their cytoplasm also has a higher concentration of K^+ compared

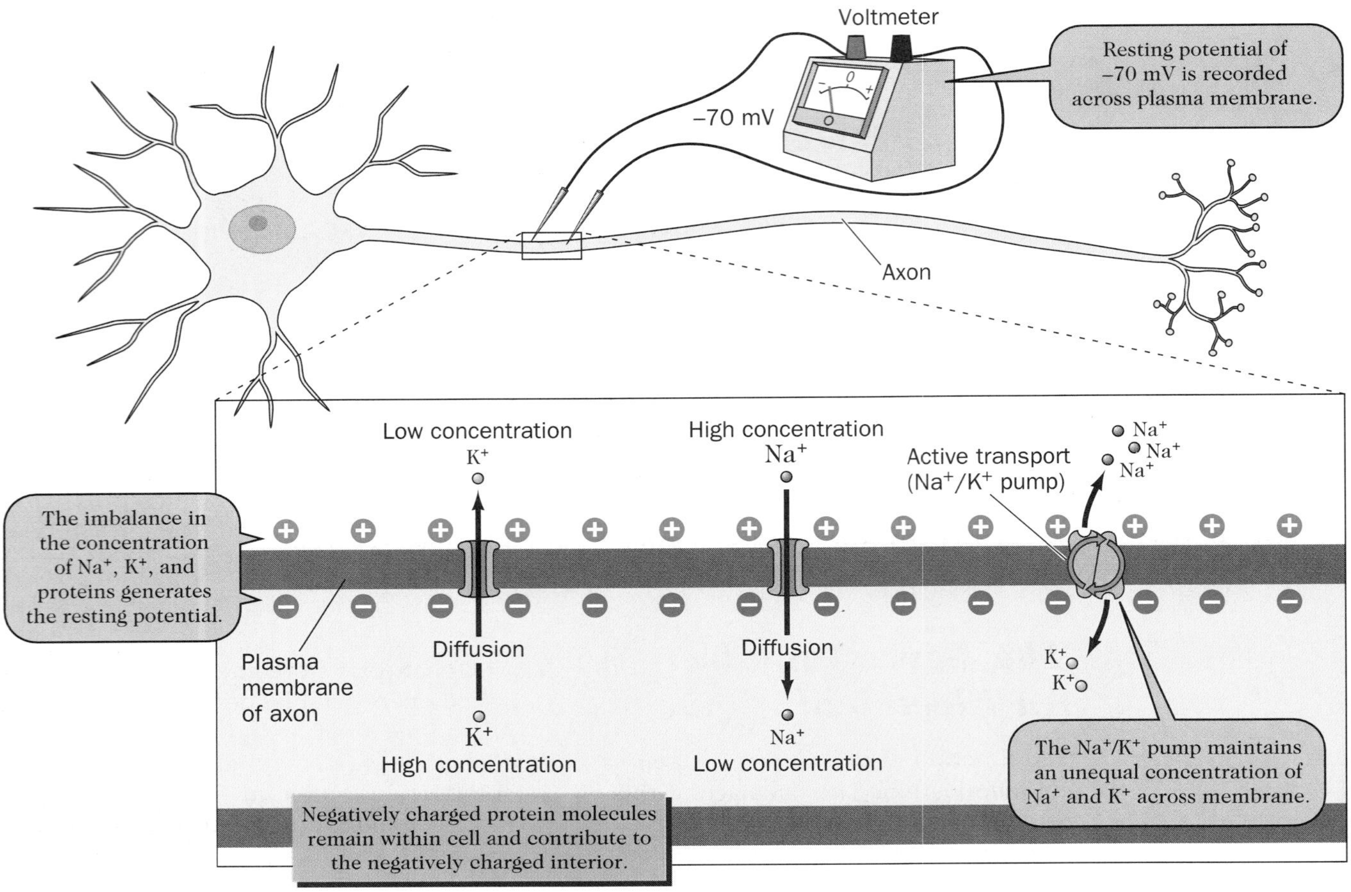
Voltmeter
−70 mV
Resting potential of −70 mV is recorded across plasma membrane.
Axon
Low concentration
K+
High concentration
Na+
Active transport (Na+/K+ pump)
Na+
Na+
Na+
The imbalance in the concentration of Na+, K+, and proteins generates the resting potential.
Diffusion
Diffusion
Plasma membrane of axon
K+
K+
K+
High concentration
Na+
Low concentration
The Na+/K+ pump maintains an unequal concentration of Na+ and K+ across membrane.
Negatively charged protein molecules remain within cell and contribute to the negatively charged interior.

FIGURE 14.5 Summary of how the diffusion and active transport of Na^+ and K^+ maintain a resting potential across the plasma membrane of a neuron.

with the outside of the cell. By contrast, Na^+ is less concentrated in the cytoplasm compared with the outside. The differences in the concentrations of these charged particles (ions and proteins) cause the interior of the cell to be negatively charged compared with the exterior. This imbalance of charges creates a voltage, or electrical potential, across the plasma membrane which we refer to as the **resting potential** of the cell. In an unstimulated resting neuron, this resting potential is about −70 millivolts (0.070 volt). The negative sign indicates that the interior of the cell is negative with respect to the outside (see Figure 14.5).

As you know from Chapter 2, the plasma membrane has protein channels through which certain ions can move. Although the negatively charged proteins are too large to diffuse across the plasma membrane, the smaller K^+ and Na^+ can do so. K^+ tends to diffuse out of the cell, while Na^+ tends to diffuse into the cell. If these movements were not compensated, the imbalance of these ions between the inside and the outside would be diminished and the resting potential would disappear.

To maintain the unequal distribution of Na^+ and K^+, the neuron uses chemical energy in the form of ATP to actively transport Na^+ out of the cell and move K^+ in. The active transport of these ions, called the **sodium-potassium pump** (or Na^+/K^+ pump), compensates for the effects of diffusion (see Figure 14.5). This Na^+/K^+ pump is a group of proteins in the plasma membrane. In each action of the pump, *three* Na^+ are moved out of the cell and *two* K^+ are moved into the cell. The Na^+/K^+ pump thus operates to maintain the ion imbalance and resting potential of neuron.

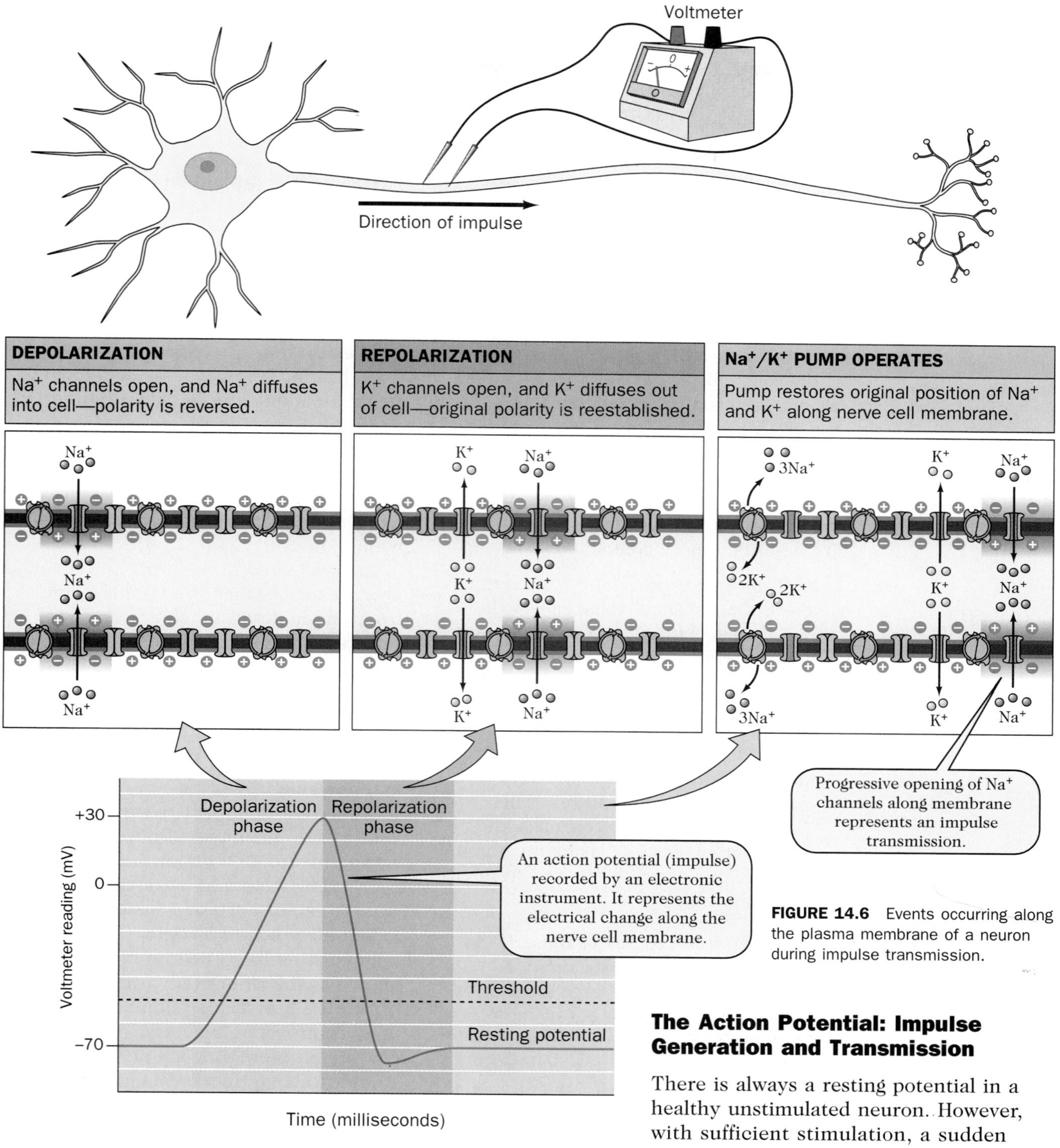
Voltmeter
Direction of impulse
DEPOLARIZATION
Na+ channels open, and Na+ diffuses into cell—polarity is reversed.
REPOLARIZATION
K+ channels open, and K+ diffuses out of cell—original polarity is reestablished.
Na+/K+ PUMP OPERATES
Pump restores original position of Na+ and K+ along nerve cell membrane.
Na+
K+
3Na+
2K+
Progressive opening of Na+ channels along membrane represents an impulse transmission.
+30
0
–70
Voltmeter reading (mV)
Depolarization phase
Repolarization phase
An action potential (impulse) recorded by an electronic instrument. It represents the electrical change along the nerve cell membrane.
Threshold
Resting potential
Time (milliseconds)

FIGURE 14.6 Events occurring along the plasma membrane of a neuron during impulse transmission.

> A neuron's exterior is positively charged because of many sodium ions; its interior is negatively charged because of its many proteins. This creates a resting potential of about −70 millivolts which is maintained by the sodium-potassium pump.

The Action Potential: Impulse Generation and Transmission

There is always a resting potential in a healthy unstimulated neuron. However, with sufficient stimulation, a sudden change in the resting potential occurs. This electrical change is called an **action potential**, the technical term for an *impulse* (Figure 14.6). An action potential may be produced by stimulation from another neuron. To produce an action potential, the stimulus must be above a minimum strength called the *threshold*.

Generating an action potential. When an action potential is produced, the neuron's resting potential rapidly changes from −70 millivolts (mV) to +30 mV (Figure 14.6). This is called *depolarization*. Then, just as suddenly, the potential changes again and goes from +30 mV back to −70 mV, the resting potential. This is called *repolarization*.

This entire depolarization and repolarization process takes place in about 2 milliseconds (0.002 second). This is not a very long time, but it is sufficient for three coordinated events to occur: (1) Na^+ rushes into the cell, (2) K^+ moves out of the cell, and (3) the Na^+ and K^+ imbalance is restored by the action of the Na^+/K^+ pump. These are the events that generate the action potential. Concentration gradients drive the first two events, while ATP provides the energy for the third. In what follows, these three events are described more completely.

1. *Sodium moves into the cell.* A stimulus above the threshold value causes the Na^+ channels in the plasma membrane to open temporarily. As a result, Na^+ rapidly diffuses into the cell from the outside, moving from a region of greater concentration to a region of lesser concentration. This large influx of positive ions causes *depolarization* from the original −70 mV to +30 mV. The abrupt change in voltage causes the Na^+ channels in the plasma membrane to close, preventing more Na^+ from entering the cell.

2. *Potassium moves out of the cell.* As the Na^+ channels close, many K^+ channels open, and a large number of potassium ions diffuse out of the cell. This loss of K^+ produces *repolarization*. As the inside of the cell once again becomes more negative, many of the potassium channels close, limiting further loss of K^+.

3. *Action of the sodium-potassium pump.* To reestablish the resting concentrations of Na^+ and K^+, the sodium-potassium pump actively transports Na^+ out of the cell and moves K^+ in. Until the resting potential is reestablished, the neuron is said to be *refractory* and cannot be depolarized by another stimulus. This refractory period lasts about 1.0 millisecond.

Transmitting an action potential. To transmit a message, an action potential must move along the axon. How does this take place? Neuroscientists have found that depolarization of one small section of the membrane opens Na^+ channels in adjacent sections of the membrane. Therefore, as the Na^+ rushes in at one area, the adjacent area becomes depolarized. Depolarizations are quickly followed by repolarizations and the action of the sodium-potassium pump. Thus, a nerve impulse is actually a series of depolarizations and repolarizations that rapidly move along the surface of the neuron membrane (see Figure 14.6).

In myelinated nerves, this movement is even faster because the impulse jumps from one node of Ranvier to the next. This happens because the depolarization at one node establishes an electrical field with an adjacent node, opening its Na^+ channels (Figure 14.7). Depolarization and an impulse follow.

Some anesthetics, such as lidocaine and procaine (Novocain), prevent impulse transmission from pain receptors by preventing Na^+ channels from opening in sensory neurons. Without the influx of Na^+, depolarization does not take place and the impulse is not transmitted.

The all-or-none principle. In responding to a stimulus, a neuron obeys what is known as the all-or-none principle. This principle states that a stimulus either will produce an action potential (impulse) or will not. Increasing the strength of the stimulus above the threshold does not increase the intensity of the impulse. Any stimulus below the threshold will not produce an impulse. This is analogous to pulling the trigger of a gun. Without a minimum pressure (threshold) on the trigger, the gun will not fire. However, any trigger pressure above the minimum produces the same effect: the gun fires.

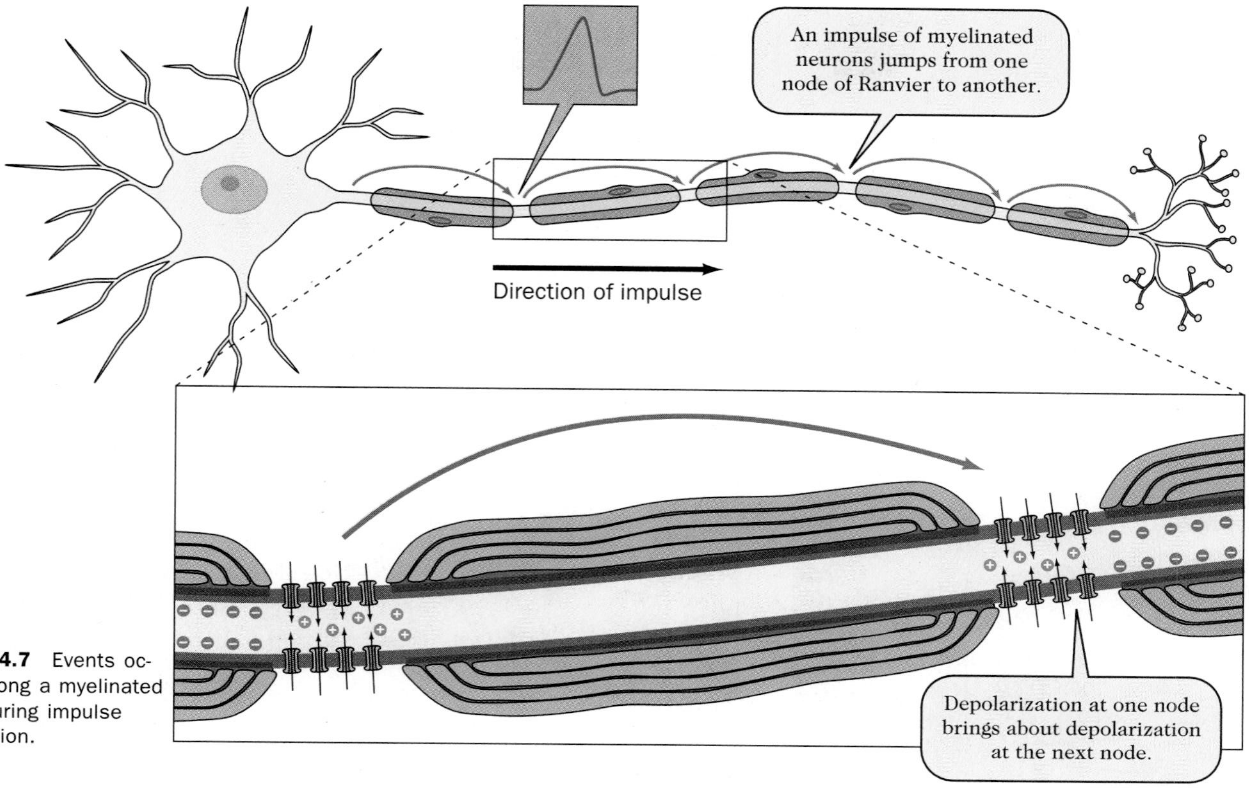

An impulse of myelinated neurons jumps from one node of Ranvier to another.
Direction of impulse
Depolarization at one node brings about depolarization at the next node.

FIGURE 14.7 Events occurring along a myelinated neuron during impulse transmission.

> A threshold stimulus initiates an impulse, or action potential. During depolarization sodium ions diffuse into the cell, followed by repolarization when potassium ions diffuse out of the cell. The Na^+/K^+ pump restores the original concentrations of these ions.

The Synapse: Impulse Transmission between Cells

Impulses are transmitted from one nerve cell to another or to a muscle or gland by moving across a junction known as a **synapse** (***sin**-aps*). Synapses exist between a neuron, called a *presynaptic cell,* and another neuron, muscle, or gland cell, called a *postsynaptic cell*.

All synapses have several common components, as shown in Figure 14.8. The *presynaptic membrane* is the plasma membrane of the axon knob. The *postsynaptic membrane* is the plasma membrane of the next neuron's dendrite or cell body, a muscle cell, or a gland cell. Between the pre- and postsynaptic membranes is a very small fluid-filled gap called the *synaptic cleft*. This cleft is only about 0.02 micrometer (about one-millionth of an inch) wide, and, therefore, chemical substances diffuse across it rapidly.

Within the axon knob of the neuron are many vesicles called *synaptic vesicles*. These vesicles contain a chemical known as a **neurotransmitter,** which is a chemical signal used to stimulate the postsynaptic cell. Upon release from the synaptic vesicle, neurotransmitters rapidly diffuse across the synaptic cleft to interact with the postsynaptic membrane.

Although we know of about 50 different neurotransmitter molecules operating in different parts of the nervous system, an individual neuron stores and releases only one type of neurotransmitter. The first neurotransmitter discovered was **acetylcholine (ACh)** (*as-ee-til-**co**-lean*). It is released by certain neurons of the CNS and by neurons of the PNS that synapse with skeletal muscle cells (Chapter 5).

All neurotransmitters operate in essentially the same fashion (Figure 14.8):

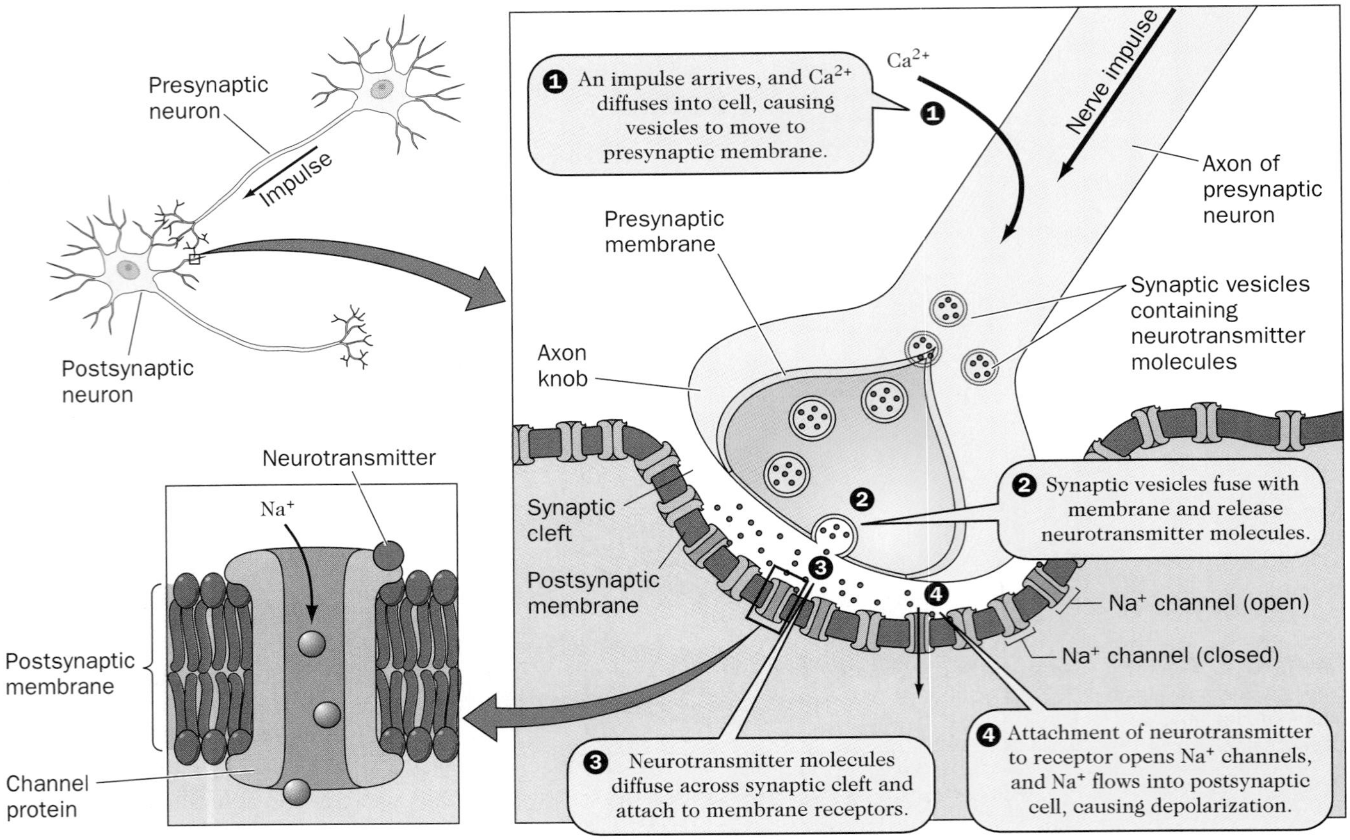
Presynaptic neuron
Impulse
Postsynaptic neuron
Neurotransmitter
Na+
Postsynaptic membrane
Channel protein
1 An impulse arrives, and Ca2+ diffuses into cell, causing vesicles to move to presynaptic membrane.
Ca2+
Nerve impulse
Axon of presynaptic neuron
Presynaptic membrane
Synaptic vesicles containing neurotransmitter molecules
Axon knob
Synaptic cleft
Postsynaptic membrane
2 Synaptic vesicles fuse with membrane and release neurotransmitter molecules.
Na+ channel (open)
Na+ channel (closed)
3 Neurotransmitter molecules diffuse across synaptic cleft and attach to membrane receptors.
4 Attachment of neurotransmitter to receptor opens Na+ channels, and Na+ flows into postsynaptic cell, causing depolarization.

FIGURE 14.8 Summary of the chemical events at a synapse.

(1) When an impulse arrives at the axon knob, it opens Ca^{2+} channels in the presynaptic membrane. As a result, Ca^{2+} diffuses into the neuron. (2) The Ca^{2+} causes as many as 100 synaptic vesicles to fuse with the presynaptic membrane, spilling their neurotransmitter molecules into the synaptic cleft. (3) The released neurotransmitter molecules diffuse across the synaptic cleft and bind to specific receptors in the postsynaptic membrane. Depending on the type of neurotransmitter, this binding may excite or inhibit the postsynaptic cell.

There are two categories of neurotransmitters: excitatory and inhibitory. When an **excitatory neurotransmitter** is released, the postsynaptic cell becomes depolarized and an impulse is generated. This means that the message has been carried one more step toward its destination. Acetylcholine and dopamine are examples of excitatory neurotransmitters (Table 14.2).

When an **inhibitory neurotransmitter** is released, it prevents the production of an impulse by causing the postsynaptic membrane to become hyperpolarized. A neuron is hyperpolarized when the cell's interior becomes even more negative in relation to its exterior, for example, when the resting potential increases from −70 to −90 millivolts (mV). This can occur when a neurotransmitter opens Cl^- channels and this negative ion diffuses into the cell. Gamma-aminobutyric acid (GABA) is an example of an inhibitory neurotransmitter (Table 14.2).

Turning off neurotransmitters. To prevent continued stimulation of a postsynaptic neuron (or a muscle or gland), the effect of the neurotransmitter must be terminated. Depending on the type of neurotransmitter, this is accomplished in

TABLE 14.2 A Few Common Neurotransmitters

Neurotransmitter	Principal Location of Release	Function
Acetylcholine	Neuromuscular junctions, areas of the brain, ANS	Usually excitatory
Norepinephrine	Areas of brain and spinal cord, ANS	Excitatory or inhibitory depending on receptors; involved in emotional responses
Dopamine	Areas of brain	Usually excitatory; involved in emotions and moods
Serotonin	Areas of brain	Usually inhibitory; involved in emotions, moods, and sleep
Gamma-aminobutyric acid	Areas of brain	Inhibitory; may play a role in pain perception
Endorphins	Areas of brain	Usually inhibitory; act like opiates to block pain

different ways. Some neurotransmitters are inactivated by an enzyme. For example, ACh is inactivated by an enzyme called *acetylcholinesterase* (*as-ee-til-co-lin-**ess**-ter-ace*). Other neurotransmitters, such as norepinephrine and dopamine, are transported back into the presynaptic axon and used again.

Synapses are targets for toxins and drugs. The function of synapses can be altered by a variety of substances from the environment. These substances include toxic chemicals produced by bacteria and plants and a variety of drugs. For example, the *botulism toxin* produced by bacteria prevents the release of ACh at neuromuscular junctions and prevents skeletal muscle contraction.

Curare is a poison traditionally placed on arrows by the native people of the Amazon rain forest. Like botulism toxin, it acts on the neuromuscular junction, but curare binds to ACh receptors on the postsynaptic membrane. In modern medicine, curare is used to produce temporary respiratory paralysis during surgery on the lungs.

Drugs such as alcohol, cocaine, valium, and amphetamines have various effects on the mind, body, and emotions, but they all act on the synapse in some way. We will describe their actions in Chapter 15 and in Spotlight on Health: Six (page 442).

> A synapse is a junction between neurons or between neurons and their effectors. A neurotransmitter crosses the synapse and attaches to receptors on the postsynaptic cell. Depending on the neurotransmitter, it may excite or inhibit the postsynaptic cell.

Summation and Integration by Postsynaptic Neurons

In your nervous system, an individual neuron usually receives impulses from hundreds of excitatory and inhibitory neurons (Figure 14.9). At every moment a particular neuron receives impulses from many, if not all, of these other neurons. Whether the particular neuron becomes depolarized depends on the total number of excitatory and inhibitory stimulations it receives. This is a process called **summation**.

If a larger number of *excitatory* neurons are releasing their depolarizing neurotransmitters, the cell will depolarize and an impulse will be generated. If a larger number of *inhibitory* neurons are releasing their hyperpolarizing neurotransmitters, the cell will not depolarize and an impulse will not be generated. Since all the neurons that form synapses

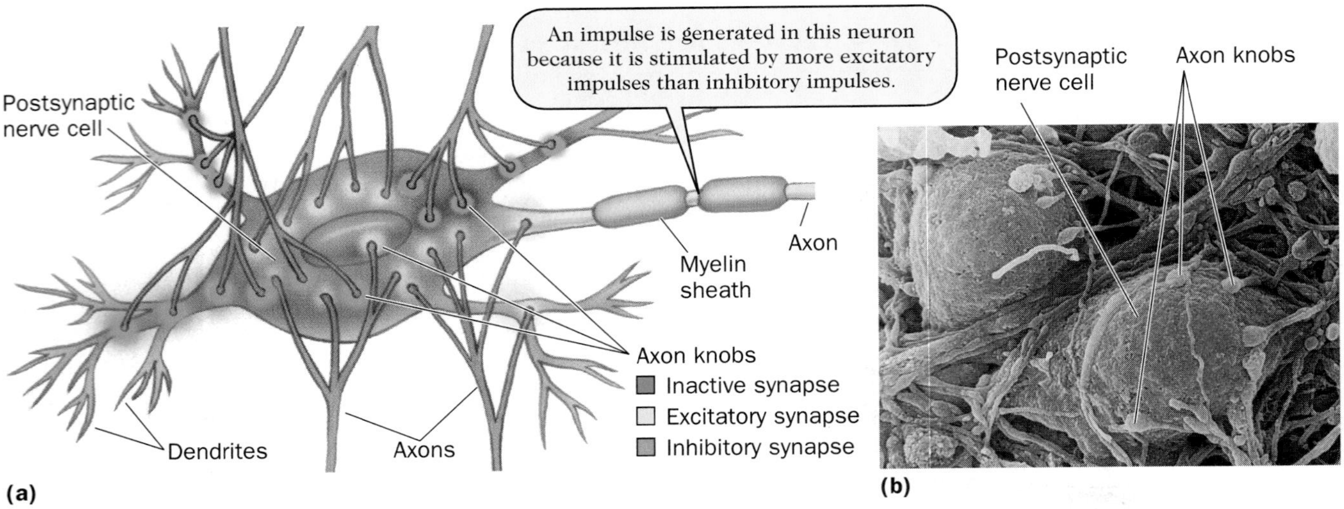
An impulse is generated in this neuron because it is stimulated by more excitatory impulses than inhibitory impulses.
Postsynaptic nerve cell
Axon
Myelin sheath
Axon knobs
Inactive synapse
Excitatory synapse
Inhibitory synapse
Dendrites
Axons
(a)
Postsynaptic nerve cell
Axon knobs
(b)

FIGURE 14.9 (a) Summation and integration of excitatory and inhibitory impulses by a postsynaptic neuron. (b) A magnified view of pre- and post-synaptic neurons.

with a particular neuron come from so many different places, the particular neuron is **integrating** an enormous amount of information. An impulse is or is not initiated in the postsynaptic cell, depending on whether it receives more excitatory or inhibitory neurotransmitters. The operation of the entire nervous system is based on this kind of decision in the billion or so cells that produce impulses.

> During summation, a nerve cell becomes stimulated through the combined effects of excitatory and inhibitory neurons. Through summation, neurons integrate sensory impulses from a variety of sources, producing an appropriate response.

The Peripheral Nervous System

Throughout the preceding chapters we have spoken of parts of the peripheral nervous system and their role in the regulation of organs and functions such as body temperature and blood pressure. Of course, the central nervous system also plays a role in many of these regulations, and the interaction of the PNS and CNS is something we want to emphasize here. Homeostasis is a direct product of both the PNS and the CNS.

Although the PNS and CNS interact, here we will focus on the PNS. What are its component parts? Where are they found? How do they operate, and how does the PNS interact with the CNS, particularly the spinal cord?

The PNS and the Spinal Cord

As we stated earlier, the PNS consists of nerves that transmit impulses to and from the CNS. It consists of two components: the somatic division and the autonomic division. In their communication with the CNS, both divisions of the PNS make use of the cranial nerves and spinal nerves. There are 12 pairs of **cranial nerves** which communicate directly with the brain. Cranial nerves are numbered with Roman numerals (I through XII) and they will be discussed further in Chapter 15.

There are 31 pairs of **spinal nerves** that communicate directly with the spinal cord along its entire length (Figure 14.10). Each spinal nerve has two branches that connect with the spinal cord: the dorsal root and the ventral root. The *dorsal root* contains sensory neurons that carry impulses from regions of the body to the spinal cord. The cell bodies of these sensory neurons are massed together in an enlarged area of the dorsal root called the *dorsal root ganglion*.

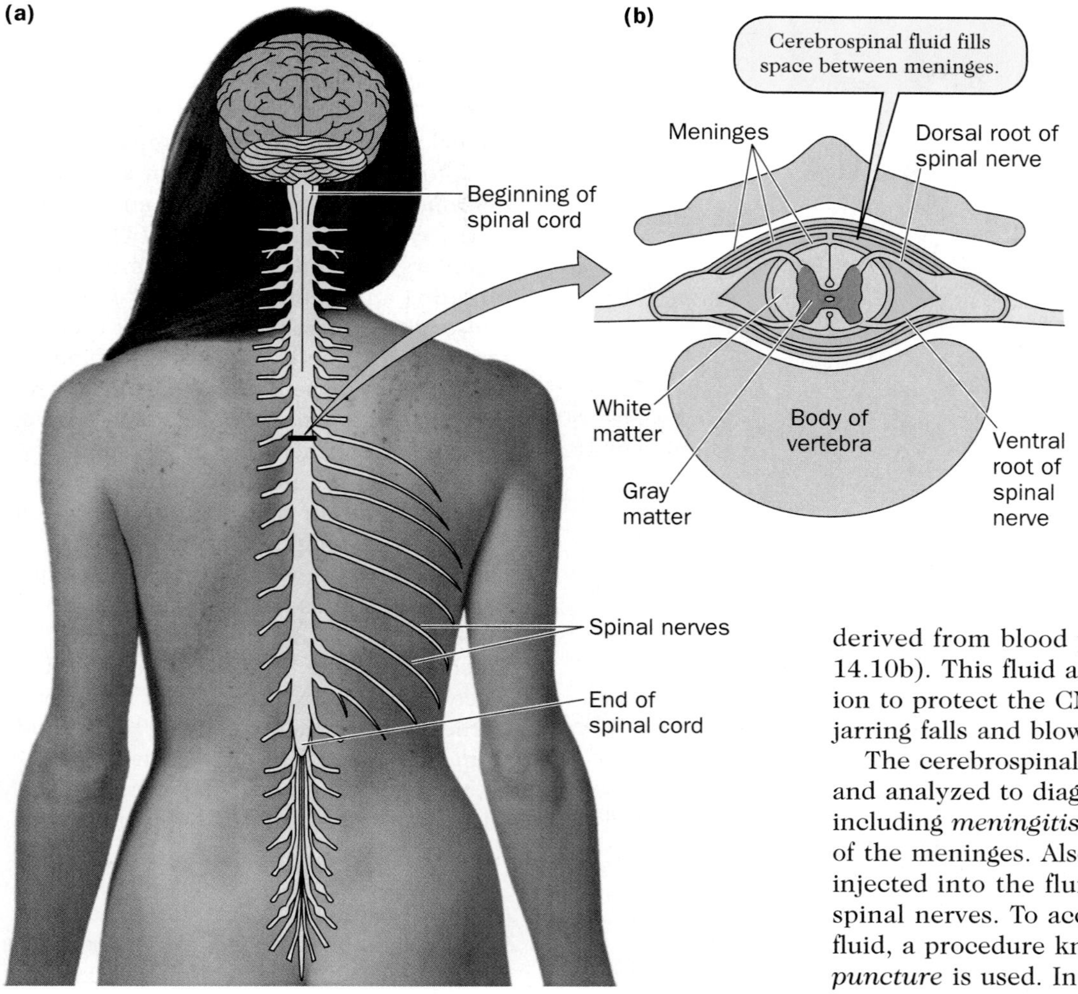
(a)
Beginning of spinal cord
Spinal nerves
End of spinal cord
(b)
Cerebrospinal fluid fills space between meninges.
Meninges
Dorsal root of spinal nerve
White matter
Gray matter
Body of vertebra
Ventral root of spinal nerve

FIGURE 14.10 The spinal cord. (a) Posterior view; (b) cross section showing dorsal and ventral roots of spinal nerve.

The *ventral root* contains motor neurons that carry impulses from the spinal cord to regions of the body. The cell bodies of these motor neurons are located in the spinal cord. To understand the operation of the spinal nerves, we need to understand the structure and functions of the spinal cord, which is considered part of the CNS.

Spinal cord structures. The spinal cord is an integrating structure that links the PNS with the brain. Housed within and protected by the vertebrae, the spinal cord is a compact structure that contains both myelinated and unmyelinated neurons. It is about the diameter of your little finger and extends from the base of the skull to the region of the first lumbar vertebra; it does not extend into the sacral region of the spine (Figure 14.10a). We will briefly examine the spinal cord's protective structures, fluid, and internal organization of neurons.

Surrounding the spinal cord are a series of three thin tissue membranes called the **meninges** (*men-**in**-geez*). Between two of these membranes is a space filled with **cerebrospinal fluid**, a clear fluid derived from blood plasma (Figure 14.10b). This fluid acts as a watery cushion to protect the CNS from injury during jarring falls and blows.

The cerebrospinal fluid can be removed and analyzed to diagnose certain diseases, including *meningitis*, a serious infection of the meninges. Also, drugs can be injected into the fluid to anesthetize the spinal nerves. To access the cerebrospinal fluid, a procedure known as a *lumbar puncture* is used. In this procedure a hypodermic needle is inserted between the third and fourth lumbar vertebrae. At this site, the inserted needle cannot damage the spinal cord but has access to the fluid (Figure 14.11).

Viewed in cross section, the internal structure of the spinal cord consists of two distinct areas called the gray matter and the white matter (see Figure 14.10b). The *gray matter* consists of unmyelinated axons arranged in an H-shaped area in the center of the cord. Within the gray matter, sensory neurons from the dorsal root form synapses with the motor neurons of the ventral root. Within the gray matter, sensory and motor neurons also form synapses with neurons that carry messages up the spinal cord to the brain. The neurons communicating with the brain have myelinated axons which are located in the *white matter* surrounding the gray matter.

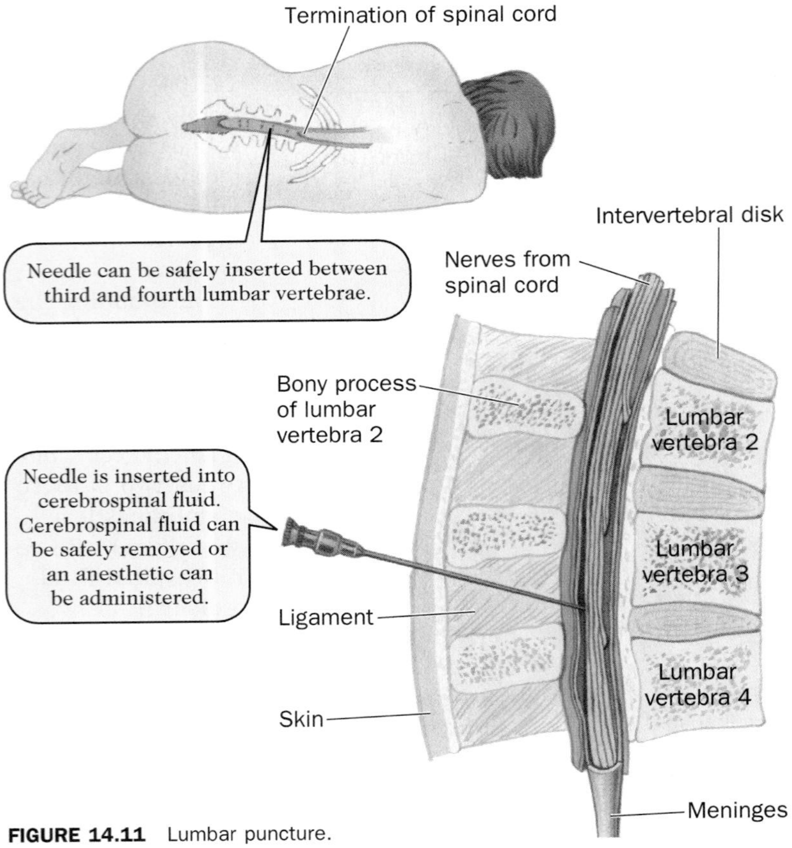
Termination of spinal cord
Needle can be safely inserted between third and fourth lumbar vertebrae.
Intervertebral disk
Nerves from spinal cord
Bony process of lumbar vertebra 2
Lumbar vertebra 2
Needle is inserted into cerebrospinal fluid. Cerebrospinal fluid can be safely removed or an anesthetic can be administered.
Lumbar vertebra 3
Ligament
Lumbar vertebra 4
Skin
Meninges

FIGURE 14.11 Lumbar puncture.

> The PNS consists of 12 pairs of cranial nerves and 31 pairs of spinal nerves. The spinal cord links the spinal nerves to the brain. Three protective meninges cover the spinal cord. Cerebrospinal fluid cushions the spinal cord.

The Somatic Division of the PNS

The **somatic division** of the PNS communicates with the skeletal muscles, the skin, and their sensory receptors. It includes both sensory and motor neurons. Sensory neurons transmit impulses to the spinal cord from sensory receptors throughout the body. Motor neurons transmit impulses from the spinal cord to the skeletal muscles.

The somatic division controls skeletal muscles by means of **somatic reflexes**. These reflexes are unlearned responses to external stimuli. They occur automatically and quickly and are not dependent on the brain. Quickly removing your hand from a sharp tack is an example of a somatic reflex.

Somatic reflexes operate because of neuron arrangements called **reflex arcs.** A typical reflex arc has five components: a receptor, a sensory neuron, an association neuron, a motor neuron, and an effector. In the case of the sharp tack: (1) *Pain receptors* in the dermis are stimulated, initiating an impulse (Figure 14.12a). (2) *Sensory neurons* attached to the receptors transmit the impulse to the spinal cord. The impulse enters the spinal cord through the dorsal root of a spinal nerve. (3) *Association neurons* in the spinal cord transfer the impulse from sensory neurons to motor neurons. (4) *Motor neurons* carry the impulse out of the spinal cord and into the ventral root and spinal nerve. The impulse is then conducted to the effectors. (5) *Effectors*, the skeletal muscles, receive the impulse and contract. In this case, the skeletal muscles of the arm contract, removing your hand from the tack. This entire sequence of events operates so fast that you are usually not aware of what is happening until the effector has acted.

Although a reflex arc operates without decision making by your brain, the brain is informed about what is happening. This occurs because the association neurons also synapse with other neurons that send impulses to the brain. Reflex information transmitted to the brain permits awareness and the development of more complicated responses to the initial stimulus, such as removing the tack, caring for the wound, and seeking medical attention.

The reflex response to the sharp tack is an example of a *withdrawal reflex*, but there are other somatic reflexes, such as the *stretch reflexes* (Figure 14.12b). Receptors for these reflexes detect stretching in a skeletal muscle, and the reflex response is a muscle contraction. The operation of stretch reflexes permits us to stand upright and move effectively without falling or thinking much about the process. An example of the stretch reflex is the knee-jerk reflex in which the slight stretching of thigh muscles causes their sudden contraction (see Figure 14.12b).

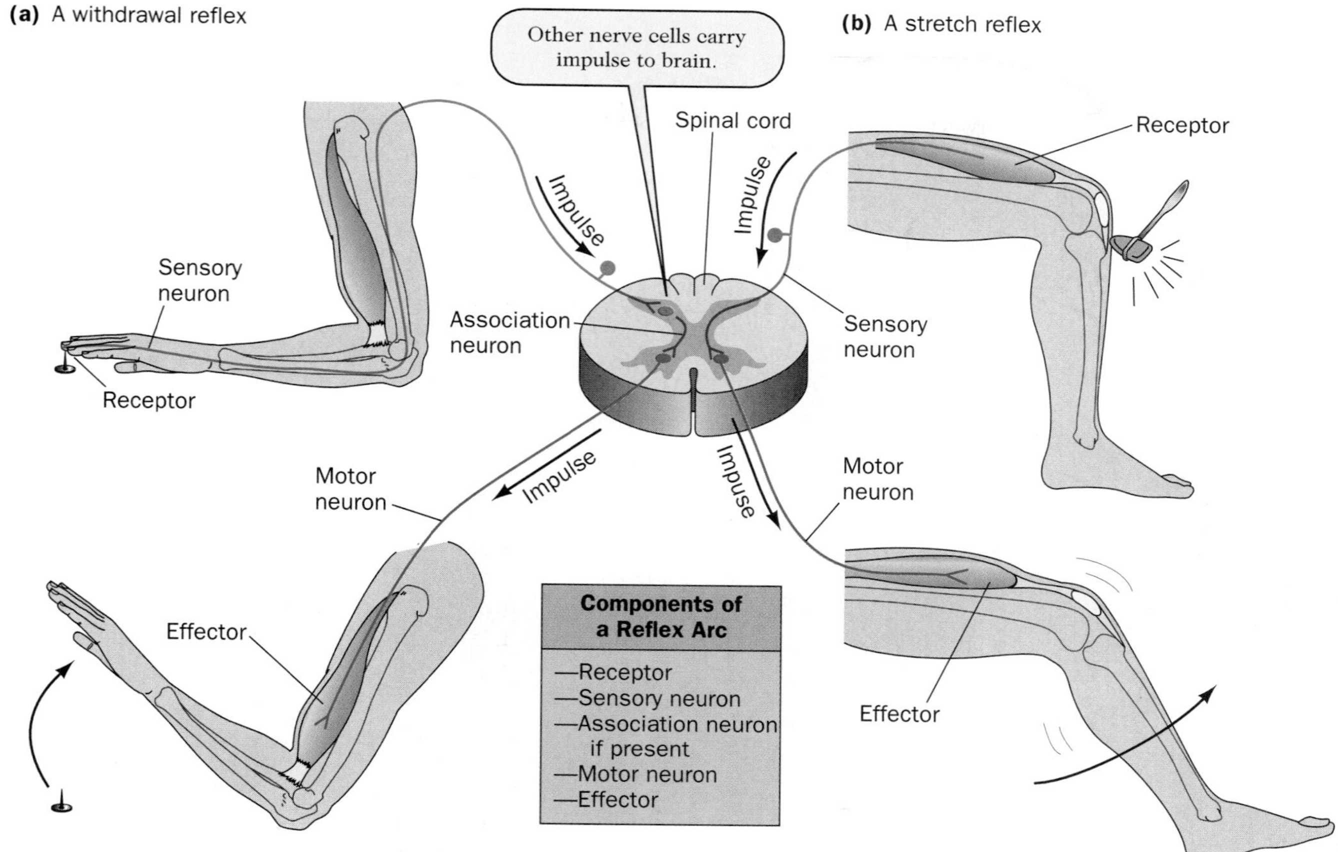
(a) A withdrawal reflex
Other nerve cells carry impulse to brain.
(b) A stretch reflex
Spinal cord
Receptor
Impulse
Impulse
Sensory neuron
Association neuron
Sensory neuron
Receptor
Motor neuron
Impulse
Impuse
Motor neuron
Effector
Components of a Reflex Arc
—Receptor
—Sensory neuron
—Association neuron if present
—Motor neuron
—Effector
Effector

FIGURE 14.12 Examples of two reflex arcs and their associated components.

You may have experienced this reflex during a medical examination. Testing this reflex indicates the condition of the spinal cord and its associated neurons.

> A reflex is a rapid, unlearned response to an external stimulus. Reflex arcs include a receptor, a sensory neuron, an association neuron, a motor neuron, and an effector.

The Autonomic Division of the PNS

The **autonomic division** of the PNS consists of motor neurons that relay impulses from the CNS to the internal organs of the body. (It is sometimes referred to as the *autonomic nervous system* or ANS.) The autonomic division generally responds to sensory information carried to the CNS by sensory nerves of the somatic division. In the preceding chapters we often encountered the operation of the autonomic division in the control of smooth muscle, cardiac muscle, and glands. It functions automatically, without conscious effort, and is essential to homeostasis.

In contrast to the somatic division, which uses only one motor neuron to communicate between the CNS and the effector organ, the autonomic division uses two motor neurons. As we will see, the second motor neurons establish the antagonistic (opposing) functions of the two components of the autonomic division: sympathetic and parasympathetic (Figure 14.13). Generally, internal organs receive motor impulses from both.

The **sympathetic division** originates from the thoracic and lumbar regions of the spinal cord. The first of the two motor neurons of the sympathetic division comes from the gray matter of the spinal cord and synapses with the second motor neuron outside the CNS. The second neuron then conducts the impulse to the internal effector organ.

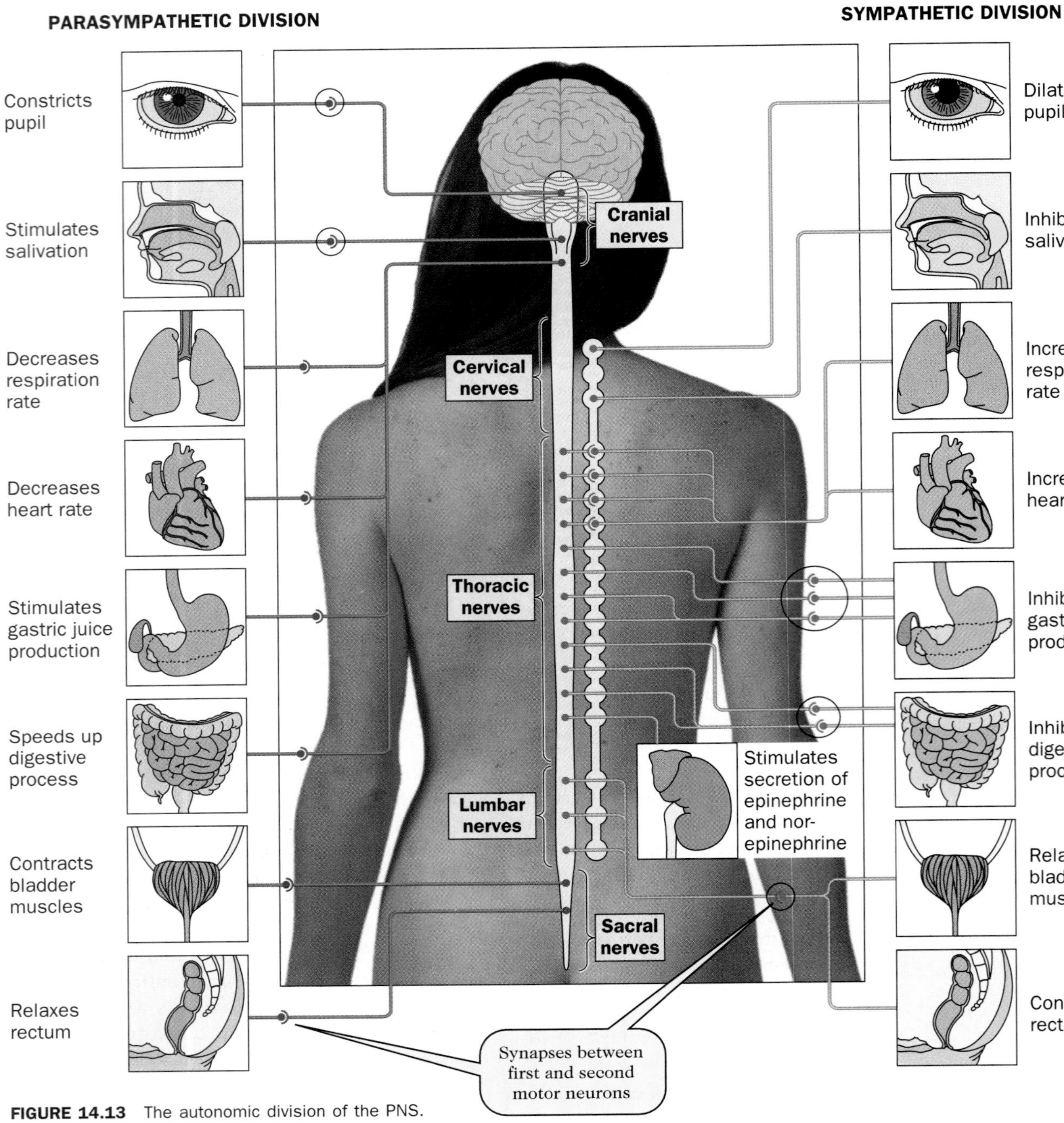
PARASYMPATHETIC DIVISION
SYMPATHETIC DIVISION
Constricts pupil
Stimulates salivation
Decreases respiration rate
Decreases heart rate
Stimulates gastric juice production
Speeds up digestive process
Contracts bladder muscles
Relaxes rectum
Cranial nerves
Cervical nerves
Thoracic nerves
Lumbar nerves
Sacral nerves
Dilates pupils
Inhibits salivation
Increases respiration rate
Increases heart rate
Inhibits gastric juice production
Inhibits digestive process
Stimulates secretion of epinephrine and nor-epinephrine
Relaxes bladder muscles
Contracts rectum
Synapses between first and second motor neurons

FIGURE 14.13 The autonomic division of the PNS.

Functionally, the sympathetic division stimulates our internal organs for real or perceived emergencies: the so called *fight-or-flight responses*. For example, when impulses are received from the sympathetic division, the heart rate and force of contraction increase, blood vessels of skeletal muscles dilate, and the respiration rate increases. These are all homeostatic adjustments that prepare the body for defensive action or flight to safety. Organs that do not contribute directly to these goals are slowed so they take less energy. For example, the digestive activity of the stomach and intestines is inhibited and blood is shunted away from the skin, producing the pale color of fear.

The **parasympathetic division** originates from the lowest part of the brain, which is called the *brain stem*, and the sacral region of the spinal cord (see Figure 14.13). The first of its two motor neurons comes from either the brain stem or the sacral region of the spinal cord and synapses with the second motor neuron outside the CNS. The second neuron conducts the impulse to the internal effector organ.

Functionally, the parasympathetic division transmits impulses that have a relaxing or restorative effect. One of its functions is to calm you down from the fight-or-flight response. For example, during periods of relaxation, your parasympathetic division decreases the heart rate and force of contraction and decreases the respiratory rate. It signals for constriction of the blood vessels to your skeletal muscles, shunting more blood to the skin, stomach, and intestines and restoring their circulation levels.

How can sympathetic and parasympathetic neurons have opposite effects on our internal organs? The answer lies with the type of neurotransmitter released by the second motor neuron. The sympathetic division releases norepinephrine, while the parasympathetic division releases acetylcholine. The postsynaptic membranes of our internal organs have receptor proteins that respond differently to these two neurotransmitters.

> The autonomic division uses two motor neurons to carry impulses to the internal organs of the body. The sympathetic component prepares us for emergencies. The parasympathetic component has a restorative effect on the body.

The Senses: Experiencing the Environment

Using specialized receptor cells, our bodies monitor both our external and internal environments. Sensory receptors that detect changes in the external environment are located close to the body surface. They give us the senses of touch, pressure, pain, heat, cold, smell, taste, vision, hearing, and balance. Our external senses permit us to anticipate the dangers as well as enjoy the pleasures of being alive.

In the preceding chapters we often spoke of the internal sensory receptors located throughout our internal organs, blood vessels, skeletal muscles, tendons, and joints. Although we are usually not consciously aware of their activities, they are essential to homeostasis because they monitor blood pressure, internal body temperature, oxygen (O_2) and carbon dioxide (CO_2) concentrations, and digestive activities. Internal sensors also provide information about movement, skeletal muscle tension, and the position of our limbs.

In what follows, we will first describe some of the simpler sensory receptors and then examine our special senses of smell, taste, vision, hearing, and balance.

Sensory Receptors: Dendrites or Modified Dendrites

When stimulated, our sensory receptors initiate impulses. The human body contains many types of receptors for sensing different types of stimuli. Figure 14.14 shows various types of sensory receptors of the skin. Although receptors may vary in structure and complexity, they share some general characteristics.

Some receptors are more complex than others, but all are constructed from the dendrites of sensory neurons. The simplest receptors, such as those which detect pain or monitor blood pressure, are unspecialized dendrites. The most complex receptors are structurally modified dendrites contained in specialized structures, for example, the receptors for the special senses. The different types of sensory receptors and their functions are presented in Table 14.3.

When a receptor receives a threshold stimulus, depolarization occurs. This depolarization generates an action potential (impulse) which is transmitted along a sensory neuron to the spinal cord and brain for a response. From an alphabet of on-off impulses, your brain creates coher-

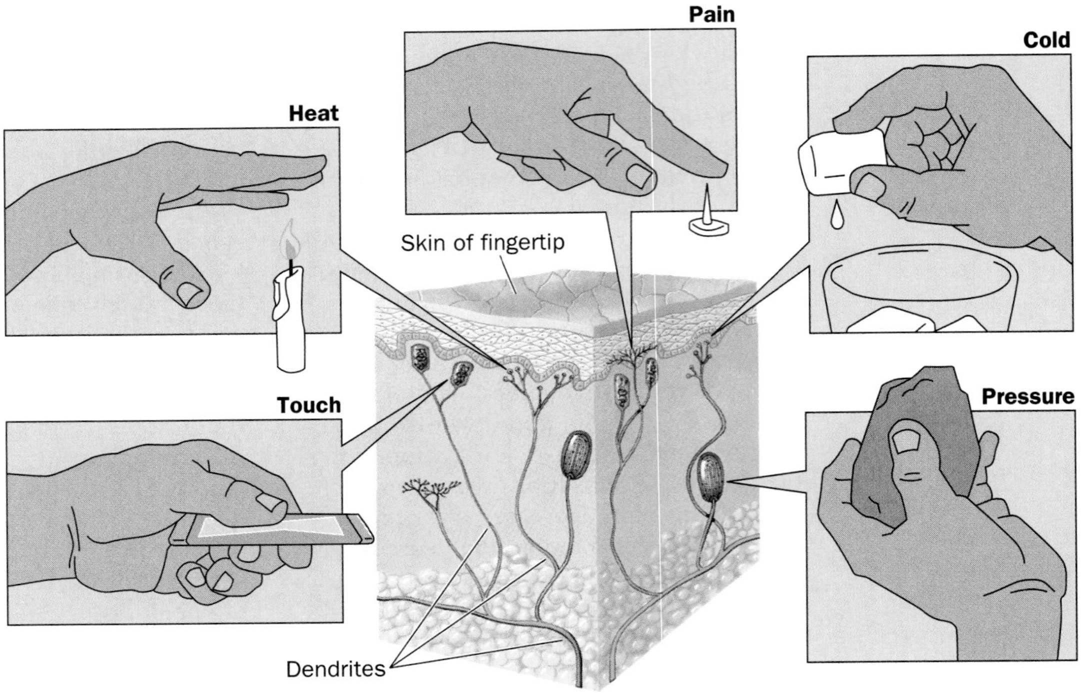

Pain
Cold
Heat
Skin of fingertip
Touch
Pressure
Dendrites

FIGURE 14.14 The sensory receptors of the skin.

ent meaning. It interprets the impulses produced by sensory receptors. For example, when photoreceptors in the eye are stimulated, the sensory neurons transmit impulses to a specific region of the brain for interpretation and we "see" light, colors, and images.

TABLE 14.3 Sensory Receptors

Sensory receptor	Stimulus	Location
General Senses		
Pressure	Pressing on tissue	Skin
Temperature	Heat	Skin
Touch	Pressing on tissue	Skin
Pain	Varied	Skin and internal organs
Stretch	Movement of tissue	Skeletal muscle and tendons
Chemical	CO_2, O_2	Blood vessels
Special Senses		
Sight	Light	Eye
Sound	Sound waves	Ear
Balance	Movement	Ear
Taste	Chemicals	Tongue
Smell	Chemicals	Nasal cavity

Frequent stimulation produces adaptation. Many sensory receptors cease to respond when they are stimulated continuously. This is called **adaptation.** Some receptors adapt more rapidly than others do. The pressure receptors of the skin represent a commonly experienced example. After a few minutes, you may no longer feel your hat even though it felt a little snug when you first put it on. Other receptors, such as pain receptors, adapt slowly or not at all. Rapid and complete adaptation of these pain receptors would affect our ability to survive, since they alert us to possible tissue damage resulting from an injury or disease.

Sensory receptors can be unspecialized dendrites or structurally modified dendrites. Many receptors adapt to continued stimulation and eventually stop responding.

Pain and Stretch Receptors

Found throughout the skin and the internal organs, **pain receptors** respond to excessive pressure, chemicals, and temperature extremes. In response to one of these stimuli, a pain receptor will generate impulses that are transmitted to the brain, where they are interpreted as pain. An injured site produces impulses, but we feel pain in the brain.

The pain associated with the internal organs is often perceived to be coming from the skin or from skeletal muscles remote from the internal organ. This is called *referred pain*, and it occurs because the impulses from internal pain receptors travel in the same spinal nerves that carry impulses from pain receptors of the skin and skeletal muscles. For this reason, the pain of a heart attack is often felt as a pain in the left shoulder and arm.

Skeletal muscles contain **stretch receptors** that inform us of the amount of stretching and tension of our skeletal muscles (see Figure 14.12b). These receptors have structures called *muscle spindles* which are located between skeletal muscle cells. Each spindle consists of several specialized muscle cells wrapped with the dendrites of a sensory neuron. When the muscle cells are stretched, the dendrites are stimulated and the sensory neuron transmits an impulse to the CNS. The CNS processes this information and sends motor impulses, causing muscle contraction which counters the stretching. It is through this mechanism that we are able to maintain an erect posture.

The tendons that attach muscle to bone contain sensory receptors that monitor muscle tension and help prevent tendon and muscle damage from excessive muscle contraction. Our skeletal joints also have sensory receptors that help the CNS monitor the position and movement of our limbs.

> Pain receptors are scattered throughout the skin and internal tissues. Stretch receptors wrap around specialized muscle cells and inform us of skeletal muscle stretching and tension.

Smelling and Tasting: Detecting Chemicals

Although not as capable as some other animals, humans can distinguish a wide variety of odors and tastes. Smell receptors are located in the nasal cavity, and taste receptors are concentrated on the tongue. They permit us to detect the chemicals in our food and in the air we breathe.

The sense of smell. Our sense of smell depends on **olfactory receptors** situated among the epithelial cells lining the roof of the nasal cavity (Figure 14.15). We each have about 10 million of these receptor cells, and they occupy an area of about 5 square centimeters (about 1 square inch).

One end of an olfactory receptor cell forms a cluster of eight or more *olfactory hairs* that project into the mucus layer of the nasal cavity. Olfactory hairs are cellular projections that contain receptor sites for airborne chemicals that have dissolved in mucus. When a chemical binds to a receptor site, an impulse is produced and transmitted to a structure called the *olfactory bulb,* where the receptor synapses

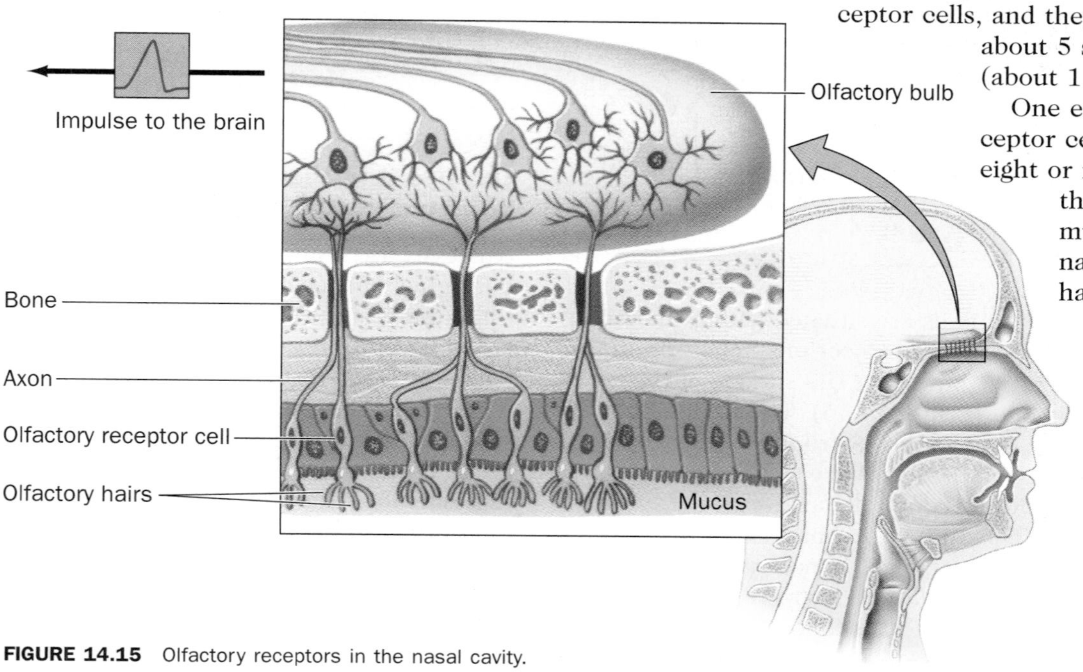
Impulse to the brain
Olfactory bulb
Bone
Axon
Olfactory receptor cell
Olfactory hairs
Mucus

FIGURE 14.15 Olfactory receptors in the nasal cavity.

with other neurons. Neurons from the olfactory bulb conduct the impulses to the brain for interpretation.

Although we can distinguish about 10,000 odors, we have only about 1000 kinds of receptor sites distributed among our 10 million olfactory cells. How a large number of odors can be detected by only 1000 different kinds of receptors sites is unknown. Perhaps each recognition site can bind with more than one type of odor molecule.

The sense of taste. Our sense of taste depends on **taste receptors** located in structures called **taste buds**. These taste buds are situated within numerous small projections called *papillae* that cover the surface of the tongue (Figure 14.16). There are about 5000 taste buds distributed along the tip, sides, and back of the tongue; few are found in the middle area.

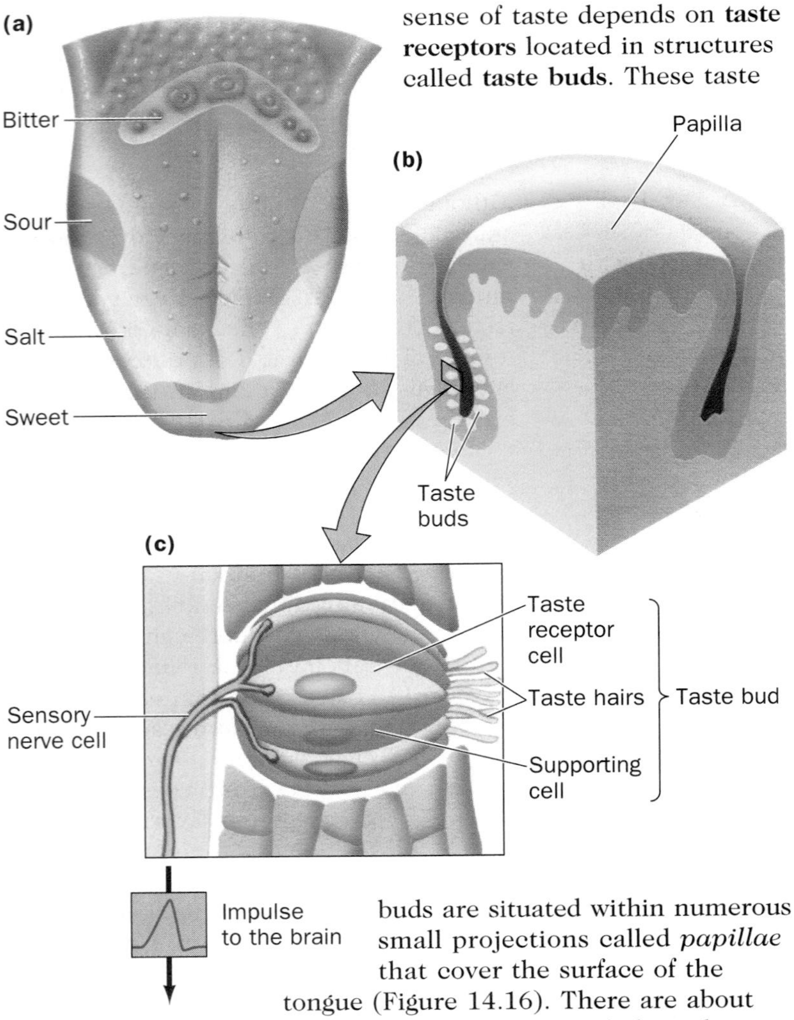
(a)
Bitter
Sour
Salt
Sweet
(b)
Papilla
Taste buds
(c)
Taste receptor cell
Taste hairs
Taste bud
Supporting cell
Sensory nerve cell
Impulse to the brain

FIGURE 14.16 Location and structure of taste buds. (a) Location of the four primary taste categories. (b) Taste buds within papilla. (c) Taste receptors within taste bud.

The exposed plasma membrane of each taste receptor cell consists of microvilli called *taste hairs*. These taste hairs contain receptor sites that bind to chemicals dissolved in saliva. This binding produces an impulse that is transmitted to sensory neurons at the base of each taste bud. Through synapses with additional neurons, such impulses reach the brain, where interpretation takes place and we have the experience of taste.

Our tastes are usually divided into four primary categories: sweet, sour, salty, and bitter (see Figure 14.16a). The front area of the tongue contains some taste receptors sensitive to sweet substances and others sensitive to salty substances. Receptors on each side of the tongue are more sensitive to sour substances, and those in the back are more sensitive to bitter substances. These categories are not absolute. Although a taste receptor cell may be more sensitive to one taste, it also may be capable of responding to another. These multiple sensitivities of our taste receptors partially explain why we are able to perceive variations of taste that go far beyond four simple categories.

Much of what we perceive as taste actually consists of our sense of smell. As food enters the mouth, easily vaporized molecules reach the olfactory receptors in the nasal cavity. The relationship between taste and smell explains why heated foods are often tastier than cold foods.

> The nasal cavity contains olfactory receptors that detect odor molecules dissolved in mucus. Gustatory receptors on the tongue detect molecules dissolved in saliva.

The Eye and Vision: Windows to the World

Our eyes are complex structures that house highly specialized receptor cells that are sensitive to light and are called **visual receptors**. In addition to sensing light, our eyes regulate the amount of light that enters and focus incoming light rays on the visual receptors so that clear images are formed. We will first describe the delicate structures of the eye and then discuss how they operate to produce an image.

The structure of the eye. Each eyeball is a roughly spherical structure about 2.5 centimeters (1 inch) in diameter that

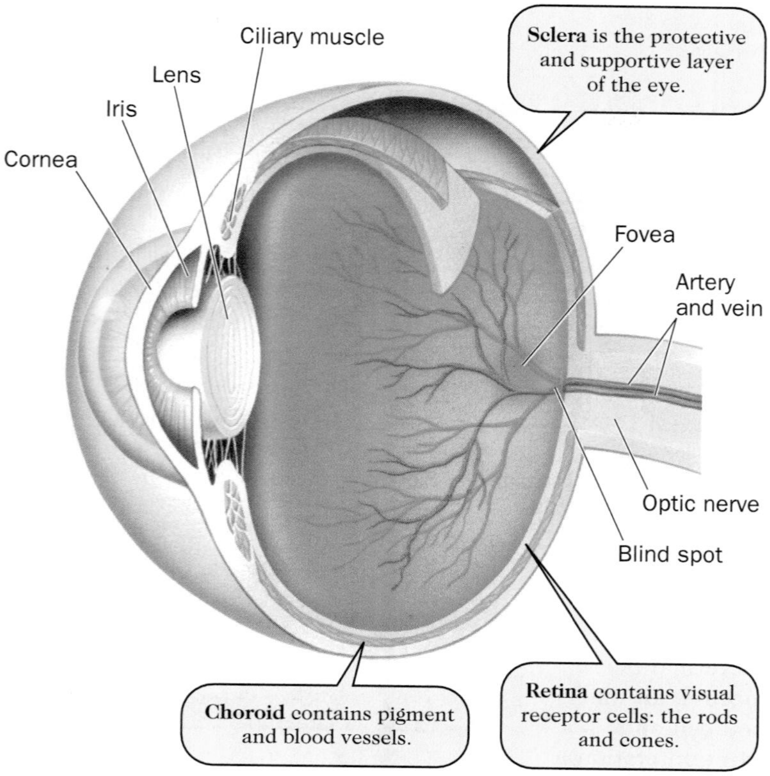
Ciliary muscle
Lens
Iris
Cornea
Sclera is the protective and supportive layer of the eye.
Fovea
Artery and vein
Optic nerve
Blind spot
Choroid contains pigment and blood vessels.
Retina contains visual receptor cells: the rods and cones.

FIGURE 14.17 Structures of the eye.

contains two fluid-filled chambers. Each eyeball is organized into three specialized layers: sclera, choroid, and retina (Figure 14.17).

The **sclera** (***skler**-ah*) is a tough layer of white connective tissue that helps maintain the shape of the eye, protects it from injury, and anchors the skeletal muscles that move the eye in the skull. The visible "white of the eye" is a portion of the sclera. At the front of the eye the sclera forms a transparent structure called the **cornea** (***core**-nee-ah*) through which light reaches the interior.

Just beneath the sclera is the **choroid** (***core**-oid*) layer. This layer is darkly pigmented to absorb scattered light and ensure that a clear image is formed on the retina. It also contains numerous blood vessels that nourish the tissues of the eye. Toward the front of the eye, the choroid changes into two structures: the iris and the ciliary body. The **iris** contains our eye color pigments and smooth muscles. These tiny smooth muscles change the diameter of an opening called the *pupil*, regulating the amount of light that reaches the interior and the retina. The smooth muscles of the iris are controlled by the autonomic division of the PNS.

The **ciliary body** contains smooth muscle and is connected to the lens by delicate ligaments. Contraction of these muscles alters the tension on the ligaments, which in turn change the shape of the lens. This permits images of both near and distant objects to be focused on the retina. The **lens** is a transparent and flexible structure made of protein. It divides the eye into two chambers that are filled with colorless transparent fluids. The chamber in front of the lens is filled with a watery fluid called the *aqueous humor* (***a**-kwee-us **hue**-mur*). The large chamber in back of the lens is filled with a gel-like fluid called the *vitreous humor* (***vit**-tree-us **hue**-mur*), which helps give shape to the eye.

The third and innermost layer is the **retina.** It contains the visual receptor cells. The retina has three layers of nerve cells: cones and rods, bipolar cells, and ganglion cells (Figure 14.18). The cones and rods are the visual receptor cells and are positioned next to the choroid layer. Impulses from the cone and rod cells are conducted by bipolar cells to the ganglionic cells, which transmit them out of the eye along the **optic nerve**.

Rods and cones: responding to light. The two types of visual receptor cells are named for their outer segments, which are either rod-shaped or cone-shaped (see Figure 14.18). **Rods** function in dim light, do not give rise to detailed images, and are not sensitive to color. **Cones** function in bright light, give rise to detailed images, and are sensitive to color. Within the rods and cones are the light-absorbing pigment molecules. When their special pigment molecules absorb light, they undergo a chemical change that generates an impulse. This impulse travels from the visual receptors to the optic nerve, which transmits it to the brain for interpretation.

In dim light the 120 million rod cells in each retina are activated. They contain a pigment known as **rhodopsin** (*row-**dop**-sin*) consisting of a protein called *opsin* and a derivative of vitamin A called *retinal*. Rod cells allow us to see in dim

FIGURE 14.18 The retina and its sensory nerve cells.

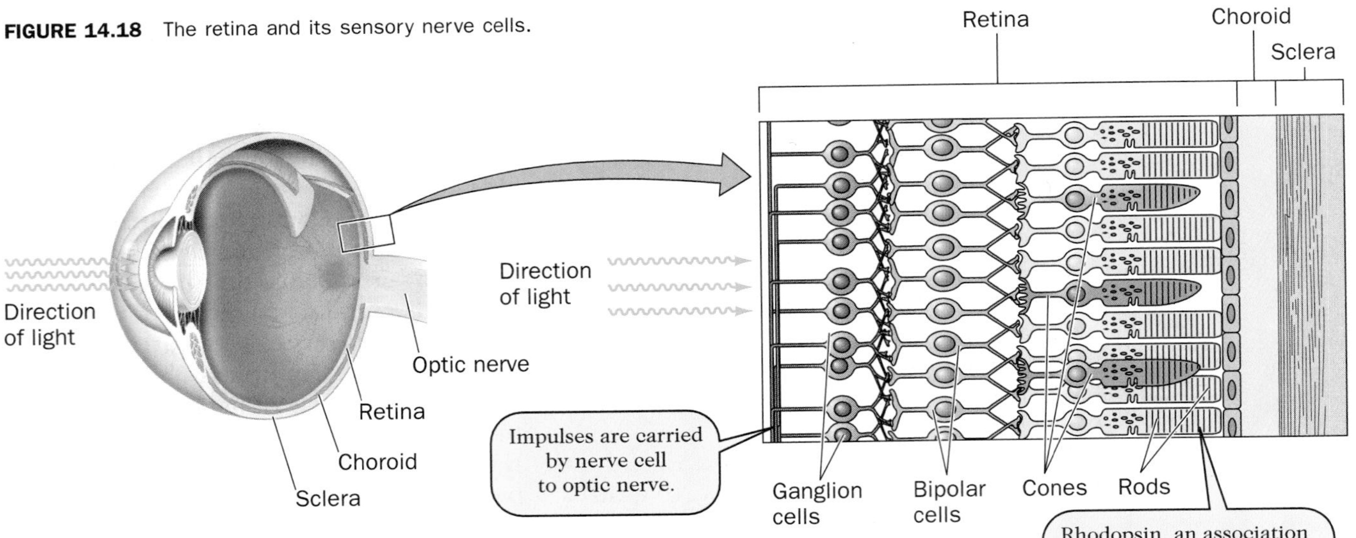
Retina
Choroid
Sclera
Direction of light
Direction of light
Optic nerve
Retina
Choroid
Sclera
Impulses are carried by nerve cell to optic nerve.
Ganglion cells
Bipolar cells
Cones
Rods

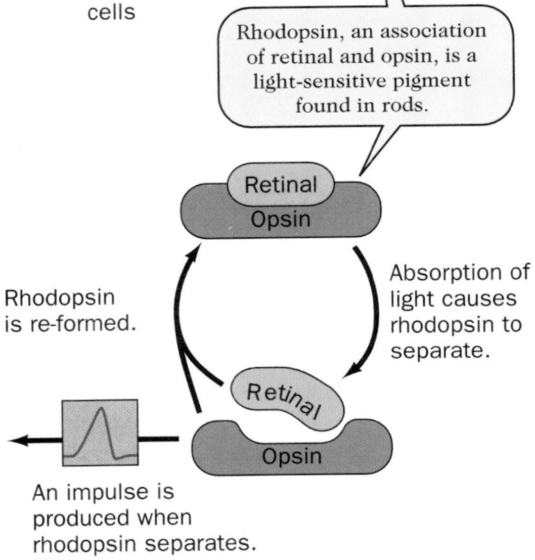
Rhodopsin, an association of retinal and opsin, is a light-sensitive pigment found in rods.
Retinal
Opsin
Absorption of light causes rhodopsin to separate.
Rhodopsin is re-formed.
Retinal
Opsin
An impulse is produced when rhodopsin separates.

light because rhodopsin is extremely sensitive to light. Even very small quantities of light cause rhodopsin to split into its two components: opsin and retinal (see Figure 14.18). When this occurs, an impulse is generated in the rod cell and transmitted to the brain for interpretation.

Seeing when the light is dim. In dim light, rhodopsin is re-formed as quickly as it is split apart. However, as the amount of light increases, rhodopsin is split apart faster than it can be re-formed. As a result, the amount of rhodopsin in the rod cells diminishes, and our rod cells produce fewer impulses. This explains why a person does not see well for a period of time after entering a dimly lighted room on a sunny day. In the bright sun, most of the rhodopsin is broken down, and until sufficient amounts are re-formed, rod cells are less capable of producing impulses. As more rhodopsin is re-formed, your ability to see in the dim light improves.

Seeing colors. In bright light, the 6 million cone cells in each retina are activated. There are three types of cone cells, and each contains a slightly different pigment: blue cones, green cones, and red cones. Like rhodopsin, each pigment consists of a protein and retinal. Retinal is the same for all photopigments, but each type of cone has a different type of opsin, making it sensitive to blue, red, or green light.

Our perception and discrimination of colors occur through the combined stimulation of the three different types of cone cells. When stimulated by different colors of light, the pigments break apart and an impulse is produced. Impulses from the different cones are transmitted to the brain, where they are interpreted as colors.

Sensory neurons from rods and cones converge and exit from the eye, forming the optic nerve. At the site where these neurons converge, there are no rods or cones, and an image focused there will not be "seen." This area is appropriately called the *blind spot* (see Figure 14.17).

Color vision deficiency. Color vision deficiency is an inherited condition in which a person is unable to distinguish certain colors. Color vision deficiency results when one or more types of cones lack their pigment or when the pigment occurs in low concentrations. Red or green deficiency is the most common condition. A person with red color blindness does not produce the red-absorbing pigment and so is unable to distinguish red colors.

Among whites of European origin, a green or red deficiency affects about 8 percent of males and 1 percent of females. The occurrence of this condition is lower among Asians and Native Americans and much lower among African-Americans.

Detailed images: visual acuity. Compared to rods, cones are less sensitive to light, and so it takes more light to stimulate them. However, when cones are stimulated, they produce more detailed images, and this is called visual acuity. A small area of the retina known as the *fovea centralis* (***foe-vee-ah cen-tral-is***) contains only cone cells and is the area of the retina that produces the sharpest vision (see Figure 14.17). When you look directly at an object, it is focused on the fovea. When the image is illuminated with sufficient light to stimulate the cones, you see a detailed image with maximum color differentiation.

Focusing light to form images. Light reflected from an object can be focused by the lens to form an image of that object on the retina. As shown in Figure 14.19a, the curvature of the cornea and the lens causes light rays from an object more than 20 feet away to converge on the retina, forming a clear image. However, if the object is closer than 20 feet, the light rays do not converge on the retina, and a clear image is not

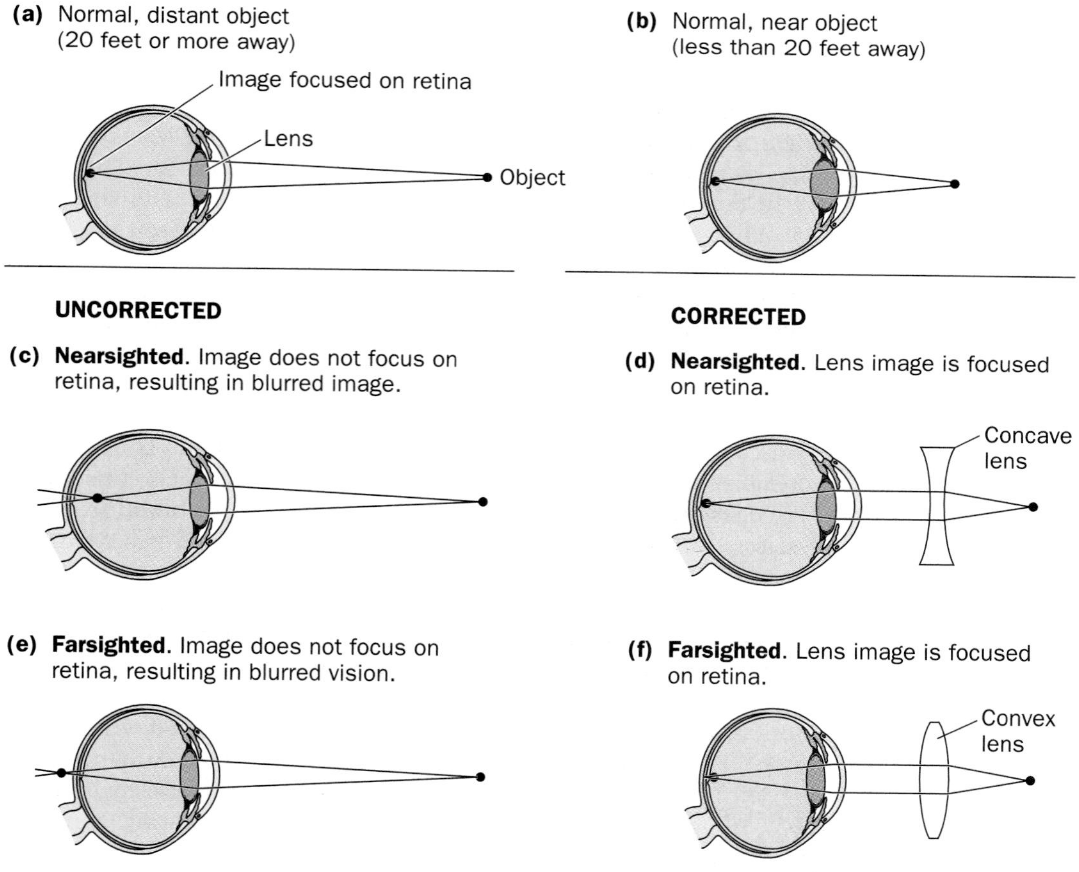

(a) Normal, distant object (20 feet or more away)
Image focused on retina
Lens
Object
(b) Normal, near object (less than 20 feet away)
UNCORRECTED
(c) Nearsighted. Image does not focus on retina, resulting in blurred image.
CORRECTED
(d) Nearsighted. Lens image is focused on retina.
Concave lens
(e) Farsighted. Image does not focus on retina, resulting in blurred vision.
(f) Farsighted. Lens image is focused on retina.
Convex lens

FIGURE 14.19 Normal and abnormal conditions of the eye. (a and b) Normal condition permitting focusing on distant or near objects. (c–f) Nearsighted and farsighted conditions and corrective lenses.

formed. To focus an image of a near object on the retina, the lens must change shape, becoming more round (Figure 14.19b). The lens shape is changed by contraction of muscles in the ciliary body. Long periods of close-up work such as reading can cause eyestrain because prolonged contraction fatigues the ciliary muscle cells.

To focus on objects 20 feet or more away, the ciliary smooth muscle relaxes and the lens assumes a more flattened shape. However, changes in the shape of the eyeball may make it impossible for the lens to focus light properly.

Changes in eyeball shape. When an eyeball is too short or too long, the image cannot be focused on the retina, resulting in nearsightedness or farsightedness (Figure 14.19c–f). *Nearsightedness* (myopia) occurs when the eyeball is too long. As a result, images of distant objects are focused in front of the retina, producing blurred images. As the term implies, nearsighted people can see near objects clearly but distant objects are blurred. The condition can be corrected by wearing concave-shaped lenses that bend the light rays so that they focus on the retina.

When the eyeball is too short, *farsightedness* (hyperopia) occurs. Images of near objects are focused behind the retina. Farsighted people can see distant objects clearly but near objects are blurred. This condition can be corrected by wearing convex-shaped lenses that bend light rays so that the image is focused on the retina.

Surface irregularities on the cornea scatter some of the light entering the eye. As a result, not all the light rays focus at the same point on the retina and the image appears blurred, causing what is known as *astigmatism*. The problem is corrected by wearing a special lens to compensate for corneal irregularities.

> The eye consists of three layers: sclera, choroid, and retina. The retina contains the visual receptor cells, the rods and cones. The lens focuses the image on the retina. Upon absorbing light, the rhodopsin in rods splits into retinal and opsin, generating a nerve impulse.

The Ear and Hearing: Detecting Vibrations

While you can shut your eyes, your ears are always open. Every waking moment they are undefended against the screaming of sirens or the song of a chickadee. Before examining the ear and how we hear, we need to consider the nature of sound. What is it that we hear?

Sounds are produced when molecules in the air are made to vibrate. For example, clapping the hands or blowing a horn produces pressure changes that alternately compress and decompress the air. A series of these pressure changes, called *sound waves*, are transmitted through the air in a fashion similar to the movement of ripples across a still pond. Sound waves cause sensitive structures of the ear to vibrate, and those vibrations are converted to nerve impulses by the receptors in the ear.

Each ear has three parts: the outer ear, middle ear, and inner ear. The outer ear and middle ear are designed to transfer sound waves from our environment to the inner ear. The inner ear has two functions. It receives sound waves and converts them into electrical impulses. It also maintains your balance.

Structures for detecting sound. The **outer ear** consists of a fleshy structure called the *pinna* (***pin**-ah*). From the pinna, a narrow canal called the *auditory canal* extends inward about 2.5 centimeters (about 1 inch). The pinna and auditory canal direct sound waves to the **tympanic membrane** (*tim-**pan**-ick*), or eardrum which is a partition between the outer ear and the middle ear (Figure 14.20). The sound waves cause the eardrum to vibrate, and those vibrations are transferred to the middle ear.

The **middle ear** is an air-filled chamber situated within the temporal bone of the skull. Attached to the eardrum and spanning the chamber of the middle ear are three tiny bones that are attached to each other: the *hammer* (malleus), the *anvil* (incus), and the *stirrup* (stapes). By their movement, these bones transmit mechanical vibrations from the eardrum to a small oval-shaped membrane of the inner ear called the *oval window*.

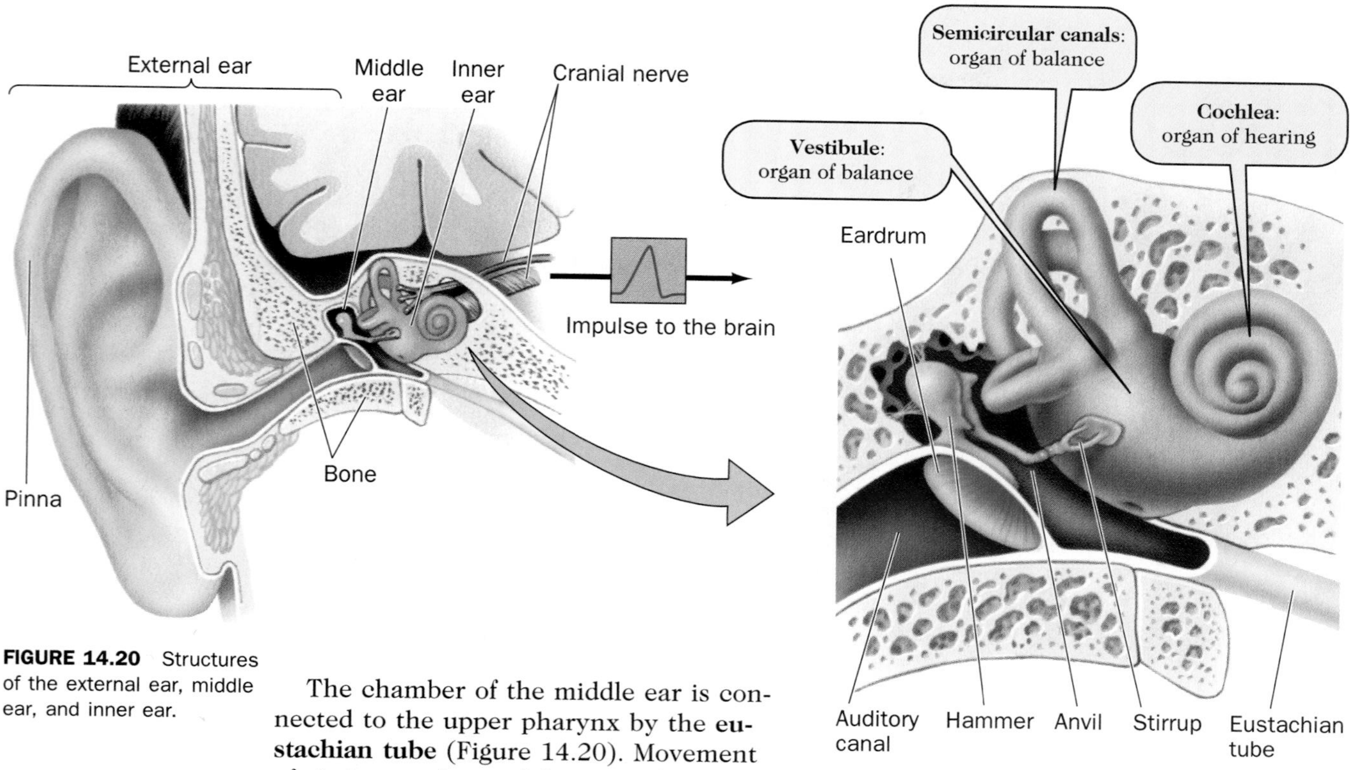
External ear
Middle ear
Inner ear
Cranial nerve
Impulse to the brain
Pinna
Bone
Semicircular canals: organ of balance
Cochlea: organ of hearing
Vestibule: organ of balance
Eardrum
Auditory canal
Hammer
Anvil
Stirrup
Eustachian tube

FIGURE 14.20 Structures of the external ear, middle ear, and inner ear.

The chamber of the middle ear is connected to the upper pharynx by the **eustachian tube** (Figure 14.20). Movement of air into and out of the eustachian tube permits the equalization of air pressure on either side of the eardrum during changes in altitude. If this is not done, unequal air pressures cause the eardrum to bulge, distorting sound. The "popping" of our ears during rapid altitude changes is due to this equalization of air pressure.

The **inner ear** structure associated with hearing is the **cochlea** (***coke**-lee-ah*), which is located within the temporal bone. The cochlea consists of three fluid-filled canals arranged in a spiral: the vestibular canal, tympanic canal, and cochlear duct (Figure 14.21).

The *vestibular canal* and the *tympanic canal* are continuous with each other in the upper cochlea or apex. Their fluid content receives the mechanical vibrations from the oval window and transmits them along the length of the canals to a thin membrane called the *round window*. The round window bulges in synchrony with the oval window to relieve pressure in the cochlea.

The **cochlear duct** is a separate fluid-filled compartment between the other two canals (Figure 14.21). Within the cochlear duct is the **organ of Corti**, whose highly specialized structures convert vibrations into nerve impulses. The *basilar membrane* of the cochlear duct supports some 16,000 specialized sensory cells called *hair cells*. These cells have modified microvilli that project into the fluid of the cochlear duct. Extending above and just touching the microvilli is a structure called the *tectorial membrane* (*tech-**tore**-ee-ul*). When vibrations cause the basilar membrane to move up and down, the microvilli are compressed against the tectorial membrane, bending the microvilli and causing an impulse in the hair cell. A summary of the events in hearing can be found in Figure 14.22. *Cranial nerve VIII* conducts the impulses to the brain for interpretation.

Perception of pitch. Pitch refers to the frequency or number of sound vibrations occurring in 1 second. The greater the frequency, the higher the pitch of the sound, and vice versa. Our ears are designed to distinguish frequencies from 20 to 20,000 vibrations per second. This is often described in *hertz (Hz) units* where 1 Hz is equal to 1 vibration per second (20,000 vibrations per second equals 20,000 Hz).

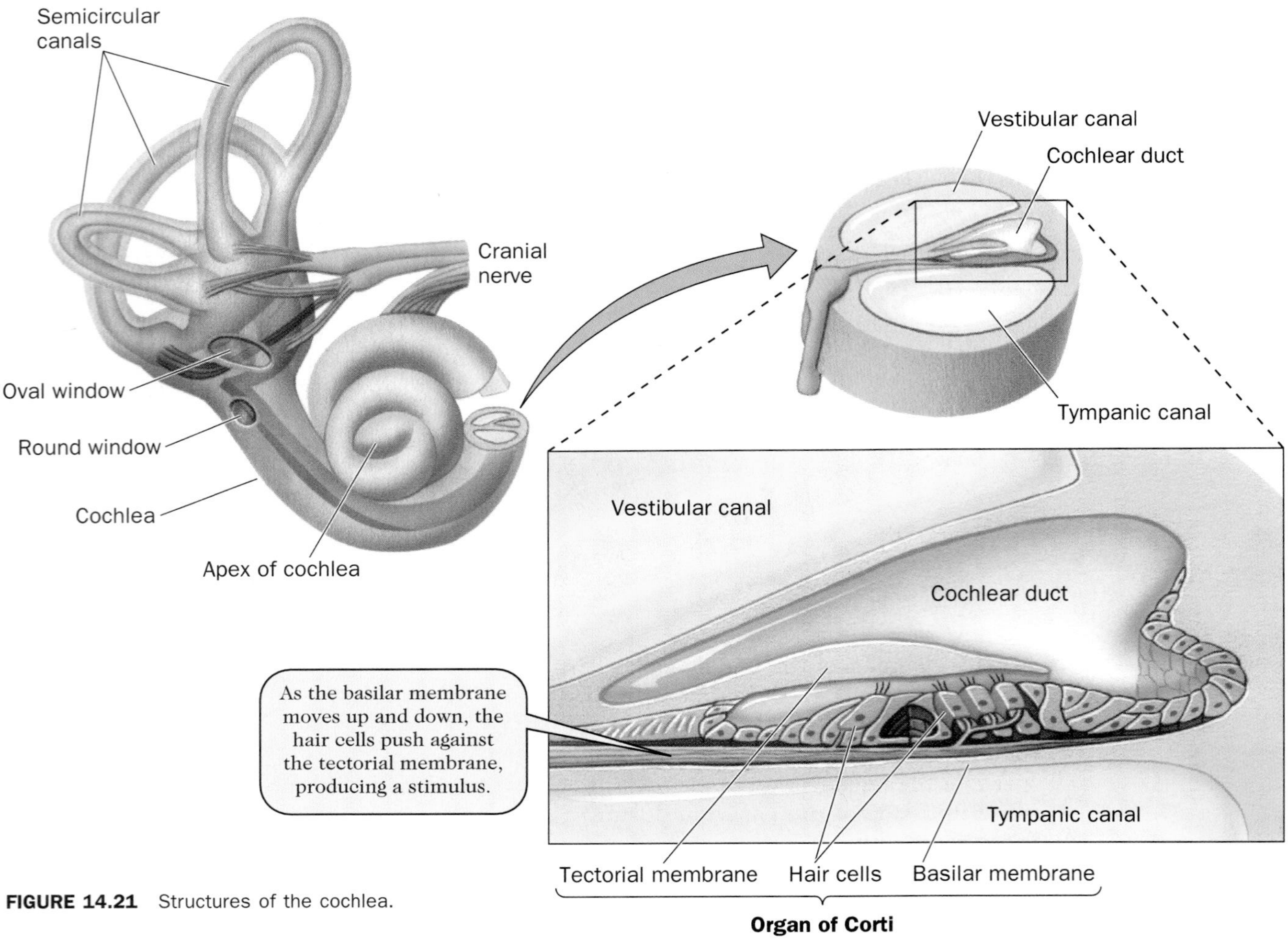
Semicircular canals
Cranial nerve
Oval window
Round window
Cochlea
Apex of cochlea
Vestibular canal
Cochlear duct
Tympanic canal
Vestibular canal
Cochlear duct
As the basilar membrane moves up and down, the hair cells push against the tectorial membrane, producing a stimulus.
Tympanic canal
Tectorial membrane
Hair cells
Basilar membrane
Organ of Corti

FIGURE 14.21 Structures of the cochlea.

How does the ear differentiate pitch? The basilar membrane progressively changes in width and flexibility from the base to the tip of the cochlear duct. At the base, the basilar membrane is narrow and rather stiff. Toward the tip, it becomes wider and more flexible. Different sound frequencies or pitches cause different sections of the basilar membrane to vibrate and stimulate hair cells. As a result, impulses are produced that travel to an area of the brain where they are interpreted as a sound with a particular pitch.

Perception of loudness. Our ears can detect the intensity or loudness of sound. Loudness is measured in decibels (db). The louder the sound, the larger the vibrations transmitted to the hair cells in the organ of Corti. As more hair cells are excited, more sensory neurons transmit impulses to the brain, which interprets them as loudness. Long unprotected exposure to sounds above 80 db can permanently damage or destroy hair cells and result in deafness (Table 14.4).

> Sound vibrations travel from the outer ear to the middle ear and inner ear. The middle ear is an air chamber that contains three bones linking the eardrum to the cochlea of the inner ear. The cochlea contains the organ of Corti, whose hair cells convert sound waves into impulses.

The Ear and Balance: Detecting Gravity and Motion

Hundreds of times each day our bodies move and then rest. Effectively making these changes requires a sense of position and balance. We must be aware of the po-

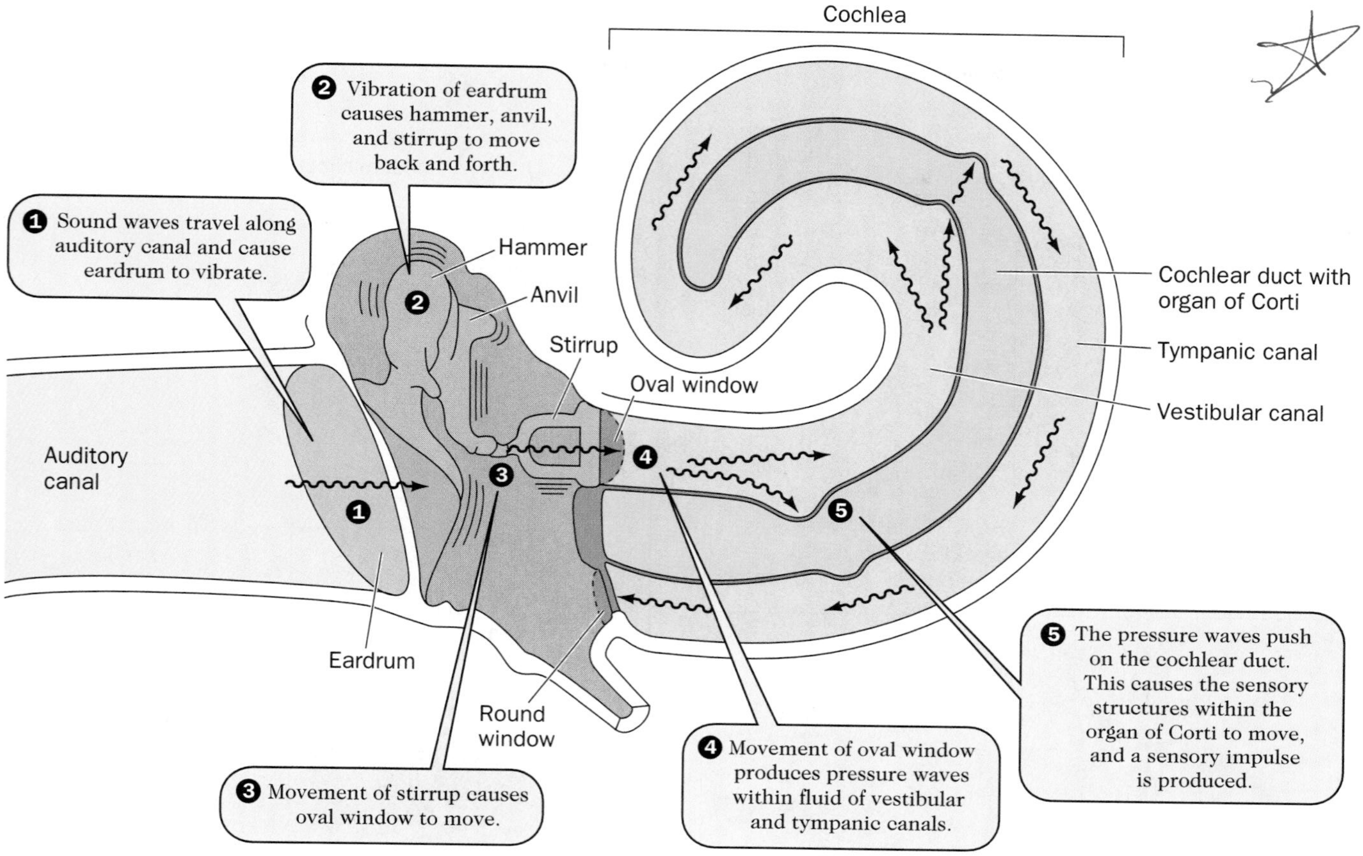
Cochlea
❷ Vibration of eardrum causes hammer, anvil, and stirrup to move back and forth.
❶ Sound waves travel along auditory canal and cause eardrum to vibrate.
Hammer
Anvil
Stirrup
Oval window
Cochlear duct with organ of Corti
Tympanic canal
Vestibular canal
Auditory canal
Eardrum
Round window
❸ Movement of stirrup causes oval window to move.
❹ Movement of oval window produces pressure waves within fluid of vestibular and tympanic canals.
❺ The pressure waves push on the cochlear duct. This causes the sensory structures within the organ of Corti to move, and a sensory impulse is produced.

FIGURE 14.22 Summary of how sound wave vibrations affect the organ of Corti. The cochlea has been partially uncoiled.

TABLE 14.4 Decibel Levels Associated with Common Sounds

Possible Hearing Damage	Type of Sound	Decibels
Harmful to hearing	Jet takeoff	150
	Shotgun blast	140
	Jackhammer	120
Possible hearing loss	Live rock band	110
	Car horn nearby	100
	Power mower	100
No hearing loss	Vacuum cleaner	70
	Normal conversation	60
	Country quiet	50
	Watch ticking	20
	Leaves rustling	10
	Absolute silence	0

sition of the head and limbs in order to activate the skeletal muscles that maintain our balance when we move. There are two types of balance (also called equilibrium): static and dynamic.

Static balance. The sensory apparatus for static balance is found in two fluid filled chambers of the inner ear called the **utricle** (***you***-*trik-el*) and **saccule** (***sack***- *yule*) (Figure 14.23). They detect the position of the head with respect to gravity when the body is stationary; they also detect linear movements (acceleration and deceleration) such as those experienced in a moving vehicle. The utricle and saccule contain thousands of sensory *hair cells* with elongated microvilli that project into a jellylike material. Resting on the upper surface of the jellylike material are many tiny crystals of calcium carbonate called *otoliths* (***oh***-*toe-liths*).

When the head is tilted, the weight of the otoliths causes the jellylike mass to shift position, bending the microvilli and

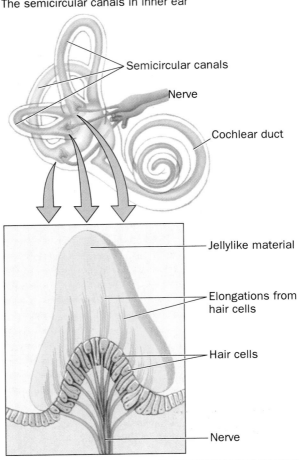

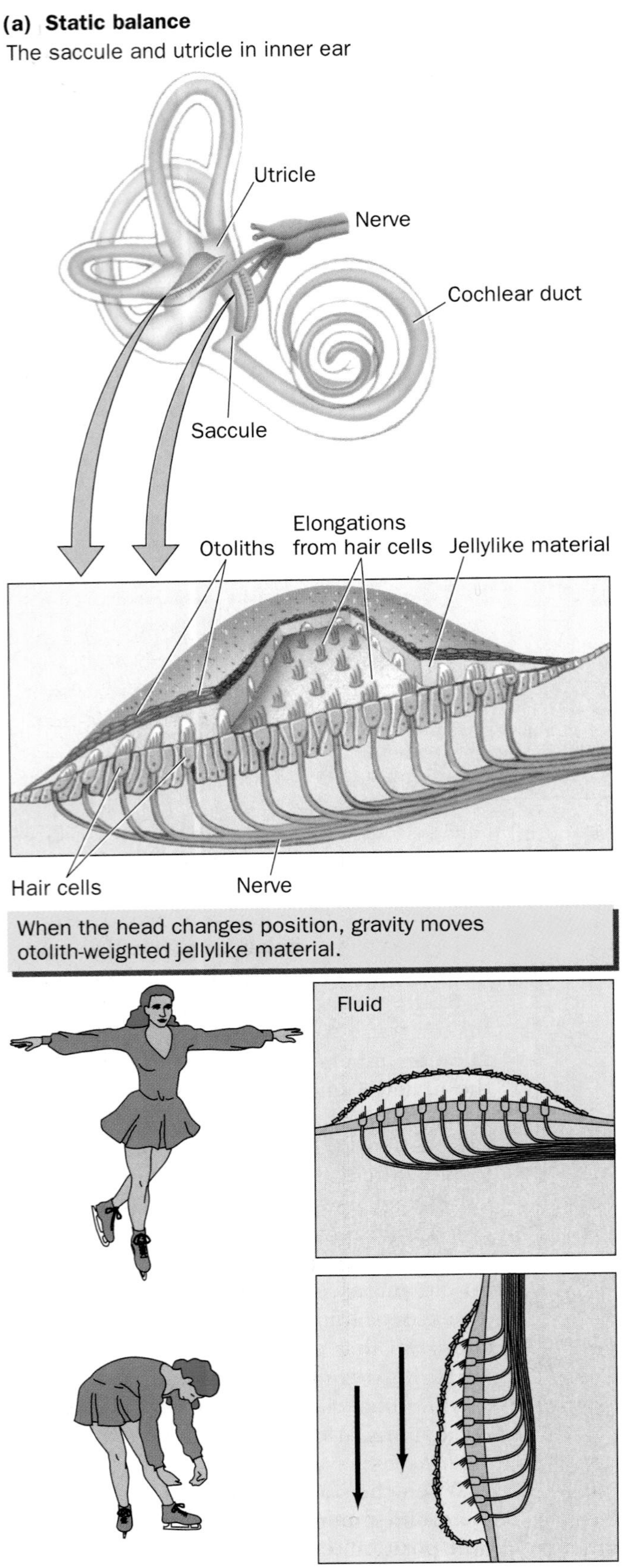
(a) Static balance
The saccule and utricle in inner ear
Utricle
Nerve
Cochlear duct
Saccule
Otoliths
Elongations from hair cells
Jellylike material
Hair cells
Nerve
When the head changes position, gravity moves otolith-weighted jellylike material.
Fluid
(b) Dynamic balance
The semicircular canals in inner ear
Semicircular canals
Nerve
Cochlear duct
Jellylike material
Elongations from hair cells
Hair cells
Nerve
When the head and body move, the jellylike material also moves.
Fluid

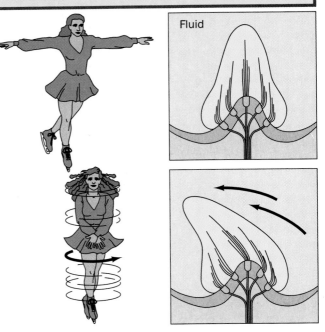

FIGURE 14.23 Sensory structures for static and dynamic balance.

causing an impulse (Figure 14.23). The impulse is transmitted along cranial nerve VIII to the brain, where the signals are interpreted. During acceleration and deceleration, the gelatinous masses move, initiating impulses that permit the brain to sense the changes in speed.

Dynamic balance. The sensory apparatus for dynamic balance is located in three bony fluid-filled chambers of the inner ear called the **semicircular canals** (Figure 14.23). These canals detect rotational or angular position changes of the head when the body is moving. The three canals are oriented at right angles to one another in three different planes. At the base of each canal is a mass of sensory *hair cells* with elongated microvilli projecting into a jellylike material.

Rotational or angular movements cause the fluid in the semicircular canals to move. This movement pushes on the jellylike mass, bending the microvilli and producing an impulse (Figure 14.23). Cranial nerve VIII conducts the impulses to the brain, where signals from all three semicircular canals are integrated. Impulses from the brain to the skeletal muscles permit corrective actions to maintain or adjust body position and balance.

> The structures for maintaining static and dynamic balance are in the inner ear. Static balance detects the position of the head when the body is stationary. Dynamic balance detects rotational and angular position changes of the head.

Diseases and Disorders

There are many serious disorders of the nervous system. Because it regulates other body systems, dysfunctions of the nervous system often cause problems in other organs. For example, an infection of spinal cord neurons can debilitate certain skeletal muscles.

Besides its susceptibility to viral and bacterial infections, the nervous system is particularly sensitive to toxic materials such as lead, alcohol, and components of tobacco smoke. Deficiencies in vitamins B_1 (thiamine) and B_{12} also have particularly severe consequences for the nervous system. In what follows, we will briefly describe selected disorders and diseases of the PNS, showing how they alter its normal functions as well as those of other body systems.

Multiple Sclerosis: A Myelin Disorder

Multiple sclerosis (MS) is a progressive disorder that destroys the myelin sheath around neurons in the brain and spinal cord. The neurons continue to function, but without the myelin covering, impulse conduction is slowed and short circuits occur. As a result, motor neurons of the PNS do not receive proper signals, resulting in muscle weakness, uncoordinated movements, and imbalance. The internal organs controlled by the autonomic division of the PNS can also be affected, resulting in problems associated with vision, peristalsis of the stomach and intestines, and urinary bladder control.

The cause of MS is unknown, but one hypothesis states that a viral infection leads to an autoimmune response (Chapter 13). The virus may alter the membrane of the myelin sheath so that the immune system attacks the sheath as if it were foreign, destroying oligodendrocytes and impairing nerve transmission.

Polio: A Spinal Cord Disease

Any disorder that disturbs the integrity of the spinal cord will ultimately interfere with the processing of nerve signals. The polio virus infects and destroys motor neurons in the gray matter of the spinal cord. Since these motor neurons transmit impulses to skeletal muscles, any reduction in their numbers weakens or paralyzes the skeletal muscles they control. The loss of motor function is permanent because nerve cells of the CNS do not regenerate. However, patients do not lose the sense of feeling in the area of paralysis because the virus does not destroy sensory neurons.

The polio virus is transmitted between humans primarily by contaminated food and water. The virus enters the digestive tract and travels by way of the bloodstream to the spinal cord. Vaccines are available that are nearly 100 percent effective in preventing polio.

Disorders of the Eye

The eye is a complex structure, and its many disorders result from this complexity. Earlier in this chapter we discussed changes in the shape of the eyeball. Here we will focus on excess eye fluid, loss of lens transparency, and loss of lens flexibility.

Excessive fluid: glaucoma. The aqueous humor of the anterior eye chamber is continuously produced and reabsorbed. However, if too much fluid is produced or not enough is reabsorbed, fluid pressure in the eye increases, compressing the blood vessels that supply the retina and its neurons. As a result, neurons of the retina die. The condition is called *glaucoma (glaw-**koh**-mah),* and it is a common cause of blindness in the elderly. Glaucoma can be treated with medication or surgery.

Loss of lens transparency. A *cataract* is a condition in which the lens has lost its transparency. If it is severe enough, a cataract results in blindness because it prevents light rays from entering the eye. The problem can be corrected by surgically removing the affected lens and inserting an artificial lens. Cataracts commonly develop later in life.

Loss of lens flexibility. The lens slowly loses its flexibility with age. At some point enough flexibility is lost that the shape of the lens cannot be adjusted for near vision. This condition is called *presbyopia (prez-bee-**oh**-pea-ah).* Whereas a younger adult can focus on reading material only 10 centimeters (4 inches) away, a 50-year-old may need to hold the same reading material at arm's length to focus. Presbyopia can be corrected with reading glasses or with bifocals for those already wearing glasses.

Disorders of the Ear

There are many disorders of the ear. We'll briefly consider infections and hearing loss. Balance disorders can occur because of problems with the sensory structures of the vestibule and semicircular canals.

Middle ear infections (otitis media). *Otitis media* is a more common problem among children than among adults because a child's eustachian tubes are shorter and more horizontally positioned. The condition occurs when infectious bacteria migrate from the throat, up the eustachian tube, and into the middle ear chamber. As a result of the infection, large amounts of fluid and pus accumulate in the chamber, causing the eardrum to bulge outward. This is painful and is interpreted as an earache. Prolonged infections can damage the middle ear bones, resulting in hearing impairment.

Hearing loss. Hearing disorders and deafness can have several causes. Damage to the eardrum or middle ear bones results in *conduction deafness*. As a result, sound vibrations are not effectively conducted to the inner ear. For example, arthritis of the middle ear bones may impair their operation. Hearing aides are tiny electrical devices that amplify sound and help overcome some deficiencies of the eardrum or middle ear bones.

Damage to the hair cell receptors within the cochlea of the inner ear results in *nerve deafness*. Cochlea hair cells can be permanently damaged from prolonged exposure to noises 80 db or above (see Table 14.4). Since destroyed hair cells are not replaced, continued losses eventually lead to hearing impairment. Prevention includes avoiding exposure to sounds above 80 db or using ear protection devices such as earplugs and ear covers.

To partially remedy severe nerve deafness, *cochlear implants* have been developed. A cochlear implant consists of a tiny microprocessor that converts sound vibrations into electrical signals. The microprocessor is implanted under the skin behind the pinna, and its electrodes are inserted along the organ of Corti to carry electrical signals to the hair cells. Cochlear implants are still rather new and produce crude sounds, but they allow nerve deaf individuals to hear.

Life Span Changes

By the second month of development, the general organization of both the CNS and PNS has been established. During the final 6 months of fetal development, nerve cells of the PNS grow and extend into developing tissues. Nerve synapses are formed, broken, and re-formed until the appropriate connections are established with muscles, glands, and other nerve cells. After the establishment of permanent synapses, the process of myelination begins.

In the 2 years after birth, the number of nerve cells increases, particularly in the brain. This is why proper nutrition during prenatal development and early infancy is so important in preventing mental deficiencies (Chapter 7). Myelination also continues during infancy and childhood. The improved coordination skills that occur during early childhood are a result of both brain development and continued myelination of nerve cells in the CNS and PNS.

Less dramatic changes of the nervous system, particularly those occurring in the brain, take place throughout our lives. Any time you learn something new, a change has taken place somewhere in your brain. Learning can continue throughout life. All things considered, your nervous system may be your most durable and dependable organ system if it is properly cared for.

Late in the life span, a variety of changes occur in the nervous system. As with other body systems, some neurons die. Like muscle cells, mature nerve cells do not undergo further cell division, and therefore, any cells that are lost are not replaced. The long-term consequence of such permanent cell loss is most pronounced in the brain.

As we age, changes also occur in the PNS. There is a gradual loss of myelin, and this slows impulse transmission. There is also a gradual decrease in the synthesis and release of neurotransmitters as well as a decrease in the number of the receptor sites to which they attach in the postsynaptic membrane. Such losses may contribute to glandular dysfunction, muscle weakness, impaired coordination, and poor reflexes among older adults. However, the precise significance of these changes varies greatly among individuals.

A loss of neurotransmitter receptors from the autonomic division has a wide range of homeostatic effects. For example, older adults often have difficulty regulating body temperature in cool and very warm environments. Also, there is a decreased ability to regulate heart rate and blood pressure when one stands up from a seated position or during exercise.

Life span changes also affect our senses. Earlier in this chapter we mentioned the formation of cataracts of the lenses, increased fluid pressure in the eye, and decreased flexibility of the lens. Loss of muscle cells from the iris reduces our control of the amount of light entering the eye. Loss of visual receptor cells may reduce both night vision and visual acuity.

Hearing loss (conduction deafness) occurs when the eardrum and the middle ear bones decline in their ability to conduct mechanical vibration to the inner ear. The eardrum becomes less rigid, while the ear bones become more rigid. Within the cochlea, hair cells and sensory neurons can be damaged, reducing the impulses that reach the brain (nerve deafness).

Chapter Summary

The Nervous System: An Overview

- The central nervous system (CNS) includes the brain and spinal cord.
- The peripheral nervous system (PNS) includes sensory and motor nerve cells that communicate between the CNS and the tissues. The PNS is divided into the somatic and autonomic divisions.
- Nerves are bundles of nerve cell extensions. Sensory nerves conduct impulses to the CNS. Motor nerves conduct impulses away from the CNS. Mixed nerves contain both sensory and motor nerve cells.

Cells of the Nervous System

- Nerve cells, or neurons, transmit impulses. Neuroglia cells support and protect neurons.
- A neuron has three components: cell body, dendrites, and axon.
- The myelin sheath of PNS nerves is formed by Schwann cells. In the CNS it is formed by oligodendrocytes. Myelinated neurons have nodes of Ranvier that speed impulse transmission.

The Nerve Impulse: An Electrical Signal

- The resting membrane potential of a neuron results from an uneven distribution of ions across its plasma membrane. The outside of the cell is more positively charged because of its many sodium ions; the inside is more negatively charged because of its many negatively charged proteins.
- An action potential (or impulse) develops when a stimulus alters the resting membrane potential.
- Impulse transmission between neurons occurs across a gap called a synapse. Chemical signals called neurotransmitters carry signals across the synapse.

The Peripheral Nervous System

- The meninges and cerebrospinal fluid surround and protect the brain and spinal cord.
- The somatic division of the PNS communicates with the skeletal muscles, the skin, and their sensory receptors. A reflex arc consists of: sensory receptor, sensory neuron, association neuron, motor neuron, and effector.
- The autonomic division of the PNS consists of motor nerves that transmit impulses to internal organs. There are two components: sympathetic and parasympathetic.

The Senses: Experiencing the Environment

- Olfactory receptors in the upper nasal cavity are stimulated by chemicals in the air to produce smell.
- Taste receptors in the taste buds of the tongue are stimulated by chemicals to produce taste.
- The eye consists of three layers: sclera, choroid, and retina. Rod and cone cells in the retina convert light into impulses.
- The outer ear, middle ear, and inner ear structures produce hearing. The organ of Corti in the inner ear converts sound vibrations into impulses.
- Static balance detection occurs in the utricle and saccule of the inner ear; dynamic balance detection occurs in the semicircular canal of the inner ear.

Diseases and Disorders

- Multiple sclerosis destroys the myelin sheath of CNS nerves.
- Polio is caused by a virus which destroys motor nerves in the spinal cord.
- Glaucoma causes blindness because of too much fluid pressure inside the eye.
- Cataracts result from reduced lens transparency.
- Presbyopia results from reduced lens flexibility.
- Otitis media is an infection of the middle ear chamber.
- There are two types of deafness: conduction deafness and nerve deafness.

Life Span Changes

- During the first year or two after birth, nerve cell numbers continue to increase.
- During the life span, some nervous system functions may be affected because of nerve cell losses. Learning continues throughout life.

Selected Key Terms

action potential (p. 367)
autonomic division (p. 375)
axon (p. 364)
cell body (p. 363)
central nervous system (p. 361)
cones (p. 381)
dendrite (p. 364)
dynamic balance (p. 389)
effector (p. 361)
motor nerve (p. 362)
myelin sheath (p. 364)
neurotransmitter (p. 369)
organ of Corti (p. 385)
peripheral nervous system (p. 361, 372)
resting potential (p. 366)
rods (p. 381)
sensory nerve (p. 362)
sensory receptor (p. 361)
static balance (p. 387)
synapse (p. 369)

Review Activities

1. Distinguish between the structures and functions of the cell body, dendrite, and axon.
2. Describe how the myelin sheath is formed around PNS neurons.
3. Explain how the ion imbalance across a nerve cell membrane develops. What changes must occur when an action potential is generated?
4. Using acetylcholine as an example, describe the following events: neurotransmitter release, movement across synapse, stimulation of next neuron, and termination of activity.
5. Describe the pathway of a reflex arc and explain how a withdrawal reflex operates.
6. Distinguish between somatic division and autonomic division functions.
7. How do the sympathetic and parasympathetic divisions differ structurally and functionally?
8. Describe how taste and smell are interrelated.

9. Trace the path of light from the cornea to the retina.
10. Explain how the ciliary body, ligaments, and lens function to focus an image of a near object on the retina.
11. Explain why it takes time to see well after entering a dimly lighted room.
12. Describe the function of rods and cones.
13. Trace the pathway of sound waves from the outer ear, through the middle ear, and to the cochlea of the inner ear. Explain how the organ of Corti converts sound vibration into electrical impulses.
14. Explain how the ear distinguishes between pitch and loudness.
15. Distinguish between static and dynamic balance.

Self-Quiz

Matching Exercise

___ 1. The division of the PNS that stimulates smooth muscle, heart muscle, and glands
___ 2. The nerve cell extension that conducts impulses away from the cell body
___ 3. Gaps between Schwann cells that are important in impulse transmission
___ 4. The ion most concentrated outside a resting neuron
___ 5. The technical term for an impulse
___ 6. The neurotransmitter from motor neurons that stimulates skeletal muscle
___ 7. A spinal nerve root in which only sensory neurons are found
___ 8. The subdivision of the autonomic division in which motor nerves originate from the brainstem and sacral region of the spinal cord
___ 9. The receptors in skeletal muscles that inform us about the amount of tension
___ 10. The visual receptor cells that detect color

A. Parasympathetic
B. Action potential
C. Autonomic
D. Axon
E. Cones
F. Sodium
G. Nodes of Ranvier
H. Stretch receptors
I. Acetylcholine
J. Dorsal

Answers to Self-Quiz

1. C; 2. D; 3. G; 4. F; 5. B; 6. I; 7. J; 8. A; 9. H; 10. E

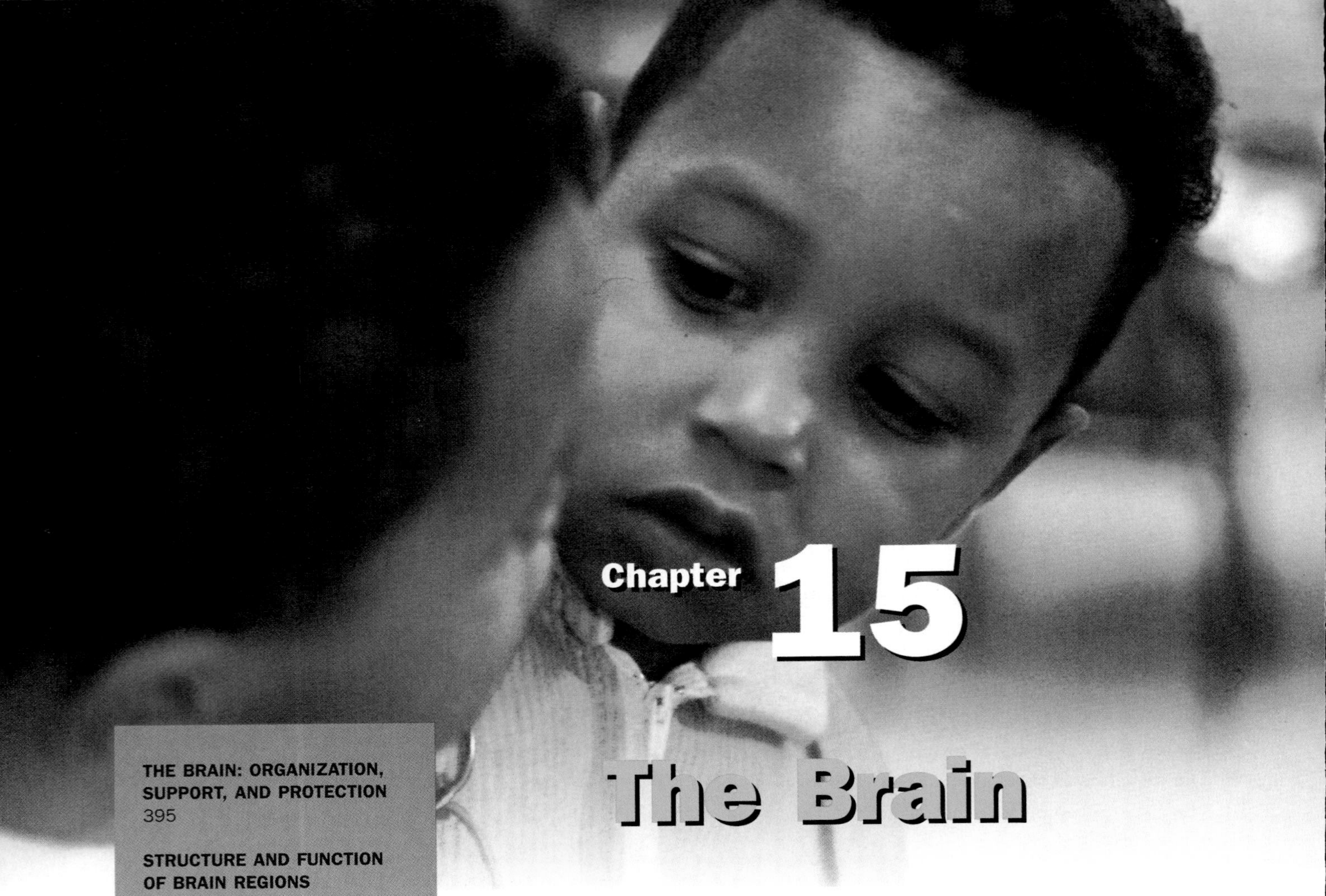

Chapter 15
The Brain

"*I* want to know myself," I said, gazing at the familiar image in the mirror. The image answered, "Which self: physical, mental, or spiritual?"

These simple sentences address the mysteries of the brain and the body, the mind and the soul, person and personality. The brain is the seat of desire ("I want"). It is the location for learning, memory, and knowledge ("to know"). It is the center for conscious awareness of our body, its image, and its environment ("myself"). The brain describes, distinguishes, compares, evaluates, reflects, and dreams. These mental capacities are often referred to as the "higher functions" of the brain, but the brain also has other functions.

Little goes on in the body without the direct or indirect involvement of the brain. By processing signals from our sensory receptors, the brain coordinates and controls the "voluntary actions" of our skeletal muscles.

The brain is also involved in coordinating and regulating the normal functions of our internal organs, and in doing so it is essential to body homeostasis. Breathing and digesting, coughing and sneezing, heartbeat and blood pressure—all are controlled by the brain without our being aware of its activity. These functions are often said to be involuntary, unconscious, or automatic.

In this chapter, we will examine the brain and describe how it is organized, covered, protected, and bathed. We will explore its different parts, the human functions associated with them, and the nature of "brain waves," sleep, dreaming, learning, and memory. In addition, we will examine differences between male and female brains and the possible role of the brain in gender identity. Of course, we will close with diseases, disorders, and life span changes of the brain.

The Brain: Organization, Support, and Protection

Soft and wrinkled, an adult brain weighs about 1300 grams (about 3 pounds), but this varies considerably among individuals. Male brains are usually 10 to 17 percent larger than female brains of the same age, probably because of the larger average body size of the male. However, male and female brains have the same total number of neurons, although specific brain regions may differ in this regard. Significantly, there appears to be no direct relationship between intelligence and brain size whether comparisons are based on gender or race.

The brain has billions of neurons (estimates range from 25 billion to 100 billion). Since each neuron may have up to 10,000 synapses, this means that there are trillions of interconnections among neurons. The brain is certainly the body's most complex organ.

For the convenience of this discussion, the human brain can be divided into three basic regions: the hindbrain, the midbrain, and the forebrain (Figure 15.1). It

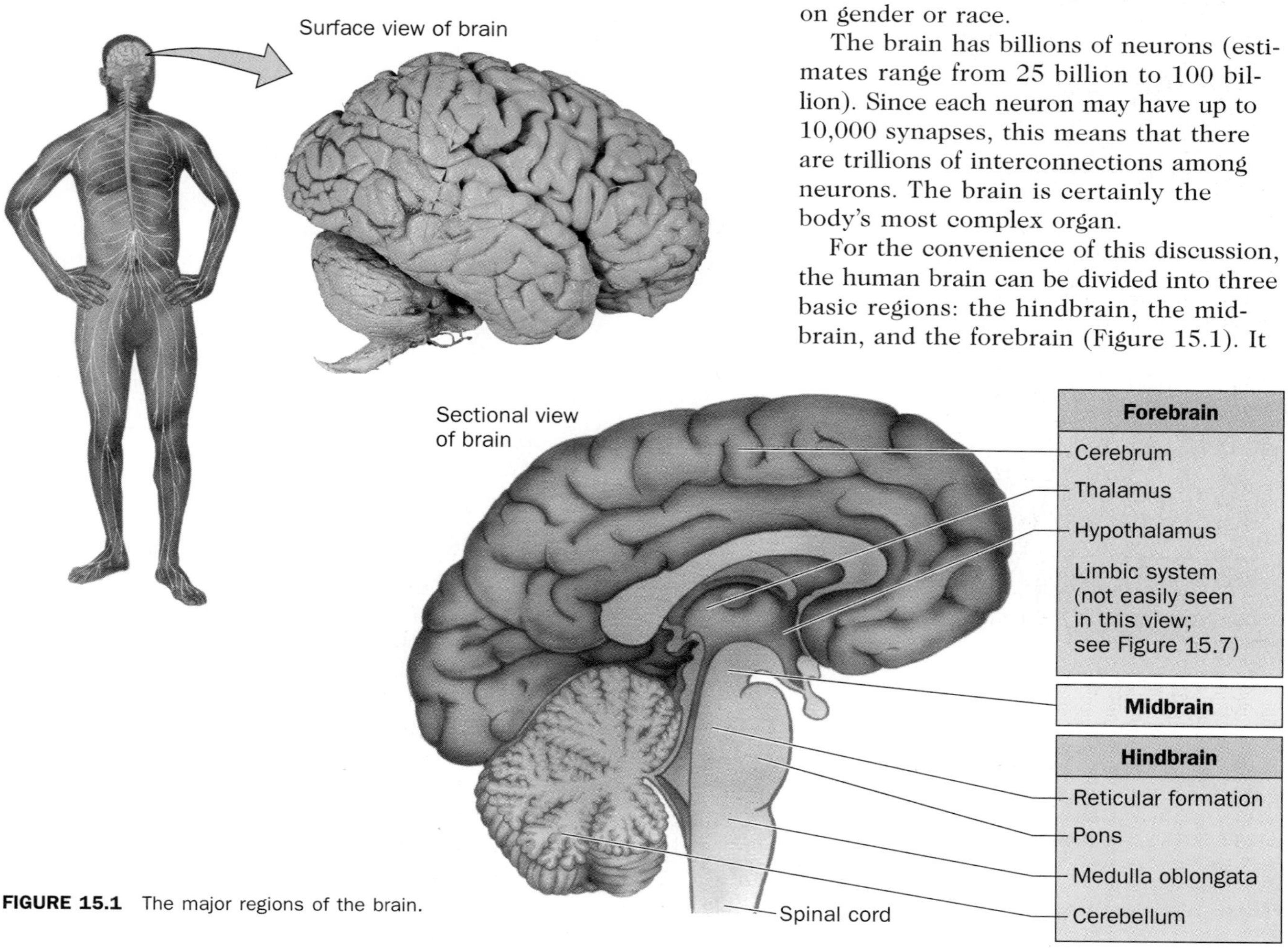
Surface view of brain
Sectional view of brain
Forebrain
Cerebrum
Thalamus
Hypothalamus
Limbic system (not easily seen in this view; see Figure 15.7)
Midbrain
Hindbrain
Reticular formation
Pons
Medulla oblongata
Cerebellum
Spinal cord

FIGURE 15.1 The major regions of the brain.

is not unreasonable to say that we actually have three brains. While each of these areas has specialized functions, they are in communication with each other and usually operate as a single integrated unit. The hindbrain, midbrain, and forebrain are connected to the entire body by the spinal cord, the spinal nerves, and the cranial nerves.

The 12 pairs of **cranial nerves** connect the brain directly to the muscles, glands, and sensory receptors of the head and neck (Figure 15.2). One pair—the *vagus nerves*—has branches that also connect with the heart, lungs, and digestive tract. Thirty-one pairs of **spinal nerves** connect the spinal cord to the muscles, glands, sensory receptors, and organs of the trunk and appendages. Ascending (sensory) and descending (motor) columns of neurons in the **spinal cord** connect the brain to the spinal nerves.

Protective Coverings and Cerebrospinal Fluid

An adult brain is nearly 75 percent water and has a gelatinlike consistency. Without adequate support and protection, it would slump to the bottom of the cranium and be easily damaged by normal body movements, bumps, and falls. The hard, flat

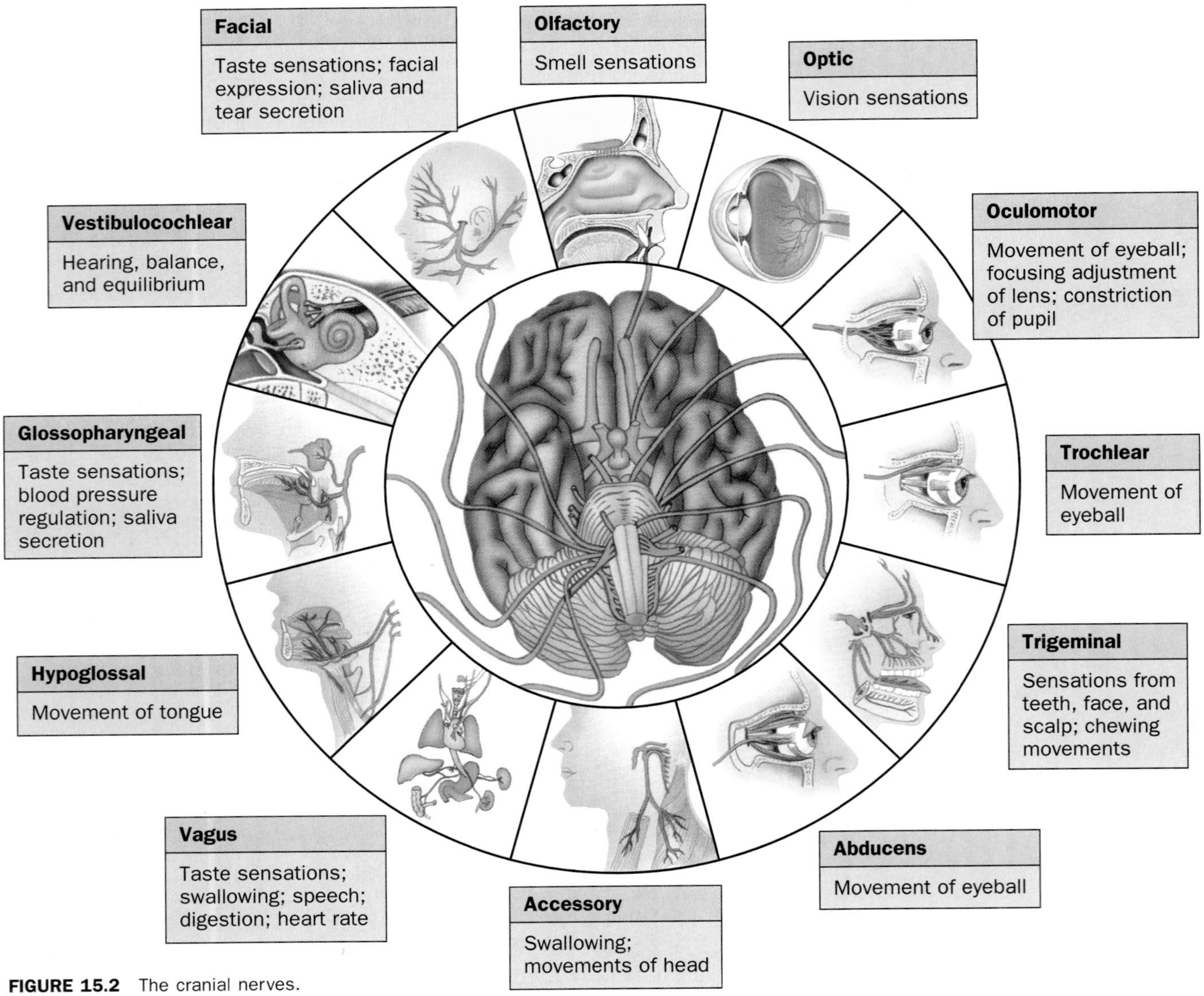

Facial
Taste sensations; facial expression; saliva and tear secretion
Olfactory
Smell sensations
Optic
Vision sensations
Oculomotor
Movement of eyeball; focusing adjustment of lens; constriction of pupil
Trochlear
Movement of eyeball
Trigeminal
Sensations from teeth, face, and scalp; chewing movements
Abducens
Movement of eyeball
Accessory
Swallowing; movements of head
Vagus
Taste sensations; swallowing; speech; digestion; heart rate
Hypoglossal
Movement of tongue
Glossopharyngeal
Taste sensations; blood pressure regulation; saliva secretion
Vestibulocochlear
Hearing, balance, and equilibrium

FIGURE 15.2 The cranial nerves.

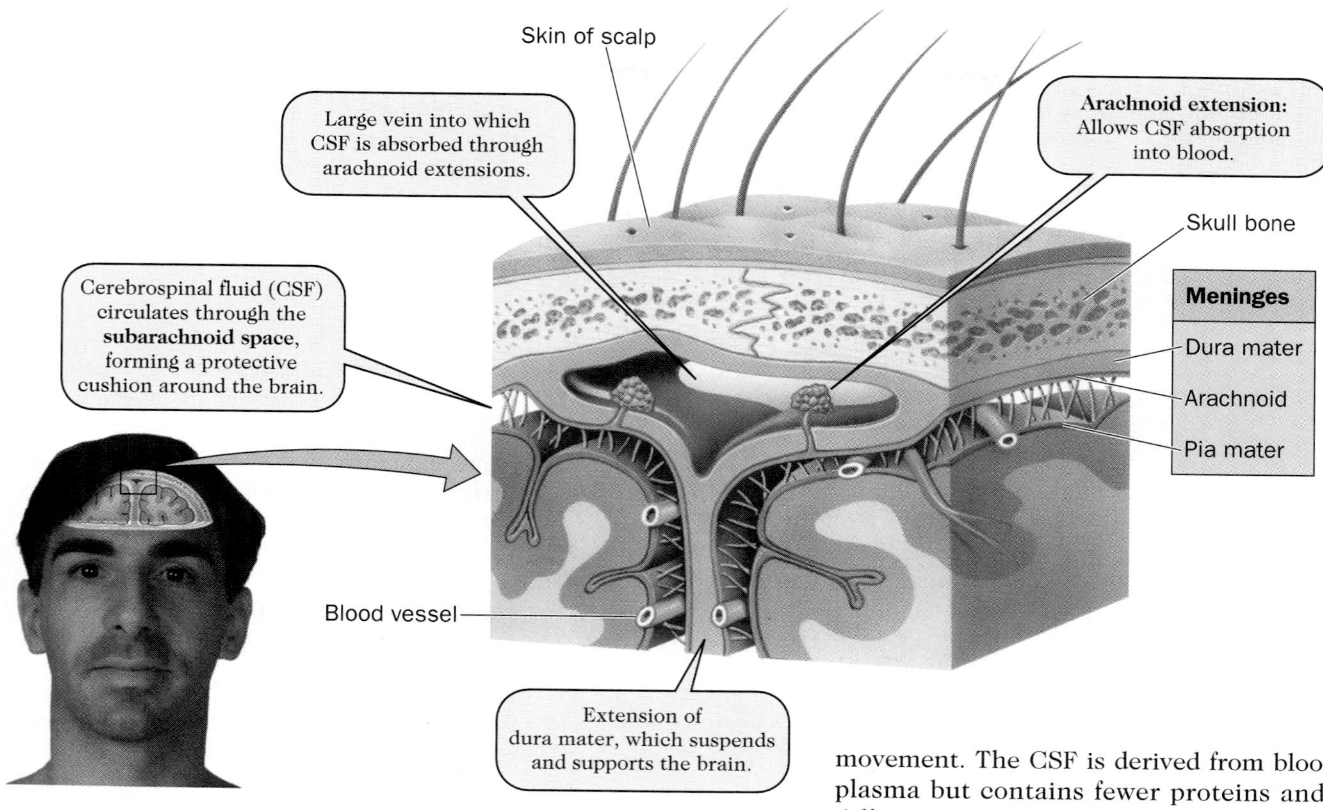
Skin of scalp
Large vein into which CSF is absorbed through arachnoid extensions.
Arachnoid extension: Allows CSF absorption into blood.
Skull bone
Cerebrospinal fluid (CSF) circulates through the subarachnoid space, forming a protective cushion around the brain.
Meninges
Dura mater
Arachnoid
Pia mater
Blood vessel
Extension of dura mater, which suspends and supports the brain.

FIGURE 15.3 The supportive and protective meninges.

bones of the cranium provide protection from most blows to the head, but more protection is needed. Directly covering the brain are the same three layers of the **meninges** which surround the spinal cord: the dura mater, the arachnoid, and the pia mater (Figure 15.3).

The thickest, toughest, and outermost of the three layers is the *dura mater*. Folds of the dura pass between the major lobes of the brain and attach to the bones of the cranium to provide suspension and support. The innermost layer is the *pia mater*. It is very thin and adheres to the brain surface, following all its contours.

Between the dura mater and the pia mater lies the middle layer: the arachnoid. Its weblike structure is intimately associated with blood vessels and allows the circulation of cerebrospinal fluid in the *sub-arachnoid space*, a compartment between the arachnoid and the pia mater.

The **cerebrospinal fluid (CSF)** in the subarachnoid space cushions the brain against the vibrations and jolts of daily movement. The CSF is derived from blood plasma but contains fewer proteins and different concentrations of ions. It provides a suitable chemical environment for the operation of brain tissue and is a vehicle for exchanging nutrients and waste products with the blood. Because of the role of the brain in the regulation and control of all body functions, the chemical composition of the CSF is essential to body homeostasis.

> The meninges consist of three layers of tissue which surround, support, and protect the brain. The subarachnoid space contains the cerebrospinal fluid, which cushions the brain and maintains a suitable environment for brain function.

The Ventricles and Circulation of the Cerebrospinal Fluid

The entire brain is constructed around four inner chambers, or **ventricles**, which are continuous with each other and with the central canal of the spinal cord (Figure 15.4). They contain about 150 milliliters (about $\frac{2}{3}$ cup) of CSF, which circulates and is replaced several times daily.

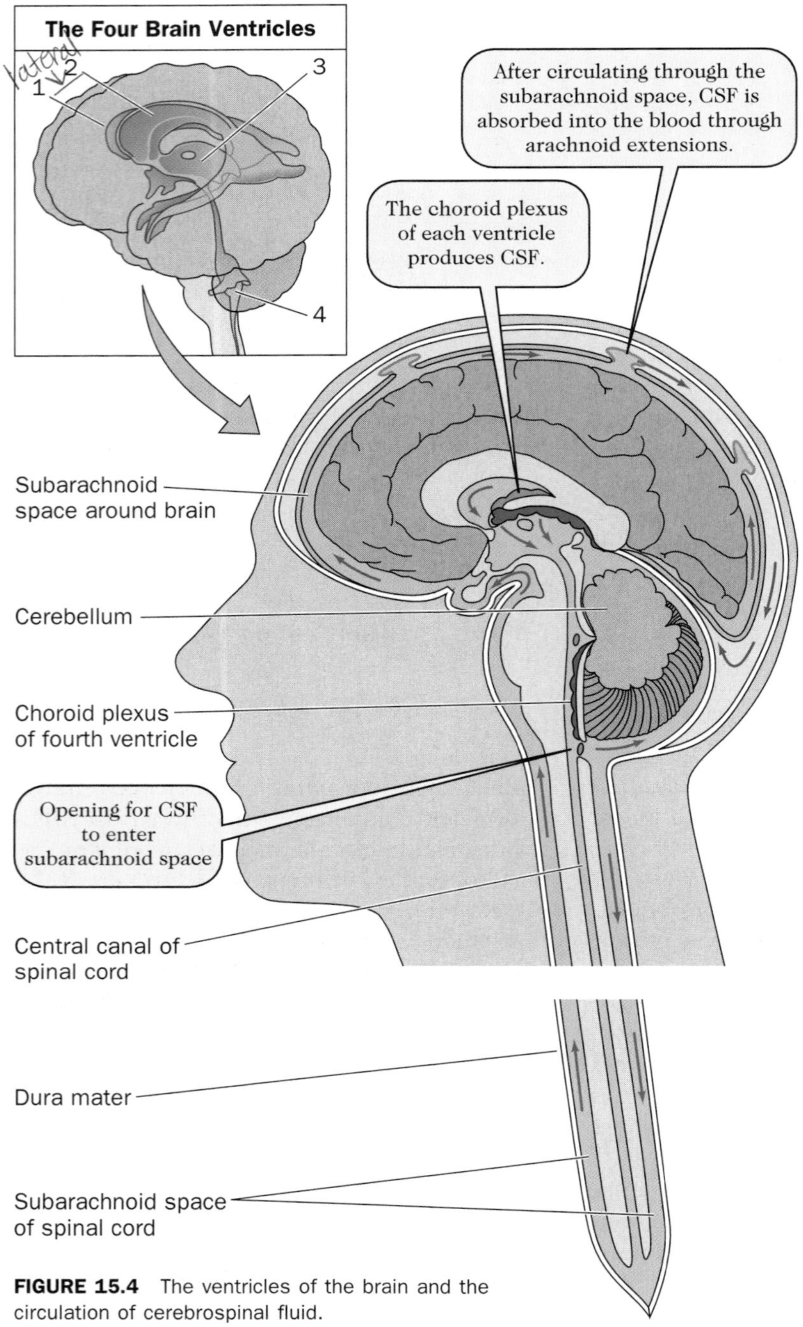

The Four Brain Ventricles
1
2
3
4
After circulating through the subarachnoid space, CSF is absorbed into the blood through arachnoid extensions.
The choroid plexus of each ventricle produces CSF.
Subarachnoid space around brain
Cerebellum
Choroid plexus of fourth ventricle
Opening for CSF to enter subarachnoid space
Central canal of spinal cord
Dura mater
Subarachnoid space of spinal cord

FIGURE 15.4 The ventricles of the brain and the circulation of cerebrospinal fluid.

CSF is produced by blood filtration through the walls of specialized capillaries called the *choroid plexus* (***kore**-oyd*) in the roof of each ventricle (Figure 15.4). From the ventricles, CSF flows slowly out into the subarachnoid space around the brain and spinal cord. This circulation is aided by the beating of cilia on the epithelial cells that line the ventricles. As excess CSF accumulates, it is reabsorbed into the blood through the arachnoid.

If flow and reabsorption of CSF are prevented by inflammation or a tumor, the CSF accumulates and puts pressure on the brain. This condition is called *hydrocephalus* and may lead to permanent brain damage if it is not corrected by surgically draining the fluid.

> The four ventricles are inner chambers of the brain. The CSF is produced by filtration of blood through capillaries in the ventricles. It circulates through the ventricles and around the brain and spinal cord before it is reabsorbed into the blood.

Blood: Supplying the Brain's Need for Glucose and O_2

Although it represents only about 2 percent of your body weight, your brain receives about 20 percent of the blood pumped from the heart and uses about 20 percent of the oxygen (O_2) available when your body is at rest. When a particular region of the brain is active, it receives increased blood flow and thus more O_2 and nutrients, particularly glucose.

Brain cells cannot live long without O_2. Any interruption of blood flow and O_2 delivery that lasts more than about 3 to 5 minutes can be fatal to brain cells. Because brain cells cannot regenerate, the death of brain neurons means a loss of brain function. Brain cells also need a constant supply of glucose. Unlike other cells in the body, brain cells use only glucose to provide energy for all their cellular functions. For this reason, a low concentration of glucose in the blood (hypoglycemia) may cause headache, dizziness, confusion, and even loss of consciousness.

The major arteries supplying oxygenated blood to the brain unite at its base to form a circular structure from which smaller arteries branch out to supply all the brain regions. Ruptures of small arteries or their blockage by blood clots are the major causes of strokes. Strokes deprive the brain of blood, O_2, and glucose and can be seriously debilitating and even fatal.

Brain capillaries are much less permeable than are most capillaries in the body. These reduced-permeability capillaries form what is called the **blood-brain barrier.** The blood-brain barrier rigidly controls the environment of your brain cells and protects them from many dangerous chemicals that get into your bloodstream. However, this barrier also makes it difficult to combat brain infections with standard antibiotics, which can't move from the blood into the infected brain.

A temporary reduction in blood flow to the brain can cause a brief loss of consciousness called *fainting*. This may happen if you rise too rapidly from a reclining position, because gravity causes blood to remain in the lower parts of your body. In addition, some people faint upon receiving an emotional stimulus such as the news of a death or the sight of blood. This happens when autonomic reflexes reduce the heart rate and dilate blood vessels, briefly lowering blood pressure, blood flow, and O_2 delivery to the brain.

Dilation or constriction of cerebral blood vessels may cause our most frequent form of pain, the common *headache*. Although there are no pain receptors in the brain tissue, they are found in the walls of the larger arteries and veins. Changes in blood pressure stimulate the receptors, and we quickly recognize the familiar ache and pounding sensation.

> Brain cells quickly die in the absence of O_2 and glucose. Blood clots or ruptures in an artery interrupt blood flow to the brain, causing cell death and loss of brain function. The blood-brain barrier restricts the diffusion of chemicals into the brain.

Structure and Function of Brain Regions

As we noted earlier in this chapter, the three regions of the brain are associated with specific roles that assist in the control and integration of body functions. The **hindbrain** sits like a cap on the spinal cord. It contains groups of nerve cells called centers that control breathing and heart rate, awaken us from sleep, maintain consciousness, and coordinate movement. The **midbrain** coordinates head movement with visual and auditory sensations.

The **forebrain** contains distinct regions which receive and interpret sensations, initiate movement, and store memories. These regions are also associated with emotions, learned behavior, conscious thought, decision making, and control of body temperature and water content. Only rarely do the regions of the forebrain operate independently. Instead, they are connected by neurons to other regions of the forebrain and to the hindbrain and midbrain so that they can work cooperatively to support integrated functioning.

The Hindbrain: Ancient and Effective

The hindbrain has sometimes been called the "reptilian" brain because it responds in a rather rigid and conventional way to regulate the basic survival activities of the body. Major areas in the hindbrain region include the medulla, the cerebellum, the pons, and the reticular formation.

The medulla: control and coordination of internal organs. The **medulla** (*muh-dull-uh*) is connected to the spinal cord at the base of the skull (Figure 15.5). All the sensory and motor nerves from the spinal cord and four of the cranial nerves are monitored by the medulla. It is also the site where motor nerves from one side of the forebrain cross over to serve the opposite side of the body. Because of this crossing over, each side of the brain controls the motor activities on the opposite side of the body.

Within the medulla are several centers which control internal organs and functions. The *cardiac control center* and the *respiratory control center* both receive sensory information about the levels of O_2, carbon dioxide (CO_2), and acidity in the blood. These centers respond by sending out messages that control the heart rate and the rate of breathing. A

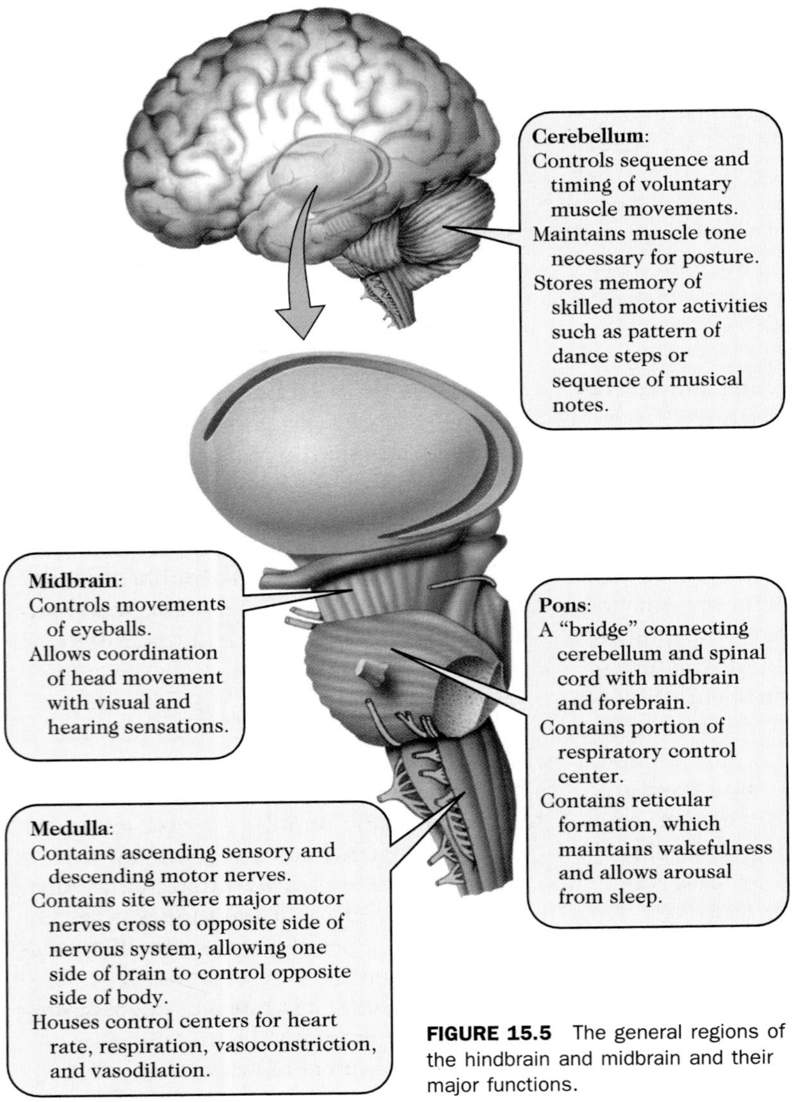
Cerebellum:
Controls sequence and timing of voluntary muscle movements.
Maintains muscle tone necessary for posture.
Stores memory of skilled motor activities such as pattern of dance steps or sequence of musical notes.
Midbrain:
Controls movements of eyeballs.
Allows coordination of head movement with visual and hearing sensations.
Pons:
A "bridge" connecting cerebellum and spinal cord with midbrain and forebrain.
Contains portion of respiratory control center.
Contains reticular formation, which maintains wakefulness and allows arousal from sleep.
Medulla:
Contains ascending sensory and descending motor nerves.
Contains site where major motor nerves cross to opposite side of nervous system, allowing one side of brain to control opposite side of body.
Houses control centers for heart rate, respiration, vasoconstriction, and vasodilation.

FIGURE 15.5 The general regions of the hindbrain and midbrain and their major functions.

vasomotor center helps regulate the diameter of arteries, controlling the distribution of blood in the body. Other centers in the medulla make possible the activities associated with coughing, sneezing, swallowing, hiccuping, and vomiting.

The cerebellum: coordination of body movements. The second largest component of the entire brain—the **cerebellum**—is located behind the medulla and is divided into two hemispheres (Figure 15.5). It helps maintain muscle tone and coordinates the sequence and timing of muscle contraction and relaxation, assuring that two antagonistic muscles do not contract at the same time. The cerebellum also stores memories of sequences of skilled movements, such as those used in a dance routine. Excessive alcohol consumption interferes with normal cerebellar operation and results in uncoordinated movements and staggering.

The cerebellum also receives sensory information from a wide variety of receptors: stretch receptors in tendons and joints, balance and motion receptors in the inner ear, and visual receptors in the retina of the eye. Indeed, we may experience nausea when this sensory information overwhelms the cerebellum, as happens in motion sickness and during viral infection of the inner ear.

The pons: connections to other regions. The third part of the hindbrain is the **pons,** which is located between the medulla and the midbrain (Figure 15.5). A portion of the pons contributes to the respiratory control center of the medulla, but the pons functions mostly as a bridge. It connects the major motor and sensory nerves from the spinal cord, medulla, and cerebellum with the midbrain and forebrain.

The reticular formation: a role in waking and sleeping. Covering portions of both the pons and the medulla is a diffuse network of neurons called the **reticular formation** (Figure 15.5). This network has both motor and sensory functions. It helps maintain muscle tone as well as alertness and wakefulness. While some portions of the reticular formation cause us to fall asleep, a sensory portion called the *reticular activating system* (*RAS*) helps us wake up. It responds to the ring of an alarm clock, the light of dawn, the smell of smoke, unusual sounds, and the absence of familiar sounds. Sleep-inducing drugs such as the barbiturates are believed to act on the reticular formation.

Injury or infection of the RAS often results in a loss of consciousness called *coma*. There are different degrees of coma, some light and others deep. In a light coma some reflexes continue to operate, but those reflexes may be lost in a deeper coma. Fortunately, a coma will sometimes lighten or even suddenly end after many years of unconsciousness.

The hindbrain consists of the medulla, cerebellum, pons, and reticular formation. The medulla controls internal organs. The cerebellum coordinates movements. The pons connects the three major parts of the brain. The reticular formation wakes us up and maintains alertness.

The Midbrain: Movement of the Eyes and Head

Located above the pons and joining the hindbrain and forebrain is the **midbrain** (Figure 15.5). This is the smallest of the three major parts of the brain. By means of the cranial nerves, it controls eye movements and pupil size. It also coordinates head movement with sight and sound, causing us to turn toward a bright flash or a loud noise. Sites in the midbrain connect to specific regions in the forebrain, coordinating our unconscious muscle movements and assuring that they are smooth and orderly.

A deficiency of the neurotransmitter *dopamine* in the midbrain has been associated with the involuntary shaking of *Parkinson's disease*. Parkinson's disease usually appears in persons over age 55 and progresses from involuntary hand tremors to shaking of the legs and head. It may eventually result in muscle paralysis, starting with the facial muscles. Treatment with the drug *L-dopa* has proved effective in relieving the tremors associated with the disease.

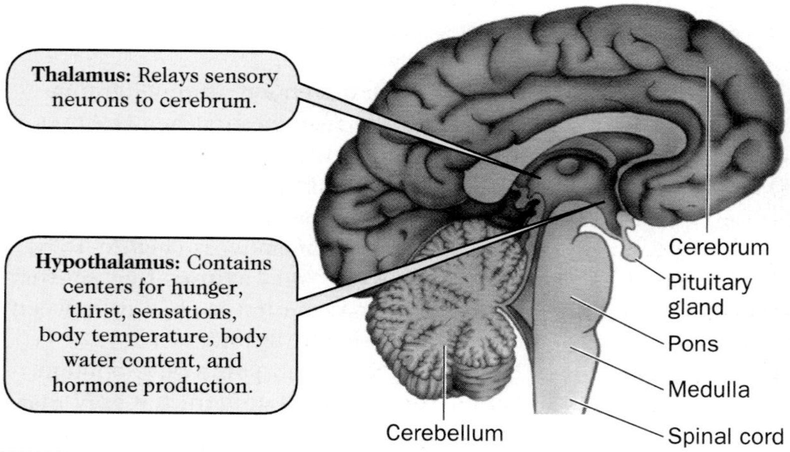
Thalamus: Relays sensory neurons to cerebrum.
Hypothalamus: Contains centers for hunger, thirst, sensations, body temperature, body water content, and hormone production.
Cerebrum
Pituitary gland
Pons
Medulla
Spinal cord
Cerebellum

FIGURE 15.6 The components of the forebrain and their functions.

The midbrain couples head movements with sight and sound sensations. It also helps maintain smooth and balanced movements through coordination of muscle groups.

The Forebrain: Unconscious Regulation and Conscious Awareness

The forebrain includes the thalamus, hypothalamus, limbic system, and cerebrum. In addition to integrating the functions of many internal organs and maintaining homeostasis, the cerebrum of the forebrain is associated with some of the mental activities we consider to be most human: language and abstract thinking.

The thalamus and hypothalamus: monitoring internal and external sensations. The portions of the forebrain closest to the midbrain include the thalamus and hypothalamus (Figure 15.6). The **thalamus** serves as the primary relay station for sensations. It receives impulses from nearly all incoming sensory neurons and transmits those impulses to specific regions in the cerebrum, where they are interpreted. As impulses from pain, temperature, and pressure receptors reach the thalamus, we become aware of a stimulus but are not aware of where it is coming from. The localization of a stimulus is done by the cerebrum.

The **hypothalamus** lies between the thalamus and the pituitary gland. The hypothalamus is one of the body's major regulators of homeostasis. It receives sensory input on sight, sound, taste, smell, temperature, blood glucose concentration, blood osmotic pressure, hunger, and thirst. Through its connections with the autonomic division of the PNS, the hypothalamus also helps regulate heart rate, blood pressure, and contractions of the urinary bladder. The hypothalamus is connected to the pituitary gland and exerts control over its hormonal activities.

The hypothalamus also controls food intake. A *hunger center* helps us know when it is time to eat, and a *satiety center* informs us when we are full. A *thirst center* controls water balance by letting us know when body fluids have become too concentrated so that we need to drink fluid. A *temperature control center* helps keep body temperature within narrow limits, causing us to shiver or sweat. The hypothalamus is a key contributor to the homeostasis we identified in previous chapters.

> The thalamus relays sensory impulses from the cranial and spinal nerves to the cerebrum for interpretation. The hypothalamus controls body temperature, water content, hunger, feeding behavior, blood sugar, heart rate, and blood pressure.

The limbic system: emotions and memories. The limbic system is actually formed by portions of the thalamus and hypothalamus and regions of the cerebrum (Figure 15.7). The **limbic system** links the conscious mental activities of the cerebrum to unconsciously controlled activities carried out by the midbrain and hindbrain. The limbic system plays an important role in the formation of strong emotions such as affection, pleasure, sexual desire, fear, pain, and anger. It appears to be a center where information is fixed into memory. Because the limbic system involves both emotion and memory, emotionally charged events are frequently recalled with vivid detail.

The sensory pathways for smell also pass through the limbic system. An odor may trigger pleasant or unpleasant memories. Long-forgotten memories of a person or place may tumble into consciousness in response to a particular odor.

The cerebrum: language, decisions, awareness, and movement. The cerebrum is the largest portion of the brain. It deals with higher brain functions that distinguish humans from all the other animals: symbolic communication, language, thought, calculation, decision making, and conscious awareness of emotions. It is the site of what we sometimes call voluntary actions and control.

As seen in Figure 15.8, the cerebrum folds over and encloses most of the other brain regions. It consists of two hemispheres divided by a deep longitudinal groove. The upper layer of the cerebrum, only 2 to 4 millimeters thick, is called the **cerebral cortex.** This layer represents only 25 percent of the brain volume but contains nearly 75 percent of the brain neurons. The many cell bodies and abundant blood vessels in the cortex give it a slightly gray-pink color, and it is often called the *gray matter*. Beneath the cortex is the *white matter,* which is made up of dendrites and abundant myelinated axons.

Although a deep longitudinal groove separates the two cerebral hemispheres, they are not entirely independent. A large group of neurons called the **corpus callosum** spans the lower part of the groove and links the hemispheres (Figure 15.8). The corpus callosum assures that sensory and motor information is shared between both sides of your brain.

During both embryonic development and early postnatal growth, the gray matter of the cerebrum grows more rapidly than does the white matter. This causes the surface of the cortex to become ex-

FIGURE 15.7 Location of the limbic system in the forebrain.

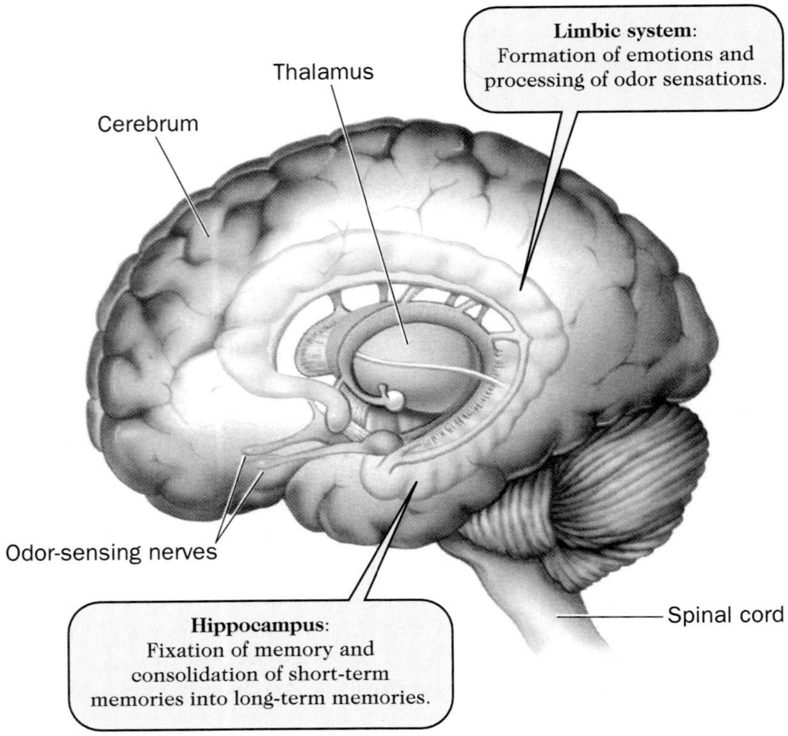
Limbic system:
Formation of emotions and processing of odor sensations.
Thalamus
Cerebrum
Odor-sensing nerves
Spinal cord
Hippocampus:
Fixation of memory and consolidation of short-term memories into long-term memories.

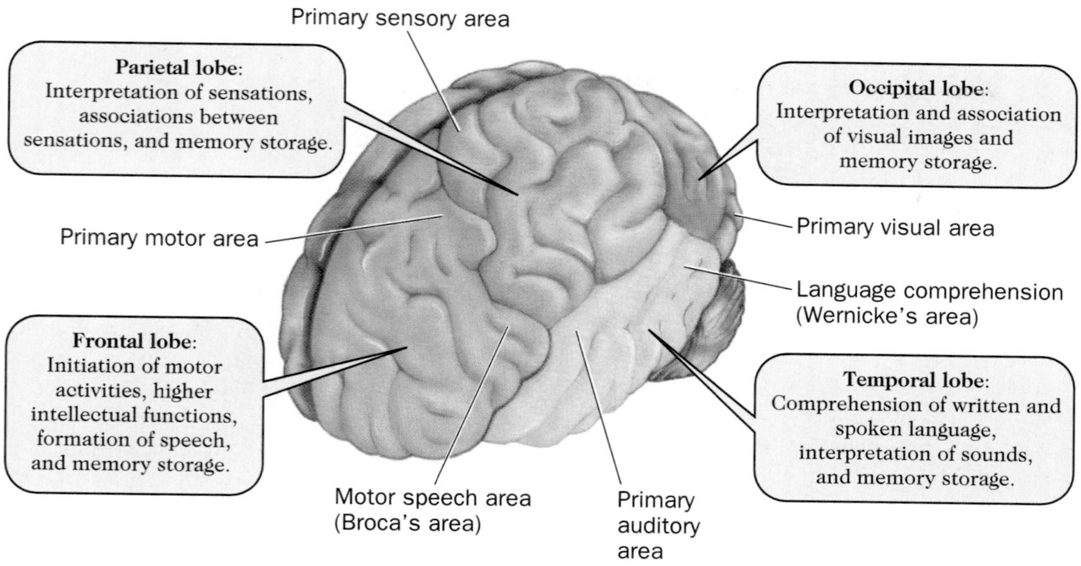

Primary sensory area
Parietal lobe: Interpretation of sensations, associations between sensations, and memory storage.
Occipital lobe: Interpretation and association of visual images and memory storage.
Primary motor area
Primary visual area
Language comprehension (Wernicke's area)
Frontal lobe: Initiation of motor activities, higher intellectual functions, formation of speech, and memory storage.
Temporal lobe: Comprehension of written and spoken language, interpretation of sounds, and memory storage.
Motor speech area (Broca's area)
Primary auditory area

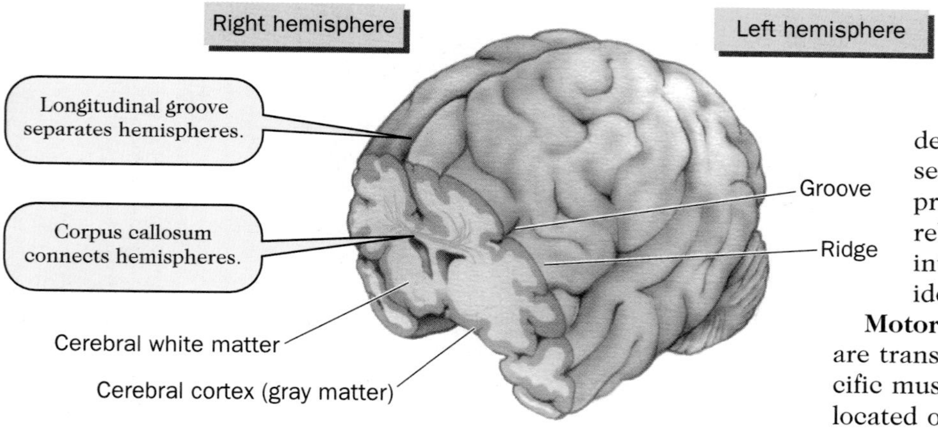

Right hemisphere
Left hemisphere
Longitudinal groove separates hemispheres.
Corpus callosum connects hemispheres.
Groove
Ridge
Cerebral white matter
Cerebral cortex (gray matter)

FIGURE 15.8 The cerebrum with its lobes, functions, grooves, and ridges.

tensively folded, forming ridges and grooves. The larger grooves divide the cerebrum into sections called **lobes.** There are five lobes—the *frontal, parietal, occipital, temporal,* and the *insula*—though only the first four can be seen from the surface (Figure 15.8). Distributed within the lobes are specialized areas for specific sensory, motor, and "mental" activities.

Sensory areas which receive and interpret impulses from sensory neurons are located in specific lobes. For example, the sensory areas for sight are in the occipital lobe, while those for hearing, smell, and taste are in the temporal lobe. A **primary sensory area** on a prominent ridge in the parietal lobe contains segments devoted to all the major body regions (Figure 15.9).

The most sensitive body regions, such as fingertips, lips, and genitalia, involve more of the primary sensory area. Less sensitive body regions cover smaller areas and have fewer neurons. This primary sensory area is where sensory information from your body is interpreted and where you become conscious of the location of a sensation. If you stub your right big toe against the leg of your desk, the sensory impulses will be sent to the big toe region of the primary sensory area of the left cerebral hemisphere. You will quickly interpret the sensation as pain and identify its location.

Motor areas initiate impulses which are transmitted by motor nerves to specific muscles. The **primary motor area** is located on a ridge in the frontal lobe and is depicted in Figure 15.9. The movement of each body region is correlated with a specific segment of that ridge. Body parts that are capable of precise and intricate movements, such as the fingers and face, span a large area and utilize many neurons. Because of the crossing of motor neurons in the medulla, the portions of the ridge in each hemisphere control the motor activities in the opposite side of the body.

Association areas help integrate sensory and motor information and conduct higher mental activities such as the reasoning, decision making, memory storage, judgment, personality traits, and intelligence we associate with the mind. While the sensory and motor areas are found in both hemispheres, some of the association areas are localized in only one cerebral hemisphere.

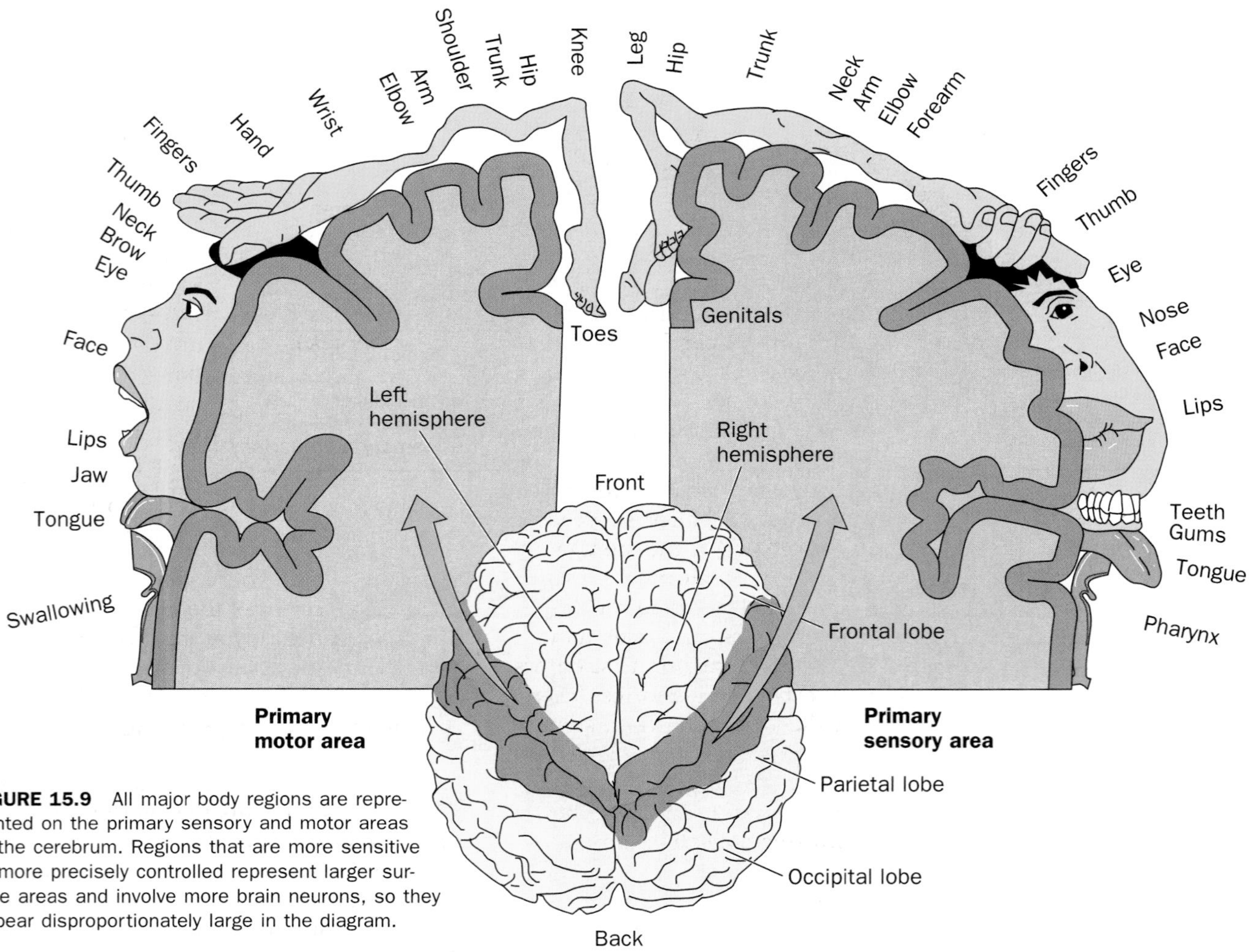
Fingers
Hand
Wrist
Elbow
Arm
Shoulder
Trunk
Hip
Knee
Thumb
Neck
Brow
Eye
Face
Lips
Jaw
Tongue
Swallowing
Toes
Left hemisphere
Leg
Hip
Trunk
Neck
Arm
Elbow
Forearm
Fingers
Thumb
Eye
Nose
Face
Lips
Teeth
Gums
Tongue
Pharynx
Genitals
Right hemisphere
Front
Frontal lobe
Parietal lobe
Occipital lobe
Back
Primary motor area
Primary sensory area

FIGURE 15.9 All major body regions are represented on the primary sensory and motor areas of the cerebrum. Regions that are more sensitive or more precisely controlled represent larger surface areas and involve more brain neurons, so they appear disproportionately large in the diagram.

> The cerebral cortex interprets sensory information and is the site for language, memory, and abstract thought, body awareness, and coordination of skeletal muscles. The cerebrum contains specialized sensory, motor, and association areas. The corpus callosum connects the cerebral hemispheres.

A cooperative network. Only rarely does a single functional area in the cerebral cortex work independently of the others. Most mental activities require the cooperative action of several areas. Imagine that you are attempting to describe a textbook picture of the brain to a friend who has not seen it. The image is assembled on the primary visual area of your occipital lobes. Impulses then pass through an association area to another area of your cortex, *Wernicke's area,* where you select the words you will use in your description (see Figure 15.8). Impulses then pass to another area of your cortex, *Broca's area,* where muscle groups are selected to vocalize the words in your description. Broca's area activates the facial regions of the primary motor area, which in turn send impulses to the muscles of the face, lips, tongue, and diaphragm to accomplish speech. All these areas are linked together by neurons so that they can interact in a cooperative network.

> Specialized regions of the cortex cooperate through a network of neurons to accomplish complex responses. Wernicke's area selects words, while Broca's area selects muscles for speech.

Learning and Memory

Humans do a lot of learning, and as you know, learning requires memory. **Learning** refers to any change in behavior that is caused by experience. It involves gaining new information or discovering new applications for old information. **Memory** involves acquiring, storing, and retrieving information. Both processes begin in the limbic system of the forebrain, but information is stored in the cerebrum or in some cases the cerebellum of the hindbrain. There are actually two stages to memory storage: short-term memory and long-term memory.

The retention of information for a few seconds or hours is called *short-term memory*. Not surprisingly, the storage space for short-term memory is limited. *Long-term memory* is retained for days, years, or even a lifetime and uses more storage space. All information first enters the brain as short-term memories. Then, in a process called *consolidation*, some short-term memories are transferred to long-term storage. The *hippocampus* region of the limbic system seems to be associated with this consolidation (see Figure 15.7).

The formation of memories appears to require changes in the physical makeup of the brain. These changes may include (1) the assembly of proteins which are stored in neurons, (2) the forming of new connections or synapses between neurons, or (3) an increase in the effectiveness of neurotransmitter communication between neurons that are already connected. Recent research has suggested that a single memory may be stored over an entire network of neurons and synapses, with their rhythmic firing producing a three-dimensional hologramlike effect for that memory.

All regions of the brain—forebrain, midbrain, and hindbrain—are capable of storing memory, including the sensory and motor regions. However, specific kinds of memory appear to be stored in specific regions. For example, the memories of the sights, sounds, tastes, and odors associated with your last vacation will be stored in the regions of the cerebrum most closely related to those individual sensations.

Frequent use or repetition of information facilitates consolidation, as does associating information in short-term memory with items that are already in long-term storage. Students needing long-term memory of technical terms and data frequently associate the information with nonsense rhymes. These and other memory tricks are called mnemonic devices (*nih-mon-ick*).

> Learning requires memory. The two stages of memory are short-term and long-term memory. All parts of the brain are capable of storing memories, but the hippocampus aids in the consolidation of long-term memories.

Two Brains in One: Differences between the Cerebral Hemispheres

Like about 10 percent of the population, I am left-handed. This means that for many skilled motor activities, my right cerebral hemisphere dominates my left hemisphere, and I prefer to use my left hand or foot. Compared to right-handed people, lefties have more narrow parietal and occipital lobes in our right cerebral hemispheres and narrower frontal lobes in our left cerebral hemispheres.

Functionally, the **left cerebral hemisphere** appears to control logical, sequential processes which can be expressed with words or numbers (Figure 15.10). As you examine this text, you are using your left hemisphere to read and comprehend the language and ideas presented here. The so-called left brain does mathematics and science, and you rely on it more when you speak or write.

Young children seem to have roughly equal speech potential in both hemispheres, but after age 8 the Broca's area of the left hemisphere becomes dominant over the same area in the right hemisphere. In adults, an injury to the left side of the cerebrum causes language, speech, and reading deficiencies. How-

Hola!
Hello!
Moshi moshi!
Konnichiwa!
Bonjour!
Spoken language
Right hand coordination
R
Left hand control
L
Music awareness
Art awareness
E=½Iω²+u
Number skills
Written language
Three-dimensional forms
Scientific skills
Reasoning
Imagination
Insight

FIGURE 15.10 Functional differences between the left and right cerebral hemispheres.

ever, this does not mean that there is no role for the right hemisphere in adult speech. Recent studies have shown that blood flow increases in both hemispheres when we speak.

The **right cerebral hemisphere** appears to be the artistic and musical side of the brain (Figure 15.10). It helps you distinguish visual and tonal information. Sounds, especially the sounds of wind, rain, ocean waves, and bird calls, are interpreted by the right hemisphere. Visual patterns, the shapes of objects, and their locations in space are interpreted by this side of the brain. With the right hemisphere we appreciate the "big picture," or the context of an event, and we recognize the faces of friends and family members.

Individual variations. We differ in the degree to which we rely on the two halves of our brain. Are you more left-brained or right-brained? Would you rather read a book or watch a movie? Are you better at repairing a toaster or arranging flowers in a vase? Do you excel at Trivial Pursuit or Pictionary? Would you rather do a crossword puzzle or sketch a picture? If you selected all the first choices, you may have a left brain tendency. If you preferred the second choices, you may have a right brain tendency.

> The two cerebral hemispheres are not functionally or structurally equal. The right hemisphere is more "artistic and musical," while the left hemisphere is the "engineering and debating" side of the brain.

Genders of the Brain: Development of Male and Female Differences

In the musical comedy *My Fair Lady,* Professor Henry Higgins laments in frustration, "Why can't a woman be more like a man?" Both men and women have often expressed such sentiment about the opposite sex. Gender-related differences in attitudes, capacities, interests, and behaviors have long been observed, documented, and discussed, often leading to controversy.

Female Brains Differ from Male Brains

What does it mean to have a *female brain*? Women seem to be better than men at seeing the context in which an event unfolds. Women are better at sensing the emotional messages in conversations, gestures, and facial expressions. In general, women speak and read at an earlier age. They learn languages more easily and usually outperform men in tests of grammar and spelling. Women see better at night than men do and have more acute senses of smell, taste, and hearing. From an early age, girls exhibit better fine motor control of the hands, usually displaying better penmanship than boys.

Women seem to use both sides of the brain more equally than men do. In a study that examined brains after death, female brains had a larger corpus callosum than did male brains. As you know, the corpus callosum is a bundle of nerves connecting the two hemispheres. Its larger size may indicate an increased number of connections between the two sides of the female brain. Measurement of electrical activity in living brains also indicates that women draw more equally on both hemispheres for verbal and visual activities and emotional responses and in predicting the three-dimensional shapes of objects.

What are the functional traits of the *male brain*? On average, men excel at solving mathematical problems, mentally rotating objects in space, and predicting three-dimensional shapes from two-dimensional projections. Boys exhibit better hand-eye coordination and more precise control of movements that use large groups of muscles. Men see more clearly in bright light than women do but tend to have reduced peripheral vision. They have a better sense of perspective, or the relative positions of close and distant objects. Men, especially those who are right-handed, use the left hemisphere almost exclusively for verbal activities and the right for visual and spatial tasks and emotional responses.

Do these differences arise from unalterable cellular and molecular differences between male and female brains, or are they learned from the social environment? The general answer is that both biological organization and the environment play a role.

Research suggests that these differences may begin to be established early in prenatal development as an embryo is exposed to hormones from both the mother and its own developing reproductive organs.

> Males and females show different behaviors, attitudes, and talents that are based on different brain organization. Men seem to use one hemisphere more than the other. Women tend to use both hemispheres more equally.

The Origins of Male and Female Gender Differences

For purpose of reproduction, humans exhibit two distinct biological sexes: male and female. In addition to biological gender, however, there are less obvious and more personal, subjective expressions of gender differences: gender identity and sexual orientation. Although current research is not entirely clear or consistent on these points, these differences appear to be associated with differences in brain structure.

Biological gender is established at the moment of conception after a sperm fertilizes an egg. As will be discussed further in Chapter 18, it is the genetic contribution of the sperm that determines whether an embryo develops as a male or a female.

The reproductive organs of both sexes begin as unspecialized clusters of tissue in the embryo. The development of this tissue is first influenced by the female sex hormone *estrogen,* which is produced by the mother's body. A female embryo responds to her mother's estrogen, and this unspecialized tissue develops into ovaries, which become part of the embryo's female reproductive anatomy and the source of her eggs. The developing ovaries produce additional estrogen that influences the development of female reproductive organs and the female brain.

In a male embryo, testes develop and produce the sex hormone *testosterone*. The secretion of testosterone influences the development of both the male reproductive organs and the male brain. While a developing female's brain is washed with estrogen, a developing male's brain receives a massive dose of his own testosterone, overcoming his mother's estrogen.

It seems clear that both testosterone and estrogen play a role in the development of different neural networks in male and female brains. However, they may also play additional roles. It has been hypothesized that these two hormones act on specific centers of the developing human brain to separately determine gender identity and sexual orientation. If additional research supports this hypothesis, it will be clear that your biological gender, gender identity, and sexual orientation are largely determined before you draw your first breath.

> Male and female sexual differences are determined at conception. Female reproductive organs develop in response to both maternal and embryonic estrogen. Male reproductive organs develop in response to testosterone. These hormones also influence brain development.

The Genders of the Brain

Gender identity is different from but is certainly influenced by biological gender. It is a personal, subjective, psychological recognition that "I am male" or "I am female." Biological gender and gender identity are both different from *sexual orientation,* which is the attraction to members of the opposite sex or the same sex. Some initial research suggests that sexual orientation and gender identity have separate centers in the brain. The development of these centers may or may not coincide with the development of biological gender.

The origin of gender variations. Sometimes a gender identity exists in association with the opposite biological gender. For example, a male body may have a female identity. This can produce extreme personal anguish in an individual as social and cultural forces conflict with a contrary sense of gender identity. How does such gender conflict come about?

Some researchers suggest that occasionally a male embryo may produce just enough testosterone to start the development of male reproductive organs but too little additional hormone to instill a male pattern in the brain. The embryo's brain is then influenced more strongly in its development by the mother's estrogen levels. To a greater or lesser degree, a female brain and a female gender identity form in a male body.

Such ambiguities can also exist in a female embryo. Such an embryo may be exposed to high testosterone levels accidentally produced by her own developing adrenal glands or those of her mother. She may develop a male brain pattern and a male gender identity in a female body.

A similar course of events may take place in the hypothesized brain center for sexual orientation. This brain center may develop at a slightly different time than does the center for gender identity and may require a certain dose of hormone to yield a heterosexual orientation. If these conditions are not met, a homosexual orientation may result. Of course, there is not one homosexual type or one heterosexual type. Postnatal learning and social conditions can greatly influence how either sexual orientation is expressed. The different possible routes for the development of biological gender, gender identity, and sexual orientation can create very complex issues for the affected individual as well as his or her family and friends.

> Exposure to male and female hormones during the initial stages of brain development may separately determine gender identity and sexual orientation. Too much or too little testosterone may result in a gender identity or sexual orientation different from the biological gender.

Brain Chemistry: Neurotransmitters in the Brain

As you learned in Chapter 14, neurons communicate with one another at synapses through the release of neurotransmitters. Some neurotransmitters have an excitatory effect, and others have an inhibitory effect. At least 50 different neurotransmitters are associated with neurons in the brain. However, only one neurotransmitter is secreted by each neuron.

Acetylcholine is a neurotransmitter that may produce either excitation or inhibition, depending on the kind of receptor it binds to in a synapse. It is used by the motor nerves ("voluntary" nerves) that begin at the primary motor area of the cerebrum, pass through the hindbrain and spinal cord, and terminate at skeletal muscles. Acetylcholine is also used in the limbic system, and the loss of neurons that make acetylcholine seems to contribute to the disabling symptoms of Alzheimer's disease, which we'll discuss later in this chapter.

Gamma-aminobutyric acid (GABA) is an inhibitory transmitter used in the midbrain and forebrain. Antianxiety drugs such as diazepam (Valium) exert a calming effect on the brain by increasing the effectiveness of GABA in the limbic system, thalamus, and hypothalamus.

Other neurotransmitters of the brain include norepinephrine, dopamine, serotonin, enkephalins, and endorphins. **Norepinephrine** is widely used in the sympathetic portion of the PNS and in the brain. It is associated with wakefulness, dream formation, and mood regulation.

Dopamine has already been mentioned in association with the *midbrain* and Parkinson's disease. Overproduction of dopamine has been associated with the mental disorder called *schizophrenia,* which is characterized by hallucinations, confusion, delusions, uncontrolled movements, and strange behaviors. Medications such as chlorpromazine (Thorazine) block the action of dopamine and reduce some schizophrenic symptoms. Other drugs, such as the amphetamines (called "speed" on the street), enhance the activity of dopamine in brain cells, producing the experience of high energy, vigilance, restlessness, and racing thoughts.

Serotonin is believed to be involved in controlling moods and states of conscious awareness. Its function in the brain is disrupted by drugs such as lysergic acid diethylamide (LSD), which produce visual and auditory hallucinations, fear, panic, and dizziness.

Enkephalins (*en-**kef**-uh-lins*) and **endorphins** (*en-**door**-fins*) are neurotransmitters with very potent painkilling effects that are released by certain neurons in the thalamus, hypothalamus, and limbic system. Their actions are even more powerful than those of morphine and heroin, which utilize the same receptor attachment sites on brain neurons.

> Neurotransmitters may have either excitatory or inhibitory effects, depending on the receptor and neuron to which they bind.

Observing the Living Brain

How do we know which regions of the brain perform specific functions? The study of dead and preserved brains offers few clues to the functions performed by specific brain structures. In the past, the study of brain-injured individuals, often soldiers or accident victims, was the primary means of locating brain function. Researchers noted the site of injury and correlated it with functions or capacities that had been lost. The establishment of some correlation between structure and function was also possible during brain surgery. However, it is now possible to observe the living brain without invading it by cutting.

Forming Images of the Brain with Radiation

Today, noninvasive technologies allow us to learn much about the brain without relying solely on accident victims or surgery. Medical imaging procedures such as CT, PET, and MRI scans now let us peer into the brain and observe its anatomy, metabolic activities, or blood flow (Figure 15.11). These procedures all use some form of harmless radiation that is manipulated to form a revealing image of an organ or tissue.

Computerized tomography (*CT* or *CAT*) uses x-rays and computer technology to show cross-sectional views of the body. *Positron emission tomography* (*PET*) can reveal the metabolic activity of tissues as they consume radioactive molecules such as glucose prepared with isotopes of carbon. *Magnetic resonance imaging* (*MRI*) uses an electromagnet and radiowaves to detect tumors and other abnormal structures.

> Imaging techniques such as CT, PET, and MRI scans allow observation of brain structure and function in a living brain without surgery or dissection.

Brain Waves: Patterns of Electrical Activity in the Brain

We can also watch the brain at work by making recordings of the electrical activity associated with different mental functions, even sleep. As you recall, nerve impulses are electrical. Although the actual electrical change in a single neuron is very small, when large numbers of cells in specific brain regions are active together, the total change is large enough to be detected on the surface of the scalp.

The instrument used to detect and record brain electrical activity is called an *electroencephalograph* (*ee-**lek**-trow-en-seph-uh-low-graf*). Electrodes are

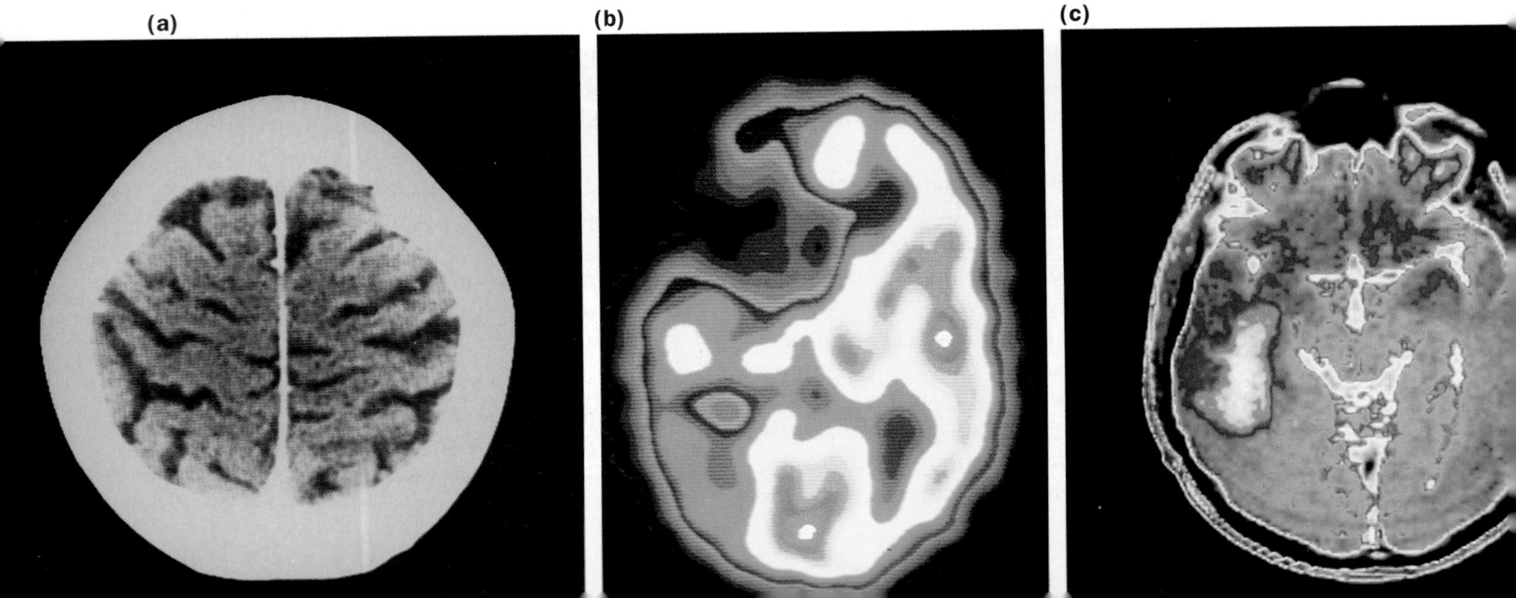

FIGURE 15.11 Medical imaging techniques. (a) A CT scan uses x-rays and a computer to assemble cross-sectional views. (b) In a PET scan, areas with high metabolic activity are reddish and those with low activity are blue. (c) The MRI technique uses a magnetic field and radiowaves to create a visual image. Tumors are seen especially well on MRI scans.

(a) A recording of electrical activity in the brain is made with surface electrodes and an electroencephalograph

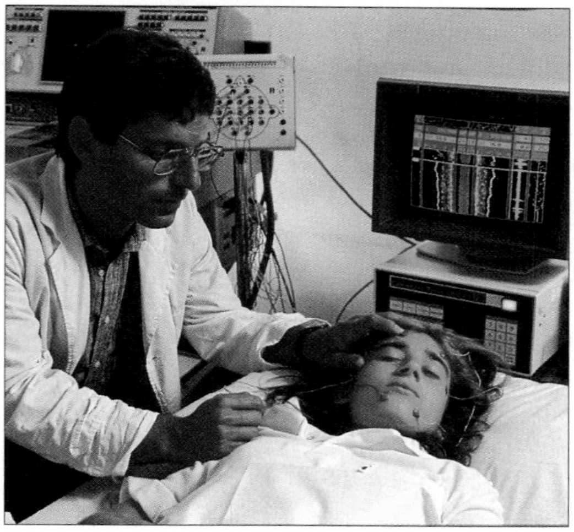

(b) Alpha, beta, delta, and theta waves as they are recorded from awake or sleeping individuals

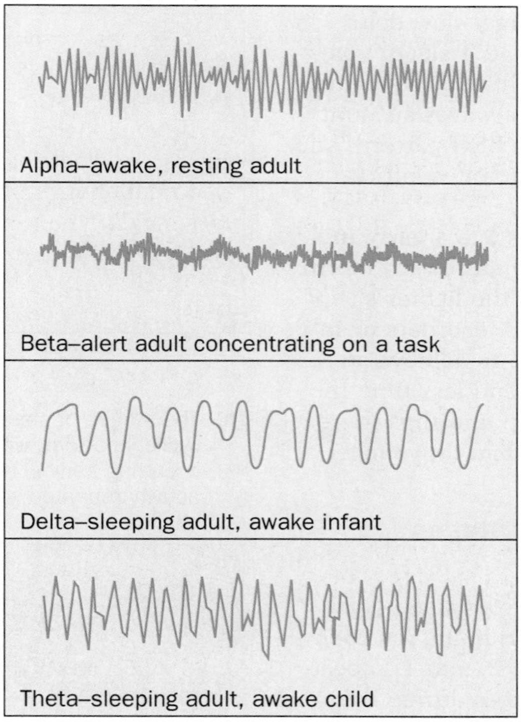
Alpha–awake, resting adult
Beta–alert adult concentrating on a task
Delta–sleeping adult, awake infant
Theta–sleeping adult, awake child

FIGURE 15.12 Brain waves.

placed on the scalp, and the tiny electrical signals they detect are amplified and displayed on a computer monitor, forming a record of brain activity known as an *electroencephalogram* (*EEG*) (Figure 15.12). EEG recordings occur as rhythmic waves that are produced when groups of neurons are active. The patterns and combinations of waves change with activity, behaviors, wakefulness, sleep, disease, and injury. The absence of electrical activity, often described as a "flat EEG," is used clinically as a sign of brain death.

There are four characteristic brain wave patterns. *Alpha waves* occur at a rate of 8 to 12 cycles per second (cps). They are recorded from calm, resting, awake individuals who have their eyes closed. Alpha waves disappear entirely during sleep. *Beta waves* (18 to 25 cps) are observed when an individual is alert and is receiving sensory information or conducting mental tasks such as reading and adding or subtracting numbers. As you read and study this page, your brain is producing beta waves. *Theta waves* (3 to 7 cps) are common in young children and sleeping adults. *Delta waves* (1 to 5 cps) are recorded from sleeping adults and awake infants. They are a sign of brain damage when recorded from an awake adult.

The EEG records electrical activity in the brain. Four basic wave patterns are associated with different levels of consciousness and activity. EEG patterns can be used to recognize and diagnose some brain and sleep disorders.

Sleep and Dreaming

Are you beginning to fall asleep while reading this chapter? Are you considering taking a nap to "let your brain rest"? Well, a nap may be a good idea, but it will not rest your brain. While some other organs may be less active while you sleep, your brain does not get much rest. With sleep, the pattern of electrical activity in the brain changes but is not reduced. Neither is blood flow or O_2 utilization reduced during sleep.

The actual function of sleep and the dreaming which often occurs are still mysteries and topics of debate. Although we may not yet understand the function of sleep, we are able to describe four distinct stages based on EEG recordings

(Figure 15.13a). **Stage 1 sleep** is a time of gentle dozing during which you may experience a floating sensation as you drift into and out of sleep. Your heart rate and breathing slow down, and your eyes slowly roll from side to side. Your brain emits slow alpha waves which are gradually replaced by theta waves. Stage 1 sleep lasts about 5 minutes or less.

During **stage 2 sleep**, your eyes remain still and you exhibit little movement. Your EEG pattern shows short sequences of "sleep spindles" (12 to 14 cps) and large, high-voltage "K complex" waves. You may dream during this stage, but you will rarely remember the dream content. In total, you spend about half your night's sleep in stage 2.

Stage 3 sleep is a time of restful sleep during which your heart rate, blood pressure, and breathing decrease and your muscles progressively relax. Nearly half of your brain waves are large, slow delta waves. As you reach **stage 4 sleep**, you pass into deep sleep from which you are not easily aroused. Delta waves account for half or more of your EEG pattern, and your heart and breathing rates fall to their lowest levels.

Typically, you reach stage 4 early in your night's sleep and then toward morning spend more time in the lighter sleep stages. People with sleep disorders or insomnia are often unable to achieve stage 4 sleep. Older adults spend less time in stage 3 and stage 4 sleep and thus more time in stages 1 and 2 than do younger adults.

Cycling through the Stages of Sleep

Normally you progress from stage 1 to stage 4 sleep in about an hour. You then return through stages 3, 2, and 1, repeating this cycle several times during the night or sleep period (Figure 15.13b). When you reenter stage 1 during each cycle, there is a period of intense, vivid dreaming and active, jerky eye movements called, not surprisingly, rapid eye movement sleep.

During **rapid eye movement sleep (REM)**, breathing and heart rate may in-

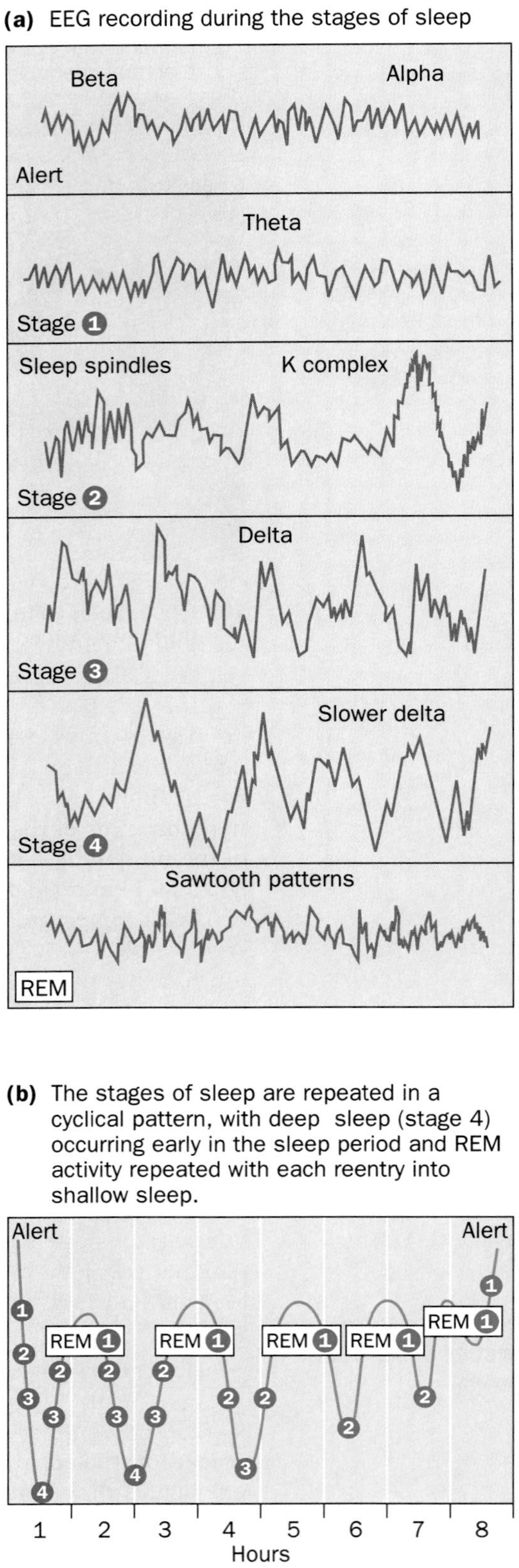
(a) EEG recording during the stages of sleep
Beta
Alpha
Alert
Theta
Stage 1
Sleep spindles
K complex
Stage 2
Delta
Stage 3
Slower delta
Stage 4
Sawtooth patterns
REM
(b) The stages of sleep are repeated in a cyclical pattern, with deep sleep (stage 4) occurring early in the sleep period and REM activity repeated with each reentry into shallow sleep.
Alert
Alert
REM 1
1 2 3 4 5 6 7 8
Hours

FIGURE 15.13 Sleep patterns.

crease. The muscles of the body are quite relaxed, and little movement occurs. Infants may spend 50 percent of their sleep time in REM sleep, while adults spend only about 15 percent. Each episode of REM sleep lasts about 10 to 15 minutes. REM sleep seems to be the time for memory consolidation. Sharp "sawtooth" waves are evident on the EEG and are actually quite similar to the alert EEG pattern.

Our pattern of waking and sleeping is one of the most familiar biological rhythms. Individual requirements for sleep may vary from 2 or 3 hours to 9 or 10 hours daily. Whatever the true function of sleep, we all require at least brief periods of it interspersed between periods of activity. You will probably spend about one-third of your life sleeping.

When our accustomed pattern of sleeping is disrupted by changes in the shift assignment at work, a jet plane trip across several time zones, all-night cramming for an exam, or a weekend party, we quickly notice the lethargy and tiredness that result until we adjust to a new pattern. The hypothalamus of the forebrain, the medulla of the hindbrain, and the pineal endocrine gland all help regulate our *sleep-wake cycle*. Body temperature cycles overlap a bit with the sleep-wake cycle. We typically fall asleep as we cool down in the evening and wake up as we begin to warm up early in the morning.

> EEG patterns distinguish four stages of sleep. We pass from stage 1 to stage 4 and then return in reverse order and repeat the cycle frequently. REM sleep occurs when we re-enter stage 1, and its EEG resembles that for the awake state; there are vivid, remembered dreams.

Diseases and Disorders

The spectrum of brain disease and symptoms reflects the anatomy and function of the brain. Some disorders are localized to particular centers of the brain and affect a single brain function, such as speech. However, because the brain controls all body regions, most brain disorders have multiple effects and complicated symptoms.

Some brain disorders are congenital and are caused by defective chromosomes or genes, while others result from lack of O_2, traumatic injury, infections, or tumors. Certain chronic diseases, such as epilepsy and Alzheimer's disease, produce significant changes in brain structures and operation, but their causes are obscure. Other chronic disorders termed "mental illnesses," such as depression and schizophrenia, are less clearly associated with brain structure changes but instead are associated with more subtle chemical changes.

Infectious Diseases of the Brain

While the meninges can become infected by a variety of bacteria, viruses, fungi, and protozoa, the blood-brain barrier usually prevents larger infective organisms from reaching brain tissue. For this reason, only the smallest infectious agents—viruses—are usually responsible for brain infections such as encephalitis and rabies.

Encephalitis is caused by a variety of viruses transmitted to humans by the bites of arthropod insects such as ticks and mosquitoes. The symptoms of encephalitis include headache, fever, muscle pain, sleepiness, convulsions, and coma. As with other viral infections, treatment is usually directed toward alleviating the symptoms.

Rabies is primarily a viral disease of mammals such as skunks, foxes, raccoons, bats, dogs, and cats. The disease may even spread to humans through the bite, scratch, or lick of an infected animal. The existence of vaccines for animals has minimized cases of human rabies in most industrialized countries, but rabies remains a major threat in developing nations.

The rabies virus spreads from the site of injury through sensory neurons to the spinal cord and brain. The virus multiplies in the brain, causing cell death and severe symptoms, including hyperactivity, agitation, disorientation, light sensitivity, painful swallowing, paralysis, and coma.

Treatment includes a series of vaccine injections.

Traumatic Cerebral Injury

The bones of the skull, the meninges, and the cerebrospinal fluid all help protect the brain against moderate traumas. However, a severe blow to the head may cause small blood vessels to rupture, leading to a hemorrhage or bruise. Bruising of the brain can result in the accumulation of blood between the bony cranium and the soft brain tissue. This increased pressure deforms the soft brain, causing altered functions and perhaps a brief period of unconsciousness called a *concussion*.

Cerebral Blood Flow and Strokes

Because of its great need for O_2 and glucose, brain tissue is extremely intolerant of reduced blood flow, and cell death occurs quickly without O_2. The delivery of O_2 also can be impaired by a loss of elasticity and hardening of the wall of an artery or by a buildup of fatty deposits.

The plugging or rupturing of a blood vessel supplying a brain region and the resulting area of cell death and loss of function are referred to as a *stroke* or *cerebrovascular accident* (*CVA*). High blood pressure and the abnormal production of blood clots in the heart or lungs increase the risk of stroke. Because of the crossing of motor neurons in the brain, a CVA in one cerebral hemisphere causes paralysis on the opposite side of the body. Speech functions are often lost when a stroke occurs in the left hemisphere. CVAs in the hindbrain are often fatal because of the damage to important respiratory and cardiac control centers. Recovery from a nonfatal stroke depends on the site and size of the affected region and the ability of surrounding brain areas to compensate for the loss of function (Figure 15.14).

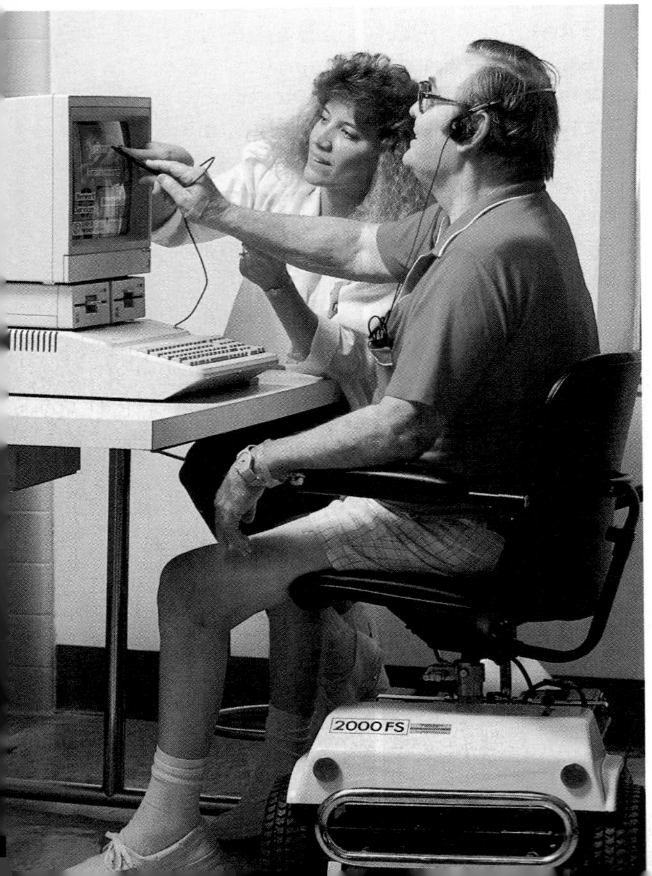
2000 FS

FIGURE 15.14 Occupational therapists aid in the physical and emotional recovery of stroke patients.

Brain Tumors

Primary brain tumors originating from the tissues of the brain are less common than *secondary brain tumors*, which are caused by the colonization of the brain by cancer cells metastasizing from tumors in other body regions. Roughly half of all brain tumors are benign, but their growth may distort adjacent brain areas and profoundly alter their functions. Radiation or chemotherapy treatments are often used to destroy such tumors or limit their growth.

The symptoms associated with brain neoplasms vary with their size and specific location. Headache arising from increased pressure around or within the brain is a common initial symptom, but tumors also may interfere with vision, speech, hearing, emotion, muscle coordination, and any of our higher mental activities.

Chronic Brain Disorders

Chronic brain disorders range from multiple sclerosis and epilepsy, which show dramatic changes in motor coordination, to conditions called *organic brain syndromes*. These syndromes are disorders of thought, emotion, and behavior which are associated with a structural change or degeneration in the brain, as in Alzheimer's disease.

Epilepsy (**ep**-uh-lep-see) afflicts about 1 in every 200 people in the United States. It is characterized by uncoordinated muscle spasms called convulsions or seizures that last a few seconds or many minutes. These convulsions result from sudden discharges of nerve impulses by clusters of neurons which disrupt the normal transmission of impulses through the brain. As a consequence, an affected person may see images or hear sounds that are not actually present, make unusual facial gestures or sounds, stare blankly into space, lose consciousness, lose urinary or bowel control, or convulse violently.

There are many causes of epilepsy, including brain tumors, infections, metabolic disturbances, alcoholic toxicity, and traumatic head injury. Characteristic EEG patterns help identify several forms of the disorder. Prescribed drugs such as *dilantin* inhibit or decrease neuronal activity, and epileptics can usually lead normal and productive lives without fear of seizures.

Age-Related Disorders

Severe memory loss and a decline in reasoning ability and attention span, often with some degree of belligerent behavior, are symptoms of *senile dementia* (*see-nile dee-men-she-uh*), a condition affecting about 10 percent of American adults over the age of 65. An early-onset form of dementia with distinct behavioral characteristics and changes in brain structure is known as *Alzheimer's disease*. Currently, over 4 million American adults are affected and this disease contributes to over 100,000 deaths each year. This makes Alzheimer's disease the fourth leading cause of death among the elderly. The disease is highly publicized because of the dramatic and debilitating changes it produces.

The first symptoms of Alzheimer's disease include forgetfulness and disorientation, and there is a progressive decline in reading and computational skills. The later stages are marked by confusion, hyperactivity, speech loss, and failure to recognize family members and friends. Patients often walk aimlessly and display extremes of emotion ranging from belligerent paranoia to joyful bliss. The final stages are characterized by increased sleepiness and coma. Death usually results from an associated condition such as pneumonia.

Upon examination after death, the brains of Alzheimer's patients show enlarged ventricles and a loss of as much as 30 percent of their estimated normal weight. There is extensive loss of brain cells, and the cerebral cortex becomes extremely thin, with narrower ridges, wider grooves, and larger ventricles. A PET scan of the living brain of an Alzheimer's patient shows reduced metabolic activity in the frontal and temporal lobes of the cerebrum, probably from a decrease in the number of available neurons (Figure 15.15). In the limbic system, where memories are consolidated, abnormal neurons contain tangled webs of protein filaments and numerous deposits of a material called *amyloid* (*am-muh-loyd*). There is also a significant decrease in the enzyme which produces the neurotransmitter acetylcholine, which is used by neurons in the limbic system.

Just what causes Alzheimer's disease is unknown, but environmental toxins, slow-acting viruses, decreased blood flow in the brain, altered cholesterol transport proteins, and gene mutations are among the leading candidates. Currently there is no treatment available for the disorder, though the progression of symptoms can be stalled by drugs which enhance the effectiveness of acetylcholine.

The identification of alternative forms of certain genes on chromosomes 14 and

FIGURE 15.15 A PET scan reveals the metabolic changes associated with brain disorders. (a) A normal adult brain. (b) The brain of an Alzheimer's disease patient. Note the reduced metabolic activity (blue and green colors) in the brain of the Alzheimer's patient.

(a) Normal brain **(b)** Alzheimer's brain

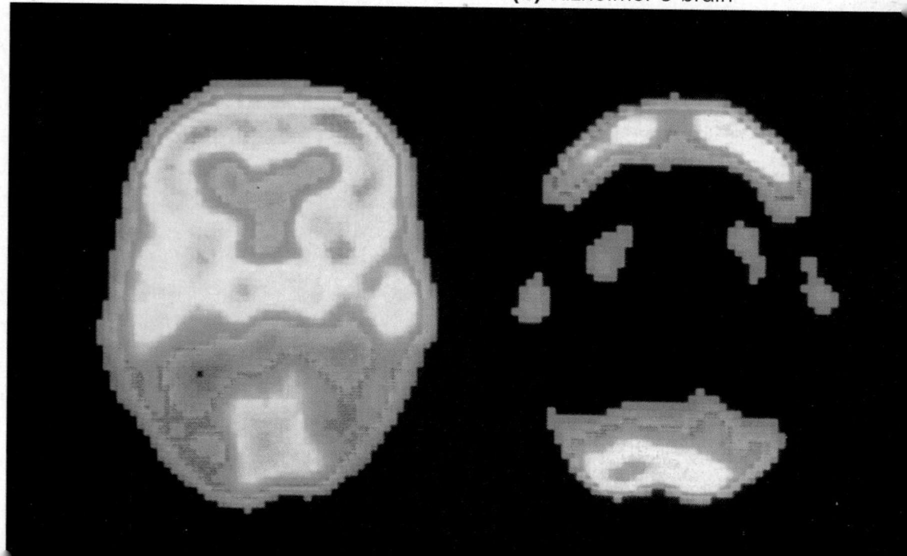

21 may eventually make possible diagnostic tests which can be administered early in life to determine who is at increased risk of developing Alzheimer's disease in old age. However, without more effective methods of treatment, such an early warning may be almost as disruptive for individuals and families as the disease itself is.

Life Span Changes

It is during the prenatal period and the first 18 months after birth that all of our brain cells are produced. The neurons are small at first but rapidly grow in size and in the number of synaptic connections they make with other neurons. The cerebrum of a newborn baby has fewer ridges and grooves than does an adult cerebrum. At birth, the brain represents about 12 percent of total body weight. It doubles in weight during the first year and reaches about 90 percent of its adult weight by age 6.

Reduction in Brain Size

Unfortunately, neurons are unable to divide by mitosis after they have been produced. Thus, as brain cells are lost from normal wear and tear, disease, or injury, they are not replaced. Estimates of the number of cells lost range as high as 100,000 per day in an adult. As cells are lost, the brain decreases in weight. A 10 percent loss in brain weight over a normal lifetime is not unusual. Both gray and white matter are lost, but loss is higher in certain brain regions. Areas of the cerebral cortex, especially the frontal and temporal lobes, may undergo a 50 percent cell loss by age 90. With the loss of neurons, the ridges become narrower and the grooves become wider. The ventricles increase in size to compensate for the reduced cellular volume of an older brain.

Accumulation of Aging Pigments

As the brain ages, cells accumulate a fatty yellow pigment called *lipofuscin*. The significance of this accumulation is still a matter of debate. Does this pigment cause aging-related changes in the brain or merely result from these changes? Future research will have to answer this question.

Learning and Memory Later in the Life Span

As we age, the number of neurons available to perform higher mental functions decreases. It will take us longer to recognize the size and shape of objects, judge distances, respond to sensations, and make decisions. Our movements will be affected by changes in the nervous system and the muscles and bones. Our verbal fluency and ability to learn and store new information will also decline, but to a lesser degree. However, not all mental functions and not all individuals are affected equally.

Many people continue to enjoy intellectual pursuits and work at mentally challenging activities until the end of life. In fact, there are definite advantages to pursuing lifelong learning. A "use it or lose it" philosophy must apply to the preservation of brain function.

Age-related loss of memory is more severe for short-term than for long-term memory. It may be the transfer or consolidation of memory that is most affected by age, making it easier to remember old memories than it is to form new ones. Postmortem studies suggest a correlation between loss of brain weight and memory loss: Those with severe memory loss show at least 20 percent reduction in brain weight. Again, memory lasts longer in those who practice making memories. People who continue to engage in challenging mental activity after retirement retain their memory capacity longer.

Chapter Summary

The Brain: Organization, Support, and Protection

- The brain can be divided into three general regions: hindbrain, midbrain, and forebrain.
- Cranial nerves (12 pairs) connect the brain to structures in the head, neck, and trunk.
- Spinal nerves (31 pairs) connect the spinal cord to the trunk and appendages.
- Three meninges protect the brain: the dura mater (outermost), arachnoid, and pia mater.
- CSF circulates through the ventricles and the subarachnoid space to cushion the brain.

Structure and Function of Brain Regions

- The hindbrain includes the medulla (cardiac, respiratory, and vasomotor control); cerebellum (muscle tone, balance, and the timing of muscle actions); pons (respiratory control; connection of hindbrain to other brain areas); and reticular formation (muscle tone; wakefulness).
- The midbrain coordinates eye movement, pupil size, and head movement.
- The forebrain includes the thalamus (relays sensory information); hypothalamus (body temperature, water balance, thirst, and hunger); limbic system (fixes memories, relays smell sensations, and forms emotions); and cerebrum (interprets sensations and assists in language, decision making, voluntary motor actions, and memory).

Learning and Memory

- Learning is a behavior change resulting from experience.
- Memory is acquiring, storing, and retrieving information.
- The hippocampus consolidates short-term memories into long-term memories.

Two Brains in One: Differences between the Cerebral Hemispheres

- The left hemisphere is used for language, speech, and mathematical calculation.
- The right cerebral hemisphere recognizes faces and interprets sounds, patterns, and shapes.

Genders of the Brain: Development of Male and Female Differences

- Biological gender is inherited.
- Gender identity may develop as hormones influence brain development.
- Sexual orientation develops separately from biological gender and gender identity.

Brain Chemistry: Neurotransmitters in the Brain

- Neurons in the brain communicate by means of excitatory or inhibitory neurotransmitters.
- Enkephalins and endorphins are natural painkillers.

Observing the Living Brain

- CT, PET, and MRI scans allow observation of living brains.
- Electroencephalograms (EEGs) are recordings of brain electrical activity.

Sleep and Dreaming

- Sleep occurs in five repeated stages: stages 1 through 4 and REM (rapid eye movement) sleep.
- The hypothalamus, medulla, and pineal gland help regulate the sleep-wake cycle.

Diseases and Disorders

- Encephalitis and rabies are caused by viruses.
- Traumatic injury may bruise the brain and cause concussion, a period of unconsciousness.
- A stroke is a ruptured or blocked blood vessel in the brain.
- Tumors arise in brain tissue or form as cancer cells metastasize from other body areas.
- Epilepsy is characterized by seizures, muscle spasms, and sometimes loss of consciousness.
- Alzheimer's disease is marked by brain weight loss, neuron filament tangles, and amyloid deposits.

Life Span Changes

- As the brain grows, brain cells increase in size and form new connections.
- With age, brain cell numbers are reduced.
- Lipofuscin accumulates in brain cells with age.
- Age-related memory loss affects short-term memory more than long-term memory.

Selected Key Terms

Alzheimer's disease (p. 415)
blood-brain barrier (p. 399)
cerebellum (p. 400)
cerebrospinal fluid (p. 397)
cerebrovascular accident (p. 414)
cerebrum (p. 402)
corpus callosum (p. 402)
electroencephalogram (p. 411)
hypothalamus (p. 401)
limbic system (p. 402)
medulla (p. 399)
meninges (p. 397)
midbrain (p. 401)
pons (p. 400)
thalamus (p. 401)

Review Activities

1. Explain the brain's relationship to the cranial and spinal nerves.
2. List the three meninges and describe how they protect the brain.
3. Describe where the CSF is produced and how it circulates through and around the brain.
4. What is the significance of the blood-brain barrier?
5. List the specialized regions of the hindbrain and summarize the functions of each one.
6. What are the functions of the midbrain?
7. List the specialized regions of the forebrain and summarize the functions of each one.
8. Explain how the arrangement of the primary sensory and motor areas of the cerebrum makes it possible to precisely locate sensations and control muscle activity.
9. Compare the responsibilities of Wernicke's and Broca's areas of the cerebrum.
10. How and where are memories stored in the brain? Which brain region assists in memory consolidation?
11. Compare the different responsibilities of the left and right cerebral hemispheres. Which structure physically connects the two hemispheres for information sharing?
12. Summarize the involvement of testosterone and estrogen in gender development and gender identity.
13. List several excitatory and inhibitory neurotransmitters used in the brain.
14. List and describe the four major patterns of brain waves.
15. List and describe the stages of sleep. Explain the nightly cycling of these stages.

Self-Quiz

Matching Exercise

___ 1. Site of respiratory and cardiac control centers	A. Reticular formation
___ 2. Area that controls muscle tone, balance, and muscle sequencing	B. Thalamus
___ 3. "Bridge" that connects hindbrain, midbrain, and forebrain	C. Cerebellum
___ 4. Area responsible for arousal and wakefulness	D. Limbic system
___ 5. Area responsible for control of head and eye movement	E. Medulla
___ 6. Structure that relays sensations to cerebrum	F. Corpus callosum
___ 7. Brain structure that controls body temperature, thirst, and hunger	G. Midbrain
___ 8. Structure that controls smell, memory consolidation, and emotion	H. Cerebrum
___ 9. Part of the brain that controls language, reasoning, and decision making	I. Pons
___ 10. Structure that connects cerebral hemispheres	J. Hypothalamus

Answers to Self-Quiz

1. E; 2. C; 3. I; 4. A; 5. G; 6. B; 7. J; 8. D; 9. H; 10. F

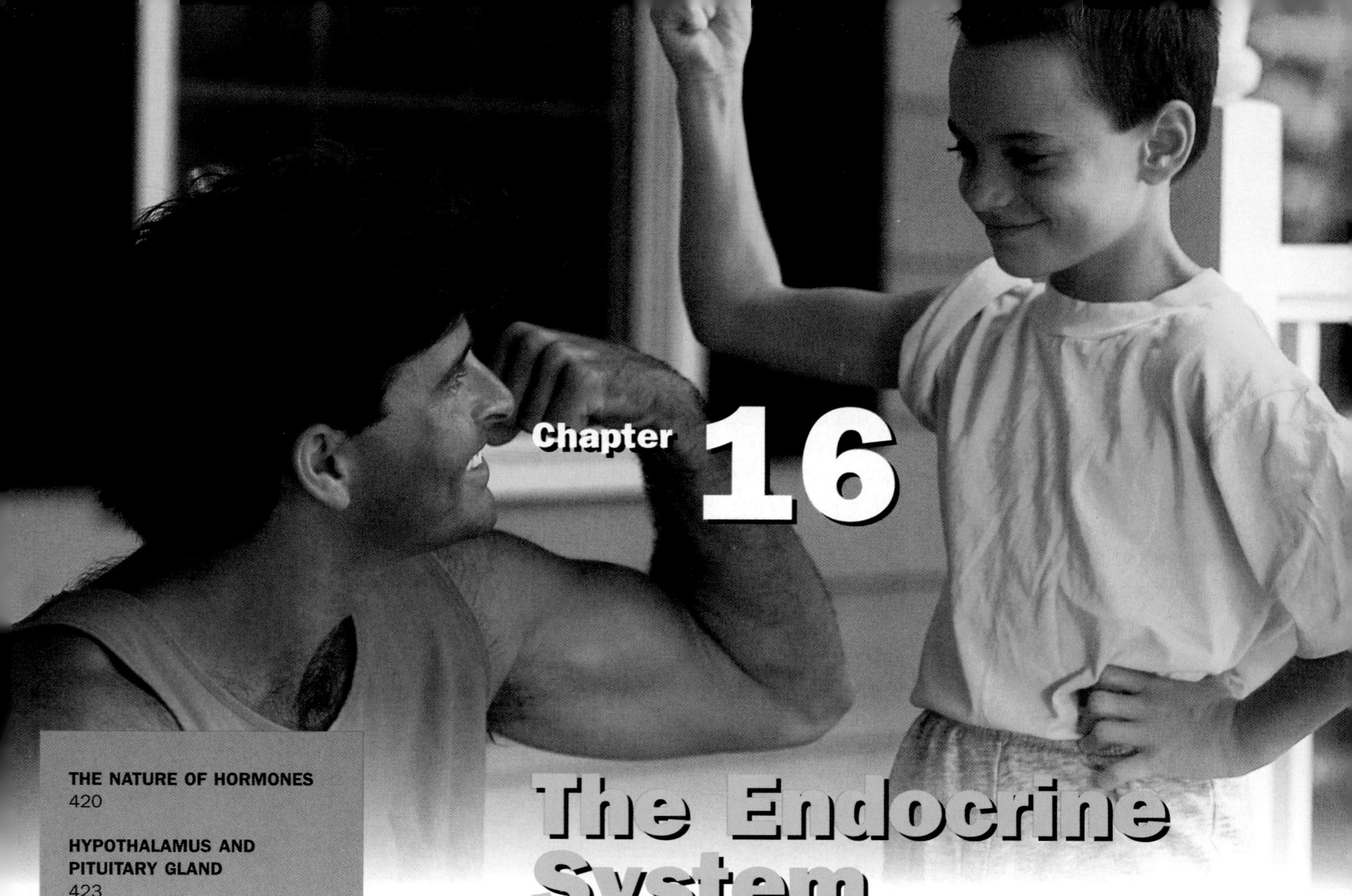

Chapter 16

The Endocrine System

Compare yourself now with the way you were in your preadolescent years. Obvious changes have occurred. Not only have you grown, you have matured sexually. You have developed characteristics that identify you as an adult and allow you to reproduce. These enormous changes in your body were orchestrated by the endocrine system and its hormones. **Hormones** are chemical substances secreted by glands or tissues of the endocrine system. The word "hormone" comes from a Greek word meaning "to activate," and hormones regulate body processes by activating and inhibiting certain cellular functions in specific tissues and organs.

In addition to long-term changes in body form and reproductive capacity, hormones govern day-to-day and even moment-to-moment changes. They control the amount of glucose in the blood and help coordinate the release of digestive secretions from the stomach and intestines. In the face of outside danger, hormones help prepare us for confrontation or flight. Responding to internal body conditions, they regulate body water, blood pressure, and the production of red blood cells. Over 50 hormones and hormonelike substances operate in the human body, regulating a variety of body functions.

Although hormones differ chemically, they all have four characteristics in common. (1) Hormones are manufactured and secreted in tiny amounts by specialized cells or ductless glands. (2) Upon release, a hormone enters the blood and is transported throughout the body. (3) Hormones produce specific effects on specific *target cells*. (4) The target cells are usually distant from the site of a hormone's secretion.

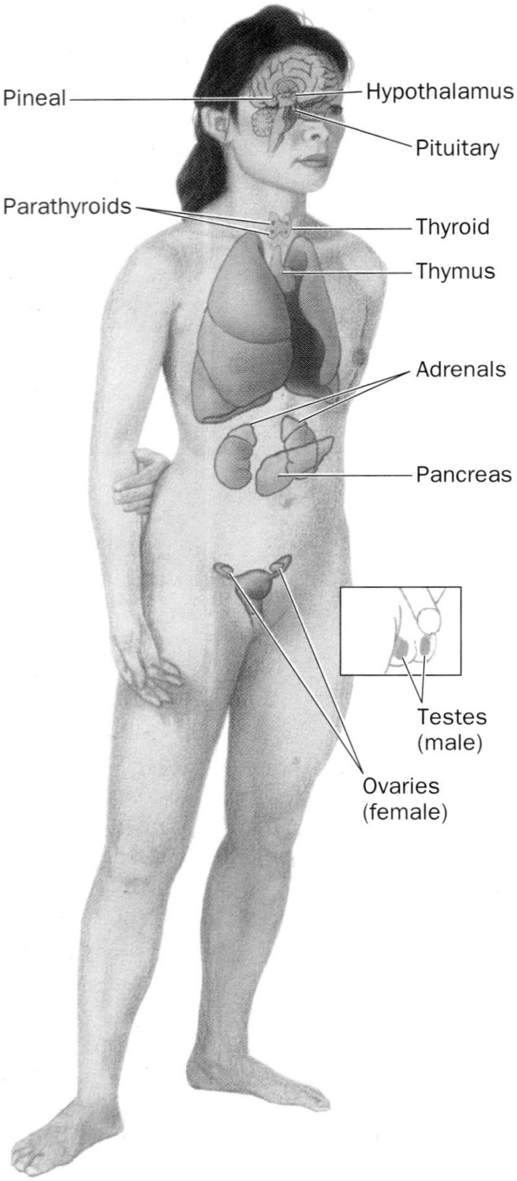

Pineal
Hypothalamus
Pituitary
Parathyroids
Thyroid
Thymus
Adrenals
Pancreas
Testes (male)
Ovaries (female)

FIGURE 16.1 Location of the major endocrine glands.

Traditionally, the **endocrine system** has been considered to consist of 10 glands, some of which may also have other functions (Figure 16.1 and Table 16.1). For example, the ovaries and testes produce eggs and sperm for reproduction but are also the source of powerful sex hormones. To the traditional list of endocrine glands we can add organs such as the stomach, small intestine, kidneys, and heart, which have been discovered to secrete hormones.

Of course, the neurons of the nervous system also communicate with the glands and organs of the body. The functions of the nervous and endocrine systems often complement each other as they continuously adjust the activity of body cells. The nervous system and its sensory receptors are involved in monitoring both the external and internal environments, while the endocrine system responds to the internal environment and signals from the nervous system.

In this chapter, we will describe how hormones alter the activity of target cells. We will then discuss the glands of the endocrine system, identifying the hormones they secrete. In the section on diseases and disorders, the focus will be on diabetes mellitus and stress. We will close the chapter with some reflections on how the endocrine system changes over the life span.

The Nature of Hormones

The body's 50 different hormones all have specific effects. Before examining individual hormones, let's consider some of their general characteristics and modes of functioning. What are hormones? If a hormone circulates throughout the body, why does it affect only its target cells? How does a hormone influence a target cell's function? How is the release of hormones regulated so that their effects occur at appropriate times?

Two Categories of Hormones and Their Actions

As hormones circulate in the body, they leave blood capillaries, enter tissue spaces, and come into contact with body cells. However, a hormone influences only those target cells which contain a specific receptor for that hormone. Depending on the hormone, the target cell's receptor is located either on the surface of the plasma membrane or in the cell. To organize our discussion about hormone action, it is convenient to group hormones into two principal categories: nonsteroid hormones and steroid hormones.

TABLE 16.1 The Major Endocrine Glands and Their Hormones

Endocrine Gland	Hormone Secreted	Principal Action
Hypothalamus	Releasing and inhibiting hormones	Controls release of anterior pituitary hormones
Pituitary		
Posterior pituitary	Oxytocin	Stimulates smooth muscle contraction of uterus and mammary gland
	Antidiuretic hormone (ADH)	Regulates water reabsorption
Anterior pituitary	Growth hormone (GH)	Influences metabolism and growth
	Prolactin	Stimulates mammary gland development
	Follicle stimulating hormone (FSH)	Stimulates development of ovary and testes
	Luteinizing hormone (LH)	Stimulates production of sex hormones
	Adrenocorticotropic hormone (ACTH)	Stimulates adrenal cortex
	Thyroid stimulating hormone (TSH)	Stimulates thyroid gland
Thyroid gland	Thyroxine	Regulates metabolism
	Calcitonin	Lowers blood calcium levels
Parathyroid glands	Parathyroid	Raises blood calcium levels
Pancreas	Insulin	Lowers blood glucose levels
	Glucagon	Raises blood glucose levels
Adrenal glands		
Medulla	Epinephrine and norepinephrine	Prepares body for quick action
Cortex	Glucocorticoids	Influences carbohydrate metabolism, inflammation
	Aldosterone	Regulates Na^+, K^+, and water concentrations
	Androgens and estrogens	Some effects in females
Testes	Testosterone	Sperm production, sex characteristics, maintenance of reproductive organs
Ovary	Estrogen and progesterone	Egg development, sex characteristics, menstrual cycle
Pineal	Melatonin	Regulates many body rhythms
Thymus	Thymosin	Stimulates T-cell lymphocyte development

Nonsteroid hormones and second messengers. Nonsteroid hormones are proteins, small chains of amino acids, or a modified amino acid (tyrosine). They include most of our hormones, including those from the hypothalamus, pituitary, and pancreas. Nonsteroid hormones are water-soluble and cannot diffuse across cellular membranes. They bind with protein receptors protruding from the plasma membrane of the target cell (Figure 16.2a).

Once binding occurs, most nonsteroid hormones have a similar mode of action. They activate an enzyme in the cell that increases the production of another chemical referred to as a **second messenger** (the first messenger is the hormone). We will describe one such second messenger, called *cyclic AMP* (cyclic adenosine monophosphate), which is produced from ATP. Cyclic AMP activates specific enzymes in the cytoplasm. These activated enzymes perform cellular functions that ultimately alter cell activity. The first and second messengers function rapidly and produce their effects on target cells within seconds or minutes.

Steroid hormones and gene activation. The steroid hormones are all lipids derived from cholesterol. They are secreted from the ovaries (estrogen, progesterone), testes (testosterone), and adrenal cortex (cortisol, aldosterone). Steroid hormones do not bind with receptors on plasma membranes. These hormones are lipid-soluble and diffuse freely across the membrane lipid bilayer and into the cell

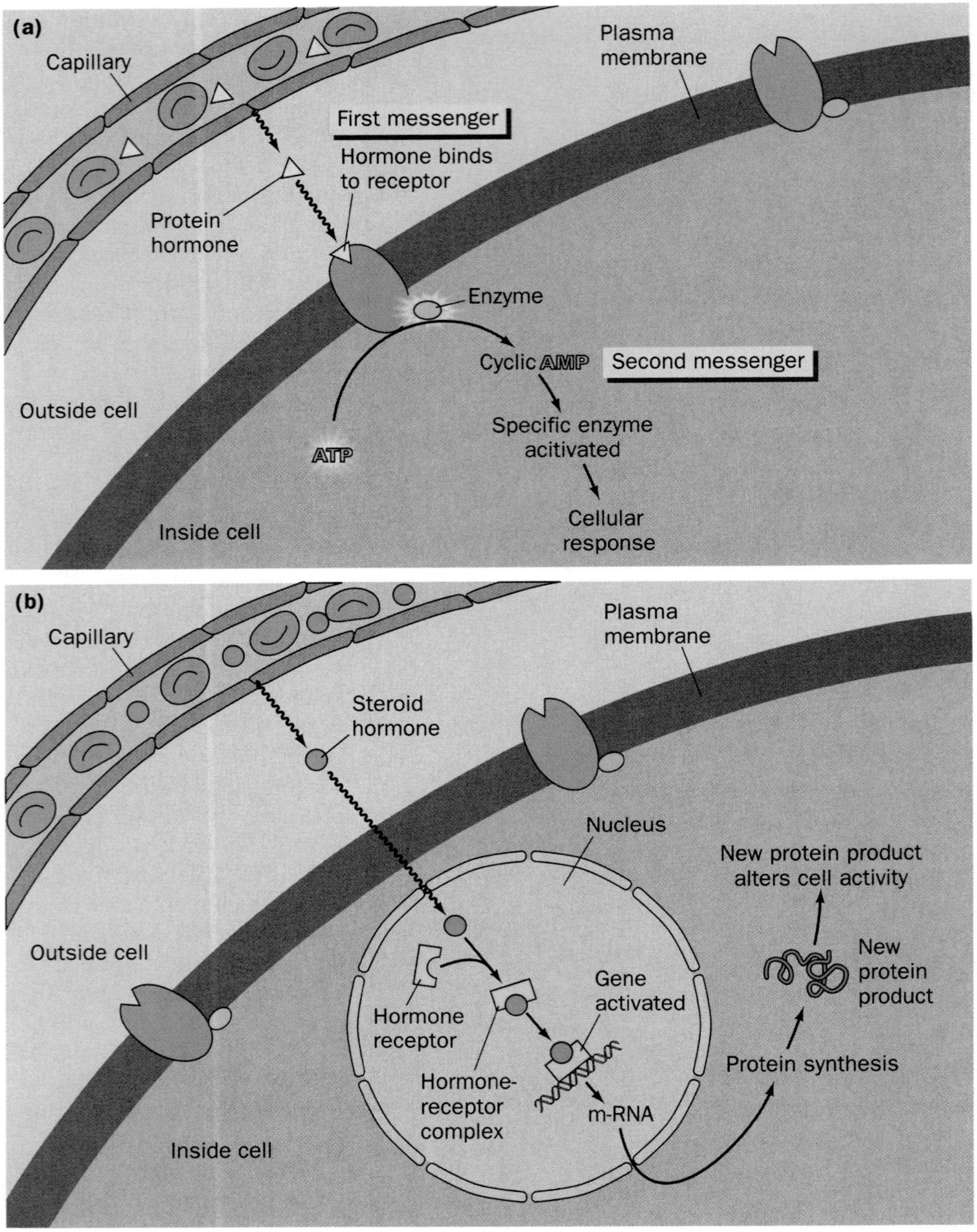
(a)
Capillary
Plasma membrane
First messenger
Hormone binds to receptor
Protein hormone
Enzyme
Cyclic AMP
Second messenger
Outside cell
Specific enzyme acitivated
ATP
Inside cell
Cellular response
(b)
Capillary
Plasma membrane
Steroid hormone
Nucleus
New protein product alters cell activity
Outside cell
New protein product
Gene activated
Hormone receptor
Protein synthesis
Hormone-receptor complex
m-RNA
Inside cell

FIGURE 16.2 (a) How nonsteroid hormones influence target cells. (b) How steroid hormones influence target cells.

(Figure 16.2b). Once it is in the cell, the hormone combines with a receptor molecule in the cytoplasm or nucleus. The hormone-receptor complex then binds to a specific site on DNA and initiates the transcription of specific genes. As a result, new proteins are synthesized in the cell.

The newly synthesized proteins may function as enzymes or structural proteins, but however they function, they produce changes in the target cell that are characteristic of a specific hormone. The response of target cells to a steroid hormone may require several hours or several days depending on the time needed for protein synthesis and for the new protein to produce changes in the cell.

> Nonsteroid hormones bind to receptors on the surface of the target cell, and activate second messengers such as cyclic AMP. Steroid hormones move across the plasma membrane, binding to receptors inside the cell and activating genes.

Regulating Hormone Secretion

Because hormones alter target cell functions, it is important that appropriate amounts of a hormone be secreted. Too much or too little of a circulating hormone can lead to abnormal functions of target cells and eventually cause endocrine disorders.

The synthesis and release of hormones are regulated by a **negative feedback loop** (Chapter 4). In negative feedback, information about the effects of a hormone on its target cells is returned (fed back) to the endocrine gland. The feedback of this information has a depressing (negative) effect on the further release of the hormone. This is a self-regulating mechanism which ensures that appropriate amounts of a hormone are released to maintain homeostasis.

Let us briefly describe an example of negative feedback using the regulation of blood calcium (Ca^{2+}) concentration by the parathyroid glands (Figure 16.3). When blood Ca^{2+} becomes low, the parathyroid glands release **parathyroid hormone.** Parathyroid hormone stimulates target cells, such as bone cells, to release Ca^{2+} into the blood. As blood Ca^{2+} increases above normal levels, this is detected by the parathyroid glands and secretion of the hormone is reduced.

It is through such negative feedback loops that our endocrine glands are regulated. We will describe other examples of negative feedback throughout this chapter.

> A negative feedback loop is used to regulate the production of hormones. It is a self-regulating mechanism that maintains homeostasis of the body by continuously adjusting hormone concentrations.

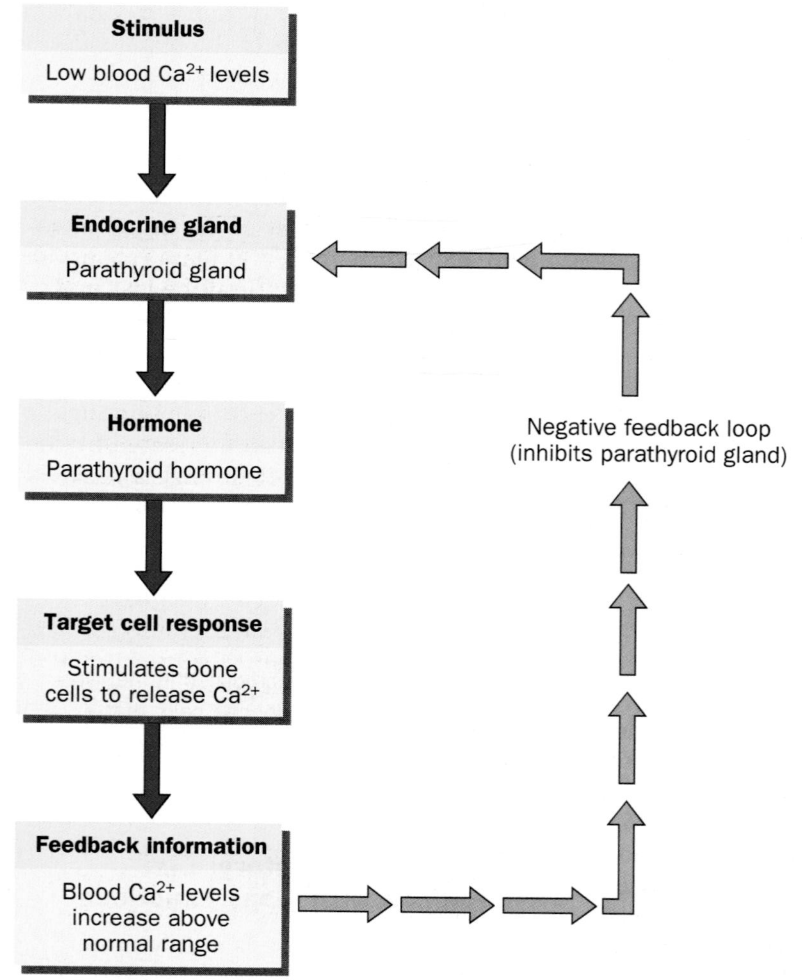

Stimulus
Low blood Ca2+ levels
Endocrine gland
Parathyroid gland
Hormone
Parathyroid hormone
Target cell response
Stimulates bone cells to release Ca2+
Feedback information
Blood Ca2+ levels increase above normal range
Negative feedback loop (inhibits parathyroid gland)

FIGURE 16.3 A negative feedback loop associated with the regulation of hormone release. Regulation of parathyroid hormone is shown here.

Hypothalamus and Pituitary Gland

The **hypothalamus** is an important regulatory center of the brain. It continuously monitors the composition of the blood that flows through it, controlling the body's water balance, temperature, and hunger. It also monitors the autonomic nervous system (ANS) and the pituitary gland. The **pituitary** (*pit-too-ih-tare-ee*) is a pea-sized endocrine gland attached by a stalk to the underside of the hypothalamus and located at a position just posterior to the eyes (Figure 16.4). The pituitary is sometimes called the "master gland" of the body because it regulates so many of the other endocrine glands.

Although the pituitary appears to be one gland, it is divided into two different parts: the anterior pituitary and the posterior pituitary. The release of hormones from both parts of the pituitary depends entirely on nerve cells in the hypothalamus. This is a good example of how the function of one body system (endocrine system) is integrated with that of another body system (nervous system).

In the next section, we will describe the posterior and anterior lobes of the pituitary, the hormones they release, the roles of those hormones in the body, and the way their release is regulated by the hypothalamus.

The Posterior Pituitary: Two Hormones with Several Functions

The posterior pituitary releases two hormones: oxytocin and antidiuretic hormone. These hormones are actually synthesized in the hypothalamus by specialized nerve cells called *neuroendocrine cells* (Figure 16.4). The neuroendocrine cells extend into the posterior pituitary, where the hormones are secreted into the blood.

Oxytocin (*ox-see-tow-sin*) is released at the end of pregnancy and causes the smooth muscle of the uterus to contract, forcing the infant from the uterus during childbirth. Oxytocin also causes the smooth muscle of the mammary glands to contract, forcing out milk for the infant. The secretion of oxytocin during breast-feeding illustrates the functional relation-

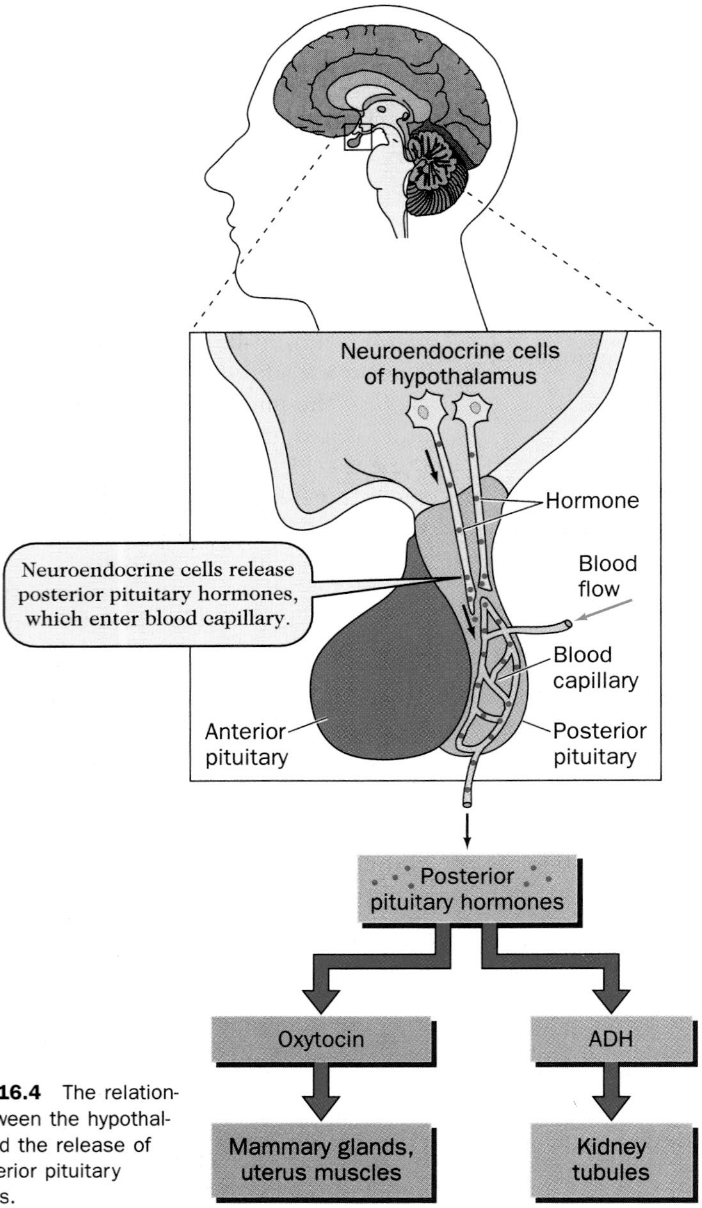

Neuroendocrine cells of hypothalamus
Hormone
Neuroendocrine cells release posterior pituitary hormones, which enter blood capillary.
Blood flow
Blood capillary
Anterior pituitary
Posterior pituitary
Posterior pituitary hormones
Oxytocin
ADH
Mammary glands, uterus muscles
Kidney tubules

FIGURE 16.4 The relationship between the hypothalamus and the release of the posterior pituitary hormones.

ship between the posterior pituitary and the hypothalamus.

An infant's suckling on the mother's breast stimulates sensory nerves in the nipple (Figure 16.5). These sensory neurons carry impulses to the hypothalamus, where the impulses are transmitted to the neuroendocrine cells that produce and store oxytocin. Stimulation of these neuroendocrine cells causes the release of oxytocin into the posterior pituitary from which it enters the blood. Upon reaching breast tissue, oxytocin encounters cells with receptors to which it binds. This binding initiates smooth muscle contraction, resulting in the release of milk.

Antidiuretic hormone (ADH) (*an-tie-die-you-**ret**-ick*) regulates water balance in the body. This hormone helps prevent dehydration by stimulating collecting ducts in the kidney to absorb more water during urine formation (Chapter 11). As a result, water is retained in the body as less water is eliminated with the urine.

The hypothalamus contains nerve cells that detect low water concentrations in the blood. When this situation occurs, those nerve cells stimulate neuroendocrine cells of the hypothalamus to release ADH into the posterior pituitary. Once it is released, ADH enters the blood and is carried to the kidneys, where it activates the reabsorption of water, increasing the water concentration in the blood. When normal water concentration is reestablished, the nerve cells of the hypothalamus are no longer stimulated and the neuroendocrine cells stop releasing ADH. This reduction in ADH secretion is an example of negative feedback.

Insufficient release of ADH allows too much water to be lost in the urine. In severe conditions, up to 30 liters (about 30 quarts) of watery urine can be lost in a day. This condition is called *diabetes insipidus* (*die-ah-**bee**-tees in-**sip**-ih-dus*) and, if left untreated, leads to death from dehydration and the loss of important ions. The condition can be treated by the administration of ADH through a nasal spray. Reduced release of ADH is also caused by alcohol, which explains why the consumption of alcohol increases urine formation.

> Oxytocin and ADH are produced in the hypothalamus by neuroendocrine cells that extend into the posterior pituitary and, upon stimulation, release their hormones.

The Anterior Pituitary: Six Hormones with Many Functions

The anterior pituitary is a little more complex than the posterior pituitary. The anterior pituitary secretes six major hor-

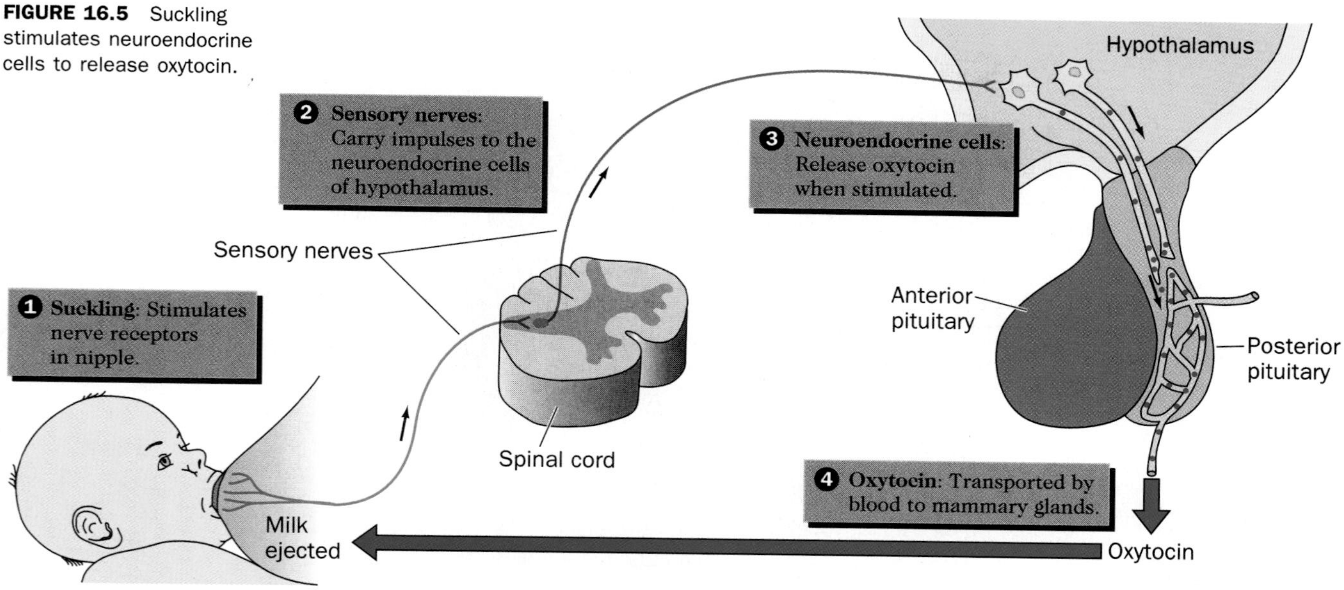

2 Sensory nerves: Carry impulses to the neuroendocrine cells of hypothalamus.
Hypothalamus
3 Neuroendocrine cells: Release oxytocin when stimulated.
Sensory nerves
1 Suckling: Stimulates nerve receptors in nipple.
Anterior pituitary
Posterior pituitary
Spinal cord
4 Oxytocin: Transported by blood to mammary glands.
Milk ejected
Oxytocin

FIGURE 16.5 Suckling stimulates neuroendocrine cells to release oxytocin.

mones; growth hormone, prolactin, follicle stimulating hormone, luteinizing hormone, adrenocorticotropic hormone, and thyroid stimulating hormone (Table 16.1). Their release is regulated by releasing hormones and inhibiting hormones that are secreted by the neuroendocrine cells of the hypothalamus (Figure 16.6). *Releasing hormones* stimulate the release of anterior pituitary hormones, while *inhibiting hormones* suppress the release of some anterior pituitary hormones. Releasing and inhibiting hormones are transported from the hypothalamus directly to the anterior pituitary through a series of capillaries called the *pituitary portal system* (Figure 16.6).

Growth hormone (GH). This hormone from the anterior pituitary influences the metabolism of cells in ways that promote the growth of body tissues. GH stimulates protein synthesis by increasing the rate at which amino acids are transported into cells. Fat utilization is encouraged by GH and, therefore, cellular energy is obtained primarily from fat instead of carbohydrates. As a result, less liver glycogen has to be broken down to glucose. The release of GH is controlled by both releasing and inhibiting hormones from the hypothalamus.

Although GH is released throughout life, its effects are most dramatic during childhood development. During childhood GH promotes an increase in the size of skeletal muscles and bone growth. As a result, a child's muscle mass increases along with an increase in bone length.

The release of too much or too little GH during childhood development influences an individual's height. Too little GH results in a relatively short individual described as a *pituitary dwarf* (Figure 16.7a). Such individuals may be no more than 1.2 meters (4 feet) in height but are normal in body proportions. Young individuals who are shorter than normal because of insufficient GH production can now be treated. If the condition is diagnosed while bone growth is still occurring, normal height can be attained through the periodic administration of GH. By contrast, the release of too much GH during development increases bone growth so that the individual becomes abnormally tall, a condition called *gigantism* (Figure 16.7b).

In adults, when too much GH is secreted by the anterior pituitary, *acromegaly* (*ak-row-**meg**-ah-lee*) can develop. In this condition, bones in the face, hands, and feet enlarge and connective tissues of the body thicken (Figure 16.8). Long bones do not increase in length because their growth regions are inactive in adults. If this condition is left

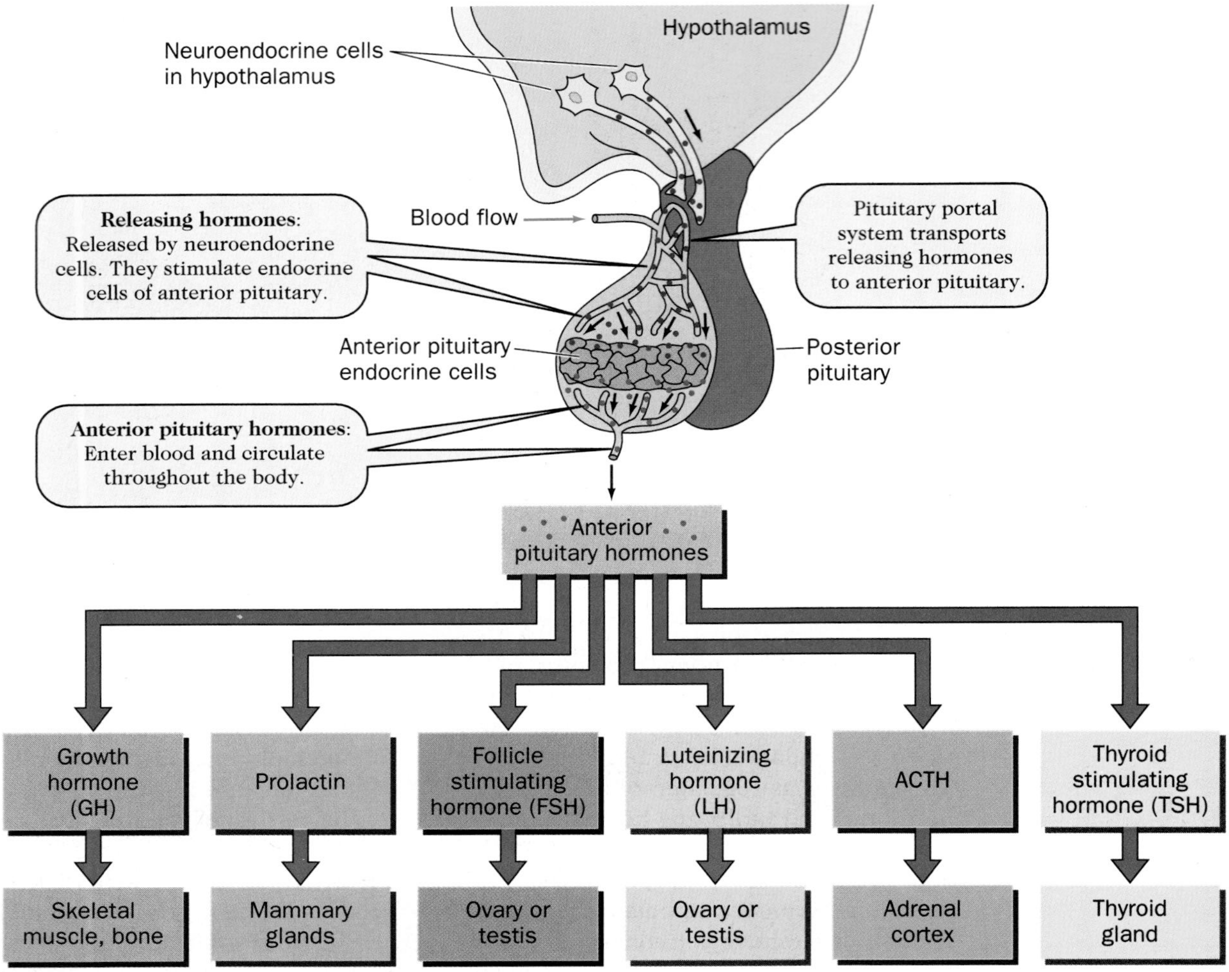
Hypothalamus
Neuroendocrine cells in hypothalamus
Releasing hormones: Released by neuroendocrine cells. They stimulate endocrine cells of anterior pituitary.
Blood flow
Pituitary portal system transports releasing hormones to anterior pituitary.
Anterior pituitary endocrine cells
Posterior pituitary
Anterior pituitary hormones: Enter blood and circulate throughout the body.
Anterior pituitary hormones
Growth hormone (GH)
Prolactin
Follicle stimulating hormone (FSH)
Luteinizing hormone (LH)
ACTH
Thyroid stimulating hormone (TSH)
Skeletal muscle, bone
Mammary glands
Ovary or testis
Ovary or testis
Adrenal cortex
Thyroid gland

FIGURE 16.6 The relationship between releasing hormones from the hypothalamus and the secretion of anterior pituitary hormones.

untreated, changes in body features may be substantial and permanent. This condition is often caused by a tumor in the hypothalamus or pituitary.

Prolactin. This hormone stimulates development of mammary gland cells and milk production. The development of mammary gland cells also requires stimulation by other hormones. During pregnancy, large amounts of estrogen and progesterone in the blood stimulate the hypothalamus to secrete more *prolactin-releasing hormone.* This stimulates the anterior pituitary to release more prolactin. Increased prolactin levels stimulate growth of the mammary glands and milk production. After birth, when estrogen and progesterone in the blood decline, suckling stimulates the continued secretion of *prolactin-releasing hormone,* and as a result, milk production continues.

In females who are not breast-feeding, more *prolactin-inhibiting hormone* is secreted by the hypothalamus, and so milk production does not occur. The function of prolactin in males is unknown.

Gonadotropic hormones. The gonadotropic hormones (*go-nah-doe-**trow**-pick*) include two hormones that stimulate the growth, development, and function of the gonads, that is, the reproductive organs. **Follicle stimulating hormone (FSH)** stimulates the development of eggs in a female's ovaries and that of sperm in a male's testes. **Luteinizing hormone (LH)** (***lew**-tea-in-eye-zing*) stimulates egg development, estrogen produc-

FIGURE 16.7 (a) A pituitary dwarf. (b) A pituitary giant.

(a) (b)

tion in the ovary, and ovulation (egg release) in females. In males, LH stimulates the cells of the testes to produce the male hormone testosterone.

The anterior pituitary does not release gonadotropins until puberty (about age 11 for females and age 13 for males). It is during puberty that the ovaries and testes begin to develop, secreting either estrogen or testosterone into the blood. The actions of these two hormones produce the body changes associated with sexual maturity.

Adrenocorticotropic hormone (ACTH). Adrenocorticotropic hormone (*ah-dree-no-core-tick-oh-**trow**-pick*) from the anterior pituitary stimulates the cortex, or outer layer, of the adrenal gland to release a group of hormones called glucocorticoids. Glucocorticoids are steroid hormones that help the body respond to stress-related conditions. The release of ACTH is controlled by a releasing hormone from the hypothalamus through a negative feedback loop that involves the concentration of glucocorticoids in the blood.

Thyroid stimulating hormone (TSH). This anterior pituitary hormone stimulates the thyroid gland to produce and release its hormones. The secretion of TSH is stimulated by *TSH-releasing hormone* from the hypothalamus. We will discuss TSH further in the following section.

> The anterior pituitary secretes GH, prolactin, FSH, LH, ACTH, and TSH. Their release is regulated by specific releasing or inhibiting hormones secreted by the hypothalamus and carried to the anterior pituitary by the pituitary portal system.

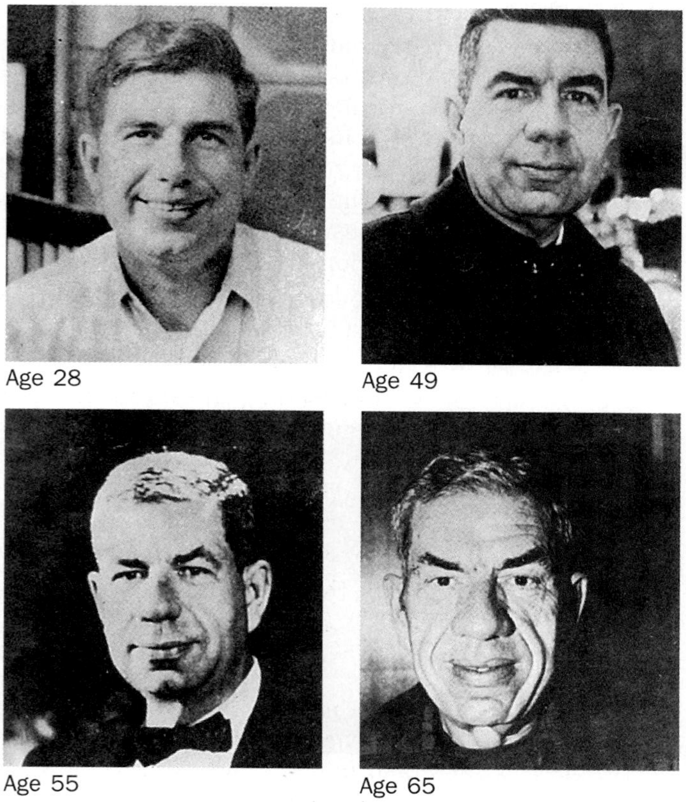
Age 28
Age 49
Age 55
Age 65

FIGURE 16.8 The progressive development of acromegaly in the face.

Thyroid and Parathyroid Glands

The thyroid and parathyroid glands are two distinct endocrine glands. We have grouped them together because they are located in the same site in the neck and because both release hormones that help regulate Ca^{2+} in the body. The thyroid gland consists of two lobes positioned in front of the trachea just below the larynx (Figure 16.9). The parathyroid glands consist of four small, almost inconspicuous tissue masses that are attached to the back of the thyroid gland.

The Thyroid Gland

The thyroid gland produces two hormones: thyroxine and calcitonin. Each performs different functions in the body. Thyroxine regulates cell metabolism; calcitonin regulates blood calcium levels.

Thyroxine. The hormone thyroxine (*thigh-**rock**-sin*) is synthesized by the thyroid gland from two molecules of the amino acid tyrosine to which are added atoms of iodine (I). Although it is a nonsteroid hormone, thyroxine does not act by attaching to receptors on the plasma membrane. Instead, it behaves like a steroid hormone and enters target cells, where it activates specific genes that code for the enzymes used in cell energy metabolism. Thyroxine regulates ATP production in almost all body cells.

As levels of thyroxine increase, the tissues of the body increase their rate of glucose breakdown and more ATP is produced. This in turn increases the body's basal metabolic rate (BMR) and the production of body heat. Decreasing thyroxine has the opposite effect.

The release of thyroxine from the thyroid gland is regulated by two hormones: *TSH-releasing hormone* from the hypothalamus and *thyroid stimulating hormone* (*TSH*) from the anterior pituitary. Conditions that increase energy requirements, such as pregnancy and cold temperatures, stimulate an increase of thyroxine through a series of steps (Figure 16.10): (1) Prolonged exposure to cold temperature, for example, stimulates the hypothalamus to release TSH-releasing hormone. (2) This stimulates the anterior pituitary to release TSH. (3) Increasing levels of TSH stimulate the thyroid gland to manufacture and secrete more thyroxine. (4) In the blood, thyroxine circulates throughout the body and increases cellular energy metabolism, maintaining normal body temperature.

Insufficient production of thyroxine in adults leads to a condition called *myxedema* (*mix-eh-**dee**-ma*). Symptoms include a low BMR, low body temperature, fluid accumulation under the skin (edema), lethargy, and weight gain. Although most common in elderly women, myxedema can occur in both sexes at all ages. In children, insufficient secretion of thyroxine may slow body growth, delay sexual maturity, and alter brain functioning. In both children and adults, the condition can be treated by taking thyroxine pills.

Since iodine is needed to chemically synthesize thyroxine, a lack of dietary iodine prevents its synthesis. Low levels of thyroxine in the blood stimulate the

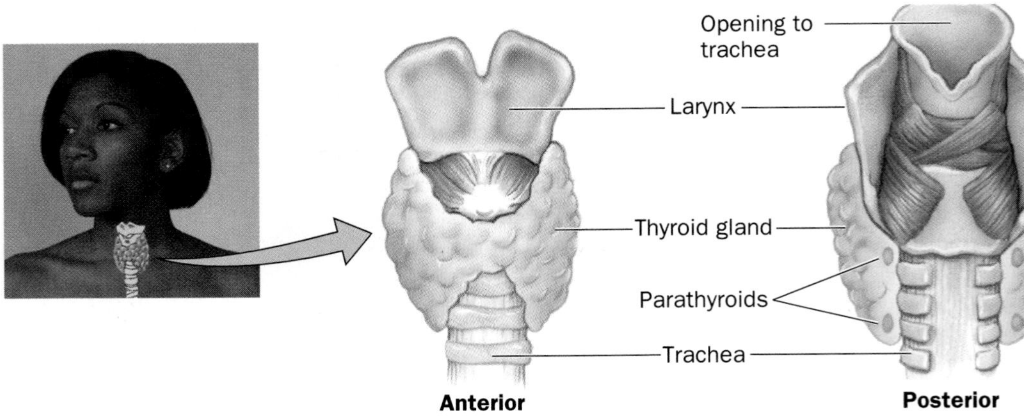
Opening to trachea
Larynx
Thyroid gland
Parathyroids
Trachea
Anterior
Posterior

FIGURE 16.9 Location of the thyroid and parathyroid glands.

FIGURE 16.10 The release and regulation of thyroxine from the thyroid gland.

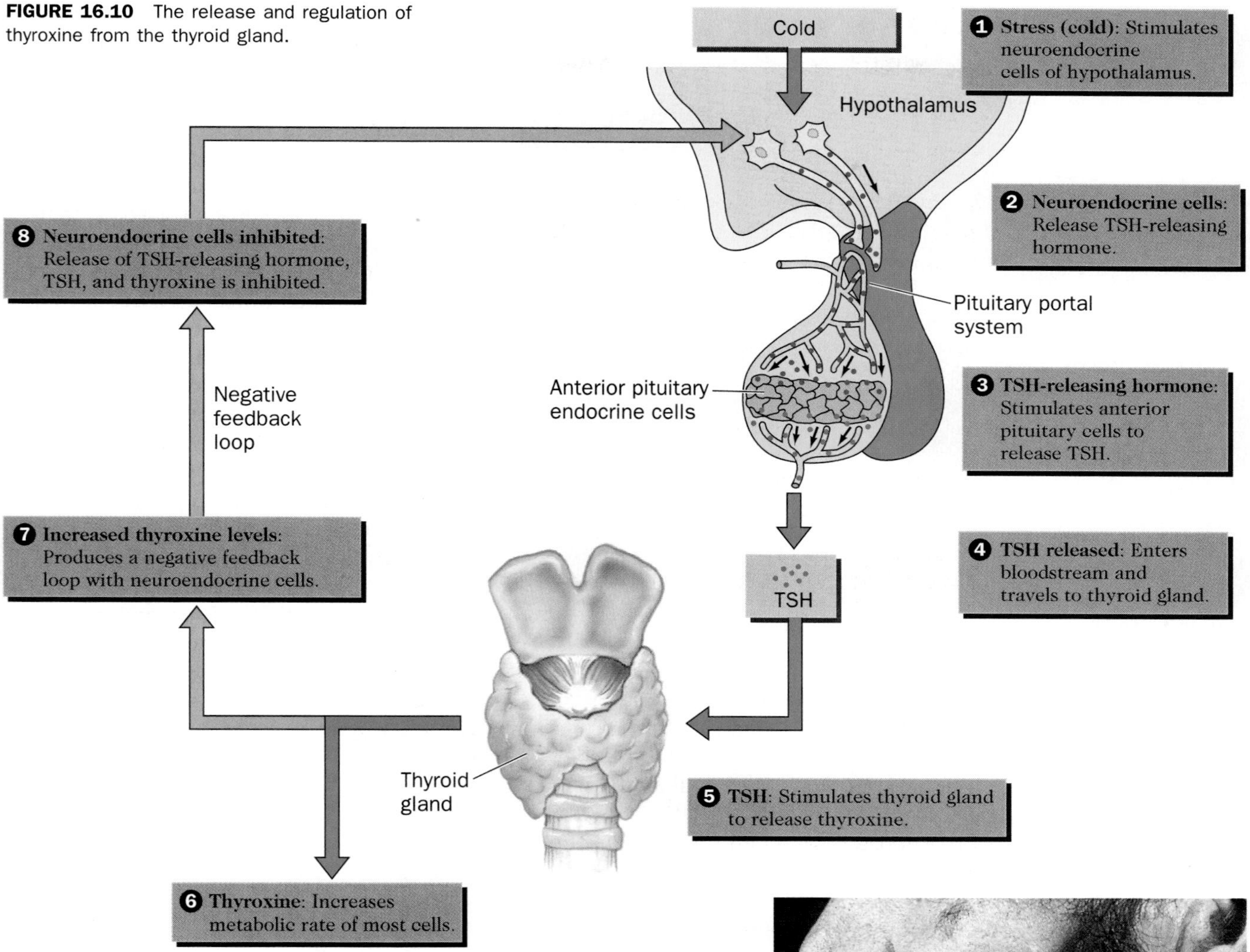
Cold
1 Stress (cold): Stimulates neuroendocrine cells of hypothalamus.
Hypothalamus
2 Neuroendocrine cells: Release TSH-releasing hormone.
Pituitary portal system
Anterior pituitary endocrine cells
3 TSH-releasing hormone: Stimulates anterior pituitary cells to release TSH.
TSH
4 TSH released: Enters bloodstream and travels to thyroid gland.
5 TSH: Stimulates thyroid gland to release thyroxine.
Thyroid gland
6 Thyroxine: Increases metabolic rate of most cells.
7 Increased thyroxine levels: Produces a negative feedback loop with neuroendocrine cells.
Negative feedback loop
8 Neuroendocrine cells inhibited: Release of TSH-releasing hormone, TSH, and thyroxine is inhibited.

secretion of TSH from the anterior pituitary. TSH causes the thyroid gland to enlarge as it attempts to synthesize more thyroxine. This condition is called *goiter* (*goy-ter*) and often results in a visually enlarged thyroid gland (Figure 16.11). Goiter is less common today because of the availability of iodized table salt.

Oversecretion of thyroxine elevates the metabolic rate of tissues, increasing heat production, physical activity, and weight loss. *Graves' disease* is the most common cause of oversecretion of thyroxine in adults. It is an autoimmune disorder in which a person's antibodies attach to thyroid gland cells and stimulate their growth, increasing hormone release. Often Graves' disease results in fluid accumulation behind the eyes, causing them to protrude (Figure 16.12).

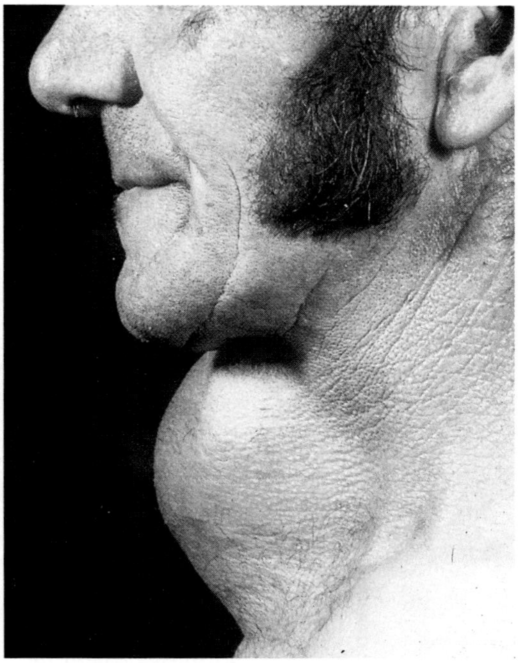

FIGURE 16.11 An endemic goiter resulting from an enlarged thyroid gland.

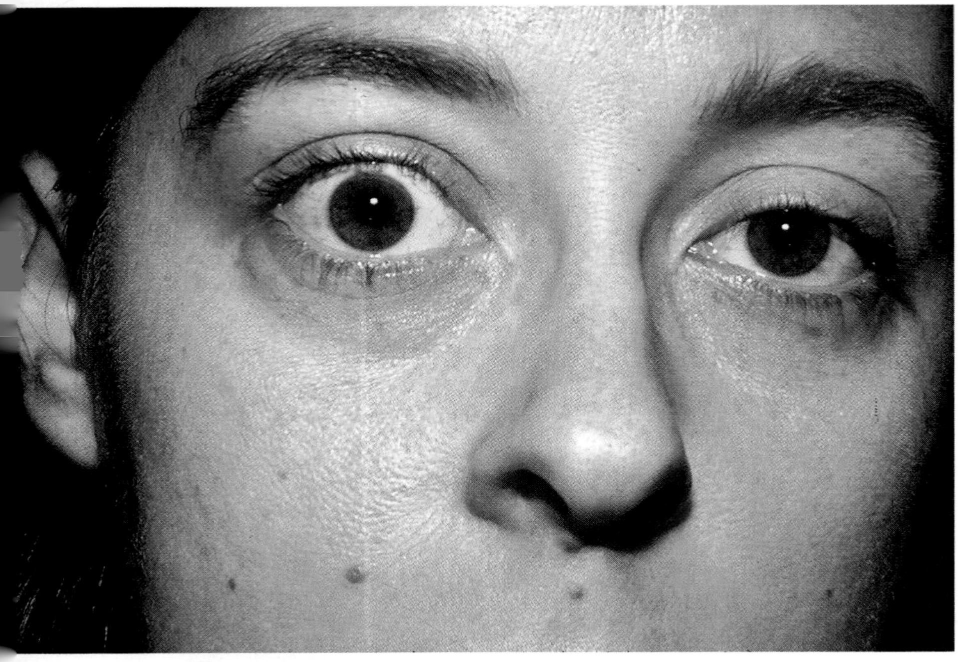

FIGURE 16.12 Protruding eyes resulting from Graves' disease.

Calcitonin. The hormone calcitonin (*cal-sih-**tone**-in*) is secreted by different cells of the thyroid gland. Calcitonin functions to lower the amount of Ca^{2+} circulating in the blood. It does this by decreasing the release of Ca^{2+} from bone (inhibits osteoclasts) and increasing the deposition of Ca^{2+} in bone (stimulates osteoblasts).

Calcitonin release is regulated through a negative feedback loop in response to the amount of Ca^{2+} in the blood. High blood calcium stimulates calcitonin release, and low blood calcium inhibits its release. Calcitonin works in conjunction with the parathyroid hormone to regulate blood calcium levels.

> The thyroid gland produces thyroxine and calcitonin. Thyroxine regulates energy metabolism. Calcitonin lowers blood calcium.

The Parathyroid Glands

The parathyroid glands release **parathyroid hormone (PTH).** PTH increases the amount of Ca^{2+} circulating in the blood. It accomplishes this by increasing three activities: calcium absorption from the intestine, calcium reabsorption by the kidneys during urine formation, and calcium removal from bone tissue (stimulates osteoclasts).

The amount of Ca^{2+} in the blood is regulated by both calcitonin (decreases Ca^{2+}) and PTH (increases Ca^{2+}). However, PTH is more important than calcitonin in maintaining Ca^{2+} levels in the blood. The secretion of PTH is stimulated by low blood Ca^{2+}. Its release is regulated through a negative feedback loop involving high blood calcium levels, as was described earlier in this chapter.

The maintenance of normal blood Ca^{2+} is important for proper functioning of muscle and nerve tissue. You will recall that Ca^{2+} is required for skeletal muscle contraction and impulse conduction between synapses (Chapters 5 and 14). Insufficient PTH leading to low blood calcium can cause muscle spasms. By contrast, excessive PTH can cause bones to weaken as calcium is continuously removed from the bone matrix.

> The four small parathyroid glands release parathyroid hormone, which increases blood calcium. Parathyroid hormone release is stimulated by low blood calcium levels.

The Pancreas

The pancreas is positioned near the stomach in the abdominal cavity and functions as both an exocrine and an endocrine gland. As an exocrine gland it produces and releases digestive enzymes into the small intestine. As an endocrine gland it produces two main hormones: insulin and glucagon. Both of these hormones regulate the amount of glucose circulating in the bloodstream. Insulin decreases blood glucose, while glucagon increases it. Because glucose is the primary energy fuel for most cells, these two hormones are intimately associated with the body's energy needs.

Insulin and glucagon are produced by more than a million small clusters of cells called *islets of Langerhans* (***eye**-lets of **lang**-er-hanz*) scattered throughout the pancreas. In the islets, *beta cells* secrete

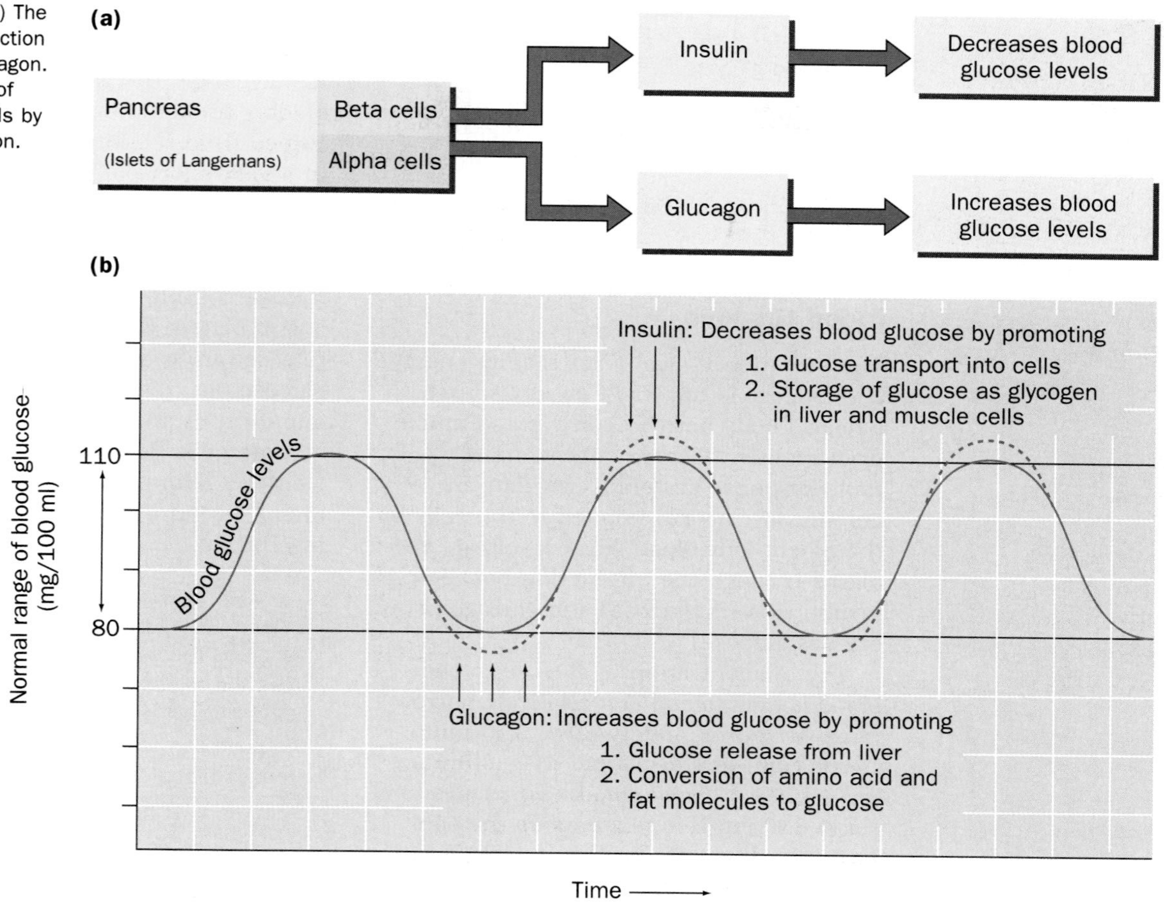
(a)
Pancreas
(Islets of Langerhans)
Beta cells
Alpha cells
Insulin
Glucagon
Decreases blood glucose levels
Increases blood glucose levels
(b)
Insulin: Decreases blood glucose by promoting
1. Glucose transport into cells
2. Storage of glucose as glycogen in liver and muscle cells
110
80
Blood glucose levels
Normal range of blood glucose (mg/100 ml)
Glucagon: Increases blood glucose by promoting
1. Glucose release from liver
2. Conversion of amino acid and fat molecules to glucose
Time

FIGURE 16.13 (a) The production and function of insulin and glucagon. (b) The regulation of blood glucose levels by insulin and glucagon.

insulin and *alpha cells* secrete glucagon (Figure 16.13a).

Insulin and Glucagon: Regulating Blood Glucose

In the discussion that follows, recall that glycogen is a storage form of glucose. When glycogen is broken down, glucose is released.

Insulin: lowering blood glucose. Insulin is released from the pancreas when blood glucose levels become elevated, as may occur after a meal (Figure 16.13b). Circulating in the blood, it acts in different ways on different cells. In muscle cells, adipose cells, and other connective tissue cells, insulin enhances the transport of glucose across the plasma membrane into cells.

Other actions of insulin include promoting the conversion of glucose to glycogen in the liver, promoting the entry of amino acid and fat molecules into cells from the blood, and stimulating protein synthesis. The importance of insulin can be appreciated by realizing that *diabetes mellitus* is caused by a lack of insulin or a decrease in its activity. We will discuss this disease in the section on diseases and disorders.

Glucagon: raising blood glucose. Glucagon (***glue-kah-gone***) is a protein hormone that is released from the pancreas when blood glucose levels are low, as may occur between meals. It functions to raise blood glucose by acting exclusively on the liver (Figure 16.13b). Fundamentally, its effects are opposite to those of insulin. Glucagon accomplishes its task by first stimulating the liver to convert glycogen to individual glucose units. However, after all the glycogen has been used, glucagon plays a second role. Under starvation conditions, glucagon stimulates the conversion of fats and amino acids (from the breakdown of muscle protein) to glucose.

Insulin from beta cells of the pancreas lowers blood glucose by enhancing glucose transport into cells. Glucagon from alpha cells of the pancreas raises blood glucose by stimulating the liver to break down glycogen.

Homeostatic Regulation of Blood Glucose

The activities of our bodies require energy, and glucose is our main energy source. Because body activity can vary so much from moment to moment, we need a rapid response system to regulate the amount of glucose supplied to the cells by the blood. Our blood sugar levels are regulated through the interaction between insulin (lowers glucose) and glucagon (increases glucose).

The concentration of glucose in the blood is maintained at a relatively narrow range of approximately 80 to 110 milligrams (mg) of glucose per 100 milliliters (ml) of blood. Blood glucose concentrations above 110 mg/100 ml stimulate the pancreas to release insulin (Figure 16.13b). When the blood glucose concentration returns to normal, the release of insulin is turned off. When the blood glucose concentration falls below 80 mg/100 ml, this stimulates the pancreas to release glucagon (Figure 16.13b). As a result, blood glucose levels rise as more glucose enters the bloodstream. When the blood glucose concentration returns to normal, the release of glucagon is turned off.

How does this negative feedback loop work during a typical day? To explain, let us start at the point where you begin your day—upon awakening. Having slept 8 hours, you've had nothing to eat during this period. All carbohydrates consumed before sleeping have been digested and absorbed.

After a night's sleep, no glucose is available from the small intestine and your blood glucose is low. Glucagon has been stimulating the liver to convert its stores of glycogen to glucose, which is then released into the blood.

Now it's time for breakfast. Let's say you eat a breakfast rich in carbohydrates: a glass of orange juice, coffee with sugar, and a stack of hotcakes with syrup. During digestion, the carbohydrates are quickly broken down to glucose and absorbed from the intestine into the blood. In a very short period of time the glucose level in the blood increases.

Now there is too much glucose in the blood. As glucose levels rise, glucagon release is inhibited and insulin release is stimulated. Enough insulin will be released from the pancreas to transport glucose out of the blood and store it in one form or another in muscle, liver, or adipose cells. It is through this negative feedback loop that your blood sugar remains within a stable range of 80 to 110 mg/100 ml.

If blood glucose falls to 80 mg/100 ml, the pancreas releases glucagon, which stimulates the release of glucose from the liver. If blood glucose rises above 110 mg/100 ml, the pancreas releases insulin, which decreases blood glucose to a normal range.

The Adrenal Glands

Your adrenal glands are positioned above each of your kidneys (*ad,* "near"; *renal,* "kidney"). Each adrenal gland is composed of two distinct parts: an inner core called the *medulla* and an outer covering called the *cortex* (Figure 16.14). Although they are located in the same gland, the adrenal medulla and adrenal cortex secrete very different hormones and function independently of each other.

Adrenal Medulla

Tissues of the adrenal medulla produce two hormones: **epinephrine (adrenaline)** and **norepinephrine (noradrenaline).** Actually, the adrenal medulla functions as part of the autonomic nervous system (ANS). During perceived or real dangers, nerve cells from the sympathetic division of the ANS directly stimulate adrenal medulla cells, causing secretion of epinephrine and norepinephrine into the bloodstream (Figure 16.15).

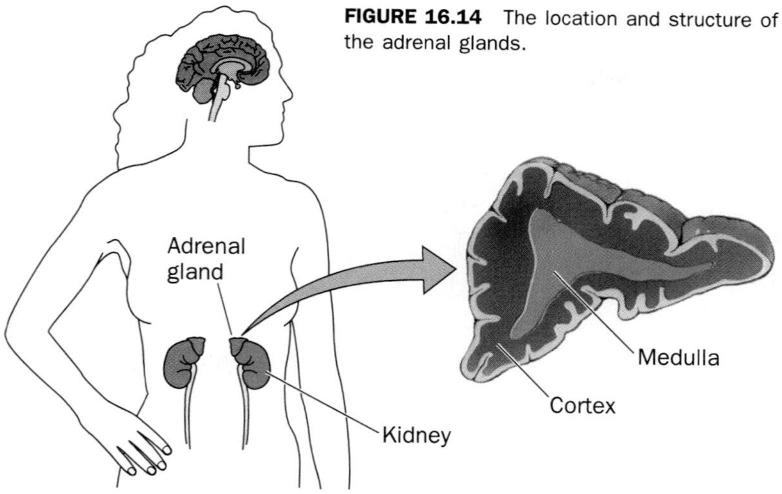
Adrenal gland
Medulla
Cortex
Kidney

FIGURE 16.14 The location and structure of the adrenal glands.

Recall that the sympathetic division acts to mobilize the body for emergency or quick action—the "fight-or-flight" response (Chapter 14). It increases the heart rate and the respiration rate, decreases digestive activity, dilates the blood vessels of the skeletal muscles, and constricts those of the skin. The hormones norepinephrine and epinephrine have the same effect on the body. However, they also increase the amount of glucose in the blood by stimulating the breakdown of glycogen in the liver. The presence of glucose in the blood assures that there is energy to fuel either fight or flight.

You may ask why the functions of the endocrine system complement those of the nervous system in stimulating fight or flight. Compared with neurotransmitters, hormones are distributed to the entire body by the blood, and they are released in larger amounts. Therefore, as hormones, norepinephrine and epinephrine affect the body about 10 times longer and reinforce, enhance, and prolong the fight-or-flight effects of the sympathetic division.

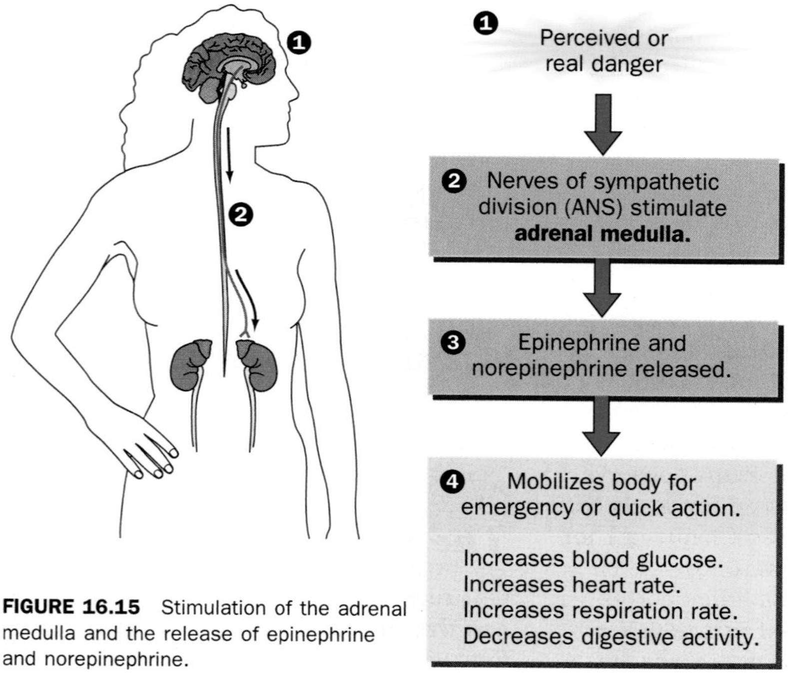
1
2
1 Perceived or real danger
2 Nerves of sympathetic division (ANS) stimulate adrenal medulla.
3 Epinephrine and norepinephrine released.
4 Mobilizes body for emergency or quick action.
Increases blood glucose.
Increases heart rate.
Increases respiration rate.
Decreases digestive activity.

FIGURE 16.15 Stimulation of the adrenal medulla and the release of epinephrine and norepinephrine.

> The inner portion of the adrenal gland—the adrenal medulla—releases epinephrine and norepinephrine when stimulated by the sympathetic division of the ANS. Both hormones prepare the body for emergency or quick actions.

Adrenal Cortex

The **adrenal cortex** produces two main groups of hormones: glucocorticoids and mineralocorticoids. Small amounts of sex hormones are also produced by the adrenal cortex. The hormones released by the adrenal cortex are steroid hormones that are derived from cholesterol and are not soluble in water.

Glucocorticoids. The glucocorticoids (*glue-koh-**core**-tih-koyds*) are a family of steroid hormones. The most prevalent hormone in this category is **cortisol,** also called hydrocortisone, which constitutes about 95 percent of the secreted glucocorticoids.

As their name implies, glucocorticoids regulate the amount of blood glucose. They also help the body respond to inflammation and stress. Released during long periods between eating (starvation conditions), glucocorticoids increase blood glucose levels by promoting the breakdown of muscle protein to amino acids. In the liver, these amino acids are converted to glucose and released into the blood. Glucocorticoids also promote the utilization of fats instead of glucose for energy.

Inflammation also increases the secretion of glucocorticoids. In general, glucocorticoids reduce inflammation by decreasing leakage from capillaries, suppressing histamine release, and reducing fever.

Stressful conditions such as infection, physical injury, surgery, and emotional conflicts can cause the adrenal cortex to secrete as much as 20 times the usual amount of cortisol. As a result, nutrients such as glucose and amino acids become more available to support the increased maintenance and repair needs of the body. We'll elaborate on stress later in the section on diseases and disorders.

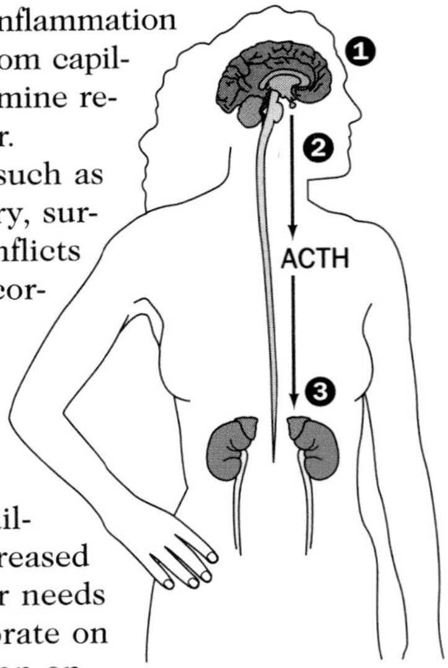
1
2
ACTH
3

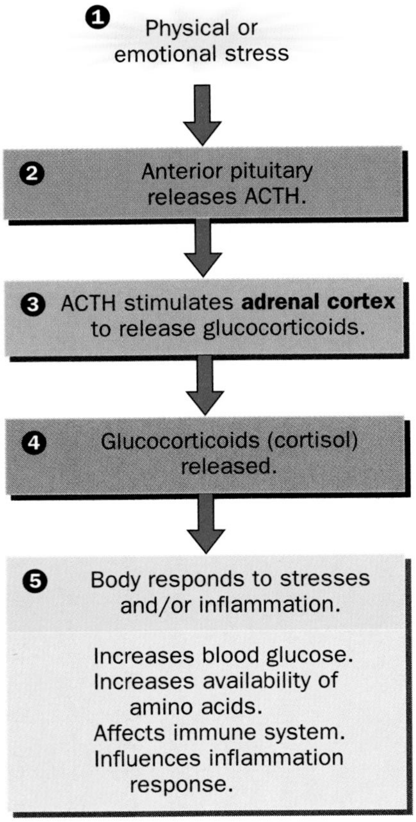
1 Physical or emotional stress
2 Anterior pituitary releases ACTH.
3 ACTH stimulates **adrenal cortex** to release glucocorticoids.
4 Glucocorticoids (cortisol) released.
5 Body responds to stresses and/or inflammation.
Increases blood glucose.
Increases availability of amino acids.
Affects immune system.
Influences inflammation response.

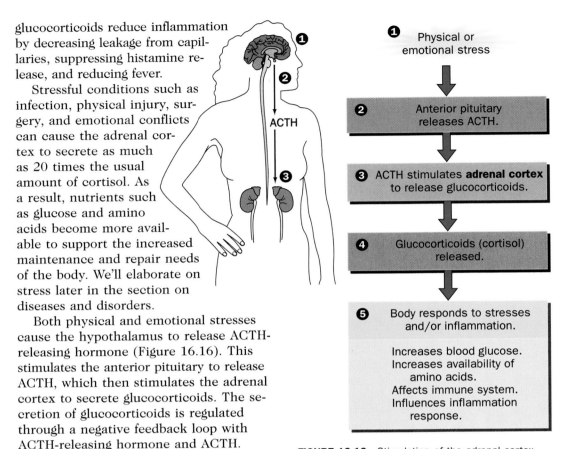

FIGURE 16.16 Stimulation of the adrenal cortex and the release of glucocorticoid hormones.

Both physical and emotional stresses cause the hypothalamus to release ACTH-releasing hormone (Figure 16.16). This stimulates the anterior pituitary to release ACTH, which then stimulates the adrenal cortex to secrete glucocorticoids. The secretion of glucocorticoids is regulated through a negative feedback loop with ACTH-releasing hormone and ACTH.

An overabundance of glucocorticoids causes *Cushing's disease.* This disease may result from an adrenal gland tumor that causes overproduction of hormones. However, it may also result when large therapeutic doses of glucocorticoid drugs (such as prednisone) are administered to control the chronic inflammation that accompanies disorders such as arthritis and asthma.

Mineralocorticoids. The mineralocorticoids (*min-er-al-oh-**core**-tih-koyds*) are a family of related steroid hormones, the most abundant and potent of which is **aldosterone** (*al-**doss**-ter-own*). Aldosterone regulates the concentration of sodium (Na^+) and potassium (K^+) in the body, enhancing the reabsorption of Na^+ and the excretion of K^+ from the kidney tubules. Aldosterone also helps conserve water because Na^+ reabsorption leads to the retention of water. See Chapter 11 for more details about aldosterone.

Sex hormones. The adrenal cortex secretes small amounts of estrogens (female sex hormones) and androgens (male sex hormones). The amount of estrogen released is insignificant in adult males and females, but the amount of androgen does affect females. In females, androgen promotes the growth of pubic and armpit hair and may also contribute to the sex drive. In males, the effect of adrenal cortex androgens is negligible because the majority of androgen is released by the testes.

> The adrenal cortex releases glucocorticoids, mineralocorticoids, and sex hormones. Glucocorticoids regulate carbohydrate metabolism, inflammation, and stress. Mineralocorticoids regulate ions and water in the body.

The Ovaries and Testes

Our reproductive glands—the male testes and the female ovaries—have a dual function. They not only produce sex cells (sperm and ova) but also produce sex hormones that control the development of sexual characteristics and the maintenance of reproductive structures.

Male Sex Hormones

Male sex hormones are called *androgens.* The primary androgen produced by males is **testosterone.** Almost all the testosterone in adult males is produced by the testes.

During fetal development, the testes produce testosterone, which causes the development of the male reproductive structures: the penis, scrotum, testes, and their associated glands. After birth and until puberty, very little, if any, testosterone is produced. During puberty (about age 13 in males), testosterone production by the testes is initiated by the release of luteinizing hormone (LH) from the anterior pituitary.

LH stimulates specific cells of the testes to produce testosterone. Testosterone stimulates sperm production, the development of male secondary sex characteristics such as body hair, larger muscles, deeper voice, and the sex drive. After puberty, the continued release of LH and testosterone maintains a male's secondary sex characteristics and sex drive.

Female Sex Hormones

Mature females produce two sex hormones: **estrogen** and **progesterone.** Both are produced in the ovary by specialized cells associated with the production of egg cells. The small amounts of estrogen released by the adrenal cortex between birth and puberty seem to have few, if any, effects.

In females, puberty occurs somewhat earlier (about age 11) than it does in males. It is during puberty that the anterior pituitary begins releasing LH and FSH. Together, LH and FSH stimulate the ovaries to produce estrogen and progesterone. Estrogen is primarily responsible for triggering the changes associated with the development of secondary sexual characteristics such as breast development, contour changes in the pelvis, and the distribution of fat under the skin. Progesterone is important in the monthly preparation of tissue in the uterus for pregnancy.

During their reproductive years, females release LH and FSH from the anterior pituitary, as well as estrogen and progesterone from the ovary in a monthly cycle known as the **menstrual cycle.** This monthly hormonal cycle is described more fully in Chapter 17.

> Testosterone is produced by the testes of mature males. Its production is stimulated by LH. Estrogen and progesterone are produced by the ovaries of mature females. Their secretion is stimulated by LH and FSH.

Other Endocrine Tissues and Their Secretions

The glands described above are not your body's only hormone-secreting tissues. In the preceding chapters you encountered organs with specialized hormone-secreting cells. Cells of the digestive system release several hormones (gastrin, secretin, cholecystokinin) that help regulate digestive processes. Kidney cells release a hormone (erythropoietin) that regulates red blood cell production and active vitamin D which can be considered a hormone. The thymus gland releases a hormone (thymosin) that aids in the development of T-cell lymphocytes.

In what follows, we'll describe the secretions of the pineal gland of the brain. Then we'll turn to hormonelike substances called prostaglandins that are made and released by almost all tissues of the body.

Pineal Gland of the Brain

The **pineal gland** (***pine**-ee-ul*) is a pea-sized structure found in the roof of the brain's third ventricle (see Figure 16.1). Its name comes from its resemblance to a pinecone. As an endocrine gland it produces the hormone **melatonin** (*mel-ah-**tone**-in*), whose release is regulated by the daily cycle of light and dark called the *diurnal cycle.* Melatonin is secreted primarily during darkness; very little is produced during daylight hours.

How does the pineal gland differentiate between daylight and darkness? It appears that visual pathways from the retina not only carry impulses to the visual centers of the brain but also connect with the nerve cells that innervate the pineal gland. During daylight, the pineal gland receives visual impulses from the retina that inhibit the release of melatonin. During darkness, visual impulses are much reduced and the release of melatonin is increased.

The diurnal release of melatonin influences our daily rhythms of sleep, hunger, and body temperature. Melatonin also may be involved in timing the onset of puberty by restricting the release of gonadotropic releasing hormone from the hypothalamus.

Melatonin also appears to affect mood and behavior. For example, some individuals experience more depression and mood swings during the shorter days of winter compared with the longer days of summer. The shorter days of winter allow excessive melatonin to be released from the pineal gland, which in an unknown way affects mood. Researchers have confirmed that such seasonal mood changes are real and have called the condition *seasonal affective disorder* (*SAD*). Daily exposure to several hours of very bright light successfully decreases melatonin levels and relieves SAD in at least 90 percent of patients (Figure 16.17). The longer days of summer also reduce the amount of melatonin released.

FIGURE 16.17 A patient with seasonal affective disorder receiving light therapy.

> The pineal gland produces melatonin during darkness. Melatonin influences daily rhythms such as sleep, hunger, and temperature.

Prostaglandins: Hormonelike Chemicals from Our Cells

The name prostaglandins (*pros-tah-**glan**-dins*) was given to these substances because at one time they were believed to be produced only by the male's prostate gland. Today we know that they are produced by nearly all body cells of both sexes. **Prostaglandins** are a family of modified fatty acid molecules. They are said to be "hormonelike," rather than true hormones, because they do not fulfill two of the defining characteristics of hormones: They are not released into the blood, and they do not act far from their sites of secretion.

Once they are released, prostaglandins attach to receptors on the plasma membrane of their target cells and stimulate or inhibit second messengers such as cyclic AMP. About 16 different prostaglandins have been discovered, and they have an array of activities throughout the body. Most of their functions can be placed in two basic categories. First, they alter the contraction of smooth muscles in our internal organs including reproductive, vascular, respiratory, digestive, and excretory organs. Second, they increase inflammation, causing pain and fever. Drugs such as aspirin and ibuprofen (Advil, Motrin) reduce pain and inflammation because they inhibit the synthesis of a specific type of prostaglandin associated with the inflammation response.

> Prostaglandins are hormonelike molecules produced by many body cells. They do not enter the blood but act on cells near where they are released. Prostaglandins regulate smooth muscle contraction or inflammation.

Diseases and Disorders

Because the endocrine system and its secretions control so many body functions, an endocrine disorder may have widespread effects on the body. During our preceding discussions we described several endocrine disorders and diseases. In this section we will focus on two topics: diabetes mellitus and how the endocrine system responds to stress.

Diabetes Mellitus

Diabetes mellitus (***mel**-eye-tus*) is caused by insufficient secretion of insulin. Without insulin, there is inadequate transport of glucose from the blood into body cells. Consequently, the blood shows abnormally high amounts of glucose. The concentration of glucose in the blood of diabetics can exceed 800 mg/100 ml of blood; the normal range is only 80 to 110 mg/100 ml.

In addition to high blood glucose, the common symptoms include (1) glucose in the urine, (2) increased urination and thirst, and (3) weight loss because fat rather than glucose is used to supply energy. If untreated, diabetes mellitus can lead to the destruction of capillaries and small blood vessels.

There are two distinct forms of diabetes mellitus: type I and type II. *Type I diabetes* is also known as juvenile-onset diabetes mellitus or insulin-dependent diabetes mellitus. This form develops primarily during childhood (ages 10 to 16), when the beta cells of the pancreas stop producing insulin. As a result, these patients must control their blood sugar with daily injections of manufactured insulin and a special diet. Together, insulin injections and diet avoid large fluctuations in blood glucose, which can be life-threatening.

Type I diabetes afflicts about 2 of every 1000 people in the United States (about 10 percent of all diabetics have this type). While it runs in families, not everyone inheriting the genes develops the disease. It is believed to be an autoimmune reaction that destroys beta cells.

Type II diabetes is also known as adult-onset or non-insulin-dependent diabetes mellitus. It occurs predominantly in adults over age 40. Patients with type II diabetes usually secrete sufficient amounts of insulin, but their target cells lack adequate membrane receptors for that hormone. For this reason, treatment does not require insulin injections, and the disease is usually controlled through diet, exercise, and weight loss.

Type II diabetes is the most common form of diabetes, afflicting about 2 of every 100 people in the United States (approximately 90 percent of all diabetics have this type). Type II diabetes also has an inherited component and runs in families. In contrast to type I diabetes, many more people with family histories of type II diabetes actually develop the disease. Obesity and a diet high in carbohydrates appear to increase a person's chances of developing this disorder.

Because complications from diabetes affect blood vessels and peripheral nerves, heart attacks occur twice as often in diabetics as in nondiabetics. Kidney disease occurs 17 times as often, blindness 25 times as often, and gangrene 10 times as often. Whether such complications occur often depends on factors such as the type of diabetes, the age of onset, the type of therapy, and genetic factors.

Stress: Its Effect on the Body

When we speak of the stress in our lives, we are usually referring to our emotional reactions to speaking in public, taking final exams, being late, or having an argument. However, stress also occurs because of physical conditions such as extreme heat and cold, starvation, injury, illness, and surgery.

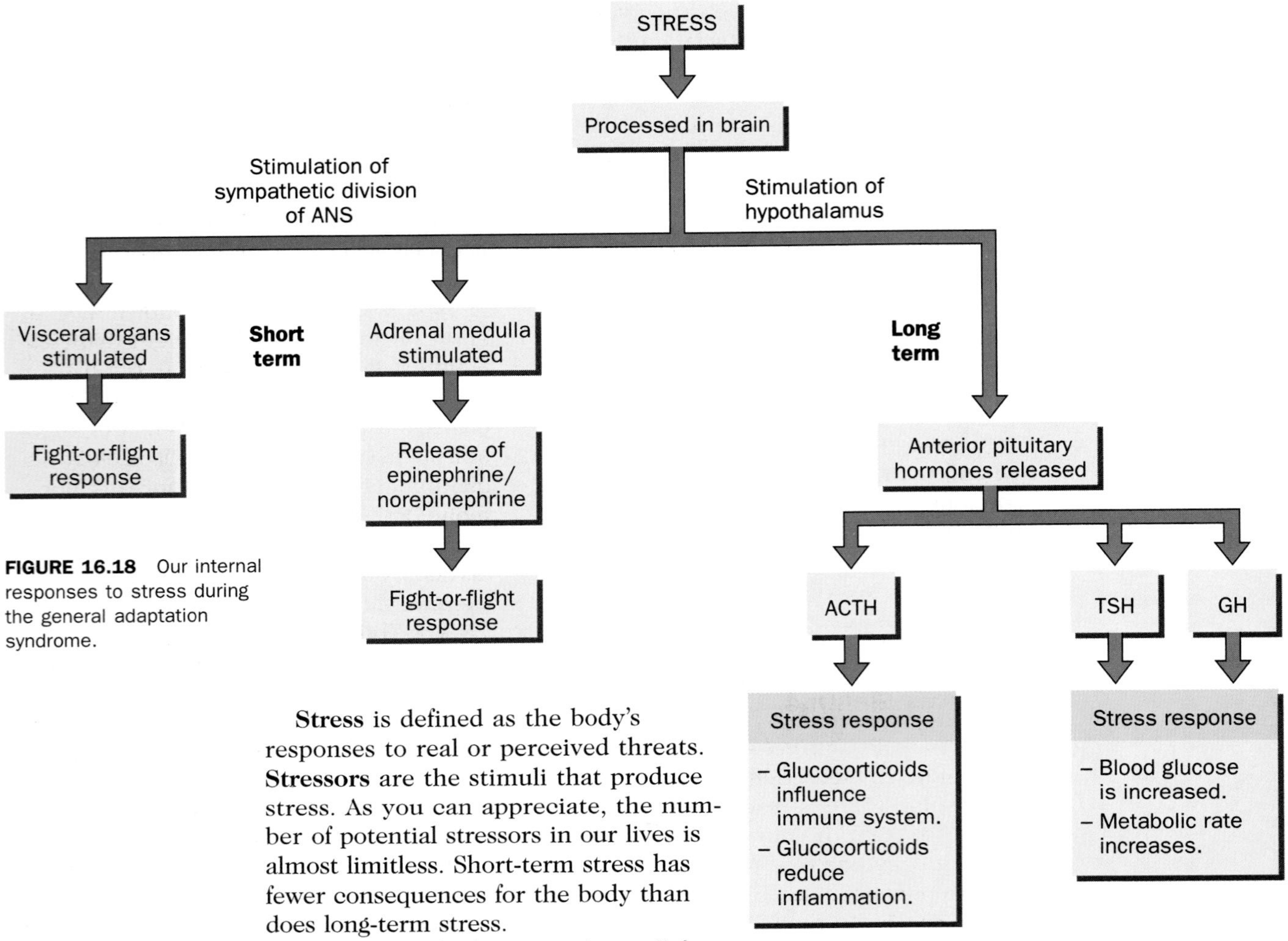
STRESS
Processed in brain
Stimulation of sympathetic division of ANS
Stimulation of hypothalamus
Short term
Long term
Visceral organs stimulated
Fight-or-flight response
Adrenal medulla stimulated
Release of epinephrine/ norepinephrine
Fight-or-flight response
Anterior pituitary hormones released
ACTH
TSH
GH
Stress response
– Glucocorticoids influence immune system.
– Glucocorticoids reduce inflammation.
Stress response
– Blood glucose is increased.
– Metabolic rate increases.

FIGURE 16.18 Our internal responses to stress during the general adaptation syndrome.

Stress is defined as the body's responses to real or perceived threats. **Stressors** are the stimuli that produce stress. As you can appreciate, the number of potential stressors in our lives is almost limitless. Short-term stress has fewer consequences for the body than does long-term stress.

The human body responds to all forms of stress in a similar manner. During stress, the nervous system and endocrine system initiate internal adjustments called the stress response or *general adaptation syndrome* (*GAS*). The GAS is traditionally separated into three stages: (1) the fight-or-flight response, (2) the resistance response, and (3) the exhaustive response.

The fight-or-flight response. You will recall from Chapter 14 that the perception of a threatening situation initiates a series of actions by the sympathetic division of the autonomic nervous system called the *fight-or-flight response.* These nerves also stimulate the adrenal medulla to release the hormones epinephrine and norepinephrine into the blood (Figure 16.18).

As a result of this combined nervous and endocrine activity, your body is prepared for rapid action. Your heart rate, blood pressure, and respiration rate increase. More blood flows to your brain and skeletal muscles, and less flows to the skin and digestive system. Glucose is released by the liver into the blood to fuel increased body action. These changes profoundly alter the homeostasis of the body, creating a "supercharged" situation. Such short-term stress can be useful and valuable.

When the perceived danger quickly passes, nerve impulses to the hypothalamus and to the adrenal medulla cease and the amount of epinephrine and norepinephrine in the blood rapidly declines. The body returns to its former more serene homeostasis. However, what happens if the perceived danger does not leave quickly?

The resistance response. Some stress is long-lasting. In modern lives, anxiety, unemployment, divorce, and conflicts at work and with family members, friends, or lovers may last for days, weeks, and even years. Your body adjusts to such long-term or chronic stress through sustained activity of the hypothalamus, the anterior pituitary, and the adrenal cortex.

With long-term stress, the hypothalamus secretes its releasing hormones, which stimulate the anterior pituitary to release ACTH, TSH, and GH (see Figure 16.18). These hormones maintain our blood glucose levels by converting stored fat and muscle protein to glucose. They also increase the rate of ATP production to fuel all cellular activities. ACTH increases cortisol secretion by the adrenal cortex, reducing the inflammation response and decreasing swelling, irritation, and pain in an injured area.

The increased cortisol can be particularly dangerous because it slows down the immune response. When we live under long-term physical or emotional stress, some research suggests that we are more susceptible to infections, cancer, and autoimmune diseases. The relationships between the nervous system, the endocrine system, and the immune system are being researched. Will we someday discover that our thoughts directly affect our health?

The exhaustive response. Long-term stress can eventually deplete our energy reserves, reduce our ability to fight infection, and exhaust our organs. For example, prolonged stimulation of the adrenal cortex may result in its exhaustion and a loss of its ability to control the concentration of essential ions. Too many sodium ions (Na^+) are retained, while too many potassium ions (K^+) are lost. Ion imbalances affect all body systems and cause a decline in health and eventual death.

Stress and Substance Abuse

In our fast-paced and hectic lives, most of us are challenged by chronic stress. It is one of the factors that can lead to the abuse of both prescription and illegal drugs (so-called recreational drugs). While tranquilizers and other therapeutic drugs have a place in medical treatment, their overuse can lead to addiction. For this and other reasons, it is important to find alternative ways to manage chronic stress, such as counseling, support groups, spiritual development, exercise, meditation, and yoga.

Drug or alcohol abuse often begins as a misguided attempt to self-medicate chronic stress. Alcohol, nicotine, caffeine, and illegal drugs such as marijuana, amphetamines, cocaine, and heroin may be used in a misdirected effort to get relief from stress. If stress is chronic and unmanaged, the psychological power of these drugs may overwhelm your best intentions to control their use and you may become dependent on them. Inevitably, addiction to any type of substance eventually creates more stress, not less. Further discussion of substance abuse is found in Spotlight on Health: Six (page 442).

Life Span Changes

Over the entire life span, there are substantial changes in some endocrine glands and their secretions, while others show few if any changes. As you learned in Chapter 15, the secretion of sex hormones from the embryonic gonads plays an important role in the development of both the anatomic and attitudinal characteristics of males and females. In addition, other hormones appear to be essential in regulating the normal development of the embryo.

After birth, the male and female gonads do not secrete hormones until puberty. During a female's reproductive years, the release of estrogen and progesterone continues on a monthly cycle (Chapter 17). At the end of a female's reproductive years, a time called *menopause* (usually between ages 45 to 55), egg development and the menstrual cycle stop, along with the ovarian secretion of estrogen and progesterone. For a woman, the beginning of menopause is accompanied by pronounced physical and psychological changes. Nothing quite so dramatic occurs in males, but after age 40 there is a gradual decline in testosterone secretion.

During menopause, the reduction of estrogen in the blood causes a number of changes, including alterations in secondary sex characteristic (decreased armpit and pubic hair, reduced breast size), reduction of sebaceous secretions in the skin, and an increased loss of calcium from bone. Continued loss of calcium weakens bones and often leads to a condition called *osteoporosis* (Chapter 5).

The testes and ovaries are not the only glands that change over the life span. The parathyroid glands increase in size until about age 30 in males and age 50 in females. Afterward, these glands retain their size, and there is no change in the amount of hormone secreted over the remaining life span. A similar pattern is observed in the adrenal glands and the endocrine tissue of the pancreas.

However, other endocrine glands decline over the life span. After puberty, the thymus gland and its secretions gradually decline, until by age 60 all secretion stops. Recall that the thymus gland plays an important role in establishing your specific immune defenses early in the life span (Chapter 13). The decline of the thymus may be linked to decreased effectiveness of the immune defenses after age 60.

Although the size of the pituitary does not change over the life span, by age 60 its blood supply decreases and there is a decline in growth hormone. However, in normal healthy people, there is no decline in the pituitary's secretion of thyroid stimulating hormone, ACTH, and antidiuretic hormone.

As in the pituitary, certain functions of the thyroid decrease while others do not. There appears to be no change in thyroxine secretion over the life span. However, there is a decline in calcitonin secretion, and this may eventually contribute to loss of calcium from bone and osteoporosis in the elderly.

Chapter Summary

The Nature of Hormones

- Nonsteroid hormones bind to protein receptors on the plasma membranes of target cells and activate specific enzymes through second messengers.
- Steroid hormones enter target cells, bind to receptors, and activate genes.
- Hormone concentrations in the body are regulated by a negative feedback loop.

Hypothalamus and Pituitary Gland

- The posterior pituitary produces and releases oxytocin and ADH. These hormones are synthesized by neuroendocrine cells in the hypothalamus and are released upon nerve stimulation.
- The anterior pituitary produces and releases GH, prolactin, FSH, LH, ACTH, and TSH. Their release is regulated by releasing or inhibiting hormones from the hypothalamus.

Thyroid and Parathyroid Glands

- The thyroid gland releases thyroxine and calcitonin. Thyroxine increases cell metabolism; calcitonin lowers Ca^{2+} levels in the blood.
- The parathyroid glands release parathyroid hormone, which increases Ca^{2+} levels in the blood.

The Pancreas

- The beta cells of the pancreas release insulin, which lowers blood glucose levels; the alpha cells release glucagon, which increases blood glucose levels.
- When glucose levels rise above 110 mg/100 ml, insulin is released. When glucose levels fall below 80 mg/100 ml, glucagon is released.

The Adrenal Glands

- The adrenal medulla releases epinephrine and norepinephrine when stimulated by the ANS.
- The adrenal cortex releases glucocorticoids and mineralocorticoids plus some sex hormones.

The Ovaries and Testes

- During and after male puberty, most testosterone is produced by the testes.
- The ovary of a sexually mature female produces estrogen and progesterone.

Other Endocrine Tissues and Their Secretions

- The pineal gland in the brain secretes the hormone melatonin, which influences the daily rhythms of sleep, hunger, and body temperature.
- The prostaglandins are a family of hormonelike substances that either increase smooth muscle contraction or increase inflammation, causing pain and fever.

Diseases and Disorders

- The symptoms of diabetes mellitus include blood glucose levels above 110 mg/100 ml of blood, glucose in urine, increased thirst and urination, and weight loss. Type I diabetes occurs primarily in children, and type II occurs primarily in adults.

- Stress is the body's response to real or perceived threats. The general adaptation syndrome (GAS) includes the fight-or-flight response, the resistance response, and exhaustive response.

Life Span Changes

- After birth, the male and female gonads do not release hormones until puberty.
- Over the life span, many endocrine gland decrease in size, without significant decreases in activity.

Selected Key Terms

anterior pituitary (p. 424)
general adaptation syndrome (p. 438)
hormone (p. 419)
negative feedback loop (p. 422)
neuroendocrine cells (p. 423)
nonsteroid hormones (p. 421)
posterior pituitary (p. 423)
releasing hormones (p. 425)
second messengers (p. 421)
steroid hormones (p. 421)

Review Activities

1. What are the four characteristics of a hormone?
2. Identify the 10 endocrine glands and briefly describe their locations and functions in the body.
3. Describe the chemical and functional differences between nonsteroid and steroid hormones.
4. Describe the regulation of hormone secretion by a negative feedback loop, using parathyroid hormone and blood Ca^{2+} levels as an example.
5. Using the release of oxytocin during suckling, describe how the nervous system, the neurosecretory cells of the hypothalamus, and the posterior pituitary gland function to release milk from the mammary glands.
6. Identify the six major hormones of the anterior pituitary and briefly describe their functions.
7. Briefly explain the cause of the following diseases: pituitary dwarfism, acromegaly, goiter, Graves' disease, myxedema, and Cushing's disease.
8. Distinguish between the functions of calcitonin and those of parathyroid hormone.
9. Identify the two main hormones of the pancreas, state which islet of Langerhan cells secrete them, and explain how each influences the blood glucose level.
10. Distinguish between the hormones released from the adrenal medulla and those released from adrenal cortex.
11. Describe the function of testosterone in adult males and that of estrogen and progesterone in adult females.
12. Where is the pineal gland located, what hormones does it secrete, and what is its function?
13. What is seasonal affective disorder (SAD), and how is melatonin involved?
14. Why are prostaglandins described as hormonelike substances?
15. Describe the relationship between prostaglandins and drugs such as asprin and ibuprofen.

Self-Quiz

Matching Exercise

___ 1. Category of hormones that use second messengers
___ 2. An example of a second messenger
___ 3. Area of brain where neuroendocrine cells that help regulate the pituitary gland are located
___ 4. Pituitary hormone that prevents dehydration
___ 5. Anterior pituitary hormone that stimulates the testes to produce sperm
___ 6. Hormone that functions to lower blood Ca^{2+} levels
___ 7. Hormone that functions to lower blood glucose levels
___ 8. Hormone that functions to regulate Na^{+} and K^{+} levels in the body
___ 9. Hormone produced by the ovary in mature females
___ 10. Hormone released primarily during darkness

A. Aldosterone
B. ADH
C. Insulin
D. Nonsteroid
E. Melatonin
F. Cyclic AMP
G. Estrogen
H. Hypothalamus
I. FSH
J. Calcitonin

Answers to Self-Quiz

1. D; 2. F; 3. H; 4. B; 5. I; 6. J; 7. C; 8. A; 9. G; 10. E

SPOTLIGHT ON HEALTH: SIX

Alcohol and Drug Use among College Students

Alcohol and other drugs on campus—a harmless rite of passage or a serious concern? According to data from the National Clearinghouse for Alcohol and Drug Information (NCADI), 90 percent of college students drink alcoholic beverages. Nearly half are binge drinkers, consuming five or more drinks at one sitting. Most drink to be social, but about half admit they drink to become drunk and a third say they drink because of problems, stress, or boredom. Every year, American college students consume over 430 million gallons of alcohol-laced beverages, enough to fill 3500 Olympic-size swimming pools. Some college students begin drinking as preteens, and many continue their consumption patterns into later adulthood.

Alcohol consumption exacts an alarming toll on college students. Students who drink more frequently die in traffic accidents, receive citations for driving under the influence/driving while intoxicated (DUI/DWI), damage personal or public property, argue or fight, attempt suicide, experience or commit sexual abuse, miss classes, receive failing grades, suffer academic probation, or drop out of college than do those who are not drinkers. Most campus rapes occur when the victim, the assailant, or both have been drinking. Rates of unplanned pregnancies and sexually transmitted diseases, including AIDS, parallel rates of alcohol consumption.

Alcohol is certainly the most commonly used drug on college campuses, but there are others. The drugs most likely to be abused by college-age students are listed in Table 1. A recent Core Alcohol and Drug Survey conducted with funding from the U.S. Department of Education indicated that depending on the institution surveyed, about 10 percent of students use marijuana at least once a year, 3 percent use it monthly, and 1 percent use it on a daily basis. About 6 percent use cocaine at least annually, and nearly 5 percent use hallucinogens. Less than 5 percent of students report tranquilizer or inhalant use, and less than 1 percent use narcotics or steroids. Nearly all students who report alcohol or other drug use admit they do so to influence mood or behavior. Many seek relief from the pressure and stress of college studies, especially when they are combined with outside work and family responsibilities.

The frequent use of any mood-altering drug may cause a dependence on the feelings of well-being or euphoria induced by the drug. This is called *psychological dependence,* and it is distinct from physical dependence. *Physical dependence* develops as brain cells become accustomed to the presence of the drug, and painful symptoms occur if the drug is withheld or unavailable. With habitual use, larger quantities are needed to produce the desired sensations. Such *drug tolerance* develops because the liver and kidney become more efficient in detoxifying and removing drugs from the bloodstream.

Dependence on alcohol and other drugs can be explained in part by their influence on the manufacture, release, and action of neurotransmitters in the brain. Alcohol has a depressant effect on the brain by acting on gamma-aminobutyric acid (GABA) receptors and increasing the inhibitory effects of GABA. Recent research suggests that alcohol does not actually kill nerve cells in the brain, as is popu-

TABLE 1 Classes of Frequently Abused Drugs			
Group	Psychological Effects	Physiological Actions	Examples
Hallucinogens	Hallucination; sensory distortion; euphoria; mood swings; paranoia; anesthesia	Stimulate CNS; dilate pupils; increase heart rate and blood pressure; tremors; increase appetite	LSD, marijuana, mescaline, peyote, phencyclidine (PCP) psilocybin
Depressants	Initial euphoria followed by depression; disorientation; anesthesia	Depress CNS; impair vision and judgment; increase reaction time; slur speech	Alcohol, barbiturates, meprobamate, methaqualone
Stimulants	Increased alertness; decreased fatigue; hyperactivity; mood swings	Stimulate sympathetic system; dilate pupils; increase heart rate; insomnia; appetite loss	Amphetamine; caffeine, cocaine, nicotine
Narcotic analgesics	Eurphoria; anesthesia; decreased sex drive	Depress CNS; inhibit pain circuits; inhibit reflexes; constrict pupils	Codeine, heroin, methadone, morphine, opium
Tranquilizers	Reduced anxiety; calmness; sleep	Depress CNS; relax skeletal muscle; reduce REM sleep	Benzodiazepines, beta blockers, Valium
Inhalants	Giddiness; decreased inhibitions; floating sensations; drowsiness; amnesia; anxiety	Eye irritation; light sensitivity; blurred vision; cardiac irregularities	Amyl nitrite, glue, lighter fluid, nitrous oxide, paint thinner
Steroids	Mood swings	Mimic male sex hormones; decrease sperm count; impotence; menstrual irregularities	Nandrolone, oxandrolone, oxymetholone, stanozolol

larly believed, but instead disconnects them, interrupting the flow of neurotransmitters and information in key brain circuits. "Crack," a modified form of cocaine, causes the massive release of norepinephrine and dopamine in the brain. These neurotransmitters stimulate the sympathetic branch of the autonomic nervous system, causing short-lived feelings of confidence and power.

Are you concerned about your own level of dependence on alcohol and other drugs? You can get a clue about your risk for developing alcohol dependency by answering the questions in Table 2 (page 444). You can modify the table to assess your risk for dependence on other drugs by slightly altering the questions.

TABLE 2 Are You Becoming an Alcohol-Dependent Person?

Answer Yes or No
1. Do you ever wake up the "morning after" a party unable to remember what happened the night before even though friends are certain you did not pass out?
2. Do you sometimes need a drink in the morning to "get your motor running"?
3. Do you sometimes miss work or scheduled appointments because of alcohol consumption?
4. Do you seem to crave a drink after a quarrel or disappointment at home or in the office to "settle your nerves"?
5. Has alcohol ever been the cause of a family or work-related quarrel?
6. Do you feel deprived if you cannot have a drink every day or at a specific time each day?
7. Do you "go out drinking" instead of having a drink when you go out?
8. Do you feel uncomfortable at a party if alcohol is not available, or do you go looking for your own refill at a party if the host is slow in making the rounds?
9. Do you pour yourself a larger amount than you serve to guests, or do you find yourself sneaking a second drink before the others have finished their first?
10. Do you lie about how much you drink when asked or feel irritated, frightened, or guilty when others question you about the amount you drink?
11. Has another family member ever been treated for alcohol dependence?
12. Have you ever been arrested for driving under the influence (DUI) or driving while intoxicated (DWI)?
A *yes* answer to any of these questions can be considered a warning sign. Two *yes* answers suggest that you may already be dependent on alcohol. Three or more *yes* answers indicate a serious problem that may require professional guidance.

PART SEVEN

Reproduction, Development, and Heredity

CHAPTER 17

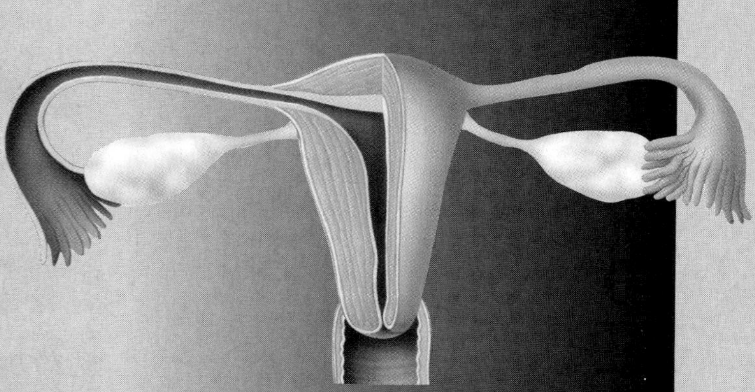

Human Reproduction

CHAPTER 18

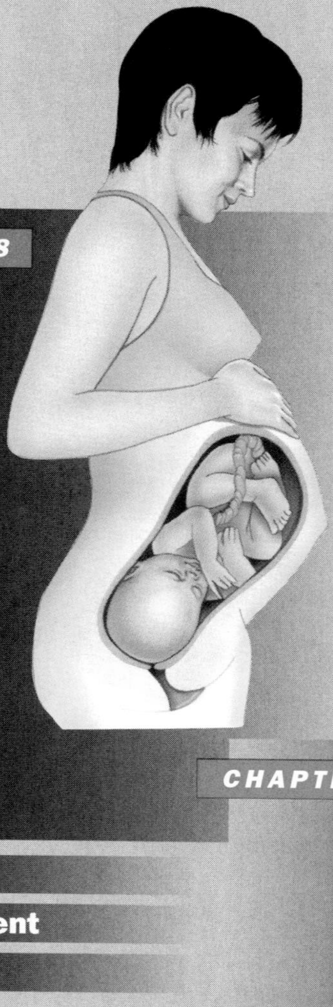

Human Development and Birth

CHAPTER 19

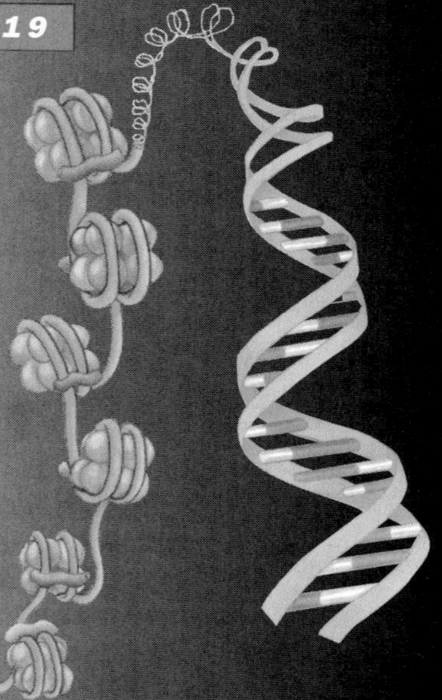

Human Genetics

Chapter 17

Human Reproduction

When does 1 plus 1 give 3 or 4 or more? The traditional nursery rhyme says "mommy and daddy, and baby makes three," but in reproducing, a man and a woman may produce several children. **Reproduction** continues the human species and contributes to the growth of the human population. Reproductive capacity lies in the specialized cells—the sperm and the egg—which carry genetic information. When these cells unite at **fertilization**, their genes initiate and direct the development of a new individual.

The story of reproduction actually consists of several stories. It includes the story of the production of the sperm and the egg and their delivery to the site of fertilization. It includes the changes in our bodies that make reproduction possible. It is a story of constancy and change, a story of homeostasis and long-lasting modifications that produce a new human life. At the personal level, it is a story of passion and pleasure, commitment and nurturing. It creates families and the emotional forces that can tear them apart.

In this chapter, we will describe the male testes and female ovaries and the production and delivery of **gametes**, the sperm and egg. We will also explore how hormones and nerves coordinate the many processes of reproduction and consider methods for both enhancing and limiting fertility. In these explorations, you will discover that modern reproductive technology complicates the nursery rhyme's concept of mommy and daddy. With gamete donors, artificial fertilization, and surrogate mothers, the lines of parenthood have become biologically and legally fuzzy.

In this chapter we consider only part of the story of reproduction. In Chapter 18, we take up the rest of the story: fertilization and the development of a new human life. Sexually transmitted diseases of the reproductive organs are discussed in Spotlight on Health: Seven (page 526).

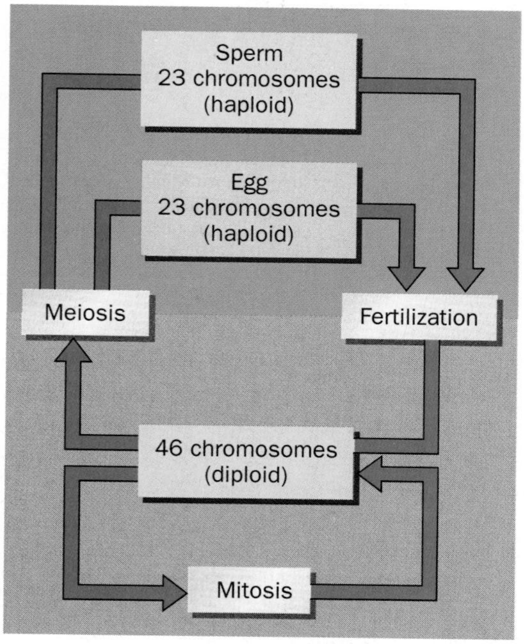
Sperm
23 chromosomes
(haploid)
Egg
23 chromosomes
(haploid)
Meiosis
Fertilization
46 chromosomes
(diploid)
Mitosis

FIGURE 17.1 Comparison of mitosis and meiosis.

Meiosis: The Production of Gametes

Before we examine the male and female reproductive systems, it is essential to understand the process of cell division called **meiosis** (*my-oh-sis*), which produces sperm and egg cells (Figure 17.1). The cellular products and processes of meiosis differ from those of mitosis, the type of cell division process that produces all other body tissues and organs. Mitosis produces two progeny cells that are genetically identical to each other and to the parent cell (Chapter 2). These cells all have 23 pairs of chromosomes (46 chromosomes total) and are said to be **diploid**. However, with meiosis, things are different.

Meiosis is a series of events that reduce the number of chromosomes by half. Meiosis produces four progeny cells, and each of those cells has only 23 chromosomes. The progeny cells are said to be **haploid**. These haploid cells are the gametes (sperm and egg) that will unite upon fertilization to reestablish the diploid condition (46 chromosomes).

To accomplish a reduction in the number of chromosomes, meiosis includes two nuclear and cytoplasmic divisions: meiosis I and meiosis II (Figure 17.2). **Meiosis I** begins with a diploid nucleus that has replicated its DNA. The chromosomes condense much as they do in mitosis, but unlike mitosis, there is an aligning and pairing of chromosomes during prophase I.

The two chromosomes that align with each other are called **homologues**, and each paired set is a **homologous pair**. During this pairing, an exchange of genetic material between homologous chromosomes called **crossing over** takes place. This exchange process increases genetic diversity in the human population and helps explain why even closely related people are unique in appearance. We will discuss genetics in Chapter 19.

During anaphase I, the paired chromosomes are separated from each other and move to opposite poles of the cell. The nuclear membrane re-forms around the chromosomes, which lose their condensed appearance. Eventually, two cells are

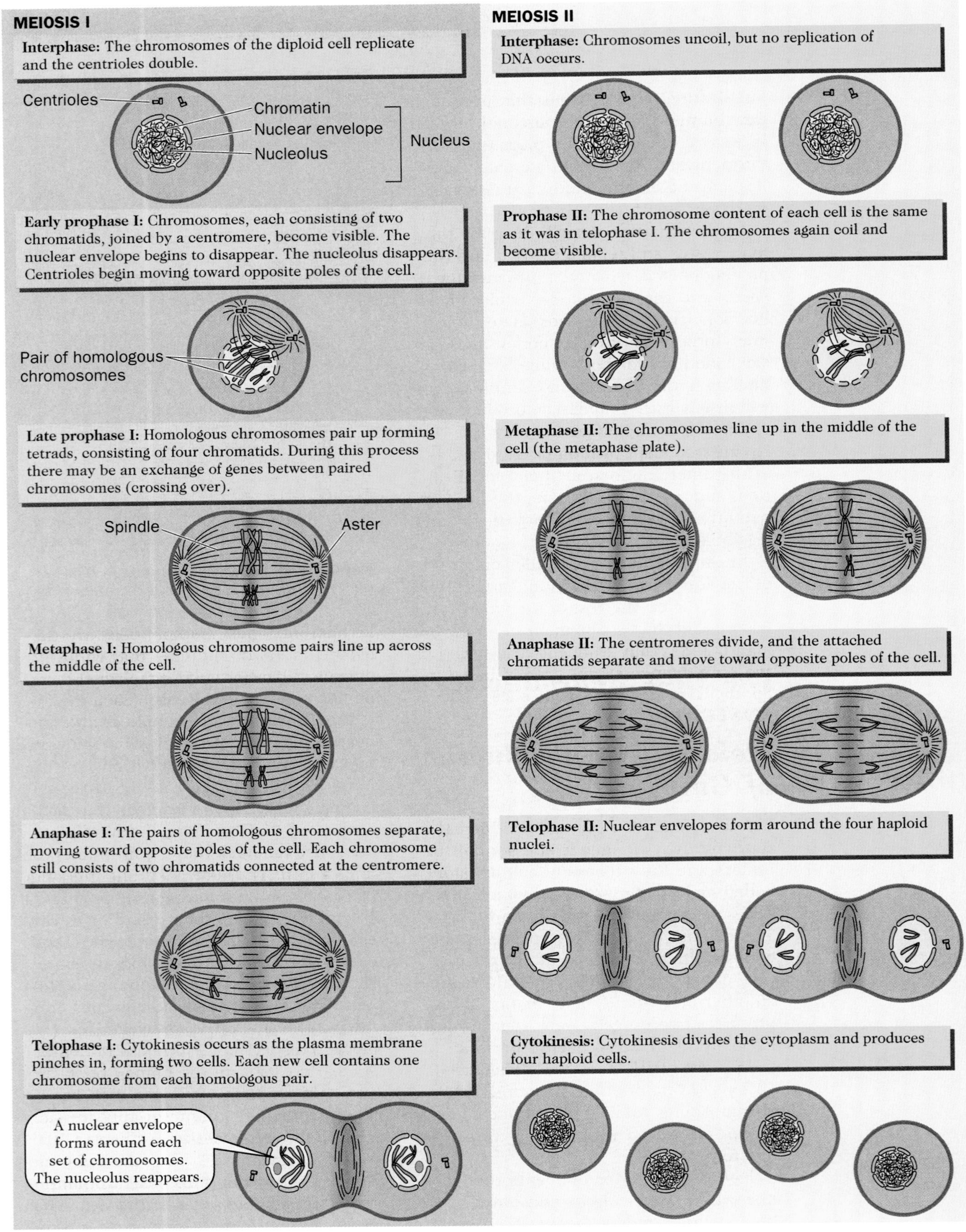
MEIOSIS I
Interphase: The chromosomes of the diploid cell replicate and the centrioles double.
Centrioles
Chromatin
Nuclear envelope
Nucleolus
Nucleus
Early prophase I: Chromosomes, each consisting of two chromatids, joined by a centromere, become visible. The nuclear envelope begins to disappear. The nucleolus disappears. Centrioles begin moving toward opposite poles of the cell.
Pair of homologous chromosomes
Late prophase I: Homologous chromosomes pair up forming tetrads, consisting of four chromatids. During this process there may be an exchange of genes between paired chromosomes (crossing over).
Spindle
Aster
Metaphase I: Homologous chromosome pairs line up across the middle of the cell.
Anaphase I: The pairs of homologous chromosomes separate, moving toward opposite poles of the cell. Each chromosome still consists of two chromatids connected at the centromere.
Telophase I: Cytokinesis occurs as the plasma membrane pinches in, forming two cells. Each new cell contains one chromosome from each homologous pair.
A nuclear envelope forms around each set of chromosomes. The nucleolus reappears.
MEIOSIS II
Interphase: Chromosomes uncoil, but no replication of DNA occurs.
Prophase II: The chromosome content of each cell is the same as it was in telophase I. The chromosomes again coil and become visible.
Metaphase II: The chromosomes line up in the middle of the cell (the metaphase plate).
Anaphase II: The centromeres divide, and the attached chromatids separate and move toward opposite poles of the cell.
Telophase II: Nuclear envelopes form around the four haploid nuclei.
Cytokinesis: Cytokinesis divides the cytoplasm and produces four haploid cells.

FIGURE 17.2 Meiotic cell division consists of meiosis I and meiosis II.

formed from meiosis I. Between meiosis I and meiosis II, there is no DNA synthesis. These two cells both undergo another division, meiosis II.

Meiosis II is essentially a form of mitosis, but the chromosomes do not pair (the homologues are now in separate cells). Each chromosome that is visible at metaphase II consists of two identical chromatids, just as in mitosis. During anaphase II, these chromatids are pulled apart and migrate to separate cells formed at the end of telophase II. As a result of both meiosis I and meiosis II, four haploid cells are formed.

> Meiosis produces the gametes (sperm and eggs). In diploid body cells, there are 23 pairs of chromosomes (46 total chromosomes). Meiosis separates homologous chromosomes into separate haploid gametes, each of which contains 23 chromosomes.

The Male Reproductive System

The reproductive system of the male does two things. First, it produces sperm and delivers them to the female to fertilize an egg. Second, it produces hormones. These hormones determine the secondary sex characteristics of the male body, help determine the functional organization of the brain, and initiate sperm formation.

The organs of male reproduction include a pair of gonads called testes or testicles for producing sperm, a series of ducts and tubules for storing and transporting sperm, accessory glands that produce secretions, and an erectile organ—the penis—to accomplish the transfer of sperm to the female (Figure 17.3a, b).

The Testes: Producing Sperm and Hormones

In the embryo, the **testes** (singular, *testis*) develop along the posterior wall of the abdomen. Shortly before birth they descend into a fleshy sac known as the **scrotum**, which hangs away from the body wall between the thighs. The penis and scrotum together constitute the male's *external genitalia*.

Temperatures in the scrotum are about 2 to 3°C cooler than those in the abdomen, and this is important for the survival of the temperature-sensitive sperm. In the wall of the scrotum, there is a layer of smooth muscle, the *dartos* (***darr***-*toes*) muscle, which responds to changing temperatures. In the cold, this muscle contracts and the testes are drawn closer to the abdomen for warmth, while heat causes the muscle to relax and the testes descend further for cooling.

Internally, the testes are subdivided into roughly 250 compartments, each of which contains tightly coiled tubules, the **seminiferous tubules** (*sem-uh-**niff**-fur-us*), where sperm are produced. In a magnified cross section of a seminiferous tubule, we can see the stages of meiosis and the development of haploid sperm cells (Figure 17.3c). Along the outer wall of the tubules are specialized cells called **spermatogonia** (*spur-**mat**-oh-**go**-knee-uh*) (singular, *spermatogonium*), which have not yet undergone meiosis. Along the innermost surface of the tubule are the maturing sperm with their flagella extending into the lumen of the tubule.

Surrounding the spermatogonia and the developing sperm are large cells called **Sertoli cells** (*sir-**toe**-lee*) (Figure 17.3c). These cells provide protection, support, and nourishment to the developing sperm. Because they are haploid, sperm differ genetically from all other body cells. They also have different surface antigens. Sertoli cells provide a barrier that protects sperm from an attack by the male's immune system.

In addition to the developing sperm cells and Sertoli cells, the testes contain **interstitial cells** (*in-ter-**stish**-ul*) which lie between seminiferous tubules (Figure 17.3c). During puberty, when they are stimulated by luteinizing hormone (LH) from the pituitary gland, these cells begin to produce the male sex hormone *testosterone*. Testosterone initiates and maintains sperm formation and provides a man's sex drive. It also promotes the development of all male secondary sex characteristics as well as male behavior.

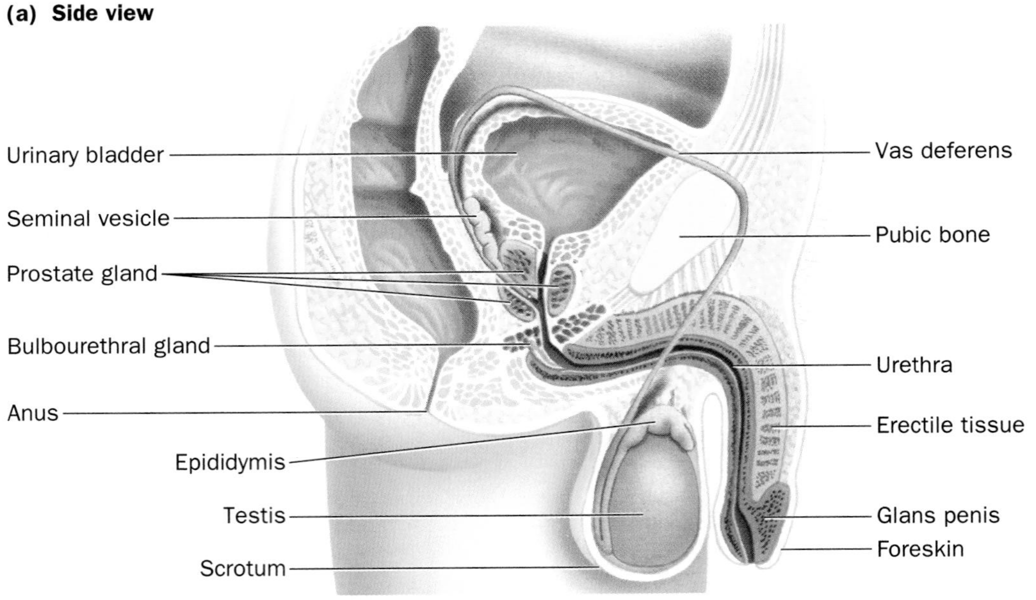
(a) Side view
Urinary bladder
Seminal vesicle
Prostate gland
Bulbourethral gland
Anus
Epididymis
Testis
Scrotum
Vas deferens
Pubic bone
Urethra
Erectile tissue
Glans penis
Foreskin

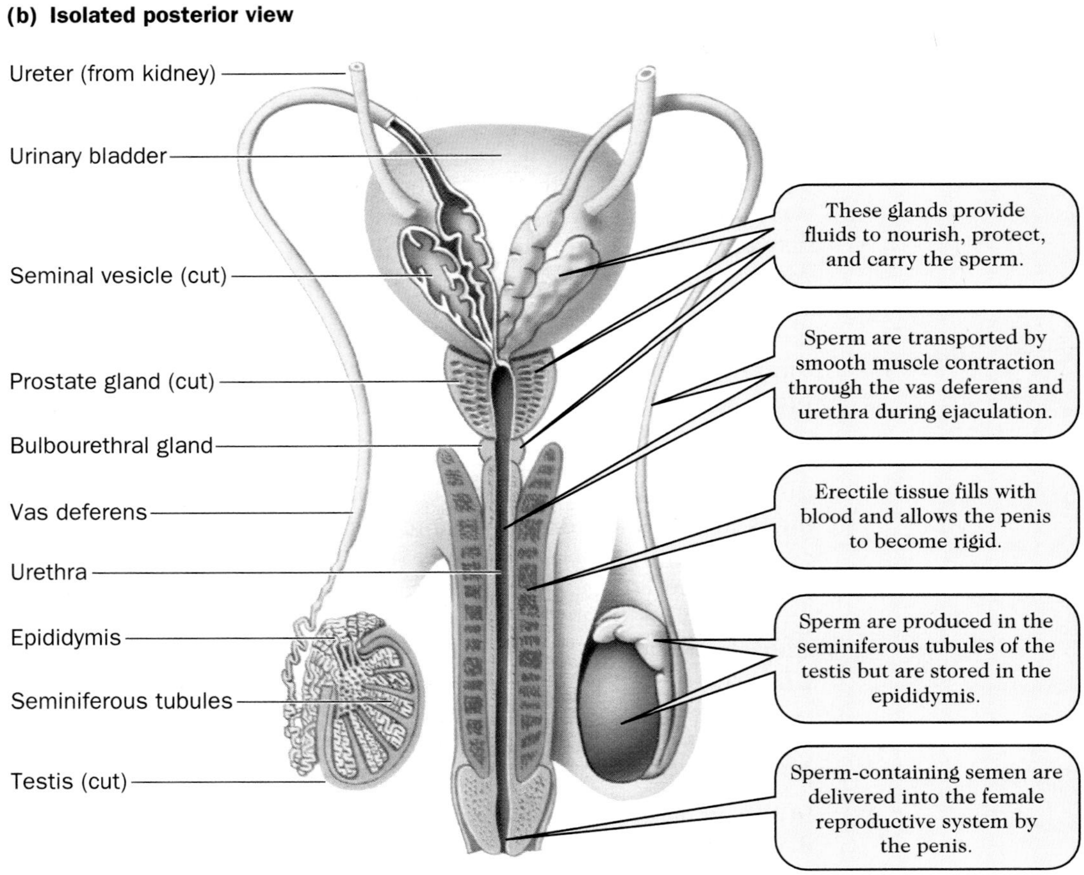
(b) Isolated posterior view
Ureter (from kidney)
Urinary bladder
Seminal vesicle (cut)
Prostate gland (cut)
Bulbourethral gland
Vas deferens
Urethra
Epididymis
Seminiferous tubules
Testis (cut)
These glands provide fluids to nourish, protect, and carry the sperm.
Sperm are transported by smooth muscle contraction through the vas deferens and urethra during ejaculation.
Erectile tissue fills with blood and allows the penis to become rigid.
Sperm are produced in the seminiferous tubules of the testis but are stored in the epididymis.
Sperm-containing semen are delivered into the female reproductive system by the penis.

FIGURE 17.3 The male reproductive system.

(c) Sectioned view of testis showing seminiferous tubules and epididymis

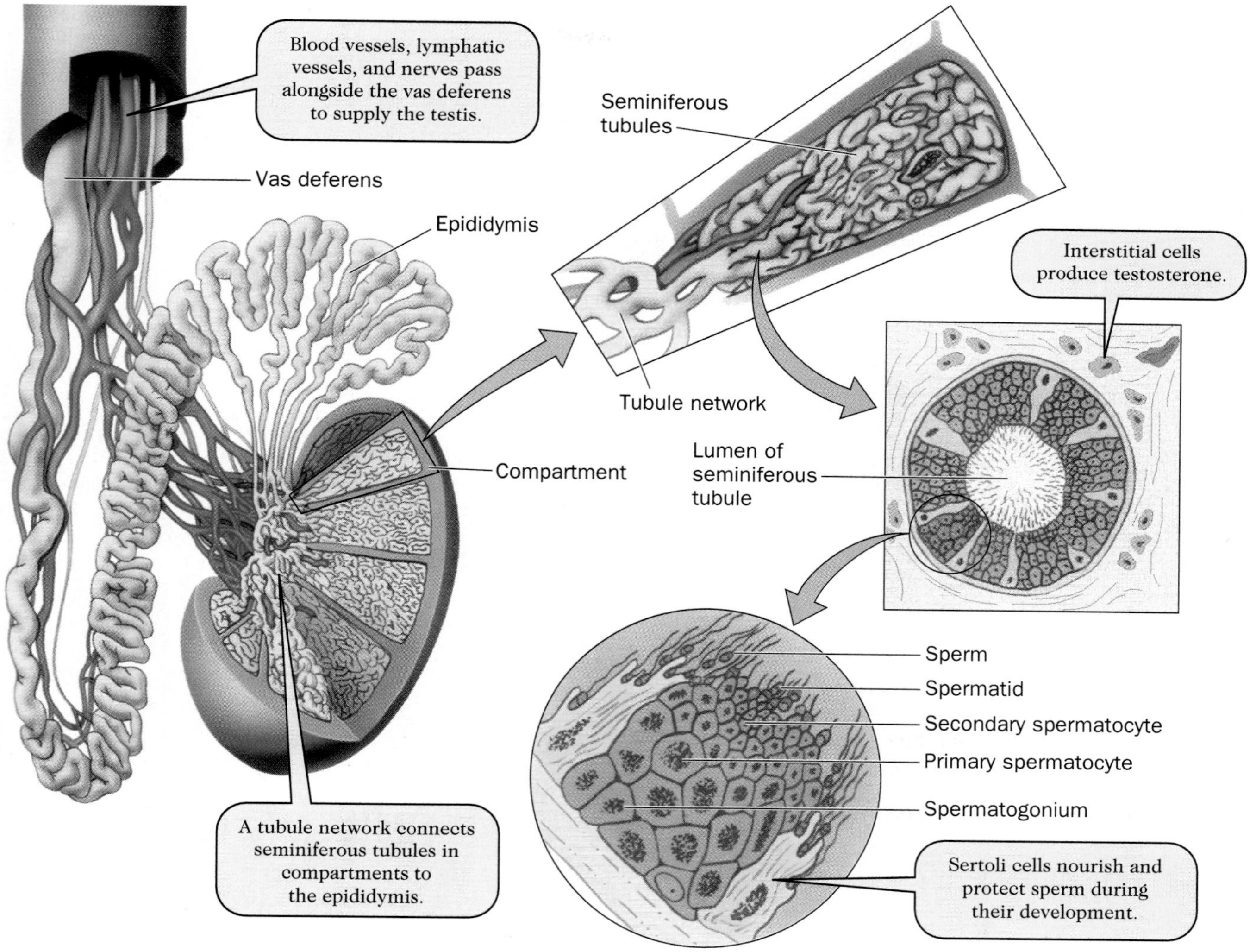
Blood vessels, lymphatic vessels, and nerves pass alongside the vas deferens to supply the testis.
Vas deferens
Epididymis
Seminiferous tubules
Tubule network
Compartment
Interstitial cells produce testosterone.
Lumen of seminiferous tubule
Sperm
Spermatid
Secondary spermatocyte
Primary spermatocyte
Spermatogonium
A tubule network connects seminiferous tubules in compartments to the epididymis.
Sertoli cells nourish and protect sperm during their development.

> Sperm are produced in the seminiferous tubules of the testes. Sertoli cells support, nourish, and protect developing sperm. Interstitial cells produce testosterone. Interstitial cells are stimulated by LH.

Spermatogenesis: Producing Sperm by Meiosis

Having identified the site of sperm development, let us now take a closer look at sperm formation, which is called **spermatogenesis** (*spur-mat-oh-jen-uh-sis*) (Figure 17.4). This process begins with a diploid cell (46 chromosomes), the **spermatogonium**, which divides by mitosis. This yields the **primary spermatocyte**, which undergoes meiosis I, producing **secondary spermatocytes**. Meiosis II then produces haploid **spermatids** (23 chromosomes). By the end of meiosis, each primary spermatocyte produces four spermatids. In association with a Sertoli cell, each spermatid develops into a mature sperm. **Sperm** consist mostly of a large nucleus, many mitochondria for generating ATP, and a long flagellum to propel the sperm toward the egg (Figure 17.5). The developmental stages of spermatogenesis are summarized in Table 17.1.

After puberty, millions of sperm mature each day. A male continues to make sperm from puberty until death, though the numbers of living sperm are reduced with age. It takes about 10 weeks for a

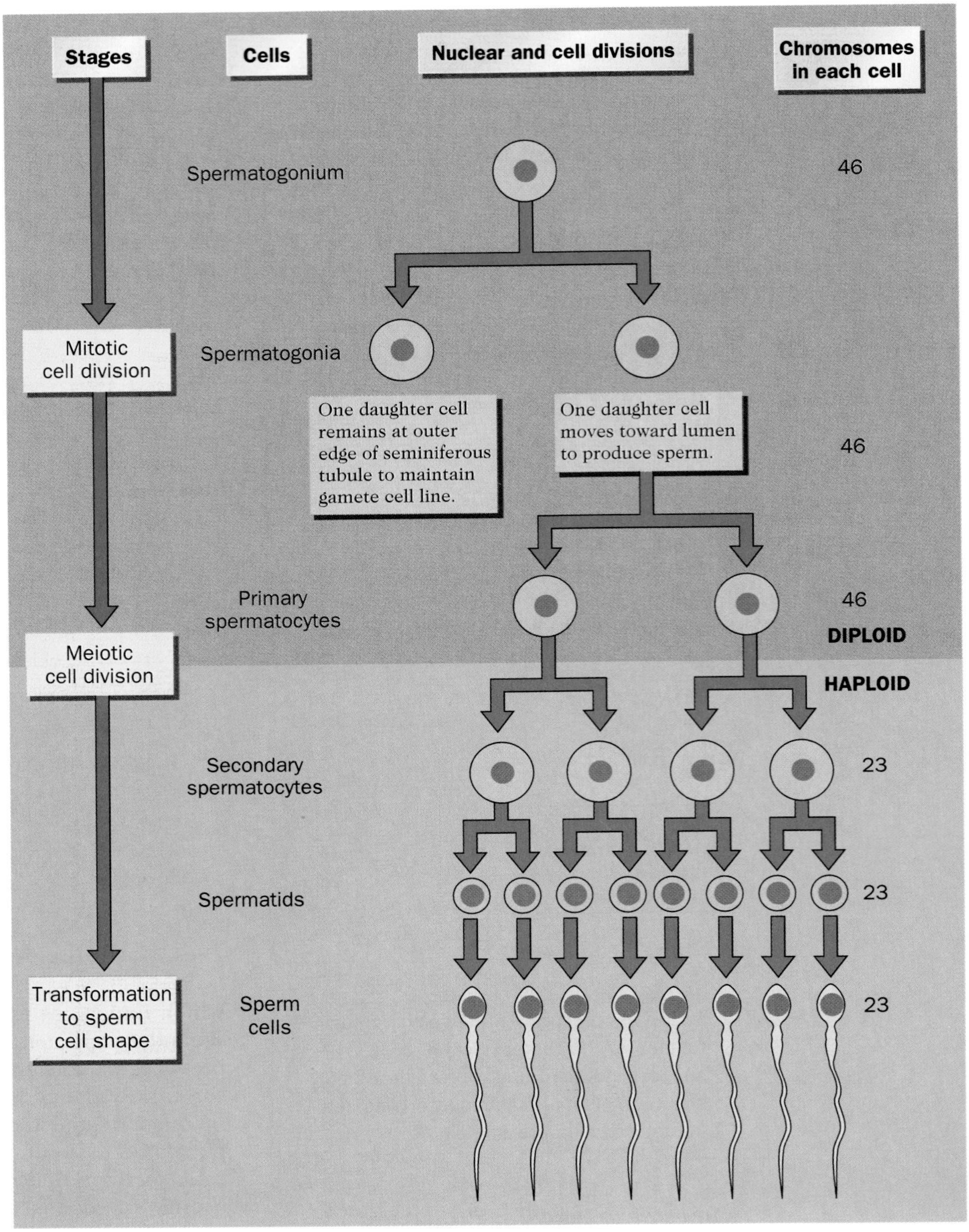
Stages
Cells
Nuclear and cell divisions
Chromosomes in each cell
Spermatogonium
46
Mitotic cell division
Spermatogonia
One daughter cell remains at outer edge of seminiferous tubule to maintain gamete cell line.
One daughter cell moves toward lumen to produce sperm.
46
Meiotic cell division
Primary spermatocytes
46
DIPLOID
HAPLOID
Secondary spermatocytes
23
Spermatids
23
Transformation to sperm cell shape
Sperm cells
23

FIGURE 17.4 The stages of spermatogenesis.

sperm to completely mature and be capable of fertilization. However, spermatogenesis is not failure proof, and a normal male produces about 10 percent defective sperm that are unable to fertilize because they are damaged or incomplete. Often flagella are missing.

During *ejaculation*, sperm are ejected from the body. Each ejaculation releases a volume of fluid containing up to 300 million sperm, although only one sperm cell is needed for the fertilization of an egg. Such large numbers are released because sperm face many obstacles in their

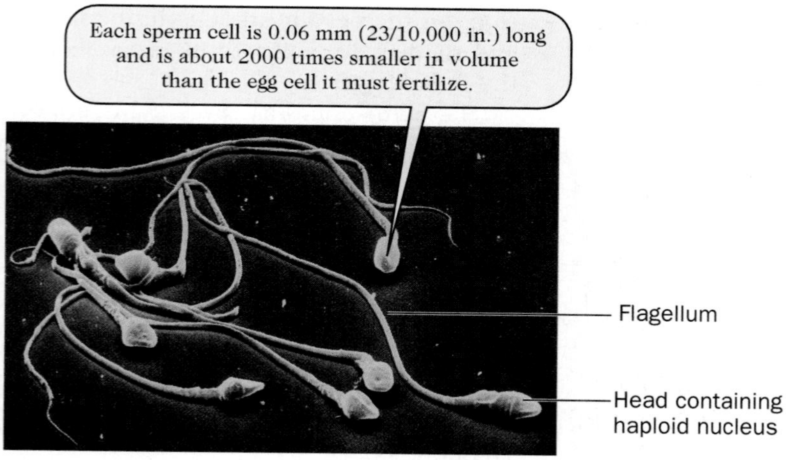
Each sperm cell is 0.06 mm (23/10,000 in.) long and is about 2000 times smaller in volume than the egg cell it must fertilize.
Flagellum
Head containing haploid nucleus

FIGURE 17.5 Sperm cells.

journey through the female system, and millions die. In fact, a male releasing fewer than 60 million sperm is usually considered infertile, as it is unlikely that any single sperm will survive the journey to ensure fertilization.

> Spermatogenesis involves mitosis, meiosis, and the development of sperm. The diploid primary spermatocyte undergoes meiosis to yield four haploid spermatids that mature into sperm. Sperm production begins at puberty and continues until death.

Transporting Sperm: Tubes, Glands, and Secretions

Maturing sperm are washed down the seminiferous tubules by secretions from Sertoli cells and begin a long journey that may end in fertilization. From the tubules, the sperm enter the epididymis, then the vas deferens, and eventually the urethra (see Figure 17.3a). Along the way, three accessory glands—the seminal vesicles, the prostate, and the bulbourethral glands—contribute secretions. Together with the sperm, these secretions form **semen**.

In each testis, the **epididymis** (*epp-puh-**did**-duh-mis*) is a long coiled tube where sperm can be stored while they mature (see Figure 17.3b, c). If sperm are not released from the epididymis they degenerate and their molecules are reabsorbed into the body. The wall of the epididymis contains smooth muscle which can contract in a rhythmic fashion to propel the sperm into the vas deferens during ejaculation.

Joining the epididymis, the **vas deferens** (***vas deaf**-fur-enz*) passes out of the testes and into the pelvic cavity posterior to the urinary bladder. Behind the bladder, each vas deferens merges with a duct from a **seminal vesicle** (***sem**-muh-nul*) (see Figure 17.3b). The two seminal vesicles produce a sugary, alkaline fluid (pH 7.2 to 7.6) which buffers the sperm against the acids in the female reproductive tract. The sugar, called fructose, is used for ATP formation by the sperm to power its swimming movements.

Next, each vas deferens merges with the urethra from the urinary bladder (Figure 17.3). The site where these tubes merge is surrounded by the **prostate gland** (***pros**-state*), which contributes a milky, alkaline fluid to the semen. In men over age 50 this gland often enlarges, narrowing the urethra and causing problems with urination. The prostate is also prone to malignant tumors.

TABLE 17.1 The Development of the Male Gamete by Spermatogenesis

Developmental Stage	Description
Spermatogonium	A diploid cell which divides by mitosis to produce primary spermatocytes
Primary spermatocyte	A diploid cell which undergoes meiosis I to produce secondary spermatocytes
Secondary spermatocyte	A haploid cell which undergoes meiosis II to produce spermatids
Spermatid	A haploid cell which develops into a sperm without further cell division
Sperm	The mature male gamete released at ejaculation

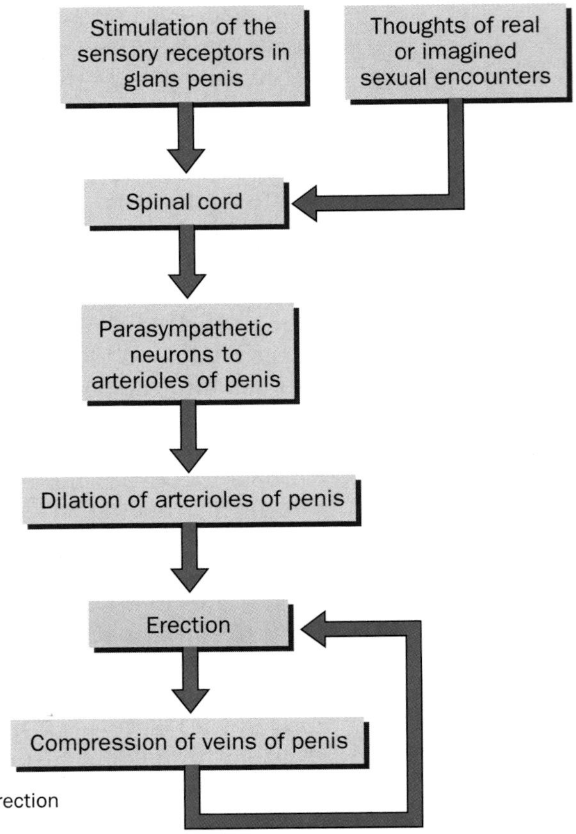
Stimulation of the sensory receptors in glans penis
Thoughts of real or imagined sexual encounters
Spinal cord
Parasympathetic neurons to arterioles of penis
Dilation of arterioles of penis
Erection
Compression of veins of penis

FIGURE 17.6 The erection reflex.

The third pair of accessory glands is the **bulbourethral glands** (***bulb**-boh-you-ree-thral*) which lie at the base of the penis (Figure 17.3). The bulbourethral glands secrete into the urethra a slippery, alkaline fluid which lubricates the urethral canal. Upon ejaculation, a series of highly coordinated events forcefully ejects about 2 to 5 milliliters (a teaspoon or less) of sperm-containing semen through the urethra of the erect penis.

> Sperm complete their maturation in the epididymis. Upon ejaculation, sperm pass into the vas deferens and urethra. Secretions of the seminal vesicles, prostate gland, and bulbourethral glands help create semen.

The Penis: Delivering the Sperm

The **penis** is the organ used during sexual intercourse to introduce sperm into the female reproductive tract. The internal structure of the penis permits **erection**, an increase in its length, diameter, and rigidity, to facilitate its entry into the female system.

The penis contains three cylinders of spongy **erectile tissue**, one of which surrounds the urethra (see Figure 17.3b). Upon sexual stimulation, an autonomic reflex causes arterioles to dilate and increase blood flow into the erectile tissues (Figure 17.6). Veins draining blood from the penis are squeezed closed as the fluid pressure in the erectile tissue increases. This traps blood in the erectile tissues, increasing and sustaining the erection.

The tip of the penis is expanded into an acorn-shaped structure called the head or *glans penis*. At birth, the glans penis is covered by a fleshy **foreskin**. A surgical procedure called *circumcision* may be performed to remove the foreskin.

> The penis contains three cylinders of erectile tissue which fill with blood upon sexual stimulation, allowing erection. The enlarged head of the penis is called the glans penis which is surrounded by the foreskin.

Male Sexual Response, Ejaculation, and Orgasm

So far we have described the essential male structures for sexual intercourse and the ejaculation process by which semen is expelled out of an erect penis. Now we will examine the **sexual response**, a series of events that coordinate erection, secretions, and muscular contractions.

Sexual response consists of four phases. (1) *Excitement* is a time of increased sexual awareness that leads to erection and increased pleasurable sensations from the penis. (2) *Plateau* is a period of intense arousal and increased breathing and heart rates, blood pressure, and muscle tension. (3) The *orgasm* phase is relatively brief in men and accomplishes ejaculation, which consists of emission and expulsion. During *emission*, sympathetic nerve impulses cause contraction of the smooth muscle in the epididymis, seminal vesicles, prostate, and bulbourethral glands and move semen into the urethra. The internal urethral sphincter at the base of the bladder closes to prevent semen from entering the bladder and urine from entering the urethra. During *expulsion*, the skeletal muscles surrounding the base of the penis contract in a rhythmic fashion to force the sperm-containing semen out through the urethra in several spurts. As expulsion continues, the male experiences an increase in breathing and heart rate and emotions of intense pleasure. These physical and emotional responses constitute the male orgasm. (4) *Resolution* is the period when the erection subsides and cannot be achieved again for a period of time. It may last several minutes to several hours.

> The male sexual response consists of four phases: excitement, plateau, orgasm, and resolution. Excitement is marked by erection. Plateau is marked by increasing arousal. Orgasm includes ejaculation. During resolution, the erection subsides.

The Female Reproductive System

While the male reproductive system produces and delivers reproductive cells, the female system is more complicated in that it also nurtures and delivers a new individual. The female reproductive system has five major functions: (1) It produces and transports the female gametes. (2) It receives sperm from the male. (3) It provides a site for fertilization. (4) It supports, protects, and nourishes the development of the fertilized egg, embryo, and fetus. (5) It provides a birth canal for the newborn infant. In addition to these reproductive functions, the female produces milk from mammary glands to nourish the newborn child.

The primary organs that accomplish these tasks are a pair of ovaries, a pair of uterine tubes, the uterus, and the vagina (Figure 17.7).

The Ovaries: Producing Follicles, Ova, and Hormones

In the female, the paired **ovaries** produce the female gametes, the eggs, or **ova** (singular, *ovum*), and the sex hormones estrogen and progesterone. The ovaries are about the size and shape of a small plum. They lie on the lateral walls of the pelvic cavity and are held in position by supportive ligaments.

Each ovary is covered with a layer of **germinal epithelium**. Within this covering there are small saclike structures called *ovarian follicles* (Figure 17.8). These follicles pass through different stages of development: primary follicles, secondary follicles, and graafian follicles. The small **primary follicles** consist of a single layer of cells and contain an immature ovum. The follicle cells support and protect the developing ovum and also produce the hormone estrogen, which maintains a woman's secondary sex characteristics.

The larger **secondary follicle** develops from the primary follicle and supports the continued maturation of the ovum. The next stage is the **graafian follicle** (***graph-ee-un***), which has several layers of cells and a fluid-filled cavity surrounding the developing ovum. The immature ovum is released from the graafian follicle in the process called *ovulation*. Ovulation occurs about once a month during a woman's childbearing years. After ovulation, the follicle structure changes into a hormone-secreting structure, the *corpus luteum*, before degrading.

The development of the follicles and the release of an immature ovum at ovulation are both initiated and controlled by follicle stimulating hormone (FSH) and luteinizing hormone (LH) from the pituitary gland (Chapter 16). The ovum will become completely mature only if it is fertilized by a sperm.

> The ovum is the female gamete. It is produced in a follicle within the ovary. Follicles undergo a process of maturation and development under the influence of pituitary hormones. Ovulation is the release of an immature ovum from the graafian follicle.

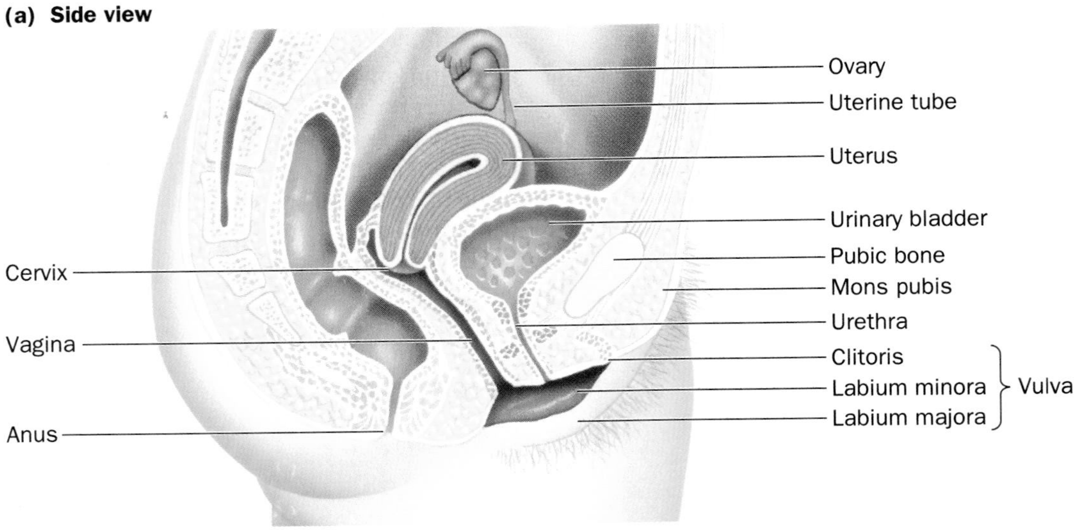
(a) Side view
Ovary
Uterine tube
Uterus
Urinary bladder
Pubic bone
Mons pubis
Urethra
Clitoris
Labium minora
Labium majora
Vulva
Cervix
Vagina
Anus

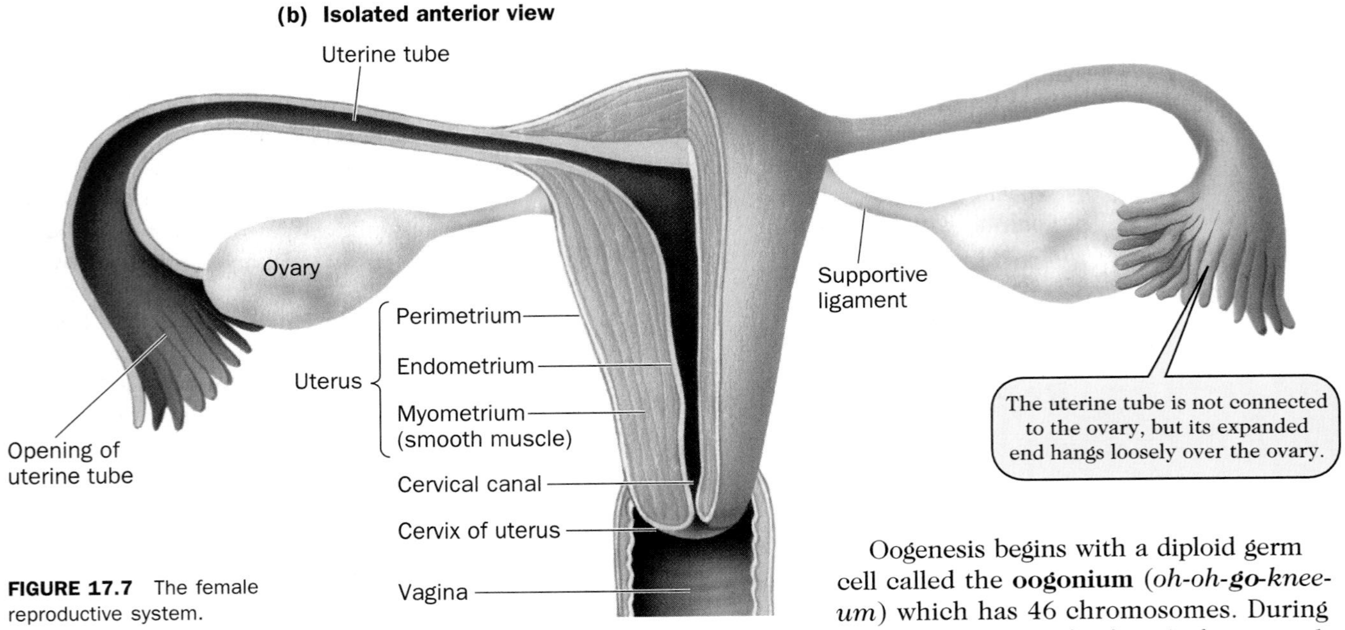
(b) Isolated anterior view
Uterine tube
Ovary
Supportive ligament
Opening of uterine tube
Uterus
Perimetrium
Endometrium
Myometrium (smooth muscle)
Cervical canal
Cervix of uterus
Vagina
The uterine tube is not connected to the ovary, but its expanded end hangs loosely over the ovary.

FIGURE 17.7 The female reproductive system.

Oogenesis: Producing a Mature Ovum by Meiosis

In females, the version of meiosis that produces ova is called **oogenesis** (*oh-oh-jen-uh-sis*) (Figure 17.9). It is initiated and partially completed in the follicles of the ovary. Unlike spermatogenesis, which begins at puberty, oogenesis begins before birth, although it is not completed until many years later. In fact, the final stages are completed only after fertilization occurs.

Oogenesis begins with a diploid germ cell called the **oogonium** (*oh-oh-go-knee-um*) which has 46 chromosomes. During the development of a female fetus, each oogonium divides by mitosis, producing **primary oocytes**, which are found in the primary follicles.

At the time of birth a female has about 500,000 primary oocytes. During her life she will make no more. In fact, most of them will degenerate before puberty. Only about 500 will develop sufficiently to be released by ovulation.

As oogenesis continues, a diploid primary oocyte undergoes meiosis I, producing a functional cell called a **secondary oocyte** and a small nonfunctional cell

(a) Sectioned view showing primary, secondary, and graafian follicles in which oocytes form through meiosis

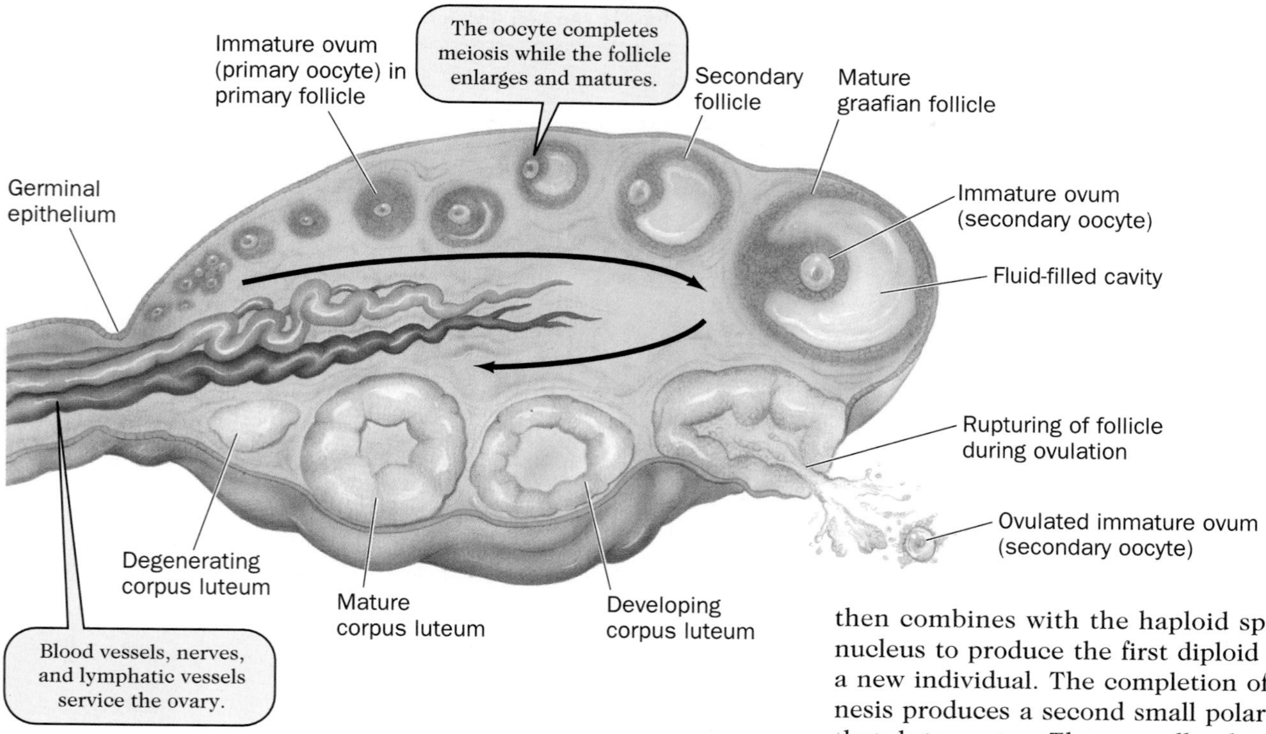
Immature ovum (primary oocyte) in primary follicle
The oocyte completes meiosis while the follicle enlarges and matures.
Secondary follicle
Mature graafian follicle
Germinal epithelium
Immature ovum (secondary oocyte)
Fluid-filled cavity
Rupturing of follicle during ovulation
Ovulated immature ovum (secondary oocyte)
Degenerating corpus luteum
Mature corpus luteum
Developing corpus luteum
Blood vessels, nerves, and lymphatic vessels service the ovary.

(b) Microscopic view of an ovulated oocyte

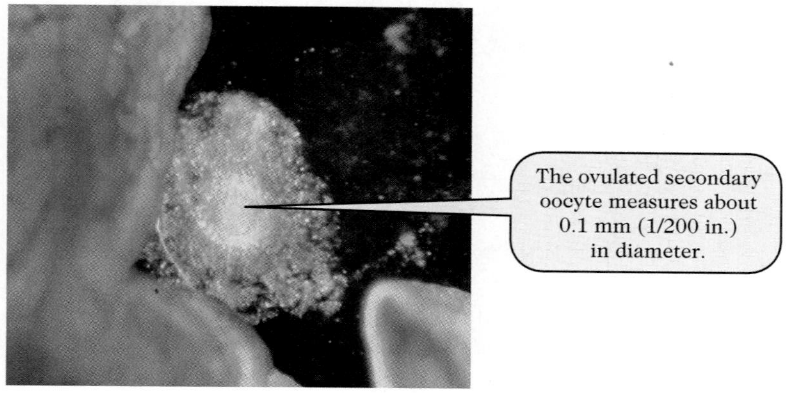
The ovulated secondary oocyte measures about 0.1 mm (1/200 in.) in diameter.

FIGURE 17.8 The ovary.

known as a **polar body**. The polar body degenerates immediately or after going through meiosis II. The secondary oocyte begins meiosis II but stops midway in the process at metaphase II. It is this secondary oocyte that is the immature ovum released from the ovary at ovulation.

Only if the secondary oocyte is penetrated by a sperm cell will it complete meiosis II, producing a mature ovum whose haploid nucleus (23 chromosomes) then combines with the haploid sperm nucleus to produce the first diploid cell of a new individual. The completion of oogenesis produces a second small polar body that degenerates. These small polar bodies contain little cytoplasm. They allow the necessary reduction in chromosome number by meiosis while permitting the ovum to retain most of its cytoplasm and stored food. Upon fertilization, these stored nutrients are used to sustain the early stages of development (Chapter 18). The stages of oogenesis are summarized in Table 17.2.

> Oogenesis involves meiosis, in which a diploid primary oocyte forms a haploid ovum and two haploid polar bodies. The secondary oocyte is released at ovulation. Meiosis II is completed only upon fertilization.

The Uterine Tubes: Site of Fertilization

The two **uterine tubes**, sometimes referred to as **oviducts** or **fallopian tubes**, convey the secondary oocyte to the uterus (see Figure 17.7b). One end of each tube connects to the uterus, while the other end forms a funnel-shaped structure that hangs loosely over the ovary.

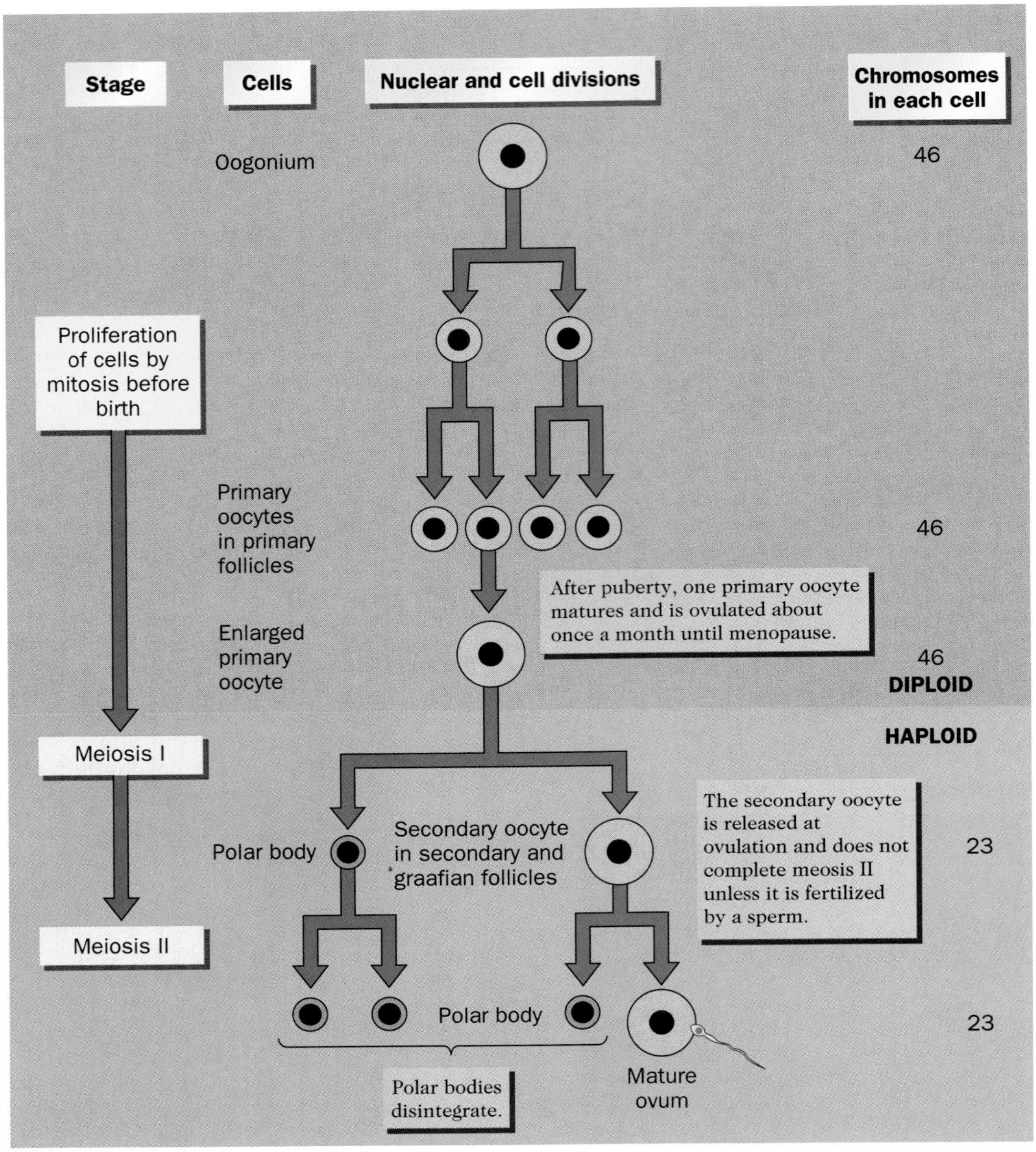

Stage
Cells
Nuclear and cell divisions
Chromosomes in each cell
Oogonium
46
Proliferation of cells by mitosis before birth
Primary oocytes in primary follicles
46
After puberty, one primary oocyte matures and is ovulated about once a month until menopause.
Enlarged primary oocyte
46
DIPLOID
HAPLOID
Meiosis I
Polar body
Secondary oocyte in secondary and graafian follicles
The secondary oocyte is released at ovulation and does not complete meosis II unless it is fertilized by a sperm.
23
Meiosis II
Polar body
23
Polar bodies disintegrate.
Mature ovum

FIGURE 17.9 The stages of oogenesis.

The secondary oocyte, which is released at ovulation, is "trapped" by the open end of the uterine tube. When an oocyte enters the uterine tube, its passage toward the uterus is facilitated by the action of cilia, which line the tube, and the contractions of smooth muscle in the tube wall.

A secondary oocyte can live only about 24 hours after ovulation. After ejaculation, sperm live only about 24 to 48 hours in the woman, and so for fertilization to take place, intercourse must occur fairly close to ovulation. If intercourse has occurred prior to ovulation, fertilization usually occurs in the upper third of the uterine tube (the distance traveled by the oocyte in a day). In humans, the only easily observed indication of ovulation is a slight rise in body temperature about a day after ovulation.

The Uterus: Site of Embryonic Development

The **uterus** is a hollow, muscular organ about the size and shape of a pear. It is here that the fertilized ovum grows

TABLE 17.2 The Development of the Female Gamete by Oogenesis

Developmental Stages	Description
Oogonium	A diploid cell which divides by mitosis to produce primary oocytes before birth
Primary oocyte	A diploid cell in a primary follicle which undergoes meiosis I to form a secondary oocyte and a polar body
Secondary oocyte	A haploid cell in a secondary follicle or graafian follicle which undergoes meiosis II and stops at metaphase II; the stage released at ovulation
Polar body	A nonfunctional cell resulting from meiosis I and II that degenerates without developing further
Ovum	The mature female gamete, produced when a secondary oocyte is fertilized and completes meiosis II

and develops (see Figure 17.7). The two uterine tubes enter on either side of its broad upper portion, and supportive ligaments attach the uterus to the pelvic wall, ovaries, and external genitalia. The narrow inferior end of the uterus, which is called the **cervix**, connects to the vagina. The entrance into the uterus, the **cervical canal**, is plugged with mucus secretions.

The wall of the uterus contains three distinct layers (see Figure 17.7b). The thin outer covering, or **perimetrium** (*per-ee-***me***-tree-um*), is actually a continuation of the peritoneal membrane. The thicker middle layer, or **myometrium**, contains smooth muscle. During pregnancy these cells increase in size, and during childbirth their contractions expel the infant into the world.

The inner lining of the uterus is called the **endometrium**. This layer contains abundant blood vessels and glands to nourish and protect the embryo. If fertilization occurs, the early embryo burrows into this layer in a process called *implantation* (Chapter 18). If fertilization and implantation do not occur, portions of this layer are shed and replaced on a monthly basis during the menstrual cycle, which we will discuss later in this chapter.

> Fertilization usually occurs in a uterine tube. The myometrium of the uterus has smooth muscle whose contractions expel the infant during birth. The endometrium of the uterus nourishes and protects an implanted embryo.

The Vagina: The Organ of Intercourse and Birth

The **vagina** is a flexible muscular tube extending about 8 to 10 centimeters (3 to 4 inches) from the cervix to the body's exterior (see Figure 17.7a). The vagina receives the penis during sexual intercourse and serves as the birth canal. Sperm are deposited in the vagina and must swim by flagellar motion through the cervix and uterus and into the uterine tubes.

The walls of the vagina, which are lined with epithelium, are normally collapsed and touching but can expand considerably during intercourse and childbirth. The acidity of the vagina (pH 3.5 to 4) limits microbial growth and is hostile to sperm. The alkaline pH of semen neutralizes this acidity.

A fold of mucous membrane may completely or partially cover the external opening of the vagina. This tissue, which is called the *hymen*, can be ruptured by initial sexual intercourse, strenuous physical activity, or the insertion of tampons to absorb the menstrual flow. At the exterior, the vagina opens into a **vestibule** which is enclosed by the external genitalia. Secretions from vestibular glands help lubricate the vagina.

> The vagina receives the penis during sexual intercourse and serves as the birth canal. Sperm deposited in the vagina must pass through the cervix to enter the uterus and uterine tubes.

The Vulva: The External Genitalia

The female external genitalia are collectively called the **vulva**. The vulva includes two sets of fleshy folds called labia (***lay-bee-uh***), the mons pubis, and the clitoris (Figure 17.10).

The **labia majora** are the large outer folds of flesh which enclose the vestibule. The skin of these labia is pigmented and contains abundant adipose tissue. It also contains smooth muscle fibers, glands, and sensory receptors. Anterior to the labia majora is a mound of fatty skin called the *mons pubis*. After puberty, the mons and the labia majora are covered with coarse hair.

Enclosed by the labia majora are the smaller **labia minora**. These labia are flaps of smooth hairless skin that enclose the vestibule. At their anterior margin, they merge to produce a hood over an erectile organ, the **clitoris** (***klit-uh-ris***). The clitoris contains two cylinders of spongy erectile tissue which fill with blood and enlarge upon sexual stimulation. This organ is abundantly supplied with sensory nerves and is the site of intense sexual feeling.

In the female body, unlike the male, the urethra does not participate in reproduction. Instead, it exits separately into the vestibule. This separation of the urinary and reproductive tracts in the female helps minimize the spread of disease between these two systems. This is particularly important during pregnancy.

> The vagina receives the penis and serves as the birth canal. The clitoris can achieve erection upon sexual stimulation. The female urethra passes separately from the bladder into the vestibule, which is surrounded by two sets of labia.

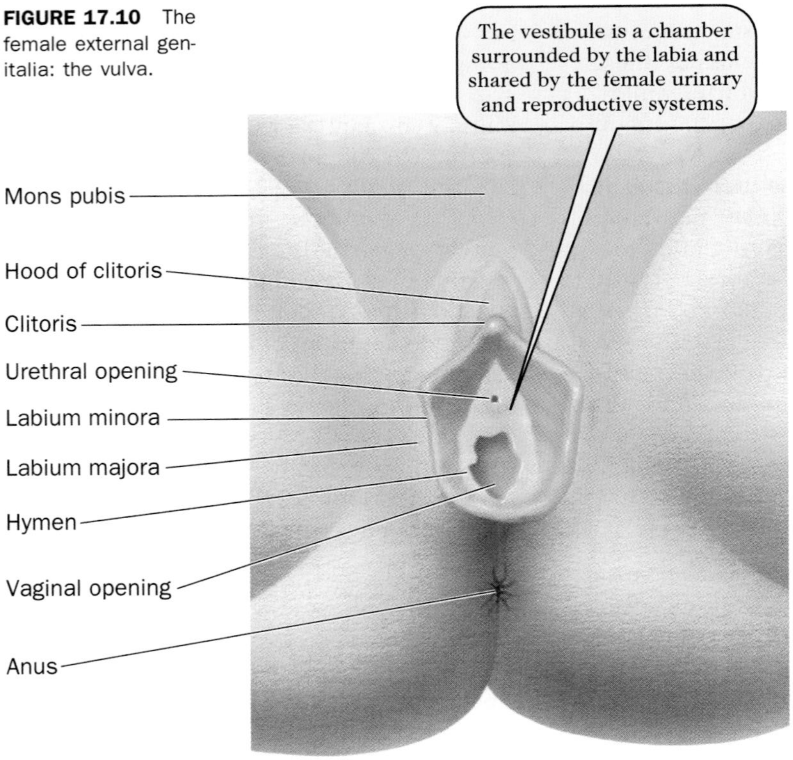

The vestibule is a chamber surrounded by the labia and shared by the female urinary and reproductive systems.
Mons pubis
Hood of clitoris
Clitoris
Urethral opening
Labium minora
Labium majora
Hymen
Vaginal opening
Anus

FIGURE 17.10 The female external genitalia: the vulva.

The Mammary Glands: Nourishing the Newborn

The **mammary glands** are modified sweat glands and are not directly connected with the rest of the female reproductive system (Figure 17.11). However, at puberty their development coincides with changes in the reproductive system. For this reason and because they nourish the newborn infant, we will discuss them here.

The mammary glands lie between the pectoralis muscles and the skin. These glands, which are typical of all mammals, are modified for milk production in a process called **lactation**. The mammary glands begin preparing for lactation during pregnancy but do not secrete milk until after childbirth.

Each breast contains a projecting **nipple** which is surrounded by a pigmented area, the **areola**. Smooth muscle tissue in the nipple causes it to become erect in the cold, during sexual stimulation, and during the nursing of an infant. Glands in the areola lubricate the nipple to keep it soft for comfortable nursing.

Internally, each breast contains many lobes of mammary gland tissue consisting of gland cells which produce and secrete milk. These cells are organized into sacs and surrounded by modified gland cells that are capable of limited contractions. Secreted milk flows through a series of narrow, branching ducts to reach a **lact-**

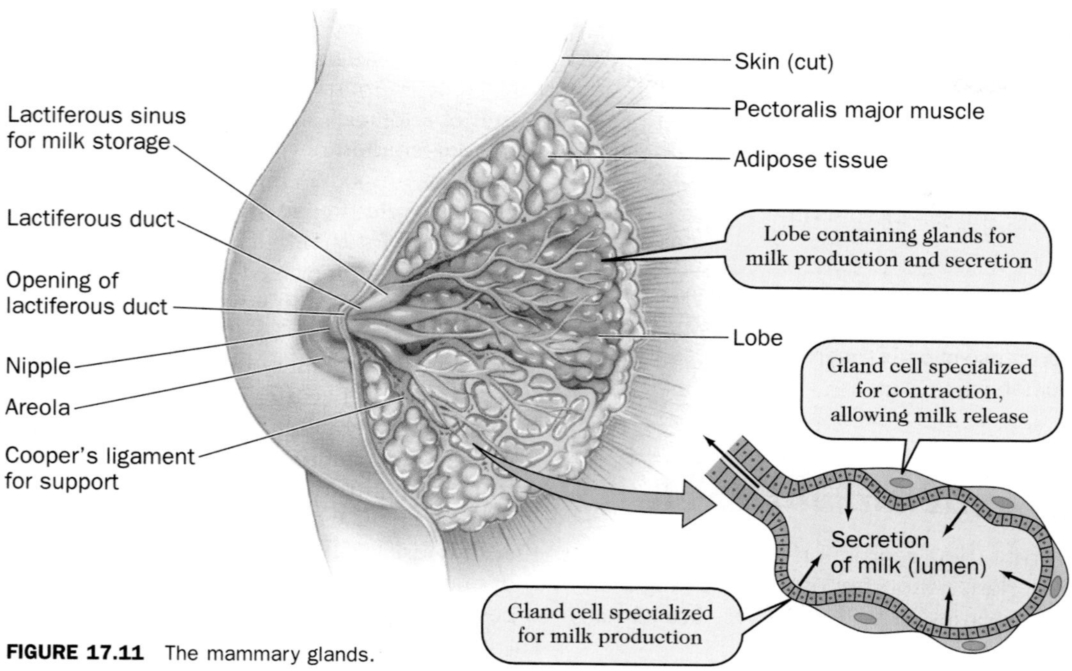

Skin (cut)
Pectoralis major muscle
Adipose tissue
Lactiferous sinus for milk storage
Lactiferous duct
Lobe containing glands for milk production and secretion
Opening of lactiferous duct
Lobe
Nipple
Areola
Gland cell specialized for contraction, allowing milk release
Cooper's ligament for support
Secretion of milk (lumen)
Gland cell specialized for milk production

FIGURE 17.11 The mammary glands.

iferous sinus where it can be briefly stored. Weak contractions in the modified epithelial cells cause the release, or "letdown," of milk. The released milk flows through a **lactiferous duct** to reach the outside through a pore in the nipple.

Except during actual lactation, the mammary glands are small and contribute little to the size of the breast. Breast size is determined largely by the amount of adipose tissue present and is not an accurate indication of the potential for milk production. The only internal support for the breasts is provided by narrow bands of connective tissue called *Cooper's ligaments*.

Although both males and females have mammary glands, they remain inactive in males. Estrogen initiates the enlargement of the breasts at puberty, but both estrogen and progesterone are needed to prepare the mammary glands for lactation. Prolactin encourages milk formation during lactation and oxytocin allows the release of milk during nursing (Chapter 16).

> Milk is produced and secreted by the mammary glands. The hormones estrogen, progesterone, prolactin, and oxytocin all play roles in lactation.

Female Sexual Response and Orgasm

Females experience the same four phases of sexual response that males experience: excitement, plateau, orgasm, and resolution. The *excitement phase* in women is triggered by tactile, visual, or psychological stimulation. The clitoris and breasts are especially sensitive to physical stimulation, which initiates spinal reflexes that dilate blood vessels in the external genitalia, the breasts, and the wall of the vagina. Increased blood flow in these regions causes the labia to swell and the clitoris and nipples to become erect. Lubrication for sexual intercourse is provided by the glands at the vestibule and by male secretions.

During the *plateau phase*, circulatory, respiratory, and muscular responses allow accumulation of blood in the wall of the vagina, causing it to swell further. As stimulation continues, women enter the orgasm phase.

The *orgasm phase* is characterized by rhythmic muscular contractions in the vaginal wall and sensations of intense physical pleasure. The experience of orgasm is highly variable, with barely perceptible sensations in some women and repeated waves of sensation in others. There is no exact female equivalent of male ejaculation.

During the *resolution phase*, breathing and heart rate return to normal and erectile tissues empty. However, women unlike men, can immediately continue sexual activity. With continued sexual stimulation, some women are able to cycle between the plateau and orgasm phases and achieve multiple orgasms.

The primary hormone influencing a woman's sex drive is testosterone, which she produces in very small quantities in her adrenal glands—only a tiny fraction of the amount produced in a man's testes. The testosterone level in the blood peaks at about the same time ovulation occurs, and some women have an increased desire for sex at this time.

Women experience the same stages of sexual response (excitement, plateau, orgasm, and resolution) as men, but express them differently.

The Menstrual Cycle

During their reproductive years women experience monthly cycles of changes in their ovaries and uterus. There are actually two overlapping cycles that cover about 28 days: the ovarian cycle and the uterine cycle. The **ovarian cycle** includes the maturation of a follicle, oogenesis, ovulation, and the degradation of the follicle. This cycle develops an oocyte for fertilization. The **uterine cycle** prepares the endometrium of the uterus for the implantation and development of the fertilized ovum and initiates shedding of the lining if implantation does not occur. Both cycles are controlled by hormones, as depicted in Figure 17.12.

The cycles begin at puberty, are interrupted only by pregnancy, and cease at the time of menopause. Their durations vary from woman to woman and even in the same woman. Women who exercise very strenuously or have very low or high levels of body fat may experience a temporary cessation of these cycles, a condition

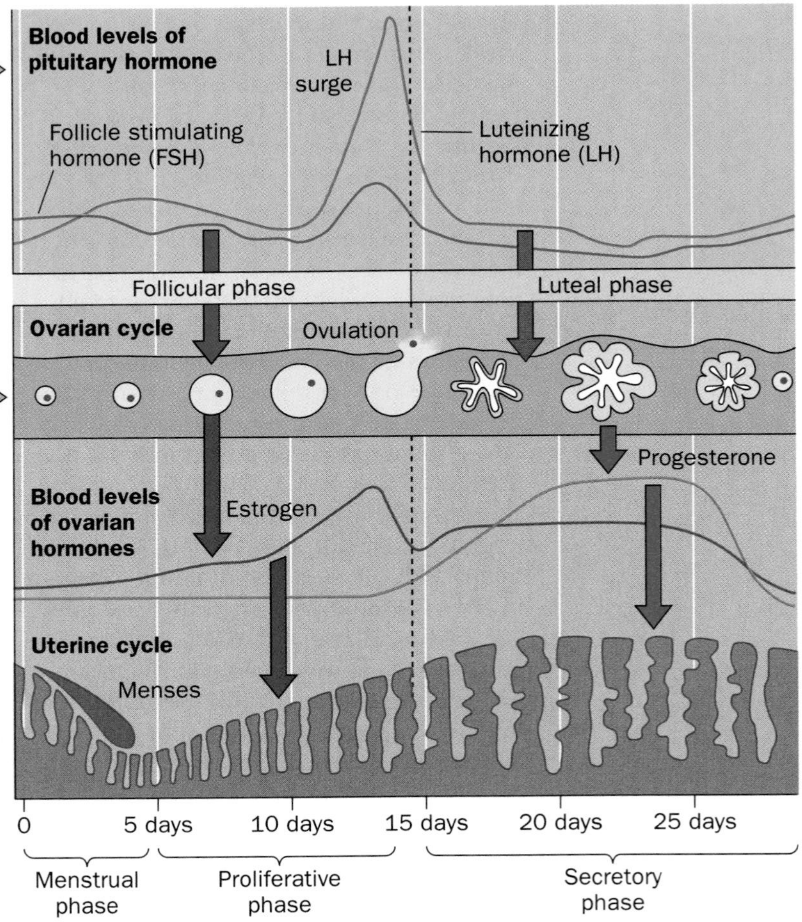
Blood levels of pituitary hormone
LH surge
Follicle stimulating hormone (FSH)
Luteinizing hormone (LH)
Follicular phase
Luteal phase
Ovarian cycle
Ovulation
Progesterone
Blood levels of ovarian hormones
Estrogen
Uterine cycle
Menses
0
5 days
10 days
15 days
20 days
25 days
Menstrual phase
Proliferative phase
Secretory phase

FIGURE 17.12 The ovarian and uterine cycles. In the follicular phase, FSH causes an ovarian follicle to develop. The developing follicle releases estrogen, causing growth of the endometrium. A surge in LH production stimulates ovulation and corpus luteum formation. In the luteal phase, progesterone promotes further thickening of the endometrium. Menstruation occurs when the corpus luteum degenerates and progesterone production declines.

called *amenorrhea* (*aa-men-uh-***ree***-uh*). The ovarian and uterine cycles are often collectively called the **menstrual cycle**.

The Ovarian Cycle

The ovaries of adolescent and adult women alternate between two phases, each lasting roughly 2 weeks. Days 1 through 14 are the **follicular phase**, which is initiated by the release of follicle stimulating hormone (FSH) from the anterior pituitary. During this phase, a primary follicle begins to enlarge and a primary oocyte begins the process of meiosis. While it is growing, the follicle secretes estrogen and smaller quantities of progesterone into the blood. By day 14, a graafian follicle has matured and pushed up against the ovary surface like a blister.

The **luteal phase** begins at around day 14, when a surge in luteinizing hormone (LH) released by the anterior pituitary causes rupture of the graafian follicle and ovulation. A home test kit (Q-Test) is available to detect this surge in LH concentration and predicts ovulation. The test measures excess LH in a woman's urine and is useful for couples who want to increase their chance of achieving fertilization and pregnancy.

After the follicle ruptures, it collapses and is transformed into a **corpus luteum**. The corpus luteum slowly increases in size, and its cells release a large quantity of progesterone and smaller amounts of estrogen into the blood. The progesterone acts on the endometrium of the uterus, preparing it to receive a fertilized ovum.

If there is no fertilization, the corpus luteum degenerates and is reabsorbed by the ovary, marking the end of the luteal phase. The decline in progesterone initiates the shedding of the endometrium in a process called **menses** or **menstruation**. However, if fertilization does occur, the corpus luteum persists and secretes progesterone for about 6 months, until another structure, the *placenta*, produces sufficient progesterone. Then the corpus luteum degenerates.

How does the follicle in the ovary know that fertilization has occurred so that the corpus luteum and its progesterone production can be maintained? The answer involves another hormone. The developing embryo implanted in the endometrium secretes a hormone called *human chorionic gonadotropin* (*HCG*), which signals the ovary to sustain the corpus luteum. This hormone is also the basis for home test kits which indicate pregnancy by detecting HCG in a urine sample (First Response, e.p.t., FACT Plus, Precise).

> The ovarian cycle has a follicular phase and a luteal phase. FSH starts the follicular phase, in which an oocyte undergoes meiosis and estrogen is secreted. LH initiates ovulation and the luteal phase, in which a corpus luteum forms and secretes progesterone.

The Uterine Cycle

The series of changes in the lining of the uterus constitutes the **uterine cycle**. These changes are initiated by cyclic fluctuations in the amounts of estrogen and progesterone secreted by the ovary during the ovarian cycle. The uterine cycle consists of three phases: the menstrual phase, the proliferative phase, and the secretory phase (Figure 17.12).

The **menstrual phase**, the most dramatic and obvious set of events, occupies the first week of the uterine cycle. It follows the decline of the corpus luteum and the abrupt decline in the amount of progesterone and estrogen in the blood.

When deprived of these two hormones, blood vessels in the endometrium lining of the uterus begin to close off, reducing the blood flow and oxygen (O_2) delivery to the endometrial cells. The cells in the outer layer of the endometrium begin to die and are shed in patches, along with blood from degenerating blood vessels. The tissue debris and blood wash into the uterine cavity, through the cervix, and into the vagina, from which they are discharged from the body. The discharge is called the *menstrual flow*.

The entire uterine lining is shed in this fashion over a 5- to 7-day period. Only a basal layer of the endometrium remains for regeneration during the remainder of the uterine cycle. Usually, the total blood loss resulting from menstrual flow is about 50 to 150 milliliters (about 2 to 5 ounces). Contractions of the smooth muscle in the uterine wall help expel the endometrium, producing the often painful cramps experienced during menstruation.

Very painful cramps are called *dysmenorrhea* (*dis-men-uh-***ree***-uh*) and are caused by the presence of prostaglandins in the uterus. Over-the-counter medications which contain ibuprofen are often used to inhibit prostaglandin production and ease menstrual cramping.

The **proliferative phase** occupies the second week of the uterine cycle. During this time a new follicle is developing in the ovary, and its secretion of estrogen causes the endometrium to grow by means of mitosis. New epithelial cells, glands, and blood vessels are produced as the endometrium thickens (Figure 17.12).

The **secretory phase** begins at ovulation and coincides with the growth and decline of the corpus luteum in the ovary, thus occupying the last 2 weeks of the uterine cycle. The increased secretion of progesterone during the expansion of the corpus luteum causes more growth of the endometrium, especially extension of blood vessels and enlargement of the uterine glands. The endometrium becomes thick and spongy as it retains water, electrolytes, and glycogen in preparation for nourishing an embryo. In the absence of fertilization, the corpus luteum and its production of progesterone decline. The decline in progesterone initiates the menstrual phase described above, and the entire uterine cycle is repeated.

Premenstrual Syndrome

Premenstrual syndrome (*PMS*) involves a cluster of physical and emotional symptoms that may appear in some women during the luteal phase of the ovarian cycle and then disappear with menstruation. Over 90 percent of women experience PMS at some time in their lives. The typical physical symptoms include swollen and tender breasts, water retention, headache, backache, and fatigue. The emotional symptoms include lethargy, depression, panic attacks, irritability, and even violent hostility.

The causes of PMS are unknown but may be related to the influence of fluctuating levels of reproductive hormones on the nervous system. Medications to reduce water retention and pain offer some relief from the physical symptoms. Personal management of the emotional symptoms can be achieved through education, therapy, relaxation techniques, stress reduction, and, if all else fails, medication.

> During the proliferative phase, estrogen stimulates the growth of the endometrium. In the secretory phase, progesterone enhances blood vessel and glandular growth. If there is no fertilization the menstrual phase begins, and the endometrium is shed.

Alternatives for Birth Control

Some women attempt to increase their opportunities for fertilization by monitoring LH levels and body temperature to match ovulation with intercourse. Other women try to decrease the probability of pregnancy by using some form of birth control. Some of the alternatives for birth control also protect against sexually transmitted diseases (STDs) (Figure 17.13).

Abstinence

Some women and men abstain from sexual activity, choosing celibacy as a philosophical or religious approach to life. Others avoid sexual contact only during certain portions of their lives or reproductive cycles. Total abstinence from vaginal sexual intercourse is the only completely failureproof way for a woman to avoid pregnancy.

Artificial Methods of Birth Control

Artificial methods of birth control include surgical sterilization, hormone manipulation, intrauterine devices (IUDs), chemicals, and barriers (Figure 17.13). Their effectiveness in preventing pregnancy varies (Table 17.3).

Sterilization. In male sterilization, which is called *vasectomy*, a small incision is made in the wall of the scrotum, the vas deferens is located and tied at two

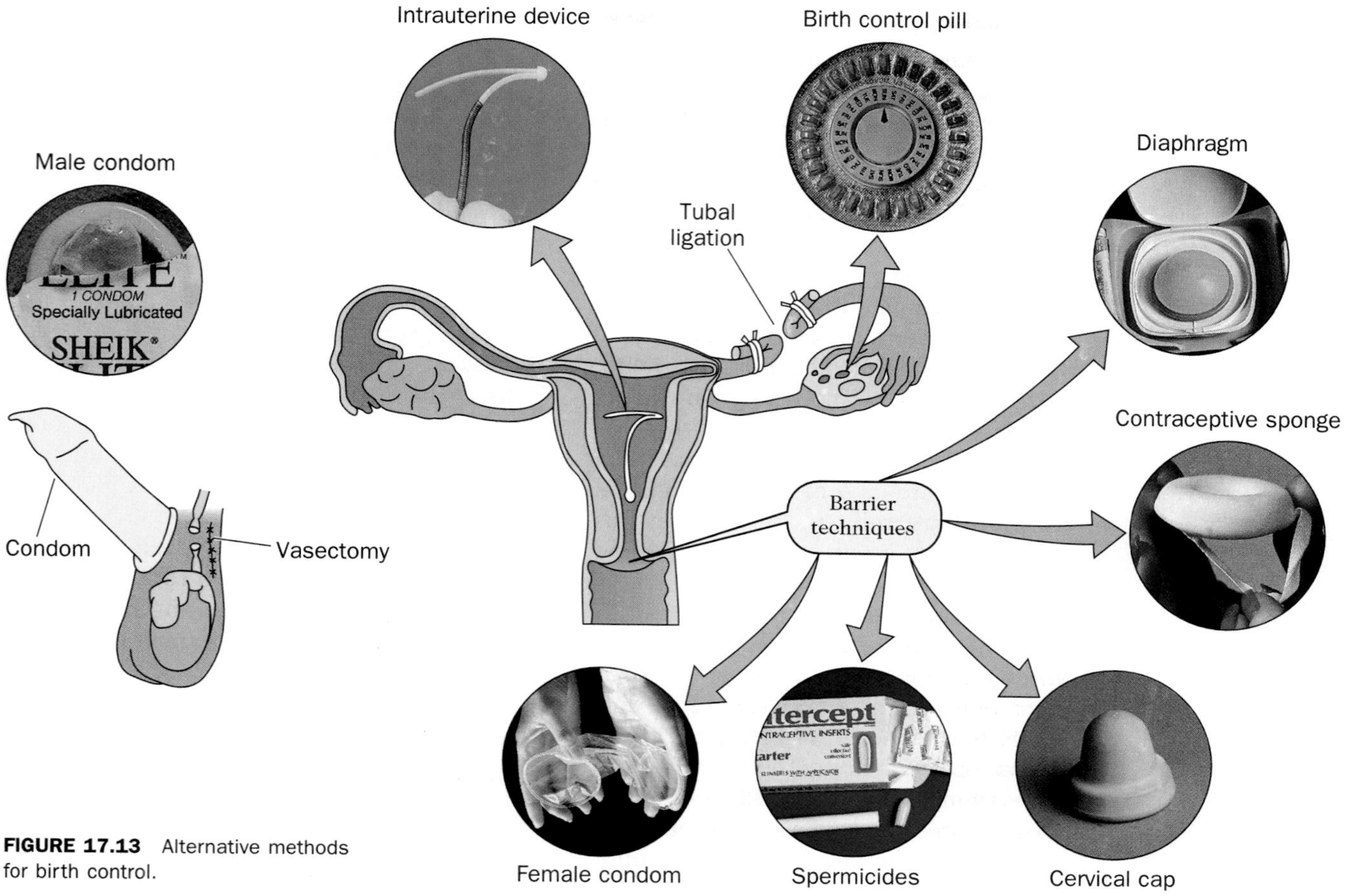
Intrauterine device
Birth control pill
Male condom
Diaphragm
Tubal ligation
1 CONDOM
Specially Lubricated
SHEIK
Contraceptive sponge
Condom
Vasectomy
Barrier techniques
Female condom
Spermicides
Cervical cap

FIGURE 17.13 Alternative methods for birth control.

TABLE 17.3 Effectiveness of Birth Control Methods		
	Percentage of Women Experiencing an Unplanned Pregnancy in First Year of Continuous Use	
Method	Typical Use*	Perfect Use†
No method	85	85
Withdrawal	4	18
Fertility awareness rhythm	20	<10
Spermicide	21	<6
Cervical cap	18	<10
Diaphragm	18	6
Female condom	21	5
Male condom	12	<3
Depo-Provera	<1	<1
Norplant	<1	<1
Sterilization	<1	<1

* Typical first-time use when not all the rules of use are observed.
† Perfect use when all the rules of use are carefully observed.

sites, and an intervening section is cut out (Figure 17.13). This prevents sperm from reaching the urethra and being ejaculated in the semen. The man continues to produce testosterone and sperm, and his secondary sex characteristics and sex drive are not reduced.

In female sterilization, or *tubal ligation*, an incision is made in the abdominal wall, the uterine tubes are tied at two points, and the intervening section is cut away (Figure 17.13). The oocyte is thus unable to reach the uterus, and sperm are unable to reach the oocyte. To avoid an external incision (and scar), a new technique called *hysteroscopy* (*hiss-tuh-**rah**-skop-pea*) involves the insertion of a tiny flexible telescope through the vagina and uterus and into the uterine tubes. An electrical current is used to seal the tubes.

Hormones. Since hormones control the production of oocytes, artificial manipulation of hormone levels can provide an effective means of reducing fertility for women. Oral contraceptives, or *birth control pills*, are used by nearly one-third of American women at some time in their lives. Various combinations of synthetic progesterone and estrogen provide about a dozen different formulations of birth control pills. These block the release of FSH and LH from the pituitary and prevent follicle development and ovulation, but offer no protection against STDs.

Birth control pills have both beneficial and undesirable side effects. On the positive side many women find that they reduce menstrual flow and cramps. There also appears to be some protection against ovarian cysts and cancer of the uterus and ovary. However, other women experience water retention, high blood pressure, and headaches. In women over age 35 who smoke, blood clots may form. Because of the potential for side effects, a woman using oral contraceptives should consult regularly with her health care provider.

Daily pill taking can be tedious, and so some women have opted for the *Norplant system*, which is a slow-release form of progesterone that can be inserted under the skin and is effective for about 5 years. An injectable derivative of progesterone, Depo-Provera, is also available. Each injection provides effective contraception for a 3-month period.

Intrauterine Devices. *Intrauterine devices (IUDs)* are small pieces of plastic, copper, or stainless steel which are inserted by a health care provider into the uterus. IUDs act by creating a mild, chronic inflammation which damages sperm and prevents implantation. IUDs may cause cramping and excessive menstrual bleeding in some women. They provide no protection against STDs.

Chemicals. Sperm-killing chemicals called *spermicides* are available as foams, jellies, creams, suppositories, and douches. These products are inserted into the vagina shortly before intercourse. They may kill some STD agents. A common ingredient, *nonoxynol-9*, has been found to be effective against the AIDS virus.

Barriers. Barrier methods include condoms, diaphragms, and cervical caps. Properly used barrier methods prevent sperm from gaining access to the uterus and uterine tubes. *Condoms* can be worn by both males and females as inexpensive, disposable barriers. Both types of condoms can be used in conjunction with spermicides to increase their effectiveness. Latex condoms are effective in reducing disease transmission, including transmission of the virus that causes AIDS.

The most effective male condom is a latex sheath which encloses the penis and prevents the ejaculated semen from contacting the vagina. Most condoms are available in a prelubricated form to increase comfort and sensitivity. The female condom is formed from two flexible rings connected by a polyurethane sheath. One ring fits over the cervix and the other fits over the external genitalia to line the entire vaginal region, preventing sperm from entering the uterus and reducing disease transmission.

Some women use a *diaphragm*, which is a rubber dome that fits over the entire cervix. The diaphragm is used along with a spermicide to block sperm from entering the uterus but must be left in place for some time after intercourse. Diaphragms must be fitted individually to the user. By themselves, they do not protect a woman against STDs.

Small latex or plastic *cervical caps* fit over the tip of the cervix. Like a diaphragm, they should be used with a spermicide, and they must be fitted to each user. They can be worn comfortably for up to 48 hours. They do not protect a woman against STDs.

Alternatives: The Rhythm Method and Withdrawal

For those who find abstinence, sterilization, hormones, barriers, or chemicals unacceptable, the *rhythm* or *fertility awareness method (FAM)* provides an alternative. FAM is actually a group of practices that take advantage of the short life spans of the sperm and oocyte and the limited period during which fertilization is possible after ovulation.

Using FAM as a birth control technique involves monitoring body temperature to watch for the slight rise in temperature (several tenths of a degree) that usually follows ovulation. After 3 days of elevated temperature, there is a reduced opportunity for fertilization. A women also may monitor other changes, including alterations in her cervical mucus (thin and watery near ovulation, thick and sticky after ovulation), slight one-sided pain in the lower abdomen, softening of the cervix, and breast tenderness. Careful calendar charting of these physical changes is essential to the effectiveness of FAM in preventing pregnancy. To prevent pregnancy, a woman should avoid intercourse for an 8-day period in the middle of her menstrual cycle: 5 days before ovulation through 3 days after ovulation.

The least effective alternative is *withdrawal* of the penis before ejaculation. This takes great control by the male and trust by the female. Before ejaculation occurs, secretions may flush some sperm out of the vas deferens, creating the possibility of fertilization.

Abortion

In cases of rape and incest or when other methods of birth control fail, an unplanned pregnancy may occur. In these cases, some women may choose an *abortion*, which is the termination and induced expulsion of a developing embryo. Methods for abortion include vacuum suction of the embryo from the uterus, surgical scraping of the uterine lining, infusion of a strong salt solution to cause rejection and the use of drugs or chemicals.

A new drug, *RU 486 (mifepristone)*, blocks the normal action of progesterone and causes a loss of the endometrium, carrying away an implanted embryo. It can be used up to 5 weeks after fertilization has occurred. Certain *prostaglandins* have also been studied as a means of "morning-after" birth control because of their ability to cause shedding of the endometrium and contractions of the uterus.

Abortion is currently the focus of heated debate, physical confrontation, and legal wrangling between those who support an embryo's "right to life" and those who want to protect a woman's "right to choose" her own reproductive alternatives. Some people prefer to emphasize abstinence or the use of contraceptives to prevent fertilization, while others see a need for safe and readily available intervention at any point in the reproductive cycle.

> Next to complete sexual abstinence or surgical sterilization, the barrier protection provided by male or female condoms, in conjunction with a spermicidal agent, offers the best protection against both pregnancy and STDs.

Diseases and Disorders

There are four general types of diseases and disorders of the reproductive systems in males and females: (1) fertility problems, which cause difficulties in becoming pregnant, (2) complications of pregnancy, (3) cancers and tumors of the reproductive system, and (4) sexually transmitted diseases (STDs). In this section we will focus on fertility problems and cancers. The complications of pregnancy are briefly discussed in Chapter 18, and STDs are treated in Spotlight on Health: Seven (page 526).

Infertility

Many couples who want children have difficulty achieving pregnancy. A couple are considered infertile if they fail to conceive a child after a year of frequent intercourse. In about 40 percent of cases of infertility there is a physical problem with the male, in another 40 percent it is the female, and in 20 percent it is both partners.

Male infertility is usually traced to a lower sperm count or an abundance of abnormal sperm. Recall that a man who ejaculates fewer than 60 million sperm is considered infertile. Sometimes the sperm count is low because a man produces too little testosterone. In other cases a man's immune system may mistakenly attack his sperm. Male fertility can also be reduced by exposure to radiation, high temperature, drugs such as anabolic steroids, and by diseases like mumps and gonorrhea.

Some males can increase their sperm counts by taking hormone injections. Others use the services of a fertility clinic to pool, preserve, and concentrate several semen samples. The accumulated semen is then artificially inserted into the female partner.

Female infertility can have many causes. Irregular or infrequent menstrual cycles make it difficult for intercourse to coincide with ovulation. Low hormone production by the pituitary gland may prevent new follicles and oocytes from being activated, limiting ovulation. Some infertile women produce too much prolactin, the hormone which stimulates milk production but also inhibits ovulation in nursing mothers.

Because of malformation or inflammation, the uterine tubes may prevent the passage of either sperm or oocytes. Some infertile women produce an overly thick cervical mucus that prevents sperm penetration, or their vaginal secretions are so acidic that they quickly destroy sperm. About 20 percent of infertile women develop benign tumors of the uterus called *fibroids*, which can interfere with the implantation of a fertilized egg.

In *endometriosis*, there is abnormal growth of endometrial tissue on the outer surface of the uterus, the uterine tubes, or even the bladder. Normal hormone cycles induce changes in this abnormally placed tissue. These changes cause pain and may interfere with conception or pregnancy. Endometriosis can sometimes be corrected with surgery, hormone therapy, or drugs which prevent menstruation by inhibiting the release of FSH and LH.

Age also affects fertility in females. Women age 40 to 50 are more likely to release oocytes with genetic damage. This damage occurs because the oocytes have had more years in which to be exposed to harmful substances and pathogens. Embryos from genetically damaged oocytes may not complete development.

In some cases, there is normal fertilization of a healthy oocyte but the embryo is not sustained in the uterus, and a spontaneous abortion, or *miscarriage*, occurs. Miscarriage is defined as the loss of a fetus before the twenty-second week or before it can survive outside the uterus. It is estimated that up to one-third of all pregnancies end in miscarriage, with most miscarriages occurring in the first 3 months. For a couple who want children, miscarriage can cause even more anxiety and frustration than infertility does.

Infertility or repeated miscarriages may persuade some couples to consider *adoption*. For others, advances in medical and reproductive technology have made alternative methods for enhancing fertility available.

Alternatives for Enhancing Fertility

Artificial insemination (sperm donation) is a technique that has been borrowed from agriculture. It increases fertility in cases where a male partner is infertile. It also permits a single female to achieve pregnancy. Sperm from a volunteer donor are inserted into the vagina or uterus of the female recipient near her normal time of ovulation. Although the name of the donor is withheld, information on the physical characteristics, abilities, educational level, achievements, and preferences of donors is given to the female to help her select a donor who best matches her own characteristics and interests.

Surrogate motherhood (uterus donation) is used when a male has a normal sperm count but his female partner is unable to maintain a pregnancy. A surrogate or stand-in mother may then donate or rent her uterus and perhaps her ovum. The surrogate is artificially inseminated by sperm donated from the male partner. The surrogate must agree legally to yield the baby for adoption at the end of the pregnancy.

In the early 1990s a woman bore her own triplet grandchildren by serving as a surrogate for her daughter, who had been born without a uterus. Other women view the situation as a moneymaking proposition, simply renting a uterus for a fee (Figure 17.14).

In vitro fertilization (IVF) can be used to overcome blockages in the uterine tubes which interfere with sperm or oocyte

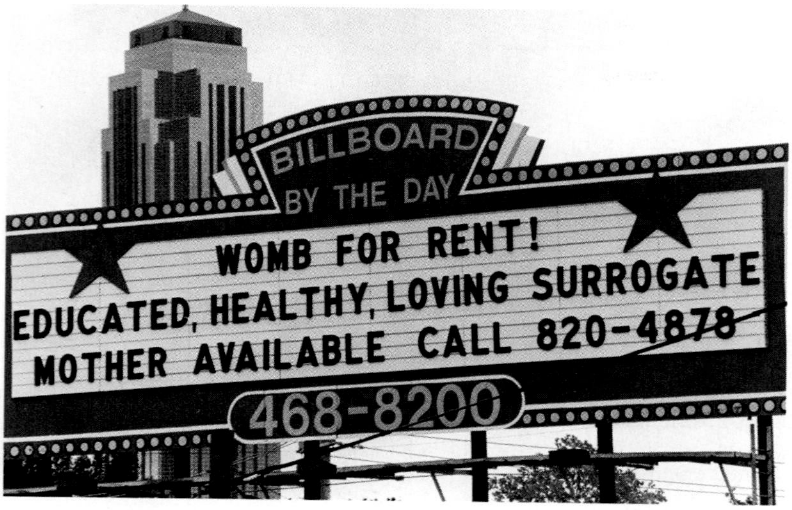
BILLBOARD
BY THE DAY
WOMB FOR RENT!
EDUCATED, HEALTHY, LOVING SURROGATE
MOTHER AVAILABLE CALL 820-4878
468-8200

FIGURE 17.14 A sign of the times: a billboard advertising the services of a surrogate mother.

transport. In this technique, sperm fertilize oocytes outside the body in a glass dish, and the resulting embryo is allowed to develop for 1 or 2 days. It is then implanted into the uterus of a female partner or a surrogate mother. While sperm are easy to collect, collecting oocytes offers more of a challenge.

In one technique, a laparoscope (lighted tube) and a suction tube are inserted through an incision in the abdominal wall to harvest oocytes directly from the ovary. Sperm and oocytes are combined in a laboratory vessel for fertilization, and several embryos are permitted to develop. Only one or two are implanted. The remaining embryos are maintained for later use by freezing them rapidly in liquid nitrogen. The success rate for each attempt at IVF is only 10 to 15 percent.

Gamete intrafallopian transfer (GIFT) is a method that was designed to improve the success and lower the cost of IVF. In GIFT, oocyte development is stimulated by hormones and the mature oocytes are removed from the ovary. Then the oocytes and the donor's sperm are immediately placed in the uterine tubes, below any obstruction, so that relatively normal fertilization can occur. The success rate is nearly 40 percent.

New experimental techniques allow the direct introduction of an individually selected sperm into a particular oocyte with the aid of tiny glass tubes. It is also possible to select the sex of the offspring by manipulating the slight differences in motility and pH tolerance of male-determining and female-determining sperm. The actual parenthood of an embryo can become quite complex if multiple alternatives are used. There may be a sperm donor parent, a female oocyte donor parent, a surrogate mother, and an adoptive mother and father.

Cancer of the Reproductive Organs

Breast cancer is the second most common form of cancer in women, trailing only lung cancer. It occurs in 10 percent of all women and is most common in women approaching or passing menopause. Breast cancer is dangerous and deadly because the cells metastasize while the tumors are still quite small and often unnoticed. Monthly self-examinations, annual exams by a physician, and mammograms (breast x-rays) for women over age 50 can help detect these cancers before they can spread. The procedure for breast self-examination is shown in Figure 17.15.

Cervical cancer occurs in only 3 percent of women under age 80, with most new cases appearing in women between the ages of 40 and 60. The cancer begins as structural changes in the cells of the cervix. The early stages of these changes can often be detected by microscopic examination of routine Pap smears, which are scrapings of cervical tissue. Treatment of cervical cancer generally requires surgical removal of the cervix and perhaps also the uterus coupled with chemotherapy or radiation therapy.

Prostate cancer (carcinoma) is the second most common form of cancer in men (skin cancer is the most common) and the third leading cause of cancer-related death in males (behind lung and colorectal cancer). However, it rarely occurs before age 50. The cause of prostate cancer is not known, but there may be a genetic predisposition. Recent research suggests that a high-fat diet may increase the risk of prostate cancer.

Except for enlargement of the prostate, there are relatively few symptoms of the cancer until it is well developed. Cancer-

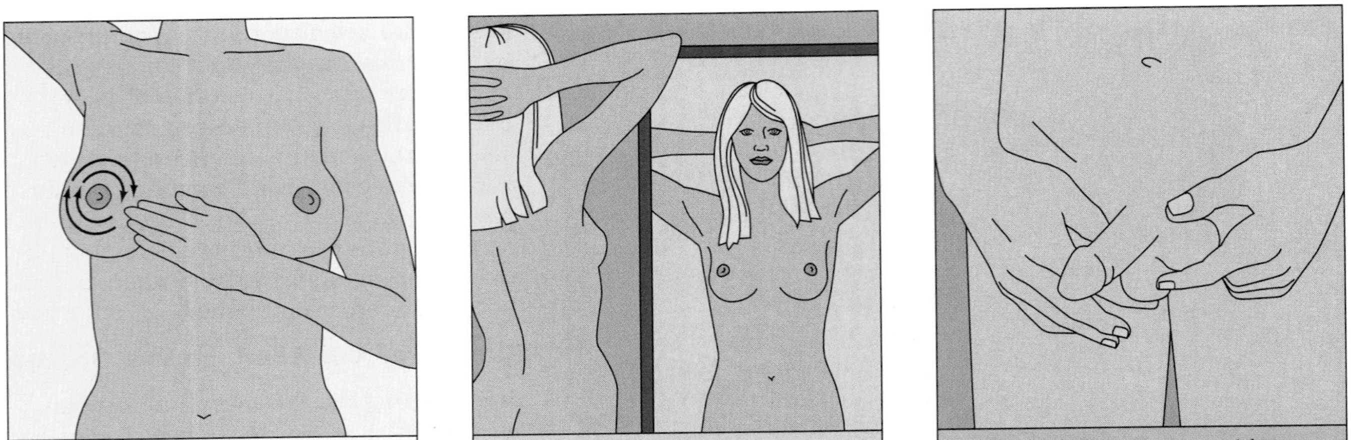

Breasts should be examined for lumps or other changes about a week after each period. To check right breast, place right hand behind head and press fingers of left hand against breast, moving in a slow spiral from the nipple outward. Check right armpit carefully. Repeat process for left breast and armpit.

Examine the appearance of the breasts in a mirror after a shower. Stand with hands on head, arms stretched above head, arms on hips, and turned toward each side. Look for changes in nipples, redness, swelling, discharge, or unusual dimpling of skin. Report changes to a health care professional.

The testes should be examined for lumps, swelling, pain, or tenderness once a month after a shower or bath. Examine by rolling each testis between thumb and fingers. Also check shaft and glans of penis for changes. Report changes to a health care professional.

FIGURE 17.15 Monthly self-examination of breasts and testicles.

ous enlargement of the gland usually occurs along its outer edges rather than around the urethra. Therefore, the cancer does not generally restrict urine flow as does benign enlargement.

Prostate enlargement can be detected by a physician probing the rectum with a gloved finger and pressing the rectal wall against the gland. All men over age 50 should receive a yearly rectal examination of the prostate. A new blood screening test for prostate specific antigen (PSA) may allow early detection even before glandular enlargement occurs. As with all cancers, early detection increases the effectiveness of treatment. Prostate cancer is treated by surgical removal, radiation therapy, and medications that reduce the secretion of testosterone, which accelerates most prostate cancers.

Testicular cancer is the most common form of cancer in younger adult males (ages 29 to 35) but is rare in older males. It is most common in men whose testes descended slowly or never descended into the scrotum. In most cases, a small, hard painless lump is noticed on the front or side of the testis. These lumps and or other swellings can best be detected during a self-examination after a shower or bath, when the scrotum is relaxed (Figure 17.15). Testicular cancer is treated with surgery and radiation therapy.

Life Span Changes

While changes in the reproductive system occur throughout the life span of both males and females, the most dramatic changes occur during youth and late adulthood. In males, puberty begins later than it does in females, and males do not see as dramatic a cessation of their reproductive capacity as females do later in life. While reproduction is diminished for couples after age 45, enjoyment of sexuality can continue to an advanced age in reasonably healthy people.

In the Male

Before puberty begins, the seminiferous tubules of young males are inactive and there is no lumen. During puberty (about age 13), a lumen develops and spermatogenesis begins. Among the earliest signs of puberty is a gradual enlargement of the testes and penis and the appearance of pubic hair. The production of testosterone by the interstitial cells of the testes accelerates general growth and enhances the development of other male secondary sex characteristics. Enlargement of the seminal vesicles and prostate gland soon occurs. The first ejaculations of semen and sperm begin about a year after penis enlargement, often as *nocturnal emissions* during erotic dreams.

Rather than a rapid decline in reproductive capacity, men usually experience gradual changes. The erectile tissue becomes a bit less elastic and responsive, increasing the time needed for an erection and reducing the rigidity of the erection. Older men may require a longer period of frictional exposure in the vagina to achieve orgasm. The secretion of testosterone is gradually reduced, and this reduces the overall muscle mass of the body and may even thin the vocal cords, raising the pitch of the voice slightly. Sperm production also decreases after age 50 or so, with about one-third less sperm produced than is the case in a young adult. The seminal vesicles decrease in size and activity. While the prostate also becomes less active, it is typically enlarged.

In the Female

By birth, a female has made all of her primary oocytes. However, the ovaries remain small until puberty (about age 11), when pituitary hormones initiate their enlargement and the maturation of primary follicles. By adulthood, the ovaries will have enlarged about 30 times. The uterus is rather large at birth (as a result of stimulation by the mother's hormones) and gradually declines until puberty, when it again enlarges. In young females, the first signs of puberty are enlargement of the breasts and the appearance of pubic hair. These signs are followed by menses.

The first menstrual cycle, called **menarche** (*men-ark-key*), usually occurs between ages 12 and 13 years, generally after a critical weight of 50 kilograms (110 pounds) has been achieved. Over the last 200 years the age of menarche has gradually declined, especially in the more developed nations. The causes of this change are not entirely clear, but better nutrition may play a role.

Menopause marks the end of the reproductive phase of a woman's life, since ovulation and menstruation cease. The age at menopause is variable but usually falls between 45 and 55. As menopause approaches, there are fewer primary follicles left in the ovary to produce estrogen and progesterone or respond to pituitary hormones. Without the negative feedback inhibition provided by estrogen and progesterone, the pituitary releases increasing amounts of FSH and LH in an unsuccessful attempt to stimulate follicular activity.

High levels of FSH and LH stimulate the adrenal glands to release androgens and cortisol, which, together with lower estrogen levels, may cause depression, anxiety, irritability, and fatigue in some women. Often a woman experiences unexpected flushing and warming of the skin ("hot flashes") as blood vessels uncontrollably dilate and constrict in response to varying hormone levels. Eventually the body adapts to the reduced levels of estrogen and establishes a new homeostasis, but some changes continue.

After menopause, the ovaries decline in weight and the ligaments, fat tissue, and glands of the breast degenerate, causing the breasts to shrink and sag. The lining of the vagina becomes thinner and less elastic. This, coupled with reduced vaginal secretions, may make intercourse somewhat painful. Estrogen creams and sterile lubricants can reduce the discomfort.

The reduction of estrogen coupled with reduced calcium absorption by the intestine may cause *osteoporosis*, making bones brittle and more easily broken (Chapter 5). The loss of estrogen also causes the voice to drop in pitch, some facial hair to appear, and pubic hair to become more coarse and sparse.

Each age has its benefits, however, and the freedom from menstruation and pregnancy that comes with menopause allows many women to find new enjoyment in sexual activity. With affectionate and caring sensitivity to each other's needs, an elderly couple can continue to enjoy a rich and satisfying sexual relationship.

Chapter Summary

Meiosis: The Production of Gametes

- Meiosis produces haploid gametes–the sperm and egg.
- During meiosis I, homologous chromosomes pair, align, and exchange segments by crossing over.

The Male Reproductive System

- Men produce sperm and the hormones that create male sexual characteristics and brain organization.
- The testes lie in the scrotum.
- Interstitial cells secrete testosterone which initiates sperm formation (spermatogenesis).
- During spermatogenesis, spermatogonia divide by mitosis to produce primary spermatocytes. Primary spermatocytes undergo meiosis I to produce secondary spermatocytes, which undergo meiosis II to form haploid spermatids, which mature into sperm.
- Sperm are stored in the epididymis until their release from the body during ejaculation. At ejaculation, sperm are expelled through the urethra.
- Semen contains sperm and secretions from the accessory glands.
- The penis contains erectile tissue.
- The four stages of male sexual response are excitement, plateau, expulsion, and resolution.

The Female Reproductive System

- Women produce ova, provide a site for fertilization and embryonic development, and produce the hormones that create female sexual characteristics and brain organization.
- Ova form in the ovaries (oogenesis). The follicles in which they develop produce estrogen.
- In oogenesis, an oogonium divides by mitosis, producing a primary oocyte. A primary oocyte undergoes meiosis I, forming a secondary oocyte and a polar body. The secondary oocyte forms an ovum and a second polar body if it is fertilized.
- At ovulation, secondary oocytes are released from the ovary. Fertilization usually occurs in the uterine tube.
- The three layers in the uterine wall are the outer perimetrium, middle myometrium (smooth muscle layer), and inner endometrium.
- The vulva or external genitalia includes the mons, labia, and clitoris.
- Milk production (lactation) is controlled by estrogen, progesterone, prolactin, and oxytocin.
- Women experience the same stages of sexual response that men experience but express them differently.

The Menstrual Cycle

- The menstrual cycle includes the ovarian cycle and the uterine cycle.
- The ovarian cycle includes a follicular phase (follicle development, estrogen secretion, and oocyte meiosis) and a luteal phase (ovulation, corpus luteum formation, and progesterone secretion).
- If fertilization occurs, the corpus luteum continues progesterone secretion during pregnancy. HGC from the embryo communicates pregnancy to the ovary.
- The uterine cycle includes a menstrual phase (shedding of endometrium), a proliferative phase (growth of endometrium), and a secretory phase (enlargement of uterine glands).

Alternatives for Birth Control

- Complete abstinence from sexual intercourse is the only failureproof way to avoid pregnancy.
- Vasectomy and tubal ligation produce obstacles for sperm and oocyte passage.
- Birth control pills (estrogen and progesterone) block the release of FSH and LH from the pituitary.
- IUDs create a mild inflammation in the uterus to damage sperm and prevent embryo implantation.
- Spermicides make barrier devices more effective.

Diseases and Disorders

- Artificial insemination involves the insertion of donor sperm into a female's reproductive tract.
- In vitro fertilization involves fertilization in a laboratory vessel and embryo implantation in a woman's uterus. GIFT introduces the sperm and egg into the uterine tubes.
- Breast and cervical cancer are more common in older women, while prostate cancer occurs most frequently in older males.

Life Span Changes

- In boys, spermatogenesis begins at puberty.
- Testosterone secretion, sperm production, and sexual responsiveness decline gradually with age.
- Girls have all their primary oocytes at birth; oogenesis begins at puberty.
- The first mestrual cycle is called menarche; the end of menstrual cycles is called menopause.

Selected Key Terms

interstitial cells (p. 449)
meiosis (p. 447)
menopause (p. 471)
menstrual cycle (p. 462)
oogenesis (p. 456)
ovarian follicles (p. 455)
ovary (p. 455)
ovulation (p. 455)
ovum (p. 455)
seminiferous tubules (p. 449)
sexual response (pp. 454, 461)
sperm (p. 451)
spermatogenesis (p. 451)
testis (p. 449)

Review Activities

1. Compare the results of mitosis and meiosis.
2. Why is it important for the testes to pass into the scrotum instead of remaining in the abdominal cavity?
3. Describe the functions of Sertoli cells and interstitial cells.
4. Summarize the process of spermatogenesis.
5. Trace the pathway followed by the sperm through the male system from point of formation to point of release.
6. Compare the products of the seminal vesicles, the prostate gland, and the bulbourethral glands. What are the functions of semen?
7. Explain how the male system is structured to release sperm and how the female system is structured to receive sperm. How is the clitoris different from the penis?
8. Compare the stages of sexual response as experienced by males and females.
9. Summarize the process of oogenesis. List the follicle stages and those for the oocyte.
10. Trace the pathway followed by the oocyte through the female system from point of formation to point of implantation, if it is fertilized. Where does fertilization normally occur?
11. Describe the structure and function of the mammary glands.
12. List the phases and events which occur during the ovarian and uterine cycles. How do they correlate with pituitary or ovarian hormones?
13. Compare the alternatives for birth control with regard to their effectiveness in preventing both pregnancy and STDs.
14. List and describe some of the alternatives for circumventing male or female infertility.
15. Compare the lifetime production of gametes by males and females. When does production begin and end, and in what quantities are the gametes produced?

Self-Quiz

Matching Exercise

___ 1. Site of spermatogenesis	A. Secondary oocyte
___ 2. Site of oogenesis	B. Epididymis
___ 3. Cell resulting from spermatogenesis	C. Ovarian follicles
___ 4. Cell resulting from oogenesis	D. Vas deferens
___ 5. Cell released at ovulation	E. Ovum
___ 6. Site of fertilization	F. Spermatid
___ 7. Nonfunctional cell produced during oogenesis	G. Mammary gland
___ 8. Storage site for sperm	H. Uterine tube
___ 9. Male tube severed during vasectomy	I. Seminiferous tubules
___ 10. Site of lactation	J. Polar body

Answers to Self-Quiz

1. I; 2. C; 3. F; 4. E; 5. A; 6. H; 7. J; 8. B; 9. D; 10. G

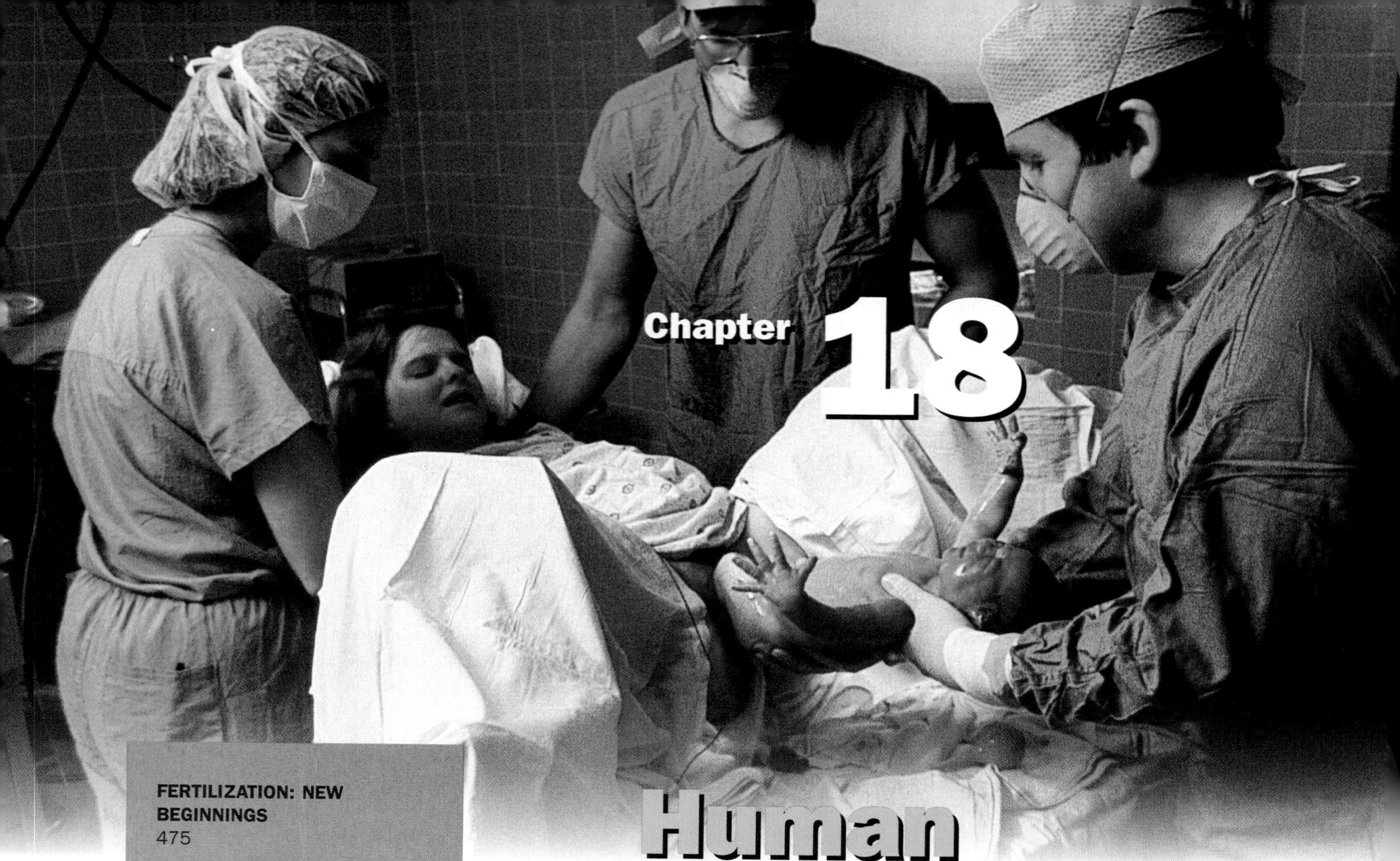

Chapter 18

Human Development and Birth

It is difficult to imagine now how I must have felt then—forming, growing, and changing as I lay head down in my mother's womb. I've always felt emotionally close to her, but never since have I been as physically close as I was then. How can I capture the sensations I felt then with the words I use now? There must have been warmth and wetness and darkness. I'm sure there were sounds—certainly the thump of her heart and the swoosh of her blood in nearby vessels. Did I struggle against the increasing pressure as birth approached? Was there pain? How did that first gasp of air feel as it inflated my lungs? Was I startled by the brightness and coolness of the delivery room? I wish I could remember!

This is a chapter about our forgotten beginnings. You will learn about sperm and oocytes, about fertilization and concep-

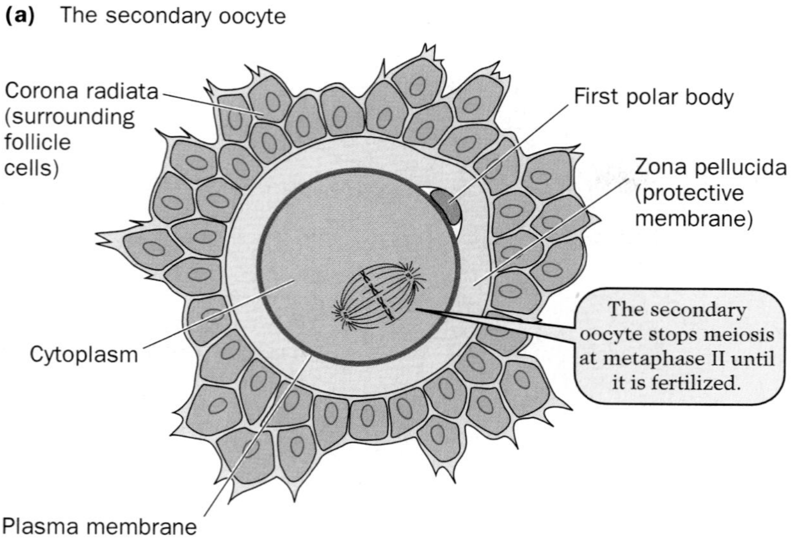
(a) The secondary oocyte
Corona radiata (surrounding follicle cells)
First polar body
Zona pellucida (protective membrane)
The secondary oocyte stops meiosis at metaphase II until it is fertilized.
Cytoplasm
Plasma membrane

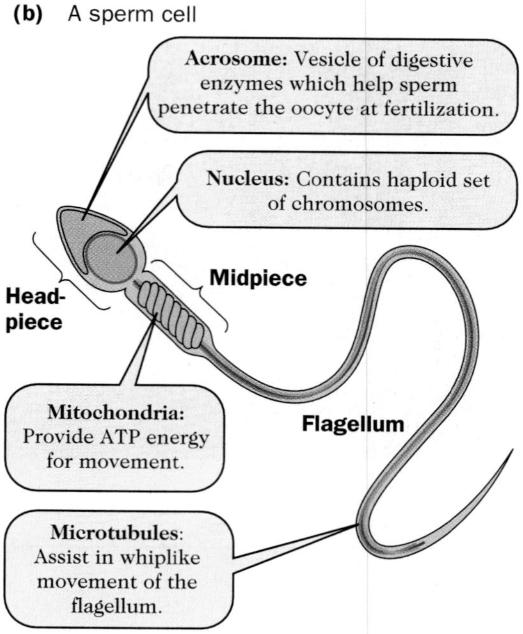
(b) A sperm cell
Acrosome: Vesicle of digestive enzymes which help sperm penetrate the oocyte at fertilization.
Nucleus: Contains haploid set of chromosomes.
Midpiece
Head-piece
Mitochondria: Provide ATP energy for movement.
Flagellum
Microtubules: Assist in whiplike movement of the flagellum.

FIGURE 18.1 Human gametes.

tion, and about embryonic and fetal development. You will discover the changes your mother experienced during pregnancy and the stages of your birth. There were obstacles and hazards that we will discuss as we proceed.

Fertilization: New Beginnings

You had a beginning when your mother's secondary oocyte was penetrated by your father's sperm. What had to happen for those two cells to meet? What followed fertilization? We'll answer these questions in the sections that follow.

The Oocyte and the Sperm

When it is released from the ovary, the secondary oocyte is surrounded by a protective sphere of cells called the **corona radiata** which is derived from the follicle in the ovary (Figure 18.1a). The size of the corona radiata and its irregular surface make it easier for the entire assembly to be "trapped" by the uterine tube.

As you learned in Chapter 17, the unequal divisions during oogenesis give most of the cytoplasm to the oocyte, instead of the polar bodies. In addition to cellular organelles, the cytoplasm of the oocyte contains stored nutrients. If the oocyte is fertilized, these nutrients will support its life for about 2 weeks before it establishes a physical relationship with the mother and obtains her nutrients.

The streamlined sperm are 2000 times smaller in volume than the oocyte. Each sperm consists of three parts: a headpiece, midpiece, and a tail or flagellum (Figure 18.1b). The **headpiece** consists mostly of a haploid nucleus and an **acrosome,** a vesiclelike cap that contains digestive enzymes. The **midpiece** contains many mitochondria, providing ATP energy for the whiplike motion of the **flagellum**.

The motion of the flagellum propels sperm at a speed of about 5 millimeters (about $\frac{1}{4}$ inch) per minute, allowing some to travel from the vagina to the uterine tube in under 30 minutes. While this journey may be rapid, it includes many hazards which kill, damage, or deform most of the approximately 300 million sperm ejaculated.

The Hazards of Being a Sperm

The first hazard to sperm after ejaculation is the environment within the vagina. In an adult woman the vagina normally has an abundant population of bacteria called lactobacilli. The lactobacilli pro-

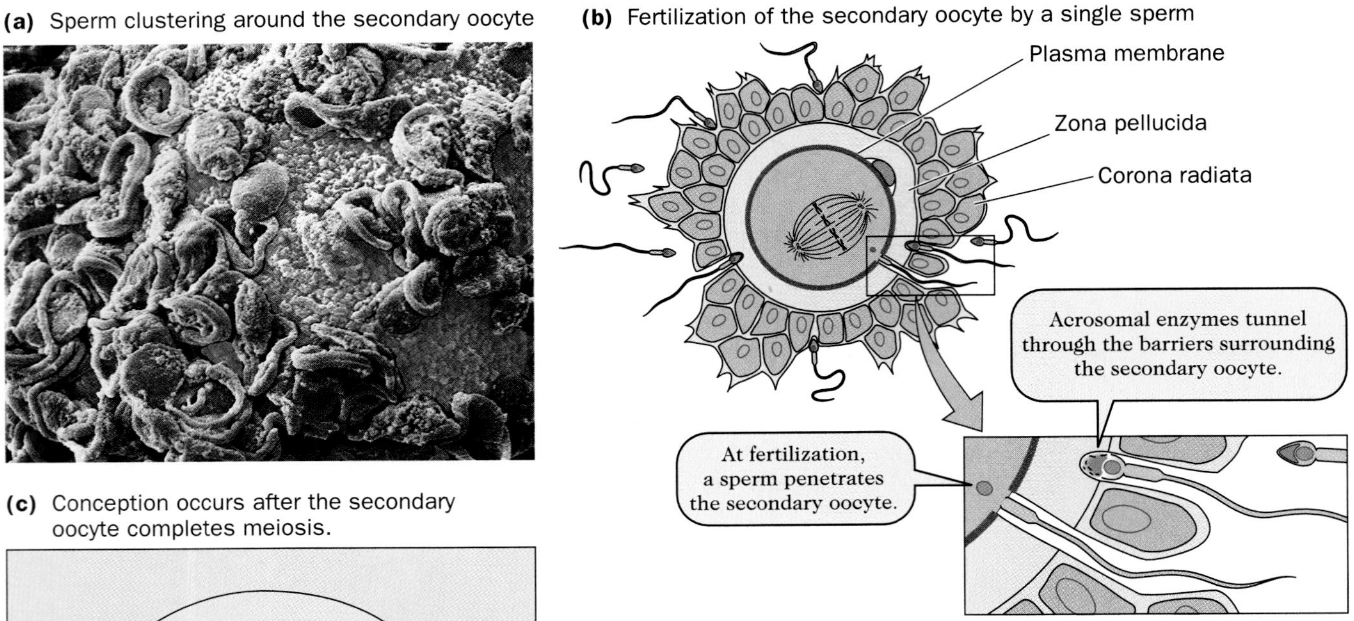
(a) Sperm clustering around the secondary oocyte
(b) Fertilization of the secondary oocyte by a single sperm
Plasma membrane
Zona pellucida
Corona radiata
Acrosomal enzymes tunnel through the barriers surrounding the secondary oocyte.
At fertilization, a sperm penetrates the secondary oocyte.

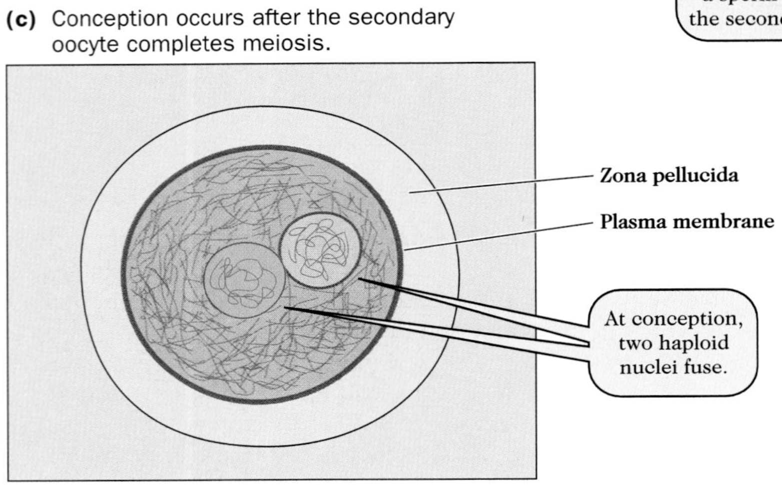
(c) Conception occurs after the secondary oocyte completes meiosis.
Zona pellucida
Plasma membrane
At conception, two haploid nuclei fuse.

FIGURE 18.2 Fertilization and conception.

duce small quantities of lactic acid which keep the vagina slightly acidic (about pH 5.0). The alkaline secretions in the semen neutralize this acidity, and this helps protect the sperm. Vaginal bacteria also may attack sperm directly, reducing their numbers or damaging their capacity to swim.

The cervix presents another hazard. Throughout most of the menstrual cycle, the cervical canal is plugged with sticky mucus which most sperm cannot penetrate. However, the pituitary hormones that stimulate ovulation also cause this mucus plug to become thin and watery, allowing sperm to enter the uterus. The broad expanse of the uterus though, offers additional challenges. Some sperm get lost among the endometrial cells. Others are killed by white blood cells traveling over the uterine lining. Yet, some sperm survive to face additional challenges.

There are two uterine tubes, and by chance, roughly half the remaining sperm ascend each tube, although only one tube will receive an ovulated oocyte. Within the uterine tubes the motion of cilia creates a downward current of fluid against which the sperm must swim. Accidents continue to happen, and many sperm die en route.

By the time the sperm reach the upper one-third of the uterine tube, where fertilization is most likely to occur, their numbers are drastically reduced. Of the 300 million or so sperm ejaculated, only dozens survive to cluster around the corona radiata that protects the oocyte (Figure 18.2a).

When the heads of the sperm make contact with the corona radiata, their acrosomes secrete the stored digestive enzymes, creating a channel which permits sperm to move toward the surface of the oocyte. Soon a single sperm will penetrate the oocyte, and fertilization will occur.

> Sperm consist of a head piece, a midpiece, and a flagellum. Only a few sperm reach the oocyte. Enzymes from the acrosome of the sperm digest a channel through the corona radiata.

Fertilization: A Single Sperm Penetrates the Oocyte

As the acrosomal enzymes digest a channel through the corona radiata, one sperm slips through and binds to special receptors on a protective covering around the oocyte known as the **zona pellucida** (*zone-uh pell-lew-sid-uh*) (Figure 18.2b). The specificity of these receptors present a barrier to fertilization between different species.

Upon binding to a single sperm, the zona pellucida undergoes a rapid chemical change, making it impenetrable to other sperm. This normally prevents fertilization of the oocyte by more than one sperm.

Fertilization is accomplished when the head of a single sperm is drawn through the plasma membrane and into the cytoplasm of the oocyte (see Figure 18.2b). The flagellum and midpiece of the sperm remain outside in the zona pellucida, where they degenerate along with the remaining sperm clustering around the corona radiata.

Upon fertilization, the oocyte changes. Within a few hours it completes meiosis II, producing an ovum with 23 chromosomes (Chapter 17). **Conception** occurs when the haploid nucleus of the ovum fuses with the haploid nucleus of the sperm to establish a diploid cell with 46 chromosomes (Figure 18.2c). This first diploid cell is called a **zygote** (see Figure 18.4).

The sperm's only contribution to the zygote is its haploid nucleus. None of the sperm's organelles enter the oocyte. All the organelles of the next generation come from the cytoplasm of the mother's oocyte.

In addition to providing a diploid set of chromosomes, fertilization has two other immediate consequences: it determines the gender of a zygote and stimulates cell division by mitosis. Gender is determined by the sex chromosomes, designated X and Y. Sperm carry either an X or Y chromosome. All ova carry an X chromosome. An X sperm combining with an X ovum produces an XX, or female, zygote. A Y sperm and an X ovum produce an XY, or male, zygote.

The Y chromosome carries a gene called the **sex-determining region** which causes the formation of testes later in development. Lacking a Y chromosome, a female will later form ovaries instead of testes.

After fertilization, the zygote begins a series of mitotic cell divisions. The resulting cells eventually form all the tissues of the new individual and the supportive membranes of the embryo.

> Changes in the zona pellucida prevent entry by more than one sperm. After fertilization, the oocyte completes meiosis, and its nucleus fuses with the sperm nucleus, forming a zygote. Conception establishes the diploid condition, determines gender, and initiates cell division.

Twins: Formed by Unusual Events

One in every 90 births results in twins, while triplets are more rare (1 in 8000). There are two very different processes that can lead to the formation of multiple embryos: multiple ovulation and splitting of an embryo. Occasionally, a female ovulates more than one oocyte. If they are all fertilized, these oocytes produce multiple zygotes. Since these zygotes are products of different sperm and oocytes, they are not necessarily the same gender and have other genetic differences. They produce **fraternal twins** (Figure 18.3a).

Though less common, more than two separate embryos sometimes develop simultaneously. Multiple embryos are particularly common when the mother has used ovulation-stimulating hormones to increase her chances of getting pregnant. The tendency toward multiple ovulations may have a genetic basis, since fraternal twins "run in families."

Identical twins are produced from a single zygote (Figure 18.3b). Initially the zygote undergoes normal cell division, creating more cells. However, at some point early in development the cells separate into two groups. When both groups of cells go on to form a complete embryo, identical twins are produced. Because identical twins come from the same sperm and ovum, they have identical genes, have the same general appearance, and are of the same gender.

During early development, sometimes there is an incomplete separation of groups of cells. The resulting embryos and newborn infants remain attached at some point in their bodies. These are **conjoined (Siamese) twins** (Figure 18.3c). Conjoined twins often can be surgically separated after birth if there is only minimal sharing of skeletal structures or organs. If they share brain regions, a heart, or a digestive tract, they usually must remain attached throughout life.

> Fraternal twins result when separate oocytes are fertilized by different sperm. Identical twins develop when cells from a single zygote separate early in development. Incomplete separation of identical twins produces conjoined twins.

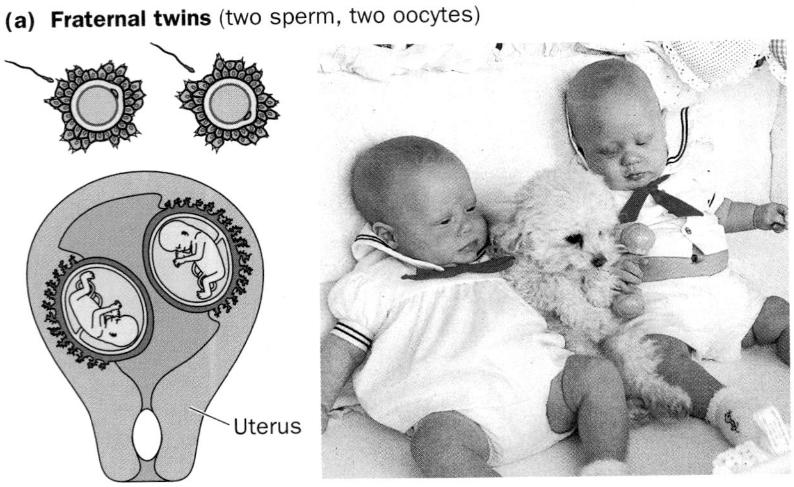
(a) Fraternal twins (two sperm, two oocytes)
Uterus

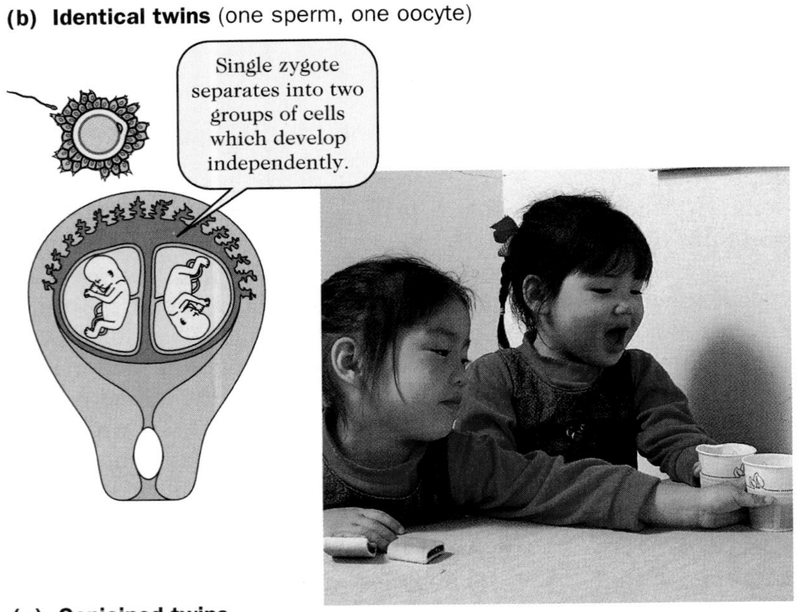
(b) Identical twins (one sperm, one oocyte)
Single zygote separates into two groups of cells which develop independently.

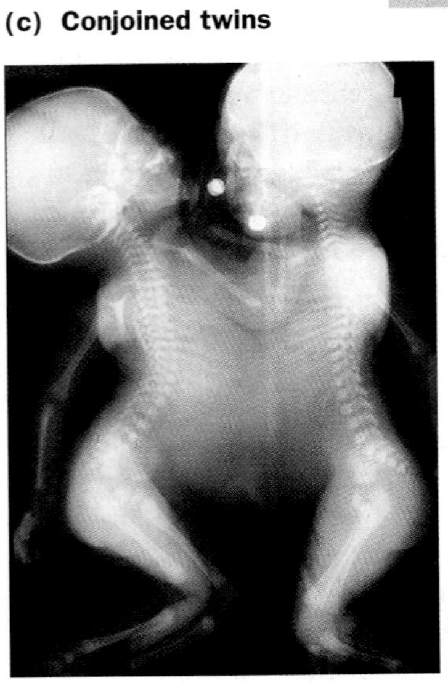
(c) Conjoined twins

FIGURE 18.3 Twins.

Human Development in the Uterus: An Overview

The development of the human body is a lengthy and complex process. It extends well into the second decade after birth. In this chapter we will focus on the development that takes place in the uterus after fertilization and before birth. Before describing some of the major features of this development, we will identify its fundamental processes and stages.

The Processes of Development

There are four distinct processes which contribute to early development: cleavage, growth, morphogenesis, and differentiation. **Cleavage** is the initial process. The first mitotic cell divisions "cleave" the components of the original zygote into separate cells. The first cell divides to form 2 cells. Then the 2 divide to form 4. The 4 form 8, the 8 form 16, and so on, with the cells doubling in number with each division (Figure 18.4).

The cells formed from these early divisions are called *blastomeres*. Since these cell divisions are very rapid, blastomeres do not grow during their very brief inter-

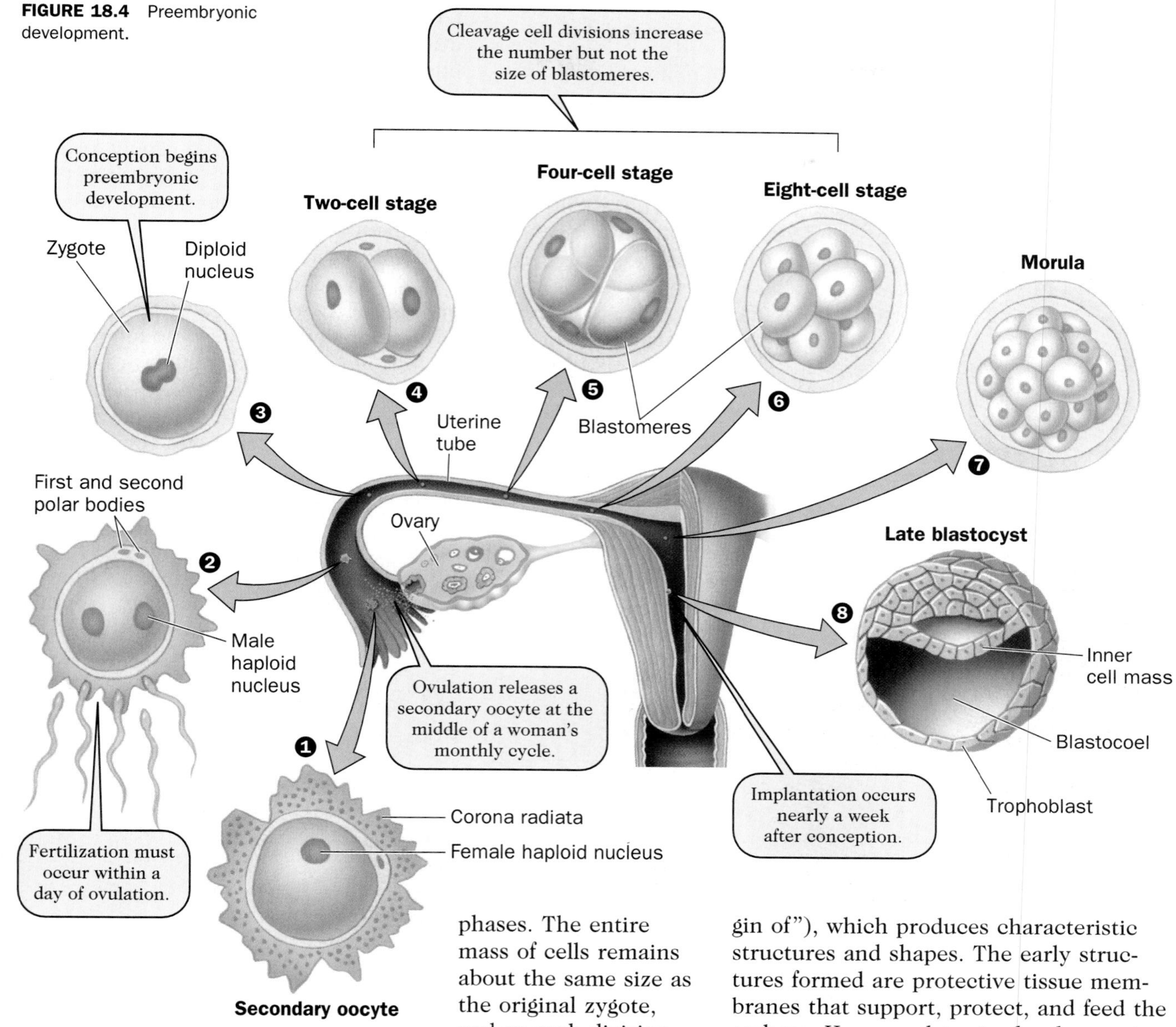
Cleavage cell divisions increase the number but not the size of blastomeres.
Conception begins preembryonic development.
Zygote
Diploid nucleus
Two-cell stage
Four-cell stage
Eight-cell stage
Morula
3
4
5
6
7
Uterine tube
Blastomeres
First and second polar bodies
Ovary
2
Male haploid nucleus
Late blastocyst
8
Inner cell mass
Blastocoel
Trophoblast
Ovulation releases a secondary oocyte at the middle of a woman's monthly cycle.
1
Implantation occurs nearly a week after conception.
Corona radiata
Female haploid nucleus
Fertilization must occur within a day of ovulation.
Secondary oocyte

FIGURE 18.4 Preembryonic development.

phases. The entire mass of cells remains about the same size as the original zygote, and so each division produces smaller cells. Early development is marked by an increase in cell number, not cell size.

Eventually the rate of cell division slows, and there is a longer interphase period. As a result, the cells grow larger in size. Overall, **growth** is a product of an increase in cell size and cell number. Before birth, growth is dramatic and we increase in length over 5000-fold. After birth, we increase in length only about $3\frac{1}{2}$ times during our roughly 20-year growth phase.

As cells proliferate and grow in size, the embryo undergoes **morphogenesis** (*morpho*, "form or shape"; *genesis*, "origin of"), which produces characteristic structures and shapes. The early structures formed are protective tissue membranes that support, protect, and feed the embryo. However, later in development, morphogenesis produces the head, limbs, and internal organs.

As development proceeds, certain groups of cells gradually become more specialized in their structures and functions. This process is called **differentiation**. Although all the cells produced by mitosis have the same chromosomes and genes, differentiation involves turning on and turning off some of those genes at specific times. Through this process, active genes code for the synthesis of specific proteins that determine the fate and function of the cells (Chapter 2). After cells differentiate, they do not return to their earlier unspecialized state.

The Stages of Development

After fertilization, development in the uterus can be divided into three stages: preembryonic development, embryonic development, and fetal development. The first stage—**preembryonic development**—lasts about 2 weeks after fertilization (Figure 18.4). In addition to the early cleavages, it is during this period that the new cells must establish a physical relationship with the mother's uterus.

Preembryonic development is a dangerous time. Nearly 90 percent of all zygotes fail to survive this stage. During this vulnerable period, the mother is typically unaware that her pregnancy is beginning and may continue hazardous occupational exposures, behaviors, or consumption patterns.

The **embryonic development** stage lasts about 6 weeks, from the third week through the eighth week (Figure 18.5). It begins at about the time the new cells establish a nourishing relationship with the uterus. Rapid cell proliferation, growth, morphogenesis, and cell differentiation follow, establishing all the organ systems. However, these systems are not yet fully operative, and the embryo is only about 2.5 centimeters (1 inch) in length at the end of this stage. Further growth and development characterize the final stage, the fetal stage.

After 8 weeks of development, the embryo is called a **fetus**, and the stage of **fetal development** occupies the final and longest period of development. It lasts from week 9, when long bones begin to calcify, until birth. During the entire fetal stage, the organ systems are completed and the fetus grows to an average weight of about 3.2 kilograms (7 pounds) and a length of 52 centimeters (20 inches). Birth occurs approximately 38 weeks after fertilization or about 40 weeks from the beginning of the mother's last monthly period.

FIGURE 18.5 Embryonic development: formation of primary germinal layers.

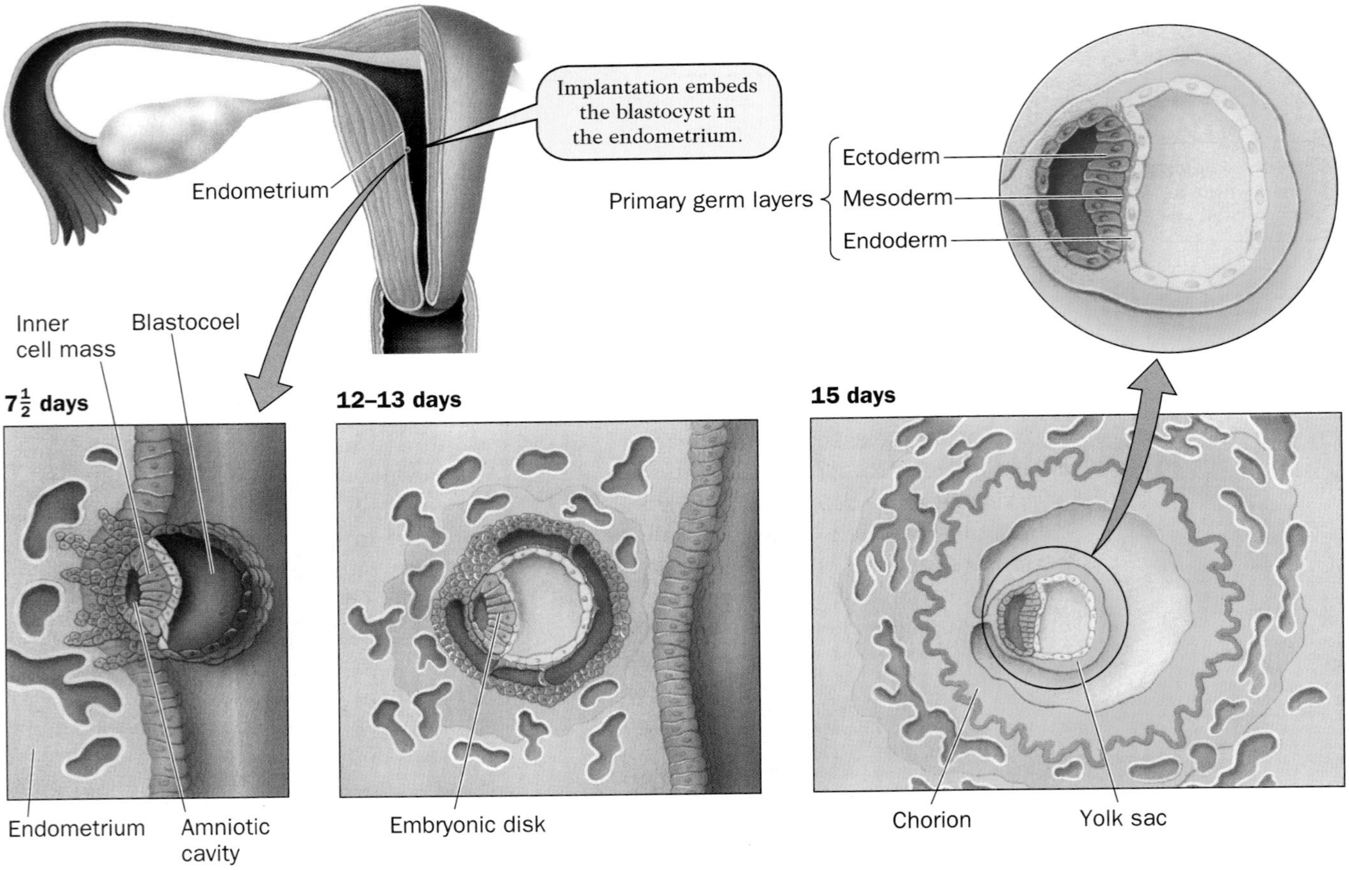
Implantation embeds the blastocyst in the endometrium.
Endometrium
Primary germ layers
Ectoderm
Mesoderm
Endoderm
Inner cell mass
Blastocoel
$7\frac{1}{2}$ days
12–13 days
15 days
Endometrium
Amniotic cavity
Embryonic disk
Chorion
Yolk sac

The processes of development include cleavage, growth, morphogenesis, and cell differentiation. There are three stages of development: preembryonic stage, embryonic development, and the fetal stage.

Preembryonic Development: The First 2 Weeks

The preembryonic stage begins with fertilization and ends with the formation of the three germinal layers: the ectoderm, endoderm, and mesoderm (Chapter 3). Immediately after fertilization, cleavages produce a solid ball of 32 cells in about 72 hours (day 3). This structure is called the *morula* (***more**-you-luh*), a term that means "little mulberry," because of the clustered appearance of the blastomeres (see Figure 18.4). The morula continues the journey down the uterine tube. After entering the uterus, it floats free for 3 or 4 days before attaching to the endometrium in a process called **implantation**.

Before implantation, the morula undergoes changes. Some of its cells begin to separate from one another, forming an internal fluid-filled cavity called the *blastocoel* (***blast**-oh-seal*). The entire structure containing this cavity is known as the **blastocyst**. If you look closely at Figure 18.4, you can see that the blastocyst consists of two distinct groups of cells. The outer ring of cells, which is just one cell layer thick, is called the **trophoblast**. The trophoblast soon begins to participate in the formation of tissue membranes that support, protect, and nourish the developing embryo. The other group of cells is the **inner cell mass** along one side of the blastocoel. The inner cell mass later forms the embryo proper.

Cleavage transforms the zygote into a solid ball of cells, the morula. The morula develops into the blastocyst, which consists of a trophoblast, an inner cell mass, and a fluid-filled blastocoel.

Implantation: Attaching to the Endometrium

After the blastocyst forms, it settles against the endometrium of the uterus, and the trophoblast secretes enzymes which digest a path into the endometrium (Figure 18.5). This is the beginning of implantation. As implantation continues, cells from the trophoblast begin to grow into the endometrium, anchoring the blastocyst and establishing a basis for exchanging materials with the mother by diffusion.

By day 12, the blastocyst is completely embedded in the endometrium and the mother is approaching the end of her normal menstrual cycle. The trophoblast cells secrete the hormone human chorionic gonadotrophin (HCG), which signals the corpus luteum in the ovary to sustain its secretion of progesterone (Chapter 17). By maintaining progesterone production, HCG prevents the onset of menses and the shedding of the endometrium with its implanted blastocyst. If the woman does not experience her next period, she will probably begin to suspect pregnancy. The high level of HCG is also thought to cause the frequent bouts of nausea (morning sickness) many women experience during the earliest stages of pregnancy.

By the end of the second week, implantation is complete and a region of the inner cell mass called the **embryonic disk** has formed two of the three germinal layers: the ectoderm and endoderm. A little later the third germinal layer, the mesoderm, forms between the other two (Chapter 3). From these three layers, all the tissues in the body will be formed (Table 18.1).

The formation of a second cavity called the **amniotic cavity** (*am-knee-**ah**-tick*) marks the end of the preembryonic period and the beginning of the embryonic period (Figure 18.5). At this stage you were smaller than the period at the end of this sentence.

The blastocyst implants, and its trophoblast secretes HCG, which maintains the corpus luteum and its secretion of progesterone,

TABLE 18.1 Adult Tissues and Organs Derived from the Primary Germinal Layers

Ectoderm	Mesoderm	Endoderm
Epidermis of skin	Dermis of skin	Mucosa of digestive tract
Hair, nails, sweat glands	All connective tissue	Alveoli of lungs
Enamel of teeth	Bone marrow	Tonsils
Lining of nasal cavity, anus	All muscle tissue	Liver and pancreas
Nervous system	Kidneys and ureters	Thyroid and parathyroid glands
Cornea, lens, retina of eye	Testes and ovaries	Lining of urinary bladder
Adrenal medulla	Adrenal cortex	Lining of urethra, vagina
Posterior pituitary gland	Lining of blood vessels	Anterior pituitary gland
Portions of all sense organs	Lymphatic tissue, vessels	Thymus gland

preventing menses. By the end of the second week the three germinal layers and the amniotic cavity have begun to form.

Abnormal Implantation: Hazards to Development

Sometimes implantation and early development begin at an abnormal site, causing what is known as an **ectopic pregnancy** (*eck-top-pick*). The blastocyst may be implanted in one of the uterine tubes (tubal pregnancy). More rarely, it may not even reach the uterine tube, passing instead into the abdominal cavity (abdominal pregnancy) and implanting and developing on the wall of the intestine or bladder or the outer surface of the uterus.

Ectopic pregnancies occur about once in every 100 pregnancies. They may cause internal bleeding, threatening the life of the mother. In spite of these dangers, some misplaced embryos do survive without defects but of course cannot be born in the usual fashion. They must be removed surgically as soon as they can survive in a hospital incubator (around the seventh month).

Embryonic Development: Weeks 3 through 8

The appearance of the amniotic cavity and the three germinal layers mark the beginning of the embryonic stage of development. During the embryonic phase all the organ systems begin to form. At the outset of this phase the extraembryonic membranes develop, protecting the embryo and forming the placenta.

Extraembryonic Membranes for Support and Protection

During early embryonic development, four special tissue membranes form to support, protect, and nourish the developing embryo. These are called the **extraembryonic membranes** because they are not part of the body of the embryo and are discarded after birth. They include the amnion, the yolk sac, the allantois, and the chorion (Figure 18.6).

The epithelial wall around the amniotic cavity forms the **amnion**, commonly known as the "bag of waters." The salty fluid contained in the amniotic cavity is called the *amniotic fluid*. It is formed by filtration from the mother's blood and provides a protective cushion. The amniotic fluid eventually receives urine when the embryo's kidneys begin to function.

By the beginning of the third week a **yolk sac** has formed. In humans, the yolk sac is the site of blood cell formation until that job is taken over by the developing liver and later the bone marrow. Parts of the yolk sac will develop into the intestines and the reproductive system. An extension of the yolk sac forms the **allantois** (*ah-lan-toe-iss*), which is a temporary structure during development. It helps form umbilical blood vessels and part of the bladder but degenerates during the second month of development.

The fourth extraembryonic membrane, the **chorion** (*core-ee-on*) is formed from the trophoblast. The chorion and amnion grow to enclose and protect the entire embryo and later the fetus. In addition, the chorion plays an important role in the transfer of nutrients, gases, and wastes between the embryo and the mother.

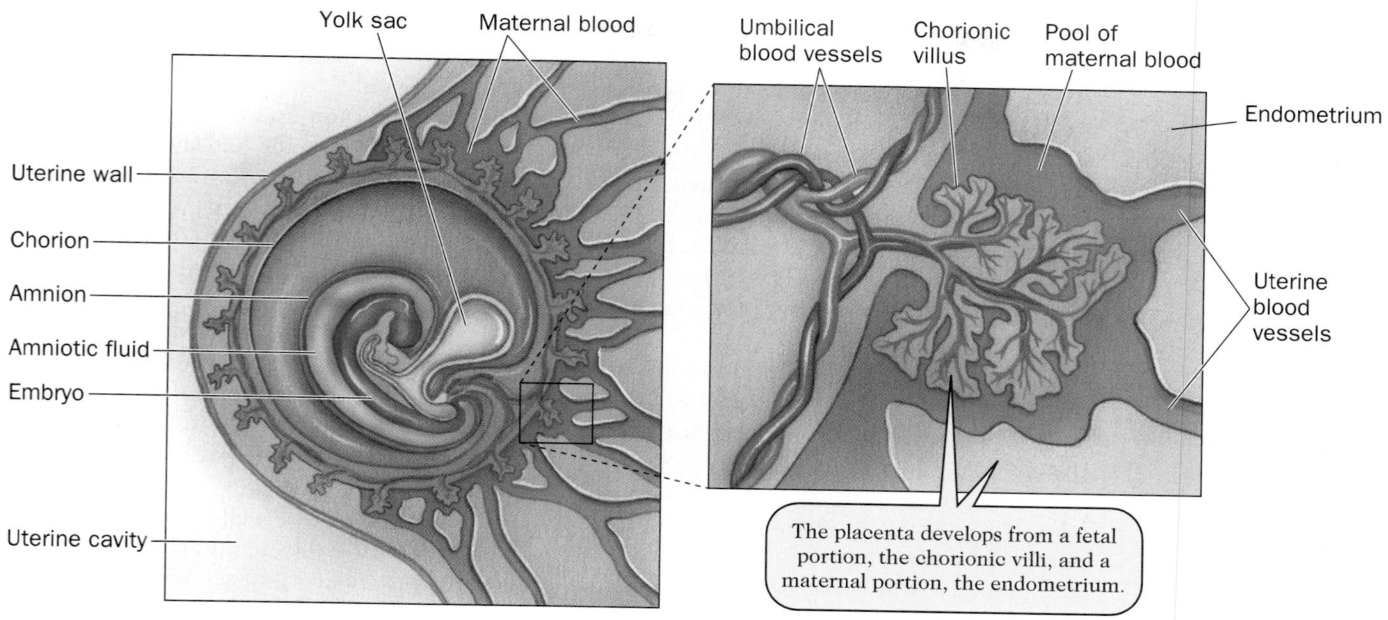

Yolk sac
Maternal blood
Uterine wall
Chorion
Amnion
Amniotic fluid
Embryo
Uterine cavity
Umbilical blood vessels
Chorionic villus
Pool of maternal blood
Endometrium
Uterine blood vessels
The placenta develops from a fetal portion, the chorionic villi, and a maternal portion, the endometrium.

FIGURE 18.6 Embryonic development: formation of the placenta.

To accomplish these transfers, the chorion establishes a close association with the endometrium. As it forms, the chorion releases digestive enzymes which eat away cavities within the endometrium near the embedded embryo. Digested capillaries fill these cavities with blood, and fingerlike extensions of the chorion called **chorionic villi** grow out into the blood-filled cavities. The chorionic villi contain the embryo's capillaries covered with a thin layer of epithelium from the chorion. Embryonic blood has now been placed very close to maternal blood.

> The embryo is suspended in amniotic fluid. The yolk sac forms blood cells. The allantois helps form the bladder. The chorion and its chorionic villi establish an exchange between the embryo's blood and the mother's blood.

The Placenta: Exchanging Nutrients and Secreting Hormones

Initially, the chorion provides the association with the mother's circulatory system, but it further develops into a more complex structure called the **placenta**. The placenta, which is shown in Figure 18.6, is the structure formed by this close association of embryonic (chorion) and maternal (endometrium) tissues. There is no mixing of embryonic and maternal blood in the placenta because they are always separated by epithelial layers. However, they are close enough for the efficient exchange of wastes, nutrients, and respiratory gases by diffusion. The placenta becomes functional during the fifth week of development, just about the time when the embryonic heart begins to beat and pump blood through the embryo.

The embryo is connected to the placenta by the long **umbilical cord**, which contains one *umbilical vein* and two *umbilical arteries* surrounded by connective tissues. The umbilical vein carries oxygenated, nutrient-rich blood from the placenta into the embryo, and the umbilical arteries return deoxygenated, nutrient-poor blood from the embryo to the placenta.

The placenta allows the developing embryo to release wastes and gain oxygen (O_2) and nutrients from its mother's circulatory system. This permits the embryo to take advantage of the work being done by the mother's digestive, respiratory, and urinary systems. The placenta also works a bit like a filter, keeping most of the mother's waste materials out of the embryonic circulation.

Unfortunately, the placenta is not a perfect filter. The increasing number of cocaine-addicted, alcohol-damaged, and

AIDS-infected infants provides evidence that damaging agents can cross the placental barrier. The old advice to pregnant women, "Don't consume anything you would not feed your newborn baby," is both true and relevant. Additional advice would be to avoid behaviors that expose the mother to sexually transmitted diseases (STDs) because STDs also can be transmitted to the embryo, fetus, or newborn.

The placenta is also an endocrine organ, secreting HCG, progesterone, and estrogen. HCG sustains progesterone production by the corpus luteum in the mother's ovary. Progesterone promotes the development of blood vessels and glands in the endometrium. Eventually, the placenta will make its own progesterone, and by about 6 months into the pregnancy, the corpus luteum will begin to degenerate.

Progesterone from the placenta also prevents contractions of the muscular myometrium layer of the uterus that might prematurely expel the embryo. In addition, it promotes the formation of a thick mucus plug in the cervical canal which keeps vaginal bacteria and contaminants from reaching the interior of the uterus. The estrogen produced by the placenta promotes the growth of myometrium as the embryo and then the fetus expand in the uterus.

> The placenta contains two separate capillary networks—one from the mother's endometrium and one from the chorion. Nutrients, gases, and wastes are exchanged between these vessels. The placenta produces HCG, progesterone, and estrogen.

Survival in a "Hostile" Womb

Since we don't remember the interior of our mother's uterus (our immature brains could not form memories), we tend to romanticize our experience there and describe it as a warm, supportive, nurturing environment. However, that is not the whole truth.

In your mother's uterus you behaved quite a bit like a parasite, burrowing into her tissues and stealing her nutrients. Her immune system should have attacked you just as it would any other foreign organism. After all, you had different genes, and your cell surfaces had different proteins which could function as antigens to initiate an immune attack. Yet this attack usually does not occur. The trophoblast helps prevent an immune response to the embryo.

It appears that chemical agents released by the trophoblast inhibit the action of the mother's *interleukin-2*. This substance normally activates a woman's cytotoxic T-cell and helper T-cell lymphocytes. Blocking of helper T cells prevents the recruitment of B-cell lymphocytes and their antibodies as a response to embryonic antigens (Chapter 13). In other words, the embryo appears to inhibit the mother's immune response, thus securing its parasitic existence until birth. Also, the separation of embryonic and maternal blood in the placenta reduces the recognition of other embryonic antigens by maternal white blood cells.

> Chemical agents from the trophoblast prevent an immune response and an attack by the mother's body on the embryo.

The Embryo: Growth and Development Week by Week

Week 3. The trophoblast expands into the chorion, which begins the development of the placenta, while the allantois forms the first umbilical blood vessels. The yolk sac begins the formation of blood cells.

Mesodermal cells form a transitional structure, the *notochord*, which provides support for the embryo until the vertebral column forms (Figure 18.7). The notochord also is important because it initiates the formation of the nervous system by causing the overlying ectoderm to thicken and fold into the *neural tube*. The anterior end of the neural tube becomes the brain, while the posterior end forms the spinal cord.

Blocks of mesoderm called *somites* form alongside the neural tube. Portions

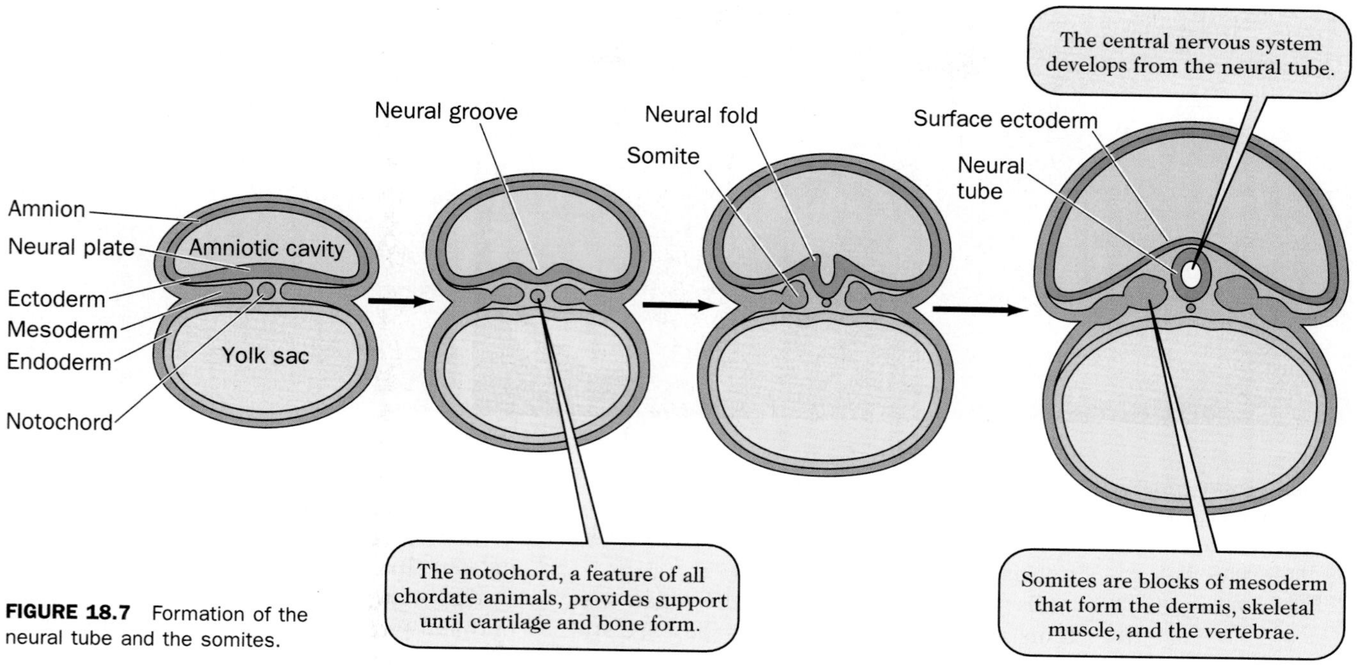
The central nervous system develops from the neural tube.
Neural groove
Neural fold
Somite
Surface ectoderm
Neural tube
Amnion
Neural plate
Amniotic cavity
Ectoderm
Mesoderm
Endoderm
Yolk sac
Notochord
The notochord, a feature of all chordate animals, provides support until cartilage and bone form.
Somites are blocks of mesoderm that form the dermis, skeletal muscle, and the vertebrae.

FIGURE 18.7 Formation of the neural tube and the somites.

of each somite form the dermis of the skin, the skeletal muscles, and the vertebrae. Other mesodermal cells construct two tubes that fuse, fold, and subdivide to form the heart. By this stage in your development, you have grown to the size of the letter "e."

Week 4. The heart begins to beat, propelling blood cells through an expanding network of blood vessels. The jaws, ears, and eyes begin their development during this period. Internally, the tongue, diaphragm, and brain begin to appear.

At the surface of the embryo, paddle-like buds appear that will form the arms and legs (Figure 18.8a). The arms and hands will grow more rapidly than will the legs and feet. In a sense, we grow from the head down. The upper body regions mature before those of the lower body do, and even after birth we grip with our hands before we walk.

Week 5. The head is growing rapidly as the brain expands into all its major functional regions, and the face is recognizably human. Even a nose is present. On the limb buds, the hands and feet are beginning to appear, though fingers and toes are not yet present. Internally, the lymphatic system and the kidneys are taking shape.

Week 6. The upper limbs are quite recognizable, with elbows, wrists, and even fingers appearing (Figure 18.8b). Inside the limbs, cartilage formation begins as a prelude to bone formation. The external ear is taking shape, and the hard palate forms as the two halves of the upper jaw begin to fuse.

Week 7. The fingers are fully differentiated and distinct from the thumb. The knee and ankle are beginning to appear in the lower leg. The toes are evident but are still connected together by a layer of skin. Teeth are beginning to form in the jaws, and the liver is starting to take over the production of blood cells from the yolk sac.

Week 8. The head is clearly separated from the trunk by the neck. Toes are distinct. Eyelids have developed and have temporarily covered the eyes. Bone formation begins as the cartilage which formed the early skeleton is replaced by bone tissue (Figure 18.8c). The skeletal muscles are contracting spontaneously, though the contractions are too gentle to be felt by the mother. The digestive tract has differentiated into its major regions, and a mouth and an anus are present. The external genitalia are appearing but cannot be distinguished as male or

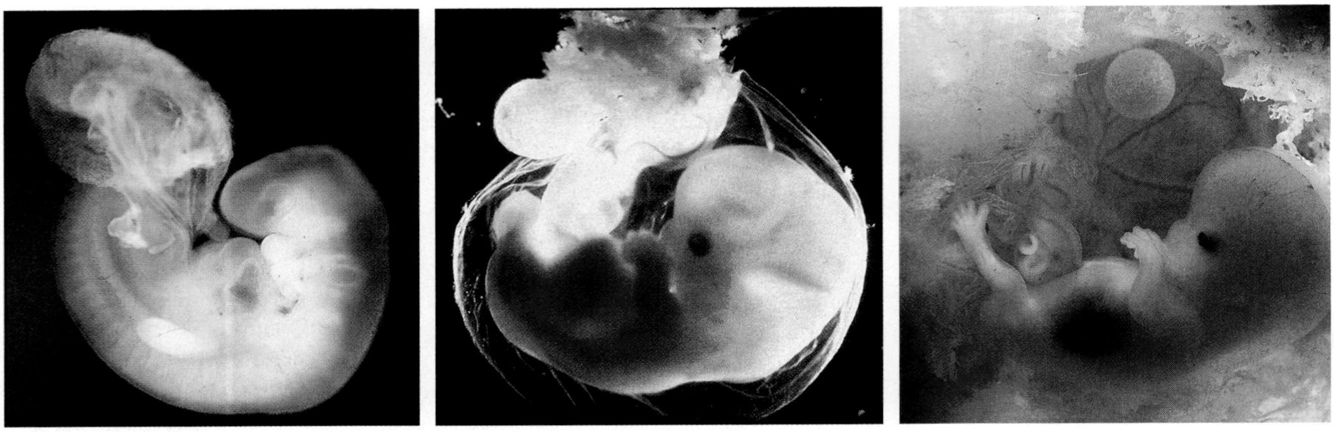

(a) (b) (c)

FIGURE 18.8 (a) The embryo at 4 weeks. (b) The embryo at 5 to 6 weeks. (c) The embryo at 8 weeks.

female. The gonads, however, are differentiating into ovaries or testes.

By the end of week 8 all the organ systems are established, though they are still incomplete and unable to sustain independent life. Among all the systems, the circulatory system is the first to become fully functional, transporting nutrients and wastes to and from the placenta. An average embryo is now nearly 2.5 centimeters (1 inch) in length and 1 gram (0.04 ounce) in weight.

> By the end of the embryonic phase of development all the organ systems are established, but only the circulatory system is fully functional. The sexes are just beginning to differentiate internally.

What Can Go Wrong?

Among the preembryos which survive to implantation, 15 to 25 percent are lost during the embryonic phase as congenital defects lead to miscarriages. Most miscarriages occur quite early in pregnancy and go unnoticed by the mother.

The 6 weeks of embryonic development are a time of rapid cell division, cell differentiation, and morphogenesis. Cells are differentiating into epithelial cells, connective tissue cells, muscle cells, or nerve cells. For all these cells to operate and for the embryo to develop normally, cellular events must occur at the right time and place. With so much going on, it is not surprising that most birth defects originate in this phase of development.

For instance, the vertebrae and skin may fail to grow around the spinal cord, leaving it exposed to potential injury during birth in a condition called *spina bifida*. The edges of the upper palate in the mouth may fail to grow together, causing *cleft palate* and leaving the mouth and nasal cavities connected instead of separated. There may be holes in the septum that separates the two ventricles of the heart.

We don't know what causes all these defects, but genes, pathogens, radiation, environmental chemicals from pollution, smoking, drinking, and drugs have all been implicated. Even a pregnant woman's use of prescription medication such as sedatives and antibiotics can affect her developing embryo. In addition, some viruses, such as rubella, which causes German measles, can cross the placenta from an infected mother to her embryo, causing deafness, blindness, retardation, and even death. Much damage can be done before a woman even knows she's pregnant, and so if pregnancy is possible, a woman should avoid the known dangers.

Diagnosing embryonic defects. The most common method of inspecting embryonic tissue for chromosomal abnormalities and the defective products of mutant genes is called **chorionic villi sampling (CVS)**. It is performed by inserting a thin, flexible tube through the vagina and into the uterus (Figure 18.9). With the help of an ultrasound scanner,

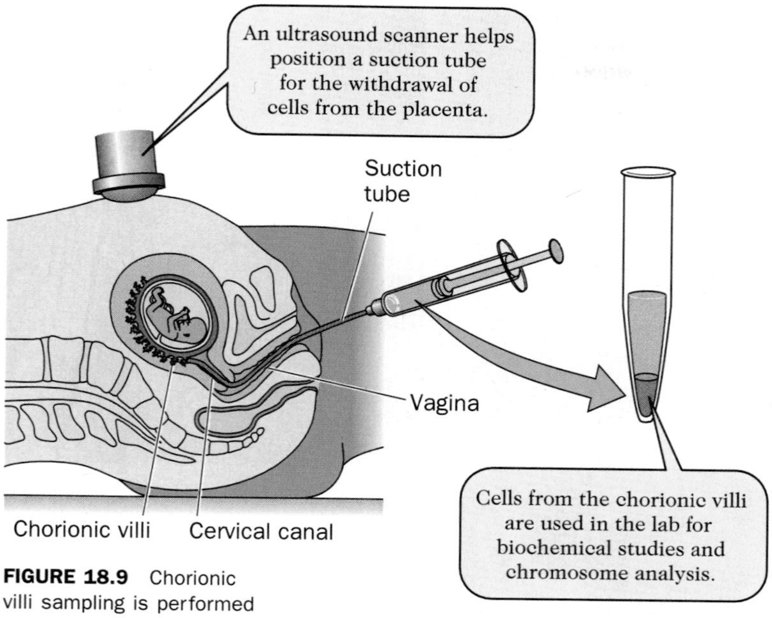
An ultrasound scanner helps position a suction tube for the withdrawal of cells from the placenta.
Suction tube
Vagina
Chorionic villi
Cervical canal
Cells from the chorionic villi are used in the lab for biochemical studies and chromosome analysis.

FIGURE 18.9 Chorionic villi sampling is performed by removing embryonic cells from the placenta.

the tip of the tube extracts tissue from the chorionic villi for examination of chromosomes and chemical analysis. Because there is a slight risk of injury to the embryo, the test usually is not performed unless there is a family history of genetic defects.

Testing can be carried out as early as 5 weeks into pregnancy. This early identification of potential problems gives the mother ample time to decide whether to continue or terminate her pregnancy. Genetic counseling centers are available at most large hospitals to help the mother interpret the results of CVS and provide information on potential consequences for the newborn and the family.

> Defective development of the embryo may be caused by genes, toxic substances, and pathogens. CVS examination of embryonic tissue for abnormal chromosomes and chemicals permits the identification of some defects in the embryo.

Fetal Development: Weeks 9 through 40

The period of fetal development that occurs between week 9 and week 40 is a time when the formation of organ systems is completed.

The Fetus: Changes Month by Month

Month 3. The external genitalia are developed well enough that if we could peek into the uterus, we would recognize the fetus as being male or female. The kidneys begin to function, producing urine that is released into the amniotic fluid. The upper limbs have just about reached their birth length, but the legs and feet are still forming. Fingerprints are developing. The liver has begun to produce bile, and the spleen is participating in blood cell formation. The end of the third month marks the completion of the first third, or the *first trimester,* of the entire developmental process (Figure 18.10a).

Month 4. Morphogenesis continues as the eyes and ears move to their final location. In the ovaries, the follicles are being produced. In the skeleton, the bone marrow is participating in blood cell formation, along with the liver and spleen (Figure 18.10b).

Month 5. By this point the nervous system is able to coordinate skeletal muscles, permitting reflex movements such as thumb sucking, kicking, and fist making. These movements, which are called *quickening*, may be felt by the mother. The skin of the fetus is covered with thin, silky hair called *lanugo*, and a fatty paste keeps the skin from absorbing too much water from the amniotic fluid. In females the uterus forms, while in males the testes begin their descent toward the scrotum. The pancreas begins the secretion of insulin and glucagon. A 5-month fetus can be seen in Figure 18.10c.

Month 6. The fetal heartbeat can be heard through a stethoscope. Fingernails are visible. The alveoli of the lungs have formed. However, the lungs are one of the last organs to become fully functional. The sixth month ends the *second trimester*, and by this point the fetus has moved from a head-up to a head-down orientation.

Month 7. The fine lanugo hair covering the early fetus is lost and replaced by coarser hair. Eyelids show occasional blinking. The legs have finally reached their birth length, and toenails are visible.

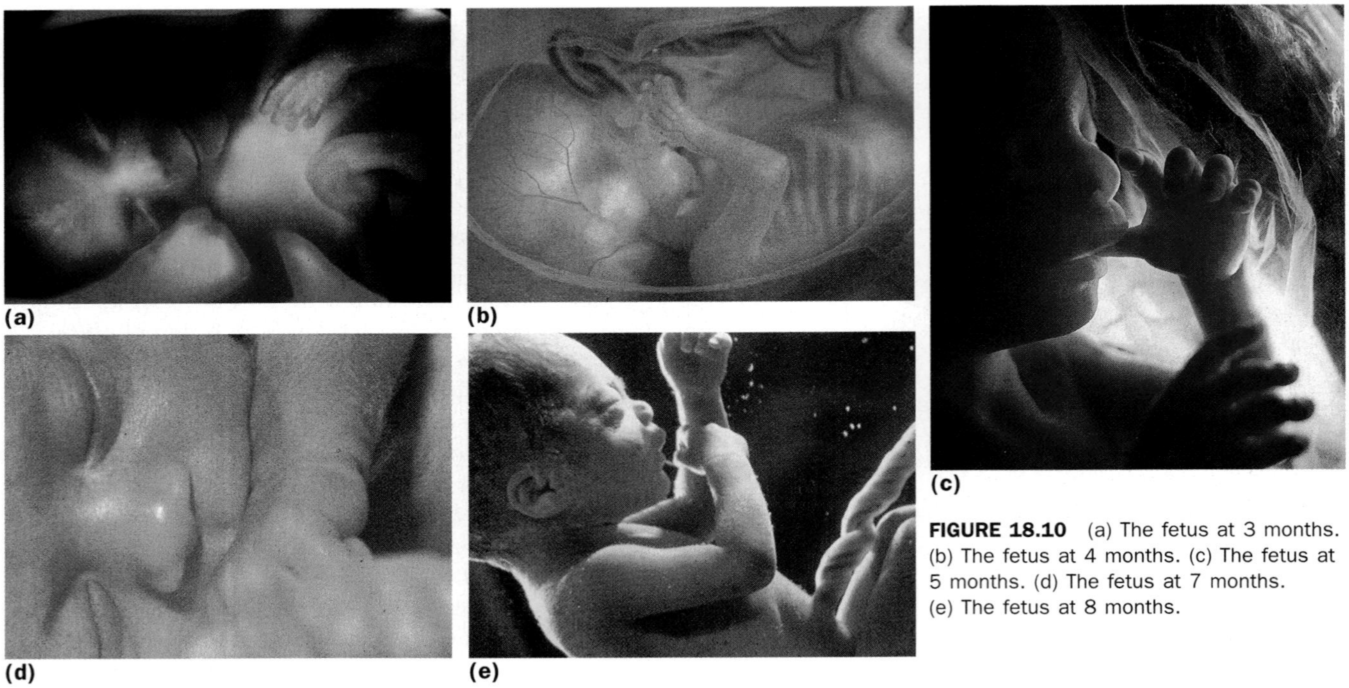

FIGURE 18.10 (a) The fetus at 3 months. (b) The fetus at 4 months. (c) The fetus at 5 months. (d) The fetus at 7 months. (e) The fetus at 8 months.

However, the wrinkled skin lacks an undercoating of fat tissue and appears quite translucent and reddish because of the many capillaries that show through to the surface (Figure 18.10d).

Months 8 and 9. During the last 2 months of development, the fetus gains weight rapidly. It forms rounded contours as a layer of fat accumulates below the fetal skin. The skin appears less red as fat intervenes between blood vessels and the skin's surface (Figure 18.10e).

By the end of 40 weeks the organ systems have completed their development. The last two systems to mature are the digestive and respiratory systems, since these systems do not fully function until birth. In the male fetus the testes have fully descended into the scrotum. If you were of average size, you weighed about 3.2 kilograms (7 pounds) and measured nearly 50 centimeters (20 inches) in length at birth.

> During fetal development, months 3 through 9, the organ systems grow in size and mature sufficiently to support life outside the uterus. The sexes become fully differentiated in appearance.

What Can Go Wrong?

A major problem faced by the fetus is obtaining enough nutrients to support its growth. Sometimes there are too many nutrients available. For example, fetuses developing in untreated diabetic mothers gain extra fat by converting the very abundant blood sugar of the mother into storage lipids. By contrast, when there are too few nutrients, the fetus wins the competition for scarce resources and maternal muscles and bones are broken down to provide for fetal growth.

The effects of heavy alcohol consumption (more than five drinks daily) during pregnancy produce *fetal alcohol syndrome*, in which a newborn exhibits a low birthweight, facial deformities, poor muscle coordination, and learning disabilities. Even moderate drinking (two drinks daily) can result in small babies who learn slowly. There is no safe level of alcohol consumption as far as fetal development is concerned. Beer and wine have the same effect as distilled liquors ("hard alcohol").

The fetus is not free from the risk of infection. Viruses may be passed through the placenta from an infected mother, causing devastating infections and even fetal death. Some infections can even be

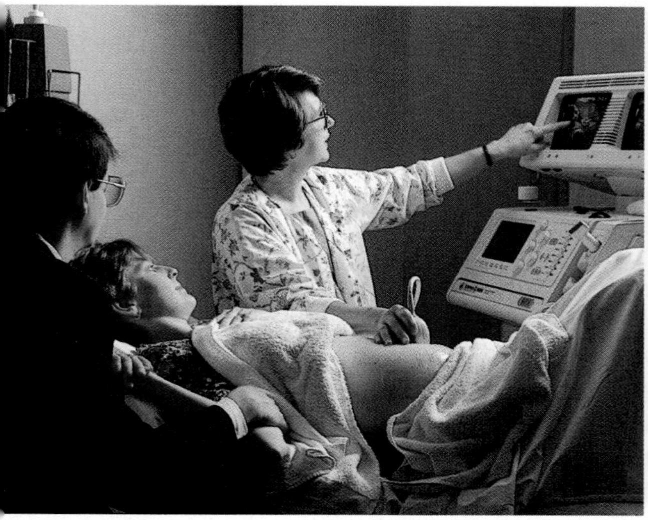

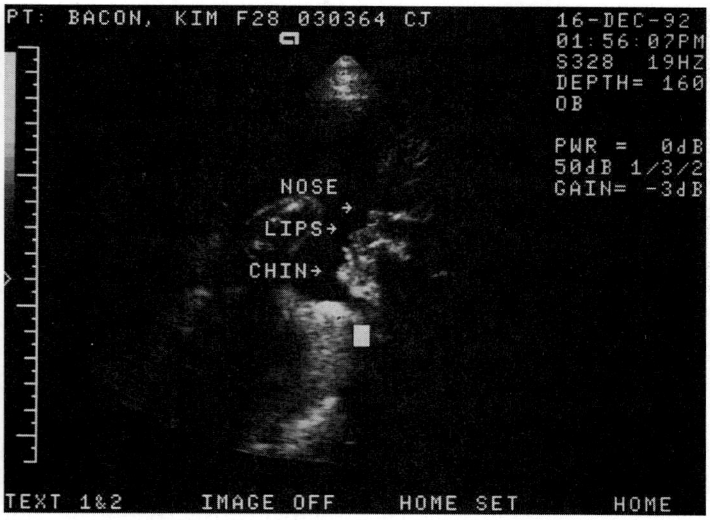
PT: BACON, KIM F28 030364 CJ
16-DEC-92
01:56:07PM
S328 19HZ
DEPTH= 160
OB
PWR = 0dB
50dB 1/3/2
GAIN= -3dB
NOSE
LIPS
CHIN
TEXT 1&2
IMAGE OFF
HOME SET
HOME

FIGURE 18.11 An ultrasound scan provides a video of the developing fetus.

passed to the fetus during its birth. The AIDS virus (HIV), the herpes virus, and the gonorrhea bacteria all can be contracted during birth, leading to life-threatening infections during the early months after birth.

Diagnosing fetal defects. The most useful tools for assessing fetal development and defects are ultrasound (sonography), amniocentesis, and alpha-fetoprotein screening.

An **ultrasound scan** provides a moving picture of a fetus in the uterus (Figure 18.11). Ultrasound technology involves the passage of high-frequency sound waves through the abdominal and uterine walls. The waves that rebound from hard tissues in the fetus are computer-analyzed to create an image. A video picture of the fetal head, vertebral column, heart, or even genitalia is displayed on a television screen for interpretation. Ultrasound scans permit an accurate estimate of the due date if the time of fertilization is uncertain.

Amniocentesis (*am-knee-oh-sen-**tea**-sus*) involves the insertion of a long needle through the abdominal and uterine walls and into the amniotic cavity (Figure 18.12). A small sample of amniotic fluid is removed, along with skin and lung cells that have been sloughed off into the fluid. In the laboratory, those cells are allowed to grow until there is a sufficient number for biochemical tests and inspection of chromosomes. Unfortunately, this takes several weeks. An additional disadvantage is the slight risk of injury to or infection of the mother and fetus.

Alpha-fetoprotein (AFP) screening is performed on a pregnant woman's blood between weeks 16 and 18. AFP normally is produced in small amounts by a developing fetus. If excessive AFP is released by the fetus, it diffuses through the placenta into the mother's circulation, from which it can be detected by the screening test. An increased amount of AFP in the maternal blood may indicate twins or may indicate a neural tube defect such as spina bifida.

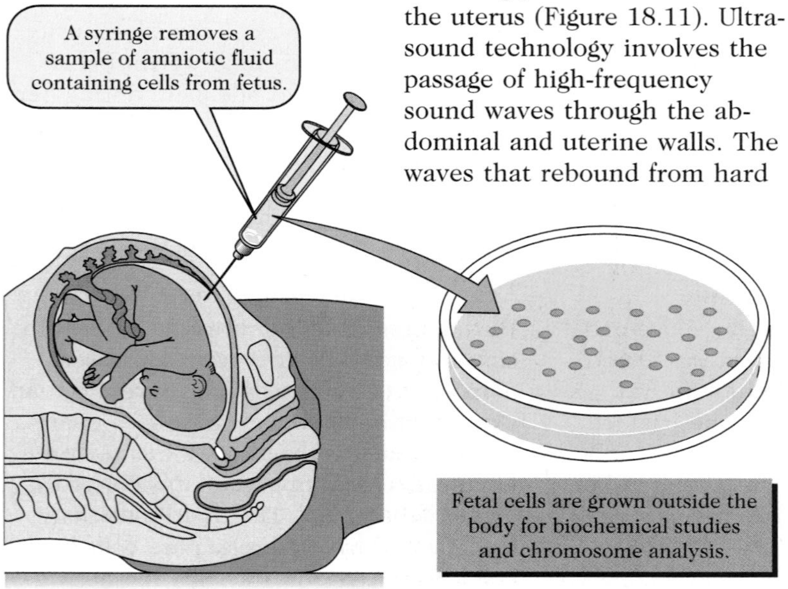
A syringe removes a sample of amniotic fluid containing cells from fetus.
Fetal cells are grown outside the body for biochemical studies and chromosome analysis.

FIGURE 18.12 Amniocentesis is performed when sufficient amniotic fluid is present, at around 14 to 15 weeks of pregnancy.

Ultrasound images reveal defects. Amniocentesis removes amniotic fluid and cells for chromosomal analysis and chemical testing. Excessive AFP in the maternal blood may indicate fetal defects.

Maternal Changes during Pregnancy

The fetus hasn't been the only one changing through 9 months of pregnancy. Its mother has also undergone significant changes in shape and function in support of the pregnancy. Nearly all the mother's body systems are affected, along with her movement, sleep, and general comfort.

The Uterus: Its Growth Affects Other Organs

Normally the size of a clenched fist, the uterus expands in size and weight about 20-fold during pregnancy. In the early stages of pregnancy this growth is due to an increase in the number of smooth muscle cells. Later in pregnancy, those cells increase in size. This expansion of the uterus can cause much discomfort to a pregnant woman.

As it grows, the uterus pushes the surrounding intestines, stomach, and liver upward against the diaphragm, putting pressure on the rib cage and widening the thoracic cavity. Pressure against the stomach may force gastric juices into the esophagus, causing heartburn. The mother may feel the need to urinate more frequently as the urinary bladder is compressed by the uterus. Compression of the colon may cause constipation.

Confined by the diaphragm above and the pelvic girdle below, the uterus eventually pushes outward and alters the woman's distribution of weight and balance while she is standing or walking. To avoid tipping forward, the woman may arch her back, exaggerating the lumbar curvature of the vertebral column. This may place more pressure on her spinal nerves, causing considerable pain.

Nutrition and Weight: Eating for Two

During pregnancy a woman gains weight from several sources. The added weight is due to the fetus, the increased size of the uterus, the placenta, the retention of water, and increased adipose tissue.

Age-old advice says that a pregnant woman must eat for two, but it is not simply a matter of increasing food intake. She needs only an additional 300 calories per day to support adequate fetal growth. High-quality food with all the essential amino acids, minerals, and vitamins is needed to provide the embryo and fetus with the proper raw materials.

Severe malnutrition during the first 3 months of pregnancy doubles the risk of bearing a child with serious defects in the nervous system. The B vitamin *folic acid* appears to be especially crucial to nervous system development during this early stage of growth, and deficiencies of this vitamin are indicated in defects such as *spina bifida* and *hydrocephalus* (accumulation of fluid in the brain).

Mothers who experience starvation in the last 6 months of pregnancy are more likely to give birth to underweight babies who suffer a high death rate in the first year of life. Studies of Dutch mothers who experienced starvation during World War II indicate that if a low-birthweight female infant survives to sexual maturity, she is likely to bear underweight babies even if nutrition was adequate during her pregnancy. Hunger imperils not only the current generation but generations yet to be conceived.

Mammary Glands: Secretions for the Newborn

During pregnancy the breasts increase in size and weight, partly from increased adipose tissue but mostly as a result of an expansion of glandular tissue in preparation for milk production. Milk is not actually produced until several days after the birth of the infant. The first fluids that are released by the breasts are called **colostrum**. Colostrum is rich in IgA antibodies which helps protect the newborn's digestive tract during the establishment of its normal flora.

Cardiovascular and Respiratory Systems: Delivering More

During pregnancy the blood volume of the mother increases about 40 percent, allowing her to send sufficient blood to the placenta. With a larger volume of blood to move around, her heart must work harder. Cardiac output rises by 20 to 30 percent and the heart rate increases by 10 to 15 percent during pregnancy, mostly in the later stages.

Pregnancy also may cause changes in the blood vessels and blood flow. Veins near the surface of the legs may lose their elasticity and become permanently stretched. Such damaged veins are called *varicose veins*. The damage occurs because the uterus presses against blood vessels and, as a result, blood has difficulty returning to the heart and tends to pool in the legs, stretching the veins.

The respiratory system also changes during pregnancy. Since the fetus is removing an increasing amount of O_2 from her blood, the mother's respiratory rate increases. Toward the end of pregnancy she may suffer shortness of breath as the expansion of the uterus hinders movement of the diaphragm.

The Skeleton: A Source of Calcium

If the mother's diet is deficient in calcium (Ca^{2+}), her bones and teeth may be partially broken down to provide calcium for the fetus. Of course, this may weaken her skeleton and damage her dental health.

As birth approaches, a hormone called *relaxin* is secreted by the placenta and ovaries, causing a loosening of connective tissues in the joints of the pelvic girdle. This loosening makes it easier for the pelvis to flex and permit passage of the fetal head and shoulders through the *birth canal* from the cervix to the vaginal opening.

Exercise for a Healthy Pregnancy

Moderate exercise during pregnancy benefits the mother physically and psychologically and does not harm the fetus. Exercise improves the capacity for O_2 delivery, increases muscular strength, and minimizes calcium loss from the skeleton. Appropriate exercises are those which use large groups of muscles, do not require delicate balance, and do not greatly increase body temperature.

If you are pregnant, warm up carefully, exercise on shock-absorbing surfaces, avoid bouncing or jarring motions, and cool down gradually afterward. Hot tubs and saunas may be tempting but should be avoided because they increase the temperature of the fetus, which may contribute to spinal cord defects.

Swimming is an ideal exercise because the water cools the body and bears some of the additional body weight, preventing compression of major blood vessels by the uterus. A fit and healthy mother is more likely to experience a smooth delivery and have a rapid recovery.

> During pregnancy, the uterus enlarges, crowding other organs and altering some of their normal functions. The mammary glands prepare for milk secretion. The heart rate and respiratory rate increase. The mother's skeleton may lose calcium.

Birth: An End and a Beginning

Birth is the final event of pregnancy. It defines a new way of life for both the mother and the child. Normally birth is accomplished through a series of events called *labor* that move the fetus from the uterus through the birth canal.

Labor appears to be initiated by hormones from both the mother and the fetus. As the fetal pituitary gland matures, it secretes the hormone ACTH into the fetal circulation. ACTH stimulates the fetal adrenal gland to secrete steroid hormones that act on the placenta, and the placenta produces prostaglandins. Prostaglandins cause contractions in the smooth muscle of the uterus (Figure 18.13).

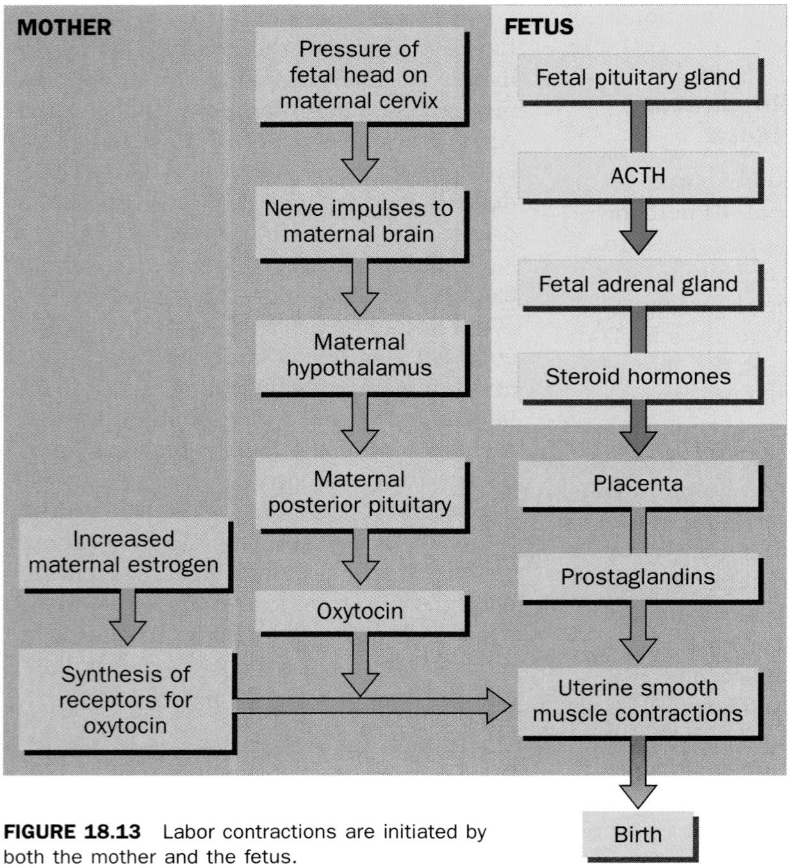
MOTHER
FETUS
Pressure of fetal head on maternal cervix
Nerve impulses to maternal brain
Maternal hypothalamus
Maternal posterior pituitary
Oxytocin
Increased maternal estrogen
Synthesis of receptors for oxytocin
Fetal pituitary gland
ACTH
Fetal adrenal gland
Steroid hormones
Placenta
Prostaglandins
Uterine smooth muscle contractions
Birth

FIGURE 18.13 Labor contractions are initiated by both the mother and the fetus.

In the last weeks of pregnancy estrogen in the mother's blood begins to increase, while progesterone declines. The increased estrogen causes the synthesis of oxytocin receptors on the smooth muscle cells in the myometrium of the uterus. With the appearance of these receptors, the oxytocin present in the mother's blood causes the uterine wall to begin slow, rhythmic contractions that are called *false labor* because they are often mistaken by a first-time mother as the beginning of labor.

As oxytocin from the mother increases, *true labor* begins with powerful, painful contractions. A positive feedback loop is initiated as these contractions cause the release of more oxytocin. More oxytocin causes even stronger contractions, and so on.

At the beginning of labor, uterine contractions last 30 seconds and occur at intervals of 15 to 30 minutes. By the end of labor, they are up to 90 seconds in duration and occur every 2 minutes or less. Labor lasts longest for the first birth (an average of 24 hours), while subsequent births may take less time. Labor can be divided into 3 stages: dilation of the cervix, expulsion, and afterbirth (Figure 18.14).

Stages of Labor

Stage 1: dilation of the cervix. The first contractions of the uterus press the fetal head against the cervical canal. Because of these contractions, fetal pressure, and the hormone relaxin, the cervix begins to soften and widen as it is drawn up and merges with the uterus. Eventually the opening will widen to a diameter of 10 centimeters (4 inches), enough to accommodate the head of the fetus.

The pressure of the head against the cervix is usually enough to tear the amnion and release the amniotic fluid through the vagina. When this happens, it is said that the mother's "water has broken," and this is a sure sign that labor is proceeding. Dilation is the longest phase of labor. It may take 6 to 12 hours or even more for a first birth, depending on the size of the fetus.

Stage 2: expulsion. Expulsion lasts from full dilation until the actual delivery of the newborn. The woman experiences frequent, long contractions and has an intense urge to push the newborn out into the world. The entire body of the fetus must pass through the mother's birth canal (the uterus, vagina, and surrounding pelvic girdle).

The appearance of the infant's head at the labia around the vagina is called "crowning." As soon as the infant's face emerges, a birth attendant removes mucus from the infant's nose and mouth to permit the first breath. As delivery of the head proceeds and tissues stretch, the mother may receive a surgical incision called an *episiotomy* (*eh-**peas**-ee-**ott**-tuh-me*). This makes more room for the baby and prevents more extensive and dangerous tissue tearing. After the head emerges, it is followed by first one shoulder and

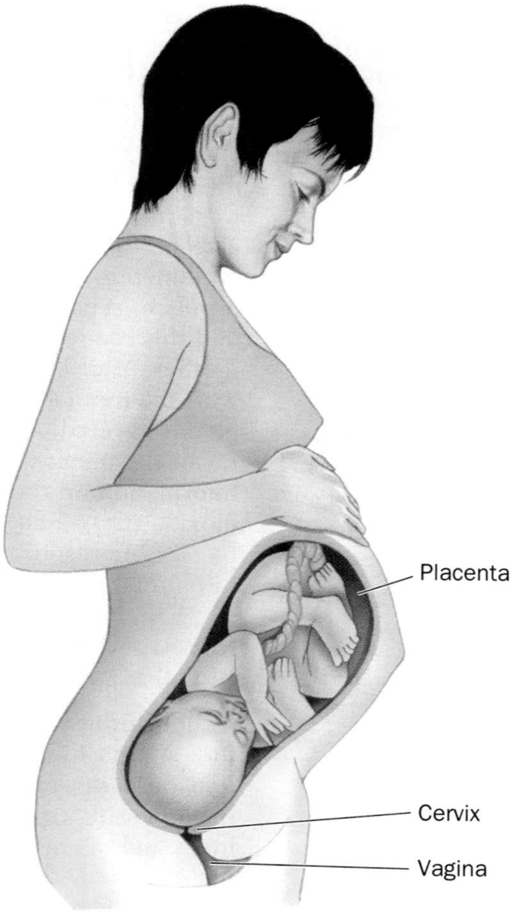
Placenta
Cervix
Vagina

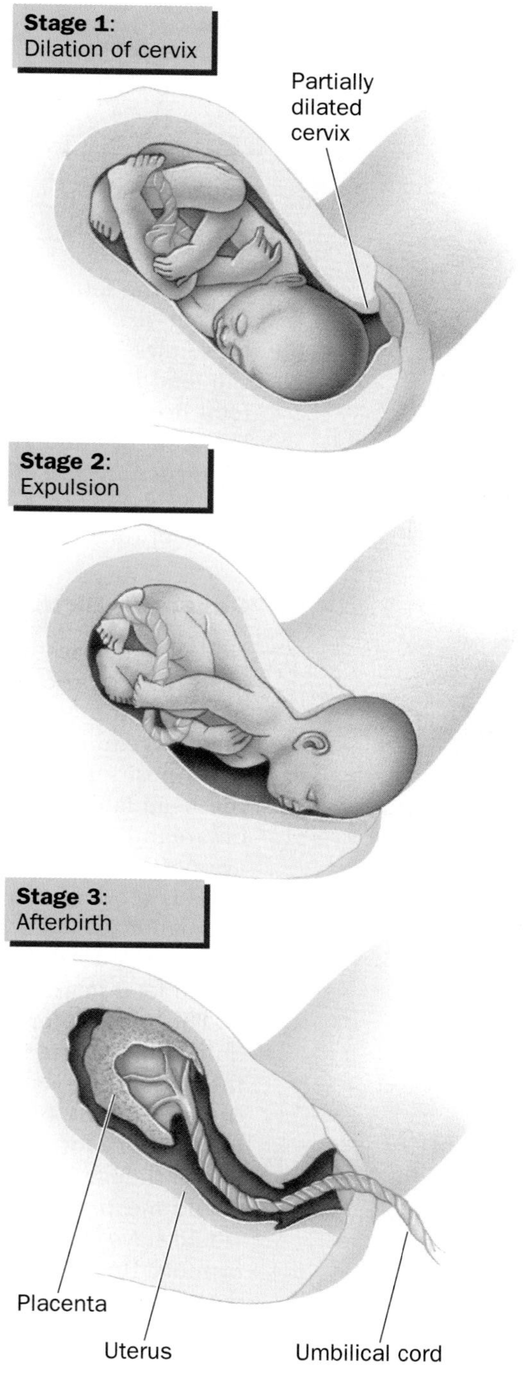
Stage 1:
Dilation of cervix
Partially
dilated
cervix
Stage 2:
Expulsion
Stage 3:
Afterbirth
Placenta
Uterus
Umbilical cord

FIGURE 18.14 The stages of labor.

then the other, after which the rest of the body emerges rapidly.

When the entire body is outside, the umbilical cord is gently extended, clamped tightly to stop blood flow, and cut. The expulsion stage usually lasts for under an hour for a first birth and even less for subsequent births.

Stage 3: afterbirth. During the afterbirth stage, the mother continues to have strong contractions which further detach the placenta from the lining of the uterus. When the placenta is completely detached, it is expelled through the birth canal along with the remnants of the other extraembryonic membranes and the umbilical cord, which are collectively called the afterbirth. The *afterbirth* is usually delivered 15 to 30 minutes after the baby, and there may be some loss of blood.

Electronic Fetal Monitoring

To provide information on the status of the fetus during labor and delivery, **electronic fetal monitoring** is often performed (Figure 18.15). Electrodes are attached to the mother's abdomen or directly to the

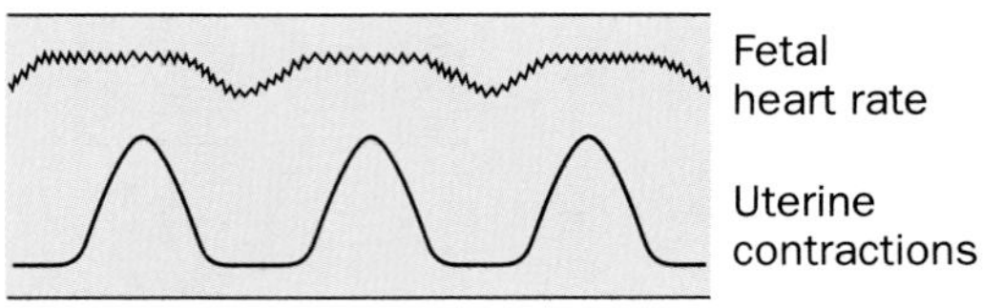
Printout
Fetal heart rate
Uterine contractions

FIGURE 18.15 Electronic fetal monitoring.

skin of the fetus through the vagina and the cervical canal. Amplified signals from these electrodes provide information on the fetal heartbeat, a good indicator of fetal health. The normal fetal heart rate is between 120 and 160 beats per minute. Abnormal increases or decreases are signs of distress which may require surgical delivery of the child.

Cesarean Birth: Surgical Birth

Not all deliveries begin normally or proceed through the birth canal. A surgical birth called a **cesarean birth**, or **C-section**, may be required if the fetus has an unusual position in the uterus. Normally the head is down and emerges first. A cesarean birth may be required if the legs are positioned next to the cervix, if the fetus is too large for the birth canal, if a long labor exhausts the mother, or if there are signs of fetal distress such as a change in the fetal heart rate.

The surgery for a cesarean birth requires anesthesia and an incision through the abdominal wall and uterus to remove the infant. The operation requires hospital recovery for the mother. In the past, it was thought that one cesarean birth would mean that all subsequent babies would have to be delivered in the same fashion. However, it has been found that with medical supervision, about 65 percent of cesarean mothers can later have normal vaginal deliveries.

> In stage 1 of a normal birth uterine contractions break the amnion and cause the cervix to widen. Stage 2 is the expulsion of the fetus through the birth canal. In stage 3, the placenta separates from the uterine wall.

The First Breath and Blood Flow

The newborn's first breath requires an enormous effort and usually is accompanied by a loud cry. Before birth, the airways had a very small diameter and the alveoli of the lungs were collapsed and nonfunctional. During labor, the placenta begins to separate from the endometrium, and this reduces gas exchange with the mother. After birth, the fetal connection with the mother's circulation is completely severed, and the respiratory center in the infant's brain receives sensory information on the increasing carbon dioxide (CO_2) and low O_2 in the infant's blood. The respiratory center responds by initiating breathing.

Rerouting the Circulation of Blood after Birth

Although the cardiovascular system is the first system to become functional in the fetus, it is not fully developed until after birth. During embryonic and fetal life, the lungs and digestive systems are active but not as active as they will be after birth, because the fetus is not eating, digesting, or breathing for itself. O_2 and nutrients are obtained by fetal blood circulating through the placenta and the umbilical arteries and veins. In the fetus, three special bypass structures shunt blood away from the liver and the lungs: the ductus venosus, the foramen ovale, and the ductus arteriosus (Figure 18.16).

In the fetus, the umbilical vein carrying O_2 and nutrients from the placenta connects to the hepatic portal vein, which normally carries blood from the digestive tract to the liver. However, in the fetus, a bypass called the **ductus venosus** connects the hepatic portal vein to the inferior vena cava, directing incoming blood away from the liver and toward the heart. Only enough blood passes to the liver to support the growth of its cells.

A second bypass occurs as blood in the inferior vena cava enters the right atrium. A hole called the **foramen ovale** (*for-ray-mun oh-val-lay*) is present in the septum between the right atrium and left atrium,

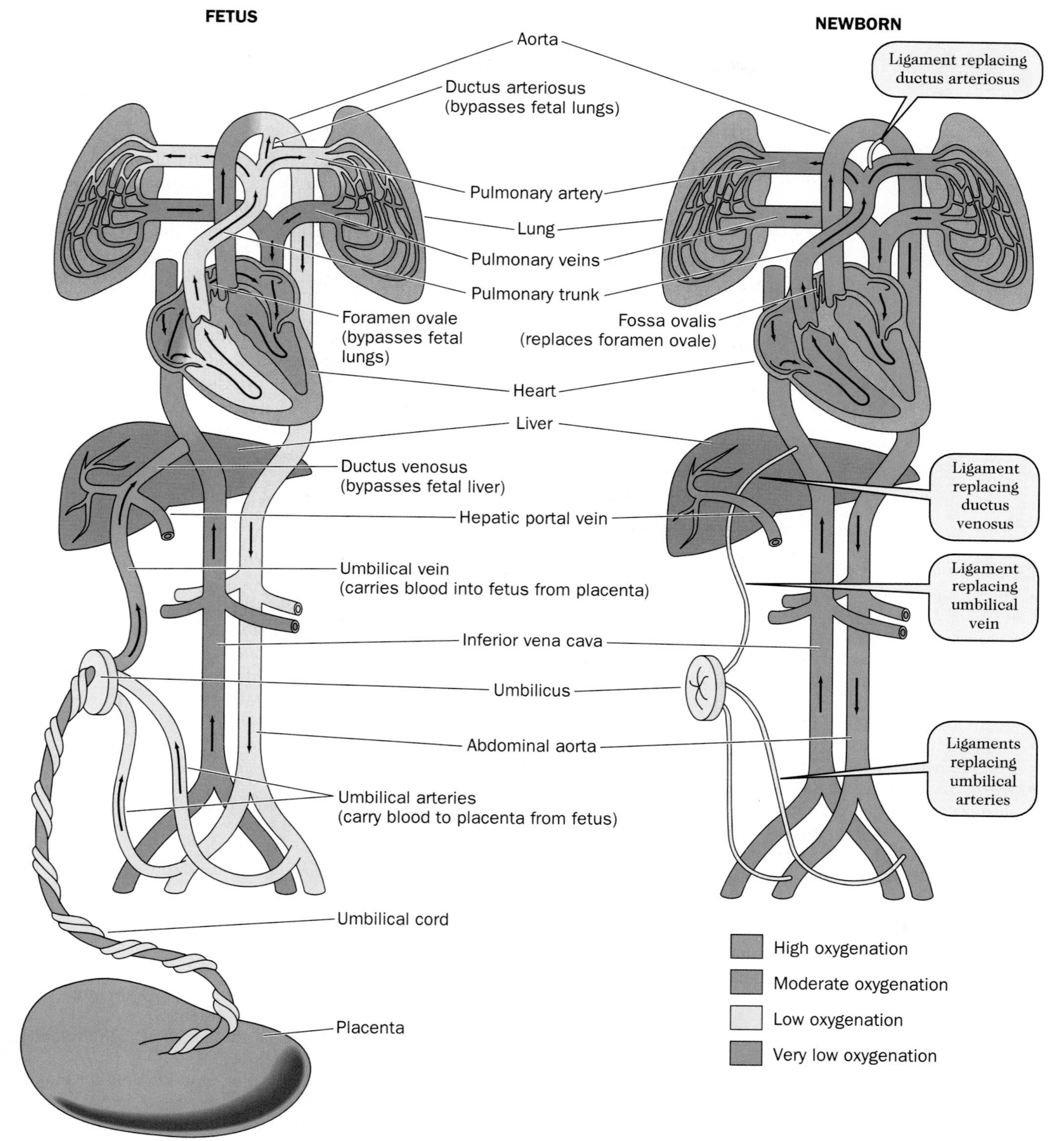

FETUS
NEWBORN
Aorta
Ductus arteriosus
(bypasses fetal lungs)
Ligament replacing ductus arteriosus
Pulmonary artery
Lung
Pulmonary veins
Pulmonary trunk
Foramen ovale
(bypasses fetal lungs)
Fossa ovalis
(replaces foramen ovale)
Heart
Liver
Ductus venosus
(bypasses fetal liver)
Ligament replacing ductus venosus
Hepatic portal vein
Umbilical vein
(carries blood into fetus from placenta)
Ligament replacing umbilical vein
Inferior vena cava
Umbilicus
Abdominal aorta
Ligaments replacing umbilical arteries
Umbilical arteries
(carry blood to placenta from fetus)
Umbilical cord
Placenta
High oxygenation
Moderate oxygenation
Low oxygenation
Very low oxygenation

FIGURE 18.16 Before birth, the lungs and liver are not fully functional and special bypasses are incorporated into the fetal circulatory system to redirect blood flow around them. At birth, these bypasses must close, as must the umbilical veins and arteries, to allow the newborn to breathe and function independently.

permitting most of the blood to rush over to the left atrium. The blood passes into the left ventricle and then into the large aorta for distribution to body tissues. This bypass avoids sending all the blood through the lungs. Only a little blood enters the right ventricle and is pumped into the pulmonary trunk and lungs.

A third bypass is encountered in the pulmonary trunk. The **ductus arteriosus** (***duck**-tus are-**teer**-ee-oh-sus*) is a short vessel that routes most of the blood in the pulmonary trunk away from the lungs and into the aorta for distribution to the body. Just enough blood passes to the lungs to provide for their growth.

These three bypasses allow the efficient delivery of O_2 and nutrients to growing tissues. However, at birth, they must close for blood to pass to the lungs and liver. As the infant takes its first breaths, the flow and pressure of blood in its cardiovascular system are altered. The ductus venosus and ductus arteriosus close and eventually are replaced by ligaments, as are the umbilical veins and arteries. The foramen ovale closes, and its edges fuse together. The blood pressure in the left atrium rises and helps seal the foramen. Only a shallow depression persists in the adult heart, marking the former location of this opening.

> In the embryo and fetus three bypasses limit blood flow to the liver and lungs. The ductus venosus bypasses the liver, while the foramen ovale and the ductus arteriosus limit blood flow through the lungs. At birth, these structures close.

Chapter Summary

Fertilization: New Beginnings

- The secondary oocyte is surrounded by the corona radiata.
- Sperm have a headpiece, a midpiece, and a flagellum.
- Sperm release acrosomal enzymes to digest a channel through the corona radiata for fertilization.
- Upon fertilization the secondary oocyte completes meiosis II, producing the ovum.
- Conception occurs when the sperm nucleus combines with the ovum nucleus to produce the zygote.
- Sperm establish zygote gender by adding an X or a Y chromosome to the X chromosome in the ovum.
- Twins form when separate oocyte are fertilized (fraternal) or when a single zygote separates into groups of cells which each form embryos (identical).

Human Development in the Uterus: An Overview

- The four processes of early development are cleavage, growth, morphogenesis, and differentiation.
- Cleavage (mitotic) cell divisions divide the zygote into cells called blastomeres.
- Morphogenesis is the change in shape of the embryo.
- Differentiation is the specialization of cell structure and function as some genes are turned on and others off.
- The stages of development are preembryonic, embryonic, and fetal development.

Preembryonic Development: The First 2 Weeks

- Preembryonic development forms the morula and the blastocyst. The blastocyst is the stage of development at implantation in the uterus.
- Implantation at an abnormal site causes an ectopic pregnancy.

Embryonic Development: Weeks 3 through 8

- Embryonic development includes the formation of extraembryonic membranes and organ systems.
- The wall of the amniotic cavity is the amnion and the fluid within the cavity is the amniotic fluid.
- The yolk sac starts blood cell formation and forms part of the intestines and reproductive system.
- The allantois forms umbilical blood vessels and portions of the urinary bladder.
- The chorion (chorionic villi) forms the embryonic portion of the placenta.
- The placenta permits exchange of nutrients, respiratory gases, and wastes between embryonic and maternal blood vessels.
- During embryonic development, all organ systems are established.
- Chorionic villi sampling removes tissues from the placenta for genetic analysis.

Fetal Development: Weeks 9 through 40

- Fetal development includes growth in body size and weight and maturation of organ systems.
- Ultrasound scans allow visual inspection of the developing fetus for defects or state of growth.
- Amniocentesis allows removal of amniotic fluid for biochemical or genetic analysis.
- AFP screening can indicate twins or structural defects.

Maternal Changes during Pregnancy

- Expansion of the uterus makes breathing, urination, defecation, and movement difficult.
- Malnutrition in the mother affects the fetal nervous system and causes low birth weight.
- The mammary glands form milk after birth; their first, antibody-rich fluid is the colostrum.

- The mother's blood volume, cardiac output, and respiratory rate increase during pregnancy.

Birth: An End and a Beginning

- Labor, initiated by fetal and maternal hormones, moves the fetus from the uterus through the birth canal.
- Stage 1 labor includes dilation of the cervix, tearing of the amnion, and release of amniotic fluid.
- Stage 2 labor involves expulsion and delivery of the newborn by powerful uterine contractions.
- Stage 3 labor includes delivery of the placenta, extraembryonic membranes, and the umbilical cord.
- Cesarean birth is the surgical removal of the fetus.

The First Breath and Blood Flow

- The newborn begins breathing as the placenta separates and the vessels in the umbilical cord close.
- Three circulatory bypasses in the fetus—the ductus venosus, foramen ovale, and ductus arteriosus—allow blood to circumvent the liver and lungs before birth.

Selected Key Terms

amniocentesis (p. 489)
amnion (p. 482)
blastocyst (p. 481)
chorionic villi sampling (p. 486)
conception (p. 477)
corona radiata (p. 475)
differentiation (p. 479)
embryonic development (p. 482)
fertilization (p. 477)
fetal development (p. 487)
implantation (p. 481)
placenta (p. 483)
preembryonic development (p. 481)
umbilical cord (p. 483)
zona pellucida (p. 477)

Review Activities

1. Sketch and label a sperm cell
2. List some of the obstacles in the female reproductive tract which reduce the number of sperm available for fertilization.
3. Summarize the events which occur at fertilization. What are three direct consequences of fertilization?
4. Explain the differences in the ways identical, fraternal, and conjoined twins are formed.
5. Distinguish between the processes of cleavage, growth, morphogenesis, and differentiation.
6. Summarize the events and development stages which occur between fertilization and implantation.
7. List the extraembryonic membranes and their functions.
8. Summarize the changes that occur in the embryo during embryonic development.
9. Explain the significance of the notochord, neural tube, and somites.
10. Compare the techniques of chorionic villi sampling and amniocentesis.
11. Summarize the changes that occur in the fetus during fetal development.
12. Summarize the changes that occur in the mother during pregnancy.
13. List the stages of labor and the major events that occur in each stage. Which maternal hormones are associated with labor?
14. How does maternal health influence fetal health? Cite several examples.
15. List the three circulatory bypasses in the fetus and give their locations.

Self-Quiz

Matching Exercise

___ 1. Term for early mitotic cell divisions
___ 2. Stage of development at implantation
___ 3. Site where the embryo proper first begins to develop
___ 4. Extraembryonic membrane that helps form placenta
___ 5. Extraembryonic membrane directly surrounding embryo
___ 6. Extraembryonic membrane that begins forming blood cells
___ 7. Extraembryonic membrane forming bladder
___ 8. Vessel carrying blood from placenta into fetus
___ 9. Vessel carrying blood from fetus to placenta
___ 10. Remnants of placenta, umbilical cord, membranes

A. Yolk sac
B. Inner cell mass
C. Amnion
D. Blastocyst
E. Allantois
F. Umbilical artery
G. Cleavage
H. Afterbirth
I. Umbilical vein
J. Chorion

Answers to Self-Quiz

1. G; 2. D; 3. B; 4. J; 5. C; 6. A; 7. E; 8. I; 9. F; 10. H

Chapter 19

Human Genetics and DNA Technology

"Be all you can be. It's in your genes." Our genes represent our inheritance, our potential, and our limitations. They form a biological blueprint for our lives as human beings and as individuals. In the preceding chapters we often mentioned the role of genes both in normal development and functioning and in diseases and disorders. What do we really know about genes? It turns out that we know a great deal.

You've been hearing about genes since Chapter 1, where we described DNA, and Chapter 2, where we described the role of DNA in protein synthesis. You know that DNA is the hereditary material that forms the genes and is passed from generation to generation by the sperm and ova. DNA controls a cell by coding for proteins.

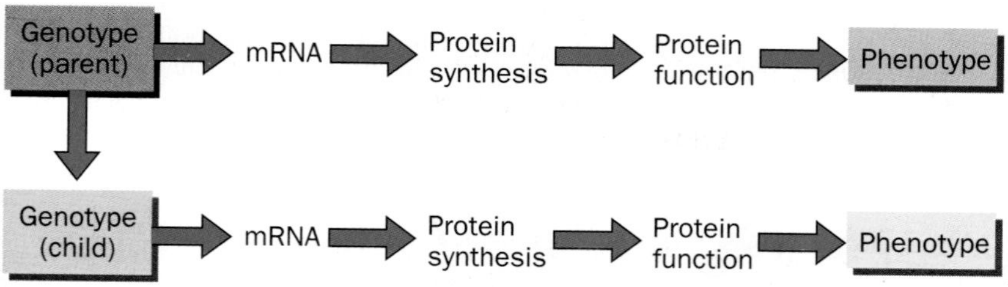
Genotype (parent)
mRNA
Protein synthesis
Protein function
Phenotype
Genotype (child)
mRNA
Protein synthesis
Protein function
Phenotype

FIGURE 19.1 Genes (genotypes) are passed from parents to their children. Genes determine the phenotype or appearance of a parent or child.

Proteins directly and indirectly produce our characteristics or traits. Traits which can be observed, measured, or chemically analyzed represent the **phenotype** of an individual. The genes that are responsible for the phenotype comprise the **genotype** (Figure 19.1). For example, brown eyes and the blood type AB are both phenotypes, while the genes coding for those conditions are part of the genotype. As we will discover, there are more genes in the genotype than there are traits in the phenotype. We do not use all the genes we inherit, but we can pass them on to the next generation through reproduction. The total genetic material you inherited from your parents is also referred to as your **genome.**

In this chapter, we will focus first on the chromosomes that bear the genes and how things can go wrong with their structure or number. Then we will look at particular genes, how they operate and interact, and how patterns of inheritance permit us to calculate the probability of a particular phenotype appearing in children born to parents with known or suspected genotypes. We will end the chapter by examining the brave new world of genetic engineering and biotechnology.

Human Chromosomes

Chromosomes carry genes. You have 46 chromosomes in each of your diploid body cells. There is no particular significance to the number 46—a potato has 48 chromosomes, cows have 60, and there are 78 in a dog. The number of chromosomes signifies the number of units into which the genome is packaged for distribution during mitosis and meiosis.

There may be up to 100,000 genes in the genome of a human cell, and they are distributed among the 46 chromosomes. In this section, we will examine the natural pairing and structure of chromosomes and their laboratory display for the diagnosis of certain disorders. Then we will turn to the role of particular chromosomes in the determination of gender and abnormalities.

The Pairing and Structure of Chromosomes

Our 46 chromosomes exist as pairs, which means that you have 23 pairs of chromosomes in your diploid cells. One set of 23, the *maternal set,* was inherited from your mother's ovum and a *paternal set* of 23 was inherited from your father's sperm. Each pair of similar chromosomes (one maternal and one paternal) is known as a **homologous pair,** and each chromosome in a pair is known as a **homologue** (Chapter 17). Homologous pairs are identical in length, and in the banding patterns produced by laboratory staining of cells during mitosis. These characteristics permit each homologous pair to be distinguished from the other pairs.

A typical human chromosome contains 60 percent protein and 40 percent DNA by weight. The protein plays packaging and regulating roles, and the DNA is the hereditary material. The DNA of a single chromosome may contain 200 million nucleotide pairs and would be about 6 centimeters (over 2 inches) long if it were drawn out straight. In other words, a single DNA molecule is thousands of times longer than the cell itself, and so it must be folded and coiled to fit into the nucleus with all the other chromosomes. This is quite a packaging accomplishment. A helpful analogy is to consider the nucleus to be the size of a basketball and to use the entire basketball court to represent the chromosomal material that must be folded to fit inside it.

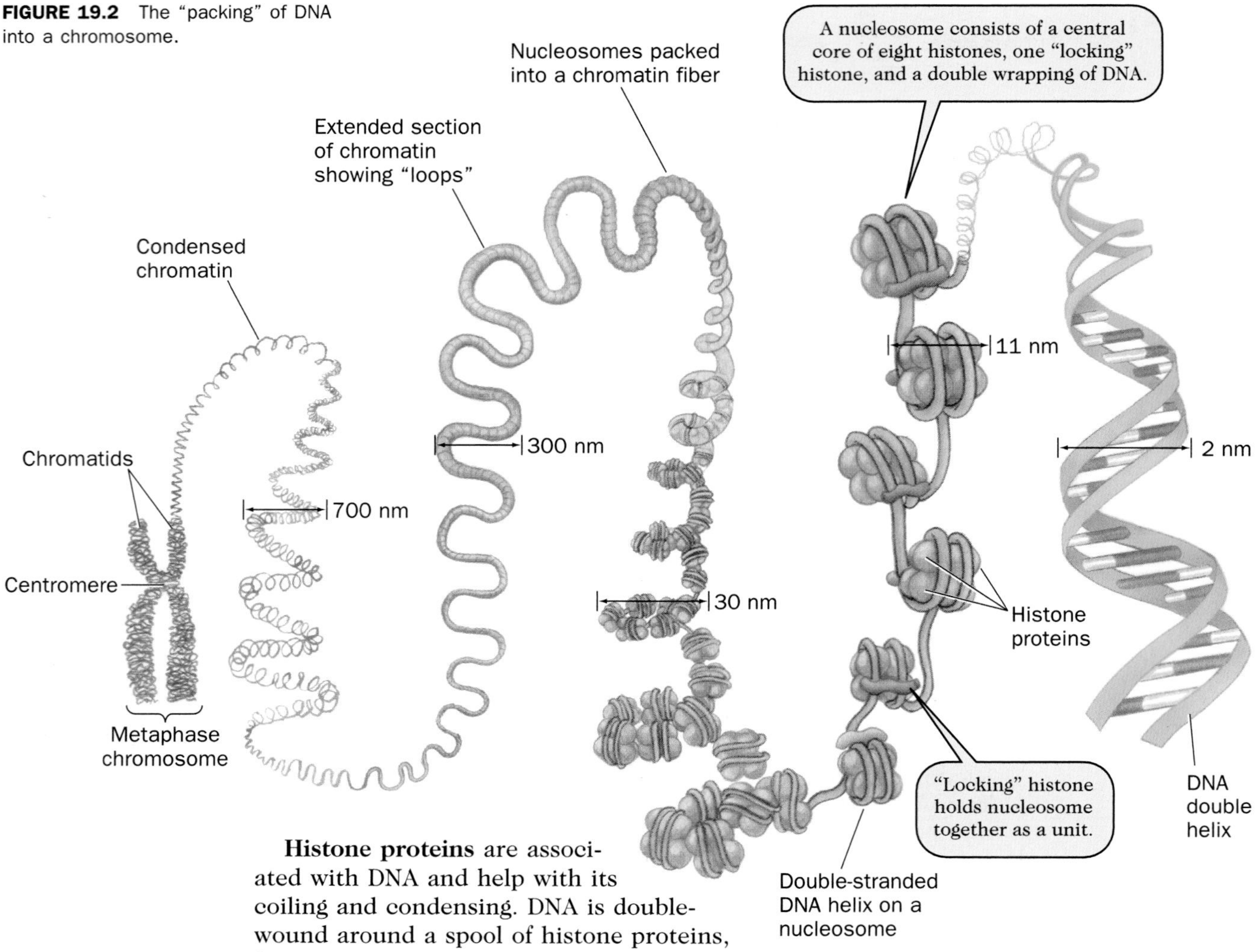

Nucleosomes packed into a chromatin fiber
A nucleosome consists of a central core of eight histones, one "locking" histone, and a double wrapping of DNA.
Extended section of chromatin showing "loops"
Condensed chromatin
Chromatids
Centromere
700 nm
300 nm
30 nm
11 nm
2 nm
Histone proteins
Metaphase chromosome
"Locking" histone holds nucleosome together as a unit.
DNA double helix
Double-stranded DNA helix on a nucleosome

FIGURE 19.2 The "packing" of DNA into a chromosome.

Histone proteins are associated with DNA and help with its coiling and condensing. DNA is double-wound around a spool of histone proteins, forming a structure called a **nucleosome** (Figure 19.2). To complete the condensation of the chromosome, nucleosomes are tightly packed together. Then the packed segments are looped, and the loops again are tightly packed, producing the characteristic chromosomal shapes.

During the interphase portion of the cell cycle, the chromosomes are elongated and difficult to see under the microscope (see Figure 2.13). When it is in this elongated form, called **chromatin** (***crome-uh-tin***), the genes (DNA) can be replicated or transcribed. Replication prepares a cell for mitosis or meiosis, while transcription is the first step in protein synthesis (Chapter 2). At the close of interphase, as the cell prepares to divide, its chromatin becomes coiled and condensed, forming chromosomes for efficient distribution to progeny cells.

All 23 pairs of human chromosomes have characteristic lengths and banding patterns. DNA is wrapped around histone proteins, forming nucleosomes. During interphase, chromatin is replicated or transcribed.

The Human Karyotype: Displaying Our Chromosomes

For the study and early diagnosis of certain birth defects, it is necessary to identify, display, and photograph all the chromosomes. Such a display is called a **karyotype,** and it is prepared by the procedures depicted in Figure 19.3. Metaphase of mitosis provides the best opportunity for observing individual chromosomes and preparing a karyotype. During metaphase, each chromosome consists of two genetically identical **chromatids** joined at a region called a **centromere.** In addition to length and banding, the different positions of the centromeres are used to identify homologous pairs of chromosomes.

Photos made at metaphase are enlarged, and the image of each chromosome is cut out. Then each image is placed with its homologue, and the 23 pairs are arranged by length. Many laboratories now use video screens and computer manipulation of images instead of cameras, photos, and cutting, but the principles are the same. The presence of extra chromosomes, missing chromosomes, or abnormal chromosomes can be detected in a karyotype.

Of the 23 homologous pairs of human chromosomes, 22 pairs are similar in appearance in men and women and are called **autosomes.** The traits associated with the genes on these chromosomes are called *autosomal traits.* The twenty-third pair are called the **sex chromosomes** because they are different in males and females. The sex chromosomes have genes that determine gender as well as genes for traits called **sex-linked traits.**

Recall that sperm carry either an X or a Y chromosome. However, ova carry only an X chromosome. When an X-carrying sperm fertilizes an oocyte, an XX, or female, zygote is produced. When a Y-carrying sperm fertilizes an oocyte, an XY, or male, zygote is produced. The X chromosome carries many genes, but the Y is small and carries very few.

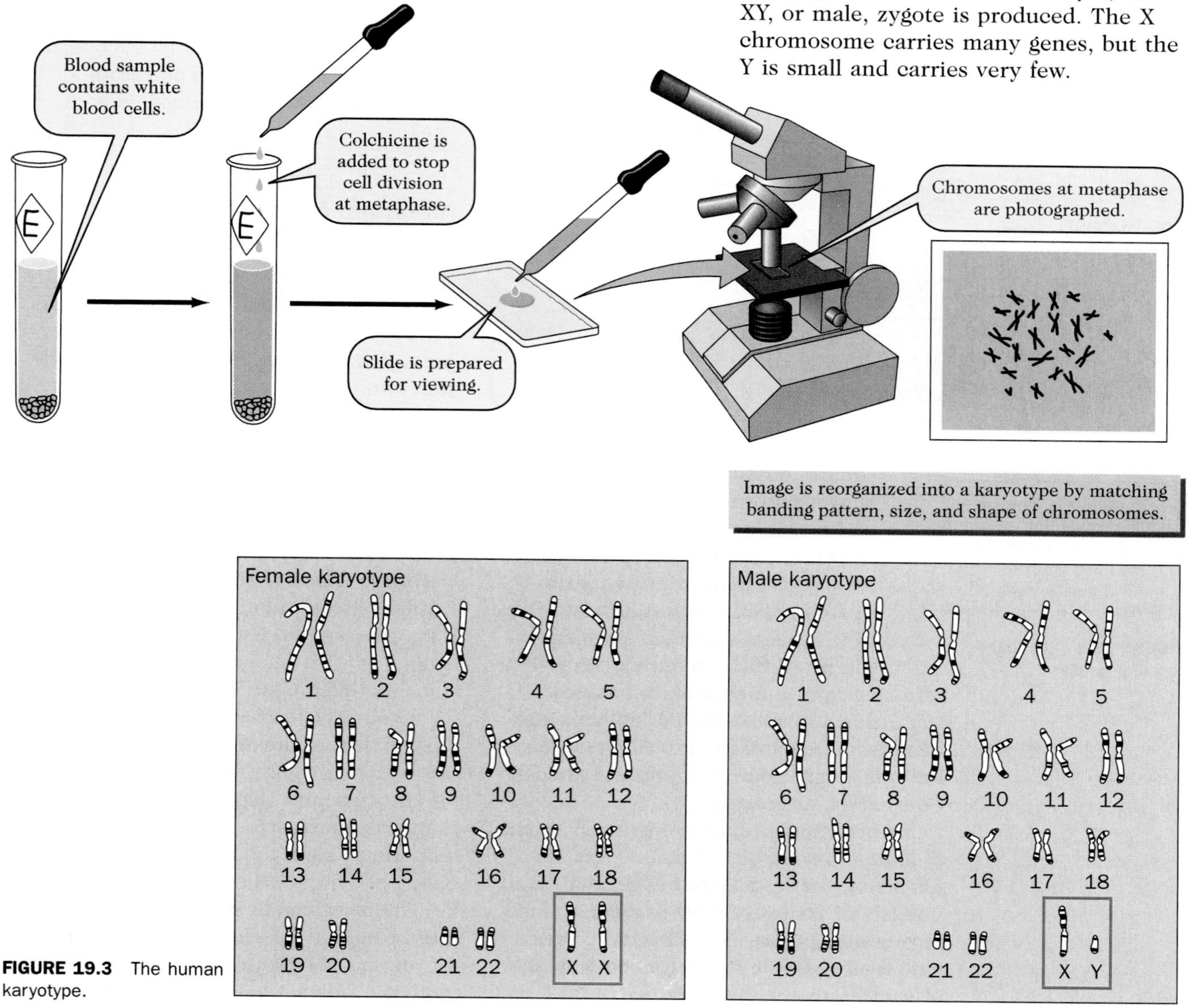
Blood sample contains white blood cells.
Colchicine is added to stop cell division at metaphase.
Slide is prepared for viewing.
Chromosomes at metaphase are photographed.
Image is reorganized into a karyotype by matching banding pattern, size, and shape of chromosomes.
Female karyotype
1 2 3 4 5
6 7 8 9 10 11 12
13 14 15 16 17 18
19 20 21 22 X X
Male karyotype
1 2 3 4 5
6 7 8 9 10 11 12
13 14 15 16 17 18
19 20 21 22 X Y

FIGURE 19.3 The human karyotype.

A karyotype shows 23 homologous pairs of chromosomes. Chromosome pairs 1 through 22 are the autosomes. Pair 23 are the sex chromosomes, which carry gender-determining genes. Males inherit an XY combination; females, an XX combination.

Inheritance of Gender: Males and Females

The karyotypes for normal males and females are displayed in Figure 19.3. A male is any individual with at least one Y chromosome in his karyotype. Recall from Chapter 18 that the genes of the *sex-determining region of the Y chromosome* (*SRY*) cause the formation of the male gonad.

The inheritance of gender can be visualized in a simple diagram called a **Punnett square** (Figure 19.4). In a Punnett square, possible types of sperm are shown on one side of the square and possible types of oocytes are shown on another side. The interior of the square represents the possible results of fertilization. Punnett squares represent the genotypic possibilities for the potential offspring from a mating. From these squares, the phenotypic possibilities can be determined and the probabilities of inheriting a particular genotype and phenotype can be calculated.

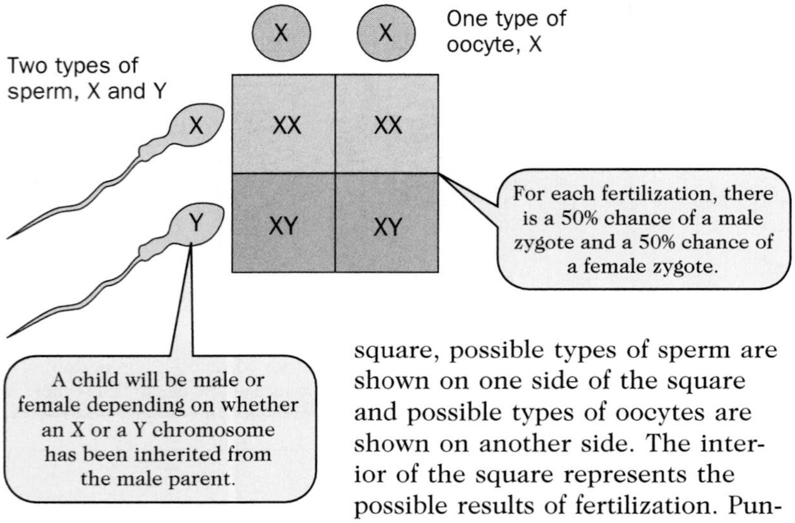

X
X
One type of oocyte, X
Two types of sperm, X and Y
X
XX
XX
Y
XY
XY
For each fertilization, there is a 50% chance of a male zygote and a 50% chance of a female zygote.
A child will be male or female depending on whether an X or a Y chromosome has been inherited from the male parent.

FIGURE 19.4 The inheritance of gender.

The Punnett square in Figure 19.4 emphasizes two important facts. First, the fertilizing sperm determines gender. Second, there is always a 50 percent chance of producing a boy or a girl in any individual fertilization as long as both kinds of sperm are equally healthy.

Because a normal woman has twice as many X chromosomes as a normal man, you might suspect that she would produce twice as many gene products for all the traits associated with her X chromosomes. To prevent this from happening, one X chromosome is inactivated in each diploid female cell. An inactive X chromosome can be seen microscopically as a dark-staining body called a **Barr body** along one edge of the nuclear membrane during interphase, when the cell is not dividing. A normal female has a single Barr body in each of her cells, while a normal male, with only one X chromosome, has none.

Punnett squares represent the genotypic and phenotypic possibilities for the offspring of a mating. Females have one active X chromosome and one Barr body in each cell.

Abnormal Numbers of Sex Chromosomes

Some men and women inherit too many X or Y chromosomes from their parents. This may result from a mistake in spermatogenesis or oogenesis called **nondisjunction.** During meiosis, homologous chromosome pairs normally separate at anaphase I and sister chromatids separate at anaphase II (Chapter 17). Nondisjunction represents a failure of these pairs to separate at either anaphase I or anaphase II. Consequently, the haploid gametes receive too many or too few copies of a particular chromosome and are abnormal. This can happen with any of the autosomes or with the sex chromosomes as shown in Figure 19.5a.

When gametes resulting from nondisjunction are involved in a fertilization, the zygote contains an abnormal number of chromosomes (Figure 19.5b). When this happens, the zygote may live but develops abnormally, and the individual shows a group of symptoms termed a syndrome.

The **metafemale syndrome (XXX)** occurs when an XX gamete (either a sperm or an oocyte) combines at fertilization with an X gamete. Metafemales are female

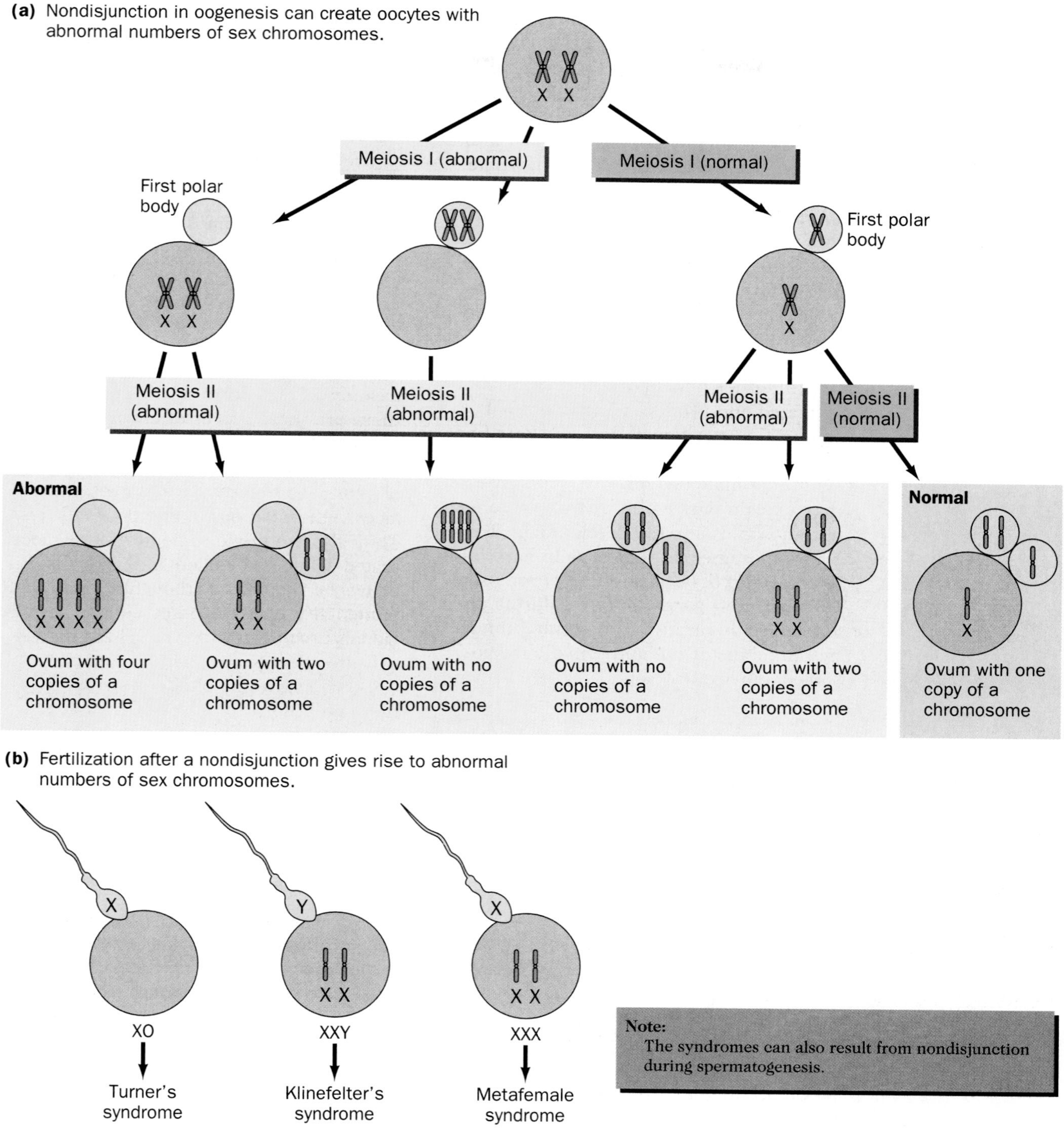
(a) Nondisjunction in oogenesis can create oocytes with abnormal numbers of sex chromosomes.
X X
Meiosis I (abnormal)
Meiosis I (normal)
First polar body
X X
First polar body
X
Meiosis II (abnormal)
Meiosis II (abnormal)
Meiosis II (abnormal)
Meiosis II (normal)
Abormal
X X X X
Ovum with four copies of a chromosome
X X
Ovum with two copies of a chromosome
Ovum with no copies of a chromosome
Ovum with no copies of a chromosome
X X
Ovum with two copies of a chromosome
Normal
X
Ovum with one copy of a chromosome
(b) Fertilization after a nondisjunction gives rise to abnormal numbers of sex chromosomes.
X
XO
Turner's syndrome
Y
X X
XXY
Klinefelter's syndrome
X
X X
XXX
Metafemale syndrome
Note:
The syndromes can also result from nondisjunction during spermatogenesis.

FIGURE 19.5 Nondisjunction is a failure in chromosome separation during meiosis.

but are often sterile and may have a reduced mental capacity. This syndrome occurs in about 1 in 1250 female births. When an X gamete combines with a gamete lacking X or Y chromosomes, **Turner's syndrome (XO)** results. The individual is female but sterile and is usually short in stature and has some mental impairment. This syndrome occurs in only 1 in every 2000 female births.

When an XX gamete combines with a Y gamete or an XY gamete combines with an X, the resulting condition is called **Klinefelter's syndrome (XXY).** The affected individual is male, but he is usually tall, has underdeveloped genitalia and enlarged breasts, and may exhibit learning disabilities. Klinefelter's syndrome occurs at a rate of 1 in 1000 male births. **XYY syndrome** males result from a nondisjunction in spermatogenesis that produces YY sperm that fertilize a normal oocyte. These males are usually quite tall, often have below-normal intelligence, and may exhibit aggressive behavior. The incidence is about 1 in every 1000 male births.

When an extra X chromosome is present in either a male or a female, it is inactivated and produces a Barr body. For example, a metafemale has two Barr bodies in each diploid cell and a Klinefelter's syndrome male has one. This inactivation may explain why embryos with extra X chromosomes survive development. Extra Y chromosomes are not usually harmful to the embryo, because the Y chromosome carries so few genes. Survival during development is rare when there are abnormal numbers of the autosomal chromosomes, our next topic.

> Nondisjunction is a failure to separate homologous chromosomes during anaphase I or II of spermatogenesis or oogenesis. The gametes will have too many or too few chromosomes. Zygotes from these gametes may live but develop characteristic syndromes.

Abnormal Numbers of Autosomal Chromosomes

Nondisjunction also can lead to abnormal numbers of autosomal chromosomes. In a **trisomy** there are three copies of a particular chromosome, while in a **monosomy** there is only one chromosome instead of the normal two. Having too many or too few autosomal chromosomes is usually lethal for the developing embryo, and a miscarriage results. Only a few embryos with such conditions survive beyond birth, and they often have serious mental and physical defects.

The most common trisomy is called **trisomy 21,** or **Down syndrome,** in which there are three copies of chromosome 21 (Figure 19.6). Affected individuals are short in stature and have a special fold of skin over slightly slanted eyes, flat faces with broad noses, and poor reflexes. They also are prone to heart disease, accelerated aging, and leukemia. Subnormal intelligence is common, though a wide range of mental capacity is observed among individuals with this syndrome. These individuals are affectionate, friendly, and cheerful.

Down syndrome children are more frequently born to older mothers. The chance of having an affected child is 1 in 3000 for mothers under 30 and 1 in 9 for those over 48. Recent research suggests that younger mothers conceive an equal number of affected embryos but that most fail to survive until birth for unknown reasons.

> Down syndrome results when three copies of chromosome 21 are inherited, a condition called trisomy 21. Down syndrome children are more frequently born to older mothers.

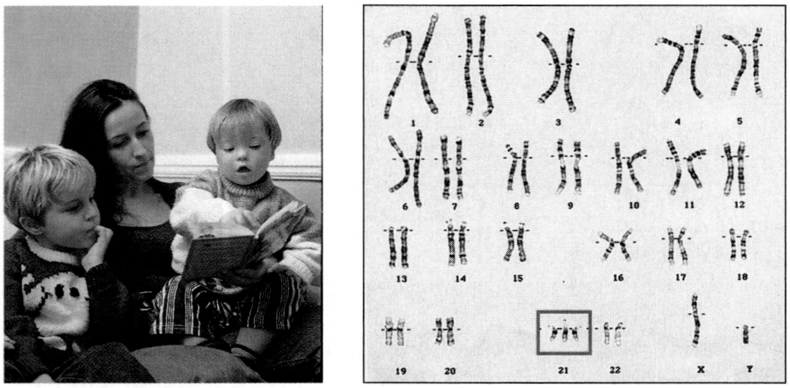
1 2 3 4 5
6 7 8 9 10 11 12
13 14 15 16 17 18
19 20 21 22 X Y

(a) A child with characteristic facial features

(b) A karyotype showing trisomy 21

FIGURE 19.6 Down syndrome.

Abnormal Chromosome Structures

Sometimes the correct number of chromosomes is inherited but the chromosomes are structurally abnormal because portions have been duplicated, rearranged, or lost (Figure 19.7). These alterations often result from chromosomal breakage after exposure to radiation or harmful chemicals.

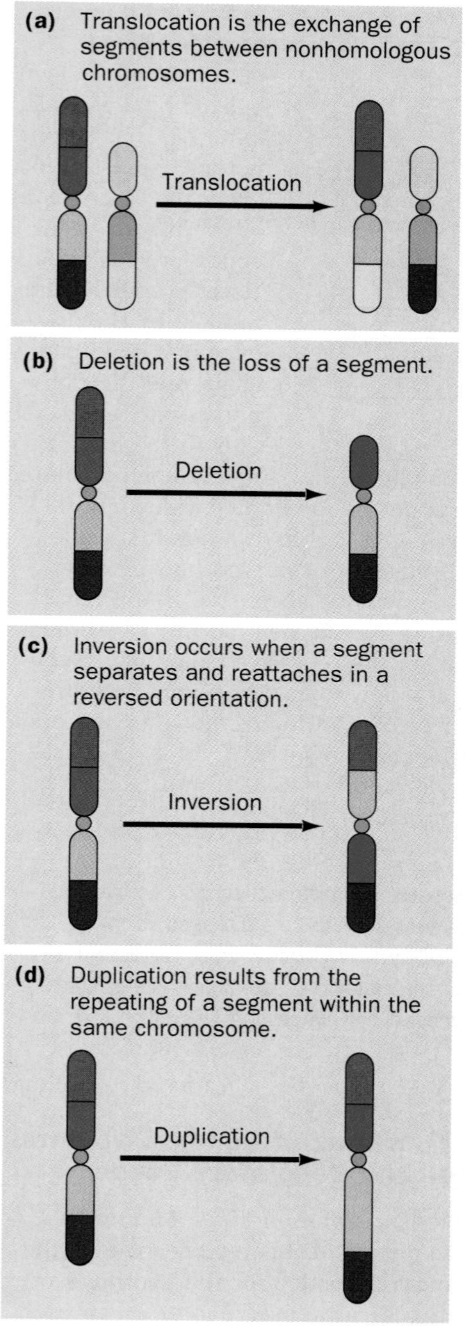
(a) Translocation is the exchange of segments between nonhomologous chromosomes.
Translocation
(b) Deletion is the loss of a segment.
Deletion
(c) Inversion occurs when a segment separates and reattaches in a reversed orientation.
Inversion
(d) Duplication results from the repeating of a segment within the same chromosome.
Duplication

FIGURE 19.7 Abnormal chromosome structures are caused by rearrangements of DNA segments.

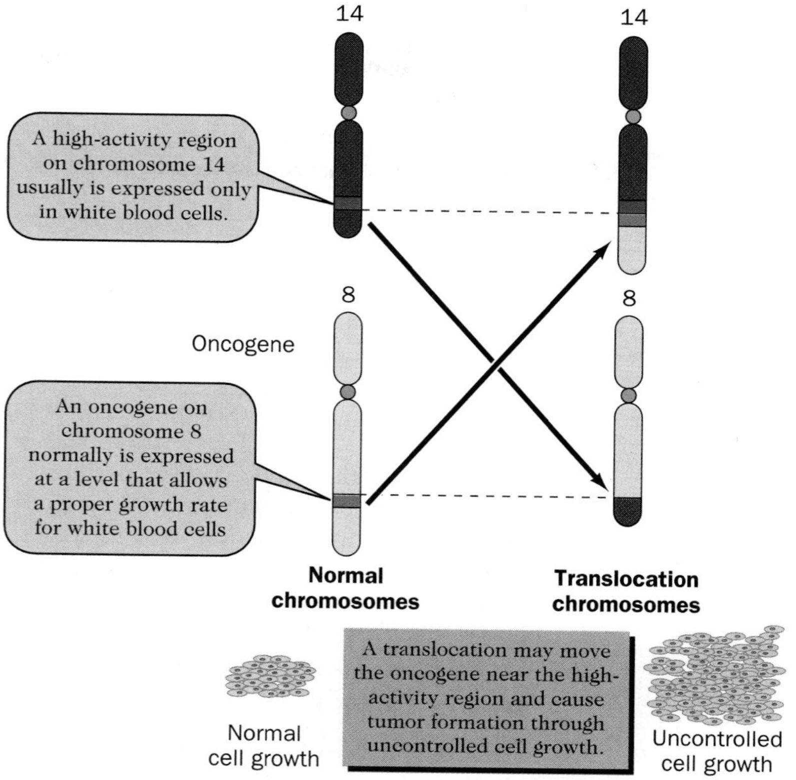
14
14
A high-activity region on chromosome 14 usually is expressed only in white blood cells.
8
8
Oncogene
An oncogene on chromosome 8 normally is expressed at a level that allows a proper growth rate for white blood cells
Normal chromosomes
Translocation chromosomes
Normal cell growth
A translocation may move the oncogene near the high-activity region and cause tumor formation through uncontrolled cell growth.
Uncontrolled cell growth

FIGURE 19.8 Activation of an oncogene by translocation.

A **duplication** results when one portion of a chromosome is repeated, resulting in extra copies of certain genes. **Translocation** occurs when nonhomologous chromosomes trade portions and groups of genes are moved to unusual locations. Translocation may increase the risk of developing certain cancers by locating potential oncogenes next to other genes which enhance their activity (Figure 19.8). An **inversion** results when a section of a chromosome separates and reattaches in a reversed direction. A **deletion** is a complete loss of a chromosome segment.

A deletion from chromosome 5 causes the condition called *cri du chat* ("cry of the cat"). The most striking features include a small head with widely spaced eyes, severe mental impairment, and a mewing cry that sounds like that of a distressed kitten. This disorder is rare, occurring in only 1 in 50,000 births.

Deletions and duplications have more severe and damaging effects than do translocations and inversions. They somewhat duplicate the conditions of trisomy or monosomy, causing the presence of too many or too few genes.

> Segments of chromosomes may be accidentally altered by duplication, translocation, inversion, and deletion.

Human Genes

Genes are the fundamental units of heredity that are passed from generation to generation. Genes reside on chromosomes, and each gene has a specific location called a **locus** on a particular chromosome (Fig-

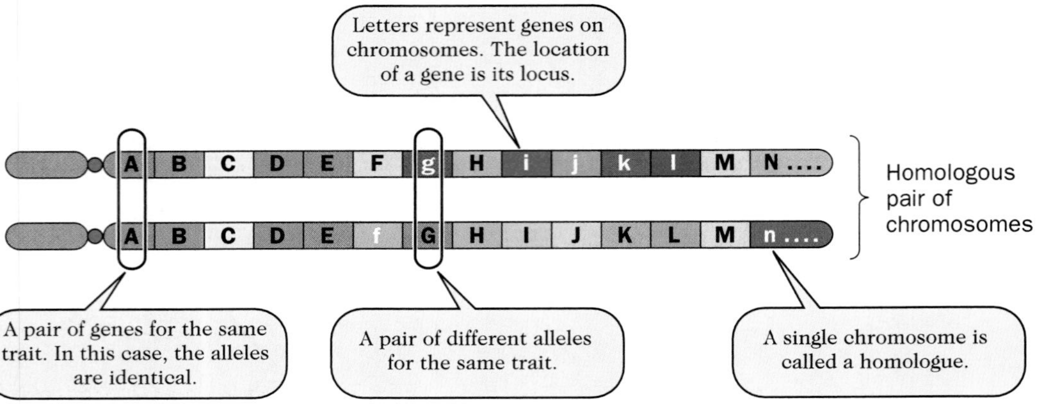
Letters represent genes on chromosomes. The location of a gene is its locus.
A B C D E F g H i j k l M N....
A B C D E f G H I J K L M n....
Homologous pair of chromosomes
A pair of genes for the same trait. In this case, the alleles are identical.
A pair of different alleles for the same trait.
A single chromosome is called a homologue.

FIGURE 19.9 Chromosome pairs and their genes.

ure 19.9). In this section, we will explore the nature of paired genes, how they interact, and how they function. We will see how changes in a gene can alter the development and the functioning of the adult body.

Genes Come in Pairs on Homologous Chromosomes

In diploid cells, homologous chromosomes come in pairs and so do genes, with one gene on each homologue. Each pair of genes controls a particular trait through the synthesis of a specific protein. We can symbolize a gene by a letter, for example, A, and a pair of genes by two letters, such as AA. Genes exist in different forms called **alleles** (Figure 19.9). For example, A, a, a′, and a″ all symbolize alleles of the same gene. Each allele is responsible for a different phenotypic expression of a trait. Although three or more alleles for a particular gene may exist in the human population, only two alleles for each trait can be present in the cells of a normal individual. The alleles for the same trait occupy the same locus, one on the maternal homologue and one on the paternal homologue.

The pair of alleles present may be identical or different. If they are identical, as in AA and aa, you are said to be **homozygous** for the trait controlled by the gene. If they are different, as in Aa, Aa′, and aa′, you are said to be **heterozygous.** The different alleles have different relationships to each other. Some alleles are dominant, while others are recessive.

Dominant and recessive alleles. A **dominant allele,** symbolized by a capital letter (A), has its phenotype expressed in either the homozygous or the heterozygous condition (AA or Aa). **Recessive alleles** are masked by the presence of a dominant allele, and their phenotypes are expressed only in the homozygous condition (aa). For example, a pointed hairline on the forehead ("widow's peak") is a trait that is due to a dominant allele (W). A straight or curved hairline results from a recessive allele (w). If you have the latter phenotype, you must be homozygous for the recessive gene, ww. If you have a widow's peak, you may be either heterozygous or homozygous for the dominant allele, Ww or WW. Table 19.1 lists some common dominant and recessive traits found in humans.

> Alleles are the different forms of a gene. Dominant alleles are expressed when homozygous or heterozygous. Recessive alleles are masked by dominant alleles and expressed only when homozygous.

The Functions of Genes: Structural Genes and Regulatory Genes

A gene is a region of DNA. The nucleotide sequence of this region encodes the information needed for the synthesis of a particular protein (Chapter 2). In the preceding chapters we encountered many of these proteins: enzymes, hormones, receptor proteins, muscle proteins, collagen, hemoglobin, and so on. The genes that code for these proteins are called **structural genes,** but there are other genes called **regulatory genes** whose proteins control the transcription of structural genes. Regulatory genes are very important in turning on and off certain

TABLE 19.1 Some Dominant and Recessive Human Genetic Traits

Dominant Trait	Recessive Trait
Normal color vision	Color blindness
Normal night vision	Night blindness
Long eyelashes	Short eyelashes
Dark hair	Light hair
Premature baldness (male)	Normal hair
Normal pigmentation	Albinism
Dimples in cheeks	No dimples in cheeks
Free earlobes	Attached earlobes
Ability to curl tongue	Inability to curl tongue
Convex nose bridge	Concave or straight nose bridge
Achondroplasia (dwarfism)	Normal height
Polydactylism (extra fingers or toes)	Normal fingers or toes
Normal arches in feet	Flat feet
Normal sugar metabolism	Diabetes mellitus
Blood groups A, B, AB	Blood group O
Normal red blood cells	Sickle-cell trait
Normal blood clotting	Hemophilia
Rh antigen (Rh positive)	No Rh antigen (Rh negative)
Migraine headaches	No migraine headaches
Widow's peak hairline	Straight hairline
Wet earwax	Dry earwax
Broad lips	Thin lips
Syndactyly (webbing of fingers or toes)	No webbing of fingers or toes

genes during development and in responding to changing conditions in an adult.

One way regulatory genes operate is by coding for a **regulatory protein** which binds to DNA, preventing the transcription of structural genes (Figure 19.10). When this happens, the structural gene is turned off. The regulatory protein also may be capable of binding with other types of molecules instead of with DNA. If these molecules enter a cell they bind to the regulatory protein, preventing its bonding to DNA. When this happens, the structural gene is free to be transcribed into mRNA and the mRNA can be translated into a protein. The gene is "turned on."

For example, steroid hormones function to initiate gene activity by binding with the protein product of a regulatory gene (Chapter 16). The regulation of gene activity is essential to the processes of development and birth described in Chapter 18. Recall how increased estrogen secretion late in pregnancy causes the synthesis of protein receptors for oxytocin on cells of the myometrium.

Although all body cells arise by mitosis and have identical genes, some of these genes are active and others are inactive. A muscle cell develops and functions because its genes for the proteins actin and myosin are activated. These genes are not active in cells that perform other functions. The orchestrated regulation of genes at specific times and places contributes significantly to normal development.

> Structural genes code for the synthesis of protein molecules. Regulatory genes control the transcription of structural genes, thus regulating protein synthesis.

Abnormal Genes: Mutations

When you consider all the biological functions of proteins, it is clear that their synthesis must occur with great accuracy. A mistake in the amino acid sequence of a protein can alter the way that particular protein functions. If a mistake occurs during the synthesis of a single protein molecule, the consequences are not very great, because each protein molecule has a limited life and will soon wear out and be replaced.

However, if a mistake occurs in a gene, the consequences are greater. All the mRNA molecules transcribed from that gene will copy the mistake. When the mRNA is translated during protein synthesis, all the protein molecules produced will be altered. A change in a gene that produces altered proteins and an abnormal phenotype is called a **mutation.** If a mutation occurs in gametes, the alter-

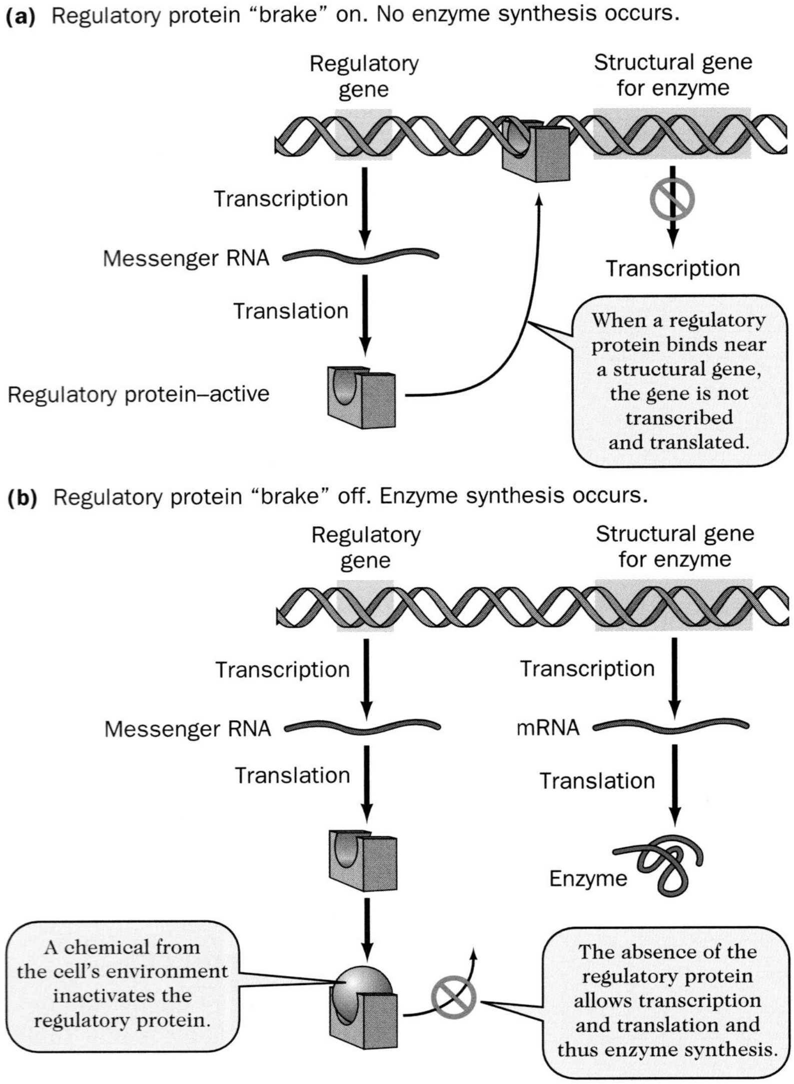
(a) Regulatory protein "brake" on. No enzyme synthesis occurs.
Regulatory gene
Structural gene for enzyme
Transcription
Messenger RNA
Translation
Regulatory protein–active
Transcription
When a regulatory protein binds near a structural gene, the gene is not transcribed and translated.
(b) Regulatory protein "brake" off. Enzyme synthesis occurs.
Regulatory gene
Structural gene for enzyme
Transcription
Messenger RNA
Translation
Transcription
mRNA
Translation
Enzyme
A chemical from the cell's environment inactivates the regulatory protein.
The absence of the regulatory protein allows transcription and translation and thus enzyme synthesis.

FIGURE 19.10 Regulation of structural genes allows a cell to control its metabolism by regulating the abundance of its enzymes.

ation can be passed on to children and future generations (Table 19.2). Two examples can illustrate the impact of mutations.

Sickle-cell anemia. *Sickle-cell anemia* is a disease of the red blood cells that is caused by a mutation in one of the structural genes for hemoglobin. In the United States, about 150 in every 100,000 children of African descent suffer from sickle-cell anemia. The defective gene causes the substitution of an incorrect amino acid in each of the two beta proteins of the hemoglobin molecule.

As the blood carries red cells to active body tissues where there is low oxygen (O_2) concentration, the defective hemoglobin causes the red blood cells to change shape, forming an unusual "sickle shape" (Figure 19.11). These abnormally shaped cells may fragment or clump together, blocking capillaries and reducing blood flow and O_2 delivery to muscles and all the internal organs. Without adequate O_2, all these organs are damaged. There is no cure, but an acute crisis may be treated with a blood transfusion to temporarily replace the sickled cells with normal red blood cells.

Normal beta proteins are produced by an allele which we'll designate S. People who are homozygous for this dominant allele, SS, have normal hemoglobin and show no signs of the disease even during strenuous activity. The full effects of the

TABLE 19.2 The Probabilities of Hereditary Diseases in Particular Ethnic Groups and Nationalities

Hereditary Disorder	Ethnic Group with Highest Risk	Probability That Individual Is a Carrier (Heterozygous)	Probability That Individual's Child Will Inherit Disease
Sickle-cell anemia	African Americans	1 in 10	About 1 in 400
Beta-thalassemia	Italian Americans	1 in 10	About 1 in 400
	Greek Americans		
Tay-Sachs disease	Jews (Ashkenazim)	1 in 30	About 1 in 4000
Adult lactose intolerance	Asians	Almost all	Nearly 100%
	African Americans	Most	About 7 in 10
Phenylketonuria	No ethnic differential	1 in 80	About 1 in 20,000
Cystic fibrosis	No ethnic differential	1 in 25	About 1 in 2500

FIGURE 19.11 Sickle-cell anemia.

anemia are seen only in people who are homozygous for the recessive allele, ss.

Heterozygous individuals, Ss, have a milder form of the disorder called *sickle-cell trait.* They produce both the normal beta protein (from allele S) and the abnormal beta protein (from the mutant allele s). These persons usually don't show symptoms unless they experience O_2 depletion at high altitudes or as a result of strenuous sustained exercise. However, under certain conditions, there are actually some benefits to the heterozygous state. If you are heterozygotic, you are less likely to become infected by the protozoan that causes malaria. Worldwide, malaria is a major disease affecting 300 million people, primarily in the tropics. Thus, for the entire human population, the benefits of the heterozygous state are very important.

A gene for cancer. Recently scientists have discovered a mutation in a regulatory gene associated with a form of cancer called *hereditary nonpolyposis colorectal cancer* (*HNPCC*). This regulatory gene is located on chromosome 2. People who inherit the mutant gene for HNPCC have an 80 percent chance of developing colon cancer as well as stomach and uterine cancer, usually before age 50.

The normal form of the gene codes for a protein that identifies incorrectly replicated DNA so that it can be repaired. The mutant form of the gene permits these mistakes to go unrecognized and uncorrected. The mistakes in the nucleotide sequence of DNA produce altered proteins that transform normal cells into cancer cells. A blood test is being developed to allow screening of individuals in affected families and thus early medical intervention and treatment.

We all carry mutations. Spontaneous mistakes in DNA structure occur at a very low rate, only one error in every billion nucleotides, yet we probably all carry a half dozen or so mutations that have accumulated in our family lines over time. The severity of their impact on our healthy functioning depends on which genes and proteins are affected.

The rate of mutation is increased by exposure to radiation and harmful chemicals in our living or working environment. Not all mutations are bad; some result in new, advantageous capabilities and may in part lead to the gradual emergence of a new species. That is the basis of evolution.

> A mutation is an alteration in the nucleotide sequence of a gene, coding for an abnormal protein. Mutations may occur spontaneously or result from exposure to environmental factors. They may cause disease or provide the raw material for evolutionary change.

Genes and the Environment

Factors from outside the cell may influence the expression of genes in an individual, modifying the phenotype that is produced by a specific genotype. In a sense, a particular genotype establishes the potential limits for a given trait. The phenotype that is actually expressed may fall short of that potential if unfavorable conditions are encountered during development. For instance, you may have inherited genes for tallness, but unless all the essential amino acids, minerals, and vitamins were available from the environment (your nutrition) during your crucial growth stages, you may not be tall. Additional examples are found in fetal exposure to alcohol and antibiotics (Chapter 18). Both of these substances, if present at certain times in sufficient quantities, can alter the normal expression of genes, producing defects in development.

The Human Genome Project

We think the human genome may contain as many as 100,000 genes. To date, over 4000 genes have been identified and more than 1500 have been mapped to their locations on specific chromosomes. Begun in 1989, an intensive worldwide scientific effort called the **Human Genome**

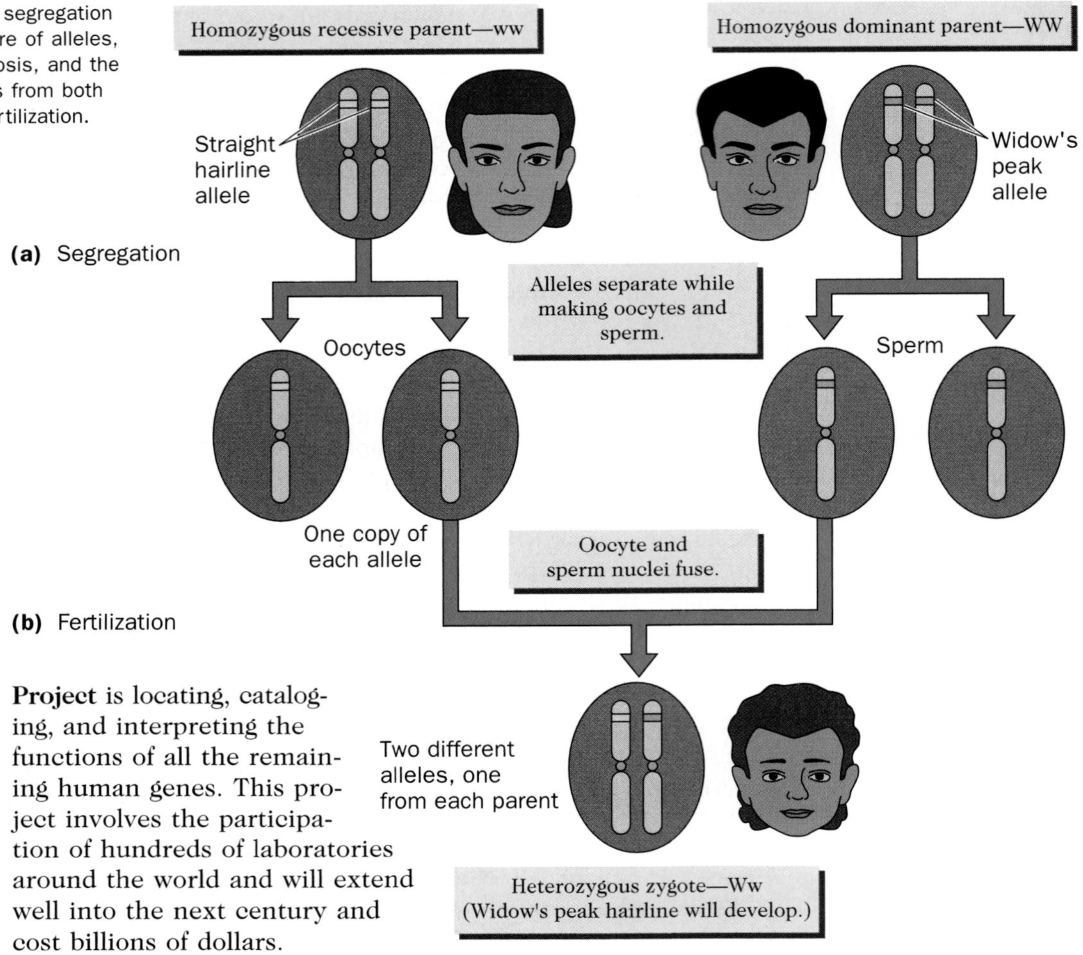
Homozygous recessive parent—ww
Homozygous dominant parent—WW
Straight hairline allele
Widow's peak allele
(a) Segregation
Alleles separate while making oocytes and sperm.
Oocytes
Sperm
One copy of each allele
Oocyte and sperm nuclei fuse.
(b) Fertilization
Two different alleles, one from each parent
Heterozygous zygote—Ww (Widow's peak hairline will develop.)

FIGURE 19.12 The law of segregation describes the discrete nature of alleles, their separation during meiosis, and the recombination of the alleles from both parents in a zygote after fertilization.

Project is locating, cataloging, and interpreting the functions of all the remaining human genes. This project involves the participation of hundreds of laboratories around the world and will extend well into the next century and cost billions of dollars.

Considerable success has already been reported in locating genes responsible for human disorders. A gene designated BRCA1 that is associated with about 5 percent of inherited breast cancers has been located on chromosome 17. A different region on chromosome 17 has been associated with psoriasis, an inflammatory skin disorder. A mutant form of a gene on chromosome 4 has been linked to abnormal bone and cartilage growth and a form of dwarfism.

A defective gene on chromosome 9 is suspected of causing malignant melanoma, a cancer of the skin's pigment-forming cells. A similar mutant tumor suppressor gene on chromosome 3 has been linked to two-thirds of all kidney cancers. A gene on chromosome 12 may determine which women suffer osteoporosis by producing defective vitamin D receptors. Researchers also have reported a gene in mice that is associated with obesity and have suggested that a human counterpart may exist.

As more of this kind of information accumulates, the potential for intervention in the individual genotype will be enormous. The opportunity for the correction of defective genes and the prevention of genetically based disorders will be greatly enhanced.

> Environmental factors can interact with the genome to alter or limit the expression of genes. The Human Genome Project is an effort to identify all the human genes.

Patterns of Inheritance

The basic rules of inheritance were first described by **Gregor Mendel**, an Augustinian monk who conducted breeding experiments with garden peas in the 1850s. Mendel knew nothing about mitosis, meiosis, chromosomes, or DNA. However, he concluded that the traits whose inheritance he studied in peas were controlled by a pair of "factors," what we now call genes. He realized that some genes are dominant and others are recessive because he observed traits that skipped a generation and then reemerged. This could happen only if the genes were discrete and did not blend or mix. Mendel proposed two laws which have guided genetic research for more than 100 years. These laws do not apply to all forms of inheritance, but the exceptions only prove their value.

Mendel's Rules of Inheritance

Mendel's first rule is the **law of segregation.** This law states that pairs of alleles separate from each other during the formation of gametes (Figure 19.12). We now know that this happens during anaphase I of meiosis when the paired homologous chromosomes separate. The different alleles of a pair have equal chances of appearing in the gametes.

Mendel's second rule is the **law of independent assortment.** This law states that when two distinct traits are inherited, the distribution of alleles for one trait into oocytes or sperm does not affect the distribution of the alleles for other traits (Figure 19.13). We now know that this rule applies only if the genes are not located on the same chromosome. Today we are aware that if genes are located near one another on the same chromosome, they are **linked.** Linked genes are inherited together and do not assort independently.

> A pair of alleles separate from one another as their homologous chromosomes separate during gamete formation. Alleles located on nonhomologous chromosomes are inherited independently of one another.

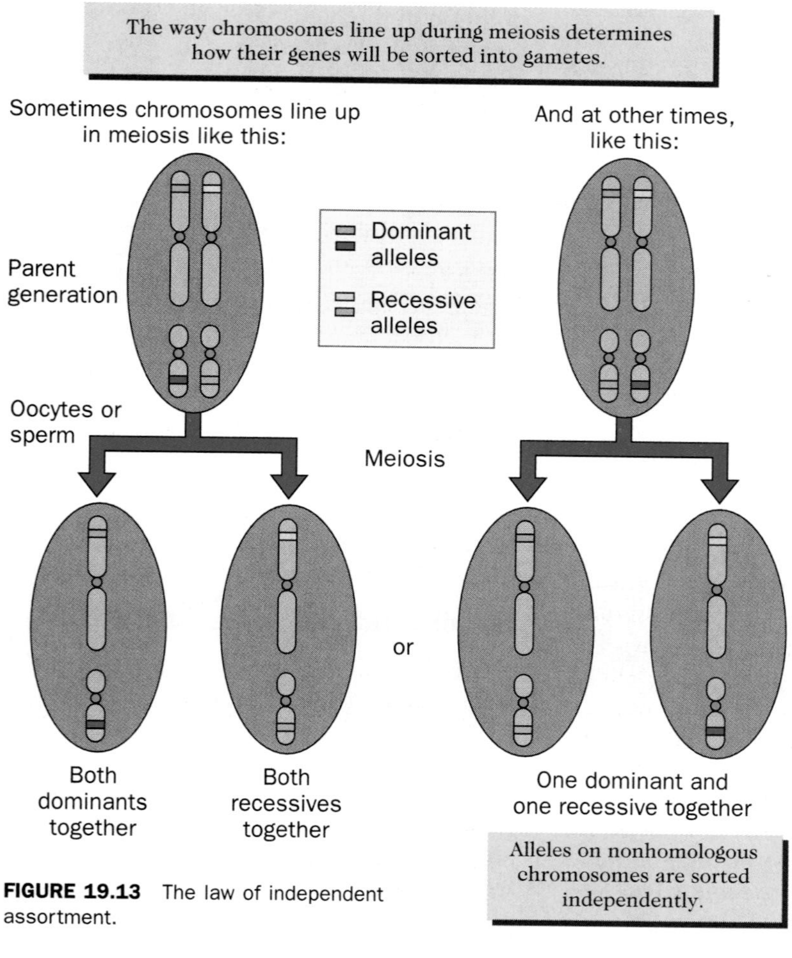
The way chromosomes line up during meiosis determines how their genes will be sorted into gametes.
Sometimes chromosomes line up in meiosis like this:
And at other times, like this:
Parent generation
Dominant alleles
Recessive alleles
Oocytes or sperm
Meiosis
or
Both dominants together
Both recessives together
One dominant and one recessive together
Alleles on nonhomologous chromosomes are sorted independently.

FIGURE 19.13 The law of independent assortment.

Using Punnett Squares to Predict Genotypes and Phenotypes

A Punnett square can be used to illustrate the laws of segregation and independent assortment and to determine the probability that certain genotypes will appear in children. For example, let us use the Punnett square to investigate the inheritance of attached or free earlobes.

Free earlobes are due to a dominant allele we'll designate E. Attached earlobes are recessive and will be represented by e. If you have attached earlobes, you must be ee, or homozygous recessive. However, if your earlobes hang freely, you may be either homozygous dominant, EE, or heterozygous, Ee.

Figure 19.14 shows earlobe inheritance expressed in a Punnett square. The sides of the square show the alleles present in the parental gametes, and the boxes in the square represent all the possible allele combinations in the zygotes. By counting the number of boxes with the same genotypes or phenotypes, we can determine the ratios and calculate probabilities. The ratio of free earlobes to attached earlobes is 3:1, and the probability of having an attached earlobe child when

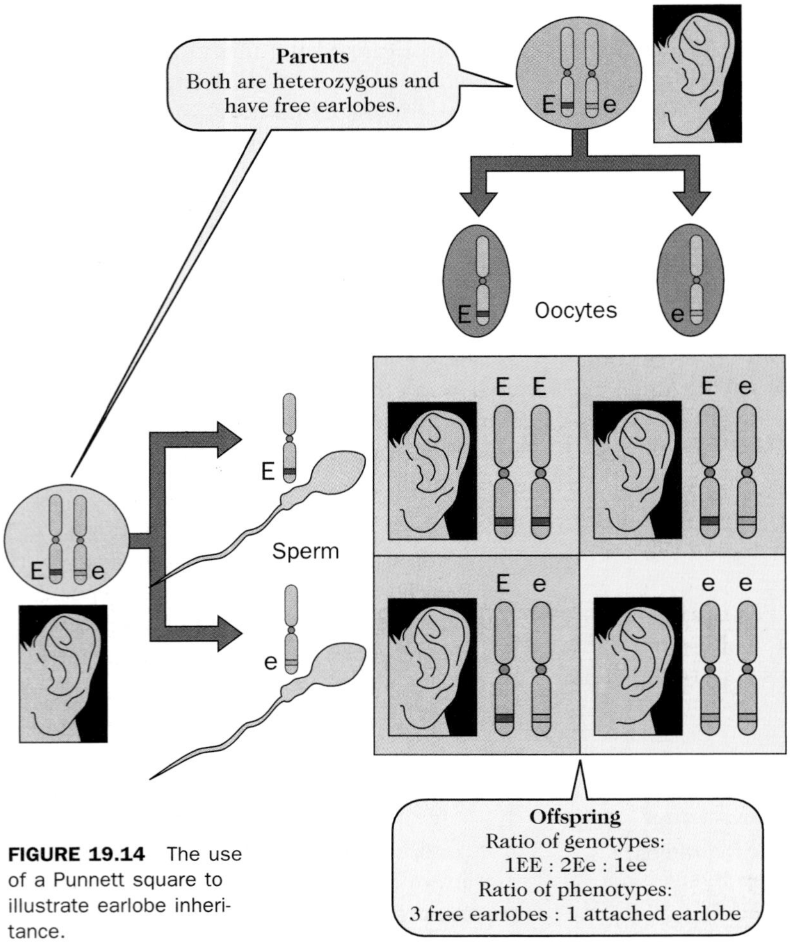
Parents
Both are heterozygous and have free earlobes.
E e
Oocytes
E
e
Sperm
E E
E e
E e
e e
Offspring
Ratio of genotypes:
1EE : 2Ee : 1ee
Ratio of phenotypes:
3 free earlobes : 1 attached earlobe

FIGURE 19.14 The use of a Punnett square to illustrate earlobe inheritance.

two heterozygotes mate is 25 percent, or one in four, as you will learn in the next section.

> A Punnett square presents all the possible genotypes and phenotypes resulting from a mating between known genotypes. From a Punnett square, ratios can be determined and probabilities can be calculated for both genotypes and phenotypes.

Probability: The Chances a Child Will Have a Certain Phenotype

Probability is a mathematical expression (fraction or percentage) of the chance that a particular event will occur. In genetics, probability expresses the chance that a particular genotype or phenotype will appear in a child of parents who have a known or suspected genotype.

To determine the probability that a child will receive a particular combination of alleles, we multiply the probabilities of inheriting each individual allele. When both parents are heterozygous, Ee, for example, the chance of an E allele appearing in a parental gamete is 1 in 2, or 1/2, which can also be expressed 1:2 (Figure 19.14). The chance of an e appearing is also 1/2. The probability of a child receiving the genotype EE is $1/2 \times 1/2 = 1/4$. The probability of inheriting the homozygous recessive condition, ee, is also $1/2 \times 1/2 = 1/4$. What about heterozygous children?

There are two ways in which a heterozygous child can be produced (Figure 19.14), and so the probability of an Ee combination is $1/2 \times 1/2 = 1/4$, $2 \times 1/4 = 1/2$. In this mating, there are three potential genotypes, and their total probability of occurrence should equal 1 $(1/4 + 1/4 + 1/2 = 1)$.

We can determine the probability of a child obtaining a particular phenotype by counting all the genotypes in a Punnett square that produce that phenotype. In Figure 19.14, out of four potential allele combinations, three genotypes yield the dominant phenotype (EE plus two Ee). There is thus a 3/4 chance of a child having free earlobes if his or her parents are heterozygous for that condition. There is only a 1 in 4, or 1/4, chance of a child receiving attached earlobes.

> Probability expresses the chance that a particular genotype or phenotype will appear in a child of parents with a known or suspected genotype.

Monohybrid and Dihybrid Matings

The mating illustrated in Figure 19.14 takes into consideration only one pair of alleles in the parents and their children and is called a *monohybrid mating.* When two traits are considered in a single mating, it is called a *dihybrid mating*. Figure 19.15 presents a dihybrid (two traits) mating involving dimpled cheeks and tongue-rolling ability when both par-

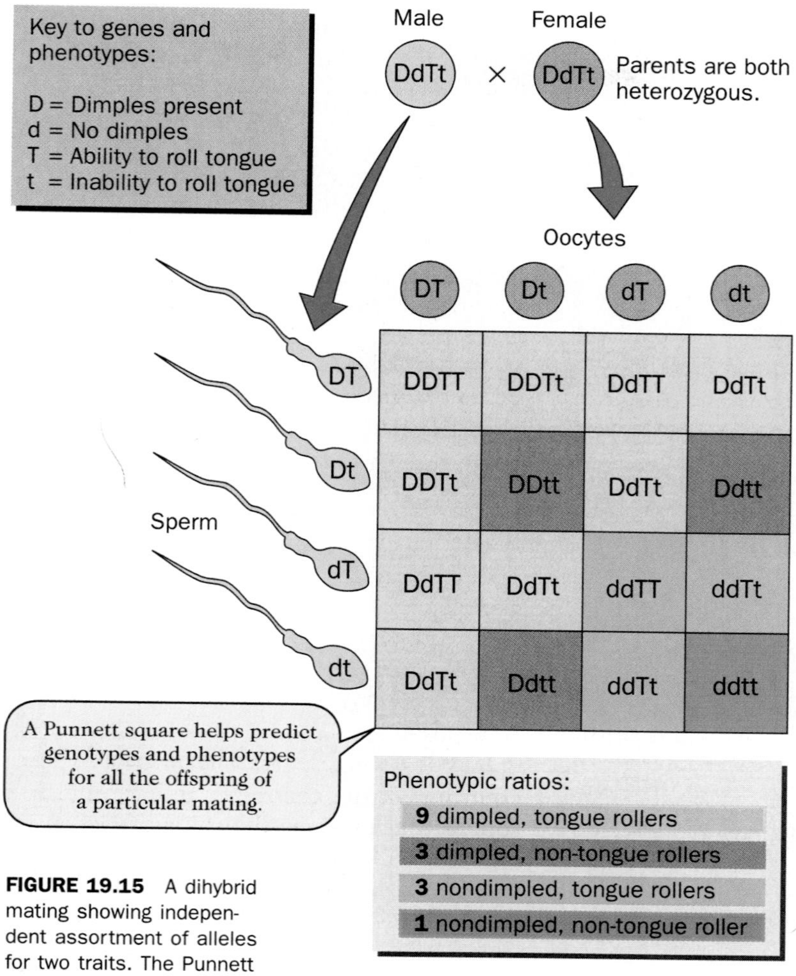
Key to genes and phenotypes:
D = Dimples present
d = No dimples
T = Ability to roll tongue
t = Inability to roll tongue
Male
Female
DdTt × DdTt
Parents are both heterozygous.
Oocytes
DT Dt dT dt
Sperm
DT Dt dT dt
DDTT DDTt DdTT DdTt
DDTt DDtt DdTt Ddtt
DdTT DdTt ddTT ddTt
DdTt Ddtt ddTt ddtt
A Punnett square helps predict genotypes and phenotypes for all the offspring of a particular mating.
Phenotypic ratios:
9 dimpled, tongue rollers
3 dimpled, non-tongue rollers
3 nondimpled, tongue rollers
1 nondimpled, non-tongue roller

FIGURE 19.15 A dihybrid mating showing independent assortment of alleles for two traits. The Punnett square shows all the possible genotypes and phenotypes for the offspring.

ents are heterozygous. A dihybrid mating like this is a good illustration of Mendel's law of independent assortment. The dimple and tongue-rolling genes are inherited independently of each other.

Determining probability for a dihybrid mating is very similar to determining probability for a monohybrid mating: You multiply the probabilities for the separate alleles to determine the probability of alleles appearing together. For example, when one trait is considered, the probability of heterozygote parents having a heterozygote child is 1/2. (Figure this out using a Punnett square.) When two traits are considered, as in Figure 19.15, the probability of having a heterozygote child is $1/2 \times 1/2 = 1/4$ (or 4 in 16 in the example). The chance of obtaining a double homozygous recessive or dominant child is $1/4 \times 1/4 = 1/16$.

For practice, examine Figure 19.15 again and list all the genotypes that result in a dimpled child who is not able to roll the tongue. How many are there? What is the probability that such a child will result from a mating of double heterozygous parents? (Answers: DDtt, Ddtt; 3; 3/16.)

A monohybrid mating deals with the alleles for a single trait. A dihybrid mating is concerned with the allele combinations for two traits.

Testcrosses and Pedigrees: Which Genes Are Present?

It is often difficult or impossible to determine an individual's genotype when only the phenotype can be observed directly. We cannot test directly for genotype, as we can for phenotype. Therefore, it is difficult to distinguish between homozygous dominant and heterozygous individuals.

With agricultural or research animals, this difficulty can be overcome by mating an individual with an unknown genotype to a known homozygous recessive individual. Such a mating is called a **testcross.** Analysis of the progeny from a testcross permits you to determine the unknown genotype (Figure 19.16a). However, for ethical and scientific reasons, this method cannot be applied to humans. From a scientific point of view, humans do not produce enough offspring from a single mating for conclusions to be drawn. For these reasons, pedigree analysis is used with humans.

Pedigree analysis. It has long been obvious that certain disorders and diseases tend to "run in families." By tracing the inheritance of those conditions through a family tree or pedigree, we can often better understand the nature of the defect. Until the more sophisticated tools of DNA technology were developed, family trees offered the only means of characterizing the inheritance of defective genes.

A **pedigree** is a family history extending over several generations. It identifies the genders, mating relationships, and specific phenotypic traits of each individual. By ex-

FIGURE 19.16 Determining genotypes. (a) A testcross. (b) A pedigree. (c) Pedigree practice. Can you write in the appropriate genotype?

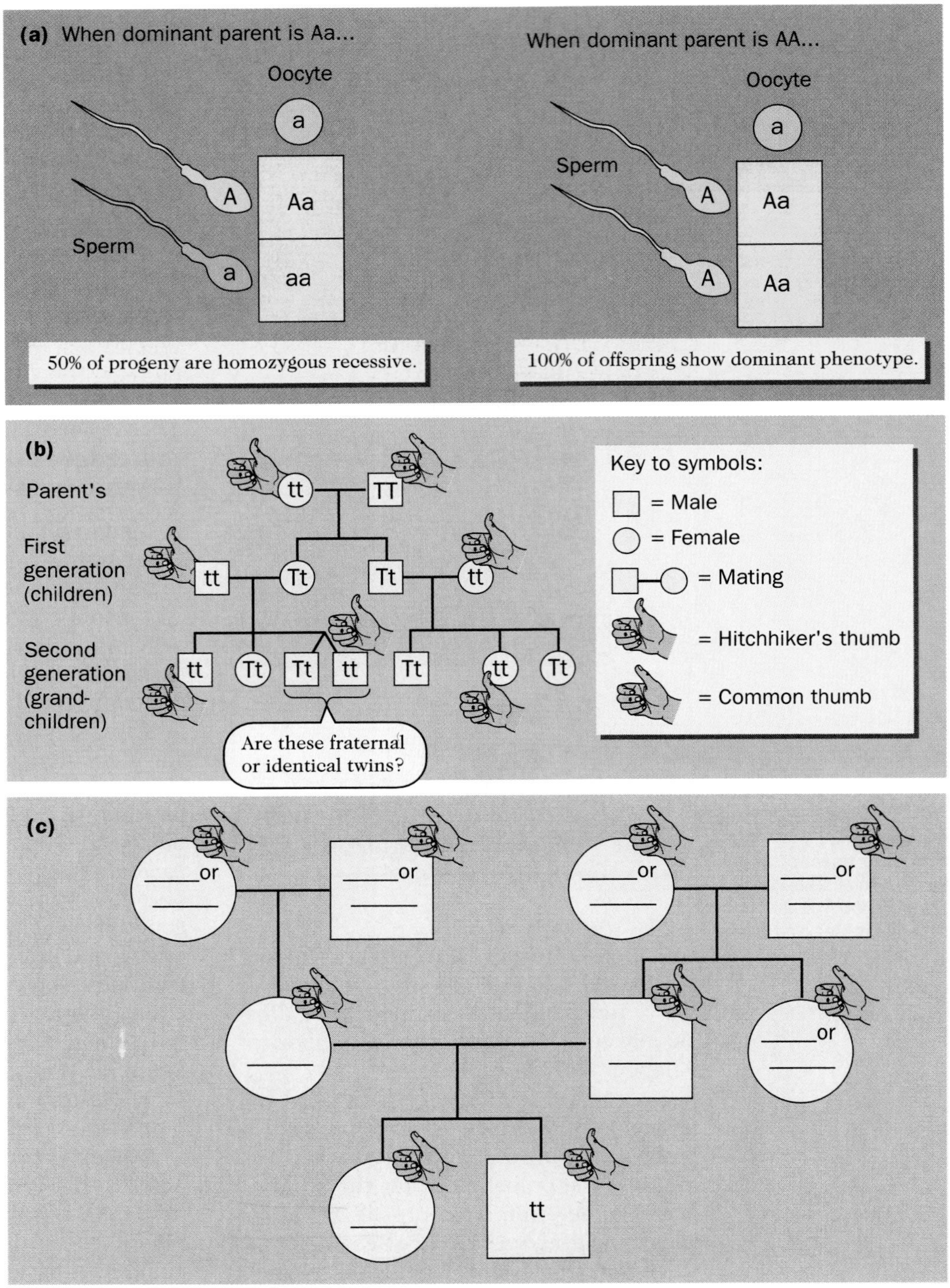
(a) When dominant parent is Aa...
Oocyte
a
A
Aa
Sperm
a
aa
50% of progeny are homozygous recessive.
When dominant parent is AA...
Oocyte
a
Sperm
A
Aa
A
Aa
100% of offspring show dominant phenotype.
(b)
Parent's
tt
TT
First generation (children)
tt
Tt
Tt
tt
Second generation (grand-children)
tt
Tt
Tt
tt
Tt
tt
Tt
Are these fraternal or identical twins?
Key to symbols:
= Male
= Female
= Mating
= Hitchhiker's thumb
= Common thumb
(c)
or
or
or
or
or
tt

amining several generations in a pedigree, a geneticist often can determine the genotype of the individuals (Figure 19.16b). From pedigrees it is also possible to decide whether a trait is inherited as a dominant or recessive allele and whether the allele is carried on the X chromosome. When this has been established, it is possible to predict the probability that a particular marriage will produce an affected child.

Testcrosses are used to determine homozygous dominant or heterozygous animals. In humans, pedigree analysis can determine the dominant, recessive, or sex-linked nature of an allele.

Polygenic Inheritance: Many Genes, One Trait

So far we have been discussing alleles that occupy the same locus on a chromosome. However, some human traits, such as skin color and hair and eye color, are controlled by several separate genes at different loci. For skin color, there are three and possibly four separate genes, each with two alleles. These genes control the amount and distribution of melanin pigment in the skin (Chapter 4). The effects of these alleles are additive, and this type of inheritance is said to be **polygenic** (multiple gene) inheritance.

Figure 19.17 shows the inheritance of skin color over three generations, assuming three genes of control and using colored blocks to represent the combinations of alleles. The dominant alleles promote dark skin color, and the recessive alleles are expressed as light coloration. These are actually trihybrid crosses, and in the last generation shown there is only a 1:64 chance of a completely homozygous dominant or recessive individual.

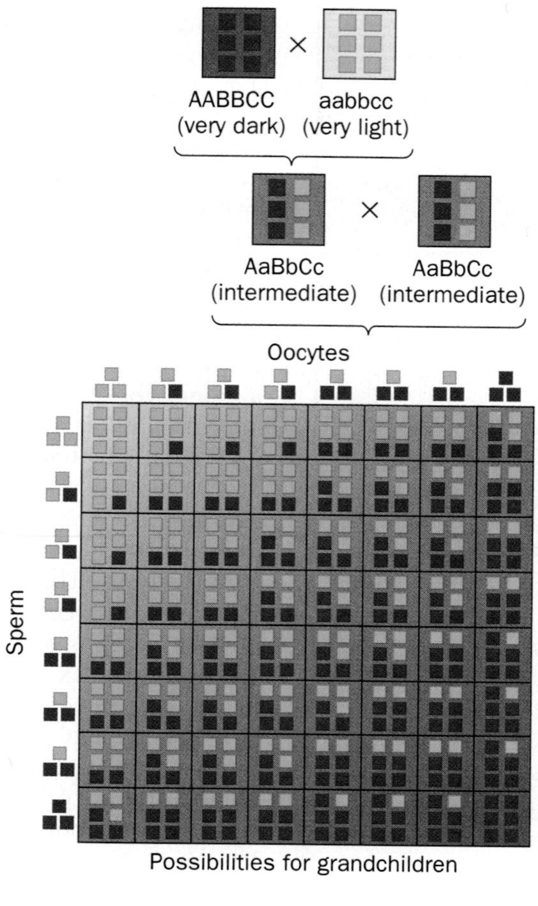
AABBCC
(very dark)
aabbcc
(very light)
AaBbCc
(intermediate)
AaBbCc
(intermediate)
Oocytes
Sperm
Possibilities for grandchildren

FIGURE 19.17 The polygenic inheritance of skin color.

Sex-Linked Inheritance: Traits That Travel with Gender

Sex-linked genes are genes carried on the sex chromosomes. When we speak of sex linkage, it is generally in reference to genes on the X chromosome, since there are only a few genes on the Y chromosome. These genes influence a wide variety of important traits, including vision, hearing, blood clotting, sweat gland function, muscle control, and formation of the hard palate.

Since a woman has two X chromosomes, her inheritance follows the typical dominant and recessive pattern we have already discussed. If she is heterozygous, the expression of a recessive allele is masked by that of the dominant allele. In contrast, a male has only one X chromosome and the Y chromosome that determines his gender.

A male does not have an opportunity to be heterozygous for genes carried on the X chromosome, which are called **X-linked genes.** He expresses whichever allele, dominant or recessive, he inherited on the X chromosome from his mother. A male shows the X-linked traits associated with his mother's family line, not those of his father's line. He will display X-linked traits observed in the maternal grandfather, for example, not in the paternal grandfather. We will discuss some sex-linked disorders that may be inherited in this fashion later in this chapter.

> A male expresses the traits of whichever alleles are present on his single X chromosome. He cannot be heterozygous for those traits, as can a female.

Genetic Disorders

Much of the reason for studying the human genome centers on the need to understand, treat, and possibly correct genetic disorders associated with defective, missing, or overly abundant genes. There are over 5000 known genetic disorders.

Autosomal Dominant Disorders

Autosomal dominant disorders can be passed on to children when only one parent exhibits the symptoms of a condition (Figure 19.18a). Since many autosomal dominant disorders are not manifested until adulthood, an affected person often begins a family before being diagnosed.

Huntington's disease (*HD*) is an autosomal dominant neurological disorder that affects 1 in 10,000 adults in the United States. The symptoms usually do not occur until age 35 to 45. HD is caused by a mutation of a gene on chromosome 4 that apparently causes the production of an abnormal protein. This defective protein damages brain cells and leads to progressive loss of motor control and mental instability.

Another autosomal dominant disorder is *hypercholesterolemia* (*high-purr-**coal**-est-er-ol-**lee**-me-uh*), a condition in which abnormally high levels of cholesterol are produced, leading ultimately to heart disease. Individuals who are homozygous (HH) or heterozygous (Hh) for this condition often suffer heart attacks at an early age.

Neurofibromatosis (*NF*) is a disorder associated with benign tumors called *neurofibromas* which accumulate under the skin and in muscles. This disorder, which occurs in 1 in every 3000 people in the United States, is associated with a defective tumor suppressor gene on chromosome 17. Extreme forms of the disease lead to skeletal deformities, vision and hearing loss, and learning disabilities.

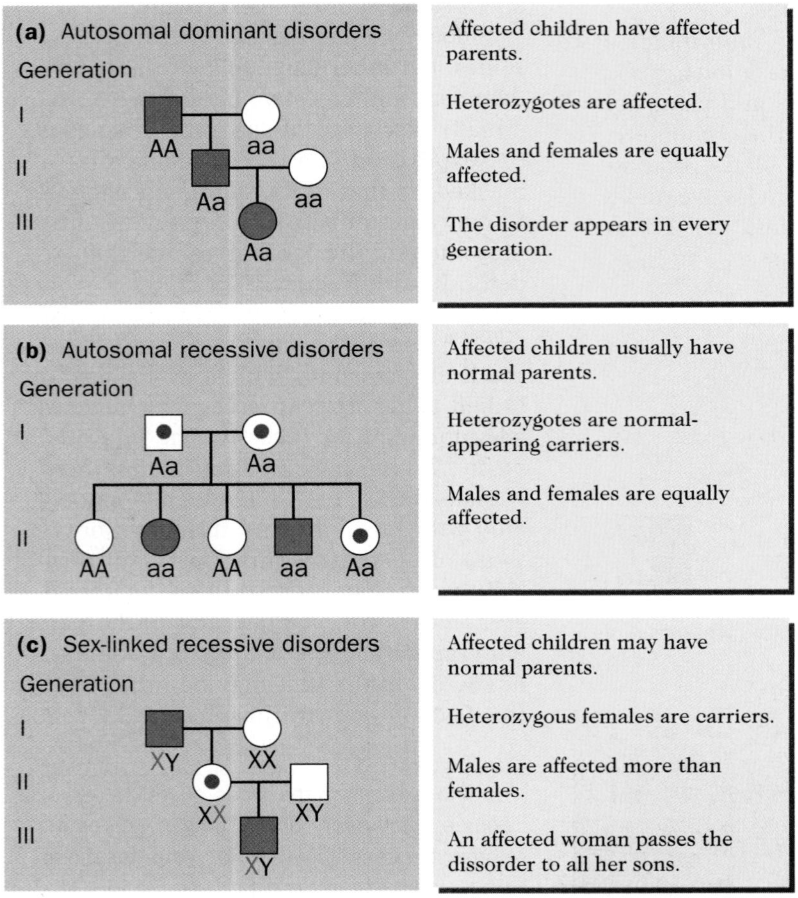

(a) Autosomal dominant disorders
Generation
I
II
III
AA
aa
Aa
aa
Aa
Affected children have affected parents.
Heterozygotes are affected.
Males and females are equally affected.
The disorder appears in every generation.
(b) Autosomal recessive disorders
Generation
I
II
Aa
Aa
AA aa AA aa Aa
Affected children usually have normal parents.
Heterozygotes are normal-appearing carriers.
Males and females are equally affected.
(c) Sex-linked recessive disorders
Generation
I
II
III
XY
XX
XX
XY
XY
Affected children may have normal parents.
Heterozygous females are carriers.
Males are affected more than females.
An affected woman passes the dissorder to all her sons.

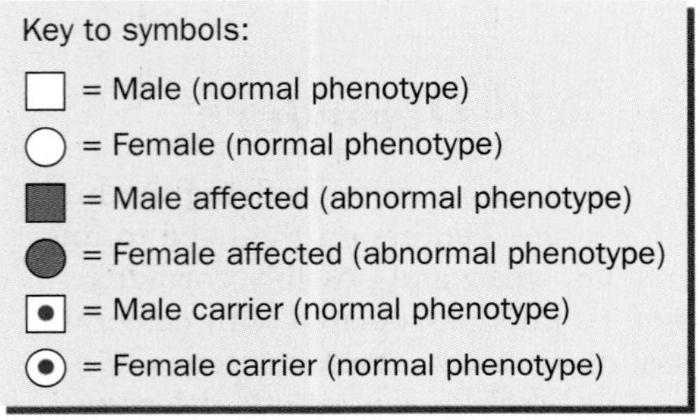

Key to symbols:
= Male (normal phenotype)
= Female (normal phenotype)
= Male affected (abnormal phenotype)
= Female affected (abnormal phenotype)
= Male carrier (normal phenotype)
= Female carrier (normal phenotype)

FIGURE 19.18 Inheritance pedigrees for common types of genetic disorders.

Autosomal Recessive Disorders

With the autosomal recessive pattern of inheritance, heterozygous parents with normal phenotypes each pass an abnormal allele to a child (Figure 19.18b). Each heterozygous parent is a "carrier" of a defective gene. In other words, two healthy parents can have an afflicted child. This pattern usually follows that of a monohybrid cross between heterozygotes and usually has a 1/4 probability of producing an affected child.

The most common autosomal recessive disorder is *cystic fibrosis* (*CF*), which occurs in 1 in every 2000 children in the United States. There are over 30,000 CF patients in the United States today. The defective gene is on chromosome 7, and new screening exams can help identify the nearly 1 in 20 adult carriers of the disorder. The mutant allele results in the production of excessively thick mucus that clogs respiratory passageways (see Chapter 10). An important new treatment utilizes a genetically engineered enzyme,

Pulmozyme, to clean up some of the debris in the lungs and thin the mucus secretions.

Phenylketonuria (*fen-nel-key-toe-**new**-ree-uh*), *PKU,* is caused by a defective gene on chromosome 12. Affected individuals are unable to synthesize an enzyme which metabolizes phenylalanine, an amino acid. Consequently, a toxic molecule, phenylketone, accumulates in the blood and damages brain cells. For most affected children and adults, avoidance of phenylalanine in the diet can help minimize the symptoms of the disorder. This is why a warning about phenylalanine content (found in the sweetening agent aspartame) is printed on the label of diet sodas containing aspartame.

Recessive disorders are more common when mates are selected from a small group of people or close relatives because there are more opportunities to generate homozygous recessives. Matings between close cousins, brothers and sisters, and even uncles and nieces are called *consanguineous matings* (*con-sang-**gwin**-ee-us*). Most cultures have strict taboos against such matings, possibly because of the resulting disorders.

FIGURE 19.19 X-linked inheritance of red-green color blindness.

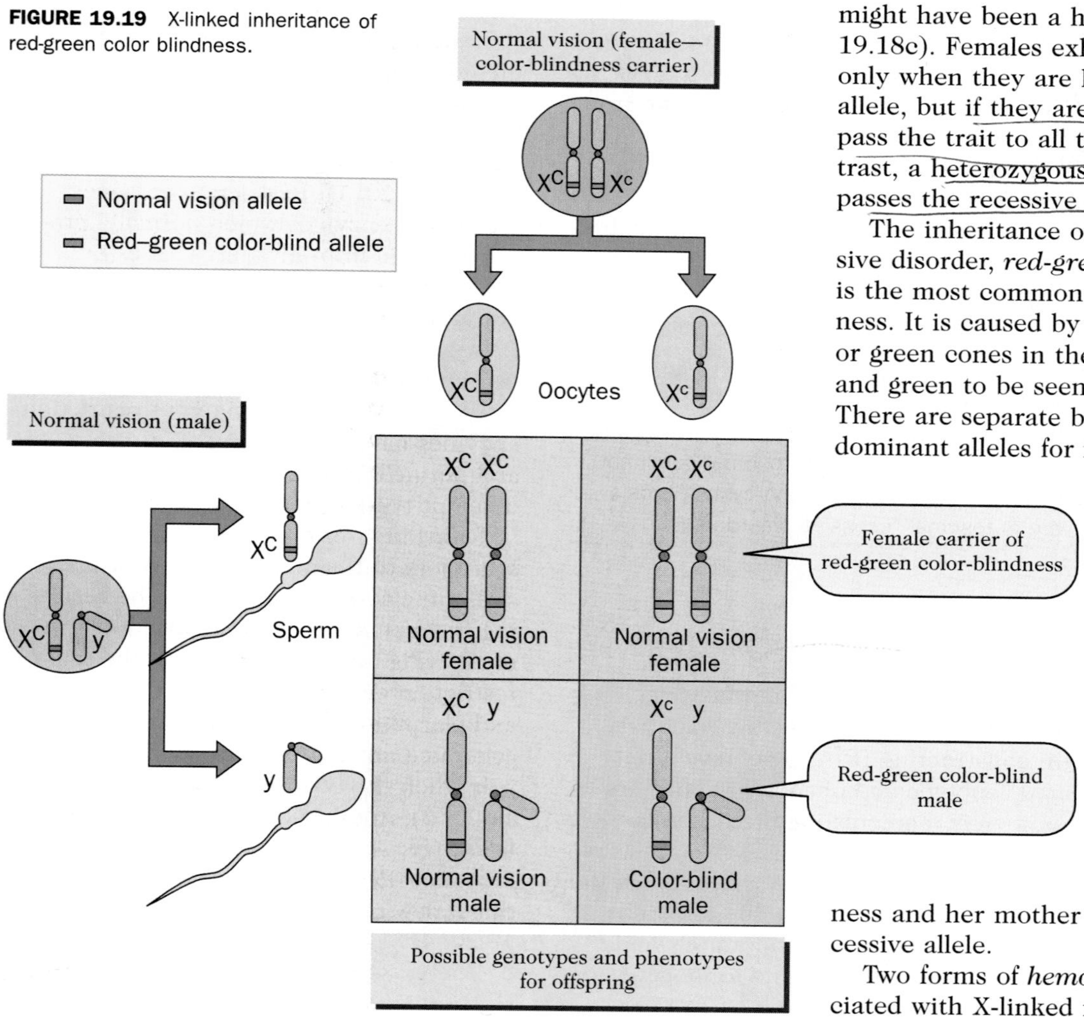
Normal vision (female—color-blindness carrier)
Normal vision allele
Red–green color-blind allele
X^C
X^c
Oocytes
Normal vision (male)
X^C
Sperm
y
X^C X^C
Normal vision female
X^C X^c
Normal vision female
Female carrier of red-green color-blindness
X^C y
Normal vision male
X^c y
Color-blind male
Red-green color-blind male
Possible genotypes and phenotypes for offspring

Sex-Linked Recessive Disorders

Males express all the genes on their X and Y chromosomes. As we noted earlier in this chapter, the X chromosome is loaded with genes but the Y chromosome has few genes. Other than testes formation, the only Y-linked trait identified so far is a condition of excessively hairy ears. A male affected by an X-linked disorder inherited it from his mother, who might have been a heterozygote (Figure 19.18c). Females exhibit X-linked traits only when they are homozygous for the allele, but if they are homozygous, they pass the trait to all their sons. By contrast, a heterozygous female theoretically passes the recessive gene to half her sons.

The inheritance of an X-linked recessive disorder, *red-green color blindness,* is the most common form of color blindness. It is caused by a deficiency of red or green cones in the retina, causing red and green to be seen as the same color. There are separate but closely linked dominant alleles for forming each type of cone and a recessive allele that prevents normal cone formation (Figure 19.19). About 8 percent of males and 0.5 percent of females in the United States are affected. A female will be affected only if her father suffered red-green color blindness and her mother also carried the recessive allele.

Two forms of *hemophilia* are associated with X-linked recessive genes. The defective genes cause a loss of clotting factors and potentially life-threatening bleeding. This disease rarely occurs in females, because until recently, affected males generally did not live long enough to reproduce and pass on the trait. Even if a woman did inherit a double dose of the affected alleles, she would have little chance of surviving puberty and her first menstrual periods. Today males suffering from hemophilia can receive treatment through the replacement of the missing clotting factors. In the 1970s and 1980s many hemophiliacs were exposed to the AIDS virus in the process of their treatment.

Other X-linked recessive disorders include cleft palate condition, night blindness, and a form of muscular dystrophy as well as *severe combined immunodeficiency disease* (*SCID*) (Chapter 13). SCID is caused by a defective gene which codes for part of a T-cell lymphocyte receptor for interleukin-2. Babies with SCID have a low T cell count and may die from minor infections. Researchers believe that SCID can be cured by inserting normal genes into lymphocyte-forming cells shortly after birth. Gene therapies such as this are among the most exciting developments in the rapidly blossoming field of DNA technology.

> Hemophilia, color blindness, and SCID are inherited as sex-linked disorders, while Huntington's disease is an autosomal dominant condition and both PKU and cystic fibrosis are autosomal recessive disorders.

DNA Technology

For the first time in human history, and not without controversy, scientists are constructing DNA to achieve a desired genotype and phenotype. They are not merely manipulating the genotype through selective breeding. New species of microorganisms are being assembled in the laboratory. It is even possible to patent a microbe created in the laboratory and own the commercial rights to its gene products.

DNA bearing the genes for desirable products is inserted into fast-growing cells which rapidly make large quantities of the substance. **DNA technology** was made possible by the invention of techniques for splicing together DNA from different sources, even different species.

Recombinant DNA and Gene Cloning: Cut, Paste, and Clone

DNA that is formed by combining genes or gene fragments from different kinds of cells is called **recombinant DNA.** The molecular tools used in constructing recombinant DNA are provided by bacteria. A common bacterium used in recombinant DNA research is *Escherichia coli* (abbreviated *E. coli*), an organism which naturally grows in the human intestine but can be easily cultivated in the laboratory.

In addition to the bacteria, the tools include plasmids and restriction enzymes. **Plasmids** are small loops of DNA in the bacterial cytoplasm, and they are separate and distinct from the bacterial genome (Figure 19.20, 1). **Restriction enzymes** are special enzymes which normally protect bacteria from invading viruses by chopping up their DNA. These enzymes are very useful in making recombinant DNA because they cut the DNA only at certain nucleotide sequences (Figure 19.20, 2). Hundreds of different restriction enzymes have been identified, isolated, and purified for commercial use in making different types of recombinant DNA.

Restriction enzymes are used by researchers to fragment any DNA at specific sites (nucleotide sequences). The resulting fragments are called **restriction fragments** (Figure 19.20, 2). When the same restriction enzyme is used on human DNA and plasmid DNA, the single-strand regions generated are complementary and base pair when the two DNAs are mixed (Figure 19.20, 3). Another bacterial enzyme, **DNA ligase** (***lye***-*gayce*), is used to bond fragments of DNA together, permanently inserting a new gene into a plasmid (Figure 19.20, 4).

The plasmid bearing the unusual gene (recombinant DNA) can be mixed with

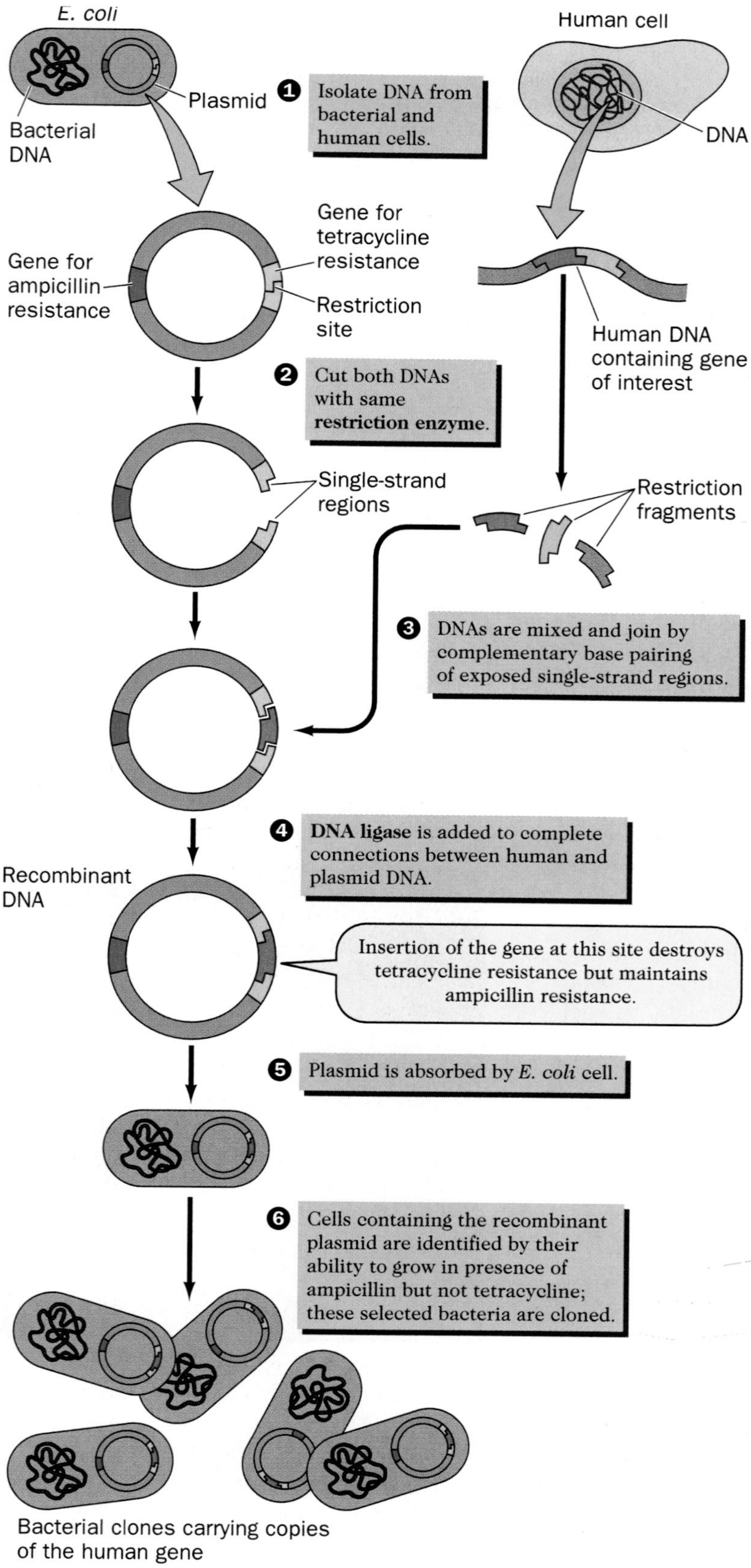
E. coli
Human cell
Plasmid
Bacterial DNA
DNA
1 Isolate DNA from bacterial and human cells.
Gene for tetracycline resistance
Gene for ampicillin resistance
Restriction site
Human DNA containing gene of interest
2 Cut both DNAs with same restriction enzyme.
Single-strand regions
Restriction fragments
3 DNAs are mixed and join by complementary base pairing of exposed single-strand regions.
4 DNA ligase is added to complete connections between human and plasmid DNA.
Recombinant DNA
Insertion of the gene at this site destroys tetracycline resistance but maintains ampicillin resistance.
5 Plasmid is absorbed by E. coli cell.
6 Cells containing the recombinant plasmid are identified by their ability to grow in presence of ampicillin but not tetracycline; these selected bacteria are cloned.
Bacterial clones carrying copies of the human gene

FIGURE 19.20 Cloning of a human gene in a bacterial cell.

and taken up by bacterial cells (Figure 19.20, 5). The recombinant DNA of the plasmid is duplicated every time the cellular DNA is duplicated before cell division. Recall from Chapter 12 how rapidly bacteria proliferate. This proliferation generates a clone of cells bearing the recombinant DNA. The copying of a gene in this fashion is called **gene cloning** (Figure 19.20, 6). However, what we are really interested in are the gene products: proteins such as hormones and enzymes or the products produced by enzymes in the bacterial cell.

Recombinant bacteria may be useful in a number of different ways: (1) They may synthesize and release the product of the inserted gene into their growth medium, from which it can be isolated and purified. (2) They may be released into the environment to carry out a biological or chemical process using the inserted gene. (3) The proliferation of trillions of identical bacterial cells can make available sufficient quantities of the gene for conventional chemical study. (4) Quantities of the gene may be available for transfer to other organisms, providing them with the desired capabilities.

> Restriction enzymes cut DNA at specific sites, creating restriction fragments. Plasmids are small loops of bacterial DNA. Plasmids and human restriction fragments can be linked together, forming recombinant DNA in bacteria.

Selecting DNA Fragments and Genes of Interest

Recombinant DNA technology may sound simple from this brief overview, but there are a few obstacles and some valuable tricks of the trade. One obstacle is the actual isolation of the DNA fragments or genes of interest that are released when a chromosome is chopped by restriction enzymes. There are so many fragments that finding the one you want is a little like finding a needle in a haystack. Two techniques are of value here: gel electrophoresis and DNA probes.

FIGURE 19.21 Gel electrophoresis of DNA.

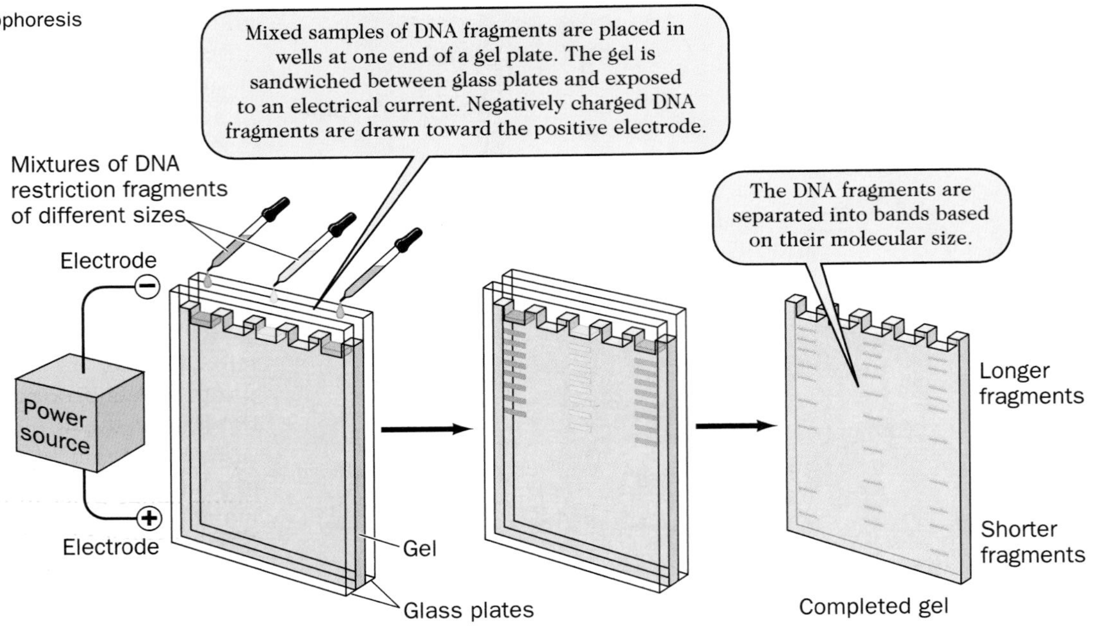
Mixed samples of DNA fragments are placed in wells at one end of a gel plate. The gel is sandwiched between glass plates and exposed to an electrical current. Negatively charged DNA fragments are drawn toward the positive electrode.
Mixtures of DNA restriction fragments of different sizes
Electrode
Power source
Electrode
Gel
Glass plates
The DNA fragments are separated into bands based on their molecular size.
Longer fragments
Shorter fragments
Completed gel

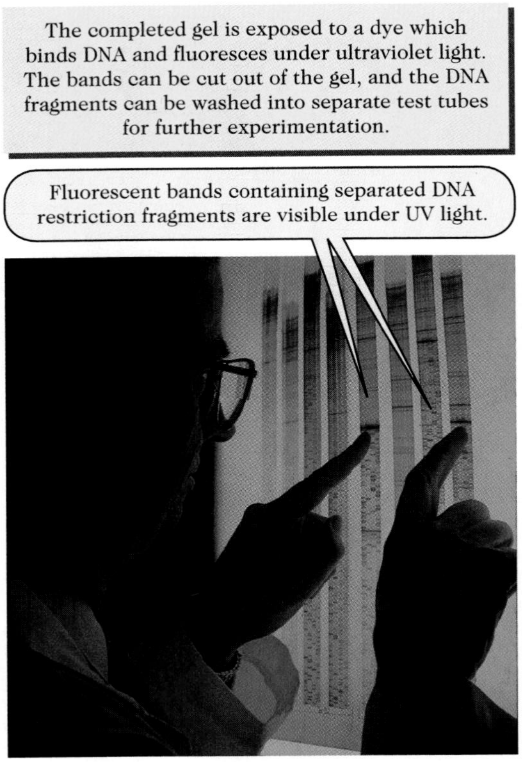
The completed gel is exposed to a dye which binds DNA and fluoresces under ultraviolet light. The bands can be cut out of the gel, and the DNA fragments can be washed into separate test tubes for further experimentation.
Fluorescent bands containing separated DNA restriction fragments are visible under UV light.

Gel electrophoresis (*ee-lek-troh-for-ee-sis*) separates DNA fragments by causing them to migrate through a porous gel under the influence of an electrical field. Differently sized fragments of DNA contain different amounts of negative charges. Therefore, these fragments migrate to different locations in the gel when they are placed in an electrical field (Figure 19.21). Such migration separates different fragments.

When separation is complete, the gel can be bathed with dyes or fluorescent chemicals to identify DNA fragments as bands on the gel. A band containing a fragment of interest can be cut out of the gel, and the DNA can be washed from the band for use in gene cloning and recombinant DNA formation.

DNA probes are single-stranded nucleotide sequences that are complementary to and thus able to bind with a desired DNA fragment or gene. DNA probes can be synthesized in a test tube by linking either DNA or RNA nucleotides in a specified sequence. Radioactive atoms are often incorporated in the probe to help locate it after it binds to a complementary DNA fragment. After the electrophoretic separation described above, a DNA probe can help locate the precise fragment of interest (the needle in the haystack).

Gel electrophoresis allows the separation of different sized DNA fragments. A DNA probe has a nucleotide sequence complementary to a desired DNA fragment and allows that fragment to be extracted from a mixture.

Polymerase Chain Reaction: Getting a Lot from a Little

Only tiny amounts of DNA can be isolated by the techniques described above. For research purposes, larger quantities are needed. Additional copies of DNA can be made by the gene cloning technique described above. However, there is now an easier and faster method: the **polymerase chain reaction (PCR).** PCR is a technique by which a DNA fragment can be copied repeatedly (amplified) in a test tube. Millions of copies can be made in just a few hours. The technique uses bacterial enzymes and essentially duplicates the process of DNA replication in a piece of automated laboratory equipment.

In fact, the primary enzyme used is **DNA polymerase,** the enzyme which makes complementary copies of DNA strands before cell division (Chapter 2). PCR, which is named for this enzyme, has been used to amplify tiny amounts of DNA obtained from fossils, embryonic cells, and even tissue or semen samples from crime victims. It was even featured, in a somewhat exaggerated fashion, in the book and movie *Jurassic Park* as the technique for copying dinosaur DNA.

As we continue to unearth the remains of the earliest humans, PCR will allow us to copy preserved DNA fragments so that they can be compared with similar fragments from living humans and we can learn just how much we have changed as a species during our time on Earth.

PCR is an automated means of repeatedly copying DNA fragments to provide enough specimens for experimentation and research. The copies can be made more rapidly than is the case with traditional gene cloning.

Restriction Fragment Length Polymorphism Analysis

As you learned earlier in this chapter, some genes exist in different forms called alleles. Alleles code for variations of the protein product of a gene. The fact that alleles must then differ subtly in their nucleotide sequences was confirmed by observation of the difference in electrophoresis bands that are produced when different alleles are cut with the same restriction enzyme (Figure 19.22).

The different sized restriction fragments have been dubbed **restriction fragment length polymorphisms,** or **RFLPs** (pronounced "Rifflips"), and their presence can be detected with the use of DNA probes. An *autoradiograph* is produced when the radioactive DNA probe exposes x-ray film and shows the location of a RFLP.

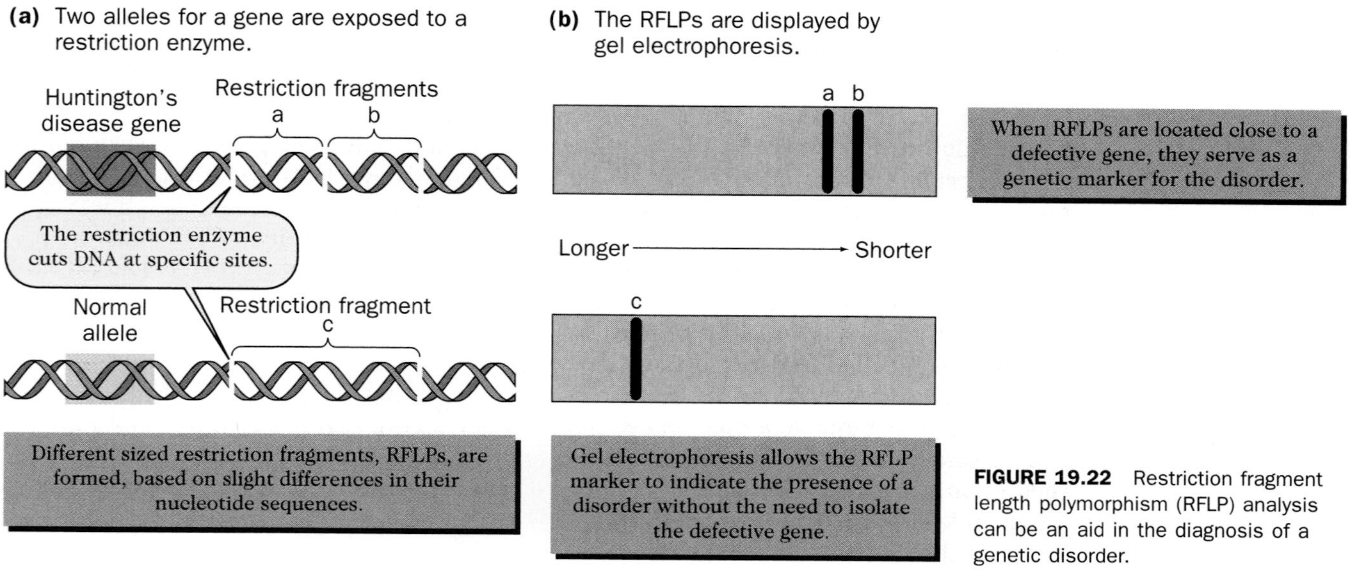
(a) Two alleles for a gene are exposed to a restriction enzyme.
Huntington's disease gene
Restriction fragments
a
b
The restriction enzyme cuts DNA at specific sites.
Normal allele
Restriction fragment
c
Different sized restriction fragments, RFLPs, are formed, based on slight differences in their nucleotide sequences.
(b) The RFLPs are displayed by gel electrophoresis.
a b
When RFLPs are located close to a defective gene, they serve as a genetic marker for the disorder.
Longer
Shorter
c
Gel electrophoresis allows the RFLP marker to indicate the presence of a disorder without the need to isolate the defective gene.

FIGURE 19.22 Restriction fragment length polymorphism (RFLP) analysis can be an aid in the diagnosis of a genetic disorder.

RFLPs can serve as genetic markers when they are always present in association with defective disease-causing alleles. Such is the case with a RFLP found on chromosome 4 in people suffering from Huntington's disease. This RFLP is thought to lie in the vicinity of the defective gene that causes the disease.

Early in life, RFLP analysis can predict with high accuracy (99 percent) the embryos or people who will later develop debilitating genetic disorders. The current genetic tests for sickle-cell anemia, cystic fibrosis, PKU, and Duchenne muscular dystrophy are all based on RFLP analysis.

> A restriction enzyme acting on a specific allele generates a variety of restriction fragments (RFLPs) which can be separated and identified by using gel electrophoresis techniques. Normal and defective alleles can be distinguished by comparison of their RFLPs.

Applications of DNA Technology

DNA technology is giving us a brave new world of opportunities, possibilities, and risks. The products we consume and our health care are being revolutionized. There are some clear benefits, but the personal and social impact may be negative as well as positive.

Agriculture: DNA for Production and Resistance

Among agricultural animals, the rate of weight gain has been increased by genetic engineering and biotechnology. For example, the gene for the protein *bovine growth hormone* (*BGH*) was identified and isolated from cows (bovines) and then cloned and inserted into *E. coli.* Rapid bacterial growth is accompanied by the production of large quantities of BGH, which is harvested, purified, and injected into dairy cows, increasing milk production. It also helps beef cattle gain weight more rapidly.

The genes for growth hormones also have been inserted directly into the oocytes or embryos of hogs, sheep, and rainbow trout to boost the rate of food production. Animals such as these, which have obtained genes from other species, are called *transgenic animals.*

Transgenic plants also are important in modern agriculture. Wheat, cotton, and soybean plants have received genes that make them more resistant to the poisons used for weed control. In 1994 the first genetically altered food, a tomato, was approved for sale by the U.S. Food and Drug Administration (FDA). These tomatoes ripen more slowly, have a longer shelf life, and are slower to spoil. Corn, cotton, and potato plants have been made more resistant to insect damage through the incorporation of a gene that codes for an insect-killing protein.

In the future, DNA technology may increase the food value of corn and wheat, ensuring that they produce all the essential amino acids required by humans. It may even be possible to reduce the need for expensive and toxic nitrogen fertilizers by genetically engineering nitrogen fixation by crops such as corn and wheat.

Medicine: DNA for Diagnosis and Treatment

In medicine, biotechnology may be used for both diagnosis and treatment. To date, DNA technology has permitted the diagnosis of over 200 human genetic disorders, primarily through the use of RFLP marker analysis and DNA probes. Genetic testing of fetal tissue obtained by chorionic villi testing or amniocentesis permits early identification of genetic defects (Chapter 18).

Genetic counseling centers at major hospitals provide information and counseling on genetic disorders. To those at risk, genetic counselors offer many helpful services: (1) They provide general information about hereditary disorders. (2) They help interpret the results of genetic testing. (3) They calculate the probability of having an affected child. (4) They identify family-planning and reproductive alternatives. (5) They provide referrals to community groups for emotional, technical, and financial support.

The identification of defective alleles creates an opportunity for **gene therapy**

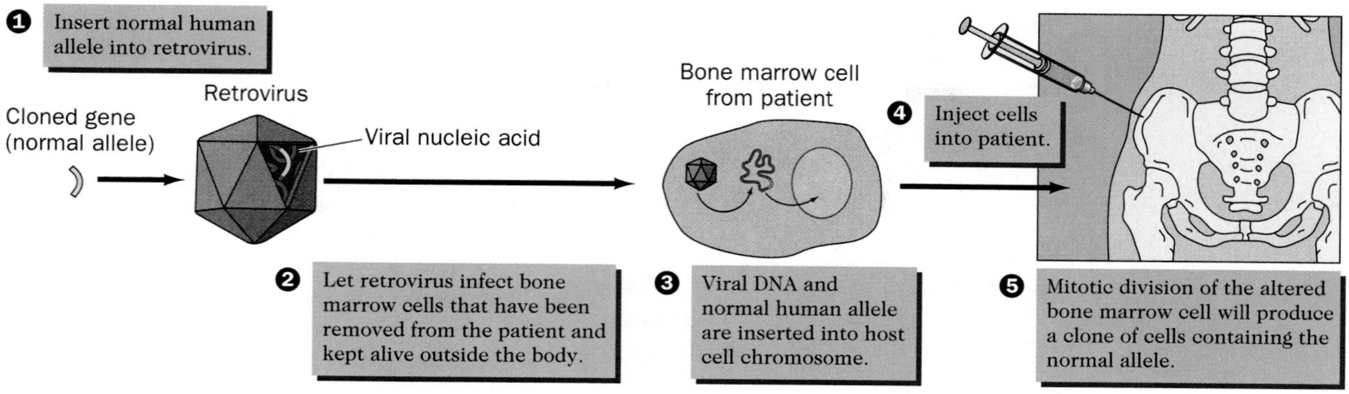
1 Insert normal human allele into retrovirus.
Retrovirus
Cloned gene (normal allele)
Viral nucleic acid
Bone marrow cell from patient
4 Inject cells into patient.
2 Let retrovirus infect bone marrow cells that have been removed from the patient and kept alive outside the body.
3 Viral DNA and normal human allele are inserted into host cell chromosome.
5 Mitotic division of the altered bone marrow cell will produce a clone of cells containing the normal allele.

FIGURE 19.23 Gene therapy. Bone marrow cells with a defective gene are removed from a person with a genetic disorder. The cells are infected with a harmless retrovirus carrying a normal allele for the defective gene. When the altered bone marrow cells are returned to the patient, all their mitotic daughter cells express the normal allele and the genetic disorder is reversed.

and the correction of an abnormal genotype. One of these techniques uses a virus (retrovirus) known for its ability to integrate DNA into human chromosomes. The virus is used to insert a cloned copy of a normal allele into human cells grown outside the body (Figure 19.23). These cells then can be injected into a patient with the hope that they will replace genetically defective cells. Gene therapy like this is best suited for disorders associated with a single gene and its protein product.

Recombinant DNA technology also is helping in the production of vaccines against infectious disease agents, especially viruses. If the gene which codes for a viral antigen can be isolated, it can be cloned, and the antigen can be produced in very large quantities for use in a vaccine. Even genes essential to infection can be altered to produce a live but harmless organism for use as a vaccine. In general, immunity is more effective when it is produced against a complete pathogen than when it is produced against a single isolated antigen.

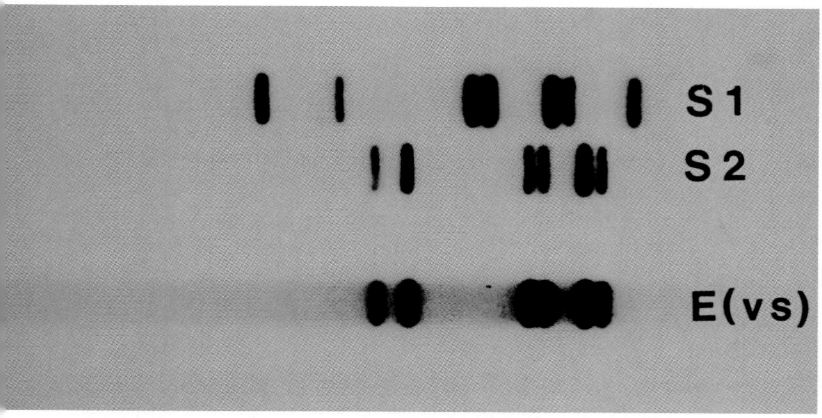
S1
S2
E(vs)

FIGURE 19.24 A DNA fingerprint (RFLP) from a murder case. RFLP analysis of the bloodstain (E) indicates that the blood on the defendant's clothes came from the victim (S2), not from the defendant (S1).

Human hormones have been made more readily available and less expensive through the use of DNA technology. *Human insulin* produced by bacteria from a cloned human gene is gradually replacing insulin extracted from the pancreas of pigs and cows. Genetic engineering has produced *human growth hormone* (enhances growth), *erythropoietin* (stimulates red blood cell production), *interferon* (combats viral infection and perhaps cancer), *interleukins* (activates immune cells), *tissue plasminogen activator* (dissolves blood clots), and *thrombopoietin* (stimulates platelet formation).

Criminal Investigation, Prosecution, and Defense

Telltale fingerprints often identify a criminal. In the 1990s **DNA fingerprints** are becoming increasingly important in criminal investigation, prosecution, and defense. A form of RFLP analysis can be done on samples such as hair, blood, skin, and semen obtained from a crime scene, matching the samples with a victim or a suspect (Figure 19.24). To date, such evidence has not always been accepted as proof of guilt but has been used to eliminate suspects. It also has been used to release innocent persons from prison.

Perils among the Promises: Ethics and Safety

DNA technology holds tremendous promise for agriculture and medicine. However, issues of ethics and safety have clouded these lofty visions. Could gene-splicing projects create a superbacterium that might cause widespread infectious disease? Federal regulations are already in place to ensure laboratory safety and prevent the escape of recombinant organisms. However, might recombinant organisms be intentionally released as a new form of terrorism or biological warfare? Could disease and herbicide-resistant plants cause environmental damage as they overgrow native plants?

As techniques evolve for testing an individual's genome, when and on whom should genetic testing be done? More important, who should have access to the results of genetic testing? Might you be denied employment or insurance because your genetic profile shows the presence of certain genes considered "defective"?

To control medical costs, will the carriers of genetic disorders be denied the opportunity to bear children or be required to pay higher insurance premiums if they have affected children? Do humans have the right to produce and patent new transgenic species?

There are many such questions and not a great deal of factual information on which to base the answers. Competing social values are also involved. Both public and private groups are struggling to raise questions, provoke discussion, provide answers, and establish policies and regulations before problems arise. DNA technology has tremendous potential, but there is obviously a need for informed public and legislative discussion and prudent regulations as we progress into this brave new world.

Chapter Summary

Human Chromosomes

- The phenotype includes traits which can be observed, measured, or analyzed.
- The genotype includes the genes for the traits of the phenotype.
- Each pair of similar chromosomes (one maternal and one paternal) form a homologous pair.
- Histone proteins act as spools on which DNA is wrapped.
- A karyotype is a photographic display of chromosomes.
- Chromosome pairs 1 to 22 are called autosomes and chromosome pair 23 is the sex chromosomes.
- In a woman, one X chromosome is inactivated in each cell forming a Barr body.
- Nondisjunction is a failure of homologous chromosomes or chromatid pairs to separate during meiosis I or II.
- Abnormal chromosomes result from duplication, translocation, inversion, or deletion of chromosome segments.

Human Genes

- Alleles are different forms of the same gene, creating different phenotypic versions of a trait. Alleles may be the same (homozygous) or different (heterozygous). Dominant alleles are expressed in the homozygous or heterozygous condition; recessive alleles are expressed in the homozygous condition.
- Genes coding for proteins are structural genes; genes controlling structural genes are regulatory genes.
- A mutation is a change in a gene that causes altered proteins to be produced.

Patterns of Inheritance

- The law of segregation describes the independent separation of homologous pairs of chromosomes during meiosis.
- The law of independent assortment describes the independent inheritance of alleles located on separate chromosomes.
- To determine the probability of inheriting two alleles together, multiply the probabilities of inheriting each separately.
- A monohybrid mating involves a single pair of alleles; a dihybrid mating involves two allele pairs.
- A pedigree is a genetic family tree showing the gender and phenotype of related individuals.
- Polygenic inheritance occurs when several separate genes control a single phenotypic trait.
- Sex-linked genes are on the sex chromosomes. A woman is homozygous or heterozygous for X-linked genes; a man expresses the alleles he inherits on his X chromosome.

Genetic Disorders

- Autosomal dominant disorders include Huntington's disease, hypercholesterolemia, and neurofibromatosis.
- Autosomal recessive disorders include cystic fibrosis and phenylketonuria.
- Sex-linked disorders include color blindness, hemophilia, and severe combined immunodeficiency disorders.

DNA Technology

- Recombinant DNA combines genes from different kinds of cells.
- Plasmids are loops of bacterial DNA. Restriction enzymes cut DNA at specific sites to form restriction fragments. DNA ligase can bond DNA fragments into a plasmid, which can then be inserted into bacteria and replicated.
- Gel electrophoresis separates different sized DNA fragments. DNA probes are complementary nucleotide

sequences that can bind to and locate desired DNA fragments.
- PCR allows a desired DNA fragment to be copied repeatedly.
- RFLPs are variations in the length of restriction fragments due to differences in nucleotide sequences.

Applications of DNA Technology

- Transgenic organisms have received genes from other species.
- Gene therapy corrects genetic disorders by directly replacing defective genes.
- Vaccines and hormones have been made more available and less expensive through DNA technology.
- RFLP analysis is performed on body tissue or fluid samples to match suspect and victim in a crime.

Selected Key Terms

alleles (p. 506)
autosomes (p. 501)
Barr body (p. 502)
gene cloning (p. 519)
genotype (p. 499)
heterozygous (p. 506)
homozygous (p. 506)
karyotype (p. 500)
mutation (p. 507)
nondisjunction (p. 502)
phenotype (p. 499)
plasmid (p. 518)
Punnett square (p. 502)
recombinant DNA (p. 518)
sex chromosomes (p. 501)

Review Activities

1. Explain the difference between genotype and phenotype. Give an example of each.
2. Explain how chromosomes are packaged to fit into the nucleus of a cell.
3. What is a karyotype? How are autosomes different from sex chromosomes?
4. Explain how gender is determined genetically. What is a Barr body?
5. Summarize the chromosome combinations and symptoms of metafemale, Turner's, Klinefelter's, and Down syndromes.
6. Summarize the differences between duplication, translocation, inversion, and deletion in chromosomes.
7. Using letters to represent alleles, demonstrate the difference between homozygous dominant, heterozygous, and homozygous recessive conditions.
8. Describe the difference between structural and regulatory genes.
9. What is a mutation? Give an example.
10. Summarize the law of segregation and the law of independent assortment.
11. Use a Punnett square to solve the following problem. The ability to roll the tongue into a U shape is dominant (T) over the inability to roll the tongue (t). Determine the probability that two individuals who are heterozygous for tongue rolling will have a non-tongue-rolling child.
12. Define the following terms: nondisjunction, testcross, pedigree, polygenic inheritance, X-linked genes.
13. Summarize the pattern of inheritance (autosomal dominant, recessive, etc.) and symptoms for Huntington's disease, hypercholesterolemia, neurofibromatosis, cystic fibrosis, PKU, color blindness, and hemophilia.
14. Explain how plasmids, restriction enzymes, and DNA ligase are used to form recombinant DNA.
15. Explain how gel electrophoresis, DNA probes, PCR, and RFLP analysis are used in DNA technology.

Self-Quiz

Matching Exercise

___ 1. Proteins that form a nucleosome with DNA	A. Sex chromosomes
___ 2. Photographic display of human chromosomes	B. Barr body
___ 3. Chromosome pairs 1 through 22	C. Karyotype
___ 4. Chromosome pair 23	D. Pedigree
___ 5. Inactive X chromosome in female cells	E. Histones
___ 6. Alteration in a structural gene causing a defective protein	F. Ligase
___ 7. Genetic history of a family	G. Restriction
___ 8. Enzyme that cuts DNA into fragments	H. Autosomes
___ 9. Enzyme that connects DNA fragments	I. PCR
___ 10. Process for making copies of DNA fragments	J. Mutation

Answers to Self-Quiz

1. E; 2. C; 3. H; 4. A; 5. B; 6. J; 7. D; 8. G; 9. F; 10. I

SPOTLIGHT ON HEALTH: SEVEN

Sexually Transmitted Diseases

Sexually transmitted diseases (STDs) are infectious diseases that are passed between humans during unprotected vaginal, anal, and oral sexual contact. They usually affect the reproductive system but may spread to other body systems. If STDs are not treated, they endanger general health and fertility and may be life-threatening. Some are incurable.

STDs have become an increasingly serious threat to public health for two reasons. First, changing patterns of sexual behavior make sexual contact more likely even at an early age. Second, many infectious organisms have become resistant to antibiotics that once were very effective treatments. In general, STDs are easier to avoid than to treat and cure. Abstinence, monogamy, and the use of condoms are the best ways to avoid infection.

For the most part, STD pathogens are fragile and do not survive long periods of exposure to air. For transmission, these pathogens require the moist environments and close contact of a sexual encounter. Our discussion of STDs will be based on the kinds of infectious agents responsible for those diseases: bacteria, viruses, fungi, protozoa, and arthropods.

STDs Caused by Bacteria

Among the more common bacterial STDs are gonorrhea, syphilis, and chlamydiosis.

Chlamydiosis: The Most Common Bacterial STD

Chlamydiosis is caused by the bacterium *Chlamydia trachomatis* (*cla-**mid**-ee-uh track-oh-**mat**-iss*) (Figure 1). Sometimes referred to as **nongonococcal urethritis (NGU)**, chlamydiosis (*clah-mid-ee-**oh**-sis*) is the most rapidly increasing STD in the United States with over 4 million cases annually.

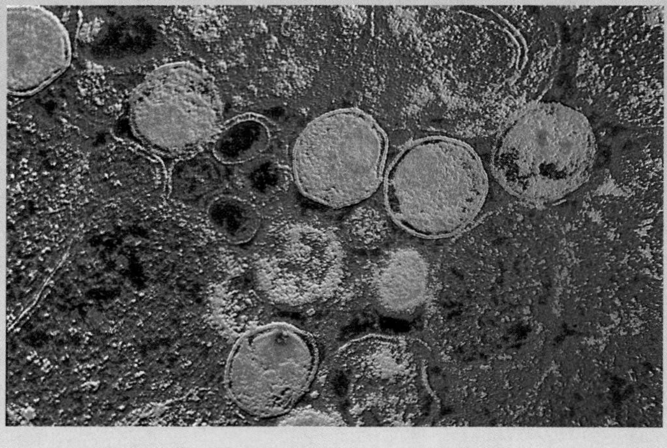

FIGURE 1 *Chlamydia trachomatis,* the pathogen for chlamydiosis.

In men, there may be a thick puslike discharge from the penis and painful urination. Women, however, may not show symptoms until the infection has spread to the cervix, uterus, and uterine tubes. Without proper diagnosis and treatment, women are particularly vulnerable to **pelvic inflammatory disease (PID)**, which can cause sterility. The symptoms in males and females are so similar to those of gonorrhea that this disorder has often been misdiagnosed as gonorrhea and incorrectly treated.

Antibiotics such as tetracycline, erythromycin, and rifampin are usually effective in the treatment of chlamydial infections. However, penicillin, the drug most often prescribed for gonorrhea, is not effective, underscoring the need for a correct diagnosis.

Gonorrhea: A Common STD

Gonorrhea is caused by the bacterial pathogen *Neisseria gonorrhoeae* (*nye-**seer**-ee-ah gon-or-**ree**-ee*) (Figure 2). About 700,000 cases of gonorrhea are reported each year. The symptoms of gonorrhea are

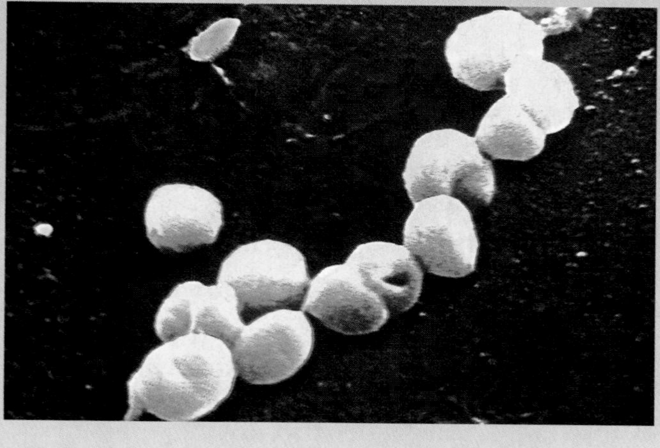

FIGURE 2 *Neisseria gonorrhoeae*, the pathogen for gonorrhea.

often more obvious in males than in females and include a discharge from the penis similar to that caused by *Chlamydia* and painful urination. In women, infection of the vagina or uterus may go unnoticed and eventually spread through the uterine tubes to the lower part of the abdominal cavity, resulting in PID and possible sterility. During birth, infants may contract gonorrheal eye infections as they pass through an infected mother's birth canal. If they are not treated, such infections can cause blindness in an infant.

For 40 or 50 years, infections of gonorrhea have been treated effectively with penicillin or spectinomycin, but newly evolved strains are resistant to both antibiotics.

Syphilis: Several Phases

Syphilis is caused by *Treponema pallidum* (***trep**-oh-**knee**-mah **pal**-lid-dum*), a slender, spiral-shaped bacterium (Figure 3). After infection, the untreated disease progresses slowly through several distinct stages. After several weeks of incubation, a hard, dry sore called a **chancre** (***shang**-kur*), which is filled with bacteria, appears at the site of infection. The appearance of a chancre marks the beginning of the **primary stage,** which lasts for about a month. Chancres are painless and may not be noticed. They disappear without treatment, leading an infected person to believe that complete healing has occurred. Unfortunately, this is not the case.

After a period without external symptoms, the **secondary stage** of syphilis begins. Its symptoms often include a diffuse skin rash and white patches on the tongue, cheeks, and gums which teem with bacteria and allow the disease to be spread by kissing. The afflicted person usually has a fever, fatigue, loss of appetite, and bone pain. This stage may last as long as a year and leads to the latent stage, in which the symptoms subside entirely. However, without treatment, about 30 percent of those who are infected proceed to the tertiary stage.

During the **tertiary stage** of syphilis, the bacteria spread to internal organs and systems, damaging the heart and injuring the brain; this can lead to paralysis and mental instability. Large disfiguring sores called **gummas** may appear on the skin in this stage. The organism can cross the placenta of an infected mother or be transmitted to a newborn during birth. A solution of silver nitrate or antibiotics is washed into the eyes of newborns as a precaution against both syphilis and gonorrhea infections. Treatment of syphilis is possible, at least in the early stages, with penicillin, tetracycline, or erythromycin.

STDs Caused by Viruses

There are four major STDs caused by viruses: genital herpes, genital warts, AIDS, and hepatitis B. Viruses are not living cells and have none of the cellular machinery usually attacked by antibiotics. When viruses enter a human cell, they take over its protein synthesis machinery, assembling hundreds of new viral particles. These new viruses may remain in the cell or be released immediately to infect other cells. After entering a cell, viruses are hidden from the body's natural defenses—phagocytosis and antibodies—often making complete eradication impossible. Thus, once you have been infected, the virus can remain with you for life.

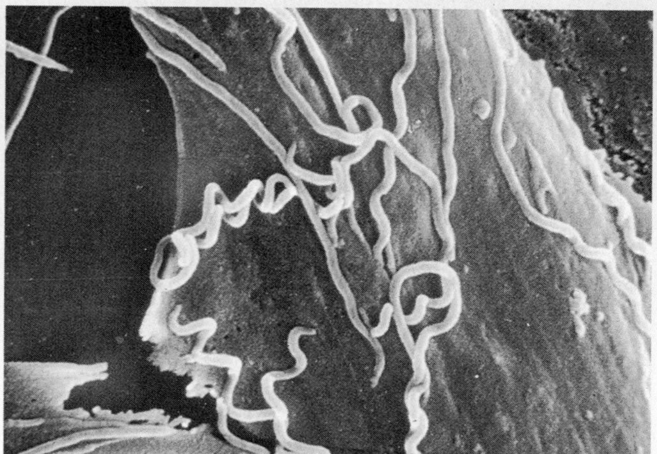

FIGURE 3 *Treponema pallidum*, the pathogen for syphilis.

Genital Herpes: Infection for a Lifetime

Herpes simplex type 2 viruses have infected over 40 million people in the United States, and there are approximately 500,000 new cases each year. The primary symptoms of genital herpes are periodically recurring watery blisters on the genitalia or buttocks which may be accompanied by fever, painful urination, and swollen lymph nodes in the groin. The blisters rupture and heal within a few weeks after each outbreak. Upon rupture, the fluid that is released contains viruses that can spread the infection. Nearly all body regions, including the mouth and hands, can become infected if contacted by broken blisters.

Between episodes of blisters, the virus lies dormant in spinal nerve cells. Before an outbreak, the virus descends through sensory neurons to reach the skin surface. Release of the virus and repeated outbreaks of blisters can be triggered by stress, menstruation, nutritional changes, and exposure to sunlight.

The infection also can be passed during birth as an infant contacts blisters in the birth canal of its infected mother. Congenital herpes infections of the eyes can cause blindness if they are not treated. Prevention is achieved through the avoidance of sexual activity during outbreak episodes or through the use of condoms. The drug Acyclovir interferes with the replication of the virus and can prevent or reduce the severity of new outbreaks.

Genital Warts: Annoyances That May Lead to Cancer

Genital warts are caused by the **human papillomavirus (HPV).** Public health authorities believe that as much as 15 percent of the U.S. population may be infected by this virus. The warts vary in appearance from tiny bumps to large, spreading masses which can appear on the penis, the labia, around the anus, in the vagina, and on the cervix. The disease spreads through contact with warts on the genitalia of an infected person. They may cause or contribute to cancers of the penis and cervix.

Warts can be removed with the drug Condylox, freezing with liquid nitrogen, cauterization with an electrical current, laser surgery, or treatment with alpha$_{n3}$-interferon (Alferon). However, there is a tendency for genital warts to recur.

AIDS: Debilitating and Deadly

AIDS is caused by **human immunodeficiency virus (HIV).** HIV infection is transmitted by contact with the blood or body fluids (semen, breast milk, and vaginal fluids) of an infected person or by birth when the mother is infected. While the virus can be transmitted sexually, it does not attack the reproductive system directly. Instead, HIV infects and destroys the cells needed for the immune response, making a person more susceptible to life-threatening secondary infections such as pneumonia. For more information on AIDS, see Spotlight on Health: Five (page 354).

Hepatitis B May Lead to Cirrhosis or Liver Cancer

Hepatitis B is caused by the **hepatitis B virus (HBV).** HBV is transmitted by blood, body fluids, and sexual contact in the same fashion as HIV. However, it is less fragile and 100 times more contagious than HIV and can withstand environmental exposure that would destroy HIV. HBV multiplies in the liver and bone marrow and often leads to a chronic (carrier) infections. The symptoms of infection include fatigue, nausea, abdominal pain, arthritis, and a yellowing of the skin (jaundice) caused by liver damage. There are approximately 300,000 new cases each year in the United States, with about 10 percent of these patients becoming chronic carriers of the virus. The infection can be fatal by itself or can lead to cirrhosis (*sir-row-sis*) of the liver (similar to alcoholic liver destruction) and even liver cancer.

Unlike most STDs, there is a vaccine against HBV. Some public health officials believe that all college students should be vaccinated against HBV to reduce sexually transmitted cases. Federal law already requires that health care students and workers be vaccinated.

STDs Caused by Fungi, Protozoa, and Arthropods

In addition to bacteria and viruses, STDs are caused by yeasts (a form of fungus), single-cell eukaryotes (protozoa), and lice (arthropods).

Vaginal Yeast Infections

Vaginal candidiasis (*can-did-eye-uh-sis*) is caused by infection with the yeast *Candida albicans.* This yeast is part of our normal flora and is found in the mouth, colon, and vagina, where its growth normally is limited by competition from other organisms in the normal flora and the body's immune defenses. If these factors are disrupted, the yeast population may grow out of control. Women with yeast infections experience painful inflammation of the vagina, often with a thick, cheesy discharge. Men may develop a painful inflammation of the urethra through sexual contact with an infected woman. A chronic low level of infection which may cause recurring bladder infections is common in many women. Treatment involves the use of antibiotics such as nystatin, clotrimazole, and miconazole or over-the-counter preparations such as Femstat, Monistat, Mycelex, and Vagistat.

Trichomonas: An Infective Protozoan

Trichomoniasis (***trih-koh-moh-nye-uh-sis***) is caused by the single-celled protozoan *Trichomonas vaginalis.* In women, this disease usually is characterized by genital itching and a foul-smelling vaginal discharge. In men, there are often no symptoms, but occasionally the disease spreads to the seminal vesicles and prostate gland and causes painful swelling. Antibiotic treatment is successful with the use of metronidazole.

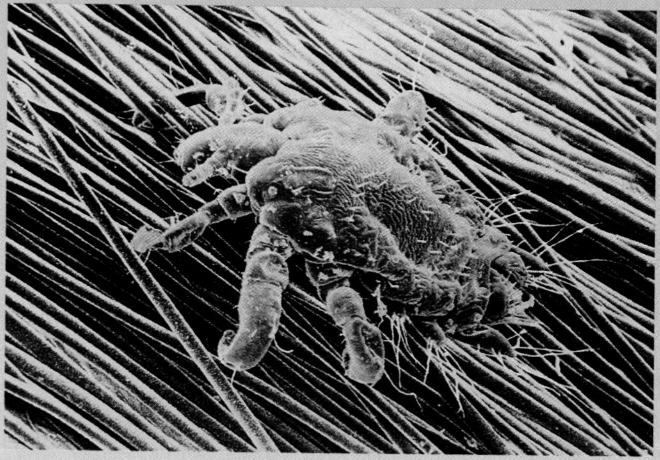

FIGURE 4 *Phthirus pubis,* the pubic louse.

Pubic Lice

Pubic lice ("crabs") are infestations of the pubic hair by the parasitic louse *Phthirus pubis* (***thir-russ pew-bus***) (Figure 4). Adult lice grip on to pubic hairs with their claws and pierce the skin with their mouth parts to draw a meal of blood. The human host may experience painful itching and reddened patches of skin. Female lice lay eggs near the base of pubic hairs, and the young hatch within a few days to expand the infestation. The lice are passed between humans by intimate contact or by sharing clothing, sheets, or blankets. Self-treatment of the infestation is possible with the use of medicated shampoos such as Kwell.

PART EIGHT

Living with Nature

CHAPTER 20

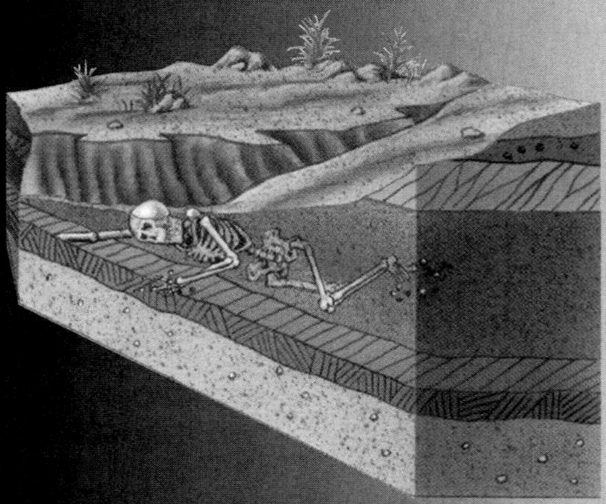

Evolution and Human Evolution

CHAPTER 21

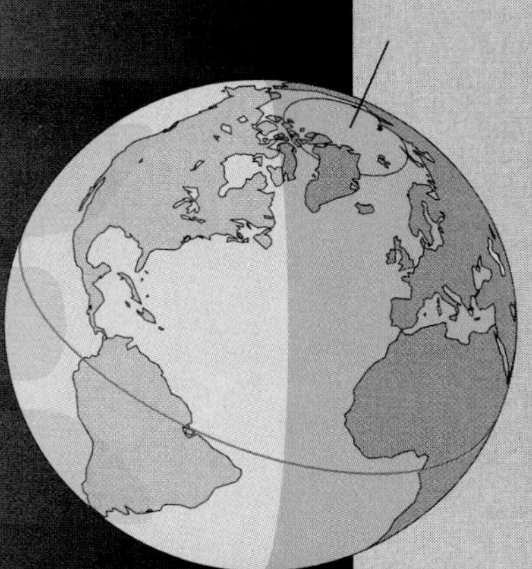

Ecology: The Interactions of Matter, Energy, and Life

CHAPTER 22

Human Ecology and Environmental Impact

Chapter 20

Evolution and Human Evolution

It is the nature of life to change. After reading the preceding chapters, you are familiar with patterns of change in living human systems. We have spoken about changes in molecules, cell structures, and genes. We have described the changes of growth and development as well as those needed to maintain homeostasis.

In considering these changes, our frame of reference has been the individual cell, tissue, organ system, or organism. Now we must expand our frame of reference and examine the individual as part of a population of similar individuals. We will be asking, how does a population change? Such changes constitute evolution.

Evolution can be defined as a change in a population over time. By *change,* we are referring to inherited alterations that result in phenotypic changes. By *population,* we are referring to a group of similar organisms that freely and successfully interbreed. By *time,* we are usually referring to geologic time, which measures change over millions of years, but in some situations time is measured in generations. Our expanded frame of reference will eventually include the interactions of a population with the environment and with other organisms.

We will begin this examination by considering the contributions of Charles Darwin (1809–1882) and then summarize the facts that support the theory of evolution and describe the mechanisms by which evolution operates. Then we will discuss the origin and early history of life on Earth. We will conclude with a presentation of human evolution and the proposed human lineage.

Charles Darwin and the Science of Evolution

The diversity of organisms on Earth is enormous. Think of the variety of birds. From the tiny, fast-flying hummingbird to the large, flightless ostrich, there are about 9000 different species of birds on the planet today. There are 1.5 million named species of animals, plants, fungi, Protista, and Monera, and there may be as many as 30 million more that have not been identified. A **species** is a group of similar organisms that have the capacity to produce living progeny in nature. In some cases different species appear to be very different; in other cases the differences may not be easily observed. How do species originate? Do they change over time? Do they become extinct?

Such questions have been discussed and debated for centuries. However, it was not until 1859, when Charles Darwin published *The Origin of Species,* that a meaningful scientific explanation was proposed to explain the mechanism of biological diversity. Darwin's book was based on scientific observations and data collected during 5 years spent exploring the coastal areas of South America as a naturalist aboard the British survey ship HMS *Beagle* (1831–1836).

Darwin's Theory: Change through Selection

On this voyage the young Darwin (age 22) observed how living forms are similar and different from the preserved remains (fossils) of animals and plants. During the voyage Darwin came to believe that the similarities between living and dead animals provided evidence for a common lineage. In other words, some forms of life had evolved from earlier and now extinct forms of life.

In addition to the diversity observed between species, Darwin also was aware of variation within a species. Although they resemble one another and can interbreed, the individual members of a species are not identical—they show individual **inherited variation.** Indeed, by controlling (selecting) the mating of domestic animals, humans can vastly change the size and appearance of progeny. By carrying out selective breeding, humans have created different breeds of dogs and pigeons, as well as agricultural plants and other animals (Figure 20.1). These activities showed Darwin the power of selection. But how could selection operate in nature?

The answer to this question became clear when Darwin realized that many

FIGURE 20.1 Selective breeding has resulted in distinct breeds of dogs with different sizes and characteristics.

(a) A body decomposes unless covered by sediment.

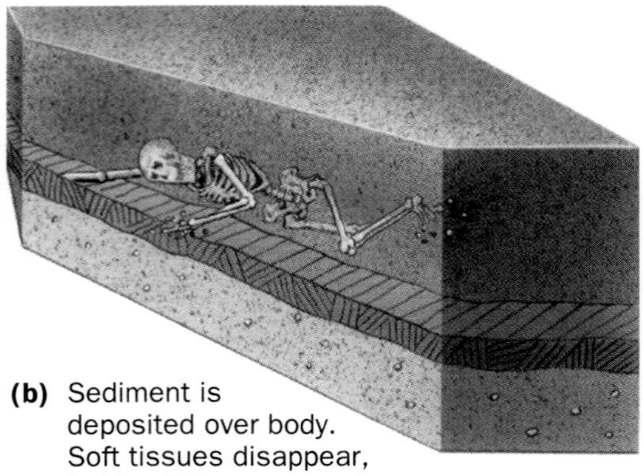

(b) Sediment is deposited over body. Soft tissues disappear, but hard tissues remain.

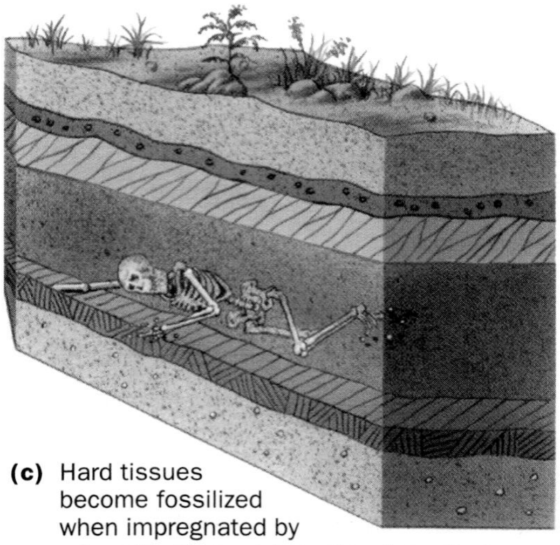

(c) Hard tissues become fossilized when impregnated by minerals. More sediment is deposited.

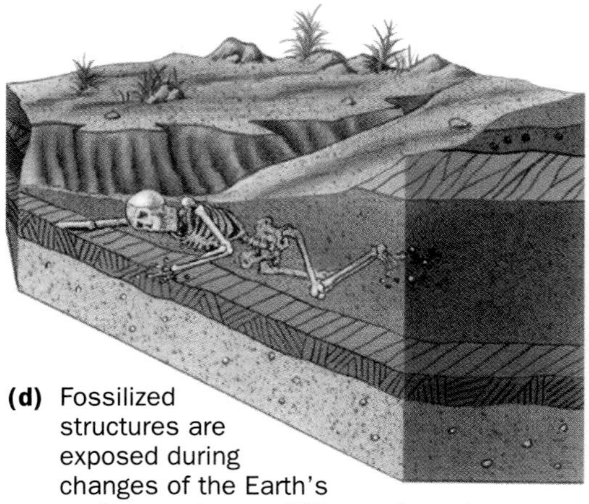

(d) Fossilized structures are exposed during changes of the Earth's surface such as uplifting and erosion.

FIGURE 20.2 The sequence of events during fossil formation.

more progeny are produced than actually survive. The progeny that survive are those with inherited traits that best adapt them to their environment and allow them to reproduce successfully. This is the theory of natural selection Darwin proposed in his book. As this process operates over long periods of time, new species result as inherited traits accumulate and populations become isolated from one another because of migration or geologic events.

Darwin also applied his ideas of evolution to humans, and this initiated a debate between science and religion which in some quarters continues today. Darwin's theory was in conflict with literal interpretations of the Bible's account of creation in 6 days and the belief that humans originated from Adam and Eve. The debate between science and religion occurs because of a misinterpretation about the roles of science and religion.

Religion is based on personal experience and/or belief in a supernatural power or creator. Science, by contrast, uses facts about the world and universe to answer questions (hypotheses) using the scientific method (see the Prologue). Science describes what has been discovered about our physical environment; it does not preclude religious faith in supernatural phenomena. In fact, virtually all biologists today accept the theory of evolution as an explanation of biological diversity, yet many also have strong religious beliefs.

From his research, Darwin concluded that living species are descended from similar extinct species. Inherited traits that help an organism adapt to its environment and allow for successful reproduction formed the basis of Darwin's theory of natural selection.

Evidence for Evolution

Evidence for evolution comes from several scientific disciplines. It comes from the study of rocks and fossils and from comparisons of the structure and chemistry of living organisms. In this section, we will summarize the kinds of evidence obtained from six different sources: fossils, comparative anatomy, vestigial structures, comparative embryology, comparative biochemistry, and biogeography.

Fossils: Snapshots of the Past

Fossils are the remnants of organisms that lived in the past. Most often, fossils develop when layers of sediment cover a dead organism before it decomposes totally (Figure 20.2). The layers of sediments can be produced by the silt of a river, lake, or sea, or they can come from windblown sand and volcanic ash. Once they are covered by sediment, bone and hard tissues become mineralized, forming a fossil.

Over thousands or millions of years, sedimentary deposits form layers or strata of geologic material. As portions of the Earth's surface are lifted up and as oceans and lakes recede, these strata become visible and accessible for excavation by scientists in search of fossils (Figure 20.3).

Sedimentary layers are organized in chronological order, with older layers covered by more recent ones. This permits a natural ordering of fossils from the oldest to the youngest.

The age of fossils can be determined by measuring the decay of certain radioactive isotopes present in the fossils. As radioisotopes decay, they change into

FIGURE 20.3 Excavation of fossils. Fossils often are found near the surface of the Earth where geologic processes have exposed long-buried strata.

other isotopes at a constant rate (Chapter 1). By measuring the amounts of each isotope present, scientists can determine the amount of time that has passed since the isotope was actively incorporated into the organism while it was alive.

By sampling fossils from successive sedimentary layers, we can observe a series of structural changes over millions of years. Examination of the strata also reveals the extinction of some forms and the expansion of others. Such information supports the concept that populations of organisms change over time and that new forms of life arise from preexisting forms.

A Common Origin for Similar Structures

Different groups of related organisms share similar anatomic features that appear to result from an evolutionary descent from a common ancestor. In a sense, the anatomy of today's organisms retains traces of past ancestors.

FIGURE 20.4 Comparing bone anatomy in the forelimbs of different vertebrate groups.

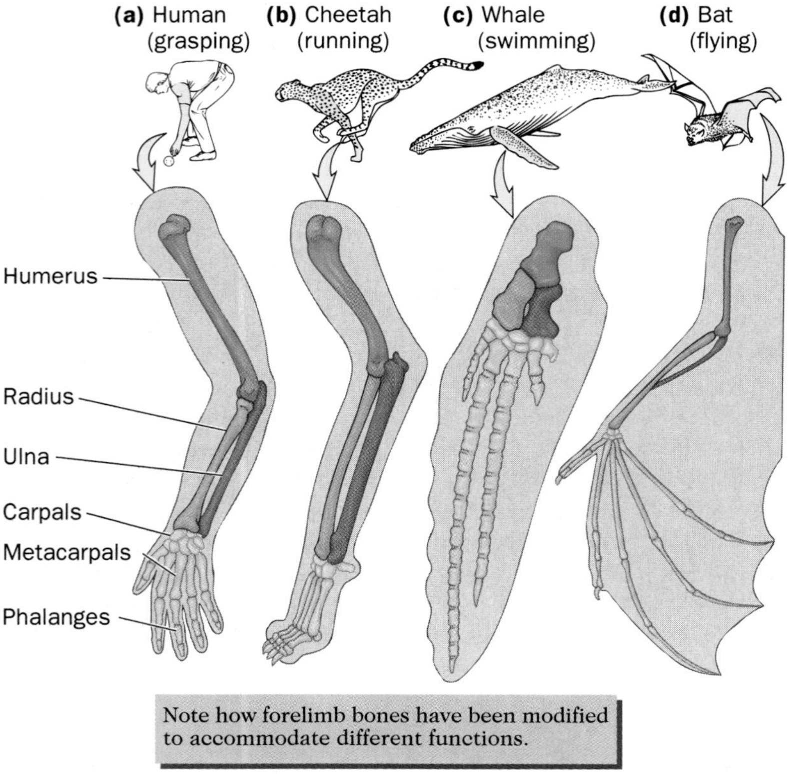
(a) Human (grasping)
(b) Cheetah (running)
(c) Whale (swimming)
(d) Bat (flying)
Humerus
Radius
Ulna
Carpals
Metacarpals
Phalanges
Note how forelimb bones have been modified to accommodate different functions.

For example, consider the forelimbs of several vertebrates (animals with a backbone). Although animals such as bats, whales, cheetahs, and humans differ in many ways, they all have forelimbs. Even though their forelimbs have different functions, each shares a similar arrangement of bones, muscles, and nerves. Figure 20.4 shows that the forelimbs of all these animals contain an upper bone (the humerus) that is connected to two lower bones (the ulna and radius), a group of wrist bones (carpals), and digits (metacarpals and phalanges).

These similarities are interpreted as evidence of descent from a common ancestor. Although wings, flippers, and arms are adapted to different ways of life, their internal anatomy is remarkably similar. They represent structural variations derived from ancestors that had the same basic structure.

Vestigial Structures: Ancestral Remnants

Vestigial structures are poorly developed structures of marginal use that are consistently found in some species. However, in other related species the same structures are fully developed and have specific functions. Therefore, scientists have concluded that vestigial structures are the remnants of anatomic structures that at one time had important functions for ancestors. Humans contain vestigial structures. The coccyx (our tailbone) is considered a remnant of tail vertebrae. In fact, human babies occasionally are born with additional coccyx bones that form a small "tail," which usually is surgically removed. Muscles that move the outer ears are still present in humans, but they are of no functional use to us. The human appendix is considered to be a vestigial structure representing the remains of a larger cecum found in many mammals.

Examples are present in other animals as well. Whales are mammals whose ancestors returned to the sea after living on land. Those ancestors made use of rear legs for terrestrial locomotion. Although these structures are not used in whales, whales have vestigial hip and leg bones

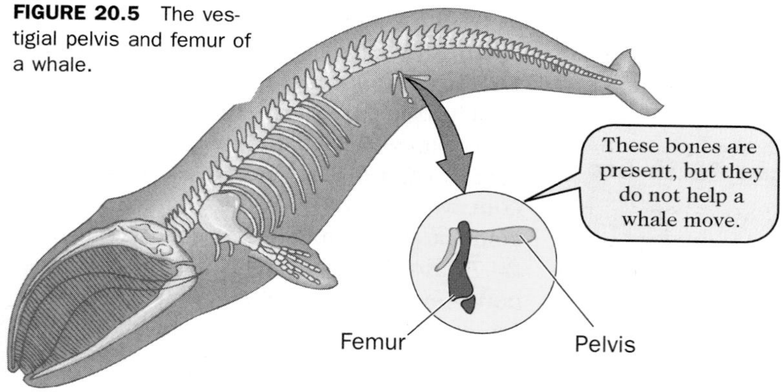
These bones are present, but they do not help a whale move.
Femur
Pelvis

FIGURE 20.5 The vestigial pelvis and femur of a whale.

that are indicative of their ancestors and mark the course of their evolution (Figure 20.5).

Comparing Development: New Uses for "Old" Structures

Adult fish, reptiles, birds, and humans are very different from one another. However, their early embryos are all remarkably similar (Figure 20.6). As development progresses, embryological structures such as gill arches and tailbones are lost or are modified into other structures. In fish, gill arches develop into gills. However, in other vertebrates that evolved from ancestral fish, gill arches either regress or become modified. For example, in humans, portions of the gill arches form the eustachian tube which connects the middle ear to the throat. The tail represents another example. The pronounced tail of early vertebrate embryos develops to varying degrees in all vertebrates except humans and a few monkeys and apes.

FIGURE 20.6 Structural similarities of early embryos from four vertebrate groups.

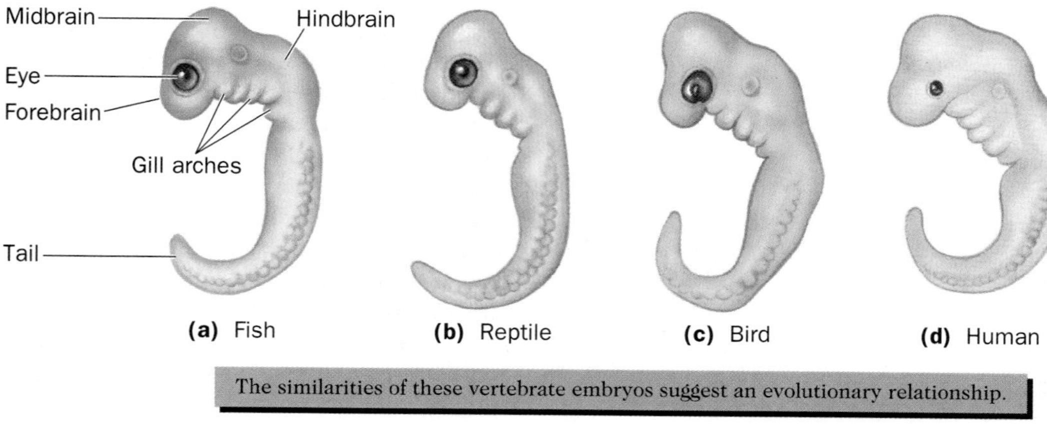
Midbrain
Hindbrain
Eye
Forebrain
Gill arches
Tail
(a) Fish
(b) Reptile
(c) Bird
(d) Human
The similarities of these vertebrate embryos suggest an evolutionary relationship.

Why are there such similarities in the early development of these diverse groups of vertebrates? Why develop a set of gill arches, for example, only to have them regress or become modified? The simplest answer is that there is an evolutionary relationship between these vertebrates. The more recently evolved groups, including humans, have retained some of the developmental processes of their ancestors.

Comparing the Sequences of Proteins and DNA

The biochemical similarity of all forms of life provides direct and compelling evidence for their evolutionary relatedness. All life-forms use lipids for membranes and proteins for enzymes. All living organisms use DNA as the hereditary material and RNA for protein synthesis. Furthermore, protein synthesis always uses the same genetic code (Chapter 2). The analysis of the sequences of proteins and nucleic acids offers rich opportunities to measure the extent of relatedness.

For example, almost all vertebrate animals use hemoglobin to transport oxygen (O_2). However, the amino acid sequence of the hemoglobin protein varies progressively among the vertebrate groups. The extent of similarity is considered evidence of how closely related different groups or species are. When analyzed, human hemoglobin is more similar to that of chimpanzees than to that of horses and more similar to that of horses than to that of chickens. Why is the protein sequence of human hemoglobin more similar to that of chimps than to that of horses? This occurs because humans are more closely related to chimps than they are to horses.

Comparison of the DNA from different groups and species offers another way to measure changes from a common ancestor. When the nucleotide sequences for different species are compared, the number of changes provides a measure of how great the divergence from a common ancestor has been.

Biogeography: Where on Earth Are the Animals?

Biogeography is the study of the distribution of plants and animals over the entire Earth. Biogeography asks, Why are zebras found on the grasslands of Africa but not on similar grasslands in North America and South America? Why do pouched mammals called marsupials, such as the kangaroo and koala, predominate in Australia but not in similar environments elsewhere? Marsupials are mammals in which the young are born underdeveloped and complete their development in an external pouch on the mother's body.

FIGURE 20.7 Movement of the Earth's continents during the last 200 million years.

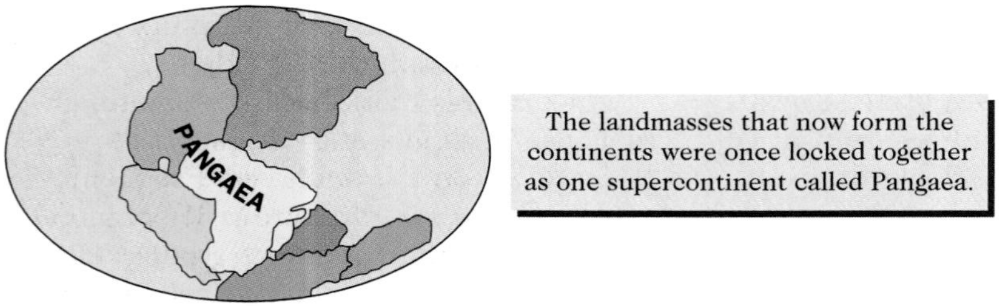
PANGAEA
The landmasses that now form the continents were once locked together as one supercontinent called Pangaea.

(a) 200 million years ago

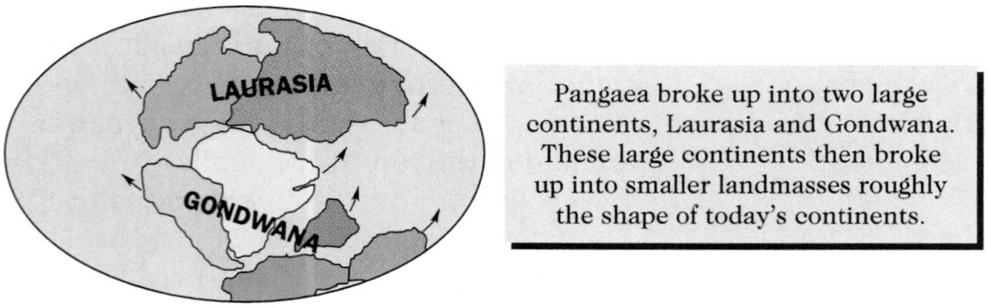
LAURASIA
GONDWANA
Pangaea broke up into two large continents, Laurasia and Gondwana. These large continents then broke up into smaller landmasses roughly the shape of today's continents.

(b) 135 million years ago

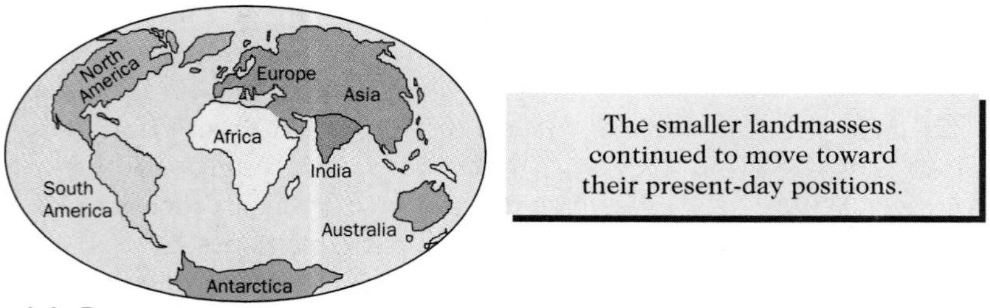
North America
Europe
Asia
Africa
India
South America
Australia
Antarctica
The smaller landmasses continued to move toward their present-day positions.

(c) Present

Geologic studies over the last 25 years have established what is known as **continental drift**; that is, the Earth's continents are located on large plates that slowly move over the Earth's surface. About 200 million years ago the major continents were locked together in one giant landmass called *Pangaea* (Figure 20.7). From that time to the present, the continental plates have been slowly moving apart to their present positions.

This movement separated continents and isolated certain groups of animals from each other. For example, Australia separated from the other continents about 65 million years ago. This isolated its population of marsupials before mammals that form placentas could occupy the landmass that was to become Australia. Isolated from competition with placental mammals, the marsupials of Australia paralleled the evolution of placental mammals on other continents. Thus there are marsupials that correspond to placental mammals such as wolves, mice, and squirrels (Figure 20.8). The only placental mammals present in Australia are the few that have been introduced by humans.

> Anatomic features of organisms, vestigial structures, embryonic development, differences in protein and nucleic acid sequences, and biogeographical isolation are used as evidence of an evolutionary relationship.

Mechanisms of Evolution

Earlier in this chapter we defined evolution as an inherited change in an interbreeding population of animals or plants. In this section, we will describe how such changes take place. At the outset, it is important to emphasize two concepts. First, to be significant for evolution, changes in the phenotype must be inherited changes. That is, they must result from a genetic change. Second, evolution is a phenomenon that works on populations over generations. In other words, populations evolve over time, whereas individuals do not.

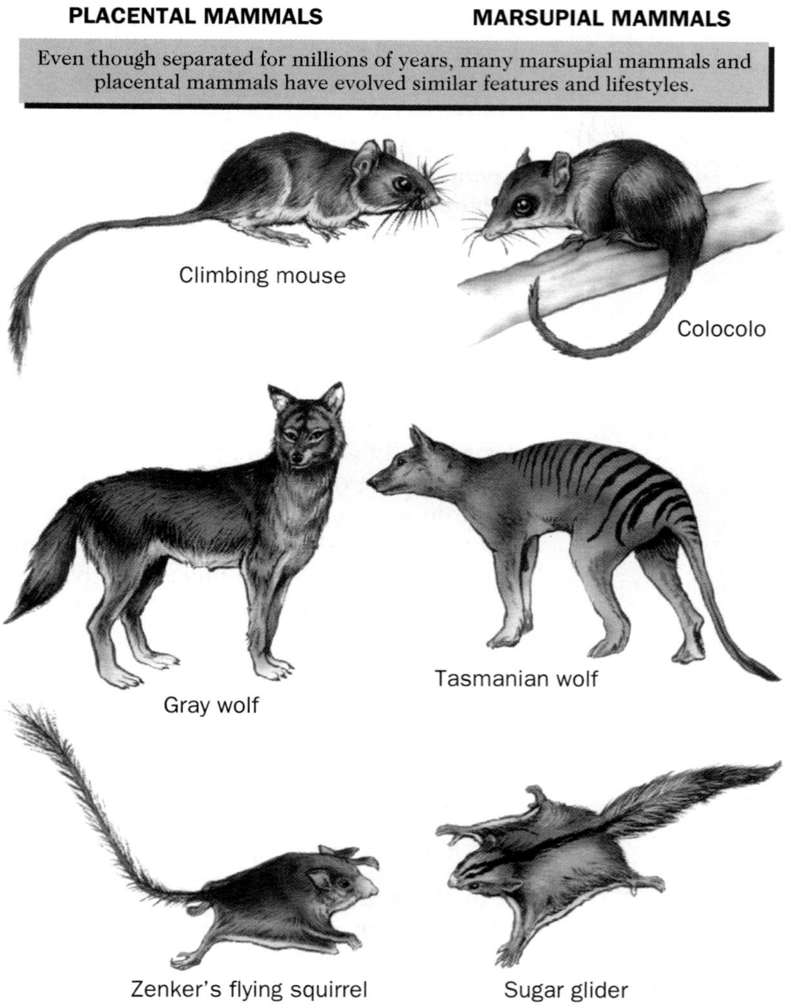
PLACENTAL MAMMALS
MARSUPIAL MAMMALS
Even though separated for millions of years, many marsupial mammals and placental mammals have evolved similar features and lifestyles.
Climbing mouse
Colocolo
Gray wolf
Tasmanian wolf
Zenker's flying squirrel
Sugar glider

FIGURE 20.8 Similarities between selected placental mammals and marsupial mammals.

The genetic composition of a population is defined as a **gene pool.** In any gene pool certain genes occur more frequently than do others. Evolution is the change in the frequency of genes in a gene pool. There are four major biological mechanisms that alter the gene frequencies of populations: natural selection, mutation, migration, and genetic drift.

Natural Selection

Natural selection is a mechanism that maintains genes favorable to the life of individuals within a population. What do the bacterial diseases gonorrhea, syphilis, and tuberculosis have in common? Through the operation of natural selection, the bacteria causing these diseases have become resistant to certain antibiotics (Chapter 12). Initially, only a small number of cells in the population of each bacterial species contained the gene for antibiotic resistance. As a particular antibiotic was used over time, it killed nonresistant cells. However, bacteria with the gene for antibiotic resistance survived and reproduced, increasing the frequency of this gene in the gene pool. In a span of a dozen or so years, a change in gene frequency occurred which resulted in the majority of bacteria having the antibiotic-resistant gene.

Charles Darwin coined the term **"natural selection"** to explain how the inherited traits of organisms within a population change over time. Fundamentally, natural selection favors the genes and phenotypic characteristics that offer a survival and reproductive advantage. Organisms without such advantages produce fewer offspring, and over generations they and their genes are slowly eliminated from the population.

The end result of natural selection is better *adaptation* to the local *environment.* The term "environment" encompasses all the external factors that impinge on an organism. These include physical factors such as temperature, pH, and sunlight as well as biological factors such as other organisms competing for food and shelter. The term "adaptation" concerns the phenotype and occurs when the genes inherited from the gene pool improve an individuals' chance to survive and reproduce in its environment.

Adaptations include diverse characteristics such as coat color, running speed, body covering, ability to conserve water, care for the young, and efficient use of energy. For example, organisms with genes for more effective camouflage may escape predation, increasing their chances of survival and reproduction (Figure 20.9). Similarly, organisms with genes that allow them to utilize an alternative nutrient may survive and reproduce during periods when the usual food is scarce. In both of these examples the organisms have inherited advantageous genes (natural selection) and are more capable of surviving condi-

Poorly adapted: Dark-colored moths are more susceptible to bird predator because they are not camouflaged.
Better adapted: Lighter-colored moths escape bird predator because they are camouflaged by light-colored growth on tree trunk.

FIGURE 20.9 An example of adaptation through camouflage.

tions of environmental change (adaptation).

The process of natural selection and adaptation is sometimes described as "survival of the fittest." In this context, "fittest" does not refer to the biggest, meanest, or toughest. Instead, it refers to those in the population that survive long enough to reproduce and leave viable offspring that reproduce. In this way, favorable genetic traits are transferred from generation to generation.

> Natural selection is a process in which certain organisms within a population have a selective advantage for survival because they have genes which better adapt them to their environment. The better adapted organisms are reproductively more successful.

Mutation: The Source of Genetic Variation

Natural selection acts on a population of various phenotypes. However, phenotypic variation results from genetic variation (Chapter 19). **Mutation** is the raw material for natural selection because it produces genetic variation. Mutations are random alterations in the DNA of genes. Altered genes in turn cause phenotypic changes (structural, physiological, or behavioral).

Mutations occur randomly as a result of enviornmental exposure to toxic chemicals or radiation, such as ultraviolet light. Because mutations are random, we cannot predict precisely where and when a particular mutation will occur. In addition, most mutations have damaging, not beneficial, effects. Damaging mutations usually are eliminated from populations because the organisms that carry them do not survive as long or leave as many progeny compared with other members of the species with more adaptive genes. However, the occasional beneficial mutation is the basis for natural selection and evolution.

> Mutations are random changes in DNA. Mutation is the raw material for evolution because it produces genetic variation.

Migration of Animals and Their Genes

Migration refers to the movement of individual organisms and their genes from one population to another. After migration and reproduction, the genes of the migrating individual become part of the gene pool of the new population. This movement (flow) of genes from one population to another is called **gene flow.** The beneficial result of gene flow is that a population can gain new genes and show increased genetic variation.

As an example, consider two hypothetical ponds separated by a distance sufficient to prevent frogs from easily migrating between them. As time passes, random mutations will have occurred in each frog population. Because mutations are random, the mutated genes will differ in the two populations. Now, a single frog migrates from one pond to the other. This could occur in a number of ways, such as through its own mobility, by flooding, or even by a child who carries the frog from one pond to the other. If the migrating frog successfully breeds, it will have introduced new alleles (genetic variation) into the population.

> Migration of an organism from one population to another can introduce new genes if reproduction is successful.

Genetic Drift: Small Populations, Large Changes

Genetic drift is a random change in the gene frequency of small populations caused by deaths or migration and isolation. In small populations the loss of a few individuals can make a big difference. We will describe two randomly occurring situations that most often lead to genetic drift: the bottleneck effect and the founder effect.

The bottleneck effect. Disasters such as fire, earthquake, a depleted food supply, and the outbreak of disease can drastically reduce the number of organisms in a population. A **bottleneck effect** occurs when a surviving population is very small (perhaps as few as a dozen individuals) and its gene pool is not representative of the original population. Some genes (alleles) found in the original population may not be present, while others may be much more plentiful than they were in the original population. As a result, since the surviving population has less genetic variability than did the original population, this can be dangerous for long-term survival.

A bottleneck effect might have occurred with the cheetah during the last ice age some 10,000 years ago. Today there are only about 15,000 cheetahs alive, and genetic studies have determined that these animals have little genetic variation. They are all surprisingly similar in their genotypes. How could this happen?

It has been hypothesized that the small numbers of cheetahs which survived the ice age were not genetically representative of the original population but contributed the genes found in today's population. In other words, that ice age produced a bottleneck effect, and the cheetahs that survived were genetically very similar (Figure 20.10). This is a serious problem for cheetahs today, because they are producing fewer offspring than expected, and there is concern that they may become extinct.

The founder effect. When a small number of organisms migrate from a large population and become established as a new population in another location, a **founder effect** can occur. As was described above, the gene pool of the smaller migrating group usually is not representative of the larger population. Some alleles may be absent, while others may be under- or overrepresented. As a consequence, when individuals reproduce and the "founding" population increases in size, its gene frequencies are different from those of the original population.

The Amish population in eastern Pennsylvania provides a clear example of the founder effect. Because of religious persecution, about 200 members of Amish families in Switzerland migrated to Pennsylvania between 1720 and 1770. These

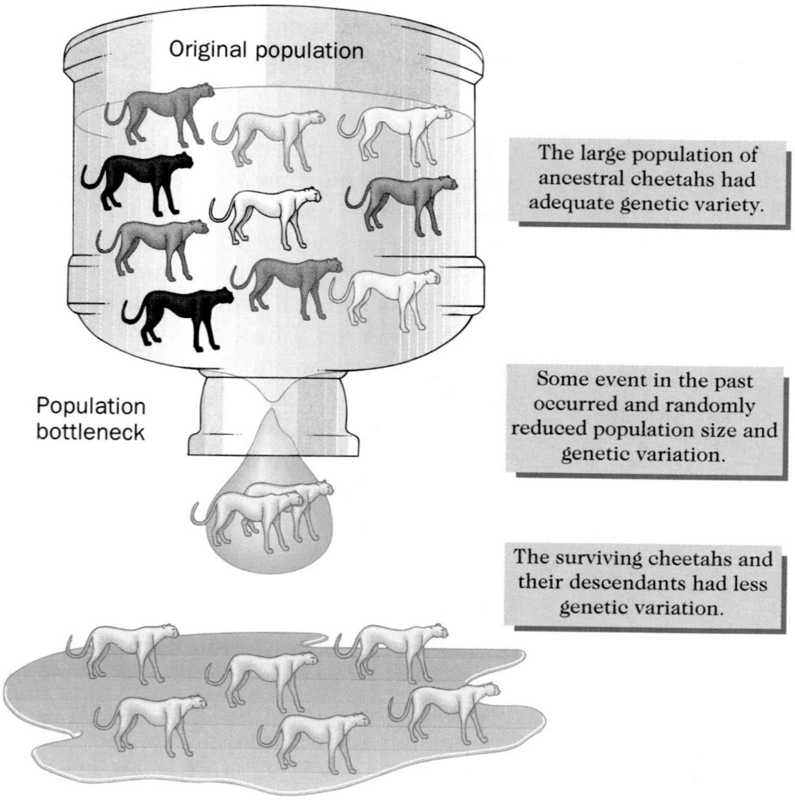
Original population
The large population of ancestral cheetahs had adequate genetic variety.
Population bottleneck
Some event in the past occurred and randomly reduced population size and genetic variation.
The surviving cheetahs and their descendants had less genetic variation.

FIGURE 20.10 The bottleneck effect occurred with the cheetah in the past.

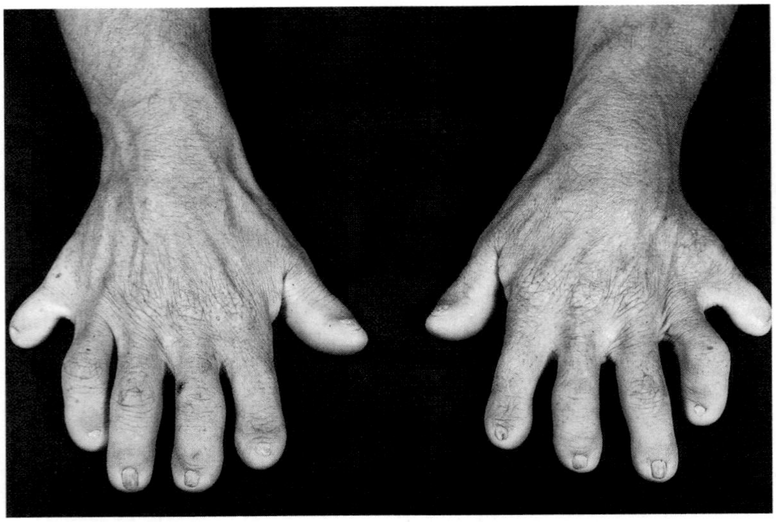

FIGURE 20.11 The hands of an adult with Ellis–van Creveld syndrome. Note the extra fingers on each hand.

200 individuals represent the founding population which produced the 12,000 Amish now living in Pennsylvania.

Genetic studies have determined that one of the founding couples carried a defective gene (allele) to America. The defective gene causes short arms and legs and extra fingers, a condition called the Ellis–van Creveld syndrome (Figure 20.11). Because of their religious beliefs, the Amish have remained reproductively isolated from the rest of society, and they have a higher incidence of this genetic defect. The frequency of the defect in the Amish population is 1 in 14, compared with 1 in 1000 in the general population.

> Small populations can experience a random event that alters the gene frequency for the entire population and future generations. Two examples of genetic drift are the bottleneck effect, involving natural disasters, and the founder effect, involving migration.

The Origin and Early History of Life on Earth

The Earth is believed to be about 4.6 billion (4,600,000,000) years old. It is hypothesized that the planet was formed from a swirling cloud of debris, dust, and gases that at one time encircled the sun. As gravitational forces compacted those materials, heat was released, melting the interior. Volcanoes and cracks in the crust permitted molten rock and gases to escape, forming the Earth's early atmosphere. These gases included carbon dioxide (CO_2), hydrogen (H_2), water vapor (H_2O), nitrogen (N_2), methane (CH_4), and ammonia (NH_3). Note that oxygen gas (O_2) was not present, and this was extremely important, as you will soon learn.

As the Earth cooled, water in the atmosphere condensed and fell as rain. When it encountered hot rocks, the water evaporated back into the atmosphere and then fell again as rain. For millions of years this cycle was repeated. Eventually the Earth cooled enough for liquid water to accumulate on the surface, forming oceans which contained soluble minerals eroded from rocks (Figure 20.12). It is within this ancient setting that life probably began on Earth about 4 billion years ago. The oldest fossils resemble bacteria and have been dated to be about 3.5 billion years old. In other words, after the Earth's formation, close to a billion years passed before the first living organisms were advanced enough to leave a fossil trace.

How did life on Earth originate? Of course, we cannot know for sure. There is no evidence that the formation of life from nonliving matter occurs under the current environmental conditions on Earth, and so such a process cannot be directly observed. However, by re-creating the early conditions on Earth, scientists can offer "rational speculation." In the discussion that follows, we will present one scientific explanation of how life might have originated from a mixture of molecules.

The Raw Materials for Life

During its early history the Earth was bombarded by energy in different forms: heat from the molten interior, lightning from clouds in the atmosphere, the impact of large meteors, and the sun's light. Such energy could have driven chemical reactions between CO_2, H_2, N_2, CH_4, NH_3,

FIGURE 20.12 How the Earth might have looked 4 billion years ago.

and H_2O in the atmosphere and oceans, producing complex organic molecules, the raw materials of life. The absence of O_2 from the early atmosphere permitted the accumulation of these complex molecules. Oxygen gas is such a reactive molecule that it would have broken down the early organic molecules, preventing their accumulation.

The possibility of producing organic molecules characteristic of living systems through such reactions was established by a series of experiments begun by Stanley Miller in the 1950s. In these experiments, gaseous mixtures were confined in an apparatus and subjected to energy from electrical discharges which resembled lightning (Figure 20.13). After a period of operation, analysis of the contents revealed a variety of amino acids, several sugar molecules, lipids, and nucleotides.

If something like this occurred over a billion years, the organic molecules formed would accumulate in pools, creating what can be described as an "organic soup." This soup contained the fundamental building blocks of all the living systems we encounter on the Earth (Chapter 1). The availability of these precursor molecules set the stage for a series of developments that ultimately produced the first living cell (Table 20.1).

During Earth's early history, chemical reactions between CO_2, H_2, N_2, CH_4, NH_3, and H_2O could have produced the organic molecules characteristic of living cells: amino acids, lipids, sugars, and nucleotides.

FIGURE 20.13 The apparatus used by Stanley Miller to simulate the conditions of early Earth.

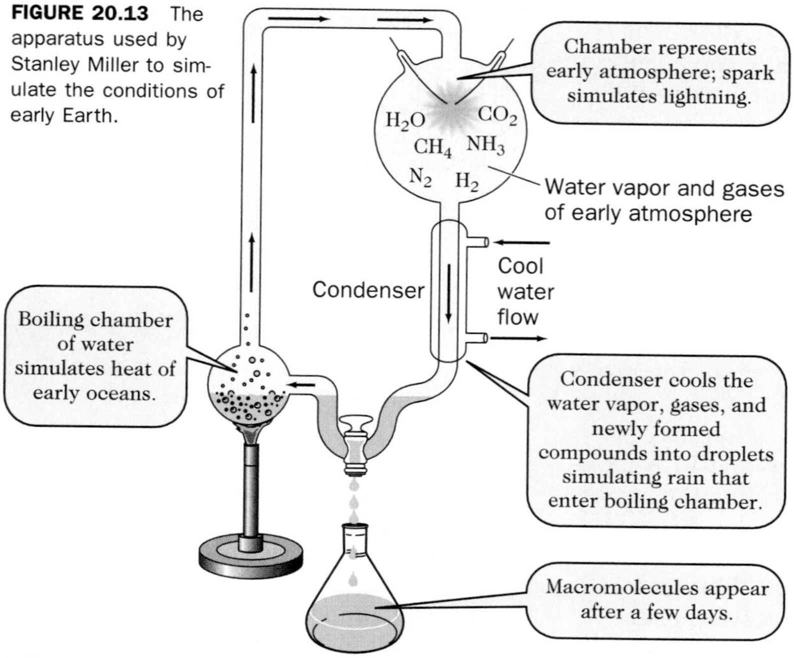

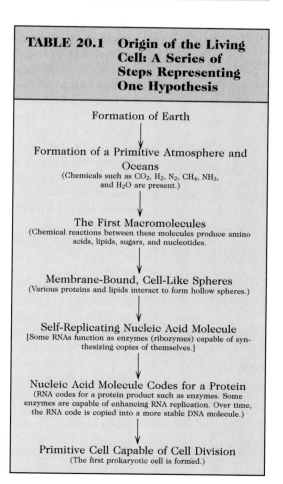

Chamber represents early atmosphere; spark simulates lightning.
H2O
CO2
CH4 NH3
N2 H2
Water vapor and gases of early atmosphere
Condenser
Cool water flow
Boiling chamber of water simulates heat of early oceans.
Condenser cools the water vapor, gases, and newly formed compounds into droplets simulating rain that enter boiling chamber.
Macromolecules appear after a few days.

TABLE 20.1 Origin of the Living Cell: A Series of Steps Representing One Hypothesis

Formation of Earth

↓

Formation of a Primitive Atmosphere and Oceans
(Chemicals such as CO_2, H_2, N_2, CH_4, NH_3, and H_2O are present.)

↓

The First Macromolecules
(Chemical reactions between these molecules produce amino acids, lipids, sugars, and nucleotides.)

↓

Membrane-Bound, Cell-Like Spheres
(Various proteins and lipids interact to form hollow spheres.)

↓

Self-Replicating Nucleic Acid Molecule
[Some RNAs function as enzymes (ribozymes) capable of synthesizing copies of themselves.]

↓

Nucleic Acid Molecule Codes for a Protein
(RNA codes for a protein product such as enzymes. Some enzymes are capable of enhancing RNA replication. Over time, the RNA code is copied into a more stable DNA molecule.)

↓

Primitive Cell Capable of Cell Division
(The first prokaryotic cell is formed.)

Macromolecules and the First Cells

Over the course of nearly a billion years organic molecules formed and accumulated in pools, creating an organic soup. Evaporation of water concentrated the soup on clay surfaces known to catalyze (speed) chemical reactions. Catalysis was as essential to the origin of life as it is to the function of cells.

When these conditions are re-created in the laboratory, polypeptides and nucleic acids are formed. Of course, these proteins are not the highly regular and functional molecules that cells make, nor do these nucleic acids consistently carry coded messages as they do in a cell. Instead, the soup and the clay were just the means of linking small molecules together in a random or nearly random order. These processes provided the raw materials from which future refinements would arise.

The origin of a membrane. One of the most important events in the chemical origin of life was the origin of a membranelike structure. Laboratory evidence shows that in water, proteins and lipids spontaneously form into microscopic hollow spheres that have characteristics of the plasma membrane of a living cell. Such spheres swell and shrink in response to changes in osmotic conditions and split in half as their volume enlarges. Within the interior of these hollow spheres, molecules could accumulate, and conditions might favor chemical reactions that could not take place in the organic soup outside.

Which evolved first, nucleic acids or proteins? We know that both nucleic acid and protein molecules are essential to living systems. How did the functional relationship between nucleic acid and proteins originate? Which evolved first, the nucleic acid and its code or the protein product of the code?

One answer to this question may be provided by the discovery that RNA molecules are capable of catalyzing the synthesis of other RNA molecules. Before this discovery was made, it was believed that only proteins functioned as biological catalysts (enzymes). The discovery of catalytic RNA, called a *ribozyme,* has led to the suggestion that RNA became abundant before DNA or specific proteins did.

At some point nucleic acid molecules began directing the synthesis of proteins. This might have happened when a protein acted as an enzyme and enhanced the replication of RNA. This association conferred a survival advantage and produced an abundance of both the RNA and the protein. If these events occurred within a membrane-bound structure, that structure might have evolved into the first cell, using RNA as its hereditary material, as do RNA viruses today.

DNA for heredity: a "late" arrival. Eventually, the more stable DNA molecule took over the role of the hereditary material. Then RNA became involved in directing protein synthesis, and the all-important function of catalyzing chemical reactions was taken over by proteins (enzymes).

The evolution of DNA (for heredity), RNA (for protein synthesis), and proteins (as enzymes) must have occurred very early in the evolution of cells, because all cells show this pattern. Of course no one knows exactly what occurred, and there is no agreement among scientists regarding the most likely events. All that can be said is that research findings show that given a time frame of hundreds of millions of years, such reactions could have taken place on the early Earth.

> Proteins and lipids can form membrane-bounded hollow spheres with cell-like properties. RNA probably preceded both the synthesis of specific proteins and the use of DNA as the hereditary material.

The Earliest Cells: The Prokaryotes

The earliest cells were probably simple self-replicating systems involving DNA and resembling present-day prokaryotes (Chapter 2). Like all cells, they required raw materials and energy. However, the earliest cells probably could not synthesize most of the molecules they required. Instead they absorbed sugars, nucleotides, lipids, and amino acids from their organic soup environment. Because there was no

O_2 available, the first cells lived off this soup by *anaerobic metabolism,* which consists of chemical reactions that do not require O_2.

Harvesting energy from sunlight. About a billion years after the earliest cells proliferated, something happened that forever changed both life and the Earth. Through a series of mutations, some early cells acquired the ability to harvest the abundant energy of sunlight to manufacture complex carbon molecules (organic molecules) from CO_2 and H_2O. This manufacturing process is called **photosynthesis.** It proved to be a great advantage because these cells were able to manufacture their organic molecules from the plentiful supply of CO_2 and H_2O. A by-product of photosynthesis is O_2, which began to increase in the atmosphere.

The production of O_2 can be dated because O_2 reacts with iron ore, converting it to iron oxide, or rust. Analysis of rock and sediment samples shows that iron oxide started accumulating about 2.5 billion years ago, which must have been about the time when photosynthesis evolved. This is about 1 billion years after the earliest known fossil cells.

O_2 changes the Earth and life. The accumulation of reactive O_2 in the atmosphere destroyed the energy-rich molecules in the soup and attacked anaerobic cells. Some of those cells evolved the capacity not only to detoxify O_2 but to use it to extract energy from organic molecules, a process called *aerobic metabolism.*

Cells using aerobic metabolism have an advantage over anaerobic cells, because much more energy is generated during aerobic metabolism than during anaerobic metabolism (Chapter 7). With the evolution of aerobic metabolism, the stage was set for the diversification of prokaryotic cells and the origin of eukaryotic cells and multicellular organisms.

Eukaryotic Cells and Multicellular Organisms

Eukaryotic cells originated from prokaryotic cells about 1.5 billion years ago, as indicated by fossils. Since then, they have become the dominant cell type on Earth, displaying the most diversity and giving rise to multicellular organisms. Multicellular animals and plants appeared some 670 million years ago. These organisms clearly had great advantages over single-celled organisms, because the fossil record shows a continual evolution and expansion of complex multicellular organisms composed of eukaryotic cells.

What were some of the selective advantages of multicellular forms? For one thing, multicellular organisms expanded their source of food as they became able to prey on unicellular forms. In addition, multicellular systems permitted specialization and division of labor among cells. This increased efficiency permitted wider exploitation of places to live and things to eat. Figure 20.14 presents a geologic time scale that illustrates the evolution from precells to complex multicellular forms.

> The first cells were anaerobic and prokaryotic. Some prokaryotic cells evolved photosynthesis. A by-product of photosynthesis, O_2, began accumulating in the environment. Prokaryotic cells evolved that used O_2. Eukaryotic cells evolved from prokaryotes.

Human Evolution

The human species appeared on Earth in the same way as all other living species, through the process of evolution. The evolutionary path that ultimately led to humans began with the earliest primates some 65 million years ago. It was not until about 4 million years ago that the first members of the human family begin to appear in the fossil record. We are a relatively new species, for our evolution has occurred only over the last 0.1 percent of the Earth's history. The question then arises: What is the evidence that supports human evolution?

Much of the evidence for human evolution is derived from the fossil record. Although the number of fossils available for studying human evolution is large, these fossils are not sufficiently complete to finalize an understanding of our ancestry. Therefore, there are scientific disagree-

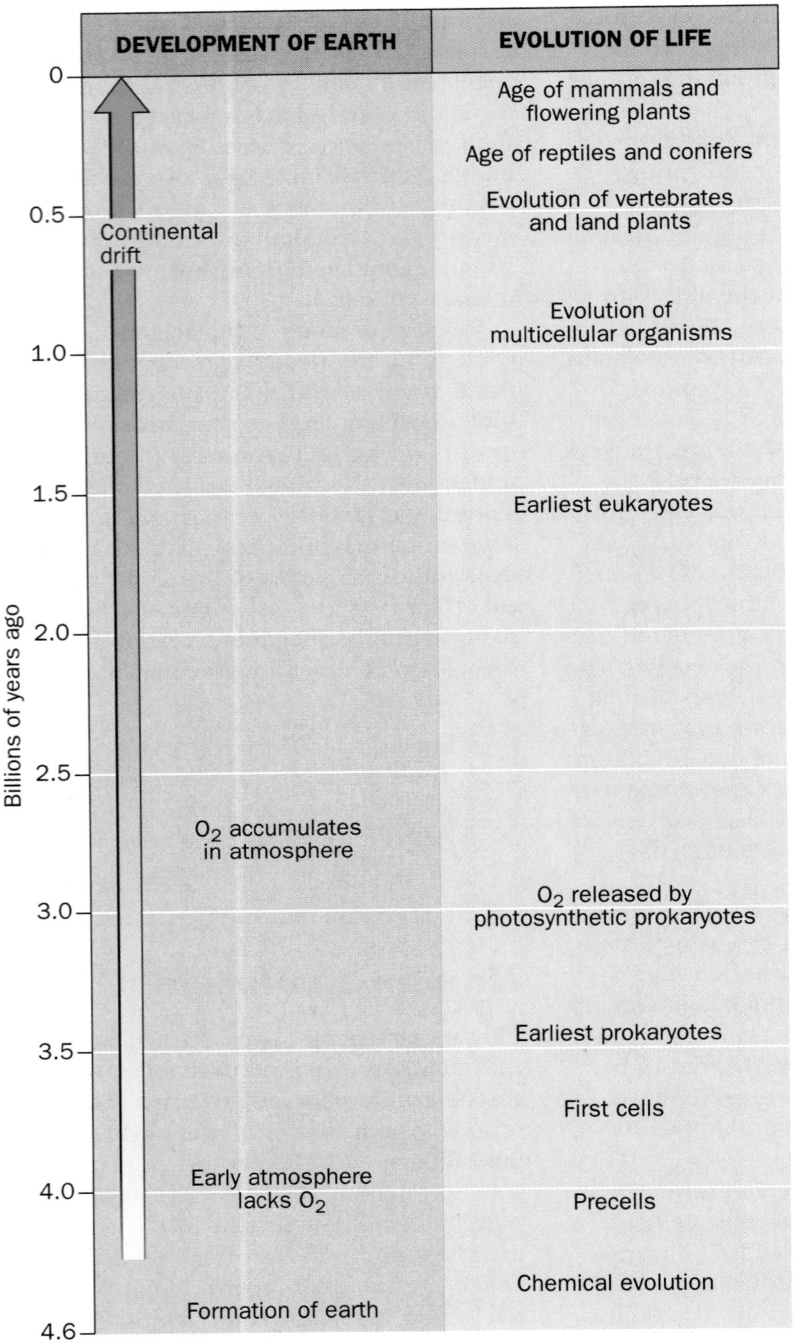

DEVELOPMENT OF EARTH
EVOLUTION OF LIFE
Billions of years ago
0
0.5
1.0
1.5
2.0
2.5
3.0
3.5
4.0
4.6
Continental drift
O_2 accumulates in atmosphere
Early atmosphere lacks O_2
Formation of earth
Age of mammals and flowering plants
Age of reptiles and conifers
Evolution of vertebrates and land plants
Evolution of multicellular organisms
Earliest eukaryotes
O_2 released by photosynthetic prokaryotes
Earliest prokaryotes
First cells
Precells
Chemical evolution

FIGURE 20.14 A summary of important events during the evolution of life since the Earth's beginning.

ments about the interpretation of some fossil evidence. The resolution of such disagreements will come about as additional fossils are discovered.

Such disagreements, however, should not be interpreted to mean that science is questioning whether human evolution occurred. Indeed, scientific thinking universally accepts the idea that humans evolved from apelike ancestors. These disagreements are actually what science and the scientific method are all about, that is, proposing hypotheses based on observations. At this time the scientific data are not complete enough for a universal consensus about how humans evolved.

We will begin this section with a brief explanation of how organisms are classified and named. Then we will describe our primate origins. We will conclude the section by tracing the evolution of our ancestry.

Classifying and Naming Organisms

Taxonomy is the branch of biology that classifies living things into groups on the basis of similarities in structure, genetics, biochemistry, embryology, and behavior. If two living things have many similarities, an interpretation is made that they are related by descent from a common ancestor. The science that studies relatedness on the basis of taxonomy is called *phylogeny.* It is phylogeny that tells us that bats are more closely related to monkeys than to birds. Taxonomy and phylogeny are supportive pursuits. The goal of taxonomy is to create classification groups that accurately reflect relatedness (phylogeny).

Seven major categories are used in taxonomy: kingdom, phylum, class, order, family, genus, and species. We have defined a **species** as a group of similar organisms that can interbreed in nature and produce offspring. A **genus** represents a group of related species, and a **family** is a group of related genera (plural of *genus*). Table 20.2 discusses the seven taxonomic categories as they apply to humans.

An organism's scientific name is represented by its genus and species name and appears in italics. For example, the genus name assigned to humans is *Homo* (Latin for "man") and the species name is *sapiens* (Latin for "wise"). Our scientific name therefore is *Homo sapiens,* abbreviated *H. sapiens.*

TABLE 20.2 The Taxonomic Classification of Modern Humans

Taxonomic Category	Scientific Name	Characteristics and Examples of Organisms in Each Taxonomic Category
Kingdom	Animalia	Many-celled organisms, eukaryotic, with a structurally complex anatomy. Includes all animals, such as sponges, jellyfish, worms, insects, clams, sea urchins, fish, amphibians, reptiles, birds, and mammals.
Phylum	Chordata	Animals with a backbone and nerve cord. Includes sea squirts, fish, amphibians, reptiles, birds, and mammals.
Class	Mammalia	Chordates with mammary glands and hair during all or part of the life cycle. Includes mice, cats, dogs, horses, kangaroo, whales, apes, and humans.
Order	Primates	Mammals with five digits, flat fingernails, and stereoscopic vision. Includes lemurs, monkeys, apes, and humans.
Family	Hominidae	Primates with an upright stance and bipedal locomotion and an enlarged brain. Includes modern humans, our human ancestors, and *Australopithecus.*
Genus	*Homo*	Hominidae with an enlarged brain, skull anatomy similar to that of modern humans, and social and cultural characteristics. Includes modern humans and our immediate ancestors.
Species	*sapiens*	*Homo* with large brains and complex social and cultural characteristics. Includes modern humans.

Taxonomy is the area of science which categorizes organisms into groups that show evolutionary relationships. The seven taxonomic categories are kingdom, phylum, class, order, family, genus, and species.

Our Primate Ancestors

Humans are placed into the taxonomic class **Mammalia.** All mammals exhibit two defining characteristics: *body hair* during all or part of life and *mammary glands* that produce milk to nourish their offspring. One group of mammals, the **primates,** is a taxonomic order that includes humans, apes, and monkeys.

As a taxonomic order, all primates share the following characteristics: (1) There are five digits on each hand, with a thumb opposable to the other four. (2) The four limbs can rotate in bony sockets. (3) There are flat fingernails and toenails instead of claws or hooves. (4) The eyes are forward-facing, which leads to three-dimensional (stereoscopic) vision and good perception of distance.

All these characteristics evolved as adaptations to life in trees, which is called an *arboreal lifestyle.* (Humans have fully adapted to life on the ground.) An opposable thumb, nails on the fingers and toes, and easily rotated limbs all improve an organism's ability to grasp branches and swing from branch to branch. Three-dimensional vision results in an increase in depth perception. This becomes important for judging distances while moving through trees as well as for capturing prey.

Primates are divided into two taxonomic groups: prosimians and anthropoids (Figure 20.15). Present-day **prosimians** (meaning premonkeys) are tree-dwelling primates ranging in size from that of a mouse to that of a cat. The **anthropoids** include old world and new world monkeys, apes, and humans. The fossil record indicates that monkeys were the first anthropoids to evolve from prosimians some 40 million years ago in Africa (Figure 20.15). The evolution of apes and humans, collectively called **hominoids** (meaning "humanlike"), is a more recent event which we'll discuss next.

Humans are primates. Primates have five digits with one digit opposable, limbs that easily rotate, fingernails and toenails, and three-dimensional vision. Anthropoids are primates that include monkeys and hominoids.

The Hominoids: Apes and Humans

Apes and humans are closely related. Apes, including gibbons, orangutans, gorillas, and chimpanzees, evolved from the old world monkeys of Africa approximately

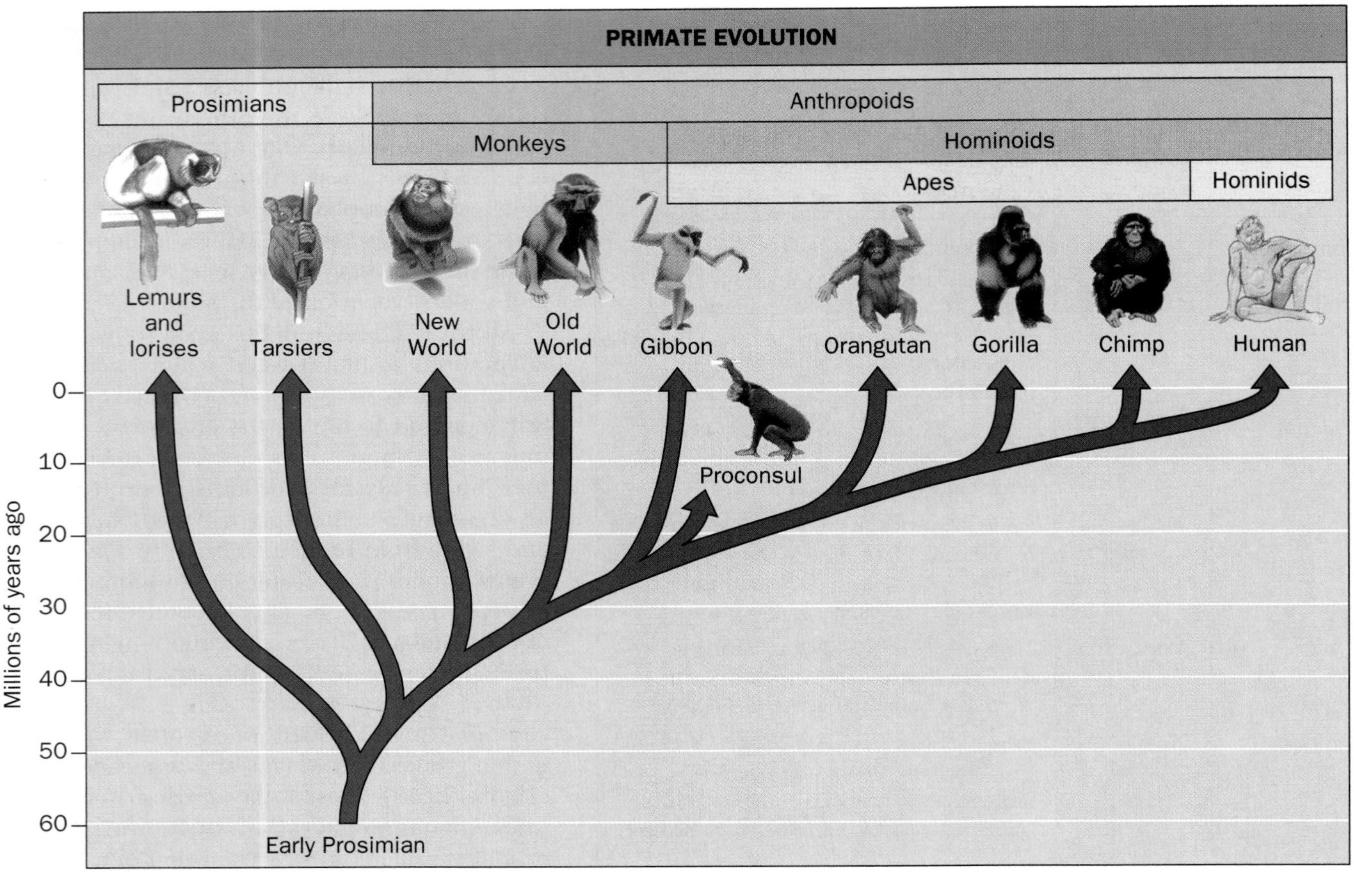
PRIMATE EVOLUTION
Prosimians
Anthropoids
Monkeys
Hominoids
Apes
Hominids
Lemurs and lorises
Tarsiers
New World
Old World
Gibbon
Orangutan
Gorilla
Chimp
Human
Proconsul
Early Prosimian
Millions of years ago
0
10
20
30
40
50
60

FIGURE 20.15 The primates and their evolutionary relationships.

25 million years ago. Compared with monkeys, apes usually are bigger, their brains are larger, they lack tails, and they exhibit more complex behaviors.

About 25 million years ago, the hominoids diverged from the old world monkey as is illustrated by an arboreal primate called *Proconsul* (Figure 20.15). Fossil evidence and DNA comparisons of apes show that the gibbon diverged first, followed by the orangutan, the gorilla, and then the chimpanzee. DNA comparisons between apes shows a progressive difference in DNA as one proceeds from gibbon, to orangutan, to gorilla, and finally to the chimpanzees. Since DNA differences result from mutations that occur over time, the progressive differences are interpreted to mean that the gibbon was the first of the modern apes to evolve and the chimpanzee was the last.

Similar molecular studies have compared the DNA of humans and modern apes. Such studies show that humans are most closely related to the chimpanzee. For example, there is only a 1.6 percent difference between the DNA nucleotide sequences of humans and chimpanzees but a 4 percent difference between those of humans and gorillas. From the fossil record and molecular studies, it appears that the evolutionary path leading to modern humans began when our early ancestors separated from the ape lineage some 4 million to 5 million years ago. We and our human ancestors are collectively known as **hominids.** The emergence and evolution of hominids are described next.

> Apes and humans are collectively called hominoids. DNA studies show that humans are more closely related to chimpanzees. Evidence suggests that human ancestors—hominids—separated from the ape lineage 4 million to 5 million years ago.

The Earliest Hominids: *Australopithecus*

Two important characteristics are used to separate our ancestors from the apes: walking upright (bipedal locomotion) and enlargement of the brain. Evidence shows that walking upright occurred first, followed by a gradual increase in brain size.

The oldest hominid fossil discovered thus far is *Australopithecus afarensis* (*aw-stray-low-**pith**-ih-cuss*). A popularly known example of *A. afarensis* is a specimen found in the Afar Desert of Ethiopia in 1974. Its nickname is "Lucy," after the popular Beatles song of 1974, "Lucy in the Sky with Diamonds." The fossil skeleton of Lucy was 40 percent complete and was dated to be about 3.5 million years old (Figure 20.16).

Other fossils of *A. afarensis* have been discovered since 1974, and they share three characteristics: (1) They walked upright, as determined by pelvis and leg bone anatomy. (2) Their faces and skulls were apelike with a cranial capacity about 450 cubic centimeters (cc). This is about the cranial capacity of a chimpanzee. In contrast, the human cranial capacity is a little less than 1400 cc. (3) They were between 3.5 and 4.5 feet tall. Further evidence that hominids could walk upright some 3.5 million years ago comes from hominid footprints discovered in volcanic ash that had hardened into place about that long ago.

From the fossil evidence, it appears that *A. afarensis* gave rise to two lineages: one continuing the genus *Australopithecus* and one leading to the genus *Homo*. Three examples of australopithecine evolution are *A. africanus, A. robustus,* and *A. boisei.* All were apelike, walked upright, and had a cranial capacity of approximately 500 cc. At present the fossil evidence for understanding their evolutionary relationship is incomplete. Consequently, several different hypotheses have been proposed to explain their evolutionary path. Figure 20.17 presents one possible evolutionary relationship among these species. Australopithecines disappeared about 1.5 million years ago, and since then the only living hominid has been *Homo*.

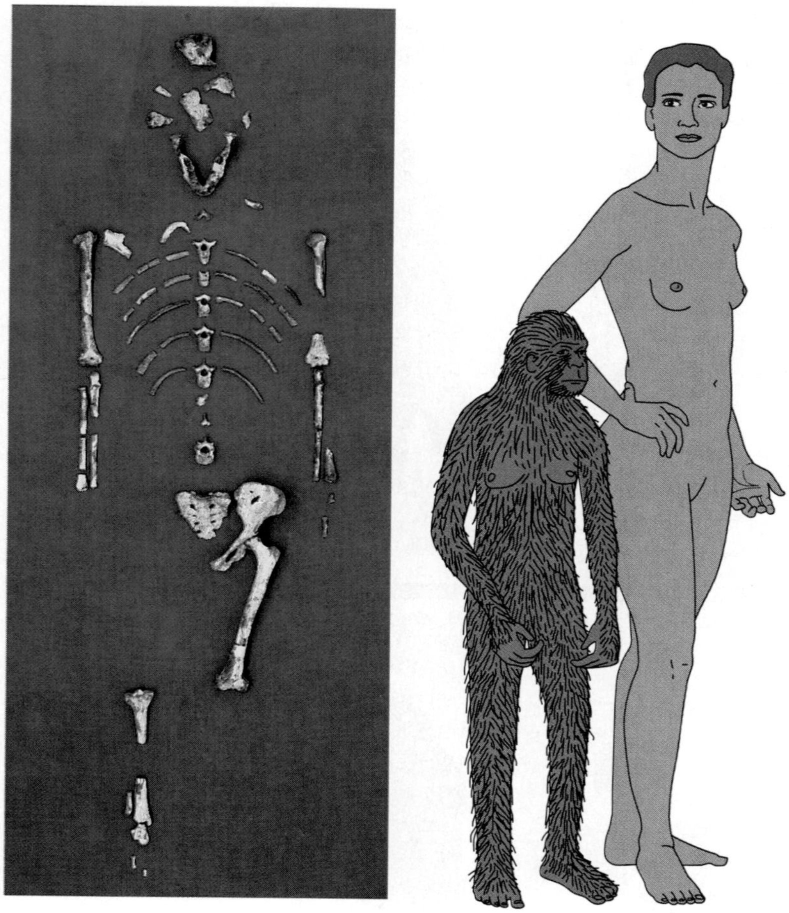

(a) About 40 percent of Lucy's skeleton was preserved.

(b) Lucy was a female, as determined by her pelvis. She was about 3 feet 7 inches tall and weighed about 60 pounds. She died in her late teens or early twenties (her wisdom teeth had come through).

FIGURE 20.16 (a) The fossil remains of Lucy. (b) A drawing of how Lucy might have looked compared with a modern human female.

> In contrast to apes, humans walk upright and have larger brains. The oldest hominid discovered thus far is *Australopithecus afarensis.* The fossil of an *A. afarensis* called "Lucy" lived about 3.5 million years ago.

The Earliest Human: *Homo habilis*

Early fossils of the first human, ***Homo habilis,*** have been dated to be 2 million years old. It appears that *H. habilis* coexisted with the australopithecines in Africa for nearly 1 million years (see Figure 20.17). Both groups walked upright on two legs and had apelike facial features, but *H. habilis* was slightly taller (5 feet) and had a slightly larger brain (700 cc) (Figure 20.18). From the structure of their teeth, it appears *H. habilis* could eat meat in addition to the plants that formed the diet of the australopithecines.

An increase in brain size appears to be linked with an increase in intelligence, for it is among the fossils of *H. habilis* that the first stone tools are found. These were simple tools crudely fashioned from large stones or sharp stone chips. The tools were used for butchering animals as well as cutting and chopping plants and meat (*Homo habilis* means "handy man"). From the type and size of their tools and from animal remains at fossil sites, it is speculated that *H. habilis* hunted only small animals. Any large animals obtained were probably scavenged from the kills of larger predators.

The earliest human. *Homo habilis,* existed 2 million years ago. The increased brain size of *H. habilis* is linked with increased intelligence and the use of primitive tools.

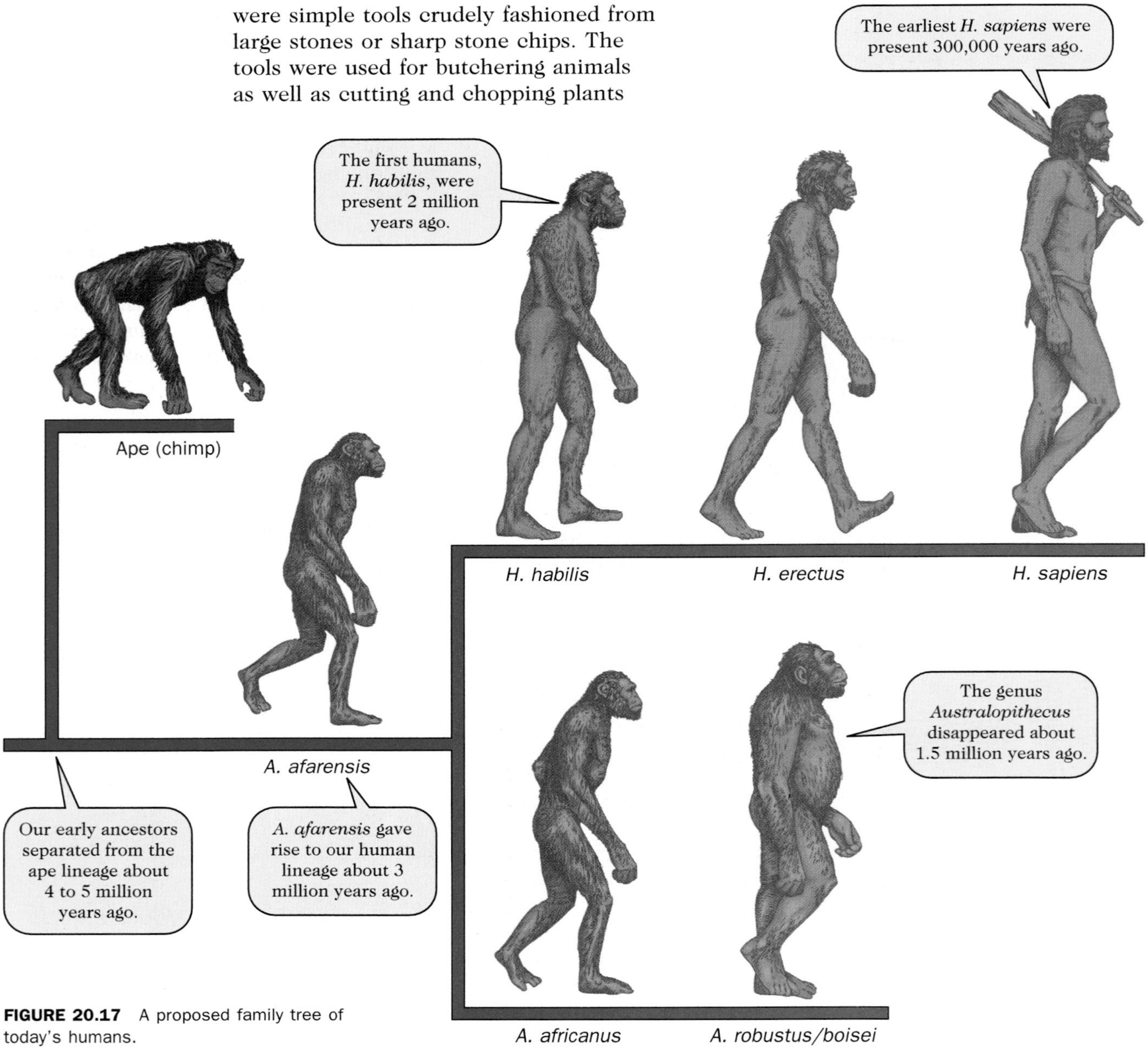
The earliest *H. sapiens* were present 300,000 years ago.
The first humans, *H. habilis*, were present 2 million years ago.
Ape (chimp)
H. habilis
H. erectus
H. sapiens
A. afarensis
The genus *Australopithecus* disappeared about 1.5 million years ago.
Our early ancestors separated from the ape lineage about 4 to 5 million years ago.
A. afarensis gave rise to our human lineage about 3 million years ago.
A. africanus
A. robustus/boisei

FIGURE 20.17 A proposed family tree of today's humans.

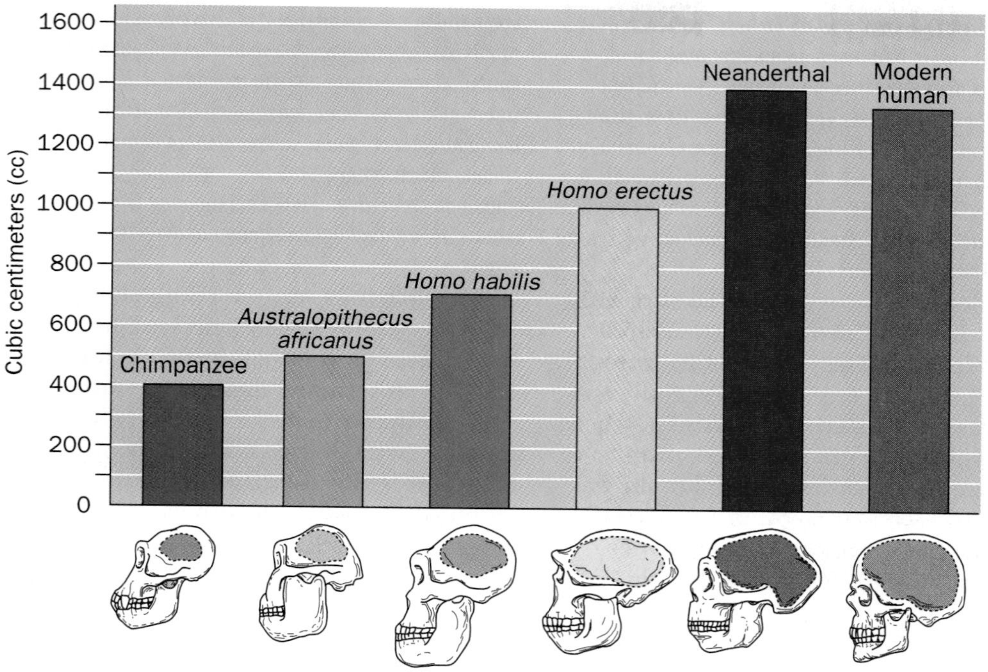
1600
1400
1200
1000
800
600
400
200
0
Cubic centimeters (cc)
Chimpanzee
Australopithecus africanus
Homo habilis
Homo erectus
Neanderthal
Modern human

FIGURE 20.18 Comparing the skull shape and brain size of chimpanzees, our human ancestors, and modern humans.

Our Immediate Ancestor: *Homo erectus*

Homo erectus (meaning "upright man") appeared as descendants of *H. habilis* about 1.5 million years ago (see Figure 20.17). They originated in Africa and ultimately spread into Asia and Europe. Fossil remains show that these humans had a brain size averaging 1000 cc, had facial features that were still rather apelike, were taller than *H. habilis,* and made more sophisticated tools.

Skull remains indicate that the brain centers devoted to more complex thought were developing; however, *H. erectus* probably were not capable of verbal communication through language. However, *Homo erectus* did form social groups of 20 to 50 individuals and lived in shelters such as huts and caves.

Fossil evidence indicates that they hunted large animals for food and clothing. This probably would have required group cooperation. Evidence from shelters shows that *H. erectus* used fire about 500,000 years ago. However, it is difficult to know what purposes fire served. Perhaps it was used for cooking, for keeping warm, for defense, or for all these purposes.

For the first time there were social groups that cooperated and interacted together in activities such as hunting, using fire, and sharing food and shelter. This indicates that *H. erectus* was beginning to develop culture. **Culture** includes knowledge, beliefs, myths, art, patterns of hunting and feeding, morals, laws, and the material expressions of those things (such as tools, buildings, and clothing). The beginning of culture means that knowledge and skills were transferred between individuals and passed from generation to generation. Many of the social and behavioral advances of *H. erectus* were no doubt instrumental in their successful migration from Africa to Asia and Europe, where the climate was colder and the environment less hospitable.

> *Homo erectus* descended from *H. habilis* about 1.5 million years ago. *H. erectus* had a larger brain. They formed social groups, used fire, and developed culture, which contributed to their successful migration from Africa to Europe and Asia.

The Human Species: *Homo sapiens*

The earliest fossils having features modern enough to be identified as ***Homo sapiens*** (meaning "wise humans") have been dated to be over 300,000 years old and probably evolved from *H. erectus* (see Figure 20.17). These early humans had features similar to those of *H. erectus,* and so they looked somewhat different from modern humans. Therefore, our species, *Homo sapiens,* is often separated into two groups: the early, or archaic, humans and the modern humans.

Archaic humans. Early humans migrated throughout a large part of the old world. Most of our fossil information about archaic humans centers on one group of *Homo sapiens* called **Neanderthals**. Fossils of this group were first discovered in the Neander Valley of Germany, but additional fossil discoveries show that they lived in the Middle East and parts of Asia from about 130,000 to 35,000 years ago.

Neanderthals were relatively short with a sturdy and muscular build. Compared with modern humans, their skulls were massive. Their faces projected slightly, and their foreheads sloped backward. They had heavy eyebrow ridges and a less pronounced chin. The Neanderthal brain was anatomically similar to ours, but it was slightly larger, averaging about 1400 cc (see Figure 20.18). It may be that this large brain size was related to increased muscle mass and control rather than intelligence.

These early humans lived in small groups and exhibited more advanced behavioral and cultural characteristics. As toolmakers, Neanderthals were skilled. In addition to stone weapons for killing animals, they made stone tools for domestic uses such as scrapping animal hides. It is thought that they wore the prepared hides as clothing.

FIGURE 20.19 Neanderthals burying a deceased clan member. Note the ceremonial use of flowers, tools, and weapons.

Neanderthals also buried their dead, and their burial sites reveal further cultural advances. For example, some fossilized skeletons were of relatively old individuals, while other skeletons show bone fractures that had healed before death. Furthermore, analysis of burial sites shows a ritualistic pattern in which the dead were buried with weapons, food, and flowers (Figure 20.19). The accumulation of such findings at burial sites indicates that Neanderthals not only cared for the elderly and injured but may have developed abstract thought such as belief in an afterlife.

Modern humans. Our understanding of how modern humans originated is unsettled and debated by scientists. Two hypotheses have been proposed to explain the origin of modern humans: the multiregional and the replacement hypotheses. The **multiregional hypothesis** proposes that *H. erectus* migrated from Africa about 1 million years ago and spread to different regions of the world. In each geographic region, they gradually evolved into modern humans. Within each geographic region modern humans developed into different races in Asia (Orientals), Europe (Caucasians), Australia (Aborigines), Africa (Africans), and so forth. Although they were somewhat geographically isolated, interbreeding and gene flow occurred between the different racial populations.

The **replacement hypothesis**, also known as the **"Out of Africa" model**, proposes that modern humans first evolved in Africa about 200,000 years ago. According to this hypothesis, modern humans began migrating out of Africa about 100,000 years ago and spread throughout the world. They coexisted with Neanderthals in Europe, whose ancestors had previously migrated out of Africa. Neanderthals, however, became extinct about 35,000 years ago. They were replaced by modern humans who presumably outcompeted them. The human races of today developed from the modern humans who migrated out of Africa some 100,000 years ago.

Analysis of the DNA in human mitochondria supports the replacement hypothesis. Mitochondrial DNA is inherited entirely from the mother and does not undergo the normal recombinations associated with nuclear DNA. For this and other reasons, analysis of human mitochondrial DNA may be a guide to the ancestry of human populations. Research comparing the variation in mitochondrial DNA from humans throughout the world concluded that the mitochondria of all living humans can be traced to a female who was a member of a group that lived in Africa about 200,000 years ago. Although there is debate among scientists about the validity of this conclusion, the press has popularized this research and calls this single female "Mitochondrial Eve."

However modern humans evolved, fossils with modern features existed in Europe about 40,000 to 30,000 years ago. These modern humans are commonly called **Cro-Magnons,** after a location in southern France where their fossils were first found. They were anatomically similar to us. Their brain size (1330 cc) and their skull shape, jaw, teeth, and face were almost indistinguishable from ours. If groomed and dressed in modern attire, they would be unnoticed by most of us on any city street.

Archaeological evidence from the caves in which they lived show that Cro-Magnons were advancing culturally as well. Not only were their tools and weapons made from stone, they used stone tools to make other tools and weapons from bone, antlers, ivory, and wood. The sizes of the groups in which they lived increased, and the groups became socially more complex. They gathered and ate grains, berries, roots, and other plant material. Groups of Cro-Magnons also hunted large animals, and this probably involved some form of verbal communication.

Art, another form of communication, first appears with the Cro-Magnon, who carved small figurines and drew on cave walls. Their paintings were large and colorful and depicted animals as well as hunting scenes (Figure 20.20). It is speculated that such paintings were done for ritualistic purposes such as to assure hunting success. Burial sites became more elaborate. They not only buried their dead with tools, weapons, and ornaments but used pigments to decorate the dead. Such burial activity suggests that Cro-Magnons were developing a religion.

FIGURE 20.20 Photograph of a cave painting by Cro-Magnons.

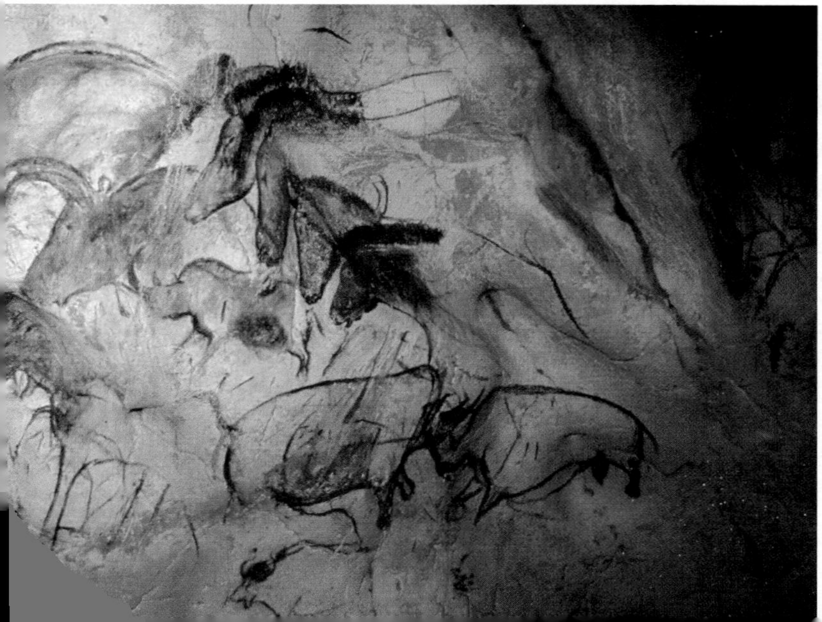

> *Homo sapiens* probably evolved from *H. erectus.* The origin of modern humans is explained by the multiregional hypothesis or the replacement hypothesis. The first modern humans, such as the Cro-Magnon, were culturally complex, used language and art, and had burial rituals.

The Beginnings of Civilization

As modern humans became more successful, they continued to migrate over the entire Earth. Modern humans learned how to plant and cultivate food crops and began to domesticate wild animals for food and clothing. Slowly, human groups changed from a society of hunters and gatherers to a society based on agriculture. *Agricultural societies* are known to have existed 10,000 years ago. They were more sedentary because of the need to care for crops and domesticated animals.

As a result, permanent settlements developed with larger numbers of people living and interacting together.

As a species, modern humans have been very successful. About 10,000 years ago the world population of humans is estimated to have been approximately 10 million. Today the world's population is approximately 5.6 billion. This increase has occurred primarily within the last 300 years because of improved nutrition, sanitation, and health care. However, such large numbers of humans have taxed Earth's resources. In Chapters 21 and 22 we'll examine the Earth's ecology and its deterioration in the face of pollution, erosion, extinction of animal and plant species, global warming, ozone depletion, and starvation.

> About 10,000 years ago modern humans changed from a hunter-gatherer society to an agricultural society. Crops were cultivated, animals were domesticated, and permanent residences were established.

Chapter Summary

Charles Darwin and the Science of Evolution

- Charles Darwin proposed the mechanism called natural selection to help explain biological diversity.
- Natural selection states that a population of organisms can change over time because some individuals have inheritable traits that better adapt them to their environment, allowing them to produce more offspring.

Evidence for Evolution

- Fossils develop when minerals are deposited in bone and other hard tissues of dead organisms. They can be used to view a series of structural changes in extinct organisms.
- Organisms that appear to be different may have similar anatomic features that show descent from a common ancestor.
- Vestigial structures are poorly developed structures of marginal use that represent remnants of fully developed and functional structures in related organisms.
- Early embryos of different vertebrates show structural similarities that represent an evolutionary relationship.
- Similarities in protein and DNA sequences show an evolutionary relationship.
- Geologic changes can help explain variations in plant and animal distribution.

Mechanisms of Evolution

- With natural selection, genes favorable to a population of organisms are maintained through inherited traits that increase adaptation and reproductive success.
- Gene alterations through mutation are the raw material for natural selection.
- New genes can be introduced into a population as organisms from one population migrate to another population and successfully reproduce.
- Genetic drift is a random change in the gene pool of small populations resulting from death, migration, or isolation.

The Origin and Early History of Life on Earth

- The Earth is estimated to be 4.6 billion years old.
- The original atmosphere contained chemical compounds that reacted together to form biological molecules.
- The earliest cells were prokaryotic and absorbed molecules from the organic soup. Cells developed that produced their organic molecules by photosynthesis. This process released gaseous O_2.
- Cells then developed which used O_2 to extract more energy from organic molecules.

Human Evolution

- Primates include humans, apes, and monkeys. They share the following characteristics: a five-digit hand with an opposable thumb, limbs that rotate easily, flat fingernails and toenails, and stereoscopic vision.
- The anthropoids are a group of primates that include humans, apes, and old world and new world monkeys.
- Apes and humans are collectively called hominoids; they diverged from monkeys about 25 million years ago.
- We and our human ancestors are known as hominids. Fossils of the oldest hominid, *Australopithecus afarensis,* have been dated to be 3.5 million years old. *Australopithecus* became extinct about 1.5 million years ago.
- The first humans were *Homo habilis.* Their fossil remains are dated to be 2 million years old.
- *Homo erectus,* a descendent of *H. habilis,* appeared about 1.5 million years ago.
- *Homo sapiens* probably evolved from *H. erectus.* The earliest fossils of *H. sapiens* have been dated to be about 300,000 years old. The origin of modern humans is explained by the multiregional hypothesis or the replacement hypothesis.

Selected Key Terms

adaptation (p. 539)
anthropoid (p. 547)
Australopithecus (p. 549)
Cro-Magnon (p. 553)
evolution (p. 532)
hominid (p. 548)
hominoid (p. 547)
Homo erectus (p. 551)
Homo habilis (p. 549)
Homo sapiens (p. 551)
natural selection (p. 539)
Neanderthal (p. 552)
primate (p. 547)
species (pp. 533, 546)
taxonomy (p. 546)

Review Activities

1. Evolution is defined as a change in a population over time. Describe what is meant by the terms "change," "population," and "time" in this definition.
2. Identify the title of the most important book written by Charles Darwin, the date it was published, and what Darwin proposed to explain biological diversity.
3. In the text, six sources of evidence were presented that support the concept of evolution. Describe at least four of them.
4. Define natural selection in one sentence.
5. What are mutations, and how are they important to the process of natural selection?
6. What is genetic drift? Identify several natural factors that might cause it to occur.
7. Briefly describe how the first cells might have developed from macromolecules of the early Earth. Include in your description the development of the plasma membrane and the relationship between nucleic acids and proteins.
8. Which occurred first, cells that could utilize gaseous O_2 or cells that could utilize gaseous CO_2 in photosynthesis? Defend your answer.
9. Distinguish between anthropoids, hominoids, and hominids.
10. What is the importance of *Australopithecus afarensis* to the evolution of modern humans?
11. List three characteristic differences between *Homo habilis* and *Homo erectus.*
12. What is a Neanderthal, where did Neanderthals live, and how long ago? Describe some of their cultural advances.
13. What is a Cro-Magnon, where did Cro-Magnons live, and how long ago? Describe some of their cultural advances.
14. In regard to the origin of modern humans, distinguish between the multiregional and replacement hypotheses.
15. Which hypothesis about the origin of modern humans is supported by "Mitochondrial Eve"?

Self-Quiz

Matching Exercise

___ 1. A group of similar organisms that can successfully reproduce	A. Hominoids
___ 2. The study of animal and plant distribution on Earth	B. Gene pool
___ 3. Term for the total number of genes within a population	C. *Homo erectus*
___ 4. The mechanism that maintains favorable genes within a population	D. Species
___ 5. Term for the transfer of genes between separate populations	E. Natural selection
___ 6. Term for random changes in a population's gene frequency caused by chance events	F. Biogeography
___ 7. Collective term that includes apes and humans	G. Gene flow
___ 8. The earliest hominid fossil	H. *Homo habilis*
___ 9. The first human fossil	I. Genetic drift
___ 10. Our immediate ancestor that evolved into *Homo sapiens*	J. *Australopithecus afarensis*

Answers to Self-Quiz

1. D; 2. F; 3. B; 4. E; 5. G; 6. I; 7. A; 8. J; 9. H; 10. C

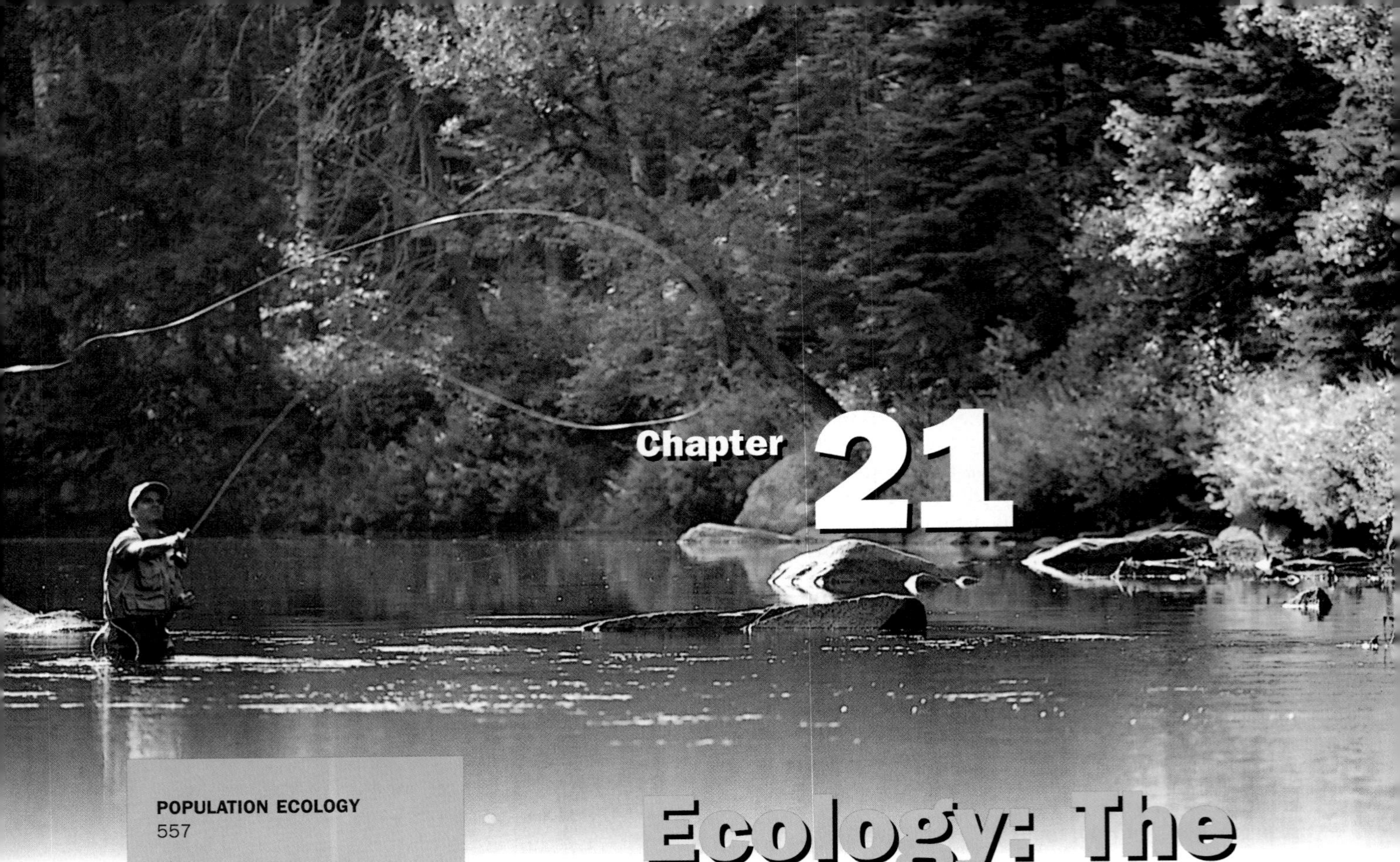

Chapter 21

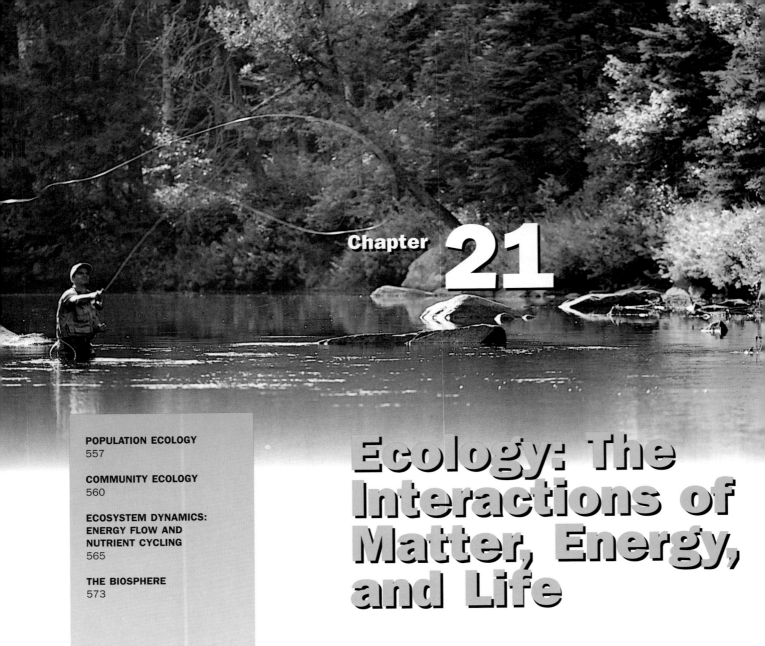

Ecology: The Interactions of Matter, Energy, and Life

We are not alone in the world. Humans do not live in isolation. We interact continuously with other living things and with our physical surroundings. Interactions with others and with the environment influence the distribution and abundance of all the species on Earth, including humans, and form the basis of the field of study called **ecology.** As we discuss basic ecological principles in this chapter, we will move beyond the individual organism and examine interactions at four increasingly higher levels of organization: populations, communities, ecosystems, and the biosphere.

A **population** is an interacting group of individuals, all of the same species, that inhabit a particular geographic area at one time. A population of students attends your college. A pink bilberry population inhabits the heathlands of northern Great Britain, and a crown of thorns starfish population exists along the Great Barrier Reef of Australia. On human skin there is a population of golden staphylococci.

A **community** contains populations of two or more species that live and interact within the confines of the same geographic area. Heathers such as the bilberry dominate the heathland community, but there are also lichens, mosses, and insects, some of which pollinate the bilberry. The crown of thorns starfish feeds on the corals which form and stabilize the Great Barrier Reef community and provide shelter for over a thousand species of reef fishes. Communities may be of any size—as large as an entire forest or as small as a single rotting apple.

An **ecosystem** is formed by a community of plant and animal populations and the physical aspects of the environment in which they live. For instance, to live in the heathland ecosystem, species must interact with each other but also must contend with cool temperatures, abundant fog, and acidic, poorly drained, nutrient-deficient soil. The Great Barrier Reef presents a brightly lit, highly saline environment with waves that pound and threaten to dislodge the clinging and creeping inhabitants.

The **biosphere** is the portion of Earth inhabited by life. It is the sum of all the ecosystems in fresh and marine waters and on land. Extending from the highest altitude to which organisms can fly, to the greatest depths of the ocean, the biosphere is influenced by global and seasonal patterns of weather and by geologic and atmospheric events.

In this chapter, we will explore the dynamics of populations and communities and the distribution of ecosystems in the biosphere. In Chapter 22, we will focus on the ecology of humans. In both chapters, we will examine the impact of humans on other organisms and the environment. No single species has as great an influence on its surroundings or as great an opportunity to shape its own destiny as do humans.

Population Ecology

Population ecology focuses on distinct populations of organisms and asks the questions: Where do different species live? How are they distributed? How large is the population, and how does it change over time?

Habitats: Where Populations Live

Populations are formed when members of the same species live together. The places where they live, obtain shelter, and find nutrients are called **habitats.** Within habitats, different patterns of distribution may be found: uniform distribution, random distribution, and clumped distribution (Figure 21.1). A **uniform distribution** is found in areas where individuals compete for limited resources such as water, especially when one species releases chemical factors that inhibit the growth of other species. A uniform distribution is also typical of plantings made for human agricultural purposes to allow easy care and harvesting, as in an orchard of fruit trees. A **random distribution** is seen when nutrients, shelter, and water are equally available throughout a habitat.

The most common distribution is a **clumped distribution,** since local geographic features such as hills, valleys, streams, caves, and lava flows create an uneven distribution of resources, concentrating them more in some areas than in others. Humans exhibit all three patterns because they live in housing tracts (uniform), on homesteads (random), and in apartment buildings (clumped).

Uniform
Random
Clumped

FIGURE 21.1 Population distribution patterns.

> A habitat is the portion of a community occupied by a population of a particular species. Within their habitat, the members of a population may be clumped together or dispersed in a random or uniform pattern.

Range: Limits to a Population's Distribution

Organisms are not equally distributed on Earth. Each species has a characteristic **geographic range** of distribution that is determined by its ability to tolerate physical environmental conditions, competition with other species, and geographic barriers. Species can be absent from a region because they are unable to tolerate its temperature, pressure, pH, mineral content, or availability of water and food.

For example, the water tupelo is a tree found in the swamps and floodplains of the southern United States. It grows best under the waterlogged conditions associated with prolonged flooding and is not found in regions that lack standing water. The ring-necked pheasant, which was introduced into the United States from Europe, survives well in most regions of the country but is not found in the southern states because this bird's eggs do not develop properly in the warmer temperatures of that region. These examples illustrate the concept known as the *law of tolerance,* which states that the growth of an organism is limited by an essential factor that is absent, in short supply, or present in harmful excess.

In areas where ranges overlap, there is competition between populations. Animals compete for food and for breeding or nesting sites. Plants compete for sunlight, water, minerals, and even pollinating insects. Sometimes competition between populations may prevent a population from occupying all of its potential range. Along the eastern slopes of the Sierra Nevada Range in California, the "least" chipmunk is capable of living at all elevations, but is forced into the sagebrush habitat at the base of the mountains by the very aggressive behavior of the "yellow pine" chipmunk, which occupies only the middle to lower elevations.

Mountain ranges, deserts, and seas present obstacles to species that are not strong fliers, crawlers, or swimmers. Humans may accidentally or intentionally expand the range of a species by transporting it across such barriers. The Africanized honeybee, often called the "killer" bee because of its aggressive behavior, was transported to Brazil in 1956 in the hope of developing a hardy tropical hybrid with a high rate of honey production. The bees accidentally escaped captivity and have gradually spread northward at over 100 kilometers (63 miles) per year. They reached Panama in 1982, Mexico in 1987, Texas in 1990, and Arizona in 1993.

> The range of a species is determined by its ability to tolerate the extremes of its physical environment and competition with other species as well as its ability to surmount geologic barriers.

Population Growth: Increasing Numbers

Populations change size through birth, death, and migration. When the gains (birth plus immigration) exceed the losses (death plus emigration), the population increases in size. In addition to the availability of suitable space and nutrients, a population's capacity for growth is determined by five factors: (1) the number of offspring produced by an individual member at one time, (2) the length of time required for offspring to reach sexual maturity, (3) the average length of life for population members, (4) the sex distribution, which is the ratio of reproductively active males to females, and (5) the age distribution of the population, since the very young and the very old often are not reproductively active.

The age distribution in a population is best illustrated with a survivorship curve. **Survivorship** refers to the percentage of a population that is alive at a given age. Three basic patterns for survivorship are seen in the graphs in Figure 21.2. These graphs are characterized by the time when deaths occur in the life cycle: late loss, constant loss, and early loss.

When populations colonize a new habitat that has abundant resources, they grow rapidly (Figure 21.3). Eventually, as the population increases, competition for food and shelter increases. Crowded conditions increase the rate of disease trans-

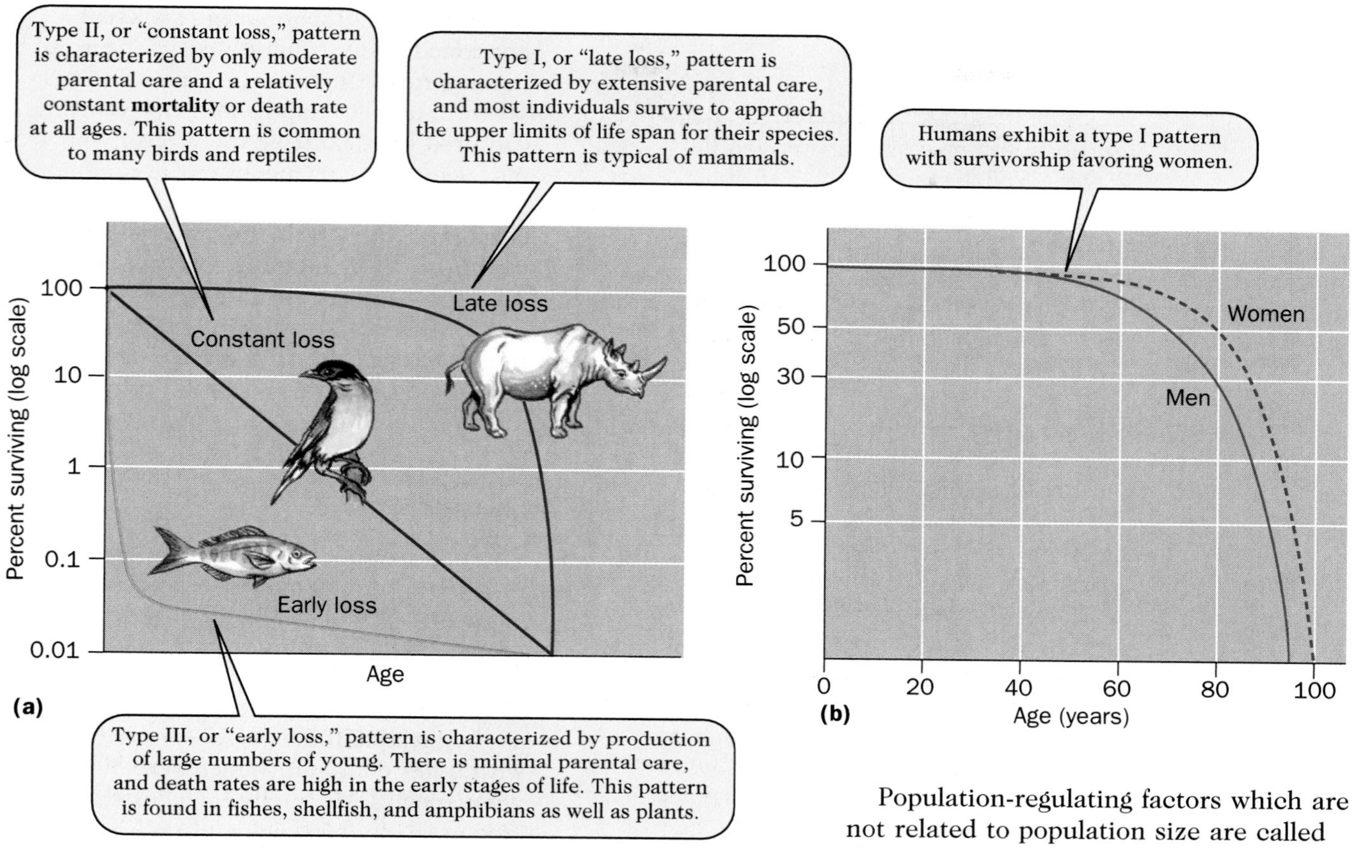

Type II, or "constant loss," pattern is characterized by only moderate parental care and a relatively constant **mortality** or death rate at all ages. This pattern is common to many birds and reptiles.
Type I, or "late loss," pattern is characterized by extensive parental care, and most individuals survive to approach the upper limits of life span for their species. This pattern is typical of mammals.
Humans exhibit a type I pattern with survivorship favoring women.
100
10
1
0.1
0.01
Percent surviving (log scale)
Constant loss
Late loss
Early loss
Age
(a)
Type III, or "early loss," pattern is characterized by production of large numbers of young. There is minimal parental care, and death rates are high in the early stages of life. This pattern is found in fishes, shellfish, and amphibians as well as plants.
100
50
30
10
5
Percent surviving (log scale)
Women
Men
0
20
40
60
80
100
(b)
Age (years)

FIGURE 21.2 Survivorship curves.

mission between individuals, survivability decreases, and the death rate increases. Over time, the population size oscillates above and below a stable level called the carrying capacity. The **carrying capacity** refers to the number of individuals that can be supported permanently by the resources available in a habitat.

Each habitat has its own *population-regulating factors* which determine the actual size of the carrying capacity for its component populations. The regulating factors which intensify as the size of a population increases are termed **density-dependent** ("density" refers to the population per unit area). They tend to increase the death rate and decrease the reproductive rate and include primarily biological factors such as competition for food, nesting sites, and light. Seedling mortality caused by shading in a mature evergreen forest is a density-dependent factor. Disease and parasitic infestations also are included because they are more likely to occur in a crowded population.

Population-regulating factors which are not related to population size are called **density-independent.** They include weather, climate, and catastrophic events such as fire, earthquake, flood, and volcanic eruption. In the Cascade Range in the state of Washington, freezing winter weather or a volcanic eruption such as that of Mount St. Helens in 1980 can cause insect mortality. These are both density-independent factors that regulate population size.

Human activity also may have an impact on the size of natural populations. The clearing of tropical rain forests for human agriculture devastates the bird, mammal, amphibian, reptile, and insect populations as well as the tree species. Hunting and destruction of hardwood forests that resulted from expanded farming in the 1800s extinguished the passenger pigeon population in the Ohio River valley. Today the expansion of farming and the presence of poachers seeking marketable ivory and other trophies threaten elephant and rhinoceros populations in Africa. As the human population has increased, the populations of many animals and plants have suffered,

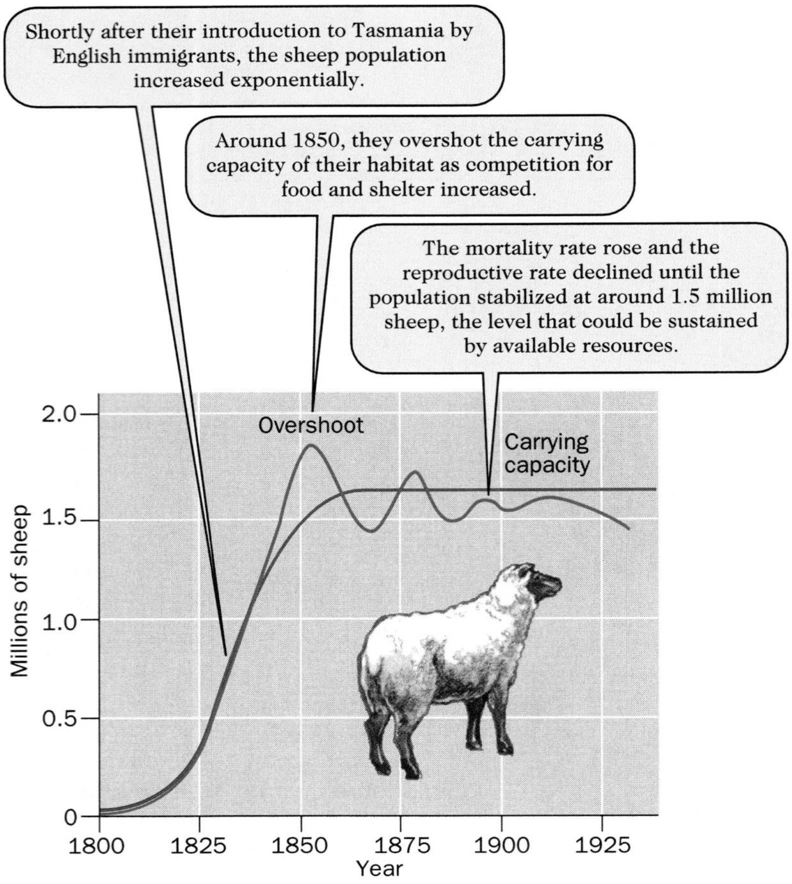
Shortly after their introduction to Tasmania by English immigrants, the sheep population increased exponentially.
Around 1850, they overshot the carrying capacity of their habitat as competition for food and shelter increased.
The mortality rate rose and the reproductive rate declined until the population stabilized at around 1.5 million sheep, the level that could be sustained by available resources.
Overshoot
Carrying capacity
Millions of sheep
2.0
1.5
1.0
0.5
0
1800
1825
1850
1875
1900
1925
Year

FIGURE 21.3 History of the sheep population in Tasmania.

but some have benefited. Rats and cockroaches have increased in abundance along with the number of human dwellings. Ravens, pigeons, and opossums have adapted to the human presence and thrive in and around many American cities and parks.

> Population growth is influenced by the sex distribution, age distribution, age of sexual maturity, length of life, and the number of offspring. Carrying capacity is influenced by factors that are dependent or independent of population density.

Community Ecology

A city park, though an idealized and not truly natural habitat, is a good place to observe the interactions between populations that are typical of many communities. You may observe carp in a small pond feeding on aquatic insects, bees pollinating roses, hummingbirds sipping nectar from fuchsias, and a spider's web with captured flies. Looking elsewhere, you may see ants carrying away the remnants of a discarded hamburger, a pile of dirt signaling a mole tunneling after earthworms, a rabbit nibbling on grass under a maple tree, or a squirrel carrying a peanut to its nest. Every animal or plant in the park, even the strolling humans, occupies a position in the community and fulfills a functional role.

Niche: The Functional Role of an Organism

The way in which an organism "fits and functions" in its community is described as its **niche** (*nitch*). The description of a niche includes the full range of environmental conditions under which an organism feeds, grows, and reproduces as well as the lifestyle it adopts as it interacts with other community members. For plant species, the niche includes the degree of shade tolerance and requirements for soil moisture, pH, and mineral content.

The niche of an animal species includes its choice of food, its time and place of feeding, and the form and location of its nest or burrow. In a natural community different species may occupy similar niches or overlap in their use of niches, but no two can jointly occupy the same niche. Competition may force species to specialize and occupy a smaller niche than they are capable of inhabiting.

> An organism's niche is described by the role that organism plays in its habitat and the conditions under which it lives, feeds, and breeds. No two species can occupy exactly the same niche in the same habitat.

Competition for Community Resources

In the forests of New England there are several species of insect-eating warblers. Each bird is potentially capable of catching insects at any site in a tree from the ground to the crown and from the trunk to the tips of the outermost branches.

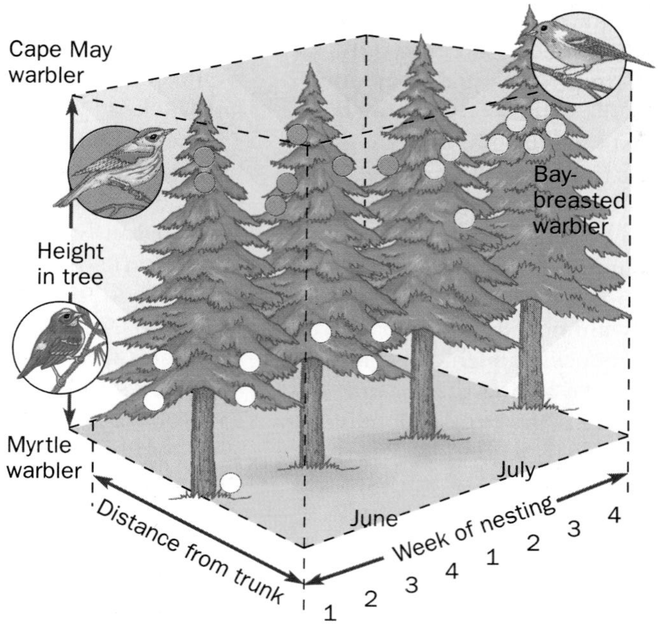
Cape May warbler
Height in tree
Myrtle warbler
Bay-breasted warbler
Distance from trunk
June
July
Week of nesting
1 2 3 4 1 2 3 4

In New England, three species of insect-catching and tree-nesting warblers compete for available resources. Each species is capable of feeding or nesting anywhere on a given tree. By feeding at different places and nesting at different times, the species occupy only a portion of their potential niche but successfully share the available resources.

FIGURE 21.4 Competition for limited resources.

These birds also are capable of nesting anywhere in the tree in June and July. When several different species of warbler occupy a particular tree, **interspecific competition** (between species) causes each species to adapt to a particular feeding and nesting site and time, resulting in a partitioning of the available resources (Figure 21.4). Myrtle warblers find insects on the ground and in the lower branches and build nests in early summer. Cape May warblers feed near the crown and tips of the upper branches, building their nests in early to middle summer. Bay-breasted warblers feed along the trunk and midlevel branches and are the last to construct nests in midsummer.

Intense interspecific competition for limited resources may lead to the **competitive exclusion** of a species from a community when it is less efficient in the utilization of a resource than its competitor is. Figure 21.5 shows how competition for bacterial food between two species of the microscopic protozoan *Paramecium* can lead to the exclusion of the less efficient species. A similar exclusion may be occurring in the deserts of southern California as wild burros outgraze bighorn sheep and aggressively defend water holes.

Predation: The Drama of Predator and Prey

Competition between populations in many communities may extend to **predation**, in which one organism—the **preda-**

FIGURE 21.5 Competitive exclusion of a *Paramecium* species.

When a large species (*Paramecium caudatum*) and a small species (*Paramecium aurelia*) are cultured separately, they exhibit similar patterns of growth. However, when they are cultured together, the smaller species is more efficient at catching bacterial food, and its success dooms the larger species.

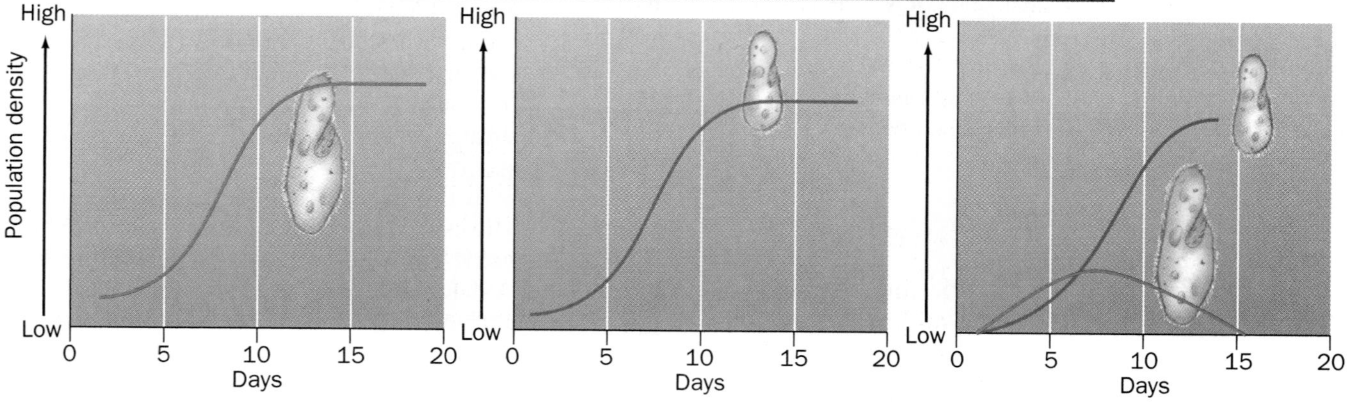
Population density
High
Low
0 5 10 15 20
Days
High
Low
0 5 10 15 20
Days
High
Low
0 5 10 15 20
Days

tor—consumes another—the **prey**—as its primary source of food. Predation may take many different forms and does not always involve the killing of the prey species. For instance, *herbivores* are animals which eat the leaves, stems, seeds, or fruits of plants, generally not killing but only damaging the plants. *Carnivores* are animals which typically kill and eat herbivores or other carnivores. *Parasites* infest plants or animals and obtain nutrition at their expense, though generally not to the extent of killing them. *Cannibalism* occurs when the prey individual is of the same species as the predatory individual.

Since the lifestyles of predator and prey are closely linked, their populations may exhibit "boom or bust" oscillations as the abundance of one influences the abundance of the other. A famous example involves the changes in the snowshoe hare (prey) and Canadian lynx (predator) populations revealed by the trapping records of the Hudson's Bay Company in Canada (Figure 21.6). The hare population rises when burrows and food are abundant and falls when the lynx population rises in response to an increase in the supply of its accustomed food (the hare). The cycling of the lynx population appears to be directly related to food supply, while the hare population oscillates in accordance with predator numbers as well as the availability of suitable food and shelter.

FIGURE 21.6 Synchronization of predator and prey populations.

Historical trapping records of the Hudson's Bay Company can be used to show a synchronous cycling of Canadian lynx and snowshoe hare populations. Oscillations in the predatory lynx population are closely linked to the increase or decline in the hare population.

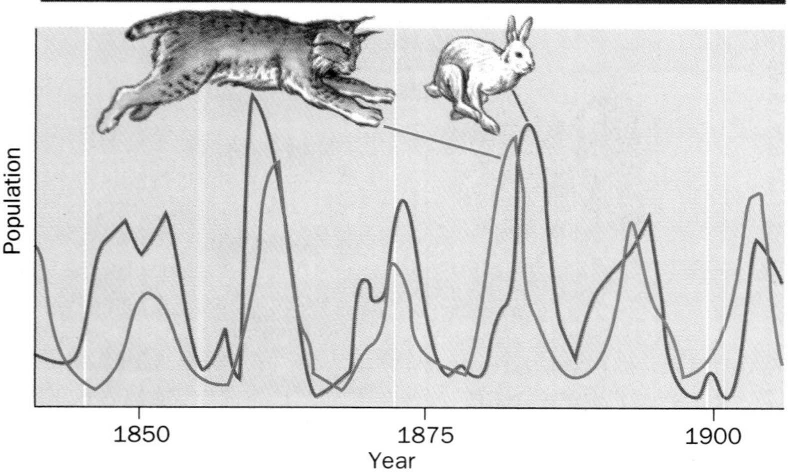
Population
1850
1875
1900
Year

The presence of predators may influence the diversity of prey species in a community. Studies along the coast of Washington State revealed that when the ocher starfish was removed from test sections of the intertidal zone, the diversity of chitons, limpets, snails, mussels, and barnacles in tide pools was reduced from 15 species to 8. The mussel, a favorite food of the starfish, dominated the simpler community. The presence of intense and efficient predation appeared to support the larger species diversity in the intertidal zone.

Avoiding predation. To avoid being eaten, prey species have evolved defensive mechanisms. Some, such as the hare, have powerful hind legs to outrun the larger lynx. The ground-nesting ring-necked pheasant uses a rapid run or short bursts of flight to escape from foxes, coyotes, and other predators. Others rely on trickery and deceit through *camouflage*, a combination of colored shapes and behavior that allows them to blend in with their surroundings and avoid predation.

Many plants and animals produce noxious or poisonous chemical substances to discourage pests and predators. The presence of a poison alone is not fully protective if a prey must first be eaten to discourage its predator. Many poisonous organisms, therefore, exhibit *warning coloration* consisting of bright yellow, red, or orange pigmentation to announce their toxicity and warn off predators. For example, after hatching from their eggs, baby red-eared slider turtles combine green, black, and yellow warning coloration with sharp claws and aggressive behavior (Figure 21.7). If one is swallowed by a largemouth bass, it claws and scratches so ferociously, that it is quickly spit out. The bass's second encounter with the hatchling will trigger avoidance. The coloration disappears as soon as the turtle grows too large to be eaten by fish.

A few nonpoisonous species have evolved *mimicry*, which is a color pattern that mimics that of a poisonous species. The similarity of coloration between the

FIGURE 21.7 The warning coloration of this hatchling red-eared slider turtle warns predatory fish that it is a difficult meal to swallow.

nontoxic viceroy butterfly and the noxious monarch butterfly prevents the viceroy from being eaten by a hungry bird (Figure 21.8).

Disrupting predator-prey balances. Humans often intentionally or accidentally disrupt the predator-prey balance in natural communities. Before 1829, marine lampreys (a predatory fish with a circular sucking mouth and rasping teeth) were prevented from entering the Great Lakes of the northeastern United States by a natural barrier, Niagara Falls. However, after construction of the Welland Canal around Niagara Falls, lampreys gradually gained access to the lakes, and the presence of this new predator devastated the lake trout population. The trout fishery has slowly returned as humans have instituted lamprey control measures and plantings of hatchery-raised trout.

The introduction of a species, however, is sometimes useful in pest control. On the west side of the Cascade Range in Oregon, Washington, and British Columbia a noxious weed, tansy ragwort, often poisons cattle and horses. Herbicidal spraying and hand removal have not provided sufficient control. However, the widespread release of the herbivorous cinnabar moth, whose larvae (immature feeding form) consume the tansy weed, has provided an important additional control measure.

> Predators influence the size and diversity of prey populations. To avoid predation, animals use camouflage coloration, noxious chemicals, or aggressive behavior. Some animals mimic the coloration of toxic animals to avoid predation.

Symbiosis: Living Together

Interactions between community members are not always an eat-or-be-eaten proposition. When different species live together with coordinated life cycles for singular or mutual benefit, the relationship is called **symbiosis.** Symbiotic relationships include parasitism, commensalism, and mutualism.

In **parasitism,** one partner in an association benefits at the expense of the other partner. Parasites infest their *hosts.* A parasite extracts nutrients from the host's blood, body fluids, or tissues. The life cycles of parasites and hosts are closely coordinated, and a series of different hosts may be involved in sustaining and transmitting a single parasite.

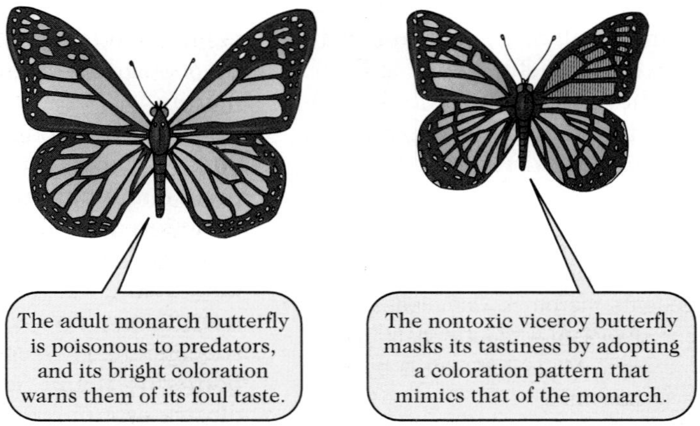
The adult monarch butterfly is poisonous to predators, and its bright coloration warns them of its foul taste.
The nontoxic viceroy butterfly masks its tastiness by adopting a coloration pattern that mimics that of the monarch.

FIGURE 21.8 Mimicry.

Internal human parasites such as intestinal tapeworms and hookworms are often highly specialized with suckers or hooks for holding on. These parasites are incapable of an independent existence. External parasites such as mites, ticks, fleas, and lice can survive temporary separation from a host.

In **commensalism,** one partner in the association benefits and the other partner neither benefits nor is harmed greatly. When barnacles grow on the thick skin of a Pacific gray whale, they are able to filter feed larger volumes of water than is the case when they grow on an exposed rock or dock piling. Cattle egrets form a similar "hitchhiking" association with cape buffalo in Africa. The egret rides on the buffalo's back and preys on insects kicked up as the large beast moves through grasses and brush. Mice may be living commensally with you in your home. The bacteria on your skin represent a form of commensalism. This commensalism can shift to parasitism, however, if the skin is injured and the bacteria (staphylococci and streptococci) enter your body.

In **mutualism** both partners benefit from the relationship. Some soil fungi enter into a mutualistic association called a *mycorrhiza* (***my-koh-rise-uh***) with the roots of trees and shrubs. The fungus benefits the plants by extending the surface area of their roots and assisting in the absorption of water and minerals from the soil. In exchange, the fungus receives some of the food manufactured by the plant and stored in its roots.

The definition of mutualism can be extended to include the relationship of humans to common household pets such as dogs, cats, and small birds. The animals derive an obvious benefit of food and shelter, and the humans gain the psychological comfort of companionship and perhaps protection.

> Symbiosis is a physical association between two organisms. Parasitism harms one organism and benefits the other. Commensalism benefits one organism without benefiting or harming the other. Mutualism allows both organisms to benefit.

Succession: Community Changes

The interactions between animal and plant populations in a community lend a degree of stability to the entire community. Unfortunately, that stability frequently is disrupted by catastrophic physical events. Forest and range fires burn away protective vegetation. Geologic uplift forces new mountains up through the Earth's surface. Earthquakes split and cleave a valley floor. Hurricanes and floods scour the land and change the course of rivers. Humans clear forests for agriculture and fill swamps for habitation. The progression of changes as a newly formed habitat is colonized or a damaged region recovers is called **succession.** Change is the only constant factor over time.

As an Alaskan glacier grows, it moves the landforms in its path. Later, the recession of the glacier exposes gravel and rock devoid of soil and fertilizing nitrogen. As shown in Figure 21.9, such newly exposed land is colonized by living organisms. This colonization of bare, lifeless land is called **primary succession.** Primary succession refers to a series of different plant and animal communities that follow one another over time. Each prepares the way for the one that follows.

More widespread is a process called secondary succession. You may have noticed this process if over several years you have watched the natural changes in an empty city lot after a building has been torn down. Or perhaps you have observed the changes in pastures on an abandoned farm or a hillside that has been clear-cut by logging. **Secondary succession** involves a series of community changes that occur when a previously occupied habitat is disturbed and invaded by populations from outside the community.

Humans are often responsible for secondary succession. We may alter a natural community for agriculture or construction and then abandon the land when the soil is depleted. The abandoned land then goes through secondary succession. Nature also can cause secondary succession. The recovery of communities bordering Mount St. Helens after its explosive eruption in 1980 is a good exam-

FIGURE 21.9 Primary succession after the retreat of an Alaskan glacier.

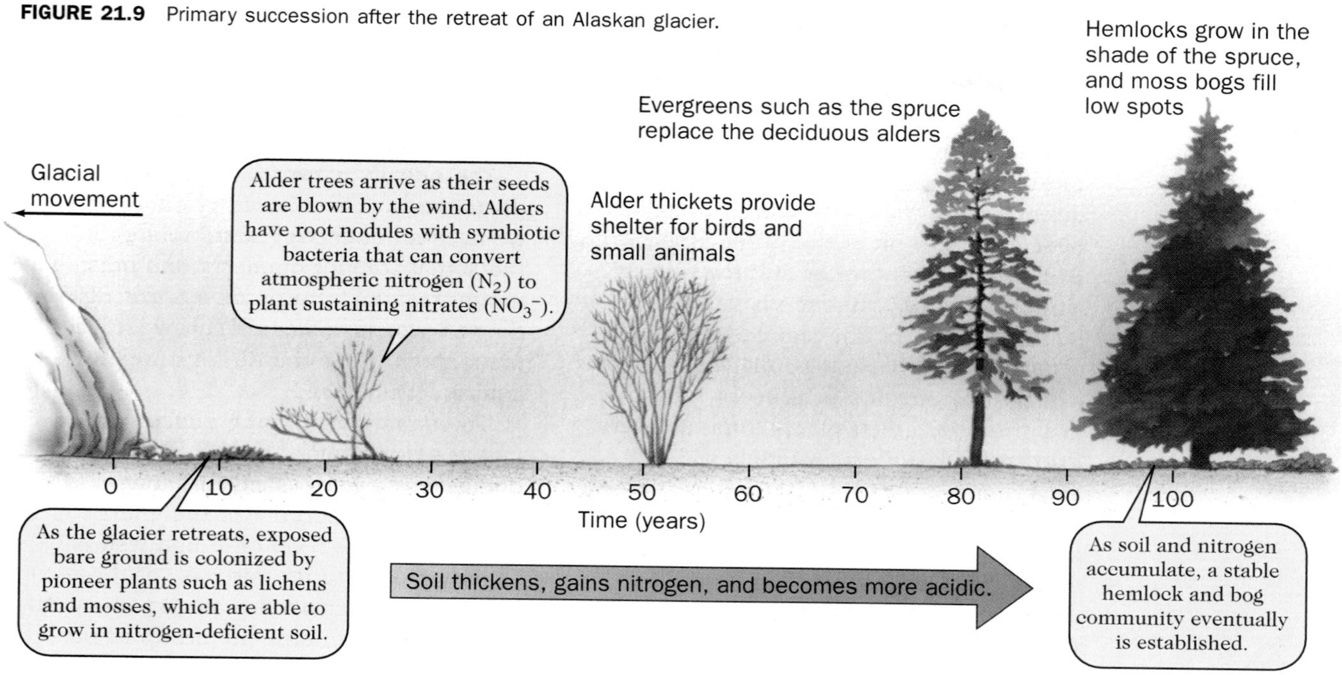
Hemlocks grow in the shade of the spruce, and moss bogs fill low spots
Evergreens such as the spruce replace the deciduous alders
Glacial movement
Alder trees arrive as their seeds are blown by the wind. Alders have root nodules with symbiotic bacteria that can convert atmospheric nitrogen (N_2) to plant sustaining nitrates (NO_3^-).
Alder thickets provide shelter for birds and small animals
0
10
20
30
40
50
60
70
80
90
100
Time (years)
As the glacier retreats, exposed bare ground is colonized by pioneer plants such as lichens and mosses, which are able to grow in nitrogen-deficient soil.
Soil thickens, gains nitrogen, and becomes more acidic.
As soil and nitrogen accumulate, a stable hemlock and bog community eventually is established.

ple. Blueberry shrubs and dogwood trees have replaced the *pioneer species* that first colonized the ash-covered ground. Douglas fir seedlings in time will produce new evergreen forests to replace those which once stood on the slopes of the mountain.

> The sequential changes in the species composition of a community are called succession. Primary succession occurs when bare land is colonized. Secondary succession occurs when a community is altered, allowing invasion by other populations.

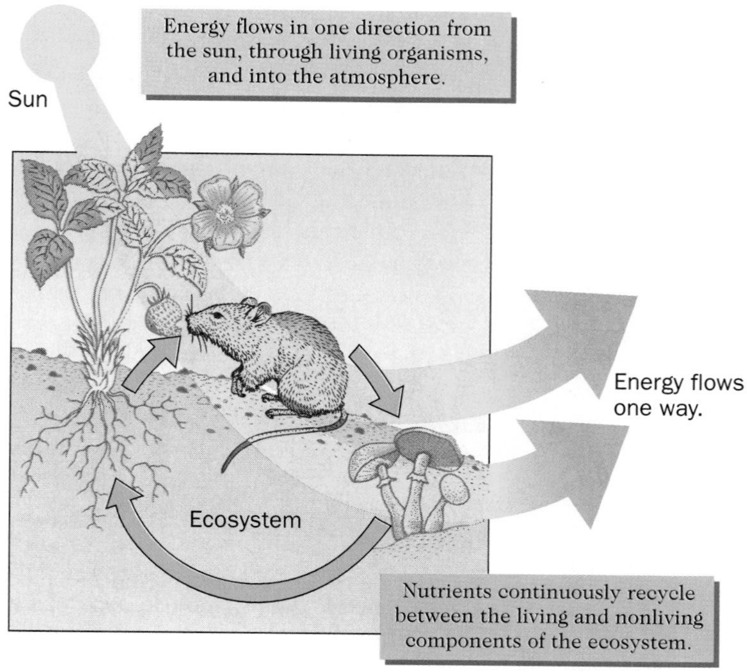
Energy flows in one direction from the sun, through living organisms, and into the atmosphere.
Sun
Energy flows one way.
Ecosystem
Nutrients continuously recycle between the living and nonliving components of the ecosystem.

FIGURE 21.10 The flow of energy through an ecosystem.

Ecosystem Dynamics: Energy Flow and Nutrient Cycling

A community interacting with its physical environment forms an ecosystem. The successional changes observed in a community over time are made possible by a flow of energy and a cycling of nutrients. It has become quite popular in human communities to recycle wastes in order to conserve resources. Natural communities have been conducting such recycling since the evolution of photosynthesis (Chapter 20).

The Flow of Matter and Energy

Energy enters terrestrial and most aquatic ecosystems in the form of sunlight (Figure 21.10). Light energy is transformed to chemical energy as it passes through the living members of a community and eventually is lost back to the atmosphere as heat energy. Note that energy flows in only one direction through an ecosystem. Figure 21.11 shows that only about 2 percent of incoming light energy is bound up in living organisms. The rest is reflected directly from the Earth's surface or lost as heat. If you think of the proportion of the Earth covered by water and ice (nearly three-quarters), you can visualize the reflective nature of our planet.

Organisms that use sunlight for photosynthesis are called **producers.** These are primarily green plants, but some are Protista and Monera. Animals which eat producers or each other are called **consumers.** Plant-eating herbivores are known as *primary consumers,* herbivore-eating carnivores are called *secondary consumers,* and carnivore-eating carnivores are termed *tertiary consumers.*

Both producers and consumers grow and metabolize, releasing carbon dioxide (CO_2) and organic wastes; ultimately they die. The wastes and dead tissues of plants and animals are consumed by **decomposers,** which play the very important role of recycling minerals and nutrients back to producers (see Figure 21.10). Bacteria, fungi, worms, nematodes, and some insects serve as decomposers in many ecosystems.

The food web: who eats whom. The flow of energy and the cycling of nutrients through the various populations in a natural community produce a **food web**, which is shown in a simplified form in Figure 21.12. Some animals are of course capable of feeding at different levels in the food web, consuming a variety of plant and animal species. That is certainly the case with humans.

The energy pyramid. A food web helps us understand who is eating whom, but we need an **energy pyramid** to appreciate the actual amount of energy and organic matter passing from producers through the various consumers. Figure 21.13 depicts an energy pyramid for a freshwater aquatic ecosystem.

The organic material produced in animal or plant tissues through the consumption of food energy is called **biomass**. Only about 10 percent of the energy available at one level of a food web is transformed into the biomass of the next higher level. Herbivores receive about 10 percent of the energy bound up in producer biomass. Warm-blooded carnivores receive 10 percent or less of the energy locked up in herbivore biomass.

Actually, the 10 percent figure is only a convenient approximation, since eco-

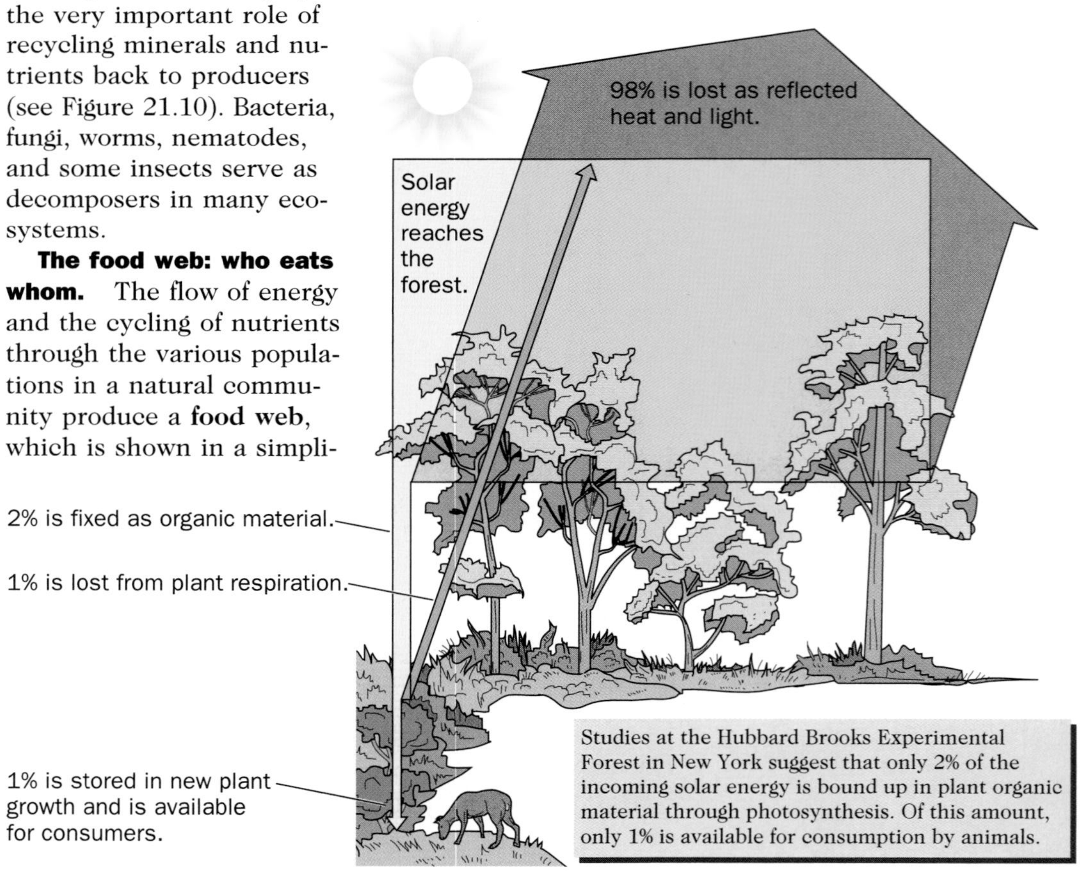
98% is lost as reflected heat and light.
Solar energy reaches the forest.
2% is fixed as organic material.
1% is lost from plant respiration.
1% is stored in new plant growth and is available for consumers.
Studies at the Hubbard Brooks Experimental Forest in New York suggest that only 2% of the incoming solar energy is bound up in plant organic material through photosynthesis. Of this amount, only 1% is available for consumption by animals.

FIGURE 21.11 Energy utilization in a forest community.

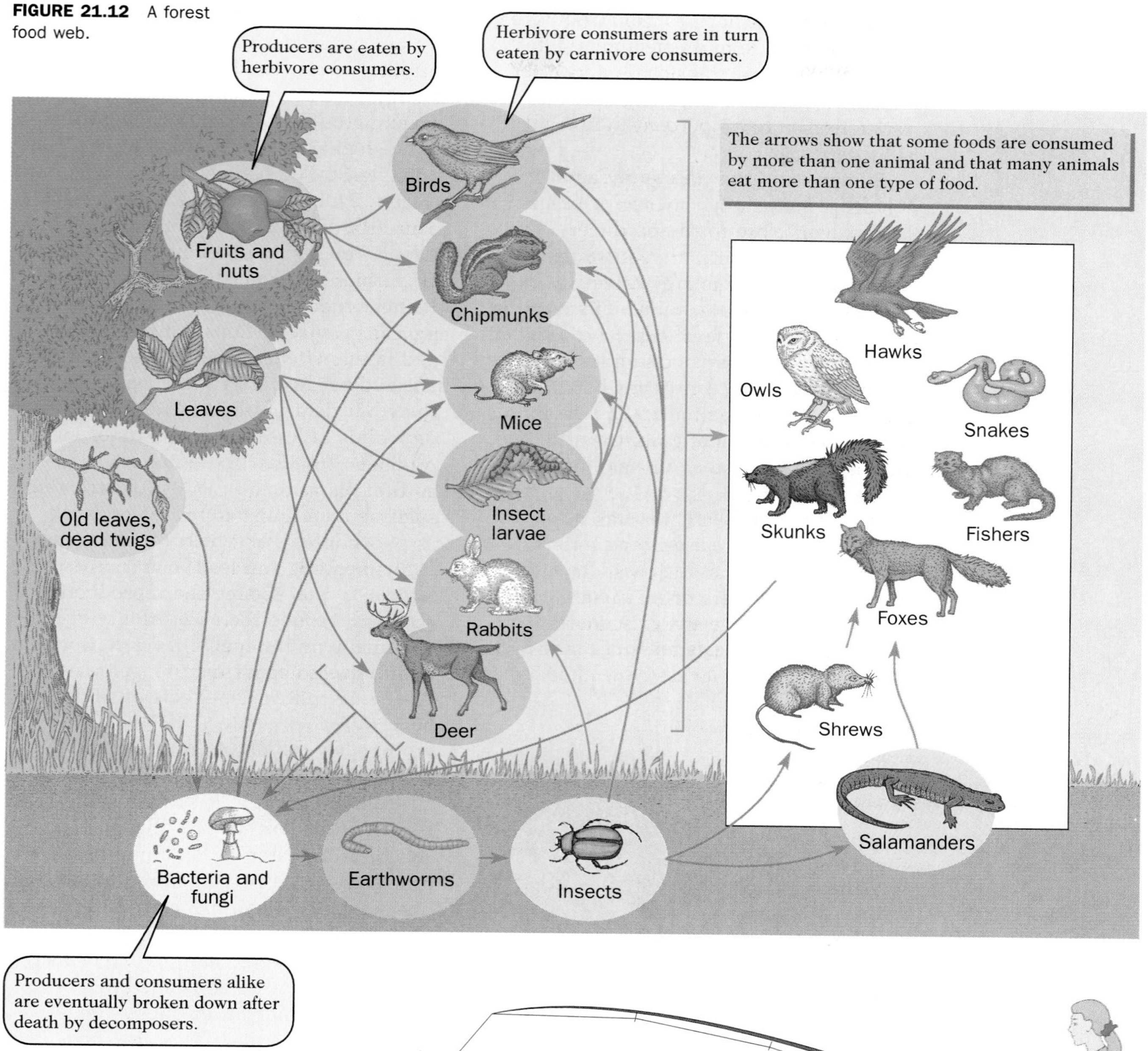
Producers are eaten by herbivore consumers.
Herbivore consumers are in turn eaten by carnivore consumers.
The arrows show that some foods are consumed by more than one animal and that many animals eat more than one type of food.
Fruits and nuts
Leaves
Old leaves, dead twigs
Birds
Chipmunks
Mice
Insect larvae
Rabbits
Deer
Hawks
Owls
Snakes
Skunks
Fishers
Foxes
Shrews
Salamanders
Bacteria and fungi
Earthworms
Insects
Producers and consumers alike are eventually broken down after death by decomposers.

FIGURE 21.12 A forest food web.

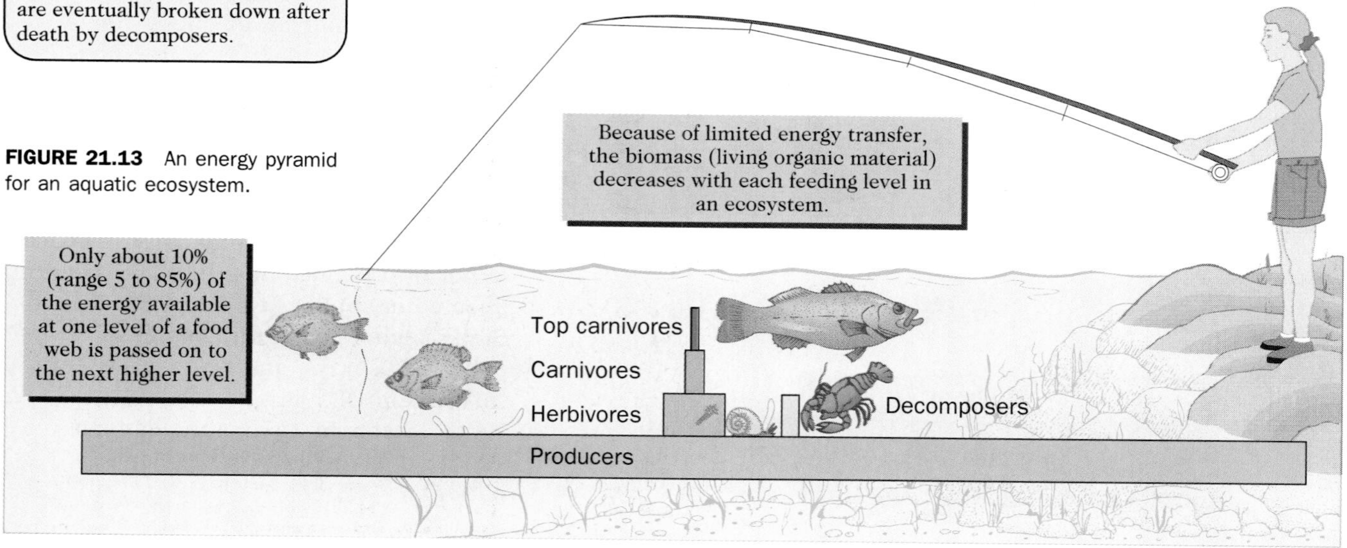
Because of limited energy transfer, the biomass (living organic material) decreases with each feeding level in an ecosystem.
Only about 10% (range 5 to 85%) of the energy available at one level of a food web is passed on to the next higher level.
Top carnivores
Carnivores
Herbivores
Decomposers
Producers

FIGURE 21.13 An energy pyramid for an aquatic ecosystem.

systems vary in their levels of photosynthetic production and consumers differ in their efficiency at transforming food energy into biomass. The actual transfer varies from 5 to 85 percent in different food webs.

Humans and the energy pyramid. When we eat as a herbivore (or vegetarian), we eat close to the producer level (low on the pyramid) and consume a larger share of the energy originally captured by photosynthesis. When we eat as a second- or third-level carnivore (higher on the pyramid), we obtain a lesser share of that energy. We can gain 1 kilogram (kg) of weight by eating 10 kg of grain, or we can feed 100 kg of grain to cattle and then eat 10 kg of steak or hamburger.

> Only 2 percent of sunlight energy is converted to chemical energy by organisms. Chemical energy passes through the food web as producers are eaten by consumers. Only a portion of the energy at one level of the food web is transferred to the next level to form biomass.

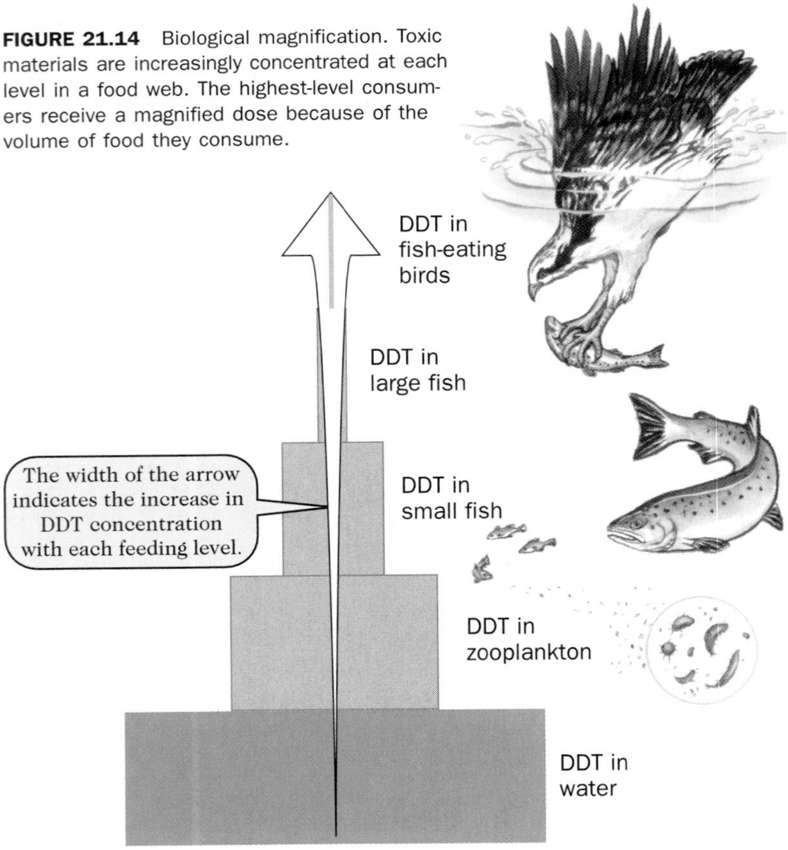

DDT in fish-eating birds
DDT in large fish
The width of the arrow indicates the increase in DDT concentration with each feeding level.
DDT in small fish
DDT in zooplankton
DDT in water

FIGURE 21.14 Biological magnification. Toxic materials are increasingly concentrated at each level in a food web. The highest-level consumers receive a magnified dose because of the volume of food they consume.

Concentrating Toxins in Food Webs

Another lesson can be learned from the energy pyramid. There is a tendency for toxic materials to become concentrated in the higher levels of a food web, a phenomenon called *biological magnification* (Figure 21.14). Magnification occurs because the biomass at any given level in a food web is produced by eating a much larger biomass from the level below. Higher-order consumers such as eagles, pelicans, ospreys, lions, killer whales, and humans tend to be the most seriously affected because they are consuming the equivalent of hundreds to thousands of pounds of lower-level consumer and producer biomass. Toxic materials such as DDT (a pesticide), PCBs (electrical insulating material), radioisotopes (such as phosphorus), and heavy metals (especially mercury and lead) can increase the mortality rate, reduce the reproductive rate, and reduce the survivability of the young among the higher-level consumers in an affected ecosystem.

The magnification effects of DDT have been well documented. DDT sprayed on marshes in Long Island through the 1950s for mosquito control accumulated in sediments, where it was consumed by debris feeders such as shrimp. The shrimp were eaten in large numbers by minnows. Since DDT is stored in fat, and not detoxified, each fish stored DDT from the shrimp it consumed during its life. The minnows were eaten by larger fish, which in turn were consumed by cormorants and merganser ducks. Some birds received lethal doses and died. Others, with sublethal doses, produced eggs with soft shells which were crushed by the parental birds in the nest. Bird populations dropped precipitously and took many years to return to their previous levels after DDT spraying was ended. Other bird populations that have been affected in similar fashion in the United States include peregrine falcons, brown pelicans, ospreys, and bald eagles. For this reason DDT was banned in the United States in 1973, although it is still used extensively in developing nations.

Toxic materials accumulate in higher-level consumers because of the large amount of lower-level organisms they must eat to grow and reproduce.

Cycling of Water, Carbon, and Minerals

The thermonuclear reactions that occur in the sun will provide the ecosystems of Earth with a continuous input of energy for many generations to come. Unlike energy, however, there is not a continued input of carbon, hydrogen, nitrogen, oxygen, phosphorus, and sulfur (CHNOPS), the materials used in constructing biological molecules (Chapter 1). With a finite amount of each material present in the atmosphere, soil, water, and living organisms, the materials must be recycled continuously through ecosystems. Since recycling loops involve geologic events, weather, and living organisms, they are called **biogeochemical cycles.** These cycles are important to the health and productivity of communities, but can be influenced by human activity.

Water cycling. Water is fundamentally important to life, serving as the universal solvent in cells and tissue fluids. Water forms about 70 percent of living material and covers about 75 percent of the Earth's surface. In the water cycle, the heat of the sun powers the evaporation of water from oceans, lakes, streams, and organisms. Then water condenses and falls as rain, snow, or hail. A complete description of the water cycle can be found in Figure 21.15.

At any time, only about 2 percent of the water on Earth is frozen in ice or held in the fluids of plants and animals. The rest is water vapor and liquid water, which are cycled continuously between the atmosphere and Earth. Humans interfere with the cycling of water by logging and not replanting forests. This increases the amount of water that does not penetrate the soil but instead runs off. Humans also divert natural runoff by constructing canals and dams. Human industrial activity has the potential to warm the atmosphere, perhaps melting glaciers and raising the sea level. This could cause salt water to invade and damage coastal soils and supplies of underground freshwater.

Water passes from a liquid to a gas during evaporation and respiration. It returns to a liquid through condensation and precipitation. On Earth and in organisms, water is the most plentiful solvent.

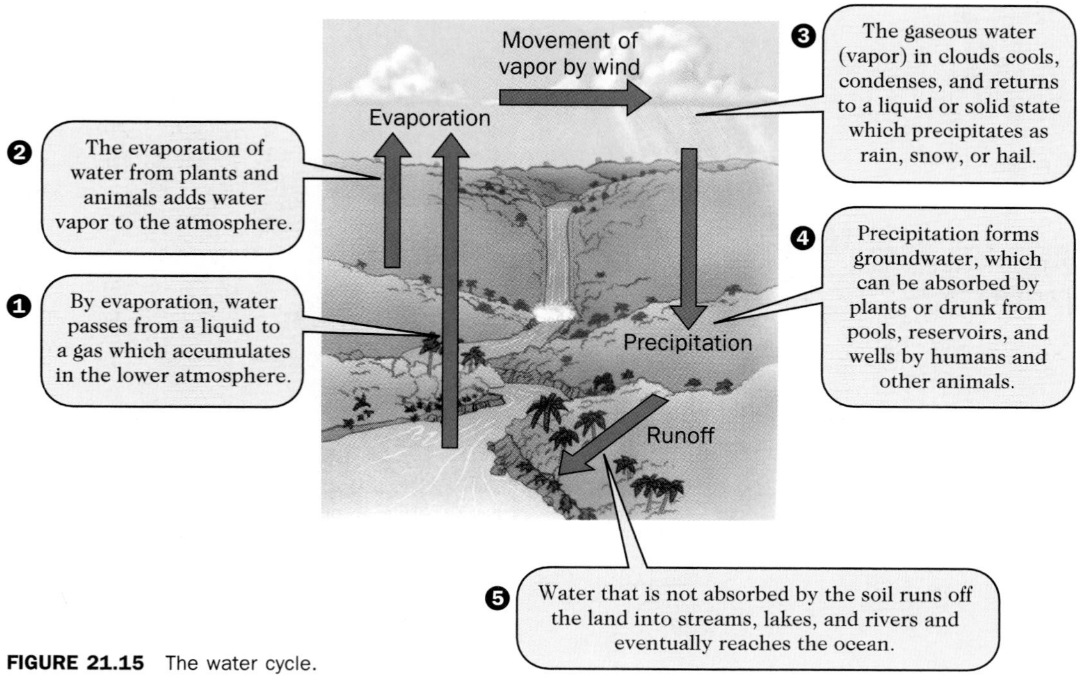
Movement of vapor by wind
Evaporation
Precipitation
Runoff
❶ By evaporation, water passes from a liquid to a gas which accumulates in the lower atmosphere.
❷ The evaporation of water from plants and animals adds water vapor to the atmosphere.
❸ The gaseous water (vapor) in clouds cools, condenses, and returns to a liquid or solid state which precipitates as rain, snow, or hail.
❹ Precipitation forms groundwater, which can be absorbed by plants or drunk from pools, reservoirs, and wells by humans and other animals.
❺ Water that is not absorbed by the soil runs off the land into streams, lakes, and rivers and eventually reaches the ocean.

FIGURE 21.15 The water cycle.

Carbon cycling. Carbon atoms continuously cycle through organisms and the atmosphere. You and I are only temporarily borrowing carbon to form our tissues. When we die and decay, the carbon will be reused in other organisms. Carbon in the form of carbon dioxide gas is removed from the atmosphere by producers and is chemically altered to form sugars and other organic molecules during photosynthesis. Figure 21.16 describes the cycling of carbon through producers, consumers, and decomposers.

When some organic matter is not immediately decomposed, it may form fossil fuel deposits of coal, oil, and natural gas. The burning of fossil fuels in human factories and automobiles contributes to an excessive accumulation of CO_2 in the atmosphere, a condition called the *greenhouse effect,* which we will explore in Chapter 22.

Carbon dioxide is used by plants to form sugars, which are consumed by herbivores, carnivores, and decomposers. Respiration returns CO_2 to the atmosphere.

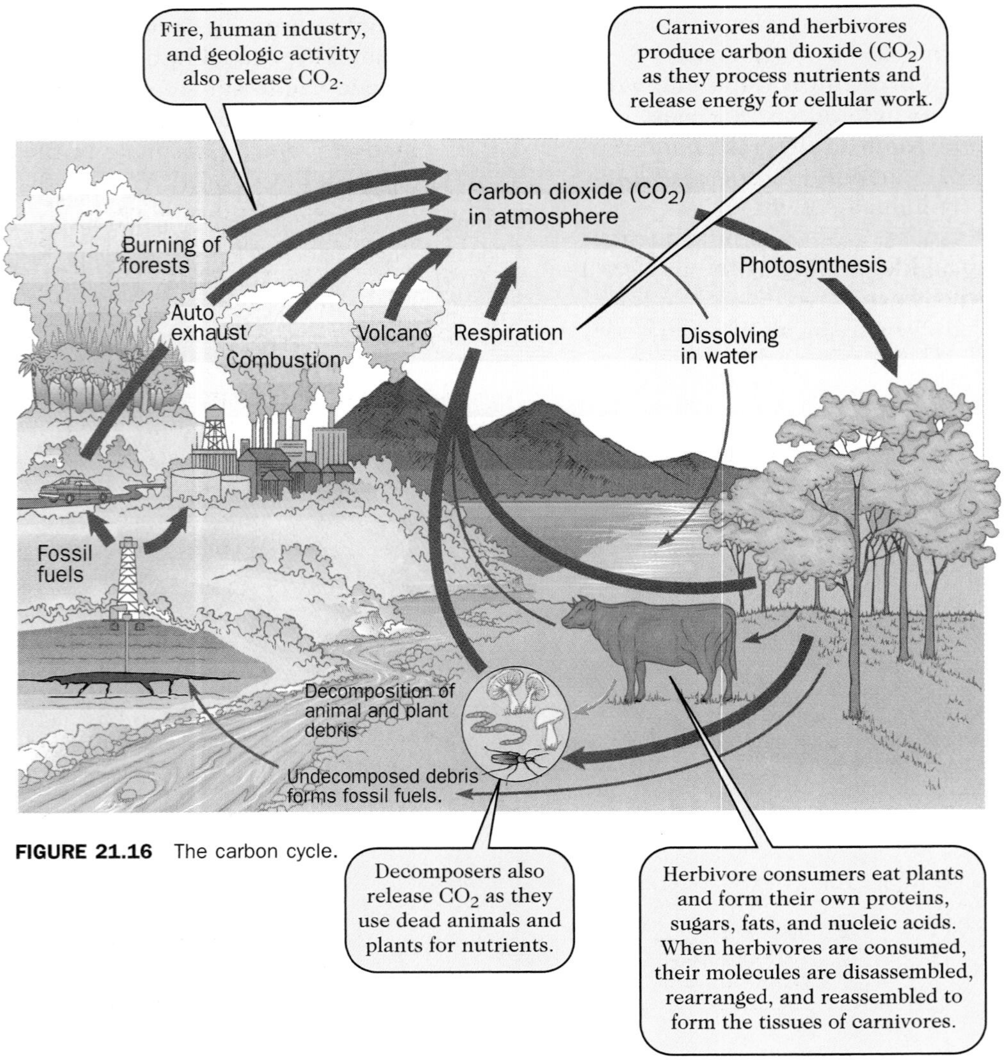

Fire, human industry, and geologic activity also release CO_2.
Carnivores and herbivores produce carbon dioxide (CO_2) as they process nutrients and release energy for cellular work.
Carbon dioxide (CO_2) in atmosphere
Burning of forests
Auto exhaust
Combustion
Volcano
Respiration
Photosynthesis
Dissolving in water
Fossil fuels
Decomposition of animal and plant debris
Undecomposed debris forms fossil fuels.
Decomposers also release CO_2 as they use dead animals and plants for nutrients.
Herbivore consumers eat plants and form their own proteins, sugars, fats, and nucleic acids. When herbivores are consumed, their molecules are disassembled, rearranged, and reassembled to form the tissues of carnivores.

FIGURE 21.16 The carbon cycle.

Nitrogen cycling. Nitrogen gas (N_2) is the most abundant of all the gases in the Earth's atmosphere, constituting nearly 80 percent of the outside air we breathe. Atoms of nitrogen are essential to life, but plants and animals lack the enzymes needed to convert N_2 to a form that can be incorporated into biological molecules. Fortunately, enzymes that do this are found in certain groups of free-living and symbiotic bacteria. The symbiotic bacteria are found in special root structures called *nodules* in terrestrial plants such as legumes (peas), alfalfa, and alder trees.

In a complex process called **nitrogen fixation,** N_2 is converted by bacteria into a form that can be used by plants to make amino acids and the nucleotides for DNA and RNA (Figure 21.17). Animals then obtain all their nitrogen by consuming and digesting the proteins and nucleic acids of plants or other animals. But what happens during decomposition? N_2 is returned to the atmosphere in a process called **denitrification,** which takes place when other soil bacteria act on animal wastes and dead plants and animals. At any given time, only about 0.03 percent of the available nitrogen is fixed into a usable biological form.

Nitrogen fertilizers versus crop rotation. Many

farmers take advantage of the nitrogen-fixing action of bacteria when they practice *crop rotation.* They plant crops such as peas or alfalfa one year to enrich the nitrogen content of the soil, and in the second year they plant corn or wheat, crops which remove nitrogen from the soil as they grow.

Farmers who do not rotate crops and repeatedly plant only corn or wheat deplete the soil of nitrogen, reducing its capacity to grow plants. To avoid such depletion, many farmers apply large amounts of expensive, commercially produced high-nitrogen fertilizers to the soil. This can lead to further problems. The excess nitrogen becomes a pollutant when it washes into streams and lakes, where it upsets food webs by poisoning animals and causing excessive growth of aquatic plants that may clog and deplete ponds and lakes.

Gaseous nitrogen compounds also can pollute. In the United States alone, as much as 25 million tons per year of gaseous nitrogen oxides is released into the atmosphere through automobile and jet plane exhausts, power-generating plants, and factories. The nitrogen oxides contribute to the formation of *acid rain,* which we'll discuss further in Chapter 22.

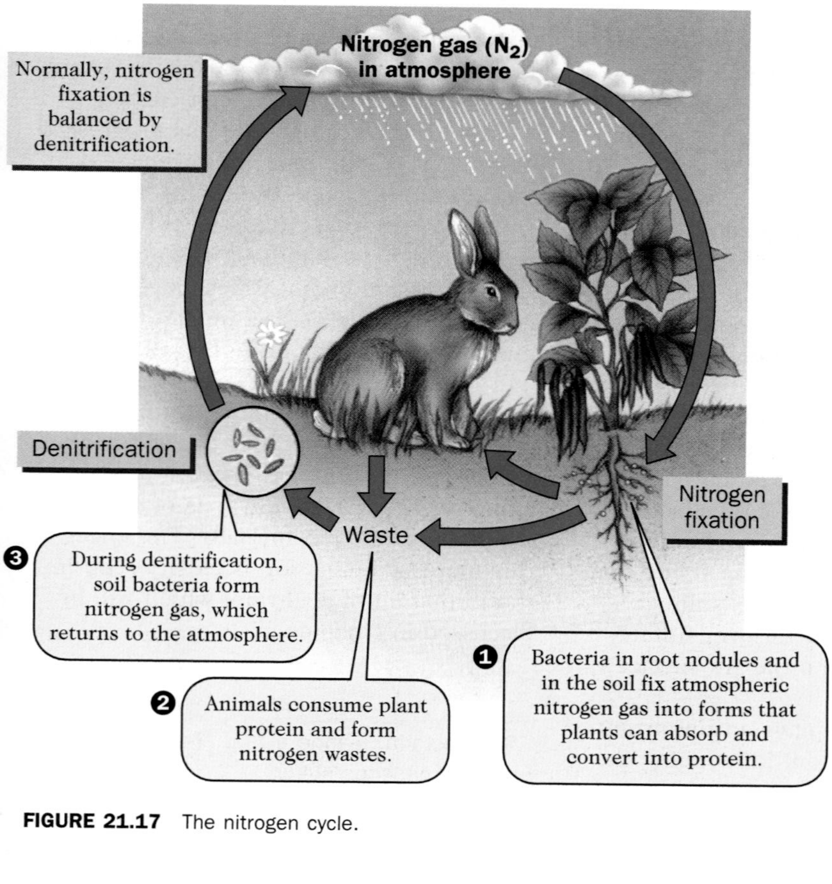

Nitrogen gas (N_2) in atmosphere
Normally, nitrogen fixation is balanced by denitrification.
Denitrification
Waste
Nitrogen fixation
3 During denitrification, soil bacteria form nitrogen gas, which returns to the atmosphere.
2 Animals consume plant protein and form nitrogen wastes.
1 Bacteria in root nodules and in the soil fix atmospheric nitrogen gas into forms that plants can absorb and convert into protein.

FIGURE 21.17 The nitrogen cycle.

> Nitrogen fixation by certain bacteria converts the gas N_2 to a form that can be used by plants. Animals obtain nitrogen by eating plants and other animals. Denitrification by soil bacteria returns nitrogen to N_2.

Phosphorus cycling. There are no ecologically significant phosphorus-containing gases. Therefore, the cycling of phosphorus lacks the gaseous phase of the previously described cycles, and this cycle operates more slowly (Figure 21.18). Phosphorus in the form of phosphate (PO_4^{3-}) is removed from the soil and incorporated into membranes, nucleic acids, and ATP by plants. When animals consume plants, they obtain phosphate and incorporate it into similar structures and molecules. When plants and animals die, decomposers return phosphate to the soil, where it can be recycled. The weathering of rock by wind and rain also returns phosphate to the soil, but this is an even slower process. Because of the slow cycling of phosphorus, it is the single nutrient that limits the proliferation of plants in many ecosystems.

Human use of phosphate-containing detergents and fertilizers pollutes the soil with excess phosphates. As with nitrogen, the excess phosphates cause extensive growth of aquatic plants and algae, causing dense blooms which turn water green. Masses of dead algal cells accumulate in the sediments, and their decomposition removes much of the available oxygen (O_2) from the water. The absence of O_2 slows the decomposition of dead plants, emits foul odors, and reduces the number of insects, fish, amphibians, and waterfowl present.

> Phosphate is removed from the soil by plants and enters the food web. Decay liberates phosphate from dead tissues, returning it to the soil. Phosphates erode from the soil and are slow to be recycled. Phosphate pollution damages ecosystems.

FIGURE 21.18 The phosphorus cycle.

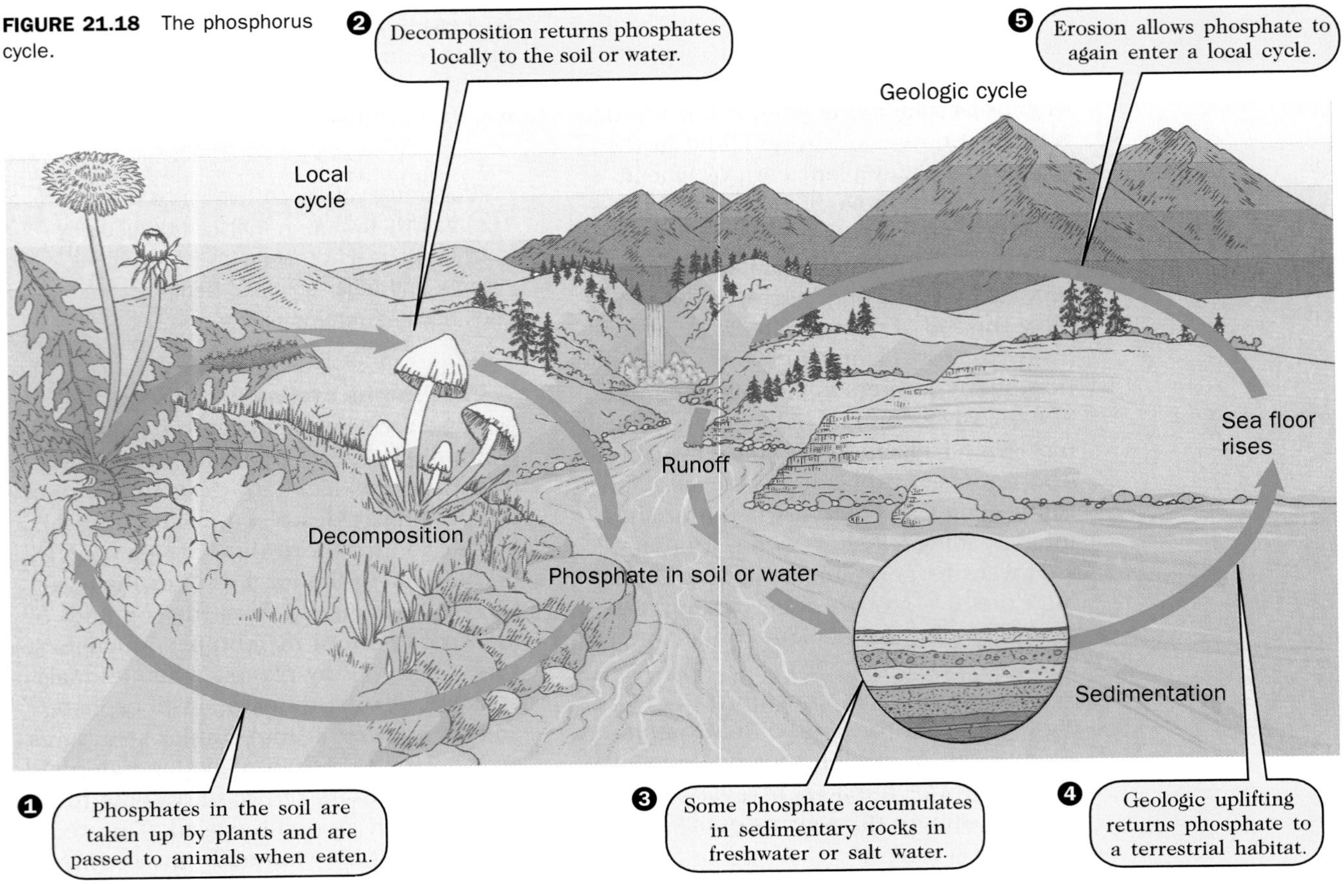
❷ Decomposition returns phosphates locally to the soil or water.
❺ Erosion allows phosphate to again enter a local cycle.
Geologic cycle
Local cycle
Runoff
Sea floor rises
Decomposition
Phosphate in soil or water
Sedimentation
❶ Phosphates in the soil are taken up by plants and are passed to animals when eaten.
❸ Some phosphate accumulates in sedimentary rocks in freshwater or salt water.
❹ Geologic uplifting returns phosphate to a terrestrial habitat.

Sulfur cycling. Sulfur in the form of sulfate (SO_4^{2-}) is used by animals and plants to construct amino acids and proteins. Sulfates are released during decomposition and are converted by bacteria in soils and sediments to the gases sulfur dioxide (SO_2) and hydrogen sulfide (H_2S), which enter the atmosphere. You can identify the activity of these bacteria when you smell a rotten egg odor in the air over a saltwater swamp or a mudflat exposed at low tide.

About half the gaseous sulfur released by living organisms is produced by coastal marine algae in the form of dimethyl sulfide. Volcanoes are a nonbiological source of atmospheric sulfur, accounting for about 10 to 20 percent of the gaseous sulfur that is present. In the atmosphere, sulfur molecules react with O_2 to form sulfate, which is returned to the Earth's surface by precipitation.

Atmospheric sulfur plays an important role in climate formation, influencing the weather experienced in various regions. Water condenses around particles of dimethyl sulfide to form clouds. Some researchers suspect that the global warming associated with high levels of carbon dioxide in the atmosphere has progressed more slowly than expected because of increased cloud cover resulting from an increase in dimethyl sulfide levels.

Humans have significantly increased sulfur emission to the atmosphere by burning sulfur-containing coal in factories, by mining and smelting iron ores which contain sulfur, and by electrical power generation and petroleum refining. Human activities now release $1\frac{1}{2}$ times the sulfur emitted by natural sources. The United States alone releases 30 million tons of sulfur dioxide into the air annually. The effect of this increased emission is clearly seen in the formation of acid rain, which will be discussed in Chapter 22.

Soil bacteria, marine algae, and volcanoes release gaseous sulfur. Sulfur is released from organisms during decomposition.

The Biosphere

All the living communities inhabiting all the ecosystems make up the Earth's **biosphere,** the thin film of life enveloping our planet. Organisms are strongly affected by weather and climate. *Weather* includes the conditions of temperature, humidity, wind, cloud cover, and precipitation that occur in a given locale over a period of hours or days. Weather conditions affect the activity, feeding, and reproduction of individual organisms. *Climate* refers to seasonal weather patterns that prevail over years or even centuries in a particular locality. Climate influences the distribution and abundance of entire species. What are the causes of climate?

Global Patterns of Climate

The world's climates are created primarily by the exposure of an orbiting and rotating planet to solar energy. The major climatic zones are created by the circulation of air and water masses powered, respectively, by the heat of the sunlight and the rotation of the Earth on its axis. Climatic zones in turn influence the distribution of plants and animals.

The origin of the seasons. In Figure 21.19 you can see that the Earth is tilted on its axis about 23.5 degrees off vertical. During the *summer*, the northern hemisphere tips toward the sun, absorbs more solar energy, and experiences warmer weather and longer periods of daylight. In the *winter,* it tips away from the sun, causing shorter days and cooler temperatures. The seasons are just the reverse for the southern hemisphere.

The sun lies over the equator at all times of the year, and its direct, intense light is concentrated on a relatively small surface area. Thus, at the equator days are warmer, brighter, and equally long from season to season. As the distance (latitude) from the equator increases, light strikes the Earth at an angle. This means that solar energy is distributed over a broader surface area, creating colder conditions and more seasonal variation. At the poles, winters are long and cold with periods of prolonged darkness and summers are short and cool with periods of continuous light.

Patterns of atmospheric circulation. The intense light falling at the equator (0 degrees latitude) heats the overlying air and reduces its density, causing it to expand and rise. The sun's heat also evaporates water from the Earth's surface, and this water vapor is drawn upward with the rising air. As the hot, moist air rises, it begins to cool and move away from the equator toward the north and south. Water condenses in the cooling air and falls as rain over a broad band

FIGURE 21.19 The rotation of a tilted Earth around the sun and the variation in light intensity with latitude produces the four annual seasons.

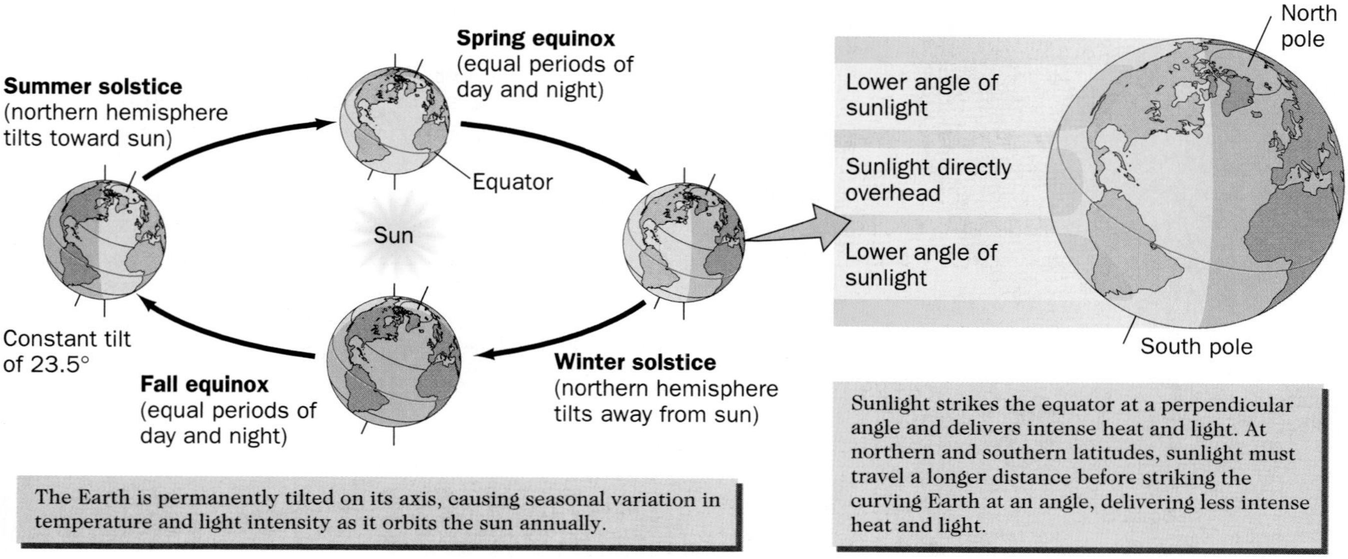
Spring equinox
(equal periods of day and night)
Summer solstice
(northern hemisphere tilts toward sun)
Equator
Sun
Constant tilt of 23.5°
Fall equinox
(equal periods of day and night)
Winter solstice
(northern hemisphere tilts away from sun)
North pole
Lower angle of sunlight
Sunlight directly overhead
Lower angle of sunlight
South pole

The Earth is permanently tilted on its axis, causing seasonal variation in temperature and light intensity as it orbits the sun annually.

Sunlight strikes the equator at a perpendicular angle and delivers intense heat and light. At northern and southern latitudes, sunlight must travel a longer distance before striking the curving Earth at an angle, delivering less intense heat and light.

neighboring the equator, creating the *tropics,* which are among the warmest, brightest, and wettest regions on Earth.

At about 30 degrees north and south latitude, the cooler, drier air begins to descend. As it nears the Earth's surface, it is warmed by compression and reflected solar energy. When it reaches the surface, it is quite warm and dry and flows back toward the equator. This warm dry air creates the Earth's major *deserts* near these latitudes.

Circling the globe, there are a total of six air circulation "coils" consisting of rising warm moist air and descending cool dry air (Figure 21.20). The Earth's rotation causes the ascending and descending masses of air to move through the circulation coils in a spiraling fashion. This movement forms the prevailing *winds* at each latitude and determines the direction in which weather fronts normally move. The direction of wind movement, together with the Earth's rotation, also powers the circulation of ocean currents.

Ocean currents and land temperatures. The surface waters of the world's oceans are heated by the sun and set in motion by blowing winds, creating **currents.** Currents move in roughly circular patterns called *gyres,* influenced by the shape of ocean basins and the presence of continents. The rotation of the Earth on its axis causes gyres to circulate clockwise in the northern hemisphere and counterclockwise in the southern hemisphere (Figure 21.21).

Since water heats and cools more slowly than does air, the range of temperatures in the oceans is not as extreme as that in the air over land. The interior of a continent typically has more variable and extreme temperatures than do coastal areas. For example, in the western Atlantic, the Gulf Stream warms the eastern seaboard of the United States and Canada, causing a warmer, moister, and more temperate climate than is found in the interior. The Gulf Stream also flows to the east, where it warms western Europe.

Continents modify climate. The shape of continents, especially the presence of mountain ranges, influences the flow of air and precipita-

FIGURE 21.20 The Earth's climatic zones.

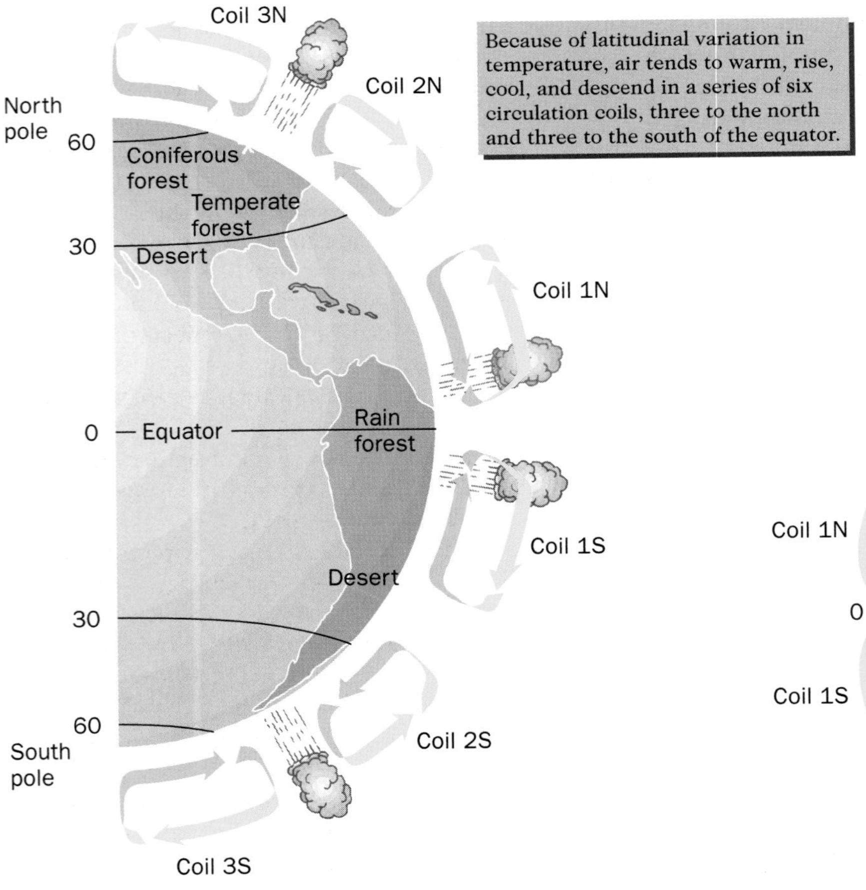
Coil 3N
Coil 2N
North pole
60
Coniferous forest
Temperate forest
30
Desert
Coil 1N
0
Equator
Rain forest
Coil 1S
Desert
30
60
South pole
Coil 2S
Coil 3S
Because of latitudinal variation in temperature, air tends to warm, rise, cool, and descend in a series of six circulation coils, three to the north and three to the south of the equator.

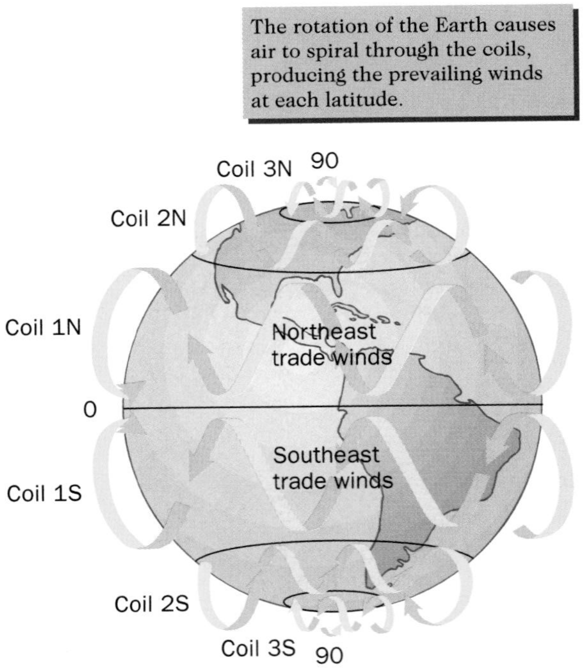
The rotation of the Earth causes air to spiral through the coils, producing the prevailing winds at each latitude.
Coil 3N
90
Coil 2N
Coil 1N
Northeast trade winds
0
Southeast trade winds
Coil 1S
Coil 2S
Coil 3S
90

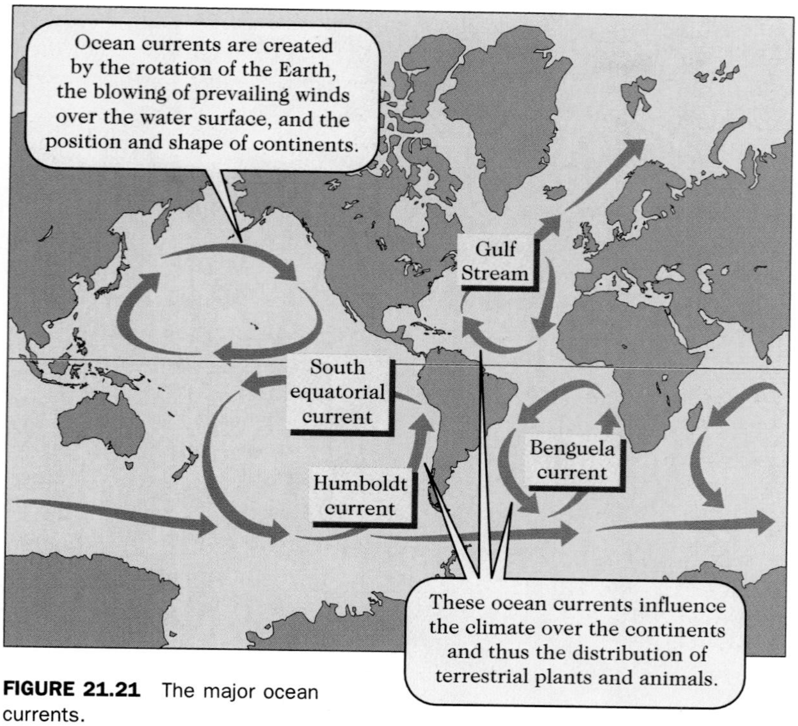
Ocean currents are created by the rotation of the Earth, the blowing of prevailing winds over the water surface, and the position and shape of continents.
Gulf Stream
South equatorial current
Benguela current
Humboldt current
These ocean currents influence the climate over the continents and thus the distribution of terrestrial plants and animals.

FIGURE 21.21 The major ocean currents.

tion and thus modifies the formation of climate. As moisture-laden air flows in from the ocean and encounters mountains, it rises and cools. Since cool air holds less moisture, rain falls on the ocean-facing side of the mountains. As cool, dry air descends on the far side of the mountains, it becomes warmer and absorbs moisture from the land, producing a dry region called a *rain shadow*. The Mojave Desert of California, for instance, is found on the dry eastern side of the southern end of the Sierra Nevada Range. Without mountains, climatic zones would be found in broad uniform bands that would coordinate more precisely with latitude.

Solar heating and the Earth's rotation create patterns of air and water circulation and the ocean currents. Circulating air coils drive the winds and move weather. Mountains influence air and water circulation, creating local precipitation patterns.

Terrestrial Biomes: Covering 25 Percent of the Earth

The climate that prevails over a geographic region establishes the conditions in an ecosystem to which plant and animal populations must adapt. The term **"biome"** is used to designate a terrestrial or aquatic region with characteristic environmental conditions that is occupied by distinctive communities of organisms.

The world's eight major terrestrial biomes are mapped in Figure 21.22. Their close correlation with patterns of air and moisture circulation that create climatic zones can be clearly seen. Terrestrial biomes typically are named for their predominant forms of vegetation. Since plants cannot move away from the weather, they reflect adaptations to the climatic conditions (especially rain-fall and temperature) that prevail in a region. As producers, plants also influence the types of consumers and decomposers with which they coexist. In reality, biomes do not have sharp and finite borders as they appear to have on a map, but instead merge gradually with one another. The species composition in similar biomes varies from continent to continent.

Animals and plants in the deserts of North America and Africa may resemble one another because they occupy similar niches, but they actually belong to different taxonomic groups. In most biomes today, species composition also is strongly influenced by human activity and habitation.

Tropical forests: few nutrients in the soil. Tropical forests contain more different species of animals and plants than do all the other terrestrial biomes combined (Figure 21.23). Tropical forests are found near the equator in South America, Africa, and Asia. Temperatures are consistently warm, averaging 23°C (74°F). In areas where there is abundant rainfall, over 250 centimeters (100 inches) annually, rain forests occur.

With a full 12 hours of sunlight each day, considerable solar energy strikes the upper canopy of the forests, but the taller trees shade those growing closer to the ground. Only about 2 percent of the incoming light reaches the forest floor. Plants that are successful in this habitat have very broad leaves (for absorbing sunlight) or reach into the upper canopy by growing high on trees or growing as long spindly vines.

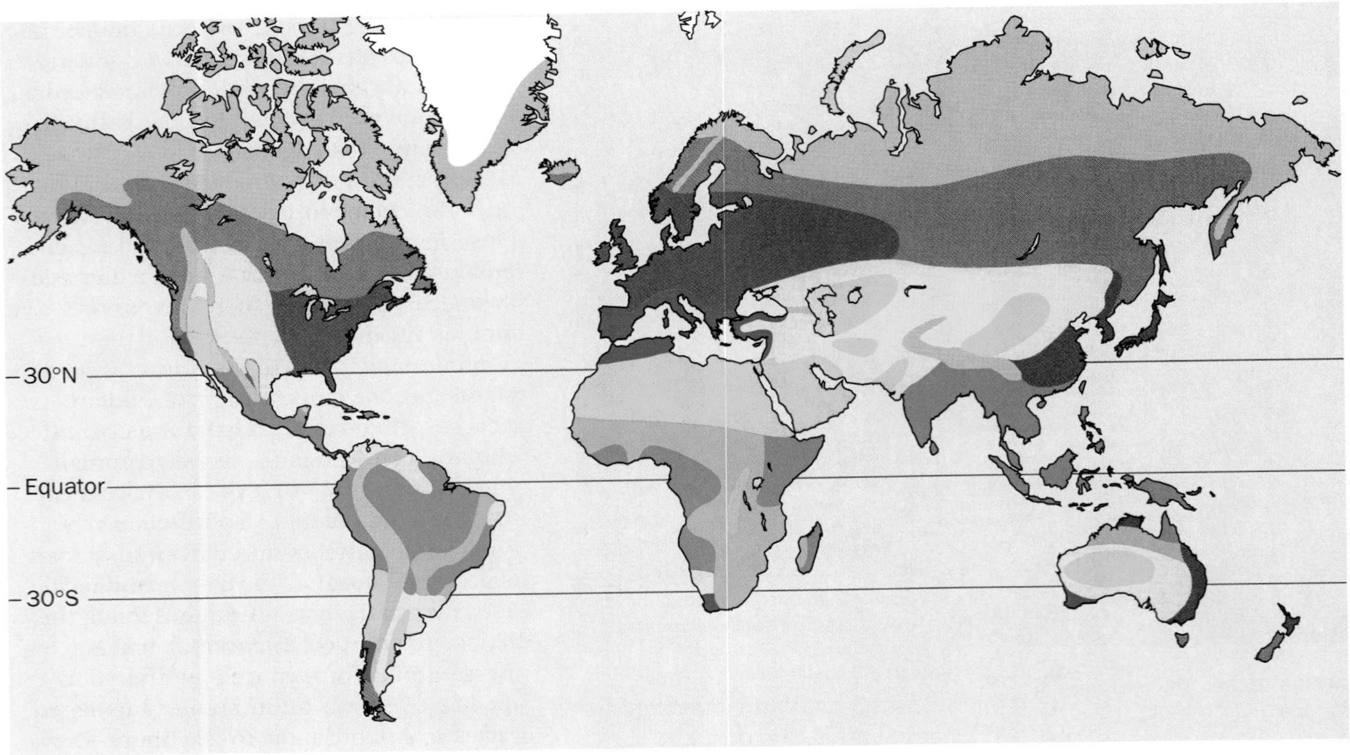
30°N
Equator
30°S
Tropical forest
Savanna
Desert
Grassland
Chaparral
Deciduous forest
Coniferous forest
Tundra

FIGURE 21.22 Map of the terrestrial biomes.

Rain forest soils are low in mineral nutrients because of three factors: (1) rapid decomposition of fallen leaves by abundant insect and microbial life, (2) quick absorption of minerals by the diverse vegetation, and (3) the leaching action by the frequent rain. The low nutrient levels frustrate the humans who cut down rain forests in an effort to create agricultural land. When rain forests are logged and converted to crop growth or cattle grazing, there are sufficient nutrients to support only a few years of sustained use. When the soil is exhausted of mineral nutrients, the farmers abandon it, clearing more forest land in a flawed attempt to create a productive agricultural region. The rapid destruction of rain forests has many consequences, which we will discuss in Chapter 22.

FIGURE 21.23 A tropical forest.

Savannas: dry, nutrient-poor grasslands. Savanna is dry grassland with scattered trees (Figure 21.24). It is common in subtropical regions of South America, Africa, and Australia. There are typically three seasons in the savanna: cool and dry, hot and dry, and warm and wet. The soils are porous and well drained and hold only small amounts of organic and mineral nutrients. The savanna is home to some of the world's largest grazing herbivores, including buffalo, giraffes, kangaroos, zebras, elephants,

FIGURE 21.24 The savanna.

and antelopes on different continents. There are also large predatory consumers such as lions, leopards, hyenas, and dingoes. Humans affect the savanna by harvesting the large mammals for food, pelts, and ivory and by grazing cattle for food or leather production and by farming.

Deserts: water limits life. Deserts have the lowest annual rainfall among the terrestrial biomes, generally totaling less than 30 centimeters (12 inches) per year. Water is the major factor controlling the diversity and abundance of plants and animals (Figure 21.25). Deserts frequently have wide fluctuations in daily temperatures, with very hot afternoons and rather cool nights. Plants are often widely spaced and, like the cacti, are modified to maximize water storage (thick, fleshy stems) and minimize water loss (thin, spinelike leaves). Some plants complete their life cycles in the brief period when water is available. For example, desert wildflowers bloom in spectacular displays after late winter rains in the southwestern deserts of the United States.

FIGURE 21.25 A desert in the southwestern United States.

Desert animals also must be specialized for water conservation. The abundant reptiles (snakes and lizards) are coated with waterproof scales. Small mammals (mice and gerbils) burrow away from the harsh sun and feed at night, when it is cooler. Many derive water from only food and as a by-product of their metabolism.

Human activity is expanding many of the world's deserts because of overgrazing and trampling of pastureland and poor use of irrigation. When properly irrigated, however, desert lands are highly productive, as exemplified by the wheat fields and vineyards of eastern Washington State, which are kept moist by water diverted from the Columbia River.

Grasslands: moderate rain, high productivity. Grasslands receive more rain than deserts but less than forests (Figure 21.26). Included in this biome are the *prairies* of the central United States, the *pampas* of Argentina, the *steppes* of central Asia, and the *veldt* of central and southern Africa. The dozens of grasses common to this habitat must be capable of withstanding frequent droughts, wildfire, and the grazing of large herbivores. Their deep roots give them access to water and allow quick regeneration of damaged aerial portions.

Herbivores common to grasslands include bison, antelopes, gazelles, wild horses, and burrowing rodents. Lions, coyotes, wolves, and hawks are common predators in grasslands. Human activity has converted much of the native grassland to agricultural use. In the southwestern United States overgrazing by cattle has reduced the growth of grasses and encouraged encroachment by sagebrush, a species more common in cool deserts.

FIGURE 21.26 A prairie grassland in the United States.

cracking open the tough seeds of chaparral plants and allowing their germination.

Browsing mammals such as deer and small fruit-eating rodents and birds are common animals, as are lizards and snakes. Since the chaparral borders coastal regions, it is frequently populated, altered, and destroyed by human residential and recreational use. In this habitat, careless humans ignite fires more frequently than do natural sources such as lightning. The brushfires, coupled with extensive trampling and clearing of vegetation, promote erosion and mudslides.

Deciduous forests lose their leaves. Deciduous forests are populated by broad-leaved trees which lose their leaves and become dormant in winter and then sprout new leaves each spring (Figure 21.28). The slowly decomposing leaf litter provides an abundant supply of nutrients to the soil, making this biome a productive agricultural region when trees are cleared, as has been done in much of the central and eastern United States. This biome experiences hot summers and very cold winters, with abundant rain throughout the year. Some of the common trees include maple, oak, birch, hickory, and beech. Insects, spiders, birds, shrews, bobcats, foxes, wolves, cougars, and bears all historically found homes in these forests. Among these animals, the large predators have been extensively displaced by humans.

Coniferous forests: the evergreens. Coniferous forests extend as a broad band of fir, pine, spruce, redwood, and hemlock across the northern United States, Canada, northern Europe, and Asia (Figure 21.29). The trees are well adapted for cold winters during which little liquid water may be available. Their drooping branches shed snow and ice, and their waxy, needle-shaped leaves reduce water loss by evaporation. Since their leaves are retained through the winter, they are available for photosynthesis as soon as the growing season begins. In areas disturbed by fire or human activity, deciduous trees such as alder, birch, aspen, and willow are common.

Small burrowing mammals survive the winters under a blanket of insulating snow that prevents deep freezing of the

Chaparral is coastal, and fires are frequent. Chaparral (***shap-uh-ral***) is a coastal biome that borders deserts and grasslands (Figure 21.27). This biome is characterized by hot, dry summers and cool, wet winters. Shrubs and trees which retain their leathery leaves throughout the year are the predominant types of vegetation. Sage, manzanita, eucalyptus, and scrub oak are common examples. These plants are resistant to the frequent fires which sweep through chaparral. However, fire may play a positive role by

FIGURE 21.27 The chaparral.

FIGURE 21.28 A deciduous forest in the eastern United States.

soil. Insects, seed-eating birds, and furry mammals such as rabbits, deer, moose, elk, bears, lynx, and wolverines frequent these forests. Though much of this biome is still underpopulated by humans, its southern borders have been logged extensively to provide lumber for construction. Massive harvesting of trees from hundreds of acres at a time creates the "clear-cuts" that dot the landscape of the Pacific northwest, causing soil erosion and loss of minerals from areas that are not quickly replanted.

Tundra: land of permafrost. Tundra lies between the northernmost coniferous forests and the polar ice caps. A vast, treeless plain, the tundra experiences very severe climatic conditions (Figure 21.30). There are dark, cold, and windy winters and bright, cool summers. The growing season lasts only a few weeks. Tundra receives less rain than do some

FIGURE 21.29 A coniferous forest in the western United States.

FIGURE 21.30 The tundra.

deserts, and during the winter water is frozen and unavailable to plants and animals.

The cold temperatures maintain a permanently frozen layer of subsoil called **permafrost** 1 meter (3.3 feet) or less below the soil surface. The permafrost prevents drainage of melting snow and ice during the summer thaw and turns the tundra into a very soggy marsh. The permafrost layer also prevents the deep root penetration needed for tree growth. As a result, dwarf shrubs, grasses, mosses, and lichens are the most successful types of vegetation.

Animals that survive on the tundra must be very well insulated against the cold or be able to burrow. The tundra is populated by large herds of herbivorous mammals insulated against the cold by hair, fat, and a warm and stable internal temperature. Caribou, reindeer, and musk oxen graze on the abundant mosses and lichens. Wolves prey on young, old, and ill herbivores, while arctic foxes and snowy owls feed on smaller mammals such as hares and lemmings. During the summer, flocks of migratory birds arrive on the tundra, many of which feed on the clouds of mosquitos and other insects which have matured in pools of standing water.

The tundra is a fragile environment that is easily damaged by human activity. Oil wells, pipelines, and military radar installations are the primary forms of human presence on the tundra. The cold temperatures dictate slow growth and repair of plants, allowing a single wheel rut from a motor vehicle to scar the tundra for many years.

The tundra biome also can be found above the tree line near the tops of tall mountains at all latitudes. In fact, altitude mimics latitude in the distribution of biomes, since precipitation and temperature vary with elevation just as they do with latitude (Figure 21.31).

As you climb a tall tropical mountain, you stroll through a tropical rain forest at the base and then ascend through a deciduous and coniferous forest to reach the tundra, which lies just below the permament ice cap at the mountain's summit.

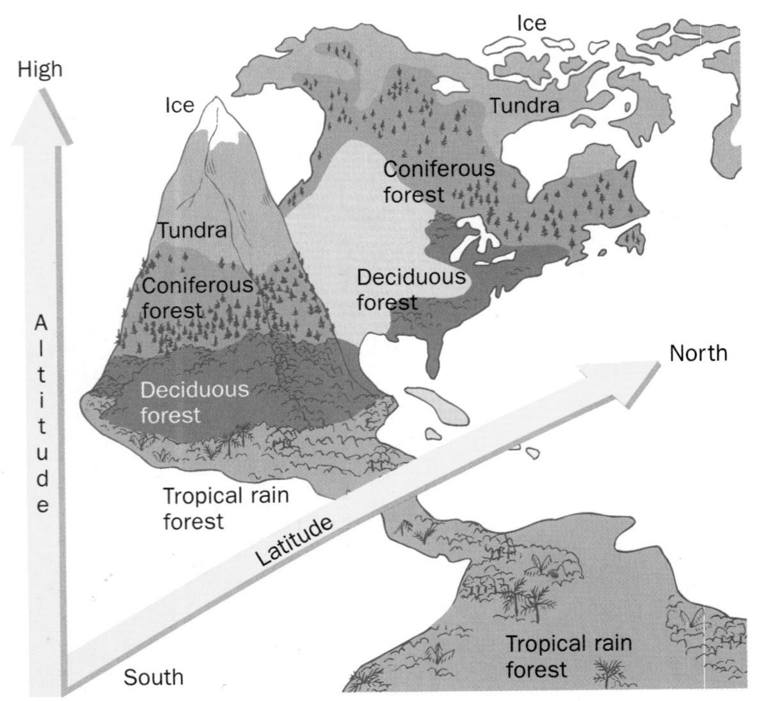
Ice
High
Ice
Tundra
Coniferous forest
Tundra
Coniferous forest
Deciduous forest
Altitude
North
Deciduous forest
Tropical rain forest
Latitude
Tropical rain forest
South
Low

As you ascend the mountain, the biomes are arranged in the same order in which they are found when you travel from the equator toward either the north pole or the south pole.

FIGURE 21.31 Altitude mimics latitude in the distribution of biomes.

Biomes are regions where climate creates conditions to which characteristic groups of plants and animals become adapted. The eight terrestrial biomes are named for their predominant forms of vegetation. In the distribution of terrestrial biomes, altitude mimics latitude.

Aquatic Biomes: Covering 75 Percent of the Earth

Life on Earth began in water and developed there for nearly 3 billion years before the land was colonized by plants and animals. Today, three-quarters of the Earth is covered with water. Aquatic organisms are buffered against rapid temperature changes by the heat-holding capacity of water. They are strongly influenced by decreasing light penetration and increasing pressure that occur with greater depth and by the salt content of their habitat.

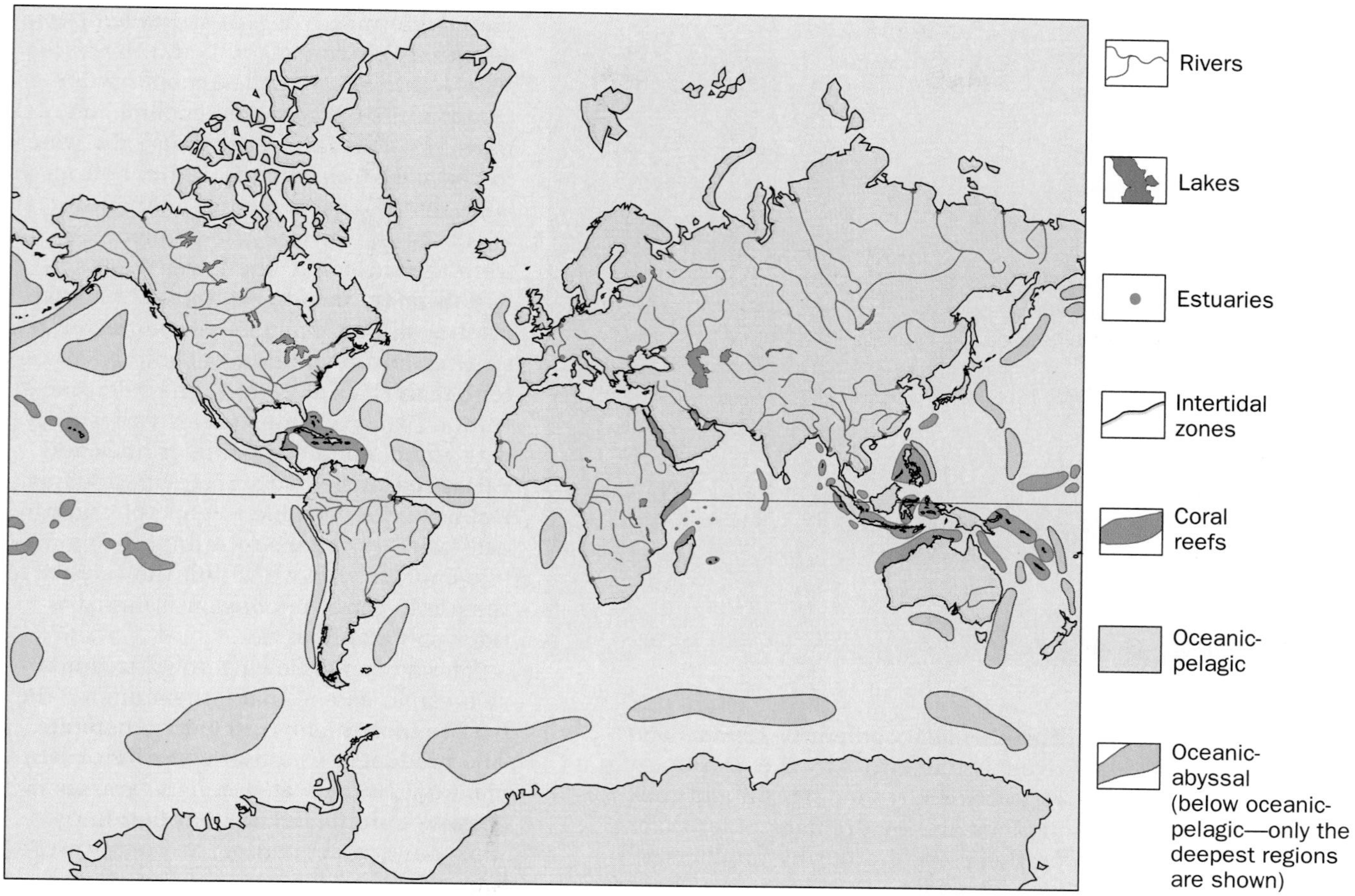
Rivers
Lakes
Estuaries
Intertidal zones
Coral reefs
Oceanic-pelagic
Oceanic-abyssal (below oceanic-pelagic—only the deepest regions are shown)

FIGURE 21.32 The major aquatic biomes.

FIGURE 21.33 A river biome.

In aquatic habitats, there is also a tendency for both organic and mineral nutrients to settle toward the bottom. This creates a separation of light energy and nutrients which influences the distribution of plants and animals. Plants must locate nearer the surface, where light is abundant, while animals can be distributed throughout the habitat, from surface to bottom. Aquatic biomes (Figure 21.32) are named on the basis of their physical and chemical properties or their location. Freshwater biomes, with less than 1 percent salt content, are closely linked to terrestrial biomes. The freshwater biomes include streams, rivers, ponds, and lakes. They represent less than 1 percent of the area of the Earth's surface.

Streams and rivers: flowing freshwaters. Streams and rivers are formed by runoff of water from mountains and hills. At some places this combines with the water from melting glaciers (Figure 21.33).

FIGURE 21.34 A pond biome.

The chemical content of streams and rivers is influenced by the erosion of minerals from surrounding soil and rock formations and by drainage of fertilizers, pesticides, and other by-products of human agriculture and industry. Their O_2 content increases with the turbulence of flow, and their temperature is determined by altitude and latitude.

Organisms that dwell in this biome must contend with the forceful flow of water by swimming strongly enough to maintain their position, attaching firmly to rocks, or seeking shelter under rocks or in burrows. Bordering many streams and rivers are soggy, seasonally flooded swamps, bogs, and marshes collectively called **inland wetlands.**

Ponds and lakes: standing freshwaters. Ponds and lakes form when runoff collects in a land-locked basin (Figure 21.34). Since some light is absorbed by water and the materials within it, light intensity decreases with depth. This physical phenomenon influences the presence of plants and the amount of photosynthetic production. Rooted plants are found along the shore and in shallow water, while floating plants occur in the deeper open water. Floating microscopic life is called **plankton.** The microscopic producers (mostly algae) are called *phytoplankton* (*phyto,* "plant"), and the microscopic animals are called *zooplankton* (*zoo,* "animal"). Seasonal cooling and warming of surface water cause the water to circulate from the top to the bottom in lakes and ponds, bringing sediment nutrients to the surface and bringing O_2 from the surface to the lower depths.

Estuaries: mixing regions. Estuaries are transitional habitats where freshwater streams and rivers flow out into a saltwater ocean (Figure 21.35). The salt concentration (salinity) in estuaries varies daily with the coming and going of tides and varies seasonally with upstream flooding. Animals must be able to cope physiologically with variations in salinity or be mobile enough to move within the estuary to remain in water of constant salinity as tides ebb and flow.

The nutrients flowing downstream enrich estuaries and make them among the world's most highly productive habitats. The producers in estuaries include phytoplankton in the water, aquatic grasses in shallow waters, and salt-tolerant marsh

FIGURE 21.35 An estuary.

plants along the shore. Estuaries are important feeding and breeding grounds for fish. The surrounding marshes, called **coastal wetlands**, frequently provide refuge for migratory birds. Fish and birds attract human recreational, commercial, and residential activity. Estuaries also receive all the polluting chemicals that have been collected during the downstream flow of rivers. Over one-half of the coastal and inland wetland areas in the United States have been destroyed by human activity.

Intertidal zones: land washed by tides. The intertidal zone biome includes the portion of the ocean shore washed over by the tides (Figure 21.36). Animals and plants may experience alternating air exposure and water coverage with the tidal cycle and often are exposed to heavy wave action. Many organisms show adaptations such as firm attachment (tube feet of starfish and holdfasts of seaweeds), protective shells (crabs, clams, and snails), or a burrowing existence (worms and other soft-bodied animals).

Sandy beaches are best for burrowing, while rocky beaches offer somewhat sheltered tide pools or rock crevices. Beaches and tidepools also attract humans, and the intertidal zones are frequently the focus of human food gathering, recreation, and residence.

Oceanic-pelagic biomes: open ocean. The oceanic-pelagic (*peh-**lah**-jik*) biome, the deep open water of the ocean, is home to floating plankton and swimming marine animals (Figure 21.37). Plants in this biome must be able to float near the surface to absorb adequate light for photosynthesis. The larger algae often have gas bladders to aid in flotation, and the microscopic algae raft together to increase their surface area or form oil droplets in their cytoplasm, both of which increase their buoyancy. Animals also must float or develop structures which allow active swimming, such as cilia, flagella, fins, and flukes.

Pelagic waters are typically cold and become colder with an increase in latitude or depth. The waters are well mixed by the movement of ocean currents, and there are occasional upwellings where bottom waters are pushed up to the surface, carrying with them mineral nutri-

FIGURE 21.36 The rocky intertidal zone.

FIGURE 21.37 The open ocean.

FIGURE 21.38 The ocean floor.

ents from the sediments on the ocean floor.

Oceanic-abyssal biomes: deep ocean floor. The oceanic-abyssal (*uh-**biss**-sul*) biome lies on the ocean floor at very great depth (Figure 21.38). It is a dark, cold, and foreboding realm where pressures are crushingly great. There are no plants conducting photosynthesis, but there are many bacteria, invertebrates (animals without backbones), and fishes. The bacteria are especially plentiful near deep sea *vents,* which are cracks in the ocean floor where water is superheated by the magma lying under the Earth's crust. Bacteria capable of oxidizing hydrogen sulfide as an energy source live in the warm water surrounding the vents. These bacteria act as producers in this unusual ecosystem and provide energy for crabs, tube-dwelling worms, sea anenomes, and fish. Some of these bacteria live symbiotically in the tissues of the animals, providing organic compounds in exchange for sulfides.

Coral reefs: deposited by animals. Coral reefs are the only biome formed primarily by the growth of animals (Figure 21.39). Corals are small soft-bodied animals that have a calcium carbonate skeleton outside their bodies. They are assisted in depositing their skeletons by symbiotic algae growing and conducting photosynthesis in their tissues. Reefs form by means of the gradual accumulation of coral skeletons over time. The reef provides an anchoring site, shelter, and food for organisms such as seaweeds, sponges, starfish, snails, and fishes. Along the Great Barrier Reef of Australia there are over 500 different species of corals, plants, and animals.

Although reef communities are old, they are also very fragile. Lying in shallow waters near shore, they are particularly threatened by erosion of soil as a result of road and building construction, agriculture, and logging. Soil eroded from the land may smother corals and other filter-feeding animals. Reefs also are threatened by souvenir-collecting tourists and overharvesting of fish, turtles, mollusks, and crustaceans. The removal of these higher-level predators disrupts the food web, allowing the populations of primary and secondary consumers to increase inappropriately and upset the community's balance. Large portions of Austrailian and African reefs have been damaged by an overabundance of coral-eating starfish and sea urchins.

FIGURE 21.39 A coral reef in Fiji.

Aquatic biomes cover 75 percent of the Earth's surface and include streams and rivers, lakes and ponds, saltwater estuaries, intertidal zones, oceanic-pelagic, oceanic-abyssal, and coral reefs.

Chapter Summary

Introduction

- Ecology is the study of the interactions between organisms and their environment.
- Populations are individuals of the same species inhabiting the same area. Communities are different populations, living and interacting in the same area. Ecosystems are communities interacting with their physical environment. The biosphere is the portion of Earth inhabited by life.

Population Ecology

- Habitats are the places where populations live and feed.
- Geographic range is determined by a species' physical tolerance, competition with other species, and by geographic barriers.
- The growth of a population is influenced by the age of sexual maturity, the number of offspring, the length of life, and the pattern of sex and age distribution.
- Carrying capacity is the number of individuals supported in a habitat by available resources.

Community Ecology

- Niche is the role of an organism in its habitat and the conditions under which it lives and thrives.
- Interspecific competition may cause the exclusion of a species from a community.
- Predation occurs when herbivores eat plants, carnivores eat animals, parasites infest other organisms, or cannibalism occurs between members of the same species. Camouflage, warning coloration, or mimicry help species avoid predation.
- Forms of symbiosis (species living together) include parasitism, commensalism, and mutualism.
- Succession is the progression of changes in a community as new or altered habitats are colonized.

Ecosystem Dynamics: Energy Flow and Nutrient Cycling

- Energy flows in one direction through an ecosystem.
- Producers are photosynthetic organisms. Herbivores and carnivores are consumers. Decomposers recycle nutrients and minerals back to producers.
- A food web describes the flow of energy and nutrients through the populations of a community.
- An energy pyramid shows the energy transfers between the various levels of a food web.
- Biomass is the organic material in plants and animals.
- Biological magnification is the concentration of toxic materials in the higher levels of a food web.
- Water moves from the atmosphere, through oceans or freshwater, organisms, soil, and back to the atmosphere.
- Carbon is absorbed (as CO_2) from the atmosphere by producers, consumed by animals and decomposers, and returned to the atmosphere (again as CO_2) by respiration.
- Nitrogen (as N_2) is removed from the atmosphere by nitrogen fixing bacteria, absorbed by plants, consumed by animals, released into the soil by decomposers, and returned to the atmosphere by denitrifying bacteria.
- Phosphorus cycling lacks a gaseous phase. Phosphorus is absorbed from the soil by plants, consumed by animals, and released again into the soil by decomposers. Geologic activity and the weathering of rock also release phosphorous into the soil.
- Sulfur cycles from the atmosphere into soil by precipitation, through plants, animals, and decomposers and returns to the atmosphere as a gas.

The Biosphere

- Climates result from circulation of air and water masses as the Earth revolves on its axis and orbits the sun.
- Seasons are created as the Earth tips on its axis, exposing the northern or southern hemispheres to the sun.
- Air circulates in spiraling coils of rising warm air and descending cool air which create prevailing winds.
- Ocean currents circulate in clockwise (northern hemisphere) and counterclockwise (southern hemisphere) gyres.
- Continents and mountains alter the circulation of air and oceans and influence the patterns of rainfall.
- Biomes are regions with characteristic environmental conditions and distinctive communities.
- Terrestrial biomes include: tropical forests, savannas, deserts, grasslands, chaparral, deciduous forests, coniferous forests, and the tundra.
- Aquatic biomes include: streams and rivers, ponds and lakes, estuaries, intertidal zones, oceanic-pelagic, oceanic-abyssal, and coral reefs.

Selected Key Terms

biogeochemical cycles (p. 569)
biological magnification (p. 568)
biome (p. 575)
biosphere (p. 573)
community (p. 557)
consumer (p. 566)
decomposer (p. 566)
ecosystem (p. 557)
food web (p. 566)
habitat (p. 557)
niche (p. 560)
population (p. 556)
producer (p. 566)
succession (p. 564)
symbiosis (p. 563)

Review Activities

1. Define population, community, ecosystem, and biosphere.
2. List and explain the patterns by which organisms are distributed in habitats. Give examples from your own experiences if possible.
3. What factors influence the range and growth of populations?
4. Distinguish between density-dependent and density-independent factors that influence carrying capacity. Give examples of each.
5. Distinguish between the niche and the habitat of an organism.
6. Describe three patterns of predation and list some of the defensive methods evolved by animals to avoid predation.
7. Explain the differences between parasitism, commensalism, and mutualism. List examples of each one.
8. Distinguish between primary succession and secondary succession. List examples you have observed.
9. Distinguish between producers, primary consumers, secondary consumers, tertiary consumers, and decomposers.
10. Explain why biological magnification of toxic chemicals occurs in a food web.
11. Briefly summarize the water, carbon, and nitrogen cycles. Distinguish between nitrogen fixation and denitrification.
12. Explain the forces that create climates and seasons and air and water circulation.
13. Why doesn't rain fall equally over landmasses?
14. List the terrestrial biomes and give a brief description of each one.
15. List the aquatic biomes and give a brief description of each one.

Self-Quiz

Matching Exercise

___	1. Places where populations live, shelter, and feed	A. Producers
___	2. Function of a species and the conditions under which it lives	B. Carrying capacity
___	3. Number of individuals supported by a given habitat	C. Gyres
___	4. Symbiosis in which one partner primarily benefits	D. Niche
___	5. Photosynthetic members of a food web	E. Chaparral
___	6. Organisms in food web that conduct decay	F. Commensalism
___	7. Circular patterns of movement by ocean currents	G. Estuary
___	8. Coastal terrestrial biome with low shrubs and frequent fires	H. Habitats
___	9. Terrestrial biome with highest diversity of species	I. Decomposers
___	10. Aquatic biome where saltwater and freshwater mix	J. Tropical forest

Answers to Self-Quiz

1. H; 2. D; 3. B; 4. F; 5. A; 6. I; 7. C; 8. E; 9. J; 10. G

Chapter 22

Human Ecology and Environmental Impact

We humans have always been a mobile species, spreading in all directions from our evolutionary points of origin. Indeed, humans have established residence in virtually all terrestrial habitats from the world's highest place in Nepal to its lowest, near the Dead Sea in Israel. Humans also can be found in Cherrapunji, India, the wettest spot on Earth; northern Chile, the driest region; Greenland, the coldest; and Araouane, Mali, the warmest.

What special features of humans have assisted our successful expansion across the globe? What is the historical pattern of growth for human populations? What has been the impact of human growth on the environment and resources of the Earth? These topics are the focus of this final chapter in the story of human biology.

Human Characteristics: What Makes Us Unique?

What does it mean to be human? The characteristics that help distinguish us from other animals include (1) a skeleton with easily rotated limbs that permits upright posture and walking, (2) stereoscopic (three-dimensional) vision, (3) gripping, versatile hands, (4) a conscious, decision-making brain, (5) speech and language, and (6) the development of culture. Together these attributes have helped our species extend its range throughout the biomes of the Earth, even to some of the most inhospitable habitats. Homeostatic physiology also has helped, allowing us to sweat when overheated, shiver when chilled, and produce more red blood cells when living at a higher altitude.

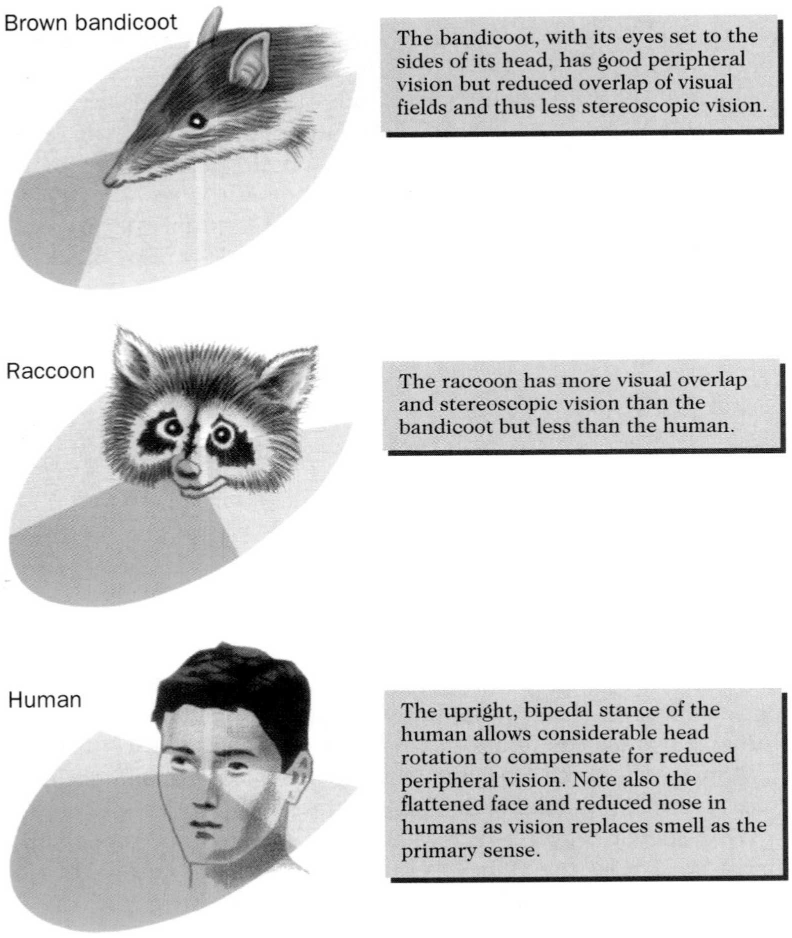
Brown bandicoot
The bandicoot, with its eyes set to the sides of its head, has good peripheral vision but reduced overlap of visual fields and thus less stereoscopic vision.
Raccoon
The raccoon has more visual overlap and stereoscopic vision than the bandicoot but less than the human.
Human
The upright, bipedal stance of the human allows considerable head rotation to compensate for reduced peripheral vision. Note also the flattened face and reduced nose in humans as vision replaces smell as the primary sense.

FIGURE 22.1 Stereoscopic vision allows depth perception and judgment of distance.

Upright Walking: Freeing the Arms and Hands

Except for flightless running birds such as ostriches, humans are the only animals to depend solely on **bipedalism,** or two-footedness, for standing, walking, and running. Human walking is a complex balancing act that requires the coordinated use of muscles in the legs, feet, hips, and back. Combined with movements of the arms and hands, bipedal locomotion also allows swimming and climbing. Long before trains, planes, automobiles, and boats and even before the use of horses and oxcarts, walking, swimming, and climbing aided human dispersal.

Standing and walking on two feet also allows us to better apply our vision to our surroundings. On level ground, a crawling human can see about 6 miles (9.6 kilometers), but a standing human can pick out low objects up to 15 miles (24 kilometers) away.

Stereoscopic Color Vision: Judging Depth and Distance

Our eyes are set well to the front of the head, allowing both to focus on the same object (Figure 22.1). This wide overlap of visual fields provides **stereoscopic** (three-dimensional) **vision,** which is indispensable for judging distances. Although our eyes have a preset focus for viewing objects 20 feet (6 meters) away, we are able to change the shape of the lens and adjust the focus for closer or more distant objects. Using six muscles and three cranial nerves, your eyes can be moved to scan your surroundings even when your head is stationary. Additionally, the cone cells of the retina provide color vision in all but the dimmest light.

Although a red-tailed hawk may see sharper detail and a housefly may register movement more quickly, your eyes offer better combinations of scanning movement, focus adjustment, and color detection. Vision has assisted human migration in many ways, from recognizing familiar landmarks to navigating by stars, maps, and instruments.

Hands and Thumbs: Carrying and Manipulating

After our eyes, our hands are perhaps the most versatile of all biological structures. Each hand has 27 bones and 25 joints that perform a total of 58 separate movements. With our hands we can greet, grasp, flick, crush, carry, caress, gesture, point, trace, turn, mold, and defend. We can even "sign" the words in this paragraph to compensate for faulty speech or hearing. Our hands are the primary instruments with which we touch, shape, and repair our world. Their evolution allowed us not only to make and use tools but also to control fire. These capacities contributed to our survival, dispersal, and further evolution.

Our thumbs are operated independently from our fingers by separate sets of muscles, providing us with a completely *opposable thumb* which is able to touch the tip of each finger. This seemingly simple capacity expands the variety of gripping and grasping movements we can perform (Figure 22.2).

As versatile as they are on their own, our hands reach their peak of usefulness when they make and manipulate tools: pencils, wrenches, zippers, buttons, brushes, remote controls, can openers, shovels, hammers, and computer keys. Tools are extensions of the human body and magnifications of its capacities. Our toolmaking ability is a product of our upright posture and free hands, our stereoscopic vision and movable eyes, and the coordination of all these factors by our most prized possession, the brain.

Consciousness: Thinking and Awareness, Past and Present

Hand and eye coordination are basic to both the simplest and the most complex human accomplishments. Recall your favorite book, painting, or building and imagine the interplay between the brain, eyes, and hands required for its production. The very fact that you can contemplate these things in their absence emphasizes the power of the human mind.

Humans are alone in our ability to observe phenomena, consider options, select a course of action based on past experience or anticipation of future events, execute that action by movement or speech, and color it with emotion. Most important, when we think, we are aware that we are thinking. We are conscious of our thoughts and actions and thus are better able to modify our behavior and surroundings for our survival. The site of these capacities

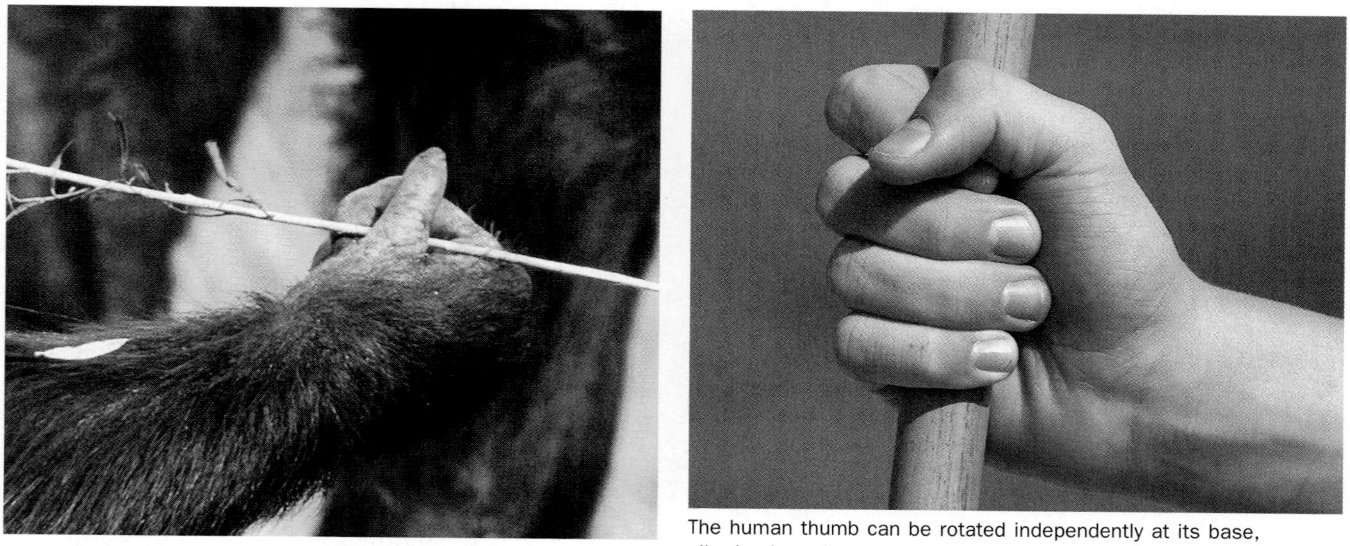

The human thumb can be rotated independently at its base, allowing it to directly oppose each finger and facilitate the grasping of objects.

FIGURE 22.2 Chimpanzee and human hands.

is the cerebrum of the brain (Chapter 15). Figure 22.3 compares the complexity and size of the human cerebrum with the cerebrums of other animals. The expanded human cerebrum, with its large association areas, also allows complex communication between individuals.

Speech and Language: Bringing the Past into the Present

Most animals can communicate with each other. Birds use song, movements, and feather displays to attract a mate. Your pet dog marks its territory and announces its gender to other dogs through drops of pheromones (external hormones) released with urine. A parrot may mimic human speech but cannot form new sentences that are based on new events. Our primate cousins, the chimpanzees, can communicate thoughts and even emotions through the visual symbols of American Sign Language, but they speak only simple words and cannot form the consonant sounds made by humans. Only humans have both a cerebrum large and complex enough to assemble words from thoughts and a larynx ("voice box") specialized enough to vocalize words and change their meaning with pitch.

The development of speech and language aided human expansion and success by allowing individuals to communicate, coordinate their actions, and cooperate. It is not always possible for one person alone to ford a river, construct a shelter, or capture dinner, and the coordination and cooperation made possible by language increase the chances of survival.

Language also permits the capacity to learn from the experiences of others. Without language, when an individual dies, experiences are lost. Language, oral or written, allows experiences and thoughts to be communicated over generations.

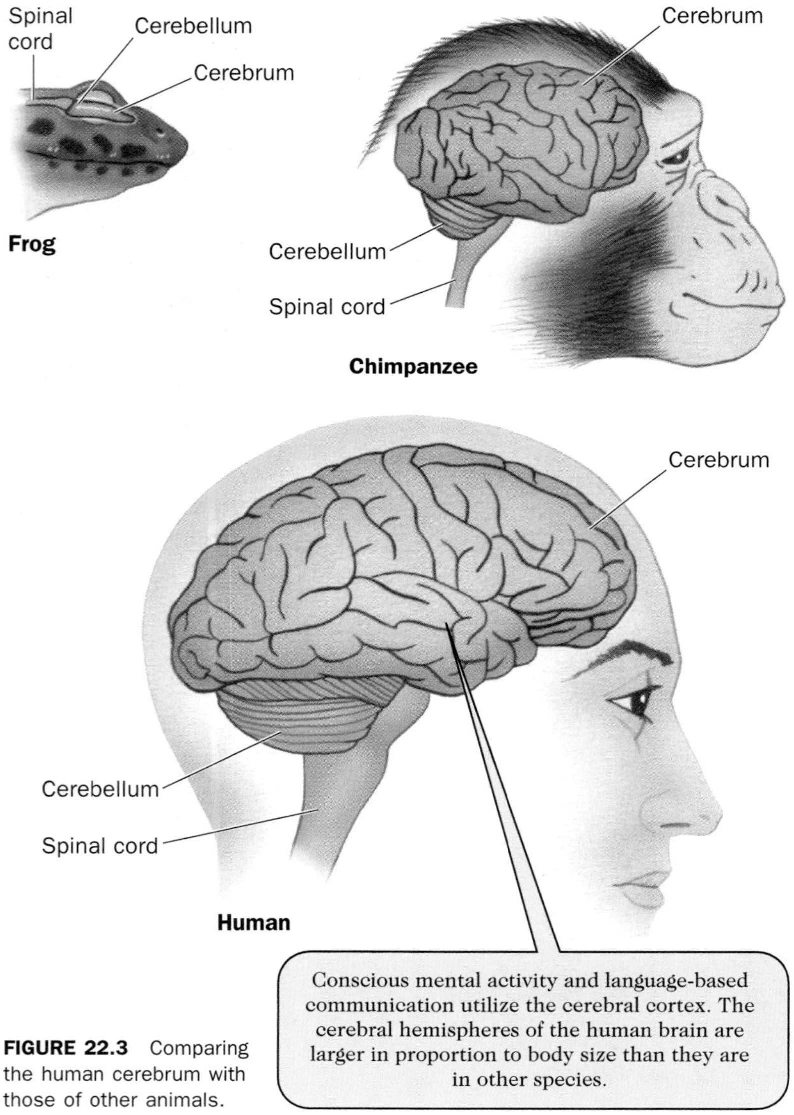
Spinal cord
Cerebellum
Cerebrum
Frog
Cerebrum
Cerebellum
Spinal cord
Chimpanzee
Cerebrum
Cerebellum
Spinal cord
Human
Conscious mental activity and language-based communication utilize the cerebral cortex. The cerebral hemispheres of the human brain are larger in proportion to body size than they are in other species.

FIGURE 22.3 Comparing the human cerebrum with those of other animals.

The Development of Culture: Beyond Biology

Language was probably the last of our unique characteristics to evolve, and oral communication preceded writing. Speech enabled early groups of humans to share beliefs, values, family histories, rituals, hunting skills, and plant lore and to teach those things to children. Speech aided survival and sharing. A shared way of life offers a sense of continuity and is the essence of **culture.** Culture establishes the rules and rituals that determine how we greet each other, the clothing and adornments we wear, our food and eating utensils, the materials with which we construct homes, the careers to which we aspire, our methods of worship, and the manner in which we raise children and bury relatives. It shapes our sense of self and justice, dictates how we play games and wage wars, and provides order, purpose, and reason in a complex world.

Globally, there are vast cultural differences between groups of humans, as can

be deduced from the pictures in Figure 22.4. Our social groupings allow richly diverse and uniquely human expressions of culture. A few animals exhibit the rudiments of culture. Chimpanzees, elephants, and macaque monkeys use natural objects as crude tools and teach simple skills to their young. Humans, however, can combine materials in different ways to fashion new tools and weapons and devise new rules for their use.

> Human success has been assisted by bipedal locomotion, stereoscopic color vision, a gripping hand and opposable thumb, a consciously thinking and remembering brain, speech and language, and the sharing of life through culture.

FIGURE 22.4 The diversity of human culture is expressed by our adornments and clothing.

The Human Population

Take a break. Close your eyes and slowly count to 60. During this 1-minute pause, 250 babies were born throughout the world and 50 people died. There will be about 2 million more humans added to the world population this week and nearly 100 million added this year. There is a human population explosion, which is increasing our numbers worldwide at an average rate of 1.6 percent per year. There are over 5.6 billion of us now, and by the year 2025 perhaps there will be as many as 8.5 billion.

History of the Human Population

Our population has not always been this large or grown this rapidly. Figure 22.5 shows that for most of human history the population was small, rarely exceeding 10 million, held in check by disease, famine, harsh climates, and stringent cultural requirements for marriage. An explosive increase in the population, however, appears to have begun in the early 1700s with the advent of the industrial revolution.

Since then, the nations of Europe, North America, Australia, and northeastern Asia have continued to develop industry-based economies. These nations are often designated **more developed countries (MDCs).** The nations of Africa, Latin America, and the remainder of Asia, home to three-fourths of the world's population, are called **less developed countries (LDCs)** because their economies have only recently begun the transition to industrialization.

Birthrates and Death Rates: Patterns of Change

Birthrates tend to be higher in agricultural societies where child labor is recruited for farming or mining and infant mortality rates are high. The rates are also higher in areas where educational and employment levels are low, especially for women.

The **total fertility rate (TFR),** the number of children borne by a woman throughout her reproductive years, averages 3.6 children per woman worldwide. MDCs have an average TFR of 1.9 (1.84 for the United States), while LDCs average 3.8 (7.94 in Kenya). TFR tends to decrease when education provides employment skills and allows career aspirations for women and society empowers women to make economic and reproductive decisions. The most powerful influence on TFR in fact is family planning. TFR is decreasing in countries where women have increased access to contraceptive technology. Today, 51 percent of married women (childbearing age) in LDCs use some form of contraception, compared with 70 percent of married women in most MDCs.

Death rates tend to decrease as nutrition is improved through better methods

FIGURE 22.5 Growth of the human population.

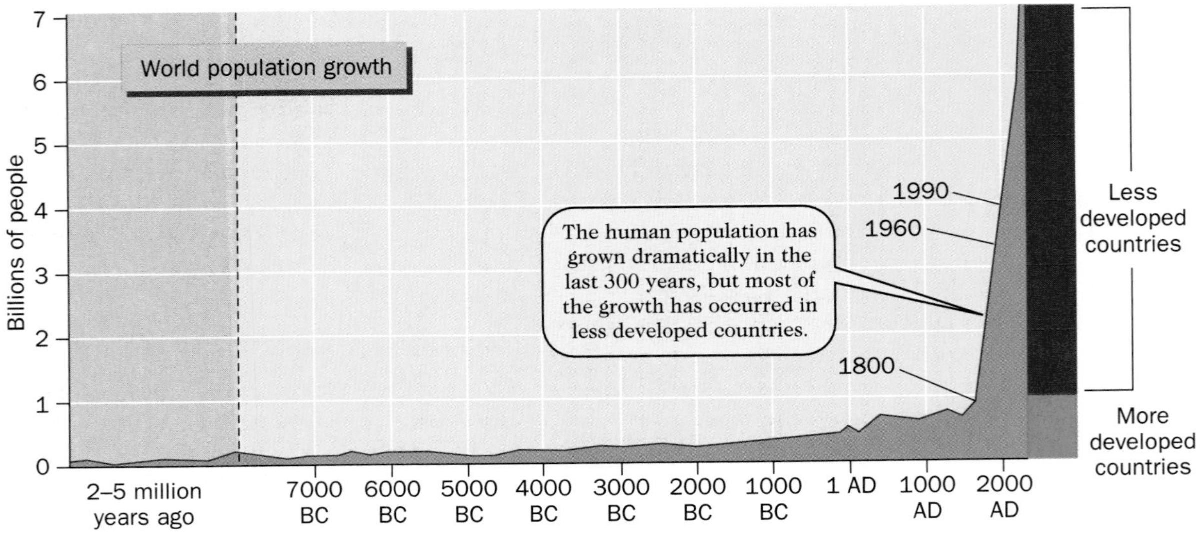
World population growth
Billions of people
0
1
2
3
4
5
6
7
The human population has grown dramatically in the last 300 years, but most of the growth has occurred in less developed countries.
1800
1960
1990
Less developed countries
More developed countries
2–5 million years ago
7000 BC
6000 BC
5000 BC
4000 BC
3000 BC
2000 BC
1000 BC
1 AD
1000 AD
2000 AD

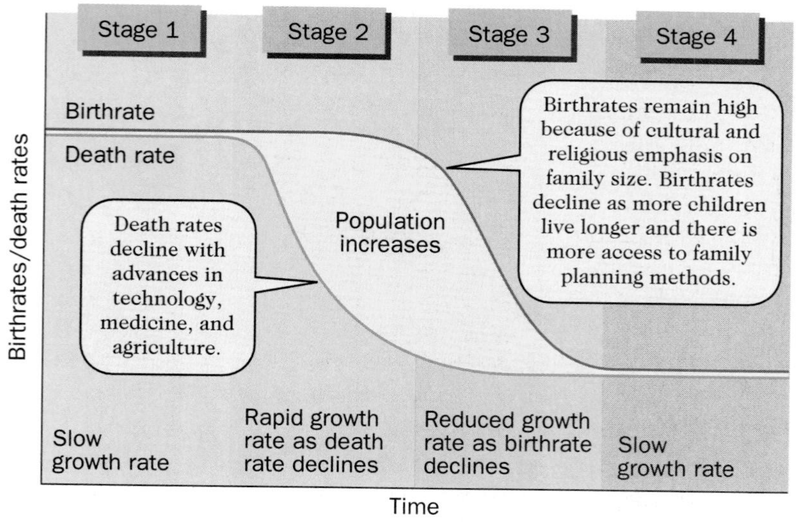

Stage 1
Stage 2
Stage 3
Stage 4
Birthrate
Death rate
Birthrates/death rates
Death rates decline with advances in technology, medicine, and agriculture.
Population increases
Birthrates remain high because of cultural and religious emphasis on family size. Birthrates decline as more children live longer and there is more access to family planning methods.
Slow growth rate
Rapid growth rate as death rate declines
Reduced growth rate as birthrate declines
Slow growth rate
Time

FIGURE 22.6 The stages of the demographic transition.

of food production and distribution. Improvements in sanitation, hygiene, and medical technology can reduce the death rate from infectious diseases. The death rate of infants is a good general index of social conditions and public health. The **infant mortality rate,** or the number of babies in a thousand who die during the first year of life, averages 77 worldwide, but is highest in LDCs, where high infant and child mortality remains a problem.

The four stages of change. Historically, improvements in nutrition, sanitation, and medicine produce changes in birthrates and death rates called the **demographic transition** (demography is the study of populations). As a population experiences technological and medical changes, its birthrates and death rates pass through a series of four stages that alter the population's rate of growth (Figure 22.6).

Stage 1 populations have a high birthrate which is more or less canceled by an equally high death rate. Most of human history has occurred at this stage, and though people were culturally encouraged to "be fruitful and multiply," disease and harsh living conditions kept the total population relatively small.

Stage 2 populations retain high birthrates because of a continued cultural emphasis on large families, but they exhibit declining death rates because of improvements in agriculture, public sanitation, personal hygiene, and access to health care. These populations increase, often straining the ability of a society to support itself with the available resources.

Stage 3 of the transition occurs as agricultural populations become increasingly industrialized and their rate of growth stabilizes as a result of declines in both birthrates and death rates.

Stage 4 occurs when the birthrate declines to the point where it once again equals the death rate. Such stage 4 populations approach what is called **zero population growth (ZPG)**.

This has been the pattern for most MDCs in Europe and North America, and today these nations have reached stage 4 and are approaching ZPG. Several have even slipped below that level and are experiencing gradual population decline. Most LDCs are in stage 2 of the transition. Increased access to antibiotics, birth control programs, and financial and agricultural aid from MDCs are helping many LDCs make the transition to stage 3 more rapidly than did the MDCs of Europe. In many LDCs birthrates are falling even in the absence of economic growth, primarily because of improvements in family planning programs and the availability of contraceptives. Other LDCs still have high annual growth rates, however, and hold the greatest potential for population increase in the next century (Figure 22.7). These LDCs are a significant source of immigration to the MDCs, swelling previously stabilized populations.

In no other single century of human history has the population doubled, yet in this century alone it will have increased *fourfold!* The doubling time for the world's population now averages 40 years. Kenya is growing the most rapidly, with a doubling time of only 17 years, while Italy's population is growing the slowest, with a doubling time of 3465 years.

Population Age Structure

The potential for future growth in a population can be represented in an **age structure diagram,** as shown in Figure 22.8. Each diagram is divided in half vertically to represent the proportion of males and

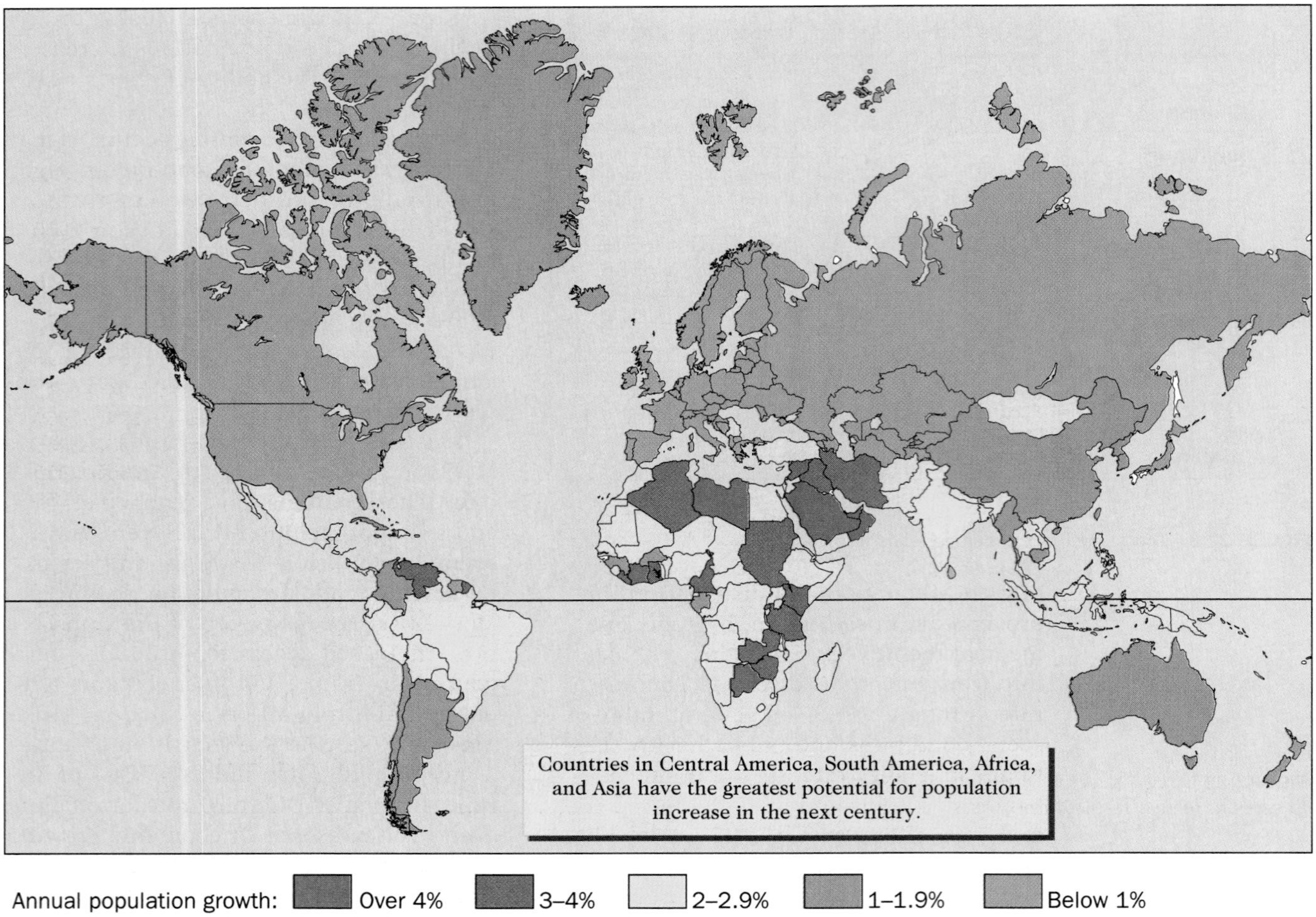

Countries in Central America, South America, Africa, and Asia have the greatest potential for population increase in the next century.
Annual population growth:
Over 4%
3–4%
2–2.9%
1–1.9%
Below 1%

FIGURE 22.7 World population growth.

females. Horizontal sections of the diagram show the size of various age groups with differing reproductive capacities. Typically, such diagrams show the prereproductive young, the reproductively active, and postreproductive older individuals.

The overall shape of the diagram represents whether a population is increasing, stabilizing, or shrinking. An expanding population appears as a pyramid, with the largest proportion of individuals in a prereproductive age grouping. Stable populations have relatively equal proportions of prereproductive and reproductive individuals. Declining populations have a prereproductive group that is smaller than either the reproductive group or the postreproductive group, implying a reduced reproductive potential for the population.

The age structure diagrams for three representative countries are shown in Figure 22.9. Among these countries, Kenya's

FIGURE 22.8 Age structure diagrams for expanding, stable, and declining populations.

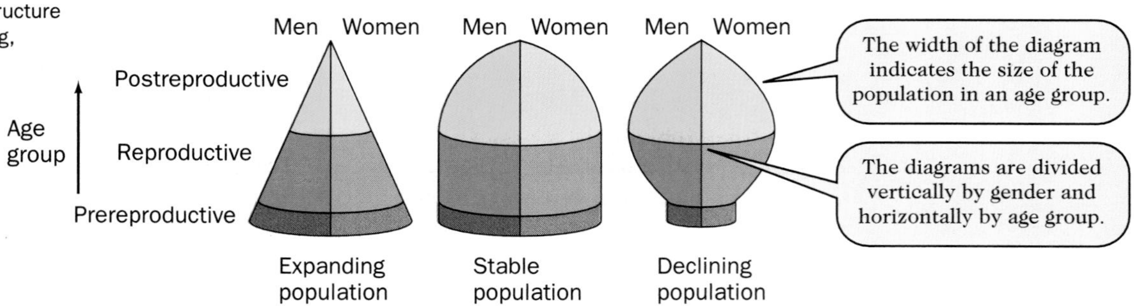

Men
Women
Men
Women
Men
Women
Age group
Postreproductive
Reproductive
Prereproductive
Expanding population
Stable population
Declining population
The width of the diagram indicates the size of the population in an age group.
The diagrams are divided vertically by gender and horizontally by age group.

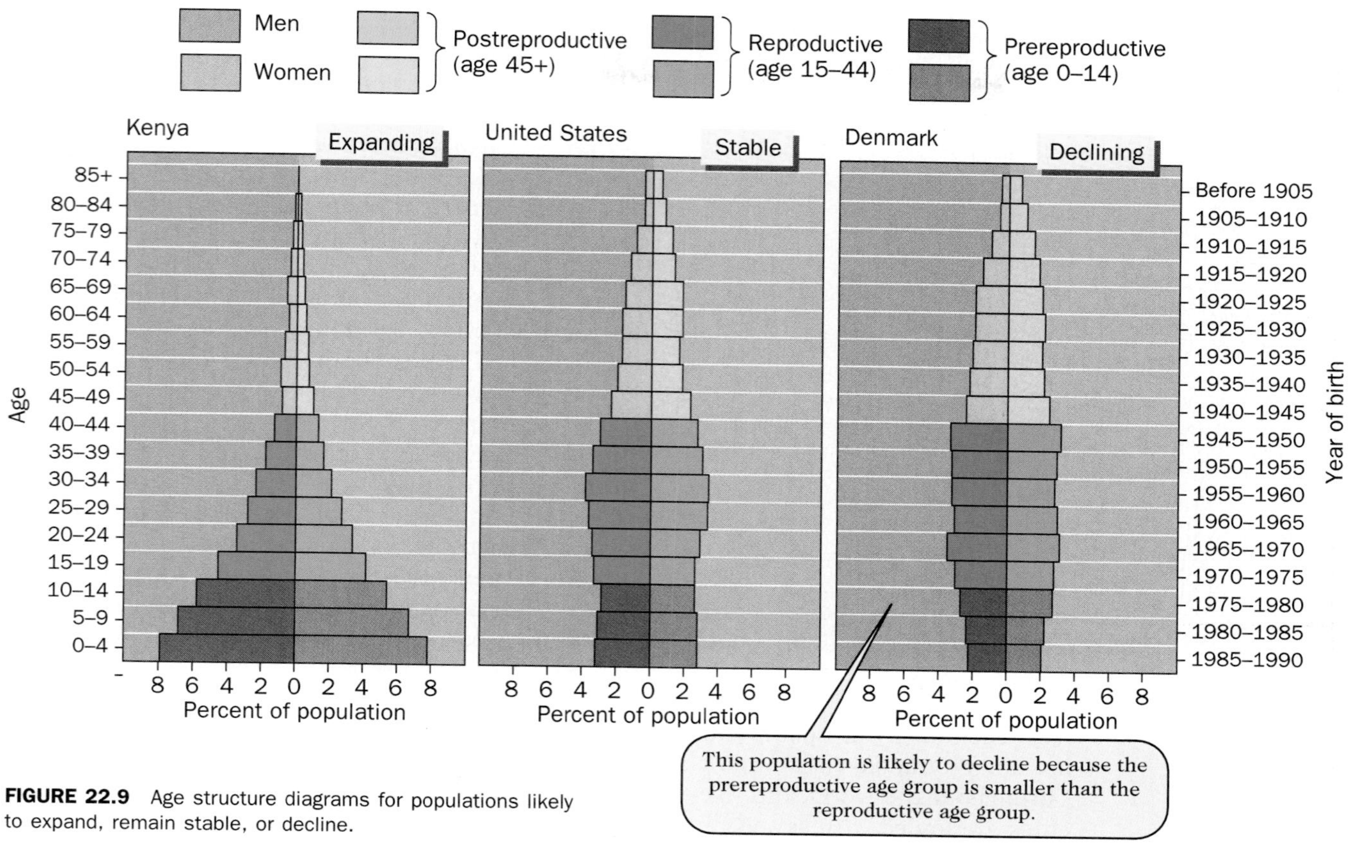
Men
Women
Postreproductive (age 45+)
Reproductive (age 15–44)
Prereproductive (age 0–14)
Kenya
Expanding
United States
Stable
Denmark
Declining
Age
85+
80–84
75–79
70–74
65–69
60–64
55–59
50–54
45–49
40–44
35–39
30–34
25–29
20–24
15–19
10–14
5–9
0–4
Before 1905
1905–1910
1910–1915
1915–1920
1920–1925
1925–1930
1930–1935
1935–1940
1940–1945
1945–1950
1950–1955
1955–1960
1960–1965
1965–1970
1970–1975
1975–1980
1980–1985
1985–1990
Year of birth
8 6 4 2 0 2 4 6 8
Percent of population
This population is likely to decline because the prereproductive age group is smaller than the reproductive age group.

FIGURE 22.9 Age structure diagrams for populations likely to expand, remain stable, or decline.

population is the most likely to expand. LDCs, such as Kenya, have a larger proportion of their populations in reproductively active or soon to be active age groups than do MDCs, and the LDCs can therefore expect a higher rate of population growth in the future.

Urbanization of the Human Population

It may be hard to grasp the magnitude of the global population increase if you live in a rural area. You are probably still surrounded by open land, though it most likely is being farmed, grazed, or logged or is undergoing residential, recreational, or commercial development. If you live in a city, though, you may more directly sense the pressure of population growth. By the end of the decade you will have a lot of company, with over 50 percent of the world population living in cities. For comparison, at the beginning of this century cities held only 15 percent of the world's population.

The largest city in the world now is Tokyo, Japan, with over 26 million inhabitants. New York City has 16 million residents but is smaller than São Paulo, Brazil, with 19.2 million. The world's cities are now gaining 1 million new inhabitants per week, an increase in population of about 4 percent per year. By the year 2000 there will be 25 major cities with over 10 million inhabitants; there are only 16 now. Most of the growth will occur in cities in the northern hemisphere and in LDCs.

Conditions in major cities, however, are not always pleasant for all the inhabitants (Figure 22.10). In Calcutta, Cairo, and Rio de Janeiro two-thirds of the people live in cardboard or tin shanties with no safe water and minimal toilet facilities. Hundreds of thousands in each city are homeless, wandering littered streets or scrambling through vast garbage dumps for scraps of food and clothing. In Shanghai, where government housing is available, residents are provided with only about 2 square meters (about 20 square

FIGURE 22.10 An urban shantytown.

feet) each for living space, about 25 times less than a middle-class American expects. Throughout the world, the rivers of major cities are fouled with sewage and industrial waste and the air is tainted with automobile and factory emissions. Crime, violence, and pollution are part of daily urban life.

Why, then, are people being drawn to cities? As desperate and deplorable as urban conditions are for the poor, they are better than rural conditions. Urban incomes in the northern hemisphere may be up to three times greater than rural incomes. Cities offer hope for employment, and electricity and drinkable water are generally more available, as are social, medical, educational, and sanitation services.

Surpassing the Earth's Carrying Capacity

Recall from Chapter 21 that the **carrying capacity** represents the number of individuals in a species that can be sustained indefinitely by the food and other resources in a given habitat. The carrying capacity is increased by certain activities and decreased by others. It increases when advances in agriculture, biotechnology, mining, and manufacturing make more resources available; it decreases when the combined effects of pollutants reduce the life-sustaining quality of the Earth's atmosphere, soil, and water.

Carrying capacity also can be altered by natural phenomena. If polar ice caps and glaciers expand or melt, ocean levels drop or rise accordingly, with consequences for human habitats and food production. Volcanoes spew light-blocking ash into the atmosphere and reduce the global level of photosynthesis, decreasing crop yields.

Estimates of the Earth's carrying capacity range from 500 million to 15 billion people. If the lower estimate is correct, the carrying capacity has already been exceeded 10-fold and a population crash is imminent. If the higher estimate is correct, the population has a few more decades before the consequences become obvious.

Currently, enough food is produced to support the existing world population. Food distribution, not food production, is of greater concern now. However, as the world population continues to increase rapidly, production will fall behind demand. Two-thirds of the world's food is now consumed by the wealthier industrialized nations. The MDCs dominate the LDCs in the world food market because of two factors. First, most LDCs lack the agricultural technology to produce adequate quantities of food to support their growing populations. Second, most LDCs are too poor to purchase sufficient food.

The world's farmers and fishers are already beginning to struggle to meet the global demand for food. Oceanic fisheries, once viewed as a limitless resource, are declining in all major fishing grounds except the Indian Ocean. Large-capacity high-technology factory vessels harvest fish more rapidly than they can be replaced naturally.

Fish farming and other forms of *aquaculture* have increased productivity but are unlikely to fill the gap left by declining stocks of wild fish. Salmon returning to the coastal streams of Washington State and British Columbia from the Northern Pacific are fewer in number, smaller in size, and less fertile than in past decades. After 20 years of increased productivity from advances in farming technology, the production of U.S. corn, western European wheat, and Asian rice has leveled, and world grain reserves are falling.

Sounding an alarm. We have been a successful species. Our unique human features and supportive technologies have allowed us to grow in number and expand our range of distribution. Our past success, however, threatens our future. Our consumption of resources and the accumulated by-products of our technologies have begun to erode the quality of the Earth's air, water, and land. Unless we slow our rates of growth, resource consumption, and waste accumulation, we will endanger our survival and the survival of the other species with which we share this planet.

> The human population has increased rapidly in the last 300 years. LDCs have had the most rapid population increase. Soon over half the human population will reside in cities. The human population is approaching the Earth's carrying capacity.

Human Impact: Air Pollution

The air you are breathing consists primarily of two gases: nitrogen (79 percent) and oxygen (20 percent). It also contains carbon dioxide (CO_2), water vapor, inert gases, and up to 2800 different chemical compounds. Many of these chemicals are considered *air pollutants,* and their presence in even tiny amounts endangers your health.

Pollutants enter the air from cigarettes, autos, factories, power plants, trains, airplanes, garbage dumps, heating plants, cooling equipment, cleaning solvents, paints, adhesives, and insulating materials. There are also natural sources of pollutants: lightning (ozone and nitrates), wildfires (CO_2 and ash), wind erosion of soil (minerals and particulates), volcanic eruptions (sulfates), leaves of trees (aroma-causing terpene hydrocarbons), ocean spray (salts and sulfates), and soil (radon gases).

Air pollution in general has intensified worldwide with the growth of cities and the expansion of industrial and transportation technology. For example, there are now 500 million registered automobiles in the world, but by 2025 there may be as many as 2 billion, with most of the increase occurring in LDCs. Along with trains and airplanes, air pollution from autos threatens human health and may be altering the climate in which we live.

CO_2, the Greenhouse Effect, and Global Warming

The Earth would be a pretty cold place to live, about $-18°C$ (0°F) on the average, if it were not for the participation of the upper atmosphere (stratosphere) in a kind of **greenhouse effect,** holding in heat and warming the land and waters. Like the glass in a greenhouse, CO_2, methane, and other gases in the atmosphere allow the passage of incoming light from the sun but prevent the outward radiation of heat from the Earth's surface (Figure 22.11).

CO_2 is produced naturally by volcanic activity, wildfires, and plant and animal respiration. Methane is released along with CO_2 by decomposers in the carbon cycle, especially in marshes, where it is called "swamp gas." The entry of these gases into the atmosphere has historically been balanced by their removal through photosynthesis or absorption into freshwater or salt water, and so the Earth has not overheated. However, as the human population has grown, its activities have threatened this delicate balance.

The burning of fossil fuels in homes, autos, and factories adds more CO_2 to the atmosphere than can be removed by photosynthesis. Many scientists are concerned that accumulation of greenhouse gases such as CO_2 will trap more heat and raise air temperatures for the entire Earth. This **global warming** may accelerate the melting of polar ice, causing a rise in ocean levels that could result in widespread flooding of human agricultural and

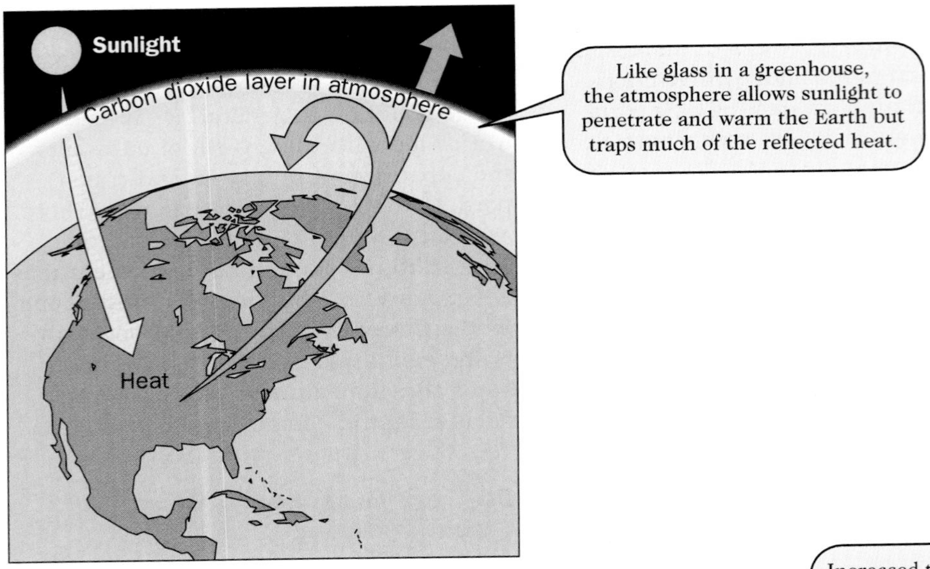
Sunlight
Carbon dioxide layer in atmosphere
Like glass in a greenhouse, the atmosphere allows sunlight to penetrate and warm the Earth but traps much of the reflected heat.
Heat

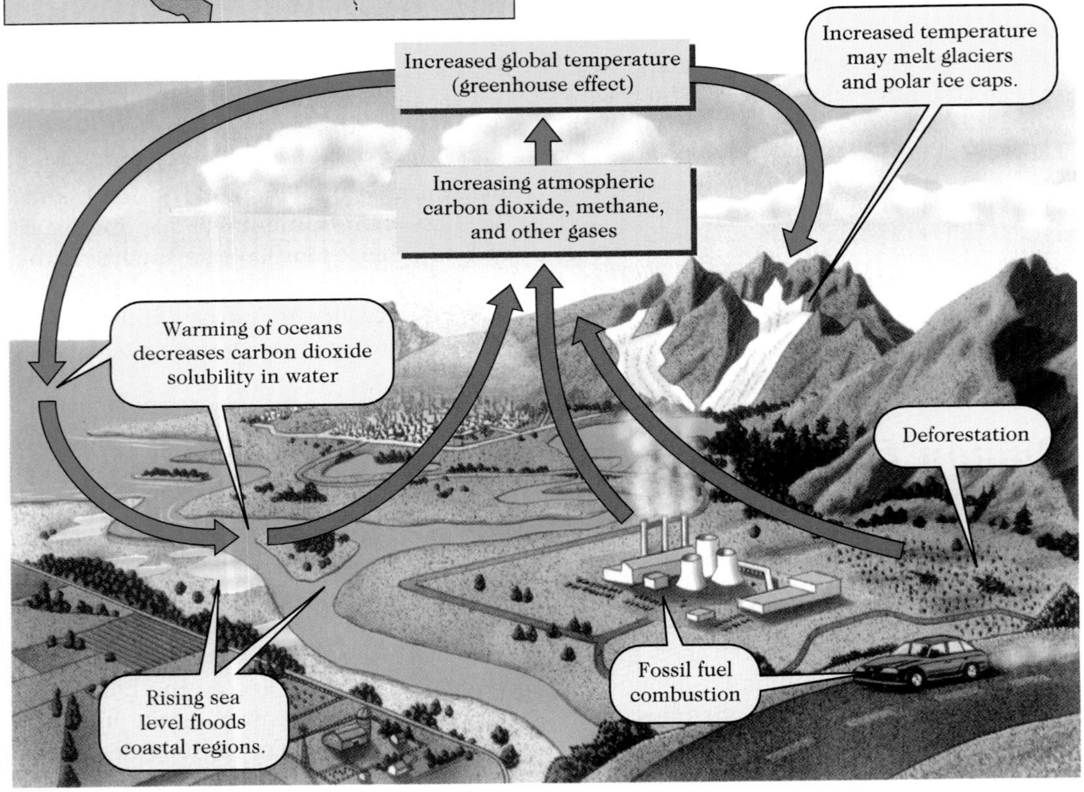
Increased global temperature (greenhouse effect)
Increased temperature may melt glaciers and polar ice caps.
Increasing atmospheric carbon dioxide, methane, and other gases
Warming of oceans decreases carbon dioxide solubility in water
Deforestation
Fossil fuel combustion
Rising sea level floods coastal regions.

FIGURE 22.11 The greenhouse effect and global warming.

urban areas. The sea level has already risen 10 to 12 centimeters (4 to 5 inches) in the last 50 years. Worst-case scenarios predict up to a 1-meter (approximately 3 feet) rise in the next 50 years. With 50 percent of the U.S. population living within 80 kilometers (50 miles) of the ocean, the cost of protective dikes, relocation of homes and industries, and loss of agricultural productivity would be crippling to this country's economy.

The causes of global warming. The concentration of CO_2 in the atmosphere normally represents about 0.04 percent of all the gases present, but in the past 200 years it has increased by 25 percent and is now increasing by 0.4 percent per year. Each gallon (about 4 liters) of gasoline burned in an automobile adds 5 pounds (2.3 kilograms) of CO_2 to the atmosphere. The cutting down of tropical rain forests (Chapter 21), the burning of the resulting brush, and the conversion of forest land to agriculture increase CO_2 as well as methane.

Methane (CH_4) is even more effective in trapping heat than CO_2 is. Atmospheric methane concentrations are currently increasing by about 1 percent per year. Methane is released by microbes in the muck of rice paddies, by termites swarming through the brush left over from a logged forest, and by decaying garbage in landfills. Over 70 million metric tons of methane are released annually by the bacteria in the rumen stomachs of domestic cattle and other grazing animals.

Chlorofluorocarbons (CFCs) are another family of gases that may contribute to the greenhouse effect. CFCs, also called *freons,* are synthetic gases used as coolants, solvents, and propellants in aerosol cans (now banned in the United States, Canada, and Scandinavia) and in the manufacture of Styrofoam. CFCs decompose very slowly in the atmosphere. When released from discarded refrigerators, air conditioners, aerosol cans, and fast-food containers, CFCs accumulate in the atmosphere and retain heat near the Earth's surface. One CFC molecule has the same greenhouse heating effect as 20,000 CO_2 molecules.

Temperature trends. Since the last ice age, the average air temperature has increased 4°C (7°F), rising 0.5°C (0.9°F) in the last century alone. It has been estimated that 25 percent of this temperature increase is due to the increase of greenhouse gases in the last 100 years. There has also been a significant increase in nighttime warming over the last 40 years in the United States, China, and eastern Europe, since greenhouse gases trap heat 24 hours a day.

Whether further increases in CO_2, methane, and CFC levels will result in continued atmospheric warming is a matter of intense scientific debate. In an attempt to control greenhouse gas emissions, 166 nations signed the Framework Convention on Climate Change agreement at the Earth Summit in Rio de Janeiro in 1992. This first world climate treaty calls for a reduction in CO_2 and CFC levels by the year 2000.

Ozone Layers: Addition and Depletion

Linked to the concern about the greenhouse effect is a concern about disruption of the protective **ozone (O_3)** layers surrounding the Earth. There are actually two ozone layers, each associated with opposite problems. The layer of ozone in the lower atmosphere (troposphere) is produced by the reaction of oxygen (O_2) with automobile exhaust, industrial emissions, and sunlight (Figure 22.12). This layer of ozone is increasing, and levels have tripled in the past century, increasing 10 to 30 times over major urban areas such as Mexico City and the Los Angeles basin. Ozone in this layer damages plants and causes respiratory distress in humans and other animals.

The other ozone layer, which is in the upper atmosphere (stratosphere), appears to be decreasing in density (5 and 8 per-

Primary pollutants interact with each other and with sunlight to form secondary pollutants.
Secondary pollutants: Nitric acid, sulfuric acid, ozone and sulfate salts
Primary pollutants: Carbon dioxide, sulfur dioxide, nitrogen oxides, hydrocarbons
Secondary pollutants plus particulates form smog.

FIGURE 22.12 Primary and secondary pollutants and the production of smog.

cent over the United States and Europe, respectively, since the late 1970s), resulting in a large "hole" over Antarctica, where the layer is naturally quite thin.

This upper ozone layer is important because it acts as a shield around the Earth to reduce the penetration of ultraviolet (UV) rays through the atmosphere. Increased UV exposure has been associated with skin cancer, cataracts, and depression of the immune system in humans. Continued depletion of the upper ozone layer might be expected to increase the incidence of these disorders. For instance, a 1 percent decrease in the ozone in this layer is projected to cause a 5 percent increase in the incidence of skin cancers. International teams of researchers are closely monitoring the large ozone hole over Antarctica for signs of expansion.

The major culprit in ozone depletion appears to be the CFCs mentioned earlier in this chapter. Chlorine (Cl) released by the slow decomposition of CFCs causes the breakdown of ozone to O_2 in the following chemical reactions:

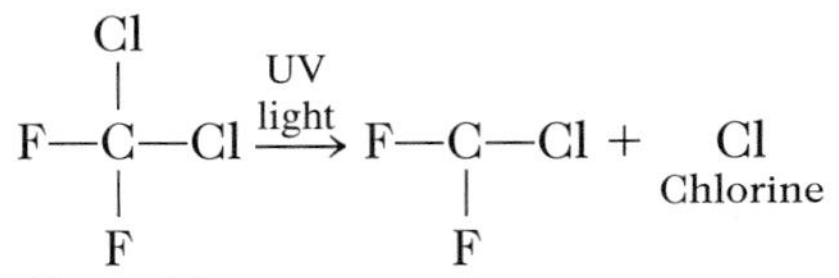
UV
light
Chlorine
Freon 12

$$\underset{\text{Chlorine}}{Cl} + \underset{\text{Ozone}}{O_3} \longrightarrow \underset{\text{Chlorine oxide}}{ClO} + \underset{\text{Oxygen}}{O_2}$$

Thankfully, global production of CFCs is falling as a result of an international agreement, the Montreal Protocol. This agreement, signed in 1987 by most industrial nations, calls for a phasing out of CFC production. However, since it takes about 15 years for CFCs to reach the stratosphere, their effect will be felt well into the next decade.

> The accumulation of CO_2, methane, and chlorofluorocarbons reduces heat loss from the Earth (the greenhouse effect). Depletion of ozone may permit excessive UV radiation to penetrate through the atmosphere.

Smog: A Stew of Pollutants

Some of the *primary pollutants,* such as nitrogen oxides (NO, NO_2, and N_2O) and sulfur dioxide (SO_2), in motor vehicle exhausts and factory emissions react with other chemicals in the air or with sunlight to form *secondary pollutants* (see Figure 22.12). These secondary pollutants, along with particulates, form hazy **smog,** which accumulates over major cities and industrial regions.

Gray smog forms when sulfur oxides and particulates are released by the burning of fossil fuels (coal and oil). Cities in cool climates such as London and New York are affected primarily during the winter months, when the demand for heating and electricity is high. **Brown smog** is more common over warm cities such as Los Angeles and Mexico City, where hydrocarbons, ozones, and nitrogen oxides from automobile exhausts accumulate during the summer months.

Conditions are particularly severe when temperature inversions in the atmosphere hold warm air and smog close to the ground. Curiously, sulfurous smog has a cooling effect on local climate because it reflects light entering the atmosphere, thus slightly countering the warming effect of greenhouse gases.

Some reduction in the smog over American cities should be achieved after 1996, since the U.S. Environmental Protection Agency (EPA) banned leaded gasoline and began requiring the use of cleaner-burning gasoline in severely affected regions. Ethanol or methanol (ethyl alcohol or methyl alcohol) will be added to gasoline to reduce automobile exhaust emissions. These reformulated gasolines contain more oxy-gen atoms, increasing the chance that carbon dioxide instead of the more toxic carbon monoxide will form in auto exhaust. The new gasolines are also expected to produce fewer hydrocarbons, nitrogen oxides, and other primary pollutants.

The EPA estimates that the cleaner fuels will reduce total U.S. production of air pollutants by 2.4 million tons annually, or about 100 fewer pounds (45.5 kilograms) per auto per year. Automobile emission standards and periodic exhaust checks are also helping reduce smog in affected cities such as Los Angeles.

The health effects of smog. Research studies suggest that acid aerosols (airborne liquid droplets) and ozone in city smog cause severe inflammation deep within the bronchial tubes of human lungs. In the United States, 1 in 4 children are at risk for health problems because they live in cities where ozone levels exceed government standards.

Even in cities where ozone levels are lower, particulate matter is produced by vehicle emissions, factory smokestacks, mining and construction activities, wood-burning stoves, and fireplaces. Such matter irritates the eyes and impairs lung function. Over time, chronic respiratory illnesses such as asthma and emphysema may result. Globally, over a fifth of the world's population (1 billion people) is exposed daily to unhealthy air.

Among the 50 states, California has the most stringent standards for particulate pollution. Currently, 45 percent of the U.S. population and 50 percent of those suffering from asthma live in areas with particulates that exceed California air quality standards. National emission and air quality standards and funding programs to limit air pollution have been established by the *Clean Air Acts.*

Indoor air pollution. Smog is not only an outdoor problem. Ozone from vehicle exhaust on busy city streets penetrates buildings and rises nearly to outside levels. Inside buildings, ozone interacts with nylon carpets, releasing potentially toxic benzene, formaldehyde, acetaldehyde, and other volatile organic chemicals. These indoor pollutants can cause dizziness, headache, coughing, sneezing, and flulike symptoms and also can exacerbate chronic respiratory illness. The EPA estimates that indoor air pollutants are more likely to cause cancer than are outdoor pollutants because they rise to higher concentrations in enclosed spaces and people generally spend more time inside buildings than outside.

Because of efforts to conserve heat, some buildings and factories are poorly ventilated, exposing residents and workers to toxic fumes from insulating or construction materials, furniture, carpets, and electrical office equipment. Such "sick" buildings may include as many as a quarter of all American offices and homes. Air pollution is not simply annoying; it is deadly. Lung diseases are the third leading cause of death in the United States and are also a leading cause of death in the LDCs.

Acid Deposition and Precipitation ("Acid Rain")

The nitrogen oxides and sulfur dioxide in smog can combine with water vapor in the air to produce nitric acid (HNO_3) and sulfuric acid (H_2SO_4), respectively. These acids precipitate out of the atmosphere as microscopic particles or dissolve in rain, snow, or fog, producing what is commonly called "acid rain." These corrosive acids often are deposited many miles from their actual source, damaging trees and agricultural crops, harming fish and waterfowl and their invertebrate prey, and dissolving historic buildings, statues, and landmarks.

The map in Figure 22.13 shows the regions of the United States that are the most strongly affected by acid deposition. Some regions are naturally protected against this corrosive action by the presence in rocks and soils of calcium carbonate (limestone), a chemical which neutralizes strong acids. The northeastern portions of the United States and Canada, however, lie on a foundation of granite and basalt that does not have this buffering action. Central European forests, notably the famous Black Forest in Bavaria, have also been severely damaged.

Acid deposition in forest soils reduces the uptake of minerals by tree roots, diminishes photosynthesis by causing the loss of leaves and needles, decreases germination and the growth of seedlings, and makes trees more susceptible to fungal diseases and insect attack. Studies in the Netherlands indicate that acid deposition leaches calcium from forest soils. Without sufficient soil calcium, forest-dwelling snails are unable to construct their shells, and the number of snails declines. Forest birds that feed primarily on snails then

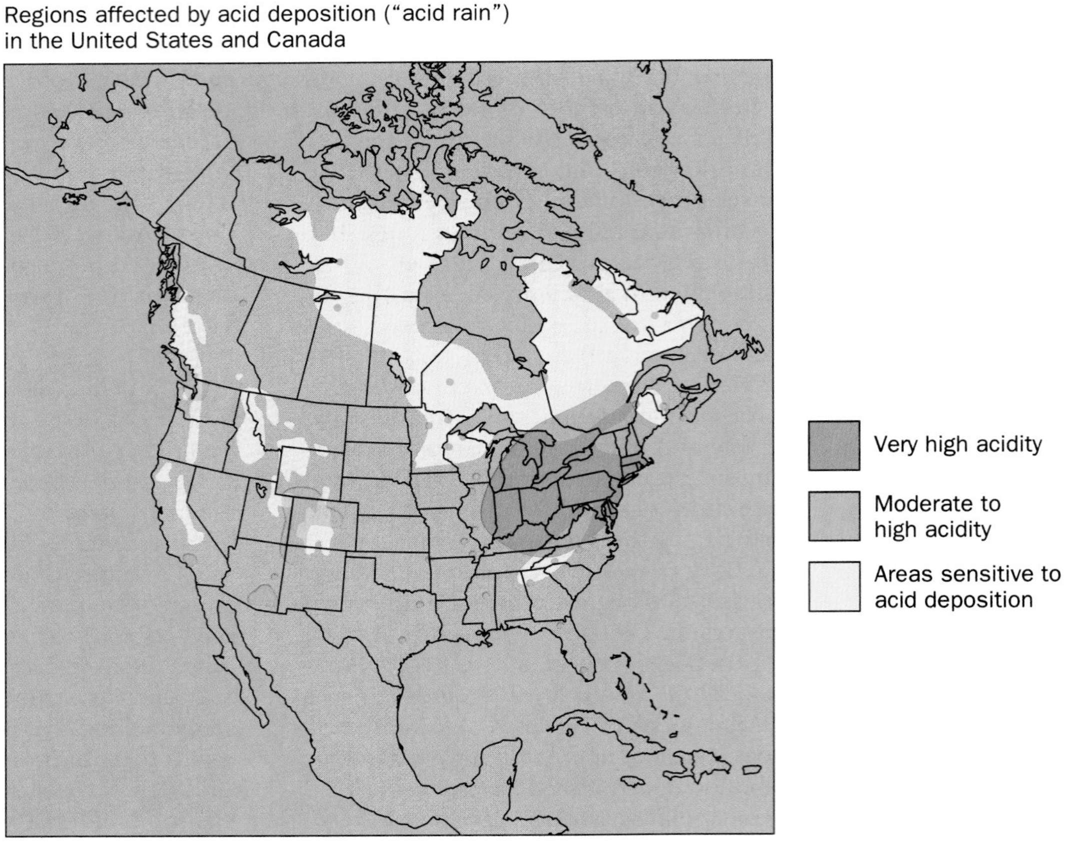
Regions affected by acid deposition ("acid rain")
in the United States and Canada
Very high acidity
Moderate to high acidity
Areas sensitive to acid deposition

Transformation in moist air to sulfuric acid and nitric acid
Nitrogen oxides and sulfur dioxide in smog
Wet acid deposition as acid droplets are dissolved in rain and snow
Dry acid deposition as particles of sulfate and nitrate salts
Acid deposition kills plants and animals and leaches minerals from the soil.
Lakes in soils low in limestone become acidic.
Lakes in soil high in limestone are buffered against acidity.

FIGURE 22.13 Acid deposition.

suffer calcium deficiency and lay soft-shelled eggs that break in the nest, reducing the bird population and affecting the entire food web.

In aquatic ecosystems, an increase in acidity causes toxic metals such as lead, nickel, mercury, cadmium, and copper to become more soluble. In their soluble form, these metals are more easily absorbed by clams, crayfish, snails, and insect larvae, which are important primary consumers. The metals are then subject to biological magnification through food webs, reaching toxic levels for higher-order consumers, including humans (Chapter 21). Terrestrial studies also raise concern. Moose grazing on plants in affected areas of Sweden, for example, show toxic accumulations of cadmium in the kidneys and liver.

> Particulates and the oxides of nitrogen and sulfur react with sunlight to form smog. Nitrogen oxides and sulfur dioxide react with water vapor to form acids that return to Earth with precipitation, damaging plants and injuring aquatic organisms.

Human Impact: Water Pollution

"Water, water everywhere and not a drop to drink" is an old refrain that describes the difficulty many humans encounter in trying to obtain a refreshing drink of water. Water, like O_2, is essential for the maintenance of human life. Although about 75 percent of the Earth is covered with water, most of it is salt water that is unfit for consumption without extensive and expensive *desalination* (salt removal).

Terrestrial animals, including humans, and plants live at the mercy of the water cycle and compete for the less than 1 percent of the freshwater lying at or close to the Earth's surface (Figure 22.14). That does not seem like very much, but it amounts to 9000 cubic kilometers (2200 cubic miles) available for use, about a third of which is currently being used by humans. Over 73 percent of the water used by humans supports world agriculture. Individuals in MDCs use more water than do those in LDCs. The average American, for instance, uses 70 times more water annually than the average Ghanian does.

Freshwater is a renewable resource (Chapter 21). However, drought is striking 20 percent of the Earth and already threatens 40 percent of the human population. Although the world population is increasing, the amount of drinkable water available does not increase and is being reduced by pollution from human activity.

Surface Freshwater Pollution

Streams, rivers, reservoirs, and lakes can become polluted by untreated human sewage, waste chemicals from factories, runoff from city streets, heat from power plants, pesticides and fertilizers from agricultural activities, sediments that erode

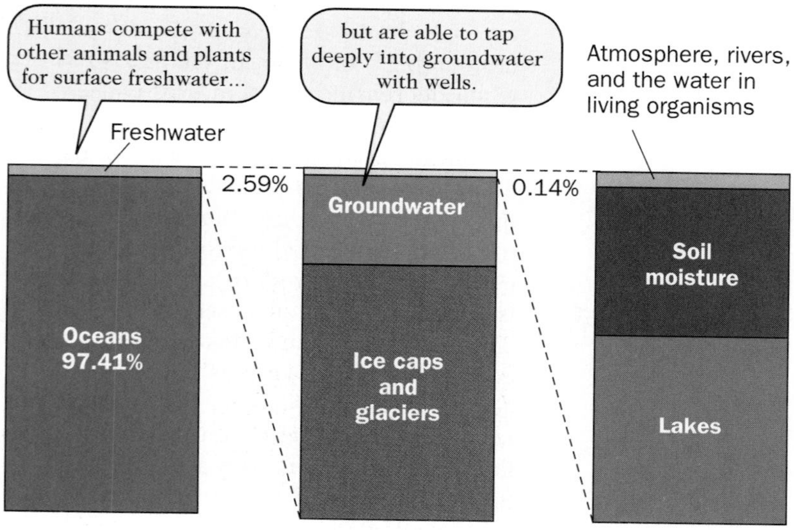
Humans compete with other animals and plants for surface freshwater...
but are able to tap deeply into groundwater with wells.
Atmosphere, rivers, and the water in living organisms
Freshwater
2.59%
0.14%
Oceans 97.41%
Groundwater
Ice caps and glaciers
Soil moisture
Lakes

FIGURE 22.14 The distribution of water on Earth.

from land, and the settling of airborne pollutants. Flowing water, such as in a rapidly moving river, dilutes pollutants more readily than does standing water, such as that in a lake, where there is a more gradual exchange of water. The precise impact of a pollutant on the aquatic inhabitants and terrestrial neighbors of a body of water depends on the chemical nature of the pollutant. We'll now consider pollution by organic and inorganic nutrients, toxins, and sediments as well as heat and pathogen pollution.

Organic nutrients. Organic nutrients are chemicals that wash away from feedlots, sewage treatment plants, paper mills, and meatpacking plants and can be metabolized by aquatic bacteria which decompose waste. As bacteria consume the organic matter, they also consume large quantities of O_2, depleting the life-sustaining quality of the water and threatening the survival of aquatic animals. **Biological oxygen demand (BOD),** the reduction of O_2 concentration in a water sample by bacterial action, can be used as a measure of water quality. A low BOD level indicates water with plenty of O_2 to sustain aquatic life. Water polluted with organic nutrients has a high BOD level.

Inorganic nutrients. Inorganic nutrients include nitrates, phosphates, and sulfates that enter freshwater as fertilizers or laundry detergents. These nutrients support plant growth but, when present in excess, cause extensive proliferation ("blooming") of algae. In fall and winter when decreasing temperatures cause aquatic plants to die and sink, bacterial decomposition results in a massive decrease in O_2 (high BOD level). This sequence of overfeeding, blooming, and dying of algae is called **eutrophication** (*you-***trow***-fuh-***kay***-shun*), a process that is part of primary succession (Chapter 21). Eutrophication of lakes hastens their conversion to dry land as bottom debris accumulates and marshes fill in. Human activities accelerate the eutrophication of lakes and reservoirs, threatening their use for recreation and aquaculture and as a municipal water supply.

Toxic pollutants. Toxic pollutants include organic chemicals such as polychlorinated biphenyls (PCBs), cleaning solvents, gasoline, oil, and pesticides, as well as inorganic acids, heavy metals, and salts. Most pollutants enter water supplies from industrial discharge, runoff from farms or city streets, sewage, and mining activities or by settling from polluted air. Generally, either they cannot be degraded to simpler chemicals by bacteria (they are not biodegradable) or they are degraded very slowly.

As these pollutants accumulate, they are toxic to fish and invertebrates as well as humans. For example, heavy metals such as mercury, lead, arsenic, and cadmium in human drinking water can cause respiratory problems, kidney failure, anemia, learning disabilities, birth defects, and cancer. Salts used on winter roads to melt snow and ice also wash into surface waters after the spring thaw and damage salt-sensitive aquatic plants and animals.

Sediments. Sediments are suspended particles of soil with bits of insoluble inorganic and organic materials. By total volume, they represent the largest form of water pollution in the United States. Sediments are formed naturally as wind and rain erode the landscape. Their formation is accelerated when logging, mining, road construction, overgrazing, and other agricultural practices loosen topsoil and allow it to be washed or blown away. Sediments decrease light penetration in water and reduce aquatic plant growth, clog the gills of fish and shellfish, cover spawning and feeding sites, and fill in lakes, reservoirs, streambeds, and shipping channels.

Heat pollution. Heat pollution occurs when power plants, steel mills, oil refineries, and paper mills use surface water to cool reactors, turbines, and other equipment and then discharge the heated water back into lakes or rivers. When several power plants discharge into the same water supply, excessive heating of the aquatic ecosystem lowers the amount of O_2 that can be dissolved in the water. At the same time, warm water increases the activity of animals and their demand for O_2. Juvenile and adult fish, amphibians, and reptiles may suffocate in warm water, their eggs fail to hatch, and the surviving animals are made more susceptible to parasites and pathogens. Heat pollution is reduced by temporarily storing heated water in open towers or ponds, allowing it to cool before it is discharged into a natural aquatic ecosystem.

Pathogen pollution. Pathogen pollution refers to the presence of disease-causing organisms in water supplies. Waterborne infectious disease is much more of a problem today in LDCs, where untreated human sewage frequently enters surface waters used for drinking, clothes washing, and bathing. Diseases such as typhoid fever and cholera (bacteria), amoebic dysentery and giardiasis (protozoa), hepatitis and polio (viruses), and schistosomiasis (parasitic worms) claim over 25,000 lives daily and incapacitate many more people. Sewage treatment plants and water purification facilities have reduced the problem in MDCs, yet as many as one-third of American rivers exhibit unacceptable numbers of bacteria according to the EPA.

Groundwater Pollution

Groundwater provides over 50 percent of the drinking water in the United States through deep wells. It forms when surface water percolates through the soil to collect in underground reservoirs. Rather than a fast-flowing underground river, groundwater is more like a slowly circulating lake.

Groundwater becomes polluted as water contaminated with organic chemicals, radioactive wastes, pesticides, fertilizers, toxic metals, and bacteria seeps into the soil from septic tanks, landfills and hazardous waste dumps, industrial waste storage tanks, farm fields, and oil wells. EPA studies have found that nearly half the public water systems relying on groundwater are contaminated with dangerous chemicals. Organic solvents such as carbon tetrachloride are the most common. Over two-thirds of rural wells also have been found to be contaminated, frequently by pesticides or nitrates from fertilizers. Human illnesses resulting from exposure to groundwater polluted by pesticides and fertilizers include nervous disorders, skin rashes, leukemia, miscarriages, low birthweights, birth defects, and infant deaths.

In the United States, the *Clean Water Act,* the *Safe Drinking Water Act,* and the *Water Quality Act* are the foundation of a national effort to ensure safe water for consumption and recreational use. These acts require the development of water quality standards, monitoring systems, waste discharge limits, and dredging guidelines. They provide grants and loans to states for the construction of sewage treatment plants, water quality monitoring stations, and regulation enforcement. Unfortunately, progress has been slow, relatively few standards have been established, funding has been lost through federal budget cuts, and gains in water quality have barely offset losses. We are in danger of poisoning ourselves.

Ocean Pollution: The Ultimate Receptacle for Too Much Waste

The ocean is the ultimate receptacle for the waste products of human habitation, industry, agriculture, and commerce. Chemical-laden runoff from farms and city streets washes downriver, through fragile estuaries, to the sea, where it adds to the raw sewage discharged offshore by underwater pipes. Garbage, sewage, medical wastes, and industrial wastes are dumped into the oceans from barges, passenger ships, and military vessels. Sediments, often laden with toxic chemicals, are dredged from rivers and harbors and dumped into the ocean. Pollutant chemicals settle from the air and fall as precipitation into the vast stretches of seawater. Fishing nets and other plastic refuse lost from commercial and recreational boats trap or injure marine birds, mammals, and reptiles.

Although garbage, sewage, and plastic dumping and ballast tank washing are now regulated in American waters, the regulations are difficult to enforce and are frequently ignored by international shipping. Even as cosmopolitan a city as Victoria, British Columbia, continues to dump untreated sewage into the Strait of Juan de Fuca at the entrance into Puget Sound.

Oil pollution. Oil pollution from both natural and human activities remains a major threat to the health of ocean ecosystems. Over 3.2 million metric tons (3.5 million tons) of oil enters the oceans annually, about half by natural seepage from oil deposits. Oil spills resulting from tanker collisions and drilling accidents receive considerable media attention. There are thousands of these incidents each year, but they account for only 20 percent of the oil fouling the sea. Over 30 percent of ocean oil pollution results from oil disposed inland and washed downriver. Most oil spills are concentrated at vulnerable habitats such as beaches and coastal wetlands.

When oil spills into the ocean, about 25 percent evaporates into the air, 50 percent is decomposed by natural bacteria, and 25 percent settles to the bottom in gooey globs. Before it can dissipate, however, oil kills algae, fish, birds, shellfish, and mammals and disrupts the food web along beaches and the ocean bottom for many years (Figure 22.15). Human cleanup efforts can minimize the impact of coastal oil spills but often merely shift the damage to inland disposal or reclamation sites.

> Freshwater may be polluted by organic and inorganic nutrients, sediments, toxins, heat, and pathogens. Groundwater may be contaminated as chemicals and microbes seep through the soil. Oceans receive pollution in the form of sewage, sediments, agricultural and industrial chemicals, and shipping debris.

FIGURE 22.15 Oil spills kill and disrupt food webs for many years.

Human Impact: Altering and Destroying Habitats

As the human family expands, there is less room for other species. The use of terrestrial and aquatic habitats for farming, mining, transportation, industry, recreation, and waste disposal leads to the loss of nesting, breeding, burrowing, and feeding or drinking sites. As a result, habitats are altered and degraded in their structure, species diversity, and carrying capacity. Humans are agents of habitat degradation, but we are also victims. As habitats become unfit for other species, their carrying capacities for humans are often reduced.

Deforestation: Losing More Than Trees

Forests historically covered nearly a third of the Earth's land surface, occupying an area fifteen times greater in size than the United States. Forests provide shelter and food to other animal species, bind soil to prevent erosion, and recycle oxygen, carbon, nitrogen, and water. They also provide humans with over 5000 commercial products, including lumber, paper, turpentine, rubber, fruits, nuts, syrups, and medicines. Over 90 percent of the people in LDCs depend on firewood and charcoal from forests as their primary source of cooking and heating fuel.

However, in MDCs, most plant-based food and the food for livestock come from cereal grains and vegetable crops. Many forestlands in the United States have been converted to farmlands and low-density home sites.

Over a third of the world's original forests have already been logged, and roughly 42.5 million more acres (17 million hectares) of forest are converted to human use each year, an area equal to the size of Washington State. One in three people in LDCs now faces a fuelwood crisis and must rely on dung patties (diverting manure from its important fertilizing role), tree leaves (reducing the humus content in soil), and seedlings or crop residues (promoting soil erosion) for heating or cooking. Replanting and conservation efforts are not keeping pace. In Africa, only 1 tree is replanted for every 30 that are logged. The net loss of forest habitats is called **deforestation.**

Tropical rain forests are among those most heavily assaulted by human activity because they occur in regions adjacent to expanding and growing human populations. Estimates of the annual loss of rain forests to ranching, farming, building, and highway construction range from 4 million to 40 million acres (1.6 million to 16 million hectares) per year. The conversion of rain forest land to farming or grazing is unfortunate, because most rain forest soils are low in nutrients and poorly support crops or animals (Chapter 21). The burning of logging wastes increases the input of CO_2 into the atmosphere, and the loss of photosynthetic plants reduces the forest's capacity to absorb atmospheric CO_2. The loss of forests alters local humidity, precipitation, and climate. Thailand, for example, has suffered so many deforestation-related floods that it has banned logging.

Desertification: Soil Erosion and Lost Productivity

As developing countries struggle to feed their growing human populations, they often resort to short-term solutions that violate sound land management practices and bring long-term disaster. Overuse of chemical fertilizers alters the structure of the soil, reducing its water-holding capacity. Overgrazing and trampling of rangeland by livestock damage the plant roots that bind soil. These practices promote soil erosion.

It may take 200 to 1000 years for an inch of topsoil to form, but it can be eroded away in just a few seasons. Topsoil is eroded when forests are lost to farms, steep hillsides are cultivated, and firewood collectors remove seedlings and brush. Over a third of the world's rain-watered cropland (825 million acres) is losing its productivity in this fashion, and billions of acres are threatened. Each year 50 million acres (20 million hectares) become too debilitated to support farming or grazing and 15 million acres (6 million hectares) become virtual desert, beyond hope of reclamation.

The conversion of previously productive land to new desert is called **desertification.** Figure 22.16 shows the areas where deserts are now expanding. The solution to desertification, however, is not merely to add more water through irrigation. Mineral salts dissolved in irrigation water are left behind in the soil when the water evaporates from arid land. Salt accumulation in the soil, called **salination,** reduces plant growth and crop productivity.

Salination threatens 100 million acres (40 million hectares) of world cropland. Much of this land lies in India and Pakistan, but Syria, Iraq, China, central Asia, and California also have extensive areas threatened by salt accumulation. Irrigation boosts food production by 3 times over rain watering, but its misuse can cause abandonment of formerly productive land. One-third of the world's food is now grown on irrigated land.

The degradation of productive cropland and rangeland through desertification and salination reduces the land's carrying capacity and thus its ability to support an increasing human population. As with deforestation, desertification changes local climate, reducing precipitation. Drought and famine and increased human suffering soon follow.

Species Extinction: We Lose When They Die

Species of animals and plants come and go on the Earth. More lived here in the past than are living here now. Extinction is part of the process of evolution. A poorly adapted species leaves too few offspring to thrive, and it disappears forever, making room for species that are better adapted. Environmental conditions change over time, and so does the list of living species.

Has the accelerating expansion of the human species and the associated habitat alteration increased the rate of extinction of other animals and plants? To clearly assess the impact of humans, we need to know how many other species are present.

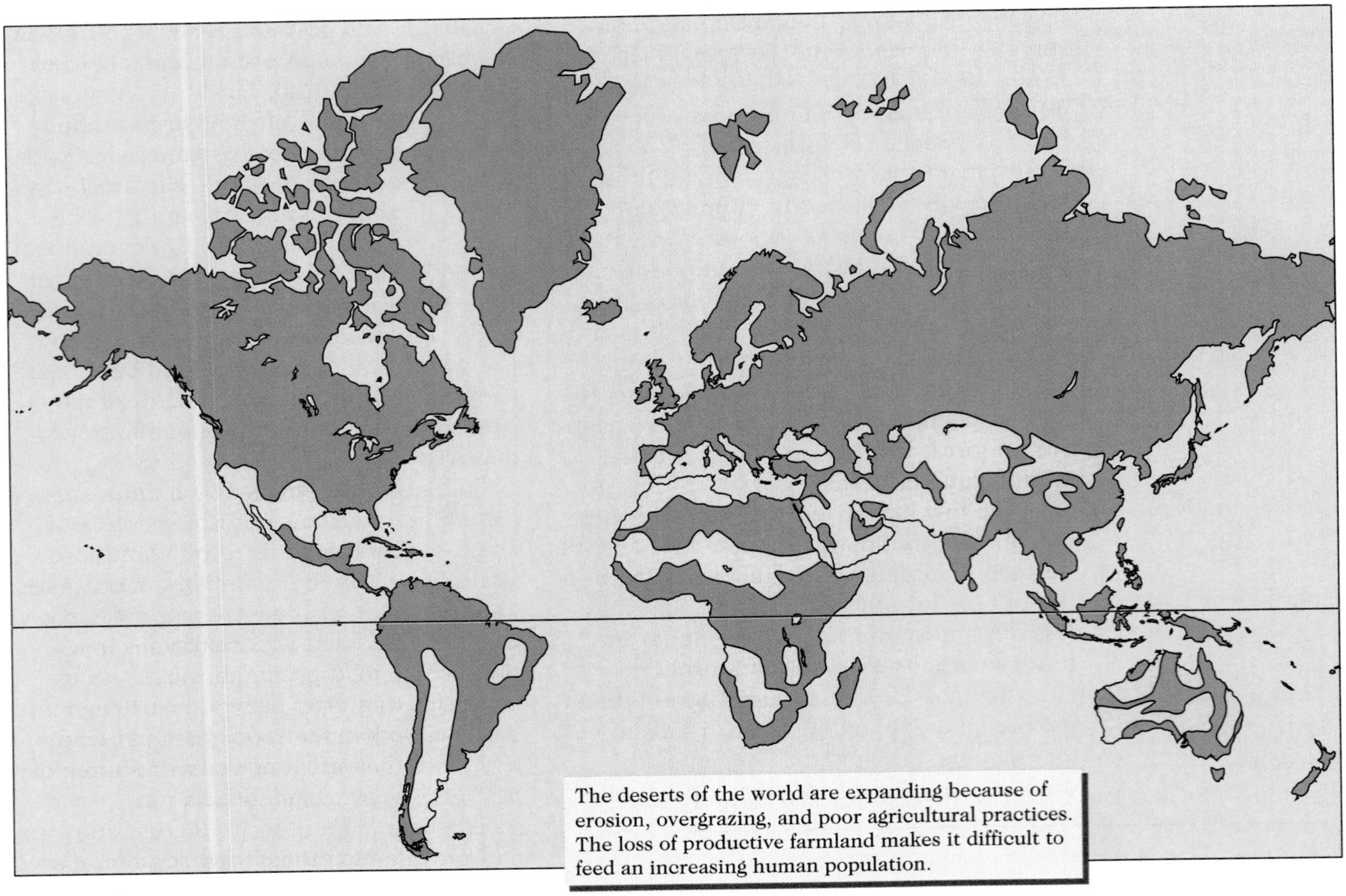
The deserts of the world are expanding because of erosion, overgrazing, and poor agricultural practices. The loss of productive farmland makes it difficult to feed an increasing human population.

FIGURE 22.16 Desertification.

Approximately 1.5 million species have already been named and cataloged by scientists, but there are many million more species on Earth. Estimates of the number of species currently being forced into extinction by human activity range from 1 to 150 a day. Some species have disappeared even before they could be adequately investigated, while others feared to be extinct have been rediscovered. However, some species have benefited from the human presence. The organisms that live in our attics, basements, lawns, sewers, alleys, barns, fields, and wharves and on our bodies have expanded their numbers and ranges right along with humans.

The awakening of national concern in the 1960s over the impact our species may have on others culminated in the *Endangered Species Act.* The act calls for the listing of endangered and threatened species and the estimation of their numbers and ranges. It also makes provisions for their protection by requiring approval of federal projects that might affect endangered species or their habitats.

There are 667 animals and plants listed as threatened or endangered in the United States with about 50 more species added each year. More of these species are associated with forest ecosystems than with any other habitat, and over 95 percent of the species listed are included because of habitat alteration or destruction. There is some good news, however. As protective regulations have increased its abundance, the bald eagle has been removed from the list of endangered species, although it is still a threatened species. Endangered gray wolves are being returned to their former ranges in Idaho and Wyoming.

Efforts which help preserve **biological diversity** have also resulted in the establishment of protective reserves in the United States and many other countries. African reserves now hold 95 percent of that continent's bird species and 75 per-

TABLE 22.1 Reducing Your Impact on the Environment

Rethink Purchases, Patronage, and Use of Spare Time

- Avoid items with plastic or Styrofoam packaging materials.
- Buy items that are made from recycled materials.
- Avoid plastic shopping bags by carrying reusable paper or canvas bags.
- Buy quality items that can be repaired instead of replaced.
- Avoid disposable plastic utensils, straws, and razors.
- Buy solar-powered or rechargeable items such as calculators and shavers.
- Purchase energy-efficient appliances and cars.
- Buy an electrical or propane barbecue instead of charcoal.
- Use cloth diapers or a diaper service instead of disposables.
- Buy food in bulk or in reusable containers.
- Patronize gas stations with vapor recovery pumps.
- Avoid aerosol sprays and use pumps whenever possible.
- Volunteer to plant trees or clean up roadways.
- Write legislators about enacting and enforcing clean air laws.
- Boycott businesses that violate clean air or water standards.
- Help seniors and others insulate or weather-strip their homes.

Reduce Consumption of Resources and Waste Production

- Walk or bicycle wherever you can do so safely.
- Use mass transit or carpooling whenever possible.
- Have your car tuned up on a regular schedule.
- Cluster errands and plan the shortest possible driving routes.
- Compost lawn and edible household wastes.
- Use hand-held mixers, can openers, or graters instead of electric appliances.
- Use energy-efficient fluorescent bulbs at home and work.
- Lower your home and office thermostats and wear warm or layered clothing.
- Improve attic insulation and door or window weather stripping in your home.
- Run only full dishwasher and washing machine loads.
- Use a kitchen timer to limit showers and install water-saving showerheads.
- Turn off the faucet while brushing your teeth, washing your face, or scrubbing dishes.
- Plant drought-tolerant groundcover instead of grass.
- Eat more fruits and vegetables and decrease meat consumption.
- Grow a home garden or purchase vegetables from farmer's markets.

Recycle by Finding New Uses for Disposable Materials

- Collect newspapers, cans, bottles, plastic containers, engine oil, clothing, and grocery bags.
- Cut off the front fold of greeting cards you receive and send them again as post-cards.
- Purchase glass or metal food containers and use them to store bulk foods.

cent of its plants. In the United States, a new 6.6 million acre (2.7 million hectares) national park has been approved to encompass and expand the existing Death Valley and Joshua Tree national monuments.

Captive breeding programs such as that for the California condor have also helped sustain species that once were on the verge of extinction. The U.S. Department of Agriculture and several international agencies are also trying to preserve plant diversity by storing seed and tissue culture specimens, especially for crop plants. We worry about the loss of other species because they provide many of our foods, medicines, and building or clothing materials; they help recycle nutrients and minerals in our habitats; and they are a continuing source of beauty and fascination.

> Deforestation and desertification threaten terrestrial habitats. Legal protection and plant and animal reserves help sustain biological diversity and protect endangered species.

The Road to Sustainable Development

The growth of the human population has profoundly affected the environment of the Earth, and the degree of human impact increases with each generation. The repair of existing environmental injury and the prevention of further habitat degradation will require changes in individual behavior and value systems. It will also require economic and political compromises by nations that are more accustomed to competing than cooperating. The environmental threats that necessitate these

TABLE 22.2 Eco-facts: Facts about the Human Impact on the Environment

- Recycling an aluminum can saves enough energy to run a TV set for 3 hours.
- Twenty times more energy is needed to make a new aluminum can than is needed to use a recycled can.
- Americans throw away enough aluminum every 3 months to rebuild the entire U.S. fleet of passenger airplanes.
- One acre (0.4 hectare) of trees is needed to absorb the CO_2 produced by driving one car 26.000 miles (42,000 kilometers) per year.
- Six acres (2.4 hectares) of trees are needed to absorb the CO_2 produced each year by a single American family of four.
- Recycling 700 paper grocery bags saves a 15-year-old tree.
- Recycling a 3-foot (1-meter) stack of newspapers saves a 20-foot (7-meter) tree.
- Newspapers can be recycled up to nine times.
- Over a lifetime, we each discard 600 times our body weight in garbage.
- Each year, an average American uses 190 pounds (86 kilograms) of plastic.
- Removing your name from junk mail lists saves $1\frac{1}{2}$ trees per year.
  Write: Mail Preference Services
  Direct Marketing Association, Inc.
  P.O. Box 9008
  Farmingdale, NY 11735-9008

changes are often far removed from the individuals and nations whose behavior must be altered. As with all potential dangers, environmental threats are subject to interpretation and debate about their severity and immediacy.

The magnitude of the social and economic change required to reverse the trend of human environmental impact is equivalent in scale to that of the industrial revolution. This "green" revolution, however, must be carefully planned and executed with international cooperation. Its goal is **sustainable development,** a balance between the demands of the human population and the supply of resources. This will allow the entire human population to live below the Earth's carrying capacity without sacrificing the needs of other species or future human generations.

What will it take to achieve sustainable development? Population control is foremost, for without it, the demands and damage created by an ever-growing populace will soon outpace advances in agriculture, energy, transportation, mining, and waste management technology. A restructuring of world debt is needed so that the prices of commodities reflect the cost of their environmental impact. All nations must share in the development of technologies that conserve resources, prevent pollution, and restore ravaged habitats.

Also important is the global adoption of an ecological ethic. The Earth is not an inexhaustible and consumable resource for human enjoyment. Humans are part of the natural world and are subject to the same ecological principles that apply to all other species. The health of any species is dependent on the health of its habitat. Healthy habitats are complexly balanced, interwoven, and diverse webs that unravel quickly when their physical, chemical, or biological threads are disrupted. In the words of Chief Sealth of the Duwamish tribe of Washington State in an 1855 letter to President Franklin Pierce: "All things are connected. Whatever befalls the Earth, befalls the children of the Earth."

What Can You Do?

The road to sustainable development begins with individual action. The three R's for living an environmentally sound or "green" life are **rethink, reduce,** and **recycle.** Table 22.1 offers several specific suggestions, while Table 22.2 presents interesting recycling facts. From newspapers to shoes to motor oil, products constructed from recycled materials are becoming increasingly popular.

Chapter Summary

Human Characteristics: What Makes Us Unique?

- Humans are distinguished from other animals by bipedal locomotion, upright posture, and versatile eyes and hands. The cerebrum permits consciousness and decision-making and, with the larynx, has allowed the development of speech and language.
- Culture is the sharing of experience and the development of rules and rituals for the conduct of life.

The Human Population

- Dramatic population growth has occurred in the last 300 years, with more growth in less developed countries (LDCs) than in more developed countries (MDCs).
- Total fertility rate (TFR) is lower in MDCs than LDCs, due to family planning and access to contraceptives. Infant mortality is lower in MDCs than LDCs.
- The demographic transition describes the sequential changes in the birthrates, death rates, and population size of an industrializing society.
- Age structure diagrams show the number of males and females in the prereproductive, reproductive, and postreproductive segments of populations.
- The carrying capacity is the number of people that can be sustained indefinitely by available resources.

Human Impact: Air Pollution

- Pollutants enter the air from human activities and industries as well as from natural sources.
- Gases like CO_2, CFCs, and methane in the atmosphere allow light penetration, but limit heat escape. Their increase due to human activity may create a green house effect and global warming.
- The lower ozone layer is produced by the interaction of O_2, pollutants, and sunlight. The upper ozone layer acts as a shield to limit UV light penetration.
- Primary pollutants react with sunlight and other chemicals to form secondary pollutants that cause smog. Gray smog (fossil fuels) occurs in cool climates and brown smog (auto exhausts) in warm climates.
- Nitrogen oxides and sulfur dioxide in smog combine with water vapor to produce acids that form acid rain, damaging plants and animals and disrupting food webs.

Human Impact: Water Pollution

- Surface waters become polluted by wastes, chemicals, pesticides and fertilizers, sediments, and airborne chemicals.
- Organic nutrients increase bacterial activity in water, which reduces oxygen concentration and thus the ability of water to support other life.
- Inorganic nutrients increase plant growth and lead to eutrophication, hastening conversion of wetlands to dry land, as part of primary succession.
- Sediments decrease plant growth, clog animal gills, and fill in waterways.
- Heat pollution interferes with animal development, and enhances opportunity for disease.
- Pathogens enter water with untreated sewage.
- Groundwater becomes polluted as chemicals and bacteria seep through the soil.
- Oceans receive wastes, chemicals, and sediments that are dumped on land and washed downstream.
- Oil pollution occurs naturally and from human activities.

Human Impact: Altering and Destroying Habitats

- Deforestation occurs as forests are converted to other uses. Desertification is increasing due to human activity and poor agricultural practices. Salination results from overirrigation of land.
- Destruction of natural habitats and conversion to other uses reduces species (biological) diversity.
- Sustainable development balances human needs and the supply of resources on Earth, allowing populations to live below carrying capacity and not sacrifice the needs of other species or future generations.
- The three R's of an environmentally sound life are rethink, reduce, recycle.

Selected Key Terms

acid deposition (p. 601)
age structure diagram (p. 593)
biological diversity (p. 608)
biological oxygen demand (p. 604)
bipedalism (p. 588)
deforestation (p. 607)
demographic transition (p. 593)
desertification (p. 607)
eutrophication (p. 604)
global warming (p. 597)
greenhouse effect (p. 597)
opposable thumb (p. 589)
ozone (p. 599)
salination (p. 607)
smog (p. 600)
sustainable development (p. 610)

Review Activities

1. List the characteristics which distinguish humans from other animals.
2. Describe the advantages provided to humans by bipedalism, stereoscopic vision, and an opposable thumb.
3. How has the human population changed over time? What might account for its recent growth? Where will most future population growth occur? Why?
4. List and describe the four stages of the demographic transition. Compare the experiences of MDCs and LDCs.
5. Sketch and label the outlines of age structure diagrams for expanding, stable, and declining populations.
6. Compare the age structure diagrams for typical MDCs, and LDCs. Explain the consequences of their shapes.
7. Is the concept of a carrying capacity appropriate for the human population? What factors might act to increase or decrease the carrying capacity?
8. Summarize the concept of a greenhouse effect. How does it relate to the potential for global warming? Which chemicals are associated with global warming? What are the long-term consequences of an increase in their atmospheric concentrations?
9. Describe the distribution of ozone in the atmosphere. Is its concentration remaining stable or changing? Explain.
10. What factors result in smog formation? How is smog related to acid rain?
11. Explain how human activities result in deforestation, desertification, and salination. What is their impact on biological diversity?
12. What is biological oxygen demand (BOD)? Why is it a useful measure of water quality?
13. Define TFR, infant mortality rate, and zero population growth. What are the influencing factors?
14. What are the potential natural and human-related sources of air pollution? Water pollution? Oil pollution?
15. Define sustainable development. What individual action might you pursue if you found it a worthy goal?

Self-Quiz

Matching Exercise

___ 1. A series of changes in birthrates and death rates	A. BOD
___ 2. Protective atmospheric layer reflecting UV light	B. Deforestation
___ 3. Combination of secondary pollutants and particulates	C. Ozone
___ 4. Reduction of O_2 concentration by bacterial action	D. Eutrophication
___ 5. Process that contributes to conversion of lakes to dry land	E. Demographic transition
___ 6. Salt accumulation in soil	F. Desertification
___ 7. Loss of forest habitat	G. Smog
___ 8. Process that converts productive land to new desert	H. Culture
___ 9. Rules, rituals, and customs	I. Greenhouse gases
___ 10. Atmospheric gases that hold in heat	J. Salination

Answers to Self-Quiz

1. E; 2. C; 3. G; 4. A; 5. D; 6. J; 7. B; 8. F; 9. H; 10. I

Health and Reproduction: Threats from Pollutants

Many of the chemicals released into water and soil from human agriculture and industry are toxins or carcinogens. Toxins kill cells, while carcinogens transform body cells into cancer cells. Some polluting chemicals, however, also act as artificial hormones when they are absorbed or ingested by animals. Certain pesticides, herbicides, detergents, and breakdown products from plastics and electrical insulation appear to act as environmental hormones, mimicking the effects of natural estrogen hormones and having a feminizing effect on wildlife and perhaps humans. In contrast, some naturally occurring chemicals may act as estrogen inhibitors, helping to protect humans against diseases such as certain cancers.

Estrogens are "female" sex hormones, a group of related chemicals that are produced in the ovary, by the placenta during pregnancy, and to a lesser degree in the adrenal cortex of men and women. In women, estrogens promote the development of secondary sex characteristics, influence brain development, assist in the maturation and release of eggs (oocytes), prepare the uterus for pregnancy and stimulate its repair during the menstrual cycle, and help prepare the mammary glands for milk production. In men, the role of estrogens is not well understood because these hormones normally are masked by the more abundant "male" sex hormones known as androgens. Abnormally high levels of estrogens in males, however, have a feminizing effect, inhibiting the growth of the testes and penis and reducing sperm production.

Estrogens exert their biological effect by binding to protein receptors in the cytoplasm of certain cells, primarily those in the reproductive system but also cells in heart, bone, and fat tissue. The estrogen-receptor complex that forms then binds to specific DNA regions in the cell's nucleus, altering the rate of cellular activity. The effects of estrogen activity are quite tissue-specific, with different results occurring in different types of tissue. Unfortunately, it appears that the estrogen receptor can bind to chemicals other than estrogen and generate an effect on cells similar to the effect of natural estrogens. It is these estrogen-mimicking or *estrogenic* chemicals, absorbed from the environment, which are now causing concern among scientists and physicians. The estrogenic chemicals do not even need to be close replicas of estrogen molecules; they merely need to be able to attach to and activate estrogen receptors.

For example, biologists have observed structural and behavioral changes in seagulls exposed to high levels of DDT, a common pesticide of the 1940s and 1950s. DDT has been banned in the United States and Europe but is still used heavily in less developed countries, especially for mosquito control. DDT can linger in the environment and in food webs for nearly 60 years after its application. Male gulls with feminized reproductive systems have been observed on Santa Barbara Island off the California coast and in New Bedford Harbor, Massachusetts. The feminized males show little interest in breeding, and the gull populations have declined in number and are dominated by females that frequently nest together.

In Florida, alligators exposed to estrogenic pesticides in a freshwater lake, Lake Apopka, have penises only one-half to one-fourth the normal size, produce abnormal levels of sex hormones (very high estrogen production by females and very little testosterone synthesis by males), and have hatchling mortality rates 10 times above the norm. Pallid sturgeon in the Mississippi River have high tissue concentrations of DDT and polychlorinated biphenyls (PCBs) and exhibit reduced fertility and altered sex organs, as do male Florida panthers. Canadian fish called white suckers that are exposed to estrogenic chemicals in paper mill discharges take longer to mature, have small reproductive organs, and exhibit reduced fertility. Wildlife may be serving as an early warning sign.

What endangers the reproduction of these wild animals also may be endangering the reproductive future of humans, particularly humans who live in industrialized regions.

In Texas, laboratory experiments with red slider turtles also indicate that certain PCBs exert their effects during embryonic development. Combinations of various PCBs caused male embryo turtles to hatch as females or develop both male and female reproductive organs, an "intersex" condition. The fact that different PCBs have a combined or synergistic action raises the concern that other hormone-mimicking chemicals, such as herbicides and pesticides, may act in combination when they are present together in an ecosystem. Their synergistic action typically is greater than their individual effects. The same may be true for other polluting chemicals as well.

Human studies of hormonelike substances obtained from the environment are focused on their involvement in alterations of the testes and genitalia of males and in female breast cancer. Fetal exposure to estrogenic chemicals may disrupt the normal pattern of development in the male reproductive tract between the seventh and fourteenth weeks of pregnancy. This disruption does not cause gender reversal but may cause subtle changes that are not manifested until later in life. This feminization effect may help explain some troubling medical observations, including data suggesting that men born to mothers who received estrogen therapy during pregnancy are the most likely to develop testicular cancer during the early adult years.

In a number of industrially developed countries, including the United States, Scotland, Great Britain, France, Denmark, Norway, and Sweden, there have been significant increases in the numbers of testicular cancers over the last 50 years (threefold to fourfold increases). In addition, the volume of semen and the number of sperm ejaculated by the average male have dramatically decreased (50 percent decline) in the same time frame. This may contribute to the lower fertility rates and population growth rates observed in these and other more developed countries (Chapter 22).

Recent British studies have noted a doubling in the incidence of cryptorchidism (*krip-**torr**-kid-iz-im*), or undescended testes, since the 1950s. Undescended testes cause sterility and increase the risk of developing testicular cancer. These studies also cite an increase in the number of males born with hypospadia (*high-poh-**spay**-dee-uh*), an opening at the base of the penis. The opening, an embryological remnant, should normally close before birth as genes on the Y chromosome and testosterone direct the masculinization of the external genitalia. Exposure to estrogenic chemicals may allow the opening to persist. A persistent opening that is not surgically corrected interferes with urination and ejaculation.

Current research is also focusing on the impact of estrogenic chemicals on nurse cells, also called Sertoli cells, in the testis (Figure 1). These cells are critical to the development of sperm and the size and descent of the testes. Indications are that the number of Sertoli cells is reduced during fetal development by exposure to estrogenic chemicals, perhaps even those from pesticides, detergents, and plastics. The lower the number of Sertoli cells, the lower the number of sperm produced. Also of interest are chemicals, such as the common fruit fungicide vinclozolin, which produce similar effects on males by directly inhibiting testosterone production instead of mimicking estrogen.

There are also concerns for women. In the last 40 years, the chance of a woman developing breast cancer has increased from 1 in 20 to 1 in 9. Only about a third of women who develop breast cancer have identifiable risk factors: family history, early first menstrual period, first pregnancy late in life, or minimal breast-feeding activity. Some medical studies indicate that the risk of breast cancer increases with the level of DDT present in the blood. DDT, like other estrogenic chemicals, is fat-soluble and may accumulate in the fat tissue of the breast, where it disrupts normal cell division regulation and triggers tumor formation.

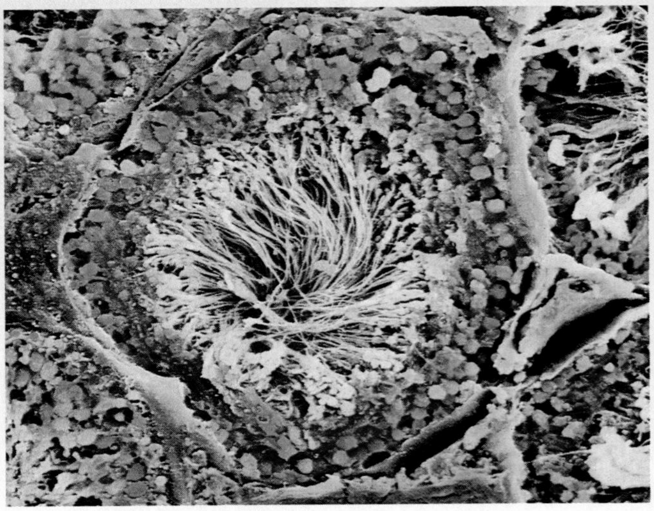

FIGURE 1 An SEM of a seminiferous tubule in the testis of a male, showing developing sperm cells. Exposure to estrogenic chemicals may lower sperm numbers by reducing the number of Sertoli cells that protect and nourish developing sperm.

Some naturally occurring environmental chemicals, however, may prevent cancer by inhibiting estrogen receptors in the body. For instance, estrogen-like chemicals in plants such as isoflavinoids and lignin, which are called *phytoestrogens,* may lower the risk of cancer when they make up a large proportion of the diet. Countries such as Japan, where traditional diets are rich in fiber and phytoestrogens, have a lower incidence of breast, prostate, and colon cancer than do countries such as the United States, where less plant material is incorporated into the diet. Lignin is abundant in flax seeds, whole wheat, and rye, while soybeans are rich in isoflavinoids. Studies with laboratory mice in the United States and Finland also indicate that animals with soy-rich diets develop prostate and mammary tumors more slowly than do animals on soy-free diets.

Prefixes and Suffixes: Clarifying the Meaning of Technical Terms

Most of these *prefixes* (meaning "to put before") and *suffixes* ("put after") are derived from Greek or Latin words. As presented below, prefixes are followed by a dash (for example, *ante-*) while suffixes are preceded by a dash (for example, *-osis*). Contemporary meanings are given, and examples are chosen from terms used in this book.

Prefix or suffix	**Meaning** (examples)
a-	**without** (anaerobic, anemia, asexual)
ante-	**before** (anterior, antecedent)
anti-	**against, opposed to** (antibiotic, antibody, antigen, anticodon)
-ase	**indicates an enzyme** (polymerase, peptidase, lipase)
auto-	**self or same** (autoimmune, autonomic)
bi-	**two or twice** (bicarbonate, biceps, binocular vision, bilateral)
bio-	**life or living** (biology, biome, biosphere, biochemistry, biopsy)
cardio-	**heart** (cardiac, myocardium, electrocardiogram)
centi-	**one hundred** (centimeter, centiliter, centigram)
cyto-, cyte-	**container or cell** (cytoplasm, cytokinesis, erythrocyte)
de-	**off or removal** (dehydration, deoxyribonucleic acid)
di-	**two, twice, or double** (disaccharide, dioxide, diploid)
eco-	**home or Earth** (ecology, ecosystem)
ecto-	**outside** (ectoderm, ectoparasite)
endo-	**within** (endoderm, endometrium, endocrine)
epi-	**over, above** (epidermis, epiglottis, epithelium, epididymis)
eu-	**true** (eukaryote)
ex-, exo-	**out of, from, beyond** (excretion, exocytosis, exocrine)
hemo-	**blood** (hemoglobin, hemophilia)
hetero-	**other, different** (heterozygote, heterosexual)
homo-, homeo-	**same** (homogeneous, homeostasis, homologous, homosexual)
hydro-, hydra-	**water or hydrogen** (hydrocarbon, dehydration, carbohydrate)

Prefix or suffix	**Meaning** (examples)
hyper-	**over, above, more than** (hypertonic, hyperactivity, hyperthyroid)
hypo-	**under, less than** (hypodermic, hypotonic, hypothalamus)
in-	**not** (indigestion, infertility, insoluble, insufficient)
inter-	**between** (intercellular, interphase, interstitial, intercourse)
intra-, intro-	**within** (intracellular, intrauterine, intravenous, introduction)
-itis	**inflammation of** (arthritis, bronchitis, dermatitis)
kilo-	**one thousand** (kilometers, kiloliters, kilograms)
-logy	**the study of** (biology, sociology, anthropology, psychology)
-lysis	**to break** (hydrolysis, analysis, catalysis)
macro-	**great or large** (macromolecule, macrophage)
meta-	**after, transfer** (metaphase, metastasis, metacarpal, metatarsal)
micro-	**small** (microbe, microscope, microbiology, microtubule)
milli-	**a thousandth** (millimeter, milliliter, milligram)
mono-	**one** (monomer, monosaccharide, monocyte, monocular)
neuro-	**nerve** (neuron, neurotransmitter, neuromuscular)
non-	**not or none** (noninfectious, nondisjunction, nonmyelinated)
-osis	**a condition** (arteriosclerosis, acidosis)
patho-	**disease** (pathogen, pathology)
-phase	**part or appearance** (metaphase, interphase)
pre-	**before** (precancerous, presynaptic membrane, prenatal)
peri-	**around** (pericardial, peripheral, peristalsis)
-phage, phago-	**to eat** (phagocytosis, macrophage)
poly-	**many** (polymer, polypeptide, polysaccharide)
post-	**behind** (posterior, postsynaptic membrane)
pro-	**before** (prokaryote, prophase, prolactin, progesterone)
prot-	**first, original** (protein, proton, Protista)
sub-	**under, below** (subatomic, subcutaneous)
trans-	**across** (translocation, translation, transport)
tri-	**three** (triangular, triceps, triglyceride, triphosphate)
uni-	**one** (unicellular, unilateral)

Glossary

- Definitions of units of measure can be found in Table 2.1.
- If you do not find a term here, check the index for the page number where a definition may be found.
- Specific diseases and disorders and their definitions can be found in the text (consult the index for page numbers).

acetylcholine (ACh) A neurotransmitter released by some nerve cells.

acetyl coenzyme A A compound containing an energy-rich 2-carbon molecule that enters the Krebs cycle in the mitochondria during ATP production.

acid A substance which in a solution donates one or more hydrogen ions (protons); a solution whose pH is less than 7.

acrosome A vesicle, located in the sperm head, that contains digestive enzymes and assists with penetration into an oocyte during fertilization.

actin A protein molecule of muscle cells that makes up the thin filaments in the sarcomere units of a myofibril.

action potential The electrical changes associated with depolarization of a nerve cell or muscle cell membrane; equivalent to an impulse.

active transport The movement of a substance across a membrane against a concentration gradient using a protein carrier and cellular energy.

adenine (A) One of the nitrogen bases found in the nucleotides of RNA and DNA.

adenosine diphosphate *See* **ADP.**

adenosine triphosphate *See* **ATP.**

ADP (adenosine diphosphate) A molecule composed of adenine, the sugar ribose, and two phosphate groups; combines with ℙ to form ATP.

adrenal gland A gland located above each kidney. Its inner portion, the adrenal medulla, releases epinephrine and norepinephrine; its outer portion, the adrenal cortex, releases several steroid hormones.

adrenocorticotropic hormone (ACTH) A hormone produced by the anterior pituitary that stimulates the cortex of the adrenal gland to release glucocorticoid hormones.

aerobic pathway The complete breakdown of glucose to CO_2, H_2O, and ATP in the mitochondria when O_2 is available.

aging The slow, progressive, and cumulative changes in the structure and function of the body that occur over the life span of an individual.

agranular leukocyte A white blood cell that contains few granules in its cytoplasm; includes lymphocytes and monocytes.

albumin A plasma protein of the blood that helps maintain the water balance between the blood and the tissues.

aldosterone A hormone released by the cortex of the adrenal gland that stimulates the kidneys to reabsorb sodium and excrete potassium.

alimentary canal The digestive tube, also called the gastrointestinal (GI) tract; includes the mouth, esophagus, stomach, small intestine, large intestine, rectum, and anus.

allantois An extraembryonic membrane which forms as an extension of the yolk sac and assists in the formation of umbilical blood vessels and the urinary bladder.

alleles The various forms of a gene which influence the same trait and occupy the same site on a chromosome.
alveolus, plural **alveoli** The microscopic air sacs in the lung where oxygen diffuses from the lung to the blood and carbon dioxide diffuses from the blood to the lung.
amino acid The monomeric unit making up proteins; consists of an amino group (—NH_2), an acid group (—COOH), and an R group. R groups have different compositions and properties, giving amino acids their distinctive characteristics.
amnion The extraembryonic membrane which directly surrounds the developing embryo and fetus, forming the amniotic cavity, a protective bag of fluid.
amoeboid movement A form of movement used by white blood cells in which portions of the cell flow forward, attach to a surface, and contract, pulling the rest of the cell forward.
amylase An enzyme from the salivary gland and pancreas that digests starch and glycogen to the sugar maltose.
anabolism The reactions of metabolism that synthesize larger molecules from smaller molecules; energy used.
anaerobic pathway The incomplete breakdown of glucose to lactic acid during glycolysis and lactic acid fermentation as a result of insufficient oxygen.
anaphase A phase of nuclear division in which chromatids (mitosis) or chromosome pairs (meiosis) separate and begin moving toward opposite poles of the cell.
antibody A protective protein produced by a plasma cell; capable of recognizing and attaching to an antigen.
anticodon In transfer RNA, three nucleotide bases that form hydrogen bonds with three complementary nucleotide bases (codon) in messenger RNA.
antidiuretic hormone (ADH) A hormone produced by the hypothalamus but released by the posterior pituitary; stimulates the kidneys to reabsorb water.
antigens Proteins or polysaccharides which allow the immune system to distinguish between "self" cells which belong in the body and "nonself" cells which are foreign to the body. Antigens bind to antibodies.
appendicular skeleton The part of the skeleton which includes the pectoral girdle and pelvic girdle and their attached limbs, the arms and legs.
aqueous humor A clear, watery fluid between the cornea and lens of the eye.
artery A blood vessel carrying blood away from the heart; its wall structure is thicker than that of veins.
atom The smallest unit of matter that participates in chemical reactions; composed of protons, neutrons, and electrons.
ATP (adenosine triphosphate) A molecule composed of adenine, the sugar ribose, and three phosphate groups. An energy shuttle, capturing energy when it is formed from ADP and a free phosphate group (ℙ) and releasing energy for cellular processes when it breaks down to ADP and ℙ.
atrioventricular (AV) node A part of the cardiac conduction system located in the lower right atrium that receives impulses from the sinoatrial node and transmits them to the ventricles.
atrioventricular (AV) valve A one-way valve located between the atria and the ventricles that prevents blood from reentering the atria during ventricular contraction.
atrium, plural **atria** One of the two upper chambers of the heart; the right chamber receives deoxygenated blood, and the left chamber receives oxygenated blood.
autoimmunity A condition in which antibodies mistakenly attack self-antigens.
autonomic nervous system (ANS) The part of the peripheral nervous system that controls the involuntary activities of the internal organs and smooth muscles.
autosomes Human chromosomes 1 through 22; determine general genetic traits but not gender.
axial skeleton The part of the skeleton which includes the skull, vertebral column, and rib cage.
axon The nerve cell structure that carries impulses away from the cell body.

bacillus, plural **bacilli** Term applied to rod-shaped bacteria.
bacterium, plural **bacteria** A prokaryotic microorganism; responsible for many human diseases.
base A substance that in solution accepts hydrogen ions (H^+); a solution whose pH is greater than 7. Also, general term for one of the components of nucleotides.
basement membrane A noncellular layer that connects an epithelium to an underlying connective tissue.
basophil A type of granular leukocyte that releases histamine and heparin, which are important in the body's response to inflammation.
B-cell lymphocytes (B cells) The white blood cells responsible for antibody production and thus antibody-mediated immunity.
benign Refers to an abnormal but noncancerous growth of cells.
biological diversity The variety of different species of organisms and the genetic variation among the individuals in a species.
biological magnification The increasing concentration of toxic materials passing through a food web from lower-level to higher-level consumers.
biomass The total organic matter in the animals and plants of a ecosystem.
biome A major terrestrial or aquatic region with characteristic environmental conditions and distinctive communities of organisms.
biosphere The entire portion of the Earth inhabited by living organisms; the total of all biomes, ecosystems, and communities.
blastocyst The stage of development at implantation; consists of a trophoblast and an inner cell mass.
blood-brain barrier A system of tight connections between epithelial cells in the walls of brain capillaries which permit only certain small molecules to enter the brain.
blood type A blood grouping based on the presence or absence of surface marker proteins (antigens) on the plasma membrane of red blood cells.

blood pressure The amount of pressure exerted by blood against the walls of blood vessels.

bond A link between atoms that holds them together in specific ways and contributes certain properties to the aggregate. *See also* **covalent bonds, ionic bonds,** and **polar covalent bonds.**

bone A connective tissue composed of bone cells embedded in hard calcium phosphate.

Bowman's capsule An epithelial structure which encloses the glomerulus and receives fluid filtered from the blood to begin urine formation.

bronchioles Narrow airway tubes that originate from the bronchi and branch extensively throughout the lung.

bronchus, plural **bronchi** An airway tube that branches from the trachea and leads into the lungs.

buffer A substance that limits the change in the pH of a solution when hydrogen ions (H^+) are added or removed.

bulbourethral glands A pair of male reproductive glands that open into the urethra at the base of the penis and release a buffering and lubricating fluid.

calcitonin A hormone secreted by the thyroid gland; lowers blood calcium levels.

Calorie The amount of energy (heat) required to raise the temperature of 1000 milliliters of water by 1° Celsius.

capillary The smallest blood vessel, one cell layer thick; the site where nutrients, waste products, and gases are exchanged between the blood and tissue cells.

carcinogen A cancer-causing agent.

cardiac cycle The sequence of heart muscle contractions and relaxations that represent one heartbeat.

cardiac muscle The striated, involuntary muscle tissue of the heart.

cardiac output The amount of blood pumped by a ventricle in 1 minute.

carrier protein A protein that transports another substance from one location to another; a protein in the plasma membrane which transports specific molecules or ions across the membrane.

carrying capacity The largest population of a species that the resources in a habitat can support over time.

cartilage A specialized type of connective tissue with a rubbery, fiber-reinforced matrix; provides support.

catabolism The reactions of metabolism that break complex compounds into simpler ones; energy released.

cell The smallest structure that shows all the characteristics of life; the fundamental unit of structure and function in all living organisms.

cell cycle The sequence of events in a eukaryotic cell; consists of interphase (growth by synthesis), mitosis (nuclear division), and cytokinesis (cell division).

cell-mediated immunity The form of specific protection against disease provided by T-cell lymphocytes.

cellular respiration Chemical reactions that break down food molecules to release energy, using oxygen and releasing carbon dioxide; involves glycolysis, the Krebs cycle, and the electron transport system.

central nervous system (CNS) The brain and spinal cord.

centromere The region of a chromosome that connects two chromatids.

cerebellum The second largest portion of the human brain; maintains muscle tone and balance, coordinates the sequence and timing of muscular activities, and stores memories of skilled movements.

cerebrospinal fluid (CSF) A protective shock-absorbing fluid formed by filtration from the blood and circulated within the ventricles and subarachnoid space of the central nervous system.

cerebrum The largest portion of the human brain; consists of two connected hemispheres divided into five lobes; the seat of conscious thought, awareness, decision making, memory storage, language, sensory interpretation, and movement.

cervix A narrow, neck-shaped portion of the uterus; projects into the vagina.

channel protein A protein that extends across the plasma membrane, forming a channel through which certain small molecules can move.

chlorofluorocarbons (CFCs) Organic molecules used in refrigerants and plastics and associated with the depletion of the upper protective layer of atmospheric ozone.

chorion The outermost extraembryonic membrane; involved in placenta formation.

choroid The middle layer of the eyeball; contains blood vessels and dark pigment for absorbing light.

chromatids The two identical copies of a chromosome produced during interphase of the cell cycle.

chromatin Tiny threadlike material consisting of DNA and proteins present in the interphase nucleus of eukaryotic cells. Chromosomes are formed as chromatin condenses early in mitosis or meiosis.

chromosome A small compact structure formed as chromatin condenses during mitosis or meiosis. Chromosomes assure the correct distribution of genes (DNA) to progeny nuclei and cells.

ciliary body The anterior part of the eye's choroid layer; changes the shape of the lens during focusing.

cilium, plural **cilia** A tiny motile structure that extends from the surface of certain cells and is composed of microtubules. Cilia move substances across a surface (mucus in the trachea) or through a tube (the human oocyte through the oviduct). *Compare* **flagellum.**

cleavage A series of mitotic cell divisions that occur after fertilization in the zygote.

clitoris A sensitive erectile organ in the female vulva at the anterior junction of the labia minora.

coccus, plural **cocci** The term applied to spherical-shaped bacteria.

cochlea A coiled fluid-filled structure of the inner ear that contains the organ of Corti with its hearing receptors.

codon In messenger RNA, three nucleotide bases that form hydrogen bonds to three complementary nucleotide bases (anticodon) in transfer RNA.

collagen A type of fiber that provides support and reinforcement in the matrix of connective tissues.
collecting duct A tubule in the renal pyramid of the kidney which collects and concentrates glomerular filtrate released from the distal convoluted tubules of connecting nephrons.
colon The part of the large intestine that extends from the cecum to the rectum.
commensalism A symbiotic relationship in which one partner benefits and the other neither benefits nor is harmed.
community All the populations of living organisms that inhabit a region.
compact bone A dense type of bone found on the outer surface and shaft of bones; contains Haversian canals.
concentration gradient The difference in the concentration of a substance between different regions which may be separated by a membrane.
cones Visual receptor cells of the retina that detect color and give rise to detailed images.
connective tissue One of the four primary tissues from which the body is constructed; consists of cells and a nonliving extracellular matrix.
consumers Nonphotosynthetic organisms that are unable to produce their own food and must eat other organisms.
cornea The transparent front portion of the sclera which allows light to enter the eye.
corpus callosum The major connection between the two cerebral hemispheres; allows cooperation and information sharing between the hemispheres of the cerebrum.
corpus luteum The remnants of an ovarian follicle after oocyte discharge at ovulation; secretes progesterone, which prepares the lining of the uterus for pregnancy.
cortisol A glucocorticoid hormone released by the cortex of the adrenal gland that increases blood glucose levels; helps the body respond to inflammation and stress.
covalent bond A link formed between atoms when they share a pair of electrons in the outermost shell.
cranial nerves The 12 pairs of peripheral nerves that originate from the brain.
Cro-Magnon The common name for a population of modern humans who lived in Europe about 40,000 to 30,000 years ago.
crossing over The exchange of DNA segments between homologous chromosomes during prophase I of meiosis.
cyclic adenosine monophosphate (cyclic AMP) A molecule that serves as a second messenger in the endocrine system by activating specific enzymes in the cytoplasm of certain cells.
cytokinesis The process by which a single cell divides into two cells after mitosis or meiosis.
cytoplasm A region surrounded by the plasma membrane but outside the nucleus.
cytosine (C) One of the nitrogen bases in the nucleotides of RNA and DNA.
cytoskeleton A network of long, thin proteins extending throughout the cytoplasm that maintains cell shape, anchors organelles, and moves the cell; consists of microfilaments, intermediate filaments, and microtubules.
cytotoxic T-cell lymphocytes T-cell lymphocytes which use lymphokines to kill infected or abnormal cells and cells foreign to the body.

decomposers Microorganisms, such as bacteria and fungi, that break down organic molecules from dead organisms and waste products to simpler inorganic molecules which can be recycled.
deforestation The net loss of forestlands caused by harvesting trees without replanting and by converting the land to other uses.
demographic transition A theory which states that as countries become industrialized, their populations experience first a declining death rate and later a declining birthrate and are subject to growth in the intervening period.
dendrite A nerve cell process that carries impulses toward the cell body.
denitrification The return of nitrogen to a gaseous, atmospheric state (N_2) as soil bacteria decompose the proteins and nucleic acids in dead animals and plants.
deoxyribonucleic acid (DNA) The molecule that carries the hereditary information in cells. The sequence of nucleotide bases in DNA codes for the synthesis of specific proteins.
depolarization A change in the permeability of a cell's plasma membrane that allows sodium ions (Na^+) to diffuse rapidly into the cell's interior.
dermis The inner, connective tissue layer of the skin where hair follicles, glands, sensory receptors, and blood vessels are located.
desertification The net increase of desert land as a result of overgrazing, erosion, drought, or climatic change.
development The complicated processes by which permanent changes in structure, form, and function take place in cells, tissues, organs, and organisms.
diaphragm A sheet of skeletal muscle that separates the thoracic and abdominal cavities; its contraction and relaxation are necessary for breathing.
diastole The phase of the cardiac cycle in which the atria and/or ventricles relax and fill with blood.
diastolic pressure The blood pressure in an artery during ventricular relaxation; the second number in a blood pressure reading.
differentiation The process by which cells develop specialized structures or functions in tissues or organs.
diffusion The net movement of substances from a region of higher concentration to a region of lower concentration (movement down a concentration gradient).
diploid The number of chromosomes (46) in a normal body cell.
disaccharide Two monosaccharide sugars bonded together; examples include sucrose, maltose, and lactose.
distal convoluted tubule The terminal, coiled segment of a nephron; receives glomerular filtrate from the loop of Henle and passes it into a collecting duct.
DNA *See* **deoxyribonucleic acid.**
DNA fingerprint A pattern of DNA fragments that is made visible by electrophoresis when DNA is digested by

restriction enzymes and the fragments are separated by size in an electrical field.

DNA ligase An enzyme that links fragments of DNA; used in biotechnology to combine separate genes into a plasmid for transfer into a microbial cell.

DNA polymerase An enzyme that links together separate nucleotides to form a polymer of DNA.

DNA probes Single-stranded nucleotide sequences prepared to be complementary with a desired DNA fragment or gene.

dominant allele An allele that expresses its trait even when only one copy is present in a heterozygote.

ductus arteriosus An embryological connection between the pulmonary trunk and the aorta; allows most blood to bypass the nonbreathing fetal lungs.

ductus venosus An embryological connection between the hepatic portal vein and the inferior vena cava; allows most blood to bypass the fetal liver.

duodenum The first 10 inches of the small intestine after the stomach; receives secretions from the pancreas and gallbladder.

ecology The study of the interactions between organisms and their environment.

ecosystem A community of organisms that interact with each other and with the nonliving chemical and physical factors of their environment.

ectoderm The outermost germ layer in the embryo; gives rise to the epidermis of the skin, the nervous system, and portions of the sensory organs.

effector A structure, such as a muscle or gland, that produces a response when stimulated.

electrolytes Mineral ions that are in solution in the body and are capable of conducting an electric current; examples include Na^+ K^+, Cl^-, and Ca^{2+}.

electron One of the three types of particles that make up an atom; has a very tiny mass and an electrical charge of -1 and orbits the atomic nucleus.

electron transport system A series of reactions in the mitochondria that extract energy from electrons of NADH and FADH; the released energy is used to make ATP.

embryo An organism in the early stages of its development before its organ systems are present.

embryological development The phase of development between the third and eighth weeks after conception, between the establishment of a placenta and the formation of organ systems.

endocrine gland A ductless gland or organ that releases hormones directly into the blood.

endocytosis The process by which cells engulf large molecules or particles and bring them into the cell. *Compare* **exocytosis.**

endocytotic vesicle A small membrane-bounded compartment that forms from the plasma membrane during endocytosis.

endoderm The innermost germ layer in the embryo; gives rise to the digestive and respiratory systems.

endometrium The lining of the uterus, a portion of which is shed monthly during the menstrual flow; the layer of the uterus which forms part of the placenta during pregnancy.

endoplasmic reticulum (ER) An extensive network of membranes that divides the cytoplasm of cells into compartments separate from the surrounding cytoplasm. Smooth ER lacks ribosomes; rough ER has ribosomes.

endorphins Naturally occurring opiumlike chemicals in the central nervous system which exert a painkilling effect.

endospore A structure produced by some bacteria that helps them survive environmental extremes.

energy The capacity to bring about changes in matter; released by some chemical reactions and required for others.

enzyme A protein that increases the speed of a chemical reaction by briefly binding to the reactants.

eosinophil A granular leukocyte that attacks parasites and participates in allergic responses.

epidermis The outer epithelial layer of the skin.

epididymis A coiled tube adjacent to the testis in the scrotum in which sperm are stored before ejaculation.

epiglottis A cartilaginous structure that covers the opening of the windpipe during swallowing; prevents food and drink from entering the windpipe.

epinephrine (adrenaline) A hormone released by the medulla of the adrenal gland that helps prepare the body for fight-or-flight situations.

epithelium One of the four primary tissues from which the body is constructed; consists of closely packed cells which cover body surfaces, line cavities or passageways, and form glands.

erythrocyte A red blood cell; contains hemoglobin molecules that transport oxygen from the lungs to the tissues.

esophagus The section of the alimentary canal that connects the mouth to the stomach.

essential amino acids Amino acids that cannot be synthesized by the body and must be obtained from the proteins we eat.

estrogens A class of hormones which regulate the menstrual cycle, support gamete formation, influence brain development, and cause the formation of female secondary sex characteristics.

eukaryotic cell A complex cell containing many internal organelles; found in humans, other animals, plants, Protista, and fungi. *Compare* **prokaryotic cell.**

eustachian tube A narrow tube that connects the middle ear chamber with the upper pharynx; permits adjustments of air pressure in the middle ear chamber.

eutrophication The accumulation of mineral and organic nutrients in a body of water, leading to overgrowth of plants, and loss of oxygen as plant debris is decomposed; hastens conversion to dry land.

exocrine glands Glands with excretory ducts leading to a body surface or passageway.

exocytosis A process of secretion from eukaryotic cells in which membrane-bounded vesicles containing material for secretion fuse with the inner surface of the plasma membrane and empty their contents outside the cell. *Compare* **endocytosis.**

experiment A set of conditions constructed to test a hypothesis.

external respiration The exchange of carbon dioxide and oxygen between the air in the lungs and the blood.

facilitated diffusion The passive transport (diffusion) of material across a plasma membrane by a specific carrier protein. The substance moves down its concentration gradient and requires no cellular energy.

FAD Flavin adenine dinucleotide; a molecule that transports hydrogen atoms to the electron transport system.

fat-soluble vitamin A vitamin that dissolves in fat; includes vitamins A, D, E, and K.

fatty acid A long chain of carbon atoms covalently bonded to each other and to hydrogen atoms; contains an acid functional group (—COOH) at one end. Fatty acids are constituents of neutral fats and phospholipids.

feedback control system A regulatory mechanism which allows changes or imbalances in the internal or external environment to be sensed so that adjustments can be made to maintain homeostatic conditions.

fertilization The combining of haploid male and female gametes to produce a diploid zygote.

fetal development The period of development from week 9 until birth, which occurs approximately 38 weeks after conception; a time when organ systems mature, weight is gained, and independent survival becomes possible.

fetus The embryo after 9 weeks of development but before birth.

fibrin An insoluble protein fiber formed from fibrinogen during blood clotting.

fibrinogen A soluble plasma protein that is changed to fibrin during blood clotting.

flagellum, plural **flagella** A long thin structure composed of microtubles that is used to propel sperm. *Compare* **cilium.**

fluid mosaic model The general structure of all biological membranes; consists of a double layer (bilayer) of phospholipids in which float proteins, many of which span the bilayer and function as channel proteins, carrier proteins, receptor proteins, or enzymes.

follicle stimulating hormone (FSH) A hormone released by the anterior pituitary that stimulates development of the ovarian follicle in females and sperm production in males.

food web The interconnecting feeding relationships in an ecosystem.

foramen ovale An embryonic opening between the right and left atria of the heart that allows most blood to bypass the nonbreathing fetal lungs.

forebrain A structure that includes the thalamus, hypothalamus, limbic system, and cerebrum of the brain; integrates the functions of internal organs and maintains homeostasis; associated with higher mental activities such as language, abstract thinking, sensory interpretation and integration, and memory storage.

fossil The preserved remains of a previously living organism.

fovea centralis An area of the retina with a high concentration of cone cells; produces sharp, detailed images.

functional group A group of atoms that associate together and contribute specific properties to the many different molecules in which they are found, for example, $—NH_2$, —COOH, $—PO_4^{3-}$.

fungus, plural **fungi** A group of eukaryotic organisms that extract nutrients from dead or decaying organic material; includes molds and yeasts.

gallbladder A saclike structure next to the liver that stores bile until it is needed for digestion.

gametes The haploid reproductive cells of males and females: sperm and ovum, respectively.

gene A unit of hereditary information; a segment of DNA with a specific sequence of nucleotide bases that codes for RNA or a protein.

gene pool All the alleles of all the genes in a population at a given time.

gene therapy The correction of mutations or genetic disorders by replacing defective genes with normal genes.

genetic engineering The deliberate altering of the genetic makeup of an organism; the transfer of genes between organisms for the purpose of accumulating cellular products useful in medicine, agriculture, or food production.

genome The complete set of genes in an individual or species.

genotype The particular combination of genes inherited by an individual at conception.

genus A taxonomic category containing species that display similarities and have an evolutionary relationship; the genus of humans is *Homo.*

global warming A rise in average annual air temperature; associated with the greenhouse effect and potentially responsible for glacial melting and a rise in sea level.

glomerular filtrate The fluid filtered under pressure from the blood at the glomerulus in Bowman's capsule of the nephron; the fluid which begins urine formation.

glomerulus: A network of capillaries formed by the afferent arteriole and drained by the efferent arteriole; the site of glomerular filtrate formation.

glucagon A hormone released by the islets of Langerhans in the pancreas that increases glucose levels in the blood.

glucocorticoids A group of steroid hormones released by the cortex of the adrenal gland; cortisol is the most abundant glucocorticoid.

glucose A monosaccharide or simple sugar ($C_6H_{12}O_6$). Extensively used for energy, it is transported by the blood (blood sugar) and stored by the liver and muscles as glycogen.

glycogen A polysaccharide of glucose used to store energy in animal cells. Stores of glycogen are found in the liver and muscles.

glycolysis A series of anaerobic reactions in the cytoplasm that catabolize a glucose molecule, forming two pyruvic acid molecules; a net of two ATPs are produced.

Golgi apparatus (Golgi) A system of membranes that receives, processes, and packages materials manufactured by the cell; the source of lysosomes and vesicles destined for exocytosis.

gonadotropic hormones Follicle stimulating hormone and luteinizing hormone, which are released by the anterior pituitary; regulate ovaries and testes.

graafian follicle The mature ovarian follicle just before the release of an oocyte at ovulation; a source of estrogen in the female.

granular leukocytes White blood cells that contain many granules in the cytoplasm; includes neutrophils, eosinophils, and basophils.

gray matter Nerve tissue that contains unmyelinated nerve cells; located in the brain and spinal cord.

greenhouse effect An accumulation of CO_2, ozone, and other pollution-related gases in the lower atmosphere, trapping heat near the Earth's surface and leading to global warming.

growth hormone (GH) A hormone released by the anterior pituitary; promotes the growth of muscle and bone, especially in children.

growth plate A cartilage plate in the epiphysis of long bones; responsible for increases in bone length.

habitat The place where an organism or a community of organisms lives, feeds, and reproduces.

haploid A single set of 23 chromosomes; half the number of chromosomes in a normal body cell; the number of chromosomes in a human gamete.

heart A muscular organ whose contractions pump blood through the blood vessels of the body.

helper T-cell lymphocytes T-cell lymphocytes which release interleukins to activate B-cell lymphocytes; forge a link between cell-mediated immunity and antibody-mediated immunity.

hemoglobin The iron-containing protein molecule in red blood cells that transports oxygen and a small percentage of carbon dioxide.

heterozygous Possessing two different alleles for a given trait.

hindbrain The medulla, cerebellum, pons, and reticular formation of the brain; coordinates breathing, heart rate, movement, and posture.

histamine A chemical released from mast cells or injured body cells that causes vasodilation during an inflammation response.

homeostasis The capacity of the human body to maintain stable internal conditions.

hominids Modern humans and their immediate ancestors; includes the species of *Homo* and *Australopithecus*.

hominoids A group of primates that includes apes and humans.

Homo erectus A descendant of *Homo habilis* that appeared about 1.5 million years ago; migrated from Africa to Asia and Europe.

Homo habilis The first humans; their fossil remains date back to 2 million years ago.

homologous pair A pair of chromosomes which contain genes for the same traits. One homologue is inherited from the male parent, and the other from the female parent.

Homo sapiens The species to which modern humans belong.

homozygous Possessing two identical alleles for a given trait.

hormones Chemical substances secreted into the blood by glands or tissues of the endocrine system; each hormone stimulates specific target cells.

human chorionic gonadotropin (HCG) A hormone secreted by the placenta and the chorionic membrane of the embryo; causes the corpus luteum to persist in the ovary during pregnancy and continue the secretion of progesterone and estrogen.

human immunodeficiency virus (HIV) The pathogen responsible for AIDS.

hydrogen bond A weak bond formed by the attraction of a partially positively charged hydrogen atom in one molecule (or part of a molecule) for a partially negatively charged atom in another molecule (or part of a molecule).

hydrophilic "Water loving"; describes polar or charged substances that are attracted to other charged substances or polar molecules such as water.

hydrophobic "Water hating"; describes nonpolar substances with no attraction to charged substances or polar molecules such as water.

hypertonic A term that describes a solution with a solute concentration greater than that of another solution. *Compare* **hypotonic** and **isotonic.**

hypodermis A layer of loose connective tissue under the dermis of the skin; stores fat which insulates the body and cushions internal organs.

hypothalamus The region of the forebrain controlling body temperature, water balance, and the pituitary gland.

hypothesis A tentative answer to a scientific question that can be tested by an experiment.

hypotonic A term that describes a solution with a solute concentration less than that of another solution. *Compare* **hypertonic** and **isotonic.**

immune memory The memory of a specific antigen that is retained by cells of the immune system long after recovery from an infection.

immunity A specific resistance to a particular disease agent.

immunoglobulins The classification of antibodies; there are five classes: IgA, IgD, IgE, IgG, and IgM.

implantation The embedding of the blastocyst into the endometrium of the uterus.

inflammation response A defensive reaction triggered by histamine; associated with blood vessel dilation, phagocyte migration, and tissue repair; characterized by symptoms of redness, swelling, heat, and pain.

inner cell mass A mound of cells in the blastocyst which develops into the embryo.

inner ear Fluid-filled chambers in the temporal bone of the skull; includes the cochlea for hearing and the semicircular canals and vestibule for balance.

insertion The end of a muscle that attaches to the bone that moves when the muscle contracts.

interferon An antiviral chemical secreted by virus-infected cells; inhibits the assembly of virus particles in uninfected cells.

interleukin-2 A lymphokine chemical secreted by helper T-cell lymphocytes to stimulate cytotoxic T cells, B cells, and natural killer cells.

internal respiration The exchange of carbon dioxide and oxygen between the blood and tissue cells of the body.

interphase The lengthy portion of the cell cycle during which the cell grows and duplicates its DNA.

interstitial cells Testosterone-secreting cells found in clusters between the seminiferous tubules of the testis.

intervertebral disks Cartilage pads located between adjacent vertebrae.

ion An electrically charged form of certain atoms.

ionic bond An electrical attraction between ions bearing opposite charges (+ and −).

iris A pigmented eye structure between the cornea and the lens; regulates diameter of pupil.

islets of Langerhans Endocrine cells in the pancreas that release the hormones insulin and glucagon.

isotonic A term describing a solution with a solute concentration identical to that of the solute concentration of another solution. *Compare* **hypotonic** and **hypertonic.**

joint Also called an articulation; the place where a bone comes in contact with another bone or a cartilage.

karyotype A photographic display of an individual's chromosomes; arranged by length, number, and type.

keratin A protein which gives strength to the cornified layer of the epidermis, hair, and nails.

kidneys The excretory organs of the urinary system; accomplish blood filtration and urine formation and thus regulate the volume, osmotic pressure, and pH of blood and body fluids.

Krebs cycle A series of chemical reactions in the mitochondria that process the 2-carbon acetyl fragments from pyruvic acid; hydrogen atoms, ATP, and carbon dioxide are released.

lactation The production and release of milk from the mammary glands.

lacteal A lymphatic capillary in a villus of the small intestine; receives fats, cholesterol, and fat-soluble vitamins.

lactic acid fermentation The anaerobic pathway that processes pyruvic acid from glycolysis when sufficient oxygen is not available; lactic acid is produced.

large intestine The part of the alimentary canal consisting of the cecum, colon, rectum, and anal canal.

larynx A cartilage structure between the pharynx and trachea that contains the vocal cords.

law of independent assortment A law stating that when multiple traits are inherited, the distribution of the alleles for one trait into oocytes or sperm does not affect the distribution of the alleles for other traits.

law of segregation A law stating that pairs of genes separate from each other during the formation of gametes; the different alleles of a pair have an equal chance of appearing in gametes.

lens A transparent structure made of protein; located behind the iris; focuses images on the retina.

leukocytes A collective term for all the different white blood cells; defend the body against foreign agents.

life span The normal length of an individual's life; consists of embryological development, growth and maturation, and adulthood.

ligaments Tough connective tissue bands that join bones together.

limbic system The region of the forebrain which plays a role in the formation of emotions and the fixation of information into memory; the sensory pathways for smell pass through this system.

linked genes Genes on the same chromosome that are inherited together.

lipase An enzyme that breaks down fat molecules.

lipids A group of organic molecules that includes fats, cholesterol, steroid hormones, and phospholipids.

liver A large organ in the upper abdominal cavity; performs many functions, including bile production, glycogen storage, and detoxification of blood.

loop of Henle A portion of the nephron located between the proximal convoluted tubule and the distal convoluted tubule; consists of a descending limb and an ascending limb.

lumen A general term that describes any internal space separated by membranes or tissues from other spaces; the interior of a compartment. Often applied to the interior space of a tube such as a blood vessel.

luteinizing hormone (LH) A hormone released by the anterior pituitary; stimulates estrogen production and ovulation in females and testosterone production in males.

lymph Tissue fluid that has entered a lymph capillary.

lymphatic vessels Thin-walled, valve-equipped vessels which conduct lymph from the tissue spaces to large veins near the heart, where it enters the blood.

lymph node An organ in the lymphatic system where lymph is filtered; contains lymphocytes.

lymphocyte An agranular leukocyte consisting of two types: B-cell lymphocytes and T-cell lymphocytes.

lymphokines Protective chemicals secreted by T-cell lymphocytes.

lysosome A membrane-bounded cytoplasmic vesicle that contains digestive enzymes; may fuse with endocytotic vesicles to digest their contents.

macrophage A defensive cell that phagocytizes foreign materials and presents antigens to B cells and T cells as part of the immune response.

major histocompatibility complex (MHC complex) A family of specific cell surface proteins that trigger responses by T cells by displaying antigen fragments.

malignant Cancerous; describes abnormal cells which invade surrounding tissues.

Mammalia The taxonomic class to which humans are assigned; animals with hair that nourish their young with milk from mammary glands.

mast cell A specialized leukocyte that does not circulate but is found in tissue spaces, where it secretes histamine as part of the inflammation response.

matrix The nonliving component of connective tissue, consisting of a network of fibers suspended in a ground substance that may be liquid, gel-like, or solid. The internal space of a mitochondrion.

matter Something that takes up space and has mass; may exist as a solid, liquid, or gas.

medulla A general term for the central portion of an organ.

medulla oblongata The region of the brain that joins with the spinal cord; controls automatic functions such as breathing, heart rate, and blood pressure.

meiosis The process of nuclear divisions by which the number of chromosomes is reduced by half, resulting in the formation of four haploid cells.

melanin The dark pigment formed by melanocytes; found in the skin and hair.

melanocytes Cells that synthesize the pigment melanin and transfer it to epidermal cells of the skin.

memory T-cell lymphocytes T cells which retain the receptors for a specific antigen long after recovery from an infection; they transform into helper T-cell lymphocytes and mount a secondary immune response.

meninges The three membranes that enclose the brain and spinal cord; dura mater, arachnoid, and pia mater.

menopause The permanent cessation of menstrual cycles at about age 50.

menses (menstruation) In the absence of pregnancy, the periodic discharge of blood, tissue, and mucus from a sexually mature uterus.

menstrual cycle A monthly series of hormone-regulated events which prepare the lining of the uterus for pregnancy and cause its shedding to the exterior in the absence of fertilization.

mesoderm The embryonic tissue that forms the skeleton, muscles, kidneys, circulatory system, and reproductive organs. One of the three germ layers. *See* **ectoderm** and **endoderm.**

messenger RNA (mRNA) A form of RNA whose nucleotide base sequence carries information from DNA to ribosomes for the synthesis of a protein.

metabolism A general term for the chemical reactions that take place in cells; extracts usable energy from fuels and transforms raw materials into molecules that can be used by the cell. *See* **anabolism** and **catabolism.**

metaphase A phase of mitosis or meiosis in which the chromosomes are aligned at the center of the cell.

metastasis The spread of cancerous cells from one body site to another, often through the circulatory or lymphatic system.

microtubules Hollow protein fibers that form part of cilia, flagella, cytoskeletons, and spindles.

microvilli Many small projections of a cell's plasma membrane; increase the cell's surface area.

midbrain The region of the brain that controls eye movements and pupil size, coordinates head movements with sight and sound, and cooperates with the forebrain to coordinate unconscious muscle movements.

middle ear An air-filled chamber that contains the three small bones that transmit sound vibrations from the eardrum to the cochlea.

mineralocorticoids Hormones released by the cortex of the adrenal gland; including aldosterone, which regulates sodium levels in the blood.

minerals Inorganic substances such as calcium, sodium, and iron required for proper body functioning.

mitochondrion, plural **mitochondria** An organelle surrounded by two membranes that contains its own ribosomes and DNA; the site where cellular respiration extracts energy from fuel molecules, producing most of the cell's ATP.

mitosis An orderly division of the nucleus resulting in the equal distribution of chromosomes to two progeny nuclei; usually followed by cell division (cytokinesis).

molecule A form of matter consisting of two or more atoms bonded together and showing characteristic properties.

Monera The kingdom that includes prokaryotes (bacteria).

monoclonal antibodies Antibodies specific to a single antigen; produced in the laboratory by selecting a single fast-growing B cell and cloning it.

monocyte An agranular leukocyte that is phagocytic.

monomers Identical or similar units that can be bonded together to form a polymer; for example, amino acid monomers form a polypeptide or protein.

monosaccharide A simple sugar such as glucose or fructose.

monosomy A term describing the condition of a cell when one chromosome is missing.

morphogenesis The development of the characteristic structure and shape of a body part or an entire organism.

morula A stage of embryonic development characterized by a solid ball of blastomere cells.

motor nerve A nerve containing only motor nerve cells; transmits impulses from the central nervous system to effectors such as muscles and glands.

mucous membrane An epithelial and connective tissue membrane that lines passageways open to the exterior.

mucus A thick sticky fluid secreted by mucous membranes and glands that keeps surfaces moist and traps irritating or infectious particles.
mutagen A substance that causes mutations.
mutation An inherited change in the structure of a gene (DNA) that produces an identifiable change in the structure and function of the organism.
mutualism A symbiotic relationship in which both partners benefit from their association.
myelin sheath A fatty insulating sheath surrounding the axons of some neurons.
myocardium The heart muscle.
myofibrils Fiberlike units in muscle cells; composed of actin and myosin filaments; responsible for muscle cell contraction.
myosin A protein molecule of muscle cells that make up the thick filaments in the sarcomere units of a myofibril.

NAD A molecule that transports hydrogen atoms to the electron transport system.
natural killer (NK) cells A family of leukocytes that destroy tumor cells or virus-infected cells; part of the nonspecific defenses.
natural selection A process which favors individuals in a population that have inheritable traits that better adapt them to their environment, allowing them to leave more offspring.
Neanderthals Archaic humans who lived in Europe and Asia, but disappeared about 35,000 years ago.
negative feedback A feedback control system in which the action of an effector organ reduces an imbalance to restore homeostasis.
nephron The fundamental functional unit of the kidney that filters the blood; consists of Bowman's capsule, proximal convoluted tubule, loop of Henle, and distal convoluted tubule.
nerve A cordlike bundle of nerve cell extensions that carries impulses to and from the central nervous system.
neuromuscular junction The synapse between a motor neuron and a skeletal muscle cell.
neuron A nerve cell consisting of cell body, dendrites, and axon; specialized to conduct electrical impulses.
neurotransmitter A chemical released by the terminal end of an axon enabling an impulse to cross a synapse.
neutral fat *See* **triglyceride.**
neutron One of the three types of particles in the nucleus of an atom; has a mass of 1 and no electrical charge.
neutrophil A granular leukocyte that phagocytizes and destroys invading microbes.
niche The environmental conditions in which an organism feeds, grows, and reproduces, as well as the lifestyle it adopts as it interacts with other community members.
nitrogen fixation A process by which certain bacteria convert nitrogen atoms from nitrogen gas (N_2) in the atmosphere to a form absorbable and usable by plants.
nodes of Ranvier Small gaps along the myelin sheath of an axon.
nondisjunction The failure of paired chromosomes to separate properly, resulting in a gamete that carries an abnormal number of chromosomes.
nonspecific defenses Body defenses that do not involve the immune response and operate against a wide range of foreign substances and cells.
nuclear envelope The double membrane that is the outermost portion of the cell nucleus.
nucleic acid A general term for RNA and DNA.
nucleolus Within the nucleus, the site for the synthesis of ribosomal RNA (rRNA).
nucleosome A double wrapping of DNA around histone proteins to form a packing unit in the chromosome.
nucleotide A monomer out of which nucleic acids are constructed; consists of a nitrogen base, a 5-carbon sugar, and a phosphate group.
nucleus The largest organelle in eukaryotic cells; surrounded by the nuclear envelope; contains most of a cell's DNA.
nutrient A general term for substances obtained from the environment to sustain life; includes carbohydrates, fats, proteins, vitamins, minerals, and water.

occipital lobe The posterior portion of the cerebrum; receives and interprets impulses from the eyes.
oncogenes Upon mutation, genes that contribute to uncontrolled cell proliferation; sometimes carried by viruses.
oogenesis The meiotic process by which an egg is formed.
optic nerve A nerve that carries impulses from the retina of the eye to the brain.
organ A group of tissues that function in a coordinated manner.
organelle A membrane-bounded structure that carries out specialized functions in eukaryotic cells, for example, the endoplasmic reticulum, mitochondria, vesicles, and Golgi apparatus.
organ of Corti A structure in the cochlea that converts sound vibrations into nerve impulses.
organ system A group of organs with integrated functioning that maintain the body.
origin The end of a muscle that attaches to a stationary bone when the muscle contracts.
osmosis The movement of water across a selectively permeable membrane from a region of higher water concentration to a region of lower water concentration.
ovaries The two female organs that produce ova (eggs).
ovulation The release of an egg (oocyte) from the ovary.
oxytocin A hormone produced by the hypothalamus but released by the posterior pituitary; stimulates the contraction of smooth muscle in the uterus and milk release from mammary glands.
ozone (O_3) A reactive gaseous form of oxygen present in the upper and lower atmosphere where it absorbs ultraviolet light and forms smog, respectively.

ℙ A symbol used for the free phosphate group $—PO_4^{3-}$.
pacemaker *See* **sinoatrial (SA) node.**
pancreas An organ near the stomach; functions as an exocrine gland during digestion and as an endocrine gland to regulate blood glucose levels.
parasitism A symbiotic relationship in which one organism lives at the expense of another.
parasympathetic division A component of the autonomic nervous system that has a relaxing or restorative effect on the body.
parathyroid glands Four small endocrine glands that release parathyroid hormone; located in back of the thyroid gland.
parathyroid hormone (PTH) A hormone released by the parathyroid glands; raises blood calcium levels.
parietal lobe In the brain, the upper middle portion of the cerebrum that receives and processes sensory impulses.
passive transport Diffusion across the plasma membrane; does not require cellular energy.
pathogen A microorganism capable of causing disease in humans, other animals, and plants.
pectoral girdle The bones (clavicle and scapula) that attach the upper limbs to the axial skeleton.
pelvic girdle The paired pelvic bones (coxal bones) that attach the lower limbs to the axial skeleton.
penis The male organ of sexual intercourse and urination.
pepsin An active protein-digesting enzyme secreted by the stomach mucosa as inactive pepsinogen.
peptide bond The chemical bond between the amino acids in a protein.
peripheral nervous system (PNS) A part of the nervous system consisting of sensory receptors, effectors, and their nerves; transmits impulses to and from the central nervous system.
peristalsis Wavelike contractions of smooth muscle that move materials through the lumen of tubelike structures.
peritoneum A serous membrane lining the abdominal cavity and covering its organs.
peritubular capillaries A network of capillaries closely associated with the nephron into which water, ions, and nutrients are reabsorbed from glomerular filtrate.
phagocytosis A form of endocytosis in which cells such as white blood cells engulf and digest cell debris and bacteria.
pharynx The throat; extends from the nasal cavity to the voice box (larynx) and is the common passageway for food, drink, and air.
phenotype An identifiable or measurable trait associated with a particular genotype.
phospholipids A family of complex lipids that contain a hydrophilic phosphate group as well as a hydrophobic hydrocarbon; essential constituents of membranes.
pH scale A scale measuring the concentration of hydrogen ions (H^+) in a solution; values less than 7 are acidic, while those greater than 7 are alkaline.
phytoplankton Microscopic plants that float near the surface of a body of water and conduct photosynthesis.
pineal gland An endocrine gland in the brain; secretes melatonin.
pioneer species The first species to invade a newly established habitat that is undergoing primary succession.
pituitary gland An endocrine gland attached to the underside of the brain; secretes eight hormones, many of which influence other endocrine glands.
placenta A temporary organ formed from fetal and maternal tissues for the exchange of gases, nutrients, and wastes between the maternal blood and the fetal blood.
plasma The liquid component of blood; composed of water, nutrients, blood proteins, hormones, and wastes.
plasma cell A B-cell lymphocyte specialized for the production and secretion of a specific type of antibody.
plasma membrane The thin flexible structure that encloses a cell and regulates the movement of material into and out of the cell.
plasmid A small circular piece of DNA found in many bacteria that is not part of a cell's chromosome. A useful vector for the transfer of recombinant DNA in genetic engineering.
platelets Cell fragments in the blood that participate in blood clotting.
pleura The serous membrane that lines the thoracic cavity and covers the lungs.
polar body During oogenesis, a small inactive cell formed during meiosis I and meiosis II.
polymer A large chainlike molecule formed by bonding together similar smaller molecules (monomers).
polymerase chain reaction (PCR) A laboratory process that uses enzymes to produce large amounts of identical DNA from a single DNA molecule.
polypeptide A chain of amino acids linked by peptide bonds; proteins are long polypeptides. *Compare* **protein.**
polysaccharide A polymer of a monosaccharide, for example, glycogen.
pons The part of the brain that connects the cerebellum and medulla with the midbrain.
population A number of individuals of the same species living in a defined geographic area.
predation A relationship in which one species, called a predator, feeds on all or part of another species, its prey.
preembryonic development The earliest events after the fertilization of an egg, includes cleavage and blastocyst formation and implantation.
primary immune response The response of the immune system to its first encounter with an antigen; involves clonal selection and establishes memory cells. *See* **secondary immune response.**
primary oocyte A cell that will enter into meiosis to produce a single egg (ovum).
primary pollutants Chemicals that accumulate in the atmosphere at harmful concentrations as a direct result of human activity or natural events.
primary spermatocytes Cells that will enter into meiosis to produce sperm.
primates A taxonomic order in the class Mammalia; includes humans, apes, monkeys, and prosimians.

producers Photosynthetic organisms which use energy from sunlight and carbon dioxide to construct organic molecules.
progesterone A hormone produced by the ovary (corpus luteum) that maintains the endometrium of the uterus.
prokaryotic cell A small cell with none of the organelles found in eukaryotic cells; bacteria are prokaryotes.
prolactin A hormone produced by the anterior pituitary that stimulates the growth of the mammary glands and milk production.
prophase The initial phase of mitosis or meiosis in which the nuclear membrane breaks down and chromosomes appear.
prostaglandins A group of lipids that act as short-range regulatory signals; produced by many body cells.
prostate gland A gland that secretes part of the semen; surrounds the urethra where it exits the bladder and connects with the vas deferens.
protein A long chain of amino acids linked by peptide bonds; proteins have specific three-dimensional shapes that are essential to their functions.
proton One of the three types of particles that make up the nucleus of an atom; has a mass of 1 and an electrical charge of +1; a hydrogen ion (H^+).
protozoa A eukaryotic microorganism.
proximal convoluted tubule The segment of the nephron between Bowman's capsule and the loop of Henle; reabsorbs nutrients, electrolytes, and water.
pulmonary circulation The part of circulatory system that transports blood from the heart to the lungs for gas exchange and then back to the heart.
pulse Rhythmic pressure changes in an artery resulting from the heartbeat.
Punnett square A square grid that shows all the possible genotypes of the progeny resulting from a mating between organisms with known genotypes.
pus Fluid containing white blood cells, dead tissue cells, cell debris, and tissue fluid; produced after inflammation at the site of tissue damage.
pyruvic acid The 3-carbon compound that is the end product of glycolysis.

rapid eye movement (REM) sleep A stage in the sleep cycle characterized by rapid eye movements, an irregular EEG pattern, and increased O_2 consumption.
receptor A protein that has a specific binding capacity for another substance (often another protein); binding initiates a change in cell activity.
recessive allele An allele whose phenotypic expression is masked when it is in the heterozygous condition.
recombinant DNA DNA that is created in the laboratory by joining the DNA of different species or individuals.
recommended dietary allowance (RDA) The recommended daily intakes of various nutrients.
red blood cell *See* **erythrocyte.**
red bone marrow Blood cell–producing tissue found in the spaces of spongy bone.
releasing hormones Hormones produced by the hypothalamus that stimulate the anterior pituitary to release its hormones.
renal cortex The outer layer of the kidney; the location where blood is initially filtered into the nephron.
renal medulla The inner layer of the kidney; the location of the renal pyramids.
renal pelvis A cavity in the kidney that receives urine from the collecting ducts and passes it to the ureter.
renal pyramid A region of the renal medulla where the collecting ducts terminate.
renin An enzyme that converts angiotensinogen to angiotensin I, causing an increase in blood pressure through kidney retention of salt and water.
replication The process by which DNA is duplicated during the S period of interphase.
repolarization The process by which the normal resting potential of a cell membrane is reestablished.
resting potential The electrical potential (voltage) that normally exists across the plasma membrane of an unstimulated cell.
restriction fragments Fragments of DNA generated by the use of restriction enzymes.
restriction fragment length polymorphism (RFLP) The diversity of DNA fragments that can be obtained using restriction enzymes; based on individual variation in gene structure.
restriction enzymes A family of enzymes that chemically cut a DNA molecule at specific sites; used in making recombinant DNA.
reticular formation A network of interconnected neurons in the brain that function to filter sensory impulses and maintain muscle tone.
retina The light-sensitive innermost layer of the eye; contains rod and cone cells.
reverse transcriptase An enzyme which allows the assembly of DNA from an RNA template; carried by HIV and some other viruses.
Rh antigen An antigen (surface marker) on the plasma membrane of red blood cells in some individuals.
rhodopsin A pigment in rod cells that is very sensitive to light; chemically composed of opsin and retinal.
ribonucleic acid *See* **RNA.**
ribosomal RNA (rRNA) The form of RNA found in ribosomes.
ribosome A complex particle involved in the synthesis of proteins; may be found free in the cytoplasm or attached to the endoplasmic reticulum.
RNA (ribonucleic acid) One of the two categories of nucleic acid found in cells; involved in the cellular assembly of protein.
RNA polymerase The enzyme that links together separate nucleotides to form a polymer of RNA.
rods The visual receptor cells in the retina that permits people to see in dim light; do not detect color.

salination Salt accumulation in soil produced when mineral salts dissolved in irrigation water remain in the soil after the water evaporates.

salivary glands The three pairs of glands that secrete saliva into the mouth.
sarcomere The basic contractile unit of muscle cells; consists of repeating units of myosin and actin.
saturated fatty acid A fatty acid containing the maximum number of hydrogen atoms and no double bonds.
Schwann cells Cells that form the myelin sheath around axons in the peripheral nervous system.
sclera The tough, white outermost protective layer of the eye.
scrotum The sac of skin that contains the testes.
secondary immune response A rapid and large response of the immune system to an antigen that was encountered previously.
secondary oocyte The large cell formed by the first meiotic division during oogenesis; the cell ovulated and penetrated by a sperm.
secondary pollutants Chemicals formed in the atmosphere when primary pollutants react with other chemicals or with sunlight.
secondary spermatocytes Cells formed by the first meiotic division during spermatogenesis.
secretion The process by which materials are moved from the inside of a cell to the outside.
selectively permeable membrane A biological membrane that permits some materials to pass across while retaining, slowing, or controlling the passage of other materials.
semen The mixture of sperm and accessory gland secretions forced from the penis during ejaculation.
semicircular canals Three fluid-filled tubes in the inner ear; responsible for dynamic balance (equilibrium).
semilunar valves Heart valves at the entrance of the aorta and pulmonary trunk; prevent blood from moving back into the ventricles during ventricular relaxation.
seminal vesicles A pair of glands that produce secretions that become part of the semen.
seminiferous tubules Tiny coiled tubes in the testes in which sperm are produced.
senescence The decline of the body in older age, when it has a reduced capacity to defend itself, repair itself, and maintain homeostasis.
sensory nerve A nerve containing only sensory nerve cells; transmits impulses from a sensory receptor to the central nervous system.
sensory receptor The specialized cell that responds to a specific kind of stimulation: light, sound, touch, heat, cold, or pain.
serotonin A brain neurotransmitter believed to be involved in controlling moods and conscious awareness.
serous membrane A continuous tissue membrane that lines body cavities not opened to the exterior and covers the outer surfaces of internal organs.
sex chromosomes The X and Y chromosomes. Normal females have two X chromosomes, and normal males have an X chromosome and a Y chromosome.
sex-linked traits Traits whose genes are found on the X and Y sex chromosomes.
sinoatrial (SA) node Specialized cardiac muscle cells in the right atrium that initiate electrical signals for each contraction of the heart.
sinuses Air spaces lined with a mucous membrane; located in certain bones of the skull.
skeletal muscle Striated, voluntary muscle attached to the skeleton; its contractions move the skeleton.
sliding filament mechanism A series of molecular events in which actin filaments slide past myosin filaments during muscle contraction.
small intestine The primary site of digestion and absorption in the alimentary canal; located between the stomach and the large intestine.
smooth muscle Nonstriated, involuntary muscle found in many internal organs.
sodium-potassium pump A plasma membrane carrier protein that uses ATP energy to move sodium and potassium ions against a concentration gradient.
solute A substance like sugar that is dissolved in a solvent (usually water). *See also* **solution** and **solvent.**
solution A homogeneous mixture of one or more solutes in a solvent (usually water). *See also* **solute** and **solvent.**
solvent The liquid (usually water) used to dissolve and disperse a solute. *See also* **solute** and **solution.**
somatic division (somatic nervous system) A division of the peripheral nervous system that is under voluntary control and stimulates skeletal muscles.
species A group of living organisms with similar structures, functions, and appearance; organisms that can interbreed with one another, producing fertile offspring.
specific defenses Defenses against foreign substances and organisms that are based on the immune response.
sperm The male gamete; equipped with a flagellar tail for propulsion.
spermatogenesis The meiotic process by which sperm cells are formed.
spermatogonia Cells which undergo mitosis to produce primary spermatocytes.
sphincter A circular band of smooth muscle that surrounds a body tube; its contractions and relaxations control the passage of materials along the tube.
spinal cord A part of the central nervous system located in the vertebral column.
spinal nerves The 31 pairs of nerves that originate from the spinal cord.
spirillum, plural **spirilla** A term applied to curve-shaped bacteria.
spleen A lymphatic organ that stores red blood cells and lymphocytes; removes damaged red blood cells.
spongy bone Bone with thin bars and plates surrounding small spaces containing red marrow.
stem cells Unspecialized bone marrow cells that give rise to red and white blood cells.
stereoscopic vision Three-dimensional vision that results from viewing with two eyes with an overlapping field of vision.
stimulus Anything in the environment that causes living cells to respond.
stomach An expanded part of the alimentary canal between the esophagus and the small intestine; site where food is mixed with gastric juice, forming chyme.

stretch receptors Sensory receptors in a muscle that detect any stretch of the muscle and signal for contractions to keep the muscle length constant.

structural genes Genes that code for the amino acid sequence of proteins, in contrast to regulatory genes.

subarachnoid space The space within which the cerebrospinal fluid circulates around the central nervous system; located between the arachnoid and pia mater.

succession A process in which the communities of plants and animals in a habitat are replaced over time by a series of progressively more complex communities. Primary succession refers to the progressive colonization of a previously unoccupied area, while secondary succession occurs after an inhabited area is disturbed and vegetation is removed.

suppressor T-cell lymphocytes T-cell lymphocytes that function to suppress the immune response.

surfactant A lipoprotein produced by cells of the alveoli that reduces surface tension in the alveoli.

sustainable development A balance between the demands of the human population and the supply of resources so that the human population will live below the Earth's carrying capacity for humans and not sacrifice the needs of other species or future human generations.

symbiosis A relationship between two species in which one relies on the other for one or more resources (food, protection, movement).

sympathetic division A component of the autonomic nervous system that prepares the body for quick action, the fight-or-flight response.

synapse The junction between an axon and another cell such as another nerve cell, a muscle cell, or a gland cell.

synaptic cleft The small space between an axon and the cell it stimulates.

synaptic vesicles Membrane-bound structures in the axon which store neurotransmitters.

synovial joint A moveable joint with a synovial cavity between bones; synovial fluid reduces friction.

systemic circulation The network of blood vessels that circulates blood from the heart to the tissues of the body and back to the heart.

systole The contraction phase of the heart during the cardiac cycle.

systolic pressure The blood pressure in an artery during ventricular contraction; the first number in a blood pressure reading.

taxonomy The classification of organisms based on their evolutionary relationships.

T-cell lymphocytes (T cells) Lymphocytes that accomplish cellular immunity: helper cells, natural killer cells, suppressor cells, and memory cells.

telophase The terminal phase of mitosis or meiosis during which the nuclear envelope re-forms and the chromosomes become less distinct; usually followed by cytokinesis.

temporal lobe In each hemisphere of the cerebrum, the middle portion of the cerebral cortex which deals with hearing.

tendon A band of connective tissue that attaches muscle to bone.

testcross A mating of a known homozygous recessive and a phenotypically dominant individual to determine whether the latter is homozygous or heterozygous.

testis, plural testes A male sex organ; produces sperm.

testosterone The male sex hormone produced in the testes; responsible for development of secondary sexual characteristics and brain development.

thalamus The portion of the brain that routes incoming nerve impulses to regions in the cerebrum.

thrombin An enzyme involved in blood clotting; converts fibrinogen to fibrin.

thymine (T) One of the nitrogen bases in DNA.

thyroid gland An endocrine gland in the neck; produces two hormones: thyroxine and calcitonin.

thyroid stimulating hormone (TSH) A hormone released by the anterior pituitary; stimulates the thyroid gland to produce thyroxine.

thyroxine A hormone released by the thyroid gland; influences cell metabolism.

tidal volume The volume of air inhaled and exhaled during normal, quiet breathing; about 500 milliliters.

tissue A group of cells that associate together and accomplish a specific function.

tissue culture The maintenance and growth of cells outside the body under laboratory conditions.

tissue fluid A fluid derived from blood plasma; occupies the spaces between cells in all body tissues.

total fertility rate (TFR) The number of children borne by a woman during her reproductive years.

trachea The windpipe; a tube reinforced by cartilage; extends from the larynx to the bronchi.

transcription The process by which the nucleotide base sequence in DNA is copied into a strand of messenger RNA (mRNA); the first phase of protein synthesis. *See also* **translation.**

transfer RNA (tRNA) RNA capable of bonding to a specific amino acid (making loaded tRNA), which it carries to mRNA for protein synthesis.

translation The process by which the nucleotide base sequence of mRNA, along with ribosomes and loaded tRNA, is used to link amino acids into a specific sequence; the second phase of protein synthesis. *See also* **transcription.**

translocation Any process in which a chromosome segment is moved from one chromosome to another.

transverse tubules Tiny tubes that carry impulses from the plasma membrane to the interior of skeletal muscle cells.

triglyceride Fat; formed by the bonding of three fatty acid molecules to a molecule of glycerol.

troponin-tropomyosin protein complex An association of two proteins that regulate the binding of myosin heads to actin filaments during muscle contraction.

tubal ligation Surgical cutting and tying of the uterine tubes to prevent fertilization of an ovum.

tubular reabsorption A process in the kidney nephron by which substances in the filtrate move back into the blood.

tubular secretion A process in the kidney nephron by which substances are actively secreted from the blood into the filtrate.

tumor An abnormal mass of cells resulting from uncontrolled and rapid cell division; may be cancerous or benign.

tympanic membrane The eardrum; its movement transmits sound wave vibrations to the ear bones of the middle ear.

umbilical cord A long structure that carries arteries and veins from the placenta to the fetus.

universal donor A person with blood type O; capable of donating blood to people with any of the other ABO blood types (A, B, and AB).

universal recipient A person with blood type AB; capable of receiving blood from people with any of the other blood types (A, B, and O).

unsaturated fatty acid A fatty acid containing one or more double bonds and less than the maximum number of hydrogen atoms.

uracil (U) A nitrogen base found in RNA, where it substitutes for thymine.

urea The primary substance for the excretion of waste nitrogen from the body.

ureter A tube that carries urine from the kidney to the bladder.

urethra The tube through which urine passes from the urinary bladder to the outside.

urinary bladder The storage organ for urine.

uterine tube A tube extending from the ovary to the uterus through which the ovum passes and in which fertilization usually takes place.

uterus A hollow female organ that receives, retains, and nourishes a fertilized egg as it develops into an embryo and fetus.

vaccine A solution of antigens that is injected into the body to stimulate the production of antibodies and confer protection against disease without causing symptoms of the disease.

vagina A tubular muscular organ extending from the cervix to the exterior; receptacle for the penis during sexual intercourse; the birth canal.

vas deferens A tube for the passage and storage of sperm; runs from the epididymis to the penis.

vasectomy Surgical cutting and tying of the vas deferens to prevent the passage of sperm.

vasoconstriction A decrease in the diameter of blood vessels caused by contraction of the surrounding smooth muscle.

vasodilation An increase in the diameter of blood vessels caused by relaxation of the surrounding smooth muscle.

vein A blood vessel that carries blood toward the heart; its wall structure is thinner than that of an artery.

ventricle A cavity or chamber, such as the two lower chambers of the heart and the ventricles of the brain.

vertebra, plural **vertebrae** A bone that makes up the vertebral column.

villus, plural **villi** A small fingerlike projection from the mucosa of the small intestine; increases the surface area for absorption.

virus An intracellular parasite that consists of a core of nucleic acid surrounded by a protein coat.

vital capacity The maximum volume of air that can be inhaled and exhaled during deep breathing.

vitamins Nutrients obtained from the diet in tiny amounts; function as parts of enzymes or participate in enzymatic reactions.

vitreous humor The gel-like fluid in the posterior chamber of the eye.

vocal cords Two bands of elastic tissue stretched across the larynx; responsible for sound when they vibrate.

vulva The external genitalia of a female.

water-soluble vitamin A vitamin that dissolves in water; any of the B vitamins and vitamin C.

white blood cells *See* **leukocytes.**

white matter Nerve tissue composed of myelinated nerve cells, for example, the white matter of the brain and spinal cord.

X-linked genes Genes located on the X chromosome; passed to progeny that receive that chromosome.

yellow bone marrow A fatty tissue found in the hollow shaft of a long bone.

zero population growth (ZPG) A condition in which the birthrate of a population equals or falls below the death rate so that there is no net growth in population size.

zona pellucida The jellylike material surrounding the egg.

zooplankton The microscopic floating or weakly swimming animal life in a body of water.

zygote The diploid cell produced by the union of haploid egg and sperm.

Credits

Prologue

Chapter opener: Jeff Greenberg/Photo Edit; Figure P.1: Left, Cesar Paredes/Stock Market; center, Bob Daemmrich/Stock, Boston; right, David Stoecklein/F-Stock; Figure P.5: Top left, Muzz Murray/Oxford Scientific Films/Earth Scenes; top center, Lefever/Grushow/Grant Heilman; top right, Chris Marona/Photo Researchers; center, Runk/Schoenberger/Grant Heilman; bottom, Manfred Kage/Peter Arnold; Figure P.6: (a) Lennart Nilsson/Bonnier Alba; (b) David M. Phillips/Visuals Unlimited; Figure P.7: Fritz Pölking/Peter Arnold; Figure P.9: Top, Manfred Kage/Peter Arnold; bottom, Biophoto Associates/Science Source/Photo Researchers.

Chapter 1

Chapter opener: Stuart McCall/Tony Stone Images; Figure 1.1: Left, Petit Format/Science Source/Photo Researchers; center, Jerry Cooke/Photo Researchers; right, Joseph Nettis/Photo Researchers; Figure 1.2: Top, David Madison; bottom, Treë; Figure 1.6: Top, Simon Fraser/Medical Physics, RVI, Newcastle/Science Photo Library/Photo Researchers; bottom, Dept. of Nuclear Medicine, Charing Cross Hospital/Science Photo Library/Photo Researchers; Figure 1.11: Biophoto Associates/Science Source/Photo Researchers; Figure 1.14: Simon Fraser/Science Photo Library/Photo Researchers; Figure 1.16: (e) Ken Eward/Science Source/Photo Researchers; Figure 1.22: Will & Deni McIntyre/Photo Researchers.

Chapter 2

Chapter opener: Francis Leroy, Biocosmos/Science Photo Library/Photo Researchers; Figure 2.2: (a) K. G. Murti/Visuals Unlimited; (b) Dr. J. T. Beveridge, University of Guelph/BPS; Figure 2.6: (a) Biophoto Associates/Photo Researchers; (b) Dr. Tony Brain/Science Photo Library/Photo Researchers; (c) David M. Phillips/The Population Council/Science Source/Photo Researchers; Figure 2.7: Courtesy Dr. J. David Robertson, Duke University Medical Center; Figure 2.10: (a, b, c) EM: Michael Sheetz, Color Map: Kenneth Eward/BioGrafx; Figure 2.14: R. Rodewald, University of Virginia/BPS; Figure 2.20: Don Fawcett, M.D./Keith Porter/Photo Researchers; Figure 2.22: David M. Phillips/Visuals Unlimited; Figure 2.24: Jim Solliday/Biological Photo Service; far left, David M. Phillips/Visuals Unlimited.

Chapter 3

Chapter opener: Dan McCoy/Rainbow; Figure 3.7: (b) Fred E. Hossler/Visuals Unlimited; Figure 3.14: (a, b, c) Lennart Nilsson/Bonnier Alba; Figure 3.16: Eddie Adams/AP/Wide World.

Chapter 4

Chapter opener: Friend & Denny/Stock Shop; Figure 4.2: A. & F. Michler/Peter Arnold; Figure 4.3: (a) Biophoto Associates/Science Source/Photo Researchers; Figure 4.4: Daemmrich/Image Works; Figure 4.5: David Fuller/Visuals Unlimited; Figure 4.6: (c) Biophoto Associates/Photo Researchers; Figure 4.14: (a) Dr. P. Marazzi/Science Photo Library/Photo Researchers; (b) Biophoto Associates/Science Source/Photo Researchers; Figure 4.15: Top, Ken Greer/Visuals Unlimited; bottom, Eamonn McNulty/Science Photo Library/Photo Researchers; Figure 4.16: Topham/Image Works.

Chapter 5

Chapter opener: Jean-Marc Loubat/Agence Vandystadt/Photo Researchers; Figure 5.13: (b) VideoSurgery/Photo Researchers; Figure 5.25: Michael Klein/Peter Arnold.

Chapter 6

Chapter opener: Charles Krebs/Stock Market; Figure 6.4: (b) David Scharf/Peter Arnold; Figure 6.7: Fred Hossler/Visuals Unlimited; Figure 6.17: Prof. P. Motta/Dept. of Anatomy/University "La Sapienza," Rome/Science Photo Library/Photo Researchers.

Chapter 7

Chapter opener: Tom Hollyman/Photo Researchers; Figure 7.2: Alan L. Detrick/Photo Researchers; Figure 7.3: Miguel Photographer/Image Bank.

Chapter 8

Chapter opener: David Young-Wolff/Photo Edit; Figure 8.2: M. Giles/Pix*Elation/Fran Heyl Associates; Figure 8.4: David Scharf/Peter Arnold; Figure 8.9: Lennart Nilsson/Bonnier Alba; Figure 8.13: (a, b) Stanley Flegler/Visuals Unlimited.

Chapter 9

Chapter opener: David Woods/Stock Market; Figure 9.2: Lennart Nilsson/Bonnier Alba; Figure 9.12: Patrick McDonnel/Custom Medical Stock Photo; Figure 9.15: (a) Melanie Carr/Custom Medical Stock Photo.

Chapter 10

Chapter opener: Carl Roessler/FPG; Figure 10.4: Lennart Nilsson/Bonnier Alba; Figure 10.6: CNRI/Science Photo Library/Photo Researchers; Figure 10.8: From Fishman: *Pulmonary Diseases and Disorders, 2nd ed.*, p. 41, ©1988 by McGraw-Hill, Inc.; Figure 10.10: (a) SIU Biomed Comm./Custom Medical Stock Photo.

Chapter 11

Chapter opener: David Madison/Duomo; Figure 11.2: CNRI/SPL/Photo Researchers; Figure 11.7: David M. Phillips/Visuals Unlimited; Figure 11.17: SIU/Photo Researchers; Figure 11.18: (a) Hank Morgan/Photo Researchers.

Chapter 12

Chapter opener: Esbin/Anderson/Omni-Photo Communications; Figure 12.1: Dr. Tony Brain & David Parker/Science Photo Library/Photo Researchers; Figure 12.5: (a, c, d) Manfred Kage/Peter Arnold; (b) David M. Phillips/Photo Researchers; (e) David M. Phillips/Visuals Unlimited; Figure 12.7: Left, Dr. C. P. Kurtzman, NCAUR, ARS, USDA; right, from Baron, E. J., Peterson, L. R., Finegold, S. M.: *Bailey & Scott's Diagnostic Microbiology*, 9th ed., Mosby, 1994; Figure 12.13: Carroll H. Weiss/Camera MD Studios; Figure 12.14: Left, Ken Greer/Visuals Unlimited; right, Biophoto Associates/Photo Researchers.

Chapter 13

Chapter opener: J. M. Petan/SPS/Vandystadt/Photo Researchers; Figure 13.4: Lennart Nilsson/Bonnier Alba; Figure 13.5: Lennart Nilsson/Bonnier Alba.

Chapter 14

Chapter opener: Chad Slattery/Tony Stone Images; Figure 14.2: From *Tissues and Organs: A Text-Atlas of Scanning Electron Microscopy* by Richard G. Kessel and Randy H. Kardon, p. 79, ©1979 by W. H. Freeman and Company; Figure 14.3: (b) David M. Phillips/Visuals Unlimited; Figure 14.9: (b) Dr. David Scott/Phototake.

Chapter 15

Chapter opener: Jeff Isaac Greenberg/Photo Researchers; Figure 15.1: Fred Hossler/Visuals Unlimited; Figure 15.11: (a) Bruce Coleman; (b) Howard Sochurek/Medichrome/Stock Shop; (c) Howard Sochurek/Medical Images; Figure 15.12: (a) Richard T. Nowitz/Photo Researchers; Figure 15.14: Will McIntyre/Photo Researchers; Figure 15.15: (a, b) NIH/Science Source/Photo Researchers.

Chapter 16

Chapter opener: Arthur Tilley/Tony Stone Images; Figure 16.7: (a) Maureen Lambray/Sygma; (b) Bettmann; Figure 16.8: From: Isselbacher: *Harrison's Principles of Internal Medicine, 13th ed.*, p. 1900, ©1994 by McGraw-Hill, Inc.; Figure 16.11: Biophoto Associates/Photo Researchers; Figure 16.12: L. V. Bergman & Associates, Inc.; Figure 16.17: John Griffin/Medichrome/Stock Shop.

Chapter 17

Chapter opener: Myrleen Ferguson Cate/Photo Edit; Figure 17.5: David M. Phillips/Visuals Unlimited; Figure 17.8: (b) C. Edelmann/La Villette/Photo Researchers; Figure 17.13: Clockwise from upper left: John Kaprielian/Photo Researchers, SIU/Peter Arnold, Ray Ellis/Photo Researchers, Ray Ellis/Photo Researchers, SIU/Visuals Unlimited, Terraphotographics/BPS, Ray Ellis/Photo Researchers, Scott Camazine & Sue Trainor/Photo Researchers; Figure 17.14: Tim Johnson/AP/Wide World.

Chapter 18

Chapter opener: SIU/Biomedical Communications/Bruce Coleman; Figure 18.2: (a) David Scharf/Peter Arnold; Figure 18.3: (a) P. H. Curran-Miller/Stock Shop; (b) Erika Stone/Peter Arnold; (c) National Medical Slide B/Custom Medical Stock Photo; Figure 18.8: (a, c) Lennart Nilsson/Bonnier Alba; Figure 18.8: (b) Petit Format/Nestle/Photo Researchers; Figure 18.10: (a) Keith/Custom Medical Stock Photo; (b) Petit Format/Nestle/Photo Researchers; (c) Lennart Nilsson/Bonnier Alba; (d, e) Petit Format/Nestle/Science Source/Photo Researchers; Figure 18.11: Left, Will & Deni McIntyre/Photo Researchers; right, Courtesy of Jim and Kim Bacon.

Chapter 19

Chapter opener: Jon Feingersh/Stock Market; Figure 19.6: (a) Gary Parker/Science Photo Library/Photo Researchers; (b) Courtesy Cytogenetics Laboratory, Dynacare Laboratory of Pathology, Seattle, WA; Figure 19.11: David M. Phillips/Visuals Unlimited; Figure 19.21: (b) SIU/Visuals Unlimited; Figure 19.24: Leonard Lessin/Peter Arnold.

Spotlight on Health: Seven

Figures 1 & 2: CNRI/Science Photo Library/Photo Researchers; Figure 3: NIAID/NIH/Peter Arnold; Figure 4: E. Gray/Science Photo Library/Photo Researchers.

Chapter 20

Chapter opener: D. Finnin/C. Chesek/American Museum of Natural History; Figure 20.1: M. Timothy O'Keefe/Bruce

Coleman; Figure 20.3: Jonathan Blair/Woodfin Camp & Associates; Figure 20.9: Science/Visuals Unlimited; Figure 20.11: Courtesy Victor McKusick, Johns Hopkins University, Center for Medical Genetics; Figure 20.12: Chelsea Bonestell; Figure 20.16: Courtesy of Cleveland Museum of Natural History, Dept. of Physical Anthropology; Figure 20.20: Jean Clottes/Sygma.

Chapter 21

Chapter opener: Chris Hackett/Image Bank; Figure 21.7: Gary Retherford/Photo Researchers; Figure 21.23: Gary Braasch/Woodfin Camp & Associates; Figure 21.24: Stephen J. Krasemann/DRK Photo; Figure 21.25: Tom Bean/DRK Photo; Figure 21.26: George Holton/Photo Researchers; Figure 21.27: Zig Leszczynski/Earth Scenes; Figure 21.28: John Gerlach/DRK Photo; Figure 21.29: Charlie Ott/Photo Researchers; Figure 21.30: Rod Kieft/Visuals Unlimited; Figure 21.33: Ed Reschke/Peter Arnold; Figure 21.34: John Lemker/Earth Scenes; Figure 21.35: Roger Cole/Visuals Unlimited; Figure 21.36: Nancy Sefton/Photo Researchers; Figure 21.37: Carl Purcell/Photo Researchers; Figure 21.38: C. Wirsen, WHOI/Science/Visuals Unlimited; Figure 21.39: Norbert Wu/DRK Photo.

Chapter 22

Chapter opener: Mark Kelley/Stock, Boston; Figure 22.2: Left, Gerry Ellis Nature Photography; right, Richard Kolar/Animals Animals; Figure 22.4: Top row: left, Dana Hyde/Photo Researchers; right, David Austen/Woodfin Camp & Associates; Middle row: left, Lindsay Hebberd/Woodfin Camp & Associates; center, Bachmann/Image Works; right, Porterfield/Chickering/Photo Researchers; Bottom row: left, Alain Evrard/Photo Researchers; center, Bob Daemmrich/Image Works; right, Stephen R. Brown/Picture Cube; Figure 22.10: Hans Peter Huber/Tony Stone Images; Figure 22.15: Don Giruin/Bruce Coleman.

Spotlight of Health: Eight

Figure 1: Professor P. Motta, Department of Anatomy, University of Rome/Science Photo Library/Photo Researchers.

Index